Lecture Notes in Computer Science 10187

Commenced Publication in 1973
Founding and Former Series Editors:
Gerhard Goos, Juris Hartmanis, and Jan van Leeuwen

More information about this series at http://www.springer.com/series/7407

Editors
Ivan Dimov
Institute of Information and Communication
  Technologies
Bulgarian Academy of Sciences
Sofia
Bulgaria

Lubin Vulkov
Department of Applied Mathematics
  and Statistics
University of Rousse
Ruse
Bulgaria

István Faragó
Department of Applied Analysis
Eötvös Loránd University
Budapest
Hungary

ISSN 0302-9743   ISSN 1611-3349   (electronic)
Lecture Notes in Computer Science
ISBN 978-3-319-57098-3   ISBN 978-3-319-57099-0   (eBook)
DOI 10.1007/978-3-319-57099-0

Library of Congress Control Number: 2017937150

LNCS Sublibrary: SL1 – Theoretical Computer Science and General Issues

Printed on acid-free paper

This Springer imprint is published by Springer Nature
The registered company is Springer International Publishing AG
The registered company address is: Gewerbestrasse 11, 6330 Cham, Switzerland

Ivan Dimov · István Faragó
Lubin Vulkov (Eds.)

Numerical Analysis and Its Applications

6th International Conference, NAA 2016
Lozenetz, Bulgaria, June 15–22, 2016
Revised Selected Papers

 Springer

Preface

This volume of the 6th International Conference on Numerical Analysis and Applications was held at the hotel Sunset Beach, Lozenetz, Bulgaria, June 15–22, 2016. The conference was organized by the Division of Numerical Analysis and Statistics, University of Rousse Angel Kanchev, Bulgaria, in cooperation with the Department of Parallel Algorithms, Institute of Information and Communication Technologies, Bulgarian Academy of Sciences, Sofia.

The conference continued the tradition of the five previous meetings (1996, 2000, 2004 in Ruse and 2008, 2012 in Lozenetz) as a forum where scientists from leading research groups from the "East" and "West" are provided with the opportunity to meet and exchange ideas and establish research cooperations. More than 120 scientists from all over the world participated in the conference.

The main tracks comprised: Numerical Modeling; Numerical Stochastics; Numerical Approximation and Computational Geometry; Numerical Linear Algebra and Numerical Solution of Transcendental Equations; Numerical Methods for Differential Equations; High-Performance Scientific Computing;

The special topics covered at the conference were: Novel methods in computational finance based on the FP7 Marie Curie Action, project Multi-ITN STRIKE—Novel Methods in Computational Finance (grant agreement number 304617); and advanced numerical and applied studies of fractional differential equations.

A wide range of problems concerning recent achievements in numerical analysis and its applications in physics, chemistry, engineering, and economics were discussed. An extensive exchange of ideas between scientists who develop and study numerical methods and researchers who use them to solve real-life problems took place during the conference.

The keynote lectures reviewed some of the advanced achievements in the field of numerical methods and their efficient applications. The conference lectures were presented by university researchers and industry engineers including applied mathematicians as well as numerical analysis and computer experts.

The success of the conference and the present volume are due to the joint efforts of the Scientific Committee, to the local organizers, and to many colleagues from various institutions and organizations. We thank to our colleagues for their help in the organization of this conference. We especially thank to M. Koleva for her help in the preparation of this volume. We are also grateful to the organizers of the minisymposia.

The 7th International Conference on Numerical Analysis and Its Applications will be held in June 2020.

January 2017

Ivan Dimov
István Faragó
Lubin Vulkov

Organization

NAA 2016 was organized by the Division of Numerical Analysis and Statistics, University of Ruse Angel Kanchev, Bulgaria, in cooperation with the Department of Parallel Algorithms, Institute of Information and Communication Technologies, Bulgarian Academy of Sciences, Sofia.

Scientific Committee

Ivan Dimov	IICT, Bulgarian Academy of Sciences, Bulgaria
Matthias Ehrhardt	Bergische Universität Wuppertal, Germany
István Faragó	Eötvös Loránd University, Hungary
Martin Gander	Université de Genève, Switzerland
Francisko Gaspar	University of Zaragoza, Spain
Abdul Khaliq	Middle Tennessee State University, USA
Raytcho Lazarov	Texas A&M University, USA
Piotr Matus	Institute of Mathematics, NAS, Belarus
Nikolay Nefedof	Lomonosov Moscow State University, Russia
Vladimir Shaidurov	Institute of Computational Modelling SB RAS, Russia
Martin Stynes	Beijing Computational Science Research Center, China
Petr Vabishchevich	Russian Academy of Sciences, Russia
Song Wang	Curtin University, Australia

Local Organizers

Luben Vulkov (Chair)	Ruse University "Angel Kanchev", Bulgaria
Tatiana Chernogorova	Sofia University "St. Kliment Ohridski", Bulgaria
Jury Dimitrov	Ruse University "Angel Kanchev", Bulgaria
Miglena Koleva	Ruse University "Angel Kanchev", Bulgaria
Walter Mudzimbabwe	Ruse University "Angel Kanchev", Bulgaria
Radoslav Valkov	University of Antwerp, Belgium

Contents

Contributed Papers

Invited Papers

Behavior of Weak Solutions to the Boundary Value Problems for Second Order Elliptic Quasi-Linear Equation with Constant and Variable Nonlinearity Exponent in a Neighborhood of a Conical Boundary Point

Yury Alkhutov[1], Mikhail Borsuk[2]($\boxtimes$), and Sebastian Jankowski[2]

[1] A.G. and N.G. Stoletov Vladimir State University, Vladimir, Russia
[2] University of Warmia and Mazury, Olsztyn, Poland
borsuk@uwm.edu.pl

Abstract. We study the behavior near the boundary conical point of weak solutions to the Dirichlet and Robin problems for elliptic quasi-linear second-order equation with the p-Laplacian and the strong non-linearity in the right side, as well as we consider the Dirichlet problem for $p(x)$-harmonic functions. We establish the exact estimate of the solutions modulus for our problems near a conical boundary point of the type $|u(x)| = O(|x|^\varkappa)$.

1 The Dirichlet Problem with Constant Nonlinearity Exponent

Let $\mathbb{C}$ be an open cone in $\mathbb{R}^n$, $n \geq 2$ with vertex in the origin $\mathcal{O}$, B_R be the open ball with radius R centered at $\mathcal{O}$ and $\mathbb{C}_R = \mathbb{C} \cap B_R$. We assume that boundary $\partial\Omega$ of $(n-1)$-dimensional domain Ω, that cut by the cone $\mathbb{C}$ on the unit sphere centered at $\mathcal{O}$, is sufficiently smooth (the smoothness property we shall define more exactly later).

Our interest is the studying of the behavior in a neighborhood of the origin $\mathcal{O}$ of solutions to the Dirichlet problem with boundary condition on the lateral surface of the cone:

$$\begin{cases} Au = b(x, u, \nabla u) & \text{in } \mathbb{C}_{R_0}, \\ u|_{\partial\mathbb{C} \cap B_{R_0}} = 0, & 0 < R_0 \leq 1. \end{cases} \tag{DQL}$$

where

$$Au \equiv \text{div}\left(|\nabla u|^{p-2} a(x) \nabla u\right), \quad p = const > 1 \tag{1}$$

under assumptions:

(i) $a(x) = \{a_{ij}(x)\}$ is a measurable symmetric matrix that satisfies for almost all $x \in \mathbb{C}_{R_0}$ the uniform ellipticity condition

$$\nu_1 |\xi|^2 \leq \sum_{i,j=1}^n a_{ij}(x)\xi_i\xi_j \leq \nu_2 |\xi|^2 \quad \forall\, \xi \in \mathbb{R}^n, \ 0 < \nu_1 \leq \nu_2, \tag{2}$$

© Springer International Publishing AG 2017
I. Dimov et al. (Eds.): NAA 2016, LNCS 10187, pp. 3–14, 2017.
DOI: 10.1007/978-3-319-57099-0_1

(ii) the function $b(x, \eta, \xi)$ is a Caratheodory function, that is $b(x, \eta, \xi)$ for almost all $x \in \mathbb{C}_{R_0}$ is continuous with respect to $(\eta, \xi) \in \mathbb{R} \times \mathbb{R}^n$, measurable with respect to x for all $(\eta, \xi) \in \mathbb{R} \times \mathbb{R}^n$ and satisfies the following inequality:

$$|b(x, u, \nabla u)| \leq \frac{\mu}{1 + |u|} |\nabla u|^p, \quad \mu \in [0, \nu_1). \tag{3}$$

We define the functions class $W \equiv W_0^{1,p}(\mathbb{C}_R, \partial \mathbb{C} \cap B_R)$ as the Sobolev space of L^p-integrable in $\mathbb{C}_R$ functions, which have the zero trace on $\partial \mathbb{C} \cap B_R$ and all its weak derivatives of first order exist and are L^p-integrable in $\mathbb{C}_R$ for any $R \in (0, R_0)$.

By a weak solution of problem (DQL) is meant a function $u \in W$ that satisfies the integral identity

$$\int\limits_{\mathbb{C}_{R_0}} \left(|\nabla u|^{p-2} \nabla u \cdot \nabla \varphi + b(x.u, \nabla u) \varphi \right) \, dx = 0$$

for all bounded test functions $\varphi \in W$ vanishing near the spherical part of the domain $\mathbb{C}_{R_0}$ boundary.

The main goal of this section is the finding of sufficient conditions on the continuity character in $\mathbb{C}_{R_0}$ of $a_{ij}(x)$ and on the boundary $\partial \Omega$ smoothness under the fulfilment which an any solution of the considered problem (DQL) behaves in a neighborhood of the cone $\mathbb{C}$ vertex as $O(|x|^\lambda)$, where λ is the exact exponent of the rate of the tending to zero as $|x| \to 0$ for solutions of the problem

$$\begin{cases} \triangle_p v \equiv \operatorname{div}(|\nabla v|^{p-2} \nabla v) = 0, & x \in \mathbb{C}_{R_0}, \\ v(x) = 0, & x \in \partial \mathbb{C} \cap B_{R_0}; \\ 0 < R_0 \leq 1. \end{cases} \tag{p_L}$$

The behavior of solutions to the p-Laplacian problem (p_L) in a neighborhood of $\mathcal{O}$ was studied by Tolksdorf in [13]. To formulate this result we consider the spherical coordinates (r, ω) and we denote by $|\nabla_\omega u|$ the projection of the gradient ∇u on the tangent plane onto the unit sphere at the point ω. Let $\Omega = \mathbb{C} \cap S^{n-1}$, where S^{n-1} is the unit sphere centered at $\mathcal{O}$, be the domain obtained by the intersection of the cone $\mathbb{C}$ and the unit sphere. If we shall find the (p_L) solution in the form $v(r, \omega) = r^\lambda \psi(\omega)$, then we obtain for $(\lambda, \psi(\omega))$ the nonlinear eigenvalue problem

$$\begin{cases} -\operatorname{div}_\omega \left((\lambda^2 \psi^2 + |\nabla_\omega \psi|^2)^{(p-2)/2} \nabla_\omega \psi \right) = \\ \lambda \left(\lambda(p-1) + n - p \right) (\lambda^2 \psi^2 + |\nabla_\omega \psi|^2)^{(p-2)/2} \psi, & \omega \in \Omega, \\ \psi|_{\partial \Omega} = 0. \end{cases} \tag{$NEVP$}$$

P. Tolksdorf investigated in 1983 [13] the behavior near $\mathcal{O}$ of solutions to $(NEVP)$ and established that if $\partial \Omega \in C^\infty$ then there exist one and only one the least positive eigenvalue

$$\lambda > \max\{0, (p-n)/(p-1)\},$$

and the corresponding eigenfunction $\psi \in C^\infty(\overline{\Omega})$ that solves above nonlinear eigenvalue problem $(NEVP)$. Moreover,

$$\psi > 0 \quad \text{in } \Omega, \ \psi^2 + |\nabla_\omega \psi|^2 > 0 \quad \text{in } \overline{\Omega}$$

and any two positive eigenfunctions are scalar multiples of each other, if they solve problem for this λ. Next he proved that any solution of (p_L) $u(x) = O(|x|^\lambda)$ and λ is the exact exponent.

Later on, in 1989, Dobrowolski [6] noted that the condition about the boundary smoothness it is possible weaken: above results are valid if $\partial\Omega \in C^{2+\beta}$ and in this case $\psi \in C^{2+\beta}(\overline{\Omega})$.

In the case of the spherical cone $\mathbb{C}$ nonlinear eigenvalue problem $(NEVP)$ in Ω was studied for the first time in the Krol and Maz'ya work [11].

Power estimates with the exact exponent near conical point of solutions to the Dirichlet problem for uniformly elliptic second order equations

$$\sum_{i,j=1}^n \frac{\partial}{\partial x_i}\left(a_{ij}(x)\frac{\partial u}{\partial x_j}\right) = 0$$

under different assumptions about the equation coefficients was established by Verzhbinsky and Maz'ya [14] as well as by Kondratiev et al. [9].

The exact exponent is characterized by the first eigenvalue of the Dirichlet problem for the Laplace-Beltrami operator in Ω. In the Verzhbinsky - Maz'ya work leader coefficients $a_{ij}(x)$ are Hölder - continuous in a conical domain, but in the Kondratiev - Kopachek - Oleinik work they are Dini-continuous.

This question was studied in more detail by Borsuk in Chap. 5 [5]; there different other estimates for the solutions smoothness are proved and the extensive bibliography is indicated.

Main result is following statement:

Theorem 1. *Let u be a weak solution of the problem (DQL), assumptions* **(i)**– **(ii)** *are satisfied, $\partial\Omega \in C^{2+\beta}$ and λ be the least positive eigenvalue of problem $(NEVP)$. Let us assume that $M_{R_0/2} = \sup\limits_{x \in \mathbb{C}_{R_0/2}} |u(x)|$ is known (see below). In addition, suppose that $a_{ij}(x)$ satisfy the Lipschitz condition, i.e.*

$$|a_{ij}(x) - a_{ij}(y)| \leq L|x-y|; \ \forall x,y \in \mathbb{C}_{R_0}; \quad a_{ij}(0) = \delta_i^j; \ i,j = 1,...,n. \quad (4)$$

Then there exist $\varrho_0(n,p,L,\Omega) \leq R_0/2$ and constant $C_0 > 0$ depending only on $\mu, \lambda, \nu_1, \varrho_0, M_{R_0/2}$ such that

$$|u(x)| \leq C_0|x|^\lambda, \quad x \in \mathbb{C}_{\varrho_0}. \quad (5)$$

Proof. At first, we prove that under conditions (2), (3) L_∞- a priori estimate holds, i.e. we derive the value $M_{R_0/2} = \sup\limits_{x \in \mathbb{C}_{R_0/2}} |u(x)|$. Moreover, condition (3) is exact. For this we give 2 counterexamples. Namely, we show that the condition $0 \leq \mu < \nu_1$ in (3) is essential and if $\mu > \nu_1$ the solution of problem (DQL) may be unbounded in the cone vertex.

Example 1. We consider the cone of the form

$$\mathbb{C}^l = \{x: \ 0 \le \theta < l, \quad \cos\theta = x_n|x|^{-1}\}, \tag{6}$$

where $l \in (0, \pi)$ is a sufficiently close to π number. We find the solution of the Dirichlet problem

$$\triangle_p v = 0, \ \text{in} \ \mathbb{C}^l; \quad v\big|_{\partial\mathbb{C}^l} = 0 \tag{7}$$

in the form $v(x) = |x|^\lambda \psi(\theta)$, where $\psi(\theta)$ is a solution to the boundary value problem (see problem (0.3)–(0.4) [10]) for the ordinary differential equation:

$$\left((\lambda^2\psi^2(\theta) + (\psi'(\theta))^2)^{(p-2)/2}\psi'(\theta)\sin^{n-2}\theta\right)'_\theta$$
$$+\lambda(\lambda(p-1) + n - p)(\lambda^2\psi^2(\theta) + (\psi'(\theta))^2)^{(p-2)/2}\sin^{n-2}\theta\psi(\theta), \ \theta \in (0, l); \tag{DODE}$$

$$\psi'(0) = 0, \quad \psi(l) = 0.$$

Let us norm $\psi(\theta)$ by $\psi(0) = 1$. Problem $(DODE)$ was investigated by Krol' in [10] for $\lambda < 0$. In this work the existence of eigenvalue $\lambda = \lambda(l) < (p-n)/(p-1)$ and corresponding single positive eigenfunction $\psi \in C^2([0, l))$ was proved (see there Theorem 1, Sect. 1). In the same place (see Theorem 2) the asymptotics of such eigenvalues for $1 < p \le n-1$ and $l \to \pi$ was derived. From this asymptotics it follows that for given $\varepsilon > 0$ there is $l = l_\varepsilon$ such that eigenvalue $\lambda_\varepsilon = \lambda(l_\varepsilon)$ of $(DODE)$ satisfies the inequality

$$\frac{p-n}{p-1} - \varepsilon < \lambda_\varepsilon < \frac{p-n}{p-1}. \tag{8}$$

Let $(\lambda_\varepsilon, \psi_\varepsilon)$ be the solution of $(DODE)$. Then function

$$v_\varepsilon = |x|^{\lambda_\varepsilon}\psi_\varepsilon(\theta) + 1$$

is a solution of the problem

$$\triangle_p v = 0, \ \text{in} \ \mathbb{C}^{l_\varepsilon}; \quad v\big|_{\partial\mathbb{C}^{l_\varepsilon}} = 1. \tag{9}$$

Now we define the function

$$h_\varepsilon = v_\varepsilon^{\frac{p-1}{p+\mu-1}} - 1.$$

By direct calculation we obtain that $h_\varepsilon \in W^{1,p}(\mathbb{C}_R^{l_\varepsilon})$, $0 < R \le 1$, if

$$\mu > 1 + \frac{\varepsilon p(p-1)}{n-p}. \tag{10}$$

as well as

$$\triangle_p h_\varepsilon = -\frac{\mu}{1+h_\varepsilon}|\nabla h_\varepsilon|^p \ \text{in} \ \mathbb{C}_R^{l_\varepsilon}, \quad h_\varepsilon\big|_{\partial\mathbb{C}^{l_\varepsilon}\cap B_R} = 0.$$

The function h_ε is positive in $\mathbb{C}_R^{l_\varepsilon}$ and unbounded in the any neighborhood of the cone $\mathbb{C}^{l_\varepsilon}$ vertex. By virtue of the arbitrariness $\varepsilon > 0$, from (10) it follows that it is impossible choose $\mu > 1$ in (3).

Example 2. Let $n > 2$ and we consider the Dirichlet problem with Laplace operator

$$\Delta u = b(x, u, \nabla u) \quad \text{in } \mathbb{C}^l \cap B, \ u|_{\partial \mathbb{C}^l_R} = 0, \ 0 < R < 1 \tag{11}$$

in a cone of the form (6) under assumption (3) with $\mu \in [0, 1)$. Let $\nu(l)$ be the least positive eigenvalue of the Dirichlet problem for the Laplace-Beltrami operator in the domain Ω^l, cut by cone $\mathbb{C}^l$ on the unit sphere centered at origin, and ψ be corresponding to ν eigenfunction. It is well known that $\psi(\theta) > 0$ in Ω^l and $\psi \in C^2([0, \theta])$. We set

$$\lambda = \lambda(l) = \frac{1}{2}\left(2 - n - \sqrt{(n-2)^2 + 4\nu(l)}\right) \quad \Longrightarrow \quad \lambda(\lambda + n - 2) = \nu. \tag{12}$$

The function $|x|^\lambda \psi(\theta)$ satisfies in $\mathbb{C}^l$ the Laplace equation and vanishes on $\partial \mathbb{C}^l$. It is well known (see Theorem 2 Sect. 2 [10]), that $\nu \to 0$ as $l \to \pi$. Therefore, by (12), for given $\varepsilon > 0$ there is $l_\varepsilon = l(\varepsilon)$ such, that for $\lambda_\varepsilon = \lambda(l_\varepsilon)$ the inequality

$$2 - n - \varepsilon < \lambda_\varepsilon < 2 - n \tag{13}$$

holds.

Let $v_\varepsilon(x) = |x|^{\lambda_\varepsilon} \psi_\varepsilon(\theta) + 1$ with ψ_ε being corresponding to ν_ε eigenfunction. The function v_ε satisfies the Laplace equation in $\mathbb{C}^l_\varepsilon$ and is equal to 1 on $\partial \mathbb{C}^{l_\varepsilon}$.

Now we define the function

$$h_\varepsilon = v_\varepsilon^{\frac{1}{1+\mu}} - 1.$$

By direct calculation we obtain that $h_\varepsilon \in W^{1,2}(\mathbb{C}^{l_\varepsilon}_R)$, $0 < R \leq 1$, if

$$\mu > 1 + \frac{2\varepsilon}{n - 2}. \tag{14}$$

as well as

$$\Delta_p h_\varepsilon = -\frac{\mu}{1 + h_\varepsilon} |\nabla h_\varepsilon|^2 \quad \text{in } \mathbb{C}^{l_\varepsilon}_R, \quad h_\varepsilon|_{\partial \mathbb{C}^{l_\varepsilon} \cap B_R} = 0.$$

The function h_ε is positive in $\mathbb{C}^{l_\varepsilon}_R$ and unbounded in the any neighborhood of the cone $\mathbb{C}^{l_\varepsilon}$ vertex. By virtue of the arbitrariness $\varepsilon > 0$, from (14) it follows that it is impossible choose $\mu > 1$ in (3) as well as in the case $p = 2$.

From the boundedness in $\mathbb{C}_R$ for $R < R_0$ of the problem (DQL) solutions follows their the Höder continuity (see the Ladyzhenskaya and Ural'tseva book [12]: Theorem 1.1 Chap. 4, Sect. 1).

The proof of the main theorem is based on the barrier method and the using of the maximum principle. At first, we take the function

$$h = h(w) = \int_0^w \frac{d\tau}{1 + \tau^\delta}, \quad w(r, \omega) = r^\lambda \psi(\omega) + r^\gamma;$$

$$\delta = \min\{1, (2\lambda)^{-1}\}, \quad \gamma = \lambda(1 + 2\delta),$$

where (λ, ψ) is the solution of $(NEVP)$ with the smallest positive λ.

Lemma 1. *Let assumptions (2), (4) are satisfied, $\partial\Omega \in C^{2+\beta}$ and λ be the least positive eigenvalue of problem (NEVP). Then the inequality*

$$Ah \leq -C_0(n,p,L,\Omega)r^{(\lambda-1)p-\lambda(1-\delta)} \quad a.e. \text{ in } \mathbb{C}_\varrho, \quad \varrho \leq r_0(n,p,L,\Omega) \leq R_0$$

holds.

Next, we derive that

$$\inf_{\overline{\mathbb{C}} \cap S_{\varrho_0}} h(w) \geq 3^{-1}\rho_0^\gamma, \tag{15}$$

where S_ϱ is the sphere with radius ϱ centered at the origin. As the barrier we define the function

$$g(w) = \sup_{\mathbb{C}_{\varrho_0}} |f(u)| \left(\inf_{\overline{\mathbb{C}} \cap S_{\varrho_0}} h(w) \right)^{-1} h(w) \text{ with} \tag{16}$$

$$f(t) = \int_0^t \exp\left(\frac{\mu}{\nu_1(p-1)} \int_0^\tau \frac{dz}{1+|z|} \right) d\tau, \tag{17}$$

ν_1 is the ellipticity constant.

Finally, by the comparison principle, we prove that

$$|u(x)| \leq |f(u)| \leq g(w) \leq C_0|x|^\lambda, \quad x \in \mathbb{C}_{\varrho_0}.$$

2   The Dirichlet Problem with Variable Nonlinearity Exponent

In this section we describe briefly recent new results of our article [2]. We consider the $p(x)$-harmonic equation in $\mathbb{C}_{R_0}$

$$\triangle_{p(x)}u \equiv \text{div}(|\nabla u|^{p(x)-2}\nabla u) = 0$$

with the exponent $p(x)$ that is a measurable function in $\mathbb{C}_{R_0}$, separated from unit and infinity:

$$1 < p_1 \leq p(x) \leq p_2 \quad \forall x \in \mathbb{C}_{R_0}.$$

Our interest is the studying of the behavior in a neighborhood of the origin $\mathcal{O}$ of the $p(x)$-harmonic in $\mathbb{C}_{R_0}$ functions u, satisfying the homogeneous Dirichlet boundary condition on the lateral surface of the cone:

$$\begin{cases} \triangle_{p(x)}u = 0, & x \in \mathbb{C}_{R_0}, \\ u|_{\partial\mathbb{C} \cap B_{R_0}} = 0, & 0 < R_0 \leq 1/2. \end{cases} \tag{$Dp(x)$}$$

We define the functions class

$$W_{loc} = \{w : w \in W_0^{1,1}(\mathbb{C}_R, \partial\mathbb{C} \cap B_R), |\nabla w|^{p(x)} \in L^1(\mathbb{C}_R)\}$$

for any $R \in (0, R_0)$. Here $W_0^{1,1}(\mathbb{C}_R, \partial\mathbb{C} \cap B_R)$ is the Sobolev space of functions which have the zero trace on $\partial\mathbb{C} \cap B_R$ and all its weak derivatives of first order exist and are L^1-integrable in $\mathbb{C}_R$.

By a solution of our problem is meant a function $u \in W_{loc}$ that satisfies the integral identity

$$\int\limits_{\mathbb{C}_{R_0}} |\nabla u|^{p(x)-2}\nabla u \cdot \nabla\varphi \, dx = 0$$

for all test functions $\varphi \in W_{loc}$ vanishing near the spherical part of the domain $\mathbb{C}_{R_0}$ boundary.

The main goal of this section is the finding of sufficient conditions on the continuity character in $\mathbb{C}_{R_0}$ of exponent $p(x)$ and on the boundary $\partial\Omega$ smoothness under the fulfilment which an any solution of the considered problem behaves in a neighborhood of the cone $\mathbb{C}$ vertex as $O(|x|^\lambda)$, where λ is the exact exponent of the rate of the tending to zero as $|x| \to 0$ for solutions of the problem

$$\begin{cases} \triangle_{p_0} v == 0, & x \in \mathbb{C}_{R_0}, \\ v|_{\partial\mathbb{C}\cap B_{R_0}} = 0, & 0 < R_0 \le 1/2 \end{cases} \tag{Dp_0}$$

with $p_0 = p(0)$. If we shall find the (Dp_0) solution in the form $v(r,\omega) = r^\lambda\psi(\omega)$, then we obtain for $(\lambda, \psi(\omega))$ the nonlinear eigenvalue problem $(NEVP)$ with $p = p_0$.

The equation with variable exponent $p(x)$ belongs to the wide class of elliptic equations with a nonstandard growth condition and is the Euler equation for variational problems with integrand $|\nabla u|^{p(x)}/p(x)$. Such variational problems was investigated by V. Zhikov in the mid-80s. The important role in the $p(x)$-Laplacian equation theory plays the known as logarithmic condition

$$|p(x) - p(y)| \le \frac{k_0}{|\ln|x-y||} \quad \forall x,y \in \mathbb{C}_{R_0}, \ |x-y| \le 1/2,$$

that was defined by Zhikov [15]. It was introduced initially in order that to prove the density of smooth functions in the solutions space. Later, it was found that the logarithmic condition has other numerous consequences. For example, it guarantees the interior Hölder-continuity of solutions, that was established with different methods by Fan [8] and Alkhutov [1]. The behavior on the boundary of solutions to the Dirichlet problem for $p(x)$-harmonic equation with exponent p, satisfying the logarithmic condition, was studied in the Alkhutov and Krasheninnikova work [3]. In particular, they established that solutions of considered problem are Hölder continuous at the cone $\mathbb{C}$ vertex.

At first, we establish that if $p(x) \in C^0(\overline{\mathbb{C}_{R_0}})$ then $u(x)$ is bounded in $\mathbb{C}_R$ for all $R \in (0, R_0)$; therefore we set

$$M_R = \sup_{\mathbb{C}_R} |u(x)|. \tag{18}$$

We obtained the main result under the assumption that exponent p satisfies the Lipschitz condition, i.e.

$$|p(x) - p(y)| \leq L|x - y|, \quad \forall x, y \in \mathbb{C}_{R_0}. \tag{19}$$

Theorem 2. *Let $\lambda > 0$ be the eigenvalue of above $(NEVP)$ with $p = p(0) = p_0$. Let the Lipschitz condition (19) for $p(x)$ satisfy and $\partial\Omega \in C^{2+\beta}$. Then for any solution of problem $(Dp(x))$ we have*

$$|u(x)| \leq C(\lambda, \beta, \rho_0, M_{R_0/2})|x|^\lambda, \quad \forall x \in \mathbb{C}_{\rho_0},$$

where $\rho_0 = \rho_0(n, p_0, p_1, L, \Omega, M_{R_0/2}) \leq R_0/2$.

Proof. We give only the proof sketch. We apply the barrier method and use the maximum principle. In particular, if $p(x) \equiv const$ then we give the simple proof of the power estimate with the exact exponent for solutions near the cone $\mathbb{C}$ vertex, that early was obtained by Tolksdorf [13], but by more complicated way.

We set

$$h = h(w) = \int\limits_0^w \frac{d\tau}{1 + \tau^\delta}, \quad \delta = \min\{1, (2\lambda)^{-1}\},$$

where $w(r, \omega) = r^\lambda \psi(\omega) + r^\gamma = h_1(r, \omega) + h_2(r), \quad \gamma = \lambda(1 + 2\delta)$.

Lemma 2. *If $p(x)$ is Lipschitz-continuous (see (19)) and $\partial\Omega \in C^{2+\beta}$, then for $\rho \leq r_0(n, p_0, p_1, L, \Omega) \leq R_0$ we have*

$$\Delta_{p(x)} h \leq -C(n, p_0, p_1, L, \Omega) r^{(\lambda-1)p_0 - \lambda(1-\delta)}$$

almost everywhere in $\mathbb{C}_\rho$.

We obtain required result from this Lemma, using the maximum principle, because

$$\frac{1}{3} w \leq h(w) \leq w \quad \text{in } \mathbb{C}_R \, \forall R \in (0, 1).$$

The Lipschitz condition play the important role for the construction of the barrier, but it is enough burdensome. Using other methods we can weaken this condition, but only in the case $p_0 = p(0) = 2$, namely if $\partial\Omega \in C^{1+\beta}$ and exponent $p(x) \in C^\alpha(\mathbb{C}_{R_0})$, i.e.

$$|p(x) - p(y)| \leq H|x - y|^\alpha, \quad \forall x, y \in \mathbb{C}_{R_0}. \tag{20}$$

In this case the exact estimate near the cone $\mathbb{C}$ vertex for solutions of considered problem is the same as for solutions of similar problem for the Laplace equation. Let μ be the first eigenvalue of the Dirichlet problem for the Laplace-Beltrami operator in Ω. We define

$$\lambda_0 = \frac{2 - n + \sqrt{(n-2)^2 + 4\mu}}{2} \quad \Longrightarrow \quad \lambda_0(\lambda_0 + n - 2) = \mu.$$

Theorem 3. *Let (20) satisfy, $\partial\Omega \in C^{1+\beta}$ and $p(0) = 2$. Then for any solution $u(x)$ of problem $((Dp(x)))$ we have*

$$|u(x)| \leq C(n, p_1, p_2, H, \alpha, \lambda_0, \rho_0, \beta, M_{R_0/2})|x|^{\lambda_0}, \quad \forall x \in \mathbb{C}_{\rho_0},$$

where $\rho_0 = \rho_0(n, p_1, p_2, H, \alpha, \lambda_0, \beta, \Omega, M_{R_0/2}) \leq R_0/4$.

Proof. The proof is based on integro-differential inequalities (see in detail [2,4,5]) and on the pointwise estimate of the gradient modulus. The last estimate follows from the Fan results [7]. Namely, he proved that if $p(x) \in C^\alpha$, then $u(x) \in C^{1+\varkappa}(\overline{\mathbb{C}}_{R_0/4} \setminus \{\mathcal{O}\})$. It is used actually the freezing method for $p(x)$.
Putting

$$U(\rho) = \int\limits_{\mathbb{C}_\rho} |x|^{2-n}|\nabla u|^2 \, dx,$$

where u is a solution of considered problem, we find that

$$(1 - \rho^{\alpha/4})U(\rho) \leq \frac{\rho}{2\lambda_0}U'(\rho) + \rho^{1+2\lambda_0}, \quad \forall \rho \leq r_0 \leq R_0/4.$$

where r_0 depends only on n, p, Ω and $M_{R_0/2}$. Hence it follows that

$$U(\rho) \leq C\left(2\lambda_0 r_0 + r_0^{-2\lambda_0}U(r_0)\right)\rho^{2\lambda_0}, \quad \forall \rho \in (0, r_0].$$

Further, we derive the local estimate at the boundary for $|u(x)|$.

Proposition 1. *Let $\partial\Omega \in C^{1+\beta}$ and (20) satisfy. Then for any $\nu > 0$ a solution u of considered problem with $\rho \leq R_0/4$ satisfies the inequality*

$$\sup_{x \in \mathbb{C}_{\rho/2, \rho/4}} |u(x)| \leq C\left(\fint_{\mathbb{C}_\rho} |u|^{p_0} \, dx\right)^{1/p_0} + C\rho^\nu,$$

where C depends only on n, p, Ω, ν and $M_{R_0/2}$, but $\mathbb{C}_{r_1, r_2} = \mathbb{C} \cap (B_{r_1} \setminus \overline{B}_{r_2})$.

Then stated result follows from this proposition. In fact, because in our case $p_0 = 2$, it is sufficient to use the Wirtinger inequality

$$\int\limits_{\mathbb{C}_\rho} |x|^{-n}|u|^2 \, dx \leq \mu^{-1}\int\limits_{\mathbb{C}_\rho} |x|^{2-n}|\nabla u|^2 \, dx = \mu^{-1}U(\rho),$$

where μ is the first eigenvalue of the Dirichlet problem for the Laplace-Beltrami operator on Ω. The last see in detail [2,4,5].

3   The Robin Problem

Here we consider two dimensional Robin problem for the p-Laplacian operator in an angle and investigate the corresponding eigenvalue problem. Let (r, ω) be polar coordinates and

$$G_0^d = \{(r, \omega) \mid r \in (0, d); \ -\frac{\omega_0}{2} < \omega < \frac{\omega_0}{2}\} \subset \mathbb{R}^2, \quad \omega_0 \in (0, \pi);$$

be an angle with sides $\Gamma_\pm^d = \{(r, \omega) \mid r \in (0, d); \ \omega = \pm\frac{\omega_0}{2}\}$, the vertex $\mathcal{O} \in \partial G_0^d$ and $\Omega_d = \{(d, \omega) \mid -\frac{\omega_0}{2} < \omega < \frac{\omega_0}{2}\}$.

We consider the following problem:

$$\begin{cases} -\operatorname{div}(|\nabla u|^{p-2}\nabla u) + b(u, \nabla u) = f(x), & x \in G_0^d \\ |\nabla u|^{p-2}\frac{\partial u}{\partial n} + \frac{\gamma}{|x|^{p-1}}u|u|^{p-2} = 0, & x \in \Gamma_\pm^d \end{cases} \quad (RQL)$$

under assumptions:

(i) $1 < p < 2$;

(ii) $|f(x)| \le f_0|x|^\beta$, $\quad f_0 \ge 0$, $\beta > \frac{(p-1)^2}{p-1+\mu}\lambda - p$; $\quad \gamma = const > 0$.

(iii) the function $b(u, \xi)$ is differentiable with respect to the u, ξ variables in $\mathfrak{M} = \mathbb{R} \times \mathbb{R}^2$ and satisfy in $\mathfrak{M}$ the following inequalities:

$$|b(u, \xi)| \le \delta|u|^{-1}|\xi|^p + b_0|u|^{p-1}, \quad \delta \in [0, \mu), \ \mu \in [0, 1);$$

$$\sqrt{\sum_{i=1}^{2}\left|\frac{\partial b(u, \xi)}{\partial \xi_i}\right|^2} \le b_1|u|^{-1}|\xi|^{p-1}; \quad \frac{\partial b(u, \xi)}{\partial u} \ge b_2|u|^{-2}|\xi|^p;$$

$$b_0 \ge 0, \ b_1 > 0, \ b_2 > 0.$$

We define the functions class

$$\mathfrak{N}_{-1,\infty}^{1,p}(G_0^d) = \left\{u \mid u(x) \in L_\infty(G_0^d) \text{ and } \int_{G_0^d} \langle r^{-p}|u|^p + |u|^{-1}|\nabla u|^p\rangle\} \, dx < \infty\right\}.$$

It is obvious that $\mathfrak{N}_{-1,\infty}^{1,p}(G_0^d) \subset W^{1,p}(G_0^d)$.

Definition 1. *The function u is called a weak bounded solution of problem (RQL) provided that $u(x) \in \mathfrak{N}_{-\infty}^{1,p}(G_0^d)$ and satisfies the integral identity*

$$Q(u, \eta) := \int_{G_0^d} \langle|\nabla u|^{p-2}u_{x_i}\eta_{x_i} + b(u, \nabla u)\eta\rangle \, dx + \gamma\int_{\Gamma_0^d} r^{1-p}u|u|^{p-2}\eta dS +$$

$$-\int_{\Omega_d} |\nabla u|^{p-2}\frac{\partial u}{\partial r}\eta d\Omega_d = \int_{G_0^d} f\eta dx. \quad (II)$$

for all $\eta(x) \in \mathfrak{N}_{-1,\infty}^{1,p}(G_0^d)$.

Main result is the following statement:

Theorem 4. *Let u be a weak bounded solution of the problem (RQL), assumptions* **(i)**–**(iii)** *are satisfied, $M_0 = \sup\limits_{x \in G_0^d} |u(x)|$ and λ be the least positive eigenvalue of problem*

$$
\begin{cases}
\left\langle \lambda^2 \psi^2(\omega) + (p-1)\psi'^2(\omega) \right\rangle \psi''(\omega) + & (RODE) \\
\quad \langle (2p-3)\lambda^2 + (2-p)\lambda \rangle \psi(\omega)\psi'^2(\omega) \\
\quad + \langle (p-1)\lambda + 2 - p \rangle \lambda^3 \psi^3(\omega) = 0, & \omega \in (-\frac{\omega_0}{2}, \frac{\omega_0}{2}); \\
\pm \psi'(\omega)\langle \lambda^2\psi^2(\omega) + \psi'^2(\omega)\rangle^{\frac{p-2}{2}} + \gamma \left(\frac{\lambda}{\varkappa}\right)^{p-1}\psi^{p-1}(\omega) = 0, & \omega = \pm\frac{\omega_0}{2}.
\end{cases}
$$

Then there exist $d(p, \lambda, (\mu - \delta), b_0)$ and constant $C_0 > 0$ depending only on $\lambda, d, M_0, p, (\mu - \delta), b_0, f_0$, such that

$$
|u(x)| \le C_0 |x|^{\varkappa}, \quad \varkappa = \frac{p-1}{p-1+\mu}\lambda; \quad \forall x \in \overline{G_0^d}. \tag{21}
$$

Proof. We give only the proof sketch. The proof of the main theorem is based on the barrier method and on the using of the comparison principle. At first, we derive $L_\infty-$ a priori estimate of the weak bounded solutions to problem (RQL) (see also Theorem 6.5, Sect. 6.3 [4]) and thus we know M_0. As the barrier function we consider

$$
w = w(r, \omega) = r^{\varkappa}\psi^{\varkappa/\lambda}(\omega), \quad \varkappa = \frac{p-1}{p-1+\mu}\lambda \tag{22}
$$

with $(\lambda, \psi(\omega))$ being the solution to the eigenvalue problem $(RODE)$. Next we verify that $w(r, \omega) \in \mathfrak{N}^{1,p}_{-1,\infty}$. For this we prove following properties of the $(RODE)$ solutions:

$$
\lambda > 0; \quad \psi(-\omega) = \psi(\omega), \quad \psi(\omega) > 0, \quad \psi'(\omega) < 0 \quad \forall \omega \in \left(-\frac{\omega_0}{2}, \frac{\omega_0}{2}\right);
$$

$$
\psi\left(\frac{\omega_0}{2}\right) = 1, \quad 1 \le \psi(\omega) \le \exp(\omega_0, y_0), \quad y_0 = const(p, \gamma, \lambda, \mu) > 0.
$$

Further, we prove

Comparison Principle.

Let operator Q satisfy assumptions **(i)** *-* **(iii)** *and $d \lll 1$. Let functions $u, w \in \mathfrak{N}^{1,p}_{-1,\infty}(G_0^d)$ satisfy the inequality*

$$
Q(u, \eta) \le Q(w, \eta)
$$

for all non-negative $\eta \in \mathfrak{N}^{1,p}_{-1,\infty}(G_0^d)$ and also the inequality

$$
u(x) \le w(x) \quad \text{a.e. on } \Omega_d
$$

holds. Then $u(x) \le w(x)$ in G_0^d.

The proof see Sect. 6.2 [4].

Finally, we show that solution u to (RQL) and the barrier function w by (22) satisfy the comparison principle. By this the proof of main theorem is completed.

References

1. Alkhutov, Y.: The Harnack inequality and the Hölder property of solutions of nonlinear elliptic equations with nonstandard growth condition. Differ. Equ. **33**(12), 1653–1663 (1997)
2. Alkhutov, Y., Borsuk, M.V.: The behavior of solutions to the Dirichlet problem for second order elliptic equations with variable nonlinearity exponent in a neighborhood of a conical boundary point. J. Math. Sci. **210**(4), 341–370 (2015)
3. Alkhutov, Y., Krasheninnikova, O.: Continuity at boundary points of solutions of quasilinear elliptic equations with a non-standard growth condition. Izv. Math. **68**(6), 1063–1117 (2004)
4. Borsuk, M.: Transmission Problems for Elliptic Second-order Equations in Nonsmoooth Domains. Frontiers in Mathematics. Birkhäuser, Boston (2010). 218 p
5. Borsuk, M., Kondratiev, V.: Elliptic boundary value problems of second order in piecewise smooth domains. North-Holland Math. Libr. **69**, 530 (2006). Elsevier
6. Dobrowolski, M.: On quasilinear elliptic equations in domains with conical boundary points. J. reine und angew. Math. **394**, 186–195 (1989)
7. Fan, X.: Global $C^{1,\alpha}$ regularity for variable exponent elliptic equations in divergence form. J. Differ. Equ. **235**(2), 397–417 (2007)
8. Fan, X., Zhao, D.: A class of De Giorgi type and Hölder continuity. Theory Methods Appl. **36**(3), 295–318 (1999)
9. Kondrat'ev, V.A., Kopachek, I., Oleinik, O.A.: On the best Hölder exponents for generalized solutions of the Dirichlet problem for a second-order elliptic equation. Math. Sb. **131**(1), 113–125 (1986). (in Russian); English transl.: Math. USSR Sb. **59**, 113–127 (1988)
10. Krol', I.N.: The solutions of the equation div $\{|\nabla u|^{p-2} \cdot \nabla u\} = 0$ with a singularity at a boundary point, (Russian) Boundary value problems of mathematical physics, 8. Tr. Mat. Inst. Steklova **125**, 127–139 (1973)
11. Krol', I.N., Maz'ya, V.G.: The absence of the continuity and Hölder continuity of the solutions of quasilinear elliptic equations near a nonlegular boundary. (Russ.) Trudy Moskov. Mat. Obšč. **26**, 75–94 (1972)
12. Ladyzhenskaya, O.A., Ural'tseva, N.N.: Linear and Quasilinear Elliptic Equations. Academic Press, New York (1968)
13. Tolksdorf, P.: On the Dirichlet problem for quasilinear equations in domains with conical boundary points. Commun. Partial. Differ. Equ. **8**, 773–817 (1983)
14. Verzhbinsky, G.M., Maz'ya, V.G.: Asymptotic behaviour of the solutions of second order elliptic equations near the boundary. I. Siberian Math. J. **12**, 874–899 (1971). II - Siberian Math. J. **13**, 858–885 (1972)
15. Zhikov, V.V.: On Lavrentiev's phenomenon. Russ. J. Math. Phys. **13**(2), 249–269 (1994)

CVA Computing by PDE Models

Iñigo Arregui$^{(\boxtimes)}$, Beatriz Salvador, and Carlos Vázquez

Department of Mathematics, University of A Coruña, A Coruña, Spain
{inigo.arregui,beatriz.salvador,carlos.vazquez.cendon}@udc.es

Abstract. In order to incorporate the credit value adjustment (CVA) in derivative contracts, we propose a set of numerical methods to solve a nonlinear partial differential equation [2] modelling the CVA. Additionally to adequate boundary conditions proposals, characteristics methods, fixed point techniques and finite elements methods are designed and implemented. A numerical test illustrates the behavior of the model and methods.

Keywords: CVA · Modelling · Numerical methods

1 Introduction

Since 2007 crisis, when important financial entities went bankrupt, the counterparty risk has become an important ingredient in all contracts. It can be described as the risk to each party of a contract that the counterparty will not live up to its contractual obligations. Thus, the neutral risk value of a derivative must take into account the following adjustments [6,7]:

- CVA: the credit value adjustment is the amount by which the value of a security is adjusted downward because of the counterparty credit risk
- DVA: the debit value adjustment corresponds to the CVA of the bank, viewed from the point of view of its counterparty
- FVA: the funding value adjustment represents the difference between a collateralized and uncollateralized trade. Moreover FVA = FBA − FCA, where FCA is the adjustment due to existence of funding costs by the issuer, and the FBA is the adjustment due to the liquidity produced by the evolution in this value.

Thus, including counterparty risk in the pricing of derivatives represents an important change in the existent risk-free pricing models. In particular, you can formulate nonlinear partial differential equation (PDE) models which have to be mathematically analyzed and solved by means of numerical methods.

Our goal is to calculate the value of derivatives, accounting for all the associated cash flows that come from the derivative itself, the act of hedging, and the management of default risk and funding cost. We will refer to all value adjustments as XVA, which is defined by:

$$XVA = DVA - CVA - FCA + FBA = DVA - CVA + FVA.$$

© Springer International Publishing AG 2017
I. Dimov et al. (Eds.): NAA 2016, LNCS 10187, pp. 15–24, 2017.
DOI: 10.1007/978-3-319-57099-0_2

In this work, we first propose boundary conditions for a one dimensional model [2], considering constant default intensities. Then, we propose a set of numerical methods (characteristics methods, fixed point techniques and finite elements) to achieve approximations of the solutions and a numerical test is solved.

2 Mathematical Model

In a first step, following [2] we hedge the derivative with a self-financing portfolio which covers all underlying risk factors of the model. Let us assume a portfolio Π consisting of:

- $\Delta(t)$ units of the underlying S,
- $\alpha_B(t)$ units of P_B, a default risky, zero-recovery, zero-coupon bond of party B
- $\alpha_C(t)$ units of P_C, an analogous bond for the counterparty C
- γ units of cash, which is made up of a financing amount, cash needed to buy a position in C's bond and a REPO amount, such that the portfolio value at time t hedges out the value of the derivative contract to the seller.

Thus,

$$-\hat{V}_t = \Pi_t = \Delta S_t + \alpha_B P_{B_t} + \alpha_C P_{C_t} + \gamma.$$

Imposing the self-financing feature of the portfolio, we deduce:

$$d\Pi_t = \Delta dS_t + \alpha_B dP_{B_t} + \alpha_C dP_{C_t} + (r\gamma_F^+ + r_F\gamma_F^- - r\gamma_{P_C} - r_R\gamma_R)dt;$$

then, applying Ito's Lemma and eliminating all risks in the portfolio we obtain the PDE modelling the value of the derivative including the counterparty risk:

$$\begin{cases} \partial_t \hat{V} + \mathcal{A}_t \hat{V} - r\hat{V} = (\lambda_B + \lambda_C)\hat{V} + s_F M^+ \\ \qquad\qquad - \lambda_B(R_B M^- + M^+) - \lambda_C(R_C M^+ + M^-) \\ \hat{V}(T, S) = H(S), \end{cases}$$

where the parabolic differential operator $\mathcal{A}_t$ is given by:

$$\mathcal{A}_t V \equiv \frac{1}{2}\sigma^2 S^2 \frac{\partial^2 V}{\partial S^2} - r_R S \frac{\partial V}{\partial S},$$

and λ_B and λ_C are constant default intensities from the counterparties, R_B and R_C are the constant recovery rates of the two counterparties, s_F is the funding cost of the entity and M represents the Mark-to-Market value of $\hat{V}$ at default.

We consider two scenarios for the determination of the derivative Mark-to-Market value at default, namely that recovery is either on the total risky value or on the riskless value. Thus, according to the M value, two PDEs are obtained:

- if $M = \hat{V}$, we obtain the nonlinear PDE:

$$\begin{cases} \partial_t \hat{V} + \mathcal{A}_t \hat{V} - r\hat{V} = (1 - R_B)\lambda_B \hat{V}^- + (1 - R_C)\lambda_C \hat{V}^+ + s_F \hat{V}^+ \\ \hat{V}(T, S) = H(S), \end{cases}$$

– if $M = V$, the following linear PDE is obtained:

$$
\begin{cases}
\partial_t \hat{V} + \mathcal{A}_t \hat{V} - (r + \lambda_B + \lambda_C)\hat{V} \\
\qquad\qquad = -(R_B \lambda_B + \lambda_C)V^- - (R_C \lambda_C + \lambda_B)V^+ + s_F V^+ \\
\hat{V}(T, S) = H(S),
\end{cases}
$$

where $H(S)$ represents the pay–off of the derivative. European vanilla call and put options and forwards have been considered.

As we aim to compute the value of adjustment, risk derivative value is written as:

$$
\hat{V} = V + U,
$$

where U is the total value adjustment and the risk-free value, V, satisfies the classical Black–Scholes equation without counterparty risk:

$$
\begin{cases}
\partial_t V + \mathcal{A}_t V - rV = 0 \\
V(T, S) = H(S).
\end{cases}
\tag{1}
$$

Thus, the total value adjustment PDE is obtained:

– if $M = \hat{V}$, we get a final value problem governed by a nonlinear PDE:

$$
\begin{cases}
\partial_t U + \mathcal{A}_t U - rU = (1 - R_B)\lambda_B (V + U)^- \\
\qquad\qquad + (1 - R_C)\lambda_C (V + U)^+ + s_F (V + U)^+ \\
U(T, S) = 0,
\end{cases}
$$

– if $M = V$, an analogous linear problem is deduced:

$$
\begin{cases}
\partial_t U + \mathcal{A}_t U - (r + \lambda_B + \lambda_C)U = (1 - R_B)\lambda_B V^- \\
\qquad\qquad + (1 - R_C)\lambda_C V^+ + s_F V^+ \\
U(T, S) = 0.
\end{cases}
$$

In both cases, variable S belongs to the unbounded domain $[0, +\infty)$, while t lies in $[0, T]$.

By using Feynman–Kac theorem, we obtain the total value adjustment (XVA) in terms of the expected value:

– if $M = \hat{V}$,

$$
U(t, S) = - (1 - R_B) \int_t^T \lambda_B(u) D_r(t, u) \mathbb{E}_t[(V(u, S(u)) + U(u, S(u)))^-] du
$$

$$
- (1 - R_C) \int_t^T \lambda_C(u) D_r(t, u) \mathbb{E}_t[(V(u, S(u)) + U(u, S(u)))^+] du
$$

$$
- \int_t^T s_F(u) D_r(t, u) \mathbb{E}_t[(V(u, S(u)) + U(u, S(u)))^+] du,
\tag{2}
$$

– if $M = V$,

$$U(t, S) = - (1 - R_B) \int_t^T \lambda_B(u) D_{r+\lambda_B+\lambda_C}(t, u) \mathbb{E}_t [(V^-(u, S(u))] du$$

$$- (1 - R_C) \int_t^T \lambda_C(u) D_{r+\lambda_B+\lambda_C}(t, u) \mathbb{E}_t [(V^+(u, S(u))] du$$

$$- \int_t^T s_F(u) D_{r+\lambda_B+\lambda_C}(t, u) \mathbb{E}_t [(V^+(u, S(u))] du. \tag{3}$$

where the three lines of (2) (or (3)), from top to bottom, correspond to DVA, CVA and FCA, respectively.

3   Numerical Methods

In order to solve the previous models, different numerical methods are proposed. In particular, a finite elements method is used for spatial discretization; thus, the truncation to a spatial bounded domain is required and adequate boundary conditions have to be deduced.

We will focus on the nonlinear problem, although similar methods are used in the linear one.

In any case, the change of variable $\tau = T - t$ is considered in order to write the following initial condition problem:

$$\begin{cases} \dfrac{\partial U}{\partial \tau} - \dfrac{\sigma^2}{2} S^2 \dfrac{\partial^2 U}{\partial S^2} - r_R S \dfrac{\partial U}{\partial S} + rU = \\ \qquad = -(1 - R_B)\lambda_B(V + U)^- - (1 - R_C)\lambda_C(V + U)^+ - s_F(V + U)^+ \\ U(0, S) = 0. \end{cases}$$
$$\tag{4}$$

3.1   Characteristics Method

Analogously to other advection–diffusion equations, we use a characteristics method [11] for time discretization. Thus, we consider the material derivative:

$$\frac{DU}{D\tau} = \frac{\partial U}{\partial \tau} + \frac{\partial U}{\partial S} \frac{\partial S}{\partial \tau}$$

so that we can write our equation as:

$$\frac{DU}{D\tau} - \frac{\sigma^2}{2} \frac{\partial}{\partial S} \left(S^2 \frac{\partial U}{\partial S} \right) + rU = $$
$$= -(1 - R_B)\lambda_B(V + U)^- - (1 - R_C)\lambda_C(V + U)^+ - s_F(V + U)^+.$$

The use of characteristics for time discretizations leads to:

$$\frac{U^{n+1} - U^n \circ \chi^n}{\Delta \tau} - \frac{\sigma^2}{2} \frac{\partial}{\partial S} \left(S^2 \frac{\partial U^{n+1}}{\partial S} \right) + rU^{n+1} = -(1 - R_B)\lambda_B(V + U^{n+1})^-$$
$$- (1 - R_C)\lambda_C(V + U^{n+1})^+ - s_F(V + U^{n+1})^+ \tag{5}$$

where $U^n(\cdot) \approx U(\tau^n, \cdot)$ and $\chi^n \equiv \chi(S, \tau^{n+1}; \tau^n)$ satisfies the final value problem:

$$\begin{cases} \dfrac{\partial \chi}{\partial \tau} = (\sigma^2 + \gamma_s - q_s)\chi(\tau) \\ \chi(\tau^{n+1}) = S. \end{cases}$$

The solution of this problem is

$$\chi(S, \tau^{n+1}; \tau^n) = S \exp((r_R - \sigma^2)\Delta\tau),$$

so that we can evaluate $U^n \circ \chi^n$ at each step of (5).

3.2   Fixed Point Scheme

In order to solve the of nonlinear Eq. (5) at each iteration of the characteristics method, we propose a fixed point algorithm. Thus, the global scheme can be written in the following way:

Let $N > 1$, $\varepsilon > 0$, U^0 given.
For $n = 0, 1, 2, \ldots$
    Let $U^{n+1,0} = U^n$
    For $k = 0, 1, 2, \ldots$, $U^{n+1,k+1}$ is computed to satisfy:

$$(1 + r\Delta\tau)\, U^{n+1,k+1} - \frac{\sigma^2 \Delta\tau}{2} \frac{\partial}{\partial S}\left(S^2 \frac{\partial U^{n+1,k+1}}{\partial S}\right) = U^n \circ \chi^n$$

$$- \Delta\tau\left[(1 - R_B)\lambda_B(V^{n+1} + U^{n+1,k})^- \right.$$
$$\left. + (1 - R_C)\lambda_C(V^{n+1} + U^{n+1,k})^+ + s_F(V^{n+1} + U^{n+1,k})^+\right] \quad (6)$$

until $\|U^{n,k+1} - U^{n,k}\| \leq \varepsilon$.

3.3   Boundary Conditions

As previously indicated, we will use finite elements to discretize the previous equations. Thus, we need to truncate the unbounded domain $[0, +\infty)$ into a bounded one.

We will assume $S \in [0, S_\infty]$, where $S_\infty > 0$ is large enough. Let us introduce function f, defined by:

$$f(U, V) = (1 - R_B)\lambda_B(V + U)^- + (1 - R_C)\lambda_C(V + U)^+ + s_F(V + U)^+.$$

The left boundary condition is obtained just by replacing $S = 0$ in (4):

$$\partial_\tau U + rU = -f(U, V);$$

this equation is approximated by an implicit Euler method:

$$U^{n+1}(0) - U^n(0) + \Delta\tau\, r\, U^{n+1}(0) = -\Delta\tau\, f(U^{n+1}(0), V^{n+1}(0)),$$

so that the nonhomogeneous Dirichlet boundary condition is obtained for each step of the global algorithm:

$$U^{n+1,k+1}(0) = \frac{1}{1 + r\Delta\tau}\left(U^n(0) - \Delta\tau\left[(1 - R_B)\lambda_B(V^{n+1}(0) + U^{n+1,k}(0))^-\right.\right.$$
$$\left.\left. + (1 - R_C)\lambda_C(V^{n+1}(0) + U^{n+1,k}(0))^+ + s_F(V^{n+1}(0) + U^{n+1,k}(0))^+\right]\right).$$

In order to deduce the boundary condition at $S = S_\infty$, we multiply Eq. (4) by S^{-2}; thus, when S tends to infinity the following condition is obtained:

$$\lim_{S\to\infty} \frac{\partial^2 U}{\partial S^2} = 0.$$

Then, following [5], we consider a solution of the form

$$U = H_0 + H_1 S,$$

where H_0 and H_1 are constant coefficients. By introducing this expression into each fixed point iteration, two simpler ODEs are obtained and the nonhomogeneous Dirichlet condition is posed:

$$U^{n+1,k+1}(S_\infty) = H_1^{n+1,k+1}S_\infty = \frac{1}{(r + \Delta\tau)}((U^n \circ \chi^n)(S_\infty)$$
$$- \Delta\tau\left[(1 - R_B)\lambda_B(V^{n+1}(S_\infty) + U^{n+1,k}(S_\infty))^-\right.$$
$$+ (1 - R_C)\lambda_C(V^{n+1}(S_\infty) + U^{n+1,k}(S_\infty))^+$$
$$\left. + s_F(V^{n+1}(S_\infty) + U^{n+1,k}(S_\infty))^+\right]). \tag{7}$$

3.4   Finite Elements Method

We can now proceed with the spatial discretization. For this purpose, let us consider the functional spaces:

$$H^1(0, S_\infty) = \{\varphi \in L^2(0, S_\infty)/\frac{\partial\varphi}{\partial S} \in L^2(0, S_\infty)\}$$
$$W = H_0^1(0, S_\infty) = \{\varphi \in H^1(0, S_\infty)/\varphi(t, 0) = 0, \varphi(t, S_\infty) = 0\}.$$

If we multiply both members of (6) by a function $\varphi \in V$ and integrate on $[0, S_\infty]$, the variational formulation consists in finding a function $U^{n+1} \in W$ such that:

$$(1 + r\Delta\tau)\int_0^{S_\infty} U^{n+1}\varphi\, dS - \Delta\tau\int_0^{S_\infty} \frac{\partial}{\partial S}\left(\frac{\sigma^2}{2}S^2\frac{\partial U^{n+1}}{\partial S}\right)\varphi\, dS$$
$$= \int_0^{S_\infty} (U^n \circ \chi^n)(S)\varphi\, dS - \Delta\tau\int_0^{S_\infty} f(U^{n+1}, V^{n+1})\varphi\, dS, \quad \forall\varphi \in W.$$

Since W is an infinite dimension functional space, a finite dimension subspace W_h is built. For this purpose, we consider a uniform finite element mesh. Let $M > 0$ such that

$$h = \frac{S_\infty - S_0}{M + 1} > 0,$$

and $S_j = S_0 + jh$ for $j = 0, \ldots, M + 1$.

Let us define the functional spaces

$$W_h = \{\varphi_h : (0, S_\infty) \to \mathbb{R} \,/\, \varphi_h \in \mathcal{C}(0, S_\infty), \varphi_h|_{[S_j, S_{j+1}]} \in \mathcal{P}_1\}$$
$$W_{h,0} = \{\varphi_h \in W_h \,/\, \varphi_h(0) = 0, \varphi_h(S_\infty) = 0\}.$$

The discrete problem consists in finding $U_h^{n+1} \in W_h$ such that:

$$(1 + r\Delta\tau) \int_0^{S_\infty} U_h^{n+1} \varphi_h \, dS - \Delta\tau \int_0^{S_\infty} \frac{\partial}{\partial S} \left(\frac{\sigma^2}{2} S^2 \frac{\partial U_h^{n+1}}{\partial S} \right) \varphi_h \, dS$$

$$= \int_0^{S_\infty} (U_h^n \circ \chi^n)(S) \varphi_h \, dS - \Delta\tau \int_0^{S_\infty} f(U_h^{n+1}, V^{n+1}) \varphi_h \, dS,$$

forall $\varphi_h \in W_{h,0}$. Applying Green's formula and classical properties of integrals, we obtain:

$$\sum_{j=0}^{M} (1 + r\Delta\tau) \int_{S_j}^{S_{j+1}} U_h^{n+1} \varphi_h dS + \sum_{j=0}^{M} \Delta\tau \int_{S_j}^{S_{j+1}} \left(\frac{\sigma^2}{2} S^2 \frac{\partial U_h^{n+1}}{\partial S} \right) \frac{\partial \varphi_h}{\partial S} dS$$

$$= \sum_{j=0}^{M} \int_{S_j}^{S_{j+1}} (U_h^n \circ \chi^n) \varphi_h dS - \sum_{j=0}^{M} \Delta\tau \int_{S_j}^{S_{j+1}} f(U_h^{n+1}, V^{n+1}) \varphi_h dS, \qquad (8)$$

where U_h^{n+1} and φ_h^{n+1} are polynomials of degree less or equal than one on each interval $[S_j, S_{j+1}]$.

We have thus deduced a system of linear Eq. (8), that can be written as:

$$\left[(1 + r\Delta\tau) A_h^1 + \frac{\Delta\tau}{2} \sigma^2 A_h^2 \right] U_h = b_h^1 - \Delta\tau b_h^2,$$

where matrices A_h^1 and A_h^2 and vectors b_h^1 and b_h^2 are built by conveniently assembling the contributions of each element. Moreover, adequate quadrature formulae are used in order to approximate each integral. More precisely,

$$\left[A_h^j \right]^1 \approx \frac{h}{6} \begin{bmatrix} 2 & 1 \\ 1 & 2 \end{bmatrix}$$

$$\left[A_h^j \right]^2 \approx \frac{1}{h_j} \sum_{i=0}^{2} \omega_i (S_j + h_j x_i)^2 \begin{bmatrix} 1 & -1 \\ -1 & 1 \end{bmatrix}$$

$$\left[b_h^j \right]^1 \approx \frac{S_{j+1} - S_j}{2} (U^n \circ \chi^n) \left(\frac{S_{j+1} + S_j}{2} \right) \begin{pmatrix} 1 \\ 1 \end{pmatrix}$$

$$\left[b_h^j \right]^2 \approx \frac{S_{j+1} - S_j}{2} \begin{pmatrix} f(U^{n+1}(S_j), V^{n+1}(S_j)) \\ f(U^{n+1}(S_{j+1}), V^{n+1}(S_{j+1})) \end{pmatrix},$$

where Simpson, three node Gaussian, midpoint and trapezoidal formulae have been respectively used.

4   Numerical Results

In this section, one problem is simulated and the behaviour of the XVA is analyzed. We first study the error and order of convergence of the applied numerical
methods, for which we take advantage of the analytic solution of the XVA problem in particular cases. If $M = \hat{V}$ and considering funding cost, $s_F = (1-R_B)\lambda_B$,
the analytical solution is:

$$U(t, S) = -(1 - \exp(-((1 - R_B)\lambda_B + (1 - R_C)\lambda_C)(T - t))) \, V(t, S).$$

As we can observe in Table 1, the order of convergence obtained with the discrete
norm $L^\infty((0, T) \times L^2([0, S_\infty]))$ is one.

Table 1. Relative errors in norm $L^\infty((0, T) \times L^2([0, S_\infty]))$, convergence ratios and order.
Example with finite elements scheme. The input parameters are $E = 15$, $S \in [0, 4E]$,
$r = 0.03$, $r_R = 0.015$, $\sigma = 0.25$, $t \in [0, 5]$, $\lambda_B = 0.02$, $\lambda_C = 0.05$, $R_B = 0.4$ and
$R_C = 0.4$

Time step	Space step	Error	R	Order
400	50	0.02232872	-	-
800	100	0.01192059	1.87312280	0.90544548
1600	200	0.00617545	1.93031711	0.94883787
3200	400	0.00315299	1.95860211	0.96982435
6400	800	0.00160323	1.96665313	0.97574253

We show in Figs. 1 and 2 the XVA value as a percentage of the risk-free value,
V. We can see the relevance of the choice of the mark-to-market value at default
(either V or $\hat{V}$), as well as the funding cost.

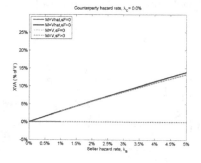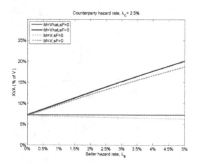

Fig. 1. CVA in the cases $M = \hat{V}$ and $M = V$ for $\lambda_C = 0\%$ and $\lambda_C = 2.5\%$ in $t = 0$

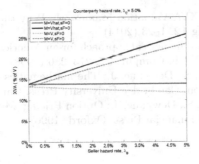

Fig. 2. CVA in the cases $M = \hat{V}$ and $M = V$ for $\lambda_C = 5\%$ in $t = 0$

5   Conclusions

In this work, we have assessed the derivative value taking into account different adjustments in the risk-free value. Models are posed when V depends on one stochastic factor, S.

To solve this problem, different numerical methods have been applied. We have analyzed one example where we can observe the relevance of considering different kinds of risks and funding cost regarding risk-free value, as well as the performance of the numerical methods.

Therefore, we can conclude the significance of taking into account different adjustments. As a result, a new valuation framework of derivatives is created, where both types of financing and counterparty risk must be considered.

References

1. Burgard, C., Kjaer, M.: In the balance. Risk **11**, 1–12 (2011)
2. Burgard, C., Kjaer, M.: PDE representations of derivatives with bilateral counterparty risk and funding costs. J. Credit Risk **7**(3), 1–19 (2011)
3. Burgard, C., Kjaer, M.: Funding costs, funding strategies. J. Credit Risk **11**, 1–17 (2013)
4. Calvo Garrido, M.C.: Mathematical analysis and numerical methods for pricing some pension plans and mortgages. Ph.D. thesis, Universidade da Coruña (2014)
5. Castillo, D., Ferreiro, A.M., García-Rodríguez, J.A., Vázquez, C.: Numerical methods to solve PDE models for pricing business companies in different regimes and implementation in GPUs. Appl. Math. Comput. **219**, 11233–11257 (2013)
6. Gregory, J.: Counterparty Credit Risk and Credit Value Adjustment. Wiley Finance, Hoboken (2012)
7. Hull, J., White, A.: Valuing derivatives: funding value adjustment and fair value. Financ. Anal. J. **3**, 1–27 (2014)
8. Huzinger, C.B.: Reviewing a framework to price a credit risky derivative post the credit crisis. Ph.D. thesis, Witwatersrand University (2014)
9. Mikosh, T.: Elementary Stochastic Calculus with Finance in View. World Scientific Press, Singapur (1998)

10. Ruiz, I.: A Complete XVA Valuation Framework. Why the "Law of One Price" is dead. iRuiz Consulting **12**, 1–23 (2014)
11. Vázquez, C.: An upwind numerical approach for an American and European option pricing model. Appl. Math. Comput. **97**, 273–286 (1998)
12. Wilmott, P., Howison, S., Dewynne, J.: The mathematics of financial derivatives. A students introduction. Cambridge University Press, Cambridge (1996)
13. Wilmott, P., Howison, S., Dewynne, J.: Option Pricing: Mathematical Models and Computation. Oxford Financial Press, Oxford (1996)

Chaotic Dynamics of Structural Members Under Regular Periodic and White Noise Excitations

J. Awrejcewicz[1,2]($\boxtimes$), A.V. Krysko[3,4], I.V. Papkova[5],
N.P. Erofeev[5], and V.A. Krysko[5]

[1] Department of Automation, Biomechanics and Mechatronics,
Lodz University of Technology, 1/15 Stefanowski Str., 90-924 Lodz, Poland
awrejcew@p.lodz.pl
[2] Institute of Vehicles, Warsaw University of Technology,
84 Narbutta Str., 02-524 Warsaw, Poland
[3] Department of Applied Mathematics and Systems Analysis,
Saratov State Technical University, 77 Politechnicheskaya Str.,
410054 Saratov, Russian Federation
[4] Cybernetic Institute, National Research Tomsk Polytechnic University,
30 Lenin Avenue, 634050 Tomsk, Russian Federation
[5] Department of Mathematics and Modeling, Saratov State Technical University,
77 Politechnicheskaya Str., 410054 Saratov, Russian Federation

Abstract. In this work we study PDEs governing beam dynamics under the Timoshenko hypotheses as well as the initial and boundary conditions which are yielded by Hamilton's variational principle. The analysed beam is subjected to both uniform transversal harmonic load and additive white Gaussian noise. The PDEs are reduced to ODEs by means of the finite difference method employing the finite differences of the second-order accuracy, and then they are solved using the 4th and 6th order Runge-Kutta methods. The numerical results are validated with the applied nodes of the beam partition. The so-called charts of the beam vibration types are constructed versus the amplitude and frequency of harmonic excitation as well as the white noise intensity.

The analysis of numerical results is carried out based on a theoretical background on non-linear dynamical systems with the help of time series, phase portraits, Poincaré maps, power spectra, Lyapunov exponents as well as using different wavelet-based studies. A few novel non-linear phenomena are detected, illustrated and discussed.

In particular, it has been detected that a transition from regular to chaotic beam vibrations without noise has been realised by the modified Ruelle-Takens-Newhouse scenario. Furthermore, it has been shown that in the studied cases, the additive white noise action has not qualitatively changed the mentioned route to chaotic dynamics.

Keywords: Non-linear dynamics · Timoshenko beam · Chaos · Bifurcations · White Gauss noise

© Springer International Publishing AG 2017
I. Dimov et al. (Eds.): NAA 2016, LNCS 10187, pp. 25–32, 2017.
DOI: 10.1007/978-3-319-57099-0_3

1   Introduction

In recent years, an increase in interest in understanding and analysis of interaction between deterministic and noisy dynamics of structural members has been observed. In particular, a special case of excitations, i.e. so-called additive white Gauss noise has been employed in numerous studies in various branches of science. In general, the employed white noise plays a crucial role while investigating vibrating structural members. White Gaussian noise is characterised by uniform spectral density and normally distributed amplitude value as well as by an adaptive action on a signal. The used term 'additive' means that the noise is summed with a given signal. On the contrary, multiplicative noise acts on a given signal as a multiplication factor.

On the other hand, either deterministic chaos or white noise is a source of difficulties in monitoring predictions of the structural members behaviour, since it is difficult to detect an exact/reliable stress-strain time-dependent relations while modelling their non-linear effects in a noisy field. Notice that non-linear problems exhibiting chaotic dynamics are highly sensitive to initial conditions, and hence any prediction of their dynamical behaviour can be difficult and often impossible, even when a periodic excitation is applied.

Investigation of an influence of stochastic noisy inputs on non-linear dynamical systems belongs to the most challenging and important research directions in the field of non-linear dynamical systems. Generally, noise implies two qualitatively different effects: it induces (i) a shift of bifurcations scenario in a deterministic system; (ii) occurrence of new types of bifurcation/phase transitions, which are associated only with noise and cannot be exhibited in the case of lack of noise.

In this paper, we show how harmonic uniform input on a beam implies occurrence of transition from regular to chaotic dynamics by the Ruelle-Takens-Newhouse scenario [1]. Notice that a modified Ruelle-Takens-Newhouse scenario exhibited by structural members has been detected and studied in reference [2]. It should be emphasised that in the problems of physics, biology, and chemistry, the noise-induced instability and phase transitions were already studied in 1987 [3]. However, in general, there are only a few results reported in the field of dynamics of structural members.

1.1   Problem Statement

We consider a beam with a rectangular transverse cross section of the length a and the height $2h$ (Fig. 1). The 2D beam is defined as follows: $\Omega = x \in [0, a]; -h \leq z \leq h, 0 \leq t \leq \infty$.

The introduced mathematical model is based on the following hypotheses: (i) although rotational inertial effects of beam elements are taken into account, only the inertial forces responsible for a transversal beam displacement are included; (ii) external forces do not change their direction during beam deformations; (iii) longitudinal beam magnitude is essentially larger than the beam transverse dimensions; (iv) geometric non-linearity is taken in the von Kármán form.

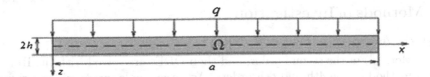

Fig. 1. The investigated beam

On the other hand, the following Timoshenko hypotheses [4] are employed: (i) transverse beam cross sections remain flat, but not perpendicular to the deformed beam axis; (ii) internal components associated with the beam cross section rotations are taken into account.

Employing Hamilton's variational principle, on a basis of the Timoshenko hypotheses, the following PDEs governing beam dynamics are obtained

$$
\begin{cases}
\dfrac{1}{3}\left\{\dfrac{\partial^2 w}{\partial x^2} + \dfrac{\partial^2 \gamma_x}{\partial x^2}\right\} + \dfrac{1}{\lambda^2}\left(L_1(w,u) + \dfrac{1}{2}L_2(w,w) + L_3(w,u) + L_2(w,w)\right) \\[2mm]
\qquad\qquad + \dfrac{1}{\lambda^2}q - \dfrac{\partial^2 w}{\partial t^2} - \varepsilon\dfrac{\partial w}{\partial t} = 0, \\[2mm]
\dfrac{\partial^2 u}{\partial x^2} + L_4(w,w) + p_x - \dfrac{\partial^2 u}{\partial t^2} = 0, \\[2mm]
\dfrac{\partial^2 \gamma_x}{\partial x^2} - 8\lambda^2\left(\dfrac{\partial w}{\partial x} + \gamma_x\right) - \dfrac{\partial^2 \gamma_x}{\partial t^2} = 0,
\end{cases}
\tag{1}
$$

where

$$
L_1(w,u) = \frac{\partial^2 w}{\partial x^2}\frac{\partial u}{\partial x}, \qquad L_2(w,w) = \frac{\partial^2 w}{\partial x^2}\left\{\frac{\partial w}{\partial x}\right\}^2,
$$

$$
L_3(w,u) = \frac{\partial w}{\partial x}\frac{\partial^2 u}{\partial x^2}, \qquad L_4(w,w) = \frac{\partial w}{\partial x}\frac{\partial^2 w}{\partial x^2}.
$$

The non-dimensional variables and parameters (with bars included) follow:

$$
\overline{w} = \frac{w}{(2h)}, \quad \overline{u} = \frac{ua}{(2h)^2}, \quad \overline{x} = \frac{x}{a}, \quad \lambda = \frac{a}{(2h)}, \quad \overline{q} = q\frac{a^4}{(2h)^4 E},
$$

$$
\overline{t} = \frac{t}{\tau}, \quad \tau = \frac{a}{k}, \quad k = \sqrt{\frac{Eg}{\gamma}}, \quad \overline{\varepsilon} = \varepsilon\frac{a}{k}.
$$

In PDEs (1) the bars are already omitted. PDEs (1) should be supplemented by boundary (here rigid clamping) and initial conditions. Both of them have the following form

$$
w(0,t) = w(1,t) = 0; \quad u(0,t) = u(1,t) = 0; \quad \gamma_x(0,t) = \gamma_x(1,t) = 0;
$$

$$
w(x,t)_{|t=0} = u(x,t)_{|t=0} = \gamma(x,t)_{|t=0}
\tag{2}
$$

$$
= \frac{\partial w(x,t)}{\partial t}\bigg|_{t=0} = \frac{\partial u(x,t)}{\partial t}\bigg|_{t=0} = \frac{\partial \gamma_x(x,t)}{\partial t}\bigg|_{t=0} = 0.
$$

2   Methods of Investigation

PDEs (1)–(2) are reduced to a Cauchy problem by means of the FDM (Finite Different Method) of the second order, and then ODEs are solved with the Runge-Kutta method of the 4th and 6th orders. As an example, we consider a rigidly clamped beam subjected to additive white noise and uniformly distributed harmonic load $q = q_0 \sin(\omega_p t)$. The results are obtained for the following fixed parameters: relative beam lengths $\lambda = a/h = 30; 50; 100$, damping factor of a surrounding medium $\varepsilon = 1$, and a time step defined by the Runge principle. The amplitude of excitation was varied within the interval $q_0 \in [0; 22000]$, whereas $\omega_p = 5$ stands for the excitation frequency. The integration interval $x \in [0; 1]$ is divided into 20, 40, and 80 parts. Owing to the Runge principle, the beam division into 40 is parts sufficient. Since the results obtained through both the 4th and 6th Runge-Kutta methods coincide, we have employed the 4th order Runge-Kutta method (time computation is twice less in comparison to the 6th order Runge-Kutta method) [5].

Additive noise has been added to the system in the form of a noisy part with a constant intensity of the form $q_{0isn} = q_{n0}(2.0 \cdot rand()/(RANDMAX + 1.0) - 1.0)$, where q_{n0} is the noise intensity, $rand()$ is a standard function of the programming language, which takes a stochastic integer number from 0 to $RANDMAX$, and $RANDMAX$ is a constant equal to 65535. As a result, the expression (without amplitude) takes a fractional value from the interval $(-1; 1)$.

In Tables 1 and 2, the following numerically obtained characteristics are reported: amplitude of harmonic excitations (a); phase portrait (b); 2D Morlet spectrum (c); and 3D Morlet spectrum (d) for the Timoshenko beams with ($w_0 = 10$) and without the white noise action. As it has been already mentioned, the Ruelle-Takens-Newhouse scenario holds when the white noise is not applied. The action of white noise accelerates a transition from periodic to chaotic vibrations. In Table 1, the already mentioned vibrational characteristics are reported for the Timoshenko beam with and without white noise, keeping the same parameters of the harmonic excitations. Notice that for $q_0 = 5$, beam vibrations are periodic (chaotic) without (with) noise.

An increase in the excitation amplitude up to $q_0 = 50$ yields two-frequency regular vibrations, whereas the same case with the noise action implies chaotic vibrations. A decrease in the amplitudes of the associated frequencies in the vicinity of the excitation frequency is observed. An increase in the excitation amplitude from $q_0 = 2 \cdot 10^3$ up to $q_0 = 8 \cdot 10^3$ implies chaotic vibrations which are similar in the presence and lack of the noise action. Furthermore, the following remarkable rule of the linear dependence of the frequencies a_2, a_3, b_1, b_2, b_3 is observed: $\omega_p - b_1 = a_1 - a_2 = a_2 - a_3 = b_1 - b_2 = a_1 - b_3 = 0.59$. In what follows we study the Timoshenko beam vibrations for $\lambda = a/h = 30; 50; 100$. The obtained numerical results have allowed to construct the vibrational characteristics with and without noise action.

In the reported examples, the vibration characteristics for $\omega_p = 5$ (linear eigenfrequency) versus both amplitude of harmonic excitation and the intensity of noise are reported. It is highly required by industry-oriented researchers to have a

Table 1. Power spectra, phase portraits, 2D and 3D wavelet spectra for different values of q_0 (without noise)

q_0	Power spectrum a)	Phase portrait b)	2D wavelet spectrum c)	3D wavelet spectrum d)
5				
50				
2000				
3000				
7500				
8000				

general picture of the change of the vibrational regime with respect to other frequencies of the excitation, which can be constructed with the help of both standard Fourier and wavelet-type analyses. For this purpose, the control plane $\{q_0, \omega_p\}$ has been divided into $9 \cdot 10^4$ pixels.

Further increase in the pixels number does not yield any changes in charts interpretations, and hence the problem has been limited to that of solving and analysing $9 \cdot 10^4$ problems for which Fourier power spectra, wavelet spectra, phase portraits, Poincaré sections as well as autocorrelation functions need to be studied.

Table 2. Power spectra, phase portraits, 2D and 3D wavelet spectra for different values of q_0 (with noise)

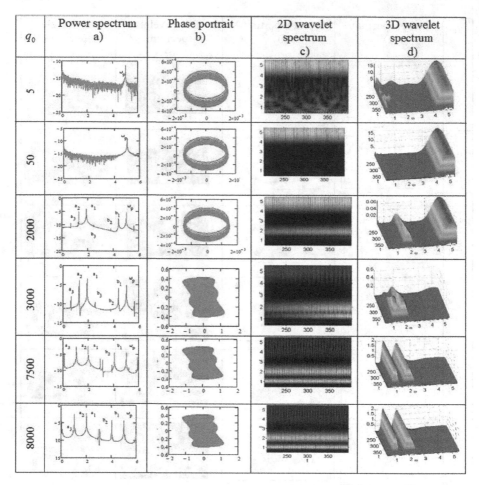

Let us study the vibrational charts with (Fig. 2b) and without (Fig. 2a) the additive white noise action. Since our numerical analysis is carried out on a given frequency interval and a given excitation amplitude, it allows to estimate a boundary between vibrations regimes.

In the case of lack of the noisy field, large zones of periodicity are exhibited which vanish in the neighborhood of high frequencies. Noisy input cancels the quasi-periodic vibrations and chaotic vibrations occur instead.

Further increase in the excitation amplitude awakes two irrational frequencies and quasi-periodic vibrations. Occurrence of a noisy input yields subharmonic vibrations.

It should be emphasised that the action of noise resulted in essential decrease in the periodic zones of vibrations. Zones of quasi-periodicity are negligible.

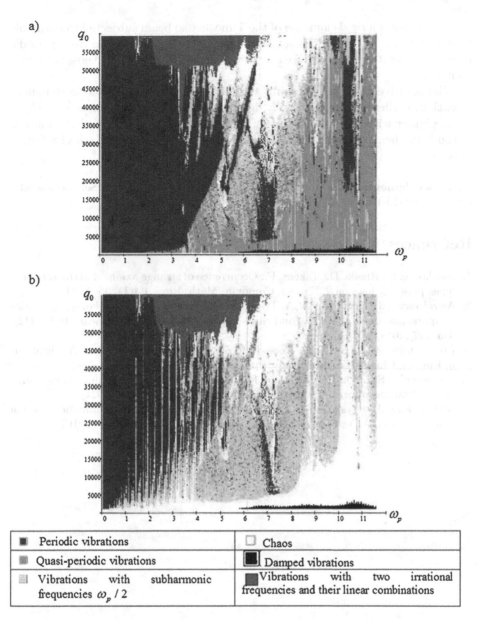

Fig. 2. Vibration charts $\{q_0, \omega_p\}$ for $w_0 = 0$ (a) and $w_0 = 10$ (b)

3   Concluding Remarks

Based on the carried out research, the following fundamental concluding remarks can be formulated:

In the case of a rigid clamping of the Timoshenko beam subjected to harmonic load with ($w_0 = 10$), and without white noise action, a transition from periodic to chaotic vibrations has been realised by the modified Ruelle-Takens-Newhouse scenario.

The additive noise action shifts the vibrating beam into chaotic dynamics essentially earlier than in the purely deterministic case (without noise).

Beginning with a threshold value of the transversal load amplitude, vibrations of the Timoshenko beam are the same in either the presence or lack of additive noise.

Acknowledgments. This work has been supported by the Russian Science Foundation (RSF-16-19-10290).

References

1. Newhouse, S., Ruelle, D., Takens, F.: Occurrence of strange axiom A attractors near quasi periodic flows on T^m, $m>3$. Commun. Math. Phys. **64**(1), 35–40 (1978)
2. Awrejcewicz, J., Krysko Jr., V.A., Papkova, I.V., Krylov, E.Y., Krysko, A.V.: Spatio-temporal non-linear dynamics and chaos in plates and shells. Nonlinear Stud. **21**(2), 313–327 (2014)
3. Horsthemke, W., Lefever, R.: Noise-Induced Transitions. Theory and Applications in Physics, Chemistry and Biology. Springer, Berlin (1984)
4. Timoshenko, S.P.: On the correction for shear of the differential equation for transverse vibrations of prismatic bars. Phil. Mag. **41**(245), 744–746 (1921)
5. Awrejcewicz, J., Krysko, V.A., Papkova, I.V., Krysko, A.V.: Deterministic Chaos in One-Dimensional Continuous Systems. World Scientific, Singapore (2016)

Convergence Order of a Finite Volume Scheme for the Time-Fractional Diffusion Equation

Abdallah Bradji[1](✉) and Jürgen Fuhrmann[2]

[1] Department of Mathematics, Faculty of Sciences,
University of Badji Mokhtar-Annaba, Annaba, Algeria
abdallah-bradji@univ-annaba.org, bradji@cmi.univ-mrs.fr
[2] Weierstrass Institute for Applied Analysis and Stochastics,
Mohrenstr. 39, 10117 Berlin, Germany
fuhrmann@wias-berlin.de
https://www.i2m.univ-amu.fr/~bradji/,
http://www.wias-berlin.de/~fuhrmann

Abstract. We consider the numerical approximation using the discrete gradient developed recently in the SUSHI method of [4] to approximate the time fractional diffusion equation in any space dimension. We derive and prove an error estimate in $\mathbb{L}^\infty(\mathbb{L}^2)$-norm.

Keywords: Time fractional diffusion equation · Non-conforming grid · SUSHI scheme · Implicit scheme · Discrete gradient

1 Problem to Be Solved and Aim of This Paper

We consider the following time fractional diffusion equation:

$$\partial_t^\alpha u(\boldsymbol{x}, t) - \Delta u(\boldsymbol{x}, t) = f(\boldsymbol{x}, t), \ (\boldsymbol{x}, t) \in \Omega \times (0, T), \tag{1}$$

where Ω is an open polygonal bounded subset in $\mathbb{R}^d$, $T > 0$, and f is a given function. Here the operator ∂_t^α is the Caputo derivative defined by:

$$\partial_t^\alpha u(\boldsymbol{x}, t) = \frac{1}{\Gamma(1-\alpha)} \int_0^t (t-s)^{-\alpha} \frac{\partial u(\boldsymbol{x}, s)}{\partial s} ds, \ 0 < \alpha < 1. \tag{2}$$

Initial condition is given by:

$$u(\boldsymbol{x}, 0) = u^0(\boldsymbol{x}), \ \boldsymbol{x} \in \Omega, \tag{3}$$

where u^0 is a given functions defined on Ω.
Homogeneous Dirichlet boundary conditions are given by

$$u(\boldsymbol{x}, t) = 0, \ (\boldsymbol{x}, t) \in \partial\Omega \times (0, T). \tag{4}$$

Fractional differential equations have been successfully used in the modeling of several processes and systems. They are used, for instance, to describe anomalous

© Springer International Publishing AG 2017
I. Dimov et al. (Eds.): NAA 2016, LNCS 10187, pp. 33–45, 2017.
DOI: 10.1007/978-3-319-57099-0_4

transport in disordered semiconductors, penetration of light beam through a turbulent medium, and penetration and acceleration of cosmic ray in the Galaxy. We refer to the monograph [8] where we find many details.

Finite volume methods have been developed to approximate different types of partial differential equations which represent a large class of problems in application. Starting from the standard volume methods which use the so-called *admissible meshes* (cf. [3]), passing by the so-called SUSHI method (cf. [4]), and then a generalization to the framework of Gradient Schemes (cf. [2]), finite volume methods have then witnessed considerable developments in the last few decades. They can be applied in arbitrary geometries and are *locally conservative*, see [3] and the references therein. One of the principles of finite volume methods is the integration of the equation to be solved on the so called *control volumes*. We use then *numerical fluxes* to approximate, using the discrete unknowns, the continuous fluxes over the boundaries of the control volumes, which appear after the integration by parts. A widely used definition of admissible finite volume meshes for viscous conservation laws can be found in [3, Definition 9.1, pp. 762–763]. However, the control volumes in the sense of [3, Definition 9.1] are open polygonal convex sets. In addition to this, for admissibility, the mesh should satisfy an orthogonality condition, that is, there exists a family of points $(x_K)_{K \in \mathscr{T}}$, such that for a given edge σ_{KL}, the line segment $x_K x_L$ is orthogonal to this edge. This condition allows to approximate the fluxes over a given edge using two point difference quotients. The construction of such meshes in general geometries is possible in many interesting cases but still linked to a number of challenging problems [5]. Consequently in many cases it is useful to drop the orthogonality condition and to assume general polyhedral control volumes, where the boundary of each control volume is a finite union of subsets of hyperplanes. Recently [4], a large class of nonconforming meshes are used in the approximation of stationary anisotropic heterogeneous diffusion equations, and some error estimates have been provided. In addition to this, a discrete gradient is developed in [4] and it satisfies the properties of stability and consistency.

The aim of the present paper is to consider a generalization of this approach to the a time fractional diffusion equation. We first construct a numerical scheme using the discrete gradient of [4] and then we derive an error estimate. We shall investigate in a future paper the case of space fractional diffusion equations.

Time fractional partial differential equations have been solved for instance using finite differences and finite elements methods, see for instance [6] and references therein.

The organization of this paper is as follows. Section 2 is devoted to the definition of some function spaces with their norms. In Sect. 3, we describe the discretizations in space and time and we provide the definition of the discrete gradient which will be used in the formulation of the finite volume scheme. In Sect. 4, we formulate our implicit scheme. Section 5 is devoted to state and prove a convergence result for the finite volume scheme presented in Sect. 4.

2   Some Preliminaries

In the investigation of non-stationary problems we shall work with functions which depend on time and have values in Banach spaces. If $\omega(\boldsymbol{x}, t)$ is a function of the space variable $\boldsymbol{x}$ and time t, then it is sometimes suitable to separate these variables and consider ω as a function $\omega(t) = \omega(\cdot, t)$ which for each t under consideration attains a value $\omega(t)$ that is a function of $\boldsymbol{x}$ and belongs to a suitable space of functions depending on $\boldsymbol{x}$.

The convergence of the finite volume scheme we want to present is analyzed using the spaces $\mathscr{C}^m\left([0, T];\ \mathscr{C}^l(\overline{\Omega})\right)$, where m and l are integers, of m-times continuously differentiable mappings of the interval $[0, T]$ with values in $\mathscr{C}^l(\overline{\Omega})$. The space $\mathscr{C}^m\left([0, T];\ \mathscr{C}^l(\overline{\Omega})\right)$ is equipped with the norm

$$
\| u \|_{\mathscr{C}^m\left([0,T];\ \mathscr{C}^l(\overline{\Omega})\right)} = \max_{j \in [\![\, 0, m\,]\!]} \left\{ \sup_{t \in [0,T]} \left\| \frac{d^j u}{dt^j}(t) \right\|_{\mathscr{C}^l(\overline{\Omega})} \right\}, \tag{5}
$$

where $\| \cdot \|_{\mathscr{C}^l(\overline{\Omega})}$ denotes the usual norm of $\mathscr{C}^l(\overline{\Omega})$.

3   Description of the Meshes and the Definition of a Discrete Gradient

We will consider, as discretization in space, the nonconforming mesh introduced in [4]. For the sake of completeness, we recall the general finite volumes mesh given in [4, Definition 2.1, p. 1012].

Definition 1 (Space discretization, cf. [4]). *Let Ω be a polyhedral open bounded subset of $\mathbb{R}^d$, where $d \in \mathbb{N} \setminus \{0\}$, and $\partial\Omega = \overline{\Omega} \setminus \Omega$ its boundary. A discretization of Ω, denoted by $\mathscr{D}$, is defined as the triplet $\mathscr{D} = (\mathscr{M}, \mathscr{E}, \mathscr{P})$, where:*

1. *$\mathscr{M}$ is a finite family of non empty connected open disjoint subsets of Ω (the "control volumes") such that $\overline{\Omega} = \cup_{K \in \mathscr{M}} \overline{K}$. For any $K \in \mathscr{M}$, let $\partial K = \overline{K} \setminus K$ be the boundary of K; let $\mathrm{m}\,(K) > 0$ denote the measure of K and h_K denote the diameter of K.*

2. *$\mathscr{E}$ is a finite family of disjoint subsets of $\overline{\Omega}$ (the "edges" of the mesh), such that, for all $\sigma \in \mathscr{E}$, σ is a non empty open subset of a hyperplane of $\mathbb{R}^d$, whose $(d-1)$-dimensional measure is strictly positive. We also assume that, for all $K \in \mathscr{M}$, there exists a subset $\mathscr{E}_K$ of $\mathscr{E}$ such that $\partial K = \cup_{\sigma \in \mathscr{E}_K} \overline{\sigma}$. For any $\sigma \in \mathscr{E}$, we denote by $\mathscr{M}_\sigma = \{K, \sigma \in \mathscr{E}_K\}$. We then assume that, for any $\sigma \in \mathscr{E}$, either $\mathscr{M}_\sigma$ has exactly one element and then $\sigma \subset \partial\Omega$ (the set of these interfaces, called boundary interfaces, denoted by $\mathscr{E}_{\text{ext}}$) or $\mathscr{M}_\sigma$ has exactly two elements (the set of these interfaces, called interior interfaces, denoted by $\mathscr{E}_{\text{int}}$). For all $\sigma \in \mathscr{E}$, we denote by $\boldsymbol{x}_\sigma$ the barycentre of σ. For all $K \in \mathscr{M}$ and $\sigma \in \mathscr{E}$, we denote by $\boldsymbol{n}_{K,\sigma}$ the unit vector normal to σ outward to K.*

3. $\mathscr{P}$ is a family of points of Ω indexed by $\mathscr{M}$, denoted by $\mathscr{P} = (\boldsymbol{x}_K)_{K \in \mathscr{M}}$, such that for all $K \in \mathscr{M}$, $\boldsymbol{x}_K \in K$ and K is assumed to be $\boldsymbol{x}_K$-star-shaped, which means that for all $\boldsymbol{x} \in K$, the property $[\boldsymbol{x}_K, \boldsymbol{x}] \subset K$ holds. Denoting by $d_{K,\sigma}$ the Euclidean distance between $\boldsymbol{x}_K$ and the hyperplane including σ, one assumes that $d_{K,\sigma} > 0$. We then denote by $\mathscr{D}_{K,\sigma}$ the cone with vertex $\boldsymbol{x}_K$ and basis σ.

The discretization of Ω is then performed using the mesh $\mathscr{D} = (\mathscr{M}, \mathscr{E}, \mathscr{P})$ described in Definition 1, whereas the time discretization is performed with a constant time step $k = \dfrac{T}{M+1}$, where $M \in \mathbb{N}^*$, and we shall denote $t_n = nk$, for $n \in [\![0, M+1]\!]$.

Throughout this paper, the letter C stands for a positive constant independent of the parameters of the space and time discretizations, the order α of the time fractional derivative, and the exact solution u.

We define the discrete space $\mathscr{X}_{\mathscr{D}}$ as the set of all $v = ((v_K)_{K \in \mathscr{M}}, (v_\sigma)_{\sigma \in \mathscr{E}})$, where $v_K, v_\sigma \in \mathbb{R}$ and $\mathscr{X}_{\mathscr{D},0} \subset \mathscr{X}_{\mathscr{D}}$ is the set of all $v \in \mathscr{X}_{\mathscr{D}}$ such that $v_\sigma = 0$ for all $\sigma \in \mathscr{E}_{\text{ext}}$. Let $H_{\mathscr{M}}(\Omega) \subset \mathbb{L}^2(\Omega)$ be the space of functions which are constant on each control volume K of the mesh $\mathscr{M}$.

For all $v \in \mathscr{X}_{\mathscr{D}}$, we denote by $\Pi_{\mathscr{M}} v \in H_{\mathscr{M}}(\Omega)$ the function defined by $\Pi_{\mathscr{M}} v(\boldsymbol{x}) = v_K$, for a.e. $\boldsymbol{x} \in K$, for all $K \in \mathscr{M}$.

In order to analyze the convergence, we need to consider the size of the discretization $\mathscr{D}$ defined by

$$h_{\mathscr{D}} = \sup \{\text{diam}(K), \ K \in \mathscr{M}\}$$

and the regularity of the mesh given by

$$\theta_{\mathscr{D}} = \max \left(\max_{\sigma \in \mathscr{E}_{\text{int}}, K, L \in \mathscr{M}} \frac{d_{K,\sigma}}{d_{L,\sigma}}, \ \max_{K \in \mathscr{M}, \sigma \in \mathscr{E}_K} \frac{h_K}{d_{K,\sigma}} \right). \tag{6}$$

The scheme we want to consider is based on the use of the discrete gradient given in [4]. For $u \in \mathscr{X}_{\mathscr{D}}$, we define, for all $K \in \mathscr{M}$

$$\nabla_{\mathscr{D}} u(\boldsymbol{x}) = \nabla_{K,\sigma} u, \quad \text{a.e. } \boldsymbol{x} \in \mathscr{D}_{K,\sigma}, \tag{7}$$

and

$$\nabla_{K,\sigma} u = \nabla_K u + \left(\frac{\sqrt{d}}{d_{K,\sigma}} (u_\sigma - u_K - \nabla_K u \cdot (\boldsymbol{x}_\sigma - \boldsymbol{x}_K)) \right) \mathbf{n}_{K,\sigma}, \tag{8}$$

where

$$\nabla_K u = \frac{1}{\text{m}(K)} \sum_{\sigma \in \mathscr{E}_K} \text{m}(\sigma) (u_\sigma - u_K) \mathbf{n}_{K,\sigma}. \tag{9}$$

We define now the following bilinear form defined on $\mathscr{X}_{\mathscr{D}} \times \mathscr{X}_{\mathscr{D}}$:

$$\langle u, v \rangle_F = \int_\Omega \nabla_{\mathscr{D}} u(\boldsymbol{x}) \cdot \nabla_{\mathscr{D}} v(\boldsymbol{x}) d\boldsymbol{x}, \quad \forall (u, v) \in \mathscr{X}_{\mathscr{D}} \times \mathscr{X}_{\mathscr{D}}. \tag{10}$$

4   Formulation of a New Implicit Scheme Using the Discrete Gradient (7)–(9)

Taking $t = t_{n+1}$ in Eq. (1) yields that

$$\partial_t^\alpha u(\boldsymbol{x}, t_{n+1}) - \Delta u(\boldsymbol{x}, t_{n+1}) = f(\boldsymbol{x}, t_{n+1}), \ \boldsymbol{x} \in \Omega. \tag{11}$$

To construct a numerical scheme, we first provide a *convenient approximation* for the term $\partial_t^\alpha u(\boldsymbol{x}, t_n)$ which appears in the left hand side (l.h.s.) of (11). Such approximation is given for instance in [7, Lemma 4.1]. For the sake of completeness and clarity, we provide this approximation along with a brief proof for the convergence. The techniques used in this proof are *almost* the same ones used in [7] with some sleight modifications. We emphasize in particular on the constants which appear in the estimates we shall obtain.

Using the definition (2) of the Caputo derivative to write

$$\partial_t^\alpha u(\boldsymbol{x}, t_{n+1}) = \frac{1}{\Gamma(1-\alpha)} \sum_{j=0}^{n} \int_{t_j}^{t_{j+1}} (t_{n+1} - s)^{-\alpha} \frac{\partial u(\boldsymbol{x}, s)}{\partial s} ds$$

$$= \frac{1}{\Gamma(1-\alpha)} \sum_{j=0}^{n} \left(\int_{t_j}^{t_{j+1}} (t_{n+1} - s)^{-\alpha} ds \right) \overline{\partial}^1 u(\boldsymbol{x}, t_{j+1}) + \mathbb{T}_1^{n+1}(\boldsymbol{x}) \tag{12}$$

where $\overline{\partial}^1$ is given by the following discrete first order derivative

$$\overline{\partial}^1 v^{j+1} = \frac{v^{j+1} - v^j}{k} \tag{13}$$

and

$$\mathbb{T}_1^{n+1}(\boldsymbol{x}) = \frac{1}{\Gamma(1-\alpha)} \sum_{j=0}^{n} \int_{t_j}^{t_{j+1}} (t_{n+1} - s)^{-\alpha} \left(\frac{\partial u(\boldsymbol{x}, s)}{\partial s} - \partial^1 u(\boldsymbol{x}, t_{j+1}) \right) ds. \tag{14}$$

The coefficients $\int_{t_j}^{t_{j+1}} (t_{n+1} - s)^{-\alpha} ds$ in (12) can be computed in a simple way

$$\int_{t_j}^{t_{j+1}} (t_{n+1} - s)^{-\alpha} ds = \frac{k^{1-\alpha}}{1-\alpha} \left((n - j + 1)^{1-\alpha} - (n - j)^{1-\alpha} \right). \tag{15}$$

Consequently, the expression (12) can then be written as

$$\partial_t^\alpha u(\boldsymbol{x}, t_{n+1}) = \frac{1}{\Gamma(2-\alpha)} \sum_{j=0}^{n} d_{j,\alpha} \overline{\partial}^\alpha u(\boldsymbol{x}, t_{n-j+1}) + \mathbb{T}_1^{n+1}(\boldsymbol{x}), \tag{16}$$

where $\overline{\partial}^\alpha$ is given by

$$\overline{\partial}^\alpha v^{j+1} = \frac{v^{j+1} - v^j}{k^\alpha} \tag{17}$$

and

$$d_{j,\alpha} = (j+1)^{1-\alpha} - j^{1-\alpha}. \tag{18}$$

Let us now estimate the remainder term $\mathbb{T}_1^{n+1}$ given by (14). To this end, we shall write $\mathbb{T}_1^{n+1}$ in some convenient way and we omit the variable x to simplify the notations. We have thanks to a Taylor expansion with integral form of the remainder

$$u(t_{j+1}) = u(t) + (t_{j+1} - t)u_s(t) + \int_t^{t_{j+1}} (t_{j+1} - r)u_{rr}(r)dr \tag{19}$$

and

$$u(t_j) = u(t) + (t_j - t)u_s(s) + \int_t^{t_j} (t_j - r)u_{rr}(r)dr. \tag{20}$$

Subtracting (20) from (19) yields that

$$u_s(s) - \bar{\partial}^1 u(t_{j+1}) = \frac{1}{k}\left(\int_s^{t_j} (t_j - r)u_{rr}(r)dr - \int_s^{t_{j+1}} (t_{j+1} - r)u_{rr}(r)dr\right). \tag{21}$$

Inserting this expansion in (14) gives

$$\mathbb{T}_1^{n+1} = \frac{1}{k\Gamma(1-\alpha)} \sum_{j=0}^n \int_{t_j}^{t_{j+1}} (t_{n+1} - s)^{-\alpha}\left(\delta_j(s) - \delta_{j+1}(s)\right)ds. \tag{22}$$

where

$$\delta_j(s) = \int_s^{t_j} (t_j - r)u_{rr}(r)dr. \tag{23}$$

Re-ordering the order of the integrals in (22) to get

$$\mathbb{T}_1^{n+1} = \frac{1}{\Gamma(2-\alpha)} \sum_{j=0}^n \int_{t_j}^{t_{j+1}} A^{n+1,j}(r)u_{rr}(r)dr, \tag{24}$$

where the function $r \mapsto A^{n+1,j}(r)$ is defined on $[t_j, t_{j+1}]$ and is given by

$$A^{n+1,j}(r) = B^{n+1}(r) - \frac{1}{k}\left(B^{n+1}(t_{j+1})(r - t_j) + B^{n+1}(t_j)(t_{j+1} - r)\right), \tag{25}$$

where $B^{n+1}(r) = (t_{n+1} - r)^{1-\alpha}$. We have, thanks to Taylor expansions

$$B^{n+1}(t_{j+1}) = B^{n+1}(r) + (t_{j+1} - r)B_r^{n+1}(r) + \int_r^{t_{j+1}} (t_{j+1} - s)B_{ss}^{n+1}(s)ds \tag{26}$$

and

$$B^{n+1}(t_j) = B^{n+1}(r) + (t_j - r)B_r^{n+1}(s) + \int_r^{t_j} (t_j - s)B_{ss}^{n+1}(s)ds. \tag{27}$$

Multiplying (26) by $r - t_j$ and (27) by $t_{j+1} - r$ and summing the results to get

$$B^{n+1}(t_{j+1})(r - t_j) + B^{n+1}(t_j)(t_{j+1} - r) = kB^{n+1}(r)$$

$$+ (r - t_j) \int_r^{t_{j+1}} (t_{j+1} - s)B_{ss}^{n+1}(s)ds + (t_{j+1} - r) \int_r^{t_j} (t_j - s)B_{ss}^{n+1}(s)ds.$$

$$(28)$$

This implies that (recall that $A^{n+1,j}(r)$ is given by (25))

$$A^{n+1,j}(r) = -\frac{r - t_j}{k} \int_r^{t_{j+1}} (t_{j+1} - s)B_{ss}^{n+1}(s)ds - \frac{t_{j+1} - r}{k} \int_r^{t_j} (t_j - s)B_{ss}^{n+1}(s)ds$$

Since $B_{ss}^{n+1}(s) = -(1 - \alpha)\alpha(t_{n+1} - s)^{-\alpha-1} < 0$, one deduces that

$$A^{n+1,j}(r) \geq 0, \quad \forall r \in (t_j, t_{j+1}). \tag{29}$$

This with (24) gives that

$$|\mathbb{T}_1^{n+1}| \leq \frac{\|u\|_{\mathscr{C}^2(0,T;\mathscr{C}(\overline{\Omega}))}}{\Gamma(2 - \alpha)} \sum_{j=0}^n \int_{t_j}^{t_{j+1}} A^{n+1,j}(r)dr. \tag{30}$$

Thanks to the triangle inequality, we have the following bound for $A^{n+1,j}(r)$, for all $r \in (t_j, t_{j+1})$

$$A^{n+1,j}(r) \leq \frac{(1 - \alpha)\alpha(r - t_j)(t_{j+1} - r)}{k}$$

$$\left(\int_r^{t_{j+1}} (t_{n+1} - s)^{-\alpha-1}ds + \int_{t_j}^r (t_{n+1} - s)^{-\alpha-1}ds \right).$$

Integrating the previous inequality over $r \in (t_j, t_{j+1})$ leads to

$$\int_{t_j}^{t_{j+1}} A^{n+1,j}(r)dr \leq \frac{(1 - \alpha)\alpha k^2}{6} \int_{t_j}^{t_{j+1}} (t_{n+1} - s)^{-\alpha-1}ds. \tag{31}$$

Gathering this last inequality with (30) yields

$$|\mathbb{T}_1^{n+1}| \leq \frac{\|u\|_{\mathscr{C}^2(0,T;\mathscr{C}(\overline{\Omega}))}}{\Gamma(2 - \alpha)} \left(\int_{t_n}^{t_{n+1}} A^{n+1,n}(r)dr + \frac{(1 - \alpha)k^{2-\alpha}}{6} \right). \tag{32}$$

But, using (25)

$$\int_{t_n}^{t_{n+1}} A^{n+1,n}(r)dr = \frac{\alpha}{2(2 - \alpha)}k^{2-\alpha}. \tag{33}$$

This with (32) imply

$$|\mathbb{T}_1^{n+1}| \leq \frac{\alpha^2 + 2}{6\Gamma(3 - \alpha)}k^{2-\alpha}\|u\|_{\mathscr{C}^2(0,T;\mathscr{C}(\overline{\Omega}))}. \tag{34}$$

From expansion (16) and estimate (34), we deduce that $\dfrac{1}{\Gamma(2-\alpha)}\displaystyle\sum_{j=0}^{n}d_{j,\alpha}$ $\overline{\partial}^{\alpha}u(t_{n-j+1})$ approximates the fractional time derivative ∂_t^{α} of u at the point t_{n+1} by order $k^{2-\alpha}$. This allows to introduce the following definition:

Definition 2 (Definition of a consistent discrete time fractional derivative of order α). *Let $\overline{\partial}^{\alpha}$ be the discrete operator given by (17) and for any $j \in [\![0, M]\!]$ let $d_{j,\alpha}$ be defined by (18). For a discrete function $(\psi^n)_{n=0}^{M+1} \in \mathbb{R}^{M+2}$, we define the discrete time fractional derivative of order α of the discrete function $(\psi^n)_{n=0}^{M+1}$ as the discrete function $\left(\overline{\partial}_{\mathbf{t}}^{\alpha}\psi^n\right)_{n=1}^{M+1}$ given by:*

$$\overline{\partial}_{\mathbf{t}}^{\alpha}\psi^n = \frac{1}{\Gamma(2-\alpha)}\sum_{j=0}^{n-1}d_{j,\alpha}\overline{\partial}^{\alpha}\psi^{n-j}. \tag{35}$$

After having obtained a suitable approximation for the time fractional derivative ∂_t^{α}, we now able to define a finite volume scheme for problem (1)–(4):

– Approximation of initial condition (3): Find $u_{\mathscr{D}}^0 \in \mathscr{X}_{\mathscr{D},0}$ such that

$$\langle u_{\mathscr{D}}^0, v\rangle_F = -\left(\Delta u^0, \Pi_{\mathscr{M}}v\right)_{\mathbb{L}^2(\Omega)}, \ \forall v \in \mathscr{X}_{\mathscr{D},0}, \tag{36}$$

– Discretization of the time factional diffusion equation: For any $n \in [\![1, M]\!]$, find $u_{\mathscr{D}}^{n+1} \in \mathscr{X}_{\mathscr{D},0}$ such that, for all $v \in \mathscr{X}_{\mathscr{D},0}$ (recall that $\overline{\partial}_{\mathbf{t}}^{\alpha}$ is the time fractional derivative of order α given by (35) of Definition 2)

$$\left(\overline{\partial}_{\mathbf{t}}^{\alpha}\Pi_{\mathscr{M}}u_{\mathscr{D}}^{n+1}, \Pi_{\mathscr{M}}v\right)_{\mathbb{L}^2(\Omega)} + \langle u_{\mathscr{D}}^{n+1}, v\rangle_F = (f(t_{n+1}), \Pi_{\mathscr{M}}v)_{\mathbb{L}^2(\Omega)}, \tag{37}$$

where $(\cdot, \cdot)_{\mathbb{L}^2(\Omega)}$ denotes the $\mathbb{L}^2(\Omega)$-inner product.

The scheme (37) can be written in the following way, since $d_{0,\alpha} = 1$

$$\left(\Pi_{\mathscr{M}}u_{\mathscr{D}}^{n+1}, \Pi_{\mathscr{M}}v\right)_{\mathbb{L}^2(\Omega)} + \gamma\langle u_{\mathscr{D}}^{n+1}, v\rangle_F = \left(\Pi_{\mathscr{M}}u_{\mathscr{D}}^n, \Pi_{\mathscr{M}}v\right)_{\mathbb{L}^2(\Omega)}$$
$$- \sum_{j=1}^{n}d_{j,\alpha}\left(\Pi_{\mathscr{M}}(u_{\mathscr{D}}^{n-j+1} - u_{\mathscr{D}}^{n-j}), \Pi_{\mathscr{M}}v\right)_{\mathbb{L}^2(\Omega)} + \gamma\left(f(t_{n+1}), \Pi_{\mathscr{M}}v\right)_{\mathbb{L}^2(\Omega)}, \tag{38}$$

where

$$\gamma = k^{\alpha}\Gamma(2-\alpha).$$

Re-ordering the sum of the right hand side of (38), we get

$$\left(\Pi_{\mathscr{M}}u_{\mathscr{D}}^{n+1}, \Pi_{\mathscr{M}}v\right)_{\mathbb{L}^2(\Omega)} + \gamma\langle u_{\mathscr{D}}^{n+1}, v\rangle_F = d_{n,\alpha}\left(\Pi_{\mathscr{M}}u_{\mathscr{D}}^0, \Pi_{\mathscr{M}}v\right)_{\mathbb{L}^2(\Omega)}$$
$$+ \gamma\left(f(t_{n+1}), \Pi_{\mathscr{M}}v\right)_{\mathbb{L}^2(\Omega)} + \sum_{j=0}^{n-1}(d_{j,\alpha} - d_{j+1,\alpha})\left(\Pi_{\mathscr{M}}u_{\mathscr{D}}^{n-j}, \Pi_{\mathscr{M}}v\right)_{\mathbb{L}^2(\Omega)}. \tag{39}$$

To compute the solution of the finite volume scheme (36) and (39), we first compute the initial solution $u_{\mathscr{D}}^0$ using (36). Equation (36) can be written in the following matrix form

$$A\xi^0 = \eta^0, \tag{40}$$

where A is a symmetric and positive definite matrix, η^0 is known. Equation (37) leads to the following linear systems, for each time step $n \in [\![0, M]\!]$

$$(M + \gamma A)\xi^{n+1} = \eta^n, \tag{41}$$

where $M + \gamma A$ is a symmetric and positive definite matrix and η^n is known from the previous steps.

5   A Convergence Result for the Finite Volume Scheme (36)–(37)

We now state one of the main results of this contribution, that is the convergence order of the finite volume scheme (36)–(37) (or equivalently (36) and (39)).

Theorem 1 (Error estimates for the finite volume scheme (36) and (39)). *Let Ω be a polyhedral open bounded subset of $\mathbb{R}^d$, where $d \in \mathbb{N} \setminus \{0\}$, and $\partial\Omega = \overline{\Omega} \setminus \Omega$ its boundary. Assume that the solution of (1)–(4) satisfies $u \in \mathscr{C}^2([0, T]; \mathscr{C}^2(\overline{\Omega}))$. Let $k = \frac{T}{M+1}$, with $M \in \mathbb{N}^*$, and denote by $t_n = nk$, for $n \in [\![0, M + 1]\!]$. Let $\mathscr{D} = (\mathscr{M}, \mathscr{E}, \mathscr{P})$ be a discretization in the sense of Definition 1. Assume that $\theta_{\mathscr{D}}$ (given by (6)) satisfies $\theta \geq \theta_{\mathscr{D}}$. Let $\nabla_{\mathscr{D}}$ be the discrete gradient given by (7)–(9) and denote by $\langle \cdot, \cdot \rangle_F$ the bilinear form defined by (10). We define the discrete time fractional derivative of order α, which is denoted by $\bar{\partial}_t^\alpha$, by (35) of Definition 2.*
Then there exists a unique solution $(u_{\mathscr{D}}^n)_{n=0}^{M+1} \in \mathscr{X}_{\mathscr{D},0}^{M+2}$ for the discrete problem (36)–(37) (or equivalently (36) and (39)) and the following $\mathbb{L}^\infty(0, T; \mathbb{L}^2(\Omega))$-estimate holds: for all $n \in [\![0, M + 1]\!]$

$$\|u(t_n) - \Pi_{\mathscr{M}}u_{\mathscr{D}}^n\|_{\mathbb{L}^2(\Omega)} \leq C\mu(k^{2-\alpha} + h_{\mathscr{D}})\|u\|_{\mathscr{C}^2([0,T];\mathscr{C}^2(\overline{\Omega}))}, \tag{42}$$

where

$$\mu = \max\left(\frac{(\alpha^2 + 2)T^\alpha}{6(2 - \alpha)(1 - \alpha)}, \frac{T}{1 - \alpha}, 1\right).$$

Proof. The uniqueness of $(u_{\mathscr{D}}^n)_{n \in [\![0,M+1]\!]}$ satisfying (36) and (39) is a consequence of the facts that matrices A and $M + \gamma A$ are positive definite, see (40) and (41).

To prove (42), we compare the solution $(u_{\mathscr{D}}^n)_{n \in [\![0,M+1]\!]}$ satisfying (36) and (39) with the solution (whose matrix involved is A which is positive definite): for any $n \in [\![0, M + 1]\!]$, find $\bar{u}_{\mathscr{D}}^n \in \mathscr{X}_{\mathscr{D},0}$ such that

$$\langle \bar{u}_{\mathscr{D}}^n, v \rangle_F = -(\Delta u(t_n), \Pi_{\mathscr{M}}v)_{\mathbb{L}^2(\Omega)}, \quad \forall v \in \mathscr{X}_{\mathscr{D},0}. \tag{43}$$

The proof of error estimate (42) can be divided into two steps.

First Step: Comparison Between $u(t_n)$ **and** $\bar{u}_{\mathscr{D}}^n$. The following $\mathbb{L}^\infty(0,T;\, \mathrm{L}^2(\Omega))$ and $\mathscr{W}^{1,\infty}(0,T;\, \mathrm{L}^2(\Omega))$ estimates are proved in [4] (see also [1] for more details):

$$\|u(t_n) - \Pi_{\mathscr{M}}\bar{u}_{\mathscr{D}}^n\|_{\mathrm{L}^2(\Omega)} \le Ch_{\mathscr{D}}\|u\|_{\mathscr{C}([0,T];\mathscr{C}^2(\overline{\Omega}))} \tag{44}$$

and

$$\|\bar{\partial}^1(u(t_n) - \Pi_{\mathscr{M}}\bar{u}_{\mathscr{D}}^n)\|_{\mathrm{L}^2(\Omega)} \le Ch_{\mathscr{D}}\|u\|_{\mathscr{C}^1([0,T];\mathscr{C}^2(\overline{\Omega}))}. \tag{45}$$

Second Step: Comparison Between $u_{\mathscr{D}}^n$ **and** $\bar{u}_{\mathscr{D}}^n$. Replacing n by $n+1$ in (43) and subtracting the result from (37) to get

$$\left(\bar{\partial}_t^{\,\alpha}\Pi_{\mathscr{M}}u_{\mathscr{D}}^{n+1}, \Pi_{\mathscr{M}}v\right)_{\mathrm{L}^2(\Omega)} + \langle\bar{e}_{\mathscr{D}}^{n+1}, v\rangle_F = (f(t_{n+1}) + \Delta u(t_{n+1}), \Pi_{\mathscr{M}}v)_{\mathrm{L}^2(\Omega)} \tag{46}$$

where

$$\bar{e}_{\mathscr{D}}^n = u_{\mathscr{D}}^n - \bar{u}_{\mathscr{D}}^n. \tag{47}$$

Taking $n = 0$ in (43), using the fact that $u(0) = u^0$, and comparing this with (36), we get the property $\bar{u}_{\mathscr{D}}^0 = u_{\mathscr{D}}^0$. This gives

$$\bar{e}_{\mathscr{D}}^0 = 0. \tag{48}$$

Substituting f by $\partial_t^\alpha u - \Delta u$ in (46) and adding $-\left(\bar{\partial}_t^{\,\alpha}\Pi_{\mathscr{M}}\bar{u}_{\mathscr{D}}^{n+1}, \Pi_{\mathscr{M}}v\right)_{\mathrm{L}^2(\Omega)}$ to both sides of the result to get

$$\left(\bar{\partial}_t^{\,\alpha}\Pi_{\mathscr{M}}\bar{e}_{\mathscr{D}}^{n+1}, \Pi_{\mathscr{M}}v\right)_{\mathrm{L}^2(\Omega)} + \langle\bar{e}_{\mathscr{D}}^{n+1}, v\rangle_F = \left(\partial_t^\alpha u(t_{n+1}) - \bar{\partial}_t^{\,\alpha}\Pi_{\mathscr{M}}\bar{u}_{\mathscr{D}}^{n+1}, \Pi_{\mathscr{M}}v\right)_{\mathrm{L}^2(\Omega)}.$$

Using the expression (16) and definition (35), the previous equation implies that

$$\left(\bar{\partial}_t^{\,\alpha}\Pi_{\mathscr{M}}\bar{e}_{\mathscr{D}}^{n+1}, \Pi_{\mathscr{M}}v\right)_{\mathrm{L}^2(\Omega)} + \langle\bar{e}_{\mathscr{D}}^{n+1}, v\rangle_F = \left(\mathbb{T}_3^{n+1}, \Pi_{\mathscr{M}}v\right)_{\mathrm{L}^2(\Omega)}, \tag{49}$$

where $\mathbb{T}_3^{n+1} = \mathbb{T}_1^{n+1} + \mathbb{T}_2^{n+1}$ with $\mathbb{T}_1^{n+1}$ is given by (14) and

$$\mathbb{T}_2^{n+1} = \bar{\partial}_t^{\,\alpha}(u(t_{n+1}) - \Pi_{\mathscr{M}}\bar{u}_{\mathscr{D}}^{n+1}). \tag{50}$$

Taking $v = \bar{e}_{\mathscr{D}}^{n+1}$ in (49) yields that

$$\left(\bar{\partial}^\alpha\Pi_{\mathscr{M}}\bar{e}_{\mathscr{D}}^{n+1}, \Pi_{\mathscr{M}}\bar{e}_{\mathscr{D}}^{n+1}\right)_{\mathrm{L}^2(\Omega)} + \Gamma(2-\alpha)\|\nabla_{\mathscr{D}}\bar{e}_{\mathscr{D}}^{n+1}\|_{\mathrm{L}^2(\Omega)^d}^2$$

$$= -\sum_{j=1}^n d_{j,\alpha}\left(\bar{\partial}^\alpha\Pi_{\mathscr{M}}\bar{e}_{\mathscr{D}}^{n-j+1}, \Pi_{\mathscr{M}}\bar{e}_{\mathscr{D}}^{n+1}\right)_{\mathrm{L}^2(\Omega)} + \Gamma(2-\alpha)\left(\mathbb{T}_3^{n+1}, \Pi_{\mathscr{M}}\bar{e}_h^{n+1}\right)_{\mathrm{L}^2(\Omega)}.$$

This gives, using the definition (17) (recall that $\gamma = k^\alpha\Gamma(2-\alpha)$)

$$\|\Pi_{\mathscr{M}}\bar{e}_{\mathscr{D}}^{n+1}\|_{\mathrm{L}^2(\Omega)}^2 + \gamma\|\nabla_{\mathscr{D}}\bar{e}_{\mathscr{D}}^{n+1}\|_{\mathrm{L}^2(\Omega)^d}^2 = \left(\Pi_{\mathscr{M}}\bar{e}_{\mathscr{D}}^n, \Pi_{\mathscr{M}}\bar{e}_{\mathscr{D}}^{n+1}\right)_{\mathrm{L}^2(\Omega)}$$

$$- \sum_{j=1}^n d_{j,\alpha}\left(\Pi_{\mathscr{M}}(\bar{e}_{\mathscr{D}}^{n-j+1} - \bar{e}_{\mathscr{D}}^{n-j}), \Pi_{\mathscr{M}}\bar{e}_{\mathscr{D}}^{n+1}\right)_{\mathrm{L}^2(\Omega)} + \gamma\left(\mathbb{T}_3^{n+1}, \Pi_{\mathscr{M}}\bar{e}_{\mathscr{D}}^{n+1}\right)_{\mathrm{L}^2(\Omega)}.$$

Re-ordering the sum of the second term on the right hand side of the previous equation and using (48), we get

$$\|\Pi_{\mathscr{M}}\bar{e}_{\mathscr{D}}^{n+1}\|_{\mathrm{L}^2(\Omega)}^2 + \gamma\|\nabla_{\mathscr{D}}\bar{e}_{\mathscr{D}}^{n+1}\|_{\mathrm{L}^2(\Omega)^d}^2 = \sum_{j=0}^{n-1}(d_{j,\alpha} - d_{j+1,\alpha})$$

$$\left(\Pi_{\mathscr{M}}\bar{e}_h^{n-j}, \Pi_{\mathscr{M}}\bar{e}_h^{n+1}\right)_{\mathrm{L}^2(\Omega)} + \gamma\left(\mathbb{T}_3^{n+1}, \Pi_{\mathscr{M}}\bar{e}_{\mathscr{D}}^{n+1}\right)_{\mathrm{L}^2(\Omega)}. \tag{51}$$

The fact that the function $s \mapsto (s+1)^{1-\alpha} - s^{1-\alpha}$ is a decreasing function for $s > 0$ implies that $d_{j,\alpha} > d_{j+1,\alpha}$ (recall that $d_{j,\alpha}$ is given by (18)) and consequently

$$d_{j,\alpha} - d_{j+1,\alpha} > 0. \tag{52}$$

Using the Cauchy Schwarz inequality, (51) implies that

$$\|\Pi_{\mathscr{M}}\bar{e}_h^{n+1}\|_{\mathrm{L}^2(\Omega)}^2 + \gamma\|\nabla_{\mathscr{D}}\bar{e}_{\mathscr{D}}^{n+1}\|_{\mathrm{L}^2(\Omega)^d}^2$$

$$\leq \left(\sum_{j=0}^{n-1}(d_{j,\alpha} - d_{j+1,\alpha})\|\Pi_{\mathscr{M}}\bar{e}_{\mathscr{D}}^{n-j}\|_{\mathrm{L}^2(\Omega)} + \gamma\|\mathbb{T}_3^{n+1}\|_{\mathrm{L}^2(\Omega)}\right)\|\Pi_{\mathscr{M}}\bar{e}_{\mathscr{D}}^{n+1}\|_{\mathrm{L}^2(\Omega)}.$$

$$\tag{53}$$

Which gives

$$\|\Pi_{\mathscr{M}}\bar{e}_h^{n+1}\|_{\mathrm{L}^2(\Omega)} \leq \sum_{j=0}^{n-1}(d_{j,\alpha} - d_{j+1,\alpha})\|\Pi_{\mathscr{M}}\bar{e}_{\mathscr{D}}^{n-j}\|_{\mathrm{L}^2(\Omega)} + \gamma\|\mathbb{T}_3^{n+1}\|_{\mathrm{L}^2(\Omega)}. \tag{54}$$

We prove by mathematical induction that there exists a positive constant denoted by θ such that for all $n \in [\![1, M+1]\!]$, we have $\|\Pi_{\mathscr{M}}\bar{e}_{\mathscr{D}}^n\|_{\mathrm{L}^2(\Omega)} \leq \theta S^n$ where

$$S^n = \max\left\{\|\mathbb{T}_3^j\|_{\mathrm{L}^2(\Omega)};\ j \in [\![1, n]\!]\right\}. \tag{55}$$

Taking $n = 0$ in (54) yields that

$$\|\Pi_{\mathscr{M}}\bar{e}_{\mathscr{D}}^1\|_{\mathrm{L}^2(\Omega)} \leq \gamma\|\mathbb{T}_3^1\|_{\mathrm{L}^2(\Omega)}. \tag{56}$$

Since $\gamma = k^\alpha \Gamma(2-\alpha) < T^\alpha \Gamma(2-\alpha)$, we choose θ such that

$$\theta \geq T^\alpha \Gamma(2-\alpha). \tag{57}$$

Assume now that $\|\Pi_{\mathscr{M}}\bar{e}_{\mathscr{D}}^j\|_{\mathrm{L}^2(\Omega)} \leq \theta S^j$ for all $j \in [\![1, n]\!]$ and prove that $\|\Pi_{\mathscr{M}}\bar{e}_{\mathscr{D}}^{n+1}\|_{\mathrm{L}^2(\Omega)}$ is bounded above by θS^{n+1}. From (54), we have (recall that $d_{j,\alpha} - d_{j+1,\alpha} > 0$ and $d_{0,\alpha} = 1$)

$$\|\Pi_{\mathscr{M}}\bar{e}_{\mathscr{D}}^{n+1}\|_{\mathrm{L}^2(\Omega)} \leq (\theta - (\theta d_{M,\alpha} - \gamma)) S^{n+1}. \tag{58}$$

It suffices then to choose θ such that $\theta d_{M,\alpha} - \gamma \geq 0$, i.e. $\theta \geq \gamma/d_{M,\alpha}$. We have

$$d_{M,\alpha} = g(M+1) - g(M) = (1-\alpha)\xi_M^{-\alpha}, \tag{59}$$

where the function g is given by $g(s) = s^{1-\alpha}$ and $\xi_M \in (M, M+1)$. Expression (59) implies that, since $k = T/(M+1)$

$$d_{M,\alpha} \geq (1-\alpha)T^{-\alpha}k^\alpha. \tag{60}$$

Therefore

$$\frac{\gamma}{d_{M,\alpha}} \leq T^\alpha \Gamma(1-\alpha).$$

From this, (57), and the fact that $\Gamma(2-\alpha) = (1-\alpha)\Gamma(1-\alpha) < \Gamma(1-\alpha)$, we are able to choose $\theta = T^\alpha\Gamma(1-\alpha)$. This leads to

$$\|\Pi_{\mathcal{M}}\bar{e}_{\mathcal{D}}^n\|_{L^2(\Omega)} \leq T^\alpha\Gamma(1-\alpha)S^n, \tag{61}$$

where S^n is given by (55).

On another hand, Gathering the definition (50) of $\mathbb{T}_2^{n+1}$ (recall that $\bar{\partial}_t^\alpha$ is given by (35)), the fact that $\bar{\partial}^\alpha = k^{1-\alpha}\bar{\partial}^1$, and estimate (45) yields that

$$\|\mathbb{T}_2^{n+1}\|_{L^2(\Omega)} \leq \frac{Ch_{\mathcal{D}}\|u\|_{\mathscr{C}^1([0,T];\mathscr{C}^2(\overline{\Omega}))}}{\Gamma(2-\alpha)} \sum_{j=0}^n k^{1-\alpha}d_{j,\alpha}. \tag{62}$$

Which implies, thanks to (15)

$$\|\mathbb{T}_2^{n+1}\|_{L^2(\Omega} \leq \frac{Ch_{\mathcal{D}}\|u\|_{\mathscr{C}^1([0,T];\mathscr{C}^2(\overline{\Omega}))}}{\Gamma(1-\alpha)} \int_0^{t_{n+1}} s^{-\alpha}ds$$

$$\leq \frac{Ch_{\mathcal{D}}\|u\|_{\mathscr{C}^1([0,T];\mathscr{C}^2(\overline{\Omega}))}}{\Gamma(2-\alpha)}T^{1-\alpha}. \tag{63}$$

Gathering estimates (61), (34), and (63) leads to (recall that $\mathbb{T}_3^{n+1} = \mathbb{T}_1^{n+1} + \mathbb{T}_2^{n+1}$)

$$\|\Pi_{\mathcal{M}}\bar{e}_{\mathcal{D}}^n\|_{L^2(\Omega)} \leq \frac{C}{1-\alpha}\max\left(\frac{(\alpha^2+2)T^\alpha}{6(2-\alpha)}, T\right)(k^{2-\alpha} + h_{\mathcal{D}})\|u\|_{\mathscr{C}^2([0,T];\mathscr{C}^2(\overline{\Omega}))}. \tag{64}$$

Gathering (64), (44), and the triangle inequality implies the desired estimate (42).

6   Conclusion

A finite volume scheme is considered to approximate a time fractional partial differential equation in any space dimension and an $\mathbb{L}^\infty(\mathbb{L}^2)$-error estimate is derived and proved.

References

1. Bradji, A., Fuhrmann, J.: Some abstract error estimates of a finite volume scheme for a nonstationary heat equation on general nonconforming multidimensional spatial meshes. Appl. Math. **58**(1), 1–38 (2013)

2. Droniou, J., Eymard, R., Gallouët, T., Herbin, R.: Gradient schemes: a generic framework for the discretization of linear, nonlinear and nonlocal elliptic and parabolic equations. Math. Models Methods Appl. Sci. **23**(13), 2395–2432 (2013)
3. Eymard, R., Gallouët, T., Herbin, R.: Finite volume methods. In: Ciarlet, P.G., Lions, J.L. (eds.) Handbook of Numerical Analysis, vol. VII, pp. 723–1020 (2000)
4. Eymard, R., Gallouët, T., Herbin, R.: Discretization of heterogeneous and anisotropic diffusion problems on general nonconforming meshes SUSHI: a scheme using stabilization and hybrid interfaces. IMA J. Numer. Anal. **30**(4), 1009–1043 (2010)
5. Gärtner, K., Si, H., Fuhrmann, J.: Boundary conforming Delaunay mesh generation. Comput. Math. Math. Phys. **50**, 38–53 (2010)
6. Jina, B., Lazarov, R., Liuc, Y., Zhou, Z.: The Galerkin finite element method for a multi-term time-fractional diffusion equation. J. Comput. Phys. **281**, 825–843 (2015)
7. Sun, Z.-Z., Wu, X.: A fully discrete difference scheme for a diffusion-wave system. Appl. Numer. Math. **56**(2), 193–209 (2006)
8. Uchaikin, V.V.: Fractional Derivatives for Physicists and Engineers. Higher Education Press/Springer, Beijing/Heidelberg (2013)

A 2nd-Order FDM for a 2D Fractional Black-Scholes Equation

W. Chen[1] and S. Wang[2(✉)]

[1] CSIRO Data61, 34 Village Street, Docklands, VIC 3008, Australia
Wen.Chen@csiro.au
[2] Department of Mathematics and Statistics, Curtin University, GPO Box U1987,
Perth 6845, Australia
Song.Wang@curtin.edu.au

Abstract. We develop a finite difference method (FDM) for a 2D fractional Black-Scholes equation arising in the optimal control problem of pricing European options on two assets under two independent geometric Lévy processes. We establish the convergence of the method by showing that the FDM is consistent, stable and monotone. We also show that the truncation error of the FDM is of 2nd order. Numerical experiments demonstrate that the method produces financially meaningful results when used for solving practical problems.

1   Introduction

In this paper we propose a 2nd-order numerical scheme for a 2D fractional Black-Scholes (fBS) equation arising in pricing options with two underlying assets [2], based the schemes in [4] for a 1D fBS equation. We prove that the developed discretization method is consistent, stable and monotone, and thus the solution generated by the numerical method converges to the exact one. Numerical experiments have been performed to demonstrate the order of convergence and usefulness of the scheme.

It is shown in [2] that the value of an option whose underlying asset price follows a geometric Lévy process is governed by a 1D fBS equation. Under the same assumptions, it is easy to show that the value U of a two-asset option (eg. Rainbow or Basket Option) which is written on two stocks whose prices S_1 and S_2 following two independent geometric Lévy processes (with zero correlation coefficient) is determined by the following 2D fBS equation:

$$\mathcal{L}U := -U_t + a_1 U_x + a_2 U_y - b_1[_{-\infty}D_x^\alpha U] - b_2[_{-\infty}D_y^\beta U] + rU = 0 \qquad (1a)$$

for $(x, y, t) \in \mathbb{R}^2 \times [0, T)$, where $x = \ln S_1$, $y = \ln S_2$, $_{-\infty}D_x^\alpha U$ and $_{-\infty}D_y^\beta U$ denote respectively the α-th and β-th derivatives of U in x and y for $\alpha, \beta \in (1, 2)$, $T > 0$ is the terminal time, $r \geq 0$ is the risk-free rate, $\sigma > 0$ is the volatility of the underlying asset prices, and $a_1 = -r - \frac{1}{2}\sigma^\alpha \sec\left(\frac{\alpha\pi}{2}\right)$, $b_1 = a_1 + r$, $a_2 = -r - \frac{1}{2}\sigma^\beta \sec\left(\frac{\beta\pi}{2}\right)$, and $b_2 = a_2 + r$. In computation, the domain $\mathbb{R}^2$ has

© Springer International Publishing AG 2017
I. Dimov et al. (Eds.): NAA 2016, LNCS 10187, pp. 46–57, 2017.
DOI: 10.1007/978-3-319-57099-0_5

to be truncated into $\Omega = (x_{\min}, x_{\max}) \times (y_{\min}, y_{\max})$ satisfying $x_{\min}, y_{\min} < 0$ and $x_{\max}, y_{\max} > 0$. We impose the following boundary and initial conditions

$$U(x, y, t) = U_0(x, y, t), (x, y) \in \partial\Omega, \ U(x, y, T) = U^*(x, y), \qquad (1b)$$

where U_0 and U^*, satisfying $U_0(x, y, T) = U^*(x, y)$ for $(x, y) \in \partial\Omega$, are known functions depending on the types of option and the strike prices K of the options. Using the aforementioned logarithmic forms, it is easy to show that $\lim_{x \to -\infty} U_x = 0$ and $\lim_{y \to -\infty} U_y = 0$ [4]. Thus, when $x_{\min}$ and $y_{\min}$ are sufficiently small, the fractional derivatives in (1a) become, up to a truncation error, the following Caputo's type

$$(x_{\min} D_x^\alpha, y_{\min} D_y^\beta)^\top V = \left(\int_{x_{\min}}^x \frac{V_{xx}(\xi, y, t)}{\Gamma_\alpha \cdot (x - \xi)^{\alpha-1}} d\xi, \int_{y_{\min}}^y \frac{V_{yy}(x, \xi, t)}{\Gamma_\beta \cdot (y - \xi)^{\beta-1}} d\xi \right)^\top,$$

where $\Gamma_u = 1/\Gamma(2 - u)$. In what following we will omit the subscripts $x_{\min}$ and $y_{\min}$ in the above derivative representations. Also, for any $\zeta = (\zeta_1, \zeta_2) \in (0, 1]^2$, we use $\nabla^\zeta U = \left(D_x^{\zeta_1} U, D_y^{\zeta_2} U \right)^\top$ to denote the ζ-th order gradient operator, where the fractional derivatives are of the Caputo type.

2   Solvability

We first reformulate (1a)–(1b) as a variational problem, and then show that the variational problem has a unique solution. Before starting this discussion, we introduce some function spaces. For any $\zeta = (\zeta_1, \zeta_2)$ and $\zeta_1, \zeta_2 \in (0, 1]$, we let $H^\zeta(\Omega) := \left\{ v : v, \nabla^\zeta v \in (L^2(\Omega))^2 \right\}$. Define $|\cdot|_\zeta$ and $\|\cdot\|_\zeta$ by $|v|_\zeta^2 = \|\nabla^\zeta v\|_{L^2(\Omega)}^2$ and $\|u\|_\zeta^2 = \|u\|_{L^2(\Omega)}^2 + |u|_\zeta^2$. Then it is easy to show that $|\cdot|_\zeta$ and $\|\cdot\|_\zeta$ are semi-norm and norm on $H^\zeta(\Omega)$ respectively. It has been shown in [7], that $H^\zeta(\Omega)$ equipped with $\|\cdot\|_\zeta$ is a Sobolev space. We also define the Sobolev space of functions the homogeneous boundary trace by $H_0^\zeta(\Omega) = \{v : v \in H^\zeta(\Omega), v|_{\partial\Omega} = 0\}$.

Without loss of generality, we assume that U_0 defined in (1b) satisfies $U_0 \in H^\gamma(\Omega)$, where $\gamma = (\alpha, \beta)$. Then, under the transformation $V = U_0 - U$, (1a) can be written as the following equation with boundary and payoff conditions:

$$\mathcal{L}V := -V_t - \nabla \cdot (B\nabla^{(\gamma-1)}V - aV) + rV = f, \qquad (2a)$$
$$V = 0 \text{ on } \partial\Omega, \ V = V^*(x, y) := U_0(x, y, T) - U^*(x, y), \qquad (2b)$$

where $a = (a_1, a_2)^\top$, $B = \text{diag}(b_1, b_2)$, $\gamma - 1 := (\alpha-1, \beta-1)$, and $f(x, y, t) = \mathcal{L}U_0$. Using the notation defined above, we pose the following problem:

Problem 1. Find $u(t) \in H_0^{\gamma/2}(\Omega)$, such that, for all $v \in H_0^{\gamma/2}(\Omega)$,

$$\left\langle -\frac{\partial u(t)}{\partial t}, v \right\rangle + A(u(t), v) = (f(t), v)$$

almost everywhere (a.e) in $(0, T)$ satisfying terminal condition (2b) a.e. in Ω, where $A(u, v) = a \langle \nabla u, v \rangle + \langle B\nabla^{(\gamma-1)}u, \nabla v \rangle + r(u, v)$ with $\langle \cdot \rangle$ denoting a duality of a pair of dual spaces.

It is easy to verify that Problem 1 is the variational problem of (2a)–(2b) (cf. [7]). From Lemma 2.1 in [4], we have shown that in the 1D case $A(\cdot,\cdot)$ is coercive and continuous. Using the lemma we now prove that $A(\cdot,\cdot)$ is also coercive and continuous, as given in the following lemma:

Lemma 1. *There exists a positive constant C, such that for any $v, w \in H_0^{\gamma/2}(\Omega)$, and $t \in (0,T)$ a.e. $A(v,v) \geq C\|v\|_{\gamma/2}^2$ and $A(v,w) \leq C\|v\|_{\gamma/2}\|w\|_{\gamma/2}$.*

The proof of this lemma, based on Lemma 2.1 in [4], is trivial and thus omitted. Using this lemma, we have the following result.

Theorem 1. *There exists a unique solution to Problem 1.*

This theorem is a consequence of Lemma 1 and Theorem 1.33 in [9], in which the unique solvability for an abstract variational inequality problem is established. The proof to Theorem 1 is thus omitted here.

3   Discretization

Numerical solution of standard BS equations has been discussed extensively in the open literature [11–14, 19, 21, 23, 24]. However, there is a very limited work available on the numerical solution of spatial fBS equations [4, 10]. Various discretization schemes have been developed for fractional DEs such as those in [8, 15–18]. In this section we will present a 2nd-order scheme for (1a), based on that in [4] for a 1D fBS equation.

For given positive integers M_x and M_y, let Ω be divided into rectangular meshes with nodes (x_i, y_j), $i = 0, .., M_x$, $j = 0, \ldots, M_y$, where $x_i = x_{\min} + ih_1$ and $y_j = y_{\min} + jh_2$ with $h_1 = (x_{\max} - x_{\min})/M_x$ and $h_2 = (y_{\max} - y_{\min})/M_y$. For a positive integer N, let $(0,T)$ be divided into N sub-intervals with the mesh points $t_n = T - n\Delta t, n = 0, 1, \ldots, N$, where $\Delta t = T/N$. The α-th partial derivative can be approximated as follows [4]:

$$D_x^\alpha V(x_i, y_j) \approx \frac{h_1^{-\alpha}}{\Gamma(2-\alpha)} \sum_{k=0}^{i+1} g_k^\alpha V_{i-k+1,j} \tag{3}$$

for any $i \in \{1, 2, \ldots, M_x - 1\}$ and $j \in \{1, 2, \ldots, M_y - 1\}$, where $V_{i-k+1,j}$ is an approximation to $V(x_{i-k+1}, y_j, t)$ and g_k^α's are given by, for $k = 3, 4, \ldots, i+1$,

$$g_0^\alpha = \frac{1}{(2-\alpha)(3-\alpha)}, \quad g_1^\alpha = \frac{2^{3-\alpha}-4}{(2-\alpha)(3-\alpha)}, \quad g_2^\alpha = \frac{3^{3-\alpha}-4\times2^{3-\alpha}+6}{(2-\alpha)(3-\alpha)}, \tag{4}$$

$$g_k^\alpha = g_0^\alpha[(k+1)^{3-\alpha} - 4k^{3-\alpha} + 6(k-1)^{3-\alpha} - 4(k-2)^{3-\alpha} + (k-3)^{3-\alpha}]. \tag{5}$$

Lemma 2. *For any $\alpha \in (1,2)$, the coefficients g_k^α, $k = 0, 1, \ldots, i+1$ satisfy:*

(1) $g_0^\alpha > 0$, $g_1^\alpha < 0$, and $g_k^\alpha > 0$ for $k = 3, 4, 5, \ldots, i+1$,
(2) there exists an $\alpha^* \in (1,2)$ such that $g_2^\alpha < 0$ when $\alpha \in (1, \alpha^*)$ and $g_2^\alpha > 0$ when $\alpha \in (\alpha^*, 2)$, and

(3) $\sum_{k=0}^{i+1} g_k^\alpha < 0$.

The proof of Lemma 2 can be found in [4]. Using (3) and its counterpart for $D_y^\beta V(x_i, y_j)$, we define the following operators:

$$(\delta_x^\alpha, \delta_y^\beta) U_{i,j}^n = \left(\frac{h_1^{-\alpha}}{\Gamma(2-\alpha)} \sum_{k=0}^{i+1} g_k^\alpha U_{i-k+1,j}^n, \frac{h_2^{-\beta}}{\Gamma(2-\beta)} \sum_{k=0}^{j+1} g_k^\beta U_{i,j-k+1}^n \right), \quad (6a)$$

$$\delta_x U_{i,j}^n = \frac{1}{2h_1}(U_{i+1,j}^n - U_{i-1,j}^n), \quad \delta_y U_{i,j}^n = \frac{1}{2h_2}(U_{i,j+1}^n - U_{i,j-1}^n), \quad (6b)$$

where $U_{k,l}^n$ denotes an approximation to $U(x_k, y_l, t_n)$. Using (6a)–(6b), we define the following scheme for (1):

$$\frac{U_{i,j}^{n+1} - U_{i,j}^n}{\Delta t} + \theta \left(a_1 \delta_x U_{i,j}^{n+1} - b_1 \delta_x^\alpha U_{i,j}^{n+1} + a_2 \delta_y U_{i,j}^{n+1} - b_2 \delta_y^\beta U_{i,j}^{n+1} + r U_{i,j}^{n+1} \right)$$

$$+ (1-\theta) \left(a_1 \delta_x U_{i,j}^n - b_1 \delta_x^\alpha U_{i,j}^n + a_2 \delta_y U_{i,j}^n - b_2 \delta_y^\beta U_{i,j}^n + r U_{i,j}^n \right) = 0 \quad (7a)$$

for $i = 1, \ldots, M_x - 1, j = 1, \ldots, M_y - 1$, and $n = 0, \ldots, N - 1$ with $\theta \in [0.5, 1]$. The boundary and payoff conditions are:

$$U_{0,j}^n = U_0(x_0, y_j, t_n), \quad U_{M_x,j}^n = U_0(x_{M_x}, y_j, t_n), \quad U_{i,0}^n = U_0(x_i, y_0, t_n), \quad (7b)$$

$$U_{i,M_y}^n = U_0(x_i, y_{M_y}, t_n), \quad U_{i,j}^N = U^*(x_i, y_j, T_N) \quad (7c)$$

for all feasible (i, j, n), To rewrite (7a) into a matrix form, we let

$$\mathbf{U}^n = (U_{1,1}^n, \ldots, U_{M_x-1,1}^n, U_{1,2}^n, \ldots, U_{M_x-1,1}^n, \ldots, U_{1,M_y-1,1}^n, \ldots U_{M_y-1,M_x-1}^n)^\top.$$

Rearranging (7a), we have

$$(\mathbf{I} + \theta\mathbf{M}) \mathbf{U}^{n+1} = (\mathbf{I} - (1-\theta)\mathbf{M}) \mathbf{U}^n + \mathbf{f}^{n+1-\theta}, \quad (8)$$

where $\mathbf{I}$ is $(M_x - 1)(M_y - 1)$ dimensional identity. The matrix $\mathbf{M}$ is a block matrix which has $(M_y - 1) \times (M_y - 1)$ blocks, and the size of each block matrix is $(M_x - 1) \times (M_x - 1)$.

$$\mathbf{M} = \begin{bmatrix} \mathbf{A}+\mathbf{B}_1 & \mathbf{B}_0 & \mathbf{0} & \cdots & \cdots & \cdots & \mathbf{0} \\ \mathbf{B}_2 & \mathbf{A}+\mathbf{B}_1 & \mathbf{B}_0 & \ddots & & & \mathbf{0} \\ \mathbf{B}_3 & \mathbf{B}_2 & \mathbf{A}+\mathbf{B}_1 & \mathbf{B}_0 & \ddots & & \vdots \\ \vdots & \ddots & \ddots & \ddots & \ddots & \ddots & \vdots \\ \mathbf{B}_{M_y-3} & & \ddots & \mathbf{B}_2 & \mathbf{A}+\mathbf{B}_1 & \mathbf{B}_0 & \mathbf{0} \\ \mathbf{B}_{M_y-2} & & & \mathbf{D}_3 & \mathbf{D}_2 & \mathbf{A}+\mathbf{B}_1 & \mathbf{B}_0 \\ \mathbf{B}_{M_y-1} & \cdots & \cdots & \cdots & \mathbf{B}_3 & \mathbf{B}_2 & \mathbf{A}+\mathbf{B}_1 \end{bmatrix}_{(M_y-1)\times(M_y-1)}, \quad (9)$$

where

$$
A_{ij} = \begin{cases} \mu_1 g_0^\alpha + \eta_1, & j = i+1 \\ \mu_1 g_1^\alpha + \frac{r}{2}\Delta t, & j = i \\ \mu_1 g_2^\alpha - \eta_1, & j = i-1 \\ \mu_1 g_k^\alpha, & j = i-k+1 \\ & k = 3, 4, \dots, i \\ 0, & \text{otherwise} \end{cases}, \quad B_j = \begin{cases} (\mu_2 g_0^\beta + \eta_2)\mathbf{I}_y, & j = 0 \\ (\mu_2 g_1^\beta + \frac{r}{2}\Delta t)\mathbf{I}_y, & j = 1 \\ (\mu_2 g_2^\beta - \eta_2)\mathbf{I}_y, & j = 2 \\ (\mu_2 g_j^\beta)\mathbf{I}_y, & j = 3, 4, \dots \\ & M_y - 1 \\ 0, & \text{otherwise} \end{cases},
$$

$$
\mu_1 = \frac{-b_1 \Delta t}{\Gamma(2-\alpha)h_1^\alpha}, \quad \eta_1 = \frac{a_1 \Delta t}{2h_1}, \quad \mu_2 = \frac{-b_2 \Delta t}{\Gamma(2-\beta)h_2^\beta}, \quad \eta_2 = \frac{a_2 \Delta t}{2h_2}, \tag{10}
$$

and $\mathbf{I}_y$ is the $(M_x - 1) \times (M_x - 1)$ identity matrix. The column vector $\mathbf{f}^{n+1-\theta} = (1 - \theta)\mathbf{f}^n + \theta \mathbf{f}^{n+1}$ is the contribution from the boundary conditions (7b)–(7c), where $\mathbf{f}^n$ and $\mathbf{f}^{n+1}$ consist of contributions of boundary values at t_n and t_{n+1} respectively. In the rest of this paper, we choose $\theta = 0.5$ which is the Crank-Nicolson method with a 2nd-order truncation error.

We comment that though the discretization method is developed for European option pricing problems, the principle developed is applicable to complementarity problems involving the fractional differential operators in (1a) governing American option valuation if a penalty method such as those in [3, 14, 20, 22, 24] is used. We will discuss this in a future paper.

4   Convergence Analysis

In this section, we show that the solution to (7) converges to the viscosity solution to (1). We start the discussion with the following theorem:

Theorem 2. *(Consistency) The finite difference scheme for (7a) is consistent with a truncation error of order $\mathcal{O}(\Delta t^2 + h_1^2 + h_2^2)$ when $\theta = 0.5$.*

Proof. In [4], we have shown that the finite difference scheme for the derivatives in x in (6) have the 2nd-order truncation error $\mathcal{O}(h_1^2)$. By symmetry, the finite difference schemes in y-direction in (6a) and (7a) have the truncation error $\mathcal{O}(h_2^2)$. It is also known that the Crank-Nicolson's scheme used in (7) has the truncation error of order $\mathcal{O}(\Delta t^2)$. Therefore, the discretization scheme (7a) has the truncation error $\mathcal{O}(\Delta t^2 + h_1^2 + h_2^2)$.

Theorem 3. *(Stability) The finite difference scheme defined by (7) is unconditionally stable.*

Proof. we use the semi-discrete Fourier transform to prove the stability of the Crank-Nicolson method with $\theta = 1/2$. From the definition, we see that all the coefficient matrices in (9) are Toeplitz matrices. Thus, each of the terms in (9) can be written as convolution of one the following vectors with a finite support $(\cdots, 0, (\mathbf{U}^k)^\top, 0, \cdots)^\top$ and $(\cdots, 0, (\mathbf{f}^{n+1/2})^\top, 0, \cdots)^\top$ for $k = n$ and $n+1$. Applying the discrete Fourier transform via the semidiscrete Fourier transform pair $U_{i,j}^n = \frac{1}{(2\pi)^2} \int_{-\pi/h_2}^{\pi/h_2} \int_{-\pi/h_1}^{\pi/h_1} e^{i(\xi_1 x_i + \xi_2 y_j)} \hat{U}^n(\xi) d\xi_1 d\xi_2$ and $\hat{U}^n(\xi_1, \xi_2) = h_2 h_1 \sum_{j=-\infty}^{\infty} \sum_{i=-\infty}^{\infty} U_{i,j}^n e^{-i(\xi_1 x_i + \xi_2 y_j)}$ to (8), or equivalently replacing $U_{i,j}^k$ and $f_{i,j}^{n+1/2}$ with $\hat{U}^k e^{(i\xi_1 h_1 + j\xi_2 h_2)i}$ and $\hat{f}^{n+1/2} e^{(i\xi_1 h_1 + j\xi_2 h_2)i}$ with $i = \sqrt{-1}$ for all admissible i, j and $k = n, n+1$, we obtain a system in $\hat{U}^{n+1}$. Solving the transformed system for $\hat{U}^{n+1}$ we have

$$\hat{U}^{n+1} = \frac{2 - \left[\bar{\eta}_1 + \mu_1 \sum_{k=0}^{i+1} g_k^\alpha e^{(1-k)\xi_1 h_1 i} + \bar{\eta}_2 + \mu_2 \sum_{k=0}^{j+1} g_k^\beta e^{(1-k)\xi_2 h_2 i} + r\Delta t\right]}{2 + \left[\bar{\eta}_1 + \mu_1 \sum_{k=0}^{i+1} g_k^\alpha e^{(1-k)\xi_1 h_1 i} + \bar{\eta}_2 + \mu_2 \sum_{k=0}^{j+1} g_k^\beta e^{(1-k)\xi_2 h_2 i} + r\Delta t\right]} \hat{U}^n$$

$$+ \frac{2\Delta t \hat{f}^{n+1/2}}{2 + \left[\bar{\eta}_1 + \mu_1 \sum_{k=0}^{i+1} g_k^\alpha e^{(1-k)\xi_1 h_1 i} + \bar{\eta}_2 + \mu_2 \sum_{k=0}^{j+1} g_k^\beta e^{(1-k)\xi_2 h_2 i} + r\Delta t\right]},$$

where $\bar{\eta}_1 = \eta_1 \left(e^{\xi_1 h_1 i} - e^{-\xi_1 h_1 i}\right)$, $\bar{\eta}_2 = \eta_2 \left(e^{\xi_2 h_2 i} - e^{-\xi_2 h_2 i}\right)$, $\xi_1 \in [-\pi/h_1, \pi/h_1]$, $\xi_2 \in [-\pi/h_2, \pi/h_2]$ and $\mu_1, \mu_2, \eta_1, \eta_2$ are defined in (10). Using Euler's formula, we rewrite the above equality as follows.

$$\hat{U}^{n+1} = \frac{1 - [(A_1 + A_2) + (B_1 + B_2)i]}{1 + [(A_1 + A_2) + (B_1 + B_2)i]} \hat{U}^n + \frac{\Delta t}{1 + [(A_1 + A_2) + (B_1 + B_2)i]} \hat{f}^{n+\frac{1}{2}},$$

where

$$A_1 = \frac{\mu_1}{2} \sum_{k=0}^{i+1} g_k^\alpha \cos((1 - k)\xi_1 h_1) + \frac{r\Delta t}{4},$$

$$B_1 = \frac{\eta_1 \sin(\xi_1 h_1)}{2} + \frac{\mu_1}{2} \sum_{k=0}^{i+1} g_k^\alpha \sin((1 - k)\xi_1 h_1),$$

and A_2 and B_2 are defined by replacing the superscript-subscript pair $(\alpha, 1)$ with $(\beta, 2)$. Taking magnitudes on both sides of the above equation, we have

$$|\hat{U}^{n+1}| = |\hat{U}^n| \sqrt{\frac{(1 - A)^2 + B^2}{(1 + A)^2 + B^2}} + |\hat{f}^{n+\frac{1}{2}}| \frac{\Delta t}{\sqrt{(1 + A)^2 + B^2}}, \qquad (11)$$

where $A = A_1 + A_2$ and $B = B_1 + B_2$. We now show that $\frac{(1-A)^2 + B^2}{(1+A)^2 + B^2} \leq 1$, or $A > 0$. Omitting the superscripts, we have from Item 3 of in Lemma 2 that $-g_1 \geq \sum_{k=0, k \neq 1}^{i+1} g_k$, with $g_k > 0$ when $k > 3$ for all $i > 3$. From the representations of g_k in (4)–(5), we have that $g_0 + g_2 > 0$. In order to estimate A_1 and A_2, we first derive the following estimate

$$\sum_{k=0}^{i+1} g_k \cos((1 - k)\xi h) = g_0 \cos(\xi h) + g_1 \cos 0 + g_2 \cos(-\xi h) + \sum_{k=3}^{i+1} g_k \cos((k - 1)\xi h)$$

$$= g_1 + (g_0 + g_2) \cos(\xi h) + \sum_{k=3}^{i+1} g_k \cos((k - 1)\xi h) \leq \sum_{k=0}^{i+1} g_k \leq 0.$$

Since $\mu_1, \mu_2 < 0$, we have the following estimations

$$\frac{\mu_1}{2} \sum_{k=0}^{i+1} g_k^\alpha \cos\left((1-k)\xi h_1\right) + \frac{r\Delta t}{4} \geq 0, \quad \frac{\mu_2}{2} \sum_{k=0}^{i+1} g_k^\beta \cos\left((1-k)\xi h_2\right) + \frac{r\Delta t}{4} \geq 0.$$

Therefore, $A_1, A_2 \geq 0$ and so $A \geq 0$. Using this result we have from (11) that, for all $\xi_i \in [-\frac{\pi}{h_i}, \frac{\pi}{h_i}]$, $i = 1, 2$,

$$\left|\hat{U}^{n+1}\right| \leq \left|\hat{U}^n\right| + \Delta t\left|\hat{f}^{n+1/2}\right| \leq \left|\hat{U}^{n-1}\right| + \Delta t\left[\left|\hat{f}^{n+1/2}\right| + \left|\hat{f}^{n-1/2}\right|\right]$$

$$\leq \cdots \leq \left|\hat{U}^0\right| + \Delta t \sum_{k=0}^{n}\left|\hat{f}^{k+1/2}\right| \leq \left|\hat{U}^0\right| + \frac{T}{N} \sum_{k=0}^{n}\left|\hat{f}^{k+1/2}\right|.$$

Using Cauchy-Schwarz inequality, we have

$$\left|\hat{U}^{n+1}\right|^2 \leq C\left(\left|\hat{U}^0\right|^2 + \frac{nT^2}{N^2} \sum_{k=0}^{n}\left|\hat{f}^{k+1/2}\right|^2\right) \leq C\left(\left|\hat{U}^0\right|^2 + \frac{1}{N} \sum_{k=0}^{n}\left|\hat{f}^{k+1/2}\right|^2\right)$$

for any $n \leq N - 1$, where C denotes a generic positive constant, independent of n and N, $\hat{U}^{n+1}, \hat{U}^0$ and $\hat{f}^{k+1/2}$ are all functions of $\xi_i \in [-\frac{\pi}{h_i}, \frac{\pi}{h_i}]$ for $i = 1, 2$. For any continuous function W on $\bar{\Omega}$, let $\|W\|_{0,h} = \left(h_2 h_1 \sum_{j=1}^{M_y-1} \sum_{i=1}^{M_x-1} |W_{i,j}|^2\right)^{1/2}$ denote the discrete L^2-norm of W. Using the properties of the discrete Fourier and its inverse transforms (particularly Parseval's equality) we have

$$\|\mathbf{U}^{j+1}\|_{0,h}^2 = \frac{1}{(2\pi)^2} \int_{-\pi/h_2}^{\pi/h_2} \int_{-\pi/h_1}^{\pi/h_1} |\hat{U}^{n+1}|^2 d\xi_1 d\xi_2$$

$$\leq \frac{1}{(2\pi)^2}\left(\int_{-\pi/h_2}^{\pi/h_2} \int_{-\pi/h_1}^{\pi/h_1} |\hat{U}^0|^2 d\xi_1 d\xi_2 + \frac{T}{N} \sum_{k=0}^{n} \int_{-\pi/h_2}^{\pi/h_2} \int_{-\pi/h_1}^{\pi/h_1} |\hat{f}^{k+1/2}|^2 d\xi_1 d\xi_2\right)$$

$$= C\left(\|\mathbf{U}^0\|_{0,h}^2 + \frac{1}{N} \sum_{k=0}^{n} \|\mathbf{f}^{k+1/2}\|_{0,h}^2\right) \leq C\left(\|\mathbf{U}^0\|_{0,h}^2 + \|\mathbf{f}\|_\infty^2\right),$$

where $\mathbf{f} = ((\mathbf{f}^0)^\top, \ldots, (\mathbf{f}^N)^\top)^\top$. Thus, we obtain $\|\mathbf{U}^{j+1}\|_{0,h} \leq C\left(\|\mathbf{U}^0\|_{0,h} + \|\mathbf{f}\|_\infty\right)$. Therefore, the numerical method is unconditionally stable.

We now show that the numerical scheme is monotone.

Theorem 4. *(Monotonicity) The discretization scheme established in (7) is monotone when $\Delta t \leq \frac{2}{r}$.*

Proof. By rearranging the discretizated equation (7a), we define a linear function $F_{i,j}^n$ of U^{n+1} and U^n as follows:

$$2F_{i,j}^{n+1}\left(U_{i,j}^{n+1}, U_{i-1,j}^{n+1}, \ldots, U_{0,j}^{n+1}, U_{i,j+1}^{n+1}, U_{i,j-1}^{n+1}, \ldots, U_{i,0}^{n+1}, U_{i+1,j}^{n}, U_{i,j}^{n},\right.$$

$$\left. U_{i-1,j}^{n}, \ldots, U_{0,j}^{n}, U_{i,j+1}^{n}, U_{i,j-1}^{n}, \ldots, U_{i,0}^{n}\right) := \left[2 + \left(\mu_1 g_1^{\alpha} + \mu_2 g_1^{\beta} + r\Delta t\right)\right] U_{i,j}^{n+1}$$

$$+ \left(\eta_1 + \mu_1 g_0^{\alpha}\right) U_{i+1,j}^{n+1} - \left(\eta_1 - \mu_1 g_2^{\alpha}\right) U_{i-1,j}^{n+1} + \mu_1 \sum_{k=3}^{i+1} g_k^{\alpha} U_{i-k+1,j}^{n+1} + \left(\eta_2 + \mu_2 g_0^{\beta}\right) U_{i,j+1}^{n+1}$$

$$- \left(\eta_2 - \mu_2 g_2^{\beta}\right) U_{i,j-1}^{n+1} + \mu_2 \sum_{k=3}^{j+1} g_k^{\beta} U_{i,j-k+1}^{n+1} - \left[2 - \left(\mu_1 g_1^{\alpha} + \mu_2 g_1^{\beta} + r\Delta t\right)\right] U_{i,j}^{n}$$

$$+ \left(\eta_1 + \mu_1 g_0^{\alpha}\right) U_{i+1,j}^{n} - \left(\eta_1 - \mu_1 g_2^{\alpha}\right) U_{i-1,j}^{n} + \mu_1 \sum_{k=3}^{i+1} g_k^{\alpha} U_{i-k+1,j}^{n}$$

$$+ \left(\eta_2 + \mu_2 g_0^{\beta}\right) U_{i,j+1}^{n} - \left(\eta_2 - \mu_2 g_2^{\beta}\right) U_{i,j-1}^{n} + \mu_2 \sum_{k=3}^{j+1} g_k^{\beta} U_{i,j-k+1}^{n}.$$

We also define the following two functions:

$$F_{i,j,+\varepsilon}^{n+1} = F_{i,j}^{n+1}\left(U_{i,j}^{n+1} + \varepsilon, U_{i+1,j}^{n+1}, U_{i-1,j}^{n+1}, \ldots, U_{0,j}^{n+1}, \ldots, U_{i,j+1}^{n}, U_{i,j-1}^{n}, \ldots, U_{i,0}^{n}\right)$$

$$F_{i,j,-\varepsilon}^{n+1} := F_{i,j}^{n+1}\left(U_{i,j}^{n+1}, U_{i+1,j}^{n+1} + \varepsilon, U_{i-1,j}^{n+1} + \varepsilon, \ldots, U_{0,j}^{n+1} + \varepsilon, \ldots, U_{i,j+1}^{n} + \varepsilon,\right.$$

$$\left. U_{i,j-1}^{n} + \varepsilon, \ldots, U_{i,0}^{n} + \varepsilon\right),$$

where $\varepsilon > 0$. It has been proved in [4] that $\left(\sum_{k=0}^{i+1} g_k^{\alpha}\right) - \frac{1}{2} g_1^{\alpha} > 0$ for $i = 1, 2, \ldots, M_x - 1$. This inequality also holds true for $\{g_k^{\beta}\}$ with $i + 1$ and M_x replaced with $j + 1$ and M_y respectively. We now use this result to prove the monotonicity of $F_{i,j}^{n+1}$. When $\Delta t \leq \frac{2}{r}$, we have from the definition of $F_{i,j}^{n+1}$ that, for any $\varepsilon > 0$ and feasible i and j,

$$F_{i,j,-\varepsilon}^{n+1} = F_{i,j}^{n+1} - \left[1 - \frac{1}{2}\left(\mu_1 g_1^{\alpha} + \mu_2 g_1^{\beta} + r\Delta t\right)\right]\varepsilon + \mu_1(g_0^{\alpha} + g_2^{\alpha})\varepsilon$$

$$+ \mu_1 \sum_{k=3}^{i+1} g_k^{\alpha} \varepsilon + \mu_2 (g_0^{\beta} + g_2^{\beta})\varepsilon + \mu_2 \sum_{k=3}^{j+1} g_k^{\beta} \varepsilon$$

$$\leq F_{i,j}^{n+1} + \mu_1 \left(\sum_{k=0}^{i+1} g_k^{\alpha} - \frac{1}{2} g_1^{\alpha}\right)\varepsilon + \mu_2 \left(\sum_{k=0}^{j+1} g_k^{\beta} - \frac{1}{2} g_1^{\beta}\right)\varepsilon - \left(1 - \frac{1}{2} r\Delta t\right)\varepsilon \leq F_{i,j}^{n+1},$$

since $\mu_1, \mu_2 < 0$. Furthermore, from Lemma 2, we know that $g_1^{\alpha} < 0$ and $g_1^{\beta} < 0$, thus we have

$$F_{i,j,+\varepsilon}^{n+1} = F_{i,j}^{n+1} + \left[1 + \frac{1}{2}\left(\mu_1 g_1^{\alpha} + \mu_2 g_1^{\beta} + r\Delta t\right)\right]\varepsilon > F_{i,j}^{n+1}.$$

Therefore, the scheme is monotone.

Combining Theorems 2, 3 and 4, we have the following convergence result.

Theorem 5. *(Convergence) Let U be the viscosity solution to (1) and $U_{h_1,h_2,\Delta t}$ be the numerical solution to (7) with spatial and time mesh size triple $(h_1, h_2, \Delta t)$. Then, $U_{h_1,h_2,\Delta t}$ converges to U as $(h_1, h_2, \Delta t) \to (0^+, 0^+, 0^+)$.*

In [1] the authors show that any finite difference scheme for a general nonlinear 2nd-order PDE which is locally consistent, stable and monotone generates a solution converging uniformly on a compact subset of $(0, T) \times \mathbb{R}$ to the unique viscosity solution of the PDE. In [5,6], Cont and Tankov extended this result to partial integro-differential equations (PIDEs). Since (1a) is an PIDE, Theorem 5 is a consequence of the results established in [1,5,6] and Theorems 2, 3 and 4.

5   Numerical Experiments

We now apply our method to the following test problem.

Example 1. Call-on-Min and Basket options: Eq. (1), with the system and market parameters $\sigma = 0.25$, $r = 0.05$, $K = 50$, $a_1 = a_2 = 0.384$, $b_1 = b_2 = 0.884$, $x_{\min} = y_{\min} = \ln(0.1)$, $x_{\max} = y_{\max} = \ln(100)$ and $T = 1$. Initial and boundary conditions can be obtained by setting $t = T$, $x = x_{\min}, x_{\max}$ or $y = y_{\min}, y_{\max}$ in the following functions.

Call-on-Min option: $U(x, y, t) = \left[\min(e^x, e^y) - Ke^{-r(T-t)}\right]_+$;

Basket option: $U(x, y, t) = \left[(e^x + e^y)/2 - Ke^{-r(T-t)}\right]_+$.

To solve this problem, we choose a uniform mesh with mesh sizes $\Delta x = \Delta y = \frac{1}{100}$ and $\Delta t = \frac{1}{100}$. The numerical solutions for these options at $t = 0$ from our method are plotted in Fig. 1 in the original independent variable $S_x = e^x$ and $S_y = e^y$. From the figures we see that these numerical solutions are qualitatively correct.

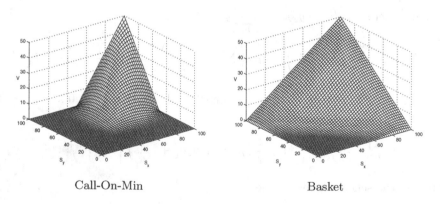

Call-On-Min   Basket

Fig. 1. Computed prices of Call-on-Min and Basket options; $\alpha = \beta = 1.5$

To see the influence of α and β on the option prices, we solve the problem for four different values of $\alpha = \beta = 1.3, 1.5, 1.7, 1.9$ and plot the differences between the numerical solutions of the standard BS equation (i.e., $\alpha = \beta = 2$) and the fractional BS equation and at $t = 0$ for Call-on-Min (Fig. 2) and Basket Option (Fig. 3). From the figures we see that the Call-on-Min and Basket options from fBS model are more expensive than their counterparts of the standard BS model. From these figures, we also see that the call prices increase as α decreases when S_1 and S_2 are greater than some critical values. This phenomenon has been observed in published results for of the 1D fBS equation [2,4] and thus our numerical results for the 2D problem are consistent with those from [2]. The figures also indicate that when α and β approach 2, the numerical solutions to the fBS equation approach to those of the BS equation.

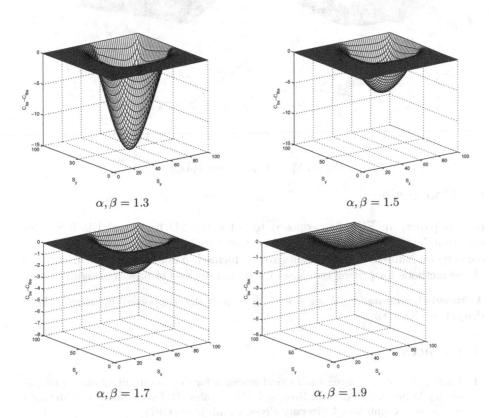

$\alpha, \beta = 1.3$ $\alpha, \beta = 1.5$

$\alpha, \beta = 1.7$ $\alpha, \beta = 1.9$

Fig. 2. $V_{bs} - V_{fbs}$ Call-on-Min option

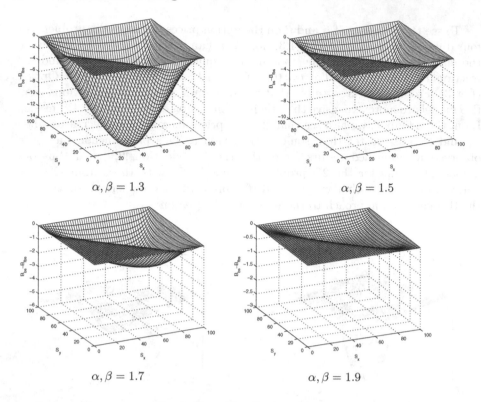

$$\alpha, \beta = 1.3 \qquad\qquad \alpha, \beta = 1.5$$

$$\alpha, \beta = 1.7 \qquad\qquad \alpha, \beta = 1.9$$

Fig. 3. $V_{bs} - V_{fbs}$ Basket option

6   Conclusion

In this paper, an FDM is proposed to solve the 2D fractional Black-Scholes equation. The discretization method is shown to be unconditionally stable and convergent. Numerical experiments are performed to demonstrate the usefulness of the methods for pricing two-asset European options of practical significance.

Acknowledgements. S. Wang's work was partially supported by the AOARD Project #15IOA095.

References

1. Barles, G.: Convergence of numerical schemes for degenerate parabolic equations arising in finance theory. In: Rogers, L.C.G., Talay, D. (eds.) Numerical Methods in Finance. Cambridge University Press, Cambridge (1997)
2. Cartea, A., del-Castillo-Negrete, D.: Fractional diffusion models of option prices in markets with jumps. Phys. A **374**, 749–763 (2007)
3. Chen, W., Wang, S.: A penalty method for a fractional order parabolic variational inequality governing American put option valuation. Comput. Math. Appl. **67**, 77–90 (2014)
4. Chen, W., Wang, S.: A finite difference method for pricing European and American options under a geometric Lévy process. J. Ind. Manag. Optim. **11**, 241–264 (2015)
5. Cont, R., Tankov, P.: Financial Modelling with Jump Processes, vol. 2. Chapman & Hall, New York (2004)

6. Cont, R., Voltchkova, E.: A finite difference scheme for option pricing in jump-diffusion and exponential Lévy models. Ecole Polytechnique Rapport Interne CMAP Working Paper No. 513 (2005)
7. Ervin, V.J., Roop, J.P.: Variational formulation for the stationary fractional advection dispersion equation. Numer. Methods Partial Differ. Equ. **22**, 558–576 (2006)
8. Ervin, V.J., Heuer, N., Roop, J.P.: Numerical approximation of a time dependent, nonlinear, space-fractional diffusion equation. SIAM J. Numer. Anal. **45**, 572–591 (2007)
9. Haslinger, J., Miettinen, M.: Finite Element Method for Hemivariational Inequalities. Kluwer Academic Publisher, Dordrecht-Boston-London (1999)
10. Koleva, M.N., Vulkov, L.G.: Numerical solution of time-fractional BlackScholes equation. Comput. Appl. Math. (2016). doi:10.1007/s40314-016-0330-z
11. Lesmana, D.C., Wang, S.: An upwind finite difference method for a nonlinear Black-Scholes equation governing European option valuation. Appl. Math. Comput. **219**, 8818–8828 (2013)
12. Lesmana, D.C., Wang, S.: Penalty approach to a nonlinear obstacle problem governing American put option valuation under transaction costs. Appl. Math. Comput. **251**, 318–330 (2015)
13. Li, W., Wang, S.: Penalty approach to the HJB equation arising in European stock option pricing with proportional transaction costs. J. Optim. Theory Appl. **143**, 279–293 (2009)
14. Li, W., Wang, S.: Pricing American options under proportional transaction costs using a penalty approach and a finite difference scheme. J. Ind. Manag. Optim. **9**, 365–389 (2013)
15. Lynch, V.E., Carreras, B.A., del-Castillo-Negrete, D., Ferreira-Mejias, K.M., Hicks, H.R.: Numerical methods for the solution of partial differential equations of fractional order. J. Comput. Phys. **192**, 406–421 (2003)
16. Meerschaert, M.M., Tadjeran, C.: Finite difference methods for two-dimensional fractional dispersion equation. J. Comput. Phys. **211**, 249–261 (2006)
17. Oldham, K.B., Spanier, J.: The Fractional Calculus. Academic Press, Cambridge (1974)
18. Tadjeran, C., Meerschaert, M.M.: A second-order accurate numerical method for the two-dimensional fractional diffusion equation. J. Comput. Phys. **220**, 813–823 (2007)
19. Wang, S.: A novel fitted finite volume method for the Black-Scholes equation governing option pricing. IMA J. Numer. Anal. **24**, 699–720 (2004)
20. Wang, S., Yang, X.Q., Teo, K.L.: Power penalty method for a linear complementarity problem arising from American option valuation. J. Optim. Theory Appl. **129**(2), 227–254 (2006)
21. Wang, S., Zhang, S., Fang, Z.: A superconvergent fitted finite volume method for BlackScholes governing European and American option valuation. Numer. Methods Partial Differ. Equ. **31**, 1190–1208 (2015)
22. Wang, S., Zhang, K.: An interior penalty method for a finite-dimensional linear complementarity problem in financial engineering. Optim. Lett. (2016). doi:10.1007/s11590-016-1050-4
23. Wilmott, P., Dewynne, J., Howison, S.: Option Pricing: Mathematical Models and Computation. Oxford Financial Press, Oxford (1993)
24. Zhang, K., Wang, S.: Pricing American bond options using a penalty method. Automatica **48**, 472–479 (2012)

Convergence of Alternant Theta-Method with Applications

István Faragó[1,2(✉)] and Zénó Farkas[2]

[1] Department of Applied Analysis and Computational Mathematics,
Eötvös Loránd University, Pázmány Péter s. 1/C., Budapest 1117, Hungary
[2] MTA-ELTE NumNet Research Group, Pázmány Péter s. 1/C.,
Budapest 1117, Hungary
faragois@cs.elte.hu

Abstract. In this work the alternant theta-method and its application is investigated. We analyze the local approximation error and the convergence of the method on the non-equidistant mesh. We define the order of convergence, as well. The main idea of this approach is the approximation of the solution of the Cauchy problems by using different numerical schemes (implicit, explicit, IMEX, one-step, multi-step etc.) with varying step-sizes. Benefits of such approximations are shown for the problems with non-smooth solutions. We show that the convergence and the error estimation can be given relatively easily for the classical θ-method in case both equidistant and non-equidistant time/space discretizations. We analyze the connection of this approach to the classical discrete Gronwall lemma. We show that the extended discrete Gronwall lemma can be successfully applied to the estimation of the convergence's rate of the alternant θ_i method. We show numerical examples for some non-linear time dependent differential equations, which have non-continuous or strongly oscillated solutions.

Keywords: Alternant theta method · Stability constant · Gronwall lemma

1 Motivation and Basics of the Alternant Theta-Method

Many scientific problems can be described by the initial value problem for first order ordinary differential equations (ODEs) of the form

$$\frac{du}{dt}(t) = f(t, u(t)), \quad t \in (0, T) \tag{1}$$

$$u(0) = u_0. \tag{2}$$

The solution of such problems plays considerable role in the mathematical modelling. As it is known under the global Lipschitz condition, i.e.,

$$|f(t, s_1) - f(t, s_2)| \le L |s_1 - s_2| \quad \text{where } (t, s_1), (t, s_2) \in \mathrm{dom}(f) \tag{3}$$

© Springer International Publishing AG 2017
I. Dimov et al. (Eds.): NAA 2016, LNCS 10187, pp. 58–69, 2017.
DOI: 10.1007/978-3-319-57099-0_6

with the Lipschitz constant $L > 0$, the problem (1) and (2) has unique solution on the domain $\text{dom}(f)$.

However, we cannot define the solutions for the majority of differential equations in analytic form, hence suitable numerical algorithms are needed for accurately approximations. The numerical integration of the problem (1) and (2) under the condition (3) is one of the most typical tasks in the numerical modelling of real-life problems.

One of our aim in this paper is to define some numerical solution at fixed points $t^{\star} \in (0, T)$ to the classical Cauchy problem (1) and (2). Let us consider the sequence of non-equidistant meshes with alternant mesh-sizes h_i of the form

$$\omega_h = \{t_0 = 0, t_n = t_{n-1} + h_n, \quad n = 1, 2, \ldots, N\}$$

and our goal is to define at the mesh-point $t^{\star} = t_N$ a suitable approximation denoted by y_N on each fixed mesh. At this point we remark that a condition should be made for time steps h_i, namely we suppose that there exists a real positive number C such that the inequality

$$h_n \leq C \cdot h, \quad \text{for all } n = 1, 2, \ldots, N-1 \text{ and for all } N = 2, 3, \ldots \quad (4)$$

holds, where $h := max_n\{h_n\}$. Condition (4) means that we demand an uniformly stepwise refinement for every time partition of the interval $(0, T)$.

The most popular and simplest methods for defining the mesh-function $y_h : \omega_h \to \mathbb{R}$ are the so-called one-step schemes, particularly, the theta-method which is notated frequently as θ-method. To proof its convergence there are many works, such as [1]. To define the more general alternant form of theta-method, we change the parameter θ at each step. These varying values are denoted by θ_n. Hence, the alternant-θ-method (in the following called ATM) can be defined as follows.

Definition 1.1. *Let us consider the sequence of parameters $\theta_n \in [0, 1]$, (n = 1, 2, \ldots, N) and the Cauchy problem defined in (1) and (2). The discrete formalization of (1) and (2) by the ATM has the following form*

$$y_n = y_{n-1} + h_n(1 - \theta_n)f(t_{n-1}, y_{n-1}) + h_n\theta_n f(t_n, y_n), \quad n = 1, 2, \ldots, N, \quad (5)$$

$$y_0 = u_0. \quad (6)$$

Hence, $\theta_n \in [0, 1]$ are fixed parameters for all $n = 1, 2, \ldots, N-1$ and it defines for $\theta_n = 0$ (for all n) explicit, otherwise implicit method. This methods are usually used for stiff systems the cases $\theta_n = 0.5$ for all n trapezoidal rule and $\theta_n = 1$ for all n backward Euler are of practical interest, for non-stiff problems we can also consider $\theta_n = 0$ for all n explicit Euler.

The main idea of this approach is the approximation of the solution of the discretized Cauchy problems (5) and (6) by using different numerical schemes (implicit, explicit, IMEX, one-step, multi-step etc.) with varying step-sizes. It has benefits e.g. for the numerical solution of problems with non-smooth solutions.

Let us define the local truncation error for the ATM. We suppose that function f is sufficiently smooth. The local truncation error $\psi_n^{(\theta)}$ for the ATM can be defined as

$$\psi_n^{(\theta)} = (1 - \theta_n)\psi_n^{(0)} + \theta_n\psi_n^{(1)} \tag{7}$$

where

$$\psi_n^{(1)} = -\frac{u(t_n) - u(t_{n-1})}{h_n} + f(t_n, u(t_n)), \qquad \psi_n^{(0)} = -\frac{u(t_n) - u(t_{n-1})}{h_n} + f(t_{n-1}, u(t_{n-1}))$$

and $u(t)$ stands for the solution of the problem (1) and (2).

By expanding $u(t_n)$ and $u(t_{n-1})$ into the Taylor series around the point $t = t_{n-1}$ and $t = t_n$, respectively, we get for error the following

$$\psi_n^{(\theta)} = \frac{h_n}{2}\left[\theta_n u''(t_n) - (1 - \theta_n)u''(t_{n-1})\right] + \frac{h_n^2}{6}\left[(1 - \theta_n)u'''(\xi_n^{(1)}) - \theta_n u'''(\xi_n^{(0)})\right] \tag{8}$$

where $\xi_n^{(1)}$ and $\xi_n^{(0)}$ are some constants defined from the Taylor series. By expanding again $u''(t_n)$ into the Taylor series around the point $t = t_{n-1}$, for the local approximation error we get the following

$$\psi_n^{(\theta)} = C_n^{(1)} h_n + C_n^{(2)} h_n^2 \tag{9}$$

where

$$C_n^{(1)} = \frac{2\theta_n - 1}{2}u''(t_{n-1}), \qquad C_n^{(2)} = \frac{1}{6}(1 - \theta_n)u'''(\xi_n^{(1)}) - \frac{1}{6}\theta_n u'''(\xi_n^{(0)}) + \frac{1}{2}u'''(\xi_n^{(2)}).$$

To define the order of the numerical method we estimate moreover the truncation error using (4) as follows

$$\left|\psi_n^{(\theta)}\right| \leq \left|C_n^{(1)}\right| h_n + \left|C_n^{(2)}\right| h_n^2 \leq \frac{2\theta_n - 1}{2}M_2 Ch + \frac{2}{3}M_3 Ch^2 \tag{10}$$

where

$$M_2 = \max_n |u''(t_n)|, \qquad M_3 = \max_n |u'''(t_n)|.$$

The order of a numerical method is defined by the local truncation error. When $\psi_n^{(\theta)}(h) = O(h^{p+1})$ for all n then the method is called consistent of order p. This means that for both Euler methods the order of consistency is equal to one, while for the "pure" trapezoidal rule the order of consistency is equal to two.

However the consistency in itself does not guarantee the convergence of a numerical method, the stability is also required. Our aim is to give an easy and elementary prove for the convergence of the general ATM. Moreover we give the expression for the stability constant of the method, as well.

The paper is organized as follows. In Sect. 2, we formulate the basic results for the ATM, proving its convergence and stability constant. Section 3 contains the connection between our proof and the classical Gronwall lemma [2] and their applications which let us to generalise the discrete form of Gronwall lemma. In Sect. 4 we show numerical examples for some non-linear time dependent differential equations, which have non-continuous or strongly oscillated solutions to proof the applicability and effectiveness of ATM. Finally we finish the paper with giving some remarks and conclusions.

2   Convergence and Stability Constant

In this section we use a sequence of meshes ω_h and we define the numerical solution at some fixed point $t^* \in (0, T)$ to the Cauchy problem (1) and (2) for the general ATM defined in (5) and (6) with

$$h_1 + h_2 + \cdots + h_N = t^*.$$

The usual way of proving the convergence of the single step θ-method is to show the zero-stability, by using its first characteristic polynomial. The proof of it can be found in [6,7]. However, the proof is complex and needs several auxalary statements.

In the sequel, we give an elementary proof of the convergence by using the following lemma.

Lemma 2.1. *Let $a_n > 0$, $b_n \geq 0$ for all $n = 1, 2, \ldots$, and s_n be such numbers that the inequalities*

$$|s_n| \leq a_n |s_{n-1}| + b_n, \qquad n = 1, 2, \ldots \tag{11}$$

hold. Then the estimate

$$|s_n| \leq \left(\prod_{l=1}^{n} a_l \right) \left[|s_0| + \sum_{j=1}^{n} b_j \left(\prod_{k=1}^{j} \frac{1}{a_k} \right) \right], \qquad n = 1, 2, \ldots \tag{12}$$

is valid.

Proof. By using induction, we can readily verify the statement. Indeed, for $n = 1$ in (12) is clearly valid. Now, under the that (12) holds for $n - 1$, from (11) we have

$$|s_n| \leq \left(\prod_{l=1}^{n} a_l \right) \left[|s_0| + \sum_{j=1}^{n-1} b_j \left(\prod_{k=1}^{j} \frac{1}{a_k} \right) \right] + b_n$$

$$= \left(\prod_{l=1}^{n} a_l \right) \left[|s_0| + \left(\sum_{j=1}^{n} b_j \prod_{k=1}^{j} \frac{1}{a_k} \right) - \left(b_n \prod_{k=1}^{n} \frac{1}{a_k} \right) \right] + b_n$$

$$= \left(\prod_{l=1}^{n} a_l \right) \left[|s_0| + \sum_{j=1}^{n} b_j \left(\prod_{k=1}^{j} \frac{1}{a_k} \right) \right] - b_n + b_n$$

which yields the statement.   □

With our notations the form of inequality (12) can be rewritten into the folowing simpler form:

$$s_n \leq |s_0| \left(\prod_{l=1}^{n} a_l \right) + \sum_{j=1}^{n} b_j \left(\prod_{l=j+1}^{n} a_l \right) \qquad \text{for all } n. \tag{13}$$

Remark 2.2. *If $a_n \equiv a$ and $b_n \equiv b$ for all $n = 1, 2, \ldots$, then, according to (12), we have*

$$|s_n| \leq a^n |s_0| + b \frac{a^n - 1}{a - 1}, \qquad n = 1, 2, \ldots \tag{14}$$

inequality (c.f. [1]).

Conclusion 2.3. *If $a_k \geq 1$, then*

$$\prod_{k=1}^{j} \frac{1}{a_k} \leq \frac{1}{a_j}$$

holds which implies the inequality

$$|s_n| \leq \left(\prod_{l=1}^{n} a_l \right) \left[|s_0| + \sum_{j=1}^{n} b_j \frac{1}{a_j} \right], \qquad n = 1, 2, \ldots \tag{15}$$

Remark 2.4. *If $a_n \equiv a \geq 1$ and $b_n \equiv b$ for all $n = 1, 2, \ldots$, then (15) implies the inequality (c.f. [1])*

$$|s_n| \leq a^n |s_0| + n a^{n-1} b, \qquad n = 0, 1, \ldots$$

In the following, we consider the global error $e_n = u(t_n) - y_n$ at the mesh-point $t = t_n$. We get the recursion in the form

$$e_n = e_{n-1} + h_n \psi_n^{(\theta)} + h_n g_n, \tag{16}$$

where $\psi_n^{(\theta)}$ is defined in (7) and

$$g_n = \theta_n \left[f(t_n, u(t_n)) - f(t_n, y_n) \right] + (1 - \theta_n) \left[f(t_{n-1}, u(t_{n-1})) - f(t_{n-1}, y_{n-1}) \right]. \tag{17}$$

Hence, using the Lipschitz property (3) and the estimations (10) and (16), we get

$$|e_n| \leq |e_{n-1}| + h_n^2 |C_n^{(1)}| + h_n^3 |C_n^{(2)}| + h_n \theta_n L |e_n| + h_n (1 - \theta_n) L |e_{n-1}| \tag{18}$$

for any $n = 1, 2, \ldots, N$. After re-arrangement of (18), we obtain

$$|e_n| \leq 1 + \frac{h_n L}{1 - h_n \theta_n L} |e_{n-1}| + \frac{h_n^2}{1 - h_n \theta_n L} \left(|C_n^{(1)}| + h_n |C_n^{(2)}| \right). \tag{19}$$

Let us denote

$$a_n = 1 + \alpha_n h_n \text{ where } \alpha_n = \frac{L}{1 - h_n \theta_n L} \tag{20}$$

and

$$b_n = \beta_n h_n^2 \text{ where } \beta_n = \frac{|C_n^{(1)}| + h_n |C_n^{(2)}|}{1 - h_n \theta_n L}. \tag{21}$$

Then, by choosing a_n and b_n according to (20) and (21), and using the inequality $1 + x \leq exp(x)$ for $x \geq 0$, Conclusion 2.3 implies the estimate

$$|e_n| \leq \left(\prod_{l=1}^{n} e^{\alpha_l h_l} \right) \cdot \left[|e_0| + \sum_{j=1}^{n} \gamma_j h_j^2 \right] \leq e^{\alpha_{max} \sum_{l=1}^{n} h_l} \left[|e_0| + \gamma_{max} \sum_{j=1}^{n} h_j^2 \right],$$
(22)

where

$$\gamma_j = \frac{\beta_j}{1 + \alpha_j h_j}, \qquad \alpha_{max} = max_j \alpha_j, \qquad \gamma_{max} = max_j \gamma_j. \tag{23}$$

Let $t^* \in (0, T)$ fixed point, $h := \frac{t^*}{n}$. Assuming (4) for all $n = 1, 2, \ldots$ we get

$$\sum_{j=1}^{n} h_j \leq C t^* h. \tag{24}$$

The next step is the estimation of α_{max} and γ_{max}. We have

$$\gamma_{max} \leq \frac{\beta_{max}}{1 + \alpha_{min}} \leq \beta_{max} \leq \frac{\frac{1}{2} max_n (2\theta_n - 1) M_2 + Ch \frac{2}{3} M_3}{1 - Ch \cdot max_n(\theta_n) L},$$

$$\alpha_{max} = \frac{L}{1 - Ch \cdot max_n \theta_n L}. \tag{25}$$

Hence, by the notation $\theta_{max} := max_n(\theta_n)$, (23),(24) and (25) implies the relation

$$|e_n| \leq exp\left(\frac{L}{1 - Ch\theta_{max}} t^* \right) (|e_0| + Cht^* \beta_{max}), \qquad n = 1, 2, \ldots . \tag{26}$$

Because $e_0 = 0$, the relation (26) results in the estimate

$$|e_n| \leq exp\left(\frac{L}{1 - Ch\theta_{max}} t^* \right) Cht^* \beta_{max}. \tag{27}$$

Theorem 2.5. *When n tends to infinity and (4) holds, then the numerical method (5) and (6) is convergent at any fixed point $t^* \in (0, T)$ in first order, assuming that $e_0 = 0$ or $e_0(h) = \mathcal{O}(h)$ is valid.*

Proof. As a consequence of assumption $h \to 0$, hence for the right side of (27) we have

$$\lim_{h \to 0} exp\left(\frac{L}{1 - Ch\theta_{max}} t^* \right) = exp(Lt^*)$$

and

$$\lim_{h \to 0} \beta_{max} = \frac{\vartheta_{max}}{2} M_2, \text{ where } \vartheta_{max} := max_n |2\theta_n - 1|$$

therefore

$$|e_n| \approx exp(Lt^*) \frac{\vartheta_{max}}{2} M_2 = Const \cdot h \tag{28}$$

which yields the first order convergence.   □

Remark 2.6. *Using the statement of Theorem 2.5 for the special case, i.e., to the pure explicit and implicit Euler schemes when for any $n = 1, 2, \ldots$ $\theta_n \equiv 0$ or $\theta_n \equiv 1$, respectively, and $h_n \equiv h$, we re-obtain the classical result.*

Consequence 2.7. *In case of $\theta_n \equiv 0.5$ for $h_n \equiv h$ for any $n = 1, 2, \ldots$ (trapezoidal formula) the convergence order is equal to two, and the stability constant is equal to $(2/3)CM_3$, since in estimation (28) the value of ϑ_{max} is equal to zero.*

3   Connection to the Discrete Version of Gronwall Lemma

In this section our aim is to find the connection between Lemma 2.1 and the classical discrete Gronwall lemma [2]. This form of discrete Gronwall lemma has many generalization and its literature is abundant. From these we mention some of them.

A typical form of discrete Gronwall lemma can be described without the initial term in the conditions. In classical form it can be described as follows. Let $a_n > 0$ and $b_n \geq 0$. The inequality

$$s_n \leq \sum_{j=1}^{n-1} a_j s_{j-1} + b_n \tag{29}$$

implies the relation

$$s_n \leq b^\star \prod_{j=1}^{n} (1 + a_j) \text{ where } b^\star := max_{1 \leq j \leq n} b_j. \tag{30}$$

According to Clark [3] the value $b^\star$ can be reduced to $min_j b_j$ for appropriate j indexes.

A generalization of (30) was developed by Holte [4]. The inequality (29) implies

$$s_n \leq \sum_{k=1}^{n} b_k a_k \prod_{j=k+1}^{n} (1 + a_k) + b_n. \tag{31}$$

The more general form of (31) was formalized by Pachpatte et al. [5]. The inequality

$$s_n \leq a_n \sum_{j=1}^{n-1} r_j s_j + b_n \tag{32}$$

implies the relation

$$s_n \leq a_n \sum_{k=0}^{n-1} r_k b_k \left(\prod_{j=k+1}^{n-1} (1 + r_j b_j) \right) + b_n. \tag{33}$$

where the sequence r_j is non-negative. The problem with these results is that they cannot be applied successfully to the convergence estimation in the above numerical methods. Namely, using the notations of (20) in (30), we get

$$\prod_{j=1}^{n}(1+a_j) = \prod_{j=1}^{n}(2+\alpha_j h_j)$$

which is not bounded if n tends to infinity. (We note that it is not surprising because the error of initial value (e_0 defined in (16)) can be arbitrarily big in this setting.)

Our aim is to apply the above inequalities with the following choice

$$b_n = \frac{h_n r_n}{1-\theta\lambda_n h_n}, \qquad a_n = \frac{1+(1-\theta)\lambda_{n-1}h_n}{1-\theta\lambda_n h_n} \tag{34}$$

where r_n and λ_n are non-negative sequences and inequalities

$$1-\theta\lambda_n h_n > 0, \qquad 1+(1-\theta)\lambda_{n-1}h_n > 0 \tag{35}$$

hold for any n.

However, this setting of coefficients isn't applicable because in the ATM the parameters θ_n usually have different values for the different n. Therefore, we can conclude, that Lemma 2.1 is some other generalization of the discrete Gronwall lemma.

Remark 3.1. *The condition $a_n > 0$ for any $n = 1, 2, \ldots$ in Lemma 2.1 can be relaxed to the condition $a_n \geq 0$, namely if inequalities (11) hold, then the estimate*

$$|s_n| \leq \prod_{l=1}^{n}|s_0| + \sum_{j=1}^{n} b_j \prod_{k=j+1}^{n} a_k \tag{36}$$

is valid which is equivalent to (12) in case of $a_n \neq 0$ for all $n = 1, 2, \ldots$. The proof can be made by induction.

4   Test Examples

In this section, we present some test examples for ordinary differential systems in order to illustrate the effectiveness of ATM.

4.1   Exponential Increasing and Slowing-Down

Consider the following non-linear ordinary differential equation with initial condition:

$$x'(t) = e^{-t} - x(t) + c$$

with the initial condition

$$x(0) = -c$$

where $c \in \mathbb{R}$ is some given constant. The exact solution of the initial value problem at t is:

$$x(t) = (t - 2c)e^{-t} + c.$$

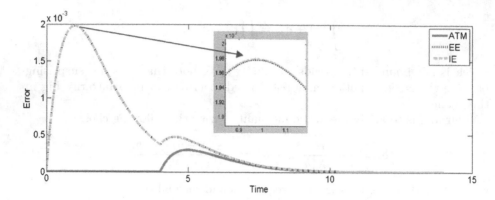

Fig. 1. Error functions in time of EE, IE and ATM methods. Used parameters are $\theta_1 = 0.5$, $\theta_2 = 0$, $h_1 = 0.001$ and $h_2 = 0.01$ on $[0, 4]$ and $[4, 8]$ intervals by each methods, respectively, and $c = 10$. (Color figure online)

This function decreases exponentially at the beginning of considered time interval and later settles. For this reason we approximate the exponential part with $\theta_1 = 0.5$ (trapezoid method) and the second part with $\theta_2 = 0$ (explicit Euler method). Figure 1 shows the error of approximated results in time by using explicit Euler (EE) (blue), implicit Euler (IE) (green) and the alternant θ_i method (ATM) (red). For time step size $h_1 = 0.001$ and $h_2 = 0.01$ are chosen on the $[0, 4]$ and $[4, 8]$ intervals by each methods, respectively. We assume furthermore that $c = 10$.

Next to same settings we investigated the results of ATM by choosing $\theta_1 = 1$ and $\theta_2 = 0$ parameters, and vice versa, namely we tested the results of alternated implicit-explicit Euler (IE-EE) and alternated explicit-implicit Euler (EE-IE). Figure 2 represents the results.

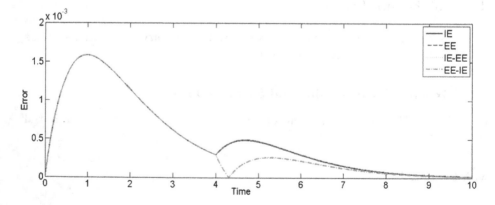

Fig. 2. Error functions in time of EE, IE, IE-EE and EE-IE methods. Used parameters are $h_1 = 0.001$ and $h_2 = 0.01$ on $[0, 4]$ and $[4, 8]$ intervals by each methods, respectively, and $c = 10$. (Color figure online)

We can conclude according to Fig. 2, that the alternated IE-EE and EE-IE methods give more accurate approximation as the simple IE or EE methods, especially on the second half of the analyzed time interval. This example illustrates well, that feasibly chosen θ_i parameters could increase the order of the numerical scheme significantly.

4.2 Discontinuous Function

Consider the following non-linear discontinuous ordinary differential equation:

$$x'(t) = \begin{cases} \frac{x(t)-c}{t} + t \cdot cos(ct) \cdot c & \text{if } 0 < t < c \\ \frac{1}{2x(t)} & \text{if } c < t \end{cases}$$

with the initial condition

$$x(0) = c$$

where $c \in \mathbb{R}$ is an appropriate constant. As one can easily check, the exact solution of the initial value problem at t is:

$$x(t) = \begin{cases} t \cdot sin(ct) + c & \text{if } 0 < t < c \\ t^{\frac{1}{2}} & \text{if } c < t. \end{cases}$$

This function waves at the beginning of considered time interval and later settles. For this reason we approximate the wave part with $\theta_1 = 0.5$ (trapezoid method) and the second part with $\theta_2 = 1$ (implicit Euler method). Figure 3 shows the error of approximated results in time by using explicit Euler (EE) (blue), implicit Euler (IE) (green) and the alternant θ_i method (ATM) (red). For time step size $h_1 = 0.001$ and $h_2 = 0.01$ are chosen on the $[0, c]$ and $[c, 10]$ intervals by each methods, respectively. We assume furthermore the $c = 4$ constant value.

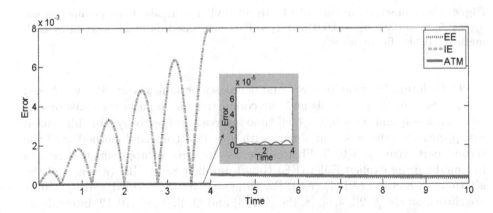

Fig. 3. Error functions in time of EE, IE and ATM methods. Used parameters are $\theta_1 = 0.5$, $\theta_2 = 1$, $h_1 = 0.001$ and $h_2 = 0.01$ on $[0, c]$ and $[c, 10]$ intervals by each methods, respectively, and $c = 4$. (Color figure online)

4.3   Periodical Exponential Increase and Decrease

Consider the following non-linear discontinuous ordinary differential equation:

$$x'(t) = \begin{cases} x(t-8k) & \text{if } 8k < t < 8k+2 \\ x(t) - e^2 & \text{if } 8k+2 < t < 8k+4 \\ -x(-t+8k-2) & \text{if } 8k+4 < t < 8k+6 \\ x(t) - e^0 & \text{if } 8k+6 < t < 8(k+1) \end{cases}$$

where $k = 0, 1, \ldots$ with the initial condition

$$x(0) = 1$$

where $c \in \mathbb{R}$ is given constant.

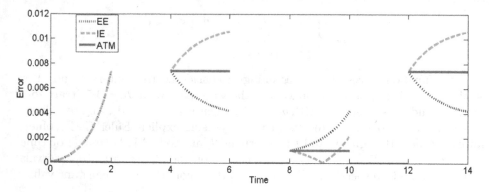

Fig. 4. Error functions in time of EE, IE and ATM methods. Used parameters are $\theta_1 = 0.5$, $\theta_2 = 0.25$, $h_1 = 0.001$ and $h_2 = 0.01$ alternated on $[0, 14]$ interval by each methods. (Color figure online)

The solution function increases and decreases periodically at $[8k, 8k+2]$ and $[8k+4, 8k+6]$ time intervals and the connective curve parts are constants at $[8k+2, 8k+4]$ and $[8k+6, 8(k+1)]$ time intervals, respectively. For this reason we approximate the exponential part with $\theta_1 = 0.5$ (trapezoid method) and the second part with $\theta_2 = 0.25$. Figure 4 shows the error of approximated results in time by using explicit Euler (EE) (blue), implicit Euler (IE) (green) and the alternant θ_i method (ATM) (red). For time step size $h_1 = 0.001$ and $h_2 = 0.01$ are chosen on the $[0, 2]$, $[4, 6]$, $[8, 10]$, $[12, 14]$ and $[2, 4]$, $[6, 8]$, $[10, 12]$ intervals by each methods, respectively.

5   Conclusion

In this work the alternant θ_i-method and its application is investigated. We analyze the local approximation error and the convergence of the method on the non-equidistant mesh. We define the order of convergence, as well. The main idea of this approach is the approximation of the solution of the Cauchy problems by using different numerical schemes (implicit, explicit, IMEX, one-step, multi-step etc.) with varying step-sizes. Benefits of such approximations are shown for the problems with non-smooth solutions. We show that the convergence and the error estimation can be given relatively easily for the classical θ-method in case both equidistant and non-equidistant time/space discretizations. We analyze the connection of this approach to the classical discrete Gronwall lemma. We show that the extended discrete Gronwall lemma can be successfully applied to the estimation of the convergence's rate of the alternant θ_i method. We show numerical examples for some non-linear time dependent differential equations, which have non-continuous or strongly oscillated solutions.

References

1. Faragó, I.: Convergence and stability constant of the theta-method. Appl. Math. **58**, 42–51 (2013)
2. Emmrich, E.: Discrete versions of Gronwall's lemma and their application to the numerical analysis of parabolic problems. Fachbereich 3 Preprint Reihe Mathematik **637**, 10–13 (1999). Berlin TU
3. Clark, D.S.: Short proof of a discrete Gronwall inequality. Discrete Appl. Math. **16**(3), 279–281 (1987)
4. Holte, J.M.: Discrete Gronwall lemma. In: MAA-NCS Meeting at the University of North Dakota (2009)
5. Pachpatte, B.G., Singare, S.M.: Discrete generalized Gronwall inequalities in three independent variables. Pacific J. Math. **82**(1), 197–210 (1979)
6. Isaacson, E., Keller, H.B.: Analysis of Numerical Methods. Wiley, New York (1966)
7. Suli, E.: Numerical Solution of Ordinary Differential Equations. Cambridge University Press, Oxford (2010)

A Numerical Study on the Compressibility of Subblocks of Schur Complement Matrices Obtained from Discretized Helmholtz Equations

Martin J. Gander[1(✉)] and Sergey Solovyev[2]

[1] Section of Mathematics, University of Geneva, Geneva, Switzerland
martin.gander@unige.ch
[2] Institute of Petroleum Geology and Geophysics SB RAS,
3 Akademika Koptyuga pr., Novosibirsk, Russia 630090
solovevsa@ipgg.sbras.ru

Abstract. The compressibility of Schur complement matrices is the essential ingredient for $\mathcal{H}$-matrix techniques, and is well understood for Laplace type problems. The Helmholtz case is more difficult: there are several theoretical results which indicate when good compression is possible with additional techniques, and in practice sometimes basic $\mathcal{H}$-matrix techniques work well. We investigate the compressibility here with extensive numerical experiments based on the SVD. We find that with growing wave number k, the ϵ-rank of blocks corresponding to a fixed size in physical space of the Green's function is always growing like $O(k^\alpha)$, with $\alpha \in [\frac{3}{4}, 1]$ in 2d and $\alpha \in [\frac{4}{3}, 2]$ in 3d.

Keywords: Helmholtz equation · Schur complements · ϵ-Rank

1 Introduction

After their introduction in the seminal paper by Hackbusch [23], $\mathcal{H}$-matrix techniques have been intensively developed over the last two decades to represent dense matrices arising from discretizations of integral equations as well as perform operations between such matrices with almost linear complexity; for a comprehensive introduction including most recent results, see [3,24]. These techniques are very much related to the fast multipole method introduced by Greengard and Rokhlin [22], which also uses the fact that the Green's functions of Laplace like problems have favorable properties; for a recent overview, see [21], and for sharp estimates for $\mathcal{H}$-matrices approximating inverses of FEM-discretized Laplace like operators, see [4].

The fast multipole method has also been studied for wave propagation phenomena [8,27], of which a typical representative is the Helmholtz equation. Although the Helmholtz equation has the deceivingly simple looking Laplacian as its principal part, the difficulty of its numerical solution is worlds apart from solving Laplace's equation; for a recent overview why standard iterative methods fail in the Helmholtz case, see [14]. Specialized methods were developed

© Springer International Publishing AG 2017
I. Dimov et al. (Eds.): NAA 2016, LNCS 10187, pp. 70–81, 2017.
DOI: 10.1007/978-3-319-57099-0_7

for the Helmholtz equation: AILU methods based on analytic incomplete LU factorizations [18,19], reinvented independently later under the name sweeping preconditioner [11,12], and optimized Schwarz methods (OSM) [16,17], which represent a unified framework for all these methods, see [20]. For an introduction to OSM, see [15].

To use $\mathcal{H}$-matrix techniques for the Helmholtz equation proves also to be more difficult: in [2], the subblocks arising are partitioned into two different types, one of which can be well treated by $\mathcal{H}^2$-matrix techniques, whereas for the other one, a multipole like expansion is needed. Almost linear complexity for the compression with a directional $\mathcal{H}^2$-matrix technique is obtained in [6], but there is no complete $\mathcal{H}^2$-matrix arithmetic available yet. For a specific geometric situation, almost linear complexity was also shown in the multipole context in [26]. Upper and lower bounds for the separability of the Green's function of the Helmholtz operator recently derived in [13] indicate however that in general the number of terms is expected to grow algebraically in the wave number. Nevertheless, even the basic $\mathcal{H}$-matrix approach was observed to work quite well in certain situations, see for example [5,25].

We study here the compressibility of off diagonal blocks of Schur complement matrices for the Helmholtz equation numerically using the SVD. We investigate the dependence of the compressibility on the wave number, mesh size, subblock selection and boundary conditions. While for the Laplace case there is a precise theoretical study for the finite difference discretizations we use [7], both in 2d and 3d, for the Helmholtz equation to the best of our knowledge this remains an open question. We measure that the ϵ-rank grows in all our numerical experiments algebraically in the wave number, and the growth is faster in 3d than in 2d.

2   Two Dimensional Study

We study the Helmholtz equation

$$(\Delta + k^2)u = f \text{ in } \Omega := (0,1)^2,$$
$$\mathcal{B}u = 0 \text{ on } \partial\Omega, \tag{1}$$

where $\mathcal{B}$ denotes a suitable boundary operator of Dirichlet and Robin type for the problem to be well posed, and we will be testing various configurations. We discretize the Helmholtz equation (1) using a standard five point finite difference discretization with mesh spacing $h := \frac{1}{n}$, which leads to a sparse linear system

$$A u = f. \tag{2}$$

If we partition the system matrix into a first block A_1 corresponding to the first line of discretization points in the y-direction, and denote the remaining diagonal block by A_2, the linear system (2) can be written in the form

$$\begin{pmatrix} A_1 & A_{12} \\ A_{21} & A_2 \end{pmatrix} \begin{pmatrix} u_1 \\ u_2 \end{pmatrix} = \begin{pmatrix} f_1 \\ f_2 \end{pmatrix}. \tag{3}$$

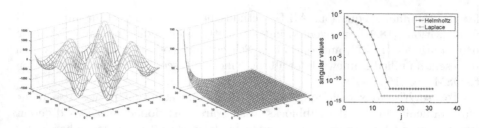

Fig. 1. Real part of the off diagonal block $S_{31,40.2}$ of the Schur complement of the Helmholtz operator (left) and the Laplace operator (middle) for Dirichlet boundary conditions, and their singular values σ_j (right).

Note that the vectors u_1 and f_1 correspond to only one line of unknowns and are thus much shorter than u_2 and f_2. We are interested in the Schur complement matrix S of the reduced system indicated by the partition in (3), i.e. when all the unknowns u_2 are eliminated from the system,

$$Su_1 = f_1 - A_{12}A_2^{-1}f_2, \quad S := A_1 - A_{12}A_2^{-1}A_{21}. \tag{4}$$

The representation of the Schur complement matrix S as an $\mathcal{H}$-matrix requires to form the so called block cluster tree, which is a partition of S into subblocks, and then to approximate these subblocks by low rank matrices. In the case of Laplace's equation, blocks on the diagonal of S or close to the diagonal can not be compressed well, since they are close to the singularity of the Green's function of Laplace's equation that lies on the diagonal. Blocks far away from the diagonal however can be very well compressed in the Laplace like cases [4]. To study the compressibility in the Helmholtz case, we apply the singular value decomposition (SVD) to the matrix subblock $S_{m,k} := S(1 : m, n - m + 1 : n)$ of the Schur complement matrix $S \in \mathbb{C}^{n \times n}$ and study the decay of the singular values as a function of m, h, and k.

We start by showing in Fig. 1 for $n = 64$ on the unit square and $k = 40.2$ the off diagonal block of the Schur complement $S_{m,k}$ of the discretized Helmholtz operator with Dirichlet boundary conditions for $m = 31$ on the left, and for comparison purposes also the corresponding Laplace case with $k = 0$ in the middle. We can clearly see that the fundamental solution visible in the Schur complement is very oscillatory for the Helmholtz operator, while it is much simpler and just decaying for the Laplace operator. This has an important influence on the decay of the singular values σ_j of these blocks, as one can see on the right: the decay in the Helmholtz case is delayed. This influences the ϵ-rank, which is defined as the smallest number r_ϵ such that $\frac{\sigma_j}{\sigma_1} < \epsilon$ for all $j > r_\epsilon$, i.e. the number of singular values we have to keep so that the approximate matrix is at most at a distance ϵ from the exact matrix in the Frobenius norm.

We next show in Table 1 the ϵ-rank of the approximate Schur complement $S_{m,k}$ for increasing k when using 10 ppw for the discretization, i.e. $k = \frac{2\pi n}{10}$, where n is the number of mesh cells in one space direction, i.e. $h = \frac{1}{n}$. We show three cases: $m = \frac{n}{2} - 1$, which means a large block of the Schur complement, about a quarter of the entire matrix, just avoiding the diagonal, $m = \frac{n}{4} - 1$ which is about half the size and now away from the diagonal, and finally $m = \frac{n}{8} - 1$, which is a relatively small block, again halved once more, and very far from the diagonal. We clearly see that the ϵ-rank is growing, and the growth is similar in the case of Robin conditions, the wave guide case where we have Robin conditions on the left and right and Dirichlet conditions at the top and bottom, and the case with Dirichlet conditions all around. For comparison purposes, we also added the ϵ-rank for the Laplacian in parentheses in the Dirichlet case. Clearly the ϵ-rank is growing for increasing wave number k, whereas it is constant for the Laplace case as soon as the block is away from the diagonal, as expected from theory [7]. Figure 2 shows the results of Table 1 graphically, and we see that the ϵ-rank is growing algebraically in k, approximately like $O(k^{\frac{3}{4}})$ when the wave number k is increasing. This seems to be independent of the choice of the block determined by the parameter m, only the constant in front of the growth is getting smaller as the block size is getting smaller and is moved further away from the diagonal. Note however that comparing in Table 1 for example the fourth line for $m = \frac{n}{2} - 1$ with the fifth line for $m = \frac{n}{4} - 1$ and sixth line for $m = \frac{n}{8} - 1$, the ranks are not growing, which indicates that with a corresponding admissibility condition, k independent compression will be possible. There will however be so many blocks then that $\mathcal{H}^2$-matrix techniques will be needed for almost linear complexity [6].

We now repeat the same experiment for the largest subblock under the condition that $k^3 h^2 = \text{const}$, which is suggested to control the pollution effect [1]. We use the same grids as in Table 1, but for a more slowly increasing wave number k. We start with the same resolution as in Table 1 for $k = 20.1$, so the first lines indicating the ϵ-rank are the same in the top row for the largest subblock in Table 1 and the new Table 2. We see again that the ϵ-rank is growing as in the 10 ppw case, and it seems the growth rate is very similar to the case in Table 1 for corresponding wave numbers k. This indicates that it is not the mesh size, but really the wave number k that dictates the growth of the ϵ-rank.

We finally test Schur complements which are not just based on the variables on one side of the domain, but include more sides. The corresponding results are shown in Table 3. We see that when a neighboring side is included in the Schur complement, the ϵ-rank is growing already a bit faster, and when opposite sides or more are included, the growth becomes $O(k)$, see also [10] for a theoretical study of the influence of geometry.

3   Three Dimensional Study

We study now the three dimensional case of the Helmholtz equation as it appears in geophysical applications, namely

Table 1. ϵ-rank in 2d for a matrix subblock $S_{m,k}$ with increasing k and three choices of m for 10 ppw (in parentheses for the Dirichlet case also the corresponding results for the Laplace operator are shown).

	Robin					Wave guide					Dirichlet				
ϵ	1e-2	1e-3	1e-4	1e-5	1e-6	1e-2	1e-3	1e-4	1e-5	1e-6	1e-2	1e-3	1e-4	1e-5	1e-6
k	Schur complement block with $m = \frac{n}{2} - 1$														
20.1	5	6	7	7	8	6	6	7	7	8	4 (3)	5 (4)	6 (5)	6 (5)	7 (6)
40.2	8	10	10	11	12	9	10	10	11	11	7 (4)	9 (5)	10 (6)	10 (6)	11 (7)
80.4	12	14	16	18	18	14	15	17	18	18	11 (4)	14 (5)	15 (6)	17 (7)	18 (8)
160.8	18	22	24	28	30	23	26	27	29	30	19 (4)	23 (6)	25 (7)	27 (8)	29 (9)
321.6	29	34	40	44	47	38	41	44	46	49	30 (4)	36 (6)	41 (7)	43 (9)	46 (10)
643.2	50	59	67	73	79	67	72	76	80	83	46 (4)	61 (6)	68 (8)	72 (9)	76 (11)
k	Schur complement block with $m = \frac{n}{4} - 1$														
20.1	3	4	4	5	5	3	4	4	4	5	3 (2)	3 (3)	3 (3)	5 (3)	5 (4)
40.2	5	5	6	6	7	5	6	6	7	7	4 (2)	5 (3)	6 (3)	6 (3)	7 (4)
80.4	6	8	9	9	10	8	9	10	10	10	7 (2)	8 (3)	8 (3)	10 (3)	10 (4)
160.8	9	12	14	16	17	13	15	16	16	17	11 (2)	14 (3)	16 (3)	16 (4)	17 (4)
321.6	15	18	22	24	27	22	24	28	28	29	21 (2)	23 (3)	25 (4)	28 (4)	29 (4)
643.2	25	32	38	42	45	42	45	48	49	51	36 (2)	43 (3)	45 (3)	49 (4)	49 (4)
k	Schur complement block with $m = \frac{n}{8} - 1$														
20.1	3	3	3	3	3	3	3	3	3	3	3 (2)	3 (2)	3 (2)	3 (3)	3 (3)
40.2	3	4	4	4	5	3	4	4	4	5	3 (2)	4 (2)	4 (3)	4 (3)	5 (3)
80.4	4	5	6	6	7	5	5	6	6	7	5 (2)	5 (2)	6 (3)	6 (3)	6 (3)
160.8	6	8	9	9	10	8	9	9	10	10	8 (2)	8 (2)	9 (3)	10 (3)	10 (3)
321.6	9	11	13	15	16	14	15	16	16	17	13 (2)	15 (2)	15 (3)	16 (3)	17 (3)
643.2	12	17	22	24	27	22	26	27	29	30	22 (2)	26 (2)	28 (3)	29 (3)	29 (3)

$$\Delta u + \frac{(2\pi\nu)^2}{V(x,y,z)^2} u = \delta(\overline{r} - \overline{r}_s) f \qquad \text{in } \Omega := (0, L)^3. \tag{5}$$

Here ν represents the frequency in Hz, $V(x, y, z)$ is a given velocity field with the velocity measured in $\frac{m}{s}$, $\overline{r}_s$ are the coordinates of a point source, whose strength is given by f, and L represents the size of the domain. We will test Dirichlet boundary conditions and perfectly matched layers (PMLs), for which we use an adaptation of the complex coordinate stretching elastic PML from [9]. We discretize the equation on a rectangular grid, and enumerate the nodes by going along the x-direction first, followed by the y-direction, and finally the z-direction. With this enumeration, the matrix subblock A_1 in (3) corresponds to the first x-y plane of discretization points, with associated Schur complement matrix S defined as in (4). We use the standard 7-point finite difference stencil, and an optimized 27-point finite difference stencil which reduces pollution and permits a reduction from 15 down to 4 ppw for equivalent results [28].

We start with a constant velocity $V(x, y, z) = 2400\,\text{m/s}$ in the cube domain with physical dimension $L = 1200\,\text{m}$. We simulate for the frequencies $\nu = 4\,\text{Hz},\ 8\,\text{Hz},\ 16\,\text{Hz}$ using the corresponding number of grid points $n = 20, 40, 80$, which implies that we are using 10 ppw in these experiments. We first show in

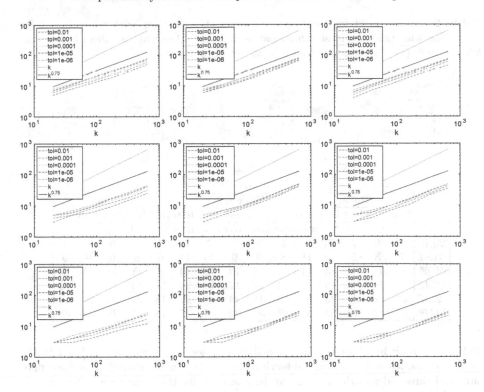

Fig. 2. Results of Table 1 shown graphically, together with two reference growth curves $O(k)$ and $O(k^{\frac{3}{4}})$.

Fig. 3 a comparison on how the singular values decay for a large off diagonal block, depending on the discretization stencil and the boundary condition. We clearly observe that increasing the frequency ν delays the decay of the singular values, and this does neither depend much on the discretization stencil, nor on the boundary condition.

We next present the ϵ-ranks of $S_{m,k}$ in Table 4, comparing the 7 point stencil to the 27 point stencil, and Dirichlet boundary conditions with PML boundary conditions using $n_{PML} = 5$ points in the PML layer. We see that the ϵ-rank grows when k increases, and the graphic visualization in Fig. 4 shows that the growth is again algebraic, this time approximately $O(k^{\frac{4}{3}})$ for k large.

To test the impact of the quality of the PML on the ϵ-rank, we show in Table 5 the ϵ-rank for an increasing depth a of the PML, for $V(x, y, z) = 2400\,\text{m/s}$ on the cube $1200\,\text{m} \times 1200\,\text{m} \times 1200\,\text{m}$ as before, for two meshes with 40 and 80 grid points in each direction and frequency $\nu = 8$ and $\nu = 16$, so we have 10 ppw. We see that improving the PML does not reduce the ϵ-rank. We also show in Table 5 the relative error in the L_1 and L_∞ norms comparing with a closed form solution, which indicates that for $n = 40$, we reach the truncation error of

Table 2. ϵ-rank in 2d for the largest matrix subblock $S_{m,k}$ with $m = \frac{n}{2} - 1$ for increasing k for the same mesh refinement as in Table 1, but wave numbers k such that $k^3 h^2 = \text{const}$ (the Laplace case would be identical to Table 1).

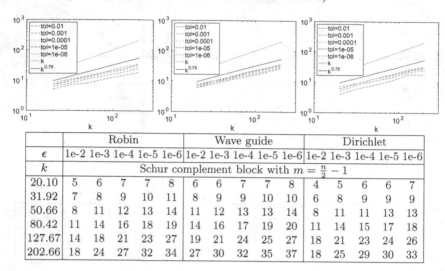

	Robin					Wave guide					Dirichlet				
ϵ	1e-2	1e-3	1e-4	1e-5	1e-6	1e-2	1e-3	1e-4	1e-5	1e-6	1e-2	1e-3	1e-4	1e-5	1e-6
k							Schur complement block with $m = \frac{n}{2} - 1$								
20.10	5	6	7	7	8	6	6	7	7	8	4	5	6	6	7
31.92	7	8	9	10	11	8	9	9	10	10	6	8	9	9	9
50.66	8	11	12	13	14	11	12	13	13	14	8	11	11	13	13
80.42	11	14	16	18	19	14	16	17	19	20	11	14	15	17	18
127.67	14	18	21	23	27	19	21	24	25	27	18	21	23	24	26
202.66	18	24	27	32	34	27	30	32	35	37	18	25	29	30	33

Table 3. ϵ-rank in 2d for the largest matrix subblock $S_{m,k}$ with $m = \frac{n}{2} - 1$ in the Dirichlet case where the matrix subblock corresponds now to two adjacent, two opposite, and three sides of the domain, for increasing k for the same mesh refinement as in Table 1.

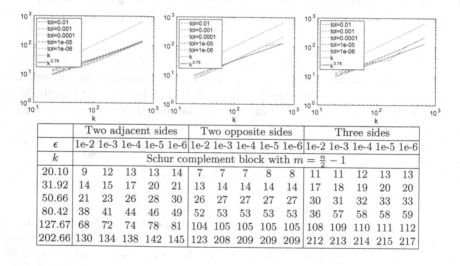

	Two adjacent sides					Two opposite sides					Three sides				
ϵ	1e-2	1e-3	1e-4	1e-5	1e-6	1e-2	1e-3	1e-4	1e-5	1e-6	1e-2	1e-3	1e-4	1e-5	1e-6
k							Schur complement block with $m = \frac{n}{2} - 1$								
20.10	9	12	13	13	14	7	7	7	8	8	11	11	12	13	13
31.92	14	15	17	20	21	13	14	14	14	14	17	18	19	20	20
50.66	21	23	26	28	30	26	27	27	27	27	30	31	32	33	33
80.42	38	41	44	46	49	52	53	53	53	53	36	57	58	58	59
127.67	68	72	74	78	81	104	105	105	105	105	108	109	110	111	112
202.66	130	134	138	142	145	123	208	209	209	209	212	213	214	215	217

the discretization at $n_{PML} = 8$, since increasing the PML depth does not reduce the error further.

We next study the behavior of the ϵ-rank in a random medium $V(x, y, z)$ using the same geometry as before. We show in Fig. 5 the decay of the singular values

Table 4. ϵ-rank in 3d for a matrix subblock $S_{m,k}$ with increasing k and three choices of m.

			7-point, Dirichlet					27-point, Dirichlet					27-point, PML				
		ϵ	1e-2	1e-3	1e-4	1e-5	1e-6	1e-2	1e-3	1e-4	1e-5	1e-6	1e-2	1e-3	1e-4	1e-5	1e-6
n	ν	k	Schur complement block with $m = \frac{(n-1)^2-1}{2}$														
20	4	0.0052	27	37	46	51	59	27	35	43	53	59	29	39	49	61	71
40	8	0.0105	6	54	68	85	110	61	78	94	112	129	58	76	97	120	140
80	16	0.0209	32	147	180	230	276	145	185	224	263	306	141	189	237	287	325
n	ν	k	Schur complement block with $m = \frac{(n-1)^2-1}{4}$														
20	4	0.0052	5	7	9	11	15	5	7	9	11	14	8	15	20	25	32
40	8	0.0105	6	13	17	21	23	16	19	23	25	27	11	16	23	29	35
80	16	0.0209	25	46	60	67	72	41	59	67	76	79	36	48	63	74	81
n	ν	k	Schur complement block with $m = \frac{(n-1)^2-1}{8}$														
20	4	0.0052	4	6	6	8	10	4	6	6	8	10	7	12	15	18	21
40	8	0.0105	5	9	9	15	17	9	13	16	18	21	8	13	15	21	24
80	16	0.0209	17	31	38	44	52	31	38	42	51	52	23	35	40	48	53

Table 5. Solution error and ϵ-rank in 3d for the largest subblock with different PML depth.

		$n=40$, $\nu = 8$, subblock 760×760							$n=80$, $\nu = 16$, subblock 3120×3120						
n_{PML}	ϵ	1e-2	1e-3	1e-4	1e-5	1e-6	L_1	L_∞	1e-2	1e-3	1e-4	1e-5	1e-6	L_1	L_∞
0		61	78	94	112	129	1.00	1.00	145	185	224	263	306	1.00	1.00
2		57	78	96	123	140	0.12	0.10	148	197	245	291	330	0.56	0.18
4		58	77	97	120	140	0.046	0.043	144	194	242	290	327	0.34	0.10
8		58	78	99	121	142	0.036	0.039	130	181	230	273	322	0.10	0.044
16		59	80	103	125	145	0.036	0.039	117	162	215	265	306	0.036	0.039

comparing the constant to the random medium case when using a fixed number

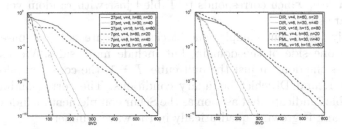

Fig. 3. Decay of the singular values in 3d, comparing the 27 and 7 point stencil with Dirichlet conditions (left) and Dirichlet and PML boundary conditions for the 27 point stencil (right) for a large matrix subblock $S_{m,k}$ corresponding to the top row in Table 4.

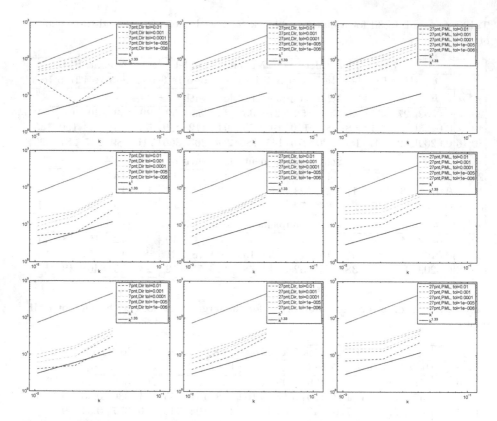

Fig. 4. Results of Table 4 shown graphically, together with two reference growth curves $O(k)$ and $O(k^{\frac{4}{3}})$.

of ppw. We see that the decay of the singular values in the random medium case is comparable to the decay of the singular values in the constant medium case, just the small ones decay a bit more slowly. This is further illustrated for the ϵ-rank in Table 6, which corresponds to Table 4 just with random velocity, and the results are similar.

We finally test Schur complement blocks which are not just based on the variables on one side of the domain, but include more sides. We use in this experiment as the domain just the unit cube, $L = 1$, the constant velocity field $V(x, y, z) = 1$, and Dirichlet boundary conditions. The corresponding results shown in Table 7 indicate that as soon as the Schur complement includes opposite sides, the ϵ-rank is growing quadratically in the wave number.

4   Conclusion

We studied numerically the compressibility of subblocks of Schur complement matrices stemming from discretized Helmholtz problems. We experimentally

Table 6. ϵ-rank in 3d for a large matrix subblock $S_{m,k}$ for a random velocity field with increasing n.

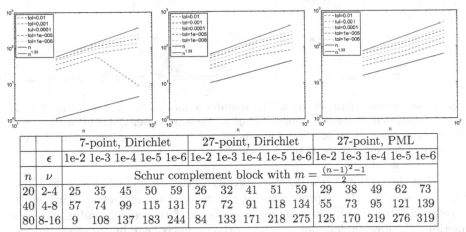

		7-point, Dirichlet					27-point, Dirichlet					27-point, PML				
ϵ		1e-2	1e-3	1e-4	1e-5	1e-6	1e-2	1e-3	1e-4	1e-5	1e-6	1e-2	1e-3	1e-4	1e-5	1e-6
n	ν	\multicolumn{15}{c}{Schur complement block with $m = \frac{(n-1)^2 - 1}{2}$}														
20	2–4	25	35	45	50	59	26	32	41	51	59	29	38	49	62	73
40	4–8	57	74	99	115	131	57	72	91	118	134	55	73	95	121	139
80	8–16	9	108	137	183	244	84	133	171	218	275	125	170	219	276	319

Table 7. ϵ-rank in 3d for a small, medium and large matrix subblock $S_{m,k}$ of a Schur complement for one side, and two opposite sides of the unit cube, for increasing k keeping 10 ppw.

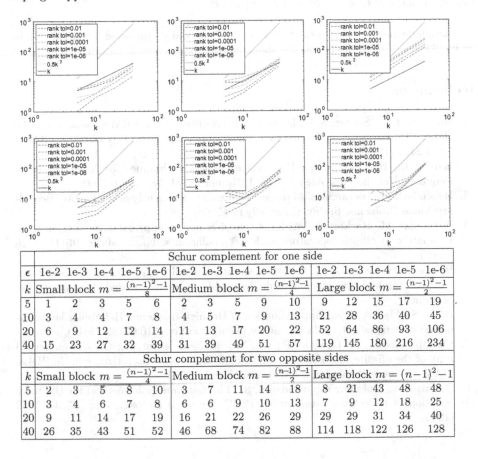

	\multicolumn{15}{c}{Schur complement for one side}														
ϵ	1e-2	1e-3	1e-4	1e-5	1e-6	1e-2	1e-3	1e-4	1e-5	1e-6	1e-2	1e-3	1e-4	1e-5	1e-6
k	\multicolumn{5}{l}{Small block $m = \frac{(n-1)^2 - 1}{8}$}					\multicolumn{5}{l}{Medium block $m = \frac{(n-1)^2 - 1}{4}$}					\multicolumn{5}{l}{Large block $m = \frac{(n-1)^2 - 1}{2}$}				
5	1	2	3	5	6	2	3	5	9	10	9	12	15	17	19
10	3	4	4	7	8	4	5	7	9	13	21	28	36	40	45
20	6	9	12	12	14	11	13	17	20	22	52	64	86	93	106
40	15	23	27	32	39	31	39	49	51	57	119	145	180	216	234
	\multicolumn{15}{c}{Schur complement for two opposite sides}														
k	\multicolumn{5}{l}{Small block $m = \frac{(n-1)^2 - 1}{4}$}					\multicolumn{5}{l}{Medium block $m = \frac{(n-1)^2 - 1}{2}$}					\multicolumn{5}{l}{Large block $m = (n-1)^2 - 1$}				
5	2	3	5	8	10	3	7	11	14	18	8	21	43	48	48
10	3	4	6	7	8	6	6	9	10	13	7	9	12	18	25
20	9	11	14	17	19	16	21	22	26	29	29	29	31	34	40
40	26	35	43	51	52	46	68	74	82	88	114	118	122	126	128

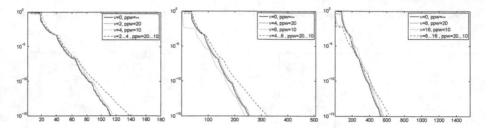

Fig. 5. Comparison of the decay of the singular values in 3d (7 point stencil with Dirichlet conditions) between a homogeneous medium and a random field varying from 2 Hz to 4 Hz with $n = 20$ number of grid points in each direction (left); 4 Hz to 8 Hz, $n = 40$ (middle) and 8 Hz to 16 Hz, $n = 80$ (right)

found that the ϵ-rank is growing algebraically in the wave number k if the block size is chosen in a fixed proportion of the Schur complement matrix, which corresponds to a fixed region in physical space of the corresponding Green's function. The growth does not depend on the boundary conditions of the underlying problem, but the choice of which variables are eliminated in forming the Schur complement. If opposite sides of the domain are kept, the worst growth is observed, namely $O(k)$ in 2d and $O(k^2)$ in 3d.

Acknowledgments. The research described was partially supported by RFBR grants 16-05-00800,17-01-00399 and the Russian Academy of Sciences Program "Arctic".

References

1. Babuska, I.M., Sauter, S.A.: Is the pollution effect of the FEM avoidable for the Helmholtz equation considering high wave numbers? SIAM J. Num. Anal. **34**(6), 2392–2423 (1997)
2. Banjai, L., Hackbusch, W.: Hierarchical matrix techniques for low- and high-frequency Helmholtz problems. IMA J. Numer. Anal. **28**, 46–79 (2008)
3. Bebendorf, M.: Hierarchical Matrices: A Means to Efficiently Solve Elliptic Boundary Value Problems. Springer, Heidelberg (2008)
4. Bebendorf, M., Hackbusch, W.: Existence of $\mathcal{H}$-matrix approximants to the inverse FE-matrix of elliptic operators with L^∞-coefficients. Num. Math. **95**(1), 1–28 (2003)
5. Betcke, T., van't Wout, E., Gélat, P.: Computationally efficient boundary element methods for high-frequency Helmholtz problems in unbounded domains (2016, preprint)
6. Börm, S., Melenk, J.M.: Approximation of the high-frequency Helmholtz kernel by nested directional interpolation. arXiv preprint arXiv:1510.07189 (2015)
7. Chandrasekaran, S., Dewilde, P., Gu, M., Somasunderam, N.: On the numerical rank of the off-diagonal blocks of Schur complements of discretized elliptic PDEs. SIAM J. Matrix Anal. Appl. **31**(5), 2261–2290 (2010)
8. Coifman, R., Rokhlin, V., Wandzura, S.: The fast multipole method for the wave equation: a pedestrian prescription. IEEE Antennas Propag. Mag. **35**(3), 7–12 (1993)

9. Collino, F., Tsogka, C.: Application of the perfectly matched layer absorbing layer model to the linear elastodynamic problem in anisotropic heterogeneous media. Geophysics **66**, 294–307 (2001)
10. Delamotte, K.: Une étude du rang du noyau de l'équation de Helmholtz: application des $\mathcal{H}$-matrices à l'EFIE. Ph.D. thesis, University Paris 13 (2016)
11. Engquist, B., Ying, L.: Sweeping preconditioner for the Helmholtz equation: hierarchical matrix representation. Comm. Pure Appl. Math. **64**(5), 697–735 (2011)
12. Engquist, B., Ying, L.: Sweeping preconditioner for the Helmholtz equation: moving perfectly matched layers. Multiscale Model. Simul. **9**(2), 686–710 (2011)
13. Engquist, B., Zhao, H.: Approximate separability of Green's function for high frequency Helmholtz equations. Technical report, DTIC Document (2014)
14. Ernst, O.G., Gander, M.J.: Why it is difficult to solve Helmholtz problems with classical iterative methods. In: Graham, I.G., Hou, T.Y., Lakkis, O., Scheichl, R. (eds.) Numerical analysis of multiscale problems, vol. 83, pp. 325–363. Springer, Heidelberg (2012)
15. Gander, M.J.: Optimized Schwarz methods. SIAM J. Numer. Anal. **44**(2), 699–731 (2006)
16. Gander, M.J., Halpern, L., Magoules, F.: An optimized Schwarz method with two-sided Robin transmission conditions for the Helmholtz equation. Int. J. Numer. Meth. Fluids **55**(2), 163–175 (2007)
17. Gander, M.J., Magoules, F., Nataf, F.: Optimized Schwarz methods without overlap for the Helmholtz equation. SIAM J. Sci. Comput. **24**(1), 38–60 (2002)
18. Gander, M.J., Nataf, F.: AILU for Helmholtz problems: a new preconditioner based on the analytic parabolic factorization. J. Comput. Acoust. **9**(04), 1499–1506 (2001)
19. Gander, M.J., Nataf, F.: An incomplete LU preconditioner for problems in acoustics. J. Comput. Acoust. **13**(03), 455–476 (2005)
20. Gander, M.J., Zhang, H.: Iterative solvers for the Helmholtz equation: factorizations, sweeping preconditioners, source transfer, single layer potentials, polarized traces, and optimized Schwarz methods (2016). Submitted
21. Greengard, L., Gueyffier, D., Martinsson, P.-G., Rokhlin, V.: Fast direct solvers for integral equations in complex three-dimensional domains. Acta Numerica **18**, 243–275 (2009)
22. Greengard, L., Rokhlin, V.: A fast algorithm for particle simulations. J. Comput. Phys. **73**(2), 325–348 (1987)
23. Hackbusch, W.: A sparse matrix arithmetic based on $\mathcal{H}$-matrices. part i: introduction to $\mathcal{H}$-matrices. Computing **62**(2), 89–108 (1999)
24. Hackbusch, W.: Hierarchical Matrices: Algorithms and Analysis, vol. 49. Springer, Heidelberg (2015)
25. Lizé, B.: Résolution directe rapide pour les éléments finis de frontière en électromagnétisme et acoustique: -Matrices. Parallélisme et applications industrielles. Ph.D. thesis, Université Paris-Nord-Paris XIII (2014)
26. Martinsson, P.G., Rokhlin, V.: A fast direct solver for scattering problems involving elongated structures. J. Comput. Phys. **221**, 288–302 (2007)
27. Rokhlin, V.: Rapid solution of integral equations of scattering theory in two dimensions. J. Comput. Phys. **86**(2), 414–439 (1990)
28. Solovyev, S., Vishnevsky, D.: A dispersion minimizing finite difference scheme and multifrontal hierarchical solver for the 3D Helmholtz equation. In: The 12th International Conference on Mathematical and Numerical Aspects of Wave Propagation (WAVES 2015), 20–24 July, pp. 428–429 (2015)

Convergence Outside the Initial Layer for a Numerical Method for the Time-Fractional Heat Equation

José Luis Gracia[1], Eugene O'Riordan[2], and Martin Stynes[3(✉)]

[1] Department of Applied Mathematics, University of Zaragoza, Zaragoza, Spain
jlgracia@unizar.es
[2] School of Mathematical Sciences, Dublin City University, Dublin, Ireland
eugene.oriordan@dcu.ie
[3] Applied and Computational Mathematics Division,
Beijing Computational Science Research Center, Beijing, China
m.stynes@csrc.ac.cn

Abstract. In this paper a fractional heat equation is considered; it has a Caputo time-fractional derivative of order δ where $0 < \delta < 1$. It is solved numerically on a uniform mesh using the classical L1 and standard three-point finite difference approximations for the time and spatial derivatives, respectively. In general the true solution exhibits a layer at the initial time $t = 0$; this reduces the global order of convergence of the finite difference method to $O(h^2 + \tau^\delta)$, where h and τ are the mesh widths in space and time, respectively. A new estimate for the L1 approximation shows that its truncation error is smaller away from $t = 0$. This motivates us to investigate if the finite difference method is more accurate away from $t = 0$. Numerical experiments with various non-smooth and incompatible initial conditions show that, away from $t = 0$, one obtains $O(h^2 + \tau)$ convergence.

Keywords: Time-fractional heat equation · Caputo fractional derivative · Initial-boundary value problem · L1 scheme · Layer region · Smooth and non-smooth data · Compatibility conditions

1 Introduction

In this paper we consider the time-fractional heat equation

$$D_t^\delta u - \frac{\partial^2 u}{\partial x^2} = 0 \tag{1a}$$

for $(x, t) \in Q := (0, \pi) \times (0, 1]$, with

$$u(0, t) = u(\pi, t) = 0 \text{ for } t \in (0, 1], \tag{1b}$$

$$u(x, 0) = \phi(x) \text{ for } x \in [0, \pi], \tag{1c}$$

© Springer International Publishing AG 2017
I. Dimov et al. (Eds.): NAA 2016, LNCS 10187, pp. 82–94, 2017.
DOI: 10.1007/978-3-319-57099-0_8

where $0 < \delta < 1$ and D_t^δ denotes the *Caputo fractional derivative*, which is defined [1] by

$$D_t^\delta g(x,t) := \left[\frac{1}{\Gamma(1-\delta)} \int_{s=0}^t (t-s)^{-\delta} \frac{\partial g}{\partial s}(x,s)\, ds \right] \quad \text{for } (x,t) \in \bar{Q}. \qquad (2)$$

For typical solutions u of (1), the temporal derivative u_t is unbounded as $t \to 0$; see, e.g., [5,8]. This weak singularity must be taken into account in any discussion of (1), for if one assumes that u_t is bounded on $\bar{Q}$, then this forces the initial condition ϕ to be identically zero, which implies that the solution $u \equiv 0$; see [7, Example 2].

In [8], assuming that the data of problem (1) satisfy certain regularity and compatibility conditions, the following estimates for its solution u were proved:

$$\left| \frac{\partial^k u}{\partial x^k}(x,t) \right| \le C, \quad \text{for } k = 0,1,2,3,4, \quad \left| \frac{\partial^\ell u}{\partial t^\ell}(x,t) \right| \le C(1 + t^{\delta-\ell}) \quad \text{for } \ell = 1,2,$$
$$(3)$$

for all $(x,t) \in [0,\pi] \times (0,1]$.

When problem (1) is approximated on a uniform mesh using the classical L1 approximation [6] for the time-fractional derivative $D_t^\delta u$ and a standard three-point scheme for $\partial^2 u/\partial x^2$, it is proved in [8] that this scheme is $O(h^2 + \tau^\delta)$ convergent nodally, where h, τ are the spatial and temporal mesh widths.

But a new sharp estimate for the truncation error of the L1 scheme shows that, while it is only $O(1)$ near $t = 0$, it is much smaller away from $t = 0$. Consequently one wonders whether the finite difference method of [8] becomes more accurate away from $t = 0$. In the present paper we investigate this question by means of numerical experiments and an improved order of convergence is indeed observed: one obtains $O(h^2 + \tau)$ convergence on subdomains of Q that are bounded away from $t = 0$, in contrast to the $O(h^2 + \tau^\delta)$ convergence in $\bar{Q}$ that was obtained in [8]. This improved accuracy away from $t = 0$ is obtained in various examples, even when the initial condition ϕ of (1c) is non-smooth or is incompatible with the other data of (1).

This phenomenon is a remarkable property of the computed solution. It was investigated in [4] in an L_2-norm setting; in the present paper we work in the discrete L_∞ norm. The discrete L_∞ analysis of the phenomenon will be considered in [3].

The structure of the present paper is as follows. In Sect. 2 we describe our difference scheme and state a new truncation error estimate for the L1 approximation. Numerical experiments in Sect. 3 verify the sharpness of this estimate and demonstrate the enhanced accuracy of the computed solution away from $t = 0$ when the data of (1) are smooth and compatible. In Sect. 4 we show that this accuracy is still obtained when the initial condition ϕ is rough or is incompatible with the other data of (1).

Notation: In this paper C denotes a generic constant that depends on the data of the boundary value problem (1) but is independent of any mesh used to solve (1) numerically. Note that C can take different values in different places.

2   The Discretisation and Truncation Error Estimates

Let M and N be positive integers. Set $h = \pi/M$ and $x_m := mh$ for $m = 0, 1, \ldots, M$. Set $\tau = 1/N$ and $t_n := n\tau$ for $n = 0, 1, \ldots, N$. Then the mesh is $\{(x_m, t_n) : m = 0, 1, \ldots, M, \ n = 0, 1, \ldots, N\}$.

The L1 approximation [6] of the Caputo fractional derivative D_t^δ is based on writing

$$D_t^\delta u(x_m, t_n) = \frac{1}{\Gamma(1-\delta)} \sum_{k=0}^{n-1} \int_{s=t_k}^{t_{k+1}} (t_n - s)^{-\delta} \frac{\partial u(x_m, s)}{\partial s}\, ds,$$

then approximating $\partial u(x_m, s)/\partial s$ by $(u(x_m, t_{k+1}) - u(x_m, t_k))/\tau$ on each time interval $[t_k, t_{k+1}]$. That is, the L1 approximation $D_N^\delta u_m^n$ is given by

$$D_N^\delta u_m^n := \frac{1}{\Gamma(1-\delta)} \sum_{k=0}^{n-1} \frac{u_m^{k+1} - u_m^k}{\tau} \int_{s=t_k}^{t_{k+1}} (t_n - s)^{-\delta}\, ds$$

$$= \frac{1}{\Gamma(2-\delta)} \sum_{k=0}^{n-1} \frac{u_m^{k+1} - u_m^k}{\tau} \left[(t_n - t_k)^{1-\delta} - (t_n - t_{k+1})^{1-\delta} \right], \qquad (4)$$

where u_m^n is the solution computed at (x_n, t_m).

In our finite difference method, the Caputo fractional derivative D_t^δ is approximated by the L1 approximation (4) and the spatial derivative u_{xx} is approximated by the standard formula

$$u_{xx}(x_m, t_n) \approx \delta_x^2 u_m^n := \frac{u_{m+1}^n - 2u_m^n + u_{m-1}^n}{h^2}.$$

Thus we approximate (1) by the discrete problem

$$D_N^\delta u_n^m - \delta_x^2 u_n^n = 0 \text{ for } 1 \le m \le M - 1, \ 1 \le n \le N, \qquad (5a)$$

$$u_0^n = u_M^n = 0 \text{ for } 0 < n \le N, \qquad (5b)$$

$$u_m^0 = \phi(x_m) \text{ for } 0 \le m \le M. \qquad (5c)$$

The discretisation (5) of (1) has been considered by several authors. Under suitable hypotheses on the data of the problem, it is shown in [8] that, taking into consideration the weak singularity indicated by the bounds (3), the solution $\{u_m^n\}$ of (5) satisfies the error estimate

$$\max_{(x_m, t_n) \in \bar{Q}} |u(x_m, t_n) - u_m^n| \le C \left(h^2 + \tau^\delta \right) \qquad (6)$$

for some constant C, and this estimate is sharp.

Remark 1. In all our numerical experiments below we shall take $M = N$, so the errors associated with the time discretisation will dominate the second-order errors associated with the spatial discretisation. For this reason we ignore the $O(h^2)$ error in our subsequent discussions and concentrate on the temporal error $O(\tau^\beta)$ for various $\beta > 0$.

Using the bounds (3) on the derivatives of u, the truncation error estimate

$$\left| D_N^\delta u(x_m, t_n) - D_t^\delta u(x_m, t_n) \right| \leq Cn^{-\delta}, \quad n = 1, 2, \ldots, N, \tag{7}$$

was derived in [8]. In [3] this estimate is sharpened as follows:

Lemma 1. *Assume that u satisfies (3). Then there exists a constant C such that for each mesh point $(x_m, t_n) \in Q$ one has*

$$\left| D_N^\delta u(x_m, t_n) - D_t^\delta u(x_m, t_n) \right| \leq Cn^{-1}.$$

Remark 2. One can write the truncation error estimate of Lemma 1 as

$$\left| D_N^\delta u(x_m, t_n) - D_t^\delta u(x_m, t_n) \right| \leq Cn^{-1} = C\tau t_n^{-1}, \tag{8}$$

i.e., the truncation error associated with the L1 approximation is first-order if $t_n \geq C_1$, where C_1 is any fixed positive constant in $(0, 1]$. This estimate motivates us to investigate if the scheme (5) provides better approximations to the solution away from $t = 0$.

Throughout the paper, numerical results will be given in the subdomain

$$\bar{Q}^* := [0, \pi] \times [0.1, 1],$$

which is a subset of Q that is bounded away from $t = 0$.

3 Numerical Experiments for Smooth and Compatible Initial-Boundary Conditions

In Sects. 3 and 4 we shall consider several test problems with various initial conditions. The errors in the solutions computed by the difference scheme (5) are estimated using the two-mesh principle [2]: on a uniform mesh with mesh steps $h/2$ and $\tau/2$, compute the numerical solution $\{z_m^n\}$ for $m = 0, 1, \ldots, 2M$ and $n = 0, 1, \ldots, 2N$ with the scheme (5). Then, calculate the two-mesh differences

$$d_{M,N}^\delta := \max_{\substack{0 \leq m \leq M, \\ 0 \leq n \leq N}} |u_m^n - z_{2m}^{2n}|; \tag{9}$$

and from these values one computes the estimated orders of convergence by

$$q_{M,N}^\delta = \log_2 \left(\frac{d_{M,N}^\delta}{d_{2M,2N}^\delta} \right). \tag{10}$$

Analogous quantities will be computed on the subdomain $\bar{Q}^*$ to estimate the maximum errors and rates of convergence on that subdomain.

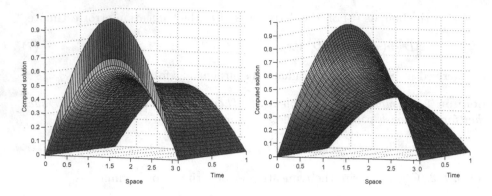

Fig. 1. Solution for $\delta = 0.2$ (left) and $\delta = 0.8$ (right).

Example 1. In (1) let the initial condition be $\phi(x) = \sin x$. One can easily verify that the solution of (1) is then $u(x,t) = E_\delta(-t^\delta)\sin x$, where $E_\delta(\cdot)$ is the Mittag-Leffler function [1], which is defined by

$$E_\delta(z) := \sum_{k=0}^{\infty} \frac{z^k}{\Gamma(\delta k + 1)}.$$

Figure 1 shows the solutions for $\delta = 0.2$ and $\delta = 0.8$. The layer at $t = 0$ is clearly visible; it is sharper for $\delta = 0.2$.

The rest of Sect. 3 presents numerical results for Example 1.

We first show that the temporal truncation error estimate of Lemma 1 is sharp. Tables 1 and 2 display for Example 1 the maximum value of the truncation error $|D_N^\delta u(x_m, t_n) - D_t^\delta u(x_m, t_n)|$ and their two-mesh orders of convergence in the domains $\bar{Q}$ and $\bar{Q}^*$. These tables indicate that the temporal truncation error of the L1 scheme does not converge in $\bar{Q}$ as N increases (this agrees with (7)) but it is first-order convergent in the subdomain $\bar{Q}^*$, in agreement with (8).

Table 1. Example 1: Temporal truncation errors in $\bar{Q}$

	N = M = 64	N = M = 128	N = M = 256	N = M = 512	N = M = 1024
$\delta = 0.2$	1.283E−001	1.325E−001	1.364E−001	1.400E−001	1.433E−001
	−0.047	−0.042	−0.037	−0.033	
$\delta = 0.4$	2.459E−001	2.498E−001	2.527E−001	2.549E−001	2.565E−001
	−0.023	−0.017	−0.012	−0.009	
$\delta = 0.6$	2.678E−001	2.660E−001	2.645E−001	2.635E−001	2.628E−001
	0.010	0.008	0.006	0.004	
$\delta = 0.8$	1.801E−001	1.757E−001	1.730E−001	1.715E−001	1.706E−001
	0.036	0.022	0.013	0.008	

Table 2. Example 1: Temporal truncation errors in the subdomain $\bar{Q}^*$

	N = M = 64	N = M = 128	N = M = 256	N = M = 512	N = M = 1024
$\delta = 0.2$	5.327E−003	2.514E 003	1.097E 003	4.824E−004	2.154E−004
	1.083	1.197	1.185	1.163	
$\delta = 0.4$	6.492E−003	2.632E−003	9.651E−004	3.564E−004	1.340E−004
	1.303	1.447	1.437	1.411	
$\delta = 0.6$	7.837E−003	3.122E−003	1.128E−003	4.108E−004	1.526E−004
	1.328	1.469	1.457	1.428	
$\delta = 0.8$	1.052E−002	4.760E−003	2.021E−003	8.664E−004	3.773E−004
	1.143	1.236	1.222	1.199	

Next, the convergence of the solution of scheme (5) is investigated. Although the exact solution of Example 1 is known, the errors and rates of convergence are computed using the two-mesh principle in accordance with the examples of Sect. 4 whose exact solutions are unknown. Table 3 displays the maximum two-mesh differences $d^\delta_{M,N}$ and orders of convergence $q^\delta_{M,N}$ for Example 1. They indicate that the method converges at the rate $O(\tau^\delta)$ in agreement with the error estimate (6) (recall that $M = N$, so temporal error dominates spatial error).

Table 3. Example 1: Maximum two-mesh differences and orders of convergence.

	N = M = 64	N = M = 128	N = M = 256	N = M = 512	N = M = 1024
$\delta = 0.2$	2.235E−002	2.072E−002	1.910E−002	1.751E−002	1.595E−002
	0.109	0.117	0.126	0.134	
$\delta = 0.4$	1.775E−002	1.402E−002	1.097E−002	8.520E−003	6.578E−003
	0.340	0.354	0.365	0.373	
$\delta = 0.6$	7.774E−003	5.172E−003	3.431E−003	2.271E−003	1.501E−003
	0.588	0.592	0.595	0.597	
$\delta = 0.8$	2.863E−003	1.624E−003	9.263E−004	5.293E−004	3.030E−004
	0.817	0.810	0.808	0.805	

Figure 2 gives the pointwise two-mesh differences at $x = \pi/2$ for $\delta = 0.2$ and the discretisation parameters $M = N = 64$ and $M = N = 128$. This plot reveals that the maximum two-mesh difference for both solutions occurs at the first interior mesh point in this example. If one computes the two-mesh differences in the subdomain $[0, \pi] \times [p, 1]$ with p a positive constant independent of N, the computed orders of convergence indicate that the method is first-order convergent. In particular, the numerical results in the subdomain $\bar{Q}^*$ for Example 1 are displayed in Table 4. The observed order of convergence for the solution of (5) is of the same order as the order of the truncation error stated in (8) (see also Table 2).

Table 4. Example 1: Maximum two-mesh differences and orders of convergence in the subdomain $\bar{Q}^*$.

	$N = M = 64$	$N = M = 128$	$N = M = 256$	$N = M = 512$	$N = M = 1024$
$\delta = 0.2$	2.023E$-$003	1.023E$-$003	4.938E$-$004	2.423E$-$004	1.210E$-$004
	0.984	1.051	1.027	1.001	
$\delta = 0.4$	3.872E$-$003	1.952E$-$003	9.406E$-$004	4.596E$-$004	2.284E$-$004
	0.989	1.053	1.033	1.009	
$\delta = 0.6$	4.371E$-$003	2.271E$-$003	1.129E$-$003	5.618E$-$004	2.814E$-$004
	0.944	1.008	1.008	0.998	
$\delta = 0.8$	2.826E$-$003	1.537E$-$003	8.142E$-$004	4.254E$-$004	2.208E$-$004
	0.878	0.917	0.937	0.946	

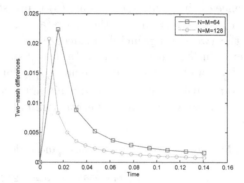

Fig. 2. Example 1: Pointwise two-mesh differences at $x = \pi/2$ with $\delta = 0.2$, $N = M = 64$ ($\square$) and $N = M = 128$ ($\circ$).

4   Numerical Experiments for Non-smooth Data

In this section it will be shown that the improvement in the order of convergence of the scheme (5) in $\bar{Q}^*$ that was observed numerically in Sect. 3 also occurs in problems where the initial condition is not smooth or corner compatibility conditions between the data of (1) are not satisfied.

Example 2. Consider now an example whose initial condition ϕ is continuous in $[0, \pi]$ but fails to be differentiable at one interior point of $(0, \pi)$. Let the initial condition be

$$\phi(x) = \begin{cases} 2x/\pi & \text{if } 0 \le x \le \pi/2, \\ 2(\pi - x)/\pi & \text{if } \pi/2 < x \le \pi, \end{cases} \tag{11}$$

which is not differentiable at $x = \pi/2$. The computed solutions for $\delta = 0.2$ and $\delta = 0.8$ are shown in Fig. 3. The maximum two-mesh differences and the orders of convergence for the scheme (5) in $\bar{Q}$ and the subdomain $\bar{Q}^*$ appear in Tables 5 and 6, respectively. The numerical results in Table 5 shows that the order of global convergence is smaller than δ (compare with the numerical results in

Table 5. Example 2: Maximum two-mesh differences and orders of convergence.

	N = M = 64	N = M = 128	N = M = 256	N = M = 512	N = M = 1024
$\delta = 0.2$	2.096E−002	1.969E 002	1.854E−002	1.743E 002	1.636E−002
	0.090	0.087	0.089	0.092	
$\delta = 0.4$	2.169E−002	1.866E−002	1.618E−002	1.407E−002	1.224E−002
	0.217	0.206	0.202	0.201	
$\delta = 0.6$	1.582E−002	1.257E−002	1.012E−002	8.194E−003	6.648E−003
	0.332	0.312	0.305	0.302	
$\delta = 0.8$	1.014E−002	7.326E−003	5.433E−003	4.077E−003	3.077E−003
	0.469	0.431	0.414	0.406	

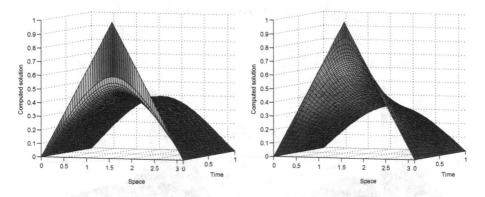

Fig. 3. Computed solution u_m^n of Example 2 when $\delta = 0.2$ (left) and $\delta = 0.8$ (right).

Table 3) — perhaps it is $\delta/2$. On the other hand, Table 6 shows again first-order convergence on the subdomain $\bar{Q}^*$.

Example 3. Consider now an example whose initial condition ϕ is discontinuous at one interior point of $(0, \pi)$. Suppose the initial condition is

$$u(x,0) = \begin{cases} 2x/\pi, & \text{if } 0 \leq x \leq \pi/2, \\ 0, & \text{if } \pi/2 < x \leq \pi, \end{cases} \tag{12}$$

which is discontinuous at $x = \pi/2$. The computed solutions for $\delta = 0.2$ and $\delta = 0.8$ are shown in Fig. 4. Observe that the solution is now more complicated: it exhibits layer regions caused by the time-fractional derivative and the discontinuity of the initial condition. Table 7 gives the maximum two-mesh differences and the orders of convergence in $\bar{Q}$; from these numerical results, the scheme does not appear to converge. Nevertheless, if we compute the two-mesh differences in the subdomain $\bar{Q}^*$ (see Table 8), then the method appears to be first-order convergent.

Table 6. Example 2: Maximum two-mesh differences and orders of convergence in the subdomain $\bar{Q}^*$.

	N = M = 64	N = M = 128	N = M = 256	N = M = 512	N = M = 1024
$\delta = 0.2$	1.881E−003	9.167E−004	4.345E−004	2.111E−004	1.049E−004
	1.037	1.077	1.041	1.008	
$\delta = 0.4$	3.581E−003	1.764E−003	8.396E−004	4.077E−004	2.020E−004
	1.022	1.071	1.042	1.013	
$\delta = 0.6$	4.378E−003	2.209E−003	1.075E−003	5.285E−004	2.632E−004
	0.987	1.039	1.024	1.006	
$\delta = 0.8$	3.861E−003	1.995E−003	1.001E−003	5.047E−004	2.562E−004
	0.952	0.995	0.988	0.978	

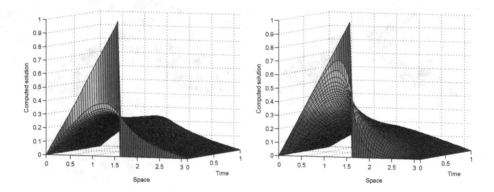

Fig. 4. Computed solution u_m^n of Example 3 when $\delta = 0.2$ (left) and $\delta = 0.8$ (right).

Table 7. Example 3: Maximum two-mesh differences and orders of convergence.

	N = M = 64	N = M = 128	N = M = 256	N = M = 512	N = M = 1024
$\delta = 0.2$	2.420E−002	2.099E−002	1.929E−002	1.828E−002	1.759E−002
	0.205	0.122	0.078	0.055	
$\delta = 0.4$	3.610E−002	3.096E−002	2.789E−002	2.593E−002	2.459E−002
	0.222	0.151	0.105	0.076	
$\delta = 0.6$	4.445E−002	3.734E−002	3.325E−002	3.063E−002	2.895E−002
	0.251	0.168	0.118	0.082	
$\delta = 0.8$	5.426E−002	4.470E−002	3.908E−002	3.545E−002	3.315E−002
	0.280	0.194	0.141	0.097	

Table 8. Example 3: Maximum two-mesh differences and orders of convergence in the subdomain $\bar{Q}^*$.

	$N = M = 64$	$N - M - 128$	$N = M = 256$	$N = M = 512$	$N - M = 1024$
$\delta = 0.2$	7.855E−003	3.925E−003	1.944E−003	9.677E−004	4.837E−004
	1.001	1.013	1.007	1.001	
$\delta = 0.4$	1.004E−002	5.023E−003	2.476E−003	1.229E−003	6.136E−004
	0.999	1.020	1.011	1.002	
$\delta = 0.6$	1.169E−002	5.876E−003	2.899E−003	1.439E−003	7.195E−004
	0.993	1.019	1.010	1.000	
$\delta = 0.8$	1.309E−002	6.612E−003	3.253E−003	1.612E−003	8.066E−004
	0.985	1.024	1.012	0.999	

Example 4. In Example 1 the initial condition satisfies the zero-order compatibility condition

$$\phi(0) = 0, \quad \phi(\pi) = 0, \tag{13}$$

and the first-order compatibility condition

$$\phi''(0) = \phi''(\pi) = 0, \tag{14}$$

(see [5]). We now consider an example that satisfies the zero-order compatibility condition (13) but not the first-order compatibility condition (14). Define the initial condition by

$$\phi(x) = \frac{4}{\pi^2} x(\pi - x). \tag{15}$$

Observe that $\phi''(0) \neq 0$ and $\phi''(\pi) \neq 0$. The computed solutions for $\delta = 0.2$ and $\delta = 0.8$ are shown in Fig. 5. Although this example does not satisfy the compatibility condition (14), we reach the same conclusions as for Example 1: The method converges with order δ in the whole domain and with first order in the subdomain $\bar{Q}^*$ (see Tables 9 and 10, respectively).

Example 5. Finally, we consider an example that fails to satisfy the zero-order compatibility condition (13). Choose

$$\phi(x) \equiv 1. \tag{16}$$

The computed solutions for $\delta = 0.2$ and $\delta = 0.8$ are shown in Fig. 6. The structure of the solution is very complicated; it has layer regions caused by the time-fractional derivative and the incompatibilities at the corners $(0,0)$ and $(\pi,0)$. The numerical results in Table 11 shows that the method is not convergent in the whole domain Q but Table 12 shows that the method is first-order convergent in the subdomain $\bar{Q}^*$.

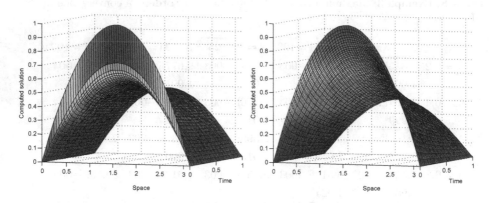

Fig. 5. Computed solution u_m^n of Example 4 when $\delta = 0.2$ (left) and $\delta = 0.8$ (right).

Table 9. Example 4: Maximum two-mesh differences and orders of convergence.

	N = M = 64	N = M = 128	N = M = 256	N = M = 512	N = M = 1024
$\delta = 0.2$	2.227E−002	2.054E−002	1.883E−002	1.714E−002	1.551E−002
	0.116	0.126	0.136	0.145	
$\delta = 0.4$	1.658E−002	1.281E−002	9.807E−003	7.462E−003	5.660E−003
	0.372	0.386	0.394	0.399	
$\delta = 0.6$	6.601E−003	4.347E−003	2.867E−003	1.891E−003	1.248E−003
	0.603	0.600	0.600	0.600	
$\delta = 0.8$	2.520E−003	1.422E−003	8.154E−004	4.680E−004	2.687E−004
	0.826	0.802	0.801	0.800	

Table 10. Example 4: Maximum two-mesh differences and orders of convergence in the subdomain $\bar{Q}^*$

	N = M = 64	N = M = 128	N = M = 256	N = M = 512	N = M = 1024
$\delta = 0.2$	2.050E−003	1.038E−003	5.013E−004	2.460E−004	1.229E−004
	0.982	1.050	1.027	1.001	
$\delta = 0.4$	3.902E−003	1.968E−003	9.490E−004	4.638E−004	2.305E−004
	0.988	1.052	1.033	1.009	
$\delta = 0.6$	4.289E−003	2.236E−003	1.116E−003	5.564E−004	2.790E−004
	0.940	1.002	1.004	0.996	
$\delta = 0.8$	2.520E−003	1.365E−003	7.325E−004	3.872E−004	2.026E−004
	0.884	0.898	0.920	0.935	

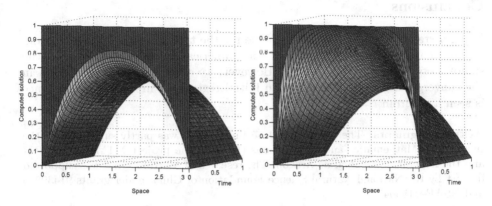

Fig. 6. Computed solution u_m^n of Example 5 when $\delta = 0.2$ (left) and $\delta = 0.8$ (right).

Table 11. Example 5: Maximum two-mesh differences and orders of convergence.

	N = M = 64	N = M = 128	N = M = 256	N = M = 512	N = M = 1024
$\delta = 0.2$	2.466E−002	2.400E−002	2.346E−002	2.305E−002	2.274E−002
	0.039	0.033	0.026	0.019	
$\delta = 0.4$	3.823E−002	3.831E−002	3.834E−002	3.836E−002	3.837E−002
	−0.003	−0.001	−0.001	−0.000	
$\delta = 0.6$	4.923E−002	4.944E−002	4.959E−002	4.965E−002	4.967E−002
	−0.006	−0.005	−0.002	−0.001	
$\delta = 0.8$	5.546E−002	5.600E−002	5.670E−002	5.691E−002	5.695E−002
	−0.014	−0.018	−0.005	−0.001	

Table 12. Example 5: Maximum two-mesh differences and orders of convergence in the subdomain $\bar{Q}^*$.

	N = M = 64	N = M = 128	N = M = 256	N = M = 512	N = M = 1024
$\delta = 0.2$	2.157E−003	1.113E−003	5.431E−004	2.679E−004	1.341E−004
	0.954	1.036	1.020	0.998	
$\delta = 0.4$	4.009E−003	2.046E−003	9.929E−004	4.868E−004	2.422E−004
	0.970	1.043	1.028	1.007	
$\delta = 0.6$	5.413E−003	2.786E−003	1.352E−003	6.623E−004	3.293E−004
	0.958	1.043	1.029	1.008	
$\delta - 0.8$	7.613E 003	3.981E−003	1.935E−003	9.462E−004	4.694E−004
	0.935	1.041	1.032	1.011	

Conclusions

Using a standard finite difference scheme for the time-fractional heat equation, we see that the convergence rate of the numerical method is affected by the smoothness of the initial condition. However, we observe that the reduction in the convergence rate is confined to an initial region and first-order convergence is seen away from $t = 0$.

Acknowledgments. The research of José Luis Gracia was partly supported by the Institute of Mathematics and Applications (IUMA), the project MTM2016-75139-R and the Diputación General de Aragón. The research of Martin Stynes was supported in part by the National Natural Science Foundation of China under grants 91430216 and NSAF-U1530401.

References

1. Diethelm, K.: The Analysis of Fractional Differential Equations. Lecture Notes in Mathematics. Springer, Berlin (2010)
2. Farrell, P.A., Hegarty, A.F., Miller, J.J.H., O'Riordan, E., Shishkin, G.I.: Robust Computational Techniques for Boundary Layers. Applied Mathematics. Chapman & Hall/CRC, Boca Raton (2000)
3. Gracia, J.L., O'Riordan, E., Stynes, M.: Convergence in positive time for a finite difference method applied to a fractional convection-diffusion equation (submitted for publication)
4. Jin, B., Lazarov, R., Zhou, Z.: An analysis of the L1 scheme for the subdiffusion equation with nonsmooth data. IMA J. Numer. Anal. **36**, 197–221 (2016)
5. Luchko, Y.: Initial-boundary-value problems for the one-dimensional time-fractional diffusion equation. Fract. Calc. Appl. Anal. **15**(1), 141–160 (2012)
6. Oldham, K.B., Spanier, J.: The Fractional Calculus. Theory and Applications of Differentiation and Integration to Arbitrary Order. With an Annotated Chronological Bibliography by Bertram Ross. Mathematics in Science and Engineering, vol. 111. Academic Press (A subsidiary of Harcourt Brace Jovanovich, Publishers), New York-London (1974)
7. Stynes, M.: Too much regularity may force too much uniqueness. Fract. Calc. Appl. Anal. **19**(6), 1554–1562 (2016)
8. Stynes, M., O'Riordan, E., Gracia, J.L.: Error analysis of a finite difference method on graded meshes for a time-fractional diffusion equation. SIAM. J. Numer. Anal. (to appear)

Multi-preconditioned Domain Decomposition Methods in the Krylov Subspaces

Valery P. Ilin[✉]

The Institute of Computational Mathematics and Mathematical Geophysics,
SB RAS, Novosibirsk State University, Novosibirsk, Russia
ilin@sscc.ru

Abstract. We consider the algebraic and geometric issues of the advanced parallel domain decomposition methods (DDMs) for solving very large non-symmetric systems of linear algebraic equations (SLAEs) that arise in the finite volume or the finite element approximation of the multi-dimensional boundary value problems on the non-structured grids. The main approaches in question for DDM include the balancing decomposition of the grid computational domain into parameterized overlapping or non-overlapping subdomains with different interface conditions on the internal boundaries. Also, we use two different sructures of the contacting the neigbour grid subdomains: with definition or without definition of the node dividers (separators) as the special grid subdomain. The proposed Schwarz parallel additive algorithms are based on the "total-flexible" multi-preconditioned semi-conjugate direction methods in the Krylov block subspaces. The acceleration of two-level iterative processes is provided by means of aggregation, or coarse grid correction, with different orders of basic functions, which realize a low - rank approximation of the original matrix. The auxiliary subsystems in subdomains are solved by direct or by the Krylov iterative methods. The parallel implementation of algorithms is based on hybrid programming with MPI-processes and multi-thread computing for the upper and the low levels of iterations, respectively. We describe some characteristic features of the computational technologies of DDMs that are realized within the framework of the library KRYLOV in the Institute of Computational Mathematics and Mathematical Geophysics, SB RAS, Novosibirsk. The technical requirements for this code are based on the absence of the program constraints on the degree of freedom and on the number of processor units. The conceptions of the creating the unified numerical envirement for DDMs are presented and discussed.

Keywords: Multi-dimensional boundary value problems · Domain decomposition · Krylov subspaces · Multi-preconditioning · Coarse grid corection · Scalable parallelism · Hybrid programming · Hierarchical memory

© Springer International Publishing AG 2017
I. Dimov et al. (Eds.): NAA 2016, LNCS 10187, pp. 95–106, 2017.
DOI: 10.1007/978-3-319-57099-0_9

1   Introduction

We consider the parallel computational methods and technologies for solving very large non-symmetric sparse SLAEs with positive definite matrices

$$Au = f, \quad A = \{a_{l,m}\} \in \mathcal{R}^{N,N}, \quad u = \{u_l\}, \quad f = \{f_l\} \in \mathcal{R}^N,$$
$$(Av, v) \geq \delta ||v||^2, \quad \delta > 0, \quad (v, w) = \sum_{i=1}^{N} v_i w_i, \quad ||v||^2 = (v, v), \tag{1}$$

which arise in finite element or finite volume approximations of the multi-dimensional boundary value problems (BVPs) on the adaptive non-structured grids. Let we have PDE

$$Lu(\mathbf{r}) = f(\mathbf{r}), \quad \mathbf{r} \in \Omega,$$
$$lu|_\Gamma = g(\mathbf{r}), \quad \mathbf{r} \in \Gamma, \quad \bar{\Omega} = \Omega \bigcup \Gamma, \tag{2}$$

where L is some linear differential operator with piece-wise smooth coefficients and l is boundary condition operator which has different types (Dirichlet, Neumann or Robin) at the different surface segments of Γ, in general. The computational domain $\bar{\Omega} = \Omega \bigcup \Gamma$ may have complicated geometry with multi-connected piece-wise boundary surfaces and contrast material properties in subdomains. We suppose that initial data of BVP (2) provide the existence of the unique solution $u(\mathbf{r})$ with the smooth enough properties, which are sufficient for validity of the numerical methods to be applied.

In recent decades, there are a lot of literature on the parallel domain decomposition methods, and we present in the reference some books and papers only [1–4]. The main approaches in question for DDM include the balancing decomposition of the grid computational domain into parameterized overlapping or non-overlapping subdomains with different interface conditions on the internal boundaries. Also, we use two different structures of the contacting the neigbour grid subdomains: with definition or without definition of the node dividers (separators) as the special (sceleton) grid subdomain. The first type decomposition (with sceleton grid subdomain) is usual for FETI approach of DDM, see [1,2], but in the second case the original matrix A has more regular block-diagonal structure. The proposed Schwarz parallel additive algorithms are based on the flexible multi-preconditioned semi-conjugate direction methods in the block Krylov subspaces. The acceleration of two - level iterative processes is provided by means of aggregation, or coarse grid correction approach, with different orders of basic functions, which realize a low - rank approximation of the original matrix. The auxiliary subsystems in subdomains are solved by direct or by the Krylov iterative methods. The parallel implementation of algorithms is based on hybrid programming with MPI-processes and multi-thread computing for the upper and the low levels of iterations, respectively.

This paper is organized as follows. In the Sect. 2 we consider the geometric issues of the different types of domain decompositions. The next section includes the description of the two level iterative processes for solving SLAEs in Krylov subspaces, on the base of multi-preconditioning approache which was proposed by C. Greif with his colleagues in [5–7]. The Sect. 4 is devoted to the parallel implementation of the algorithms, which are realized in the framework of the library KRYLOV, Institute of Computational Mathematics and Mathematical Geophysics, SB RAS, Novosibirsk. The technical requirements of this code are based on the absence of the formal program constraints on the degree of freedom and on the number of processor units. In the conclusion we discuss the efficiency of the proposed methods and technologies, as well as the conceptions of the creating the unified numerical envirenment for fast solving very large sparse SLAEs and high perfomance computing for parallel DDMs.

2   Geometric Issues of DDM

Domain decomposition approaches can be considered at the continuous level and at the discrete level. We use the second way and suppose that the original computational domain Ω is discretized already into grid computation domain Ω^h. So, in the following, DDM is implemented to the grid domains only, and upper index "h" is omitted for bravity.

Let us decompose Ω into P subdomains (with or without overlapping):

$$\Omega = \bigcup_{q=1}^{P} \Omega_q, \quad \bar{\Omega}_q = \Omega_q \bigcup \Gamma_q, \quad \Gamma_q = \bigcup_{q' \in \omega_q} \Gamma_{q,q'}, \quad \Gamma_{q,q'} = \Gamma_q \bigcap \bar{\Omega}_{q'}, \quad q' \neq q. \quad (3)$$

Here Γ_q is the boundary of Ω_q which is composed from the segments $\Gamma_{q,q'}$, $q' \in \omega_q$, and $\omega_q = \{q_1, ..., q_{M_q}\}$ is a set of M_q contacting, or conjuncted, subdomains. Formally, we can denote also by $\Omega_0 = R^d / \Omega$ the external subdomain:

$$\bar{\Omega}_0 = \Omega_0 \bigcup \Gamma, \quad \Gamma_{q,0} = \Gamma_q \bigcap \bar{\Omega}_0 = \Gamma_q \bigcap \Gamma, \quad \Gamma_q = \Gamma_q^i \bigcup \Gamma_{q,0}, \quad (4)$$

where $\Gamma_q^i = \bigcup_{q' \neq 0} \Gamma_{q,q'}$ and $\Gamma_{q,0} = \Gamma_q^e$ mean internal and external parts of the boundary of Ω_q. We define also an overlapping $\Delta_{q,q'} = \Omega_q \bigcap \Omega_{q'}$ of the neighbouring subdomains. If $\Gamma_{q,q'} = \Gamma_{q',q}$ and $\Delta_{q,q'} = 0$ then overlapping of Ω_q and $\Omega_{q'}$ is empty. In particular, we suppose in (3) that each of P subdomains has no intersection with Ω_0 ($\Omega_q \bigcap \Omega_0 = 0$).

The idea of DDM includes the definition of the sets of BVPs for all subdomains which should be equivalent to the original problem (1):

$$Lu_q(r) = f_q, \quad r \in \Omega_q, \quad l_{q,q'}(u_q)\big|_{\Gamma_{q,q'}} = g_{q,q'} \equiv l_{q',q}(u_{q'})\big|_{\Gamma_{q',q}},$$
$$q' \in \omega_q, \quad l_{q,0}u_q|_{\Gamma_{q,0}} = g_{q,0}, \quad q = 1, ..., P. \quad (5)$$

At each segment of the internal boundaries of subdomains, with operators $l_{q,q'}$ from (4), the interface conditions in the form of the Robin boundary condition are imposed:

$$\alpha_q u_q + \beta_q \frac{\partial u_q}{\partial n_q}\big|_{\Gamma_{q,q'}} = \alpha_{q'} u_{q'} + \beta_{q'} \frac{\partial u_{q'}}{\partial n_{q'}}\big|_{\Gamma_{q',q}}, \quad |\alpha_q| + |\beta_q| > 0, \quad \alpha_q \cdot \beta_q \geq 0.$$
$$(6)$$

Here $\alpha_{q'} = \alpha_q, \beta_{q'} = \beta_q$ and n_q means the outer normal to the boundary segment $\Gamma_{q,q'}$ of the subdomain Ω_q. Strictly speaking, at each part of the internal boundary $\Gamma_{q,q'}, q' \neq 0$, the pair of different coefficients $\alpha_q^{(1)}, \beta_q^{(1)}$ and $\alpha_q^{(2)}, \beta_q^{(2)}$ for the conditions of the type (5) should be given. For example, $\alpha_q^{(1)} = 1, \beta_q^{(1)} = 0$ and $\alpha_q^{(2)} = 0, \beta_q^{(2)} = 1$ correspond formally to the continuity of the solution to be sought for and its normal derivative respectively. The additive Schwarz algorithm in DDM is based on the iterative process, in which the BVPs in each q-th subdomain are solved simultaneously, and right hand sides of boundary condition in (5), (6) are taken from the previous iteration.

We implement domain decomposition in two steps. At the first one, we define subdomains Ω_q without overlapping, i.e. the contacting grid subdomains have no the joint nodes, and each node belongs to one subdomain only. Then we define the grid boundary $\Gamma_q = \Gamma_q^0$ of Ω_q, as well as the extensions of $\bar{\Omega}_q^t = \Omega_q^t \bigcup \Gamma_q^t, \Omega_q^0 = \Omega_q, t = 0, ..., \Delta$, layer by layer:

$$\Gamma_q \equiv \Gamma_q^0 = \{l' \in \hat{\omega}_l, \ l \in \Omega_q, \ l' \notin \Omega_q, \ \Omega_q^1 = \bar{\Omega}_q^0 = \Omega_q \bigcup \Gamma_q^0\},$$
$$\Gamma_q^t = \{l' \in \hat{\omega}_l, \ l \in \Omega_q^{t-1}, \ l' \in \Omega_q^{t-1}, \ \Omega_q^t = \bar{\Omega}_q^{t-1} = \Omega_q^{t-1} \bigcup \Gamma_q^{t-1}\}. \qquad (7)$$

Here Δ means the parameter of extension, or overlapping.

At the Fig. 1, we present an example of 2D grid domain decomposition with grid sceleton subdomain whose node dividers are denoted by crosses.

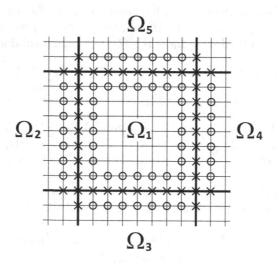

Fig. 1. Decomposition of 2-D domain with grid sceleton subdomain

The second example is presented at the Fig. 2 where we have the grid decomposition without node separators for overlapping parameter $\Delta = 3$.

Algebraic interpretation of DDM, after approximations of BVPs (5), is described by the block version of SLAEs (1):

$$A_{q,q}u_q + \sum_{r\in\hat{\omega}_q} A_{q,r}u_r = f_q, \quad q = 1, ..., P, \tag{8}$$

where $u_q, f_q \in \mathcal{R}^{N_q^{\Delta}}$ are subvectors with the components which belong to corresponding subdomain Ω_q^{Δ}, and N_q^{Δ} is the number of nodes in Ω_q^{Δ}.

In the case for Fig. 1, if the sceleton subdomain is numbered as the last one, the block matrix A has the following arrow type structure:

$$A = \{A_{q,r}\} = \begin{vmatrix} A_{1,1} & & 0 & A_{1,P+1} \\ & \ddots & & \vdots \\ 0 & & A_{P,P} & A_{P,P+1} \\ \hline A_{P+1,1} & \cdots & A_{P+1,P} & A_{P+1,P+1} \end{vmatrix}.$$

In the second case (decomposition without node dividers, Fig. 2), the matrix A has more regular block-diagonal structure.

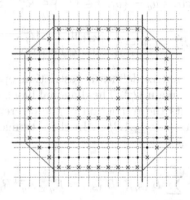

Fig. 2. Decomposition of the grid domain without dividing nodes

The implementation of the interface conditions between adjacent subdomains can be described as follows. Let the l-th node be a near-boundary one in subdomain Ω_q, see Fig. 3. Then write down the corresponding equation in the form

$$(B_{q,q}u)_l \equiv (a_{l,m} + \theta_l \sum_{m\notin\omega_q} a_{l,m})u_l + \sum_{m\in\omega_q} a_{l,m}u_m = f_l + \sum_{m\notin\omega_q} a_{l,m}(\theta_l u_l - u_m). \tag{9}$$

Here θ_l is some parameter which corresponds to different type of boundary condition at the boundary Γ_q: $\theta_l = 0$ corresponds to Dirichlet condition, $\theta_l = 1$ corresponds to Neumann condition, and $\theta_l \in (0,1)$ – to the Robin boundary condition. The diagonal blocks $B_{q,q}$ define the block-diagonal matrix

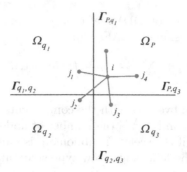

Fig. 3. The grid stencil for near boundary node

$B_s = $ block-diag $\{B_{q,q}\}$ for the additive Schwarz (AS) iterative process. In the implementation of AS we take the right hand side of (9) from the previous iteration.

The additive Schwarz iterative algorithm is defined by the corresponding preconditioning matrix B_{AS} which can be described as follows. For subdomain Ω_q^Δ with overlapping parameter Δ we define a prolongation matrix $R_{q,\Delta}^T \in \mathcal{R}^{N,N_q^\Delta}$ which extends vectors $u_q = \{u_l, \ l \in \Omega_q^\Delta\} \in \mathcal{R}^{N_q^\Delta}$ to $\mathcal{R}^N$ by the relations

$$(R_{q,\Delta}^T u_q)_l = \begin{cases} (u_q)_l \text{ if } \ l \in \Omega_q^\Delta, \\ 0 \quad \text{otherwise.} \end{cases}$$

The tranpose of this matrix defines a restriction operator which restricts vectors in $\mathcal{R}^N$ to the subdomain Ω_q^Δ. The diagonal block of the preconditioning matrix B_{AS}, which presents the restriction of the discretized BVP to the q-th subdomain, is given by $\hat{A}_q = R_{q,\Delta} A R_{q,\Delta}^T$. In these terms, the additive Schwarz preconditioner is defined as

$$B_{AS} = \sum_{q=1}^{P} B_{AS,q}, \ \ B_{AS,q} = R_{q,\Delta}^T \hat{A}_q^{-1} R_{q,\Delta}.$$

Also, we define so called restricted additive Schwarz preconditioner by considering the prolongation $R_{q,0}^T$ instead of $R_{q,\Delta}^T$, i.e.

$$B_{RAS} = \sum_{q=1}^{P} B_{RAS,q}, \ \ B_{RAS,q} = R_{q,0}^T \hat{A}_q^{-1} R_{q,\Delta}.$$

Let us remark, that B_{RAS} is non-symmetric matrix even A is symmetric one.

The second preconditioning matrix which we use for the DDM iterations in Krylov subspaces is responsible for the coarse grid correction, or aggregation approach, based on the low-rank approximation of the original matrix A. We define coarse grid, or macrogrid, Ω_c and corresponding coarse space with degree of freedom $N_c \ll N$, as well as some basic functions $w^k \in \mathcal{R}^N$, $k = 1, ..., N_c$. We suppose that the rectangular matrix $W = (w^1 ... w^{N_c}) \in \mathcal{R}^{N,N_c}$ has a full rank and define the coarse grid preconditioner B_c as follows:

$$B_c^{-1} = W \hat{A}^{-1} W^T, \quad \hat{A} = W^T A W \in \mathcal{R}^{N_c,N_c},$$

where small matrix $\hat{A}$ is low rank approximation of A, W is called restriction matrix and transposed matrix W^T is prolongation one.

In some papers, see [4,8], aggregation, or deflation, technique is applied for improvement of the initial guess for Krylov's iterative methods. Let u^{-1} be an arbitrary vector. Then we can compute the initial vectors u^0 and r^0 by the formulaes

$$u^0 = u^{-1} + B_c^{-1} r^{-1}, \quad r^{-1} = f - A u^{-1},$$
$$r^0 = f - A u^0, \quad p^0 = r^0 - B_c^{-1} r^0, \tag{10}$$

which provide the orthogonal properties

$$W^T r^0 = 0, \quad W^T A p^0 = 0, \tag{11}$$

where p^0 is convential initial direction vector in the Krylov's methods.

It is possible to choose the basic functions $w^k(\boldsymbol{r})$ from the approximation principle. If the solution u of the original problem is smooth enough, we can write

$$u = \{u_l \approx u_l^c = \sum_{k=1}^{N_c} c_k w^k(\boldsymbol{r}_l)\} \cong W \hat{u},$$

where the vector $\hat{u} = \{c_k\} \in \mathcal{R}^{N_c}$ consists of the coefficient of the expansion. If we substitute this representation in (1), we obtain the approximate equation $A W \hat{u} \approx f$. From here, we have by multiplication with W^T the both sides of this equation,

$$W^T A W \hat{u} \approx W^T f, \quad \hat{u} \cong \hat{A}^{-1} W^T f, \quad \hat{A} = W^T A W,$$
$$u \approx W \hat{u} \approx B_c^{-1} f, \quad B_c = W \hat{A}^{-1} W^T. \tag{12}$$

It is natural to use basic function $w^k(\boldsymbol{r})$ as some finite interpolation functions of the different orders. In the simplest case the functions $w^k(\boldsymbol{r})$, $k = 1, ..., N_c = P$, are choosen as unit in k-th subdomain and equal to zero in the other subdomains. It is important, that in general the coarse grid Ω_c does not depend of the domain decomposion.

Let us remark, that instead of the deflation approach (12), which is based on the multiplication of the both sides of the original SLAE with W^T, we can use multiplication with $W^T A^T$. In this case we obtain

$$W^T A^T A W \breve{u} \approx W^T A^T f, \quad \breve{u} = \breve{A}^{-1} W^T A^T f,$$
$$u \approx W \breve{u} \approx \breve{B}_c^{-1} f, \quad \breve{A} = W^T A^T A W, \quad \breve{B}_c = W \breve{A}^{-1} W,$$

and application of the formulae (10) with $\breve{B}_c$, instead of B_c, provides the initial guess with the other kind of orthogonal properties:

$$W^T A^T r^0 = 0, \quad W^T A^T A p^0 = 0.$$

3   Multi-preconditioned SCR

Now, we present the general description of the multi-preconditioned semi-conjugate residual iterative method. Let $r^0 = f - Au^0$ be an initial residual of the algebraic system, and $B_0^{(1)}, ..., B_0^{(m_0)}$ – be a set of some non-singular easily invertible preconditioning matrices. Using them, we define a rectangular matrix composed of the initial direction vectors p_k^0, $k = 1, ..., m_0$:

$$P_0 = [p_1^0 \cdots p_{m_0}^0] \in \mathcal{R}^{N,m_0}, \quad p_l^0 = (B_0^{(l)})^{-1}r^0, \tag{13}$$

which are assumed to be linearly independent.

Successive approximations u^n and the corresponding residuals r^n will be sought for with the help of the recursions

$$\begin{aligned} u^{n+1} &= u^n + P_n \bar{\alpha}_n = u^0 + P_0 \bar{\alpha}_0 + ... + P_n \bar{\alpha}_n, \\ r^{n+1} &= r^n - AP_n \bar{\alpha}_n = r^0 - AP_0 \bar{\alpha}_0 - ... - AP_n \bar{\alpha}_n. \end{aligned} \tag{14}$$

Here $\bar{\alpha}_n = (\alpha_n^1, ..., \alpha_n^{m_n})^T$ are m_n-dimensional vectors. The direction vectors p_l^n, $l = 1, ..., m_n$ forming the columns of the rectangular matrices $P_n = [P_1^n \cdots P_{m_n}^n] \in \mathcal{R}^{N,m_n}$ are defined as orthogonal ones in the sense of satisfying the relations

$$P_n^T A^T AP_k = D_{n,k} = 0 \quad \text{for} \quad k \neq n, \tag{15}$$

where $D_{n,n} = diag\{\rho_{n,l}\}$ is a symmetric positive definite matrix, because the matrices P_k have the full rank as is supposed.

Orthogonal properties (15) provide the minimization of the residual norm $||r^{n+1}||_2$ in the block Krylov subspace of the dimension M_n:

$$K_{M_n} = Span\{P_0, ..., A^{n-1}P_{n-1}\}, \quad M_n = \sum_{k=0}^{n-1} m_k, \tag{16}$$

if we define the coefficient vectors $\bar{\alpha}_n$ by the formula

$$\bar{\alpha}_n = \{\alpha_{n,l}\} = (D_{n,n}^{-1})^{-1} P_n^T A^T r^0. \tag{17}$$

For such values of $\bar{\alpha}_n$ it is easy to check that the vectors p_k^n, r_k^n satisfy to semi-conjugation condition, i.e.

$$P_k^T A^T r^{n+1} = 0, \quad k = 0, 1, ..., n. \tag{18}$$

In this case, the following relations are valid for the functionals of the residuals:

$$\begin{aligned} ||r^{n+1}||^2 &\equiv (r^{n+1}, r^{n+1}) = (r^n, r^n) - (C_n r^0, r^0) = \\ (r^0, r^0) &- (C_0 r^0, r^0) - ... - (C_n r^0, r^0), \quad C_n = P_n AD_{n,n}^{-1} A^T P_n^T. \end{aligned} \tag{19}$$

Let us remark that such properties are valid for any direction vectors p_k^n which satisfy to orthogonal condition (15). We will look for the matrices composed of the direction vectors from the recurrent relations

$$P_{n+1} = Q_{n+1} - \sum_{k=0}^{n} P_k \bar{\beta}_{k,n}, \tag{20}$$

where the auxiliary matrices

$$Q_{n+1} = [q_1^{n+1} \dots q_{m_n}^{n+1}], \quad q_l^{n+1} = (B_{n+1}^{(l)})^{-1} r^{n+1}, \quad l = 1, \dots, m_n, \tag{21}$$

are introduced, $B_{n+1}^{(l)}$ are some non-singular easy invertible preconditioning matrices and $\bar{\beta}_{k,n}$ are the coefficient vectors, which are defined after substitution of (18) into orthogonality conditions (15,) by the formula

$$\bar{\beta}_{k,n} = D_{k,k}^{-1} P_k^T A^T A Q_{n+1}. \tag{22}$$

The following statement is valid.

Theorem 1. *Let the matrices A and $B_n^{(l)}$, $n = 0, 1, \dots$; $l = 1, \dots, m_n$ be nonsingular, and the matrices P_n have a full rank. Then the iterative process (14), (17), (20)–(22) provides minimization of the residual norm $\|r^n\|$ in the block Krylov subspaces (16). Moreover, the following semi-orthogonal properties of residual vectors are valid:*

$$(A^\gamma B_{k,l}^{-1} r^n, r^k) = \begin{cases} 0, & k < n, \\ \sigma_n^{(\gamma)} = (A^\gamma B_{n,l}^{-1} r^n, r^n), & k = n. \end{cases} \tag{23}$$

Also, the coefficients $\alpha_{n,l}$ can be computed by the formula

$$\alpha_{n,l}^{(\gamma)} = (A^\gamma B_{n,l}^{-1} r^n, r^n)/\rho_{n,l}. \tag{24}$$

instead of (17).

The presented MPSCR method use the dynamic, or flexible, definition of the preconditioners $B_n^{(l)}$ and, moreover, their number m_n is variable at the different iteration steps. We propose to use the "total-flexible" variants of "coarse" preconditioning matrics B_c and Jacobi-Schwarz ones B_s. It means that at each n-th iteration we can apply several number of the different coarse grids $\Omega_{c,n}^{l'}$, $l' = 0, 1, \dots, m_n^c$, of the corresponding dimensions $N_{c,n}^{l'}$, and different number of Schwartz – type preconditioners $B_{s,n}^{l''}$, $l'' = 0, 1, \dots, m_n^s$ ($m_n^s = 0$ or $m_n^c = 0$ means no using the corresponding preconditioning at the current iterations). Let us note that application of several Schwarz preconditioners at each iteration corresponds to weighted version of domain decomposition, proposed by Greif in [7], and using several aggregation approaches at one iterative step can be interpretated, in a sense, as additive multi-grid techniques. The simplified versions of SCR where considered in [9]–[11], and in [10] it is called as Generalized Conjugate Residual (GCR) method. In a sense, SCR method is an alternative to wellknown GMRES, see [12].

The disadvantage of the considered algorithm, as well as other Krylov's methods for solving non-symmetric SLAEs, is the necessity of saving the all direction vectors what requires a lot of memory for large number of iterations. There are two main ways to avoid these circumstances. The first one consists into developing the restarts periodically: at each iteration with the numbers $n_r = r \cdot n_{rest}$, $r = 1, 2, ...$, the current residual vector is computed not by recurion (14), but from equation $r^{n_r} = f - Au^{n_r}$, and the next values $u^{n+1}, r^{n+1}, P_{n+1}$ are computed recursivaly again from the beginning. In the second way the limited orthogonalization is used, and only n_{lim} last direction matrices $P_n, P_{n-1}, ..., P_{n-n_{lim}}$ are saved and used in the formula (20). Also, for the case when we have too large deminsions of the Krylov subspaces, in [7] it was proposed to use selected multi-preconditioning when some preconditioners and corresponding direction vectors are omitted on some steps of the iterative process.

Let us remark also, that in the overlapping DDM the vectors u^n are overdetermined in the intersections of the subdomains, and we use restricted additive Schwartz approach with the preconditioner B_{RAS} in this case.

4   Parallel Algorithms and Technologies

The presented principles of the constructing the algorithms are implemented in the library Krylov [12], Institute of Computational Mathematics and Mathematical Geophysics, SB RAS, Novosibirsk, for the efficient parallel solution of the large SLAEs with sparse matrices, which are saved in the compressed sparse row format (CSR, [13]). Of course, the convertors to other convential the key approach for the automatical construction partitioning of the weighted oriented graph that presents the structure of grid set of equation, see Fig. 4. The synchronization of the distributed computing in DDMs is provided by the MPI-processes which are implemented for corresponding subdomains at the multi-core CPUs with shared memory.

The main requirements to develop a proper software are the following:

- no program formal restrictions on the degree of freedom of algebraic systems to be solved and on the number of the processor units or computational cores used; in another words, the numerical environement would be provide the scalable parallelism in the weak and/or in the strong sense;
- application of the advanced iterative methods, with the possibility of the extension of the library functionality;
- robust implementation on the base of hybrid programming of two-level computational process: using MPI tools for outer Krylov's iterations and multi-thread techniques for solving the algebraic subsystems in subdomains;
- high performance computing by using the efficient SPARSE BLAS tools [13] and communication optimization.

The library tools include automatical construction of the balancing domain decomposition, application of the different number of subdomains, size of overlapping, type of interface conditions, using various preconditioners and Krylov's algorithms, etc. The current version of the library does not use multi-GPU computer configuration yet, and corresponding development is in progress.

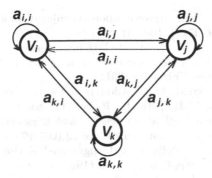

Fig. 4. The example of oriented weighted graph

5   Conclusion

At present time, there are many wellknown libraries and packages of algebraic iterative solvers for large sparse SLAEs: HYPRE, PETSc, Sparse Kit, and others,– which are available by context at the Internet. The goal of the creating the new library Krylov consists in the development of the extendable efficient envirenment for scalable parallel solving the various types of grid algebraic systems (real and complex, symmetric and non-symmetric, positive definite and non-definite, Hermitian and non-Hermitian, etc.) by advanced approaches of DDM by means of hybrid programming at the geterogenous multi-CPU, multi-core and multi-GPU computers. The program implementation is organized as Open Source adapted to the evolution of the computer architectures and platforms. This library has two-fold destinations. The first one consists in the providing the numerical tools for automatical construction of the algorithms, fast developing, validation, verification, testing, and comparative analysis of the new methods. The second aim includes the development of the high preformance code and friendly interface for the end users. In principle, this is not group project, and it is oriented to the wide cooperation of the computational algebra community.

Acknowledgements. The paper is supported by the Russian Scientific Foundation grant N 14-11-00485 and RFBR grant N 14-07-00128.

References

1. Toselli, A., Widlund, O.: Domain Decomposition Methods - Algorithms and Theory. Springer Series in Computational Mathematics, vol. 34. Springer, Heidelberg (2005)
2. Dolean, V., Jolivet, P., Nataf, F.: An Introduction to Domain Decomposition Methods: Algorithms, Theory and Parallel Implementation. SIAM, Philadelphia (2015)
3. Dubois, O., Gander, M.J., Loisel, S., St-Cyr, A., Szyld, D.: The optimized Schwarz method with a coarse grid correction. SIAM J. Sci. Comput. **34**(1), 421–458 (2012)
4. Gurieva, Y.L., Il'in, V.P.: Parallel approaches and technologies of domain decomposition methods. J. Math. Sci. **207**(5), 724–735 (2015)

5. Bridson, R., Greif, C.: A multipreconditioned conjugate gradient algorithm. SIAM J. Matrix Anal. Appl. **27**(4), 1056–1068 (2006)
6. Greif, C., Rees, T., Szyld, D.B.: MPGMRES: a generalized minimum residual method with multiple preconditioners. Technical Report 11-12-23, Department of Mathematics, Temple University (2011)
7. Greif, C., Rees, T., Szyld, D.B.: Additive Schwarz with variable weights. In: Erhel, J., Gander, M.J., Halpern, L., Pichot, G., Sassi, T., Widlund, O. (eds.) Domain Decomposition Methods in Science and Engineering XXI. LNCSE, vol. 98, pp. 779–787. Springer, Cham (2014). doi:10.1007/978-3-319-05789-7_75
8. Chapman, A., Saad, Y.: Deflated and augmented Krylov subspace techniques. Numer. Linear Algebra Appl. **4**(1), 43–66 (1997)
9. Il'in, V.P., Itskovich, E.A.: On the semi-conjugate direction methods with dynamic preconditioning. J. Appl. Ind. Math. **3**(2), 222–233 (2009)
10. Eisenstat, S.C., Elman, H.C., Schultz, M.H.: Variational iterative methods for non-symmetric systems of linear equations. SIAM J. Numer. Anal. **20**(3), 345–357 (1983)
11. Yuan, J.Y., Golub, G.H., Plemmons, R.J., Cecilio, W.A.: Semi-conjugate direction methods for real positive definite systems. BIT **44**(1), 189–207 (2004)
12. Butyugin, D.S., Gurieva, Y.L., Il'in, V.P., Perevozkin, D.V., Petukhov, A.V., Skopin, I.N.: Functionality and technologies of the algebraic solvers in the library Krylov (in Russian). Her. SUSU Ser. Comput. Math. Inf. **2**(3), 92–103 (2013)
13. Intel(R) Math Kernel Library (2013). http://sofware.intel.com/en-us/articles/intel-mkl/

Use of Asymptotics for New Dynamic Adapted Mesh Construction for Periodic Solutions with an Interior Layer of Reaction-Diffusion-Advection Equations

Dmitry Lukyanenko, Nikolay Nefedov$^{(\boxtimes)}$, Egor Nikulin, and Vladimir Volkov

Faculty of Physics, Department of Mathematics,
Lomonosov Moscow State University, Moscow 119991, Russia
nefedov@phys.msu.ru

Abstract. This paper presents the development of analytic-numerical approaches to study periodically moving fronts in singularly perturbed reaction-diffusion-advection models. We describe the results of rigorous asymptotic treatment of the problem and suggest a method to generate a dynamic adapted mesh for the numerical solution of such problems. This method based on *a priori* information. In particular, we take into account a priori estimates on the location of the transition layer, its width and structure. An example is presented to demonstrate the effectiveness of the proposed method.

Keywords: Singularly perturbed parabolic periodic problems · Interior layer · Shishkin mesh · Dynamic adapted mesh

1 Introduction

Singularly perturbed reaction-diffusion-advection problems are used in many practical applications in order to model narrow boundary and interior layers. Their numerical treatment by means of difference schemes requires meshes with a very large number of nodes. To overcome such problems we propose an effective asymptotic-numerical approach for problems with moving interior layers in nonlinear reaction-diffusion-advection equations, which based on an appropriate combination of asymptotic analysis and numerical schemes and should improve the effectiveness of numerical calculations.

This idea has been used recently for problems with stationary interior layers in [1–5] and [6–8], where special grids were used for improving of the effectiveness of calculations.

In the following we present for the numerical approximation of periodic solutions with interior layers in reaction-diffusion-advection equations an effective analytic-numerical approach which exploits and develops the asymptotic results obtained in [9] and which is based on the constructing of a *dynamic adapted mesh* (DAM). Thus, to demonstrate our approach we consider the following problem

© Springer International Publishing AG 2017
I. Dimov et al. (Eds.): NAA 2016, LNCS 10187, pp. 107–118, 2017.
DOI: 10.1007/978-3-319-57099-0_10

$$\begin{cases} \varepsilon^2 \left(\frac{\partial^2 u}{\partial x^2} - \frac{\partial u}{\partial t} \right) - \varepsilon^2 A(u,x,t) \frac{\partial u}{\partial x} - F(u,x,t) = 0, \\ (x,t) \in D := \{(x,t) \in R^2 : -1 < x < 1, t \in R\}, \\ \frac{\partial u}{\partial x}(-1,t,\varepsilon) = u^{(0)}(t), \quad \frac{\partial u}{\partial x}(1,t,\varepsilon) = u^{(1)}(t), \quad t \in R, \\ u(x,t,\varepsilon) = u(x,t+T,\varepsilon), \quad (x,t) \in \overline{D}. \end{cases} \quad (1)$$

where the parameter ε is sufficiently small, $A(u,x,t)$, $F(u,x,t)$, $u^{(0)}(t)$ and $u^{(1)}(t)$ are sufficiently smooth and T-periodic in t.

A similar problem for a reaction-diffusion equation (equations without the advection term) was considered in [10]. Note that the constructing of formal asymptotics in problem (1) is carried out in the standard way and differs from the present paper only slightly. However, the justification of the result requires the further development of our approach and is of interest for a series of problems taking advection terms into account. To justify the asymptotics thus constructed, the asymptotic method of differential inequalities is used. This method is based on known comparison theorems and develops the ideas of formal asymptotic expansions to construct upper and lower solutions in problems with internal and boundary layers (see [9,11] and references therein). In the paper, we also state the Lyapunov asymptotic stability and the local uniqueness of the periodic solution.

2   Asymptotic Analysis Results

2.1   Asymptotics Construction

We assume

- (A1) Let $A(u,x,t)$, $F(u,x,t)$, $u^{(0)}(t)$, and $u^{(1)}(t)$ be functions sufficiently smooth with respect to their arguments and T-periodic with respect to t.
- (A2) The reduced equation $F(u,x,t) = 0$ has precisely three isolated solutions T-periodic with respect to t: $\phi^{(-)}(x,t)$, $\phi^{(0)}(x,t)$, and $\phi^{(+)}(x,t)$, which are ordered, $\phi^{(-)}(x,t) < \phi^{(0)}(x,t) < \phi^{(+)}(x,t)$, $(x,t) \in \overline{D}$.
- (A3) These roots satisfy the following inequalities: $F_u(\phi^{(\pm)}(x,t),x,t) > 0$, $F_u(\phi^{(0)}(x,t),x,t) < 0$.

Introduce the function

$$I(x,t) := \int_{\phi^{(-)}(x,t)}^{\phi^{(+)}(x,t)} F(u,x,t) du$$

and assume the validity of the following conditions:

- (A4) Let the equation $I(x,t) = 0$ have an isolated solution $x_0(t)$ belonging to the interval $(0,1)$.
- (A5) The root $x_0(t)$ of the equation $I(x,t) = 0$ satisfies the inequality $\frac{\partial I}{\partial x}(x_0(t),t) < 0$ for $\forall t \in R$.

It is known [9] that under conditions above problem (1) has a solution of moving front type: in the interval $(-1, 1)$ there is some periodically moving point $x^*(t, \varepsilon)$ which is connected with a thin transition layer containing $x^*(t, \varepsilon)$ such that the solution for $x < x^*(t, \varepsilon)$ is close to the root $\phi^{(-)}(x, t)$ and for $x > x^*(t, \varepsilon)$ is close to the root $\phi^{(+)}(x, t)$.

To construct the formal asymptotic, we split problem (1) into two problems in the domains $\overline{D}^{(\mp)}$, which are considered with respect to $x^*(t, \varepsilon)$ as left and right parts of the domain $[-1, 1]$. We use the additional boundary condition in the point $x^*(t, \varepsilon) : u(x^*, t, \varepsilon) = \phi^{(0)}(x^*, t)$. We construct the asymptotic approximation of the solution in the form of a series in powers of ε

$$U^{(\pm)}(x, t, \varepsilon, x^*) = \bar{u}^{(\pm)}(x, t, \varepsilon) + Q^{(\pm)}(\tau, x^*, t, \varepsilon) + \Pi^{(\pm)}(\xi, t, \varepsilon), \qquad (2)$$

where the regular part is

$$\bar{u}^{(\pm)}(x, t, \varepsilon) = \bar{u}_0^{(\pm)}(x, t) + \varepsilon \bar{u}_1^{(\pm)}(x, t) + \ldots + \varepsilon^n \bar{u}_n^{(\pm)}(x, t) + \ldots,$$

the boundary layer in a neighborhood of $x = 0$ for $u^{(-)}$ and in a neighborhood of $x = 1$ for $u^{(+)}$ is

$$\Pi^{(\pm)}(\xi^{(\pm)}, t, \varepsilon) = \Pi_0^{(\pm)}(\xi^{(\pm)}, t, \varepsilon) + \varepsilon \Pi_1^{(\pm)}(\xi^{(\pm)}, t, \varepsilon) + \ldots + \varepsilon^n \Pi_n^{(\pm)}(\xi^{(\pm)}, t, \varepsilon) + \ldots,$$

and the part of the internal transition layer in a neighborhood of x^*, $\tau = (x - x^*)/\varepsilon$ is

$$Q^{(\pm)}(\tau, x^*, t, \varepsilon) = Q_0^{(\pm)}(\tau, x^*, t) + \varepsilon Q_1^{(\pm)}(\tau, x^*, t) + \ldots + \varepsilon^n Q_n^{(\pm)}(\tau, x^*, t) + \ldots.$$

Using the method of boundary layer functions and taking into account the specific features of a parabolic operator (see [9,12,13]), we get a sequence of problems for determination of the coefficients of the asymptotic series (2), from which, in particular, we see that $\bar{u}_0^{(-)}(x, t) = \phi^{(-)}(x, t)$, $\bar{u}_0^{(+)}(x, t) = \phi^{(+)}(x, t)$.

The regular and boundary layer terms of the asymptotic expansion (2) are constructed in the standard way. The terms $Q_0^{(\pm)}(\tau, x^*, t)$ are found from the following problems:

$$\frac{\partial^2 Q_0^{(\pm)}(\tau, x^*, t)}{\partial \tau^2} = F(\phi^{(\pm)}(x^*, t) + Q_0^{(\pm)}, x^*, t),$$
$$Q_0^{(\pm)}(0, x^*, t) + \bar{u}_0^{(\pm)}(x^*, t) = \phi^{(\pm)}(x^*, t), \qquad (3)$$
$$Q_0^{(\pm)}(\pm\infty, x^*, t) = 0.$$

It is known (see for e.q. [13]) that

$$|Q_0^{(\pm)}(\tau, x^*, t)| \le C e^{-k|\tau|}, \qquad (4)$$

where C and k are some positive constants.

The functions $Q_1^{(\pm)}$ are defined by the following problems:

$$\frac{\partial^2 Q_1^{(\pm)}}{\partial \tau^2} - \frac{\partial \tilde{F}^*}{\partial u} Q_1^{(\pm)} = r_1^{(\pm)},$$
$$Q_1^{(\pm)}(0, x^*, t) + \bar{u}_1^{(1)}(x^*, t) = 0,$$
$$Q_1^{(\pm)}(\pm\infty, x^*, t) = 0, \qquad (5)$$
$$r_1^{(\pm)}(\tau, x^*, t) := \frac{\partial Q_0^{(\pm)}}{\partial \tau}(\tau, x^*, t) \left(-\frac{\partial x^*}{\partial t} + \tilde{A}^* \right) + \tau \left(\frac{\partial \tilde{F}^*}{\partial u} \frac{\partial \bar{u}_0}{\partial x}(x^*, t) + \frac{\partial \tilde{F}^*}{\partial x} \right),$$

where the symbols "∼" and "∗" above and to the right of a function mean that the value of the function is taken at the argument $(\tilde{u}^{(\pm)}(\tau, x^*, t), x^*, t)$, and $\tilde{u}^{(\pm)}(\tau, x^*, t) := \bar{u}_0^{(\pm)}(x^*, t) + Q_0^{(\pm)}(\tau, x^*, t)$. Next terms Q_i are defined from the problems similar to (5) and have the exponential estimate of type (4).

We seek the function x^* in the form

$$x^*(t, \varepsilon) = x_0(t) + \varepsilon x_1(t) + \varepsilon^2 x_2(t) + \ldots,$$

and use C^1-matching condition of the constructed asymptotics to find x_i. We have that x_0 is defined by assumption $(A4)$ and for x_1 we have the equation

$$\frac{\partial I(x_0, t)}{\partial x} x_1 + \int_{-\infty}^{+\infty} \left(\Phi^2(\tau, x_0, t)(\tilde{A} - \frac{\partial x_0}{\partial t}) + \Phi(\tau, x_0, t)\frac{\partial \tilde{F}}{\partial x}\tau \right) d\tau = 0,$$

where $\Phi = \Phi^{(\pm)}(\tau, x^*, t) = \frac{\partial Q_0^{(\pm)}(\tau, x^*, t)}{\partial \tau}$. Next terms x_i can be found from the similar problems.

2.2   Main Result

We denote by $U_n(x, t, \varepsilon, x^*)$ the partial sums of order n of the asymptotic series constructed above in which the argument τ of the Q-functions is replaced by $\tau_n = \left(x - \sum_{i=0}^{n+2} \varepsilon^i x_i(t) \right) / \varepsilon$ and x^* by $x_n^* = \sum_{i=0}^{n+2} \varepsilon^i x_i(t)$.

Theorem 1. *If conditions $(A1)$–$(A5)$ hold, then, for sufficiently small ε, there is a solution $u(x, t, \varepsilon)$ of problem (1) which is a contrast structure of the step type, asymptotically stable, and the following bound holds:*

$$|U_n(x, t, \varepsilon) - u(x, t, \varepsilon)| < C\varepsilon^{n+1}, \quad (x, t) \in \overline{D}.$$

The proof follows [9] and we do not present it here.

The results of the asymptotic analyses is used to create the numerical approach in the Sect. 3. In particular, we use the information about the location of the interior layer, that is $x_0(t) + \varepsilon x_1(t)$, the structure of interior layer, which is defined by the problem (3), and the width of the transition layer, which is defined by the estimate for Q_0 (see (4)) and similar estimate for Q_k, we have

$$h = C|\varepsilon \ln \varepsilon|. \tag{6}$$

2.3   Example

In Sect. 3 we illustrate our approach by means of the following particular case of problem (1)

$$\begin{cases} \varepsilon^2 \left(\frac{\partial^2 u}{\partial x^2} - \frac{\partial u}{\partial t} \right) - \varepsilon^2 u \frac{\partial u}{\partial x} = (u^2 - 4)(u + (x - h_0(t))), \\ x \in (-1, 1), \ t \in R, \quad \frac{\partial u}{\partial x}(-1, t) = 0, \ \frac{\partial u}{\partial x}(1, t) = 0, \ t \in R \\ u(x, t, \varepsilon) = u(x, t + T, \varepsilon), \ x \in (-1, 1), \ t \in R \end{cases} \tag{7}$$

where $h_0(t) = \frac{\sin(t)}{2}$. For this example we have: $\phi^{(\pm)}(x,t) = \pm 2$, $\phi^{(0)}(x,t) = h_0(t) - x$, and the function $I(x,t) = \int_{-2}^{2}(u^2 - 4)(u + (x - h_0(t)))du = \frac{8}{3}(-x + h_0(t))$. Therefore, we get $x_0(t) = h_0(t)$ and it follows, that assumptions $(A1)$–$(A5)$ are satisfied. The problem for $\tilde{u}(\tau, x_0, t) = \bar{u}_0(x_0, t) + Q_0(\tau, x_0, t)$ has the form

$$\frac{\partial^2 \tilde{u}^{(\pm)}}{\partial \tau^2} = (\tilde{u}^{(\pm)})^3 - 4\tilde{u}^{(\pm)}, \ \tau \in R, \ \tilde{u}^{(\pm)}(0, x_0, t) = 0, \tilde{u}^{(\pm)}(\pm\infty, x_0, t) = \pm 2,$$

and we have $\tilde{u}^{(\pm)}(\tau, x_0, t) = 2\tanh(\sqrt{2}\tau)$, $\Phi^{(\pm)}(\tau, x_0, t) = 2\sqrt{2}\frac{1}{\cosh^2(\sqrt{2}\tau)}$.

We also have $m(t) = \int_{-\infty}^{+\infty} \Phi^2(\tau, x_0, t)d\tau = \frac{16\sqrt{2}}{3}$, $\tilde{F} = \tilde{u}(\tilde{u}^2 - 4)$, $\tilde{F}_x = \tilde{u}^2 - 4$, $\frac{\partial I(x_0, t)}{\partial x} = -\frac{8}{3}$.

Taking into account, that

$$\int_{-\infty}^{+\infty} \Phi^2(\tau, x_0, t)\tilde{A}d\tau = 0, \qquad \int_{-\infty}^{+\infty} \Phi(\tau, x_0, t)\frac{\partial \tilde{F}}{\partial x}\tau = 0,$$

the equation for x_1

$$\frac{\partial I(x_0, t)}{\partial x}x_1 + \int_{-\infty}^{+\infty}\left(\Phi^2(\tau, x_0, t)(\tilde{A} - \frac{\partial x_0}{\partial t}) + \Phi(\tau, x_0, t)\frac{\partial \tilde{F}}{\partial x}\tau\right)d\tau = 0,$$

can be rewritten as $-\frac{8}{3}x_1 = -\frac{8\sqrt{2}}{3}\cos(t)$ and therefore $x_1(t) = \sqrt{2}\cos(t)$.

3   Dynamic Adapted Mesh Construction

Using as *a priori* information the result from Sect. 2 about dependents of the transition point location in time we are able to construct so called *dynamic adapted mesh* (which is ε–independent on spatial variable x). This mesh will have in each time moment the same (or proportional) number of nodes inside and outside of the interior layer and this number will not depend on the small parameter ε.

At first, we introduce uniform mesh T_M only on t–dimension (it is also possible to use some quasiuniform mesh [16] without any changes at the further algorithm) that has number of nodes $M + 1$ (that equals to M intervals): $T_M = \{t_m, 0 \le m \le M : 0 = t_0 < t_1 < t_2 < \ldots < t_{M-1} < t_M = T\}$. The choice of the time steep τ (in the case of uniform mesh) is very important. In process of constructing of dynamic adapted mesh it is possible to obtain a situation in which local refining of the mesh on the new time layer does not have intersection (or this intersection is small enough) with local refining of the mesh on the previous time layer. In order to avoid such situation we have to check the condition on the maximal allowed time step τ—the transition point position does not have to leave the current local refining of the mesh in one time step:

$$\tau = t_{m+1} - t_m, \quad \text{where} \quad t_{m+1} = \text{argmax}_{t_m \in T_M} |x'_{tr}(t_m)|. \tag{8}$$

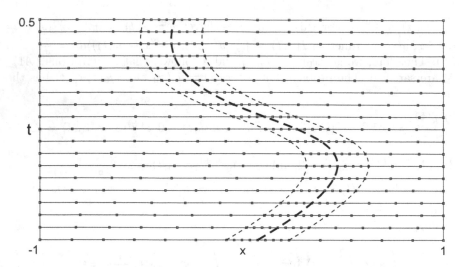

Fig. 1. Some example of the *dynamic adapted mesh* XT_M.

In the case of the considered example the time step τ must be less than $(T - 0)/|2\varepsilon \ln \varepsilon|$.

After that, for each time node m of the mesh T_M we are able to construct a piecewise uniform mesh XT_m on x–dimension that has some uniform refining in $\varepsilon \ln \varepsilon$–neighborhood of the transition point $x_{tr}(t_m)$:

$$
\begin{aligned}
XT_m = \{ \, x_n, \; 0 \leq n \leq N + KN : \\
x_n = 0 \quad \text{for} \quad n = 0, \\
x_n = x_{n-1} + h_{left} \quad \text{for} \quad n = \overline{1, N_{left}}, \\
x_n = x_{n-1} + h_{int} \quad \text{for} \quad n = \overline{N_{left} + 1, N_{left} + KN}, \\
x_n = x_{n-1} + h_{right} \quad \text{for} \quad n = \overline{N_{left} + K_{int}N + 1, N + KN} \},
\end{aligned}
$$

where N is a control parameter that determine number of intervals allocated inside the interior layer (intervals outside of the layer rationed out, so $h_{left} \cong h_{right}$),

$$
N_{left} = \left[\frac{\left(x_{tr}(t_m) - C|\varepsilon \ln \varepsilon| \right) - (-1)}{1 - (-1) - 2C|\varepsilon \ln \varepsilon|} \, N \right],
$$

$$
N_{right} = N - N_{left},
$$

$$
h_{left} = \frac{\left(x_{tr}(t_m) - C|\varepsilon \ln \varepsilon| \right) - (-1)}{N_{left}},
$$

$$
h_{right} = \frac{1 - \left(x_{tr}(t_m) + C|\varepsilon \ln \varepsilon| \right)}{N_{right}},
$$

$$h_{int} = \frac{2C|\varepsilon \ln \varepsilon|}{K_{int}N},$$

C is a control parameters that is able to adjust thickness of the interior layer ($C = 1$ on default); K is a control parameters that denote relative density of the refining inside the interior layer ($K = 1$ on default).

As a result we have constructed the *dynamic adapted mesh* (DAM) XT_M (see for example Fig. 1), which has on each time layer T_m number of intervals equals to $N + KN$. Also we want to remind that this mesh is ε-independent on spatial variable.

4 Numerical Example

Let us rewrite the Burgers equation with periodic coefficients for which we have performed asymptotic investigations in Sect. 2:

$$\begin{cases} \varepsilon^2 \left(\frac{\partial^2 u}{\partial x^2} - \frac{\partial u}{\partial t} \right) - \varepsilon^2 u \frac{\partial u}{\partial x} = \left(u^2 - 4 \right) \left(u + (x - h_0(t)) \right) & \text{for } x \in (-1, 1), \ t \in (0, 2\pi], \\ \frac{\partial u}{\partial x}(-1, t) = \alpha(t), \ \frac{\partial u}{\partial x}(1, t) = \beta(t) & \text{for } t \in (0, 2\pi], \\ u(x, 0) = u_{init}(x), \end{cases}$$
(9)

where $\alpha(t) = 0$, $\beta(t) = 0$ and $h_0(t) = \frac{1}{2} \sin t$.

As we have shown at the end of Sect. 2, all conditions for the asymptotic procedure are fulfilled for this example.

At the first step we estimate $u_{init}(x)$ by using the solution of the stationary problem (see Fig. 2):

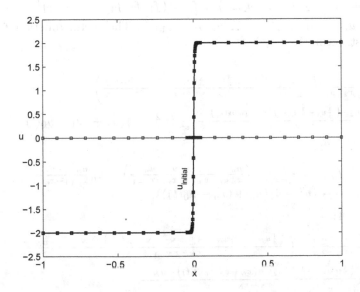

Fig. 2. The example of $u_{init}(x)$ for $\varepsilon = 10^{-2}$ (some refining of the mesh in $\varepsilon \ln \varepsilon$-neighborhood of the transition point has been performed at the initial time moment).

$$\begin{cases} \varepsilon^2 \left(\frac{\partial^2 g}{\partial x^2} - g \frac{\partial g}{\partial x} \right) = (g^2 - 4)(g + (x - h_0(0))) & \text{for } x \in (-1,1), \\ g(-1) = \alpha(0), \quad g(1) = \beta(0). \end{cases} \tag{10}$$

For the numerical solution of the Eq. (9) we apply the stiff method of lines (SMOL) [14,15] in order to reduce the PDE to a system of ODEs that can be solved by the Rosenbrock scheme with complex coefficient [14,17].

Approximating the derivatives with second order accuracy by finite-differences in (9), we obtain the following system of ODEs to determine $N - 1$ unknown functions $u_n \equiv u_n(t) \equiv u(x_n, t)$ $(n = \overline{1, N-1}$, u_0 and u_N we get from the boundary conditions):

$$\begin{cases} \frac{du_n}{dt} = \frac{2}{x_{n+1} - x_{n-1}} \left(\frac{u_{n+1} - u_n}{x_{n+1} - x_n} - \frac{u_n - u_{n-1}}{x_n - x_{n-1}} \right) - u_n \frac{u_{n+1} - u_{n-1}}{x_{n+1} - x_{n-1}} \\ \qquad - \frac{1}{\varepsilon^2} (u_n^2 - 4)(u_n + (x_n - h_0(t))), \\ u_0 = \frac{4}{3} u_1 - \frac{1}{3} u_2 - \frac{2}{3}(x_1 - x_0)\alpha(t), \\ u_N = \frac{4}{3} u_{N-1} - \frac{1}{3} u_{N-2} + \frac{2}{3}(x_N - x_{N-1})\beta(t), \\ u(x_n, 0) = u_{init}(x_n). \end{cases} \tag{11}$$

Note, that we have used approximations of the first derivatives in boundary conditions with asymmetric formula with the second order of approximation on an uniform mesh. This assumption has to do with the fact that we suppose that the moving front never achieve the boundaries (so in a neighborhood of boundary the mesh is always uniform).

This system (11) can be rewritten as

$$\begin{cases} \frac{d\boldsymbol{u}}{dt} = \boldsymbol{f}(\boldsymbol{u}, t), \\ \boldsymbol{u}(0) = \boldsymbol{u}_{init}, \end{cases} \tag{12}$$

where $\boldsymbol{u} = \begin{pmatrix} u_1 & u_2 & u_3 & \dots & u_{N-1} \end{pmatrix}^T$, $\boldsymbol{f} = \begin{pmatrix} f_1 & f_2 & f_3 & \dots & f_{N-1} \end{pmatrix}^T$ and $\boldsymbol{u}_{init} = \begin{pmatrix} u_{init}(x_1) & u_{init}(x_2) & u_{init}(x_3) & \dots & u_{init}(x_{N-1}) \end{pmatrix}^T$. The vector-function $\boldsymbol{f}$ has the following structure:

for $n = 1$:

$$f_1 = \frac{2}{x_2 - x_0} \left(\frac{u_2 - u_1}{x_2 - x_1} - \frac{u_1 - \left(\frac{4}{3} u_1 - \frac{1}{3} u_2 - \frac{2}{3}(x_1 - x_0)\alpha(t) \right)}{x_1 - x_0} \right)$$
$$- u_1 \frac{u_2 - \left(\frac{4}{3} u_1 - \frac{1}{3} u_2 - \frac{2}{3}(x_1 - x_0)\alpha(t) \right)}{x_2 - x_0} - \frac{1}{\varepsilon^2} (u_1^2 - 4)(u_1 + (x_1 - h_0(t))), \tag{13}$$

for $n = \overline{2, N-2}$:

$$f_n = \frac{2}{x_{n+1} - x_{n-1}} \left(\frac{u_{n+1} - u_n}{x_{n+1} - x_n} - \frac{u_n - u_{n-1}}{x_n - x_{n-1}} \right) - u_n \frac{u_{n+1} - u_{n-1}}{x_{n+1} - x_{n-1}}$$
$$- \frac{1}{\varepsilon^2} (u_n^2 - 4)(u_n + (x_n - h_0(t))), \tag{14}$$

for $n = N - 1$:

$$f_{N-1} = \frac{2}{x_N - x_{N-2}} \left(\frac{\left(\frac{4}{3} u_{N-1} - \frac{1}{3} u_{N-2} + \frac{2}{3}(x_N - x_{N-1})\beta(t) \right) - u_{N-1}}{x_N - x_{N-1}} - \frac{u_{N-1} - u_{N-2}}{x_{N-1} - x_{N-2}} \right)$$
$$- u_{N-1} \frac{\left(\frac{4}{3} u_{N-1} - \frac{1}{3} u_{N-2} + \frac{2}{3}(x_N - x_{N-1})\beta(t) \right) - u_{N-2}}{x_N - x_{N-2}}$$
$$- \frac{1}{\varepsilon^2} (u_{N-1}^2 - 4)(u_{N-1} + (x_{N-1} - h_0(t))). \tag{15}$$

For the numerical solution of this system of ODEs (12) we use the Rosenbrock scheme with complex coefficients (CROS1) which is monotone, stable and has the order of the accuracy $O(\tau^2)$ [14].

$$\boldsymbol{u}(t_{m+1}) = \boldsymbol{u}(t_m) + (t_{m+1} - t_m)\,Re\,\boldsymbol{w},$$
where $\boldsymbol{w}$ is the solution of the SLAE
$$\left[E - \tfrac{1+i}{2}(t_{m+1} - t_m)\boldsymbol{f}_u\Big(\boldsymbol{u}(t_m), t\Big)\right]\boldsymbol{w} = \boldsymbol{f}\left(\boldsymbol{u}(t_m), \tfrac{t_m + t_{m+1}}{2}\right). \tag{16}$$

Here, E is identity matrix, $\boldsymbol{f}_u$ is the Jacobian matrix which in the considered case has the following structure:

for $n = 1$:

$$\frac{\partial f_1}{\partial u_1} = \frac{2}{x_2 - x_0}\left(-\frac{1}{x_2 - x_1} - \frac{1}{x_1 - x_0}\right)$$
$$- \frac{u_2 - \left(\frac{4}{3}u_1 - \frac{1}{3}u_2 - \frac{2}{3}(x_1 - x_0)\alpha(t)\right)}{x_2 - x_0}$$
$$- \frac{1}{\varepsilon^2}(u_1^2 - 4 + 2u_1(u_1 + x_1 - h_0(t))),$$

$$\frac{\partial f_1}{\partial u_2} = \frac{2}{x_2 - x_0}\left(\frac{1}{x_2 - x_1}\right) - \frac{u_1}{x_2 - x_0},$$

for $n = \overline{2, N-2}$:

$$\frac{\partial f_n}{\partial u_{n-1}} = \frac{2}{x_{n+1} - x_{n-1}}\left(\frac{1}{x_n - x_{n-1}}\right) + \frac{u_n}{x_{n+1} - x_{n-1}},$$
$$\frac{\partial f_n}{\partial u_n} = \frac{2}{x_{n+1} - x_{n-1}}\left(-\frac{1}{x_{n+1} - x_n} - \frac{1}{x_n - x_{n-1}}\right)$$
$$- \frac{u_{n+1} - u_{n-1}}{x_{n+1} - x_{n-1}} - \frac{1}{\varepsilon^2}(u_n^2 - 4 + 2u_n(u_n + x_n - h_0(t))),$$
$$\frac{\partial f_n}{\partial u_{n+1}} = \frac{2}{x_{n+1} - x_{n-1}}\left(\frac{1}{x_{n+1} - x_n}\right) - \frac{u_n}{x_{n+1} - x_{n-1}},$$

for $n = N - 1$:

$$\frac{\partial f_{N-1}}{\partial u_{N-2}} = \frac{2}{x_N - x_{N-2}}\left(\frac{1}{x_{N-1} - x_{N-2}}\right) + \frac{u_{N-1}}{x_N - x_{N-2}},$$
$$\frac{\partial f_{N-1}}{\partial u_{N-1}} = \frac{2}{x_N - x_{N-2}}\left(-\frac{1}{x_N - x_{N-1}} - \frac{1}{x_{N-1} - x_{N-2}}\right)$$
$$- \frac{\left(\frac{4}{3}u_{N-1} - \frac{1}{3}u_{N-2} + \frac{2}{3}(x_N - x_{N-1})\beta(t)\right) - u_{N-2}}{x_N - x_{N-2}}$$
$$- \frac{1}{\varepsilon^2}(u_{N-1}^2 - 4 + 2u_{N-1}(u_{N-1} + x_{N-1} - h_0(t))).$$

Other components of $\boldsymbol{f}_u$ for the considered equation equal to zero.

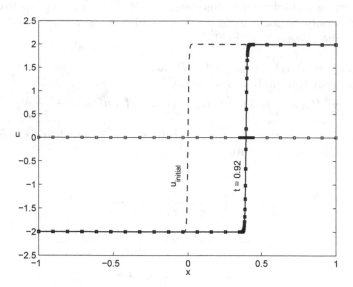

Fig. 3. The example of calculation for $\varepsilon = 10^{-2}$, $N = 20$, $C = 1$, $K = 3$, $T = 2\pi$ (some refining of the mesh in $\varepsilon \ln \varepsilon$–neighborhood of the transition point has been performed at the current time moment).

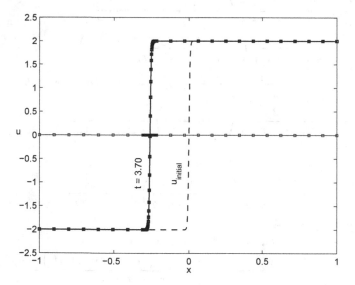

Fig. 4. The example of calculation for $\varepsilon = 10^{-2}$, $N = 20$, $C = 1$, $K = 3$, $T = 2\pi$ (some refining of the mesh in $\varepsilon \ln \varepsilon$–neighborhood of the transition point has been performed at the current time moment).

Note, that in the numerical scheme (16) we use for calculations $u(t_m)$ that has been determined on the XT_m mesh. So, after applying (16), we have to interpolate $u(t_{m+1})$ on XT_{m+1} mesh.

Some example of numerical calculations is represented on the Figs. 3 and 4.

Acknowledgements. The work was supported by RFBR (projects No. 17-01-00519, 17-01-00670, 17-01-00159, 16-01-00755 and 16-01-00437) and the Ministry of Education and Science of the Russian Federation.

References

1. Shishkin, G.: Grid approximation of a singularly perturbed quasilinear equation in the presence of a transition layer. Russ. Acad. Sci. Dokl. Math. **47**(1), 83–88 (1993)
2. O'Riordan, E., Shishkin, G.: Singularly perturbed parabolic problems with non-smooth data. J. Comput. Appl. Math. **1**, 233–245 (2004)
3. Franz, S., Kopteva, N.: Green's function estimates for a singularly perturbed convection-diffusion problem. J. Differ. Equ. **252**, 1521–1545 (2012)
4. Kopteva, N.: Numerical analysis of a 2d singularly perturbed semilinear reaction-diffusion problem. In: Margenov, S., Vulkov, L.G., Waśniewski, J. (eds.) NAA 2008. LNCS, vol. 5434, pp. 80–91. Springer, Heidelberg (2009). doi:10.1007/978-3-642-00464-3_8
5. Kopteva, N., O'Riordan, E.: Shishkin meshes in the numerical solution of singularly perturbed differential equations. Int. J. Numer. Anal. Model. **1**(1), 1–18 (2009)
6. O'Riordan, E., Quinn, J.: Numerical method for a nonlinear singularly perturbed interior layer problem. In: Clavero, C., Gracia, J., Lisbona, F. (eds.) BAIL 2010 - Boundary and Interior Layers, Computational and Asymptotic Methods. Lecture Notes in Computational Science and Engineering, vol. 81, pp. 187–195. Springer, Heidelberg (2011)
7. O'Riordan, E., Quinn, J.: Parameter-uniform numerical method for some linear and nonlinear singularly perturbed convection-diffusion boundary turning point problems. BIT Numer. Math. **51**(2), 317–337 (2011)
8. Kopteva, N., Stynes, M.: Stabilised approximation of interior-layer solutons of a singularly perturbed semilinear reaction-diffusion problem. Numer. Math. **119**, 787–810 (2011)
9. Nefedov, N., Nikulin, E.: Existence and stability of periodic contrast structures in the reaction-advection-diffusion problem. Russ. J. Math. Phys. **22**(2), 215–226 (2015)
10. Volkov, V., Nefedov, N.: Development of the asymptotic method of differential inequalities for investigation of periodic contrast structures in reaction-diffusion equations. Zh. Vychisl. Mat. Mat. Fiz. **46**(4), 615–623 (2006). Comput. Math. Math. Phys. **46**(4), 585–593 (2006)
11. Nefedov, N.: Comparison principle for reaction-diffusion-advection problems with boundary and internal layers. In: Dimov, I., Faragó, I., Vulkov, L. (eds.) NAA 2012. LNCS, vol. 8236, pp. 62–72. Springer, Heidelberg (2013). doi:10.1007/978-3-042-41515-9_6
12. Nefedov, N., Recke, L., Schnieder, K.: Existence and Asymptotic stability of periodic solutions with an interior layer of reaction-advection-diffusion equations. J. Math. Anal. Appl. **405**, 90–103 (2013)

13. Vasileva, A., Butuzov, V., Nefedov, N.: Singularly perturbed problems with boundary and internal layers. Tr. Mat. Inst. Steklova. **268**, 268–283 (2010). Proc. Steklov Inst. Math. **268**(1), 258–273 (2010)
14. Al'shin, A., Al'shina, E., Kalitkin, N., Koryagina, A.: Rosenbrock schemes with complex coefficients for stiff and differential algebraic systems. Comput. Math. Math. Phys. **46**(8), 1320–1340 (2006)
15. Hairer, E., Wanner, G.: Solving of Ordinary Differential Equations. Stiff and Differential-Algebraic Problems. Springer, Heidelberg (2002)
16. Kalitkin, N., Al'shin, A., Al'shina, E., Rogov, B.: Computations on Quasi-Uniform Grids. Fizmatlit, Moscow (2005). (in Russian)
17. Rosenbrock, H.: Some general implicit processes for the numerical solution of differential equations. Comput. J. **5**(4), 329–330 (1963)

A Mathematical Model and a Numerical Algorithm for an Asteroid-Comet Body in the Earth's Atmosphere

V. Shaydurov[✉], G. Shchepanovskaya, and M. Yakubovich

Institute of Computational Modeling of Siberian Branch of Russian Academy of Sciences, 50/44 Akademgorodok, Krasnoyarsk 660036, Russia
shaidurov04@mail.ru

Abstract. In the paper, a mathematical model is proposed for the modeling of the complex of phenomena which accompany the passage of a friable asteroid-comet body through the Earth's atmosphere: the material ablation, the dissociation of molecules, and the ionization. The model is constructed on the basis of the time-dependent Navier-Stokes equations for viscous heat-conducting gas with an additional equation for the propagation of friable lumpy-dust material in air. A numerical algorithm is proposed for solving the formulated initial boundary-value problem as the combination of the semi-Lagrangian approximation for Lagrange transport derivatives and the conforming finite element method for other terms. A numerical example illustrates these approaches.

Keywords: Time-dependent Navier-Stokes equations · Viscous heat-conducting gas · Friable asteroid-comet body · semi-Lagrangian approximation · Conforming finite element method

1 Introduction

In the paper, a mathematical model and a numerical algorithm are proposed for the modeling of the complex of phenomena which accompany the passage of a friable asteroid-comet body through the Earth's atmosphere. This model describes the processes of body ablation and the relation between two kinds of energy: the translational one (that defines the thermodynamical temperature and pressure) and energy of rotation, oscillation, electronic excitation of molecules and atoms, dissociation of molecules into atoms, radicals, and ions. For the unperturbed atmosphere, the distribution of density, pressure, and temperature in height is taken for the standard atmosphere [1] considering gravity. Moreover, we assume that the atmosphere gas consists entirely of molecular nitrogen N_2 and oxygen O_2. This is close to the ISA international standard atmosphere model wherein the content of those gases is 98.6% by weight. The proposed model is constructed on the basis of the time-dependent Navier-Stokes equations for viscous heat-conducting gas [2] with an additional equation for the motion and propagation of friable lumpy-dust material in air. The energy equation is

© Springer International Publishing AG 2017
I. Dimov et al. (Eds.): NAA 2016, LNCS 10187, pp. 119–131, 2017.
DOI: 10.1007/978-3-319-57099-0_11

modified for the relation between two its kinds: the usual energy of translational movement of molecules (which defines the local thermodynamical temperature and pressure) and the combined energy of their rotation, oscillation, dissociation into atoms and ions, electronic excitation atoms and molecules, and radiation. An asteroid-comet body is initially round and consists of a friable lumpy-dust material. This body may invade the atmosphere at a tremendous speed of 10 to 30 km/s at different angles to the Earth's surface.

The modified system of the Navier-Stokes equations is solved by the combination of the semi-Lagrangian approximation for Lagrange transport derivatives and the conforming finite element method for other terms [3–5].

For the computational purpose, we replace an unknown energy by its square root. The stencil adaptation of the semi-Lagrangian approximation along trajectories allows one to avoid the Courant-Friedrichs-Lewy restriction on the smallness of a time step in relation to a mesh-size of the space triangulation.

2   The Mathematical Formulation of the Problem

Consider the two-dimensional model of motion of a friable lumpy-dust body at a hypersonic speed. The body is assumed to be round when entering the Earth's atmosphere and is aggregated of gas and lumpy dust of significant viscosity and substantial density. In the model, this gas-dust body and the atmosphere are regarded as a single continuous gas environment with the presence of solid material therein initially concentrated and then diffused by generated streams of gas. To describe the motion of the body in the atmosphere, we use the time-dependent Navier-Stokes equations in Cartesian coordinates: the Ox axis coincides with the initial direction of the velocity vector of the cosmic body, and the Oy axis is perpendicular to Ox (Fig. 1). For scaling to dimensionless quantities, the size L_b of the body is taken for the length scale; the density ρ_a of the gas on the surface of the Earth is taken as the scale of density; the initial speed of the body u_b does as the scale of speed; L_b/u_b is taken for the time scale. And for the scale of pressure and internal energy we take the values from the condition of the unperturbed atmosphere at the Earth's surface [1].

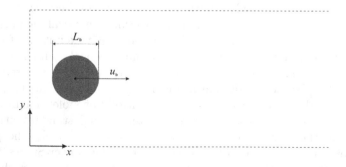

Fig. 1. A heavy lumpy-dust body Ω_b at an initial moment of computation.

For a simple illustration, we consider the two-dimensional statement of the problem. Let $\Omega = (0, H_1) \times (0, H_2)$ be the rectangular computational domain (Fig. 1) with the boundary Γ and points $\boldsymbol{x} = (x, y)$ where x is a distance along the initial direction of a body trajectory and y means the coordinate in the orthogonal direction. Since we do not consider the interaction of the body with the Earth's surface, all parts of the boundary Γ are placed at an appropriate height above the Earth's surface. Note that the Ox axis is not necessarily perpendicular to the Earth's surface and may be directed at an angle $\alpha \in (0°, 90°]$ to it. A friable lumpy-dust body Ω_b is initially placed near the left part of the boundary where the atmosphere is rarefied and moves with the speed $\boldsymbol{u}_b = (u_b, 0)$ in the Ox-direction.

We write the dimensionless Navier-Stokes equations as five equations with the unknown functions ρ_g, ρ_b, u, v, and e in the cylinder $(t, \boldsymbol{x}) \in (0, t_{fin}) \times \Omega$:

$$\frac{D\rho_g}{Dt} = 0, \tag{1}$$

$$\rho \frac{du}{dt} = -\frac{\partial P}{\partial x} + \frac{\partial \tau_{xx}}{\partial x} + \frac{\partial \tau_{xy}}{\partial y} + \mathrm{Fr}\chi\rho \cos\alpha, \tag{2}$$

$$\rho \frac{dv}{dt} = -\frac{\partial P}{\partial y} + \frac{\partial \tau_{xy}}{\partial x} + \frac{\partial \tau_{yy}}{\partial y} + \mathrm{Fr}\chi\rho \sin\alpha, \tag{3}$$

$$\rho \frac{de}{dt} + P\left(\frac{\partial u}{\partial x} + \frac{\partial v}{\partial y}\right) = -\frac{\partial q_x}{\partial x} - \frac{\partial q_y}{\partial y} + \Phi - c_{rad}e, \tag{4}$$

$$\frac{D\rho_b}{Dt} = 0. \tag{5}$$

To simplify the description, we introduce the expressions $d(\cdot)/dt$ and $D(\cdot)/Dt$ which mean the Lagrangian derivatives

$$\frac{da}{dt} \equiv \frac{\partial a}{\partial t} + u\frac{\partial a}{\partial x} + v\frac{\partial a}{\partial y} \quad \text{and} \quad \frac{Da}{Dt} \equiv \frac{\partial a}{\partial t} + \frac{\partial(ua)}{\partial x} + \frac{\partial(av)}{\partial y}.$$

Here $\rho(t, \boldsymbol{x})$ is the "full density" which combines two parts: $\rho(t, \boldsymbol{x}) = \rho_g(t, \boldsymbol{x}) + \rho_b(t, \boldsymbol{x})$ where ρ_g denotes the density of the gas component, and ρ_b denotes the density of the rigid-dust one. Besides, we use the following notations. $P(t, \boldsymbol{x})$ is the pressure; $u(t, \boldsymbol{x})$, $v(t, \boldsymbol{x})$ are the components of the velocity vector $\boldsymbol{u} = (u, v)$. Fr is the Froude number: $\mathrm{Fr} = u_b^2/gL_b$ where $g = 9.81 \, m/s^2$ is gravity. χ is the constant $G = 0.667 \cdot 10^{-10} m^3/kg \cdot c^2$ made dimensionless.

$e(t, \boldsymbol{x})$ is the specific "combined internal energy" which consists of two parts:

$$e(t, \boldsymbol{x}) = e_{tr}(t, \boldsymbol{x}) + e_{lt}(t, \boldsymbol{x}) \tag{6}$$

where e_{tr} means the specific translational energy of molecules and atoms. Note that e_{tr} defines the pressure of gas and is proportional to the thermodynamical temperature T:

$$e_{tr} = 3kT/2 \quad \text{with Boltzmann constant} \quad k = 1.38 \times 10^{-23} \, J/K. \tag{7}$$

e_{lt} means the specific energy which contains rotational and oscillation energy and the combined energy of electronic excitations of molecules and atoms as well as the losses by the dissociation of molecules into atom, radicals, and ions and energy of radiation. The time of relaxation between high-level kinds of energy is sufficiently small in comparison with a time step of our numerical algorithm. Thus, at each discrete time level, we take the local state of gas with the maximal entropy which corresponds to the state with the maximal probability. Therefore we consider e_{lt} as a function of pressure P and temperature T of the form

$$e_{lt}(t, \boldsymbol{x}) = c_{lt}(P(t, \boldsymbol{x}), T(t, \boldsymbol{x})). \tag{8}$$

The function $c_{lt}(P, T)$ is defined by the tables in [6–8] and is obtained by calculating the state with the maximal possible probability for fixed P and T.

Then we take the usual algebraic thermal equation of state

$$P = c_P \rho_g e_{tr} \tag{9}$$

with a constant c_P. Of course, this equation of state is the simplest of many known ones [2,9]. We do not discuss their usefulness here and confine ourselves to the remark that during the further refinement of the actual model parameters this equation holds an algebraic form with principal part (9).

The functions $\tau_{xx}, \tau_{xy}, \tau_{yy}$ are the components of the stress tensor

$$T = \begin{pmatrix} \tau_{xx} & \tau_{xy} \\ \tau_{xy} & \tau_{yy} \end{pmatrix} : \tau_{\mathbf{xx}} = \frac{2}{3\mathrm{Re}}\mu\left(2\frac{\partial \mathbf{u}}{\partial \mathbf{x}} - \frac{\partial \mathbf{v}}{\partial \mathbf{y}}\right),$$

$$\tau_{yy} = \frac{2}{3\mathrm{Re}}\mu\left(2\frac{\partial v}{\partial y} - \frac{\partial u}{\partial x}\right), \quad \tau_{xy} = \frac{\mu}{\mathrm{Re}}\left(\frac{\partial u}{\partial y} + \frac{\partial v}{\partial x}\right) \tag{10}$$

where μ is the viscosity coefficient consisting of two parts: $\mu = \mu_g + \mu_b$. μ_g is Sutherland's dynamical viscosity of a gas component [2]:

$$\mu_g = \mu_0 (T/T_0)^{3/2} (T_0 + T_S)/(T + T_S)\,\rho_g/\rho \tag{11}$$

where μ_0 is an experimental viscosity for the test temperature T_0; and T_S is the adjusting temperature for air [9]. μ_b is an additional viscosity providing rather high friction between rigid-dust particles:

$$\mu_b = c_b \exp(T_b/T)\rho_b/\rho \tag{12}$$

where c_b and T_b are adjusting coefficients. Again we confine ourselves to the remark that during the further refinement of the actual model parameters, equalities (11)–(12) remain (piecewise) algebraic expressions which do not change the algorithmic structure.

Re is the Reynolds number (here it is sufficiently great: $\mathrm{Re} = u_b \rho_g L_b / \mu_g$). The functions $q_x(t, \boldsymbol{x})$, $q_y(t, \boldsymbol{x})$ are heat-fluxes given by the formulae

$$q_x = -c_{\mathrm{heat}}\partial e_{tr}/\partial x, \quad q_y = -c_{\mathrm{heat}}\partial e_{tr}/\partial y$$

where

$$c_{\text{heat}} = (c_1 \mu_g \rho_g + c_2 \mu_b \rho_b)/\rho \tag{13}$$

with corresponding constants c_1, c_2.

The dissipative function $\Phi(t, \boldsymbol{x})$ is taken in the form

$$\Phi = \frac{\mu}{\text{Re}} \left[\frac{2}{3} \left(\frac{\partial u}{\partial x} \right)^2 + \frac{2}{3} \left(\frac{\partial v}{\partial y} \right)^2 + \left(\frac{\partial v}{\partial x} + \frac{\partial u}{\partial y} \right)^2 + \frac{2}{3} \left(\frac{\partial u}{\partial x} - \frac{\partial v}{\partial y} \right)^2 \right]. \tag{14}$$

The last term $c_{\text{rad}}e$ in (4) describes the energy loss by radiation with empirical coefficient c_{rad} depending on T.

To start the computations, we take the initial conditions

$$\begin{aligned}
\rho(0, \boldsymbol{x}) &= \rho_0(\boldsymbol{x}) > 0 & \text{for} \quad \boldsymbol{x} \in \Omega, \\
\boldsymbol{u}(0, \boldsymbol{x}) &= \boldsymbol{u}_0(\boldsymbol{x}) & \text{for} \quad \boldsymbol{x} \in \Omega, \\
e(0, \boldsymbol{x}) &= e_0(\boldsymbol{x}) > 0 & \text{for} \quad \boldsymbol{x} \in \Omega.
\end{aligned} \tag{15}$$

There are many ways to impose boundary conditions of different complexity on the computational boundary. We take the boundary conditions [5] constructed by analogy with "do nothing" conditions developed initially for the Navier-Stokes equations for viscous incompressible fluid [10]. Here they are formulated in the following way:

$$P\boldsymbol{n} - T\boldsymbol{n} = \mathbf{P}_{\text{out}}\boldsymbol{n} \quad \text{for} \quad (t, \boldsymbol{x}) \in [0, t_{\text{fin}}] \times \Gamma \tag{16}$$

where $P_{\text{out}}(t, \boldsymbol{x})$ is the pressure in the outer undisturbed atmosphere and $\boldsymbol{n}$ is an outward normal. For the full energy we get the boundary condition

$$\boldsymbol{q} \cdot \boldsymbol{n} = 0 \quad \text{for} \quad (t, \boldsymbol{x}) \in [0, t_{\text{fin}}] \times \Gamma \tag{17}$$

with the vector $\boldsymbol{q} = (q_x, q_y)$.

Before formulating the numerical algorithm, we rewrite all terms of Eq. (4) with one unknown function e instead of e_{tr} and e_{it} with the function $c_{\text{tr}} \geq 0$ resulting from the relations (6)–(8):

$$e_{\text{tr}} = c_{\text{tr}}^2 e \quad \text{and} \quad e_{\text{it}} = (1 - c_{\text{tr}}^2)e;$$

and then fulfil the substitution [5]

$$\varepsilon = e^{1/2} \tag{18}$$

which simplifies the implementation of the conforming finite element method. Moreover, we redefine some coefficients of Eq. (4)

$$\hat{P} = P/\varepsilon, \quad \hat{\mu} = \mu/\varepsilon, \quad \hat{c}_{\text{heat}} = c_{\text{heat}}/\varepsilon.$$

Thus, we get the equation

$$2\rho\frac{d\varepsilon}{dt} - 2\hat{c}_{\text{heat}}\left(\frac{\partial(c_{\text{tr}}\varepsilon)}{\partial x}\right)^2 - 2c_{\text{tr}}\frac{\partial}{\partial x}\left(c_{\text{heat}}\frac{\partial(c_{\text{tr}}\varepsilon)}{\partial x}\right) - 2\hat{c}_{\text{heat}}\left(\frac{\partial(c_{\text{tr}}\varepsilon)}{\partial y}\right)^2$$

$$-2c_{\text{tr}}\frac{\partial}{\partial y}\left(c_{\text{heat}}\frac{\partial(c_{\text{tr}}\varepsilon)}{\partial y}\right) = -\widehat{P}\left(\frac{\partial u}{\partial x} + \frac{\partial v}{\partial y}\right) \tag{19}$$

$$+\frac{\widehat{\mu}}{\text{Re}}\left[\frac{2}{3}\left(\frac{\partial u}{\partial x}\right)^2 + \frac{2}{3}\left(\frac{\partial v}{\partial y}\right)^2 + \left(\frac{\partial v}{\partial x} + \frac{\partial u}{\partial y}\right)^2 + \frac{2}{3}\left(\frac{\partial u}{\partial x} - \frac{\partial v}{\partial y}\right)^2\right] - c_{\text{rad}}\varepsilon.$$

3   The Finite Element Method

For the time discretization, we subdivide the time segment $[0, t_{\text{fin}}]$ into K subintervals by the points $t_k = \tau k$, $k = 0, ..., K$, with the step $\tau = t_{\text{fin}}/K$. To construct a discrete analogue of the above problem, fix time $t = t_k$ for some $k \in [1, K]$ and take into consideration Eq. (2). Replace du/dt into the right-hand side at a moment. Thus, we come to the stationary Stokes equation. Let us write it in the weak form. For this purpose, first we introduce the space $L_2(\Omega)$ of measurable functions with the inner product and the finite norm

$$(f, g)_\Omega = \int_\Omega fg\,d\Omega \quad \text{and} \quad \|f\|_0 = (f, f)_\Omega^{1/2}.$$

And then we introduce the space H^1 of all functions that belong to $L_2(\Omega)$ together with their first-order derivatives and have the finite norm

$$\|f\|_1 = \left(\|f\|_0^2 + \|\partial f/\partial x\|_0^2 + \|\partial f/\partial y\|_0^2\right)^{1/2}.$$

For the weak formulation, multiply (2) by an arbitrary function $\varphi \in H^1$, integrate over Ω, and use boundary condition (16):

$$\int_\Omega \left(P\frac{\partial\varphi}{\partial x} + \tau_{xx}\frac{\partial\varphi}{\partial x} + \tau_{xy}\frac{\partial\varphi}{\partial y}\right)d\Omega$$

$$= \int_\Gamma n_1 P_{\text{ext}}\varphi\,d\Gamma - \int_\Omega \rho\frac{du}{dt}\varphi\,d\Omega + \int_\Omega \varphi\text{Fr}\chi\rho\cos\alpha\,d\Omega. \tag{20}$$

The same operations with (16) give the equality

$$\int\limits_{\Omega} \left(P\frac{\partial \varphi}{\partial y} + \tau_{xy}\frac{\partial \varphi}{\partial x} + \tau_{yy}\frac{\partial \varphi}{\partial y} \right)\, d\Omega$$

$$= \int\limits_{\Gamma} n_2 P_{\text{ext}}\varphi\, d\Gamma - \int\limits_{\Omega} \rho\frac{dv}{dt}\varphi\, d\Omega + \int\limits_{\Omega} \varphi\, \mathrm{Fr}\chi\rho\sin\alpha\, d\Omega. \tag{21}$$

And for Eq. (19) we use boundary condition (17) and get

$$\int\limits_{\Omega} \left(2c_{\text{heat}}\frac{\partial(c_{\text{tr}}\varepsilon)}{\partial x}\frac{\partial(c_{\text{tr}}\varphi)}{\partial x} + 2c_{\text{heat}}\frac{\partial(c_{\text{tr}}\varepsilon)}{\partial y}\frac{\partial(c_{\text{tr}}\varphi)}{\partial y} \right.$$

$$\left. -2\hat{c}_{\text{heat}}\left(\frac{\partial(c_{\text{tr}}\varepsilon)}{\partial x}\right)^2 \varphi - 2\hat{c}_{\text{heat}}\left(\frac{\partial(c_{\text{tr}}\varepsilon)}{\partial y}\right)^2 \varphi \right)\, d\Omega \tag{22}$$

$$= -\int\limits_{\Omega} \left(\hat{P}\left(\frac{\partial u}{\partial x} + \frac{\partial v}{\partial y}\right) - \frac{\hat{\mu}}{\mathrm{Re}}\left[\frac{2}{3}\left(\frac{\partial u}{\partial x}\right)^2 + \frac{2}{3}\left(\frac{\partial v}{\partial y}\right)^2 + \left(\frac{\partial v}{\partial x} + \frac{\partial u}{\partial y}\right)^2 \right. \right.$$

$$\left. \left. +\frac{2}{3}\left(\frac{\partial u}{\partial x} - \frac{\partial v}{\partial y}\right)^2 \right] + c_{\text{rad}}\varepsilon + 2\rho\frac{d\varepsilon}{dt} \right)\varphi d\Omega.$$

Now we derive the integral equalities for the Galerkin method at a fixed time level $t = t_k$ with the simultaneous approximation of the derivatives du/dt using known values from the previous time level. First, in the computational domain Ω we construct a grid by two sets of the parallel lines $x_i = ih_1$, $y_j = jh_2$, $i = 0, ..., N_1$, $j = 0, ..., N_2$, with mesh-sizes $h_1 = H_1/N_1$, $h_2 = H_2/N_2$ for positive integers N_1 and N_2. Denote the set of all nodes $z_{i,j} = (x_i, y_j)$ lying in $\bar{\Omega}$ by $\bar{\Omega}_h$ and all nodes lying on Γ by Γ_h. These lines subdivide $\bar{\Omega}$ and form the triangulation $\mathfrak{I}$ consisting of the cells $\omega_{i,j} = [x_i, x_{i+1}] \times [y_j, y_{j+1}]$.

For the element-wise assembling of the finite element system, we consider a typical bilinear basis function $\varphi_{i,j}(x, y)$ of the Galerkin method

$$\varphi_{i,j}(x, y) = \begin{cases} (1 - |x - x_i|/h_1)(1 - |y - y_j|/h_2) \\ \text{for } (x, y) \in [x_{i-1}, x_{i+1}] \times [y_{j-1}, y_{j+1}], \\ \\ 0 \quad \text{otherwise.} \end{cases} \tag{23}$$

This function equals 1 at $z_{i,j}$, equals 0 at all other nodes of $\bar{\Omega}_h$, is bilinear on each cell $\omega_{i,j} \in \mathfrak{I}$, and is continuous on $\bar{\Omega}$. Its support is the rectangle $\text{supp}(\varphi_{i,j}) = ([x_{i-1}, x_{i+1}] \times [y_{j-1}, y_{j+1}]) \cap \bar{\Omega}$.

Denote the linear span of these functions by H_h:

$$H_h = span\,\{\varphi_{i,j}\}_{i,j=0}^{N_1, N_2}. \tag{24}$$

Now we are ready to derive the Galerkin formulation for problem (20)–(22). *Find functions* $u_h, v_h, \varepsilon_h \in H_h$ *which satisfy the equalities*

$$\int_{\Omega} \left(P_h \frac{\partial \varphi_h}{\partial x} + (\tau_{xx})_h \frac{\partial \varphi_h}{\partial x} + (\tau_{xy})_h \frac{\partial \varphi_h}{\partial y} \right) d\Omega$$

$$= \int_{\Gamma} n_1 P_{\text{ext}} \varphi_h \, d\Gamma - \int_{\Omega} \rho_h \left(\frac{du}{dt} \right)_h \varphi_h \, d\Omega + \int_{\Omega} \varphi_h \, \mathrm{Fr} \chi \rho_h \cos\alpha \, d\Omega,$$

(25)

$$\int_{\Omega} \left(P_h \frac{\partial \varphi_h}{\partial y} + (\tau_{xy})_h \frac{\partial \varphi_h}{\partial x} + (\tau_{yy})_h \frac{\partial \varphi_h}{\partial y} \right) d\Omega$$

$$= \int_{\Gamma} n_2 P_{\text{ext}} \varphi_h \, d\Gamma - \int_{\Omega} \rho_h \left(\frac{dv}{dt} \right)_h \varphi_h \, d\Omega + \int_{\Omega} \varphi_h \, \mathrm{Fr} \chi \rho_h \sin\alpha \, d\Omega,$$

(26)

$$\int_{\Omega} \left(2(c_{\text{heat}})_h \frac{\partial((c_{\text{tr}})_h \varepsilon_h)}{\partial x} \frac{\partial((c_{\text{tr}})_h \varphi_h)}{\partial x} + 2(c_{\text{heat}})_h \frac{\partial((c_{\text{tr}})_h \varepsilon_h)}{\partial y} \frac{\partial((c_{\text{tr}})_h \varphi_h)}{\partial y} \right) d\Omega$$

$$- \int_{\Omega} \left(2(\hat{c}_{\text{heat}})_h \left(\frac{\partial((c_{\text{tr}})_h \varepsilon_h)}{\partial x} \right)^2 \varphi_h - 2(\hat{c}_{\text{heat}})_h \left(\frac{\partial((c_{\text{tr}})_h \varepsilon_h)}{\partial y} \right)^2 \varphi_h \right) d\Omega$$

(27)

$$= - \int_{\Omega} \left((\hat{P})_h \left(\frac{\partial u_h}{\partial x} + \frac{\partial v_h}{\partial y} \right) - \frac{(\hat{\mu})_h}{\mathrm{Re}} \left[\frac{2}{3} \left(\frac{\partial u_h}{\partial x} \right)^2 + \frac{2}{3} \left(\frac{\partial v_h}{\partial y} \right)^2 + \left(\frac{\partial v_h}{\partial x} + \frac{\partial u_h}{\partial y} \right)^2 \right. \right.$$

$$\left. \left. + \frac{2}{3} \left(\frac{\partial u_h}{\partial x} - \frac{\partial v_h}{\partial y} \right)^2 \right] \right) \varphi_h d\Omega - \int_{\Omega} \left((c_{\text{rad}})_h \varepsilon_h + 2\rho_h \frac{d\varepsilon_h}{dt} \right) \varphi_h d\Omega$$

for any $\varphi_h \in H_h$. Here the expression $(f)_h$ for any function in brackets means the algebraic definition introduced above where the functions ρ, u, v, ε are replaced by $\rho_h, u_h, v_h, \varepsilon_h$.

We perform the discretization of this problem in x, y variables in 3 steps. First, we take our unknown functions in the following form:

$$u_h(t, \boldsymbol{x}) = \sum_{\boldsymbol{z}_{i,j} \in \bar{\Omega}_h} u_{h,i,j}(t) \varphi_{i,j}(\boldsymbol{x}), \quad v_h(t, \boldsymbol{x}) = \sum_{\boldsymbol{z}_{i,j} \in \bar{\Omega}_h} v_{h,i,j}(t) \varphi_{i,j}(\boldsymbol{x}),$$

$$\varepsilon_h(t, \boldsymbol{x}) = \sum_{\boldsymbol{z}_{i,j} \in \bar{\Omega}_h} \varepsilon_{h,i,j}(t) \varphi_{i,j}(\boldsymbol{x}).$$

(28)

Then we take Eqs. (25)–(27) for $\varphi = \varphi_{i,j} \, \forall i = 0, ..., N_1, \, j = 0, ..., N_2$. And finally we use the trapezoidal quadrature rule in one-dimensional integrals and their Cartesian product for integrals over $\omega_{i,j} \in \Im$:

$$\int_{\omega_{i,j}} f \, d\Omega = (f(\boldsymbol{z}_{i,j}) + f(\boldsymbol{z}_{i+1,j}) + f(\boldsymbol{z}_{i,j+1}) + f(\boldsymbol{z}_{i+1,j+1})) \, h_1 h_2 / 4.$$

Now transform the integrals in the right-hand side of (22)–(24) of the form

$$\int_{\Omega} \rho \frac{du}{dt} \varphi_{i,j} \, d\Omega.$$

Due to the quadrature rule we obtain the expression

$$\int_{\Omega} \rho \frac{du}{dt} \varphi_{i,j} \, d\Omega \big|_{t=t_k} \approx \rho(t_k, z_{i,j}) \frac{du}{dt}(t_k, z_{i,j}) \ \text{meas}\,(\text{supp}\,\varphi_{i,j})/4. \qquad (29)$$

The last factor equals $h_1 h_2$ for $z_{i,j} \in \Omega$, $h_1 h_2/2$ for $z_{i,j} \in \Gamma \setminus \{\text{its vertices}\}$, and $h_1 h_2/4$ at its four vertices.

4 The semi-Lagrangian Approximation

For the semi-Lagrangian approximation, construct the trajectory $(t, \bar{x}(t), \bar{y}(t))$ from a point $(t_k, z_{i,j})$ to the previous time level $t = t_{k-1}$ as the solution of the system of ordinary differential equations

$$\begin{cases} \bar{x}'(t) = u(t, \bar{x}(t), \bar{y}(t)), \\[2mm] \bar{y}'(t) = v(t, \bar{x}(t), \bar{y}(t)), \end{cases} \qquad t \in [t_{k-1}, t_k], \qquad (30)$$

back in time for the "initial" value at level $t = t_k$:

$$\bar{x}(t_k) = x_i, \quad \bar{y}(t_k) = y_j. \qquad (31)$$

Remark 1. Since a strong increase of temperature causes extension of air, gas flows go out of $\bar{\Omega}$ throughout the computational experiment due to short real time of observation. Therefore the points $(t_{k-1}, \bar{x}(t_{k-1}), \bar{y}(t_{k-1}))$ of the trajectory lay inside $[t_{k-1}, t_k] \times \bar{\Omega}$. Usually we use Runge-Kutta methods of order no higher than 2. This is sufficient to get a first-order final algorithm [4].

Then we approximate the second integral in the right-hand side of (25) by the difference

$$\rho(t_k, z_{i,j}) \left(u(t_k, z_{i,j}) - u\,(t_{k-1}, \bar{x}(t_{k-1}), \bar{y}(t_{k-1})) \right) \ \text{meas}\,(\text{supp}\,\varphi_{i,j})/4\tau. \qquad (32)$$

Similarly we obtain

$$\int_{\Omega} \rho \frac{dv}{dt} \varphi_{i,j} \, d\Omega \big|_{t=t_k} \approx \rho(t_k, z_{i,j}) \frac{dv}{dt}(t_k, z_{i,j}) \ \text{meas}\,(\text{supp}\,\varphi_{i,j})/4$$

$$\approx \rho(t_k, z_{i,j}) \left(v(t_k, z_{i,j}) - v\,(t_{k-1}, \bar{x}(t_{k-1}), \bar{y}(t_{k-1})) \right) \ \text{meas}\,(\text{supp}\,\varphi_{i,j})/4\tau, \qquad (33)$$

$$\int_{\Omega} \rho \frac{d\varepsilon}{dt} \varphi_{i,j} \, d\Omega \big|_{t=t_k} \approx \rho(t_k, z_{i,j}) \frac{d\varepsilon}{dt}(t_k, z_{i,j}) \ \text{meas}\,(\text{supp}\,\varphi_{i,j})/4 \qquad (34)$$

$$\approx \rho(t_k, z_{i,j}) \left(\varepsilon(t_k, z_{i,j}) - \varepsilon\,(t_{k-1}, \bar{x}(t_{k-1}), \bar{y}(t_{k-1})) \right) \ \text{meas}\,(\text{supp}\,\varphi_{i,j})/4\tau.$$

The discretization of Eqs. (1) and (5) is carried out in a slightly different way. Take a node $z_{i,j}$ and its vicinity $\Pi_{i,j} = [x_i - h_1/2, x_i + h_1/2] \times [y_j - h_2/2, y_j + h_2/2] \cap \bar{\Omega}$ with the boundary $\gamma_{i,j}$. Out of each point of $\{t_k\} \times \gamma_{i,j}$, draw the trajectory $(t, \bar{x}(t), \bar{y}(t))$ up to the intersection with the plane $t = t_{k-1}$ which is the solution of system (27) with corresponding "initial" conditions on the plane $t = t_k$. These trajectories bound a curvilinear quadrangle $Q_{i,j}$ on the plane $t = t_{k-1}$. As a result, we get the volume $V_{i,j}$, bounded from above by the face $\{t_k\} \times \Pi_{i,j}$, from below by $\{t_{k-1}\} \times Q_{i,j}$, and by four side faces consisting of the constructed trajectories (Fig. 2). Integrate Eq. (1) over volime $V_{i,j}$ and use the Gauss-Ostrogradskii equality [4]:

$$\int_{\Pi_{i,j}} \rho_g(t_k, x, y) \ d\Omega = \int_{Q_{i,j}} \rho_g(t_{k-1}, x, y) \ d\Omega. \qquad (35)$$

In [4,5] we described several ways to calculate such an integral at level $t = t_{k-1}$. Here we use the following one. First, from the integration of the left-hand side of (35) we get an algebraic expression with unknowns $\rho_{g,h}(t_{k-1}, z_{i,j})$. Second, we replace $Q_{i,j}$ by the rectangle $S_{i,j}$ with the same vertices and find the transformation $R_{i,j} : [0,1] \times [0,1] \to S_{i,j}$ with Jacobian $J_{i,j}$. Then we replace the integration of the right-hand side of (32) by the integration over $S_{i,j}$ with the help of the quadrature rule with central rectangles on an additional uniform grid. Thus, expression (35) turns into a linear algebraic equation in unknowns $\rho_{g,h}(t_{k-1}, z_{i,j})$ with a known right-hand side from the previous time level. Going over all indices, we get the system of linear algebraic equations for all unknowns $\rho_{g,h}(t_{k-1}, z_{i,j})$. Similarly we get the system of linear algebraic equations for all unknowns $\rho_{b,h}(t_k, z_{i,j})$.

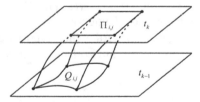

Fig. 2. The construction of the volume $V_{i,j}$.

Remark 2. In this section, the functions u, v, ε, ρ_g, and ρ_b are calculated at points between time levels t_{k-1} and t_k by the interpolation. But at level t_k they are determined only after solving the system of the finite element method. To bypass this contradiction, the iteration "with respect to nonlinearity" is applied: first the values of these functions are continued from the previous time level t_{k-1}; and at the next iteration step they are taken as the linear interpolation between the time levels (along trajectories). Note that in the numerical experiments it takes no more than three iteration steps to make the error of this approach less than the approximation error.

When combined together, these equalities form the system of *nonlinear* algebraic equations at level t_k

$$\mathbf{AU} \equiv \begin{bmatrix} A_{11} & A_{12} & A_{13} & 0 \\ A_{12}^T & A_{22} & A_{23} & 0 \\ A_{13}^T & A_{23}^T & A_{33} & 0 \\ 0 & 0 & 0 & A_{44} \end{bmatrix} \begin{bmatrix} U \\ V \\ E \\ R \end{bmatrix} = \begin{bmatrix} F_1 \\ F_2 \\ F_3 \\ F_4 \end{bmatrix} \equiv \mathbf{F} \tag{36}$$

where the vectors U, V, E, R combine the unknown numbered coefficients $\{u_{h,i,j}\}$, $\{v_{h,i,j}\}$, $\{\varepsilon_{h,i,j}\}$, $\{\rho_{g,h}(t_k, z_{i,j}), \rho_{b,h}(t_k, z_{i,j})\}$, respectively, and the symbol 0 means a matrix of corresponding size with zero entries.

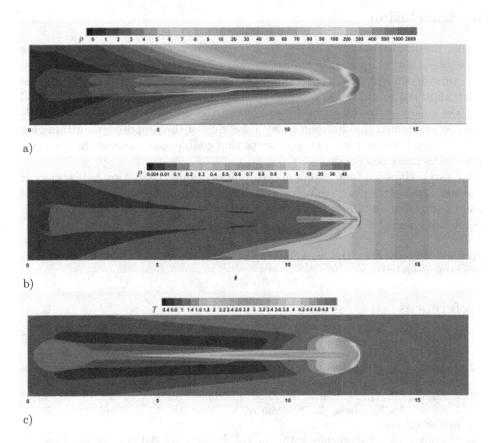

Fig. 3. The density of a friable lumpy dust body and the air (a), the pressure (b), and the temperature (c) of medium at the same instants of time.

5   A Numerical Experiment

The calculation of integrals in the right-hand sides of system (36) at each time level and at each step of iteration with respect to nonlinearity is the most tedious computational part of the method. Some analysis of accuracy, stability, and computational cost of the algorithm can be found in [4].

We take a small time step to provide reasonable accuracy in the transition mode and to accelerate convergence of iteration processes when solving system (36). For insignificant viscosity, we get matrices with strict diagonal dominance that provides fast convergence of the Jacobi and Gauss-Seidel iterative algorithms. The initial guess for the first iteration step is taken from the previous time level. But because of high velocities it is taken along the trajectories (30)–(31).

In Fig. 3(a)–(c) we demonstrate the behaviour of the dencity ρ, the pressure P, and the temperature T at the same instants of time.

6   Conclusion

Thus, the implementation of the conforming finite element method with the subsequent semi-Lagrangian approximation of the transfer derivatives reduces the time-dependent problem to the sequence of the discrete algebraic systems with improved properties in comparison with those of the finite element method for the steady-state Stokes problem. Solving the problem with great Reynolds numbers demonstrated independence of stability of the proposed algorithm on a relationship between time and space steps that enables one to avoid the Courant-Friedrichs-Lewy restriction.

Exactly the semi-Lagrangian approximation along the motion trajectories of gas and dust elementary volumes allows one to thoroughly track the ratio of solid and gaseous components of density as well as translational energy and the combined energy of other types.

Acknowledgements. The work is supported by Project 14-11-00147 of Russian Scientific Foundation.

References

1. Standard Atmosphere, International Organization for Standardization. ISO 2533:1975 (1975)
2. Johnson, R.W.: The Handbook of Fluid Dynamics. CRC Press LLC/Springer, Boca Raton/Heidelberg (1998)
3. Chen, H., Lin, Q., Shaidurov, V., Zhou, J.: Numer. Anal. Appl. **4**, 345–362 (2011)
4. Efremov, A., Karepova, E., Shaydurov, V., Vyatkin, A.: J. Appl. Math. **2014**, 610398 (2014)
5. Shaydurov, V., Shchepanovskaya, G., Yakubovich, M.: AIP Conference Proceedings, vol. 1684, pp. 020003-1-11 (2015)
6. Tables of Thermodynamical Functions of Air for the Temperature from 200 to 6000 °K and for the Pressure from 0.00001 to 100 ATM. Moscow, Computing Center of AS USSR (1962). (in Russian)

7. Tables of Thermodynamical Functions of Air for the Temperature from 6000 °K to 12000 °K and for the Pressure from 0.001 to 1000 ATM. Moscow, Academy of Sciences of USSR (1957). (in Russian)
8. Tables of Thermodynamical Functions of Air for the Temperature from 12000 °K to 20000 °K and for the Pressure from 0.001 to 1000 ATM. Moscow, Academy of Sciences of USSR (1959). (in Russian)
9. Belov, G.: Thermodynamics. Urait, Moscow (2016). (in Russian)
10. Rannacher, R.: Incompressible viscous flow. In: Encyclopedia of Computational Mechanics. Wiley, New York (2011)

On Stochastic Representation of Blow-Ups for Distributed Parameter Systems

Milan Stehlík[1,2]([⊠]) and Jozef Kiseľák[3]

[1] Department of Applied Statistics and Linz Institute of Technology,
Johannes Kepler University in Linz, Linz, Austria
milan.stehlik@jku.at

[2] Institute of Statistics, Universidad de Valparaíso, Valparaíso, Chile

[3] Institute of Mathematics, Faculty of Science,
P.J. Šafárik University in Košice, Košice, Slovakia
jozef.kiselak@upjs.sk

Abstract. This paper studies the regularity properties of stochastic representation of the stationary solution of the Fokker-Planck equation and related Dirichlet problem. When the correlation structure is exponential, we relate the representation process to the Ornstein-Uhlenbeck process. We derive some properties of blow-up identification by the singular points of the correlation structure.

Keywords: Blow-up · Fokker-Planck equation · Gaussian random field · Hilbert space · Ornstein-Uhlenbeck process · Stochastic representation

1 Introduction

Here we consider a Poisson equation, often written as $\Delta u = f$, or $\nabla^2 u = f$. Poisson equation has a broad utility in electrostatics, mechanical engineering and theoretical physics.

The Dirichlet problem for Poisson equation consists in finding a solution $u \in H_g^1(G); = \{v \in H^1(G) : v = g \text{ on } \Gamma\}$ (H^1 is a Sobolev space) on some domain G such that for $f \in L^2(G)$ and $\Gamma = \partial G$

$$- \Delta u = f, \text{ in } G, \tag{1}$$

$$u = g, \text{ in } \Gamma. \tag{2}$$

This fomulation is equivalent (in some sense) in finding of minimizer of (energy) functional $\int_G \left(\frac{1}{2}|\nabla u|^2 - uf \right) dx$ in $H_g^1(G)$. It well known that if G is open, bounded and $g \in C(\Gamma), f \in C(G)$, then exists a unique solution $u \in C^2(G) \cap C(\Gamma)$ of (1) and (2) (if moreover Γ is C^1, then $u \in C^2(\overline{G})$). We say, that solution

M. Stehlík—I thank to Professor Lyuben Valkov for his invitation.

I. Dimov et al. (Eds.): NAA 2016, LNCS 10187, pp. 132–140, 2017.
DOI: 10.1007/978-3-319-57099-0_12

to (1) and (2) is regular, if $u \in C^2(G) \cap C(G \cup \Gamma)$. Furthermore, if $u \in C^2(G)$ is harmonic, then necessarily $u \in C^\infty(G)$. More on regularity one can find e.g. in [3].

Now let G be a bounded domain in $\mathbb{R}^m$ and let u be a sufficiently regular solution of the equation

$$\frac{1}{2}\Delta u + f(\mathbf{x}) = 0$$

in G, let $u \in C(\bar{G})$ and $u = g$ on ∂G. Then the following representation for $\mathbf{x} \in G$ holds

$$u(\mathbf{x}) = \mathbb{E}\left[g(\mathbf{x} + \mathbf{W}_{\tau(\mathbf{x})}) + \int_0^{\tau(\mathbf{x})} f(\mathbf{x} + \mathbf{W}_t)dt\right], \qquad (3)$$

where $\tau(\mathbf{x}) = \inf\{t : \mathbf{x} + \mathbf{W}_t \in G\}$. This relation between the Wiener process $\mathbf{W}_t$ and the Laplace operator has been obtained by the employing of probabilistic construction of the solution of the heat equation (see [5]).

We may say that such a representation is local, in the sense, that we are computing the value of the solution at a given point of the domain. In [7] we have derived a global probability representation of the Dirichlet boundary problem solution. Therein we consider an isotropic Gaussian random field $Y_t \in \mathcal{Y} \subset \mathbb{R}^k$ with parametrized covariance function $\text{cov}(Y_s, Y_t) = c(\|s - t\|, \mathbf{r})$, measured on some compact design space $T \subset \mathbb{R}^p$, and parameter $\mathbf{r} \in G \subset \mathbb{R}^m$. Such a field is in [7] related to the D-optimality problem when the errors are correlated. Let $h(\mathbf{y}; \mathbf{r})$ be the density of the Gaussian random field Y_t. We have defined the abstract energy $\mathcal{E}(\mathbf{r}) = -\int_{\mathcal{Y}} |\nabla_{\mathbf{r}} u|^2 e^u d\mathbf{y}$, where $d\mu = e^u d\mathbf{y}$ is some (unit) mass distribution over $\mathcal{Y}$. We say abstract energy, because the integration can be employed over space of dimension k, but the operator (Laplacian) is according to m-dimensional cartesian coordinates $(r_1, \ldots, r_m)$ of the parameter $\mathbf{r}$. This can correspond to the "classical" energy $E = -\int_{\mathcal{Y}} |\nabla_{\mathbf{r}(\mathbf{y})} u|^2 d\mu$ under $m = k$, due to some diffeomorphism $\mathbf{r} = \mathbf{r}(\mathbf{y})$. Some interpretations of these abstract energies can be found in theoretical physics. This abstract energy is from the Statistical point of view the Fisher information on the parameter $\mathbf{r}$.

In [7] is given that for any Gaussian random field there exists the Dirichlet problem for Poisson equation, e.g. there exist sufficiently regular f and g so that $\ln h$ is the solution of (1), (2). Furthermore, there exists the abstract energy $\mathcal{E}(\mathbf{r})$ which is equal to the Fisher information. In [8] we have studied the blow up identification by the singular points of the correlation structure. Based on [9] we have provided a deterministic interpretation of abstract energy assuming some regularity conditions on an inverse problem operator given in [1]. We related the abstract energy to Turing's measure of conditioning (see [10]).

In the present paper, we show that for the Gaussian representant with parametrized covariance function, the limiting case $r \to 0^+$ together with $\sigma = \sqrt{2r}$ of Ornstein-Uhlenbeck process represents the blow up of the limit of the regular solutions of the sequence of Dirichlet problems for the Poisson equation.

2   Blow-Up and Ornstein-Uhlenbeck Process

It is well know that for an Itô process described by the SDE

$$d\mathbf{X}_t = \boldsymbol{\mu}(\mathbf{X}_t, t)\, dt + \boldsymbol{\sigma}(\mathbf{X}_t, t)\, d\mathbf{W}_t,$$
$$\mathbf{X}_0 = \boldsymbol{\xi}, \tag{4}$$

where $\mathbf{W}_t$ is an r-dimensional Brownian motion independent on the k-dimensional initial vector $\boldsymbol{\xi}$, drift vector $\boldsymbol{\mu}$ and diffusion tensor $D_{ij}(\mathbf{x}, t) = \sum_{l=1}^{k} \sigma_{il}(\mathbf{x}, t)\sigma_{jl}(\mathbf{x}, t)$, the Fokker-Planck equation for the probability density $h(\mathbf{x}, t)$ of the random variable $\mathbf{X}_t$ is

$$\frac{\partial h(\mathbf{x}, t)}{\partial t} = -\sum_{i=1}^{k} \frac{\partial}{\partial x_i} [\mu_i(\mathbf{x}, t)h(\mathbf{x}, t)] + \frac{1}{2}\sum_{i=1}^{k}\sum_{j=1}^{k} \frac{\partial^2}{\partial x_i\,\partial x_j} [D_{ij}(\mathbf{x}, t)h(\mathbf{x}, t)].$$

In this section we relate the stochastic representation via the Gaussian field to the linear stochastic differential equation. Let us consider the linear stochastic equation (for more details see [4]):

$$d\mathbf{X}_t = (\mathbf{A}(t)\,\mathbf{X}_t + \mathbf{a}(t))\, dt + \boldsymbol{\sigma}(t)\, d\mathbf{W}_t, \quad 0 \le t < \infty,$$
$$\mathbf{X}_0 = \boldsymbol{\xi}. \tag{5}$$

Here the $k \times k, k \times 1$ and $k \times r$ matrices $\mathbf{A}(t), \mathbf{a}(t)$ and $\boldsymbol{\sigma}(t)$ are nonrandom, measurable and locally bounded. In linear case one can obtain an explicit solution in the form

$$\mathbf{X}_t = \boldsymbol{\Phi}(t)\left[\boldsymbol{\xi} + \int_0^t \boldsymbol{\Phi}(s)^{-1}\,\mathbf{a}(s)\,ds + \int_0^t \boldsymbol{\Phi}(s)^{-1}\,\boldsymbol{\sigma}(s)\,d\mathbf{W}_s\right],$$

where $\boldsymbol{\Phi}$ is a fundamental matrix, i.e. the matrix solution of the problem $\boldsymbol{\Phi}(t)' = \mathbf{A}(t)\boldsymbol{\Phi}(t)$, $\boldsymbol{\Phi}(0) = \mathbf{I}$. The principal matrix (the state transition matrix) $\varphi(t, s) = \boldsymbol{\Phi}(t)\boldsymbol{\Phi}^{-1}(s)$ associated with $\mathbf{A}(t)$ has the following properties $\varphi(s, t) = \varphi^{-1}(t, s)$, $\varphi(t, s) = \varphi(t, \tau)\,\varphi(\tau, s)$ for $s \le \tau \le t$ and $\varphi(t, t) = \mathbf{I}$.

For $\mathbf{a}(t) \equiv 0$ and constant coefficients $\mathbf{A}(t) = \mathbf{A}$ and $\boldsymbol{\sigma}(t) = \boldsymbol{\sigma}$ we can then write

$$\mathbf{X}_t = \varphi(t, t_0)\,\boldsymbol{\xi} + \int_{t_0}^t \varphi(t, s)\,\boldsymbol{\sigma}\,d\mathbf{W}_s.$$

If $\boldsymbol{\xi} \sim \mathcal{N}(\mathbf{m}_0, \mathbf{C}_0)$ (this is a typical case for statistics, i.e. we observe process in a short time-horizon) and the process $\mathbf{W}_t$ has incremental covariance $\mathrm{cov}(d\mathbf{W}_t) = \mathbf{Q}\, dt$ then

$$\mathbf{m}(t) := \mathbb{E}[\mathbf{X}_t] = \boldsymbol{\Phi}(t)\,\mathbf{m}_0,$$

since $\frac{d\mathbf{m}(t)}{dt} = \mathbf{A}\,\mathbf{m}(t)$, $\mathbf{m}(0) = \mathbf{m}_0$, and $\mathbf{C}(t) := \mathbb{E}\left[(\mathbf{X}_t - \mathbf{m}(t))(\mathbf{X}_t - \mathbf{m}(t))^T\right]$ satisfies $\frac{d\mathbf{C}(t)}{dt} = \mathbf{A}\mathbf{C}(t) + \mathbf{C}(t)\mathbf{A}^T + \boldsymbol{\sigma}\,\mathbf{Q}\,\boldsymbol{\sigma}^T$, $\mathbf{C}(0) = \mathbf{C}_0$, which implies

$$\mathbf{C}(t) = \boldsymbol{\Phi}(t)\,\mathbf{C}_0\,\boldsymbol{\Phi}^T(t) + \int_0^t \boldsymbol{\Phi}(s)\,\boldsymbol{\sigma}\,\mathbf{Q}\,\boldsymbol{\sigma}^T\,\boldsymbol{\Phi}^T(s)ds.$$

Moreover, the vector $\mathbf{X}_t$ is a Gaussian one with mean function $\mathbf{m}(t)$ and covariance function $\mathbf{R}(s,t) := \text{cov}(\mathbf{X}_t, \mathbf{X}_s) = \varphi(s,t)\mathbf{C}(t)$ for $s \geq t$. In particular, considered system is time-invariant and this yields $\mathbf{R}(s,t) = \boldsymbol{\Phi}(|t-s|)\mathbf{C}(t)$.

2.1   1D Case

In the case $k = r = 1$, $p = 2$, $a(t) \equiv 0$, $A(t) = -\alpha$ and $\sigma(t) = \sigma > 0$ we obtain the oldest example

$$dX_t = -\alpha X_t \, dt + \sigma \, dW_t$$

of a stochastic differential equation (see [4], p. 358). The corresponding Fokker-Planck equation is

$$\frac{\partial h(x,t)}{\partial t} = \alpha \frac{\partial}{\partial x}(x\,h(x,t)) + \frac{\sigma^2}{2} \frac{\partial^2 h(x,t)}{\partial x^2},$$

whereas the fundamental stationary solution ($\partial_t h = 0$) is $h(x,t) = \sqrt{\dfrac{\alpha}{\pi\sigma^2}} e^{-\frac{\alpha x^2}{\sigma^2}}$.

Therefore, if the initial random variable X_0 has a normal distribution with mean zero and variance $\frac{\sigma^2}{2\alpha}$, then X_t is a stationary, zero-mean Gaussian process with covariance function $\rho(s,t) = \frac{\sigma^2}{2\alpha} e^{-\alpha|t-s|}$. If the representing process has exponential correlation of the form $\exp(-rd)$, then it is driven by Ornstein-Uhlenbeck process via

$$dX_t = -rX_t dt + \sqrt{2r} dW_t.$$

The Ornstein-Uhlenbeck process is stationary, Gaussian, Markovian and continuous in probability.

Now let us consider the blow-up problem introduced in [8]. Let us have a system $\{G_n, f_n\}$ where $G_n \subset G_{n+1} \subset \mathbb{R}^m$ is the increasing system of sufficiently regular domains and $f_{n+1}|G_n = f_n$ is the system of right sides for Dirichlet problem. The limiting domain is $G_\infty = \cup_{i=1}^\infty G_i$ with boundary $\Gamma_\infty = \partial G_\infty$.

If the Dirichlet problem (1), (2) is regular for all n, we are interested in whether there is a blow up at Γ_∞. Theorem 2.1 in [7] is saying that if the correlation structure of the representation process is collapsing, then there is a blow-up. Theorem 1 in [8] is giving a dimensionality interpretation of such blow-up at $\mathbf{r}^\star$ in the sense of the dimensionality loss of the stochastic representant. In more details, the representing stochastic process $\mathbf{X}_t$ is loosing the dimension, i.e. there exist a hyperplane L_1 such that $L_1 \subset \mathbb{R}^k$ and $\dim L_1 < k$ and there exists a neighborhood $U(\mathbf{r}^\star)$ of point $\mathbf{r}^\star$ such that for all $\mathbf{r} \in U(\mathbf{r}^\star) \cap G_\infty$ we have $\mathbb{P}(\mathbf{X}_t \in L_1|\mathbf{r}) < 1$ but $\mathbb{P}(\mathbf{X}_t \in L_1|\mathbf{r}^\star) = 1$.

The following theorem will provide a different interpretation via Ornstein-Uhlenbeck process. Note that we can represent the whole system $\{G_n, f_n\}$ just by one Gaussian field.

Theorem 1. *Let $\{G_n, f_n\}_{n=1}^\infty$ is the system of domains and right sides given above. Assume that the solution of Dirichlet problem (1,2) is regular for all n and $\det \Sigma(r) \to 0$ for $r \to 0^+ \in \Gamma_\infty$. Let the representant is Gaussian with exponential covariance $\exp(-rd)$. Then there is a blow-up at $r^\star = 0$ and the systems*

$\{\ln h(X_{\mathbf{t}}; r)\}_{r>0}$ and $\{h(X_{\mathbf{t}}; r)\}_{r>0}$ are stochastically unbounded. Moreover, the related Ornstein-Uhlenbeck process

$$dX_t = -rX_t dt + \sqrt{2r}dW_t$$

degenerates to the $dX_t = 0, t > 0$.

Proof. Set $\mathbf{t} = (t, s)$ and $|t - s| = d$. It is easy to check that $\det \Sigma \to 0$, for $r \to 0^+ \in G_\infty$. From continuity of the map $r \to \Sigma(r)$ we have also $\det \Sigma(0) = 0$. Now let $C > 0$ be arbitrary constant. Then

$$\mathbb{P}(\ln h(X_{\mathbf{t}}; r) > C) = \int_{\mathsf{Y}} d\mu(y),$$

where $\mathsf{Y} = \left\{y : y^T \Sigma^{-1} y < -2\ln\left(2\pi e^C (\det \Sigma)^{1/2}\right)\right\}$. Let U be the nonsingular matrix such that $U^T \Sigma^{-1} U = I$. Then after substitution $w = U^{-1}y$ we employ the polar coordinates substitution to obtain the latter integral in the form $1 - e^{-\sqrt{-2\ln(2\pi e^C (\det \Sigma)^{1/2})}}$. Thus we have for $0 < r < -\frac{1}{2d}\ln\left(1 - \frac{e^{-2C}}{4\pi^2}\right)$ the formula

$$\mathbb{P}(\ln h(X_{\mathbf{t}}; r) > C) = 1 - e^{-\sqrt{-2\ln(2\pi e^C \sqrt{1-e^{-2rd}})}}.$$

So finally $\lim_{r \to 0^+} \mathbb{P}(\ln h(X_{\mathbf{t}}; r) > C) = 1$. We have proved, that systems $\{\ln h(X_{\mathbf{t}}; r)\}_{r>0}$ and $\{h(X_{\mathbf{t}}; r)\}_{r>0}$ are stochastically unbounded. The representing process satisfies

$$dX_t = -rX_t dt + \sqrt{2r}dW_t$$

and for $r \to 0^+$ we have $dX_t = 0$. This completes the proof.

The following example is illustrating such a situation (Fig. 1).

Example 1. *Let us consider* $m = 1, k = 2, G_n = \left(\frac{1}{n}, \frac{1}{3}\right), \Gamma_n = \left\{\frac{1}{n}, \frac{1}{3}\right\}, n > 3$, $\mathcal{Y} = \mathbb{R}^2, X_{t_1} = y_1, X_{t_2} = y_2, d = t_2 - t_1 > 0$ *(here* $\{t_1, t_2\}, t_1 < t_2$, *is the design in* $T = [-1, 1]$). *Further we have* $f_n = f/G_n$ *where* $f = \exp(-rd)d^2\{-2\exp(-rd) + 2\exp(-3rd) + 2y_1^2 \exp(-3rd) + 2y_1^2 \exp(-rd) - y_1 y_2 - 6y_1 y_2 \exp(-2rd) - y_1 y_2 \exp(-4rd) + 2y_2^2 \exp(-3rd) + 2y_2^2 \exp(-rd)\}/(-1 + \exp(-2rd))^3$.

Let us consider the system of boundary functions $g_n = g/\Gamma_n, g(\Gamma_n) \subset \mathbb{R}$ *such that*

$$\lim_{n \to +\infty} g\left(\frac{1}{n}\right) = \infty$$

and

$$g\left(\frac{1}{3}\right) = -\ln 2\pi - 0.5 \ln \det \Sigma\left(\frac{1}{3}\right) + 0.5 \frac{y_1^2 - 2y_1 y_2 \exp(-\frac{d}{3}) + y_2^2}{-1 + \exp(-\frac{2}{3}d)},$$

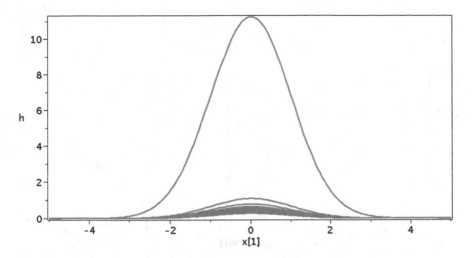

Fig. 1. Sections of graph of the density h by the plane $x_2 = x_1$ for several $r \in [0.0001, 0.2]$, $|t_1 - t_2| = d = 1$.

where $\Sigma_{i,j}(r) = \exp(-r|t_i - t_j|)$.

We obtain $d\mu(y) = (2\pi)^{-1}|\Sigma|^{-1/2}\exp(-1/2\, y^T\Sigma^{-1}y)dy_1 dy_2$, and solution

$$u = -\ln 2\pi - 0.5\ln\det\Sigma(r) + 0.5\frac{h(y_1, y_2; r)}{-1 + \exp(-2rd)}$$

where $h(y_1, y_2; r) = y_1^2 - 2y_1 y_2 \exp(-rd) + y_2^2$. Such a solution u is blowing-up for all evolutions $r \to h(y_1, y_2; r)$ in $\mathcal{Y}$ for which

$$\lim_{r\to 0^+}\frac{h(y_1, y_2; r)}{\det\Sigma(r)} < +\infty.$$

2.2   2D Case for Independent Processes

Consider now process $\mathbf{X}_t$ given by system (5) with $d = r = p = 2$, $\mathbf{a}(t) \equiv 0$ and constant coefficients $\mathbf{A} = \mathrm{diag}(-r_1, -r_2)$ and $\boldsymbol{\sigma} = \mathrm{diag}(\sqrt{2r_1}, \sqrt{2r_2})$. We also assume that $\mathbf{Q} = \mathbf{I}$ and $\boldsymbol{\xi} \sim \mathcal{N}(\mathbf{0}, \mathbf{I})$. One can then easily compute

$$\boldsymbol{\Phi}(t) = \begin{pmatrix} e^{-r_1 t} & 0 \\ 0 & e^{-r_2 t} \end{pmatrix}$$

and $\mathbf{C}(t) = \mathbf{I}$, which for $|t - s| = d$ yields covariance structure

$$\tilde{\boldsymbol{\Sigma}} := \tilde{\mathbf{R}}(d) = \begin{pmatrix} 1 & e^{-r_1 d} & 0 & 0 \\ e^{-r_1 d} & 1 & 0 & 0 \\ 0 & 0 & 1 & e^{-r_2 d} \\ 0 & 0 & e^{-r_2 d} & 1 \end{pmatrix}.$$

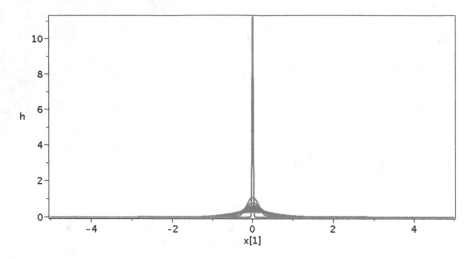

Fig. 2. Sections of graph of the density h by the plane $x_2 = 0$ for several $r \in [0.0001, 0.2]$, $|t_1 - t_2| = d = 1$.

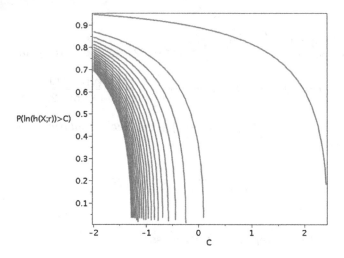

Fig. 3. $\mathbb{P}(\ln h(X_\mathbf{t}; r) > C)$ for several $r \in [0.0001, 0.2]$ depending on C, $|t_1 - t_2| = d = 1$.

Clearly $\det \tilde{\boldsymbol{\Sigma}} = (1 - e^{-2r_1 d})(1 - e^{-2r_2 d})$ and by an inverse formula we have

$$
\tilde{\boldsymbol{\Sigma}}^{-1} = \left(
\begin{array}{cc|cc}
\frac{1}{1-e^{-2r_1 d}} & -\frac{e^{-r_1 d}}{1-e^{-2r_1 d}} & 0 & 0 \\
-\frac{e^{-r_1 d}}{1-e^{-2r_1 d}} & \frac{1}{1-e^{-2r_1 d}} & 0 & 0 \\
\hline
0 & 0 & \frac{1}{1-e^{-2r_2 d}} & -\frac{e^{-r_2 d}}{1-e^{-2r_2 d}} \\
0 & 0 & -\frac{e^{-r_2 d}}{1-e^{-2r_2 d}} & \frac{1}{1-e^{-2r_2 d}}
\end{array}
\right).
$$

But obviusly $X^1_{(t,s)}$ and $X^2_{(t,s)}$ are jointly normally distributed and uncorrelated, therefore independent. More formally, $\mathbf{X}_{(t,s)}$ has density

$$h(\mathbf{x};\mathbf{r}) = \frac{1}{\sqrt{(2\pi)^4 \det \tilde{\boldsymbol{\Sigma}}}} \exp\left(-\frac{1}{2}\mathbf{x}^{\mathrm{T}}\tilde{\boldsymbol{\Sigma}}^{-1}\mathbf{x}\right) = h_{(1)}(\mathbf{x}_{(1)})\, h_{(2)}(\mathbf{x}_{(2)}),$$

where $h_{(1)}(\mathbf{x}_{(1)}) = \frac{1}{2\pi\sqrt{\det \tilde{\boldsymbol{\Sigma}}_{(1)}}} \exp\left(-\frac{1}{2}\mathbf{x}_{(1)}{}^{\mathrm{T}}\tilde{\boldsymbol{\Sigma}}_{(1)}{}^{-1}\mathbf{x}_{(1)}\right)$, $\mathbf{x}_{(1)} = (x_1, x_2)$ and $\tilde{\boldsymbol{\Sigma}}_{(1)}$ is upper block matrix of $\tilde{\boldsymbol{\Sigma}}$ and for $\mathbf{x}_{(2)} = (y_1, y_2)$, $\tilde{\boldsymbol{\Sigma}}_{(2)}$, the lower block matrix of $\tilde{\boldsymbol{\Sigma}}$, is $h_{(2)}(\mathbf{x}_{(2)}) = \frac{1}{2\pi\sqrt{\det \tilde{\boldsymbol{\Sigma}}_{(2)}}} \exp\left(-\frac{1}{2}\mathbf{x}_{(2)}{}^{\mathrm{T}}\tilde{\boldsymbol{\Sigma}}_{(2)}{}^{-1}\mathbf{x}_{(2)}\right)$. Now we can conclude that

$$\mathbb{P}(\ln h(\mathbf{X}_{(t,s)}(\mathbf{x});\mathbf{r}) > C) = \mathbb{P}(\ln h_{(1)}(X^1_{(t,s)}(\mathbf{x}_{(1)})) + \ln h_{(2)}(X^2_{(t,s)}(\mathbf{x}_{(2)})) > C)$$

$$\geq \mathbb{P}(\ln h_{(1)}(X^1_{(t,s)}(\mathbf{x}_{(1)})) > C \;\cap\; \ln h_{(2)}(X^2_{(t,s)}(\mathbf{x}_{(2)})) > C)$$

$$= \mathbb{P}(\ln h_{(1)}(X^1_{(t,s)}(\mathbf{x}_{(1)})) > C)\, \mathbb{P}(\ln h_{(2)}(X^2_{(t,s)}(\mathbf{x}_{(2)})) > C).$$

Therefore, previous ideas mentioned in one dimensional case can be applied. Notice however, that the observation here is slightly different. If for $r_1 \to r_1^*$, $r_2 \to r_2^*$ only one of the block covariance structure has zero determinant (it is clear that $\det \tilde{\boldsymbol{\Sigma}} = 0$),

$$\lim_{(r_1,r_2)\to(r_1^*,r_2^*)} \mathbb{P}(\ln h(\mathbf{X}_{(t,s)}(\mathbf{x});\mathbf{r}) > C) < 1$$

since one of the block covariances is regular. (See also Figs. 1, 2 and 3 for graphical view.)

Remark 1. *For two dimensional (dependent) process* $\mathbf{X}_t$ *given by system* (5), *whereas* $\mathbf{a}(t) \equiv 0$ *and* $\mathbf{A}(t) = \begin{pmatrix} -r_1 & 0 \\ -r_2 & 0 \end{pmatrix}$, $\sigma = \begin{pmatrix} \sigma & 0 \\ 0 & 0 \end{pmatrix}$ *with* $\mathbf{Q} = \mathbf{I}$ *and* $\xi \sim \mathcal{N}(0, \mathbf{I})$ *analogous method can be used. Notice that the process* $\mathbf{X}_t = \left(X^1_t, -r_2 \int_0^t X^1_s\, ds\right)$, *is the Ornstein-Uhlenbeck process and its time integral, which is also Gaussian, whereas the latter can be used to generate noise with a* $\frac{1}{f}$ *power spectrum (pink noise).*

3   Discussion

Typically, an interesting case of the Ornstein-Uhlenbeck process is studied, when r approaches zero with σ becomes infinite in such a way, that $r\sigma^2$ approaches a fixed constant (see [6]). Usually it is accomplished by letting $\sigma^2 = \frac{D}{2r}$, where D is a constant, called the diffusivity. This limiting case is often called the Gaussian random walk process. The correlation function $\sigma^2 e^{-rd}$ of the random walk is not defined, since $\sigma \to +\infty$ as $r \to 0^+$. However, the variogram defined as

$$\psi(\tau) = \mathbb{E}([x(t) - x(t+\tau)]^2) = 2\sigma^2(1 - e^{-r|\tau|}) = \frac{D}{r}(1 - e^{-r|\tau|})$$

does have meaning in the limit, namely $\psi(\tau) = D|\tau|$.

However, as we have shown in the present paper, also limit $r \to 0^+$ together with $\sigma = \sqrt{2r}$ gives an interesting interpretation for the blow up of the limit of the regular solutions of the sequence of Dirichlet problems for the Poisson equation.

The generalization of the Gaussian field representation is through the set probability density function with support on a finite interval. The set of probability functions is a convex subset of L^1 and it does not have a linear structure when using ordinary sum and multiplication by real constants. The crucial point is that usual distances are not invariant under relevant transformations of densities. To overcome these limitations, Aitchinson's ideas on compositional data analysis have been used in [2], generalizing perturbation and power transformation, as well as the Aitchison inner product, to operations on probability density functions with support on a finite interval. Authors showed that the set of bounded probability density functions on finite intervals is a pre-Hilbert space.

Acknowledgments. Research was supported by the Slovak Research and Development Agency under the contract No. SK-AT-2015-0019. Corresponding author was supported by Fondecyt Proyecto Regular N 1151441 and WTZ Project SK 09/2016. Jozef Kiseľák was partially supported by grant VEGA MŠ SR 1/0344/14.

References

1. Chung, C.B., Kravaris, C.: Identification of spatially discontinuous parameters in 2nd order parabolic systems by piecewise regularization. Inverse Prob. **4**, 973–994 (1988)
2. Egozcue, J.J., Diaz-Barrero, J.L., Pawlowsky-Glahn, V.: Hilbert space of probability density functions based on Aitchison geometry. Acta Math. Sin. English Ser. **22**(4), 1175–1182 (2006)
3. Evans, L.C.: Partial Differential Equations: Graduate Studies in Mathematics. American Mathematical Society, Providence (1998). 2nd edn. (2010)
4. Karatzas, I., Shreve, S.E.: Brownian Motion and Stochastic Calculus. Graduate Texts in Mathematics, vol. 113, 2nd edn. Springer, Heidelberg (1991)
5. Prokhorov, Y., Shiryaev, A.N.: Probability Theory III. Encyclopaedia of Mathematical Sciences, vol. 45. Springer, Heidelberg (1998)
6. Rybicki, G.B.: Notes on the Ornstein-Uhlenbeck Process, unpublished notes (1994)
7. Stehlík, M.: Stochastic modelling for solving PDE and D-optimality under correlation. Romai J. **1**(1), 175–182 (2005)
8. Stehlík, M. On Stochastic Representation for Dirichlet Problem Solution. MendelNet (2006). Alfaknihy, Prague (in press)
9. Uciński, D.: Optimal Measurement Methods for Distributed Parameter System Identification. CRC Press, Boca Raton (2005)
10. Walter, É., Pronzato, L.: Qualitative and quantitative experiment design for phenomenological models-A survey. Automatica **26**(2), 195–213 (1990)

A Singularly Perturbed Boundary Value Problems with Fractional Powers of Elliptic Operators

Petr N. Vabishchevich[1,2]($\boxtimes$)

[1] Nuclear Safety Institute, 52, B. Tulskaya, 115191 Moscow, Russia
vabishchevich@gmail.com
[2] North-Eastern Federal University, 58, Belinskogo, 677000 Yakutsk, Russia

Abstract. A boundary value problem for a fractional power $0 < \varepsilon < 1$ of the second-order elliptic operator is considered. The boundary value problem is singularly perturbed when $\varepsilon \to 0$. It is solved numerically using a time-dependent problem for a pseudo-parabolic equation. For the auxiliary Cauchy problem, the standard two-level schemes with weights are applied. The numerical results are presented for a model two-dimensional boundary value problem with a fractional power of an elliptic operator. Our work focuses on the solution of the boundary value problem with $0 < \varepsilon \ll 1$.

1 Introduction

Non-local applied mathematical models based on the use of fractional derivatives in time and space are actively discussed in the literature [2,11]. Many models, which are used in applied physics, biology, hydrology, and finance, involve both sub-diffusion (fractional in time) and super-diffusion (fractional in space) operators. Super-diffusion problems are treated as problems with a fractional power of an elliptic operator. For example, suppose that in a bounded domain Ω on the set of functions $u(x) = 0$, $x \in \partial\Omega$, there is defined the operator $\mathcal{A}$: $\mathcal{A}u = -\triangle u$, $x \in \Omega$. We seek the solution of the problem for the equation with the fractional power of an elliptic operator:

$$\mathcal{A}^\varepsilon u = f,$$

with $0 < \varepsilon < 1$ for a given $f(x)$, $x \in \Omega$.

To solve problems with the fractional power of an elliptic operator, we can apply finite volume or finite element methods oriented to using arbitrary domains discretized by irregular computational grids [12,15]. The computational realization is associated with the implementation of the matrix function-vector multiplication. For such problems, different approaches [7] are available. Problems of using Krylov subspace methods with the Lanczos approximation when solving systems of linear equations associated with the fractional elliptic equations are discussed, e.g., in [10]. A comparative analysis of the contour integral method,

© Springer International Publishing AG 2017
I. Dimov et al. (Eds.): NAA 2016, LNCS 10187, pp. 141–152, 2017.
DOI: 10.1007/978-3-319-57099-0_13

the extended Krylov subspace method, and the preassigned poles and interpolation nodes method for solving space-fractional reaction-diffusion equations is presented in [6]. The simplest variant is associated with the explicit construction of the solution using the known eigenvalues and eigenfunctions of the elliptic operator with diagonalization of the corresponding matrix [5,8,9]. Unfortunately, all these approaches demonstrates too high computational complexity for multidimensional problems.

We have proposed [19] a computational algorithm for solving an equation with fractional powers of elliptic operators on the basis of a transition to a pseudo-parabolic equation. For the auxiliary Cauchy problem, the standard two-level schemes are applied. The computational algorithm is simple for practical use, robust, and applicable to solving a wide class of problems. A small number of pseudo-time steps is required to reach a steady-state solution. This computational algorithm for solving equations with fractional powers of operators is promising when considering transient problems.

The boundary value problem for the fractional power of an elliptic operator is singularly perturbed when $\varepsilon \to 0$. To solve it numerically, we focus on numerical methods that are designed for classical elliptic problems of convection-diffusion-reaction [14,17]. In particular, the main features are taken into account via using locally refining grids. The standard strategy of goal-oriented error control for conforming finite element discretizations [1,3] is applied.

2   Problem Formulation

In a bounded polygonal domain $\Omega \subset R^m$, $m = 1, 2, 3$ with the Lipschitz continuous boundary $\partial\Omega$, we search the solution for the problem with a fractional power of an elliptic operator. Define the elliptic operator as

$$\mathcal{A}u = -\mathrm{div}(k(\boldsymbol{x})\mathrm{grad}\, u) \tag{1}$$

with coefficient $0 < k_1 \leq k(\boldsymbol{x}) \leq k_2$. The operator $\mathcal{A}$ is defined on the set of functions $u(\boldsymbol{x})$ that satisfy on the boundary $\partial\Omega$ the following conditions:

$$u(\boldsymbol{x}) = 0, \quad \boldsymbol{x} \in \partial\Omega. \tag{2}$$

In the Hilbert space $H = L_2(\Omega)$, we define the scalar product and norm in the standard way:

$$(u, v) = \int_\Omega u(\boldsymbol{x})v(\boldsymbol{x})d\boldsymbol{x}, \quad \|u\| = (u, u)^{1/2}.$$

For the spectral problem

$$\mathcal{A}\varphi_k = \lambda_k \varphi_k, \quad \boldsymbol{x} \in \Omega,$$

$$\varphi_k(\boldsymbol{x}) = 0, \quad \boldsymbol{x} \in \partial\Omega,$$

we have
$$\lambda_1 \le \lambda_2 \le \dots,$$
and the eigenfunctions $\varphi_k, \|\varphi_k\| = 1, k = 1, 2, \dots$ form a basis in $L_2(\Omega)$. Therefore,
$$u = \sum_{k=1}^{\infty} (u, \varphi_k)\varphi_k.$$

Let the operator $\mathcal{A}$ be defined in the following domain:

$$D(\mathcal{A}) = \{u \mid u(x) \in L_2(\Omega), \sum_{k=0}^{\infty} |(u, \varphi_k)|^2 \lambda_k < \infty\}.$$

Under these conditions the operator $\mathcal{D}$ is self-adjoint and positive defined:

$$\mathcal{A} = \mathcal{A}^* \ge \delta I, \quad \delta > 0, \tag{3}$$

where I is the identity operator in H. For δ, we have $\delta = \lambda_1$. In applications, the value of λ_1 is unknown (the spectral problem must be solved). Therefore, we assume that $\delta \le \lambda_1$ in (3). Let us assume for the fractional power of the operator $\mathcal{A}$

$$\mathcal{A}^\varepsilon u = \sum_{k=0}^{\infty} (u, \varphi_k)\lambda_k^\varepsilon \varphi_k.$$

We seek the solution of the problem with the fractional power of the operator $\mathcal{A}$. The solution $u(x)$ satisfies the equation

$$\mathcal{A}^\varepsilon u = f, \tag{4}$$

with $0 < \varepsilon < 1$ for a given $f(x)$, $x \in \Omega$.

The key issue in the study of the computational algorithm for solving the problem (4) is to establish the stability of the approximate solution with respect to small perturbations of the right-hand side in various norms. In view of (3), the solution of the problem (4) satisfies the a priori estimate

$$\|u\| \le \delta^{-\varepsilon} \|f\|, \tag{5}$$

which is valid for all $0 < \varepsilon < 1$.

The boundary value problem for the fractional power of the elliptic operator (4) demonstrates a reduced smoothness when $\varepsilon \to 0$. For the solution, we have (see, e.g., [20]) the estimate

$$\|u\|_{2\varepsilon} \le C\|f\|,$$

with $0 \le \varepsilon < 1/2$, is $\| \cdot \|_{2\varepsilon}$ is the norm in $H^{2\varepsilon}(\Omega)$. For the limiting solution, we have

$$u_0(x) = f(x), \quad x \in \Omega.$$

Thus, a singular behavior of the solution of the problem (4) appears with $\varepsilon \to 0$ and is governed by the right-hand side $f(x)$.

3   Discretization in Space

To solve numerically the problem (4), we employ finite-element approximations in space [4,18]. For (1) and (2), we define the bilinear form

$$a(u, v) = \int_\Omega k \operatorname{grad} u \operatorname{grad} v.$$

By (3), we have

$$a(u, u) \geq \delta \|u\|^2.$$

Define a subspace of finite elements $V^h \subset H_0^1(\Omega)$. Let $\boldsymbol{x}_i$, $i = 1, 2, \dots, M_h$ be triangulation points for the domain Ω. Define pyramid function $\chi_i(\boldsymbol{x}) \subset V^h$, $i = 1, 2, \dots, M_h$, where

$$\chi_i(\boldsymbol{x}_j) = \begin{cases} 1, \text{ if } i = j, \\ 0, \text{ if } i \neq j. \end{cases}$$

For $v \in V_h$, we have

$$v(\boldsymbol{x}) = \sum_{i=i}^{M_h} v_i \chi_i(\boldsymbol{x}),$$

where $v_i = v(\boldsymbol{x}_i)$, $i = 1, 2, \dots, M_h$. We have defined Lagrangian finite elements of first degree, i.e., based on the piecewise-linear approximation. We will also use Lagrangian finite elements of second degree defined in a similar way.

We define the discrete elliptic operator A as

$$(Ay, v) = a(y, v), \quad \forall \, y, v \in V^h.$$

The fractional power of the operator A is defined similarly to $\mathcal{A}^\varepsilon$. For the spectral problem

$$A\widetilde{\varphi}_k = \widetilde{\lambda}_k$$

we have

$$\widetilde{\lambda}_1 \leq \widetilde{\lambda}_2 \leq \dots \leq \widetilde{\lambda}_{M_h}, \quad \|\widetilde{\varphi}_k\| = 1, \quad k = 1, 2, \dots, M_h.$$

The domain of definition for the operator D is

$$D(A) = \{y \mid y \in V^h, \ \sum_{k=0}^{M_h} |(y, \widetilde{\varphi}_k)|^2 \widetilde{\lambda}_k < \infty\}.$$

The operator A acts on a finite dimensional space V^h defined on the domain $D(A)$ and, similarly to (3),

$$A = A^* \geq \delta I, \quad \delta > 0, \tag{6}$$

where $\delta \leq \lambda_1 \leq \widetilde{\lambda}_1$. For the fractional power of the operator A, we suppose

$$A^\varepsilon y = \sum_{k=1}^{M_h} (y, \widetilde{\varphi}_k) \widetilde{\lambda}_k^\varepsilon \widetilde{\varphi}_k.$$

For the problem (4), we put into the correspondence the operator equation for $w(t) \in V^h$:

$$A^\varepsilon w = \psi, \tag{7}$$

where $\psi = Pf$ with P denoting L_2-projection onto V^h. For the solution of the problem (6), (7), we obtain (see (5)) the estimate

$$\|w\| \le \delta^{-\varepsilon}\|\psi\|, \tag{8}$$

for all $0 < \varepsilon < 1$.

4 Singularly Perturbed Problem for a Diffusion-Reaction Equation

The object of our study is associated with the development of a computational algorithm for approximate solving the singularly perturbed problem (4). After constructing a finite element approximation, we arrive at Eq. (7). Features of the solution related to a boundary layer are investigated on a model singularly perturbed problem for an equation of diffusion-reaction. The key moment is associated with selecting adaptive computational grids (triangulations).

In view of

$$A^\varepsilon = (\exp(\ln A))^\varepsilon = I + \varepsilon \ln A + \mathcal{O}(\varepsilon^2),$$

we put the problem (7) into the correspondence with solving the equation

$$\varepsilon Au + u = \psi. \tag{9}$$

The Eq. (9) corresponds to solving the Dirichlet problem (see the condition (2)) for the diffusion-reaction equation

$$-\varepsilon \operatorname{div}(k(\boldsymbol{x})\operatorname{grad} u) + u = f(\boldsymbol{x}), \quad \boldsymbol{x} \in \Omega. \tag{10}$$

Basic computational algorithms for the singularly perturbed boundary problem (2), (10) are considered, for example, in [14, 17].

In terms of practical applications, the most interesting approach is based on an adaptation of a computational grid to peculiarities of the problem solution via a posteriori error estimates. Among main approaches, we highlight the strategy of the goal-oriented error control for conforming finite element discretizations [1,3], which is applied to approximate solving boundary value problems for elliptic equations.

The strategy of goal-oriented error control is based on choosing a calculated functional. The accuracy of its evaluation is tracked during computations. In our Dirichlet problem for the second-order elliptic equation, the solution is varied drastically near the boundary. So, it seems natural to control the accuracy of calculations for the normal derivatives of the solution (fluxes) across the boundary or a portion of it. Because of this, we put

$$G(u) = -\int_{\partial\Omega} \varepsilon k(\boldsymbol{x})(\operatorname{grad} u \cdot \boldsymbol{n})d\boldsymbol{x},$$

where n is the outward normal to the boundary. An adaptation of a finite element mesh is based on an iterative local refinement of the grid in order to evaluate the goal functional with a given accuracy η on the deriving approximate solution u_h, i.e.,

$$|G(u) - G(u_h)| \leq \eta.$$

To conduct our calculations, we used the FEniCS framework (see, e.g., [13]) developed for general engineering and scientific calculations via finite elements. Features of the goal-oriented procedure for local refinement of the computational grid are described in [16] in detain. Here, we consider only a key idea of the adaptation strategy of finite element meshes, which is associated with selecting the goal functional.

The model problem (2), (10) is considered with

$$k(x) = 1, \quad f(x) = (1 - x_1)x_2^2,$$

in the unit square ($\Omega = (0,1) \times (0,1)$). The threshold of accuracy for calculating the functional $G(u)$ is defined by the value of $\eta = 10^{-5}$. As an initial mesh, there is used the uniform grid obtained via division by 8 intervals in each direction (step 0–128 cells).

First, Lagrangian finite elements of first order have been used in our calculations. For this case, the improvement of the goal functional during the iterative procedure of adaptation is illustrated by the data presented in Table 1. Table 2 demonstrates values of the goal functional $G(u_h)$ calculated on the final computational grid, the number of vertices of this final grid and the number of adaptation steps for solving the problem at various values of the small parameter ε. These numerical results demonstrate the efficiency of the proposed strategy for goal-

Table 1. Calculation of the goal functional during adaptation steps

ε	10^{-1}		10^{-3}		10^{-5}	
Step of adaptation s	$G(u_h)$	M_h	$G(u_h)$	M_h	$G(u_h)$	M_h
0	0.087608	81	0.0056973	81	0.00006643	81
1	0.110432	97	0.0107507	98	0.00015584	95
2	0.116155	140	0.0129506	132	0.00023996	120
3	0.119766	222	0.0155597	195	0.00035644	164
4	0.122702	384	0.0175113	305	0.00050472	225
5	0.125653	694	0.0194985	466	0.00068154	349
6	0.127950	1235	0.0210232	754	0.00090839	550
7	0.128835	2179	0.0221562	1279	0.00115091	853
8	0.129542	3841	0.0229284	2132	0.00137740	1242
9	0.129940	6540	0.0234492	3753	0.00161273	1865
10	0.130149	11040	0.0237487	6626	0.00181249	2711

Table 2. Adaptation for various values of ε

ε	Goal functional $G(u_h)$	Number of vertices	Number of adaptation steps s
10^{-1}	0.130396	51868	13
10^{-2}	0.064867	72297	14
10^{-3}	0.024191	90170	15
10^{-4}	0.008061	67476	16
10^{-5}	0.002580	99003	18

oriented error control for conforming finite element discretizations applied to approximate solving singular perturbed problems of diffusion-reaction (2), (10).

Next, similar results have been obtained using Lagrangian finite elements of second order. For this case, summary data are presented in Table 3. As expected, the desired accuracy $\eta = 10^{-5}$ is reached on adaptive meshes of smaller sizes than in the case of Lagrangian finite elements of first order (see Table 2 for a comparison).

Table 3. Adaptation for Lagrangian elements of second order

ε	Goal functional $G(u_h)$	Number of vertices	Number of adaptation steps s
10^{-1}	0.130423	3574	7
10^{-2}	0.064884	5137	8
10^{-3}	0.024184	6573	9
10^{-4}	0.008076	12775	11
10^{-5}	0.002574	18501	12

5   Numerical Algorithm for the Problem with a Fractional Power

An approximate solution of the problem (7) is sought as a solution of an auxiliary pseudo-time evolutionary problem [19]. Assume that

$$y(t) = \delta^{\varepsilon}(t(A - \delta I) + \delta I)^{-\varepsilon} y(0).$$

Therefore

$$y(1) = \delta^{\varepsilon} A^{-\varepsilon} y(0)$$

and then $w = y(1)$. The function $y(t)$ satisfies the evolutionary equation

$$(tD + \delta I)\frac{dy}{dt} + \varepsilon Dy - 0, \quad 0 < t \leq 1, \tag{11}$$

where

$$D = A - \delta I.$$

By (6), we get

$$D = D^* > 0. \tag{12}$$

We supplement (11) with the initial condition

$$y(0) = \delta^{-\varepsilon}\psi. \tag{13}$$

The solution of Eq. (7) can be defined as the solution of the Cauchy problem (11)–(13) at the final pseudo-time moment $t = 1$.

For the solution of the problem (11), (13), it is possible to obtain various a priori estimates. The elementary estimate that is consistent with the estimate (8) have the form

$$\|y(t)\| \le \|y(0)\|. \tag{14}$$

To get (14), multiply scalarly Eq. (11) by $\varepsilon y + t\, dy/dt$.

To solve numerically the problem (11), (13), we use the simplest implicit two-level scheme. Let τ be a step of a uniform grid in time such that $y^n = y(t^n)$, $t^n = n\tau$, $n = 0, 1, \ldots, N$, $N\tau = 1$. Let us approximate Eq. (11) by the implicit two-level scheme

$$(t^{\sigma(n)} D + \delta I)\frac{y^{n+1} - y^n}{\tau} + \varepsilon D y^{\sigma(n)} = 0, \quad n = 0, 1, \ldots, N - 1, \tag{15}$$

$$y^0 = \delta^{-\varepsilon}\psi. \tag{16}$$

We use the notation

$$t^{\sigma(n)} = \sigma t^{n+1} + (1 - \sigma)t^n, \quad y^{\sigma(n)} = \sigma y^{n+1} + (1 - \sigma)y^n.$$

For $\sigma = 0.5$, the difference scheme (15), (16) approximates the problem (11), (12) with the second order by τ, whereas for other values of σ, we have only the first order.

Theorem 1. *For $\sigma \ge 0.5$ the difference scheme (15), (16) is unconditionally stable with respect to the initial data. The approximate solution satisfies the estimate*

$$\|y^{n+1}\| \le \|y^0\|, \quad n = 0, 1, \ldots, N - 1. \tag{17}$$

Proof. Rewrite Eq. (15) in the following form:

$$\delta\frac{y^{n+1} - y^n}{\tau} + D\left(\varepsilon y^{\sigma(n)} + t^{\sigma(n)}\frac{y^{n+1} - y^n}{\tau}\right) = 0.$$

Multiplying scalarly it by

$$\varepsilon y^{\sigma(n)} + t^{\sigma(n)}\frac{y^{n+1} - y^n}{\tau},$$

in view of (12), we arrive at

$$\left(\frac{y^{n+1} - y^n}{\tau}, y^{\sigma(n)}\right) \le 0.$$

We have

$$y^{\sigma(n)} = \left(\sigma - \frac{1}{2}\right)\tau\frac{y^{n+1} - y^n}{\tau} + \frac{1}{2}(y^{n+1} + y^n).$$

If $\sigma \geq 0.5$, then

$$\|y^{n+1}\| \leq \|y^n\|, \quad n = 0, 1, \ldots, N - 1.$$

Thus, we obtain (17).

The key point in approximate solving singularly perturbed boundary value problems is associated with mesh adaptation. In the case of solving the problem (4), we use finite element approximations and proceed to the problem (7) and then formulate the Cauchy problem (11), (13) approximated by the scheme (15), (16). In our case, singularity is associated only with spatial variables.

The decomposition of the solution of the problem (11), (13) by eigenfunctions of the operator A results in

$$y(t) = \sum_{k=1}^{N_h} a_k(t)\tilde{\varphi}_k.$$

For coefficients $a_k(t)$, we get

$$a_k(t) = (\psi, \tilde{\varphi}_k)(\delta + (\tilde{\lambda}_k - \delta)t)^{-\varepsilon}, \quad k = 1, 2, \ldots, M_h.$$

Because of this, errors in specifying the initial conditions monotonically decrease for increasing t. A similar behavior demonstrates an approximate solution of the Cauchy problem (11), (13) obtained using the fully implicit scheme with $\sigma = 1$ in (15), (16). For the Crank-Nicolson scheme (i.e., $\sigma = 0.5$ in (15), (16)), we cannot guarantee a monotone decrease of errors in time, but the error at $t = 1$ will not be more than at $t = 0$. The practical significance of such an analysis is that it provides us a simple adaptation strategy for computational grids in solving the problem (11), (13), namely, spatial mesh adaptation is conducted at the first time step of calculations.

6   Solution of a Model Problem

Below, there are presented some results of numerical solving the problem (7) for small values of ε. A computational algorithm must track a singular behavior of the solution, which is directly related to the singular behavior of the right-hand side $f(x)$. Let us consider the problem (2), (10) in the unit square $\Omega = (0, 1) \times (0, 1)$ with

$$k(x) = 1, \quad f(x) = \left(1 - x_1 - \exp\left(-\frac{x_1}{\mu}\right)\right)\left(x_2^2 - \exp\left(-\frac{1 - x_2}{\mu}\right)\right).$$

The singularity of the right-hand side (the singularity of a numerical solution of the problem with a fractional power of an elliptic operator) results from existing a boundary layer at low values of μ.

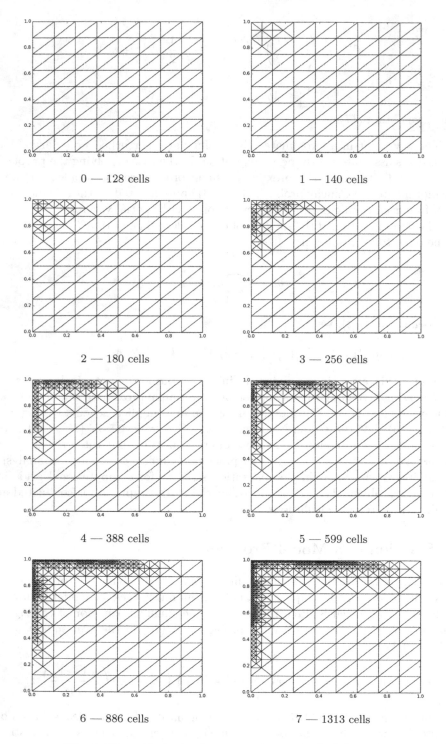

Fig. 1. The grid obtained at successive steps of adaptation

Table 4. Calculation of the goal functional during adaptation steps

ε	10^{-1}		10^{-2}		10^{-3}	
Step of adaptation s	$G(u_h; t = \tau)$	M_h	$G(u_h; t = \tau)$	M_h	$G(u_h; t - \tau)$	M_h
0	16.0955	289	21.8932	289	22.5773	289
1	24.3875	315	33.7810	315	34.9016	315
2	31.1692	399	43.8893	399	45.4226	399
3	37.0996	559	52.8249	559	54.7328	559
4	42.0854	833	60.4373	837	62.2786	834
5	45.7594	1270	66.2019	1282	68.4363	1264
6	48.6087	1849	70.4881	1885	73.2101	1889
7	50.3070	2753	73.1491	2778	75.9998	2774
8	51.2621	4067	74.6778	4120	77.5762	4125
9	51.9766	5965	75.8362	5968	78.7648	6028
10	52.2862	9201	76.3402	9235	79.2942	9261

An adaptation of the computational grid is performed during the calculation of the first time step using the two-level scheme (15), (16). For the basic variant, it is assumed that $\varepsilon = 10^{-2}$, $\mu = 10^{-2}$, the initial uniform spatial grid contains 8 intervals in each direction and the time step is $\tau = 10^{-2}$. The parameter $\theta = 2\pi^2$ corresponds the minimal eigenvalue of the elliptic operator $\mathcal{A}$. Mesh adaptation is carried out taking into account peculiarities of the right-hand side and the goal functional defined in the form

$$G(u; t = \tau) = -\int_{\partial\Omega} k(\boldsymbol{x})(\operatorname{grad} u \cdot \boldsymbol{n})d\,\boldsymbol{x}.$$

Next, the problem (2), (10) is solved using the derived grid in space and the uniform grid in time. Thus, we apply the simplest one-stage starting adaptation of the computational grid for numerical solving the unsteady problem. Lagrangian finite elements of second order are used. For time-stepping, the Crank-Nicolson ($\sigma = 0.5$ in (15)) scheme is utilized. The sequence of calculated adaptive grids is shown in Fig. 1. Note that this sequence is weakly dependent on the choice of a time step. The goal functional dynamics for different levels of adaptation is presented in Table 4. The problem is solved with different values of ε.

Acknowledgements. This work was supported by the Russian Foundation for Basic Research (projects 14-01-00785, 15-01-00026).

References

1. Ainsworth, M., Oden, J.T.: A Posteriori Error Estimation in Finite Element Analysis. Wiley, New York (2000)

2. Baleanu, D.: Fractional Calculus: Models and Numerical Methods. World Scientific, New York (2012)
3. Bangerth, W., Rannacher, R.: Adaptive Finite Element Methods for Differential Equations. Birkhäuser, Basel (2003)
4. Brenner, S.C., Scott, L.R.: The Mathematical Theory of Finite Element Methods. Springer, New York (2008)
5. Bueno-Orovio, A., Kay, D., Burrage, K.: Fourier spectral methods for fractional-in-space reaction-diffusion equations. BIT Numer. Math. **54**(4), 937–954 (2014)
6. Burrage, K., Hale, N., Kay, D.: An efficient implicit FEM scheme for fractional-in-space reaction-diffusion equations. SIAM J. Sci. Comput. **34**(4), A2145–A2172 (2012)
7. Higham, N.J.: Functions of Matrices: Theory and Computation. SIAM, Philadelphia (2008)
8. Ilic, M., Liu, F., Turner, I., Anh, V.: Numerical approximation of a fractional-in-space diffusion equation, I. Fract. Calculus Appl. Anal. **8**(3), 323–341 (2005)
9. Ilic, M., Liu, F., Turner, I., Anh, V.: Numerical approximation of a fractional-in-space diffusion equation. II with nonhomogeneous boundary conditions. Fract. Calculus Appl. Anal. **9**(4), 333–349 (2006)
10. Ilić, M., Turner, I.W., Anh, V.: A numerical solution using an adaptively preconditioned Lanczos method for a class of linear systems related with the fractional Poisson equation. Int. J. Stoch. Anal. **2008**, Article ID 104525 (2008). 26 p
11. Kilbas, A.A., Srivastava, H.M., Trujillo, J.J.: Theory and Applications of Fractional Differential Equations. North-Holland Mathematics Studies. Elsevier, Amsterdam (2006)
12. Knabner, P., Angermann, L.: Numerical Methods for Elliptic and Parabolic Partial Differential Equations. Springer, New York (2003)
13. Logg, A., Mardal, K.A., Wells, G.: Automated Solution of Differential Equations by the Finite Element Method: The FEniCS Book. Springer, Berlin (2012)
14. Miller, J.J.H., O'Riordan, E., Shishkin, G.I.: Fitted Numerical Methods for Singular Perturbation Problems: Error Estimates in the Maximum Norm for Linear Problems in One and Two Dimensions. World Scientific, New Jersey (2012)
15. Quarteroni, A., Valli, A.: Numerical Approximation of Partial Differential Equations. Springer, Berlin (1994)
16. Rognes, M.E., Logg, A.: Automated goal-oriented error control I: stationary variational problems. SIAM J. Sci. Comput. **35**(3), C173–C193 (2013)
17. Roos, H.G., Stynes, M., Tobiska, L.: Robust Numerical Methods for Singularly Perturbed Differential Equations: Convection-Diffusion-Reaction and Flow Problems. Springer, Berlin (2008)
18. Thomée, V.: Galerkin Finite Element Methods for Parabolic Problems. Springer, Berlin (2006)
19. Vabishchevich, P.N.: Numerically solving an equation for fractional powers of elliptic operators. J. Comput. Phys. **282**(1), 289–302 (2015)
20. Yagi, A.: Abstract Parabolic Evolution Equations and Their Applications. Springer, Berlin (2009)

Contributed Papers

Simulation of Surface Heating Process with Laser

Tatiana Akimenko$^{(\boxtimes)}$, Olga Gorbunova, and Valery Dunaev

Department of Robotics and Automation, Department of Physics,
Department of Missilery, Tula State University, Tula 300012, Russia
tantan72@mail.ru, oygor@mail.ru, dwa222@mail.ru

Abstract. The model of heating of surface of solid substance with infrared laser and cooling with air flow is created. From initial differential partial equations the discrete form of equation is obtained. The method of integration of system under dynamic change of energy of laser beam and constant parameters of airflow is worked out. Due to this method the initial model was transformed into a discrete form. The discrete equation is solved for the case of dynamically changing laser beam energy and constant parameters of the airflow. The result of solving of equation is graphed in 3D space: distance from center of laser beam/distance from surface of target/temperature.

1 Introduction

The mechanism of target surface heating with a laser beam is implemented in many processes [1–4]. Heating the surface by laser emission activates heat exchange process therein. When the laser beam hits the surface it heats the surface and heat propagates deep into the material by heat conductance. Practical implementation of this operation with the optimal parameters of heating and cooling is not possible without a mathematical model, which takes into account the specific conditions of the surface laser heating and cooling the surface by the convection. Therefore modeling of surface heating in such conditions is an urgent problem. The complication and urgency of this problem increases if the emission source is pulsating. The result depends on the absorption coefficient of the material at the wavelength of the laser emission, its peak power, the relative pulse duration, the heat conductivity coefficient and the heat capacity coefficient of the material [1].

The result of emission exposure depends on the intensity distribution at the focal point and its surroundings. For a laser beam operating at fundamental mode intensity distribution has a Gaussian profile [1]:

$$I(r) = \frac{I_0}{2\pi \rho^2} \exp \left(-\frac{r^2}{2\rho^2} \right), \tag{1}$$

where ρ - gaussian width parameter, I_0 - intensity in the centre of laser hot spot $r = 0$.

© Springer International Publishing AG 2017
I. Dimov et al. (Eds.): NAA 2016, LNCS 10187, pp. 155–163, 2017.
DOI: 10.1007/978-3-319-57099-0_14

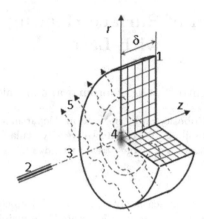

Fig. 1. The heating of target by the laser

The object of study is shown in Fig. 1, where 1- the heated target - an infinite plate thickness of which is δ; 2- the laser that generates a laser beam (3), which generates on the target surface the hot spot (4); airflow (5) uniformly blows the target.

The target is shown in cylindrical coordinates z and r. Coordinate z is directed from the center of the laser hot spot orthogonally to the surface deep into the target. Coordinate r is a radius of the circle which is constructed from the center of the laser hot spot. The target material is isotropic, i.e. its physical properties in all directions are the same.

The laser emits pulses of heat, their intensity is shown in Fig. 2. The principle of operation of the laser is described by the following equation:

$$I(r) = \begin{cases} I_{max} \text{ for } (\tau k) < t < (t_1 + \tau k), \\ I_{min} \text{ for } (t_1 + \tau k) \leqslant t \leqslant (t_2 + \tau k), \end{cases} \quad (2)$$

where $\tau = t_1 + t_2$ - pulse repetition period; t_1 - pulse duration; t_2 - spacing interval (time space between the pulses); k - the number of pulses. The relative pulse duration is defined by this relation: $t_c = \frac{t_1}{t_1+t_2}$.

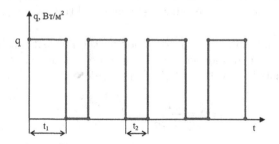

Fig. 2. The operating mode of the pulsating laser

2 Mathematical Model

The process of heating and cooling is described by the following differential equations. We use axisymmetric formulation of the problem:

- Differential equation of the heat conductivity

$$c\rho\frac{\partial T}{\partial t} = \frac{1}{r}\frac{\partial}{\partial r}\left(r\lambda_r\frac{\partial T}{\partial r}\right) + \frac{\partial}{\partial z}\left(\lambda_z\frac{\partial T}{\partial z}\right), \, \forall\, r,z \in V, \, t > 0, \tag{3}$$

where T - temperature; t - time; r, z - radial and axial coordinates; λ_r - the heat conductivity coefficient in the isotropy plane in the direction of axis r; λ_z - the heat conductivity coefficient in the direction of axis z, V - heating surface volume.

- Boundary conditions on the laser hot spot surface S

$$q_l = -\lambda_r\frac{\partial T}{\partial r}, \, \forall\, r,z \in S, \, t > 0; \tag{4}$$

- Boundary conditions of heat exchange with ambience on the surface Sa

$$\alpha\,(T_g - T) = -\lambda_r\frac{\partial T}{\partial r}, \, \forall\, r,z \in S, \, t > 0; \tag{5}$$

where α - heat exchange coefficient; T_g - ambient temperature; Sa - target surface, in the r direction it's an infinite plate, in the z direction thickness of the target is δ;

- Initial conditions

$$T = T_0, \, \forall\, r,z \in V, \, t = 0; \tag{6}$$

where T_0 - initial temperature.

The finite element method is used to solve this problem [5,6]. Its approximate variational formulation for the differential heat equation written above uses functional:

$$\Phi\,[T(r,z)] = \int\limits_v \left[\frac{1}{2}\left(\lambda_r^*\left(\frac{\partial T}{\partial r}\right)^2 + \lambda_z^*\left(\frac{\partial T}{\partial z}\right)^2\right) + q_c^*T\right] r\,dr\,dz$$

$$- \int\limits_S \left[q_l T + \alpha\left(T_g - \frac{1}{2}T\right)T\right] dS, \tag{7}$$

where $q_c^* = c^*\rho^*\frac{\partial T^*}{\partial t}$ - the heat flow, conditioned by the heat capacity of the material, index "$*$" corresponds to a temperature distribution in the space fixed in this moment.

The problem reduces to finding a function of the temperature field, satisfying the stationary value of the functional Φ for the field of research:

$$\delta\Phi\,[T\,(r,z,t^*)] = 0. \tag{8}$$

As a finite element we use a hybrid elemental annulus with a quadrangular cross-section, which is formed by the union of several triangles. A function of the temperature distribution is a linear function and the heat capacity of the element is concentrated in the nodes [5–7].

The temperature at any point of finite element is determined by nodal temperatures quantity linearly:

$$T = N_1 T + N_2 T + N_3 T, \tag{9}$$

where N_i - shape coefficient of triangular finite element calculated by the formulas:

$$N_i = \frac{1}{2S_T} \left[z_{jk} \left(r - r_k \right) - r_{jk} \left(z - z_k \right) \right], \tag{10}$$

where $S_T = \frac{1}{2} \left(r_{21} \cdot z_{32} - r_{32} \cdot z_{21} \right)$ - the area of the triangle, $i, j, k = 1, 2, 3, 1, 2$.

The heat flow which absorbed by heating element due to heat capacity of the material is expressed through the nodal heat flow:

$$q_c = \int_v c \frac{\partial T}{\partial t} dv = \int_v c \left[\sum_{i=1}^{3} N_i \cdot \frac{\partial T_i}{\partial t} \right] dv = \sum_{i=1}^{3} \frac{\partial T_i}{\partial t} \int_v c N_i dv = \sum_{i=1}^{3} q_{ci},$$

$$q_{ci} = \frac{\partial T_i}{\partial t} c_i; \quad c_i = \int_v c N_i dv, \tag{11}$$

where c_i - reduced heat capacity of the node; $v = S_T \cdot 2 \cdot \pi \cdot r_T$ - the volume of triangle finite element, r_T - centre of gravity of triangle.

Consider the original functional in the class of linear functions of temperature (9) with point sources (11). Minimization of this functional can be carried out over the nodal values [5], as they uniquely define the temperature at any point in the field of research.

Introducing into the original functional conditions (9)–(11) and differentiating the original functional with respect to the nodal temperature T_i for the field of the finite element we draw:

$$\frac{\partial \Phi}{\partial T_1} = \int \int_v \left[\lambda_z \frac{\partial T}{\partial z} \frac{\partial}{\partial T_1} \left(\frac{\partial T}{\partial z} \right) + \lambda_r \frac{\partial T}{\partial r} \frac{\partial}{\partial T_1} \left(\frac{\partial T}{\partial r} \right) \right] r dr dz + q_{c1}$$

$$= \left[(\lambda_z r_{23} r_{23} + \lambda_r z_{23} z_{23}) T_1 + (\lambda_z r_{31} r_{23} + \lambda_r z_{31} z_{23}) T_2 \right. \tag{12}$$

$$+ (\lambda_z r_{12} r_{23} + \lambda_r z_{12} z_{23}) T_3 \left] \frac{1}{4S_T} + c_1 \frac{\partial T}{\partial t}. \right.$$

Differentiating similarly T_2 and T_3, combine these equations and write the result in the following form:

$$\left\{ \frac{\partial \Phi}{\partial T} \right\} = [\lambda] \{T\} + \{c\}^T \left\{ \frac{\partial T}{\partial t} \right\}, \tag{13}$$

where $\{c\}^T$ - heat capacity matrix of the finite element; $r_{ij} = r_i - r_j$; $\{T\}$ - temperature matrix of all nodal points of the body; $[\lambda]$ - heat conduction matrix of the finite element:

$$\{c\}^T = \begin{Bmatrix} c_1 \\ c_2 \\ c_3 \end{Bmatrix}, \; \{T\} = \begin{Bmatrix} T_1 \\ T_2 \\ T_3 \end{Bmatrix}. \tag{14}$$

$$[\lambda] = \tfrac{1}{4S_T}$$
$$\cdot \begin{bmatrix} (\lambda_r z_{23} z_{23} + \lambda_z r_{23} r_{23}) ; (\lambda_r z_{31} z_{23} + \lambda_z r_{23} r_{23}) ; (\lambda_r z_{12} z_{23} + \lambda_z r_{12} r_{23}) \\ (\lambda_r z_{23} z_{31} + \lambda_z r_{23} r_{31}) ; (\lambda_r z_{31} z_{31} + \lambda_z r_{23} r_{31}) ; (\lambda_r z_{12} z_{31} + \lambda_z r_{12} r_{31}) \\ (\lambda_r z_{23} z_{12} + \lambda_z r_{23} r_{12}) ; (\lambda_r z_{31} z_{12} + \lambda_z r_{23} r_{12}) ; (\lambda_r z_{12} z_{12} + \lambda_z r_{12} r_{12}) \end{bmatrix}. \tag{15}$$

The final equation of the process of the original functional minimization on the temperature at the nodal points are obtained by combining all the derivatives (12) over all finite elements on which the field of research is sampled. Equate them to zero:

$$\sum_{i=1}^{n} \sum_{i=1}^{3} \frac{\partial \Phi}{\partial T} = \{0\}, \tag{16}$$

where n - the number of finite elements. Heat conduction matrix, heat capacity matrix and vectors of the nodal heat flow for the structure can be obtained by adding the respective members of heat conduction matrix, heat capacity matrix and finite elements heat flows. Resulting equation of the finite element method for this case takes the form:

$$[A]\{T\} = -\{C\}^T \left\{\frac{\partial T}{\partial t}\right\} + \{Q_\alpha\} + \{Q_L\}, \tag{17}$$

where $[\lambda]$, $\{C\}^T$, $\{Q_\alpha\}$, $\{Q_L\}$ - global heat conductivity matrix, heat capacity matrix and vectors of the nodal heat flow of radiant and convective heat exchange.

3 Numerical Calculations

To perform numerical calculations of the processes sampling on time was carried out in the equation. The difference Eq. (18) is solved applying an implicit difference scheme. Using the considered finite element with concentrated heat capacity allows excluding deviation in the numerical values of temperature for sharply pronounced transient heating processes and get the calculating scheme for transient temperature field, suitable for constructing the method of numerical investigation of the heat state of the investigated target. Using a finite-difference expression for the time derivative in the form $\left(T^{k+1} - T^k\right)/\Delta t$, we get:

$$[A]\{T^{k+1}\} = -\{C\}^T \left(\{T^{k+1}\} - \{T^k\}\right) \frac{1}{\Delta t}$$
$$+ \{A\}^T \left(\{T_g^{k+1}\} - \{T^{k+1}\}\right) + \{Q_\alpha^{k+1}\} + \{Q_L^{k+1}\}, \tag{18}$$

where k - an index corresponding the argument - time; $\{T^{k+1}\}$, $\{T^k\}$ - vectors of nodal temperatures in this and subsequent time points, $\{T_g\}$ - ambient temperature.

To combine in the left side of an equation all terms containing unknown quantities we pick out from the matrix $\{Q_\alpha\}$ unknown temperatures of structure nodal points:

$$\{Q_\alpha\} = \{A\}\left(\{T_G\} - \{T\}\right),\qquad(19)$$

where $\{A\}$ - a convective heat exchange vector of surface as a whole. Shifting to the left side of the Eq. (21) the terms containing the unknown $\{T^{k+1}\}$ we get:

$$\left([A] + \frac{1}{\Delta t}[C] + [A]\right)\{T^{k+1}\} = \frac{1}{\Delta t}\{C\}^T\{T^k\}$$
$$+ \{A\}^T\left(\{T_g^{k+1}\} - \{T^{k+1}\}\right) + \{Q_L^{k+1}\},\qquad(20)$$

where $[A]$ - heat conduction matrix, presented in the diagonal form.

The matrix $\{T\}$ at the initial time is determined from the initial conditions. The solution of resulting linear system of algebraic equations is carried out by the method of adjoint gradient [5,9].

4   Numerical Simulation Algorithm

On the basis of a mathematical model we developed an algorithm for solving integral differential heat conduction equation with boundary and initial conditions. The algorithm is developed for computing experiments for heat exchange investigation.

The main steps of the algorithm are: task creation and initial data creation for computational experiment; specifying of thermo-physical properties of the material: the density, the specific heat capacity, the heat conductivity coefficient; creation of the finite element grid (target geometry image with specified boundary conditions on the laser hot spot surface S and boundary conditions of heat exchange with ambience on the surface Sa, matrix of the nodes bonds of a finite element); creation the boundary and initial conditions; organization and management of calculations, the current condition saving, results output; checking the conditions of the end of the calculation. Formation of the boundary and initial conditions is the assignment of the desired values to specified subset of the nodal parameters vectors. Created data files and geometry are transferred into the main program and the process of investigated processes modeling begins. The results can be presented as graphs, diagrams, isolines or color fields. Numerical simulation of the process is carried out by means of theoretical and computational complex of computer modeling and visualization of the heat exchange processes [10].

5   Computer Simulation of the Process

In the computer simulation the impact of the laser beam on the target surface was investigated. The following characteristics were given: the material - steel;

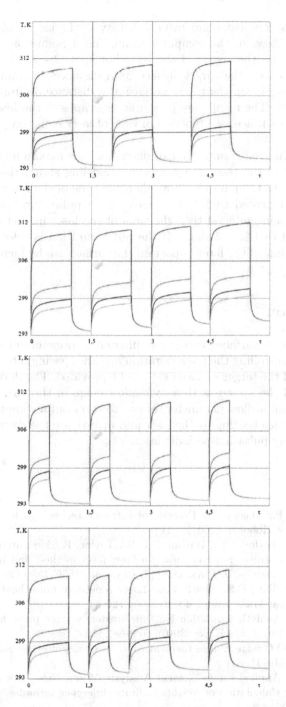

Fig. 3. The results of the computer simulation of the target surface heating by the laser beam: (a) the pulse duration is equal to the spacing interval $t_1 = t_2$; (b) the pulse duration is two times as much the spacing interval $t_1 = 2t_2$; (c) the pulse duration is two times less than the spacing interval $t_1 = 0.5t_2$; (d) the pulse duration is selected in random order

initial conditions: ambient temperature, density of the laser emission capacity, speed of the air flow. In the computer simulation 4 points on the target surface were taken: 1 - the centre of the laser hot spot; 2 - at the border of the laser hot spot; 3 - on the target surface at a distance $2r$ from the center of the laser hot spot; 4 - on the target surface at a distance $3r$ from the center of the laser hot spot. The results are presented as graphs of the heating temperature of the target surface as a function of time of laser emission pulses exposure (Fig. 3).

After analyzing the results, we can determine the maximum target heating modes. Analysis of the results of computer modeling shows that, if the pulse duration is equal to the pause interval, then the time of the laser impact on the surface can be increased to 1.2...1.5 times. If the pulse duration is two times less than the spacing interval then the time of the laser impact on the surface can be increased to 1.5...1.8 times. If the pulse duration is selected in random order, then the time of the laser impact on the surface can be increased to 2...2.5 times.

6   Conclusion

Research the target surface heating for different parameters of the laser pulse shown that by controlling the pulse duration and the spacing interval a predetermined pattern of the target surface heat can be provided. The developed method of integration of the system with a dynamic change in the energy of the laser beam and constant airflow parameters allows to carry out a computer simulation of the target surface heating by the laser and investigate the dynamics of target surface heating by pulsating-periodic laser.

References

1. Maldague, X.P.: Theory and Practice of Infrared Technology for Nondestructive Testing. Wiley, Hoboken (2001). 704 p
2. Sheich, M.A., Taylor, S.C., Hayhurst, D.R., Taylor, R.: Measurement of thermal diffusivity of isotropic materials using a laser flash method and its validation by finite element analysis. J. Phys. D Appl. Phys. 33, 1536–1550 (2000)
3. Shuja, S.Z., Yilbas, B.S., Shazli, S.Z.: Laser repetitive pulse heating influence of pulse duty. Heat Mass Transf. 43, 949–955 (2007)
4. Yilbas, B.S.: Analytical solution for time unsteady laser pulse heating of semi-infinite solid. Int. J. Mech. Sci. 39(6), 671–682 (1997)
5. Zienkiewiez, O.C.: The Finite Element Method in Engineering Science. McGraw-Hill, London (1971). 541 p
6. Segerlind, L.: Applied Finite Element Analysis. Wiley, New York (1976). 392 p
7. Fornberg, B.: Calculation of weights in finite difference formulas. SIAM Rev. 40, 685–691 (1998)
8. Babu'ka, I., Banerjee, U., Osborn, J.E.: Generalized finite element methods: main ideas, results, and perspective. Int. J. Comput. Methods 1(1), 67–103 (2004). doi:10.1142/S0219876204000083

9. Gerald, C.F., Wheatley, P.O.: Applied Numerical Analysis. Addison-Wesley Longman Inc., Boston (1997). 680 p

10. Dunaev, V.A.: Theoretical and computational complex computer modeling and visualization conjugate heat and mass transfer processes. Thermal measurements at the beginning of the XXI century. Abstracts of the Fourth International Thermophysical School, pp. 112–113. TSTU, Tambov (2001)

A Higher Order Difference Scheme for the Time Fractional Diffusion Equation with the Steklov Nonlocal Boundary Value Problem of the Second Kind

Anatoly A. Alikhanov[1]([⊠]) and Inna Z. Kodzokova[2]

[1] Institute of Applied Mathematics and Automation, Russian Academy of Sciences,
ul. Shortanova 89 a, Nalchik 360000, Russia
`aaalikhanov@gmail.com`
[2] Kabardino-Balkarian State University,
ul. Chernyshevskogo 173, Nalchik 360004, Russia

Abstract. We consider finite-difference schemes of higher order approximation for the time fractional diffusion equation with nonlocal boundary conditions containing real parameters α, β and γ. We obtain a priori estimates for the solution of the difference problem, which imply the stability and convergence of the constructed difference schemes. The obtained results are supported by the numerical calculations carried out for some test problems as well.

1 Introduction

The nonlocal boundary value problem with the boundary conditions $u(b,t) = \rho u(a,t)$, $u_x(b,t) = \sigma u_x(a,t) + \tau u(a,t)$ for the simplest equations of mathematical physics, referred to as conditions of the second class, was studied in the monograph [1] for the case with $\rho\sigma - 1 = 0$ and $\rho\tau \leq 0$. Difference schemes for a problem with $\alpha = \beta$, $\gamma = 0$ and $\nu = 1$ (the classical diffusion equation) were studied in [2]. In this case, the operator occurring in the elliptic part is self-adjoint. The self-adjointness permits one to use general theorems on the stability of two-layer difference schemes in energy spaces and consider difference schemes for equations with variable coefficients. Stability criteria for difference schemes for the heat equation with nonlocal boundary conditions were studied in [3,4]. The difference schemes considered in these papers have the specific feature that the corresponding difference operators are not self-adjoint. The method of energy inequalities was developed in [5–8] for the derivation of a priory estimates for solutions of difference schemes for the classical diffusion equation with variable coefficients in the case of nonlocal boundary conditions. Using the energy inequality method, a priory estimates for the solution of the Dirichlet and Robin boundary value problems for the fractional, variable and distributed order diffusion equation with Caputo fractional derivative have been obtained [9–14]. A priori estimates for the difference problems analyzed in [15] by using the maximum principle imply the stability and convergence of these difference schemes.

© Springer International Publishing AG 2017
I. Dimov et al. (Eds.): NAA 2016, LNCS 10187, pp. 164–171, 2017.
DOI: 10.1007/978-3-319-57099-0_15

In the case $\nu = 1$, $\alpha = 0$, $\beta = 1$, $\gamma = 0$ and the symmetric coefficients, the stability and convergence of the difference schemes in the mesh C-norm have been proved [16].

In the present paper, a difference scheme of the higher order approximation is constructed. A priori estimates for the solution of difference problem is obtained. A theorem stating that the corresponding difference scheme converges with the rate equal to the order of the approximation error is proved. The obtained results are supported by numerical calculations carried out for some test problems as well.

2 A Higher Order Difference Scheme for the Time Fractional Diffusion Equation with Nonlocal Boundary Conditions

Consider the nonlocal boundary value problem

$$\partial_{0t}^{\nu} u = k(t)u_{xx} - q(t)u + f(x,t), \quad 0 < x < 1, \quad 0 < t \leq T, \tag{1}$$

$$\begin{cases} u(0,t) = \alpha u(1,t), \\ u_x(1,t) = \beta u_x(0,t) + \gamma u(1,t) + \mu(t), \quad 0 \leq t \leq T, \end{cases} \tag{2}$$

$$u(x,0) = u_0(x), \quad 0 \leq x \leq 1, \tag{3}$$

where $\partial_{0t}^{\nu} u = \frac{1}{\Gamma(1-\nu)} \int\limits_0^t (t-s)^{-\nu} \frac{\partial u}{\partial s}(x,s)ds$ is a Caputo fractional derivative of order ν, $0 < \nu < 1$; $k(t)$ and $q(t)$ are given by sufficiently smooth functions, $\mu(t)$ and $f(x,t)$ are given functions that satisfy conditions ensuring sufficient smoothness of the solution $u(x,t)$ of the problem (1)–(3), $k(t) \geq c_1 > 0$, $q(t) \geq 0$ for all $t \in [0,T]$; α, β and γ are real numbers.

We introduce the space grid $\bar{\omega}_h = \{x_i = ih\}_{i=0}^N$, and the time grid $\bar{\omega}_\tau = \{t_j = j\tau\}_{j=0}^{j_0}$ with increments $h = 1/N$ and $\tau = T/j_0$. Set $a^{j+1} = k(t_j + \sigma\tau)$, $d^{j+1} = q(t_j + \sigma\tau)$, $\varphi_i^{j+1} = f(x_i, t_j + \sigma\tau)$, $y_i^j = y(x_i,t_j)$, $y_{\bar{x},i}^j = (y_i^j - y_{i-1}^j)/h$, $y_{x,i}^j = (y_{i+1}^j - y_i^j)/h$, $y_{\bar{x}x,i}^j = (y_{i+1}^j - 2y_i^j + y_{i-1}^j)/h^2$, $y_{t,i}^j = (y_i^{j+1} - y_i^j)/\tau$, $y_i^{(\sigma)} = \sigma y_i^{j+1} + (1-\sigma)y_i^j$, $\sigma = 1 - \nu/2$, $\mathcal{H}_h y_i = (y_{i+1} + 10y_i + y_{i-1})/12$, $i = 1,2,\ldots,N-1$.

Let us approximate the Caputo fractional derivative of order $\nu \in (0,1)$ by the $L2 - 1_\sigma$ formula [9]:

$$\Delta_{0t_{j+\sigma}}^{\nu} y_i = \frac{\tau^{1-\nu}}{\Gamma(2-\nu)} \sum_{s=0}^j c_{j-s}^{(\nu,\sigma)} y_{t,i}^s,$$

where

$$a_0^{(\nu,\sigma)} = \sigma^{1-\nu}, \quad a_l^{(\nu,\sigma)} = (l+\sigma)^{1-\nu} - (l-1+\sigma)^{1-\nu},$$

$$b_l^{(\nu,\sigma)} = \frac{1}{2-\nu} \left((l+\sigma)^{2-\nu} - (l-1+\sigma)^{2-\nu} \right)$$

$$- \frac{1}{2} \left((l+\sigma)^{1-\nu} + (l-1+\sigma)^{1-\nu} \right), \quad l \geq 1;$$

$c_0^{(\nu,\sigma)} = a_0^{(\nu,\sigma)}$, for $j = 0$; and for $j \geq 1$,

$$
c_s^{(\nu,\sigma)} = \begin{cases}
a_0^{(\nu,\sigma)} + b_1^{(\nu,\sigma)}, & s=0, \\
a_s^{(\nu,\sigma)} + b_{s+1}^{(\nu,\sigma)} - b_s^{(\nu,\sigma)}, & 1 \leq s \leq j-1, \\
a_j^{(\nu,\sigma)} - b_j^{(\nu,\sigma)}, & s = j.
\end{cases} \tag{4}
$$

Lemma 1. [9]. *For any* $\nu \in (0,1)$ *and* $u(t) \in C^3[0, t_{j+1}]$

$$
|\partial_{0t_{j+\sigma}}^\nu u - \Delta_{0t_{j+\sigma}}^\nu u| = O(\tau^{3-\nu}). \tag{5}
$$

Consider the scheme

$$
\Delta_{0t_{j+\sigma}}^\nu \mathcal{H}_h y_i = a^{j+1} y_{\bar{x}x,i}^{(\sigma)} - d^{j+1}\mathcal{H}_h y_i^{(\sigma)} + \mathcal{H}_h \varphi_i^{j+1}, \; i = \overline{1, N-1}, j = \overline{1, j_0 - 1}, \tag{6}
$$

$$
\begin{cases}
y_0^{j+1} = \alpha y_N^{j+1}, & j = 0, 1, \ldots, j_0 - 1, \\
\frac{3(1+\alpha\beta) - h\gamma}{3}\Delta_{0t_{j+\sigma}}^\nu y_N + \frac{2}{h}a^{j+1}\left(y_{\bar{x},N}^{(\sigma)} - \beta y_{x,0}^{(\sigma)} - \gamma^{j+1}y_N^{(\sigma)}\right) = \frac{2}{h}\varphi_N^{j+1},
\end{cases} \tag{7}
$$

$$
y_i^0 = u_0(x_i), \quad i = 0, 1, \ldots, N, \tag{8}
$$

where

$$
\gamma^{j+1} = \gamma - \frac{h}{2}\frac{q(t_{j+\sigma})}{k(t_{j+\sigma})}(1+\alpha\beta) + \frac{h^2}{6}\frac{q(t_{j+\sigma})}{k(t_{j+\sigma})}\gamma,
$$

$$
\varphi_N^{j+1} = k(t_{j+\sigma})\mu(t_{j+\sigma}) + \frac{h}{2}\left(\beta f(\tfrac{h}{3}, t_{j+\sigma}) + f(1 - \tfrac{h}{3}, t_{j+\sigma})\right)
$$

$$
+ \frac{h^2}{6}q(t_{j+\sigma})\mu(t_{j+\sigma}) + \frac{h^2}{6}\Delta_{0t_{j+\sigma}}^\nu\mu.
$$

The difference equation (8) approximates the differential equation (1) with the approximation order $O(\tau^2 + h^4)$ [9]. The nonlocal boundary conditions (7) have approximation order $O(\tau^2 + h^3)$.

3   Stability and Convergence

Lemma 2. [9]. *For any function* $y(t)$ *defined on the grid* $\bar{\omega}_\tau$ *one has the equality*

$$
y^{(\sigma)}\Delta_{0t_{j+\sigma}}^\nu y \geq \frac{1}{2}\Delta_{0t_{j+\sigma}}^\nu(y^2). \tag{9}
$$

Theorem 1. *If* $\alpha = \beta \neq 1$ *and* $\gamma \leq 0$, *then the difference scheme (6)–(8) is unconditionally stable and its solution satisfies the following a priori estimate:*

$$
\|[y^{j+1}]\|_0^2 \leq M\left(\|[y^0]\|_0^2 + \max_{0 \leq j \leq j_0-1}\left((\varphi_N^{j+1})^2 + \|\mathcal{H}_h\varphi^{j+1}\|_0^2\right)\right), \tag{10}
$$

where $\|[y]\|_0^2 = 0.5y_0^2 h + 0.5y_N^2 h + \sum_{i=1}^{N-1} y_i^2 h$, $M > 0$ *is a known number independent of* h *and* τ.

Proof. Taking the inner product of the Eq. (6) with $\mathcal{H}_h y^{(\sigma)}$ leads to

$$(\mathcal{H}_h y^{(\sigma)}, \Delta^\nu_{0t_{j+\sigma}} \mathcal{H}_h y) - a^{j+1}(\mathcal{H}_h y^{(\sigma)}, y^{(\sigma)}_{\bar{x}x}) + d^{j+1}(\mathcal{H}_h y^{(\sigma)}, \mathcal{H}_h y^{(\sigma)})$$
$$= (\mathcal{H}_h y^{(\sigma)}, \mathcal{H}_h \varphi^{j+1}),$$

(11)

where $(y, v) = \sum_{i=1}^{N-1} y_i v_i h$.

Using inequality (9) and Green's first difference formula, we get

$$\tfrac{1}{2} \Delta^\nu_{0t_{j+\sigma}} \|y\|_1^2 + \tfrac{2c_1}{3} \|y^{(\sigma)}_{\bar{x}}\|_0^2$$
$$\le \varepsilon \left((y^{(\sigma)}_N)^2 + \|\mathcal{H}_h y^{(\sigma)}\|_0^2 \right) + \tfrac{1}{4\varepsilon} \left((\varphi^{j+1}_N)^2 + \|\mathcal{H}_h \varphi^{j+1}\|_0^2 \right),$$

(12)

where

$$\|y\|_1^2 = \|\mathcal{H}_h y^{(\sigma)}\|_0^2 + \frac{h}{2} \left((1 + \alpha^2) - \frac{h\gamma}{3} \right) y_N^2.$$

Since $\alpha \ne 1$, then

$$(y^{(\sigma)}_N)^2 = \left(\frac{1}{1-\alpha} \sum_{i=1}^N y^{(\sigma)}_{\bar{x},i} h \right)^2 \le \frac{1}{(1-\alpha)^2} \|y^{(\sigma)}_{\bar{x}}\|_0^2.$$

(13)

The following estimation

$$\|\mathcal{H}_h y^{(\sigma)}\|_0^2 \le \|y^{(\sigma)}\|_0^2 + \tfrac{1+\alpha^2}{12} (y^{(\sigma)}_N)^2 h \le \|y_{\bar{x}}\|_0^2$$
$$+ \left(2 + \tfrac{1+\alpha^2}{12} \right) (y^{(\sigma)}_N)^2 \le \left(1 + \tfrac{25+\alpha^2}{12(1-\alpha)^2} \right) \|y_{\bar{x}}\|_0^2$$

(14)

holds.

Using (12), (13) and (14) we have the inequality

$$\tfrac{1}{2} \Delta^\nu_{0t_{j+\sigma}} \|y\|_1^2 + \left(\tfrac{2c_1}{3} - \varepsilon \left(1 + \tfrac{25+\alpha^2}{12(1-\alpha)^2} \right) \right) \|y^{(\sigma)}_{\bar{x}}\|_0^2$$
$$\le \tfrac{1}{4\varepsilon} \left((\varphi^{j+1}_N)^2 + \|\mathcal{H}_h \varphi^{j+1}\|_0^2 \right).$$

(15)

When $\varepsilon = \frac{24c_1(1-\alpha)^2}{3(12(1-\alpha)^2 + 25 + \alpha^2)} > 0$, it follows from (15) that

$$\Delta^\nu_{0t_{j+\sigma}} \|y\|_1^2 \le M_1 \left((\varphi^{j+1}_N)^2 + \|\mathcal{H}_h \varphi^{j+1}\|_0^2 \right),$$

(16)

where $M_1 > 0$ is constant independent on h, τ and T.

Rewrite the inequality (16) as follows

$$g^{j+1}_j \|y^{j+1}\|_1^2 \le \sum_{s=1}^j (g^{j+1}_s - g^{j+1}_{s-1}) \|y^s\|_1^2$$
$$+ g^{j+1}_0 \|y^0\|_1^2 + M_1 \left((\varphi^{j+1}_N)^2 + \|\mathcal{H}_h \varphi^{j+1}\|_0^2 \right),$$

(17)

where

$$g_s^{j+1} = \frac{c_{j-s}^{(\nu,\sigma)}}{\tau^\nu \Gamma(2-\nu)}, \quad 0 \le s \le j \le j_0 - 1.$$

Note that [9]

$$g_0^{j+1} = \frac{c_j^{(\nu,\sigma)}}{\tau^\nu \Gamma(2-\nu)} > \frac{1}{2t_{j+\sigma}^\nu \Gamma(1-\nu)} > \frac{1}{2T^\nu \Gamma(1-\nu)},$$

then

$$g_j^{j+1} \|y^{j+1}\|_1^2 \le \sum_{s=1}^{j} (g_s^{j+1} - g_{s-1}^{j+1}) \|y^s\|_1^2 + g_0^{j+1} E, \tag{18}$$

where

$$E = \|y^0\|_1^2 + 2T^\nu \Gamma(1-\nu) M_1 \max_{0 \le j \le j_0 - 1} \left((\varphi_N^{j+1})^2 + \|\mathcal{H}_h \varphi^{j+1}\|_0^2 \right).$$

It is obvious that at $j = 0$ the a priori estimate $\|y^1\|_1^2 \le E$ follows from (18). Let us prove that $\|y^{j+1}\|_1^2 \le E$ holds for $j = 0, 1, \ldots, j_0 - 1$ by using the mathematical induction method. For this purpose, let us assume that the a priori estimate $\|y^{j+1}\|_1^2 \le E$ takes place for all $j = 0, 1, \ldots, k-1, (k \ge 1)$.

From (18) at $j = k$ one has

$$g_k^{k+1} \|y^{k+1}\|_1^2 \le \sum_{s=1}^{k} (g_s^{k+1} - g_{s-1}^{k+1}) \|y^s\|_1^2 + g_0^{k+1} E \le g_k^{k+1} E.$$

Thus by induction we can obtain the estimate

$$\|y^{j+1}\|_1^2 \le E, \quad j = 0, 1, \ldots, j_0 - 1.$$

We have to just show that $\|y\|_1$ and $|[y]|_0$ norms are equivalent for the grid function which satisfies nonlocal boundary condition $y_0 = \alpha y_N$.

The norm equivalence follows from below inequalities

$$\frac{5}{12} |[y]|_0^2 \le \|y\|_1^2 \le \left(\frac{7}{6} - \frac{\gamma}{3} \right) |[y]|_0^2, \quad \gamma \le 0.$$

Theorem 1 is proved.

Let y_i^j is the solution of problem (6)–(8), then the function $v_i^j = \delta y_i^j + y_{N-i}^j$ at $\delta \ne \pm 1, -\alpha, \beta$ be the solution of the following nonlocal problem

$$\Delta_{0t_{j+\sigma}}^\nu \mathcal{H}_h v_i^{(\sigma)} = a^{j+1} v_{\bar{x}x,i}^{(\sigma)} - d^{j+1} \mathcal{H}_h v_i^{(\sigma)} + \mathcal{H}_h \tilde{\varphi}_i^{j+1}, \tag{19}$$

$$\begin{cases} v_0^{j+1} = \alpha_1 v_N^{j+1}, \\ \frac{3(1+\alpha_1\beta_1) - h\gamma_1}{3} \Delta_{0t_{j+\sigma}}^\nu v_N + \frac{2a^{j+1}}{h} \left(v_{\bar{x},N}^{(\sigma)} - \beta_1 v_{x,0}^{(\sigma)} - \gamma_1^{j+1} v_N^{(\sigma)} \right) = \frac{2\tilde{\varphi}_N^{j+1}}{h}, \end{cases} \tag{20}$$

$$v_i^0 = v_0(x_i), \tag{21}$$

where

$$\alpha_1 = \frac{\delta\alpha+1}{\delta+\alpha}, \quad \beta_1 = \frac{\delta^2-1}{\delta-\beta}, \quad \gamma_1 = \frac{\gamma(\delta^2-1)}{(\delta+\alpha)(\delta-\beta)},$$

$$\tilde{\varphi}_i^{j+1} = \delta\varphi_i^{j+1} + \varphi_{N-i}^{j+1}, \quad i = 1,2,\ldots,N-1, \quad \tilde{f}(x,t) = \delta f(x,t) + f(1-x,t),$$

$$\tilde{\varphi}_N^{j+1} = k(t_{j+\sigma})\mu_1(t_{j+\sigma}) + \frac{h}{2}\left(\beta_1\tilde{f}(\tfrac{h}{3},t_{j+\sigma}) + \tilde{f}(1-\tfrac{h}{3},t_{j+\sigma})\right)$$

$$+\frac{h^2}{6}q(t_{j+\sigma})\mu_1(t_{j+\sigma}) + \frac{h^2}{6}\Delta_{0t_{j+\sigma}}^{\nu}\mu_1, \quad \mu_1(t) = \frac{\delta^2-1}{\delta-\beta}\mu(t), \quad \gamma_1^{j+1} = \frac{\gamma^{j+1}(\delta^2-1)}{(\delta+\alpha)(\delta-\beta)},$$

$$v_0(x_i) = \delta u_0(x_i) + u_0(1-x_i).$$

Theorem 2. *If (1) $|\alpha| < 1, |\beta| < 1$ and $\gamma \le 0$; or (2) $|\alpha| > 1, |\beta| > 1$ and $\alpha\beta\gamma \le 0$, then the difference scheme (6)–(8) is unconditionally stable and its solution satisfies the following a priori estimate*

$$\|[y^{j+1}]\|_0^2 \le M_2\left(\|[y^0]\|_0^2 + \max_{0\le j\le j_0-1}\left((\varphi_N^{j+1})^2 + \|\mathcal{H}_h\varphi^{j+1}\|_0^2\right)\right), \qquad (22)$$

where $M_2 > 0$ is a known constant independent of h and τ.

Proof. Let us find a δ, for which the assumptions of Theorem 1 are satisfied for problem (19)–(21). The condition $\alpha_1 = \beta_1$ leads to the quadratic equation

$$\delta^2 - 2\frac{\alpha\beta - 1}{\alpha - \beta}\delta + 1 = 0,$$

which for $(\alpha^2 - 1)(\beta^2 - 1) > 0$ has two distinct real roots:

$$\delta_1 = \frac{\alpha\beta - 1 - \sqrt{(\alpha^2-1)(\beta^2-1)}}{\alpha-\beta}, \quad \delta_2 = \frac{\alpha\beta - 1 + \sqrt{(\alpha^2-1)(\beta^2-1)}}{\alpha-\beta}.$$

Take $\delta = \delta_1$ for $\alpha^2 - 1 < 0$ and $\beta^2 - 1 < 0$ and $\delta = \delta_2$ for $\alpha^2 - 1 > 0$ and $\beta^2 - 1 > 0$. This ensures the conditions $\delta \ne -\alpha, \beta$.

Let us consider the above mentioned two cases.

(1) Let $\alpha^2 - 1 < 0$, $\beta^2 - 1 < 0$ and $\delta = \delta_1$. The second assumption of Theorem 1 implies the inequality

$$\frac{\gamma(\delta^2 - 1)}{(\delta + \alpha)(\delta - \beta)} \le 0,$$

which for $\delta = \delta_1$ takes the form

$$\gamma\frac{((\sqrt{1-\alpha^2} + \sqrt{1-\beta^2})^2 + (\alpha-\beta)^2)}{(\sqrt{1-\alpha^2} + \sqrt{1-\beta^2})^2} \le 0$$

and is equivalent to the inequality $\gamma \le 0$ for $|\alpha| < 1$ and $|\beta| < 1$.

(2) If $\alpha^2 - 1 > 0$, $\beta^2 - 1 > 0$ and $\delta = \delta_2$, then the inequality $\gamma_1 \leq 0$ has the form

$$\gamma \frac{((\sqrt{\alpha^2 - 1} + \sqrt{\beta^2 - 1})^2 - (\alpha - \beta)^2)}{(\sqrt{\alpha^2 - 1} + \sqrt{\beta^2 - 1})^2} \leq 0$$

and is equivalent to the inequality $\alpha\beta\gamma \leq 0$ for $|\alpha| > 1$ and $|\beta| > 1$.

Theorem 2 is proved.

4   Numerical Results

Numerical results were obtained for the test problem in which the function

$$u(x,t) = (\alpha + 1 + \sin(\pi x) + (\alpha - 1)\cos(\pi x))(t^3 + t^2 + 1)$$

is a solution of problem (1)–(3) with coefficients $k(t) = e^t$ and $q(t) = 1 - \sin(2t)$.

Table 1 presents the accuracy and the convergence rate (CR) of the difference scheme in the norms $\|\cdot\|_0$ and $\|\cdot\|_{C(\bar{\omega}_{h\tau})}$.

Table 1. $T = 1$, $\alpha = -7$, $\beta = 4$, $\gamma = 10$, $h^3 = \tau^2$

ν	h	$\max\limits_{0 \leq j \leq j_0} \|z^n\|_0$	CR in $\|\cdot\|_0$	$\|z\|_{C(\bar{\omega}_{h\tau})}$	CR in $\|\cdot\|_{C(\bar{\omega}_{h\tau})}$
0.01	1/10	8.8584e–3		1.6665e–2	
	1/20	1.1139e–3	2.9915	2.1037e–3	2.9858
	1/40	1.3909e–4	3.0014	2.6303e–4	2.9996
	1/80	1.7363e–5	3.0020	3.3847e–5	3.0014
0.99	1/10	1.5406e–3		2.4823e–3	
	1/20	1.7659e–4	3.1250	2.8017e–4	3.1473
	1/40	2.2198e–5	2.9919	3.5654e–5	2.9741
	1/80	2.7433e–6	3.0165	4.5191e–6	2.9799

The convergence order is computed by the formula $\log_{\frac{h_1}{h_2}} \frac{e_1}{e_2}$.

Acknowledgements. This work is executed under grant of the President of the Russian Federation for the state support of young Russian scientists MK–3360.2015.1.

References

1. Steklov, V.A.: Osnovnye zadachi matematicheskoi fiziki (Main Problems of Mathematical Physics). Nauka, Moscow (1983). (in Russian)
2. Gulin, A.V., Morozova, V.A.: Family of self-adjoint nonlocal finite-difference schemes. Differ. Equ. **44**(9), 1297–1304 (2008)
3. Gulin, A.V., Morozova, V.A.: Stability of the two-parameter set of nonlocal difference schemes. Comput. Methods Appl. Math. **9**(1), 79–99 (2009)

4. Gulin, A.V., Morozova, V.A.: On a family of nonlocal difference schemes. Differ. Equ. **45**(7), 1020–1033 (2009)
5. Alikhanov, A.A.: Nonlocal boundary value problems in differential and difference settings. Differ. Equ. **44**(7), 952–959 (2008)
6. Alikhanov, A.A.: On the stability and convergence of nonlocal difference schemes. Differ. Equ. **46**(7), 949–961 (2010)
7. Alikhanov, A.A.: Stability and convergence of difference schemes approximating a two-parameter nonlocal boundary value problem. Differ. Equ. **49**(7), 796–806 (2013)
8. Alikhanov, A.A.: Stability and convergence of difference schemes approximating a nonlocal Steklov boundary value problem of the second class. Differ. Equ. **51**(1), 95–107 (2015)
9. Alikhanov, A.A.: A new difference scheme for the time fractional diffusion equation. J. Comput. Phys. **280**, 424–438 (2015)
10. Alikhanov, A.A.: A priori estimates for solutions of boundary value problems for fractional-order equations. Differ. Equ. **46**(5), 660–666 (2010)
11. Alikhanov, A.A.: Boundary value problems for the diffusion equation of the variable order in differential and difference settings. Appl. Math. Comput. **219**, 3938–3946 (2012)
12. Alikhanov, A.A.: Numerical methods of solutions of boundary value problems for the multi-term variable-distributed order diffusion equation. Appl. Math. Comput. **268**, 12–22 (2015)
13. Alikhanov, A.A.: Stability and convergence of difference schemes approximating a two-parameter nonlocal boundary value problem for time-fractional diffusion equation. Comput. Math. Modul. **26**(2), 252–272 (2015)
14. Alikhanov, A.A.: Stability and convergence of difference schemes for boundary value problems for the fractional-order diffusion equation. Comput. Math. Math. Phys. **56**(4), 561–575 (2016)
15. Shkhanukov-Lafishev, M.K., Taukenova, F.I.: Difference methods for solving boundary value problems for fractional differential equations. Comput. Math. Math. Phys. **46**(10), 1785–1795 (2006)
16. Ionkin, N.I., Makarov, V.L., Furletov, D.G.: Stability and convergence of difference schemes in Chebyshev norm for parabolic equation with nonlocal boundary condition. Mat. Model. **4**(4), 63–73 (1992). (in Russian)

Local Discontinuous Galerkin Methods for Reaction-Diffusion Systems on Unstructured Triangular Meshes

Na An[1,2], Xijun Yu[2(✉)], Chaobao Huang[2,3], and Maochang Duan[2]

[1] School of Mathematics and Statistics,
Shandong Normal University, Jinan 250014, China
[2] Laboratory of Computational Physics, Institute of Applied Physics and
Computational Mathematics, Beijing 100088, China
yuxj@iapcm.ac.cn
[3] Applied and Computational Mathematics Division,
Beijing Computational Science Research Center, Beijing 100193, China

Abstract. In this paper, on two-dimension unstructured meshes, a fully-discrete scheme is presented for the reaction-diffusion systems, which are often used as mathematical models for many biological, physical and chemical applications. By using local discontinuous Galerkin (LDG) method, the scheme can derive the numerical approximations not only for solutions but also for their gradients at the same time. In addition, the scheme employs the implicit integration factor (IIF) method for temporal discretization, which allows us to take the time-step as $\delta t = O(h_{min})$, and can be computed element by element, so that it reduces the computational cost greatly. Numerical simulations for the chlorite-iodide-malonic acid (CIMA) model demonstrate the expected behavior of the solutions, the efficiency and advantages of the proposed scheme.

Keywords: Nonlinear reaction-diffusion systems · Unstructured triangular meshes · Local discontinuous Galerkin · Implicit integration factor

1 Introduction

Since the seminal paper of Turing [11] in 1952, the reaction-diffusion systems have attracted a growing interest. The typical examples include Turing-type models such as the chloride-iodide-malonic acid (CIMA) model [6], the Schnakenberg model [10], the Gierer-Meinhardt model [5] and the Gray-Scott model [14].

The reaction-diffusion systems as mathematical models for many applications are of the following form,

$$\frac{\partial \mathbf{u}}{\partial t} = \mathbb{D}\nabla^2 \mathbf{u} + \mathbf{F}(\mathbf{u}), \tag{1}$$

with

$$\mathbf{u} = \begin{pmatrix} u \\ v \end{pmatrix}, \quad \mathbb{D} = \begin{pmatrix} D_u & 0 \\ 0 & D_v \end{pmatrix}, \quad \mathbf{F}(\mathbf{u}) = \begin{pmatrix} f(u, v) \\ g(u, v) \end{pmatrix},$$

© Springer International Publishing AG 2017
I. Dimov et al. (Eds.): NAA 2016, LNCS 10187, pp. 172–179, 2017.
DOI: 10.1007/978-3-319-57099-0_16

where u, v are two chemical concentrations, $\mathbb{D}$ is a diffusion constant matrix, and f, g describe the reaction term.

Many numerical schemes have been studied to solve the system (1) and discontinuous Galerkin (DG) methods have attracted more attention recently. The Cheng-Shu DG method was used to solve the system (1) in both [13] with the Strang type symmetrical operator splitting technique and [2] with Krylov implicit integration factor (IIF) method. And Zhang et al. [12] introduced the direct discontinuous Galerkin (DDG) method to discrete the system. Moreover, in [4], local discontinuous Galerkin (LDG) methods [3], combined with the explicit exponential time differencing method, was applied to discrete (1) in space.

In this paper, we choose to pursue the LDG method, where more general numerical fluxes than those in [4] are used, coupled with IIF methods [9] for the system (1). The method can derive the numerical approximations for both solutions and fluxes. In addition, the method relax the time-step to $\Delta t = O(h_{min})$, allows us to compute element by element and avoid solving a global system of nonlinear algebraic equations as the standard implicit schemes do, which can reduce the computational cost greatly.

The rest of this paper is organized as follows. In Sect. 2, the fully-discrete scheme is given by combining the LDG method with the IIF method. And numerical experiments for the CIMA reactive model are conducted to show formation of spatial patterns and confirm the efficiency and advantages of the proposed scheme in Sect. 3. Some conclusions are drawn in Sect. 4.

2   Construction of the Fully-Discrete Scheme

We consider the nonlinear reaction-diffusion system (1) defined on $\Omega \times [0, T]$, together with no-flux boundary conditions, where Ω is an open, bounded domain.

2.1   The LDG Method for Spatial Discretization

By introducing two auxiliary variables $\mathbf{q} = D_u \nabla u$, $\mathbf{p} = D_v \nabla u$, we write (1) into

$$\begin{cases} \mathbf{q} - D_u \nabla u = 0, & \mathbf{p} - D_v \nabla v = 0, \\ \frac{\partial u}{\partial t} - \nabla \cdot \mathbf{q} = f(u, v), & \frac{\partial v}{\partial t} - \nabla \cdot \mathbf{p} = g(u, v). \end{cases} \quad (2)$$

Let $\mathcal{T}_h = \{K\}$ be a regular triangulation of Ω with h the mesh size. $\mathcal{E}_h$ denotes the collection of all edges in $\mathcal{T}_h$. $\mathcal{E}_h^\circ$ and $\mathcal{E}_h^b$ are the sets of interior edges and boundary edges, respectively.

Define the finite element space as follows,

$$V_h = \{r_h \in L^2(\Omega) : r_h|_K \in P^1(K), \ \forall K \in \mathcal{T}_h\},$$

$$\mathbf{W}_h = \{\mathbf{w}_h \in (L^2(\Omega))^2 : \mathbf{w}_h|_K \in (P^1(K))^2, \ \forall K \in \mathcal{T}_h\},$$

where $P^1(K)$ denotes the linear polynomials on K.

Then the semi-discrete LDG formulation is: For $t \in (0, T]$, find $\mathbf{q}_h(t)$, $\mathbf{p}_h(t) \in \mathbf{W}_h$ and $u_h(t)$, $v_h(t) \in V_h$ such that, for $\forall \mathbf{w}_h^u, \mathbf{w}_h^v \in \mathbf{W}_h$ and $\forall r_h^u, r_h^v \in V_h$,

$$
\begin{cases}
\int_K \mathbf{q}_h \cdot \mathbf{w}_h^u \, d\mathbf{x} + \int_K u_h \nabla \cdot (D_u \mathbf{w}_h^u) \, d\mathbf{x} - \int_{\partial K} \widehat{u_h} \, D_u \mathbf{w}_h^u \cdot \mathbf{n}_K \, ds = 0, \\
\int_K \mathbf{p}_h \cdot \mathbf{w}_h^v \, d\mathbf{x} + \int_K v_h \nabla \cdot (D_v \mathbf{w}_h^v) \, d\mathbf{x} - \int_{\partial K} \widehat{v_h} \, D_v \mathbf{w}_h^v \cdot \mathbf{n}_K \, ds = 0, \\
\int_K \frac{\partial u_h}{\partial t} r_h^u \, d\mathbf{x} + \int_K \mathbf{q}_h \cdot \nabla r_h^u \, d\mathbf{x} - \int_{\partial K} \widehat{\mathbf{q}_h} \cdot \mathbf{n}_K \, r_h^u \, ds = \int_K f(u_h, v_h) r_h^u \, d\mathbf{x}, \\
\int_K \frac{\partial v_h}{\partial t} r_h^v \, d\mathbf{x} + \int_K \mathbf{p}_h \cdot \nabla r_h^v \, d\mathbf{x} - \int_{\partial K} \widehat{\mathbf{p}_h} \cdot \mathbf{n}_K \, r_h^v \, ds = \int_K g(u_h, v_h) r_h^v \, d\mathbf{x},
\end{cases}
\tag{3}
$$

where $\mathbf{n}_K$ is the outward unit normal vector to ∂K. The quantities $\widehat{u_h}$ and $\widehat{\mathbf{q}_h}$ are the so-called numerical fluxes and chosen as [1],

$$
\forall e \in \mathcal{E}_h^\circ, \quad
\begin{cases}
\widehat{u_h}|_e = \{u_h\} + \mathbf{C}_{12} \cdot [u_h], \\
\widehat{\mathbf{q}_h}|_e = \{\mathbf{q}_h\} - C_{11}[u_h] - \mathbf{C}_{12}[\mathbf{q}_h],
\end{cases}
\quad
\forall e \in \mathcal{E}_h^b, \quad
\begin{cases}
\widehat{u_h}|_e = u_h, \\
\widehat{\mathbf{q}_h}|_e = \mathbf{0},
\end{cases}
\tag{4}
$$

where the parameter $C_{11}(e) > 0$ and $\mathbf{C}_{12}(e)$ are both defined on every edge e. And $\widehat{v_h}$, $\widehat{\mathbf{p}_h}$ can be defined similarly to $\widehat{u_h}$ and $\widehat{\mathbf{q}_h}$.

The functions u_h, v_h and $\mathbf{q}_h = (q_{h,x}, q_{h,y})$, $\mathbf{p}_h = (p_{h,x}, p_{h,y})$ in V_h and $\mathbf{W}_h$ can be expressed by their basis functions as

$$
u_h = \sum_{i=1}^3 u_i(t)\phi_i(x,y) = \varPhi^T \overline{\mathbf{u}}, \quad v_h = \sum_{i=1}^3 v_i(t)\phi_i(x,y) = \varPhi^T \overline{\mathbf{v}},
$$
$$
q_{h,l} = \sum_{i=1}^3 q_{l,i}(t)\phi_i(x,y) = \varPhi^T \overline{\mathbf{q}_l}, \quad p_{h,l} = \sum_{i=1}^3 p_{l,i}(t)\phi_i(x,y) = \varPhi^T \overline{\mathbf{p}_l}, \quad l = x, y,
\tag{5}
$$

where $\varPhi = (\phi_1(x,y), \phi_2(x,y), \phi_3(x,y))^T$ are the basis functions, and $\overline{\mathbf{u}}$, $\overline{\mathbf{v}}$, $\overline{\mathbf{q}_l}$, $\overline{\mathbf{p}_l}$ are the corresponding degrees of freedom,

$$
\overline{\mathbf{u}} = \begin{pmatrix} u_1(t) \\ u_2(t) \\ u_3(t) \end{pmatrix}, \quad
\overline{\mathbf{v}} = \begin{pmatrix} v_1(t) \\ v_2(t) \\ v_3(t) \end{pmatrix}, \quad
\overline{\mathbf{q}_l} = \begin{pmatrix} q_{l,1}(t) \\ q_{l,2}(t) \\ q_{l,3}(t) \end{pmatrix}, \quad
\overline{\mathbf{p}_l} = \begin{pmatrix} p_{l,1}(t) \\ p_{l,2}(t) \\ p_{l,3}(t) \end{pmatrix}, \quad l = x, y.
$$

For the element K, let K_i, $i = 1, 2, 3$ denote its three adjacent elements. Then we use K_{ij}, $j = 1, 2, 3$ to denote three adjacent elements of K_i, and set $K_{ij} = K$ when $i = j$. In the following, we employ the subscript (NB) to denote the quantities belonging to the adjacent elements.

Then, substituting (4) for the interior edges into (3), and applying (5), by the same way as [7], we can obtain two separate matrix equations for the interior element K,

$$
\begin{pmatrix}
\mathbb{M} & 0 & 0 & 0 \\
0 & \mathbb{M} & 0 & 0 \\
0 & 0 & \mathbb{M} & 0 \\
0 & 0 & 0 & \mathbb{M}
\end{pmatrix}
\begin{pmatrix}
\overline{\mathbf{q}_x} \\ \overline{\mathbf{q}_y} \\ \overline{\mathbf{p}_x} \\ \overline{\mathbf{p}_y}
\end{pmatrix}
+
\begin{pmatrix}
D_u \mathbb{HH}_x & 0 \\
D_u \mathbb{HH}_y & 0 \\
0 & D_v \mathbb{HH}_x \\
0 & D_v \mathbb{HH}_y
\end{pmatrix}
\begin{pmatrix}
\overline{\mathbf{u}} \\ \overline{\mathbf{v}}
\end{pmatrix}
$$
$$
+ \sum_{i=1}^3
\begin{pmatrix}
D_u \mathbb{H}_{x,B,i} & 0 \\
D_u \mathbb{H}_{y,B,i} & 0 \\
0 & D_v \mathbb{H}_{x,B,i} \\
0 & D_v \mathbb{H}_{y,B,i}
\end{pmatrix}
\begin{pmatrix}
\overline{\mathbf{u}} \\ \overline{\mathbf{v}}
\end{pmatrix}_{(NB,i)}
=
\begin{pmatrix}
0 \\ 0 \\ 0 \\ 0
\end{pmatrix},
\tag{6}
$$

and

$$
\begin{pmatrix} \mathbb{M} & 0 \\ 0 & \mathbb{M} \end{pmatrix} \begin{pmatrix} (\overline{\mathbf{u}})_t \\ (\overline{\mathbf{v}})_t \end{pmatrix} + \begin{pmatrix} \mathbb{JJ}_x & \mathbb{JJ}_y & 0 & 0 \\ 0 & 0 & \mathbb{JJ}_x & \mathbb{JJ}_y \end{pmatrix} \begin{pmatrix} \overline{\mathbf{q}_x} \\ \overline{\mathbf{q}_y} \\ \overline{\mathbf{p}_x} \\ \overline{\mathbf{p}_y} \end{pmatrix} + \begin{pmatrix} \mathbb{G}_U & 0 \\ 0 & \mathbb{G}_U \end{pmatrix} \begin{pmatrix} \overline{\mathbf{u}} \\ \overline{\mathbf{v}} \end{pmatrix}
$$

$$
+ \sum_{i=1}^{3} \begin{pmatrix} \mathbb{J}_{x,B,i} & \mathbb{J}_{y,B,i} & 0 & 0 \\ 0 & 0 & \mathbb{J}_{x,B,i} & \mathbb{J}_{y,B,i} \end{pmatrix} \begin{pmatrix} \overline{\mathbf{q}_x} \\ \overline{\mathbf{q}_y} \\ \overline{\mathbf{p}_x} \\ \overline{\mathbf{p}_y} \end{pmatrix}_{(NB,i)} \tag{7}
$$

$$
+ \sum_{i=1}^{3} \begin{pmatrix} \mathbb{G}_{u,B,i} & 0 \\ 0 & \mathbb{G}_{u,B,i} \end{pmatrix} \begin{pmatrix} \overline{\mathbf{u}} \\ \overline{\mathbf{v}} \end{pmatrix}_{(NB,i)} = \begin{pmatrix} \mathbf{S}_u \\ \mathbf{S}_v \end{pmatrix},
$$

where $(\overline{\mathbf{u}})_t = \frac{\partial \overline{\mathbf{u}}}{\partial t}$, $(\overline{\mathbf{v}})_t = \frac{\partial \overline{\mathbf{v}}}{\partial t}$, and we use the notifications

$$
\mathbb{HH}_x = \mathbb{H}_x + \sum_{i=1}^{3} \mathbb{H}_{x,i}, \quad \mathbb{HH}_y = \mathbb{H}_y + \sum_{i=1}^{3} \mathbb{H}_{y,i},
$$

$$
\mathbb{JJ}_x = \mathbb{J}_x + \sum_{i=1}^{3} \mathbb{J}_{x,i}, \quad \mathbb{JJ}_y = \mathbb{J}_y + \sum_{i=1}^{3} \mathbb{J}_{y,i}, \quad \mathbb{G}_U = \sum_{i=1}^{3} \mathbb{G}_{u,i}. \tag{8}
$$

The 3×3 matrices are calculated as follows: for $m, n = 1, 2, 3$,

$$
\mathbb{M}_{mn} = \int_K \phi_m \phi_n d\mathbf{x}, \quad \mathbf{S}_{u,m} = \int_K f(u_h, v_h) \phi_m d\mathbf{x}, \quad \mathbf{S}_{v,m} = \int_K g(u_h, v_h) \phi_m d\mathbf{x},
$$

$$
\mathbb{J}_{l,mn} = \int_K \frac{\partial \phi_m}{\partial l} \phi_n d\mathbf{x}, \quad \mathbb{H}_{l,mn} = \mathbb{J}_{l,mn} = \int_K \frac{\partial \phi_m}{\partial l} \phi_n d\mathbf{x}, \quad l = x, y,
$$

$$
\mathbb{G}_{u,i,mn} = \int_{(\partial K)_i} C_{11} \phi_m \phi_n d\mathbf{x}, \quad \mathbb{G}_{u,B,i,mn} = -\int_{(\partial K)_i} C_{11} \phi_m (\phi_n)_{(NB,i)} d\mathbf{x}, \quad i = 1, 2, 3,
$$

$$
\mathbb{H}_{l,i,mn} = -\int_{(\partial K)_i} (\tfrac{1}{2} + C_{12}) \phi_m \phi_n n_l ds, \quad \mathbb{H}_{l,B,i,mn} = -\int_{(\partial K)_i} (\tfrac{1}{2} - C_{12}) \phi_m (\phi_n)_{(NB,i)} n_l ds,
$$

$$
\mathbb{J}_{l,i,mn} = -\int_{(\partial K)_i} (\tfrac{1}{2} - C_{12}) \phi_m \phi_n n_l ds, \quad \mathbb{J}_{l,B,i,mn} = -\int_{(\partial K)_i} (\tfrac{1}{2} + C_{12}) \phi_m (\phi_n)_{(NB,i)} n_l ds,
$$

with $\partial K = \cup_{i=1}^{3} (\partial K)_i$, $(\partial K)_i$ denoting the common edge between the element K and its adjacent element K_i. We also use the relations that $\mathbf{n} = \mathbf{n}_K = (n_x, n_y)$, $C_{12} = \mathbf{C}_{12} \cdot \mathbf{n}$ and $\mathbf{n}_{(NB)} = -\mathbf{n}$.

Remark 1. If the edge $(\partial K)_i$ sharing by K and K_i is a boundary edge, then the above matrices related to $(\partial K)_i$ can have some differences. The quantities $\mathbb{G}_{u,B,i}$, $\mathbb{H}_{l,B,i}$, $\mathbb{J}_{l,B,i}$, $l = x, y$, do not exist and should be got rid of. In addition,

$$
\mathbb{J}_{l,i,mn} = 0, \quad \mathbb{G}_{u,i,mn} = 0, \quad \mathbb{H}_{l,i,mn} = -\int_{(\partial K)_i} \phi_m \phi_n n_l ds, \quad l = x, y.
$$

We take the degrees of freedom as the values of midpoints at three edges of K, then

$$
\mathbb{M} \approx \frac{|K|}{3} \mathbb{I}, \quad \mathbf{S}_u \approx \frac{|K|}{3} \mathbf{f}(\overline{\mathbf{u}}, \overline{\mathbf{v}}), \quad \mathbf{S}_v \approx \frac{|K|}{3} \mathbf{g}(\overline{\mathbf{u}}, \overline{\mathbf{v}}),
$$

where $|K|$ is the area of the element K, $\mathbb{I}$ is the unit matrix.

Now, substituting (6) into (7), we derive a system including original variable only for the element K,

$$
\frac{d\mathbf{w}}{dt} = -\frac{3}{|K|} \left(\mathbb{N}\mathbf{w} + \sum_{i=1}^{3} \overline{\mathbb{N}}_i \mathbf{w}_{(NB,i)} + \sum_{i=1}^{3} \sum_{j=1}^{3} \widetilde{\mathbb{N}}_{ij} \mathbf{w}_{(NB,i)_{(NB,j)}} \right) + \mathbf{F}(\mathbf{w}), \tag{9}
$$

where (NB, i), $(NB, i)_{(NB,j)}$, $i, j = 1, 2, 3$ denote the quantities belonging to K_i and K_{ij}, respectively. And $\mathbf{w} = (\overline{\mathbf{u}}, \ \overline{\mathbf{v}})^T$, $\mathbf{F}(\mathbf{w}) = (\mathbf{f}(\overline{\mathbf{u}}, \overline{\mathbf{v}}), \ \mathbf{g}(\overline{\mathbf{u}}, \overline{\mathbf{v}}))^T$,

$$N = \begin{pmatrix} N_u & 0 \\ 0 & N_v \end{pmatrix}, \quad \overline{N}_i = \begin{pmatrix} \overline{N}_{i,u} & 0 \\ 0 & \overline{N}_{i,v} \end{pmatrix}, \quad \widetilde{N}_{ij} = \begin{pmatrix} \widetilde{N}_{ij,u} & 0 \\ 0 & \widetilde{N}_{ij,v} \end{pmatrix},$$

$$N_u = G_U - \tfrac{3}{|K|} D_u \left(\mathbb{JJ}_x \mathbb{HH}_x + \mathbb{JJ}_y \mathbb{HH}_y \right), \quad N_v = G_U - \tfrac{3}{|K|} D_v \left(\mathbb{JJ}_x \mathbb{HH}_x + \mathbb{JJ}_y \mathbb{HH}_y \right),$$

$$\overline{N}_{i,u} = G_{u,B,i} - \tfrac{3}{|K|} D_u \left(\mathbb{JJ}_x \mathbb{H}_{x,B,i} + \mathbb{JJ}_y \mathbb{H}_{y,B,i} \right) - \tfrac{3}{|K_i|} D_u \left(\mathbb{J}_{x,B,i} \mathbb{HH}_x^{(NB,i)} + \mathbb{J}_{y,B,i} \mathbb{HH}_y^{(NB,i)} \right),$$

$$\overline{N}_{i,v} = G_{u,B,i} - \tfrac{3}{|K|} D_v \left(\mathbb{JJ}_x \mathbb{H}_{x,B,i} + \mathbb{JJ}_y \mathbb{H}_{y,B,i} \right) - \tfrac{3}{|K_i|} D_v \left(\mathbb{J}_{x,B,i} \mathbb{HH}_x^{(NB,i)} + \mathbb{J}_{y,B,i} \mathbb{HH}_y^{(NB,i)} \right),$$

$$\widetilde{N}_{i,j,u} = -\tfrac{3}{|K_i|} D_u \left(\mathbb{J}_{x,B,i} \mathbb{H}_{x,B,j}^{(NB,i)} + \mathbb{J}_{y,B,i} \mathbb{H}_{y,B,j}^{(NB,i)} \right),$$

$$\widetilde{N}_{i,j,v} = -\tfrac{3}{|K_i|} D_v \left(\mathbb{J}_{x,B,i} \mathbb{H}_{x,B,j}^{(NB,i)} + \mathbb{J}_{y,B,i} \mathbb{H}_{y,B,j}^{(NB,i)} \right).$$

2.2 The Second Order IIF Method for Time Discretization

Assembling (9) over all of the elements in $\mathcal{T}_h$, we derive the global system,

$$\frac{d\mathbf{W}}{dt} = \mathbb{A}\mathbf{W} + \mathbf{F}(\mathbf{W}), \tag{10}$$

where $\mathbf{W} = (\mathbf{w}_1^T, \mathbf{w}_2^T, \cdots, \mathbf{w}_{N_e}^T)^T$, $\mathbf{F}(\mathbf{W}) = (\mathbf{F}(\mathbf{w}_1)^T, \mathbf{F}(\mathbf{w}_2)^T, \cdots, \mathbf{F}(\mathbf{w}_{N_e})^T)^T$, $\mathbf{w}_j$ is the degrees of freedom on K_j, $j = 1, 2, \cdots, N_e$, and N_e denotes the number of elements. The global matrix $\mathbb{A}$ is $6Ne \times 6N_e$ and sparse.

By applying the second order IIF scheme for the time evolution in (10),

$$\mathbf{W}^{n+1} = e^{\mathbb{A}\Delta t} \left(\mathbf{W}^n + \frac{\Delta t}{2} \mathbf{F}(\mathbf{W}^n) \right) + \frac{\Delta t}{2} \mathbf{F}(\mathbf{W}^{n+1}),$$

where n is the time level, $t_{n+1} = t_n + \Delta t$ and $\mathbf{W}^n = \mathbf{W}(t_n)$.

Since the vector $\mathbf{Q} = e^{\mathbb{A}\Delta t} \left(\mathbf{W}^n + \frac{\Delta t}{2} \mathbf{F}(\mathbf{W}^n) \right)$ is known at the earlier time level and can be computed by the Krylov subspace approximation shown in [2], the nonlinear system at t_{n+1} is decoupled from the diffusion with a simple form

$$\mathbf{W}^{n+1} = \mathbf{Q} + \frac{\Delta t}{2} \mathbf{F}(\mathbf{W}^{n+1}), \tag{11}$$

which can be solved element by element, and the local algebraic system on every element K_j, $j = 1, 2, \cdots, N_e$ are $\mathbf{R}(\mathbf{w}_j^{n+1}) = \mathbf{0}$, where,

$$\mathbf{R}(\mathbf{w}_j^{n+1}) = \begin{pmatrix} u_{j,1}^{n+1} - \mathbf{Q}(6(j-1)+1) - \frac{\Delta t}{2} f(u_{j,1}^{n+1}, v_{j,1}^{n+1}) \\ u_{j,2}^{n+1} - \mathbf{Q}(6(j-1)+2) - \frac{\Delta t}{2} f(u_{j,2}^{n+1}, v_{j,2}^{n+1}) \\ u_{j,3}^{n+1} - \mathbf{Q}(6(j-1)+3) - \frac{\Delta t}{2} f(u_{j,3}^{n+1}, v_{j,3}^{n+1}) \\ v_{j,1}^{n+1} - \mathbf{Q}(6(j-1)+4) - \frac{\Delta t}{2} g(u_{j,1}^{n+1}, v_{j,1}^{n+1}) \\ v_{j,2}^{n+1} - \mathbf{Q}(6(j-1)+5) - \frac{\Delta t}{2} g(u_{j,2}^{n+1}, v_{j,2}^{n+1}) \\ v_{j,3}^{n+1} - \mathbf{Q}(6(j-1)+6) - \frac{\Delta t}{2} g(u_{j,3}^{n+1}, v_{j,3}^{n+1}) \end{pmatrix}.$$

An iterative method can be applied for solving the above system. In the following numerical examples, we choose the Newton method and set the threshold value for judging Newton iteration to 10^{-13}.

3 Numerical Experiments

In this section, we come to verify the efficiency and advantages of the scheme (11) by solving the chlorite-iodide-malonic acid (CIMA) model proposed in [6].

We consider the CIMA model defined on the circular domain $\Omega = \{(x,y)|$ $(x-50)^2 + (y-50)^2 \le 50^2\}$,

$$\begin{cases} \frac{\partial u}{\partial t} - D_u \nabla^2 u = \frac{1}{\sigma}(a - u - 4\frac{uv}{1+u^2}), \\ \frac{\partial v}{\partial t} - D_v \nabla^2 v = b(u - \frac{uv}{1+u^2}), \end{cases} \tag{12}$$

with the initial conditions

$$u(\mathbf{x},0) = \frac{1}{10}rand(\mathbf{x}), \quad v(\mathbf{x},0) = \frac{1}{10}rand(\mathbf{x}),$$

where $\mathbf{x} = (x,y) \in \Omega$, and rand: $\Omega \to [0,1]$ is a random function in Fortran. We take the parameters as $\sigma = 50$, $D_u = \frac{1}{\sigma}$, $D_v = 1.07$, $a = 8.8$ and $b = 0.09$.

In the computation, we divide Ω into 16680 triangles and 8491 vertices by the Delaunay partitions. And now the length of the minimum edge is $h_{min} = 0.6545$. The time step size is taken as $\Delta t = 0.3 * h_{min}$, and the auxiliary parameters in the numerical fluxes are taken as

$$C_{11} = \frac{1}{h_e}, \quad \mathbf{C}_{12} \cdot \mathbf{n}_e = \frac{1}{2}sign(n_{e1} + n_{e2}), \quad \forall e \in \partial K \cap \mathcal{E}_h,$$

where h_e is the length of edge e.

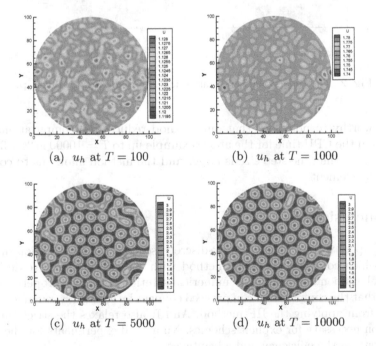

(a) u_h at $T = 100$ (b) u_h at $T = 1000$

(c) u_h at $T = 5000$ (d) u_h at $T = 10000$

Fig. 1. The time evolution of numerical solutions u_h.

Since the patterns of u_h and v_h are always of the same type, we focus on that of u_h. Figure 1 shows the pictures of u_h at $T = 100$, $T = 1000$, $T = 5000$ and $T = 10000$, respectively. We can see a H_0 hexagon pattern, which agrees well with those in [8,12]. In addition, by use of our method, we derive the approximations for fluxes as well. And it is also the same type for the two components of fluxes, and here we only present the graphs for x-direction component of $\mathbf{p}_h$ in Fig. 2.

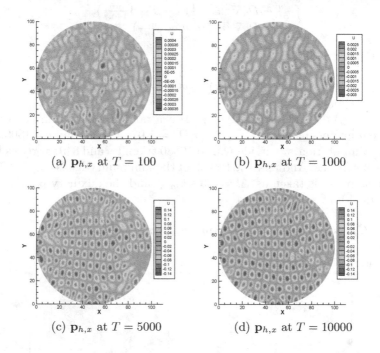

(a) $\mathbf{p}_{h,x}$ at $T = 100$ (b) $\mathbf{p}_{h,x}$ at $T = 1000$

(c) $\mathbf{p}_{h,x}$ at $T = 5000$ (d) $\mathbf{p}_{h,x}$ at $T = 10000$

Fig. 2. The time evolution of numerical approximates for fluxes $\mathbf{p}$.

It is worthy to point out that our method reduces the computational cost greatly and the CPU time for the above example up to $T = 10000$ is 26403 s. The reason is that the time-step size is larger and the method allows us to compute element by element.

4   Conclusions

In this paper, we developed a fully-discrete scheme for the reaction-diffusion system which combines the LDG method with the IIF method. We derive the numerical gradients as well as the numerical solutions. And the property of LDG method that the computation can proceed element by element is remained, which benefits from applying the IIF method. And It also relaxes the strict time-step restriction necessary for explicit schemes. Numerical experiments for the CIMA model confirm the efficiency and advantages.

Acknowledgments. This work is supported by the Natural Science Foundation of China (Grant No. 11571002), the Science and Technology Development Foundation of CAEP (Grant Nos. 2013A0202011 and 2015B0101021) and the Defense Industrial Technology Development Program (Grant No. B1520133015).

References

1. Castillo, P., Cockburn, B., Perugia, I., Schötzau, D.: An a priori error analysis of the local discontinuous Galerkin method for elliptic problems. SIAM J. Numer. Anal. **38**, 1676–1706 (2000)
2. Chen, S.Q., Zhang, Y.T.: Krylov implicit integration factor methods for spatial discretization on high dimensional unstructured meshes: application to discontnuous Galerkin methods. J. Comput. Phys. **230**, 4336–4352 (2011)
3. Cockburn, B., Shu, C.W.: The local discontinuous Galerkin method for time-dependent convection-diffusion system. SIAM J. Numer. Anal. **35**, 2440–2463 (1998)
4. Dehghan, M., Abbaszadeh, M.: Variational multiscale element free Galerkin (VMEFG) and local discontinuous Galerkin (LDG) methods for solving two-dimensional Brusselator reaction-diffusion system with and without cross-diffusion. Comput. Methods Appl. Mech. Eng. **300**, 770–797 (2016)
5. Gierer, A., Meinhardt, H.: A theory of biological pattern formation. Kybernetik **12**, 30–39 (1972)
6. Lengyel, I., Epstein, I.R.: Modelling of Turing structures in the chlorite-iodide-malonic acid-starch reaction system. Science **251**, 650–652 (1991)
7. Li, B.Q.: Discontinuous Finite Elements in Fluid Dynamics and Heat Transfer, pp. 105–156. Springer, London (2006)
8. Li, Q., Zheng, C.G., Wang, N.C., Shi, B.C.: LBGK simulations of Turing patterns in CIMA model. J. Sci. Comput. **16**, 121–134 (2001)
9. Nie, Q., Zhang, Y.T., Zhao, R.: Efficient semi-implicit schemes for stiff systems. J. Comput. Phys. **214**, 512–537 (2006)
10. Schnakenberg, J.: Simple chemical reaction systems with limit cycle behavior. J. Theoret. Biol. **81**, 389–400 (1979)
11. Turing, A.M.: The chemical basis of morphogenesis. Philos. Trans. R. Soc. Lond. B **237**, 37–72 (1952)
12. Zhang, R.P., Yu, X.J., Zhu, J., Loula, A.F.D.: Direct discontinuous Galerkin method for nonlinear reaction-diffusion systems in pattern formulation. Appl. Math. Model. **38**, 1612–1621 (2014)
13. Zhu, J.F., Zhang, Y.T., Newman, S.A.: Application of discontinuous Galerkin methods for reaction-diffusion systems in developmental biology. J. Sci. Comput. **40**, 391–418 (2009)
14. Gray, P., Scott, S.K.: Autocatalytic reactions in the isothermal, continuous stirred tank reactor: isolas and other forms of multistability. Chem. Eng. Sci. **38**, 29–43 (1983)

A Method for Linearization of a Beam Problem

A.B. Andreev[1] and M.R. Racheva[2(✉)]

[1] Institute of Information and Communication Technologies, BAS, Sofia, Bulgaria
[2] Technical University of Gabrovo, Gabrovo, Bulgaria
milena@tugab.bg

Abstract. The bending vibrations of an elastic shaft under action of compression are considered. Taking into account the beam torsion, the eigenparameter appears nonlinearly into the model fourth-order differential equation. We propose an approach for avoiding this nonlinearity using an appropriate mixed formulation and thus some variational numerical methods could be applied in a standard way.

The possibility for symmetrization of the weak mixed formulation and corresponding finite element analysis are presented. Different matrix structures related to the proposed approach and its finite element implementation are discussed. Finally, we illustrate the method by numerical example.

1 Introduction and Problem Settings

The purpose of this paper is to discuss the free vibrations of an elastic shaft with length l compressed by an axial force. The beam angular deformation causes appearance of the eigenvalue parameter λ in nonlinear form [1,2]. Thus, we are led to seek a number λ and a function $u(x) \neq 0$ so that

$$-C_1 u^{IV}(x) = \lambda u''(x) + \lambda C_2 u(x) + \lambda^2 C_3 u(x),$$

$$0 < x < l,$$

(1)

with hinged boundary conditions or simply supported boundaries

$$u(0) = u(l) = u''(0) = u''(l) = 0.$$

(2)

Here, the parameters C_1, C_2 and C_3 in (1) include:

- $C_1 > 0$ is the module of linear deformation (flexural rigidity);
- $C_2 \geq 0$ is the rigidity of the elastic base;
- $C_3 > 0$ represents the bending module of rigidity (torsional rigidity).

Mixed variational methods are very appropriate approach for solving of this type of fourth-order equations. In the mixed formulations (see [3,4]), one has two unknown fields to be approximated by Galerkin procedure and by finite element methods (FEMs) in particular. More general investigations concerning the mixed formulation for fourth-order elliptic eigenvalue problems could be found in [5] and [6]. Our main goal here is to find such a representation of the problem (1), (2) thereby to avoid the presence of λ^2.

© Springer International Publishing AG 2017
I. Dimov et al. (Eds.): NAA 2016, LNCS 10187, pp. 180–186, 2017.
DOI: 10.1007/978-3-319-57099-0_17

2   Main Results

The Eq. (1) could be written as:

$$-u^{IV} = L(\lambda, u),$$

where L is a second-order self-adjoint linear operator in $C^1(0, l) \cap H^2[0, l]$. Here and further H^m denotes the Sobolev space of order $m \geq 0$. Consequently, the eigenvalues constitute a sequence of real numbers $\lambda_1 \leq \lambda_2 \leq \lambda_3 \leq \ldots$ and the eigenfunctions u_j, $j = 1, 2, 3, \ldots$ being normalized $\|u_j\|_{0,(0,l)} = 1$.

First, we will present a different mixed formulation of the problem (1), (2). In general, such kind of presentations are not unique (see, e.g. [4,6,7]). So, we introduce a new unknown function $\sigma \in H^2(0, l)$.

Theorem 1. *The following mixed formulation is equivalent to the problem (1), (2):*

$$\left|\begin{array}{l} -\sqrt{C_1}u'' - \lambda\sqrt{C_3}u = \sigma \\[2mm] -\sqrt{C_1}\sigma'' - \lambda\sqrt{C_3}\sigma = \lambda\left(2\sqrt{C_1C_3} - 1\right)u'' - C_2\lambda u, \end{array}\right. \tag{3}$$

with boundary conditions

$$u(0) = u(l) = \sigma(0) = \sigma(l) = 0.$$

Thus, a linear eigenvalue problem in mixed form is obtained.

Proof. Consider the following system (we drop the argument x):

$$\left|\begin{array}{l} \alpha_1 u'' + \beta_1 u' + \gamma_1 u + \lambda k_1 u = \sigma \\[2mm] \alpha_2 \sigma'' + \beta_2 \sigma' + \gamma_2 \sigma + \lambda k_2 \sigma = M(\lambda, u), \end{array}\right. \tag{4}$$

where $\alpha_i, \beta_i, \gamma_i, k_i$, $i = 1; 2$ are unknown parameters and $M(\lambda, u)$ is a second-order linear operator.

We replace the first equation of (4) into the second one. So, it is easy to obtain:

$$\alpha_1\alpha_2 u^{IV} + (\alpha_1\beta_2 + \alpha_2\beta_1)u''' + (\alpha_1\gamma_2 + \alpha_2\gamma_1 + \beta_1\beta_2)u''$$

$$+ (\beta_1\gamma_2 + \beta_2\gamma_1)u' + \gamma_1\gamma_2 u + \lambda(k_1\alpha_2 + k_2\alpha_1)u'' \tag{5}$$

$$+ \lambda(k_1\beta_2 + k_2\beta_1)u' + \lambda(k_1\gamma_2 + k_2\gamma_1)u + \lambda^2 k_1 k_2 u = M(\lambda, u).$$

Now, by comparisson of (1) with (5) we get:

$$\left|\begin{array}{l} \alpha_1\alpha_2 = C_1 \\[2mm] \alpha_1\beta_2 + \alpha_2\beta_1 = 0 \\[2mm] \alpha_1\gamma_2 + \alpha_2\gamma_1 + \beta_1\beta_2 = 0 \\[2mm] \beta_1\gamma_2 + \beta_2\gamma_1 = 0 \\[2mm] \gamma_1\gamma_2 = 0. \end{array}\right. \tag{6}$$

And however:

$$\lambda u'' + \lambda C_2 u + \lambda^2 C_3 u = \lambda(k_1\alpha_2 + k_2\alpha_1)u''$$
$$\lambda(k_1\beta_2 + k_2\beta_1)u' + \lambda(k_1\gamma_2 + k_2\gamma_1)u + \lambda^2 k_1 k_2 u - M(\lambda, u). \tag{7}$$

In addition, $\alpha_i, \beta_i, \gamma_i$ and k_i, $i = 1; 2$ will be determined in such a way that (4) leads to symmetric formulation.

For that purpose we put:

$$\alpha_1 = \alpha_2 := \alpha; \beta_1 = \beta_2 := \beta; \gamma_1 = \gamma_2 := \gamma; k_1 = k_2 := k.$$

Then, (7) takes the form

$$\alpha u'' + \lambda C_2 u + \lambda^2 C_3 u = 2\lambda k\alpha u'' + \lambda^2 k^2 u - M(\lambda, u).$$

From this equality, it is easy seen that

$$k = \pm\sqrt{C_3},$$

and, moreover,

$$\alpha u'' + \lambda C_2 u = 2\lambda k\alpha u'' - M(\lambda, u),$$

which means:

$$M(\lambda, u) = \lambda(2k\alpha - 1)u'' - \lambda C_2 u.$$

Thus, we proved (3), or $(\alpha = \pm\sqrt{C_1},\ k = \pm\sqrt{C_3})$:

$$\left|\begin{array}{l} \alpha u'' + \lambda k u = \sigma \\[2mm] \alpha\sigma'' + \lambda k\sigma = \lambda(2k\alpha - 1)u'' - \lambda C_2 u. \end{array}\right. \tag{8}$$

The boundary conditions (2) are transformed in:

$$\left|\begin{array}{l} u(0) = u(l) = 0 \\[2mm] \sigma(0) = \sigma(l) = 0. \end{array}\right. \tag{9}$$

Remark 1. The differential mixed formulation from Theorem 1 is not unique. In fact, the system (3) could be written in the form:

$$\left| \begin{array}{l} \pm\sqrt{C_1}u'' \pm \lambda\sqrt{C_3}u = \sigma \\ \pm\sqrt{C_1}\sigma'' \pm \lambda\sqrt{C_3}\sigma = \lambda\left(2\sqrt{C_1 C_2} - 1\right)u'' - \lambda C_2 u. \end{array} \right.$$

Next, we will obtain the corresponding mixed variational formulation of (8), (9).

Theorem 2. *The problem (8), (9) admits symmetric variational formulation.*

Proof. Let us introduce two functional variational spaces such that

$$u \in H_0^1(0,l) = V \text{ and } \sigma \in H_0^1(0,l) = \Sigma.$$

Multiplying the first equation of (8) by a function $\psi \in \Sigma$ and integrating it on the interval $(0,l)$, we obtain:

$$\int_0^l \alpha u'' \psi'' \, dx + \lambda \int_0^l k u \psi \, dx = \int_0^l \sigma \psi \, dx.$$

Consequently, after integration by parts,

$$\alpha \int_0^l u' \psi' \, dx + \int_0^l \sigma \psi \, dx = \lambda k \int_0^l u \psi \, dx. \tag{10}$$

In the same manner, multiplying the second equation of (8) by a function $v \in V$ and integrating by parts on the interval $(0,l)$ reveals that:

$$\int_0^l \alpha \sigma'' v \, dx + \lambda \int_0^l k \sigma v \, dx = \lambda \int_0^l (2k\alpha - 1)u'' v \, dx - \lambda \int_0^l C_2 u v \, dx.$$

Hence,

$$\alpha \int_0^l \sigma' v' \, dx = \lambda k \int_0^l \sigma v \, dx - \lambda(2k\alpha - 1) \int_0^l u' v' \, dx - \lambda C_2 \int_0^l u v \, dx.$$

This equation together with (10) represent a mixed variational formulation of the problem (1), (2).

Finally, we have obtained the following mixed variational system: find $(\lambda, u, \sigma) \in \mathbf{R} \times V \times \Sigma$ such that

$$\left| \begin{array}{l} a_1(u, \psi) + b_1(\sigma, \psi) = \lambda h_2(u, \psi), \forall \psi \in \Sigma, \\ a_1(\sigma, v) = \lambda b_2(\sigma, v) - \lambda a_2(u, v) - \lambda b_3(u, v), \forall v \in V, \end{array} \right. \tag{11}$$

where

$$a_1(u, \psi) = \alpha \int_0^l u'\psi' \, dx, \, a_2(u, v) = (2k\alpha - 1) \int_0^l u'v' \, dx,$$

$$b_1(\sigma, \psi) = \int_0^l \sigma\psi \, dx, \, b_2(u, \psi) = k \int_0^l u\psi \, dx, \, b_3(u, v) = C_2 \int_0^l uv \, dx.$$

Moreover, $\alpha = \pm\sqrt{C_1}$ and $k = \pm\sqrt{C_3}$.

In conclusion, the mixed variational formulation is symmetric. It should be noted that the eigenparameter λ appears linearly in (11), which is the main advantage of the proposed method. Obviously, the mixed symmetric presentation is not unique (see Remark 1).

Let V_h and Σ_h be coupled finite element spaces such that $V_h \times \Sigma_h \subset V \times \Sigma$. The discrete variational system corresponding to (11) is: find $(\lambda_h, u_h, \sigma_h) \in \mathbf{R} \times V_h \times \Sigma_h$ such that

$$\begin{vmatrix} a_1(u_h, \psi) + b_1(\sigma_h, \psi) = \lambda_h b_2(u_h, \psi), \forall \psi \in \Sigma_h, \\ a_1(\sigma_h, v) = \lambda_h b_2(\sigma_h, v) - \lambda_h a_2(u_h, v) - \lambda_h b_3(u_h, v), \forall v \in V_h. \end{vmatrix} \quad (12)$$

From (12), the two equations could be united by the equality:

$$a_1(u_h, \psi) + a_1(\sigma_h, v) + b_1(\sigma_h, \psi) = \lambda_h \left[b_2(u_h, \psi) + b_2(\sigma_h, v) \right.$$

$$\left. -a_2(u_h, v) - b_3(u_h, v) \right], \forall (v, \psi) \in V_h \times \Sigma_h. \quad (13)$$

3   Matrix Representation of the Mixed Variational Equation

Let $\varphi_i(x)$, $i = 1, 2, \ldots, n_1$ and $\psi_j(x)$, $j = 1, 2, \ldots, n_2$, $n_1, n_2 \in \mathbf{N}$, be the shape functions of the finite element spaces V_h and Σ_h, respectively. Then u_h and σ_h have the representation

$$u_h(x) = \sum_{i=1}^{n_1} u_i \varphi_i(x) \text{ and } \sigma_h(x) = \sum_{j=1}^{n_2} \sigma_j \psi_j(x), x \in [0, l].$$

Finding the solution $(\lambda_h, u_h, \sigma_h) \in \mathbf{R} \times V_h \times \Sigma_h$ of (13) consists of resulting matrix equation solving.

If we denote $\mathcal{U} = (u_1, u_2, \ldots, u_{n_1})^T$ and $\mathcal{S} = (\sigma_1, \sigma_2, \ldots, \sigma_{n_2})^T$, then from (13) we obtain matrix equation which could be written either as

$$\begin{pmatrix} \mathcal{B}_1 & \mathcal{A}_1^T \\ \mathcal{A}_1 & \mathcal{O} \end{pmatrix} \begin{pmatrix} \mathcal{S} \\ \mathcal{U} \end{pmatrix} = \lambda \begin{pmatrix} \mathcal{O} & \mathcal{B}_2^T \\ \mathcal{B}_2 & \mathcal{A}_2 + \mathcal{B}_3 \end{pmatrix} \begin{pmatrix} \mathcal{S} \\ \mathcal{U} \end{pmatrix}, \quad (14)$$

or into the form

$$\begin{pmatrix} \mathcal{O} & \mathcal{A}_1 \\ \mathcal{A}_1^T & \mathcal{B}_1 \end{pmatrix} \begin{pmatrix} u \\ s \end{pmatrix} = \lambda \begin{pmatrix} \mathcal{A}_2 + \mathcal{B}_3 & \mathcal{B}_2 \\ \mathcal{B}_2^T & \mathcal{O} \end{pmatrix} \begin{pmatrix} u \\ s \end{pmatrix}, \qquad (15)$$

where the matrices $\mathcal{A}_s$, $s = 1; 2$ and $\mathcal{B}_k$, $k = 1; 2; 3$ correspond to the bilinear forms $a_s(\cdot, \cdot)$ and $b_k(\cdot, \cdot)$, respectively.

More precisely,

$$\mathcal{A}_1 = \left(\alpha \varphi_i' \psi_j'\right)_{i=1\,j=1}^{n_1\ \ n_2}; \mathcal{A}_2 = \left((2k\alpha - 1)\varphi_i' \varphi_j'\right)_{i,j=1}^{n_1};$$

$$\mathcal{B}_1 = \left(\psi_i \psi_j\right)_{i,j=1}^{n_2}; \mathcal{B}_2 = \left(k\varphi_i \psi_j\right)_{i=1\,j=1}^{n_1\ \ n_2}; \mathcal{B}_3 = \left(C_2 \varphi_i \varphi_j\right)_{i,j=1}^{n_1}$$

and $\mathcal{O}$ is square null matrix with dimensions $n_1 \times n_1$ and $n_2 \times n_2$ in the left hand side and right hand side of $(14), (15)$, respectively.

It is to be noted here that the global block matrices in (14) and (15) are symmetric matrices of type $(n_1 + n_2) \times (n_1 + n_2)$.

4 Numerical Example

Consider the differential equation

$$u^{IV}(x) + \lambda u''(x) + \frac{\lambda^2}{4}u(x) = 0, 0 < x < \pi,$$

with boundary conditions

$$u(0) = u(\pi) = u''(0) = u''(\pi) = 0,$$

i.e. into the Eq. (1) we take $l = \pi$ and accordingly

$$C_1 = 1; C_2 = 0; C_3 = \frac{1}{4}.$$

Coefficients selected hardly make a realistic example, but it serves as confirmation of the proposed mixed method and an illustration how does the method works. The exact eigenvalues are known, namely they are equal to $2k^2$, $k = 1, 2, 3, \ldots$.

We divide uniformly the interval $(0, \pi)$ into n subintervals, $n = 4; 8; 12; 16; 20$, so that the mesh parameter h is equal to π/h. We solve the Eq. (13) using quadratic finite elements for V_h as well as for Σ_h. It is also possible to use linear finite elements in mixed finite element methods for fourth-order problems [8], but this special case deserves to be considered separately.

In Table 1 approximations of the first three exact eigenvalues by means of the proposed mixed method solving are presented. It is well-known that standard conforming finite element methods give approximations of the eigenvalues from above, while for any mixed finite element method this is not clear in general. From Table 1 it is seen that for the concrete numerical example under consideration the mixed method approximates the exact eigenvalues assymptoticaly from below.

Table 1. Approximations $\lambda_{i,h}$ of the exact eigenvalues λ_i, $i = 1, 2, 3$ obtained after finite element implementation of the mixed method by means of quadratic FEs

n	$\lambda_{1,h}$	$\lambda_{2,h}$	$\lambda_{3,h}$
4	1.87500168327	7.50024574565	16.8777742386
8	1.88435145645	7.55633318370	17.1577066573
12	1.88895073293	7.58382506188	17.3424752329
16	1.88985592278	7.61239091796	17.5661581790
20	1.89128727333	7.63198991806	17.5900004588
	$\lambda_1 = 2$	$\lambda_2 = 8$	$\lambda_3 = 18$

Acknowledgements. This work was partially supported by the Technical University of Gabrovo under grant D1602E/2016.

References

1. Collatz, L.: Eigenwertaufgaben mit Technischen Anwendungen. Acad. Verlag, Leipzig (1963)
2. Timoshenko, S.P., Weaver, W., Young, D.H.: Vibration Problems in Engineering, 5th edn. Wiley, New York (1990)
3. Stoyan, G., Baran, A.: Eigenvalue problems. Elementary Numerical Mathematics for Programmers and Engineers. CTM, pp. 85–109. Springer, Cham (2016). doi:10.1007/978-3-319-44660-8_5
4. Brenner, S., Scott, L.R.: The Mathematical Theory for Finite Element Methods. Springer, New York (1994)
5. Racheva, M.: Mixed variational formulation for some fourth order beam problems. C. R. Acad. Bulg. Sci. Tome **64**(11), 1525–1532 (2011)
6. Andreev, A., Racheva, M.: Variational aspects of the mixed formulation for fourth-order elliptic eigenvalue problems. Math. Balk. **18**(1–2), 41–51 (2004)
7. Ciarlet, P.G.: The Finite Element Method for Elliptic Problems. North-Holland, Amsterdam (1978)
8. Ishihara, K.: A mixed finite element approximation for the buckling of plates. Numer. Math. **33**, 195–210 (1979)

Numerical Modeling of Fluid Flow in Liver Lobule Using Double Porosity Model

M. Yu. Antonov[1], A.V. Grigorev[1,2(✉)], and A.E. Kolesov[1]

[1] M.K.Ammosov North-Eastern Federal University, Yakutsk, Russia
[2] Institute of Computational Mathematics and Mathematical
Geophysics SB RAS, Novosibirsk, Russia
re5itsme@gmail.com

Abstract. Earlier in the paper of Bonfiglio et al. [2] numerical simulation of blood circulation in the liver lobule was carried out using single porosity model. Electron microscopy reveals structure of the liver lobule, which has some of the properties of fractured porous media. In this work we consider double porosity model for modeling of blood filtration in the liver lobule. A numerical algorithm based on the spatial finite element approximation and finite difference approximation in time direction using explicit-implicit computational scheme is proposed.

Liver is a vital organ of vertebrates and some other animal. One of its main purposes is filtering the blood, keeping it pure of toxins and harmful substances. The liver receives blood from two major sources: about 80% is partially deoxygenated blood from the portal vein, while the other part is oxygenated blood from the hepatic artery [6]. Blood from the portal vein is rich with nutrients extracted from food, and the liver processes these nutrients while also filtering the blood from toxins.

Before reaching the liver, the portal vein divides into right and left branches. It is further divided, forming smaller venous branches and ultimately interlobular veins. Each interlobular vein as is accompanied with a branch of hepatic artery. Blood from both sources mixes together and then passed out to the central veins through the hepatic sinusoids and perisinusoidal space. Central veins join to form the larger hepatic veins, which drain the liver blood into the inferior vena cava.

Structurally, the liver is made up of smaller blocks, which can be described in a different ways. One of the most commonly used description is "classic lobule" [7]. Each lobule has approximately hexagonal shape, with central vein in the center, surrounded by a sinusoidal space. Portal tracts are located on the angles of the hexagon, each consisting of a branch of the hepatic portal vein, a branch of the hepatic artery and the bile duct. Blood from the portal tracts flows through the sinusoidal space and empties into the central vein of each lobule.

Sinusoids are low pressure vascular channels that receive blood from terminal branches of the hepatic artery and portal vein at the periphery of lobules and deliver it into central veins. Sinusoids are lined with endothelial cells. Kupffer cells are located inside the sinusoids and function to filter viral particles from

I. Dimov et al. (Eds.): NAA 2016, LNCS 10187, pp. 187–194, 2017.
DOI: 10.1007/978-3-319-57099-0_18

blood. Blood plasma can also filter through the endothelial cells into perisinu-soidal space, which separates the sinusoids from hepatocytes.

In this work we propose mathematical model of the blood circulation in the liver lobule. Based on known lobule structure, we consider double porosity model for representation of sinusoids and sinusoidal space (an analogue of cracks and pores [1]). We use this model to find the pressure and flow distribution inside of the liver lobule. The mathematical model (Sect. 1) and results (Sect. 4) are performed in ICMMG SB RAS under sole support by RSF, grant 15-11-10024.

1   Mathematical Model

For studying the filtration of the blood in a classical lobule, we assume that liver lobules can be considered as nearly identical hexagonal cells arranged into a spatial lattice (Fig. 1). The hexagonal cross-section of the single lobule is shown on Fig. 2. We assume that the portal tracts and the central vein have cylindrical shape. Portal tract axes are located on the lobule edges, and the central vein axe is located in the center of the hexagon. The remaining volume of the lobule in our model is considered as a sinusoidal space, which is considered as porous medium and we apply Darcy's law to describe the blood flow in it. Assuming that the length of the lobule is significantly larger than its cross-section, the end effects can be neglected, and the problem is reduced to the two-dimensional case.

In our mathematical model we consider sinusoids and sinusoidal space as a continuous, homogeneous, isotropic, porous and solid permeable medium with

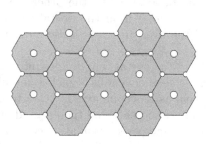

Fig. 1. View of the lobule

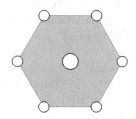

Fig. 2. Lobule scheme, consisting of region Ω, boundaries: lobule center — Γ_v, hexagon vertexes — Γ_p, hexagon edges — Γ_b

permeability k_α. Let $K_\alpha = k_\alpha/\mu$, where μ is dynamic viscosity of the blood. Hereinafter, the index $\alpha = 1$ indicates belonging to sinusoids (In our model represented as fractured part of the porous medium), and $\alpha = 2$ — to sinusoidal space (as porous part of the porous medium).

Assuming that the blood can be described as weakly compressible Newtonian fluid, we model filtration in a sinusoidal space using Darcy's law supplemented with a continuity assumption:

$$v_\alpha = -K_\alpha \nabla p_\alpha(\boldsymbol{x}), \quad c_\alpha \frac{\partial p_\alpha}{\partial t} + \nabla \cdot \rho v_\alpha + R = 0, \tag{1}$$

Here $v_\alpha(\boldsymbol{x})$ and $p_\alpha(\boldsymbol{x})$ are the flow and the pressure of the fluid in $\boldsymbol{x}$. R term represents exchange between the pores and the fractures, and built on the pressure difference $r(p_2 - p_1)$ with sign $+$ at the $\alpha = 1$, and with sign $-$ at the $\alpha = 2$. Combining these Eq. (1) we obtain:

$$c_\alpha \frac{\partial p_\alpha}{\partial t} - \nabla \cdot (K_\alpha \nabla p_\alpha(\boldsymbol{x})) + R = 0. \tag{2}$$

We consider the following boundary conditions:

$$p_1 = p_1^p, \quad \nabla p_2 \cdot \boldsymbol{n} = 0, \quad \boldsymbol{x} \in \Gamma_p, \tag{3}$$

$$p_\alpha = p_\alpha^v, \quad \nabla p_2 \cdot \boldsymbol{n} = 0, \quad \boldsymbol{x} \in \Gamma_v, \tag{4}$$

$$\nabla p_\alpha \cdot \boldsymbol{n} = 0, \quad \boldsymbol{x} \in \Gamma_b. \tag{5}$$

These boundary conditions are used for simulate the process, in which the blood from the portal tracts enter the sinusoids, and subsequently filtered through the sinusoidal space. Fluid exchange between sinusoidal capillaries and sinusoidal space is modelled using the double porosity concept. Despite its simplicity, this model gives insight into the pressure distribution in the lobule.

2   Model Problem

Let's write the two dimensional model problem in terms of dimensionless variables:

$$p_\alpha^* = \frac{p_\alpha - p_\alpha^v}{p_\alpha^p - p_\alpha^v}, \quad x^* = \frac{x}{L}. \tag{6}$$

Define boundary and initial conditions for the problem:

$$p_1^*(\boldsymbol{x}, t) = 1 - \exp(-\delta t), \quad \nabla p_2^* \cdot \boldsymbol{n} = 0, \quad \boldsymbol{x} \in \Gamma_p, \tag{7}$$

$$p_\alpha^*(\boldsymbol{x}, t) = 0, \quad \nabla p_2^* \cdot \boldsymbol{n} = 0, \quad \boldsymbol{x} \in \Gamma_v, \tag{8}$$

$$\nabla p_\alpha^* \cdot \boldsymbol{n} = 0, \quad \boldsymbol{x} \in \Gamma_b. \tag{9}$$

$$p_\alpha^*(\boldsymbol{x}, t) = 0, \quad \boldsymbol{x} \in \Omega, \quad t = 0. \tag{10}$$

The next system of equations can be used to describe the blood flow in a liver lobule:

$$c\frac{\partial u_1}{\partial t} - \operatorname{div} \operatorname{grad} u_1 + r\,(u_2 - u_1) = 0, \tag{11}$$

$$\frac{\partial u_2}{\partial t} - d \operatorname{div} \operatorname{grad} u_2 - r\,(u_2 - u_1) = 0. \tag{12}$$

Here $c = c_1/c_2$ is dimensionless capacity, $d = k_2/k_1$ is dimensionless permeability, also here we relabel $u_\alpha = p_\alpha^*$.

The model problem deals with simple geometry Ω as on Fig. 2. We apply next boundary conditions:

$$u_1(\boldsymbol{x}, t) = 1 - \exp(-\delta t), \quad \boldsymbol{x} \in \Gamma_p, \quad u_\alpha(\boldsymbol{x}, t) = 0, \quad \boldsymbol{x} \in \Gamma_v, \tag{13}$$

$$\nabla u_2 \cdot \boldsymbol{n} = 0, \quad \boldsymbol{x} \in \Gamma_p, \quad \nabla u_\alpha \cdot \boldsymbol{n} = 0, \quad \boldsymbol{x} \in \Gamma_b, \quad t \in (0, T], \tag{14}$$

We assume zero-initial condition:

$$u_\alpha(\boldsymbol{x}, 0) = 0, \quad x \in \Omega, \quad \alpha = 1, 2. \tag{15}$$

On the boundary Γ_p the pressure increases from 0 to 1 at a rate of parameter δ. This boundary condition is used to describe pressure rising at portal vein.

3   Computational Algorithm

Space discretization is performed using finite element method [3,9]. We use **Python** programming language and numerical finite element library **FEniCS** [4,8]. To perform time approximation of non-stationary filtration problem we introduce even time mesh with time step τ:

$$\overline{\omega}_\tau = \omega_\tau \cup \{T\} = \{t^n = n\tau, \quad n = 0, 1, ..., N, \quad \tau N = T\}, \tag{16}$$

We use standard Lagrange finite elements of first degree. In the Ω area triangular calculational mesh is constructed, refined in the areas with the higher expected gradients of solution.

Next splitting scheme for time discretization is used:

$$c\frac{y_1^{k+1} - y_1^k}{\tau} - \operatorname{div} \operatorname{grad} y_1^{k+1} + r\,(y_2^k - y_1^k) = 0, \tag{17}$$

$$\frac{y_2^{k+1} - y_2^k}{\tau} - d \operatorname{div} \operatorname{grad} y_2^{k+1} - r\,(y_2^k - y_1^k) = 0. \tag{18}$$

where $y^n = y(t^n)$. This kind of schemes are offered and considered in paper [5]. Using explicit-implicit scheme we decouple the problem into separate independent subproblems.

We consider variational (or weak) formulation of the problem (17, 18):

$$c \int_\Omega \frac{y_1^{k+1} - y_1^k}{\tau} v_1 dx + \int_\Omega K \operatorname{grad} y_1^{k+1} \operatorname{grad} v_1 dx + r \int_\Omega (y_2^k - y_1^k) v_1 dx = 0, \quad (19)$$

$$\int_\Omega \frac{y_2^{k+1} - y_2^k}{\tau} v_2 dx + d \int_\Omega K \operatorname{grad} y_2^{k+1} \operatorname{grad} v_2 dx - r \int_\Omega (y_2^k - y_1^k) v_2 dx = 0. \quad (20)$$

4   Results

Numerical experiments were performed on the series of meshes (283, 746, 3252, 11874 and 47496 elements) and the results are presented below. Parameters of the simulations are described in the Table 1. One of computational meshes is demonstrated on Fig. 3.

Table 1. Parameters for simulation

Description	Symbol	Value
Total time	T	1.0
Length of edge	L	1.0
Radius of central vein	R_v	0.15
Radius of portal vein	R_p	0.1
Dimensionless capacity	c	0.01
Dimensionless permeability	d	0.01
Transient flow coefficient	r	0.01
Rising rate coefficient	δ	10.0

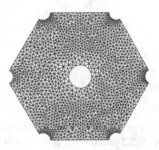

Fig. 3. Computational mesh: 1716 vertices, 3252 elements

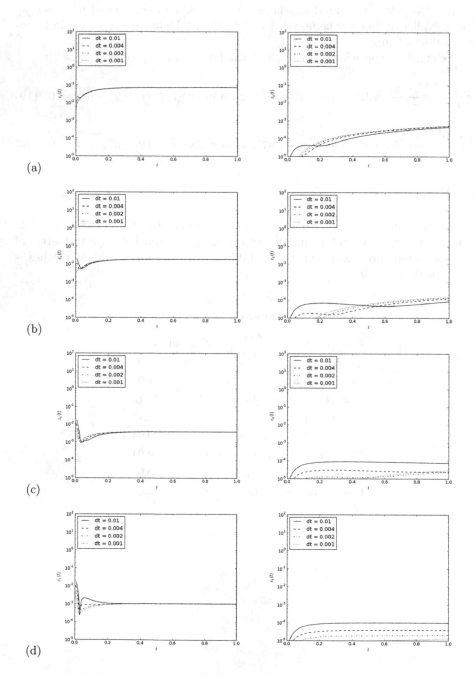

Fig. 4. Error norm u_1 and u_2: a — coarse mesh (283 elements), b — medium mesh (746 elements), c — fine mesh (3252 elements), d — extra fine mesh (11874 elements)

For convergence analysis and error estimation we consider reference solution $\bar{y}^k(\boldsymbol{x})$, which is numerical solution on the most detailed mesh (47496 elements, $\tau = 10^{-4}$). Convergence of numerical solutions is tested by comparison with the reference solution. Next simulation parameters are used $c = 0.01$, $d = 0.01$, $r = 0.01$. The estimated error norm with respect to time is shown in Fig. 4 in norm $L_2(\Omega)$.

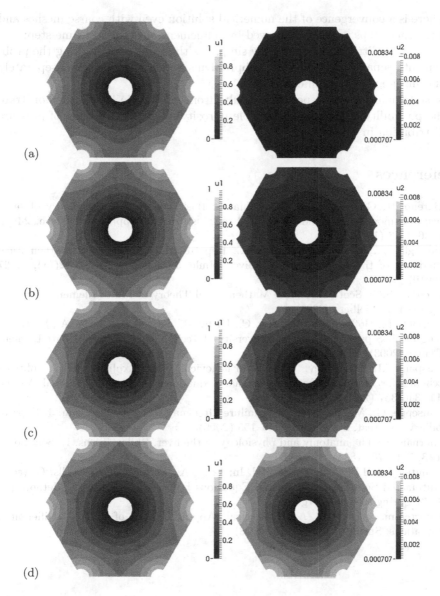

Fig. 5. Pressure distribution u_1 and u_2 at moment: a — $t = 0.25$, b — $t = 0.5$, c — $t = 0.75$, d — $t = 1$

$$\varepsilon = \|\overline{y}^k - y^k\|_{L_2(\Omega)}. \tag{21}$$

The results shown on Fig. 5 are obtained on a mesh with 3252 elements and with time step $\tau = 0.02$. The pressure distribution in the sinusoids and sinusoidal space is shown in Fig. 5.

Based on the numerical results we can make following conclusions:

- There is a convergence of the numerical solution even with coarse meshes and large time steps which is improved by refinement of mesh and time step;
- Proposed splitting scheme greatly simplifies the task by decoupling the problem into separate independent subproblems which can be solved separately and independently from each other;
- Results qualitatively repeat the results from the article [2] and demonstrate the possibility of the using of double porosity model for studying of pressure distribution in the lobule.

References

1. Barenblatt, G., Zheltov, I.P., Kochina, I.: Basic concepts in the theory of seepage of homogeneous liquids in fissured rocks [strata]. J. Appl. Math. Mech. **24**(5), 1286–1303 (1960)
2. Bonfiglio, A., Leungchavaphongse, K., Repetto, R., Siggers, J.H.: Mathematical modeling of the circulation in the liver lobule. J. Biomech. Eng. **132**(11), 11–21 (2010)
3. Brenner, S.C., Scott, L.R.: The Mathematical Theory of Finite Element Methods. Springer, Heidelberg (2008)
4. Dupont, T., Hoffman, J., Johnson, C., Kirby, R., Larson, M., Logg, A., Scott, L.: The FEniCS project. Chalmers University of Technology, Chalmers Finite Element Centre (2003)
5. Gaspar, F.H., Grigoriev, A.V., Vabishchevich, P.N.: Explicit-implicit splitting schemes for some systems of evolutionary equations. Int. J. Numer. Anal. Model. **11**, 346–357 (2014)
6. Giuseppe, G., Maddern, G.: Liver failure after major hepatic resection. J. Hepatobiliary Pancreat. Surg. **16**(2), 145–155 (2009)
7. Kiernan, F.: The anatomy and physiology of the liver. Philos. Trans. R. Soc. Lond. **123**, 711–770 (1833)
8. Langtangen, H.P.: A fenics tutorial. In: Logg, A., Mardal, K.-A., Wells, G. (eds.) Automated Solution of Differential Equations by the Finite Element Method, pp. 1–73. Springer, Heidelberg (2012)
9. Quarteroni, A., Valli, A.: Numerical Approximation of Partial Differential Equations. Springer, Heidelberg (2008)

Numerical Modelling of Ion Transport in 5-HT3 Serotonin Receptor Using Molecular Dynamics

M. Yu. Antonov[1(✉)], A.V. Popinako[2], G.A. Prokopiev[1], and A.O. Vasilyev[1]

[1] M.K. Ammosov North-Eastern Federal University, Yakutsk, Russia
mikhail@s-vfu.ru
[2] Research Center of Biotechnology RAS, Moscow, Russia

Abstract. Cation selective ligand-gated ion channels are pore-forming membrane proteins. They are responsible for generating of transmembrane voltage and action potential, playing an important role in functioning of nervous systems. Mathematical modelling of transmembrane transport in membrane and membrane/protein structures using molecular dynamics (MD) method is often associated with difficulties, because it is nearly impossible to observe spontaneous diffusion in MD experiments. In this work Molecular Dynamics (MD) and Umbrella Sampling (US) methods are used to study ion transport through 5-HT3 Serotonin receptor.

Keywords: Biophysics · Molecular dynamics · Biomembranes · Ion channels · Transmembrane transport · Serotonin receptor

1 Introduction

Cation selective ligand-gated ion channels form a large class of functional membrane proteins and are involved in regulating a variety of cellular processes. Pore-forming membrane proteins responsible, in particular, for generating of transmembrane voltage and action potential, playing an important role in functioning of nervous systems [1,2]. Dysfunctions of ion channels can lead to severe neurological diseases. At the same time studying of such objects is of considerable interest from fundamental and therapeutic points of view. Ion channels are the third largest group of targets in drugs development [3,4]. All this makes studying of the structure and the function of ion channels of considerable interest.

At the same time, these complex objects are very difficult for experimental study. Only a limited number of precise 3D atomic structures are available, while molecular mechanism of functioning is not completely clear [5,6]. Computer simulation methods have shown to be effective instrument in many different areas of science, and using computer modelling techniques is widely used for studying biological objects [7,8].

One of the popular methods here is molecular dynamics (MD) method, widely used for mathematical modelling of biological systems. Unfortunately modelling of dynamics and transmembrane transport in big systems, such as membrane and

© Springer International Publishing AG 2017
I. Dimov et al. (Eds.): NAA 2016, LNCS 10187, pp. 195–202, 2017.
DOI: 10.1007/978-3-319-57099-0_19

membrane/protein structures, is often associated with computational difficulties, therefore, of interest are those approaches and methods that we can use to get adequate results in reasonable time.

In this work molecular dynamics simulations and umbrella sampling methods are used to study of mouse serotonin 5HT-3 receptor. In 2014 high resolution (0.35 nm) structure of the 5-HT3 receptor in complex with stabilizing nanobodies was determined. Homology modelling revealed, that thansmembrane domain of the structure is highly possibly in the open state, which makes studying of the structure using modern computational methods of high interest [6].

In this study the model of the 5-HT3 receptor was built with explicit membrane and solvent. The simulations show good agreement with experimentally known macroscopic parameters, such as area per lipid, thickness and compressibility of the membrane, acyl chain order parameters and crystallographic protein structure.

2 Materials and Methods

In essence, the molecular dynamics method is based on representing the system under study as a set of material points whose interaction is described by the laws of classical mechanics. Each point is an atom or a group of atoms. The total potential energy of the system is described as a sum of partial potentials:

$$U\left(r\right) = \sum_{i,j} U_{ij}^{v}\left(b_{i,j}\right) + \sum_{i,j,k} U_{ijk}^{\theta}\left(\theta_{i,j,k}\right) + \sum_{i,j,k,l} U_{ijkl}^{\phi}\left(\phi_{i,j,k,l}\right)$$
$$+ \sum_{i\neq j}\left[U_{ij}^{c}\left(r_{i,j}\right) + U_{ij}^{vdw}\left(r_{i,j}\right)\right] \tag{1}$$

where the first three terms represent interaction between chemically bonded atoms (energies of valence bonds, valence angles and torsion angles), and the last two terms represent the electrostatic and Van der Waals forces between the pairs of atoms which are not bonded. Partial potential functions are represented with a differentiable function of atomic coordinates. The motion of atoms in the potential field is described by the system

$$\left\{ m_i \vec{\ddot{r}}_i = -\left(\frac{\partial U}{\partial x_i}, \frac{\partial U}{\partial y_i}, \frac{\partial U}{\partial z_i}\right) \right. \tag{2}$$

of second order ordinary differential equations, which is to be solved numerically. The applicable numerical methods differ in accuracy and computational complexity. The Verlet method and its variants are widely used in molecular dynamics, as a compromise between the accuracy of the procedure and the calculation speed. The coordinates of atoms in the new temporal layer are calculated as

$$r_i\left(t + \triangle t\right) = 2r_i\left(t\right) - r_i\left(t - \triangle t\right) + \frac{F_i\left(t\right)}{m_i}\triangle t^2 \tag{3}$$

$$v_i\left(t\right) = \frac{1}{2\triangle t}\left(r_i\left(t + \triangle t\right) - r_i\left(t - \triangle t\right)\right) \tag{4}$$

Umbrella Sampling Method. Umbrella sampling method is used to improve sampling of a system where ergodicity is hindered by the form of the system's energy landscape. In this approach, we apply to the system an additional potential capable of holding the ions in a potentially unfavourable region inside of the ion channel [9,10]. The modification of the original potential energy function amounts to the addition of an external potential (Fig. 1a):

$$\dot{U}(r) = U(r) - W(r), \ W(r) = k_w(\xi(r) - \xi_0)^2 \tag{5}$$

Since the form of W(r) is known, we evaluate the free energy G of the unperturbed system as

$$G(\xi) = -k_b T \cdot ln(\dot{p}(\xi)) - W(\xi) + const \tag{6}$$

where $\dot{p}(\xi)$ is the density of the probability of finding the system in a state with this value of ξ in the perturbed system, T is temperature and k_b is Boltzmann constant.

Using this approach, we can reconstruct the profile of free energy only in a small neighbourhood of ξ_0 hence, we choose a set of initial points along the reaction coordinate so that the distribution functions of the system near each initial point overlap with the distribution functions associated to the neighbouring points. There are several methods for reconstructing the original profile of free energy from the set of distributions found near each initial point. In our research we use the weighted histogram analysis method (WHAM, [11]).

Molecular Dynamics Protocol. We used Gromacs 4.6.7 [12] software for MD simulations with Gromos 53a6 [13] forcefield. As a model for eukaryotic cell membrane we used bilayers of heavy-atom model of 1-palmitoyl-2-oleoyl-sn-glycero-3-phosphatidylcholine (POPC), lipid structure were obtained from lipid structures database of the University of Calgary [14]. Stochastic dynamics with $\tau_t = 1$ ps was used for temperature control. Semi-isotropic Berendsen barostat with $\tau_p = 0.5$ ps was used for pressure coupling. Cut-off radius of 1.8 nm was used for calculating of non-bonded interactions.

POPC bilayer consisted of 498 lipid molecules. Initial surface area per molecule of the lipid was 68–72 $Å^2$. The resulting structures were hydrated by water molecules (Model TIP4P [15]) of not less than 150 molecules of water per one molecule of the lipid.

5-HT3 receptor structure was obtained from the RCSB Protein Data Bank [16] (PDB code 4PIR [6]). 3-Dimensional 5-HT3 serotonin receptor structure, obtained by X-ray analysis method is shown on Fig. 2. Because the structure is not fully determined, the ends of the missing chains were capped with ACE/NH2 residues.

For immersing the ion channel structure into the membrane we used technique described in [17]. ACE/NII2 residues spatial position was fixed with parabolic potential with the stiffness constant of $1000 \cdot kJ/(mol \cdot nm^2)$. The model system with the complete structure of the receptor, solvent and POPC bilayer contained 274,814 atoms, and is shown in Fig. 1b.

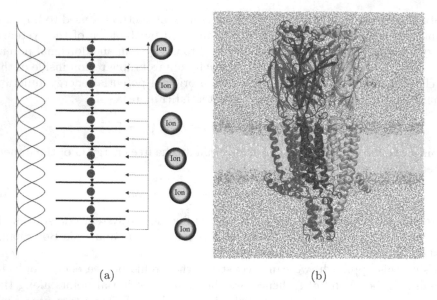

(a)   (b)

Fig. 1. Harmonic potential along reaction coordinate for restraining the ion in energetically unfavourable region(1a), The system under study (1b)

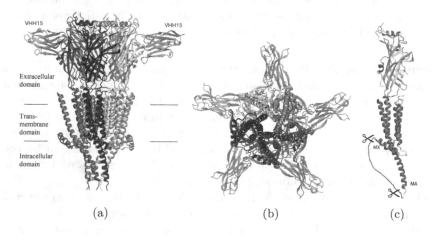

(a)   (b)   (c)

Fig. 2. X-Ray structure of 5-HT3 receptor [6]. Parallel to the membrane plane (2a). Perpendicular to the membrane plane (2b). Structure of the subunit. Chain structure between MX-MA is not determined (2c)

Molecular dynamics simulations were performed for 13 ns at 310 K with lateral pressure of −50 bar. Pressure in a perpendicular to the membrane was 1 bar in all calculations. Structure of the ion channel was fixed, which allowed the bilayer and solvent enter into interaction with the channel structure. In the last 3 ns of trajectory we applied temperature coupling of 490 K to the water molecules, which allowed water molecules to penetrate into the pore of the ion channel. Then we performed 30 ns MD simulation without restrictions on the ion channel structure. We used last 15 ns of trajectory for processing.

For umbrella sampling procedure we chose 129 starting configurations for the Na+ ion inside of the pore of the channel with the spatial step of 0.1 nm along the pore of the channel (z-axis, reaction coordinate). We used the harmonic potential with the force constant of $1000 \cdot kJ/(mol \cdot nm^2)$. The trajectories to gather statistics were 3 ns long. The total simulation time was 430 ns.

3   Results

After simulation fully hydrated model of the 5-HT3 serotonin receptor ion channel in the POPC membrane was obtained. The thickness of the membrane was 3.5 ± 0.2 nm, and the average area per lipid was 0.63 nm^2, which generally corresponds to the experimental estimates of 0.62–0.68 nm^2 for POPC bilayers [18,19]. The thickness of the membrane was determined as the average distance between atoms of phosphorus in monolayers.

Thickness map of the bilayer Fig. 3a shows no significant signs of clustering or other undesirable effects, which may occur during simulation. The average thickness of the membrane is within the acceptable values for membrane systems in the liquid crystalline phase [20].

The cell volume fluctuations (Fig. 3b) were 3174 ± 1.92 nm^3, which correspond to compressibility constant χ_T of $2.7 \cdot 10^{-10} Pa^{-1}$, which is in agreement with known experimental χ_T estimates for membranes from $1 \cdot 10^{-10}$ to $6 \cdot 10^{-10} Pa^{-1}$ [21]. Thus, the reasonable values of the compressibility factor χ_T, and Gaussian distribution pattern for volume fluctuations indicates local equilibrium conditions.

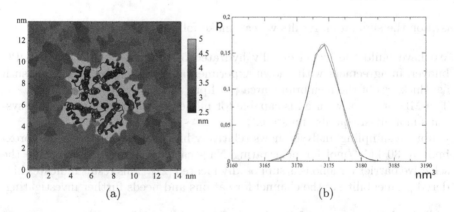

(a)   (b)

Fig. 3. Thickness distribution over the bilayer (3a). Cell volume probability density function (3b).

In vivo, 5-HT3 receptors in open state are permeable to Na+ and K+ ions, and have a negligible permeability to negatively charged ions. The calculated mean force potential (Fig. 4a) indicates potential barrier of approximately

30 kCal/mol in the intracellular domain of the ion channel, where supposedly selective filter is located. Relatively high potential barrier of this magnitude can prevent Na+ cations from extracellular space from getting inside of the channel. The histograms in (Fig. 4b) characterize the statistical covering of the conformation space along the reaction coordinate in the simulations. The histogram reveals few regions with relatively low population, which suggests that using spatial step of 0.1 nm with the force constant K_{harm} of the harmonic potential equal to $1000 \cdot kJ/(mol \cdot nm^2)$ is not enough for good statistical coverage.

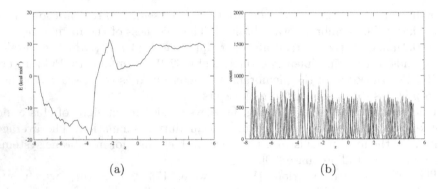

(a)   (b)

Fig. 4. Mean force potential over the channel pore (4a). Population density histograms (4b).

4 Conclusion

Based on the simulation results we can make following conclusions:

- We have build the model of fully hydrated 5-HT3 ion channel with POPC bilayer, in agreement with known experimental macroscopic parameters such as: thickness of the membrane, area per lipid and bilayer compressibility.
- The MD protocol we used is capable for modelling of protein-membrane systems in a quasi-equilibrium state.
- Umbrella sampling analysis shows relatively high potential barrier of approximately 30 kCal/mol for penetrating Na+ cations in the region, where the selective barrier for anions is supposedly located [22]. This result indicates hindered permeability of the channel for cations and needs further investigating.

All simulations were performed using "Arian Kuzmin" supercomputer center of M.K. Ammosov North-Eastern Federal University, Yakutsk, Russia. The work was supported by the Russian Foundation for Basic Research (Grant 16-34-60252) and the Ministry of Education and Science of Russian Federation.

References

1. Jegla, T.J., Zmasek, C.M., Batalov, S., Nayak, S.K.: Evolution of the human ion channel set. Comb. Chem. High Throughput Screen 12(1), 2–23 (2009)
2. Kullmann, D.M., Hanna, M.G.: Neurological disorders caused by inherited ion-channel mutations. Lancet Neurol. 1(3), 157–166 (2002)
3. Overington, J.P., Al-Lazikani, B., Hopkins, A.L.: How many drug targets are there? Nat. Rev. Drug Discov. 5(12), 993–996 (2006)
4. Hopkins, A.L., Groom, C.R.: The druggable genome. Nat. Rev. Drug Discov. 1(9), 727–730 (2002)
5. Wickenden, A., Priest, B., Erdemli, G.: Ion channel drug discovery: challenges and future directions. Future Med. Chem. 4(5), 661–679 (2012)
6. Hassaine, G., Deluz, C., Grasso, L., Wyss, R., Tol, M.B., Hovius, R., Graff, A., Stahlberg, H., Tomizaki, T., Desmyter, A., Moreau, C., Li, X.D., Poitevin, F., Vogel, H., Nury, H.: X-ray structure of the mouse serotonin 5-HT3 receptor. Nature 512, 276–281 (2014)
7. Popinako, A.V., Levtsova, O.V., Antonov, MYu., Nikolaev, I.N., Shaitan, K.V.: The structural and dynamic model of the serotonin 5-HT3 receptor. Comparative analysis of the structure of the channel domain of pentameric ligand-gated ion channels. Biofizika 56(6), 1111–1116 (2011)
8. Pischalnikova, A.V., Sokolova, O.S.: The domain and conformational organization in potassium voltage-gated ion channels. J. Neuroimmune Pharmacol. 4(1), 71–82 (2009)
9. Kastner, J.: Umbrella sampling. Wiley Interdiscip. Rev. Comput. Mol. Sci. 1(6), 932–942 (2011)
10. Torrie, G.M., Valleau, J.P.: Nonphysical sampling distributions in Monte Carlo free-energy estimation: umbrella sampling. J. Comput. Phys. 23(2), 187–199 (1977)
11. Kumar, S., Rosenberg, J.M., Bouzida, D., Swendsen, R.H., Kollman, P.A.: The weighted histogram analysis method for free-energy calculations on biomolecules. I. The method. J. Comput. Chem. 13(8), 1011–1021 (1992)
12. Hess, B., Kutzner, C., van der Spoel, D., Lindahl, E.: GROMACS 4: algorithms for highly efficient, load-balanced, and scalable molecular simulation. J. Chem. Theory Comput. 4(3), 435–447 (2008)
13. Oostenbrink, C., Villa, A., Mark, A.E., van Gunsteren, W.F.: A biomolecular force field based on the free enthalpy of hydration and solvation: the GROMOS force-field parameter sets 53A5 and 53A6. J. Comput. Chem. 25(13), 1656–1676 (2004)
14. Tieleman, D.P., Forrest, L.R., Sansom, M.S., Berendsen, H.J.: Lipid properties and the orientation of aromatic residues in OmpF, inuenza M2, and alamethicin systems: molecular dynamics simulations. Biochemistry 37(50), 17554–17561 (1998)
15. Jorgensen, W.L., Chandrasekhar, J., Madura, J.D., Impey, R.W., Klein, M.L.: Comparison of simple potential functions for simulating liquid water. J. Chem. Phys. 79(2), 926 (1983)
16. Berman, H.M.: The protein data bank: a historical perspective. Acta Crystallogr. A 64(Pt 1), 88–95 (2008)
17. Schmidt, T.H., Kandt, C.: LAMBADA and InflateGRO2: efficient membrane alignment and insertion of membrane proteins for molecular dynamics simulations. J. Chem. Inf. Model. 52(10), 2657–2669 (2012)
18. Lague, P., Zuckermann, M.J., Roux, B.: Lipid-mediated interactions between intrinsic membrane proteins: dependence on protein size and lipid composition. Biophys. J. 81(1), 276–284 (2001)

19. Nagle, J.F.: Area/lipid of bilayers from NMR. Biophys. J. **64**(5), 1476–1481 (1993)
20. Shaitan, K.V., Antonov, M.Y., Tourleigh, Y.V., Levtsova, O.V., Tereshkina, K.B., Nikolaev, I.N., Kirpichnikov, M.P.: Comparative study of molecular dynamics, diffusion, and permeability for ligands in biomembranes of different lipid composition. Biochem. Suppl. Ser. A. Membr. Cell Biol. **2**, 73–81 (2008)
21. Braganza, L.F., Worcester, D.L.: Structural changes in lipid bilayers and biological membranes caused by hydrostatic pressure. Biochemistry **25**(23), 7484–7488 (1986)
22. Popinako, A.V., Levtsova, O.V., Antonov, M.Y., Nikolaev, I.N., Shaitan, K.V.: The structural and dynamic model of the serotonin 5-HT3 receptor. Comparative analysis of the structure of the channel domain of pentameric ligand-gated ion channels. Biophysics **56**(6), 1078–1082 (2011)

Induced Dimension Reduction Method to Solve the Quadratic Eigenvalue Problem

R. Astudillo$^{(\boxtimes)}$ and M.B. van Gijzen

Delft Institute of Applied Mathematics, Delft University of Technology,
Mekelweg 4, 2628 CD Delft, The Netherlands
{R.A.Astudillo,M.B.vanGijzen}@tudelft.nl

Abstract. In this work we are interested in the numerical solution of
the Quadratic Eigenvalue Problem (QEP)

$$(\lambda^2 M + \lambda D + K)\mathbf{x} = \mathbf{0},$$

where M, D, and K are given matrices of order N. Particularly, we study
the applicability of the IDR(s) for eigenvalues to solve QEP. We present
an IDR(s) algorithm that exploits the special block structure of the linearized QEP to compute its eigenpairs. To this end we incorporate ideas
from Second Order Arnoldi method proposed in [3].

Keywords: Quadratic Eigenvalue Problem · Induced Dimension
Reduction

1 Introduction

In this work we are interested in solving the Quadratic Eigenvalue Problem
(QEP), i.e., find a subset of pairs $(\lambda, \mathbf{x})$, where $\lambda \in \mathbb{C}$ and $\mathbf{x} \in \mathbb{C}^N$ such that,

$$(\lambda^2 M + \lambda D + K)\mathbf{x} = \mathbf{0}, \tag{1}$$

where M, D, and K are (sparse) matrices of order N often referred as mass,
damping, and stiffness matrices, respectively. The QEP appears in different areas
like vibration analysis, dynamical systems, or stability of flows in fluid mechanics
(see [8] and their references within).

One of the most common options to solve the quadratic eigenvalue problem
(1) is to linearized it to an standard eigenvalue problem. First, problem (1) can
be written as a generalized eigenvalue problem, i.e.

$$C\mathbf{y} = \lambda G\mathbf{y}, \tag{2}$$

where

$$C = \begin{bmatrix} -D & -K \\ I & 0 \end{bmatrix}, \quad \text{and} \quad G = \begin{bmatrix} M & 0 \\ 0 & I \end{bmatrix}.$$

© Springer International Publishing AG 2017
I. Dimov et al. (Eds.): NAA 2016, LNCS 10187, pp. 203–211, 2017.
DOI: 10.1007/978-3-319-57099-0_20

Second, if the matrix M is not singular, (2) can be rewritten as standard eigenvalue problem,

$$A\mathbf{y} = \lambda\mathbf{y}, \tag{3}$$

with

$$A = \begin{bmatrix} -M^{-1}D & -M^{-1}K \\ I & 0 \end{bmatrix}. \tag{4}$$

It is easy to check that the eigenvalues of A and the eigenvalues of (1) are related by:

$$\mathbf{y}_i = \begin{bmatrix} \lambda_i \mathbf{x}_i \\ \mathbf{x}_i \end{bmatrix}. \tag{5}$$

Then, one can apply any eigensolver software for the standard eigenvalue (3) and obtain approximate solutions of the quadratic eigenvalue problem (1). This approach has two main disadvantages. First, it solves a standard eigenvalue problem of double the dimension of the original quadratic eigenvalue problem. Second, some properties of the matrices M, D, and K are lost during the linealization; for example, matrices M, D and K can be symmetric positive definite (SPD) matrices but the matrix A does not keep the SPD property.

To overcome the disadvantages of using the linealization (3), the authors in [3] propose a method called Second Order Arnoldi (SOAR), which is a modification of the Arnoldi method [1]. By exploiting the block structure of the matrix A, the SOAR method uses approximately half of the memory of the classical Arnoldi method applied to the problem (3). Also and more importantly, this method preserves essential structures and properties of the matrices involved.

The Arnoldi method has as main drawback its demanding computational requirements. In this contribution, we study the Induced Dimension Reduction Method for eigenvalue problem [2] to solve the Quadratic Eigenvalue Problem as an alternative to the Arnoldi method.

This document is organized as follow. Section 2 presents an introduction to the Induced Dimension Reduction method for solving linear system of equations. In Sect. 3, we present how the IDR(s) method has been adapted to solve the standard eigenvalue problem, and in Sect. 4, we present an IDR(s) to solve the Quadratic Eigenvalue problem, using ideas from SOAR. In Sect. 5, we conduct numerical experiments to illustrate the numerical behavior of the IDR(s) for QEP. Section 6 presents the conclusions and remarks of this work.

2   Induced Dimension Reduction Method - IDR(s)

IDR(s) was presented originally in [7], as a short recurrences iterative Krylov method to solve large and sparse systems of linear equations,

$$A\mathbf{x} = \mathbf{b}. \tag{6}$$

The IDR(s) method is based on the following theorem,

Theorem 1 (IDR(s) theorem). *Let A be any matrix in $\mathbb{C}^{N \times N}$, and let $P = [\mathbf{p}_1, \mathbf{p}_2, \ldots, \mathbf{p}_s]$ be an $N \times s$ matrix with s linear independent columns. Let $\{\mu_j\}$ be a sequence in $\mathbb{C}$. With $\mathcal{G}_0 \equiv \mathbb{C}^N$, define*

$$\mathcal{G}_{j+1} \equiv (A - \mu_{j+1}I)(\mathcal{G}_j \cap P^\perp) \qquad j = 1, 2 \ldots, \tag{7}$$

where $P^\perp$ represents the orthogonal complement of P. If $P^\perp$ does not contain an eigenvector of A, then, for all $j = 0, 1, 2 \ldots$, the following hold

1. *$\mathcal{G}_{j+1} \subset \mathcal{G}_j$, and*
2. *dimension$(\mathcal{G}_{j+1}) <$ dimension$(\mathcal{G}_j)$ unless $\mathcal{G}_j = \{\mathbf{0}\}$.*

Proof. See [7].

The subspaces $\mathcal{G}_j$ for $j = 0, 1, 2, \ldots$ are shrinking and nested subspaces. IDR(s) uses recurrences of size s to create an approximated solution $\mathbf{x}_{k+1}$ forcing its corresponding residual vector $\mathbf{r}_{k+1} = \mathbf{b} - A\mathbf{x}_{k+1}$ to be in the subspace $\mathcal{G}_{j+1}$. Using the fact that $\mathcal{G}_{j^*} = \{\mathbf{0}\}$ for some j^*, the residual will become zero and IDR(s) will obtain the solution of (6).

3   IDR(s) to Solve the Eigenvalue Problem

Several methods to compute a subset of eigenpairs $(\lambda_i, \mathbf{x}_i)$ of a large and sparse matrix $A \in \mathbb{C}^{N \times N}$ rely on the construction of a standard Hessenberg relation of the form,

$$AU_m = U_m H_m + \mathbf{f}\,\mathbf{e}_m^T, \tag{8}$$

where H_m is an upper Hessenberg matrix of order m (much smaller than N), $U_m = [\mathbf{u}_1, \mathbf{u}_2, \ldots, \mathbf{u}_m] \in \mathbb{C}^{N \times m}$ is a basis for the Krylov subspace $\mathcal{K}_m(A, \mathbf{u}_1)$ with $\mathbf{u}_1 \neq \mathbf{0}$, $\mathbf{f} \in \mathbb{C}^N$, and $\mathbf{e}_m$ is the m-th canonical vector. It can be proved that the eigenpairs of A can be approximated by $(\hat{\lambda}_j, U_m\hat{\mathbf{y}}_j)$, where $(\hat{\lambda}_j, \hat{\mathbf{y}}_j)$ are the eigenpairs of the smaller matrix H_m.

Two examples of well-known methods to construct a Hessenberg relation and approximate eigenpairs, are the Lanczos [5] and Arnoldi [1] method. While the Lanczos method is suitable when the coefficient matrix is symmetric, in the case of unsymmetric matrices this method might suffer from numerical instability. For this reason, the Arnoldi method is the most common option to build a Hessenberg relation for unsymmetric matrices. The Arnoldi method explicitly build an orthogonal basis for the Krylov subspace and because of this the work and storage per iteration grow with the number of iterations.

In [4] and later in [2] the IDR(s) method was adapted to build a Hessenberg relation and approximate eigenpairs of unsymmetric large matrices. The Hessenberg relations based on the IDR(s) method keep the computational work (almost) constant per iteration. Next, we review of how to obtain an IDR-Hessenberg relation.

IDR creates a vector in $\mathcal{G}_{j+1}$ with the assumption that $s + 1$ vectors are already in $\mathcal{G}_j$ namely $\{\mathbf{w}_{k-i}\}_{i=0}^s$. A new vector $\mathbf{w}_{k+1}$ in the subspace $\mathcal{G}_j$ can be written as,

$$\mathbf{w}_{k+1} = (A - \mu_{j+1} I) \left(\mathbf{w}_k - \sum_{i=1}^s \beta_i \mathbf{w}_{k-i} \right),\tag{9}$$

where the coefficient β_i are computed via the solution of the following $s \times s$ system of linear equation,

$$P^T [\mathbf{w}_{k-1}, \mathbf{w}_{k-2}, \ldots, \mathbf{w}_{k-s}]\mathbf{c} = P^T \mathbf{w}_k \quad \text{where} \quad \mathbf{c} = [\beta_1, \ldots, \beta_s]^T.$$

It is possible to rewrite (9) as,

$$A \left(\sum_{i=0}^s \beta_i \mathbf{w}_{k-i} \right) = \mathbf{w}_{k+1} - \mu_{j+1} \sum_{i=0}^s \beta_i \mathbf{w}_{k-i},\tag{10}$$

with $\beta_0 = -1$. From the equation above, the authors in [4] constructed a generalized Hessenberg relation,

$$A W_m \hat{U}_m = W_m \hat{H}_m + \mathbf{w} \mathbf{e}_m^T,\tag{11}$$

where $\hat{U}_m$ is an upper triangular matrix and $\hat{H}_m$ is an upper banded Hessenberg matrix. The banded matrix pencil $(\hat{H}_m, \hat{U}_m)$ is called the Sonneveld pencil. The eigenvalues of this pencil are divided into two sets: $\{\mu_k\}_{i=1}^t$ where t is the number of subspaces $\mathcal{G}_j$ created, and the approximations to the eigenvalues of A or Ritz values $\{\theta_k\}_{i=t}^m$. In [2], the authors construct a standard Hessenberg relation using IDR(s),

$$A W_m = W_m H_m + \mathbf{w} \mathbf{e}_m^T = W_{m+1} \bar{H}_m,\tag{12}$$

where the i−th column of the upper Hessenberg matrix H_m is defined as,

$$\mathbf{h}_i = \left(\begin{bmatrix} \begin{bmatrix} 0 \\ \vdots \\ 0 \end{bmatrix} \\ -\mu_{j+1} \begin{bmatrix} c_1 \\ \vdots \\ c_s \end{bmatrix} \\ \mu_{j+1} \\ 1 \end{bmatrix} + \sum_{\ell=1}^s c_\ell \mathbf{h}_{i-\ell} \right).\tag{13}$$

and this matrix has the same eigenvalues as the pencil $(\hat{H}_m, \hat{U}_m)$ obtained from (11) in [4]. At this point, one can apply directly IDR(s) for eigenvalues to the problem (3). Especially in applications where only the eigenvalues are needed, IDR(s) can be used as a short-recurrences method to create H_m (see Eq. (9) and (13)) and approximate the eigenvalues of (1).

4   IDR(s) to Solve the Quadratic Eigenvalue Problem

Bai and Su in [3] proposed a special version of the Arnoldi algorithm for the QEP. This so called second order Arnoldi algorithm (SOAR) exploits the structure of (3)–(5) to reduce the memory requirements for the Arnoldi method by a factor of two. In this section we examine how the ideas underlying in SOAR can be incorporated in IDR to obtain a second order IDR (SOIDR) algorithm.

4.1   Second Order IDR(s)

One can exploit the block structure of the $2N \times 2N$ matrix A in Eq. (4) for the creation a standard Hessenberg relation (12). Let us consider Eq. (12), with the matrix W_m rewritten in two block matrices of size $N \times m$ as,

$$W_m = \begin{bmatrix} W_m^{(U)} \\ W_m^{(L)} \end{bmatrix}, \tag{14}$$

Then Eq. (12) can be written as,

$$-M^{-1}DW_m^{(U)} - M^{-1}KW_m^{(L)} = W_m^{(U)}H_m + \mathbf{w}_{m+1}^{(U)}\mathbf{e}_m^T \tag{15}$$

$$W_m^{(U)} = W_m^{(L)}H_m + \mathbf{w}_{m+1}^{(L)}\mathbf{e}_m^T. \tag{16}$$

From Eq. (16), and assuming that the first column vector ($\mathbf{w}_1$) of the matrix W_m has the following pattern

$$\mathbf{w}_1 = \begin{bmatrix} \mathbf{u} \\ \mathbf{0} \end{bmatrix}, \qquad \text{with } \mathbf{u} \neq \mathbf{0} \in \mathbb{C}^N,$$

we have (using the Matlab subindex notation),

$$W_m^{(U)} = W_{m+1}^{(L)}\bar{H}_m = W_{m+1}^{(L)}(:, 2 : m + 1)\bar{H}_m(2 : m + 1, 1 : m). \tag{17}$$

The Eq. (15) can be rewritten as,

$$-M^{-1}DW_m^{(U)} - M^{-1}KW_m^{(U)}T_m = W_m^{(U)}H_m + \mathbf{w}_{m+1}^{(U)}\mathbf{e}_m^T, \tag{18}$$

where

$$T_m = \begin{bmatrix} \mathbf{0} & \bar{H}_m(2 : m, 1 : m - 1)^{-1} \\ 0 & \mathbf{0} \end{bmatrix}. \tag{19}$$

Equations (18) and (19) suggest a formula to compute the column vectors of the matrix $W_m^{(L)}$ as a linear combination of the column vector of $W_m^{(U)}$. Algorithm 1 shows a possible implementation of these ideas. This method needs only half of the memory of the classical Arnoldi method applied to the matrix A. The memory requirements of SOIDR are equivalent to those from SOAR [3].

Algorithm 1. SOIDR(s) for solving the QEP.

1: Given $s \in \mathbb{N}$, $P \in \mathbb{R}^{N \times s}$, M, D, and K.
2: Run SOAR to obtain $W \in \mathbb{C}^{n \times s + 1}$ and $H \in \mathbb{C}^{s+1 \times s}$, s.t.

$$A \begin{bmatrix} W^{(U)} \\ W^{(L)} \end{bmatrix}_s = \begin{bmatrix} W^{(U)} \\ W^{(L)} \end{bmatrix}_{s+1} \bar{H}_s.$$

3: $T_s = \bar{H}_s(2:s, 1:s-1)^{-1}$
4: **for** $i = s + 1, \ldots, m$ **do**
5:     **if** i is multiple of $s + 1$ **then**
6:         Choose the parameter μ_j for the subspace $\mathcal{G}_j$.
7:     **end if**
8:     Solve $(P^T [\mathbf{w}_{i-s}^{(U)}, \mathbf{w}_{i-s+1}^{(U)}, \ldots, \mathbf{w}_{i-1}^{(U)}]) \mathbf{c} = P^T \mathbf{w}_i^{(U)}$.
9:     $\mathbf{v} = \mathbf{w}_i^{(U)} - \sum_{\ell=1}^{s} \beta_\ell \mathbf{w}_{i-\ell}^{(U)}$.
10:     Compute the latest column of $W_i^{(L)}$ as $\mathbf{v}^{(L)}$ using (17) and (19).
11:     $\mathbf{w}_{i+1}^{(U)} = -M^{-1}(D\mathbf{v} + K\mathbf{v}^{(L)}) - \mu_j \mathbf{v}$.
12:     Create the i-th column of H according to (13).
13:     Update T_i using (19).
14:     $W_{i+1}^{(U)} = [\mathbf{w}_1^{(U)}, \mathbf{w}_2^{(U)}, \ldots, \mathbf{w}_i^{(U)}, \mathbf{w}_{i+1}^{(U)}]$.
15: **end for**
16: Compute the eigenpairs $\{(\lambda_i, \mathbf{z}_i)\}_{i=1}^{m}$ s.t. $H_m \mathbf{z}_i = \lambda_i \mathbf{z}_i$.
17: **return** $\{(\lambda_i, W_m^{(U)} \mathbf{z}_i)\}_{i=1}^{m}$.

It is important to mention that the SOIDR(s) algorithm and IDR(s) for eigenvalues generate the same Ritz values for the same input parameters. The low-memory IDR(s) for eigenvalues algorithm is a short-recurrences method that uses $2s + 2$ vectors of dimension $2N$ and it is only possible obtain approximations to the eigenvalues. While SOIDR(s) can obtain approximation of the eigenpairs, this is a more expensive algorithm. SOIDR(s) uses long-recurrences in the step (10) of Algorithm 1 and the complete upper part of W_m needs to be stored.

5   Numerical Experiments

Experiment 1: The purpose of this example is to compare the convergence of Arnoldi, SOAR, and SOIDR for the exterior eigenvalues of the QEP. The matrices M, D, and K are random sparse of order 400. Figure 1 shows a comparison between the errors of the Ritz values generated by Arnoldi, SOAR, and SOIDR(4) or 35 Ritz values after 40 matrix vector products (Fig. 2 shows the Ritz values computed).

Experiment 2: In our second experiment we measure the execution times for SOAR, SOIDR(4), and IDR(4) for eigenvalues [2]. We only compute a set of 10 the eigenvalues of the (1) and the matrices M, D and K are random matrices of size 6000×6000. Table 1 shows the CPU times for each method.

Experiment 3: This example was presented in [6], and models the propagation of sound waves in a room with five solid walls and one wall of a sound-absorbing material,

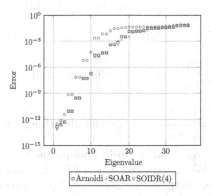

Fig. 1. *Experiment 1:* Convergence for 35 Ritz values after 40 matrix vector products. One can see a similar convergence behavior, however some Ritz values of the SOIDR(4) have a larger error.

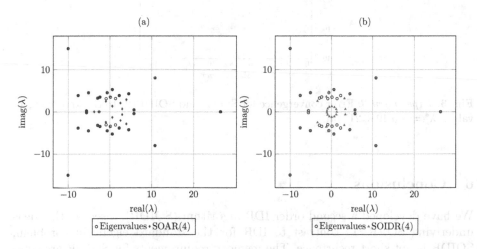

Fig. 2. *Experiment 1:* (a) Exterior eigenvalues and their approximation by SOAR. (b) Exterior eigenvalues and their approximation by SOIDR(4).

Table 1. *Experiment 2:* Execution time comparison for SOAR, SOIDR(4), and IDR(4) after 40 matrix vector products.

Method	Time [s]
SOAR	3.67
SOIDR(4)	3.78
IDR(4)	2.41

$$\frac{\lambda}{c^2}p - \triangle p = 0 \quad \text{in } [-2.0,\ 2.0]^3 \tag{20}$$

where c is the speed of sound (340 meter/second) and the boundary conditions are,

$$\frac{\partial p}{\partial n} = 0 \qquad \text{for the solid walls,} \tag{21}$$

and,

$$\frac{\partial p}{\partial n} = -\frac{\lambda}{cZ_n}p \qquad \text{for the absorbing wall.} \tag{22}$$

Selecting an impedance $Z_n = 0.2 - 1.5i0$, this problem has an analytical eigenvalue $-5.19 + 217.5i$. We discretized Eqs. (20)–(22) using finite element and obtain matrices of order 1681. Figure 3 shows the evolution of the error of the Ritz values generated by SOAR and SOIDR(2).

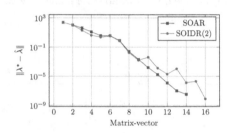

Fig. 3. *Experiment 3:* Error convergence for SOAR and SOIDR(2) to the known eigenvalues $\lambda^* = -5.19 + 217.5i$.

6   Conclusions

We have developed a second order IDR algorithm (SOIDR) based on the ideas underlying SOAR. In contrast to IDR for the standard eigenvalue problem, SOIDR is not short recurrence. The memory requirements for SOIDR are comparable to that of SOAR. Compared to SOIDR, SOAR is for our test cases the preferred method, since it exhibits a faster convergence. However, more numerical test are needed.

References

1. Arnoldi, W.E.: The principle of minimized iterations in the solution of the matrix eigenvalue problem. Q. Appl. Math. **9**, 17–29 (1951)
2. Astudillo, R., van Gijzen, M.B.: A restarted induced dimension reduction method to approximate eigenpairs of large unsymmetric matrices. J. Comput. Appl. Math. **296**, 24–35 (2016)
3. Bai, Z., Su, Y.: SOAR: a second-order arnoldi method for the solution of the quadratic eigenvalue problem. SIAM J. Matrix Anal. Appl. **26**(3), 640–659 (2005)

4. Gutknecht, M.H., Zemke, J.-P.M.: Eigenvalue computations based on IDR. SIAM J. Matrix Anal. Appl. **34**(2), 283–311 (2013)
5. Lanczos, C.: An iteration method for the solution of the eigenvalue problem of linear differential and integral operators. J. Res. Natl. Bur. Stand. **45**(4), 255–281 (1950)
6. Sleijpen, G.L.G., van der Vorst, H.A., van Gijzen, M.B.: Quadratic eigenproblems are no problem. SIAM News **29**(7), 8–19 (1996)
7. Sonneveld, P., van Gijzen, M.B.: IDR(s): a family of simple and fast algorithms for solving large nonsymmetric systems of linear equations. SIAM J. Sci. Comput. **31**(2), 1035–1062 (2008)
8. Tisseur, F., Meerbergen, K.: The quadratic eigenvalue problem. SIAM Rev. **43**(2), 235–286 (2001)

Algorithms for Numerical Simulation of Non-stationary Neutron Diffusion Problems

A.V. Avvakumov[1], V.F. Strizhov[2], P.N. Vabishchevich[2], and A.O. Vasilev[3(✉)]

[1] National Research Center Kurchatov Institute, Moscow, Russia
[2] Nuclear Safety Institute of RAS, Moscow, Russia
[3] North-Eastern Federal University, Yakutsk, Russia
haska87@gmail.com

Abstract. The paper is devoted to the issues of modelling the reactor dynamics using multi-group neutron diffusion approximation. Two schemes for the time approximation were considered, namely, an implicit and an explicit-implicit one. For the numerical solution a finite element software was developed based on the package FEniCS and the spectral problems library SLEPc. The code Gmsh is used for the mesh generation. Numerical tests were performed to analyse a regular mode of the VVER-type reactor model.

1 Introduction

The physical processes occuring in a nuclear reactor, depend on the distribution of neutron flux, a mathematical description of which is based on neutron transport equation [13]. This equation is integro-differential and the neutron flux distribution is a function of time, energy, spatial and angular variables. For practical use simplified forms of the equation are usually used in nuclear reactor calculations. The multi-group diffusion approximation [14] is most widely used to analyze the reactor. This approach is used in the majority of engineering neutronic codes.

Engineering neutronic codes are developed to model neutron transport in the multi-group diffusion approximation using, as a rule, finite-difference schemes for space variables. The diffusion equation solution is the energy-space distribution of neutron flux (including effective multiplication factor for steady-state calculation). The calculated neutronics parameters are, as a matter of fact, functionals of neutron flux density.

To increase calculational accuracy, nodal methods are widely used (see, for instance, [9]). These methods allow modelling on a rather coarse mesh (only a few points per assembly in plane and several tens of axial layers). The basic idea of a nodal method is to represent neutron flux within a calculational cell as a few-degree polynomial function or as a set of 1-D or 2-D functions. Nodal methods can be connected, in some respect, to the special variants of finite-element approximation [7]. We can note that it is more appropriate to use the standard procedures of increasing accuracy of the finite-element approximation for boundary problem solution with mesh refinement and using high-order finite elements.

© Springer International Publishing AG 2017
I. Dimov et al. (Eds.): NAA 2016, LNCS 10187, pp. 212–219, 2017.
DOI: 10.1007/978-3-319-57099-0_21

After the discretization of the neutron flux equation, we get a set of linear algebraic equations which is solved iteratively. To obtain suitable calculational accuracy for a boundary problem of a set of diffusion equations, it is necessary to use a rather fine calculational mesh (about several tens of points per assembly in plane and so many axial layers) that leads to increasing calculational time.

To model reactor dynamics, the standard methods to obtain approximate non-stationary solution are used [13,14]. The most attention should be payed to two-level weighted schemes (θ-method) [1], also the Runge-Kutta and Rosenbrok schemes are used [8]. Note a special class of methods for modelling non-stationary neutron transport within the multi-group diffusion approach. It is based on multiplicative performance of the solution as space-time factorization or quasi-static method [6]. An approximate solution is defined as a product of two functions, i.e., one of which depends on time and is connected with an amplitude, another describes space distribution (the shape function). At such approach it is difficult to control approximate solution accuracy, in particular, while calculating complicated neutron flux dynamics.

To characterize the reactor dynamic nature, a spectral parameter α is used. It is defined as the fundamental eigenvalue of the spectral problem (time-eigenvalue or α-eigenvalue problem), related to non-stationary diffusion equations [2,15]. Analogously to ordinary problems of heat transfer (see, for example, [11]) we can define a regular reactor mode. At large times a neutron flux behaviour is asymptotic. Then we can state space-time factorization of the solution with $\exp(\alpha t)$, as amplitude, and the shape function as the eigenfunction of the spectral problem.

We studied appearance of the regular mode of neutron flux on the example of modelling a two-group water-type reactor test problem. We represented two schemes with explicit and implicit approximations of terms related to neutron generation.

2   Problem Definition

Let's consider modelling neutron flux in a multi-group diffusion approximation. Neutron flux dynamics is considered within limited 2D or 3D domain Ω ($\boldsymbol{x} = \{x_1, ..., x_d\} \in \Omega$, $d = 2, 3$) with a convex boundary $\partial\Omega$. Neutron transport is described by the set of equations without taking into account delayed neutron source:

$$\frac{1}{v_g}\frac{\partial \phi_g}{\partial t} - \nabla \cdot D_g \nabla \phi_g + \Sigma_g \phi_g - \sum_{g \neq g'=1}^{G} \Sigma_{s,g' \to g}\phi_{g'}$$
$$= ((1-\beta)\chi_g + \beta\tilde{\chi}_g)\sum_{g'=1}^{G} \nu\Sigma_{fg'}\phi_{g'}, \quad g = 1, 2, ..., G. \tag{1}$$

Here $\phi_g(\boldsymbol{x}, t)$ is the neutron flux in group g at point $\boldsymbol{x}$ at time moment t, G is the number of energy groups, v_g is the effective neutron velocity of group g, $D_g(\boldsymbol{x})$ is the diffusion coefficient, $\Sigma_g(\boldsymbol{x}, t)$ is the absorption cross-section, $\Sigma_{s,g' \to g}(\boldsymbol{x}, t)$ is the cross-section for scattering a neutron from group g' to group g, χ_g ($\widetilde{\chi}_g$) is the fraction of the of the fission neutrons (delayed neutrons) in group g, $\nu \Sigma_{fg}(\boldsymbol{x}, t)$ is the generation cross-section in group g.

The albedo-type conditions are set at the boundary $\partial \Omega$:

$$D_g \frac{\partial \phi_g}{\partial n} + \gamma_g \phi_g = 0, \qquad g = 1, 2, ..., G, \tag{2}$$

where n is the outer normal to the boundary $\partial \Omega$. Let's consider the problem for Eq. (1) with boundary conditions (2), and initial conditions:

$$\phi_g(\boldsymbol{x}, 0) = \phi_g^0(\boldsymbol{x}), \qquad g = 1, 2, ..., G. \tag{3}$$

Let's write the boundary problem (1), (2), (3) in operator notation. We define the vector $\boldsymbol{\phi} = \{\phi_1, \phi_2, ..., \phi_G\}$ and the matrices

$$V = (v_{gg'}), \quad v_{gg'} = \delta_{gg'} v_g^{-1},$$

$$D = (d_{gg'}), \quad d_{gg'} = -\delta_{gg'} \nabla \cdot D_g \nabla,$$

$$S = (s_{gg'}), \quad s_{gg'} = \delta_{gg'} \Sigma_g - \Sigma_{s,g' \to g},$$

$$R = (r_{gg'}), \quad r_{gg'} = ((1 - \beta)\chi_g + \beta \widetilde{\chi}_g)\nu \Sigma_{fg'},$$

$$g, g' = 1, 2, ..., G,$$

where

$$\delta_{gg'} = \begin{cases} 1, & g = g', \\ 0, & g \neq g', \end{cases}$$

is the Kronecker delta. We shall use the set of vectors $\boldsymbol{\phi}$, whose components satisfy the boundary conditions (3). Using introduced definitions, the system of equations (1) can be written in the form of the first-order evolutionary equation:

$$V \frac{d\boldsymbol{\phi}}{dt} + (D + S)\boldsymbol{\phi} = R\boldsymbol{\phi}. \tag{4}$$

The Cauchy problem is solved for (4), when

$$\boldsymbol{\phi}(0) = \boldsymbol{\phi}^0, \tag{5}$$

where $\boldsymbol{\phi}^0 = \{\phi_1^0, \phi_2^0, ..., \phi_G^0\}$.

3   Problem Discretization

Space discretization is performed using the standard Lagrange finite elements [3]. Some details are discussed below while considering the test problem.

To perform time approximation, let's consider two schemes: the fully implicit and explicit-implicit ones. Let's τ is the step of even time mesh, that is $\boldsymbol{\phi}^n = \boldsymbol{\phi}(\boldsymbol{x}, t_n)$, where $t_n = n\tau$, $n = 0, 1, ..., N$, $N\tau = T$. We consider the fully implicit scheme of first order [12] as the basic one to obtain an approximate solution of the problem (4), (5). In this case

$$V\frac{\phi^{n+1} - \phi^n}{\tau} + (D + S)\phi^{n+1} = R\phi^{n+1}, \tag{6}$$

at the known ϕ^0. At new time level the following problem is to be solved:

$$A\phi^{n+1} = b, \tag{7}$$

where

$$A = \frac{V}{\tau} + D + S - R, \quad b = \frac{V}{\tau}\phi^n.$$

To move to new time level we should solve a set of coupled elliptic equations of second order. We can use a decoupled scheme which takes into account the solution features of the non-stationary neutron problems. The operator-matrices V, D are diagonal, and S—lower triangular matrix. Coupling of the equations is due to only the operator-matrix of neutron generation R. For this reason the explicit-implicit scheme has evident calculational prospects:

$$V\frac{\phi^{n+1} - \phi^n}{\tau} + (D + S)\phi^{n+1} = R\phi^n. \tag{8}$$

This scheme leads to problem (7) with

$$A = \frac{V}{\tau} + D + S, \quad b = \left(\frac{V}{\tau} + R\right)\phi^n.$$

While using the explicit-implicit scheme (8) we have separate independent sub-problems for neutrons of separate groups.

4   Test Problem

The test problem for reactor VVER-1000 without a reflector in a two-dimensional approximation (Ω—is the reactor core area) is considered [5]. The geometrical model of the VVER-1000 reactor core consists of a set of hexagonal assemblies and is presented in Fig. 1, where assemblies of various types are marked with various numbers. The fuel assembly pitch equals 23.6 cm. Diffusion neutronics constants in the common notations are given in Table 1. The boundary conditions (3) are used at $\gamma_g = 0.5$, $g = 1, 2$. The problem is solved at $v_1 = 1.25 \cdot 10^7$ $v_2 = 2.5 \cdot 10^5$.

To obtain approximate solutions we use the finite element methods [3] with a triangular calculational mesh. The number of triangles per one assembly κ varies from 6 to 96. The standard Lagrangian finite elements of degree $p = 1, 2, 3$ are used. The software has been developed using the engineering and scientific calculation library FEniCS [10]. The SLEPc package [4] has been used for numerical solution of the spectral problems.

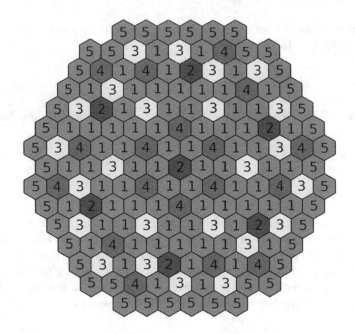

Fig. 1. Geometrcial model of the VVER-1000 reactor core

Table 1. Diffusion neutronics constants for VVER-1000 test problem

Material	1	2	3	4	5
D_1	1.38320e−0	1.38299e−0	1.39522e−0	1.39446e−0	1.39506e−0
D_2	3.86277e−1	3.89403e−1	3.86225e−1	3.87723e−1	3.84492e−1
$\Sigma_1 + \Sigma_{s,1\to2}$	2.48836e−2	2.62865e−2	2.45662e−2	2.60117e−2	2.46141e−2
Σ_2	6.73049e−2	8.10328e−2	8.44801e−1	9.89671e−2	8.93878e−2
$\Sigma_{s,1\to2}$	1.64977e−2	1.47315e−2	1.56219e−2	1.40185e−2	1.54981e−2
$\nu\Sigma_{f1}$	4.81619e−3	4.66953e−3	6.04889e−3	5.91507e−3	6.40256e−3
$\nu\Sigma_{f2}$	8.46154e−2	8.52264e−2	1.19428e−1	1.20497e−1	1.29281e−1

Let's consider the solution of the spectral problem, called Lambda Modes problem:

$$A\varphi = \lambda^{(\alpha)}V\varphi, \quad A = D + S - R. \tag{9}$$

The fundamental eigenvalue

$$\alpha = \lambda_1^{(\alpha)}$$

is called α–eigenvalues or period eigenvalues. The asymptotic behaviour (at large times) of the solution of Cauchy problem (4), (5) can be connected with the α–eigenvalue. In this regular mode, the reactor behaviour is described by the function $e^{-\alpha t}\varphi_\alpha(x)$.

In the framework of used two-group model, the spectral problem (9) can be written as:

$$-\nabla \cdot D_1 \nabla \varphi_1 + \Sigma_1 \varphi_1 + \Sigma_{s,1 \to 2} \varphi_1 - (\nu \Sigma_{f1} \varphi_1 + \nu \Sigma_{f2} \varphi_2) = \lambda^{(\alpha)} \frac{1}{v_1} \varphi_1,$$

$$-\nabla \cdot D_2 \nabla \varphi_2 + \Sigma_2 \varphi_2 - \Sigma_{s,1 \to 2} \varphi_1 = \lambda^{(\alpha)} \frac{1}{v_2} \varphi_2. \tag{10}$$

The aim is to define the fundamental eigenvalue $\alpha = \lambda_1^{(\alpha)}$: $\lambda_1^{(\alpha)} \le \lambda_2^{(\alpha)} \le ....$

The results of solution of the spectral problem (10) for the first eigenvalues $\alpha_n = \lambda_n^{(\alpha)}$, $n = 1, 2, ..., 5$ using the different grids and finite elements are shown in Table 2. The eigenvalues $\alpha_2, \alpha_3,\ \alpha_4, \alpha_5,\ \alpha_9, \alpha_{10}$ are the complex values with small imaginary parts, and the eigenvalues $\alpha_1, \alpha_6,\ \alpha_7, \alpha_8$ are the real values.

Table 2. The eigenvalues $\alpha_n = \lambda_n^{(\alpha)}$, $n = 1, 2, ..., 5$

κ	p	α_1	α_2, α_3	α_4, α_5
6	1	-105.032	$159.802 \pm 0.025510i$	$659.109 \pm 0.034667i$
	2	-139.090	$115.793 \pm 0.029186i$	$591.782 \pm 0.034667i$
	3	-140.223	$114.035 \pm 0.033814i$	$588.762 \pm 0.069025i$
24	1	-130.422	$126.984 \pm 0.034409i$	$608.734 \pm 0.070724i$
	2	-140.187	$114.089 \pm 0.033512i$	$588.849 \pm 0.068555i$
	3	-140.281	$113.887 \pm 0.033604i$	$588.415 \pm 0.068695i$
96	1	-137.704	$117.345 \pm 0.033823i$	$593.818 \pm 0.069254i$
	2	-140.284	$113.886 \pm 0.033599i$	$588.419 \pm 0.068687i$
	3	-140.308	$113.842 \pm 0.033603i$	$588.336 \pm 0.068690i$

The eigenvalues $\lambda_1^{(\alpha)} \le \lambda_2^{(\alpha)} \le ...$ are well separated. In our example, the fundamental eigenvalue is negative and therefore the main harmonic will increase, while all others will attenuate. The regular mode of the reactor is thereby defined. The value $\alpha = \lambda_1^{(\alpha)}$ determines the amplitude of neutron flux development and connects directly with the reactor period in the regular mode.

The fully implicit (6) and explicit-implicit (8) schemes are used to define approximate solution at the initial conditions:

$$\phi_1^0 = 1.0, \quad \phi_2^0 = 0.25$$

and $k = 24$, $p = 2$. Let's $T = 5 \times 10^{-3}$ and consider the fully implicit solution at $\tau = 1 \times 10^{-5}$ as the reference one.

The appearance of the regular mode of neutron flux is controlled by the proximity of the normed solution of the non-stationary problem and the fundamental eigenfunction ϕ. Let's define for $g = 1$:

$$\eta(t) = \|\overline{\phi}_1(t) - \overline{\varphi}_1\|, \quad \overline{\phi}_1(t) = \frac{\phi_1(t)}{\|\phi_1(t)\|}, \quad \overline{\varphi}_1 = \frac{\varphi_1}{\|\varphi_1\|}.$$

The appearance of the dynamic behaviour trend is evaluated by the value:

$$\theta(t) = \frac{1}{\sqrt{2}\,\|\alpha\|} \left(\left\| \frac{1}{\phi_1(t)} \frac{\partial \phi_1(t)}{\partial t} - \alpha \right\|^2 + \left\| \frac{1}{\phi_2(t)} \frac{\partial \phi_2(t)}{\partial t} - \alpha \right\|^2 \right)^{1/2}.$$

Dependence of $\eta(t)$ and $\theta(t)$ on the used time step is shown in Fig. 2 for the fully implicit scheme and in Fig. 3 for the explicit-implicit scheme. The implicit scheme (6) has significantly higher accuracy compared to the explicit-implicit scheme (8).

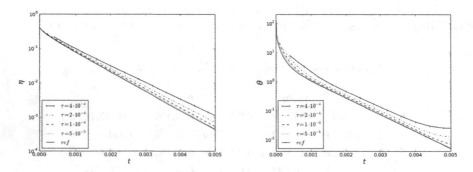

Fig. 2. The fully implicit scheme

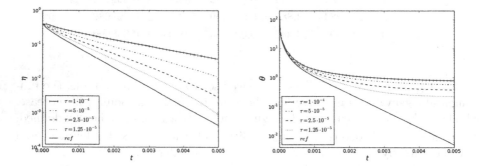

Fig. 3. The explicit-implicit scheme

Acknowledgements. This work was supported by the Russian Foundation for Basic Research (project 16-08-01215) and by the Scientific and Educational Foundation for Young Scientists of Republic of Sakha (Yakutia) 201604010207.

References

1. Ascher, U.M.: Numerical Methods for Evolutionary Differential Equations. Society for Industrial Mathematics, Philadelphia (2008)

2. Bell, G.I., Glasstone, S.: Nuclear Reactor Theory. Van Nostrand Reinhold Company, New York (1970)
3. Brenner, S.C., Scott, R.: The Mathematical Theory of Finite Element Methods. Springer, New York (2008)
4. Campos, C., Roman, J., Romero, E., Tomas, A.: SLEPc Users Manual (2013)
5. Chao, Y.A., Shatilla, Y.A.: Conformal mapping and hexagonal nodal methods-II: Implementation in the ANC-H Code. Nucl. Sci. Eng. **121**, 210–225 (1995)
6. Goluoglu, S., Dodds, H.L.: A time-dependent, three-dimensional neutron transport methodology. Nucl. Sci. Eng. **139**(3), 248–261 (2001)
7. Grossman, L.M., Hennart, J.P.: Nodal diffusion methods for space-time neutron kinetics. Prog. Nucl. Energy **49**(3), 181–216 (2007)
8. Hairer, E., Wanner, G.: Solving Ordinary Differential Equations II: Stiff and Differential-Algebraic Problems. Springer, Heidelberg (2010)
9. Lawrence, R.D.: Progress in nodal methods for the solution of the neutron diffusion and transport equations. Prog. Nucl. Energy **17**(3), 271–301 (1986)
10. Logg, A., Mardal, K., Wells, G.: Automated Solution of Differential Equations by the Finite Element Method: The FEniCS Book. Lecture Notes in Computational Science and Engineering. Springer, Heidelberg (2012). http://books.google.ru/books?id=ASWN_VRr1NQC
11. Luikov, A.: Analytical Heat Diffusion Theory. Academic Press, London (1968)
12. Samarskii, A.A.: The Theory of Difference Schemes. Marcel Dekker, New York (2001)
13. Stacey, W.M.: Nuclear Reactor Physics. Wiley, Hoboken (2007)
14. Sutton, T.M., Aviles, B.N.: Diffusion theory methods for spatial kinetics calculations. Prog. Nucl. Energy **30**(2), 119–182 (1996)
15. Verdu, G., Ginestar, D., Roman, J., Vidal, V.: 3D alpha modes of a nuclear power reactor. J. Nucl. Sci. Technol. **47**(5), 501–514 (2010)

Regularization Methods of the Continuation Problem for the Parabolic Equation

Andrey Belonosov[1,3] and Maxim Shishlenin[1,2,3(✉)]

[1] Institute of Computational Mathematics and Mathematical Geophysics,
prospect Akademika Lavrentjeva, 6, Novosibirsk 630090, Russia
`white@sscc.ru, mshishlenin@ngs.ru`
[2] Sobolev Institute of Mathematics, 4 Acad. Koptyug Avenue,
Novosibirsk 630090, Russia
[3] Novosibirsk State University, 2 Pirogova Str., Novosibirsk 630090, Russia

Abstract. We investigate the one-dimensional continuation problem (the Cauchy problem) for the parabolic equation with the data on the part of the boundary. For numerical solution we apply finite-difference scheme inversion, the singular value decomposition and the gradient method of the minimizing the goal functional. The comparative analysis of numerical methods are presented.

Keywords: Continuation problem · Parabolic equation · Numerical methods

1 Introduction

Inverse problems have numerous applications in science and engineering [11,12]. This paper is concerned with an continuation heat conduction problem, where the unknown temperature of the object is to be found based on observations (measurements) of the temperature on the part of the boundary [3,4]. The continuation problem for heat equation is ill-posed (the Cauchy problem for heat equation). We reduce the continuation problem to the inverse problem where the unknown initial data and boundary condition on unaccessible part of the boundary has to be found based on known measurements of the temperature. Therefore, the problem can be classified as a boundary inverse problem. The heat equation considered describes the evolution of temperature in a medium where the thermal conductivity is a function of spatial variables and time.

Numerical methods for continuation problems for PDEs are considered in [5–8,18,20–22]. Methods, similar to the ones of this paper, were applied in publications of the second author with coauthors [11,14–17,19].

Let us consider the continuation problem for parabolic equation

$$u_t = a(x,t)u_{xx}, \qquad x \in \mathbb{R}, \quad t > 0; \tag{1}$$

$$u|_{x=0} = f(t), \qquad u_x|_{x=0} = g(t), \qquad t > 0. \tag{2}$$

The problem is to find function $u(x,t)$ by known $f(t)$ and $g(t)$.

The continuation problem (1), (2) is ill-posed problem [10].

© Springer International Publishing AG 2017
I. Dimov et al. (Eds.): NAA 2016, LNCS 10187, pp. 220–226, 2017.
DOI: 10.1007/978-3-319-57099-0_22

2   Theoretical Results

Let us formulate results for the continuation problem for the heat equation [2]. Let in the half strip $\Pi = \{(x,t) : x \geq 0, 0 \leq t \leq 1\}$ twice continuously differentiable function $u(x,t)$ holds the following differential inequality

$$[a(x,t)u_{xx} - u_t]^2 \leq k_1 u^2 + k_2 u_x^2. \tag{3}$$

Here k_1, k_2 are positive constants, coefficient $a(x,t)$ is the continuously differentiable function in the domain Π and $a(x,t) > 0$ for $(x,t) \in \Pi$.

Then the following theoretical results can be proved [19].

Theorem 1. *If*

$$u(0,t) < e^{-\frac{1}{\eta(t)}}, \quad u_x(0,t) < e^{-\frac{1}{\eta(t)}},$$

where $\frac{\eta(t)}{t} \to 0$ *for* $t \to 0$, *then* $u(x,t) \equiv 0$ *for* $x \geq 0$.

Theorem 2. *Let*

$$0 < a_0 < a(x,t) \leq 1, \quad |a_x(x,t)| \leq 1, \quad |a_t(x,t)| \leq 1, \tag{4}$$

$$\max\{|u(x,t)|, |u_x(x,t)|, |u_t(x,t)|\} \leq \text{const} < \infty, \tag{5}$$

for $(x,t) \in \Pi$,

$$\int_0^\infty |a_t(\xi,t)| \, d\xi \leq \text{const} < \infty, \quad t \in [0,1].$$

Let a positive monotonically increasing sequence x_n *hold the condition*

$$x_n \leq \text{const} \cdot n^\mu.$$

Here $\mu < \frac{1}{2}$. *If* $u(x_n,t) = 0$ *for* $t \in [0,1]$, $n = 0,1,2,\ldots$, *then* $u \equiv 0$ *in the domain* Π.

Theorem 1 is a strengthening of the relevant approval E. M. Landis, proved in [1]. Theorem 2 gives a sufficient condition (which should impose on the sequence, x_n) uniqueness of the problem of the continuation for solutions of differential inequality (1) with an infinite sequence of intervals $\{(a, x_n, t) : 0 \leq t \leq 1\}$. Proof of Theorems 1 and 2 follows from the properties of the assessment decision of the Cauchy problem for a differential inequality (1) with data on premenopausal surface. In [4] it is proved that this estimate is exponential in nature. We are interested in its dependence on the field size and distance to the border.

Let us make some definitions

$$V = \{(x,t) : 0 \leq x \leq h, -T_0 \leq t \leq T_0\}. \tag{6}$$

Here $h > 0$, $0 < T_0 < 1$.

$$\varepsilon = \max_{t \in [-T_0,T_0]} \{|u(0,t)|, |u_x(0,t)|\}, \tag{7}$$

$$\lambda = \max_{t \in [-T_0, T_0]} \left\{ \int_0^h |a_t(\xi, t)| \, d\xi \right\} + 1, \tag{8}$$

$$M = \max_{(x,t) \in V} \{|u(x,t)|, |u_x(x,t)|, |u_t(x,t)|\} + 1. \tag{9}$$

The proofs of the theorems are followed from the following lemmas.

Lemma 1. *Let in the domain V twice continuously differentiable function $u(x,t)$ hold differential inequality (1) and coefficient $a(x,t)$ holds the condition (2) in the domain V. Let $\rho \in (0,h)$, $r = h - \rho$. Then the constants C_1, C_2 exist such that $C_i = C_i(a_0, k_1, k_2)$, $i = 1, 2$ and if*

$$\varepsilon < \exp\{-C_1(h+1)^2\lambda^2[T_0^{-1} + r^{-2} + \ln M]\},$$

then

$$|u(\rho, 0)| < \exp\{-C_2(h+1)^2\lambda^2[T_0^{-1} + r^{-2} + \ln M]r\}.$$

Lemma 2. *Let in the domain $V = \{(x,t) : 0 \le x \le h, 0 \le t \le 1\}$ the function $u(x,t) \in C_2(V)$ hold the inequality (1) and the conditions (2). Therefore the constant C_0 exists and $C_0 = C_0(a_0, k_1, k_2, h)$, such that $\max\{|u(0,t)|, |u_X(0,t)|\} < e^{-C_0/t}$, then $u(x,0) \equiv 0$ for $x \in [0,h]$.*

For the proof of the Theorem 2, we also need the following three auxiliary lemmas.

Lemma 3. *Let in the domain $V = \{(x,t) : 0 \le x \le \delta, 0 \le t \le 1\}$ function $u(x,t) \in C_2(V)$ hold the inequality (1) and the conditions (2) and*

$$\max_{(x,t) \in V} \{|(u(x,t)|, |u_x(x,t)|\} \le M.$$

If $u(0,t) = u(\delta, t) = 0$ for $t \in [0,1])$, then

$$\int_0^\delta u^2(x,t)dx \le M^2\delta \cdot \exp\left\{ -t\left[a_0\delta^{-2} - \left(\frac{k_2 + 2}{a_0} + \frac{k_1 a_0}{k_2}\right)\right]\right\} = \varepsilon.$$

Moreover if $\varepsilon < 1$, then for $(x,t) \in V$ we have

$$|u(x,t)| \le (M+1)\varepsilon^{1/4}. \tag{10}$$

Lemma 4. *Let in the domain $V = \{(x,t) : 0 \le x \le \delta, 0 \le t \le 1\}$ the function $u(x,t) \in C_2(V)$ satisfy the inequality (1) and condition (2), and*

$$\max_{(x,t) \in V} \{|(u(x,t)|, |u_x(x,t)|\} \le M.$$

If

$$\max_{(x,t) \in V} \{|(u(0,\delta)|, |u(\delta,t)|, |u(x,0)|\} \le \varepsilon,$$

then in any subdomain

$$V_\sigma = \{(x,t) : \sigma \le x \le \delta - \sigma, \sigma \le t \le 1\}, \ 0 < 2\sigma < \min(1,\delta)$$

the following inequality holds true

$$|u_x(x,t)| < C(\delta + \sigma^{-2})\varepsilon.$$

Here C is the constant which depends on a_0, k_1, k_2 and M.

Lemma 5. *Let x_n be a positive monotonically increasing sequence. If $x_n < Cn^\mu$ and $\mu < \frac{1}{2}$, then $\lim\limits_{n\to\infty} x_n(x_n + x_{n-1}) = 0$.*

3   Numerical Results

We reformulate continuation problem in the form of inverse problem.
Let us consider the direct problem:

$$u_t = a(x,t)u_{xx}, \qquad x \in (0,L), \quad t \in (0,T); \tag{11}$$

$$u|_{t=0} = \varphi(x), \qquad x \in (0,L); \tag{12}$$

$$u_x|_{x=0} = g(t), \quad u|_{x=L} = q(t), \qquad t \in (0,T). \tag{13}$$

Inverse problem is to find the unknown functions $\varphi(x)$, $g(t)$ and $q(t)$ by known additional information

$$u|_{x=0} = f(t), \qquad t \in (0,T). \tag{14}$$

Let $h_x = T/N_x$ be a mesh size with respect to x, h_t be a mesh size with respect to t and we suppose the CFL condition holds. Replace the derivatives by finite-difference analogs and move on to the solution of the direct discrete problem:

$$\frac{v_i^{k+1} - v_i^k}{h_t} = a_i^k \frac{v_{i+1}^k - 2v_i^k + v_{i-1}^k}{h_x^2}, \qquad i = \overline{1, N_x - 1}, \quad k = \overline{0, N_t - 1};$$
$$v_i^0 = \varphi_i, \qquad i = \overline{0, N_x}.$$

From the approximation of the boundary conditions with second order we obtain

$$v_0^{k+1} = v_0^k + \frac{2a_0^k h_t}{h_x^2}[v_1^k - v_0^k - h_x g^k],$$

$$v_{N_x}^{k+1} = v_{N_x}^k + \frac{2a_{N_x}^k h_t}{h_x^2}[v_{N_x-1}^k - v_{N_x}^k].$$

For numerical implementation we apply the following methods:

1. Finite-difference scheme inversion (FDSI). The continuation problem is replaced by the finite-difference analogue and then the solution is recovered inside the field region layer by layer.

$$v_{i+1}^k = 2v_i^k - v_{i-1}^k + \frac{h_x^2}{a_i^k h_t}[v_i^{k+1} - v_i^k], \qquad i = \overline{1, N_x - 1}, \quad k = \overline{i, N_t - i};$$
$$v_0^k = f^k, \qquad k = \overline{0, N_t};$$
$$v_1^k = f^k + h_x g^k + \frac{h_x^2}{a_0^k}[f']^k, \qquad k = \overline{0, N_t}.$$

2. **Singular value decomposition.** Instead of FDSI method we consider an implicit scheme Euler for the equation

$$\frac{v_i^{k+1} - v_i^k}{h_t} = a_i^{k+1} \frac{v_{i+1}^{k+1} - 2v_i^{k+1} + v_{i-1}^{k+1}}{h_x^2}, \qquad i = \overline{1, N_x - 1}, \quad k = \overline{0, N_t - 1}$$

(15)

and approximate the boundary conditions on $x = 0$:

$$v_0^k = f^0, \quad v_0^{k+1} = v_0^k + \frac{2a_0^k h_t}{h_x^2}[v_1^k - v_0^k - h_x g^k].$$

(16)

The discrete continuation problem (15), (16) is reduced to the system of linear algebraic equations [14–17]

$$Av = f.$$

Here $v = (v_0^0, \ldots, v_{N_x}^0, v_0^1, \ldots, v_{N_x}^1, \ldots, v_0^{N_t}, \ldots, v_{N_x}^{N_t})$.

3. **Gradient method.** Let us find the solution of inverse problem (10) and (11) by minimizing the cost functional [9,13]:

$$J(\boldsymbol{q}) = \int_0^T [u(0,t) - f(t)]^2 dt.$$

Here $\boldsymbol{q} = (\varphi, g, q)$.

We apply the gradient method for minimizing the cost functional

$$q_{n+1} = q_n - \alpha_n J'(\boldsymbol{q}).$$

Here q_0 is initial guess, α_n is the descent parameter, $J'(\boldsymbol{q})$ is the gradient of the functional (Figs. 1, 2 and 3).

The gradient of the functional can be calculated via the solution of the adjoint problem

$$\psi_t = -a(x,t)\psi_{xx}, \qquad x \in (0,L), \quad t \in (0,T);$$
$$\psi|_{t=T} = 0, \qquad x \in (0,L);$$
$$\psi_x|_{x=0} = 2(u(0,t) - f(t)), \quad \psi|_{x=L} = 0, \qquad t \in (0,T)$$

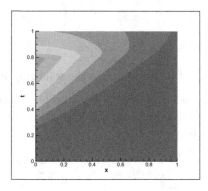

Fig. 1. Direct problem solution

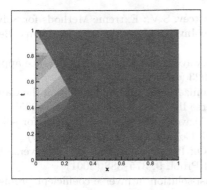 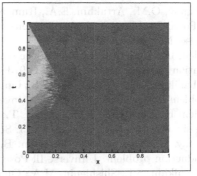

Fig. 2. The solution of the continuation problem by FDSI. Left—exact data. Right—1% noisy data

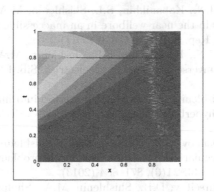

Fig. 3. The solution of the continuation problem with the regularization: add boundary conditions $u_t|_{t=0} = u_t|_{t=T} = 0$.

by formula

$$J'(q) = (\psi_x(0,t), \psi(x,0), \psi_t(x,T)).$$

Acknowledgments. The work was supported by the RFBR (grants 14-01-00208, 16-01-00755 and 16-29-15120), the Ministry of Education and Science of the Russian Federation and the Ministry of Education and Science of the Republic of Kazakhstan (project 1746/GF4).

References

1. Landis, E.M.: Some questions of the qualitative theory of elliptic and parabolic equations. Success. Math. Sci. **14**(1), 21–85 (1959). (in Russian)
2. Belonosov, A.S.: Two theorems about uniqueness of the continuation of the solutions for the parabolic equation. Math. Probl. Geophys. **5**, 30–45 (1974). (in Russian)
3. Alifanov, O.M.: Inverse Heat Conduction Problems. Springer, New York (1994)

4. Alifanov, O.M., Artukhin, E.A., Rumyantcev, S.V.: Extreme Methods for Solving Ill-Posed Problems with Applications to Inverse Heat Transfer Problems. Begell House, New York (1995)
5. Eldén, L.: Numerical solution of the sideways heat equation by difference approximation in time. Inverse Prob. **11**, 913–923 (1995)
6. Kabanikhin, S.I., Karchevsky, A.L.: Optimizational method for solving the Cauchy problem for an elliptic equation. J. Inverse Ill-Posed Probl. **3**, 21–46 (1995)
7. Eldén, L., Berntsson, F., Regińska, T.: Wavelet and Fourier methods for solving the sideways heat equation. SIAM J. Sci. Comput. **21**, 2187–2205 (2000)
8. Bastay, G., Kozlov, V.A., Turesson, B.O.: Iterative methods for an inverse heat conduction problem. J. Inverse Ill-Posed Probl. **9**, 375–388 (2001)
9. Kabanikhin, S.I., Shishlenin, M.A.: Quasi-solution in inverse coefficient problems. J. Inverse Ill-Posed Probl. **16**(7), 705–713 (2008)
10. Kabanikhin, S.I.: Inverse and Ill-Posed Problems: Theory and Applications. de Gruyter, Berlin (2011)
11. Epov, M.I., Eltsov, I.N., Kabanikhin, S.I., Shishlenin, M.A.: The recovering the boundary conditions in the near-wellbore in an inaccessible part of the boundary. Sib. Electron. Math. Rep. **8**, C.400–C.410 (2011)
12. Epov, M.I., Eltsov, I.N., Kabanikhin, S.I., Shishlenin, M.A.: The combined formulation of the two inverse problems of geoelectrics. Sib. Electron. Math. Rep. **8**, C.394–C.399 (2011)
13. Kabanikhin, S.I., Shishlenin, M.A.: Using the a priori information in the coefficient inverse problems for hyperbolic equations. Proc. Inst. Math. Mech. UrB RAS **18**(1), 147–164 (2012)
14. Kabanikhin, S.I., Gasimov, Y.S., Nurseitov, D.B., Shishlenin, M.A., Sholpanbaev, B.B., Kasenov, S.: Regularization of the continuation problem for elliptic equations. J. Inverse Ill-Posed Probl. **21**(6), 871–884 (2013)
15. Kabanikhin, S.I., Nurseitov, D.B., Shishlenin, M.A., Sholpanbaev, B.B.: Inverse problems for the ground penetrating radar. J. Inverse Ill-Posed Probl. **21**(6), 885–892 (2013)
16. Shishlenin, M.A.: Matrix method in inverse source problems. Sib. Electron. Math. Rep. **11**, C.161–C.171 (2014)
17. Kabanikhin, S.I., Shishlenin, M.A., Nurseitov, D.B., Nurseitova, A.T., Kasenov, S.E.: comparative analysis of methods for regularizing an initial boundary value problem for the Helmholtz equation. J. Appl. Math. 7 (2014). doi:10.1155/2014/786326
18. Rischette, R., Baranger, T.N., Debit, N.: Numerical analysis of an energy-like minimization method to solve a parabolic Cauchy problem with noisy data. J. Comput. Appl. Math. **271**, 206–222 (2014)
19. Belonosov, A.S., Shishlenin, M.A.: Continuation problem for the parabolic equation with the data on the part of the boundary. Sib. Electron. Math. Rep. **11**, C.22–C.34 (2014)
20. Klibanov, M.V.: Carleman estimates for the regularization of ill-posed Cauchy problems. Appl. Numer. Math. **94**, 46–74 (2015)
21. Becache, E., Bourgeois, L., Franceschini, L., Dardé, J.: Application of mixed formulations of quasi-reversibility to solve ill-posed problems for heat and wave equations: the 1D case. Inverse Probl. Imaging **9**, 971–1002 (2015)
22. Dardé, J.: Iterated quasi-reversibility method applied to elliptic and parabolic data completion problems. Inverse Probl. Imaging **10**, 379–407 (2016)

Computer Simulation of Plasma Dynamics in Open Plasma Trap

Evgeny Berendeev[1], Galina Dudnikova[2,3], Anna Efimova[1(✉)],
and Vitaly Vshivkov[1]

[1] Institute of Computational Mathematics and Mathematical Geophysics SB RAS,
Prospect Akademika Lavrentjeva, 6, Novosibirsk 630090, Russia
{berendeev,efimova,vsh}@ssd.sscc.ru
[2] Institute of Computational Technologies SB RAS, Novosibirsk, Russia
dudn@ict.nsc.ru
[3] University of Maryland, College Park, MD 20742, USA
http://www.ict.nsc.ru, http://www.umd.edu, http://www.icmmg.nsc.ru

Abstract. We present a 2D particle-in-cell (PIC) model and the corresponding parallel code for computer simulation of plasma dynamics in open plasma traps. The mathematical model includes the Boltzmann equations for ions and electrons and system of Maxwell's equations for the self-coordinate electromagnetic fields. The combination of the modified PIC-method and the Monte-Carlo methods is used to solve these equations. The problem of minimization of the plasma losses in trap with multipole magnetic walls has been investigated on the base of computer simulation.

Keywords: Particles-in-cell method · Maxwell's equations · The Boltzmann equation · The Vlasov equation · Open plasma trap

1 Introduction

We present a 2D particle-in-cell (PIC) model and the corresponding parallel code for computer simulation of plasma dynamics in open plasma traps. Open magnetic traps are one of the directions in solution of the controlled thermonuclear fusion problem. One of advantages of these systems before the closed configurations consists in possibility of injection of high energy electron beams into plasma. Let us consider two problems of nonlinear plasma dynamic in open magnetic traps. In the first one the beam and plasma parameters were chosen close to the parameters of the experiment on the GOL-3 facility (the Budker Institute of Nuclear Physics SB RAS, Novosibirsk) [1]. In the second considered problem we simulate plasma dynamics in the linear axially symmetric plasma trap with inverse magnetic mirrors with the return magnetic field (the BINP SB RAS) [2]. The problem of minimization of the plasma losses in the wide aperture passage holes in the end faces with inverse magnetic mirrors, as well as through cylindrical multipole magnetic walls of the trap is investigated on the base of

© Springer International Publishing AG 2017
I. Dimov et al. (Eds.): NAA 2016, LNCS 10187, pp. 227–234, 2017.
DOI: 10.1007/978-3-319-57099-0_23

computer simulation. A mathematical model includes the Boltzmann equation for the distribution functions for ions and electrons and system of Maxwell's equations for the self-coordinate electromagnetic fields. The combination of the modified PIC-method in the cylindrical r-z coordinates and the Monte-Carlo methods is used to solve these equations. The numerical convergence of the solution depending on calculating parameters was investigated, good compliance with available analytical decisions is obtained. The modification Euler-Lagrange domain decomposition was proposed for uniform load of computing nodes. High scalability of the offered parallel algorithm was shown by means of computing experiments on the supercomputer. Calculations are performed on the supercomputers NKS-30T (the Siberian Supercomputing Center) and the Lomonosov (the Supercomputing Center of Lomonosov Moscow State University). The numerical models presented in this work are the development and generalization of the models set out in [3–5].

2   The Problem Statement

The plasma is described by the system of the Boltzmann-Maxwell equations: (1–4):

$$\frac{\partial f_k}{\partial t} + (\overrightarrow{v}, \overrightarrow{\nabla}) f_k + q_k \left(\overrightarrow{E} + \frac{1}{c} [\overrightarrow{v} \times \overrightarrow{B}] \right) \frac{\partial f_k}{\partial \overrightarrow{p}} = St\{f_k\}, \tag{1}$$

$$rot\overrightarrow{B} = \frac{4\pi}{c} \overrightarrow{j} + \frac{1}{c} \frac{\partial \overrightarrow{E}}{\partial t}, \quad rot\overrightarrow{E} = -\frac{1}{c} \frac{\partial \overrightarrow{B}}{\partial t}, \tag{2}$$

$$div\overrightarrow{E} = 4\pi\rho, \tag{3}$$

$$div\overrightarrow{B} = 0, \tag{4}$$

$$\overrightarrow{j} = \sum_k q_k \int \overrightarrow{v} f_k(\overrightarrow{p}, \overrightarrow{r}, t) d\overrightarrow{p}, \quad \rho = \sum_k q_k \int f_k(\overrightarrow{p}, \overrightarrow{r}, t) d\overrightarrow{p}. \tag{5}$$

where f_k is the particle distribution function of the species k (beam electrons, plasma electrons and ions), $\overrightarrow{B}$ is the magnetic field, $\overrightarrow{E}$ is the electric field, c is the speed of light, v is the velocity of the particles, ρ is the electric charge density, j is the electric current density and q_k is the charge of the particle of the species k. $St\{f_k\}$ is the function describing the following physical processes:

1. ionization   of   the   hydrogen   atoms   of   the   cathode   electrons $H + e^- \rightarrow H^+ + 2e^-$;
2. ionization of the hydrogen molecules of the cathode electrons $H_2 + e^- \rightarrow H_2^+ + 2e^-$;
3. ionization of the ions H^+ hydrogen atoms $H_2 + H^+ \rightarrow H_3^+ + H$.

In the case of collisionless plasma approximation $St\{f_k\} = 0$, and the Boltzmann equation (1) turns into the Vlasov equation [6].

3   Main Equations Solutions

Based on the splitting method, the solution of the Boltzmann equation is reduced to the solution of the Vlasov equation with the corresponding corrections of the particles trajectories with allowance for physical processes using the Monte-Carlo method [6].

The PIC-method (see [6–8]) is used to solve the Vlasov equation. In this method, the plasma is simulated by a set of separate particles, each characterizing by the motion of many physical particles. The characteristics of the Vlasov equation describe trajectories of the particles. Equation (6) for these characteristics are:

$$\frac{d\vec{p}_{i,e}}{dt} = q_k(\vec{E} + [\vec{v_{i,e}}, \vec{B}]), \quad \frac{d\vec{r}_{i,e}}{dt} = \vec{v}_{i,e}, \quad \vec{p}_{i,e} = \frac{\vec{v}_{i,e}}{\sqrt{1 - \vec{v}^2_{i,e}}}. \tag{6}$$

The leap-frog scheme [7] is used to solve these equations.

For the GOL-3 simulation we use the mathematical model based on the Vlasov-Maxwell equation and Cartesian coordinates. For second problem a target has a cylindrical shape, so we consider the problem in two-dimensional formulation in cylindrical coordinates R and Z. Besides, all three components of the fields and velocities of the particles (2D3V production) are taken into account. In order to avoid singularities at the target axis a local transformation from Cartesian to cylindrical coordinates is used. It based on the Boris scheme [9].

The probability of the collision of charged particles with neutral atoms is used to simulate the process of ionization. If collision has occurred, then particles with required parameters are added in the computational domain.

The collision probability for a j-th particle possessing the velocity v_j for time Δt should be calculated on the basis of [10]:

$$P_j = 1 - \exp\left(-\Delta s_j \sigma(\varepsilon_j) n_j(r_j)\right), \tag{7}$$

where $\Delta s_j = v_j \Delta t$, $\sigma(\varepsilon_j)$ — the collision cross section, ε_j — kinetic energy of the j-th particle, n_j — the local density for particles of the corresponding type.

Maxwell's equations are solved with the Langdon-Lazinsky scheme which is described by [11]. In this scheme, the electric and magnetic fields are calculated on the grids displaced as related to each other with respect to time and space that allows to attaining second order of accuracy.

The charge and current densities, Eq. (8), accordingly, which are necessary for their solution are defined by particle velocities and coordinates:

$$\rho(r, t) = \sum_{k=1}^{K} q_k R(r, r_k(t)), \quad j(r, t) = \sum_{k=1}^{K} q_k v_k(t) R(r, r_k(t)). \tag{8}$$

Here q_k is the charge of the particle with the number k and the function $R(r, r_k(t))$ is the form-factor of the PIC-method. It characterizes the particle form and size and the charge distribution [7]. The PIC form-factor has been used for the first problem.

Satisfaction of the continuity Eq. (9) is necessary:

$$\frac{\partial \rho}{\partial t} + div_h \vec{j} = 0. \tag{9}$$

For this purpose we should calculate intermediate value of electric field E^* with not specified value of density of current. Let the required electric field E be equal $E = E^* - \nabla_h \delta\Phi$, where $\delta\Phi$ is a correction to the electrical potential. We can determine the correction $\delta\Phi$ by solving Poisson's equation:

$$\triangle_h \delta\Phi = div_h \vec{E}^* - \rho. \tag{10}$$

This method is the most resource-intensive because it is very difficult to construct a parallel algorithm based on domain decomposition. In order to avoid solving the Poisson's equation and find the correct grid values of the electric and magnetic fields we use another way. Assume that the continuity Eq. (9) is satisfied, then the Eq. (3) is met at any time, if $div_h \vec{E}^0 = \rho^0$. In works [12,13] calculation of the current density and the charge density have been satisfied in accordance with (9) for some functions R. The main condition for such accordance is symmetry of a form-factor:

$$R\big((x, y, z), (x_j(t), y_j(t), z_j(t))\big) = S(x, x_j(t)) * S(y, y_j(t)) * S(z, z_j(t)). \tag{11}$$

For the first problem of the GOL-3 trap we use PIC form-factor. For the target trap the form-factor is not symmetric in a cylindrical coordinate system, and the function R for a node number (i, l, k) is given by

$$R\big((r_i, \varphi_l, z_k), (r_j, \varphi_j, z_j)\big) = \begin{cases} \frac{1}{V_i} * \frac{r_{i+1}^2 - r_j^2}{r_{i+1}^2 - r_i^2} *, \frac{h_\varphi - |\varphi_l - \varphi_j|}{h_\varphi} * \frac{h_z - |z_k - z_j|}{h_z}, \\ \text{if } r_i < r_j < r_{i+1}, |\varphi_l - \varphi_j| < h_\varphi, |z_k - z_j| < h_z; \\ \frac{1}{V_i} * \frac{r_{i-1}^2 - r_j^2}{r_{i-1}^2 - r_i^2} * \frac{h_\varphi - |\varphi_l - \varphi_j|}{h_\varphi} * \frac{h_z - |z_k - z_j|}{h_z}, \\ \text{if } r_{i-1} < r_j < r_i, |\varphi_l - \varphi_j| < h_\varphi, |z_k - z_j| < h_z; \\ 0, \text{ otherwise} \end{cases} \tag{12}$$

where $V_i = 2\pi r_i h_r h_\varphi h_z$ is a volume of a cell. We don't consider the case $r = 0$ because of it doesn't affect on the further development of the algorithm and can be considered independently.

In initial problem definition it is necessary to consider all three components of velocity. The continuity equation should be viewed in three-dimensional space, with the subsequent elimination of dependence from φ. Let us consider the function of the particle form-factor (12) as a combination of three one-dimensional form-factors:

$$R\big((r, \varphi, z), (r_j(t), \varphi_j(t), z_j(t))\big) = \frac{1}{V} SR(r, r_j(t)) * SP(\varphi, \varphi_j(t)) * SZ(z, z_j(t)). \tag{13}$$

Assume that the particle move with constant velocity $(v_r^{m+1/2}, v_\varphi^{m+1/2}, v_z^{m+1/2})$ and doesn't leave the neighborhood $r_i < r_j < r_{i+1}$, $\varphi_l < \varphi_j < \varphi_{l+1}$, $z_k < z_j < z_{k+1}$. We write the continuity equation in the node (i, l, k) for the period of time $[t^m, t^{m+1}]$:

$$\frac{\rho_{i,l,k}^{m+1} - \rho_{i,l,k}^m}{\tau} = \frac{1}{\tau} \int_{t^m}^{t^{m+1}} \frac{d}{dt} \rho_{i,l,k}(t)\, dt$$

$$= \frac{1}{\tau} \int_{t^m}^{t^{m+1}} \frac{d}{dt} \left(\frac{1}{V_i} SR(r_i, r_j(t)) * SP(\varphi_l, \varphi_j(t)) * SZ(z_k, z_j(t)) \right) dt \tag{14}$$

Suppose that the particle moved from (r^m, φ^m, z^m) to $(r^{m+1}, \varphi^{m+1}, z^{m+1})$. We introduce the operators $\Delta f = f^{m+1} - f^m$, $\delta f = \frac{f^{m+1}+f^m}{2}$ and pass from $[t^m, t^{m+1}]$ to $[-\frac{1}{2}, \frac{1}{2}]$. In this case the particle trajectory is described as follows:

$$r(t) = \delta r + \Delta r \ \ t, \ \varphi(t) = \delta\varphi + \Delta\varphi \ \ t, \ z(t) = \delta z + \Delta z \ \ t. \tag{15}$$

Having substituted these expressions into the Eq. (14), we obtain:

$$\frac{1}{V_i} \int_{-1/2}^{1/2} \left(v_r^{m+1/2} \frac{2r(t)}{r_{i+1}^2 - r_i^2} SP(\varphi_l, \varphi_j(t)) * SZ(z_k, z_j(t)) \right) dt$$

$$+ \frac{1}{V_i} \int_{-1/2}^{1/2} \left(v_\varphi^{m+1/2} \frac{1}{h_\varphi} SR(r_i, r_j(t)) * SZ(z_k, z_j(t)) \right) dt$$

$$+ \frac{1}{V_i} \int_{-1/2}^{1/2} \left(v_z^{m+1/2} \frac{1}{h_z} SP(r_i, r_j(t)) * SR(z_k, z_j(t)) \right) dt$$

$$= \frac{2 * v_r^{m+1/2}}{V_i * (r_{i+1}^2 - r_i^2) h_\varphi h_z} * \left(\delta r * (\varphi_{l+1} - \delta\varphi)(z_{k+1} - \delta z) \right. \tag{16}$$

$$\left. + \frac{\Delta r \Delta\varphi(z_{k+1} - \delta z) + \delta z \delta\varphi \delta r + \Delta r \Delta z(\varphi_{l+1} - \delta\varphi)}{12} \right)$$

$$+ \frac{v_\varphi^{m+1/2}}{V_i * (r_{i+1}^2 - r_i^2) h_\varphi h_z} * \left((r_{i+1}^2 - \delta r * \delta r)(z_{k+1} - \delta z) \right.$$

$$\left. + \frac{\Delta r \Delta r(z_{k+1} - \delta z) + 2 * \Delta r \Delta z \delta r}{12} \right)$$

$$+ \frac{v_z^{m+1/2}}{V_i * (r_{i+1}^2 - r_i^2) h_\varphi h_z} * \left((r_{i+1}^2 - \delta r * \delta r)(\varphi_{k+1} - \delta\varphi) \right.$$

$$\left. + \frac{\Delta r \Delta r(\varphi_{k+1} - \delta\varphi) + 2 * \Delta r \Delta\varphi \delta r}{12} \right).$$

These three terms correspond to the three components of the current density (JR, JP, JZ) Since there is no current outside a cell, therefore we receive:

$$JR_{i-1/2,l,k}^{m+1/2} = JP_{i,l-1/2,k}^{m+1/2} = JZ_{i,l,k-1/2}^{m+1/2} = 0. \tag{17}$$

From (16) we obtain three current density components, for a two-dimensional case they are:

$$
\begin{aligned}
JR_{i+1/2,k}^{m+1/2} &= \frac{2 * v_r^{m+1/2} * r_i h_r}{V_i * (r_{i+1}^2 - r_i^2) h_z r_{i+1/2}} * \left(\delta r * (z_{k+1} - \delta z) + \frac{\Delta r \Delta z}{12} \right), \\
JP_{i,k}^{m+1/2} &= \frac{r_i v_\varphi^{m+1/2}}{V_i * (r_{i+1}^2 - r_i^2) h_z} * \left((r_{i+1}^2 - \delta r * \delta r)(z_{k+1} - \delta z) \right. \\
&\quad \left. + \frac{\Delta r \Delta r (z_{k+1} - \delta z) + 2 * \Delta r \Delta z \delta r}{12} \right), \\
JZ_{i,k+1/2}^{m+1/2} &= \frac{v_z^{m+1/2}}{V_i * (r_{i+1}^2 - r_i^2)} * \left((r_{i+1}^2 - \delta r * \delta r) + \frac{\Delta r \Delta r}{12} \right),
\end{aligned}
\tag{18}
$$

where $V_i = 2\pi r_i h_r h_z$ is a volume of a cell. The algorithm of the current density distribution between cells can be used when the particle moves between cells, as described in [14]. Note that Eq. (14) holds for any finite size of form-factor, therefore all the calculations obtained above should be generalized at all types of finite size of the form-factors and any coordinate system.

4   Simulation Results

From the first problem the solution convergence was obtained with the increase of the number of the particles in cell, as Fig. 1 shows. On this graph the evolution of the electric field for different values of particles in a cell (lp) is presented, where 1 - lp = 250, 2 - lp = 500, 3 - lp = 1000, 4 - lp = 2000, 5 - lp = 4000, 6 - lp = 8000. One can see that the maximum value of 250 particles differs from the one for 8000 particles twice. As follows from the figure, at least 1000 particles in the cell must be taken in order to receive acceptable solution accuracy.

For computer simulation of the second problem we need to take into account the high gradient of magnetic field that requires the step h of spatial grid no more than 0.1 mm, and so for the size of computational box of 1355 mm $\times$ 99.6 mm the selected grid nodes are 13550 $\times$ 996. Up to 10^{10} particles are used in computer simulation. To evaluate the efficiency of plasma confinement distribution of the particles was calculated to acquire inverse magnetic mirrors of a trap. Radial distribution of ion density in the field of inverse magnetic mirrors is given on Fig. 2, where 1 - H^+, 2 - H_2^+, 3 - H_3^+. The figure demonstrates that the ion density decreases rapidly along the radius, the density of ions H_2^+ and H_3^+ becomes smooth in radial direction. However, the concentration of ions of different types decreases abruptly with the radius of 25 mm. This means that only particles with high shear energy move out of a trap.

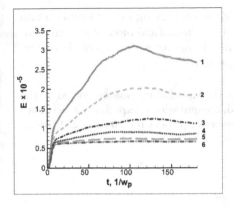

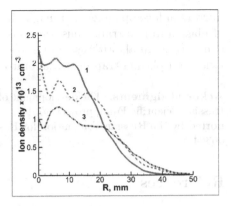

Fig. 1. The electron energy time history for different numbers of particles in a cell (lp).

Fig. 2. Radial distribution of density of ions in the field of inverse magnetic mirrors.

The average density of plasma achieved in calculations has made up to $2 * 10^{13} cm^3$. Results of computer simulation good agree with results of laboratory experiments [2].

The efficient scalable parallel algorithm with the mixed Euler-Lagrangian decomposition was developed. A group of processors is associated with each sub-domain, and particles are divided among the processors group within each sub-domain. A group of processors solves the equations for the fields only in the sub-domain. In this case, there is an exchange of the boundary field values between the groups. Also, the group should exchange with the particles, which move to other sub-domain. It is necessary to exchange current density values within the group.

5   Conclusion

In the paper the model and program for the simulation of nonlinear effects in open plasma traps are presented. The mathematical model of a target trap is developed for highly effective neutralization of powerful beam of negative ions. The model is based on the combination of the PIC-method and the Monte-Carlo method and describes the plasma column holding in the vacuum chamber by the magnetic field of complex geometry. Calculation of current density in a cylindrical coordinate system is based on the continuity equation of direct formula. This allows getting rid of global data dependencies and build scalable parallel algorithm. Analysis of influence of computational parameters on the solution is made, the optimum quantity of particles is defined.

Using of supercomputers permitted calculating plasma dynamics, and determining plasma flows at the end-trap holes. Numerical experiments have shown that the magnetic system with the weak longitudinal field and inverse magnetic mirrors in face openings in a magnetic field allows achieving rather small stream of plasma from a trap. Thus, the constructed mathematical models and efficient computational algorithms can be used for the important estimates of characteristics of a plasma trap in the laboratory experiment.

Acknowledgments. The development of numerical algorithms was supported by the Russian Scientific Fund, project no. 14-11-00485, computational experiments were supported by the Russian Foundation for Basic Research, project no. 16-01-00209, 16-31-00304.

References

1. Burdakov, A.V., Avrorov, A.P., Arzhannikov, A.V., Astrelin, V.T., et al.: Development of extended heating pulse operation mode at GOL-3. Fusion Sci. Technol. **63**(1T), 29–34 (2013)
2. Dimov, G.I., Emelev, I.S.: Experiments to study the confinement of a target plasma in a magnetic trap with inverse plugs and circular multipole walls. Tech. Phys. **59**(2), 181–189 (2014)
3. Il'in, V.P., Skopin, I.N.: About performance and intellectuality of supercomputer modeling. Program. Comput. Softw. **42**(1), 5–16 (2016)
4. Berendeev, E.A., Efimova, A.A.: Effective parallel calculations realization for modeling big problems of the plasma physics by the PIC-method. Vestnik UGATU **17**(2(55)), 112–116 (2013)
5. Efimova, A.A., Berendeev, E.A., Dudnikova, G.I., Vshivkov, V.A.: Numerical simulation of nonlinear processes in a beam-plasma system. In: AIP Conference Proceedings, vol. 1684, p. 100001 (2015)
6. Birdsall, C.K., Langdon, A.B.: Plasma Physics via Computer Simulation. Institute of Physics Publishing, Bristol (1991)
7. Berezin, Y.A., Vshivkov, V.A.: Particles Method in the Dynamics of a Rarefied Plasma. Novosibirsk, Nayka (1980)
8. Hockney, R.W., Eastwood, J.W.: Computer Simulation Using Particles. CRC Press, Boca Raton (1988)
9. Boris, J.P.: Relativistic plasma simulation - optimization of a hybrid code. In: Fourth Conference on Numerical Simulation of Plasmas, Washington, pp. 3–67 (1970)
10. Vahedi, V., Surendra, M.: A Monte Carlo collision model for the particle-in-cell method: applications to argon and oxygen discharges. Comput. Phys. Commun. **87**, 179–198 (1995)
11. Langdon, A.B., Lasinski, B.F.: Electromagnetic and relativistic plasma simulation models. Methods Comput. Phys. **16**, 327–366 (1976)
12. Esirkepov, T.Z.: Exact charge conservation scheme for particle-in-cell simulation with an arbitrary form-factor. Comput. Phys. Commun. **135**, 144–153 (2001)
13. Barthelme, R.: Conservation de la charge dans les codes PIC. C.R. Math. **341**(11), 689–694 (2005)
14. Umeda, T., Omura, Y., Tominaga, T., Matsumoto, H.: A new charge conservation method in electromagnetic particle-in-cell simulations. Comput. Phys. Commun. **156**(1), 73–85 (2003)

Note on a New High Order Piecewise Linear Finite Element Approximation for the Wave Equation in One Dimensional Space

Abdallah Bradji[✉]

Faculty of Sciences, Department of Mathematics, University of Badji Mokhtar-Annaba, Annaba, Algeria
abdallah-bradji@univ-annaba.org
https://www.i2m.univ-amu.fr/~bradji/

Abstract. We consider the piecewise linear finite element method in space for solving the one dimensional wave equation on general spatial meshes. The discretization in time is performed using the Newmark method. We show that the error between the finite element approximate solution and the piecewise linear interpolant of the exact solution is of order $(h+k)^2$ in several discrete norms. We construct a new approximation of order $(h+k)^3$. This new third order approximation can be computed using the same linear systems used to compute the finite element approximate solution with the same matrices while the right hand sides are changed. The matrices used to compute this new high-order approximation are tridiagonal and consequently the systems involving these matrices are easily to solve. The convergence analysis is performed in several discrete norms.

Keywords: Wave equation · Second order hyperbolic equations · Piecewise linear finite element approximations of high convergence order · Nonequidistant meshes · Tridiagonal matrice

1 Preliminaries and Aim of This Paper

We consider the following one dimensional wave problem:

$$u_{tt}(x,t) - u_{xx}(x,t) = f(x,t), \ (x,t) \in \mathbf{I} \times (0,T), \tag{1}$$

where $\mathbf{I} = (0,1)$, $T > 0$, and f is a given function.

Initial conditions are defined by, for given functions u^0 and u^1:

$$u(x,0) = u^0(x) \text{ and } u_t(x,0) = u^1(x), \ x \in \mathbf{I}, \tag{2}$$

Homogeneous Dirichlet boundary conditions are given by

$$u(0,t) = u(1,t) = 0, \ t \in (0,T). \tag{3}$$

Usually, when we need to get high order numerical solutions, we have either to use schemes of high order in case of finite difference and finite volume methods or to

© Springer International Publishing AG 2017
I. Dimov et al. (Eds.): NAA 2016, LNCS 10187, pp. 235–242, 2017.
DOI: 10.1007/978-3-319-57099-0_24

increase the degree of the polynomials involved in the finite element spaces in case of finite element methods. However, these two paths lead in general to linear systems with many non zero entries. The path we will fellow here is to construct high order approximations without make appeal neither to schemes of high order nor finite element spaces with high degree polynomials. We use instead low order schemes in which the matrices used to compute the approximate solutions are in general sparse and then we construct high order approximations using simple sparse matrices.

2   A Basic Scheme and Description of the Approach

Thanks to the Sobolev imbeddings, any function in $H^1(\mathbf{I})$ can be identified by a function in $\mathscr{C}(\bar{\mathbf{I}})$ and the following inequality can be easily checked:

$$\|v\|_{\mathscr{C}(\bar{\mathbf{I}})} \leq |v|_{1,\mathbf{I}}, \ \forall v \in H_0^1(\mathbf{I}). \tag{4}$$

The space discretization is performed using a *non-uniform* mesh with the points $0 < x_0 < \ldots < x_{N+1} = 1$. We define $h_i = x_{i+1} - x_i$, for all $i \in [\![0,N]\!]$. We then set $h = \max\limits_{i=0}^{N} h_i$. Let $\mathscr{V}^h$ be the piecewise linear finite element space, i.e.

$$\mathscr{V}^h = \left\{v \in \mathscr{C}(\bar{\mathbf{I}}), v|_{\bar{\mathbf{I}}_i} \in \mathscr{P}_1, \ \forall i \in [\![0,N]\!]\right\}, \tag{5}$$

and we denote by

$$\mathscr{V}_0^h = \left\{v \in \mathscr{C}(\bar{\mathbf{I}}), v|_{\mathbf{I}_i} \in \mathscr{P}_1, \ \forall i \in [\![0,N]\!]\right\} \cap H_0^1(\bar{\mathbf{I}}), \tag{6}$$

where $\mathbf{I}_i$ is the subinterval $[x_i, x_{i+1}]$ and $\mathscr{P}_1$ denotes the space of polynomials of degree less than or equal to 1.

The time discretization is performed using a constant time step $k = \frac{T}{M+1}$, where $M \in \mathbb{N} \backslash \{0\}$, and we shall denote $t_n = nk$, for $n \in [\![0, M+1]\!]$.

Throughout this paper, the letter C stands for a generic positive constant independent of the parameters of the discretization.

Let $\varphi_i \in \mathscr{V}^h$ be the usual piecewise linear basis function associated with the mesh point x_i, i.e.

$$\varphi_i(x_i) = 1 \text{ and } \varphi_i(x_j) = 0, \ \forall i \neq j. \tag{7}$$

For any function $\Psi \in \mathscr{C}(\bar{\mathbf{I}})$, we denote by $\Pi\Psi$ the piecewise linear interpolation of Ψ in the space $\mathscr{V}^h$, that is $\Pi\Psi = \sum\limits_{i=0}^{N+1} \Psi(x_i)\varphi_i$.

In order to define the finite element approximation for our problem (1)–(3), we need to define the following discrete first and second time derivatives:

$$\partial^1 v^n = \frac{v^n - v^{n-1}}{k} \quad \text{and} \quad \partial^2 v^n = \frac{v^n - 2v^{n-1} + v^{n-2}}{k^2}. \tag{8}$$

The family of finite element schemes approximating (1)–(3) we want to study in this work is based on the use of a Newmark's method as discretization in time, see for instance [2] and references therein. We define the finite element approximate solution as $(u_h^n)_{n=0}^{M+1} \in (\mathcal{V}_0^h)^{M+2}$ (see (6)) such that

$$\mathbf{a}(u_h^0, v) = -\left((u^0)_{xx}, v\right)_{\mathbb{L}^2(\mathbf{I})} = \mathbf{a}(u^0, v), \; \forall v \in \mathcal{V}_0^h, \tag{9}$$

$$\mathbf{a}(\partial^1 u_h^1, v) = -\left((\bar{u}^1)_{xx}, v\right)_{\mathbb{L}^2(\Omega)} = \mathbf{a}(\bar{u}^1, v), \; \forall v \in \mathcal{V}_0^h, \tag{10}$$

and for any $n \in [\![1, M]\!]$, find $u_h^{n+1} \in \mathcal{V}_0^h$ such that, for all $v \in \mathcal{V}_0^h$

$$(\partial^2 u_h^{n+1}, v)_{\mathbb{L}^2(\mathbf{I})} + \frac{1}{2}\mathbf{a}(u_h^{n+1} + u_h^{n-1}, v) = \frac{1}{2}(f(t_{n+1}) + f(t_{n-1}), v)_{\mathbb{L}^2(\mathbf{I})}, \tag{11}$$

where

$$\bar{u}^1 = u^1 + \frac{k}{2}((u^0)_{xx} + f(0)), \tag{12}$$

and $(\cdot, \cdot)_{\mathbb{L}^2(\mathbf{I})}$ denotes the $\mathbb{L}^2$ inner product and $\mathbf{a}(\cdot, \cdot)$ is the bilinear form:

$$\mathbf{a}(u, v) = \int_{\mathbf{I}} u_x(x) v_x(x) dx, \quad \forall v \in H^1(\mathbf{I}).$$

We denote by $|\cdot|$, the semi-norm (it is norm on $H_0^1(\mathbf{I})$):

$$|v|^2 = \mathbf{a}(v, v), \quad \forall v \in H^1(\mathbf{I}).$$

Equations (9) and (10) approximate respectively the initial conditions $u(0) = u^0$ and $u_t(0) = u^1$ which appear in (2). Equation (9) allows to compute u_h^0 as an approximation for $u(0)$, whereas (10) allows to compute u_h^1 as an approximation for $u(1)$ as shown below

$$\mathbf{a}(u_h^1, v) = -\left((u^0 + k\bar{u}^1)_{xx}, v\right)_{\mathbb{L}^2(\Omega)} = \mathbf{a}(\bar{u}^1, v), \; \forall v \in \mathcal{V}_0^h. \tag{13}$$

To compute the solution of the finite element scheme (9)–(12) (see [2, pp. 500–501]), we first compute the initial solution u_h^0 using (9) and then we use u_h^0 to compute u_h^1 using (13). Equations (9) and (13) can respectively be written in the following matrix forms

$$AU^0 = \eta^0 \text{ and } AU^1 = \eta^1, \tag{14}$$

where A is a symmetric and positive definite matrix, η^0 is known, η^1 defined in terms of U^0, and $U^n = (U_1^n, \ldots, U_N^n)^T$ with $u_h^n = \sum_{i=1}^N U_i^n \varphi_i$ where φ_i are the basis functions of $\mathcal{V}^h$.

Equation (11) leads to the following linear systems, for each time step $n \in [\![2, M+1]\!]$

$$(M + \frac{k^2}{2}A)U^n = \eta^n, \tag{15}$$

where $M + \dfrac{k^2}{2}A$ is a symmetric, positive definite matrix and η^n is known from the previous steps.

Assume that the exact solution u is satisfying $\mathscr{C}^4([0, T]; H^2(\mathbf{I}))$. Then, the following error estimates are proved in [1]:

- Discrete $\mathbb{L}^\infty(0,T;H_0^1(\mathbf{I}))$–estimate: for all $n \in [\![0,M+1]\!]$

$$|u_h^n - u(t_n)|_{1,\mathbf{I}} \leq C(h+k^2)\|u\|_{\mathscr{C}^4([0,T];H^2(\mathbf{I}))}, \tag{16}$$

- Discrete $\mathscr{W}^{1,\infty}(0,T;\mathbb{L}^2(\mathbf{I}))$–estimate: for all $n \in [\![1,M+1]\!]$

$$\|\partial^1(u_h^n - u(t_n))\|_{\mathbb{L}^2(\mathbf{I})} \leq C(h+k^2)\|u\|_{\mathscr{C}^4([0,T];H^2(\mathbf{I}))}. \tag{17}$$

We will show that the finite element solution (9)–(11) is of order two in the sense of (22)–(24) of Theorem 1 below. Our aim in the present note is to improve this order in the following sense:

1. We compute a new piecewise linear approximation $u_h^{n,1} \in \mathscr{V}_0^h$, for all $\forall n \in [\![0,M+1]\!]$ (using the piecewise linear finite element solution u_h^n defined by (9)–(11)) such that the order of the interpolation of the error between $u_h^{n,1}$ and the exact solution is $(h+k)^3$ in the discrete norms of $\mathbb{L}^\infty(H^1)$ and $\mathscr{W}^{1,\infty}(\mathbb{L}^2)$, i.e. $\|u_h^{n,1} - \Pi u^n\|_{1,\mathbf{I}}$ and $\|\partial^1(u_h^{n,1} - \Pi u(t_n))\|_{\mathbb{L}^2(\mathbf{I})}$ are of order $(h+k)^3$ uniformly in n.
2. For each $n \in [\![0,M+1]\!]$, the new approximation $u_h^{n,1} \in \mathscr{V}_0^h$ can be computed using the same linear systems (14)–(15) by keeping the matrix A, when computing $u_h^{0,1}$ and $u_h^{1,1}$, and the matrix $M + \frac{k^2}{2}A$, when computing $u_h^{n,1}$ for $n \in [\![2,M+1]\!]$, and changing only the right hand sides η^n.
3. Thanks to $u_h^{n,1} \in \mathscr{V}_0^h$, we are able to construct a new third order approximation for $u(t_n)$ (instead of its interpolant $\Pi u(t_n)$) and its first derivatives (both spatial and temporal) using $u_h^{n,1}$ without using any high order scheme. This is possible by adding some suitable piecewise third–degree polynomial to $u_h^{n,1}$ (or its temporal derivative). This will be detailed in a future paper.

To analyze the convergence, we introduce the auxiliary discrete problem: For each $n \in [\![0,M+1]\!]$, find $\bar{u}_h^n \in \mathscr{V}_0^h$ such that, for all $v \in \mathscr{V}_0^h$

$$\mathbf{a}(\bar{u}_h^n,v) = -(u_{xx}(t_n), v)_{\mathbb{L}^2(\mathbf{I})} = (u_x(t_n),v_x)_{\mathbb{L}^2(\mathbf{I})} = \mathbf{a}(u(t_n),v). \tag{18}$$

Taking $n = 0$ in (18), comparing this with (9), and using (2) to get the useful property

$$\bar{u}_h^0 = u_h^0. \tag{19}$$

Using (18) yields, since v_x is constant over each $\mathbf{I}_i$ and $u(x_i) - \Pi u(x_i) = 0$

$$\mathbf{a}(\bar{u}_h^n - \Pi u(t_n),v) = \mathbf{a}(u(t_n) - \Pi u(t_n),v) = \sum_{i=0}^N v_x(x) \int_{\mathbf{I}_i} (u(t_n)(x) - \Pi u(t_n)(x))_x \, dx$$

$$= 0. \tag{20}$$

This yields the following nice property (see also [3, p. 50]), for all $n \in [\![0,M+1]\!]$

$$\bar{u}_h^n = \Pi u(t_n). \tag{21}$$

This means that $\bar{u}_h^n$ agrees identically with $u(t_n)$ at nodes.

The following theorem provides a convergence analysis for the scheme (9)–(11):

Theorem 1 (Convergence analysis for the scheme (9)–(11): a first main result).
Let $0 < x_0 < \ldots < x_{N+1} = 1$ be an arbitrary mesh for $I = (0,1)$ and we define $h_i = x_{i+1} - x_i$. We define the mesh size as $h = \max\limits_{i=0}^{N} h_i$. Let $\mathcal{V}^h$ and $\mathcal{V}_0^h$ be the piecewise linear finite element spaces defined respectively by (5) and (6). For any function $\Psi \in \mathscr{C}(\bar{I})$, we denote by $\Pi\Psi$ the piecewise linear interpolation of Ψ in the space $\mathcal{V}^h$. Let $k = \frac{T}{M+1}$, with $M \in \mathbb{N}^*$, and denote by $t_n = nk$, for $n \in [\![0,M+1]\!]$. Assume that the solution u is satisfying $u \in \mathscr{C}^2([0,T]; \mathscr{C}^2(\bar{I}))$ (this implies that $f \in \mathscr{C}([0,T]; \mathscr{C}(\bar{I}))$ which gives sense to (11)). Then, there exists a unique solution $(u_h^n)_{n=0}^{M+1} \in (\mathcal{V}_0^h)^{M+2}$ for (9)–(11). Assume in addition that that the exact solution u is satisfying $\mathscr{C}^4([0,T]; \mathscr{C}^2(\bar{I}))$. Then, the following error estimates hold:

– *Discrete $\mathbb{L}^\infty(0,T; H_0^1(I))$–estimate: for all $n \in [\![0,M+1]\!]$*

$$|u_h^n - \Pi u^n|_{1,I} \leq C(h+k)^2 \|u\|_{\mathscr{C}^4([0,T];\, \mathscr{C}^2(\bar{I}))}, \tag{22}$$

– *Discrete $\mathbb{L}^\infty(0,T; \mathbb{L}^\infty(I))$–estimate: for all $n \in [\![0,M+1]\!]$*

$$\|u_h^n - \Pi u^n\|_{\mathscr{C}(\bar{I})} \leq C(h+k)^2 \|u\|_{\mathscr{C}^4([0,T];\, \mathscr{C}^2(\bar{I}))}, \tag{23}$$

– *Discrete $\mathscr{W}^{1,\infty}(0,T; \mathbb{L}^2(I))$–estimate: for all $n \in [\![1,M+1]\!]$*

$$\|\partial^1(u_h^n - \Pi u(t_n))\|_{\mathbb{L}^2(I)} \leq C(h+k)^2 \|u\|_{\mathscr{C}^4([0,T];\, \mathscr{C}^2(\bar{I}))}. \tag{24}$$

To prove Theorem 1, we need to use the following *a priori estimate* which has been proved in [1] in a general framework:

Lemma 1 (An a priori estimate, cf. [1]). *Under the same hypotheses of Theorem 1, assume that there exits $(\eta_h^n)_{n=0}^{M+1} \in (\mathcal{V}_0^h)^{M+2}$ such that for all $n \in [\![1,M]\!]$*

$$(\partial^2 \eta_h^{n+1}, v)_{\mathbb{L}^2(I)} + \frac{1}{2}a(\eta_h^{n+1} + \eta_h^{n-1}, v) = (\mathscr{S}^n, v)_{\mathbb{L}^2(I)}, \; \forall v \in \mathcal{V}_0^h, \tag{25}$$

where $\mathscr{S}^n \in \mathbb{L}^2(I)$, for all $n \in [\![1,M]\!]$.
Then the following estimate holds, for all $J \in [\![1,M]\!]$:

$$\|\partial^1 \eta_h^{J+1}\|_{\mathbb{L}^2(I)}^2 + |\eta_h^{J+1}|_{1,I}^2 \leq C\left(\|\partial^1 \eta_h^1\|_{\mathbb{L}^2(I)}^2 + |\eta_h^1|_{1,I}^2 + |\eta_h^0|_{1,I}^2 + (\mathscr{S})^2\right), \tag{26}$$

where $\mathscr{S} = \max\limits_{n=1}^{M} \|\mathscr{S}^n\|_{\mathbb{L}^2(I)}$.

Sketch of the proof of Theorem 1. The existence and uniqueness stem from the fact that the matrices A and $M + \frac{k^2}{2}A$ are definite positive. We first remark that error estimate (23) is a consequence of (22) thanks to inequality (4). To prove the error estimates (22) and (24), we consider the error $\bar{e}_h^n = u_h^n - \bar{u}_h^n$ (recall that $\bar{u}_h^n$ is the solution (18) whose existence and uniqueness stem also from the fact the matrix A is definite positive). The property (21) implies that $\bar{e}_h^n = u_h^n - \Pi u(t_n)$. Using some elementary calculation, we are able to justify that

$$(\partial^2 \bar{e}_h^{n+1}, v)_{\mathbb{L}^2(I)} + \frac{1}{2}\mathbf{a}(\bar{e}_h^{n+1} + \bar{e}_h^{n-1}, v) = (\mathbb{T}^n, v)_{\mathbb{L}^2(I)}, \tag{27}$$

where

$$\mathbb{T}^n = \partial^2(u(t_{n+1}) - \Pi u(t_{n+1})) + \frac{u_{tt}(t_{n+1}) + u_{tt}(t_{n-1})}{2} - \partial^2 u(t_{n+1}). \tag{28}$$

The property (19) implies that $\bar{e}_h^0 = 0$. On another hand, thanks to the choice (10) (recall that $\bar{u}^1$ is given in (12)), we have

$$\mathbf{a}(\partial^1 \bar{e}_h^1, v) = \mathbf{a}(\partial^1 u_h^1, v) - \mathbf{a}(\partial^1 \bar{u}_h^1, v) = -\left((\bar{u}^1 - \partial^1 u(t_1))_{xx}, v\right)_{\mathbb{L}^2(\mathbf{I})}, \ \forall v \in \mathcal{V}_0^h. \tag{29}$$

One remarks that $\bar{e}_h^n$ is satisfying (27)–(29) and therefore $\bar{e}_h^n$ is satisfying hypotheses of Lemma 1, one can apply Lemma 1 to obtain, for all $J \in [\![1, M]\!]$

$$\|\partial^1 \bar{e}_h^{J+1}\|_{\mathbb{L}^2(\mathbf{I})}^2 + |\bar{e}_h^{J+1}|_{1,\mathbf{I}}^2 \le C\left(\|\partial^1 \bar{e}_h^1\|_{\mathbb{L}^2(\mathbf{I})}^2 + |\bar{e}_h^1|_{1,\mathbf{I}}^2 + (\mathcal{S})^2\right), \tag{30}$$

where $\mathcal{S} = \max\limits_{n=1}^{M} \|\mathbb{T}^n\|_{\mathbb{L}^2(\mathbf{I})}$. This, together with some suitable Taylor expansions, yields the error estimates (22) and (24) of Theorem 1. ∎

3 Formulation of a New Third Order Approximation

The formulation of the new third order approximation we want to present includes some *suitable approximations* for the derivatives u_{ttxx} and u_{tttt}. Next, we provide these approximations and then we give the formulation of the third order approximation:

1. *A convenient approximation* $(w^n)_n$ *for* u_{ttxx}. Using equation (1) along with boundary conditions (3) and initial conditions (2) and acting some suitable differentiations, we find that $w = u_{ttxx}$ is satisfying:

$$w_{tt}(x,t) - w_{xx}(x,t) = F(x,t), \ (x,t) \in \mathbf{I} \times (0, T), \tag{31}$$

where $F = f_{ttxx}$. In addition to this, the initial conditions of $w = u_{ttxx}$ are given by

$$w(x,0) = G^0(x) \quad \text{and} \quad w_t(x,0) = G^1(x), \ x \in \mathbf{I}, \tag{32}$$

where

$$G^0(x) = (u^0)_{xxxx}(x) + f_{xx}(x,0) \quad \text{and} \quad G^1(x) = (u^1)_{xxxx}(x) + f_{xxt}(x,0). \tag{33}$$

The boundary conditions are given by

$$w(0,t) = -f_{tt}(0,t) \quad \text{and} \quad w(1,t) = -f_{tt}(1,t), \ t \in (0, T). \tag{34}$$

Since $w = u_{ttxx}$ satisfies Eq. (31) which is similar to (1) satisfied by u with of course different boundary and initial conditions, hence $w = u_{ttxx}$ can be approximated using the same scheme (9)–(11) and changing only the right hand sides as follows:

- Find $w_h^0 \in \mathcal{V}^h$ such that $w_h^0(0) = -f_{tt}(0,0)$, $w_h^0(1) = -f_{tt}(1,0)$, and

$$\mathbf{a}(w_h^0, v) = -\left((G^0)_{xx}, v\right)_{\mathbb{L}^2(\mathbf{I}_i)}, \quad \forall v \in \mathcal{V}_0^h \tag{35}$$

- Find $w_h^1 \in \mathcal{V}^h$ such that $w_h^1(0) = -f_{tt}(0,t_1)$, $w_h^1(1) = -f_{tt}(1,t_1)$, and

$$\mathbf{a}(\partial^1 w_h^1, v) = -\left((\bar{G}^1)_{xx}, v\right)_{\mathbb{L}^2(\mathbf{I}_i)}, \quad \forall v \in \mathcal{V}_0^h \tag{36}$$

where

$$\bar{G}^1 = G^1 + \frac{k}{2}\left((G^0)_{xx} + F(0)\right), \tag{37}$$

- We have $w_h^{n+1}(0) = -f_{tt}(0,t_{n+1})$, $w_h^{n+1}(1) = -f_{tt}(1,t_{n+1})$, and, for all $v \in \mathcal{V}_0^h$

$$\left(\partial^2 w_h^{n+1}, v\right)_{\mathbb{L}^2(\mathbf{I})} + \frac{1}{2}\mathbf{a}(w_h^{n+1} + w_h^{n-1}, v) = \frac{1}{2}\left(F(t_{n+1}) + F(t_{n-1}), v\right)_{\mathbb{L}^2(\mathbf{I})}. \tag{38}$$

2. *Approximation of $z = u_{tttt}$*. Differentiating (1) twice with respect to t gives (recall that $w = u_{ttxx}$) $z(x,t) - w(x,t) = f_{tt}(x,t)$. As an approximation z^n for the unknown function $z(t_n) = u_{tttt}(t_n)$, we suggest

$$z_h^n = w_h^n + f_{tt}(t_n). \tag{39}$$

For each $n \in [\![0, M+1]\!]$, let $s_h^n \in \mathcal{V}_0^h$ be the solution (corrector) of the following scheme:

$$s_h^0 = 0, \tag{40}$$

$$\mathbf{a}(\partial^1 s_h^1, v) = \frac{k^2}{6}\left((f_t(0) + (u^1)_{xx})_{xx}, v\right)_{\mathbb{L}^2(\mathbf{I})}, \quad \forall v \in \mathcal{V}_0^h, \tag{41}$$

and for all $n \in [\![1, M]\!]$

$$\left(\partial^2 s_h^{n+1}, v\right)_{\mathbb{L}^2(\mathbf{I})} + \frac{1}{2}\mathbf{a}\left(s_h^{n+1} + s_h^{n-1}, v\right) = \sum_{i=0}^{N}\int_{\mathbf{I}_i}\alpha_i(x)w_h^n(x_i)v(x)\,dx$$

$$+ \frac{5k^2}{12}\left(z_h^n, v\right)_{\mathbb{L}^2(\mathbf{I})}, \tag{42}$$

where α_i is given by, for all $x \in \mathbf{I}_i$

$$\alpha_i(x) = \frac{1}{2}\left((x-x_i)^2 - \Pi(x-x_i)^2\right) = \frac{1}{2}\left((x-x_i)^2 - h_i^2\varphi_{i+1}(x)\right). \tag{43}$$

Gathering (40) and (41) yields that

$$\mathbf{a}(s_h^1, v) = \frac{k^3}{6}\left((f_t(0) + \Delta u^1)_{xx}, v\right)_{\mathbb{L}^2(\mathbf{I})}, \quad \forall v \in \mathcal{V}_0^h. \tag{44}$$

We are now able to define the new third approximation for the solution of problem (1)–(3): For each $n \in [\![0, M+1]\!]$, let $u_h^{n,1} \in \mathcal{V}_0^h$ be the new approximation given by

$$u_h^{n,1} = u_h^n - s_h^n. \tag{45}$$

Theorem 2 (A new piecewise linear approximation of order three: a second main result). *In addition to the hypotheses of Theorem 1, we assume that the exact solution u is satisfying $\mathscr{C}^6([0,T]; \mathscr{C}^6(\bar{I}))$. Let α_i be defined as in (43) and $F = f_{ttxx}$. Let G^0 and $\bar{G}^1$ be the functions given respectively by (33) and (37). Then, there exist unique solutions $\left(u_h^n\right)_{n=0}^{M+1}, \left(w_h^n\right)_{n=0}^{M+1}, \left(s_h^n\right)_{n=0}^{M+1} \in \left(\mathscr{V}_0^h\right)^{M+2}$ for respectively the schemes (9)–(11), (35)–(38), and (40)–(42) where $\left(z_h^n\right)_{n=0}^{M+1} \in \left(\mathscr{V}_0^h\right)^{M+2}$ involved in the right hand side of (42) is given by (39).*

Let $\left(u_h^{n,1}\right)_{n=0}^{M+1} \in \left(\mathscr{V}_0^h\right)^{M+2}$ be given by (45). Then, the following error estimates hold:

– *Discrete $\mathbb{L}^\infty(0,T; H_0^1(I))$–estimate: for all $n \in [\![0, M+1]\!]$*

$$|u_h^{n,1} - \Pi u^n|_{1,I} \leq C(h+k)^3 \|u\|_{\mathscr{C}^6([0,T]; \mathscr{C}^4(\bar{I}))}, \tag{46}$$

– *Discrete $\mathbb{L}^\infty(0,T; \mathbb{L}^\infty(I))$–estimate: for all $n \in [\![0, M+1]\!]$*

$$\|u_h^{n,1} - \Pi u^n\|_{\mathscr{C}(\bar{I})} \leq C(h+k)^3 \|u\|_{\mathscr{C}^6([0,T]; \mathscr{C}^4(\bar{I}))}, \tag{47}$$

– *Discrete $\mathscr{W}^{1,\infty}(0,T; \mathbb{L}^2(I))$–estimate: for all $n \in [\![1, M+1]\!]$*

$$\|\partial^1\left(u_h^{n,1} - \Pi u(t_n)\right)\|_{\mathbb{L}^2(I)} \leq C(h+k)^3 \|u\|_{\mathscr{C}^6([0,T]; \mathscr{C}^4(\bar{I}))}. \tag{48}$$

Proof. The proof of Theorem 2 can be performed using the same techniques used in the proof of Theorem 1 along with some convenient Taylor expansions. In particular, we use the equality (27) along with a Taylor expansion for the term $\mathbb{T}^n$ given by (28).

4   Conclusion and a Perspective

We considered linear finite element method as discretization in space for the one dimensional wave equation on an arbitrary spatial mesh. The discretization in time is performed using a Newmark method. A *basic scheme* of order two is suggested whose matrices involved are sparse. We suggested an approach allowed to obtain a new third order piecewise linear approximation. This new high order approximation can be computed using the same matrices involved in *basic scheme*. The process we described (in brief) in this note can be repeated to obtain high order approximations with arbitrary order using the same matrices. This path can be extended to some cases in two and three dimensions. This will be detailed in a future paper.

References

1. Bradji, A., Fuhrmann, J.: Some new error estimates for finite element methods for the acoustic wave equation using the Newmark method. Math. Bohem. **139**(2), 125–136 (2014)
2. Quarteroni, A., Valli, A.: Numerical Approximation of Partial Differential Equations. Springer Series in Computational Mathematics, vol. 23. Springer, Berlin (2008)
3. Strang, G., Fix, G.J.: An Analysis of the Finite Element Method. Prentice-Hall Series in Automatic Computation, vol. XIV. Prentice-Hall, Inc., Englewood Cliffs (1973)

Short Rate as a Sum of Two CKLS-Type Processes

Zuzana Bučková[1,2]([✉]), Jana Halgašová[1,3], and Beáta Stehlíková[1]

[1] Department of Applied Mathematics and Statistics,
Comenius University, 842 48 Bratislava, Slovakia
zuzana.zikova.buckova@gmail.com
[2] Chair of Applied Mathematics/Numerical Analysis,
University of Wuppertal, Gaußstraße 20, 42119 Wuppertal, Germany
[3] ČSOB Bank, Bratislava, Slovakia

Abstract. We study the short rate model of interest rates, in which the short rate is defined as a sum of two stochastic factors. Each of these factors is modelled by a stochastic differential equation with a linear drift and the volatility proportional to a power of the factor. We show a calibration methods which - under the assumption of constant volatilities - allows us to estimate the term structure of interest rate as well as the unobserved short rate, although we are not able to recover all the parameters. We apply it to real data and show that it can provide a better fit compared to a one-factor model. A simple simulated example suggests that the method can be also applied to estimate the short rate even if the volatilities have a general form. Therefore we propose an analytical approximation formula for bond prices in such a model and derive the order of its accuracy.

1 Introduction

A discount bond is a security which pays a unit amount of money to its holder at specified time T which is called a maturity of the bond. Its price determines the interest rate for the given maturity. Short rate interest rate models are formulated in terms of a stochastic differential equation (or a system of them in multifactor models) governing the evolution of so called short rate, which is the interest rate for infinitesimal time interval. After specification of the so called market price of risk, the bond prices can be computed as solutions to a parabolic partial differential equation. Alternatively they can be formulate in the equivalent, risk neutral measure, which is sufficient to formulate the partial differential equation problem without any additional input. For more details on short rate models see, e.g. [1,6].

There are many different specifications of the short rate dynamics available in the literature. A popular model, because of its tractability, is Vasicek model [10], where the short rate follows a mean reversion process $dr = \kappa(\theta-r)dt+\sigma dw$, where w is a Wiener process and $\kappa, \theta, \sigma > 0$ are constants. Its generalization with nonconstant volatility has been proposed in [2] in the form $dr = \kappa(\theta-r)dt+\sigma r^\gamma$

© Springer International Publishing AG 2017
I. Dimov et al. (Eds.): NAA 2016, LNCS 10187, pp. 243–251, 2017.
DOI: 10.1007/978-3-319-57099-0_25

with additional parameter $\gamma > 0$, which we will refer to as a CKLS model. In addition to Vasicek model, it encompasses also other known models as special cases (we particularly note Cox-Ingersoll-Ross model [4], CIR hereafter, with $\gamma = 1/2$). Two factor models include models with stochastic volatility, convergence models modelling interest rates in a country before joining a monetary union or models where the short rate is a sum of certain factors (see [1] for a detailed treatment of different interest rate models). We study the last mentioned class of models. In particular, we are concerned with a model where the short rate r is given by $r = r_1 + r_2$ and the risk neutral dynamics of the factors r_1 and r_2 is as follows:

$$
\begin{aligned}
dr_1 &= (\alpha_1 + \beta_1 r_1)dt + \sigma_1 r_1^{\gamma_1} dw_1, \\
dr_2 &= (\alpha_2 + \beta_2 r_2)dt + \sigma_2 r_2^{\gamma_2} dw_2,
\end{aligned}
\tag{1}
$$

where the correlation between increments of Wiener processes is ρ, i.e., $\mathbb{E}(dw_1 dw_1) = \rho dt$. In particular we note that by taking $\gamma_1 > 0$ and $\gamma_2 = 0$ we are able to model negative interest rates (both instantaneous short rate and interest rates with other maturities) which were actually a reality recently in Eurozone (see historical data at [11]. This can be accomplished also by a simple one-factor Vasicek model. However, a consequence of Vasicek model is the same variance of short rate, regardless of its level. On the other hand, the real data suggest that volatilities of interest rates decrease as interest rates themselves decrease. The model with $\gamma_1 > 0$ and $\gamma_2 = 0$ has the variance dependent on the level of factor r_1.

Before using a certain model we need to calibrate it, i.e., estimate its parameters from the available data. One approach to calibration of interest rate models is based on minimizing the weighted squared differences between theoretical yields and the real market ones, see, e.g., [8,9]. Let R_{ij} be the yield observed at i-th day for j-th maturity τ_j and $R(\tau_j, r_{1j}, r_{2j})$ the yield computed from the two factor model, where r_{1i} and r_{2i} are factors of the short rate at i-th day. We denote by w_{ij} the weight of the i-th day and j-th maturity observation in the objective function. In general, we look for the values of the parameters and the decomposition of the short rate to the factors, which minimize the objective function

$$
F(r_{1i}, r_{2i}, \alpha_i, \beta_i, \gamma_i, \sigma_i) = \sum_{i=1}^{n} \sum_{j=1}^{m} w_{ij} \Big(R(\tau_j, r_{1i}, r_{2i}) - R_{ij} \Big)^2.
\tag{2}
$$

In order to solve this optimization problem, we need to evaluate the yields given by the model which is equivalent to solving the PDE for bond prices $P(\tau, r_1, r_2)$, which reads as

$$
-\frac{\partial P}{\partial \tau} + [\alpha_1 + \beta_1 r_1]\frac{\partial P}{\partial r_1} + [\alpha_2 + \beta_2 r_2]\frac{\partial P}{\partial r_2}
$$

$$
+\frac{\sigma_1^2 r_1^{2\gamma_1}}{2}\frac{\partial^2 P}{\partial r_1^2} + \frac{\sigma_2^2 r_2^{2\gamma_2}}{2}\frac{\partial^2 P}{\partial r_2^2} + \rho\sigma_1\sigma_2 r_1^{\gamma_1} r_2^{\gamma_2}\frac{\partial^2 P}{\partial r_1 \partial r_2} - (r_1 + r_2)P = 0
\tag{3}
$$

for any r_1, r_2 from their domain and any time to maturity $\tau \in [0, T)$, with initial condition $P(0, r_1, r_2) = 1$ for any r_1, r_2, see [6]. Closed form solutions are available only in special cases. For the model (1), cf. [1], it is only the Vasicek case $\gamma_1 = \gamma_2 = 0$ and the CIR case $\gamma_1 = \gamma_2 - 1/2$ but only with zero correlation $\rho = 0$ and a mixed model $\gamma_1 = 0$, $\gamma_2 = 1/2$ again with $\rho = 0$. In the remaining cases we need some approximation, which can be obtained using a certain numerical method, Monte Carlo simulation of an approximate analytical solution.

The paper is formulated as follows: In the following section we consider the uncorrelated case of the two-factor Vasicek model, i.e., the model (1) with $\gamma_1 = \gamma_2 = 0$ and $\rho = 0$, and the possibility to estimate its parameters and the short rate factors using the objective function (2). In Sect. 3 we apply this algorithm to real data and we note its advantage in fitting the market interest rates, compared to one-factor Vasicek model. Section 4 present a simulated example which shows a performance of this algorithm when estimating the short rate from a general model (1), i.e., a robustness to missspecified volatility. This motivates us to develop an analytical approximation formula for the bond prices for the model (1) and derive the order of its accuracy which we do in Sect. 5. We end the paper with concluding remarks.

2 Two-Factor Vasicek Model: Singularity and Transformation

In this section we consider the model (1) with $\gamma_1 = \gamma_2 = 0$, in which case the formulae for the bond prices are known, see for example [1]. Moreover we assume that $\rho = 0$, so the increments of the Wiener processed determining the factors of the short rate are uncorrelated. We write the bond price P as

$$\log P(\tau, r_1, r_2) = c_{01}(\tau)r_1 + c_{02}(\tau)r_2 + c_{11}(\tau)\alpha_1 + c_{12}(\tau)\alpha_2 + c_{21}(\tau)\sigma_1^2 + c_{22}(\tau)\sigma_2^2,$$

where, for $k = 1$ and $k = 2$,

$$c_{0k} = \frac{1 - e^{\beta_k \tau}}{\beta_k}, c_{1k} = \frac{1}{\beta_k}\left(\frac{1 - e^{\beta_k \tau}}{\beta_k} + \tau\right), c_{2k} = \frac{1}{2\beta_k^2}\left(\frac{1 - e^{\beta_k \tau}}{\beta_k} + \tau + \frac{(1 - e^{\beta_k \tau})^2}{2\beta_k}\right)$$

We fix the values of β_1 and β_2. Then the objective function (2) can be written as

$$F = \sum_{i=1}^{n}\sum_{j=1}^{m} \frac{w_{ij}}{\tau_j^2}\left(\log P(\tau_j, r_{1i}, r_{2i}) + R_{ij}\tau_j\right)^2$$

$$= \sum_{i=1}^{n}\sum_{j=1}^{m} \frac{w_{ij}}{\tau_j^2}\left(c_{01}(\tau_j)r_{1i} + c_{02}(\tau)r_{2i} + c_{11}(\tau_j)\alpha_1 + c_{12}(\tau_j)\alpha_2 + c_{21}(\tau_j)\sigma_1^2 + c_{22}(\tau_j)\sigma_2^2 + R_{ij}\tau_j\right)^2,$$

which can be represented as a weighted linear regression problem without intercept, with parameters $r_{1i}, r_{2i}, \alpha_1, \alpha_2, \sigma_1^2, \sigma_2^2$ to be estimated. However, the regressors are linearly dependent and hence the estimates minimizing the objective function are not uniquely determined. In the context of calibrating the yield curves, this means that different sets of parameter values and factor evolutions lead to the same optimal fit of the term structures. In particular, we have

$$-\frac{1}{\beta_2}c_{01}(\tau) + \frac{1}{\beta_2}c_{02}(\tau) + \frac{\beta_1}{\beta_2}c_{11}(\tau) = c_{12}(\tau).$$

Substituting this into the formula for the logarithm of the bond price we get

$$\log P(\tau, r_1, r_2) = c_{01}(\tau)r_1 + c_{02}(\tau)r_2 + c_{11}(\tau)\alpha_1 + c_{12}(\tau)\alpha_2 + c_{21}(\tau)\sigma_1^2 + c_{22}(\tau)\sigma_2^2$$
$$= \left(r_{1i} - \frac{\alpha_2}{\beta_2}\right)c_{01}(\tau_j) + \left(r_{2i} + \frac{\alpha_2}{\beta_2}\right)c_{02}(\tau_j)\left(\alpha_1 + \frac{\alpha_2\beta_1}{\beta_2}\right)c_{11}(\tau_j)$$
$$+ c_{21}(\tau_j)\sigma_1^2 + c_{22}(\tau_j)\sigma_2^2.$$

The objective function of the regression problem then reads as

$$F = \sum_{i=1}^{n}\sum_{j=1}^{m}\frac{w_{ij}}{\tau_j^2}\left(\left(r_{1i} - \frac{\alpha_2}{\beta_2}\right)c_{01}(\tau_j) + \left(r_{2i} + \frac{\alpha_2}{\beta_2}\right)c_{02}(\tau_j)\right.$$
$$\left. + \left(\alpha_1 + \frac{\alpha_2\beta_1}{\beta_2}\right)c_{11}(\tau_j) + c_{21}(\tau_j)\sigma_1^2 + c_{22}(\tau_j)\sigma_2^2 + R_{ij}\tau_j\right)^2, \qquad (4)$$

which is already regular. Note that we are not able to estimate all the parameters, nor the separate factors r_1 and r_2. However, the sum of the parameters corresponding to c_{01} and c_{02} is the sum of r_1 and r_2, i.e., the short rate r.

Thus, for a given pair (β_1, β_2) we find the optimal values of the regression problem above and note the attained value of the objective function. Then, we optimize for the values of β_1, β_2. For these optimal β_1, β_2 we note the coefficients corresponding to c_{01} and c_{02}. These are estimated shifted factors and their sum is the estimate of the short rate.

3 Application to Real Data

We use this algorithm to the two data sets considered in paper [5] dealing with estimating the short rate using one-factor Vasicek model: Euribor data from last quarter of 2008 and last quarter of 2011. We note that in the first case, the fit of the one-factor Vasicek was much better then in the second case.

It can be expected that in the case when already a one-factor model provides a good fit, estimating a two-factor model does not bring much change into the results. However, if the fit of a one-factor model is not satisfactory, the estimates from the two-factor model can be more substantially different. From Fig. 1 we can see that the fit of the term structures has significantly improved by adding the second factor in the last quarter of 2011.

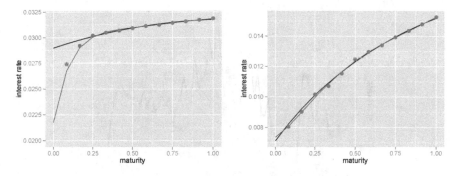

Fig. 1. Fitted yield curves using real data - a selected day in 2008 (left) and 2010 (right): blue lines show the fit from the 2-factor model, black lines from the 1-factor model, red circles are market data (Color figure online)

4   Robustness of the Short Rate Estimates

Naturally, the algorithm described in the previous section works well in case of data simulated from the two-factor Vasicek model. However, we noted the estimate of the short rate is remarkable accurate even when the volatility is misspecified. In particular, since we are able to compute exact bond prices from the two-factor CIR model with uncorrelated factors and test the algorithm on these data.

We simulate two factor CIR model with the parameters taken from [3]: $\kappa_1 = 1.8341, \theta_1 = 0.05148, \sigma_1 = 0.1543,$ $\kappa_2 = 0.005212, \theta_2 = 0.03083, \sigma_2 = 0.06689.$ We simulate daily data from one quarter (assuming 252 trading days in a year). Then, we consider market prices of risk $\lambda)1 = -0.1253, \lambda_2 = -0.06650$ from [3] and compute the term structures for maturities $1, 2, , \ldots, 12$ months for each day using the exact formulae. These data are used as inputs to estimation of the two-factor Vasicek model. A sample result, comparing the simulated short rate and its estimate is presented in Figure 2.

In spite of misspecification of the model, the terms corresponding to $\left(r_{1i} - \frac{\alpha_2}{\beta_2}\right)$ and $\left(r_{2i} + \frac{\alpha_2}{\beta_2}\right)$ indeed estimate the factors up to a constant shift. This is displayed in Fig. 3; note the vertical axis for each pair of the graphs.

5   Approximation of the Bond Prices in the CKLS Model

Based on the example in the previous section, we might want to estimate the short rate by application of the algorithms for the two-factor Vasicek model, even though we expect the volatility to have a more general form. Estimates of the short rate factors, up to an additive constant, might be a valuable results, since their knowledge greatly reduced the dimension of the optimization problem (2). However, we need to compute the bond prices in a CKLS general model - either their exact values or a sufficiently accurate approximation. Since they are

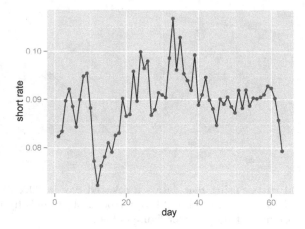

Fig. 2. Estimating short rate using data simulated from the two-factor CIR model: simulated (points) and estimated (line) short rate.

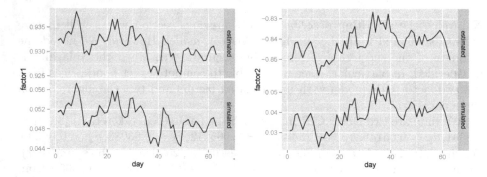

Fig. 3. Estimating factors up to an additive constant using data simulated from the two-factor CIR model.

going to be used in a calibration of a certain kind, they should be calculated quickly and without numerical problems. The aim of this section is to provide an analytical approximation formula for these bond prices and to derive order of its accuracy.

The motivation comes from the paper [7] where an approximation of bond prices for a one-factor CKLS model was proposed. Note that if the correlation in the two-factor CKLS model is zero, the bond price is equal to the sum of two terms corresponding to solutions to bond pricing PDE originating from one factor CKLS models, with factors r_1 and r_2 taking the role of a short rate. Therefore, the bond price could be approximated as a sum of the approximations corresponding to these one-factor models. They are obtained from the Vasicek bond price formula, by substituting its constant volatility by instantaneous volatility from the CKLS model. It is shown in [7] that the error of logarithm of the bond price is then $O(\tau^4)$ as $\tau \to 0^+$. We generalize this idea to the two-factor case and suggest the following approximation.

Theorem 1. *Let P^{ap} be the approximative and P^{ex} be the exact price of the bond in CKLS model. Then for $\tau \to 0^+$*

$$ln P^{ap}(\tau, r_1, r_2) - ln P^{ex}(\tau, r_1, r_2) = c_4(r_1, r_2)\tau^4 + o(\tau^4) \tag{5}$$

where coefficient c_4 is given by

$$c_4(r_1, r_2) = -\frac{1}{24r_1^2 r_2^2}\Big((2\gamma_1^2 - \gamma_1)(r_1^{4\gamma_1} r_2^2 \sigma_1^4) + (2\gamma_2^2 - \gamma_2)(r_1^2 r_2^{4\gamma_2} \sigma_2^4) \tag{6}$$

$$+ \rho\gamma_1(\gamma_1 - 1)r_1^{3\gamma_1} r_2^{\gamma_2+2}\sigma_1^3\sigma_2 + \rho\gamma_2(\gamma_2 - 1)r_1^{\gamma_1+2} r_2^{3\gamma_2}\sigma_1\sigma_2^3 \tag{7}$$

$$+ 2\gamma_2(\alpha_2 + \beta_2 r_2)(\rho\sigma_1\sigma_2 r_1^{2+\gamma_1} r_2^{1+\gamma_2} + \sigma_2^2 r_1^2 r_2^{1+2\gamma_2}) + 2\gamma_1\gamma_2\rho^2\sigma_1^2\sigma_2^2 r_1^{2\gamma_1+1} r_2^{2\gamma_2+1} \tag{8}$$

$$+ 2\gamma_1 r_1 r_2^2 \sigma_1(\alpha_1 + \beta_1 r_1)(r_1^{2\gamma_1}\sigma_1 + \rho\sigma_2 r_1^{\gamma_1} r_2^{\gamma_2})\Big). \tag{9}$$

Remark 1. From the above considerations it follows that $\log P^{ap} - \log P^{ex}$ is $O(\tau^4)$ in the case of zero correlation ρ. What needs to be done is showing that the same order of accuracy is achieved also in the case of general ρ.

Proof. Let us define function $f^{ex}(\tau, r_1, r_2) = ln P^{ex}(\tau, r_1, r_2)$, where P^{ex} is the exact solution of the Eq. (3) Then the partial differential Eq. (3) for f^{ex} is given by:

$$-\frac{\partial f^{ex}}{\partial\tau} + [\alpha_1 + \beta_1 r_1]\frac{\partial f^{ex}}{\partial r_1} + [\alpha_2 + \beta_2 r_2]\frac{\partial f^{ex}}{\partial r_2}$$

$$+ \frac{\sigma_1^2 r_1^{2\gamma_1}}{2}\left[\left(\frac{\partial f^{ex}}{\partial r_i}\right)^2 + \frac{\partial^2 f^{ex}}{\partial r_i^2}\right] + \frac{\sigma_2^2 r_2^{2\gamma_2}}{2}\left[\left(\frac{\partial f^{ex}}{\partial r_i}\right)^2 + \frac{\partial^2 f^{ex}}{\partial r_i^2}\right]$$

$$+ \rho\sigma_1\sigma_2 r_1^{\gamma_1} r_2^{\gamma_2}\left[\frac{\partial f^{ex}}{\partial r_1}\frac{\partial f^{ex}}{\partial r_2} + \frac{\partial^2 f^{ex}}{\partial r_1 \partial r_2}\right] - (r_1 + r_2) = 0.$$

For the approximation $f^{ap}(\tau, r_1, r_2) = ln P^{ap}(\tau, r_1, r_2)$ we obtain from the former PDE equation with nontrivial right-hand side $h(\tau, r_1, r_2)$:

$$-\frac{\partial f^{ap}}{\partial\tau} + [\alpha_1 + \beta_1 r_1]\frac{\partial f^{ap}}{\partial r_1} + [\alpha_2 + \beta_2 r_2]\frac{\partial f^{ap}}{\partial r_2}$$

$$+ \frac{\sigma_1^2 r_1^{2\gamma_1}}{2}\left[\left(\frac{\partial f^{ap}}{\partial r_i}\right)^2 + \frac{\partial^2 f^{ap}}{\partial r_i^2}\right] + \frac{\sigma_2^2 r_2^{2\gamma_2}}{2}\left[\left(\frac{\partial f^{ap}}{\partial r_i}\right)^2 + \frac{\partial^2 f^{ap}}{\partial r_i^2}\right]$$

$$+ \rho\sigma_1\sigma_2 r_1^{\gamma_1} r_2^{\gamma_2}\left[\frac{\partial f^{ap}}{\partial r_1}\frac{\partial f^{ap}}{\partial r_2} + \frac{\partial^2 f^{ap}}{\partial r_1 \partial r_2}\right] - (r_1 + r_2) = h(\tau, r_1, r_2).$$

In the next step we substitute to the previous equation approximation of the bond price and make a Taylor expansion of all the terms with respect to τ:

$$h(\tau, r_1, r_2) = k_3(r_1, r_2)\tau^3 + o(\tau^3),$$

where k_3 reads as

$$
\begin{aligned}
k_3(r_1,r_2) = \frac{1}{6r_1^2 r_2^2} & \Big((2\gamma_1^2 - \gamma_1)(r_1^{4\gamma_1} r_2^2 \sigma_1^4) + (2\gamma_2^2 - \gamma_2)(r_1^2 r_2^{4\gamma_2} \sigma_2^4) \\
& + \rho\gamma_1(\gamma_1 - 1)r_1^{3\gamma_1} r_2^{\gamma_2+2} \sigma_1^3 \sigma_2 + \rho\gamma_2(\gamma_2 - 1)r_1^{\gamma_1+2} r_2^{3\gamma_2} \sigma_1 \sigma_2^3 \\
& + 2\gamma_2(\alpha_2 + \beta_2 r_2)(\rho\sigma_1\sigma_2 r_1^{2+\gamma_1} r_2^{1+\gamma_2} + \sigma_2^2 r_1^2 r_2^{1+2\gamma_2}) + 2\gamma_1\gamma_2\rho^2 \sigma_1^2 \sigma_2^2 r_1^{2\gamma_1+1} r_2^{2\gamma_2+1} \\
& + 2\gamma_1 r_1 r_2^2 \sigma_1(\alpha_1 + \beta_1 r_1)(r_1^{2\gamma_1}\sigma_1 + \rho\sigma_2 r_1^{\gamma_1} r_2^{\gamma_2}) \Big).
\end{aligned}
$$

Let us consider function $g(\tau, r_1, r_2) = f^{ap} - f^{ex}$. It satisfies the equation

$$
\begin{aligned}
& -\frac{\partial g}{\partial \tau} + [\alpha_1 + \beta_1 r_1]\frac{\partial g}{\partial r_1} + [\alpha_2 + \beta_2 r_2]\frac{\partial g}{\partial r_2} + \frac{\sigma_1^2 r_1^{2\gamma_1}}{2}\left[\left(\frac{\partial^2 g}{\partial r_1^2}\right)^2 + \frac{\partial^2 g}{\partial r_1}\right] \\
& + \frac{\sigma_2^2 r_2^{2\gamma_2}}{2}\left[\left(\frac{\partial^2 g}{\partial r_2^2}\right)^2 + \frac{\partial^2 g}{\partial r_2}\right] + \rho\sigma_1\sigma_2 r_1^{\gamma_1} r_2^{\gamma_2}\left[\frac{\partial g}{\partial r_1}\frac{\partial g}{\partial r_2} + \frac{\partial^2 g}{\partial r_1 \partial r_2}\right] \\
& = h(\tau, r_1, r_2) - \sigma_1^2 r_1^{2\gamma_1}\frac{\partial f^{ex}}{\partial r_1}\frac{\partial g}{\partial r_1} - \sigma_2^2 r_2^{2\gamma_2}\frac{\partial f^{ex}}{\partial r_2}\frac{\partial g}{\partial r_2} - \rho\sigma_1\sigma_2 r_1^{\gamma_1} r_2^{\gamma_2}\left[\frac{\partial g}{\partial r_1}\frac{\partial f^{ex}}{\partial r_2} - \frac{\partial g}{\partial r_2}\frac{\partial f^{ex}}{\partial r_1}\right].
\end{aligned}
\tag{10}
$$

Taylor expansion of this equation with respect to τ is given by:

$$
g(\tau, r_1, r_2) = \sum_{i=0}^{\infty} c_i(r_1, r_2)\,\tau^i = \sum_{i=\omega}^{\infty} c_i(r_1, r_2)\tau^i,
$$

where coefficient $c_\omega(r_1, r_2)\tau^\omega$ is the first non-zero term. Thus we have $\partial_\tau g = \omega c_\omega(r_1, r_2)\tau^{\omega-1} + o(\tau^{\omega-1})$. Note that $\omega \neq 0$. Coefficient c_0 can not be the first non-zero term in the expansion, because it represents value of the function g in the maturity time of the bond and hence it equals zero (since both f^{ap} and f^{ex} are equal to 1 at maturity). Except for function $h(\tau, r_1, r_2) = k_3(r_1, r_2)\tau^3 + o(\tau^3)$, all the terms in the Eq. (10) are multiplied by at least one of the derivatives $\partial_{r_1}g$, $\partial_{r_2}g$, which are of order $O(\tau)$. Hence all the terms, except $h(\tau, r_1, r_2)$, are of the order $o(\tau^{\omega-1})$ for $\tau \to 0^+$. Equation (10) then implies

$$
-\omega c_\omega(r_1, r_2)\tau^{\omega-1} = k_3(r_1, r_2)\tau^3.
$$

We get $\omega = 4$, which means that

$$
g(\tau, r_1, r_2) = \ln P^{ap}(\tau, r_1, r_2) - \ln P^{ex}(\tau, r_1, r_2) = -\frac{1}{4}k_3(r_1, r_2)\tau^4 + o(\tau^4).
$$

Note that considering a difference of the logarithms of the bond prices is convenient because of calculation of the relative error and the differences in the term structures.

6   Conclusions

In this paper we studied a particular class of two-factor models of interest rates, in which the short rate is defined as a sum of two CKLS-type processes. We developed a method of estimating the short rate and fitting the term structures for a special Vasicek case model and showed its usefulness by applying it to fitting Euribor interest rates. An example from the simulated data, where the procedure gave a very precise estimate of the short rate even if applied to a data generated from a model with nonconstant volatilities, motivated us to propose an approximation of bond prices in such a model and prove its order of accuracy. We note that besides a precise estimate of the short rate, we have also its decomposition into the factors, but these are shifted by a constant. Still, it provides a lot of information about the process and hence our future work will be concerned with using this information together with the approximation of the bonds which we derived to obtain estimates for all the parameters of the model.

Acknowledgements. The work was supported by VEGA 1/0251/16 grant and by the European Union in the FP7-PEOPLE-2012-ITN Program under Grant Agreement Number 304617 (FP7 Marie Curie Action, Project Multi-ITN *STRIKE – Novel Methods in Computational Finance*).

References

1. Brigo, D., Mercurio, F.: Interest Rate Models - Theory and Practice: With Smile, Inflation and Credit. Springer, Berlin (2007)
2. Chan, K.C., Karolyi, G.A., Longstaff, F.A., Sanders, A.B.: An empirical comparison of alternative models of the short-term interest rate. J. Financ. **47**(3), 1209–1227 (1992)
3. Chen, R.R., Scott, L.: Maximum likelihood estimation for a multifactor equilibrium model of the term structure of interest rates. J. Fixed Income **3**(3), 14–31 (1993)
4. Cox, J.C., Ingersoll, J.E., Ross, S.A.: A theory of the term structure of interest rates. Econometrica **53**(2), 385–407 (1985)
5. Halgašová, J., Stehlíková, B., Bučková, Z.: Estimating the short rate from the term structures in the Vasicek model. Tatra Mountains Math. Publ. **61**, 1–17 (2014)
6. Kwok, Y.K.: Mathematical Models of Financial Derivatives. Springer, Heidelberg (2008)
7. Stehlíková, B.: A simple analytic approximation formula for the bond price in the Chan-Karolyi-Longstaff-Sanders model. Int. J. Numer. Anal. Model. Ser. B **4**(3), 224–234 (2013)
8. Ševčovič, D., Urbánová Csajková, A.: Calibration of one factor interest rate models. J. Electr. Eng. **55**(12/s), 46–50 (2004)
9. Ševčovič, D., Urbánová Csajková, A.: On a two-phase minmax method for parameter estimation of the Cox, Ingersoll, and Ross interest rate model. Cent. Eur. J. Oper. Res. **13**, 169–188 (2005)
10. Vasicek, O.: An equilibrium characterization of the term structure. J. Financ. Econ. **5**(2), 177–188 (1977)
11. EMMI - European Money Markets Institute. Euribor Rates. http://www.emmi-benchmarks.eu/euribor-org/euribor-rates.html

Improving the Convergence of Differential Evolution

Petr Bujok[(✉)]

Department of Computer Science, University of Ostrava,
30. dubna 22, 701 03 Ostrava, Czech Republic
petr.bujok@osu.cz

Abstract. A new variant of differential evolution (DE) algorithm with a selection of mutation strategy based on the mutant point distance (DEMD) is proposed. Three DEMD variants are compared with state-of-the-art DE variants on CEC 2015 problems at four dimension levels. The results show that one of proposed DEMD variants performs best in 35% of the problems compared to the other examined DE algorithms.

Keywords: Differential evolution · Convergence · Euclidean distance · CEC 2015 test suite · Experimental comparison

1 Differential Evolution

Differential evolution (DE) introduced by Storn and Price in [8] is a population-based evolutionary algorithm for problems with a real-valued cost function. DE is one of the most efficient optimization technique for solving a global optimization problem. A global optimization problem is defined in the search space Ω which is limited by its boundary constrains, $\Omega = \prod_{j=1}^{D}[a_j, b_j]$, $a_j < b_j$. The objective function f is defined in all $\boldsymbol{x} \in \Omega$ and the point $\boldsymbol{x}^*$ for $f(\boldsymbol{x}^*) \leq f(\boldsymbol{x}), \forall \boldsymbol{x} \in \Omega$ is the solution of the global optimization problem.

The main problem of DE consists in the control of the convergence. The convergence of DE becomes too fast and then only a local solution is found. When the convergence is too slow, fixed function evaluations are reached before the global solution is found. The speed of the convergence is dependent on the control parameters of DE - mutation scheme, crossover type and the values of F and CR. A comprehensive summary of advanced results in DE research is available in [2,6], where several kinds of mutation and crossover were listed and some adaptive or self-adaptive DE variants are described.

Performance of the several DE strategies was compared in [3]. The performance depends on the speed of convergence (higher speed causes getting stuck of P in a local minimum area) and quality of found solution (higher quality means less function-error values of the solution found by the search process). The mutation variants with best performance are *rand/1* (1), *best/2* (2), and *rand-to-best/1* (3)

$$\boldsymbol{u} = \boldsymbol{x}_{r_1} + F \cdot (\boldsymbol{x}_{r_2} - \boldsymbol{x}_{r_3}) \tag{1}$$

© Springer International Publishing AG 2017
I. Dimov et al. (Eds.): NAA 2016, LNCS 10187, pp. 252–260, 2017.
DOI: 10.1007/978-3-319-57099-0_26

$$u = x_{best} + F \cdot (x_{r_1} - x_{r_2}) + F \cdot (x_{r_3} - x_{r_4}) \tag{2}$$

$$u = x_{r_1} + F \cdot (x_{best} - x_{r_1}) + F \cdot (x_{r_2} - x_{r_3}) \tag{3}$$

where $x_{r_1}, x_{r_2}, x_{r_3}, x_{r_4}$ are mutually different points $r_1 \neq r_2 \neq r_3 \neq r_4 \neq i$ and x_{best} is best point of P.

Another study [14] is focused on the diversity of the population and avoiding a premature convergence of the algorithm. The recommended mutation strategies for increasing the population diversity are $DE/rand/1$ (1), $DE/current$-to-$rand/1$ (4), and $DE/either$-or (5). $DE/best/$ strategies are not recommended because they decrease the population diversity. The K parameter of the mutations $current$-to $rand/1$ and $either$-or is set up in two ways. An experimental study of various DE strategies [11] shows that a higher reliability is achieved by $DE/current$-to-$rand/1$ (4) if $K = rand(0,1)$ and good results also occur for $DE/either$-or (5) when $K = 0.5 \cdot (F + 1)$.

$$u = x_i + K \cdot (x_{r_1} - x_i) + F \cdot (x_{r_2} - x_{r_3}) \tag{4}$$

$$u = \begin{cases} x_{r_1} + F \cdot (x_{r_2} - x_{r_3}) & \text{if } F > rand(0,1) \\ x_{r_1} + K \cdot (x_{r_2} + x_{r_3} - 2 \cdot x_{r_1}) & \text{otherwise} \end{cases} \tag{5}$$

It was also shown that no DE strategy in the experimental comparison is able to outperform all the others in each optimization problem, which corresponds with the result of No-free-lunch theorem [13].

2   Proposed DE Variant with Selection of Mutant Vectors

The main motivation of this paper is to control the speed of the convergence in DE algorithm, namely to prefer exploration at the first stage of the search process and exploitation in the latter stage. We suppose that such a strategy prevents from getting stuck in a local minimum and supports appropriate convergence speed of the algorithm. We generate several (k_{max}) mutation vectors and select only one of them for the crossover and trial point generation. The selection is based on Euclidean distance from a chosen point in the population. The choice of the point and strategy of selection depends on the stage of the search process. A pseudo-code of newly proposed DEMD algorithm is shown in Algorithm 1. We are not able to distinguish the stages strictly a priori, which is valid generally for any optimization problem, so we propose a stochastic approach based on the ratio of ($FES/maxFES$) described in (6).

$$dist = \begin{cases} \sqrt{\sum_{j=1}^{D} (u_{k,j} - x_{i,j})^2} & k = 1,2,3, \quad \text{if } FES/maxFES < rand(0,1) \\ \sqrt{\sum_{j=1}^{D} (u_{k,j} - x_{best,j})^2} & k = 1,2,3, \quad \text{otherwise,} \end{cases} \tag{6}$$

where *FES* is the current count of function evaluations and *maxFES* maximum *FES* allowed for the run. The binomial crossover (7) is applied to the selected mutant point

$$
y_{i,j} = \begin{cases} u_{i,j}, & \text{if } rand_j(0,1) \leq CR \text{ or } j = rand_j(1, D) \\ x_{i,j}, & \text{otherwise.} \end{cases} \tag{7}
$$

Algorithm 1. Differential evolution with mutant distances control (DEMD)

1: initialize population $P = \{x_1, x_2, \ldots, x_N\}$
2: evaluate $f(x_i)$, $i = 1, 2, \ldots, N$
3: **while** stopping condition not reached **do**
4:     **for** $i = 1, 2, \ldots, N$ **do**
5:         **for** $k = 1, 2, \ldots, k_{max}$ **do**
6:             create mutant vector u_k according to kth mutation scheme
7:             compute distances of u_k according to (6)
8:         **end for**
9:         select an appropriate mutant point according to stage and selection strategy
10:        create a new trial vector y by application of the crossover (7)
11:        evaluate $f(y)$
12:        **if** $f(y) \leq f(x_i)$ **then**
13:            insert y into next generation Q
14:        **else**
15:            insert x_i into next generation Q
16:        **end if**
17:    **end for**
18:    $P \leftarrow Q$
19: **end while**

The selection of an appropriate mutant point in statement at line 9 is specified vague. The point with minimal distance from the best point is selected in the second (exploitation) stage. In the first stage, the selection of the mutant point with the maximal distance from the current point x_i prefers exploration but may deteriorate convergence. That is why the strategy selecting the mutant point with the minimal distance from the current point x_i was also considered.

The selection of the mutation strategies for DEMD is based on the results in [3] and in [14]. Two different triplets of DE mutation strategies were chosen for newly proposed DEMD algorithm, hereafter labelled DEMD1 and DEMD2. DEMD1 uses three mutations defined by (1), (2), and (3), while DEMD2 the triplet of (1), (4), and (5) mutations. Moreover, each DEMD version can use a different strategy of the mutant vector selection in the explorative stage. When the mutant point with a minimal distance from the current point in the first stage is used, the label of the version has the "min" suffix, if the mutant point with maximal distance from the current point is selected, the suffix is "max".

Compared to CoDE [12], where three trial points are also created and one is selected according the least value of the cost function, the newly proposed selection is not based on function values but on the distance from points of the population. The computational demands of this approach are smaller, especially in the case when the evaluation of function value is computationally expensive.

Table 1. Medians of function-error values for DEMD variants, $D = 10, 30$.

Problem	$D = 10$			$D = 30$		
	DEMD1min	DEMD1max	DEMD2min	DEMD1min	DEMD1max	DEMD2min
1	0	0	0	**1121.67**	4079.16	1852.86
2	0	0	0	0	0	0
3	**20.0297**	20.0369	20.0412	**20.1486**	20.1652	20.1909
4	2.98488	2.98488	2.98488	45.768	**32.2068**	40.7933
5	131.262	**35.2388**	60.761	1822.42	**1606.5**	1737.77
6	**1.41125**	5.18293	119.641	**1059.13**	1690.38	1938.85
7	0.04064	**0.02915**	0.03723	3.52222	**3.27956**	7.96049
8	**0.31759**	0.41078	16.786	**199.371**	428.318	465.95
9	**100.067**	100.133	100.314	**106.82**	106.996	108.319
10	143.235	**143.109**	143.367	**650.346**	699.658	698.024
11	**2.51798**	3.02793	2.89708	**300.852**	450.744	493.047
12	112.193	**111.794**	112.143	109.677	**109.189**	110.859
13	**0.09273**	0.09433	0.09674	0.01061	0.01061	0.01113
14	**6668.76**	6688.58	6794	**42672.9**	43883.6	44760.5
15	100	100	100	100	100	100
#best	**7**	4	0	**8**	4	0

The control parameters of F and CR in DEMD are adapted during the search process for each point x_i. The parameter of CR_i is initialized randomly from a uniform interval $(0, 1)$ and mutated during the search with a value from $(0, 1)$ with the probability of 0.1, similarly to [1]. Parameter F is adapted as follows. In the first stage, N equidistant values $F_i = i/N$, $i = 1, 2, \ldots, N$ are assigned to the points of population in random rank and modified using $F_i \leftarrow F_i + 0.1 * rand(0, 1)$ in each generation. A similar method is applied in [10]. In the exploitation phase, the value of F_i for each current point is computed as random number from the uniform interval $(0, 1)$ in each generation.

The mutation can cause that a mutant point u moves out of the domain Ω. In such a case, the values of $u_j \notin [a_j, b_j]$ are turned over into Ω by using transformation $u_j \leftarrow 2 \times a_j - u_j$ or $u_j \leftarrow 2 \times b_j - u_j$ for the violated component.

3 Experiments

Adaptive DE variants jDE [1], EPSDE [5], SaDE [7], SHADE [9], CoDE [12], and JADE [15] are used in comparison with the newly proposed DEMD variants.

Table 2. Medians of function-error values for DEMD variants, $D = 50, 100$.

Problem	$D = 50$			$D = 100$		
	DEMD1min	DEMD1max	DEMD2min	DEMD1min	DEMD1max	DEMD2min
1	**45010.8**	84218.6	61761.7	**306931**	660208	492705
2	0	0	0	0	0	3.21E-06
3	**20.1967**	20.2368	20.2713	**20.3021**	20.3711	20.4255
4	107.84	**77.6067**	112.43	321.37	**227.845**	400.967
5	3626.16	**3249.69**	3466.16	10754.3	**9769.15**	10283.1
6	**5859.76**	11801.1	9089.55	**44792.7**	72548.7	114546
7	39.945	41.7814	**15.4363**	101.309	105.935	**99.043**
8	**1630.87**	2408.32	2602.08	**15135.6**	22525.8	23207.7
9	102.206	**102.06**	102.749	**107.406**	107.616	111.064
10	**1385.76**	1523.95	1914.51	**3741.63**	4761.24	4908.87
11	**303.898**	678.555	951.091	1886	**1701.96**	2345.27
12	**116.838**	201.536	117.671	**115.686**	118.722	116.481
13	0.02584	**0.02574**	0.02915	0.06578	**0.06513**	0.0817
14	**52680.9**	52694.8	52735.3	**108887**	108905	108936
15	100	100	101.701	**101.515**	106.708	116.72
#best	**8**	4	1	**9**	4	1

These adaptive DE algorithms are considered, according to [2], the state-of-the-art DE variants. Another adaptive algorithm in this comparison is the recently published IDE [10], which has proven very good performance. Three standard DE using *DE/rand/1/bin, DE/best/2/bin,* and *DE/rand-to-best/1/bin* strategies are also included into experimental comparison and their labels in results are derived from the strategies. These algorithms are compared with three newly proposed DEMD variants: DEMD1min, DEMD1max and DEMD2min. DEMD2max variant is not included into comparison due to its bad performance in preliminary experiments.

The test suite of 15 problems was proposed for a special session on Real-Parameter Numerical Optimization, a part of Congress on Evolutionary Computation 2015 [4]. This session was intended as a competition of optimization algorithms where new variants of algorithms are introduced. All algorithms are implemented and executed in Matlab 2010a on a standard PC. An experimental setting follows the requirements given in the report [4], where 15 minimization problems are also defined.

Our tests were carried out at four levels of dimension, $D = 10, 30, 50, 100$, with 51 independent runs per each test function. The function-error value is computed as the difference between the function value of the current point and the known function value in the global minimum point. The run of the algorithm stops if the prescribed amount of function evaluation $MaxFES = D \times 10^4$ is

Table 3. Mean ranks from Friedman-rank test results for all the algorithms in comparison.

Alg., D	10	30	50	100	*Mean*
DEMD1min	**4.6**	<u>**4.4**</u>	<u>**4.2**</u>	**4.3**	4.3
JADE	<u>**4.7**</u>	**4.6**	<u>**4.2**</u>	<u>**4.2**</u>	4.4
IDE	<u>**4.4**</u>	<u>**4.7**</u>	6	5.5	5.1
DEMD1max	5.1	5.7	<u>5.3</u>	5.4	5.4
jDE	6.1	5.5	5.4	<u>4.8</u>	5.5
EPSDE	6.8	5.3	<u>5.3</u>	6	5.8
SaDE	6	6.8	6.8	6.2	6.5
CoDE	8.7	6.8	6.7	6.6	7.2
DEMD2min	7	7.6	7	7.3	7.3
rand/1	7.4	7.8	8.9	8.5	8.2
SHADE	6.3	7.3	12	7.4	8.3
randbest/1	12.9	12.8	6.8	12.7	11.3
best/2	11.1	11.9	12	12.1	11.6

Table 4. Medians of function-error values and results of Kruskal-Wallis tests, $D = 10, 30$.

Problem	$D = 10$				$D = 30$			
	DEMD1min	JADE	IDE	p	DEMD1min	JADE	IDE	p
1	**0**	**0**	<u>0.00017</u>	0.00000	1121.67	**1.31028**	<u>179030</u>	0.00000
2	0	0	0	1	**0**	**0**	<u>1.20E-06</u>	0.00000
3	**20.0297**	20.0579	20.0631	0.00000	**20.1486**	20.2779	<u>20.4352</u>	0.00000
4	2.98488	3.55754	**1.98992**	0.00000	<u>45.768</u>	**26.5575**	**23.879**	0.00000
5	131.262	52.7562	**15.182**	0.00001	1822.42	**1699.69**	1742.1	0.01542
6	<u>1.41125</u>	**0.41629**	**0.41629**	0.00002	1059.13	941.019	**538.187**	0.00000
7	0.04064	<u>0.30991</u>	**0.02958**	0.00000	**3.52222**	<u>7.87472</u>	5.64696	0.00000
8	**0.31759**	<u>0.53595</u>	**0.31345**	0.00170	199.371	219.249	**93.1529**	0.00073
9	<u>100.067</u>	**100**	100.003	0.00000	<u>106.82</u>	106.547	**105.748**	0.00000
10	143.235	143.108	**141.53**	0.03320	650.346	<u>715.256</u>	**551.086**	0.00000
11	**2.51798**	<u>2.97343</u>	**1.83843**	0.00061	**300.852**	<u>408.796</u>	400	0.00000
12	<u>112.193</u>	111.798	**111.503**	0.00000	<u>109.677</u>	108.972	**108.357**	0.00000
13	**0.09273**	**0.09273**	<u>0.09781</u>	0.00000	<u>0.01061</u>	**0.01049**	**0.01048**	0.01513
14	**6668.76**	**6670.66**	<u>6794</u>	0.00000	**42672.9**	43620	<u>44351.6</u>	0.00000
15	100	100	100	1	100	100	100	1
1st	6	5	8		5	5	7	
2nd	1	1	1		2	4	2	
3rd	3	3	3		4	3	4	

Table 5. Medians of function-error values and results of Kruskal-Wallis tests, $D = 50, 100$.

Problem	$D = 50$				$D = 100$			
	DEMD1min	JADE	EPSDE	p	DEMD1min	JADE	jDE	p
1	45010.8	**7272.18**	<u>175045</u>	0.00000	306931	**76376.1**	<u>1.43E+06</u>	0.00000
2	<u>0</u>	**0**	**0**	0.04785	<u>0</u>	**0**	**0**	0.00000
3	**20.1967**	20.3602	<u>20.8877</u>	0.00000	**20.3021**	20.4627	<u>20.6601</u>	0.00000
4	107.84	**51.7394**	<u>286.909</u>	0.00000	<u>321.37</u>	**154.44**	225.627	0.00000
5	**3626.16**	3532.94	<u>10383.8</u>	0.00000	**10754.3**	10313.6	<u>12710.9</u>	0.00000
6	<u>5859.76</u>	**2661.89**	2172.48	0.00000	44792.7	**11887.2**	<u>188422</u>	0.00000
7	**39.945**	<u>42.5838</u>	41.0566	0.00000	**101.309**	116.745	134.681	0.00000
8	<u>1630.87</u>	**1197.8**	**869.352**	0.00000	15135.6	**3835.97**	<u>79939.2</u>	0.00000
9	**102.206**	<u>102.66</u>	102.456	0.00000	107.406	<u>110.555</u>	**107.077**	0.00000
10	1385.76	<u>1760.11</u>	**1003.44**	0.00000	**3741.63**	<u>4361.36</u>	3791.2	0.00000
11	**303.898**	<u>522.613</u>	**439.034**	0.00000	1886	1259.63	**1079.89**	0.00008
12	**116.838**	116.342	<u>201.536</u>	0.00000	115.686	116.311	114.994	0.14132
13	<u>0.02584</u>	**0.02550**	**0.02516**	0.00000	<u>0.06579</u>	0.06331	**0.06175**	0.00000
14	52680.9	52678.2	**52662.7**	0.00012	108887	108881	108887	0.69090
15	100	100	100	1	101.515	101.302	**100**	0.00000
1st	6	8	7		4	6	6	
2nd	3	1	2		4	2	1	
3rd	4	4	5		3	2	5	

reached or if the minimum function-error in the population is less than 1×10^{-8}. Such an error value is considered sufficiently small for an acceptable approximation of the correct solution. The values of the function-error less than 1×10^{-8} are treated as zero in further processing. The population size of the state-of-the-art DE variants is set $N = 100$, for IDE $N = 50, 100, 200, 200$, and for remaining algorithms $N = 30$. The smaller size of P was determined based on previous experiments of standard DE. The control parameters for the standard DE are set $F = 0.8$ and $CR = 0.8$. The control parameter of DEMD is set $k_{max} = 3$. The remaining control parameters of the algorithms were set up to the values recommended by the authors in original papers.

4   Results

The medians of the function-error values for newly proposed DEMD variants are presented in Tables 1 and 2. In the last row, the count of the best median values per algorithm is given. High efficiency of DEMD1min variant is obvious (wins in $\approx 53\%$), its efficiency increases with increasing dimension D.

The overall performance of all 13 algorithms was compared using Friedman test for medians of function-error values. The null hypothesis on the equal performance of the algorithms was rejected, achieved p value for rejection was

$p < 5 \times 10^{-7}$. Mean ranks of the algorithms are presented in Table 3. Note that the algorithm winning uniquely in all the problems has the mean rank 1 and another algorithm being unique loser in all the problems has the mean rank 13. In the last column of Table 3, the average mean rank is computed for all dimensions. It is obvious that the newly proposed DEMD1min has the least average mean rank.

Based on this comparison, three algorithms with the least mean rank in each dimension are selected and compared in more detail for each problem by Kruskal-Wallis test (Tables 4 and 5). In these Tables, medians of function-error values and p values of Kruskal-Wallis tests are presented. When p value is less than the significance level 0.05, the null hypothesis on the equal performance of the three algorithms is rejected. Counts of wins included shared, unique second positions and unique last positions are summarized at the bottom of the tables. The proposed DEMD1min variant wins in 35%, reaches the second position in 17% of the test problems while it appears on the third position only in 23% of the test problems.

5   Conclusion

Three new variants of DE algorithm with mutation selection based on mutant vector distance (DEMD) are proposed and applied to CEC 2015 test problems. The performance of DEMD variants is compared with the state-of-the-art adaptive DE variants. The results of this comparison show that the best performance is achieved by DEMD1min variant. This DEMD1min variant outperforms significantly the majority of the state-of-the-art adaptive DE variants and therefore could be mentioned at least as comparable to state-of-the-art. This simple idea of mutation selection will be studied in more detail in further research.

Acknowledgments. This work was supported by University of Ostrava from the project SGS08/UVAFM/2016.

References

1. Brest, J., Greiner, S., Boškovič, B., Mernik, M., Žumer, V.: Self-adapting control parameters in differential evolution: a comparative study on numerical benchmark problems. IEEE Trans. Evol. Comput. **10**, 646–657 (2006)
2. Das, S., Suganthan, P.N.: Differential evolution: a survey of the state-of-the-art. IEEE Trans. Evol. Comput. **15**, 27–54 (2011)
3. Jeyakumar, G., Shanmugavelayutham, C.: Convergence analysis of differential evolution variants on unconstrained global optimization functions. Int. J. Artif. Intell. Appl. (IJAIA) **2**(2), 116–127 (2011)
4. Liang, J.J., Suganthan, P.N., Chen, Q.: Problem definitions and evaluation criteria for the CEC 2015 competition on learning-based real-parameter single objective optimization. Technical report, Computational Intelligence Laboratory, Zhengzhou University, Zhengzhou China and Nanyang Technological University (2014). http://www.ntu.edu.sg/home/epnsugan/

5. Mallipeddi, R., Suganthan, P.N., Pan, Q.K., Tasgetiren, M.F.: Differential evolution algorithm with ensemble of parameters and mutation strategies. Appl. Soft Comput. **11**, 1679–1696 (2011)
6. Neri, F., Tirronen, V.: Recent advances in differential evolution: a survey and experimental analysis. Artif. Intell. Rev. **33**, 61–106 (2010)
7. Qin, A.K., Huang, V.L., Suganthan, P.N.: Differential evolution algorithm with strategy adaptation for global numerical optimization. IEEE Trans. Evol. Comput. **13**(2), 398–417 (2009)
8. Storn, R., Price, K.V.: Differential evolution - a simple and efficient heuristic for global optimization over continuous spaces. J. Glob. Optim. **11**, 341–359 (1997)
9. Tanabe, R., Fukunaga, A.: Success-history based parameter adaptation for differential evolution. In: IEEE Congress on Evolutionary Computation (CEC), pp. 71–78, June 2013
10. Tang, L., Dong, Y., Liu, J.: Differential evolution with an individual-dependent mechanism. IEEE Trans. Evol. Comput. **19**(4), 560–574 (2015)
11. Tvrdík, J., Bujok, P.: A comparison of various strategies in differential evolution. In: MENDEL 2011, 17th International Conference on Soft Computing, pp. 48–55. University of Technology, Brno (2011)
12. Wang, Y., Cai, Z., Zhang, Q.: Differential evolution with composite trial vector generation strategies and control parameters. IEEE Trans. Evol. Comput. **15**, 55–66 (2011)
13. Wolpert, D.H., Macready, W.G.: No free lunch theorems for optimization. IEEE Trans. Evol. Comput. **1**, 67–82 (1997)
14. Zaharie, D.: Differential evolution: from theoretical analysis to practical insights. In: MENDEL 2012, 18th International Conference on Soft Computing, pp. 126–131. University of Technology, Brno (2012)
15. Zhang, J., Sanderson, A.C.: JADE: adaptive differential evolution with optional external archive. IEEE Trans. Evol. Comput. **13**, 945–958 (2009)

The Service-Oriented Multiagent Approach to High-Performance Scientific Computing

Igor Bychkov, Gennady Oparin, Alexander Feoktistov, Vera Bogdanova, and Ivan Sidorov$^{(\boxtimes)}$

Matrosov Institute for System Dynamics and Control Theory
of Siberian Branch of Russian Academy of Sciences,
Lermontov str. 134, 664033 Irkutsk, Russia
{bychkov,oparin,agf,bvg,ivan.sidorov}@icc.ru
http://www.idstu.irk.ru/

Abstract. The tools for intelligent management of high-performance computing in a heterogeneous distributed computing environment for solving large scientific problems are represented and the service-oriented multiagent approach to solve such problems using these tools is proposed. A purpose of our research is expansion of opportunities for management of the considered environment. Advantages of the proposed approach as compared with approaches based on use of the traditional systems for a distributed computing management are illustrated with two examples of scientific services. Experimental results show a high scalability and efficiency for calculations carried out with use of these services.

Keywords: Scientific services · High-performance computing · Agents

1 Introduction

At present, effective solving large-scale problems of mathematical modeling on the base of numerical methods demand the use of a high-performance computing (HPC) systems [1] including heterogeneous distributed computing environments (HDCE) such as Grid-systems or Cloud-infrastructures. Moreover, some problems cannot be solved without the use of HPC systems due to the high computational complexity of the problems. A modern service-oriented HDCE has some properties [2,3] that significantly complicate processes of a distributed computing management (DCM). These include the following properties: a dynamism of a HDCE; a spectrum diversity of solved problems; a shared use of scarce resources of a HDCE by users, who pursue their subjective purposes of resources exploitation; a use of the model Application-as-a-Service; a need to support a scalability of computing, fault-tolerance of HDCE nodes, and quality of services. The traditional management systems for distributed computing such as PBS [4] or GridWay [5] do not solve all problems arising from the listed properties and do not fully take into account a specificity of problems [6]. A practically significant approach to solve these problems is the intellectualization of management systems using multiagent technologies. Usually, the use of multiagent

© Springer International Publishing AG 2017
I. Dimov et al. (Eds.): NAA 2016, LNCS 10187, pp. 261–268, 2017.
DOI: 10.1007/978-3-319-57099-0_27

systems (MAS) with economic mechanisms [7] for a DCM provides improving the efficiency for computing processes of users depending on the degree of the user interest in such improvements [8,9]. A number of successful solutions for a multiagent DCM in a HDCE are known [10,11]. However, many MAS for a DCM are created in the single copy for the unique HDCE. Besides, at present developers do not have the necessary high-level tools for mass creating and using MAS [12]. The high-level tools [13] for creating services exist, but the agent development in the form of services with the orientation to some subject area remains the relevant problem today.

In this paper the tools for a multiagent DCM in a HDCE are represented and the approach to solving scientific problems with use of these tools is proposed. The main purpose of our research is to expand opportunities of a multiagent DCM in a HDCE. We consider two examples of services implemented with applying the represented approach, which illustrate high scalability and efficiency of a multiagent DCM with use these services. The services are intended for research in both classes of problems, which are very relevant at present [14,15].

2   Multiagent Management for Distributed Computing

The computation management scheme in a HDCE is represented in Fig. 1. In this scheme all systems operate with the same aggregated model of a HDCE. Unlike known models [16] that usually describe the partial aspects of a distributed computing, this model allows more fully to specify various components and processes of a HDCE including knowledge about their use. For example, the models, methods and algorithms for planning of parallel programs, job flows distribution, computing resources allocation and job scheduling as well as software-hardware infrastructure of a HDCE, subject areas for solved problems and other aspects of distributed computing can be specified with the use of this model.

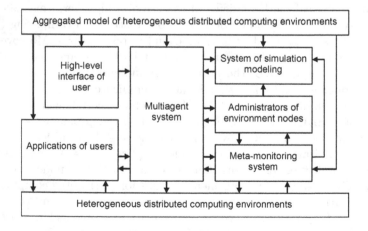

Fig. 1. Computation management scheme.

The high-level user interface is intended to define a hierarchical structure of MAS, construct agents, and configure these agents to work with the models and algorithms of a DCM in a HDCE. Capabilities of the interface are realized with use the HpcSoMaS Framework [17]. Agents are implemented as services. The hierarchical structure of MAS can include two or more functional layers for agents. At each layer agents can play a variety of roles and perform different functions. Layers differ between themselves because of the extent of agent knowledge at the concrete layer. Agents of the higher level have more knowledge as compared with agents of the lower level. However, knowledge of agents at the lower layers is not fully replicated at the higher levels. Agents of the lower levels have much unique knowledge, and agents of the higher levels can request this knowledge when it is necessary. A study of the possible changes a HDCE state and prediction of events in its nodes are carried out by the system of simulation modeling. The features concerning simulation modeling of a HDCE are considered in [17]. The meta-monitoring system is used to test HDCE nodes, to collect data about current state of HDCE nodes, to detect nodes faults, to diagnose and to partially repair these faults. In details this meta-monitoring system is described in [18]. Administrators of HDCE nodes define settings for the meta-monitoring system, simulation modeling system and multiagent system.

It is assumed that the user application is scalable. A scalability means that the problem solving time decreases in inverse proportion to number computational units used, taking into account the unit performance in a HDCE node. The application includes library applied programs for parallel solving a problem in various HDCE nodes. Such library must include program modules for the problem decomposition depending on data, pre-processing and post-processing of input and output data. In the particular case the automated methods for forming value combinations of input variables for solved problem on the basis of the specified value domains and steps of the value changes are applied for organization of multivariant calculations. The use of such methods is assigned to the MAS. Data files of users with value combinations of input variables for solved problem can also be used. The solved problems must have the ability to decompose into subproblems, which can be run independently in various nodes.

The MAS of a DCM for scientific services has a four-level structure: the first level includes user agents, agent-schedulers, agent of a problem classifying; the second level includes agent-managers; the third level includes of local agents; the fourth level includes agents for parametric adjustment of agents from the higher three levels. The MAS provides a problem formulation, problem structure element formalization and application of various schemes for problems decomposition. These functional features allow to form job flow for a scalables application and effectively allocate resources in the static or dynamic mode. A successful applying the MAS features is provides by using the problem-oriented knowledge, which has been put into aggregated model of a HDCE, and actual information about the current state of a HDCE receiving from the meta-monitoring system.

3   Experimental Results

As the first example, let us consider the scientific service for solving the problem, which is constructing a region of required dynamics in the space of two selected parameters of the regulator of a closed-loop control system. This system is described by a differential equation in which the elements of the matrix A depend continuously on parameters. Finding such regions is an important step in the problems of automated analysis and parametric synthesis of a variety of control systems of dynamic objects. Required dynamics is determined by calculating the necessary indicators and quality control checking of user restrictions (in the form of inequalities) to this parameter. Quality indicator of control is a function of the spectrum of the matrix A, and the restrictions define some region D in the plane of the complex variable $\lambda = \alpha + i\omega$. Traditional criteria of linear analysis and parametric synthesis of dynamic performance such as a stability, stability degree, oscillativity and damping of period are used as indicators.

The service provides a possibility of direct determination by user region D of the required location of the spectrum of the matrix A, which allows expanding the search of quality restrictions for transition processes in the system. As part of the parallel computing paradigm, this problem is naturally reduced to solving a set of independent subproblems (multivariant calculations) to determine the spectrum of the matrix A for changing the parameter values in the specified range $K_{min} \leq K \leq K_{max}$ and $T_{min} \leq T \leq T_{max}$ increments ΔK and ΔT respectively. These subproblems are formed on the base of the numerical grid formed by varying values of parameters K and T. A necessary condition for the solution of the original problem is the requirement to solve each subproblem. The calculation of the eigenvalues of the matrix of any density performed using the algorithms represented in [19]. The output of service represents tabular data and graphical representation of the stability region.

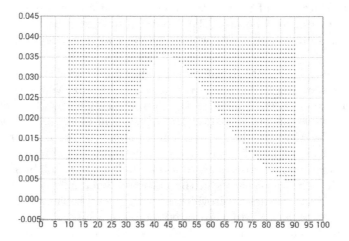

Fig. 2. The stability region of the system (1).

For example, Fig. 2 shows the stability region for the simple model system (1):

$$\begin{cases} \dot{x} = -50\,(K+1)\,z, \\ \dot{y} = x - 50\,[1.3 + K\,(T+0.01)]\,z, \\ \dot{y} = x - 50\,[1.3 + K\,(T+0.01)]\,z, \\ \dot{z} = y - 50\,(0.32 + 0.01KT)\,z, \end{cases} \tag{1}$$

$$0.005 \le T \le 0.04, 10 \le K \le 90.$$

The dimension of the matrix A, and especially the number of points of the numerical grid, in which is determined the stability of the system influences on the difficulty of solving of this problem. The Fig. 3 shows the result of the service operation for the next closed-loop control system with coefficients of the matrix A of the seventh order:

$$\begin{cases} \dot{x}_1 = Tx_7, \\ \dot{x}_2 = x_1 + (T-19)\,x_7, \\ \dot{x}_3 = x_2 + (K+2T-10)\,x_7, \\ \dot{x}_4 = x_3 + (K-T-101)\,x_7, \\ \dot{x}_5 = x_4 + (2K-8)\,x_7, \\ \dot{x}_6 = x_5 - (K+32)\,x_7, \\ \dot{x}_7 = x_6 - 2x_7, \end{cases} \tag{2}$$

$$-55 \le T \le 10, -14 \le K \le 2.$$

Two disconnected stability regions are obtained with the use of this service. This example is taken from [20]. With the help of multivariant calculations for the numerical grid formed for specified intervals of variation of K and T with steps of 0.1 (approximately 65000 pixels) stability region is constructed in a few seconds. Multivariant calculation allows using a numerical grid of a much larger size. The approach on base of the multivariant calculation provides good scalability of this problem in a HDCE for such numerical grids.

As the second example, let us consider the scientific service for solving the SAT-problem [21], which is an NP-complete. On the basis of SAT-service, the SAT-problem can be solved in parallel, and its run-time may be reduced by multivariant calculations. In this case, each independent subproblem of decomposed SAT-problem can be solved at its node of the HDCE, and the multivariant computations can be run in the mode of dynamic choice of resources. If one of subproblem obtains a solution, then the rest of subproblems are removed from running. If after the execution of all subproblems the solution is not found, then the user is informed about the completion of problem solution with the answer unsat. Additionally the service may represent tabular data and graphical representation of a speedup and efficiency of the tests set.

In the experiments, the SAT-solver hpcsat developed by authors [22] and existing SAT-solvers [23] with the open-license were used. Experiments were carried out on a number of Boolean models: the classical test problem of pigeons and holes; the random 3-SAT represented in [23], having clauses in 4.25 times

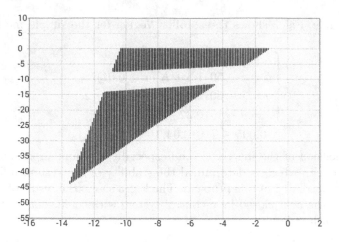

Fig. 3. The stability regions of the system (2).

as many variables; the Euler Knight's Tour problem (formulated as a problem of Boolean satisfiability). Experimental results from the tables were obtained with SAT-solver hpcsat. The speedup was estimated as the ratio of the run-time on one unit of processor cores to N units of processor cores. The unit of processor cores includes 64 cores in experiments. The efficiency of parallel solving was evaluated as the ratio of the speedup to the number of units of processor cores. Speedup and efficiency which are depended on the number of units and the dimension of the Pigeon problem is shown in Table 1. For large-scale SAT-problems, the efficiency was close to 1. The advantages of dynamic resource allocation are represented in Table 2. The average speedup and the efficiency for groups of random 3-SAT instances of the equal size solved with hpcsat represented in Table 3. In comparison with these results HordeSat showed an efficiency of less than 0.5, and its reducing with increasing the number of core from 256 to 512. All experiments were carried out in the HDCE formed on the base of the Irkutsk Supercomputer Center [24].

Table 1. The increase of the speedup and the efficiency with increase of problem size.

Problem	Cores number									
	64	160	320	640	960	64	160	320	640	960
	Speedup					Efficiency				
Pigeon12	1.00	2.22	2.96	3.48	3.81	1.00	0.89	0.59	0.35	0.25
Pigeon13	1.00	2.45	4.61	8.17	11.30	1.00	0.98	0.92	0.82	0.75
Pigeon14	1.00	2.49	4.99	9.82	14.60	1.00	1.00	1.00	0.98	0.97

Table 2. The run-time of Knights Tour Problem (static and dynamic distributing of recourse).

Problem	Variable/clause	Average time, s		Minimum time, s	
		Static	Dynamic	Static	Dynamic
knight8	4096/491024	183.0	132.0	61.0	0.6
knight9	6561/1007603	499.0	359.0	282.0	199.0
knight10	10000/1913276	3599.0	2464.0	651.0	276.0

Table 3. The average value of the efficiency for random SAT-problems.

Problem	Cores number			Problem	Cores number		
	64	256	512		64	256	512
unsat400	1.000000	0.627600	0.765741	sat370	1.000000	0.903296	0.739006
unsat410	1.000000	0.759542	0.674310	sat380	1.000000	0.982782	0.671372
unsat420	1.000000	0.650151	0.519320	sat390	1.000000	1.000000	0.988889
unsat430	1.000000	0.664335	0.557357	sat400	1.000000	1.042545	1.000558
unsat440	1.000000	0.541955	0.524159	sat410	1.000000	0.769553	0.639982

4 Conclusion

The new service-oriented approach to problem-oriented distributed computing in a HDCE with the intelligent management of computing is proposed. This approach provides the expansion of opportunities for management of a HDCE. Experimental results obtained in the real HDCE show high scalability and efficiency for calculations carried out using of the services for problems of different classes. Accounting for the subject area features of solved problems and the monitoring the current state of the used HDCE enabled achieving such results.

Acknowledgments. The research was supported by Russian Foundation of Basic Research, projects no. 15-29-07955-ofi_m and no. 16-07-00931-a, and partially supported by the Council for Grants of the President of the Russian Federation for state support of the leading scientific schools, project NSh-8081.2016.9.

References

1. Gergel, V.P., Linev, A.V.: Exaflop Performance of Supercomputers: Challenges and Trends. Vestnik of Lobachevsky State University of Nizhni Novgorod, no. 3–1, pp. 189–198 (2012). (in Russian)
2. Shamakina, A.V.: Survey on distributed computing technologies. Bull. South Ural State Univ. Ser. Comput. Math. Softw. Eng. **3**, 51–85 (2014). (in Russian)
3. Wu, L., Garg, S.K., Buyya, R.: Service level agreement (SLA) based SaaS cloud management system. In: 21st IEEE International Conference on Parallel and Distributed Systems, pp. 14–17. IEEE Press, Melbourne (2015)

4. PBS Works. http://www.pbsworks.com/
5. GridWay. http://www.gridway.org/doku.php
6. Tao, J., Kolodziej, J., Ranjan, R., Jayaraman, P.P., Buyya, R.: A note on new trends in data-aware scheduling and resource provisioning in modern HPC systems. Future Gener. Comput. Syst. **51**, 45–46 (2015)
7. Toporkov, V.V., Yemelyanov, D.M.: Economic model of scheduling and fair resource sharing in distributed computations. Program. Comput. Softw. **40**, 35–42 (2014)
8. Buyya, R., Venugopal, S.: Market-oriented grid and global grids: an introduction. In: Buyya, R., Bubendorfer, K. (eds.) Market-Oriented Grid and Utility Computing, pp. 3–27. Wiley, Hoboken (2010)
9. Perez-Gonzalez, P., Framinan, J.: A common framework and taxonomy for multicriteria scheduling problems with interfering and competing jobs: multi-agent scheduling problems. Eur. J. Oper. Res. **235**, 1–16 (2014)
10. Bogdanova, V.G., Bychkov, I.V., Korsukov, A.S., Oparin, G.A., Feoktistov, A.G.: Multiagent approach to controlling distributed computing in a cluster grid system. J. Comput. Syst. Sci. Int. **53**, 713–722 (2014)
11. Kalyaev, A.I., Kalyaev, I.A., Korovin, S.J.: Method of multiagent dispatching resources in heterogeneous cloud environments while performing flow of incoming tasks. Herald Comput. Inf. Technol. **137**, 31–40 (2015). (in Russian)
12. Kravari, K., Bassiliades, N.: A Survey of Agent Platforms. http://jasss.soc.surrey.ac.uk/18/1/11.html
13. Buyya, R., Vecchiola, C., Selvi, S.T.: Mastering Cloud Computing. Morgan Kaufmann, Burlington (2013)
14. Sokolov, V.F.: Adaptive stabilization of minimum phase plant under Lipschitz uncertainty. Autom. Remote Control **73**, 405–415 (2016)
15. Vizel, Y., Weissenbacher, G., Malik, S.: Boolean satisfiability solvers and their applications in model checking. Proc. IEEE **103**, 2021–2035 (2015)
16. Toporkov, V.V.: Models of Distributed Computing. Fizmatlit Publ, Moscow (2004). (in Russian)
17. Bychkov, I.V., Oparin, G.A., Feoktistov, A.G., Bogdanova, V.G., Pashinin, A.A.: Service-oriented multiagent control of distributed computations. Autom. Remote Control **76**, 2000–2010 (2015)
18. Bychkov, I., Oparin, G., Novopashin, A., Sidorov, I.: Agent-based approach to monitoring and control of distributed computing environment. In: Malyshkin, V. (ed.) PaCT 2015. LNCS, vol. 9251, pp. 253–257. Springer, Cham (2015). doi:10.1007/978-3-319-21909-7_24
19. Wilkinson, J.X., Reinsch, C.: Handbook for Automatic Computation. Volume II: Linear Algebra. Springer, Heidelberg (1971)
20. Gryazina, E.N., Polyak, B.T.: Multidimensional stability domain of special polynomial families. Autom. Remote Control **68**, 1608–3032 (2007)
21. Gu, J., Purdom, P.W., Franco, J., Wah, B.: Algorithms for the satisfability problem: a survay. DIMACS Ser. Discret. Math. Theoret. Comput. Sci. Am. Math. Soc. **35**, 19–151 (1997)
22. Bogdanova, V.G., Gorsky, S.A., Pashinin, A.A.: Service-oriented toolkit for solving of Boolean satisfiability problem. Fundam. Res. **2–6**, 1151–1156 (2015). (in Russian)
23. Balyo, T., Sanders, P., Sinz, C.: HordeSat: a massively parallel portfolio SAT solver. In: Heule, M., Weaver, S. (eds.) SAT 2015. LNCS, vol. 9340, pp. 156–172. Springer, Cham (2015). doi:10.1007/978-3-319-24318-4_12
24. Irkutsk Supercomputer Center. http://hpc.icc.ru

Innovative Integrators for Computing the Optimal State in LQR Problems

Petra Csomós[1]([✉]) and Hermann Mena[2,3]

[1] Eötvös Loránd University and MTA-ELTE Numerical Analysis and Large Networks Research Group, Pázmány Péter st. 1/C, Budapest 1117, Hungary
csomos@cs.elte.hu
[2] Department of Mathematics, University of Innsbruck, Innsbruck, Austria
hermann.mena@uibk.ac.at
[3] School of Mathematical Sciences and Information Technology,
Yachay Tech, Urcuqui, Ecuador
mena@yachaytech.edu.ec
http://www.cs.elte.hu/~csomos, http://homepage.uibk.ac.at/~c7021020

Abstract. We consider the numerical approximation of linear quadratic optimal control problems for partial differential equations where the dynamics is driven by a strongly continuous semigroup. For this problems, the optimal control is given in feedback form, i.e., it relies on solving the associated Riccati equation and the optimal state. We propose innovative integrators for solving the optimal state based on operator splitting procedures and exponential integrators and prove their convergence. We illustrate the performance of our approach in numerical experiments.

Keywords: Abstract LQR problems · Optimal control · Operator splitting procedures · Exponential integrators · Convergence analysis

Optimal control plays an important role in science and engineering. Particularly, the linear quadratic regulator (LQR) for partial differential equations arises naturally in many applications see e.g. [11,12]. For this problem the feedback characterization of the optimal control is given in terms of Riccati equations. Thus, the computational cost relies on solving the associated Riccati equations and computing the optimal state. Numerical methods for solving large scale Riccati equations arising from the discretization have been proposed in the literature [4]. In this paper, we propose efficient integrators based on operator splitting procedures and exponential integrators for computing the optimal state. The advantage of their use is the separate numerical treatment of the dynamics and the control part. Hence, it is unnecessary to approximate the combined operator containing the dynamics and the control. More precisely, when assuming that the dynamics is driven by a strongly continuous semigroup, operator splitting procedures and exponential integrators make it possible to determine it in the most convenient/usual way, and to treat the control part separately. Therefore, we assume that the semigroup corresponding to the dynamics can be determined or approximated efficiently, and we show how to treat the control part in order to ease the computations compared to the usual case when they are treated together. We will show that application of the innovative integrators mentioned above leads to first-order convergent method which yield more accurate and stable numerical results then the usual methods.

© Springer International Publishing AG 2017
I. Dimov et al. (Eds.): NAA 2016, LNCS 10187, pp. 269–276, 2017.
DOI: 10.1007/978-3-319-57099-0_28

1   Abstract LQR Problem

We introduce now the LQR problem in an abstract setting, for a detailed explanation we refer to [11]. We treat partial differential equations whose reformulation leads to an abstract Cauchy problem. In particular, we consider the separable Hilbert spaces $\mathcal{H}$ and $\mathcal{U}$ called state space and control space, respectively, and the linear operators $A\colon D(A) \subset \mathcal{H} \to \mathcal{H}$ and $B\colon \mathcal{U} \to \mathcal{H}$. Then the state equation is given by the initial value problem for the unknown differentiable function $x\colon [0, +\infty) \to \mathcal{H}$ as

$$\begin{cases} \frac{\mathrm{d}}{\mathrm{d}t}x(t) = Ax(t) + Bu(t), & t > 0, \\ x(0) = x_0 \in \mathcal{H} \end{cases} \tag{1}$$

where $u \in L^2(0, \infty; \mathcal{U})$ is the control function. In addition, a quadratic cost functional is assumed:

$$J(x_0, u) = \frac{1}{2} \int_0^\infty \big(\langle x(t), Qx(t) \rangle_{\mathcal{H}} + \langle u(t), Ru(t) \rangle_{\mathcal{U}} \big) \mathrm{d}t, \tag{2}$$

where $\langle \cdot, \cdot \rangle$ represents an inner product in the corresponding Hilbert space, and the operators $Q\colon \mathcal{H} \to \mathcal{H}$ and $R\colon \mathcal{U} \to \mathcal{U}$ are weight operators to be specified for each application.

Throughout the paper we suppose the following standard assumptions.

Property 1. 1. Operator $A\colon D(A) \subset \mathcal{H} \to \mathcal{H}$ is the generator of the strongly continuous semigroup $(e^{tA})_{t \geq 0}$ on $\mathcal{H}$.
2. Operator $B\colon \mathcal{U} \to \mathcal{H}$ is linear and bounded.
3. For every initial value $x_0 \in \mathcal{H}$, there exists a control function $u \in L^2(0, \infty; \mathcal{U})$ for which $J(x_0, u) < +\infty$ holds (then u is called an admissible control function).
4. Operator $Q : \mathcal{H} \to \mathcal{H}$ is positive semi-definite, and operator $R\colon \mathcal{U} \to \mathcal{U}$ is positive definite.

It was shown e.g. in [5] that under Property 1, the solution to the abstract LQR problem (1) can be obtained analogously to the finite-dimensional case, presented e.g. in [4, 8, 11], as a feedback control with the control function

$$u(t) = -R^{-1}B^*Xx(t), \quad t \geq 0, \tag{3}$$

where $X\colon \mathcal{H} \to \mathcal{H}$ represents the solution to the algebraic operator Riccati equation

$$Q + A^*X + XA - XBR^{-1}B^*X = 0, \tag{4}$$

being unique under Property 1, see e.g. in [5, Theorem V.3.1]. We note that function u defined in formula (3) minimizes the cost function (2) and is called an optimal control. By inserting the *optimal control* (3) into problem (1), we define the equation of the *optimal state* as

$$\begin{cases} \frac{\mathrm{d}}{\mathrm{d}t}x(t) = (A - BR^{-1}B^*X)x(t), & t > 0, \\ x(0) = x_0 \in \mathcal{H}. \end{cases} \tag{5}$$

The paper focuses on the numerical solution of (5). After defining a time step $\tau > 0$, we consider the approximation x_n to the exact solution $x(n\tau)$ for all $n \in \mathbb{N}$, being of the form $x_n = F(\tau)^n x_0$ with $F: [0, +\infty) \to \mathscr{L}(\mathcal{H})$.

Definition 1. *Let $F: [0, +\infty) \to \mathscr{L}(\mathcal{H})$ be a function with the following properties: (i) $F(0) = \mathrm{I}$, the identity operator in $\mathcal{H}$, (ii) the map $0 \le \tau \mapsto F(\tau)h \in \mathcal{H}$ is continuous for all $h \in \mathcal{H}$. A numerical method providing the approximate solution of the form $x_n = F(\tau)^n x_0$, $n \in \mathbb{N}$, is called convergent of order $p \ge 1$ if there exists constant $c > 0$, being independent of n, such that*

$$\|F(\tau)^n x_0 - x(t)\| \le c\tau^p$$

for all $x_0 \in \mathcal{H}$, $n \in \mathbb{N}$, and $t \in [0, T]$ for some $T \ge 0$.

We propose the application of operator splitting procedures and exponential integrators to system (5) and prove their convergence. We also discuss the strength and weakness of the methods.

2 Operator Splitting Procedures

Operator splitting procedures are useful when the equation involves the sum of operators describing different physical/biological/etc. phenomena, thus, having different nature. With their help one solves the sub-problems, corresponding to each single phenomenon, in a cycle. The sub-solutions are then connected by the initial conditions of the sub-problems, namely, each sub-problem uses the solution to the previous sub-problem as initial condition. By defining the operator $C := -BR^{-1}B^*X$, problem (5) is written as

$$\begin{cases} \frac{\mathrm{d}}{\mathrm{d}t}x(t) = (A + C)x(t), & t > 0, \\ x(0) = x_0 \in \mathcal{H}. \end{cases} \tag{6}$$

To ensure that the sub-problems possess unique solutions, we need to ensure that operators A and C are generators as well.

Lemma 1. *Under Property 1, operators $(A, D(A))$, $C \in \mathscr{L}(\mathcal{H})$, and $(A + C, D(A))$ are generators of strongly continuous semigroups.*

Proof. Operator $A: D(A) \subset \mathcal{H} \to \mathcal{H}$ generates the strongly continuous semigroup $(e^{tA})_{t \ge 0}$ on $\mathcal{H}$ by Property 1/(i). It suffices to show that operator $C = -BR^{-1}B^*X: \mathcal{H} \to \mathcal{H}$ is bounded. It was shown e.g. in [3] that operator X is bounded. Since operator $B \in \mathscr{L}(\mathcal{U}, \mathcal{H})$ by Property 1/(ii), we have that $B^* \in \mathscr{L}(\mathcal{H}, \mathcal{U})$. Similarly, since operator R is positive definite by Property 1/(iv), we immediately have $R^{-1} \in \mathscr{L}(\mathcal{U})$. Thus, operator C is a bounded linear operator on $\mathcal{H}$, and therefore generates the strongly continuous semigroup $(e^{tC})_{t \ge 0}$ by [7, Theorem I.3.5]. On the other hand, operator C is then the bounded perturbation of the generator A, therefore, the operator $A + C$ with the domain $D(A + C) = D(A)$ generates the strongly continuous semigroup $(e^{t(A+C)})_{t \ge 0}$ on $\mathcal{H}$ by [7, Theorem III.1.3].

Application of the sequential splitting, proposed by [2], leads to the sub-problems with an arbitrary fixed splitting time step $\tau > 0$:

$$
\begin{cases}
\frac{\mathrm{d}}{\mathrm{d}t} x_A^{(k)}(t) = A x_A^{(k)}(t), & t \in [(k-1)\tau, k\tau), \\
x_A((k-1)\tau) = x_C^{(k-1)}((k-1)\tau),
\end{cases}
$$
$$
\begin{cases}
\frac{\mathrm{d}}{\mathrm{d}t} x_C^{(k)}(t) = C x_C^{(k)}(t), & t \in [(k-1)\tau, k\tau), \\
x_C((k-1)\tau) = x_A^k(k\tau),
\end{cases}
\tag{7}
$$

with $x_C^{(0)}(0) = x_0$, for all $k \in \mathbb{N}$. Then the approximation x_n^{sq} to $x(n\tau)$ equals $x_n^{sq} = x_C^{(k)}(n\tau)$. Another frequently used procedure is the Strang splitting, introduced in [13,14]. In both cases, the solution is of the form $x_n^{spl} = F_{spl}(\tau)^n x_0$, $n \in \mathbb{N}$, where the index 'spl' refers to the various splittings, and the operator $F_{spl}(\tau)$ is defined as

Sequential splitting: $\qquad F_{sq}(\tau) = e^{\tau C} e^{\tau A}$, $\qquad$ (8)

Strang splitting: $\qquad F_{St}(\tau) = e^{\frac{\tau}{2} A} e^{\tau C} e^{\frac{\tau}{2} C}$. $\qquad$ (9)

Since it is difficult to approximate the semigroup operator $e^{t(A+C)}$ in practice, the advantage of applying operator splitting is that one needs the computation of the separate semigroup operators e^{tA} and e^{tC}. One can compute then the effect of the control in the physical space, and the semigroup operator e^{tA} can be computed by using an appropriate numerical method, or even explicitly (e.g. in Fourier space).

In the case when the operators are generators, we present the basic convergence result.

Theorem 1 (Chernoff's product formula, Corollary III.5.3 in [7]). *Let $\mathcal{B}$ be a Banach space, and consider the map $F \colon [0, +\infty) \to \mathscr{L}(\mathcal{B})$ which satisfies the following requirements:*

(a) $F(0) = I$, the identity operator on $\mathcal{B}$,

(b) there exist constants $M \geq 0$, $\omega \in \mathbb{R}$ such that $\|F(\tau)^n\| \leq M e^{n\omega t}$ holds for all $n \in \mathbb{N}$ and $\tau \geq 0$,

(c) there exists $D \subset X$ for which the limit $Gf := \lim_{\tau \searrow 0} \frac{1}{\tau}(F(\tau)f - f)$ exists for all $f \in D$, moreover $(\lambda - A)D$ is dense for some $\lambda > \omega$.

Then the closure $\overline{G}$ of operator G generates the strongly continuous semigroup $(S(t))_{t \geq 0}$ which is given by the limit $S(t)f = \lim_{n \to \infty} (F(\frac{t}{n}))^n f$ for all $f \in \mathcal{B}$ and uniformly for t in compact intervals.

Chernoff's product formula directly implies the convergence of the sequential and Strang splittings for abstract LQR problems.

Proposition 1. *We suppose Property 1 and that operator A generates the quasi-contractive semigroup $(S(t))_{t \geq 0}$ for which there exists $\omega_A \in \mathbb{R}$ such that $\|S(t)\| \leq e^{t\omega_A}$ holds for all $t \geq 0$. Then the sequential splitting (8) is convergent when applied to problem (5).*

Proof. We consider $F(\tau) = F_{\text{sq}}(\tau) = e^{tA}e^{tC}$. We have then $F(0) = I$. Since $(A, D(A))$ generates a quasi-contractive semigroup and operator C is bounded, the stability condition (b) reads for all $t \geq 0$ as

$$\left\|\left(e^{\frac{t}{n}A}e^{\frac{t}{n}C}\right)^n\right\| \leq \left\|e^{\frac{t}{n}A}\right\|^n \left\|e^{\frac{t}{n}C}\right\|^n \leq e^{t(\omega_A + \|C\|)},$$

that is, $M = 1$ and $\omega = \omega_A + \|C\|$. Moreover, due to the boudedness of operator C, we have $D(A + C) = D(A)$ which is dense in $\mathcal{H}$, since A is a generator. Similarly, $A + C$ is a generator with the domain $D(A)$ by Lemma 1, therefore, the set $(\lambda - (A + C))D(A)$ is dense in $\mathcal{H}$ for some/all $\lambda > \omega$.

We remark that the convergence of the Strang splitting follows from Theorem 1 for $F(\tau) = F_{\text{St}}(\tau)$ similarly. The required stability follows again from the quasi-contractivity of the semigroup $(e^{tA})_{t \geq 0}$. We note that the first-order convergence of sequential splitting (8) follows from the idea presented in [10] where the second-order convergence of the Strang splitting (9) is shown.

3   Exponential Integrators

Exponential integrators are efficient tools for solving semilinear problems like

$$\begin{cases} \frac{\mathrm{d}}{\mathrm{d}t}x(t) = Ax(t) + g(x(t)), & t > 0, \\ x(0) = x_0 \in D(A), \end{cases} \tag{10}$$

where the function $g\colon \mathcal{H} \to \mathcal{H}$ has some desirable properties. They are widely investigated in [9]. By solving this problem by the variation of constants, we arrive at

$$x(t) = e^{tA}x_0 + \int_0^t e^{(t-s)A}g(x(s))\mathrm{d}s \approx e^{tA}x_0 + \int_0^t e^{(t-s)A}g(x_0)\mathrm{d}s$$

when approximating $g(x(s)) \approx g(x_0)$ for all $s \in [0, t]$. After choosing a time step $\tau > 0$, we define the exponential Euler method for the numerical solution x_n^{exp} which approximates $x(n\tau)$ for all $n \in \mathbb{N}$ as

$$x_n^{\text{exp}} = e^{\tau A}x_{n-1}^{\text{exp}} + \int_0^\tau e^{(\tau-s)A}g(x_{n-1}^{\text{exp}})\mathrm{d}s, \quad n \in \mathbb{N}. \tag{11}$$

The advantage of applying exponential Euler method is that one only needs to compute the action of operator C on x_n^{exp} which can be efficiently done as presented in [1]. That is, in contrast to the operator splitting procedures, the computation of the semigroup operator e^{tC} is not needed.

Proposition 2. *Consider the semilinear problem* (10) *and take the linear function* $g(x) = Cx$, $x \in \mathcal{H}$, *with operator* $C \in \mathscr{L}(\mathcal{H})$. *For the sake of easier presentation, suppose contraction semigroup operators* $S(t)$ *for all* $t \geq 0$. *Then the exponential Euler method* (11) *is convergent of first order under Property 1.*

Proof. By Definition 1, we have to show that there exists a constant $c > 0$ such that the global error behaves like $\varepsilon_{n+1} = \|x(t_{n+1}) - x_{n+1}^{\exp}\| \leq c\tau$ for all $n \in \mathbb{N}$. The exact and the numerical solutions, respectively, can be written at time $t_{n+1} = (n+1)\tau$ as

$$x(t_{n+1}) = e^{\tau A}x(t_n) + \int_0^\tau e^{(\tau-s)A}Cx(t_n + s)ds,$$

$$x_{n+1}^{\exp} = e^{\tau A}x_n^{\exp} + \int_0^\tau e^{(\tau-s)A}Cx_n^{\exp}ds.$$

The global error e_{n+1} at time t_{n+1} equals then

$$e_{n+1} = x(t_{n+1}) - x_{n+1}^{\exp} = e^{\tau A}e_n + \int_0^\tau e^{(\tau-s)A}C(x(t_n + s) - x_n^{\exp})ds.$$

Since C is the bounded perturbation of the generator A, the exact solution reads as $x(t_n + s) = e^{(t_n+s)(A+C)}x_0$ and can be rewritten by using the Taylor expansion, see [7, Theorem II.1.3]:

$$e_{n+1} = e^{\tau A}e_n + \int_0^\tau e^{(\tau-s)A}Ce_nds + \int_0^\tau e^{(\tau-s)A}C + \int_0^s (A+C)x(t_n + q)dqds.$$

Since $x_0 \in D(A) = D(A+C)$, there exists $K \geq 0$ such that

$$\sup_{q\in[0,\tau]} \|(A+C)x(t_n + q)\| = \sup_{q\in[0,\tau]} \|\tfrac{d}{dt}x(t_n + q)\| = K.$$

By denoting $\|e_n\| = \varepsilon_n$ and taking into account the contractivity of the semigroup operators e^{tA}, $t \geq 0$, we have the estimate

$$\varepsilon_{n+1} \leq \varepsilon_n + \tau\|C\|\varepsilon_n + \tfrac{1}{2}\tau^2\|C\|K = (1 + \tau\|C\|)\varepsilon_n + \tfrac{1}{2}\tau^2\|C\|K.$$

We take now $a = (1 + \tau\|C\|)$ and $b = \tfrac{1}{2}\tau^2\|C\|K$ and by solving the recursion above we arrive at the estimate

$$\varepsilon_{n+1} \leq a^n(a\varepsilon_0 + b) + b\sum_{j=0}^{n-1} a^j = a^n(a\varepsilon_0 + b) + b\frac{a^n - 1}{a - 1} = a^{n+1}\varepsilon_0 + b\frac{a^{n+1} - 1}{a - 1}.$$

For $\varepsilon_0 = 0$, this leads to the following estimate on the global error:

$$\varepsilon_{n+1} \leq \tfrac{1}{2}\tau^2\|C\|K\frac{(1 + \tau\|C\|)^{n+1} - 1}{\tau\|C\|} = \tfrac{1}{2}\tau K\big((1 + \tfrac{t_{n+1}\|C\|}{n+1})^{n+1} - 1\big)$$

$$\leq \tfrac{1}{2}\tau K(e^{t_{n+1}\|C\|} - 1) = c\tau,$$

where c is independent of n itself, which proves the assertion.

4 Numerical Experiments

We consider the following partial differential equation for the unknown function $w\colon \mathbb{R}_0^+ \times \mathbb{R} \to \mathbb{R}$:

$$\begin{cases} \partial_t w(t,\xi) = -\alpha \partial_\xi w(t,\xi) - \beta w(t,\xi), & t > 0,\ \xi \in (0,1) \\ w(0,\xi) = w_0(\xi), & \xi \in (0,1) \end{cases} \tag{12}$$

with $\alpha = \beta = 10^{-3}$ and periodic boundary condition. Problem (12) is a basic model for distributed control acting over the whole spatial domain. The initial function has the form $w_0(\xi) = e^{-10^3(\xi-0.3)^2}$. Problem (12) possesses the exact solution $w(t,\xi) = e^{-\beta t} w_0(\xi - \alpha t)$, and can be formulated as an abstract LQR problem (5) for the optimal state on $\mathcal{H} = L^2(\mathbb{R})$ with $Ah := h'$, $D(A) = W^{1,2}(\mathbb{R})$, and $C = -\beta I_{\mathcal{H}}$. For our numerical experiments we choose 100 for the number of spatial grid points.

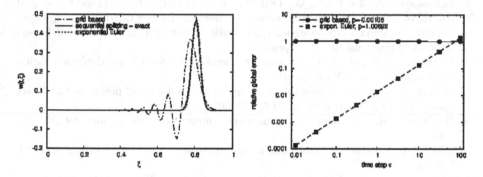

Fig. 1. Left panel: exact and numerical solutions to problem (12) at time $t = 500$ with time step $\tau = 10$. Right panel: global error of grid based method and exponential Euler method at $t = 500$ for various values of time step τ.

In the left panel of Fig. 1 the exact and the numerical solutions are shown for $\tau = 10$. The usual grid based finite difference method causes oscillations which is a well-known phenomenon being avoided in the cases when splitting or exponential integrator is applied. Since operators A and C commute, the error due to splitting vanishes (see [15]), that is, it gives the exact solution. The exponential Euler method performs very well, and by decreasing the time step τ, it provides even more accurate solution.

The right panel of Fig. 1 shows the relative global error for the usual grid based method and the exponential Euler method. The slope p of the lines in logarithmic scale gives the order of the corresponding errors. One can see the exponential Euler method's first-order convergence.

We remark that the study above can be extended to more complex examples, such as the control problem involving the one- and two-dimensional linearized shallow water equations, being capable to model flood prevention. Our corresponding results are presented in [6].

Acknowledgements. P. Csomós acknowledges the support of the National Research, Development, and Innovation Office (NKFIH) under the grant PD121117. H. Mena was supported by the project Solution of large scale Lyapunov differential equations (P 27926) founded by the Austrian Science Foundation FWF.

References

1. Al-Mohy, A.H., Higham, N.J.: Computing the action of the matrix exponential, with an application to exponential integrators. SIAM J. Sci. Comput. **33**, 488–511 (2011)
2. Bagrinovskii, K.A., Godunov, S.K.: Difference schemes for multidimensional problems (in Russian). Dokl. Akad. Nauk. USSR **115**, 431–433 (1957)
3. Balakrishnan, A.: Applied Functional Analysis. Springer, New York (1981)
4. Benner, P., Mena, H.: Numerical solution of the infinite-dimensional LQR-problem and the associated differential Riccati equations. J. Numer. Math. (to appear)
5. Bensoussan, A., Da Prato, G., Delfour, M.C., Mitter, S.K.: Representation and control of infinite dimensional systems. Birkhäuser, Boston (2007)
6. Csomós, P., Mena, H., Schwaighofer, J.: Fourier-splitting method for solving hyperbolic LQR problems (2016, preprint)
7. Engel, K.-J., Nagel, R.: One-Parameter Semigroups for Linear Evolution Equations. Springer, New York (2000)
8. Gibson, J.: The Riccati integral equation for optimal control problems in Hilbert spaces. SIAM J. Control Optim. **17**, 537–565 (1979)
9. Hochbruch, M., Ostermann, A.: Exponential integrators. Acta Numer. **19**, 209–286 (2010)
10. Jahnke, T., Lubich, C.: Error bounds for exponential operator splittings. BIT **40**, 735–744 (2000)
11. Lasiecka, I., Triggiani, R.: Control Theory for Partial Differential Equations: Continuous and Approximation Theories I, Abstract Parabolic Systems. Cambridge University Press, Cambridge (2000)
12. Lasiecka, I., Triggiani, R.: Control Theory for Partial Differential Equations: Continuous and Approximation Theories II, Abstract Hyperbolic-Like Systems over a Finite Time Horizon. Cambridge University Press, Cambridge (2000)
13. Marchuk, G.I.: Some application of splitting-up methods to the solution of mathematical physics problems. Apl. Mat. **13**, 103–132 (1968)
14. Strang, G.: On the construction and comparison of difference schemes. SIAM J. Numer. Anal. **5**, 506–517 (1968)
15. Varadarajan, V.S.: Lie Groups, Lie Algebras, and Their Representations. Springer, New York (1974)

Existence and Stability of Contrast Structures in Multidimensional Singularly Perturbed Reaction-Diffusion-Advection Problems

M.A. Davydova$^{(\boxtimes)}$ and N.N. Nefedov

Faculty of Physics, Department of Mathematics,
Lomonosov Moscow State University, Moscow 119991, Russia
m.davydova@bk.ru, nefedov@phys.msu.ru

Abstract. We consider stationary solutions with boundary and internal transition layers (contrast structures) for a nonlinear singularly perturbed equation that is referred to in applications as the stationary reaction-diffusion-advection equation. We construct an asymptotic approximation of an arbitrary-order accuracy to such solutions and prove the existence theorem. We suggest an afficient algorithm for constructing an asymptotic approximation to the localization surface of the transition layer. To justify the constructed asymptotics, we use and develop, to this class of problems, an asymptotic method of differential inequalities, which also permits one to prove the Lyapunov stability of such stationary solutions. The results can be used to create the numerical method which uses the asymptotic analyses to create space non uniform meshes to describe internal layer behavior of the solution.

Keywords: Reaction-diffusion-advection problems · Interior layer · Contrast structure

1 Statement of the Problem

Consider the boundary value problem for the nonlinear singularly perturbed equation

$$\varepsilon^2 \Delta u - f(\varepsilon \nabla u, u, x) = 0, \quad x = (x_1, ..., x_N) \in D \subset R^N, \tag{1}$$

$$u(x, \varepsilon) = g(x), \quad x \in \partial D, \tag{2}$$

where $\varepsilon > 0$ is a small parameter, the functions f, g and the boundary ∂D are assumed to be sufficiently smooth, and $\Delta = \sum\limits_{k=1}^{N} \partial^2 / \partial x_k^2$ is the Laplace operator. The notation $\varepsilon \nabla u$ implies the dependence of the function f on the arguments $\varepsilon \partial u / \partial x_k$, $k = \overline{1, N}$. We assume that the function f satisfies the standard condition of the most quadratic growth with respect to ∇u.

© Springer International Publishing AG 2017
I. Dimov et al. (Eds.): NAA 2016, LNCS 10187, pp. 277–285, 2017.
DOI: 10.1007/978-3-319-57099-0_29

The interest in Eq. (1) is caused by its numerous applications. We mark out the following particular cases of the Eq. (1):

$$f = \varepsilon(\mathbf{A}(u,x), \nabla u) + B(u,x), \tag{3}$$

$$f = A(u,x)(\varepsilon\nabla u)^2 + B(u,x), \tag{4}$$

where the components of the vector-function $\mathbf{A}(u,x) = \{A_1(u,x), ..., A_N(u,x)\}$ and the function $B(u,x)$ are sufficiently smooth in the domain $u \in U$ and $x \in D \cup \partial D$. For example, under certain generalizations, the Burgers equation, which arises in the theory of nonlinear waves, can be reduced to a one-dimensional equation of the type (1), (3), boundary value problems for an equation of the type (1), (3) appear in the description of the gas admixture transport in the surface layer of the atmosphere. In the simulation of the passage of a warm front in the solid sample with nonlinear characteristics in the case of the stabilized mode, we arrive to a boundary value problem for an Eqs. (1), (4). As one application of the problem for an Eq. (1), one can indicate the mathematical model of reaction-diffusion-advection, which appears in problems of the development of new methods of petroleum production and describes the process of in-situ combustion [1], where, in the consideration of the stabilized mode, we obtain the boundary value problem of the type (1), (3).

The study of the solutions of problem (1), (2) with internal transition layers (contrast structures) is directly related to the study of the existence of solutions with boundary layers in reaction-diffusion-advection type problems (see, for example, [2,3]). The main complexity of such problems is the description of the transition surfase in whose neighborhood the internal layer is localized. We use a more efficient method for localizing the transition surface, which permits one to develop our approach to the more complicated case of balanced advection and reaction (the so-called critical case).

2   Solutions with Boundary Layers

We assume the validity of the following condition.

Condition 1.1. *The degenerate equation $f(0,u,x) = 0$ has a solution $u = \varphi(x)$, with $f_u(0, \varphi(x), x) > 0$ for $x \in D \cup \partial D$.*

We define a sufficiently small neighborhood ∂D^δ of the surface ∂D:

$$\partial D^\delta := \left\{ x \in D : dist(x, \partial D^\delta) < \delta \right\}, \delta > 0,$$

and the local coordinates in this neighborhood

$$x \in \partial D^\delta \mapsto (y,r) \in \partial D \times [0; \delta],$$

where $y \in \partial D, \quad r = dist(x, \partial D) = dist(x,y).$

If $\mathbf{n}(y)$ is the unit internal normal to the surface ∂D at the point y, then the one-to-one correspondence between the coordinates is given by the expressions

$$x_i = y_i(\theta_1, ..., \theta_{N-1}) + r n_i(\theta_1, ..., \theta_{N-1}), \quad i - \overline{1, N}, \tag{5}$$

where $\{\theta_1, ..., \theta_{N-1}\}$ are the curvilinear coordinates on the surface ∂D, $x_i = y_i(\theta_1, ..., \theta_{N-1})$ are parametric equations of the surface ∂D and $n_i(\theta_1, ..., \theta_{N-1})$ are the direction cosines of the normal $\mathbf{n}(y)$. In this coordinate system, the derivatives $\partial u / \partial x_i$ can be calculated as follows:

$$\frac{\partial u}{\partial x_i} = d^i(r, \theta) \frac{\partial u}{\partial r} + \sum_{j=1}^{N-1} q_j^i(r, \theta) \frac{\partial u}{\partial \theta_j}, \quad i = \overline{1, N},$$

where $d^i(r, \theta)$, $q_j^i(r, \theta)$ are known functions, $\theta = (\theta_1, ..., \theta_{N-1}) \in \Theta$, Θ is the range of variation of the coordinate θ on the surface ∂D. On the surface ∂D, we define the associated system of equations

$$\frac{\partial \tilde{v}}{\partial \rho} = f(d^1(0, \theta)\tilde{v}, .., d^N(0, \theta)\tilde{v}, \tilde{u}, 0, \theta), \quad \frac{\partial \tilde{u}}{\partial \rho} = \tilde{v}, \quad 0 < \rho < +\infty, \tag{6}$$

where θ is regarded as a parameter.

Condition 1.2. *In the phase plane of system (6), the straight line $\tilde{u} = g(0, \theta)$ intersects the separatrix entering the saddle $(\varphi(0, \theta), 0)$ as $\rho \to +\infty$ and for $\theta \in \Theta$.*

We search the asymptotics of the solution of boundary layer type in the form

$$u(x, \varepsilon) = \bar{u}(x, \varepsilon) + \Pi u(\rho, \theta, \varepsilon), \tag{7}$$

where $\bar{u}(x, \varepsilon) = \varphi(x_1, ...x_N) + \varepsilon \bar{u}_1(x_1, ...x_N) + ...$ is a regular series, $\Pi u(\rho, \theta_1, ..., \theta_{N-1}, \varepsilon) = \Pi_0 u(\rho, \theta_1, ..., \theta_{N-1}) + \varepsilon \Pi_1 u(\rho, \theta_1, ..., \theta_{N-1}) + ...$ is the boundary series describing the boundary layer in a neighborhood of the boundary ∂D and $\rho = r/\varepsilon$. Further, taking into account the representation of the nonlinear operator $\varepsilon^2 \Delta - f(\varepsilon \nabla ..., ..., x)$ expressed in the variables (r, θ) and substituting the series (7) into problem (1), (2) by the boundary-function method, we obtain the equations for the terms of this series [4].

In the zeroth approximation, we obtain the nonlinear problem

$$\frac{\partial^2 \tilde{u}}{\partial \rho^2} - f\left(d^1(0, \theta) \frac{\partial \tilde{u}}{\partial \rho}, ..., d^N(0, \theta) \frac{\partial \tilde{u}}{\partial \rho}, \tilde{u}, 0, \theta\right) = 0, \tag{8}$$

$$\tilde{u}(0, \theta) = g(0, \theta), \quad \tilde{u}(+\infty, \theta) = \varphi(0, \theta),$$

where $\tilde{u}(\rho, \theta) = \varphi(0, \theta) + \Pi_0 u(\rho, \theta)$. System (6) corresponds to the equation in system (8). Condition 1.2 guarantees the existence of the solution of problem (8). For the solution of problem (8) we choose a function that varies monotonically from the value $\tilde{u} = g(0, \theta)$ for $\rho = 0$ to the value $\tilde{u} = \varphi(0, \theta)$ as $\rho \to +\infty$ for a fixed $\theta \in \Theta$.

For $n \geq 1$, we obtain the linear problems, whose solutions can be represented in the closed form [4]. The terms of the asymptotics (7) satisfy the following estimates:

$$|\Pi_n u| \leq C_n \exp[-\chi_n \rho], \; \chi_n > 0, \; C_n > 0, \; n \geq 0 .$$

The proof of the existence of a solution with the obtained asymptotics is carried out with the use of the general scheme of the asymptotic method of differential inequalities [5]. If the functions f, g and the boundary ∂D are sufficiently smooth, then, under Conditions 1.1 and 1.2, there exists a solution $u(x, \varepsilon)$ of problem (1), (2) which satisfies the uniform estimate

$$|u(x, \varepsilon) - U_n(x, \varepsilon)| \leq C\varepsilon^{n+1}, \tag{9}$$

in the domain $D \cup \partial D$, where U_n is the n-th order partial sum of the series (7) and the constant C is independent of ε.

EXAMPLES 1. Consider the Dirichlet boundary value problem for the Eqs. (1), (4). Let the degenerate equation $B(u, x) = 0$ have an isolated solution $u = \varphi(x)$, and let $B_u(\varphi(x), x) > 0$. Then, under the condition $g(0, \theta) > \varphi(0, \theta)$, the equation of the separatrix entering the saddle $(\varphi(0, \theta), 0)$ as $\rho \to +\infty$ has the form

$$\tilde{v}(\tilde{u}, \theta) = -\sqrt{\int_{\varphi(0,\theta)}^{\tilde{u}} 2B(\xi, 0, \theta) \exp\left(2\sum_{i=1}^{N} (d^i(0, \theta)^2 \int_{\xi}^{\tilde{u}} A(\eta, 0, \theta)d\eta\right) d\xi}. \tag{10}$$

For the existence of the point, at which the separatrix (10) intersects the straight line $\tilde{u} = g(0, \theta)$ (see Condition 1.2), it suffices that the following condition holds:

$$\int_{\varphi(0,\theta)}^{g(0,\theta)} B(\xi, 0, \theta) \exp\left(2\sum_{i=1}^{N} (d^i(0, \theta)^2 \int_{\xi}^{g(0,\theta)} A(\eta, 0, \theta)d\eta\right) d\xi > 0.$$

3   Contrast Structures

To state conditions under which problem (1), (2) is considered, we introduce the set of surfaces

$$\{\bar{\Omega}\} := \{\bar{\Omega} \in D : \bar{\Omega} \text{ is a sufficiently smooth closed}$$
$$\text{manifold of dimensional N-1}\}.$$

For each surface $\bar{\Omega}$ in the sufficiently small neighborhood $\bar{\Omega}^\delta$ we introduce local coordinates

$$x \in \bar{\Omega}^\delta \mapsto (y, r) \in \bar{\Omega} \times [-\delta; \delta],$$

where $y \in \bar{\Omega}$. The one-to-one correspondence between coordinates is given by analogy the expressions (5):

$$x_i = y_i(\theta_1, ..., \theta_{N-1}, \bar{\Omega}) + rn_i(\theta_1, ..., \theta_{N-1}, \bar{\Omega}), \; i = \overline{1, N},$$

where $\{\theta_1, ..., \theta_{N-1}\}$ are curvilinear coordinates on the surface $\bar{\Omega}$, $x_i = y_i(\theta_1, ..., \theta_{N-1}, \bar{\Omega})$ are parametric equations of the surface $\bar{\Omega}$ and $n_i(\theta_1, ..., \theta_{N-1}, \bar{\Omega})$ are the direction cosines of the outward normal to the surface $\bar{\Omega}$. In the each neighborhood $\bar{\Omega}^\delta$, we introduce the associated system of equations

$$\frac{\partial \tilde{v}}{\partial \xi} = f(l^1(r, \theta)\tilde{v}, .., l^N(r, \theta)\tilde{v}, \tilde{u}, r, \theta), \quad \frac{\partial \tilde{u}}{\partial \xi} = \tilde{v}, \quad -\infty < \xi < +\infty, \quad (11)$$

where r and θ are treated as parameters, $\theta = (\theta_1, ..., \theta_{N-1}) \in \bar{\Theta}$, $\bar{\Theta}$ is the domain where the coordinate θ on the surface $\bar{\Omega}$ varies, $l^1(r, \theta)...l^N(r, \theta)$ are known functions defined by analogy with the functions $d^1(r, \theta), ..., d^N(r, \theta)$.

In what follows, we consider two cases: *the case of unbalanced reaction and advection and the balanced case*. Let the following condition be satisfied.

Condition 2.1. *The degenerate equation* $f(0, u, x) = 0$ *has exactly three roots* $u = \varphi_i(x)$, $i = \overline{1, 3}$, *satisfying the conditions* $\varphi_1(x) < \varphi_2(x) < \varphi_3(x)$, $f_u(0, \varphi_i(x), x) > 0$, $i = 1, 3$, $f_u(0, \varphi_2(x), x) < 0$ *for* $x \in D \cup \partial D$.

For each surface $\bar{\Omega} \in \{\bar{\Omega}\}$, we define the function

$$H(r, \theta) \equiv \tilde{v}^+(0, r, \theta) - \tilde{v}^-(0, r, \theta), (r, \theta) \in [-\delta; \delta] \times \bar{\Theta},$$

where the $\tilde{v}^\pm(\xi, r, \theta)$ are solutions of system (11) satisfying conditions:

$$\tilde{u}^\mp(\mp\infty, r, \theta) = \varphi_i(r, \theta), \quad i = 1, 3, \tilde{v}^\mp(\mp\infty, r, \theta) = 0.$$

THE CASE OF UNBALANCED REACTION AND ADVECTION. Let the following conditions be satisfied.

Condition 2.2. *There exists a surface* $\Omega_0 \in \{\bar{\Omega}\}$ *such that* $H(0, \theta) = 0, \theta \in \Theta_0$.

Here Θ_0 is the domain where the coordinate θ on the surface Ω_0 varies. This condition defines the location of the transition surface Ω in the zero approximation.

Condition 2.3. *The derivative* $H_r(0, \theta)$ *is positive for* $\theta \in \Theta_0$.

The condition 2.2 implies that, on the surface Ω_0 of the phase plane $(\tilde{u}, \tilde{v})$, there exists a separatrix joining the saddles $(\varphi_1(0, \theta), 0)$ and $(\varphi_3(0, \theta), 0)$.

We obtain the asymptotic expansion of the contrast structure by the C^1- matching of the boundary-layer asymptotics

$$u^-(x, \varepsilon) = \bar{u}^-(x, \varepsilon) + Qu^-(\xi, \lambda^*, \theta, \varepsilon),$$
$$u^+(x, \varepsilon) = \bar{u}^+(x, \varepsilon) + \Pi u(\rho, \eta, \varepsilon) + Qu^+(\xi, \lambda^*, \theta, \varepsilon), \quad (12)$$

on the transition surface $\Omega \in \{\bar{\Omega}\}$. Here $\bar{u}^-(x, \varepsilon) = \varphi_1(x) + \varepsilon \bar{u}_1^-(x) + ...$, $u^+(x, \varepsilon) = \varphi_3(x) + \varepsilon \bar{u}_1^+(x) + ...$ are regular series, $\Pi u(\rho, \eta, \varepsilon) = \Pi_0 u(\rho, \eta) + \varepsilon \Pi_1 u(\rho, \eta) + ...$ is a boundary series describing the boundary layer in a neighborhood of the boundary ∂D, and $Qu^\pm(\xi, \lambda^*, \theta, \varepsilon) = Q_0 u^\pm(\xi, \lambda^*, \theta) + \varepsilon Q_1 u^\pm(\xi, \lambda^*, \theta) + ...$ are series describing the boundary layers localized in a neighborhood of the transition surface Ω, whose position is defined by the condition

$$u(x, \varepsilon) = \varphi_2(x), \ x \in \Omega. \tag{13}$$

Next, in accordance with the algorithms of the method of boundary functions, developed in [4] and expounded in Sect. 2, we construct boundary-layer asymptotics (12).

We seek the asymptotics of the transition surface Ω in the form of a series in powers of ε; moreover, for the leading term we take the surface Ω_0; thus the equation of the surface Ω in the local coordinate system defined in the neighborhood of the surface Ω_0 has the form

$$r = \lambda^*(\theta, \varepsilon) = \varepsilon(\lambda_1(\theta) + \varepsilon\lambda_2(\theta) + ...) \tag{14}$$

The coefficients of the expansion (14) are found from following condition, which, together with (13), is a condition of C^1−matching of the asymptotics on the surface Ω

$$\begin{aligned}
\varepsilon \left(\frac{\partial u^+}{\partial r} - \frac{\partial u^-}{\partial r} \right)\Big|_{r=\lambda^*(\theta,\varepsilon)} &= H(r,\theta)|_{r=\lambda^*(\theta,\varepsilon)} \\
+\varepsilon \left(\frac{\partial \varphi_3}{\partial r}(r,\theta) - \frac{\partial \varphi_1}{\partial r}(r,\theta) + \frac{\partial Q_1 u^+}{\partial \xi}(0,r,\theta) - \frac{\partial Q_1 u^-}{\partial \xi}(0,r,\theta) \right)\Big|_{r=\lambda^*(\theta,\varepsilon)} &+ ... \\
\equiv H(\lambda^*(\theta,\varepsilon),\theta) + \varepsilon G_1(\lambda^*(\theta,\varepsilon),\theta) + ... &= 0.
\end{aligned} \tag{15}$$

By substituting the asymptotic (14) into Eq. (15) and representing each term in (15) as an expansion in powers of ε, we obtain the uniquely solvable equation for the coeffcient $\lambda_n(\theta)$ in the asymptotic expansion (14):

$$\lambda_n(\theta)H_r(0,\theta) - \Phi_n(\theta) = 0, \ \theta \in \Theta_0,$$

where $\Phi_n(\theta)$ are known functions. By virtue of the condition 2.2, the condition (15) in the zero approximation holds.

THE CASE OF BALANCED REACTION AND ADVECTION. Let the function $H(r,\theta)$ satisfies the condition.

Condition 3.1. $H(r,\theta) = 0$ for $(r,\theta) \in [-\delta; \delta] \times \bar{\Theta}$ and for any surface in the set $\{\bar{\Omega}\}$.

This requirement marks out the balanced case. Condition 3.1 implies that, for any surface $\bar{\Omega} \in \{\bar{\Omega}\}$ on the phase plane $(\tilde{u}, \tilde{v})$, there exists a separatrix joining the saddles $(\varphi_1(r,\theta),0)$ and $(\varphi_3(r,\theta),0)$ for $\theta \in \bar{\Theta}$.

The construction of the asymptotic expansion of a solution of the type of a contrast structure in the critical case is performed by analogy with the construction in the noncritical case.

By representing each term in (15) as an expansion in powers of ε, we obtain:

$$\varepsilon G_1(0,\theta) + \varepsilon^2 \left(M(\theta)\lambda_1(\theta) - \bar{\Phi}_2(\theta) \right) + ... = 0. \tag{16}$$

Therefore, unlike the previous case, where the choice of the surface Ω_0 is performed with the use of condition 2.2, in the case of balanced reaction and advection the following condition defines the position of the surface Ω_0:

Condition 3.2. *There exists a surface* $\Omega_0 \in \{\bar{\Omega}\}$ *such that* $G_1(0, \theta) = 0$, $\theta \in \Theta_0$.

By using the Eq. (16), we obtain the linear differential equations for the functions $\lambda_n(\theta)$:

$$M(\theta)\lambda_n(\theta) - \bar{\Phi}_{n+1}(\theta) = 0, \ \theta \in \Theta_0, \tag{17}$$

where $M(\theta)$ is a linear elliptic operator of the second order with periodic coefficients, $\bar{\Phi}_{n+1}(\theta)$ are know functions.

Then, by taking into account the Eq. (17) together with the Θ_0- periodicity conditions for the functions $\lambda_n(\theta)$, we obtain the solvable differential problems under following condition:

Condition 3.3. *The operator* $M(\theta)$ *is positively invertible for* $\theta \in \Theta_0$ *on the set of positive periodic right-hand sides.*

In accordance with the asymptotic method of differential inequalities, we define an upper solution and a lower solution as the modified partial nth-order sums of the asymptotic expansion (12). The existence of upper and lower solutions implies the existence of a solution $u(x, \varepsilon)$ of problem (1), (2), which satisfies the inequality of the type (9) in the domain $D \cup \partial D$, where U_n is the nth-order partial sum of the series (12) and the constant C is independent of ε.

Remark. If we treat solutions of problem (1), (2) with the obtained asymptotics (7) and (12) as stationary solutions of the corresponding parabolic problems, then Lyapunov stability follows from well-known results (see, for example, [5]).

EXAMPLE 2. Consider the boundary value problem

$$\varepsilon^2 \Delta u - \varepsilon \left(x_1 \frac{\partial u}{\partial x_1} + x_2 \frac{\partial u}{\partial x_2} \right) - u(u-1)(u-3) = 0,$$

$$D := \left\{ (x_1, x_2) : x_1^2 + x_2^2 < 2 \right\} \subset R^2,$$

$$u(x, \varepsilon) = g(x), \ x \in \partial D := \left\{ (x_1, x_2) : x_1^2 + x_2^2 = 2 \right\}.$$

The degenerate equation

$$u(u-1)(u-3) = 0$$

has three roots $u = 0$, $u = 1$, $u = 3$.

Let the curve Ω_0 in whose ε-neighborhood the internal layer is localized be described by the equations

$$\Omega_0 : x_1 = R_0 \cos \theta, \ x_2 = R_0 \sin \theta, 0 \le \theta \le 2\pi, \quad R_0 = const$$

in some polar coordinate system with a pole inside the domain D. In accordance with our algorithm in the zeroth approximation we obtain

$$\frac{\partial \tilde{v}^\pm}{\partial \xi} = R_0 \tilde{v}^\pm + \tilde{u}^\pm(\tilde{u}^\pm - 1)(\tilde{u}^\pm - 3), \quad \frac{\partial \tilde{u}^\pm}{\partial \xi} = \tilde{v}^\pm,$$
$$\tilde{u}^-(-\infty) = 0, \quad \tilde{u}^\pm(0) = 1, \quad \tilde{u}^+(+\infty) = 3, \quad \tilde{v}^\pm(\pm\infty) = 0, \tag{18}$$

where $\tilde{u}^-(\xi) = Q_0 u^-(\xi)$, $\tilde{u}^+(\xi) = 3 + Q_0 u^+(\xi)$. It is known that, under the equality $R_0 = \frac{1}{\sqrt{2}}$, there exists a separatrix joining the saddles $(0;0)$ and $(3;0)$ on the phase plane of the system (18) in case of $\tilde{v} > 0$. The separatrix can be represented in the form:

$$\tilde{v} = -\sqrt{\frac{1}{2}}\tilde{u}(\tilde{u} - 3). \tag{19}$$

The local coordinates (r, θ) in a neighborhood of the curve Ω_0 are related to the Cartesian coordinates by the formulas:

$$x_1 = \left(\frac{1}{\sqrt{2}} + r\right)\cos\theta, \quad x_2 = \left(\frac{1}{\sqrt{2}} + r\right)\sin\theta.$$

The equation of the transition curve Ω in the local coordinate system defined near the curve Ω_0 has the form $r = \varepsilon\lambda_1 + O(\varepsilon^2)$, where the coefficient λ_1 is defined by the equality

$$\lambda_1 = -\left\{2 + \left[\int_{-\infty}^{+\infty} e^{-\frac{7\xi}{\sqrt{2}}}\left(1 + 2e^{-\frac{3\xi}{\sqrt{2}}}\right)^{-4} d\xi\right]^{-1}\left[\int_{-\infty}^{+\infty} e^{-\frac{7\xi}{\sqrt{2}}}\left(1 + 2e^{-\frac{3\xi}{\sqrt{2}}}\right)^{-4}\xi d\xi\right]\right\}$$

By using the Eq. (19), we obtain the function describing the internal transition layer in the zeroth approximation:

$$u(x_1, x_2) = \frac{3}{1 + 2\exp\left(\frac{3}{2\varepsilon}\left(1 + \sqrt{2}\varepsilon\lambda_1 - \sqrt{2(x_1^2 + x_2^2)}\right)\right)}.$$

The results can be used to create the numerical method which uses the asymptotic analyses to create space non uniform meshes to describe internal layer behavior of the solution [6,7].

By numerical analyzes of our results, one can readily see that the smaller ε the more satisfactory our algorithms.

The research was supported by the Russian Foundation for Basic Research (project no. 16-01-00437).

References

1. Volkov, V.T., Grachev, N.E., Dmitriev, A.V., Nefedov, N.N.: Front formation and dynamics in a reaction-diffusion-advection model. Math. Model. **22**(8), 109–118 (2010)

2. Nefedov, N.N., Davydova, M.A.: Contrast structures in singularly perturbed quasi-linear reaction-diffusion-advection equations. Differ. Uravn. **49**(4), 715–733 (2013). Differ. Equations 49 (4), 688–706

3. Levashova, N.T., Nefedov, N.N., Yagremtsev, A.V.: Contrast structures in the reaction-diffusion-advection equations in the case of balanced advection. Zh. Vychisl. Mat. i Mat. Fiz. **53**(1), 365–376 (2013). Comput. Math. and Math. Phys. 53 (1), 273–283

4. Davydova, M.A.: Existence and stability of solutions with boundary layers multidimensional singularly perturbed reaction-diffusion-advection problems. Math. Notes **98**(6), 45–55 (2015)

5. Nefedov, N.: Comparison principle for reaction-diffusion-advection problems with boundary and internal layers. In: Dimov, I., Faragó, I., Vulkov, L. (eds.) NAA 2012. LNCS, vol. 8236, pp. 62–72. Springer, Heidelberg (2013). doi:10.1007/978-3-642-41515-9_6

6. Lukyanenko, D.V., Volkov, V.T., Nefedov, N.N., Recke, L., Schneider, K.: Analytic-numerical approach to solving singularly perturbed parabolic equations with the use of dynamic adapted meshes. Model. Anal. Inf. Syst. **23**(3), 334–341 (2016)

7. Volkov, V., Nefedov, N.: Asymptotic-numerical investigation of generation and motion of fronts in phase transition models. In: Dimov, I., Faragó, I., Vulkov, L. (eds.) NAA 2012. LNCS, vol. 8236, pp. 524–531. Springer, Heidelberg (2013). doi:10.1007/978-3-642-41515-9_60

A Comparison of Numerical Techniques for the FEM for the Stokes Problem for Incompressible Flow

Ekaterina Dementyeva[1,2](✉) and Evgeniya Karepova[1,2]

[1] Institute of Computational Modelling of SB RAS,
660036 Akademgorodok, Krasnoyarsk, Russia
[2] IM&CS, Siberian Federal University, Krasnoyarsk, Russia
{e.v.dementyeva,e.d.karepova}@icm.krasn.ru
http://icm.krasn.ru

Abstract. In this paper the two-dimensional Stokes equations are considered for a viscous incompressible fluid in a channel. To construct a discrete problem, we use the Taylor – Hood finite elements. When solving the discrete problem, we are interested in the comparison the stabilized biconjugate gradient method, the Arrow – Hurwicz algorithm, and the Uzawa methods. Moreover, we investigate a new modification of the Uzawa algorithm. The numerical analysis shows that the new algorithm is competitive with the Uzawa and gradient methods.

Keywords: Stokes problem · FEM · Uzawa method · BiCGstab

1 Introduction

Nowadays the finite element method (FEM) is widely used for the construction of a discrete analogue of an initial-boundary value problem for the Stokes and Navier–Stokes equations including the case of a computational domain of a complex shape with an unstructured non-uniform fine grid [1, 2]. At the same time, the FEM is a compute- and resource-intensive approach. In the case of a fine grid the storage of a global matrix of the FEM system and its postprocessing can involve difficulties due to high requirements upon computing power. Therefore, we are interested in a numerical technology for the solution of a discrete problem which enables one to avoid the assembly and storage of blocks of the global matrix. Moreover, the algorithm should not require a direct inversion of the matrix.

In this paper the two-dimensional Stokes equations for a viscous incompressible fluid in a channel are considered as a model problem [3]. We use the FEM with biquadratic finite elements on rectangles for the components of a velocity vector and with bilinear elements for a pressure [2]. These mixed finite elements satisfy the Ladyzhenskaya – Babuska – Brezzi condition which ensures stability of a pressure [4].

© Springer International Publishing AG 2017
I. Dimov et al. (Eds.): NAA 2016, LNCS 10187, pp. 286–293, 2017.
DOI: 10.1007/978-3-319-57099-0_30

As a result, we have a saddle point problem with a symmetric indefinite ill-conditioned matrix. For such a problem, the Krylov subspace iterative methods and Uzawa-type algorithms are widely used [5–7]. To improve the order of convergence, a suitable preconditioner can be applied. However, the choice of a preconditioner is strongly problem-dependent [5,6]. Besides, a preconditioner, as a rule, increases computing time per one iteration loop and requires extensive storage resources. Therefore, here we do not take into account preconditioners.

For the solution of the discrete problem, we compare well-known approaches and a new modification of the Uzawa-type method. In the new approach, the Uzawa algorithm is used in combination with the Richardson iterative method. At first, the pressure is calculated without the explicit computation of the Schur matrix (contrary to the most of known methods). Then the velocity is calculated using the pressure. Hence the velocity satisfies well the divergence condition, in addition, the direct inversion of a matrix, which is the bottleneck of the Uzawa methods, is not used.

2   The Differential Problem

Let $\Omega = (0, H_1) \times (0, H_2)$, H_1, $H_2 > 0$ be a rectangular domain with the boundary Γ which includes the left inlet part $\Gamma_{in} = \{(x,y) : x = 0, \ 0 < y < H_2\}$, the right outlet part $\Gamma_{out} = \{(x,y) : x = H_1, \ 0 < y < H_2\}$, and two rigid sides $\Gamma_{rigid} = \{(x,y) : 0 \le x \le H_1, \ y = 0, H_2\}$ of the channel. We consider a model of the flow, so that its domain is the rectangle $D = (0, H_1) \times (0, H_2)$ in the space of variables (x, y).

For the unknown vector function $\mathbf{u}(x, y) = (u(x,y), v(x,y))$ and the function $p = p(x,y)$ we consider the linear Stokes equations

$$-\nu \Delta \mathbf{u} + \nabla p = \mathbf{f} \quad \text{in} \quad \Omega, \tag{1}$$

$$\nabla \cdot \mathbf{u} = 0 \quad \text{in} \quad \Omega. \tag{2}$$

Here $\mathbf{u}$ is the velocity vector-function; p is the pressure; $\mathbf{f}(x,y) = (f_1(x,y), f_2(x,y))$ is a given vector-function of body forces; ν is the kinematic viscosity coefficient (inversely proportional to the Reynolds number Re); and Δ, ∇, $\nabla \cdot$ are the Laplace, gradient, and divergence operators, respectively.

At the inlet Γ_{in}, a flow is assumed to be given:

$$\mathbf{u}(0, y) = \mathbf{u}_{in}(y) \quad \forall y \in \Gamma_{in}. \tag{3}$$

On the rigid boundary Γ_{rigid} we apply the no-slip condition:

$$\mathbf{u}(x, y) = (0, 0) \quad \forall (x, y) \in \Gamma_{rigid}. \tag{4}$$

To close the computational domain at the outlet Γ_{out}, we put the "do nothing" boundary condition [1,8], i.e., the boundary condition "without significant impact":

$$- \nu \partial_n \mathbf{u} + p\mathbf{n} = p_{ext}\mathbf{n} \quad \forall (x, y) \in \Gamma_{out}, \tag{5}$$

where $\mathbf{n} = (n_x(x,y), n_y(x,y))$ is an outward normal to Γ, $\partial_n \mathbf{u}$ is a derivative of the velocity vector in the direction $\mathbf{n}$, and $p_{ext}(y)$ is an external pressure.

3   The Discrete Problem

Let $\mathbf{H}_0^1 = H_0^1 \times H_0^1$ be the vector space where H_0^1 is the Sobolev space of functions that belong to $L^2(\Omega)$ together with their first-order derivatives and vanish on $\Gamma_{rigid} \cup \Gamma_{in}$. $\mathbf{L}^2 = L^2(\Omega) \times L^2(\Omega)$ is the vector space with the related inner product:

$$(f,g) = \int_\Omega fg\,d\Omega, \quad \|f\|_0^2 = (f,f), \quad \|\tilde{f}\|_1^2 = \left(\left\| \frac{\partial \tilde{f}}{\partial x} \right\|_0^2 + \left\| \frac{\partial \tilde{f}}{\partial y} \right\|_0^2 \right)$$

$$(\mathbf{f},\mathbf{g}) = \int_\Omega f_1 g_1 + f_2 g_2\,d\Omega, \quad \|\mathbf{f}\|_0^2 = (\mathbf{f},\mathbf{f}), \quad \|\tilde{\mathbf{f}}\|_1^2 = \left(\|\tilde{f}_1\|_1^2 + \|\tilde{f}_2\|_1^2 \right)$$

for $f, g \in L^2(\Omega)$, $\tilde{f} \in H_0^1$ and for vector functions $\mathbf{f} = (f_1, f_2)$, $\mathbf{g} = (g_1, g_2) \in \mathbf{L}^2$, $\tilde{\mathbf{f}} = (\tilde{f}_1, \tilde{f}_2) \in \mathbf{H}_0^1$.

To construct a discrete problem, consider two uniform rectangular grids on $\bar{\Omega}$:

$$\bar{\Omega}_h := \{\mathbf{z}_{i,j} = (x_i, y_j), \ x_i = ih_x, \ y_j = jh_y, \ i = 0, \ldots, N_x, \ j = 0, \ldots N_y\},$$
$$\bar{\Omega}_{h/2} := \{\tilde{\mathbf{z}}_{i,j} = (\tilde{x}_i, \tilde{y}_j), \ \tilde{x}_i = ih_x/2, \ \tilde{y}_j = jh_y/2, i = 0, \ldots, 2N_x, \ j = 0, \ldots 2N_y\}$$

with mesh-sizes $h_x = H_1/N_x$, $h_y = H_2/N_y$ in the x- and y-directions, respectively, and $N_x, N_y \geq 2$. Denote $\Omega_{h/2} = \bar{\Omega}_{h/2} \setminus (\Gamma_{in} \cup \Gamma_{rigid})$. Let $\mathcal{T}_h = \{\omega_{i,j}\}_{i,j=0}^{N_x-1, \ N_y-1}$ be the uniform rectangular triangulation of $\bar{\Omega}$ with elementary cells

$$\omega_{i,j} = [x_i, x_{i+1}] \times [y_j, y_{j+1}], \ i = 0, \ldots, N_x - 1, \ j = 0, \ldots N_y - 1.$$

For each node $\mathbf{z}_{i,j} \in \bar{\Omega}_h$ we introduce the basis function $\varphi_{i,j}$ which equals 1 at $\mathbf{z}_{i,j}$, equals 0 at the other nodes of $\bar{\Omega}_h$, and is bilinear on each cell $\omega_{i,j} \in \mathcal{T}_h$. Denote the span of these functions by $P^h = span\{\varphi_{i,j}\}_{i,j=0}^{N_x, N_y}$.

For each node $\tilde{\mathbf{z}}_{i,j} \in \bar{\Omega}_{h/2}$ we introduce the basis function $\psi_{i,j}$ which equals 1 at $\tilde{\mathbf{z}}_{i,j}$, equals 0 at the other nodes of $\bar{\Omega}_{h/2}$, and is biquadratic on each cell $\omega_{i,j} \in \mathcal{T}_h$. Denote the span of these functions by $H_u^h = span\{\psi_{i,j}\}_{i,j=0}^{2N_x, 2N_y}$.

Consider the Q_2 finite element space [2]:

$$\mathbf{W}^h := \left\{ \mathbf{w}_h = (w_x^h, w_y^h) : \mathbf{w}_h \in H_u^h \times H_u^h; \ \mathbf{w}_h = (0,0) \text{ on } \Gamma_{in} \cup \Gamma_{rigid} \right\} \subset \mathbf{H}_0^1.$$

Let $\mathbf{u}_{h,0}(x,y) \in H_u^h \times H_u^h$ be a smooth vector function which equals $(0,0)$ on Γ_{rigid} and coincides with the piecewise quadratic interpolant of $\mathbf{u}_{in}(y)$ on Γ_{in}; let also $\mathbf{v}_h \in \mathbf{W}^h$. We state the following Galerkin problem based on Stokes problem (1)–(5): find $\mathbf{u}_h = \mathbf{u}_{h,0} + \mathbf{v}_h$ and $p_h \in P^h$ such that

$$\nu((\nabla \mathbf{u}_h, \nabla \mathbf{w}_h)) - (p_h, \nabla \cdot \mathbf{w}_h) = (\mathbf{f}_h, \mathbf{w}_h) - \int_{\Gamma_{out}} p_{ext} \mathbf{n} \cdot \mathbf{w}_h \, dy \quad \forall \mathbf{w}_h \in \mathbf{W}^h, (6)$$

$$(\nabla \cdot \mathbf{u}_h, q_h) = 0 \quad \forall q_h \in P^h. \tag{7}$$

Here we use the notation $((\nabla \mathbf{u}_h, \nabla \mathbf{w}_h)) = \left(\dfrac{\partial \mathbf{u}_h}{\partial x}, \dfrac{\partial \mathbf{w}_h}{\partial x}\right) + \left(\dfrac{\partial \mathbf{u}_h}{\partial y}, \dfrac{\partial \mathbf{w}_h}{\partial y}\right)$.

The $Q_2 - Q_1$ Taylor – Hood finite element spaces satisfy the Ladyzhenskaya-Babuska-Brezzi condition [2,8]. Hence, problem (6)–(7) has a unique solution $(\mathbf{u}_h, p_h) \in \mathbf{W}^h \times P^h$ [4].

We use Simpson's rule and its Cartesian product of it to approximate the integrals over Γ_{out} and over Ω in (6)–(7) with lumping effect [3].

Then the system of linear equations corresponding to (6)–(7) can be derived in the form

$$A_h \mathbf{U}_h + B_h{}^T \mathbf{P}_h = \mathbf{F}_h, \tag{8}$$

$$B_h \mathbf{U}_h = 0 \tag{9}$$

where $A_h \mathbf{U}_h$ is the discretization of the viscous terms, $B_h \mathbf{U}_h$ denotes the discretization of the negative divergence of $\mathbf{U}_h$, $B_h{}^T \mathbf{P}_h$ is the discretization of the gradient of the pressure. The right-hand side vector $\mathbf{F}_h$ involves all contributions of the source terms, as well as the boundary integrals taking into account the contribution of the prescribed boundary conditions.

4   The Numerical Techniques

To solve saddle-point system (8)–(9), we consider a new approach that is a combination of an Uzawa-type and a Richardson iterative methods [6,9].

The scheme of the Uzawa – Richardson iterative algorithm

1. Put $n = 0$. Choose $\mathbf{U}_h^{(0)}$, $\mathbf{P}_h^{(0)}$. Choose $\tau_u \in (0, 2\lambda_u^{-1})$, $\tau_p \in (0, 2\lambda_p^{-2})$.
2. Put $\mathbf{r}^{(0)} = A_h \mathbf{U}_h^{(0)} + B_h^T \mathbf{P}_h^{(0)} - \mathbf{F}_h$; $\mathbf{q}^{(0)} = B_h \mathbf{U}_h^{(0)}$.
3. While $(\|\mathbf{res}^{(n)}\|_0^{h,\mathbf{U}} < \varepsilon)$, do

    3.1. $\mathbf{P}_h^{(n+1)} = \mathbf{P}_h^{(n)} - \tau_p(B_h \mathbf{r}^{(n)} - \tau_u^{-1} \mathbf{q}^{(n)})$;

    3.2. $\mathbf{U}_h^{(n+1)} = \mathbf{U}_h^{(n)} - \tau_u(A_h \mathbf{U}_h^{(n)} + B_h^T \mathbf{P}_h^{(n+1)} - \mathbf{F}_h)$;

    3.3. $\mathbf{r}^{(n+1)} = A_h \mathbf{U}_h^{(n+1)} + B_h^T \mathbf{P}_h^{(n+1)} - \mathbf{F}_h$;

    3.4. $\mathbf{q}^{(n+1)} = B_h \mathbf{U}_h^{(n+1)}$;

    3.5. $n = n + 1$.

Herein n is a number of an iteration step; λ_u, λ_p are the maximal eigenvalues of matrices A_h and B_h respectively, they are evaluated using Gershgorin's theorem; $\mathbf{res}^{(n)} = (\mathbf{r}^{(n)}, \mathbf{q}^{(n)})$ is a residual vector for iteration step n; $\| \cdot \|_0^{h,\mathbf{U}}$ is a discrete analogue of the norm $\| \cdot \|_0$ for a vector-function when the integrals are approximated by the Cartesian product of Simpson's rule; ε is a given accuracy.

At each iteration step of the iterative algorithm, first the pressure is calculated without explicit determination of the Schur matrix (contrary to the most of known methods), then the velocity is calculated using the known pressure.

As a result, the velocity satisfies well the divergence (continuity) condition. The convergence of the Uzawa – Richardson iterative algorithm can be proved using an approach like in [6].

We compare the Uzawa – Richardson algorithm with two well-known methods: the Arrow-Hurwicz algorithm of the Uzawa method and the stabilized biconjugate gradient method (BiCGstab). The Arrow-Hurwicz algorithm was developed for solving saddle-point problems. At the same time, BiCGstab is efficient when solving a problem for a sparse large-scale system obtained by the FEM. Both algorithms are used in the form described in [7]. The iterative parameters in the Arrow-Hurwicz algorithm are chosen according to [10].

Notice some features of our implementation of the algorithms.

First, there is no need to store the global matrix of the FEM system explicitly. A residual and any product of a matrix by a vector are assembled element-by-element with the use of local stiffness matrices which are stored in the program.

Second, to align the Euclidean norms of the matrix rows and corresponding columns with respect to one, we determine the coefficients of scalability for the matrices A_h, B_h, and $B_h{}^T$, and make a transformation of system (8)–(9). As a rule, the scalability improves the condition number of a matrix as well as the calculation accuracy [11]. As a result, we get the following scaled system:

$$DA_h\mathbf{U}_h + DB_h{}^T\mathbf{P}_h = D\mathbf{F}_h, \tag{10}$$
$$GB_h\mathbf{U}_h = 0. \tag{11}$$

Herein $D = \{d_{i,i}\}$, $G = \{g_{k,k}\}$ are the diagonal matrices of scalability, their elements can be found in sequence from the relations:

$$\sum_{j=0}^{2N_x(2N_y-1)-1} a_{i,j}^2 d_{i,i}^2 = 1, \quad i = 0, ..., 2N_x(2N_y - 1) - 1,$$

$$\sum_{l=0}^{2N_x(2N_y-1)-1} d_{l,l}^2 b_{k,l}^2 g_{k,k}^2 = 1, \quad k = 0, ..., (N_x + 1)(N_y + 1) - 1$$

where $a_{i,j}$ and $b_{k,l}$ are the elements of the matrices A_h and B_h, respectively. Notice that, to determine the scalability coefficients, we do not need to know the whole global matrix of the FEM system. It is sufficient to assemble only some "typical" rows that correspond to every type of the nodes of a finite element. For Q_2-Q_1 elements there are four typical nodes: at an angle of Q_2- and Q_1-element, at the midpoint of Q_2-element, and at the midpoint of an edge of Q_2-element. Then, using only these rows, we can calculate the scalability coefficients for the whole system. This procedure allow one to decrease memory footprint.

To verify the proposed iterative algorithm and to review techniques, we test the following model problem. In the closed domain $\Omega = [0, 10]^2$ we consider the Poiseuille flow for problem (1)–(2) with a parabolic inflow boundary condition $u_{in}(0, y) = -0.5Re(y^2 - 10y)$ and external outflow pressure $p_{ext}(10, y) = -5$ in (5). The exact solution of this problem has the form:

$$u_{ex}(x, y) = -0.5Re(y^2 - 10y), \quad v_{ex}(x, y) = 0, \quad p_{ex}(x, y) = -x + 5.$$

Note, the $Q_2 - Q_1$ finite elements are exact with respect to quadratic $u_{ex}(x, y)$, $v_{ex}(x, y)$, and linear $p_{ex}(x, y)$ functions.

Hereinafter we use the notations $\mathbf{V}_h = (u_h, v_h, p_h)$, $\mathbf{V}_{ex} = (u_{ex}, v_{cx}, p_{ex})$ for a numerical and an exact solution, respectively; $\| \cdot \|_0^{h,U}$ and $\| \cdot \|_0^{h,p}$ are discrete analogues of the norm $\| \cdot \|_0$ for a scalar-function when the integrals are approximated by the Cartesian product of Simpson's rule and by the trapezoid rule, respectively; we also consider the compound norm $\left(\| \cdot \|_0^h \right)^2 = \left(\| \cdot \|_0^{h,U} \right)^2 + \left(\| \cdot \|_0^{h,U} \right)^2 + \left(\| \cdot \|_0^{h,p} \right)^2$. In all numerical experiments, an iterative algorithm is stopped when the norm $\| \cdot \|_0^h$ of the residual vector reaches the desired accuracy $\varepsilon = 10^{-6}$.

We start with the studies of scalability. In Table 1 the convergence results for the Poiseuille flow by the Uzawa – Richardson iterative algorithm are shown when the original and the scaled matrices are used.

In this experiment $Re = 100$. The numerical results show that the transformation allows one to decrease the number of iteration steps and to improve the accuracy of computation by an order. We use the scaled system in all further numerical experiments.

Table 1. The discrete norms of numerical error for the Poiseuille flow with the original $(-)$ and the scaled $(+)$ matrix.

$N_x \times N_y$	Scalability	Iterations	$\|\mathbf{V}_h - \mathbf{V}_{ex}\|_0^h$	$\|u_h - u_{ex}\|_0^{h,U}$	$\|v_h - v_{ex}\|_0^{h,U}$	$\|p_h - p_{ex}\|_0^{h,p}$
16×16	$-$	5703	$4.6E - 04$	$4.6E - 04$	$2.4E - 05$	$1.6E - 05$
	$+$	5059	$3.9E - 05$	$3.9E - 05$	$2.0E - 06$	$7.1E - 07$
32×32	$-$	18604	$4.8E - 04$	$4.7E - 04$	$1.5E - 05$	$3.2E - 05$
	$+$	16526	$3.8E - 05$	$3.8E - 05$	$1.2E - 06$	$2.6E - 06$
64×64	$-$	59992	$5.4E - 04$	$5.4E - 04$	$1.0E - 05$	$3.2E - 06$
	$+$	53151	$4.5E - 05$	$4.5E - 05$	$8.8E - 07$	$1.7E - 06$
128×128	$-$	191512	$5.6E - 04$	$5.5E - 04$	$2.7E - 05$	$1.1E - 05$
	$+$	168947	$5.0E - 05$	$5.0E - 05$	$3.9E - 06$	$4.9E - 06$

Table 2 shows the number of iteration steps and the execution time which is needed to solve the model problem by the Arrow – Hurwitz, the Uzawa – Richardson, and the BiCGstab methods. As the execution time for the Arrow – Hurwitz method is close to that for the Uzawa – Richardson method, only the order of the execution time in seconds is presented in Table 2.

With increasing the Reynolds number, the investigated parameters do not increase considerably for the Uzawa type methods. The number of iteration steps needed for the Uzawa – Richardson algorithm is less. At the same time, the BiCGstab demonstrates the best results when the Reynolds number is up to 100. For $Re = 1000$ the number of iteration steps of the BiCGstab increases considerably. Moreover, in the case of 128×128 grid, the BiCGstab does not converge to the solution unlike other methods.

We consider the numerical error of a solution of the model problem for the discussed methods (Table 3).

Table 2. The cost of the iterative methods

Mesh	Arrow-Hurwitz		Uzawa-Richardson		BiCGstab	
$N_x \times N_y$	Iterations	Time(s)	Iterations	Time(s)	Iterations	Time(s)
$Re = 1$						
32×32	18199	$O(10)$	14395	$O(10)$	1483	$O(1)$
64×64	53955	$O(10^2)$	46557	$O(10^2)$	5558	$O(10)$
128×128	158411	$O(10^3)$	148727	$O(10^3)$	21066	$O(10^2)$
$Re = 100$						
32×32	20633	$O(10)$	16526	$O(10)$	2121	$O(1)$
64×64	60867	$O(10^2)$	53151	$O(10^2)$	5026	$O(10)$
128×128	177324	$O(10^3)$	168947	$O(10^3)$	17105	$O(10^2)$
$Re = 1000$						
32×32	22969	$O(10)$	18295	$O(10)$	12463	$O(10)$
64×64	67714	$O(10^2)$	59034	$O(10^2)$	38066	$O(10^2)$
128×128	197110	$O(10^3)$	186963	$O(10^3)$	—	—

Table 3. The norms of the numerical error of a solution of the model problem by the iterative methods

$N_x \times N_y$	Method	$\|\mathbf{V}_h - \mathbf{V}_{ex}\|_0^h$	$\|u_h - u_{ex}\|_0^{h,U}$	$\|v_h - v_{ex}\|_0^{h,U}$	$\|p_h - p_{ex}\|_0^{h,p}$
$Re = 1$					
64×64	Arrow-Hurwics	$9.20E-04$	$2.57E-07$	$3.20E-08$	$9.20E-04$
	Uzawa-Richardson	$9.26E-04$	$3.94E-07$	$5.37E-08$	$9.26E-04$
	BiCGstab	$1.17E-04$	$4.81E-05$	$7.54E-05$	$7.47E-05$
128×128	Arrow-Hurwics	$1.83E-03$	$2.79E-07$	$4.82E-08$	$1.83E-03$
	Uzawa-Richardson	$1.83E-03$	$1.30E-06$	$3.07E-07$	$1.83E-03$
	BiCGstab	$2.02E-04$	$9.42E-05$	$1.44E-04$	$1.06E-04$
$Re = 100$					
64×64	Arrow-Hurwics	$3.62E-05$	$3.62E-05$	$6.10E-07$	$3.64E-07$
	Uzawa-Richardson	$4.56E-05$	$4.56E-05$	$8.81E-07$	$1.74E-06$
	BiCGstab	$4.23E-05$	$2.80E-05$	$3.18E-05$	$1.07E-06$
128×128	Arrow-Hurwics	$4.15E-05$	$4.14E-05$	$6.74E-07$	$1.62E-06$
	Uzawa-Richardson	$5.14E-05$	$5.10E-05$	$3.92E-06$	$4.90E-06$
	BiCGstab	$4.96E-05$	$2.62E-05$	$4.21E-05$	$1.49E-06$
$Re = 1000$					
64×64	Arrow-Hurwics	$4.28E-05$	$4.28E-05$	$7.16E-07$	$1.87E-07$
	Uzawa-Richardson	$5.71E-05$	$5.71E-05$	$1.07E-06$	$2.43E-07$
	BiCGstab	$1.16E-04$	$6.48E-05$	$9.57E-05$	$1.41E-06$
128×128	Arrow-Hurwics	$5.06E-05$	$5.06E-05$	$4.54E-07$	$2.48E-07$
	Uzawa-Richardson	$6.98E-05$	$6.96E-05$	$4.60E-06$	$7.05E-07$
	BiCGstab	no convergence	—	—	—

When $Re = 1$, the velocity is calculated more exactly then the pressure with the Uzawa type methods. In this case the BiCGstab gives the best result and a similar error order both for the velocity and the pressure. For $Re = 100$, the considered methods yield the same error in the compound norm. The BiCGstab gives a less accurate solution for the second (zero) components of the velocity. For coarser meshes, the pressure is calculated with minimal error order with the Arrow-Hurwics method, but for the 128×128 grid the results of the methods become the same. For $Re = 1000$, the Uzawa type methods demonstrate the advantage in accuracy of calculation over the BiCGstab method.

In summary, the stabilized biconjugate gradient method is efficient for the Reynolds number up to 100. It provides good convergence rate and calculation accuracy. However, for a large Reynolds number, the execution time of BiCGstab increases considerably with decreasing efficiency. Besides, the BiCGstab has no theoretical proof and does not ensure the convergence to an exact solution. In this sense, the considered methods of the Uzawa type are more robust. They are better in the time cost and accuracy for $Re = 1000$.

The numerical experiments show that the proposed Uzawa – Richardson iterative algorithm is competitive with the considered methods. It converges faster (with respect to the number of iteration steps) than the Arrow – Hurwics method. Besides, it is as good as the last one in accuracy of calculations.

Acknowledgements. The work is supported by RFBR (Project 14-01-00296).

References

1. Heywood, J.G., Rannacher, R.: Finite element approximation of the nonstationary Navier – Stokes problem. I. Regularity of solutions and second order error estimates for spatial discretization. SIAM J. Numer. Anal. **19**, 275–311 (1982)
2. Bercovier, M., Pironneau, O.: Error estimates for finite element method solution of the Stokes problem in the primitive variable. Numer. Math. **35**, 211–224 (1979)
3. Dementyeva, E., Karepova, E., Shaidurov, V.: The semi-Lagrangian approximation in the finite element method for the Navier-Stokes equations. In: AIP Conference Proceedings, vol. 1684, pp. 090009-1–090009-8 (2015)
4. Brezzi, F., Fortin, M.: Mixed and Hybrid finite element methods. Springer, New York (1991)
5. Benzi, M., Golub, G., Liesen, J.: Numerical solution of saddle point problems. Acta Numer. **14**, 1–137 (2005)
6. Bychenkov, Y., Chizhonkov, E.: Iterative methods for solving saddle point problems. BINOM, Knowledge Laboratory, Moscow (2010)
7. Saad, Y.: Iterative Methods for Sparse Linear Systems, 2nd edn. Society for Industrial and Applied Mathematics, Philadelphia (2003)
8. ur Rehman, M., Vuik, C., Segal, G.: A comparison of preconditioners for incompressible Navier-Stokes solvers. Int. J. Numer. Meth. Fluids **57**, 1731–1751 (2007)
9. Zulehner, W.: Analysis of iterative methods for saddle-point problems: a unified approach. Math. Comput. **71**, 479–505 (2002)
10. Astrakhantsev, G.P.: Analysis of algorithms of the Arrow – Hurwicz type. Zh. Vychisl. Mat. Mat. Fiz. **41**, 17–28 (2001)
11. Wilkinson, J.H., Reinsch, C.: Handbook for Automatic Computation: Linear Algebra. Springer, Heidelberg (1976)

Numerical Modeling of Coupled Problems of External Aerothermodynamics and Internal Heat-and-Mass Transfer in High-Speed Vehicle Composite Constructions

Yury Dimitrienko, Mikhail Koryakov, and Andrey Zakharov[✉]

Bauman Moscow State Technical University,
2-ya Baumanskay st., 5, Moscow, Russia
{dimit,mkoryakov,azaharov}@bmstu.ru
http://www.bmstu.ru/~fn11/english/echif.htm

Abstract. A coupled problem statement for aerogasdynamics, internal heat and mass transfer and thermal strength of heat shield structures of hypersonic vehicles is formulated. A method for numerical solving of the problem is suggested which is based on the iterative solution of the three types of detached problems: a gasdynamics problem for viscous heat-conducting flows, internal heat and mass transfer into constructions of hypersonic vehicles and thermoelasticity of shell constructions. An example of the numerical solution of the coupled problem of aerogasdynamics and thermal strength of elements of heat shield structures of an advanced vehicle is given. It is shown that due to the high temperatures of the aerodynamic heating of the structure made of a polymer composite material there can appear a polymer phase thermodecomposition and intensive internal gas generation into the structure of the material.

Keywords: Coupled simulation · Computational fluid dynamics · Hypersonic flows · Aerothermodynamics · Thermomechanics · Polymer composites · Thermodecomposition structures · Heat shield · Parallel processing

1 Introduction

Flying at a hypersonic speed in the Earth's atmosphere causes an increase in the temperature of the air up to 2,000 K and above in a blunt nose. For this reason, there is the high heat flux (0.5 MW/m^2 and above) to the aircraft surface. The polymer composite materials (glass-, carbon- and organo-plastics) are often used for high-temperature protection of aircraft elements. During operation the mechanical and thermal decomposition of these materials occurs. The proper selection of a heat-resistant material is an important task in the design phase of an aircraft.

The problem of analyzing the behavior of the heat-resistant material should be solved together with the problem of gas dynamics in order to take into account

© Springer International Publishing AG 2017
I. Dimov et al. (Eds.): NAA 2016, LNCS 10187, pp. 294–301, 2017.
DOI: 10.1007/978-3-319-57099-0_31

the mutual influence of the incoming gas flow and thermal decomposition products of the composite materials. Existing commercial software packages do not allow to simulate the coupled problem, so the specialized software need to be developed.

2 Mathematical Formulation

The general formulation of the coupled problem of aerothermodynamics and thermomechanics consists of the three systems of equations

- the Navier-Stokes equations of an external gas flow;
- the internal heat and mass transfer equations;
- the equations of thermoelasticity of a shell.

2.1 System of Gasdynamics Equation

The system of equations of a viscous heat-conducting compressible gas flow consists of the continuity equation, momentum equations and energy equation [1].

The boundary conditions on the solid surface, which is the interface of the gas and solid domains, are as follows

$$v = 0, \qquad -\lambda \nabla \theta \cdot n + \varepsilon_g \sigma \theta_{max}^4 = -\lambda_s \nabla \theta_s \cdot n + \varepsilon_s \sigma \theta_s^4, \qquad \theta_s = \theta,$$

where v is the velocity vector, λ and λ_s are the thermal conductivity of the gas and solid surface respectively, θ is the gas temperature, θ_{max} is the maximum temperature into the boundary layer, θ_s is the temperature of the solid surface, $\nabla \theta_s$ is the temperature gradient on the solid wall from the construction, ε and ε_s are the emissivity of the heated gas and solid surface respectively and σ is the Stefan–Boltzmann constant.

2.2 System of Equations of Internal Heat and Mass Transfer

We consider the typical aircraft shell which is made of a heat-resistant composite material consisting of a polymer master with heat-resistant filaments. There are physical and chemical processes of thermodecomposition in such composite master under high temperatures of aerodynamic heating. In these processes the gaseous products of thermal decomposition are generated, then they are accumulated in the pores of the material and filtered into the outer gas flow, as well as a new solid phase is formed. It is the phase of pyrolytic master which has significantly lower elastic-strength properties than the original polymer phase. The four-phase model to describe the internal heat and mass transfer and deformations of such composite is proposed in [2]. This model consists of the equation of change of mass of polymer master phase, the equation of filtration of gaseous products of thermodestruction in pores of the composite material and the heat transfer equation in the thermodestruction composite

$$\rho_b \frac{\partial \varphi_b}{\partial t} = -J, \qquad \frac{\partial \rho_g \varphi_g}{\partial t} + \boldsymbol{\nabla} \cdot \rho_g \varphi_g \boldsymbol{v} = J\Gamma, \tag{1}$$

$$\rho c \frac{\partial \theta}{\partial t} = -\boldsymbol{\nabla} \cdot \boldsymbol{q} - c_g \boldsymbol{\nabla}\theta \cdot \rho_g \varphi_g \boldsymbol{v} - J\triangle e^0, \tag{2}$$

where φ_b, φ_g are the volume concentrations of the phase of the initial polymer master and gas phase; ρ_b is the density of the phase of the initial polymer master which is assumed to be constant; ρ_g is the average pore density of the gas phase (variable); c_g is the specific heat of the gas phase at constant volume, ρ and c are the density and specific heat of the composite as a whole, $\boldsymbol{q}$ is the heat flux vector, θ is the composite temperature for all phases in common; $\boldsymbol{v}$ is the velocity vector of the gas phase in the pores; $\triangle e^0$ is the specific heat of the master thermodecomposition; J is the mass velocity of the master thermodecomposition and Γ is the gas-producing factor of the master.

The Eqs. (1)–(2) are added the relations between the heat flux vector $\boldsymbol{q}$, velocity vector of the gas in pores $\boldsymbol{v}$ with the temperature gradient $\boldsymbol{\nabla}\theta$ and pressure gradient $\boldsymbol{\nabla}p$ using the Fourier and Darcy laws, as well as the Arrhenius relation for the mass velocity of the master thermodecomposition J and Mendeleev-Clapeyron equation for the pore pressure of the gas phase p

$$\boldsymbol{q} = -\boldsymbol{\Lambda} \cdot \boldsymbol{\nabla}\theta, \qquad \rho_g \varphi_g \boldsymbol{v} = -\boldsymbol{K} \cdot \boldsymbol{\nabla}p, \qquad J = J_0 e^{-E_A/R\theta}, \qquad p = \rho_g \frac{R}{\mu_g}\theta,$$

where J_0 is the pre-exponential factor, E_A is the activation energy of the thermodecomposition process, μ_g is the molecular weight of the gas phase. And $\boldsymbol{\Lambda}$ is the thermal conductivity tensor, $\boldsymbol{K}$ is the permeability tensor of the composite. They depend on the phase concentration.

The conditions for the Eqs. (1)–(2) on the heated surface of the construction are as follows: $p = p_e$, $\theta = \theta_e$, where p_e, θ_e are the pressure and temperature of the flow on the surface.

The boundary conditions of the tightness and thermal insulation are specified on the rest of the composite surface: $\boldsymbol{n} \cdot \boldsymbol{\nabla}p = 0$, $\boldsymbol{n} \cdot \boldsymbol{\Lambda} \cdot \boldsymbol{\nabla}\theta = 0$.

2.3   System of Equations of Thermoelasticity

The system of equations in curvilinear coordinates $Oq_1q_2q_3$ associated with the middle surface of an thermoelastic shell structure consists of [2]

– the equilibrium equations of the shell

$$\frac{\partial A_\beta T_{\alpha\alpha}}{\partial q_\alpha} + \frac{\partial A_\alpha T_{\alpha\beta}}{\partial q_\beta} - \frac{\partial A_\beta}{\partial q_\alpha}T_{\beta\beta} + \frac{\partial A_\alpha}{\partial q_\beta}T_{\alpha\beta} + A_\beta\left(A_\alpha k_\alpha Q_\alpha - \frac{\partial P_g}{\partial q_\alpha}\right) = 0, \tag{3}$$

$$\frac{\partial A_\beta M_{\alpha\alpha}}{\partial q_\alpha} + \frac{\partial A_\alpha M_{\alpha\beta}}{\partial q_\beta} - \frac{\partial A_\beta}{\partial q_\alpha}M_{\beta\beta} + \frac{\partial A_\alpha}{\partial q_\beta}M_{\alpha\beta} - A_\beta\left(A_\alpha Q_\alpha - \frac{\partial P_g}{\partial q_\alpha}\right) = 0, \tag{4}$$

$$-A_1 A_2(k_1 T_{11} + k_2 T_{22} + p_e) + \frac{\partial A_2 Q_1}{\partial q_1} + \frac{\partial A_1 Q_2}{\partial q_2} - (k_1 + k_2)A_1 A_2 \varphi_g P_g = 0; \tag{5}$$

– the kinematic relations

$$e_{\alpha\alpha} = \frac{1}{A_\alpha}\frac{\partial U_\alpha}{\partial q_\alpha} + \frac{1}{A_1 A_2}\frac{\partial A_\alpha}{\partial q_\beta}U_\beta + k_\alpha W, \quad 2e_{\alpha 3} = \frac{1}{A_\alpha}\frac{\partial W}{\partial q_\alpha} + \gamma_\alpha - k_\alpha U_\alpha, \quad (6)$$

$$2e_{12} = \frac{1}{A_2}\frac{\partial U_1}{\partial q_2} + \frac{1}{A_1}\frac{\partial U_2}{\partial q_1} - \frac{1}{A_1 A_2}\left(\frac{\partial A_1}{\partial q_2}U_1 + \frac{\partial A_2}{\partial q_1}U_2\right), \quad (7)$$

$$\kappa_{\alpha\alpha} = \frac{1}{A_\alpha}\frac{\partial \gamma_\alpha}{\partial q_\alpha} + \frac{1}{A_1 A_2}\frac{\partial A_\alpha}{\partial q_\beta}\gamma_\beta, \qquad 2\kappa_{\alpha 3} = -k_\alpha\gamma_\alpha, \quad (8)$$

$$2\kappa_{12} = \frac{1}{A_2}\frac{\partial \gamma_1}{\partial q_2} + \frac{1}{A_1}\frac{\partial \gamma_2}{\partial q_1} - \frac{1}{A_1 A_2}\left(\frac{\partial A_1}{\partial q_2}\gamma_1 + \frac{\partial A_2}{\partial q_1}\gamma_2\right); \quad (9)$$

– the defining relationships of thermoelasticity shell

$$T_{\alpha\alpha} = \sum_{\beta=1}^{2}(C_{\alpha\beta}e_{\beta\beta} + N_{\alpha\beta}\kappa_{\beta\beta}) - P_{g\alpha} - \hat{T}_\alpha, \quad T_{12} = 2(C_{66}e_{12} + N_{66}\kappa_{12}), \quad (10)$$

$$M_{\alpha\alpha} = \sum_{\beta=1}^{2}(N_{\alpha\beta}e_{\beta\beta} + D_{\alpha\beta}\kappa_{\beta\beta}) - M_{g\alpha} - \hat{M}_\alpha, \quad (11)$$

$$M_{12} = 2(N_{66}e_{12} + D_{66}\kappa_{12}), \qquad Q_\alpha = \bar{C}_{\alpha+3,\alpha+3}e_{\alpha 3}; \quad (12)$$

where $T_{\alpha\alpha}$, $T_{\alpha\beta}$, $M_{\alpha\alpha}$, $M_{\alpha\beta}$ are the forces and moments in the shell; Q_α are the shear forces; $e_{\alpha\alpha}$, $e_{\alpha 3}$, e_{12} are the deformations of the middle surface; $\kappa_{\alpha\alpha}$, $\kappa_{\alpha 3}$, κ_{12} are the curvatures of the middle surface; U_α, γ_α, W are the displacement, angles of curvature and deflection of the middle surface; A_α, k_α are the parameters of the first quadratic form and principal curvatures of the middle surface and P_g, M_g are the forces and moments of the pore pressure, $\alpha, \beta = 1, 2$; $\alpha \neq \beta$.

The forces and moments of the thermal stresses $\hat{T}_\alpha$, $\hat{M}_\alpha$, depending on the thermal deformations $\hat{\varepsilon}_\alpha$ for the shell are introduced as follows

$$\hat{T}_\alpha = \sum_{\beta=1}^{3}C_{\alpha\beta}\hat{\varepsilon}_\beta^{(0)}, \qquad \hat{M}_\alpha = \sum_{\beta=1}^{3}C_{\alpha\beta}\hat{\varepsilon}_\beta^{(1)}; \qquad \gamma = 1, 2, 3;$$

$$\hat{\varepsilon}_\beta^{(j)} = \int_{-h/2}^{h/2} a_{\theta 1}\hat{\varepsilon}_\beta q_3^j\, dq_3, \qquad \hat{\varepsilon}_3^{(j)} = \int_{-h/2}^{h/2} a_{\theta 2}\hat{\varepsilon}_3 q_3^j\, dq_3; \qquad j = 0, 1; \quad \beta = 1, 2;$$

$$\hat{\varepsilon}_\gamma = (\alpha_f\varphi_f B_\gamma + \alpha_b\varphi_b\Omega_\gamma)(\theta - \theta_0) + \alpha_p\Omega_\gamma\int_0^t (\theta(t) - \theta(\tau))\,\hat{\varphi}_p\, d\tau - \beta_p\varphi_p\Omega_\gamma,$$

where $\alpha_f, \alpha_b, \alpha_p$ are the coefficients of the thermal expansion of the filament, polymer and pyrolytic phases of the master, β_p is the shrinkage ratio, B_γ, Ω_γ are the coefficients depending on the location of the filaments in the composite [2]. Due to the softening of the polymer master and its thermodecomposition, stiffnesses of the shell are changing when heated. This change for orthotropic composite shells are realized using 2 functions $a_{\theta 1}$, $a_{\theta 2}$ [2].

Deformations $\varepsilon_{\alpha\beta}$ and stresses $\sigma_{\alpha\beta}$ in the shell are calculated using the following formulas

$$\varepsilon_{\alpha\beta} = e_{\alpha\beta} + q_3\kappa_{\alpha\beta}; \quad \alpha, \beta = 1, 2; \quad \varepsilon_{33} = 0, \quad \varepsilon_{\alpha3} = e_{\alpha3}; \quad \alpha = 1, 2;$$

$$\sigma_{\alpha\alpha} = -\hat{f}_\alpha p + a_{\theta1}\sum_{\beta=1}^{3} C_{\alpha\beta}\left(\varepsilon_{\beta\beta} + q_3\kappa_{\beta\beta} - \hat{\varepsilon}_\beta\right), \quad \sigma_{12} = a_{\theta1}C_{66}(\varepsilon_{12} + q_3\kappa_{12}).$$

The transverse normal stress σ_{33} and shear interlayer stresses $\sigma_{\alpha3}$ have the following formulas

$$\sigma_{33} = 6\eta\left(\frac{p_1 + p_2}{2} - \frac{P_{g1}}{h} + \frac{1}{h}C_{31}\left(a_{\theta1}^{(0)}e_{11} + a_{\theta1}^{(1)}\kappa_{11} - \hat{\varepsilon}_1^{(0)}\right)\right. \tag{13}$$

$$+ \frac{1}{h}C_{32}^0\left(a_{\theta1}^{(0)}e_{22} + a_{\theta1}^{(1)}\kappa_{22} - \hat{\varepsilon}_2^{(0)}\right) - \frac{1}{h}C_{33}^0\varepsilon_3^{(0)}\right)$$

$$+ (p_2 - p_1)\,\xi\,(q_3) + \frac{p_1 - p_2}{2} + \varphi_g p,$$

$$\sigma_{13} = \frac{12\eta\,(q_3)}{h}C_{44}^{(0)}e_{13}a_{\theta2}^{(0)}; \quad \sigma_{23} = \frac{12\eta\,(q_3)}{h}C_{55}^{(0)}e_{23}a_{\theta2}^{(0)}, \tag{14}$$

$$\xi\,(q_3) = \frac{1}{2} - \frac{q_3}{h}, \quad \eta\,(q_3) = \frac{1}{4} - \left(\frac{q_3}{h}\right)^2,$$

where p_1, p_2 are the pressures on the external surfaces of the shell.

3 Computational Method

A step-by-step method with two time steps Δt_1 and Δt_2 is used for the numerical solution of the coupled problem. The time step Δt_1 is used for changing the boundary conditions for the Navier-Stokes equations and the equations of internal heat and mass transfer. The solution is carried out in 4 stages for each step Δt_1.

Stage 1. The numerical solution of the equations of internal heat and mass transfer (1)–(2) is carried out by the numerical finite-difference method using the method of linearization and the implicit difference scheme which are described in [2]. The temperature of the solid surface θ_w interacting with the incoming flow is taken from the previous time step Δt_1.

Stage 2. The numerical integration of the Navier-Stokes equations is carried out over the time steps Δt_2 to establish flow. We use the finite volume method based on elements centered on the grid point (vertex-centered volume) [3].

Stage 3. The solution of the equations of thermoelasticity for a shell (3)–(12) is carried out by means of the finite-element method which is described in [2]. The input data for this task are pressures on the outer p_1 and inner p_2 surface of the shell, which are calculated after the solution of the Navier-Stokes equations, as well as the distributions of the temperature θ, volume concentrations of the phases and the pore pressure p which are calculated by the solving of the equations of internal heat and mass transfer (1)–(2) for the current time step.

Stage 4. The calculation of thermal stresses in the shell is carried out by means of formulas (13)–(14).

4   Results and Discussion

A numerical solution of the coupled problem was calculated for a hypersonic flow ($M = 6$) around the nose of a typical vehicle model [4] flying at the altitude of 15 km.

Figure 1 shows the distribution of the pressure and temperature of the flow near the body. The temperature reaches 1,600 K at the stagnation point on the nose of the vehicle and decreases monotonically as the distance from the stagnation point, but it remains rather high value.

The results of numerical calculations of the fields of internal heat and mass transfer in the shell element of the hypersonic aircraft are shown in Fig. 2. The curves show the distributions along the shell thickness (the dimensionless coordinate $x = 0.5 + q_3/h$ is introduced) at a control point on the bottom surface at the distance of $30\,r$ from the stagnation point, where r is the blunted body radius at the stagnation point. Different colors correspond to the 7 different time points $t_1, \ldots, t_7$ of the aircraft flight.

Figure 3 shows the distribution of the thermal transversal deformation $\hat{\varepsilon}_3$ and transversal stress σ_{33}. The thermal deformation increases in the domain of high temperature and decreases after passing through the conventional point of the beginning of the intensive thermal decomposition due to the formation of pyrolytic residue. The thermal transversal deformation $\hat{\varepsilon}_3$ becomes negative at a certain moment, because there is the shrinkage of the matrix. The emergence of shrinkage causes the formation of shrinkage stresses σ_{33} in the heated subsurface shell layer. The graph of the transversal stresses σ_{33} shows that the peak of the positive (tensile) stress σ_{33} occurs in the subsurface layer due to the increase of the pore pressure and the formation of shrinkage deformation of the shell after the initiation of thermal decomposition of the polymer. The transversal stresses are reached values of 0.13 GPa at the bottom of the shell at the time t_6. It is significantly higher than the breaking point of the composite's shell in the transversal direction. The destruction of the bundle type may occur in this part of the shell in which upper layers of the composite's fabric are delaminated from the rest of the material. A similar effect may occur at the top of the shell to the time t_7, as the transversal stresses rise to 0.06 GPa. It is less than the values at the bottom of the shell, but also comparable to the breaking point of the composite's shell in the transversal direction (see also [5]).

Figure 4 shows the efficiency of parallelization using MPI and MSU supercomputer SKIF 'Tchebychev' for solving one of the problems of gas dynamics. The mesh consisted about 16 million nodes. 5,000 time steps were calculated. Efficiency closed to linear was saved up to 256 cores then it started to decrease.

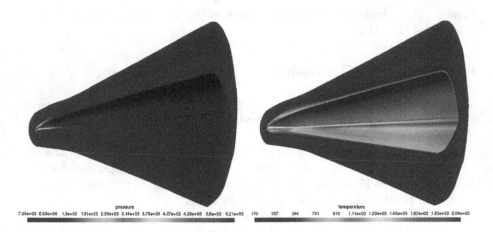

pressure
7.05e+03 6.85e+04 1.3e+05 1.91e+05 2.53e+05 3.14e+05 3.76e+05 4.37e+05 4.98e+05 5.6e+05 6.21e+05   170   357   544   731   918   1.11e+03 1.29e+03 1.48e+03 1.67e+03 1.85e+03 2.04e+03

Fig. 1. Distributions of the parameters of the gas flow near the surface of hypersonic vehicle: (left) pressure p (Pa); (right) temperature θ (K)

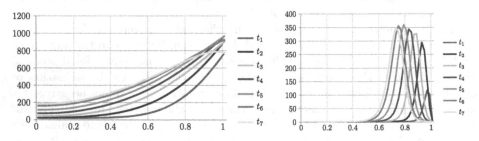

Fig. 2. Distributions of the parameters along the shell thickness: (left) temperature gradient $\theta - \theta_0$ (°C); (right) pore pressure p (atm)

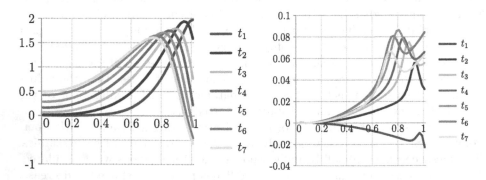

Fig. 3. Distributions of the parameters along the shell thickness: (left) thermal transversal deformation $\hat{\varepsilon}_3$; (right) transversal stress σ_{33} (GPa)

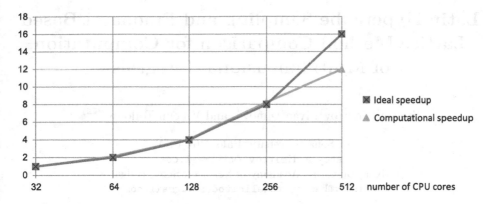

Fig. 4. Speedup ratio and parallel efficiency of the aerogasdynamics problem

Acknowledgements. Work was funded by a grant of the President of the Russian Federation research project no. MK-3007.2015.8. The reported study was supported by the Supercomputing Center of Lomonosov Moscow State University [6].

References

1. Dimitrienko, Y., Koryakov, M., Zakharov, A.: Application of finite difference TVD methods in hypersonic aerodynamics. In: Dimov, I., Faragó, I., Vulkov, L. (eds.) FDM 2014. LNCS, vol. 9045, pp. 161–168. Springer, Cham (2015). doi:10.1007/978-3-319-20239-6_15
2. Dimitrienko, Y.: Thermomechanics of Composites Under High Temperatures. Kluwer Academic Publishers, Boston (1999)
3. Barth, T.J.: Aspects of unstructured grids and finite-volume solvers for the Euler and Navier-Stokes equations. VKI Lecture Series, Belgium, Von Karman Institute for Fluid Dynamics, 1994–1905 (1994)
4. http://www.boeing.com/features/2014/05/bds-aiaa-x51-05-02-14.page
5. Dimitrienko, Y., Koryakov, M., Zakharov, A.: Computational modeling of conjugated aerodynamic and thermomechanical processes in composite structures of high-speed aircraft. Appl. Math. Sci. **9**(98), 4873–4880 (2015)
6. Sadovnichy, V., Tikhonravov, A., Voevodin, V., Opanasenko, V.: "Lomonosov": supercomputing at Moscow State University. In: Contemporary High Performance Computing: From Petascale Toward Exascale, pp. 283–307. Chapman & Hall/CRC Computational Science, CRC Press, Boca Raton (2013)

Latin Hypercube Sampling and Fibonacci Based Lattice Method Comparison for Computation of Multidimensional Integrals

Stoyan Dimitrov[1], Ivan Dimov[2], and Venelin Todorov[2(✉)]

[1] Sofia University, Palo Alto, USA
stoyan.dimitrov.fmi@gmail.com
[2] IICT, Bulgarian Academy of Sciences, Sofia, Bulgaria
ivdimov@bas.bg, venelinltodorov@gmail.com

Abstract. We perform computational investigations to compare the performance of Latin hypercube sampling (LHS method) and a particular QMC lattice rule based on generalized Fibonacci numbers (FIBO method) for integration of smooth functions of various dimensions. The two methods have not been compared before and both are generally recommended in case of smooth integrands. The numerical results suggests that the FIBO method is better than LHS method for low-dimensional integrals, while LHS outperforms FIBO when the integrand dimension is higher. The Sobol nets, which performance is given as a reference, are outperformed by at least one of the two discussed methods, in any of the considered examples.

1 Introduction

The year 1949 is generally regarded as the official birthday of the Monte Carlo method when the paper of Metropolis and Ulam [8] was published, although some authors point to earlier dates. Ermakov [3], for example, notes that a solution of a problem by the Monte Carlo method is contained in the Old Testament. Some of the most popular are the ordinary Crude MC, variance reduction methods such as Stratified sampling, Importance Sampling and Adaptive algorithms, as well as techniques using quasi-random sequences like those of Sobol, Holton and others. An additional and relatively new type of quasi - sequences are those of lattice type, which are appropriate in case of smooth and periodic integrands. Our aim is to compare the performance of a particular variant of the latter method, namely rank-1 lattice sets with generating vector based on the generalized Fibonacci sequence (FIBO), with a widely used type of stratified sampling called Latin Hypercube Sampling (LHS). Both methods are especially recommended for evaluation of integrals of smooth functions, which motivates our study. We aim to answer the question which of the two methods should be preferred depending on the dimensionality of the integral considered. A comparison of the relative errors obtained by the two methods for the same number of sample points is presented together with comparison of these errors for equal

© Springer International Publishing AG 2017
I. Dimov et al. (Eds.): NAA 2016, LNCS 10187, pp. 302–310, 2017.
DOI: 10.1007/978-3-319-57099-0_32

amount of time. Comparisons between LHS and QMC methods were done in the paper of Kucherenko et al. [6] and those of Wang and Sloan [14]. A systematic overview of Monte Carlo methods for various classes of problems may be found in the monograph [1]. The Latin Hypercube Sampling method was first published by McKay et al. [7]. Good sources providing concise explanations of the LHS method are the books of Owen [10] and Kroese et al. [5]. The monographs of Sloan and Kachoyan [12], Niederreiter [9] and Sloan and Joe [11] provide comprehensive expositions of the theory of integration lattices.

The layout of the rest of this paper is as follows. The next section describes the Latin Hypercube Sampling (LHS). Section 3 discusses the settings of the lattice rules technique and the concrete sequences that we implemented together with results concerning its discrepancy. The last section examines the comparison between the two algorithms in terms of minimization of the error when computing integrals of four different smooth functions of dimensions 4, 7, 25 and 30 respectively. Here, information concerning the time efficiency is also provided. Finally, some conclusions were made concerning the advantages and disadvantages of the two algorithms studied.

2 Latin Hypercube Sampling

Throughout this paper we will consider the problem of approximating the multiple integral

$$\int_{G_s} f(\mathbf{x})d\mathbf{x} = \int_0^1 d\mathbf{x}^{(1)} \int_0^1 d\mathbf{x}^{(2)} \dots \int_0^1 d\mathbf{x}^{(s)} f(x^{(1)}, x^{(2)}, \dots, x^{(s)}), \quad (1)$$

where $G_s = [0,1)^s$, $x = (x^{(1)}, \dots, x^{(s)}) \in [0,1)^s$ and the function f is integrable in G_s.

2.1 Stratified Sampling

One of the main problems of some widely used MC methods like Importance sampling is that a sample that is very close to another does not provide much new information about the function being integrated. One powerful variance-reduction technique that addresses this problem is called stratified sampling. Stratified sampling works by splitting up the original integral into a sum of integrals over sub-domains. In its simplest form, stratified sampling divides the domain G_s into N sub-domains (or stratas) and places a random sample within each of these intervals. A quantity of interest is the variance of the obtained approximation, considered as a random variable. It can be shown that stratified sampling can never result in higher variance than pure random sampling [13]. If $N = 1$, we have random sampling over the entire sample space.

2.2 LHS

Latin Hypercube Sampling (LHS) is a type of stratified sampling. If we wish to ensure that each of the input variables x_i has all portions of its distribution represented by input values, we can divide the range of each x_i, in our case the interval $[0, 1]$, into M strata of equal marginal probability $\frac{1}{M}$, and sample once from each stratum. In the case of integral approximation we must simply divide the interval $[0, 1]$ into M disjoint intervals, each of length $\frac{1}{M}$ and to sample one point from each of them. Let this sample be X_{kj}, for dimensions $k = 1, \ldots, s$, $j = 1, \ldots, M$. Those of them having first index k ($k = 1, \ldots, s$) are the different components for the k-th dimension of the random points that are used for the integral's approximation. These components are matched at random. Thus the maximum number of combinations for a Latin Hypercube of M divisions and s variables (i.e., dimensions) can be computed by the formula $(M!)^{s-1}$. In the context of statistical sampling, a square grid containing sample positions is a Latin square if (and only if) there is only one sample in each row and each column. Thus following the described algorithm, we obtain a set of points with positions forming a Latin square. Note that this sampling scheme does not require more samples for more dimensions (variables); this independence is one of the main advantages of LHS scheme. Below are given examples of random, stratified and latin hypercube samplings with 16 points ($s = 2$, $M = 4$ in the LHS case, the figure was taken from [4]).

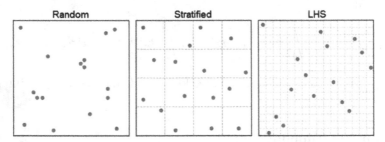

There aren't many theoretical facts demonstrating the usefulness of LHS. However, we include below a weaker variant of a theorem proved in [7]:

Theorem 1. *If $Y = f(X_1, X_2, \ldots X_s)$ is monotonic in each of its arguments then $Var(T_L) \leq Var(T_R)$, where T_L is the approximation of $\int_{G_s} f(\mathbf{x})d\mathbf{x}$ obtained by the LHS method and T_R is the approximation of the same integral obtained by random sampling.*

3 Lattice Sets

Let us take the point set $X = \{x_i \mid i = 1, 2, \ldots N\}$ in G_s. If $x_i = (x_i^{(1)}, x_i^{(2)}, \ldots, x_i^{(s)})$ and $J(v) = [0, v_1) \times [0, v_2) \times \ldots \times [0, v_s)$ represent a box, then the discrepancy of the point set is defined as the largest difference between the proportion of points in a box and the volume of this box:

$$D(X) := \sup_{0 \le v_j \le 1} \| \frac{\#\{x_i \in J(v)\}}{N} - \prod_{j=1}^{s} v_j \| . \tag{2}$$

3.1 Latices and Lattice Rules

Quasi-Monte Carlo methods rely on sets of points that are evenly spread over the integration domain i.e. these sets have small discrepancies. One effective way of obtaining such sets is to use the lattice point method.

The most widely used types of lattice sets are the following:

$$x_n = \left\{ \frac{n}{N} a \right\}, \ n = 0, \dots, N-1$$

Here $a = (a_1, \dots, a_n)$ is the generating vector of dimensionality s which has no common divisor with N. Every component of the vector is replaced with its fractional part.

A lattice S is an infinite set of points in s with the following three properties [12]:

1. If x and x' belong to S, then so do $x + x'$ and $x - x'$.
2. S contains s linearly independent points.
3. There exists a sphere centered at 0 that contains no points of the lattice other than 0 itself.

A multiple-integration lattice is then defined to be a lattice that contains $\mathbb{Z}^s$ as a sub-lattice. By a "lattice rule" then, we shall mean a rule of the form $I_N(f) = \frac{1}{N} \sum_{j=0}^{N-1} f(x_j)$, in which $x_0, \dots, x_{N-1}$ are all the points of a multiple-integration lattice that lie in G^s.

Let n be an integer, $n \ge 2$ and $a = (a_1, a_2, \dots a_s)$ be an integer vector modulo n. A set of the form [11]

$$P_n = \left\{ \left\{ \frac{ak}{n} \right\} = \left(\left\{ \frac{a_1 k}{n} \right\}, \dots, \left\{ \frac{a_s k}{n} \right\} \right) \mid k = 1, \dots, n \right\} \tag{3}$$

is called a lattice point set, where $\{x\}$ denotes the fractional part of x. Here, the vector a is the generator of the set. We are considering only sets of the form (1), which are also called node sets of rank-1 lattices. The formula describing these lattices is very simple and it could be implemented easily. However, finding a good value of a, such that the points in the set are evenly distributed over the unit cube is the difficult part here.

Indeed, the choice of a good generating vector, which leads to small errors is not trivial. Complicated methods from theory of numbers are widely used, for example Zaremba's index or error of the worst function. Now, we will focus on one concrete vector with good properties. We will refer to the lattice method using this vector as Flbo lattice method or FIBO method. Consider the generating vector below for some natural number n:

$$a = (1, F_n(2), \dots, F_n(s)) \tag{4}$$

Here, $F_n(j) = F_{n+j-1} - F_{n+j-2} - \ldots - F_n$, where F_i are the corresponding generalized Fibonacci number of dimensionality s, i.e.:

$$F_{l+s} = F_l + F_{l+1} + \ldots + F_{l+s-1}, l = 0, 1, \ldots \tag{5}$$

with initial conditions:

$$F_0 = F_1 = \ldots = F_{s-2} = 0, F_{s-1} = 1, \tag{6}$$

for $l = 0, 1, \ldots$.

After simplifying, one can see that the generating vector above is:

$$a = (1, F_{n-1} + F_{n-2} + \ldots + F_{n-s+1}, \ldots, F_{n-1} + F_{n-2}, F_{n-1}) \tag{7}$$

We have the following estimation for the discrepancy of the set obtained by using this vector and F_n number of points:

Theorem 2. *The set*

$$\left(\left\{ \frac{1}{F_n} k \right\}, \left\{ \frac{F_n(2)}{F_n} k \right\}, \ldots, \left\{ \frac{F_n(s)}{F_n} k \right\} \right), \quad 1 \le k \le F_n,$$

has discrepancy $D(F_n) = \mathcal{O}(F_n^{-\frac{1}{2} - \frac{1}{2^{s+1} \cdot \log 2} - \frac{1}{2^{2s+3}}})$.

In [11] is pointed that Fibonacci lattice sets are very appropriate for integration of smooth functions. Apart from the low discrepancy, the biggest advantage of the FIBO algorithm is its time efficiency. The number of calculation required to obtain the generating vector is asymptotically $\mathcal{O}(n) = \mathcal{O}(\log F_n)$. Once we have this vector, the generation of a new point requires constant number of operations. Thus, since we have to generate F_n points, to obtain a lattice set of the described kind consisting of F_n points, $\mathcal{O}(F_n)$ number of operations are required.

4   Numerical Examples

Here are given some results of numerical experiments. We considered four different examples of 4, 7, 25 and 30 dimensional integrals, respectively, for which we have computed their exact values.

Example 1. s= 4.

$$\int_{[0,1]^4} x_1 x_2^2 e^{x_1 x_2} \sin(x_3) \cos(x_4) \approx 0.108975. \tag{8}$$

Example 2. s= 7.

$$\int_{[0,1]^7} e^{1 - \sum_{i=1}^{3} \sin(\frac{\pi}{2} . x_i)} . arcsin(sin(1) + \frac{\sum_{j=1}^{7} x_j}{200}) \approx 0.7515. \tag{9}$$

Example 3. s= 25.

$$\int_{[0,1]^{25}} \frac{4x_1 x_3^2 e^{2x_1 x_3}}{(1 + x_2 + x_4)^2} e^{x_5 + \cdots + x_{20}} x_{21} \ldots x_{25} \approx 108.808. \tag{10}$$

Example 4. s= 30.

$$\int_{[0,1]^{30}} \frac{4x_1 x_3^2 e^{2x_1 x_3}}{(1 + x_2 + x_4)^2} e^{x_5 + \cdots + x_{20}} x_{21} \ldots x_{30} \approx 3.244. \tag{11}$$

The results are given in the tables below. Each table contains information about the MC approach which is applied, the obtained relative error, the needed CP-time and the number of points. Note that when the FIBO method is tested, the number of sampled points are always Generalized Fibonacci numbers of the corresponding dimensionality. In addition, the widely used Sobol nets' method performance is shown as a reference. We have used CPU Intel Core i5-2410M @ 2.30 GHz and MATLAB (Fig. 1) (Tables 1 and 2).

Table 1. Comparison of the relative errors for the 4 dimensional integral (8)

# of points	LHS	Time, s	SOBOL	Time, s	FIBO	Time, s
1490	4.59e−3	0.009	9.46e−4	0.43	1.01e−3	0.009
10671	5.61e−4	0.09	5.28e−4	1.4	8.59e−5	0.01
20569	4.42e−4	0.14	3.52e−5	4.32	3.89e−5	0.03
39648	2.67e−4	0.26	2.68e−5	7.77	3.01e−5	0.07
147312	1.40e−5	0.91	2.29e−6	23.7	3.71e−6	0.34
547337	3.93e−6	3.16	1.40e−7	76	9.94e−7	1.21

Table 2. Comparison of the relative errors for the 4 dimensional integral 8 and equal execution times

Time, s	LHS	SOBOL	FIBO
0.1	2.18e−4	4.07e−4	1.44e−5
1	3.32e−4	3.54e−5	8.01e−6
1	4.23e−5	5.26e−5	5.38e−7
5	3.48e−5	6.50e−6	3.77e−6
10	7.10e−5	6.55o−6	2.67e−8

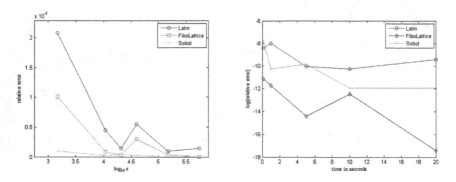

Fig. 1. Relative errors and computational times for the 4 dimensional integral

5 Conclusion

A particular 1-rank lattice Monte Carlo algorithm and a variance reduction method called Latin Hypercube Sampling are presented and discussed. We apply the two methods to 4, 7, 25 and 30 dimensional integrals of smooth functions. It was a question of interest which of these methods outperforms the other for lower and higher dimensions since they have not been compared. The tables with the obtained errors are shown. It can be seen that FIBO method produces better results for lower dimensions, but when the integrand is of dimension higher than 20 (as in Examples 3 and 4), the LHS method outperforms the FIBO method. This confirms the conclusions made in [14] regarding the comparison of MC and QMC methods in general. The reference method using the quasi-random method of Sobol characterizes with sufficiently good accuracy regardless of the dimension. It is worth mentioning that the results obtained by latin hypercube method are the best for higher dimensions and much better than the results obtained by Karaivanova and Dimov in [2] for the same integrals with the adaptive and importance separation techniques. Moreover, we received that each of the presented methods is more accurate than the method using the QMC Sobol sequence, for certain dimensions (Tables 3, 4, 5 and 6).

Table 3. The relative errors for the 7 dimensional integral (9)

# of points	Latin	Time, s	Sobol	Time, s	Lattice	Time, s
2000	7.90e−3	0.72	5.45e−4	1.24	2.57e−2	0.23
7936	1.76e−3	2.28	2.28e−4	3.68	1.19e−3	0.87
15808	9.58e−4	3.89	1.65e−4	5.26	2.81e−4	1.73
62725	7.55e−4	11.64	1.18e−4	20.3	2.78e−5	3.41
124946	2.22e−4	28.41	8.47e−5	35.4	6.87e−5	6.90
495776	7.27e−5	69.3	7.30e−5	148	8.86e−6	28.83

Table 4. Algorithm comparison of the rel.erros for the 25 dimensional integral and equal execution times

Time, s	Latin	Sobol	Lattice
1	7.24e−4	9.51e−2	2.11e−1
5	4.16e−5	5.76e−2	1.61e−1
10	2.18e−5	2.71e−2	9.58e−2
20	8.13e−6	8.28e−3	7.87e−2

Table 5. The relative errors for the 30 dimensional integral (11)

# of points	LHS	Time, s	SOBOL	Time, s	FIBO	Time, s
1024	2.01e−2	0.03	1.18e−1	0.42	8.81e−1	0.02
16384	6.53e−3	0.69	8.40e−2	4.5	6.19e−1	0.14
131072	1.35e−3	2.54	1.18e−2	30.2	2.78e−1	1.16
1048576	2.11e−4	13.9	9.20e−3	168	9.86e−2	8.61

Table 6. Comparison of the rel.errors for the 30 dimensional integral and equal execution times

Time, s	Latin	Sobol	Lattice
1	4.38e−3	1.01e−1	2.38e−1
5	8.16e−4	7.76e−2	1.81e−1
10	3.11e−4	5.71e−2	9.48e−2
20	8.63e−5	1.28e−2	7.87e−2

Acknowledgements. This work was supported by the Program for career development of young scientists, BAS, Grant No. DFNP-91/04.05.2016, by the Bulgarian National Science Fund under grant DFNI I02-20/2014, and the financial funds allocated to the Sofia University St. Kl. Ohridski, grant No. 197/2016.

References

1. Dimov, I.: Monte Carlo Methods for Applied Scientists, New Jersey, London, Singapore, World Scientific, 291 p. (2008). ISBN-10 981–02-2329-3
2. Dimov, I., Karaivanova, A.: Error analysis of an adaptive Monte Carlo method for numerical integration. Math. Comput. Simul. **47**, 201–213 (1998)
3. Ermakov, S.M.: Monte Carlo Methods and Mixed Problems. Nauka, Moscow (1985)
4. Jarosz, W.: Efficient Monte Carlo Methods for Light Transport in Scattering Media, Ph.D. dissertation, UCSD (2008)
5. Kroese, D.P., Taimre, T., Botev, Z.: Handbook of Monte Carlo Methods. Wiley Series in Probability and Statistics. Wiley, Hoboken (2011)
6. Kucherenko, S., Albrecht, D., Saltelli, A.: Exploring multi-dimensional spaces: a Comparison of Latin Hypercube and Quasi Monte Carlo Sampling Techniques, arXiv preprint arXiv:1505.02350 (2015)

7. McKay, M.D., Beckman, R.J., Conover, W.J.: A comparison of three methods for selecting values of input variables in the analysis of output from a computer code. Technometrics **21**(2), 239–245 (1979)
8. Metropolis, N., Ulam, S.: The Monte Carlo method. J. Am. Stat. Assoc. **44**(247), 335–341 (1949)
9. Niederreiter, H., Talay, D.: Monte Carlo and Quasi-Monte Carlo Methods, 2010. Springer, Heidelberg (2012)
10. Owen, A.: Monte Carlo theory, methods and examples (2013)
11. Sloan, I.H., Joe, S.: Lattice Methods for Multiple Integration. Oxford University Press, Oxford (1994)
12. Sloan, I.H., Kachoyan, P.J.: Lattice methods for multiple integration: theory, error analysis and examples. J. Numer. Anal. **24**, 116–128 (1987)
13. Vose, D.: The pros and cons of Latin Hypercube sampling (2014)
14. Wang, X., Sloan, I.H.: Low discrepancy sequences in high dimensions: how well are their projections distributed? J. Comput. Appl. Math. **213**(2), 366–386 (2008)

Non-singular Model for Evaporating Sessile Droplets

Stanislav Z. Dunin and Oleg V. Nagornov[✉]

National Research Nuclear University MEPhI, Kashirskoe shosse,
31, 115409 Moscow, Russia
nagornov@yandex.ru

Abstract. Evaporation of liquid droplets on solid substrates is used in many technological processes. Complexity and interdependence of the physical processes inside droplet and their mathematical descriptions, phase transitioning, energy and mass balance, points of singularity and thermocappilar effects force researchers to simplify models and to use numerical methods. We derive solution that has not non-singular heat flux at the droplet edge and describes the vapor concentration out of droplet and temperature at the surface. It allowed us to find out the Marangonni force and calculate non-singular velocity field that can change its direction in the stagnation points.

Keywords: Evaporation · Droplets · Heat and mass transfer

1 Introduction

Parameters of evaporating droplet have been studied both experimentally and theoretically [1–15]. Due to complex processes of evaporation and heat exchange of droplet, gas and substrate many authors simplify models. The known models contain the singularity of local heat flow near the droplet boundary similar to approach developed by Deegan et al. [1–3]. These models suggest constant concentration of vapor at the droplet surface. The vapor flow at the plane gas-substrate boundary equals to zero and concentration far from the droplet is c_∞. In such statement the local heat flow is singular at the droplet edge while this feature is integrable. It is non-physical peculiarity and some authors pointed out that they have been excluded [14,16,17]. For example, it is possible to introduce artificial multiplier that reduces the heat flow to zero at the droplet edge [17]. Semenov et al. [14] carried out calculations in range without the droplet edge. They took into account the Deryagin forces that allowed them to find out non-singular solution.

We exclude the singular component of local heat flow, and find out the solution of the steady-state droplet evaporation.

© Springer International Publishing AG 2017
I. Dimov et al. (Eds.): NAA 2016, LNCS 10187, pp. 311–316, 2017.
DOI: 10.1007/978-3-319-57099-0_33

2   Mathematical Statement of Problem

We consider a drop of partially wetting liquid resting on flat solid layer. Initial contact angle is $\theta_c < 90°$. We assume drop with pinned contact line, so during evaporation process the contact angle decreases to keep the same wetted area (Fig. 1).

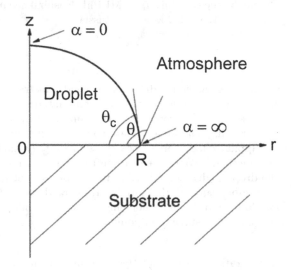

Fig. 1. Droplet in gas phase at substrate.

The initial parameters of the drop are radius R, height h and contact angle θ_c. In small drops due to Bond number $B_0 = \rho g R h_0 / \sigma = 0.03 \div 0.4$ the surface tension effects dominate gravitational effects and drops can be considered as spherical cap shape (Fig. 1).

We dont take into account the last phase of evaporation, where wetted area rapidly decreases. It is a subject for further investigations.

We also consider the Peclet number to be small $Pe << 1$, so the thermal convection is negligible and the thermal diffusivity time to be much less than life time of the drop: $t << R^2 k_{L,s}^{-1} = t_{L,s}$ where R is radius of the drop, k_s and k_L are thermal conductivities of substrate and liquid, respectively.

In gas phase we consider no thermal convection (i.e. $Pe << 1$), so concentration field also satisfies Laplaces equation $\triangle c(\alpha, \theta) = 0$.

In this geometry the problem can be solved in toroidal coordinates (α, θ), where $\theta = 0$ and $\theta = \theta_c$ represent the drop base and surface respectively. We consider no normal concentration flow through the substrate-gas boundary. The general solution of Laplaces equation for concentration can be expressed as:

$$c(\alpha, \theta) = c_\infty + \{2(\cosh \alpha + \cos \theta)\}^{\frac{1}{2}}$$

$$\cdot \int_0^\infty d\tau \, P_{-\frac{1}{2}+i\tau}(\cosh \alpha) B_D(\tau) \frac{\cosh \tau \theta_c}{\cosh \tau \pi} \frac{\cosh \tau(\pi - \theta)}{\cosh \tau(\pi - \theta_c)}, \quad \theta_c < \theta < \pi \quad (1)$$

where c is a vapor concentration, c_∞ are ambient temperature and vapor concentration, respectively, and θ_c is the contact angle. In (1) $P_{-\frac{1}{2}+i\tau}(\cosh\alpha)$ is the Miller-Fock function.

At the droplet surface the Klapeyron-Klausius equation connects concentration of saturated vapor and temperature $c_{sat}(\alpha,\theta_c)\exp(\frac{G}{kT(\alpha,\theta_c)}) = const$, where c_{sat} is the concentration of saturated vapor. Out of the droplet the vapor concentration is c_∞.

The mass and the edge angle changes during quasi-steady-state evaporation of the droplet according to equation [9]:

$$-\dot{M}(t) = J_D(\theta_c) = 4\pi(c_{s\infty} - c_\infty)f_D(\theta_c);$$

$$\dot{\theta}_c(t) = -\frac{\pi}{\rho R^2}(1 + \cos^2(\theta_c))f_D(\theta_c);$$

$$f_D(\theta_c) = \int_0^\infty d\tau \, B_D(\tau)\frac{\cosh\tau\theta_c}{\cosh\tau\pi}\frac{1}{\cosh\tau(\pi - \theta_c)}. \tag{2}$$

To define nuclei B_D we assume some conditions for concentration and its derivative at the gas-droplet interface. The local heat flow at the gas-droplet boundary is

$$\frac{R}{D}j(\alpha,\theta_c) = -(\cosh\alpha + \cos\theta_c)\nabla_\theta c(\alpha,\theta_c) = \frac{\sin\theta_c}{2}\{[2(\cosh\alpha + \cos\theta_c)]^{\frac{1}{2}}$$

$$\cdot\int_0^\infty d\tau \, P_{-\frac{1}{2}+i\tau}(\cosh\alpha)\frac{\cosh\tau\theta_c}{\cosh\tau\pi}B_D(\tau) + \sqrt{2}(\cosh\alpha + \cos\theta_c)^{\frac{3}{2}}$$

$$\cdot\int_0^\infty d\tau \, P_{-\frac{1}{2}+i\tau}(\cosh\alpha)\frac{\cosh\tau\theta_c}{\cosh\tau\pi}B_D(\tau)\tau\frac{\sinh\tau(\pi - \theta)}{\cosh\tau(\pi - \theta_c)}\} \tag{3}$$

The singularity of local heat flow can arise from the second term B_D in the figure brackets (3). Let us denote

$$\psi_0(\theta_c) = \{\sqrt{2}(\cosh\alpha + \cos\theta_c)\}^{\frac{3}{2}}$$

$$\cdot\int_0^\infty d\tau \, P_{-\frac{1}{2}+i\tau}(\cosh\alpha)\frac{\cosh\tau\theta_c}{\cosh\tau\pi}B_D(\tau)\tau\frac{\sinh\tau(\pi - \theta)}{\cosh\tau(\pi - \theta_c)} \tag{4}$$

Then B_D can be determined as solution of the integral equation (4)

$$\sqrt{2}B_D\frac{\cosh\tau\theta_c}{\cosh\tau\pi}\frac{\tanh\tau(\pi - \theta_c)}{\tanh\tau\pi}$$

$$= \int_0^\infty d\alpha \, \psi_0(\alpha,\theta_c)P_{-\frac{1}{2}+i\tau}(\cosh\alpha)\frac{\sinh\alpha}{(\cosh\alpha + \cos\theta_c)^{3/2}} \tag{5}$$

In this paper we consider simplification $\psi_0(\alpha,\theta_c) = \psi_0(\theta_c)$. General case will be subject of future study. Then nuclei is

$$B_D(\tau) = 2\psi_0(\theta_c)\frac{\tanh\tau\theta_c}{\sin\theta_c}\frac{\cosh\tau(\pi - \theta)}{\sinh\tau(\pi - \theta_c)} \tag{6}$$

Concentration in the droplet is calculated:

$$c(\alpha, \theta) = c_\infty + \frac{2\psi_0(\theta_c)}{\sin\theta_c}\{2(\cosh\alpha + \cos\theta)\}^{\frac{1}{2}}$$

$$\cdot \int_0^\infty d\tau\, P_{-\frac{1}{2}+i\tau}(\cosh\alpha)\frac{\sinh\tau\theta_c}{\cosh\tau\pi}\frac{\cosh\tau(\pi-\theta)}{\sinh\tau(\pi-\theta_c)} \quad \theta_c < \theta \le \pi$$

(7)

Function ψ_0 is expressed through concentration at the droplet edge $c(\infty, \theta) = c_{tr}$: $(c_{tr} - c_\infty)(\pi - \theta_c)\frac{\sin\theta_c}{2\theta_c} = \psi_0(\theta_c)$, where $c_{tr} = c(\infty, \theta)$. Deegan et al.'s model [1–3] assumes that $c(\alpha, \theta_c) \equiv c_{tr} = const$ while our model admits variable concentration at the droplet surface.

Calculation of integral (7) can be done in the complex plane τ using the residual. The main input in the integreal gives the poles near real axis τ. Functions $P_{-\frac{1}{2}+i\tau}(\cosh\alpha)$ can be expressed via hypergeometric function F and gamma function Γ as:

$$P_{-\frac{1}{2}+i\tau}(\cosh\alpha) = \frac{\Gamma(i\tau)}{\sqrt{\pi}\Gamma(0.5 + i\tau)}(2\cosh\alpha)^{i\tau}$$

$$\cdot F(3/4 - i\tau, 1/4 - i\tau, 1 + i\tau, 1/\cosh^2\alpha) + CC$$

(8)

where CC is complex conjugate function. Substitution of this formulae under integral allows us to extend integration by τ fom $-\infty$ to $+\infty$. At $\alpha \gg 1$, F is close to 1, then concentration can be written as follows

$$c(\alpha, \theta) \approx c_\infty + \frac{2\psi_0(\theta_c)}{\sin\theta_c}$$

$$\cdot \int_{-\infty}^\infty d\tau\, \frac{\sinh\tau\theta_c}{\cosh\tau\pi}\frac{\cosh\tau(\pi-\theta)}{\sinh\tau(\pi-\theta_c)}\frac{\Gamma(i\tau)}{\sqrt{\pi}\Gamma(0.5 + i\tau)}(2\cosh\alpha)^{i\tau}$$

(9)

In complex plane τ in upper half-plane $\mathrm{Im}\tau \ge 0$ the contour of integration along real axis can be closed by hemi-circle of great radius according to the Jordan lemma. The main input from poles results from those situated on imagine axis $\tau_D(\pi - \theta_c) = \pi$; $0 \le \theta_c \le \pi/2$; $\theta_c \le \theta \le \pi$; $1 \le \tau_D \le 2$.

Finally, we get

$$c(\alpha, \theta) \approx c_{tr} - \frac{2\psi_0(\theta_c)}{\sin\theta_c}\frac{\sin\tau_D\theta_c}{\cos\tau_D\pi}(1 - (\frac{r}{R})^2)^{-\tau_D}$$

(10)

Let us note that for Deegan-Lebedev model $\tau_{D-L}(\pi - \theta_c) = \pi/2$; $0 \le \theta_c \le \pi/2$; $\theta_c \le \theta \le \pi$; $0.5 \le \tau_{D-L} \le 1$, that results in singular local heat flow at the droplet edge. It is in a contradiction to numerious experimental data [9, 11, 12].

Local mass flow of liquid from the droplet surface is

$$j(\alpha, \theta_c) = 1 + \{2(\cosh\alpha + \cos\theta_c)\}^{\frac{1}{2}}$$

$$\cdot \int_0^\infty d\tau\, P_{-\frac{1}{2}+i\tau}(\cosh\alpha)\frac{\sinh\tau\theta_c}{\cosh\tau\pi}\frac{\cosh\tau(\pi-\theta_c)}{\sinh\tau(\pi-\theta_c)}\psi_0$$

(11)

In our model the temperature at the droplet surface is not constant and it is connected with concentration according to the Klapeyron-Klausius relationship. Then in linear approach $c(\alpha, \theta_c) - c_{s\infty} = c'_{s\infty T}(T_L(\alpha, \theta_c) - T_\infty)$, $c'_{s\infty T} = dc_{s\infty}/dT$, the temperature at the surface is

$$(T_L(\alpha, \theta_c) - T_\infty)c'_{s\infty T} = c_\infty - c_{s\infty} + \frac{2\psi_0(\theta_c)}{\sin\theta_c}\{2(\cosh\alpha + \cos\theta_c)\}^{\frac{1}{2}}$$
$$\cdot \int_0^\infty d\tau\, P_{-\frac{1}{2}+i\tau}(\cosh\alpha)\frac{\sinh\tau\theta_c}{\cosh\tau\pi}\frac{\cosh\tau(\pi - \theta_c)}{\sinh\tau(\pi - \theta_c)} \quad (12)$$

In the vicinity of the droplet edge the temperature decreases as follows

$$T_L(\alpha, \theta_c) \approx T_{tr} - (c_{tr} - c_\infty)(\pi - \theta_c)\frac{\sin\tau_D\theta_c}{\theta_c c'_{s\infty T}}(1 - (\frac{r}{R})^2)^{\tau_D} \quad (13)$$

Based on the temperature field in the droplet it is possible to calculate the velocities inside liquid arised from the Marangoni force $F_M(\alpha, \theta_c) = \sigma'_T \nabla_{\tau_g} T$.

3 Conclusions

The mathematical model for evaporation of sessile liquid droplet is developed. The exact solutions for the concentration are found out. The local heat and mass flows from the droplet have not singular behaviour. The calculated temperature at the droplet surface allows for calculating the Marangoni force and mass velocity field.

References

1. Deegan, R.D., Bakajin, O., Dupont, T.F., Huber, G., Nagel, S.R., Witten, T.A.: Capillary flow as the cause of ring stains from dried liquid drops. Nature **389**(6653), 827–829 (1997)
2. Deegan, R.D., Bakajin, O., Dupont, T.F., Huber, G., Nagel, S.R., Witten, T.A.: Contact line deposits in an evaporating drop. Phys. Rev. E **62**, 756–765 (2000)
3. Deegan, R.D.: Pattern formation in drying drops. Phys. Rev. E **61**, 475–485 (1998)
4. Ristenpart, W.D., Kim, P.G., Domingues, C., Wan, J., Stone, H.A.: Influence of substrate conductivity on circulation reversal in evaporating drops. Phys. Rev. Lett. **99**, 234502 (2007)
5. Hu, H., Larson, R.G.: Analysis of the effects of Marangoni stresses on the microflow in an evaporating sessile droplet. Langmuir **21**, 3972–3980 (2005)
6. Hu, H., Larson, R.G.: Evaporation of a sessile droplet on a substrate. J. Phys. Chem. B **106**, 1334–1344 (2002)
7. Hu, H., Larson, R.G.: Analysis of the micro fluid flow in an evaporating sessile droplet. Langmuir **21**, 3963–3971 (2005)
8. Hu, H., Larson, R.G.: Marangoni effect reverses coffee-ring depositions. J. Phys. Chem. B **110**, 7090–7094 (2006)
9. Dunin, S.Z., Nagornov, O.V., Starostin, N.V., Trifonenkov, V.P.: Analytical solution for evaporating sessile drops on solid substrates. In: Recent Advances in Applied Mathematics, Modelling and Simulation, pp. 252–255 (2014)

10. Popov, Y.O.: Evaporative deposition patterns: spatial dimensions of the deposit. Phys. Rev. E **71**, 036313 (2005)
11. Dunn, G.J., Wilson, S.K., Duffy, B.R., David, S., Sefiane, K.: The strong influence of substrate conductivity on droplet evaporation. J. Fluid Mech. **623**, 329–351 (2009)
12. Sefiane, K., Wilson, S.K., David, S., Dunn, G.J., Duffy, B.R.: On the effect of the atmosphere on the evaporation of sessile droplet of water. Phys. Fluids **21**, 062101 (2009)
13. David, S., Sefiane, K., Tadrist, L.: Experimental investigation of the effect of thermal properties of the substrate in the wetting and evaporation of sessile drops. Colloids Surf. A: Physicochem. Eng. Aspects **298**, 108–114 (2007)
14. Semenov, S., Starov, V.M., Rubio, R.G., Agogo, H., Velarde, M.G.: Evaporation of sessile water droplets: universal behaviour in presence of contact angle hysteresis. Colloids Surf. Aspects **391**, 135–144 (2011)
15. Saada, M.A., Chikh, S., Tadrist, L.: Evaporation of a sessile drop with pinned or receding contact line on a substrate with different thermophysical properties. Int. J. Heat Mass Transf. **58**, 197–208 (2013)
16. Semenov, S., Starov, V.M., Rubiob, R.G., Velarde, M.G.: Instantaneous distribution of fluxes in the course of evaporation of sessile liquid droplets: computer simulations. Colloids Surf. A: Physicochem. Eng. Aspects **372**, 127–134 (2010)
17. Fischer, B.J.: Particle convection in an evaporating colloidal droplet. Langmuir **18**, 60–67 (2002)
18. Larson, R.G.: Transport and deposition patterns in drying sessile droplets. AIChE J. **60**, 1538–1571 (2014)

Combinatorial Modeling Approach to Find Rational Ways of Energy Development with Regard to Energy Security Requirements

Alexey Edelev[1] and Ivan Sidorov[2(✉)]

[1] ESI SB RAS, Lermonotva str, 130, Irkutsk, Russia
alexedelev@gmail.com
[2] ISDCT SB RAS, Lermonotva str, 134, Irkutsk, Russia
yvan.sidorov@gmail.com

Abstract. This paper describes the model, methods and tools to find rational ways of the energy development with regard to energy security requirements. As known, energy security is directly related to the uninterrupted energy supply. It is important to choose rational ways of the energy development with ensuring energy security in the future. A number of the specific external condition combinations of energy sector operation and development taking into account uncertainties and other factors leads to a huge possible energy sector states set. Therefore it cannot be processed in reasonable time. To overcome this issue an approach of combinatorial modeling is applied to manage the growing size of the energy sector states set.

Keywords: Energy security · Combinatorial modeling · Distributed computing

1 Introduction

Energy security (ES) of a country can be defined as the state of protection of its citizens, society, and national economy from a shortage in the provision of substantiated energy demands [1–4]. The actuality of studying ES problems is currently very high and is substantiated by a number of threats to normal fuel and energy supply such as equipment depreciation, social tension, great tension of energy facilities, etc. [5].

To monitor the ES status the appropriate indicators are used to represent the most important factors of the operation and potential development of the energy sector [6]. These indicators values must adequately characterize the composition and depth of ES threats to analyze negative tendencies rise or fading.

It is important to choose the rational ways of the energy development to ensure ES in the future. This problem can be formulated as follows: for a given set of the possible energy development scenarios it is necessary to find a rational development path and optimal set of possible measures to ensure the required level of ES subject to the realization of projected emergencies. The conformity

© Springer International Publishing AG 2017
I. Dimov et al. (Eds.): NAA 2016, LNCS 10187, pp. 317–324, 2017.
DOI: 10.1007/978-3-319-57099-0_34

between the energy development options and ES requirements is controlled by means of ES indicators.

A number of the specific combinations of the external energy development and functionality conditions, accounting uncertainties and other factors can lead to the huge amount of the possible energy sector states. Hence, it might not be addressed in a reasonable time. To overcome this issue an approach of combinatorial modeling is applied to manage the energy sector states set size.

2   Combinatorial Modeling of Energy Development in Terms of Energy Security

A set of possible energy development ways is represented as a directed graph. A graph node corresponds to the state and an arc is transition between states.

At first step a reference graph with one node for each time moments is created (Fig. 1). The reference graph describes the basic energy development scenario which information is used to create new states.

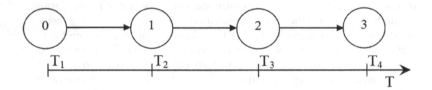

Fig. 1. Basic scenario of energy sector development

At second step the energy infrastructure is separated into several components by territorial or industrial criterion. For each component a graph is built to describe the energy facility parameters changes during the considered time period.

At third step the reference graph data and the information of different component graphs belonging to the same moment in time are combined resulting in a set of possible energy sector states for each moment in time. The created states (nodes) are linked by transitions (arcs) to form an energy development graph. All paths in this graph begin from the common initial node.

At fourth step the invalid nodes are found in the energy development graph. Each state is evaluated by the multistage constraint system. The mathematical energy sector model is the first stage to determine the admissibility of the state. If the model solution exists, then the state is evaluated by means of ES indicators which values are calculated on basis of the solution data and then are compared with their thresholds.

The main task of the last step is to find optimal (rational) path with regard to ES requirements.

A sample energy development graph for 3 moments in time is shown on Fig. 2. It consists of nine nodes. Four nodes are invalid as they do not ensure the required ES level. Dotted circles denote invalid nodes. The nodes and arcs with double line width show optimal energy development path.

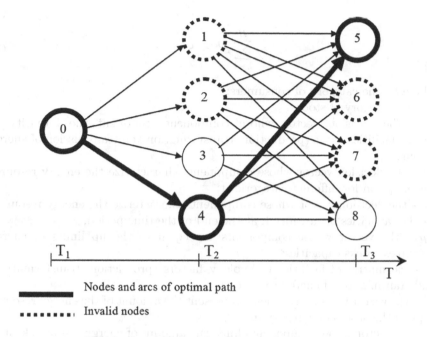

Nodes and arcs of optimal path

Invalid nodes

Fig. 2. A sample graph of energy development

3   Energy Sector Model

The energy sector model is used to evaluate the state of energy sector in a certain time period. By principles of the energy facilities relationships and properties description it is close to such models as MARKAL [7–9], MESSAGE [10,11], EFOMENV [12,13], TIMES [14] and others. The model allows:

– considering whole energy sector from the production of energy resources to final their consumption in the various economic sectors including all stages of energy transformation;
– investigating energy technological and territorial structure development;
– taking into account environmental constraints.

The energy sector in the model includes natural gas, coal, oil supply and power systems. In addition main consumers of fuel and energy are sorted by category.

The energy sector model is the following linear programming problem:

$$S_H + AX - \sum_{t=1}^{T} y^t - \sum_{h=1}^{H} S_k^h = 0 \tag{1}$$

$$0 \le X \le D \tag{2}$$

$$0 \le Y^t \le R^t \tag{3}$$

$$0 \le S_k^h \le S^h \tag{4}$$

$$\sum_{h=1}^{H} S^h \le S \tag{5}$$

- where t - the category of consumers;
- h - the category of stocks;
- X - the decision vector whose components represent the intensity of energy facilities usage (production, transformation and transmission of energy resources);
- Y^t - the decision vector whose components characterize the energy resource consumption for different categories t;
- S_k^h - the decision vector whose components characterize the energy resources stocks capacities of category h at the end of the time period;
- S_H - the vector whose components are equal to the up limits of energy resources stocks capacities;
- A - the matrix of facilities technology factors (production, transformation) and transmission of energy resources;
- S^h - the vector whose components represent the amount of the energy resource kept in the stocks of category h;
- S - the vector whose components limit the amount of energy resource kept in the stocks;
- D - the vector that determines technically possible capacities of production, transformation and transmission facilities;
- R^t - the vector that defines energy resources demands of the category t consumers.

The objective function is as follows:

$$(C, X) + \sum_{t=1}^{T} (r^t, g^t) + \sum_{h=1}^{H} (q^h, \bar{S}^h - S_k^h) \to min \tag{6}$$

The first component of this objective function reflects the operation costs of the energy sector. The vector C contains unit functioning costs for the existing, reconstructed, upgraded and newly built production, transformation and transmission facilities. The second component represents the losses due to the energy resources deficit for the different consumer categories. The energy resources deficit g^t of the category t is equal to the difference between R^t and Y^t. Vector r^t consists of the components called "specific losses" for consumer category t. The third component is similar to the second, and corresponds to the losses due to the energy resources deficit in the stocks. The ratio of the coefficients q^h reflects the preference of the certain stocks.

4   Distributed Software Package "Corrective"

The methodology to find rational ways of energy development with regard to ES requirements is implemented in the distributed software package (DSP) Corrective [15]. Distributed calculations are organized using DISCOMP toolkit [16], which focuses on the automation of the development and implementation of DSP in heterogeneous distributed computing environments (DCE). DSP Corrective consists of the following modules:

- module m_1 to design basic scenario of energy sector development to study,
- module m_2 to create energy sector development graph,
- module m_3 to evaluate possible state of energy sector (node of development graph), and
- module m_4 to support expert analysis of energy sector development paths.

The scheme of information and logical links between modules of software package Corrective is shown in Fig. 3 in the form of bipartite directed graph where modules m_1, m_2, m_3, m_4 are black ovals.

The main aim of the module m_1 is to read information from database A and to transform it into the basic scenario of energy sector development B. List of input parameters A includes the following:

- feasible energy development strategies to investigate;
- ES indicators;
- emergencies that may occur during the considered time period, and
- activity to ensure ES with their technical and economic parameters.

Module m_2 implements methods of combinatorial modeling. After creating a graph, each node is completely independent from the other in terms of performing calculations. Total computation time can be significantly reduced with the help of distributed computing technology. This is achieved by dividing set of N nodes into groups of smaller size and processing them parallel in a distributed computing environment.

The activity to ensure ES may include building special energy facilities, the change of structure and properties of existing energy facilities, redundant supply of energy and fuels and advancing the energy sources diversification for consumers. All input parameters except for ES indicators are implemented as change of basic development scenario data according to certain laws or rules.

A key part of m_2 is the model generator, which creates a new possible state of energy sector. The Generator is controlled by a set of rules that transform input parameters data into the components of energy sector model. The researcher has the ability to change these rules.

The activity to ensure ES may include building special energy facilities, the change of structure and properties of existing energy facilities, redundant supply of energy and fuels and advancing the energy sources diversification for consumers. All input parameters except for ES indicators are implemented as change of basic development scenario data according to certain laws or rules.

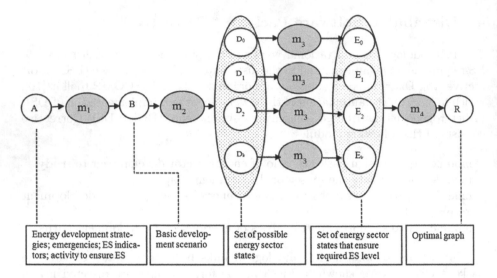

Fig. 3. The logical scheme of distributed software package Corrective

A key part of m_2 is the model generator, which creates a new possible state of energy sector. The Generator is controlled by a set of rules that transform input parameters data into the components of energy sector model. The researcher has the ability to change these rules.

A set of possible energy sector states D is described in the DISCOMP toolkit as a structured data type a parallel list. Each element of the parallel list D is transmitted to the input of module m_3. Module m_3 is used to test the validity of each possible state of energy sector of set D with a multistage constraint system. At the first stage, the linear programming problem solver finds the solution for the energy sector model. If the solution to the problem exists, then m_3 validates ES level by indicative analysis. If a state is valid in terms of the ES requirements, then it is added to the set E. The calculations in this step are independent of one another and run parallel on the available DCE nodes.

Module m_4 assists the researcher to identify the optimal energy sector development path based on set E and to compare the found optimal path with others.

The entire sequence of Corrective user actions, from assigning the optimal parameters of the energy sector model generator to solving linear programming problems and solution analysis, is automated. All modules of the DSP version of the Corrective are cross-platform and can operate under different operating systems in a heterogeneous DCE. This possibility allows using the full potential of the available computing resources to increase the efficiency of the package.

5   Results

Russia was divided into 9 regions and it results in an energy development graph where the amount of nodes is equal to $1 + N * 272$, where N is the number of time moments. An average calculation time of one energy sector state on one processor core equals 10 s. Calculation of all the states on one processor core can take more than 1015 years.

To carry out a computational experiment on an energy development graph involving 20000 states per one time moment a DCE was organized, including a high-performance nodes of HPC-cluster "Academician V.M. Matrosov" [17]. Software and hardware specifications of the DCE are:

- Software: Linux 3.12; Lua 5.1.5; lpsolve 5.5; DISCOMP toolkit 1.2;
- 160 cores AMD Opteron 6276 2,3 GHz;
- 8 GB RAM per processor core.

The total time of calculation in this DCE for 20000 states was less than 20 min.

6   Conclusion

Approach suggested above is proven to be applicable for Russia where the basic parts to build countrys energy sector development graph are regions and their energy facilities, but also for. For the similar energy sectors of other countries the distributed computing technology is potentially applicable.

Acknowledgments. The research was supported by Russian Foundation of Basic Research, projects no. 15-29-07955-ofi_m and no. 16-07-00931-a, and partially supported by the Council for Grants of the President of the Russian Federation for state support of the leading scientific schools, project NSh-8081.2016.9.

References

1. Voropai, N.I., Klimenko, S.M., Krivorutsky, L.D., Pyatkova, N.I., Rabchuk, V.I., Senderov, S.M., Trufanov, V.V., Cheltsov, M.B.: Comprehensive substantiation of the adaptive development of energy systems in terms of changing external conditions. Int. J. Glob. Energy Issue **20**(4), 416–424 (2003)
2. Senderov, S.: Energy security of the largest asia pacific countries: main trends. Int. J. Energy Power **1**(1), 1–6 (2012)
3. Ang, B.W., Choong, W.L., Ng, T.S.: Energy security: definitions, dimensions and indexes. Renew. Sustain. Energy Rev. **42**, 1077–1093 (2015)
4. Winzer, C.: Conceptualizing energy security. Energy Policy **46**, 36–48 (2012)
5. Mansson, A., Bengt, J., Lars, J.: Assessing energy security: an overview of commonly used methodologies. Energy **73**, 1–14 (2014)
6. Kruyt, B.: Indicators for energy security. Energy Policy **37**(6), 2166–2181 (2009)
7. Fishbone, L.G., Abilock, H.: MARKAL, a linear-programming model for energy systems analysis: technical description of the BNL version. Energy Res. **5**, 353–375 (1981)
8. Rostamihozori, N.: Development of energy and emission control strategies for Iran. Ph.D. dissertation. Karlsruhe (2002)
9. Beeck, N.: Classification of Energy Models. Tilburg University (1999)

10. Golodnikov, A., Gritsevskii, A., Messner, S.: A stochastic Version of the Dynamic Linear Programming Model MESSEGE III. WP-95-94. IIASA, Luxemburg (1995)
11. Messner, S., Strubegger, M.: Users Guide for MESSAGE III. WP-95-69. IIASA Luxemburg (1995)
12. Broek, V., Oostvoorn, M.F., Harmelen, T., Arkel, W.: The EC Energy and Environmental Model EFOM-ENV Specified in GAMS. The case of the Netherlands. ECN-report, no. ECN-C-92-003. Petten (1992)
13. The EC Energy and Environmental model EFOM-ENV/GAMS. Netherlands Energy Research Foundation ECN (1991)
14. Loulou, R., Labriet, M.: ETSAP-TIAM: the TIMES integrated assessment model Part I: model structure. Comput. Manag. Sci. **5**(1), 7–40 (2008)
15. Edelev, A.V.: Corrective software to study long-term development of energy sector of Vietnam. Softw. Syst. **4**, 211–216 (2014)
16. Oparin, G.A., Feoktistov, A.G., Sidorov, I.A.: Developing and applying of distributed software packages. Softw. Syst. **2**, 108–111 (2010)
17. Irkutsk Supercomputer Center. http://hpc.icc.ru

A Conservative Semi-Lagrangian Method for the Advection Problem

Alexandr Efremov[1], Evgeniya Karepova[1,2(✉)], and Vladimir Shaidurov[1]

[1] Institute of Computational Modelling of SB RAS Akademgorodok,
660036 Krasnoyarsk, Russia
e.d.karepova@icm.krasn.ru
[2] IM&CS, Siberian Federal University, Krasnoyarsk, Russia

Abstract. In the paper, a new discrete analogue of an initial-boundary value problem is presented for the two-dimensional advection equation arising from a scalar time-dependent hyperbolic conservation law. At each time level, an approximate solution is found as a bilinear function on a uniform rectangular grid. For the presented scheme, a discrete analogue of the local integral balance equation is valid between two neighboring time levels. The numerical experiments are discussed for a solution with strong gradients.

Keywords: Semi-lagrangian approach · Advection equation · Hyperbolic conservation law

1 Introduction

Nowadays a semi-Lagrangian approach to the approximation of the advection terms is intensively developed in many fluid dynamics applications [1–4]. The main feature of the initial semi-Lagrangian approaches consists in the approximation of the advection terms as the whole directional (Lagrangian) derivative in the motion direction (see [5] and the references therein).

In this paper, an initial-boundary value problem is considered for the two-dimensional advection equation arising from a scalar time-dependent hyperbolic conservation law. We start with the integral balance equality between two neighboring time levels. To construct a discrete problem, the finite volume method is used with the approximation of each integral in this balance equality.

Then we study the behavior of a numerical solution with strong gradients.

2 The Differential Problem

Let $D = (0,1) \times (0,1)$ be the unit square with the boundary Γ. In the closed domain $[0,T] \times \overline{D}$ consider the two-dimensional advection equation in the form

$$\frac{\partial \rho}{\partial t} + \nabla \cdot (\mathbf{U}\rho) = 0. \tag{1}$$

© Springer International Publishing AG 2017
I. Dimov et al. (Eds.): NAA 2016, LNCS 10187, pp. 325–333, 2017.
DOI: 10.1007/978-3-319-57099-0_35

Here $\rho(t, x, y)$ is an unknown function (such as a density or a concentration); $u(t, x, y)$, and $v(t, x, y)$ are the known components of a velocity vector $\mathbf{U} = (u, v)$.

Consider three types of the behavior of velocity $\mathbf{U}(t, x, y)$ on the boundary $\Gamma := \Gamma_{\text{in}} \cup \Gamma_{\text{out}} \cup \Gamma_{\text{rigid}}$ at any instant $t \in [0, T]$ [6]. At the inlet boundary $\Gamma_{\text{in}} = \{(0, y) : 0 \leq y \leq 1\}$ we suppose that

$$\mathbf{U} \cdot \mathbf{n} < 0 \quad \forall (t, x, y) \in [0, T] \times \Gamma_{\text{in}} \tag{2}$$

where $\mathbf{n} = (n_x(x, y), n_y(x, y))$ is the outward normal to Γ. Then at the outlet boundary $\Gamma_{\text{out}} = \{(1, y) : 0 \leq y \leq 1\}$ we suppose that the stream goes out:

$$\mathbf{U} \cdot \mathbf{n} > 0 \quad \forall (t, x, y) \in [0, T] \times \Gamma_{\text{out}}. \tag{3}$$

Finally, at the rigid boundary $\Gamma_{\text{rigid}} = \{(x, y) : x \in [0, 1], \ y = 0, 1\}$ we impose the no-slip condition

$$\mathbf{U} = (0, 0) \quad \forall (t, x, y) \in [0, T] \times \Gamma_{\text{rigid}}. \tag{4}$$

For the unknown function ρ the following initial and boundary conditions are specified:

$$\rho(0, x, y) = \rho_{\text{init}}(x, y) \quad \forall (x, y) \in \overline{D}, \tag{5}$$

$$\rho(t, 0, y) = \rho_{\text{in}}(t, y) \quad \forall (t, y) \in [0, T] \times \Gamma_{\text{in}}. \tag{6}$$

3 The Numerical Scheme

To construct the semi-Lagrangian method for problem (1), (5)–(6) on the time segment $[0, T]$, we introduce $K + 1$ time levels $t_k = k\tau$, $k = 0, 1, \ldots, K$ for time step $\tau = T/K$. In $\overline{D}$ we construct a uniform grid $\overline{D}_h$ for mesh-sizes $h_x = 1/N_x$ and $h_y = 1/N_y$ in the x- and y-directions, respectively, for integers $N_x, N_y \geq 2$:

$$\overline{D}_h = \{z_{i,j} = (x_i, y_j) : x_i = ih_x, y_j = jh_y, \ i = 0, \ldots, N_x, \ j = 0, \ldots N_y\}.$$

Denote the set of the corner points of $\overline{D}_h$ by S_h. Put $\Gamma_h := (\overline{D}_h \setminus S_h) \cap \Gamma$ and $D_h := \overline{D}_h \setminus \Gamma_{\text{in}}$. Let $\mathcal{T}_h = \{\omega_{i,j}\}_{i,j=0}^{N_x-1, \ N_y-1}$ be the uniform rectangular triangulation of $\overline{D}_h$ with elementary cells

$$\omega_{i,j} = [x_i, x_{i+1}] \times [y_j, y_{j+1}], \quad i = 0, \ldots, N_x - 1, \ j = 0, \ldots N_y - 1.$$

For each node $z_{i,j} \in \overline{D}_h$ we introduce the basis function $\varphi_{i,j}$ which equals 1 at $z_{i,j}$, equals 0 at all other nodes of $\overline{D}_h$, and is bilinear on each cell $\omega_{i,j} \in \mathcal{T}_h$. Denote the span of these functions by

$$H^h = span\{\varphi_{i,j}\}_{i,j=0}^{N_x, N_y}.$$

At each time level t_k, $k = 0, 1, \ldots, K$ we find an approximate solution of problem (1), (5)–(6) as a bilinear function $\rho_h^k(x, y) \in H^h$ which satisfies the initial condition

$$\rho_h^0(x_i, y_j) = \rho_{\text{init}}(x_i, y_j) \quad \forall (x_i, y_j) \in \overline{D}_h \tag{7}$$

and the boundary condition

$$\rho_h^k(0, y_j) = \rho_{\text{in}}(t_k, y_j) \quad \forall (0, y_j) \in \overline{D}_h \cap \Gamma_{\text{in}}, \ k = 0, 1, \ldots, K. \tag{8}$$

Define the rectangular neighbourhood $\Omega_{i,j} := [x_i - h_x/2, x_i + h_x/2] \times [y_j - h_y/2, y_j + h_y/2] \cap \overline{D}_h$ for any node $z_{i,j} \in \overline{D}_h$ (Fig. 1a).

At time level t_k for each point $(\widetilde{x}, \widetilde{y}) \in \Omega_{i,j}$ we construct the trajectory down to the plane $t = t_{k-1}$. It is determined as a solution of the Cauchy problem for the system of ordinary differential equations

$$\hat{x}'(t) = u(t, \hat{x}(t), \hat{y}(t)), \quad \hat{y}'(t) = v(t, \hat{x}(t), \hat{y}(t)) \tag{9}$$

backward in time on $[t_{k-1}, t_k]$ with the "initial" condition

$$\hat{x}(t_k) = \widetilde{x}, \quad \hat{y}(t_k) = \widetilde{y}, \quad (\widetilde{x}, \widetilde{y}) \in \Omega_{i,j}. \tag{10}$$

Denote the solution of this problem by $\bar{x}(t; \widetilde{x}, \widetilde{y})$, $\bar{y}(t; \widetilde{x}, \widetilde{y})$. Thus, on the plane $t = t_{k-1}$ we get a curvilinear quadrangle (upstream cell) $Q_{i,j}^{k-1}$ (Fig. 1b). If $\Omega_{i,j}$ is located near the inflow boundary Γ_{in}, some set of trajectories may reach the plane $x = 0$. In this case, we get an additional curvilinear polygon on this plane. Denote it by $I_{i,j}^{k-1}$ (Fig. 1c). Generally speaking, any of $Q_{i,j}^{k-1}$ and $I_{i,j}^{k-1}$ may be a triangular, quadrangular, pentagonal, or *empty* curvilinear polygon.

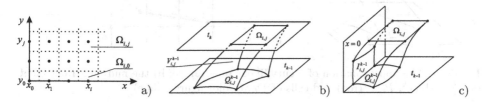

Fig. 1. The construction of curvilinear quadrangles: a) the neighbourhood $\Omega_{i,j}$ of an inner node and the neighbourhood $\Omega_{i,0}$ of a boundary node; b) an inner quadrangle $Q_{i,j}^{k-1}$; c) the case of a boundary quadrangle $I_{i,j}^{k-1}$

Let $V_{i,j}^{k-1}$ be a volume bounded by the following surfaces: $\Omega_{i,j}$ at the level $t = t_k$, $Q_{i,j}^{k-1}$ at the level $t = t_{k-1}$, possibly $I_{i,j}^{k-1}$ on the plane $x = 0$, and the trajectories that issue out of all boundary points of $\Omega_{i,j}$ at the level $t = t_k$.

Integrating (1) over $V_{i,j}^{k-1}$ with the help of (2)–(6) and the divergence theorem, we arrive at the following equality [7–9].

Statement 1. *For a smooth solution of problem* (1), (5)–(6) *under condition* (2)–(4) *we have the equality*

$$\int_{\Omega_{i,j}} \rho(t_k, x, y)\, d\Omega = \int_{Q_{i,j}^{k-1}} \rho(t_{k-1}, x, y)\, dQ + \int_{I_{i,j}^{k-1}} \rho_{\text{in}}(t, y) u(t, 0, y)\, dI, \tag{11}$$

$$i = 1, \ldots, N_x, \ j = 1, \ldots, N_y.$$

If either of $Q_{i,j}^{k-1}$, $I_{i,j}^{k-1}$ is empty, then the integral over an empty domain is supposed to equal zero.

To construct the numerical scheme, we approximate each term of equality (11).

1. To compute the first integral in the right-hand side of (11), we replace the exact solution $\rho(t_{k-1}, x, y)$ by the numerical one $\rho_h^{k-1}(x, y)$.

Moreover, we approximate the curvilinear polygon $Q_{i,j}^{k-1}$ by the straight-sided polygon $P_{i,j}^{k-1}$ with the same 4 vertices $B_n = \left(\bar{x}(t_{k-1}; \tilde{x}_n, \tilde{y}_n), \bar{y}(t_{k-1}; \tilde{x}_n, \tilde{y}_n) \right)$ $n = 1, \ldots, 4$, obtained by the intersection of the plane $t = t_{k-1}$ with characteristic trajectories which issue out of the vertices $A_n = (t_k, \tilde{x}_n, \tilde{y}_n)$ of the rectangle $\Omega_{i,j}$ at the time level t_k (Fig. 2a). To compute the coordinates of B_n, we solve Cauchy problem (9) backward in time on $[t_{k-1}, t_k]$ with corresponding "initial" conditions by the fourth-order Runge–Kutta method [9]. In principle, it is possible to use a second-order Runge–Kutta method. Due to small number of steps, it gives the third-order accuracy.

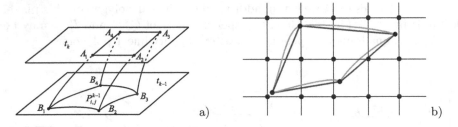

Fig. 2. a) The approximation of a curvilinear quadrangle; b) the mutual arrangement of $Q_{i,j}^{k-1}$ (blue), $P_{i,j}^{k-1}$ (red) and cells of D_h. (Color figure online)

In the general case, the vertices B_n of $P_{i,j}^{k-1}$ are irregularly spaced with respect to the nodes of $\overline{D}_h$ (Fig. 2b). To compute an integral over $P_{i,j}^{k-1}$, we use an isoparametric transformation of unit the square $[0, 1] \times [0, 1]$ into $P_{i,j}^{k-1}$. Denote the coordinates of a point from the unit square by (ξ, η). Then the corresponding coordinates (x^P, y^P) of a point from $P_{i,j}^{k-1}$ are

$$
\begin{aligned}
x^P(\xi, \eta) = &\, \bar{x}(t_{k-1}; \tilde{x}_1, \tilde{y}_1) + (\bar{x}(t_{k-1}; \tilde{x}_2, \tilde{y}_2) - \bar{x}(t_{k-1}; \tilde{x}_1, \tilde{y}_1)) \xi \\
& + (\bar{x}(t_{k-1}; \tilde{x}_4, \tilde{y}_4) - \bar{x}(t_{k-1}; \tilde{x}_1, \tilde{y}_1)) \eta + (\bar{x}(t_{k-1}; \tilde{x}_1, \tilde{y}_1) \\
& - \bar{x}(t_{k-1}; \tilde{x}_2, \tilde{y}_2) + \bar{x}(t_{k-1}; \tilde{x}_3, \tilde{y}_3) - \bar{x}(t_{k-1}; \tilde{x}_4, \tilde{y}_4)) \xi\eta,
\end{aligned} \tag{12}
$$

$$
\begin{aligned}
y^P(\xi, \eta) = &\, \bar{y}(t_{k-1}; \tilde{x}_1, \tilde{y}_1) + (\bar{y}(t_{k-1}; \tilde{x}_2, \tilde{y}_2) - \bar{y}(t_{k-1}; \tilde{x}_1, \tilde{y}_1)) \xi \\
& + (\bar{y}(t_{k-1}; \tilde{x}_4, \tilde{y}_4) - \bar{y}(t_{k-1}; \tilde{x}_1, \tilde{y}_1)) \eta + (\bar{y}(t_{k-1}; \tilde{x}_1, \tilde{y}_1) \\
& - \bar{y}(t_{k-1}; \tilde{x}_2, \tilde{y}_2) + \bar{y}(t_{k-1}; \tilde{x}_3, \tilde{y}_3) - \bar{y}(t_{k-1}; \tilde{x}_4, \tilde{y}_4)) \xi\eta.
\end{aligned} \tag{13}
$$

Thus, we obtain the approximations

$$\int_{Q_{i,j}^{k-1}} \rho(t_{k-1}, x, y) \, dQ \approx \int_{P_{i,j}^{k-1}} \rho_h^{k-1}(x, y) \, dQ \tag{14}$$

$$\approx \int_{[0,1]\times[0,1]} \rho_h^{k-1}(x^P(\xi, \eta), y^P(\xi, \eta)) J_{x,y}^{k-1}(\xi, \eta) \, d\xi d\eta$$

with the Jacobian $J_{x,y}^{k-1}$ of transformation (12)–(13).

We calculate the last integral in (14) by the Cartesian product of the composite rectangle midpoint rule. To compute a value of $\rho_h^{k-1}(x^P(\xi, \eta), y^P(\xi, \eta))$, we must determine the elementary cell $\omega_{a,b}$ which contains the point $(x^P(\xi, \eta), y^P(\xi, \eta))$.

2. Consider the uniform rectangular triangulation of $[0, T] \times [0, 1]$ with elementary cells

$$\omega_{r,q}^\tau = [t_r, t_{r+1}] \times [y_q, y_{q+1}], \quad r = 0, \dots, K-1, \quad q = 0, \dots N_y - 1.$$

For each node (t_r, y_q) of the grid we introduce the basis function $\psi_{r,q}^y(t, y)$ which equals 1 at (t_r, y_q), equals 0 at the other nodes of the grid, and is bilinear on each cell $\omega_{r,q}^\tau$.

To approximate $\rho_{\text{in}}(t, x, y)u(t, x, y)$ in the second integral in the right-hand side of (11), we use the corresponding bilinear interpolants.

Thus, we obtain the approximation

$$\int_{I_{i,j}^{k-1}} \rho_{\text{in}}(t, y)u(t, 0, y) \, dI \approx \int_{L_{i,j}^{k-1}} (\rho_{\text{in}}(t, y)u(t, 0, y))_\tau^I \, dI \tag{15}$$

for the straight-sided polygon $L_{i,j}^{k-1}$ which has the same vertices as $I_{i,j}^{k-1}$. The integrals in the right-hand side of (15) can be calculated in the same way like (14) using the Cartesian product of the composite rectangle midpoint rule.

3. To compute the integral in the left-hand side of (11), we replace the exact solution $\rho(t_k, x, y)$ by the approximate one $\rho_h^k(x, y) \in H^h$. Thus,

$$\int_{\Omega_{i,j}} \rho(t_k, x, y) \, d\Omega \approx \int_{\Omega_{i,j}\cap\omega_{i,j}} \rho_h^k(x, y) \, d\Omega + \int_{\Omega_{i,j}\cap\omega_{i-1,j}} \rho_h^k(x, y) \, d\Omega$$

$$+ \int_{\Omega_{i,j}\cap\omega_{i,j-1}} \rho_h^k(x, y) \, d\Omega + \int_{\Omega_{i,j}\cap\omega_{i-1,j-1}} \rho_h^k(x, y) \, d\Omega. \tag{16}$$

To compute integrals in the right-hand side of (16), we use the Cartesian product of the trapezoidal rule, which is exact for any bilinear function on $\omega_{i,j}$. As a result, we get the following approximation of the left-hand side of (11)

$$\int\limits_{\Omega_{i,j}} \rho(t_k, x, y)\, d\Omega \approx \int\limits_{\Omega_{i,j}} \rho_h^k(x, y)\, d\Omega \tag{17}$$

$$= \frac{h_x h_y}{64} \Big(36\rho_{h,i,j}^k + 6 \left(\rho_{h,i-1,j}^k + \rho_{h,i+1,j}^k + \rho_{h,i,j-1}^k + \rho_{h,i,j+1}^k \right)$$

$$+ \rho_{h,i-1,j-1}^k + \rho_{h,i-1,j+1}^k + \rho_{h,i+1,j-1}^k + \rho_{h,i+1,j+1}^k \Big).$$

Hereinafter we use the notations $f_{h,i,j}^k := f_h^k(x_i, y_j)$ for any $f_h(x, y)$ defined at this point.

Discrete problem 1. *For each time level* t_k, $k = 1, \ldots, K$, *find* $\rho_h^k(x, y) \in H^h$ *satisfying initial condition (7), boundary condition (8), and the following system of the linear algebraic equations*

$$36\rho_{h,i,j}^k + 6 \left(\rho_{h,i-1,j}^k + \rho_{h,i+1,j}^k + \rho_{h,i,j-1}^k + \rho_{h,i,j+1}^k \right) + \rho_{h,i-1,j-1}^k + \rho_{h,i-1,j+1}^k$$

$$+ \rho_{h,i+1,j-1}^k + \rho_{h,i+1,j+1}^k = \frac{64}{h_x h_y} \Phi_{i,j}^k, \quad (x_i, y_j) \in D_h;$$

$$18\rho_{h,i,0}^k + 6\rho_{h,i,1}^k + 3 \left(\rho_{h,i-1,0}^k + \rho_{h,i+1,0}^k \right) + \rho_{h,i-1,1}^k + \rho_{h,i+1,1}^k = \frac{64}{h_x h_y} \Phi_{i,0}^k,$$

$$18\rho_{h,i,N_y}^k + 6\rho_{h,i,N_y-1}^k + 3 \left(\rho_{h,i-1,N_y}^k + \rho_{h,i+1,N_y}^k \right) + \rho_{h,i-1,N_y-1}^k$$

$$+ \rho_{h,i+1,N_y-1}^k = \frac{64}{h_x h_y} \Phi_{i,N_y}^k, \quad i = 1, \ldots, N_x - 1;$$

$$18\rho_{h,N_x,j}^k + 6\rho_{h,N_x-1,j}^k + 3 \left(\rho_{h,N_x,j-1}^k + \rho_{h,N_x,j+1}^k \right) + \rho_{h,N_x-1,j-1}^k$$

$$+ \rho_{h,N_x-1,j+1}^k = \frac{64}{h_x h_y} \Phi_{N_x,j}^k, \quad j = 1, \ldots, N_y - 1;$$

$$9\rho_{h,N_x,0}^k + 3 \left(\rho_{h,N_x-1,0}^k + \rho_{h,N_x,1}^k \right) + \rho_{h,N_x-1,1}^k = \frac{64}{h_x h_y} \Phi_{N_x,0}^k,$$

$$9\rho_{h,N_x,N_y}^k + 3 \left(\rho_{h,N_x,N_y-1}^k + \rho_{h,N_x-1,N_y}^k \right) + \rho_{h,N_x-1,N_y-1}^k = \frac{64}{h_x h_y} \Phi_{N_x,N_y}^k.$$

Here $\Phi_{i,j}^k$ is an approximation of two integrals

$$\int\limits_{[0,1]\times[0,1]} \rho_h^{k-1}(x^P(\xi,\eta), y^P(\xi,\eta)) J_{x,y}^{k-1}(\xi,\eta)\, d\xi d\eta$$

$$+ \int\limits_{[0,1]\times[0,1]} \left(\rho_{\mathrm{in}}(t^L(\xi,\eta), y^L(\xi,\eta)) u(t^L(\xi,\eta), 0, y^L(\xi,\eta)) \right)_\tau^I J_{t,y}^{k-1}(\xi,\eta)\, d\xi d\eta$$

where $J_{x,y}^{k-1}$, $J_{t,y}^{k-1}$ are the Jacobians of the transformations of the unit square $[0,1] \times [0,1]$ into $P_{i,j}^{k-1}$, $I_{i,j}^{k-1}$, respectively, which are similar to (12)–(13).

4 Numerical Experiments

To verify the proposed algorithm, in the closed domain $\overline{D} = [0, 1] \times [0, 1]$ we consider problem (1) on the rotation of a unit cylinder of radius $r = 0.1$

$$\rho(t, x, y) = \begin{cases} 1 & (x - x_0)^2 + (y - y_0)^2 < 0.01, \\ 0 & \text{otherwise} \end{cases} \tag{18}$$

in the field of constant angular velocity $w = 1$. Here the centre $(x_0(t), y_0(t))$ of the cylinder rotates together with medium around the point $(0.5, 0.5)$; the initial location is $x_0(0) = 0.5$, $y_0(0) = 0.7$; $u(t, x, y) = 0.5 - y$, $v(t, x, y) = 0.5 + x$ are the components of the velocity.

Table 1 presents the convergence results in the discrete norm

$$\|f_h\|_1^h = \frac{1}{(N + 1)^2} \sum_{i,j=0}^{N} |f_{h,i,j}|, \quad N_x = N_y = N$$

of a function f_h defined on a uniform square grid $\overline{D}_h$. The numerical solution $\rho_h^K(x, y)$ after two complete revolutions $(K\tau = 4\pi)$ and the absolute value of the error $|\rho_h^K(x, y) - \rho(T, x, y)|$ are represented in Fig. 3 (a) – (b). The quantities

$$M = \frac{1}{(N + 1)^2} \sum_{i,j=0}^{N} \rho_h^k(x_i, y_j)/\|\rho\|_1^h, \quad M_1 = \|\rho_h^k\|_1^h/\|\rho\|_1^h$$

depending on the number k of a time instant are demonstrated in Fig. 4 (a). The first quantity M ("mass") is conserved with time. The second quantity M_1 can be considered as the measure of a change of the oscillation amplitude in the vicinity of strong gradients. M_1 increases slowly with time. The numerical experiments show that the oscillation decreases with improving accuracy of the calculation of the integrals in (14)–(15) (Fig. 4 (b)). The blue curve corresponds to one of cross sections ρ_h, when cylinder (18) moves from $(0, 0)$ to $(1, 1)$ with the constant unit velocity and for $\tau = h$. In this case the integrals near the boundary of the cylinder are calculated accurately and the oscillation is eliminated.

Table 1. The convergence result for the test problem

$N \times N$	h	K	$e_h = \left\|\rho_h^K - \rho\|_{t=t_K}\right\|_1^h$	e_h/e_{2h}	$\log(e_h/e_{2h})$
50×50	0.02	100	$6.67E - 03$		
100×100	0.01	200	$3.70E - 03$	1.80	0.852
200×200	0.005	400	$1.96E - 03$	1.89	0.915
400×400	0.0025	800	$1.01E - 03$	1.94	0.958
800×800	0.00125	1600	$5.99E - 04$	1.68	0.753

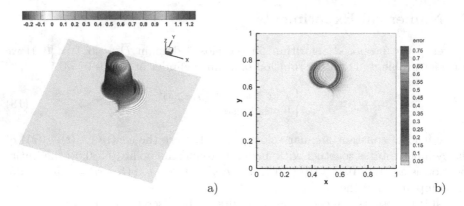

Fig. 3. The rotation of a unit cylinder, $N_x = N_y = 400$, $T = 4\pi$: a) the numerical solution; b) the absolute value of an error

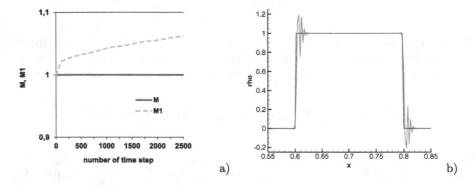

Fig. 4. The numerical solution for $N_x = N_y = 400$: a) $M(k)$ (blue curve) and $M_1(k)$ (red curve); b) a fragment of the cross section $y = 0.7$ of ρ_h for different accuracy of the calculation of the integrals (14): $\tau = h$ (blue curve), $\tau = h/1.01$ (red curve). (Color figure online)

Acknowledgements. The work is supported by RFBR (Project 14-01-00296).

References

1. Dupont, T., Liu, Y.: Back and forth error compensation and correction methods for semi-Lagrangian schemes with application to level set interface computations. Math. Comp. **76**(258), 647–668 (2007)
2. Efremov, A.A., Karepova, E.D., Vyatkin, A.V.: Some features of the CUDA implementation of the semi-Lagrangian method for the advection problem. In: AIP Conference Proceedings, vol. 1684, 090003-1-10 (2015)
3. Selle, A., Fedkiw, R., Kim, B., Liu, Y., Rossignac, J.: An unconditionally stable MacCormack method. J. Sci. Comp. **35**(2), 350–371 (2008)

4. Zerroukat, M., Wood, N., Staniforth, A.: Application of the parabolic spline method (PSM) to a multidimensional conservative semi-lagrangian transport scheme (SLICE). J. Comput. Phys. **225**(1), 935–948 (2007)
5. Lentine, M., Grétarsson, J.T., Fedkiw, R.: An unconditionally stable fully conservative semi-lagrangian method. J. Comput. Phys. **230**(8), 2857–2879 (2011)
6. Griebel, M., Dornseifer, T., Neunhoeffer, T.: Numerical Simulation in Fluid Dynamics: A Practical Introduction. SIAM, Philadelphia (1998)
7. Anderson, J.D.: Computational Fluid Dynamics: The Basics with Applications. McGraw-Hill, New York (1995)
8. Morton, K.W.: Numerical Solution of Convection-Diffusion Problems. Chapman & Hall, London (1996)
9. Efremov, A.A., Karepova, E.D., Shaydurov, V.V., Vyatkin, A.V.: A computational realization of a semi-lagrangian method for solving the advection equation. J. Appl. Math. **2014**, 610398 (2014)

Fast Meshless Techniques Based on the Regularized Method of Fundamental Solutions

Csaba Gáspár[✉]

Széchenyi István University,
Egyetem tér 1, Györ 9026, Hungary
gasparcs@sze.hu

Abstract. A regularized method of fundamental solutions is presented. The method can handle Neumann and mixed boundary conditions as well without using a desingularization technique. Instead, the approach combines the regularized method of fundamental solutions with traditional finite difference techniques based on some inner collocation points located in the vicinity of the boundary collocation points. Nevertheless, the resulting method remains meshless. The method avoids the problem of singularity and can be embedded in a natural multi-level context. The method is illustrated via a numerical example.

Keywords: Meshless methods · Method of fundamental solutions · Regularization · Finite difference schemes · Multi-level methods

1 Introduction

In the last decade, meshless methods have become quite popular for solving various types of partial differential equations. The reason of popularity is that these methods require neither domain nor boundary grid or mesh structure, which is an essential and often quite complicated task when applying a standard finite difference or finite element method. Meshless methods need a finite set of points only, without any structure.

A special meshless method is the Method of Fundamental Solutions (MFS, see. e.g. [5]), which is, in addition to the above properties, a boundary-only technique; that is, it requires some (unstructured) points along the boundary but not inside the domain.

Assume, for simplicity, that the original problem to be solved is the 2D Laplace equation supplied with mixed boundary conditions:

$$\Delta u = 0 \text{ in } \Omega, \quad u|_{\Gamma_D} = u_0, \quad \frac{\partial u}{\partial n}|_{\Gamma_N} = v_0 \tag{1}$$

Here Ω denotes a two-dimensional, bounded and sufficiently regular domain, the boundary of which is decomposed into a Dirichlet part Γ_D and a Neumann part Γ_N. It is well known that (1) has a unique solution in a proper Sobolev space.

© Springer International Publishing AG 2017
I. Dimov et al. (Eds.): NAA 2016, LNCS 10187, pp. 334–341, 2017.
DOI: 10.1007/978-3-319-57099-0_36

In its original form, the MFS provides the approximate solution of (1) in the following form:

$$u(x) := \sum_{j=1}^{N} \alpha_j \Phi(x - \tilde{x}_j), \tag{2}$$

where Φ is a fundamental solution of the Laplacian, i.e. $\Phi(x) = \frac{1}{2\pi} \log \|x\|$ (here $\|.\|$ denotes the Euclidean norm in $\mathbb{R}^2$). The points $\tilde{x}_1$, $\tilde{x}_2$, $\tilde{x}_N$ are predefined external points (*source points*). Thus, the function u defined by (2) exactly satisfies the Laplace equation in Ω. The a priori unknown coefficients α_1, α_2,... α_N can be obtained by enforcing the boundary conditions at some predefined *boundary collocation points* x_1, x_2,...,x_N:

$$\sum_{j=1}^{N} \alpha_j \Phi(x_k - \tilde{x}_j) = u_0(x_k) \quad (\text{if } x_k \in \Gamma_D)$$
$$\sum_{j=1}^{N} \alpha_j \frac{\partial \Phi}{\partial n_k}(x_k - \tilde{x}_j) = v_0(x_k) \quad (\text{if } x_k \in \Gamma_N) \tag{3}$$

where n_k denotes the outward normal unit vector at the point x_k.

The method is simple, easily programmable and essentially dimension-in dependent. It leads to much smaller linear systems than the traditional domain type methods, and has excellent convergence properties [7]. However, the matrix of the resulting linear system is non-selfadjoint, fully populated and often extremely ill-conditioned, especially when the source points are located far from the boundary. On the other hand, if they are too close to the boundary, numerical singularities are generated, which destroys the accuracy.

A general technique to overcome these difficulties is to allow the source points and the boundary collocation points to coincide. Using nonsingular solutions instead of the fundamental solution, the problem of singularity can be avoided [1] (but the problem of ill-conditioned matrices remains the case). The picture is similar, if fundamental solutions concentrated to straight lines are used [3]. If classical fundamental solutions are used, the diagonal entries of the matrix appearing in (3) become singular and their proper evaluation requires special regularization tools. Such a regularization can be, for instance, the use of approximate fundamental solutions which have no singularity at the origin. This approach works well in the case of pure Dirichlet problems. However, Neumann or mixed boundary conditions require more sophisticated (desingularization) tools [8,10], often based on the solution of some auxiliary problem [2,4,6].

In this paper, we present a simple technique to handle the Neumann boundary condition without desingularization techniques. The main idea is to combine the regularized MFS approach with a classical finite difference discretization of the Neumann boundary condition based on some fictitious gridpoints in the interior of the domain. The condition number of the system remains moderate. The system can be efficiently solved in an iterative way as well; in every iteration step, a pure Dirichlet problem has to be solved. The computational efficiency can be increased further by embedding the method in a natural multi-level context.

2   A Regularization of the MFS

A regularized fundamental solution is an approximate fundamental solution which has no singularity at the origin. The simplest way is the use of truncation. From now on, denote by Φ the truncated fundamental solution:

$$\Phi(r) := \begin{cases} \frac{1}{2\pi} \log(c\|x\|) & \text{if } c\|x\| > 1 \\ 0 & \text{if } c\|x\| \leq 1 \end{cases} \tag{4}$$

where $c > 0$ is a predefined scaling constant.

Define the source points to be identical to the boundary collocation points: $\tilde{x}_j := x_j$ $(j = 1, ..., N)$. Then the approximate solution of (1) has the form:

$$u(x) := \sum_{j=1}^{N} \alpha_j \Phi(x - x_j). \tag{5}$$

In case of pure Dirichlet problem (when $\Gamma_N = \emptyset$), the coefficients $\alpha_1, ..., \alpha_N$ can be obtained by solving the linear system:

$$\sum_{j=1}^{N} \alpha_j \Phi(x_k - x_j) = u_0(x_k) \quad (k = 1, 2, ..., N) \tag{6}$$

Due to the truncation, the diagonal entries of the matrix of (6) are well-defined i.e. not singularity problem arises. For pure Dirichlet problems, the approach works well, provided that the scaling constant c is defined to be inversely proportional to the characteristic distance of the boundary collocation points, see [3,4]. However, for Neumann or mixed problems, the method fails to work, due to the stronger singularity of the normal derivative of the fundamental solution at the origin.

Remark: Another regularization technique is the use of the nearly radial function $\log \frac{r_1 + r_2 + \sqrt{(r_1 + r_2)^2 - a^2}}{a}$, where $r_1 := \|x + x_0\|$, $r_2 := \|x - x_0\|$, $x_0 := (\frac{a}{2}, 0)$. This function is continuous everywhere in $\mathbb{R}^2$ and harmonic outside of the segment $[-x_0, x_0]$ and vanishes on this segment. The length of the segment should be proportional to the characteristic distance of the boundary collocation points. See [3] for details.

2.1   Approximation of Neumann Boundary Condition by Inner Layers

The normal derivative of the solution can be approximated by simple formulas based on standard Taylor series expansions taken in the outward normal direction. Let $x \in \Gamma$ be a boundary point and denote by n the outward normal unit vector at x. Let $\bar{x} := x - \delta \cdot n$, $\bar{\bar{x}} := x - 2\delta \cdot n$ be inner points, where $\delta > 0$ is a predefined constant. If the function u is sufficiently smooth, we have:

$$u(\overline{x}) = u(x) - \frac{\partial u}{\partial n}(x) \cdot \delta + \frac{1}{2}\frac{\partial^2 u}{\partial n^2}(x) \cdot \delta^2 + \mathcal{O}(\delta^3) \tag{7}$$

$$u(\overline{\overline{x}}) = u(x) - \frac{\partial u}{\partial n}(x) \cdot 2\delta + \frac{\partial^2 u}{\partial n^2}(x) \cdot 2\delta^2 + \mathcal{O}(\delta^3) \tag{8}$$

Hence we obtain the trivial one-layer, first-order scheme (*Scheme 1*):

$$\frac{\partial u}{\partial n}(x) = \frac{u(x) - u(\overline{x})}{\delta} + \mathcal{O}(\delta) \tag{9}$$

and a second-order, two-layer scheme, the distances of the layers from the boundary are δ and 2δ, respectively (*Scheme 2*):

$$\frac{\partial u}{\partial n}(x) = \frac{3u(x) - 4u(\overline{x}) + u(\overline{\overline{x}})}{2\delta} + \mathcal{O}(\delta^2) \tag{10}$$

Another second-order but one-layer scheme can be constructed if the function u is harmonic. This enables us to use (7) only, by approximating $\frac{\partial^2 u}{\partial n^2}(x) = -\frac{\partial^2 u}{\partial e^2}(x)$ by a standard second-order central 3-point scheme based on the points

$$\overline{x}^W := \overline{x} - \delta \cdot e, \qquad \overline{x}, \qquad \overline{x}^E := \overline{x} + \delta \cdot e$$

where e denotes the tangential unit vector at x. This leads to *Scheme 3*:

$$\frac{\partial u}{\partial n}(x) = \frac{2u(x) - u(\overline{x}^W) - u(\overline{x}^E)}{2\delta} + \mathcal{O}(\delta^2) \tag{11}$$

The values of the function u at the points $x, \overline{x}, \overline{\overline{x}}, \overline{x}^W, \overline{x}^E$, can be expressed in the form (5). (Note that the layer distance δ should be proportional to the characteristic distance of the boundary collocation points.) Thus, the approximate solution of (1) has again the form (5), and the coefficients $\alpha_1,...,\alpha_N$ can be determined by solving the following linear system. When $x_k \in \Gamma_D$, the corresponding equation is always

$$\sum_{j=1}^{N} \alpha_j \Phi(x_k - x_j) = u_0(x_k) \tag{12}$$

Along the Neumann boundary, i.e. when $x_k \in \Gamma_N$:

Scheme 1 (first-order, one-layer approximation):

$$\sum_{j=1}^{N} \alpha_j \frac{\Phi(x_k - x_j) - \Phi(\overline{x}_k - x_j)}{\delta} = v_k \tag{13}$$

Scheme 2 (second-order, two-layer approximation):

$$\sum_{j=1}^{N} \alpha_j \frac{3\Phi(x_k - x_j) - 4\Phi(\overline{x}_k - x_j) + \Phi(\overline{\overline{x}}_k - x_j)}{2\delta} = v_k \tag{14}$$

Scheme 3 (second-order, one-layer approximation):

$$\sum_{j=1}^{N} \alpha_j \frac{2\Phi(x_k - x_j) - \Phi(\overline{x}_k^W - x_j) - \Phi(\overline{x}_k^E - x_j)}{2\delta} = v_k \qquad (15)$$

Here $\overline{x}_k := x_k - \delta \cdot n_k$, $\overline{x}_k^W := \overline{x}_k - \delta \cdot e_k$, $\overline{x}_k^E := \overline{x}_k + \delta \cdot e_k$, and e_k is a unit vector which is orthogonal to n_k.

Remark: When applying Scheme 3, it is possible to use a stepsize in the direction of e_k which is different from δ i.e. the stepsize in the direction of n_k. This results in a little change in Scheme 3.

2.2 Error Estimations

Assume, for simplicity, that Ω is a half-stripe: $\Omega := (-\pi, \pi) \times (0, +\infty)$ with periodic boundary conditions along the half-lines $\{(-\pi, y) : y > 0\}$ and $\{(\pi, y) : y > 0\}$ (the general case can be deduced to this one by suitable coordinate transforms). Consider the Laplace equation in Ω supplied by pure Neumann condition $\frac{\partial u}{\partial n} = v$, where we assume that $\int_\Gamma v \, d\Gamma = 0$, so that the Neumann problem has a solution (which is unique in the Sobolev space $H^1(\Omega)$, since $H^1(\Omega)$ does not contain the constant functions). Using the more familiar notations x, y for the space variables, let us express the Neumann boundary condition v in terms of complex Fourier series:

$$v(x) = \sum_k \beta_k e^{ikx} \qquad (16)$$

Note that the compatibility condition $\int_\Gamma v \, d\Gamma = 0$ implies that $\beta_0 = 0$. The exact solution of the Neumann problem is as follows:

$$u^*(x, y) = \sum_{k \neq 0} \frac{\beta_k}{|k|} e^{-|k|y} e^{ikx} \qquad (17)$$

Denote by u the approximate solution of the Neumann problem, where the Neumann boundary condition is approximated by inner layer as indicated in the previous subsection. Then u has the form

$$u(x, y) = \sum_{k \neq 0} \alpha_k e^{-|k|y} e^{ikx} \qquad (18)$$

Simple calculations show that

$$\alpha_k = \frac{\delta}{1 - e^{-|k|\delta}} \cdot \beta_k \qquad (Scheme\ 1)$$

$$\alpha_k = \frac{2\delta}{3 - 4e^{-|k|\delta} + e^{-2|k|\delta}} \cdot \beta_k \qquad (Scheme\ 2)$$

$$\alpha_k = \frac{\delta}{1 - (\cos k\delta) \cdot e^{-|k|\delta}} \cdot \beta_k \qquad (Scheme\ 3)$$

Thus, the error of the approximations along the boundary Γ can easily be expressed in terms of Fourier series, yielding:

$$u(x,0) - u^*(x,0) = \sum_{k \neq 0} c_k \cdot \frac{\beta_k}{|k|} e^{ikx} \tag{19}$$

where

$$c_k = \frac{|k|\delta}{1 - e^{-|k|\delta}} - 1 \qquad (Scheme\ 1)$$

$$c_k = \frac{2|k|\delta}{3 - 4e^{-|k|\delta} + e^{-2|k|\delta}} - 1 \qquad (Scheme\ 2)$$

$$c_k = \frac{|k|\delta}{1 - (\cos k\delta) \cdot e^{-|k|\delta}} - 1 \qquad (Scheme\ 3)$$

An elementary analysis of the functions defined by

$$f(t) := \frac{t}{1 - e^{-t}} - 1 \qquad (Scheme\ 1)$$

$$f(t) := \frac{2t}{3 - 4e^{-t} + e^{-2t}} - 1 \qquad (Scheme\ 2)$$

$$f(t) := \frac{t}{1 - \cos t \cdot e^{-t}} - 1 \qquad (Scheme\ 3)$$

shows that, with a proper constant $C > 0$, $f(t)^2 \leq Ct^2$ (in Scheme 1), and $f(t)^2 \leq Ct^4$ in Schemes 2 and 3, respectively. This implies that the errors in Sobolev norms can be estimated as follows:

$$\|u - u^*\|_{H^s(\Gamma)} \leq \begin{cases} C \cdot \delta \cdot \|v\|_{H^s(\Gamma)} & (Scheme\ 1) \\ \\ C \cdot \delta^2 \cdot \|v\|_{H^{s+1}(\Gamma)} & (Schemes\ 2,\ 3) \end{cases} \tag{20}$$

for all $s > 0$.

2.3   Improvement by Multi-level Tools

Instead of solving system (12)–(13) (resp. (12)–(14) or (12)–(15)) directly, it is possible to apply simple iterative methods. In fact, the above defined schemes can convert the original mixed problem to a pure Dirichlet problem. More precisely, if $x \in \Gamma_N$, then the Dirichlet data $u(x)$ can be expressed by the Neumann data $v(x)$ as follows:

$$u(x) = u(\overline{x}) + \delta \cdot v(x) + \mathcal{O}(\delta^2) \qquad (Scheme\ 1)$$

$$u(x) = \frac{4}{3}u(\overline{x}) - \frac{1}{3}u(\overline{\overline{x}}) + \frac{2}{3}\delta \cdot v(x) + \mathcal{O}(\delta^3) \qquad (Scheme\ 2)$$

$$u(x) = \frac{1}{2}u(\overline{x}^W) + \frac{1}{2}u(\overline{x}^E) + \delta \cdot v(x) + \mathcal{O}(\delta^3) \qquad (Scheme\ 3)$$

If the corresponding pure Dirichlet problem is solved by the regularized MFS, i.e. the approximate solution is expressed in the form (5), then the above expressions define iterative methods to update the Dirichlet data u_k at Neumann boundary Γ_N:

$$u_k^{improved} = u(\overline{x}_k) + \delta \cdot v_k \qquad (Scheme\ 1)$$

$$u_k^{improved} = \frac{4}{3}u(\overline{x}_k) - \frac{1}{3}u(\overline{\overline{x}}_k) + \frac{2}{3}\delta \cdot v_k \qquad (Scheme\ 2)$$

$$u_k^{improved} = \frac{1}{2}u(\overline{x}_k^W) + \frac{1}{2}u(\overline{x}_k^E) + \delta \cdot v_k \qquad (Scheme\ 3)$$

A detailed analysis shows that though the above iterations converge slowly, they significantly reduce the high-frequency components of the error *independently of N*, i.e. they can be used as smoothing iterations of boundary multi-level methods (for multi-level methods, see e.g. [9]). At each level, only pure Dirichlet problems have to be solved, which makes the computational cost moderate. Details are omitted.

3   A Numerical Example

To illustrate the method, consider the problem (1) defined in the unit circle Ω with the test solution

$$u(x, y) = \cos \pi x \cdot \sinh \pi y \qquad (21)$$

In this test example, let the boundary conditions of the problem (1) be essentially of Neumann type. More precisely, in each discretisation of Γ, exactly one boundary collocation point is of Dirichlet type (to ensure the solvability and uniqueness), and the remaining points are of Neumann type. We have applied the regularized MFS with the truncated fundamental solution (4) and the scaling constant $c := N$, where N is the number of the (equally spaced) boundary collocation points. The Neumann boundary condition was approximated by the Schemes 1, 2 and 3, as defined in the previous section. The discrete relative L_2-errors have been calculated along Γ for different values of N. At the same time, the corresponding condition numbers of the matrix of the system (12)–(15) have been calculated as well. The results are summarized in Table 1. The calculated errors are in accordance with the results that Scheme 1 is of first order, while Scheme 2 and Scheme 3 are of second order. Note that the condition numbers remain moderate, however, they slightly increase with N and seem to be (approximately) proportional to $N^{5/4}$ in all cases.

4   Summary and Conclusions

A regularized version of the MFS has been presented. The method is based on truncated fundamental solutions, which circumvents the problem of singularity, when the source and collocation points coincide. The Neumann boundary condition is approximated by inner layers based on inner collocation points. This results in first- and second-order approximations of the Neumann boundary conditions. No additional desingularization technique is needed. The method can be modified in such a way that the original mixed problem is converted to a pure Dirichlet problem, and the Dirichlet data along the Neumann boundary

Table 1. Relative L_2-errors of the approximate solution of Schemes 1,2,3 (%) and the corresponding condition numbers.

N	16	32	64	128	256
Scheme 1:					
Relative L_2-error	62.84	26.16	12.09	5.839	2.889
Condition number	27.44	68.05	1.62.6	378.4	863.0
Scheme 2:					
Relative L_2-error	22.36	5.603	1.528	0.4537	0.1860
Condition number	19.51	48.71	116.7	271.9	620.5
Scheme 3:					
Relative L_2-error	33.54	6.676	1.659	0.4481	0.1465
Condition number	23.68	57.67	137.2	318.8	726.7

are updated iteratively. This iteration significantly reduces the high-frequency error components, therefore it can be used as a smoothing procedure of a natural multi-level method.

References

1. Chen, W., Shen, L.J., Shen, Z.J., Yuan, G.W.: Boundary knot method for Poisson equations. Eng. Anal. Boundary Elem. **29**, 756–760 (2005)
2. Chen, W., Wang, F.Z.: A method of fundamental solutions without fictitious boundary. Eng. Anal. Boundary Elem. **34**, 530–532 (2010)
3. Gáspár, C.: Some variants of the method of fundamental solutions: regularization using radial and nearly radial basis functions. Central Eur. J. Math. **11**(8), 1429–1440 (2013)
4. Gáspár, C.: A regularized multi-level technique for solving potential problems by the method of fundamental solutions. Eng. Anal. Boundary Elem. **57**, 66–71 (2015)
5. Golberg, M.A.: The method of fundamental solutions for Poisson's equation. Eng. Anal. Boundary Elem. **16**, 205–213 (1995)
6. Gu, Y., Chen, W., Zhang, J.: Investigation on near-boundary solutions by singular boundary method. Eng. Anal. Boundary Elem. **36**, 1173–1182 (2012)
7. Li, X.: On convergence of the method of fundamental solutions for solving the Dirichlet problem of Poisson's equation. Adv. Comput. Math. **23**, 265–277 (2005)
8. Šarler, B.: Solution of potential flow problems by the modified method of fundamental solutions: formulations with the single layer and the double layer fundamental solutions. Eng. Anal. Boundary Elem. **33**, 1374–1382 (2009)
9. Stüben, K., Trottenberg, U.: Multigrid methods: fundamental algorithms, model problem analysis and applications. In: GDM-Studien, vol. 96, Birlinghoven, Germany (1984)
10. Young, D.L., Chen, K.H., Lee, C.W.: Novel meshless method for solving the potential problems with arbitrary domain. J. Comput. Phys. **209**, 290–321 (2005)

Parallel Computations for Solving 3D Helmholtz Problem by Using Direct Solver with Low-Rank Approximation and HSS Technique

Boris Glinskiy[1], Nikolay Kuchin[1], Victor Kostin[2], and Sergey Solovyev[2(✉)]

[1] Siberian Supercomputer Center, Institute of Computational Mathematics and Mathematical Geophysics SB RAS, 6 Lavrentyeva pr., Novosibirsk, Russia 630090
{gbm,kuchin}@sscc.ru
[2] Institute of Petroleum Geology and Geophysics SB RAS,
3 Akademika Koptyuga pr., Novosibirsk, Russia 630090
{kostinvi,solovevsa}@ipgg.sbras.ru

Abstract. The modern methods of processing the geophysical data, such as Reverse Time Migration (RTM) and Full Waveform Inversion (FWI) require solving series of forward problems where the main step is solution of Systems of Linear Algebraic Equations (SLAE) of big size. For big sizes, it is time and memory consuming problem.

In this paper, we present a parallel direct algorithm to solve boundary value problems for 3D Helmholtz equation discretized with help of finite differences. The memory consumption has been resolved due to Nested Dissection approach, low-rank approximation technique and HSS format. OpenMP parallelization is based on standard BLAS and LAPACK functionality. For MPI parallelization, we propose a novel algorithm that uses dynamical distribution of the elimination tree nodes across cluster nodes. Numerical experiments show performance benefits of the proposed cluster algorithm compared to the not parallel version and demonstrate significant memory advantages over direct solvers without low-rank approximation.

1 An Algorithm of Solving Linear Algebraic Systems with a Sparse Matrix

The objective of this paper is solving the system of linear algebraic equations (SLAE):

$$AU = F \tag{1}$$

with a sparse $N \times N$-matrix and K right-hand sides, which are columns of $N \times K$-matrix F. The solution is the $N \times K$-matrix U. Such systems arise from the numerical approximation of a boundary value problem of partial elliptic differential equations. Particulary, all the numerical experiments discussed in this paper are provided for the Helmholtz equation:

$$\triangle u + \frac{(2\pi\nu)^2}{V^2} u = \delta(\overline{r} - \overline{r}_s)f, \tag{2}$$

© Springer International Publishing AG 2017
I. Dimov et al. (Eds.): NAA 2016, LNCS 10187, pp. 342–349, 2017.
DOI: 10.1007/978-3-319-57099-0_37

in the 3D domain with the radiation boundary condition. The computational domain is restricted by a parallelepiped with Perfectly Matching Layer (PML) and the Dirichlet boundary condition in the external domain. The PML condition significantly decreases the reflections from the external boundaries [5]. In Eq. (2) ν is the frequency, V is the velocity, f is the source, r_s is the coordinate of the source.

The finite difference approximation of (2) on 27-point stencil on the parallelepipedal uniform grid provides the second order approximation with low numerical dispersion [3]. As a result, we obtain the system of linear equations (SLAE) (1), where: the right-hand side is discretization of the right-hand side of Eq. (2), unknowns u of the SLAE being the approximation of the function u (2).

This SLAE has one right-hand side. However, usually it is needed to solve problem (2) with sources, located on different coordinates (in geophysical problems, there can be more than 1000), hence vectors of the right-hand sides can be collected into the matrix F. Such an approach is useful for direct solvers because factorization of the matrix can be carried out at a time for all right-hand sides. Let us note that using the PML gives an symmetric matrix with complex coefficients, so the memory consumption and FLOPS for solving SLAE are twice as large as in a real case.

The solution algorithm of the symmetric SLAE (1) based on LDL^t-decomposition, is described in the [1]. In this paper, we assume the matrix A to be non-singular and the matrix D to be non-singular as well. Thus, the matrices L, D, L^t can be inverted.

Let us recall that matrix A is sparse, but the matrix L is dense (the number of nonzero entries is significantly large than in A). To decrease the fill-in of the matrix L the reordering columns and rows of the initial matrix A is used [2].

2   The Algorithm of LDL^t-Decomposition

The L-factor structure of a sparse matrix arises from the finite difference approximation of the Helmholtz boundary value problem (the Nested Dissection reordering has been applied), is shown in Fig. 1. The elimination tree is presented in the same figure: an elimination tree-node is a panel of the matrix L, constructed from the diagonal and sub-diagonal blocks.

The last diagonal blocks of L are bigger and are denser than the first diagonal blocks. The indicator of the density in the figure is the shadow: the denser – the darker. The white color marks the zero block; the black color – the dense block.

The elimination tree completely determines the data dependence in the computational process. For example, the computation of the L-panel, corresponding to node 13 (Fig. 1) can be started after completing the computation of nodes 9 and 10. The last nodes should wait the completing computations of nodes 1, 2, 3 and 4. Node 13 should receive data both from 9, 10 and from descendants 1, 2, 3, 4.

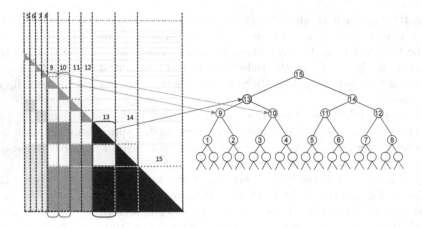

Fig. 1. The pattern of L (left) and their elimination tree (right). (Color figure online)

To additional decreasing memory, the Low-Rank approximation of the off-diagonal blocks of L is used; the diagonal blocks are compressed by using the Hierarchically Semiseparable Structure format (HSS) [6] (Fig. 2).

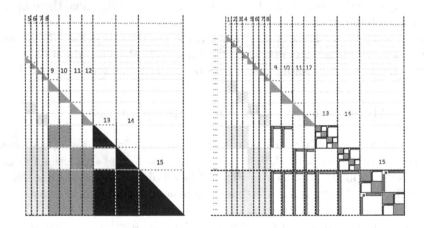

Fig. 2. The pattern of L-factor (left) and its low-rank approximation (right).

The result of the factorization is the approximation $\tilde{L}$ of the matrix L. The solution $\tilde{x}$ of the System of Linear Algebraic Equations (SLAE) $\tilde{L}\tilde{D}\tilde{L}^t\tilde{x} = b$ approximates the solution of the initial system. The error of such an approximation depends on a low-rank approximation error of each off-diagonal block and HSS compressing error of diagonal blocks.

3   Parallel Computations

The parallel version of the solver includes: the cluster nodes scheduling to factor-
ize the elimination of tree nodes; the optimized algorithm of send/receive com-
munications between cluster nodes. Also we should take into account the fact
that the size of the free RAM of the cluster is sufficient to perform computations.

The L-factor blocks can be large and do not fit RAM. Such blocks are separated by the not very large blocks. Hence, the binary tree becomes semi-binary,
where each node has one or two children Fig. 3.

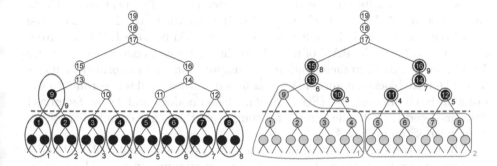

Fig. 3. Data distribution among cluster nodes at the first step (left) and at the second
factorization step (right). (Color figure online)

Let us suppose that before carrying out factorization a sparse matrix A has
been stored at one cluster node (0) and the reordering step has been performed.
This limitation is essential because of the number of non-zero entries of the initial
matrix is significantly less than nonzero elements of L-factor, so a free RAM of
one cluster node is sufficient for saving A. The description of the algorithm would
be more intricate without this limitation, so in this paper we will use it. The first
functionality of 0-node is to save the initial matrix. The cluster nodes receive
these data from 0-node on the fly. Moreover, 0-node saves the results of the
symbolic factorization, i.e. an additional step before carrying out the numerical
factorization to obtain the data structure of L.

After conducting the symbolic factorization, the positions of all non-zero
elements of L become known, the total memory to save all L-factors is known as
well. The approach of a low-rank approximation and HSS structure, described in
the previous section, decreases the memory consumption, but this compression
ration cannot be estimated a priori at the symbolic factorization step.

Let us describe the cluster algorithm proposed. If we map the elimination
tree on to cluster nodes, so us computational tasks be essentially determined
by edges of the tree. Some computational node received data (a group of the
initial matrix A) from node 0 and results of its descendants (parts of L-factors
for computing the Schur complement). Tasks are dynamically assigned to cluster
nodes, as described below.

At the first step each cluster node has a set of A-matrix columns, correspond-ing to a certain sub-tree below the level marked as a red dash line in the Fig. 3. If the number of cluster nodes exceeds the number of the sub-trees, then the nodes of tree, located above this line, are distributed among other cluster nodes.

Let us have 9 cluster nodes, marked as a red circle in the Fig. 3. Block-columns of the initial matrix, belonging to subtrees 1...8 are distributed between cluster nodes. The 9-th cluster node saves the 9-th node of the elimination tree. We suppose that a free RAM of each node is sufficient to store these data, temporary arrays for low-rank compressing, factorization and results of factorization.

After carrying out factorization and low-rank approximation, sub-trees and nodes 1–9, the results are stored at 1–9 cluster nodes. The next step collects these data in RAM of some cluster nodes. It is shown in the Fig. 3 (left): the data from nodes 1, 2, 3, 4, 9 are collected in the RAM of node 1, the data from nodes 5, 6, 7, 8—in the RAM of node 2. At the next steps, computational nodes 1 and 2 do not perform factorization, all computations are performed by nodes 3–9. The next block-columns of A are assigned to these nodes according to the order of the elimination tree (Fig. 3, right). Cluster nodes 1 and 2 store factorized data and send them to the parents for computing the Schur complement of the remainder of the matrix A.

The next steps of the algorithm are the same as above. Computational nodes take tree-nodes for factorization according to an elimination tree. These nodes receive data from node 0 (block columns of A) and data needed for computing the Schur complement from the children, save the parts of L-factor. After computing the Schur complement, it is compressed by the HSS/low-rank technique and collected at some cluster nodes, so RAM of other nodes is cleaned and is ready to continue the factorization process. In this case the factorization is carried out from bottom to top, from left to right. The number of cluster nodes, assigned to the computation, is decreased, whereas the number of nodes to save data becomes larger and large (the nodes, which computed L-factor at the previous step save the results obtained now).

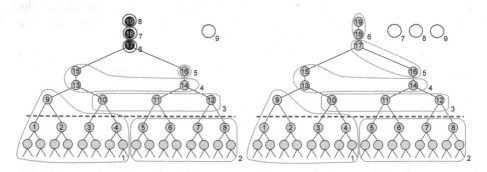

Fig. 4. Data distribution among cluster nodes at the third step (left) and at the last factorization step (right).

The next steps of the algorithm are presented in Fig. 4. The factorized tree-nodes are marked as grey, and as black – in the process of factorization. The notations like those in the Fig. 3, where factorization has not started yet for the nodes marked as white. The green contour means cluster nodes, saving factorized data (the part of L); the red contour – cluster nodes for factorization. At the end of the factorization process columns of the matrix L are distributed among cluster nodes. However some nodes can be free and do not contain the factorization data (Fig. 4, right).

At the solution step (inversion LDL^t) the right-hand side b (the matrix B in the common case) is distributed among cluster nodes according to L-distribution. Let us neglect the details of the solution step, just say that it consists of the three items:

- Forward step (inversion L);
- Diagonal step (inversion D);
- Backward step (inversion L^t),

where there is data exchange among cluster nodes at the first and at the second steps. As a result, the solution is distributed among cluster nodes like in the right-hand side b.

4 Numerical Experiments

Numerical tests demonstrate performances and memory efficiency of the algorithm proposed. We consider a homogeneous medium with the wave propagation velocities $V_p = 2500$ m/s and the source delta function. The numerical solution is compared with the analytical solution of the Helmholtz equation, which can be computed due to the model in question.

The computational domain is a cube. The Helmholtz equation is discretized on the $256 \times 256 \times 256$ uniform mesh. The ε-rank of the low-rank approximation off-diagonal blocks is 10^{-6}, the number of OMP-threads is 4 at each cluster node.

The aim of the test is demonstrating advantages of the cluster version under the one-node version (not using the MPI parallelism). The one-node version was tested on the server with 512 GB RAM. The cluster version was run at 32 cluster nodes with the same total size of the RAM. The description of computational resources is shown in Table 1. The server and the cluster have different processors, so we propose the performance comparison as the "Processor-performance ratio". We compute this ratio experimentally by running various small tests both on the server and at one cluster node. So, for estimating the speedup of the cluster version $\frac{17198}{7715} \approx 2.3$ it should be corrected by 2.2 performance advantage 3 times of the server processor against the cluster one.

Let us note that using a low-rank approximation decreases the memory consumption (about 240 GB) as compared with the full-rank approach (about 800 GB).

The second test demonstrates the memory efficiency of the proposed MPI algorithm for solving large problems. Testing the algorithm was done on the

Table 1. One-node and cluster versions computational resources.

	Server 512 G RAM	Cluster NSK-30 32 × 16 G RAM
Processor Intel(R) Xeon(R)	E5-2690 v2, 3.00 GHz AVX	E5540, 2.53 GHz SSE4.2
Processor performance ratio	x2.2	x1.0
Factorization time	17 198 s	7 715 s

heterogenous cluster NSK-30T of the Siberian Supercomputer Center [8], which consists of the one powerful node with eight processors Intel(R) Xeon(R) CPU E7-4870 @ 2.40 GHz, 1000 GB RAM and four nodes with two processors Intel(R) Xeon(R) CPU X5675 @ 3.07 GHz, 96 GB RAM. Such a configuration uses 10% RAM for the internal needs, for example, the OS, thus the size of all free RAM is about 1250 GB. The computational domain is 12 km × 12 km × 6 km parallelepiped with homogenous velocity model (v = 2400 m/s). The size of the grid is 30 m in each direction, the frequency being 4 Hz. Taking into account the PML layer (10 grid points), the computational grid is 460 × 460 × 230, so the total number of unknowns of the SLAE is about $5 \cdot 10^7$. The source (the delta function of the right-hand side, RHS) is located in the upper corner of the domain. The cluster version of the algorithm proposed uses 860 GB RAM, that is less by a factor 4 even less than the direct solvers in the full-rank arithmetic (3500 GB). The factorization step was completed in 14 h, the solution step (LDL^t inversion) takes 3 min per one RHS.

The 0-node with large memory (1000 GB) saves results, while other four nodes perform the factorization process. As a result, L-factor is stored at the 0-node, the solution step is performed at the same node as well.

The relative error of the numerical solution compared with the analytic one is about 1% in C norm. The relative residual $||F - AU||/||F|| \simeq 10^{-5}$ is in C norm as well.

5   Conclusion

A new cluster version of the SLAE solver has been developed. It is aimed at solving the partial differential equations (PDEs) and is based on the matrix reordering algorithms, LDL^t-factorization, the low-rank approximation approach and the HSS-structure of diagonal blocks. The tasks of the factorization process are effectively redistributed due to using the proposed dynamic parallelism. In particular, the cluster nodes with large free RAM are used to save computed data, whereas the factorization process is performed at the nodes with a smaller size of RAM.

The numerical experiments have demonstrated a high quality of the algorithm proposed and its speedup in comparison with a one-node version. Moreover, we have shown advantages of the low-rank approximation approach in handling with large size problems which cannot be resolved by direct solvers in the full-rank arithmetic.

Acknowledgements. The research described was partially supported by RFBR grants 14-05-00049, 14-05-93090, 16-05-00800 and the Russian Academy of Sciences Program "Arctic". All computations were performed on cluster NSK-30T of Siberian Supercomputer Center.

References

1. Solovyev, S.A.: Application of the low-rank approximation technique in the gauss elimination method for sparse linear systems. Vychisl. Metody Programm. **15**, 441–460 (2014). (in Russian)
2. George, A.: Nested dissection of a regular finite elementmesh. SIAM J. Numer. Anal. **10**(2), 345–363 (1973)
3. Solovyev, S.A.: Multifrontal hierarchically semi-separable solver for 3D Helmholtz problem using 27-point finite difference scheme. In: Expanded Abstracts of 77th EAGE Conference and Exhibition 77, p. We P4 09 (2015)
4. Dewilde, P., Gu, M., Somasunderam, N., Chandrasekaran, S.: On the numerical rank of the off-diagonal blocks of schur complements of discretized elliptic PDEs. SIAM J. Matrix Anal. Appl. **31**(5), 2261–2290 (2010)
5. Collino, F., Tsogka, C.: Application of the perfectly matched layer absorbing layer model to the linear elastodynamic problem in anisotropic heterogeneous media. Geophysics **66**, 294–307 (2001)
6. Xia, J.: Robust and efficient multifrontal solver for large discretized PDEs. In: Berry, M.W., Gallivan, K.A., Gallopoulos, E., Grama, A., Philippe, B., Saad, Y., Saied, F. (eds.) High-Performance Scientific Computing, pp. 199–217. Springer, Heidelberg (2012). doi:10.1007/978-1-4471-2437-5_10
7. Tordeux, S., Solovyev, S.A.: An efficient truncated SVD of large matrices based on the low-rank approximation for inverse geophysical problems. Sib. Electron. Math. Rep. **12**, 592–609 (2015)
8. Kuchin, N., Glinsky, B., Chernykh, I., et al.: Control and managing the HPC cluster in siberian supercomputer center. In: Proceedings of the International Conference Russian Supercomputing Days, Moscow, Russia, 28–29 September 2015 (Mosk. Gos. Univ., Moscow 2015), pp. 667–675 (2015)

ADI Schemes for 2D Subdiffusion Equation

Sandra Živanović[✉] and Boško S. Jovanović

Faculty of Mathematics, University of Belgrade,
Studentski Trg 16, 11000 Belgrade, Serbia
{sandra,bosko}@matf.bg.ac.rs

Abstract. An additive and modified factorized finite-difference scheme for an initial-boundary value problem for a two-dimensional subdiffusion equation are proposed. Its stability and convergence are investigated.

Keywords: Fractional derivatives · Subdiffusion · ADI Schemes · Stability · Convergence rate

1 Introduction

The partial differential equations of fractional order begun to play an important role in particular in modeling of the so called anomalous phenomena and in the theory of the complex systems during the last few decades. The subject of a huge number of papers and a number of monographs (see [6,9,10]) are different applications of the fractional differential equations in physics, mechanics, chemistry, technique, engineering, biology and astrophysics.

The time fractional diffusion equation (TFDE) is obtained from the classical diffusion equation by replacing the first order time derivative by fractional derivative of order α with $\alpha \in (0,1)$. The TFDE have been investigated in analytical and numerical frames. But, analytic solutions of the most fractional differential equations cannot be obtained explicitly. So many authors proposed discrete numerical approximations and discussed about its stability and convergence (see [1,3,7]).

2 Fractional Derivatives and Subdiffusion Equation

The most popular two definitions of fractional derivative are the Rimann–Liouville and the Caputo definitions.

The left Riemann-Liouville fractional derivative of order α is defined as

$$D^{\alpha}_{a+} u(t) = \frac{1}{\Gamma(n-\alpha)} \frac{d^n}{dt^n} \int_a^t \frac{u(s)}{(t-s)^{\alpha+1-n}} \, ds, \quad t \geq a,$$

where $n-1 \leq \alpha < n, n \in N$ and $\Gamma(\cdot)$ is the Gamma function.

© Springer International Publishing AG 2017
I. Dimov et al. (Eds.): NAA 2016, LNCS 10187, pp. 350–358, 2017.
DOI: 10.1007/978-3-319-57099-0_38

By interchanging the derivative and integral operators in the previous definition one obtains so called the left Caputo fractional derivative

$$^{C}D_{a+}^{\alpha}u(t) = \frac{1}{\Gamma(n-\alpha)} \int_{a}^{t} \frac{u^{(n)}(s)}{(t-s)^{\alpha+1-n}} \, ds \quad t \geq u.$$

In particular, $D_{a+}^{\alpha}u(t) = {}^{C}D_{a+}^{\alpha}u(t)$ if $u(a) = u'(a) = \cdots = u^{(n-1)}(a) = 0$. Analogously one defines the right derivatives. We use some special spaces and norms and inner products in them. For $\alpha > 0$ we set

$$|u|_{C_{+}^{\alpha}[a,b]} = \|D_{a+}^{\alpha}u\|_{C[a,b]}, \qquad |u|_{C_{-}^{\alpha}[a,b]} = \|D_{b-}^{\alpha}u\|_{C[a,b]},$$

$$\|u\|_{C_{\pm}^{\alpha}[a,b]} = \left(\|u\|_{C^{[\alpha]^{-}}[a,b]}^{2} + |u|_{C_{\pm}^{\alpha}[a,b]}^{2} \right)^{1/2},$$

$$|u|_{H_{+}^{\alpha}(a,b)} = \|D_{a+}^{\alpha}u\|_{L^{2}(a,b)}, \qquad |u|_{H_{-}^{\alpha}(a,b)} = \|D_{b-}^{\alpha}u\|_{L^{2}(a,b)}$$

and

$$\|u\|_{H_{\pm}^{\alpha}(a,b)} = \left(\|u\|_{H^{[\alpha]^{-}}(a,b)}^{2} + |u|_{H_{\pm}^{\alpha}(a,b)}^{2} \right)^{1/2},$$

where $[\alpha]^{-}$ denotes the largest integer $< \alpha$. Then we define $C_{\pm}^{\alpha}[a,b]$ as the space of functions $u \in C^{[\alpha]^{-}}[a,b]$ with finite norm $\|u\|_{C_{\pm}^{\alpha}[a,b]}$. The space $H_{\pm}^{\alpha}(a,b)$ is defined analogously, while the space $\dot{H}_{\pm}^{\alpha}(a,b)$ is defined as the closure of $\dot{C}^{\infty}(a,b) = C_{0}^{\infty}(a,b)$ with the respect to the norm $\| \cdot \|_{H_{\pm}^{\alpha}(a,b)}$.

Let $0 < \alpha < 1$, $\Omega = (0,1) \times (0,1)$ and $Q = \Omega \times (0,T)$. We shall consider the next equation

$$D_{t,0+}^{\alpha}u + \mathcal{L}u = f(x,t), \quad x = (x_1, x_2) \in \Omega, \quad t \in (0,T), \tag{1}$$

subject to boundary and initial conditions

$$u(x,t) = 0, \quad x \in \partial\Omega, \quad t \in (0,T), \tag{2}$$

$$u(x,0) = 0, \quad x \in \bar{\Omega}, \tag{3}$$

where $\mathcal{L}u = -\sum_{i=1}^{2} \frac{\partial}{\partial x_i} \left(a_i \frac{\partial u}{\partial x_i} \right)$. We also assume, unless otherwise stated, that $a_i = a(x_1, x_2, t)$ are differentiable functions which satisfy $0 < c_0 < a_i < c_1$, $i = 1, 2$.

By multiplying (1) with a test function v and after partial integration, we obtain the following weak formulation of the problem (1)–(3):

Find $u \in \dot{H}^{1,\alpha/2}(Q) = L^2((0,T), \dot{H}^1(\Omega)) \cap \dot{H}^{\alpha/2}((0,T), L^2(\Omega))$ such that

$$a(u,v) = f(v), \quad \forall v \in \dot{H}^{1,\alpha/2}(Q),$$

where $a(u,v) = \left(D_{t,0+}^{\alpha/2}u, D_{t,T-}^{\alpha/2}v \right)_{L^2(Q)} + \left(a_1 \frac{\partial u}{\partial x_1}, \frac{\partial v}{\partial x_1} \right)_{L^2(Q)} + \left(a_2 \frac{\partial u}{\partial x_2}, \frac{\partial v}{\partial x_2} \right)_{L^2(Q)}$ and $f(v) = (f, v)_{L^2(Q)}$.

Theorem 1 (see [2,8]). *Let $f \in L^2(Q)$, $a_i \in L^\infty(Q)$, and let $0 < c_0 < a_i < c_1$. Then the problem (1)–(3) is well posed in $\dot{H}^{1,\alpha/2}(Q)$ and its solution satisfy the a priori estimate*

$$\|u\|_{H^{1,\alpha/2}(Q)} \leq C\|f\|_{L^2(Q)}.$$

From Theorem 1 it immediately follows the weaker a priori estimate

$$\|u\|_{B^{1,\alpha/2}(Q)} \leq C\|f\|_{L^2(Q)},$$

where tne norm $\|\cdot\|_{B^{1,\alpha/2}(Q)}$ is defined by

$$\|u\|^2_{B^{1,\alpha/2}(Q)} = \int_0^T \left[(T-t)^{-\alpha}\|u(\cdot,t)\|^2_{L^2(\Omega)} + \|u(\cdot,t)\|^2_{H^1(\Omega)} \right] dt.$$

3   Approximation

We define the uniform mesh $\bar{Q}_{h\tau} = \bar\omega_h \times \bar\omega_\tau$ where $\bar\omega_h = \{x = (n_1 h, n_2 h) : n_1, n_2 = 0, 1, ..., N; h = 1/N\}$ and $\bar\omega_\tau = \{t_k = k\tau : k = 0, 1, ..., M; \tau = T/M\}$. We also define $\omega_h = \bar\omega_h \cap \Omega$, $\omega_{1h} = \bar\omega_h \cap ((0,1] \times (0,1))$, $\omega_{2h} = \bar\omega_h \cap ((0,1) \times (0,1])$, $\omega_\tau = \bar\omega_\tau \cap (0,T)$, $\omega_\tau^- = \bar\omega_\tau \cap [0,T)$, $\omega_\tau^+ = \bar\omega_\tau \cap (0,T]$ and $\gamma_h = \bar\omega_h \setminus \omega_h$. We shall use standard notation from the theory of the finite difference schemes (see [11]):

$$v = v(x,t), \quad \hat{v} = v(x,t+\tau), \quad \check{v} = v(x,t-\tau), \quad v^k = v(x,t_k), \quad x \in \bar\omega_h,$$

$$v_{x_i} = v(x + he_i, t) - v(x,t)/h = v_{\bar{x}_i}(x + he_i, t), \quad i = 1, 2,$$

$$v_t = v(x,t+\tau) - v(x,t)/\tau = v_{\bar{t}}(x, t+\tau) = \hat{v}_{\bar{t}},$$

where e_i denotes the unit vector of the axis $0x_i$.

For a function u defined on $\bar{Q}$ which satisfies a homogeneous initial condtion, we use two discrete approximations of fractional derivative. The first is

$$^C D^\alpha_{t,0+,\tau} u^k = \frac{\tau^{1-\alpha}}{\Gamma(2-\alpha)} \sum_{l=0}^{k-1} d_{k-l} u_t^l, \tag{4}$$

where $d_{k-l} = (k-l)^{1-\alpha} - (k-l-1)^{1-\alpha}$, $\quad 0 \leq l < k \leq M$, and the second is

$$D^\alpha_{t,0+,\tau} u^k = \frac{1}{\tau\Gamma(1-\alpha)} \left(b_0 u^k + \sum_{l=1}^{k-1} [b_{k-l} - b_{k-l-1}] u^l \right) \tag{5}$$

where $b_{k-l} = \frac{\tau^{1-\alpha}}{1-\alpha} d_{k-l+1}$. Note that in our case $^C D^\alpha_{t,0+,\tau} u^k = D^\alpha_{t,0+,\tau} u^k$.

Lemma 1 [12]. *Suppose that $\alpha \in (0,1)$, $u \in C^2([0,t], C(\bar\Omega))$ and $t \in \omega_\tau^+$. Then*

$$|^C D^\alpha_{t,0+} u - {}^C D^\alpha_{t,0+,\tau} u| \leq \frac{\tau^{2-\alpha}}{1-\alpha} \left[\frac{1-\alpha}{12} + \frac{2^{2-\alpha}}{2-\alpha} - (1 + 2^{-\alpha}) \right] \max_{\bar{Q}_t} \left| \frac{\partial^2 u}{\partial t^2} \right|,$$

where denoted $Q_t = (0,t) \times \Omega$.

Lemma 2 [5]. *For* $0 < \alpha < 1$ *and any function* $v(t)$ *defined on* $\bar{\omega}_\tau$ *which satisfies* $v(0) = 0$ *the following inequality is valid*

$$v^k (D^\alpha_{t,0+,\tau} v)^k \geq \frac{1}{2} \left(D^\alpha_{t,0+,\tau} (v^2) \right)^k + \frac{\tau^{2-\alpha}(1 - 2^{-\alpha})}{\Gamma(2 - \alpha)} (v_t^{k-1})^2. \tag{6}$$

We shall use the following inner products and norms.

$$(v, w)_h = (v, w)_{L^2(\omega_h)} = h^2 \sum_{x \in \omega_h} vw, \quad \|v\|_h = \|v\|_{L^2(\omega_h)} = (v, v)_h^{1/2},$$

$$(v, w)_{ih} = (v, w)_{L^2(\omega_{ih})} = h^2 \sum_{x \in \omega_{ih}} vw, \quad \|v\|_{ih} = \|v\|_{L^2(\omega_{ih})} = (v, v)_{ih}^{1/2}, \quad i = 0, 1, 2,$$

$$|v|^2_{H^1(\omega_h)} = \sum_{i=1}^{2} \|v_{\bar{x}_i}\|^2_{ih}, \quad \|v\|^2_{H^1(\omega_h)} = |v|^2_{H^1(\omega_h)} + \|v\|^2_h,$$

$$\|v\|^2_{L^2(Q_{h\tau})} = \tau \sum_{k=1}^{M} \|v^k\|^2_h, \quad \|v\|^2_{L^2(Q_{ih\tau})} = \tau \sum_{k=1}^{M} \|v^k\|^2_{ih}, \quad i = 0, 1, 2,$$

3.1   Additive Scheme

Let $M = 2m$. We approximate (1)–(3) with the following additive difference scheme:

$$D^\alpha_{t,0+,\tau} v^{2k-1} - 2(\tilde{a}_1 v_{\bar{x}_1})^{2k-1}_{x_1} = \bar{f}^{2k-1}, \quad x \in \omega_h, \quad k = 1, 2, \ldots, m \tag{7}$$

$$D^\alpha_{t,0+,\tau} v^{2k} - 2(\tilde{a}_2 v_{\bar{x}_2})^{2k}_{x_2} = \bar{f}^{2k}, \quad x \in \omega_h, \quad k = 1, 2, \ldots, m \tag{8}$$

$$v(x, t) = 0, \quad (x, t) \in \gamma_h \times \omega_h^+, \tag{9}$$

$$v(x, 0) = 0, \quad x \in \bar{\omega}_h, \tag{10}$$

where $\tilde{a}_i(x, t) = a_i(x - 0.5he_i, t)$. The idea of this difference scheme is to reduce two dimensional problem to one dimensional. On each time level we have to solve a sequence of linear systems with three-diagonal matrices. Because of that, this scheme is economical.

When the right-hand side f is a continuous function, we set $\bar{f} = f$, otherwise we must use some averaged value, for example $\bar{f} = T_1 T_2 f$, where T_i are Steklov averaging operators: $T_i f(x, t) = \int\limits_{-1/2}^{1/2} f(x + hse_i, t) \, ds, \quad i = 1, 2.$

Here, we define the following norm

$$\|v\|_{B^{1,\alpha/2}(Q_{h\tau})} = \left[\tau \sum_{k=1}^{2m} \left(D^\alpha_{t,0+,\tau} (\|v\|^2_h) \right)^k + \tau \sum_{k=1}^{m} \left(\|v^{2k-1}_{\bar{x}_1}\|^2_{L^2(\omega_{1h})} + \|v^{2k}_{\bar{x}_2}\|^2_{L^2(\omega_{2h})} \right) \right]^{1/2}.$$

Theorem 2. *Let $0 < \alpha < 1$ and $f \in L^2(Q)$. Let also a_i be continuous functions. Then the difference scheme (7)–(10) is absolutely stable and its solution satisfies the following a priori estimate*

$$\|v\|_{B^{1,\alpha/2}(Q_{h\tau})} \leq C\|\bar{f}\|_{L^2(Q_{h\tau})}. \tag{11}$$

Proof. Taking the inner products of Eqs. (7) and (8) with v^{2k-1} and v^{2k}, respectively, we obtain

$$\left(v^{2k-1}, D^\alpha_{t,0+,\tau}v^{2k-1}\right)_h + 2\left(\tilde{a}_1 v^{2k-1}_{\bar{x}_1}, v^{2k-1}_{\bar{x}_1}\right)_h = \left(\bar{f}^{2k-1}, v^{2k-1}\right)_h,$$

$$\left(v^{2k}, D^\alpha_{t,0+,\tau}v^{2k}\right)_h + 2\left(\tilde{a}_2 v^{2k}_{\bar{x}_2}, v^{2k}_{\bar{x}_2}\right)_h = \left(\bar{f}^{2k}, v^{2k}\right)_h.$$

From here, applying Lemma 2, it follows that

$$\frac{1}{2}\left(D^\alpha_{t,0+,\tau}\left(\|v\|^2_h\right)\right)^{2k-1} + 2c_0\left\|v^{2k-1}_{\bar{x}_1}\right\|^2_{1h} \leq \varepsilon\left\|v^{2k-1}\right\|^2_h + \frac{1}{4\varepsilon}\left\|\bar{f}^{2k-1}\right\|^2_h,$$

$$\frac{1}{2}\left(D^\alpha_{t,0+,\tau}\left(\|v\|^2_h\right)\right)^{2k} + 2c_0\left\|v^{2k}_{\bar{x}_2}\right\|^2_{2h} \leq \varepsilon\left\|v^{2k}\right\|^2_h + \frac{1}{4\varepsilon}\left\|\bar{f}^{2k}\right\|^2_h.$$

By using the Poincaré inequality on the right-hand side and for $\varepsilon = 8c_0$, we obtain

$$\frac{1}{2}\left(D^\alpha_{t,0+,\tau}\left(\|v\|^2_h\right)\right)^{2k-1} + c_0\left\|v^{2k-1}_{\bar{x}_1}\right\|^2_{1h} \leq \frac{1}{32c_0}\left\|\bar{f}^{2k-1}\right\|^2_h,$$

$$\frac{1}{2}\left(D^\alpha_{t,0+,\tau}\left(\|v\|^2_h\right)\right)^{2k} + c_0\left\|v^{2k}_{\bar{x}_2}\right\|^2_{2h} \leq \frac{1}{32c_0}\left\|\bar{f}^{2k}\right\|^2_h,$$

Multiplying with $32c_0\tau$ and summing these two inequalities through $k = 1,\ldots,m$, the proof is completed where $C = (16c_0 \min\{1, 2c_0\})^{-1/2}$.   □

Let us analyze the error and the accuracy of the proposed scheme. Let u be the solution of the initial-boundary-value problem (1)–(3) and v the solution of the difference problem (7)–(10) with $\bar{f} = T_1 T_2 f$, then the error $z = u - v$ is defined on the mesh $\bar{\omega}_h \times \bar{\omega}_\tau$. Putting $v = u - z$ into (7)–(10) it follows that error satisfies

$$D^\alpha_{t,0+,\tau}z^{2k-1} - 2(\tilde{a}_1 z_{\bar{x}_1})^{2k-1}_{x_1} = \psi^{2k-1}_1, \quad x \in \omega_h, \quad k = 1, 2, \ldots, m, \tag{12}$$

$$D^\alpha_{t,0+,\tau}z^{2k} - 2(\tilde{a}_2 z_{\bar{x}_2})^{2k}_{x_2} = \psi^{2k}_2, \quad x \in \omega_h, \quad k = 1, 2, \ldots, m, \tag{13}$$

$$z = 0, \quad x \in \gamma_h, \quad t \in \bar{\omega}_\tau, \tag{14}$$

$$z^0 = z(x, 0) = 0, \quad x \in \omega_h. \tag{15}$$

where

$$\psi^{2k-1}_1 = \xi^{2k-1} + 2\eta^{2k-1}_1 + \chi^{2k-1}, \qquad \psi^{2k}_2 = \xi^{2k} + 2\eta^{2k}_2 - \chi^{2k}.$$

Here we denoted

$$\xi = D^\alpha_{t,0+,\tau}u - T_1 T_2 D^\alpha_{t,0+}u,$$

$$\eta_i = T_1 T_2 \left(\frac{\partial}{\partial x_i} \left(a_i \frac{\partial u}{\partial x_i} \right) \right) - (\tilde{a}_i u_{\bar{x}_i})_{x_i}, \quad i = 1, 2,$$

$$\chi = T_1 T_2 \left(\frac{\partial}{\partial x_2} \left(a_2 \frac{\partial u}{\partial x_2} \right) - \frac{\partial}{\partial x_1} \left(a_1 \frac{\partial u}{\partial x_1} \right) \right).$$

Further, using the properties of Steklov averaging operators, we obtain $\eta_1 = \zeta_{1,x_1}$ and $\eta_2 = \zeta_{2,x_2}$ where

$$\zeta_i = T_{3-i} \left(a_i \frac{\partial u}{\partial x_i} \right) \Big|_{(x - 0.5he_i, t)} - a_i(x - 0.5he_i, t) \, u_{\bar{x}_i}(x, t), \quad i = 1, 2.$$

Lemma 3 [5]. *Additive difference scheme (12)–(15) is absolutely stable and the following a priori estimate holds*

$$\|z\|_{B^{1,\alpha/2}(Q_{h\tau})} \leq C \left(\tau \sum_{k=1}^{2m} \|\xi^k\|_h^2 + \tau \sum_{k=1}^{m} \|\zeta_1^{2k-1}\|_{1h}^2 + \tau \sum_{k=1}^{m} \|\zeta_2^{2k}\|_{2h}^2 \right.$$
$$\left. + \tau^3 \sum_{k=1}^{m} \|\chi_t^{2k-1}\|_h^2 + \tau^{1+\alpha} \sum_{k=1}^{m} \|\chi^{2k}\|_h^2 \right)^{1/2}.$$
(16)

Proof. The proof is similar to the proof of Lemma 6.1 in [5].   □

Theorem 3. *Let the solution u of initial-boundary value problem (1)–(3) belong to the space $C^2([0,T], C(\bar{\Omega})) \cap C^1([0,T], H^2(\Omega)) \cap C([0,T], H^3(\Omega))$ and $a_i \in C([0,T], H^2(\Omega))$, $i = 1, 2$. Then the solution v of finite difference scheme (7)–(10) with $\bar{f} = T_1 T_2 f$ converges to u and the following convergence rate estimate holds:*

$$\|u - v\|_{B^{1,\alpha/2}(Q_{h\tau})} = O(h^2 + \tau^{\alpha/2}).$$

Proof. We must estimate the terms on the right-hand side of the last inequality. Let us set $\xi = \xi_1 + \xi_2$, where

$$\xi_1 = D_{t,0+,\tau}^{\alpha} u - D_{t,0+}^{\alpha} u, \qquad \xi_2 = D_{t,0+}^{\alpha} u - T_1 T_2 D_{t,0+}^{\alpha} u = D_{t,0+}^{\alpha}(u - T_1 T_2 u).$$

From Lemma 1 and from integral representation of $u - T_1 T_2 u$, one can obtains

$$\left(\tau \sum_{k=1}^{2m} \|\xi_1^k\|_h^2 \right)^{1/2} \leq C\tau^{2-\alpha} \|u\|_{C^2([0,T], C(\bar{\Omega}))}.$$
(17)

$$\left(\tau \sum_{k=1}^{2m} \|\xi_2^k\|_h^2 \right)^{1/2} \leq Ch^2 \|u\|_{C_+^{\alpha}([0,T], H^2(\Omega))}.$$
(18)

The terms ζ_i, $i = 1, 2$, we decompose in the next way

$$\zeta_i = \zeta_{i1} + \zeta_{i2} + \zeta_{i3}, \qquad \text{where}$$

$$\zeta_{i1} = T_{3-i}\left(a_i \frac{\partial u}{\partial x_i}\right)\Big|_{(x-0.5he_i,t)} - T_{3-i}\left(a_i\right)\Big|_{(x-0.5he_i,t)} T_{3-i}\left(\frac{\partial u}{\partial x_i}\right)\Big|_{(x-0.5he_i,t)},$$

$$\zeta_{i2} = T_{3-i}\left(a_i\right)\Big|_{(x-0.5he_i,t)}\left[T_{3-i}\left(\frac{\partial u}{\partial x_i}\right)\Big|_{(x-0.5he_i,t)} - u_{\bar{x}_i}(x,t)\right],$$

$$\zeta_{i3} = \left[T_{3-i}\left(a_i\right)\Big|_{(x-0.5he_i,t)} - a_i(x - 0.5he_i, t)\right]u_{\bar{x}_i}(x,t).$$

These terms we can estimate like in [2]

$$\|\zeta_{ij}\|_{L^2(Q_{ih\tau})} \leq Ch^2 \|a_i\|_{C([0,T],H^2(\Omega))} \|u\|_{C([0,T],H^3(\Omega))}, \quad i = 1, 2, \; j = 1, 2, 3,$$
$$(19)$$

while the terms χ and χ_t are bounded and them we can estimate directly. $\qquad \square$

3.2   The Modified Factorized Scheme

The factorized difference scheme for a 2D fractional in time subdiffusion equation is proposed in [2,4]. By multiplying discretized fractional derivative with operator $(I+\theta\tau^\alpha A_1)(I+\theta\tau^\alpha A_2)$, where $A_i v = -v_{x_i\bar{x}_i}$, and θ positive parameter, we add a small error term, but achieved an economical scheme. In [13], when the operator $\mathcal{L}$ with constant coefficients does not contain mixed partial derivatives, the authors used the part of this factorized operator and obtained better convergence rate in time direction. Here, we consider (1)–(3), where $a_i = a_i(x_i)$. The operator $\mathcal{L}$ we approximate with $L_h = L_{h,1} + L_{h,2}$, where $L_{h,i}v = -(\tilde{a}_i v_{\bar{x}_i})_{x_i}$ and $\tilde{a}_i = \tilde{a}_i(x_i) = a_i(x_i - 0.5h)$. The problem (1)–(3) we approximate with the following scheme

$$\left(I + \mu^2 L_{h,1}L_{h,2}\right) D^\alpha_{t,0+,\tau} v^k + L_h v^k = \bar{f}^k, \quad x \in \omega_h, \tag{20}$$

where $\mu = \Gamma(2 - \alpha)\tau^\alpha$, subject to homogeneous boundary and initial conditions (9), (10). By using formula (5), it follows from (20)

$$\left(I + \mu L_{h,1}\right)\left(I + \mu L_{h,2}\right)v^k = -\sum_{l=1}^{k-1}[d_{k-l+1} - d_{k-l}]\left(v^l + \mu^2 L_{h,1}L_{h,2}v^l\right) + \mu\bar{f}^k.$$

This scheme is also economical. To compute the solution on time level k, we have to solve two systems of linear equations with three-diagonal matrices. Now, we define the norm $\| \cdot \|_{B^{1,\alpha/2}(Q_{h\tau})}$ on the following way

$$\|v\|^2_{B^{1,\alpha/2}(Q_{h\tau})} = \tau \sum_{k=1}^{M}\left[\left(D^\alpha_{t,0+,\tau}\left(\|v\|^2_h\right)\right)^k + |v^k|^2_{H^1(\omega_h)}\right].$$

Theorem 4. *Let $0 < \alpha < 1$ and $a_i \in C[0,1]$. Then the difference scheme (20), (9), (10) is absolutely stable and its solution satisfies the a apriori estimate*

$$\|v\|_{B^{1,\alpha/2}(Q_{h\tau})} \leq C\|\bar{f}\|_{L^2(Q_{h\tau})}. \tag{21}$$

Proof. If we denote $B = I + \mu^2 L_{h,1} L_{h,2}$, it is obviously that $B > I$ and because of that $\|v\|_B^2 := (Bv,v)_h \geq \|v\|_h^2$. Multiplying (20) with v^k, we obtain

$$\frac{1}{2} \left(D_{t,0+,\tau}^\alpha(\|v\|_h^2)\right)^k + c_0|v^k|_{H^1(\omega_h)}^2 \leq \varepsilon\|v^k\|^2 + \frac{1}{4\varepsilon}\|\bar{f}^k\|_h^2.$$

Like in the proof of the Theorem 2, for $\varepsilon = 8c_0$, we show that (21) is valid, where $C = (16c_0 \min\{1, c_0\})^{-1/2}$. $\qquad\square$

If u is the solution of the initial-boundary-value problem (1)–(3) and v the solution of the difference problem (20), (9), (10) with $\bar{f} = T_1 T_2 f$, then the error $z = u - v$ satisfies

$$\left(I + \mu^2 L_{h,1} L_{h,2}\right) D_{t,0+,\tau}^\alpha z^k + L_h z^k = \psi^k, \quad x \in \omega_h, \tag{22}$$

where $\psi^k = \xi^k + \frac{1}{2}\left(\zeta_{1,x_1}^k + \zeta_{2,x_2}^k\right) + \nu_{1,x_1}^k + \nu_{2,x_2}^k$. The terms ξ and ζ_i are already defined and $\nu_i = \frac{1}{2}\mu^2 L_{h,3-i} D_{t,0+,\tau}^\alpha(\tilde{a}_i u_{\bar{x}_i}^k)$. The next assertion holds.

Lemma 4. *Difference scheme (22),(14),(15) is apsolutely stable and the following a priori estimate is valid*

$$\|z\|_{B^{1,\alpha/2}(Q_{h\tau})} \leq C \left(\tau \sum_{k=1}^M \left(\|\xi^k\|_h^2 + \sum_{i=1}^2 \left(\|\zeta_i^k\|_{ih}^2 + \|\nu_i^k\|_{ih}^2\right)\right)\right)^{1/2}. \tag{23}$$

Now, for the convergence rate of the proposed scheme, we have to estimate the terms on the right-hand side in (23). It can be seen that the temporal convergence rate is better here than in additive scheme.

Theorem 5. *Let the solution u of the initial-boundary value problem (1)–(3) belong to the space $C^2([0,T], C(\bar{\Omega})) \cap C^1([0,T], H^3(\Omega))$ and $a_i \in C^1[0,1]$. Then the solution v of the difference scheme (20), (14), (15) with $\bar{f} = T_1 T_2 f$ convergens to the u and the following convergence rate estimate holds*

$$\|u - v\|_{B^{1,\alpha/2}(Q_{h\tau})} = O(h^2 + \tau^{\min\{2\alpha, 2-\alpha\}}).$$

Proof. The result follows from (17)–(19) and from

$$\left(\tau \sum_{k=1}^M \|\nu_i\|_{ih}^2\right)^{1/2} \leq C\tau^{2\alpha} \max_{j=1,2} \|a_j\|_{C^1[0,1]} \|u\|_{C^1([0,T], H^3(\Omega))}.$$

$\qquad\square$

Acknowledgement. This research was supported by Ministry of Education, Science and Technological Development of Republic of Serbia thorough the project No. 174015.

References

1. Chen, M., Deng, W.: A second-order numerical method for two-dimensional two-sided space fractional convection diffusion equation. Appl. Math. Model. **38**, 3244–3259 (2014)
2. Delić, A., Hodžić, S., Jovanović, B.: Factorized difference scheme for two-dimensional subdiffusion equation in nonhomogeneous media. Publ. Inst. Math. **99**, 1–13 (2016)
3. Delić, A., Jovanović, B.S.: Numerical approximation of an interface problem for fractional in time diffusion equation. Appl. Math. Comput. **229**, 467–479 (2014)
4. Hodžić, S.: Factorized difference scheme for 2D fractional in time diffusion equation. Appl. Anal. Discrete Math. **9**, 199–208 (2015)
5. Hodžić-Živanovic, S., Jovanović, B.S.: Additive difference scheme for two-dimensional fractional in time diffusion equation. Filomat **31**, 217–226 (2017)
6. Kilbas, A., Srivastava, H., Trujillo, J.: Theory and Applications of Fractional Differential Equations. North-Holland Mathematics Studies, vol. 204. Elsevier Science and Technology, Boston (2006)
7. Li, L., Xu, D.: Alternating direction implicit-Euler method for the two-dimensional fractional evolution equation. J. Comput. Phys. **236**, 157–168 (2013)
8. Li, X., Xu, C.: A space-time spectral method for the time fractional diffusion equation. SIAM J. Numer. Anal. **47**, 2108–2131 (2009)
9. Oldham, B.K., Spanier, J.: The Fractional Calculus. Academic Press, New York (1974)
10. Podlubny, I.: Fractional Differential Equations. Academic Press, San Diego (1999)
11. Samarskiĭ, A.A.: Theory of Difference Schemes. Pure and Applied Mathematics, vol. 240. Marcel Dekker Inc, New York City (2001)
12. Sun, Z.Z., Wu, X.N.: A fully discrete difference scheme for a diffusion-wave system. Appl. Numer. Math. **56**, 193–209 (2006)
13. Yao, W., Sun, J., Wu, B., Shi, S.: Numerical simulation of a class of a fractional subdiffusion equation via the alternating direction implicit method. Numer. Methods Partial Differ. Eqn. **32**, 531–547 (2016)

Evolution of Copulas in Discrete Processes with Application to a Numerical Modeling of Dependence Relation Between Exchange Rates

Naoyuki Ishimura[1(✉)] and Yasukazu Yoshizawa[2]

[1] Faculty of Commerce, Chuo University, Tokyo 192-0393, Japan
naoyuki@tamacc.chuo-u.ac.jp
[2] Tokio Marine & Nichido Life Insurance Co., Ltd., Tokyo 100-8050, Japan
y.yoshizawa4416@gmail.com

Abstract. Copulas are known to provide a flexible tool for analyzing the dependence structure between random events. Here we apply the newly introduced notion of evolution of copulas to real data of exchange rates so that we ensure the quality of practically employing our theory. Results show that our algorithm provides a prospective handy method in computational finance.

1 Introduction

Copulas are well employed tool for analyzing the possibly nonlinear dependence structure among random events [1,5]. However, it is recognized that the copula method is not suitable to time-dependent relations. We have introduced, on the other hand, the concept of evolution of copulas both in continuous and discrete senses [3,10], which assumes that the copula itself evolves according to the time-variable.

Here we apply our evolution of copulas to numerical modeling in computational finance; in particular, following the exposition by Yoshizawa [9], we analyze the dependence relation model between euro (EUR)-Japanese yen (JPY) exchange rates and Swiss franc (CHF)-JPY exchange rates. We consider events such that the direction of change are almost stable. We compare the computed value of Kendall's tau under discrete evolution to that of the moving averages of the empirical copulas. Kendall's tau is one of popular measures of association. Results show that the discrete evolution copulas approximate fairy well the smoothed transition of empirical copulas from the viewpoint of Kendall's tau.

2 Preliminary

2.1 Copulas

We first recall the definition of copula and the fundamental theorem due to Sklar [8] for completeness, which are formulated as follows in the case of a bivariate joint distribution.

© Springer International Publishing AG 2017
I. Dimov et al. (Eds.): NAA 2016, LNCS 10187, pp. 359–366, 2017.
DOI: 10.1007/978-3-319-57099-0_39

Definition 1. *A function C defined on $I^2 := [0,1] \times [0,1]$ and valued in I is called a copula if the following conditions are fulfilled.*

(i) *For every $(u,v) \in I^2$,*

$$C(u,0) = C(0,v) = 0, \quad C(u,1) = u \quad and \quad C(1,v) = v. \tag{1}$$

(ii) *For every $(u_i, v_i) \in I^2$ $(i = 1,2)$ with $u_1 \le u_2$ and $v_1 \le v_2$,*

$$C(u_1,v_1) - C(u_1,v_2) - C(u_2,v_1) + C(u_2,v_2) \ge 0. \tag{2}$$

The requirement (2) is referred to as *the 2-increasing condition*. We also note that a copula is continuous by its definition.

Theorem 1 (Sklar's theorem). *Let H be a bivariate joint distribution function with marginal distribution functions F and G; that is,*

$$\lim_{x \to \infty} H(x,y) = G(y), \qquad \lim_{y \to \infty} H(x,y) = F(x).$$

Then there exists a copula, which is uniquely determined on $RanF \times RanG$, such that

$$H(x,y) = C(F(x), G(y)). \tag{3}$$

Conversely, if C is a copula and F and G are distribution functions, then the function H defined by (3) is a bivariate joint distribution function with margins F and G.

The so-called rank correlations are a sort of dependence measures. The population version of Kendall's tau and Spearman's rho, which are denoted by τ and ρ, respectively, are typical and important examples. These τ and ρ are known to be represented in terms of copulas. Stated explicitly, let X and Y be continuous random variables whose copula is C. Then we have

$$\tau_{X,Y} = \tau_C = 4 \iint_{I^2} C(u,v) dC(u,v) - 1,$$

$$\rho_{X,Y} = \rho_C = 12 \iint_{I^2} C(u,v) \, dudv - 3 = 12 \iint_{I^2} (C(u,v) - uv) \, dudv. \tag{4}$$

2.2   Empirical Copulas

Copulas for sample data sets are also know, which are expressed in terms of the so-called empirical copulas and the corresponding empirical copula frequency function. Here we recall the basic ingredients according to Sect. 5.6 of Nelsen [6].

Below throughout the paper, for $N > 1$ and $0 < h \ll 1$, we put

$$\Delta u = \Delta v := \frac{1}{N}, \qquad \Delta t := h, \qquad \lambda := \frac{\Delta t}{(\Delta u)^2} = \frac{\Delta t}{(\Delta v)^2} = hN^2,$$

$$u_i := i\Delta u = \frac{i}{N}, \qquad v_j := j\Delta v = \frac{j}{N} \qquad (i,j = 0,1,\cdots,N). \tag{5}$$

Definition 2. *Let* $\{(x_k, y_k)\}_{k=1,2,\cdots,N}$ *denote a sample size* N *from a continuous bivariate distribution and let* $x_{(i)}, y_{(j)}$ $(i \leq i, j \leq N)$ *denote order statistics from the sample. The empirical copula is the function* C_N *given by*

$$C_N(u_i, v_j) = \frac{number\ of\ pairs\ (x, y)\ in\ the\ sample\ with\ x \leq x_{(i)}, y \leq y_{(j)}}{N}, \quad (6)$$

and the corresponding empirical copula frequency c_N *is given by*

$$c_N(u_i, v_j) = \begin{cases} 1/N & if\ (x_{(i)}, y_{(j)})\ is\ an\ element\ of\ the\ sample, \\ 0 & otherwise. \end{cases}$$

It is to be noted that C_N and c_N are related with the next formulas.

$$C_N(u_i, v_j) = \sum_{p=1}^{i} \sum_{q=1}^{j} c_N(u_p, v_q),$$

$$c_N(u_i, v_j) = C_N(u_i, v_j) - C_N(u_{i-1}, v_j) - C_N(u_i, v_{j-1}) + C_N(u_{i-1}, v_{j-1}).$$

The sample versions of Kendall's tau t is then given by

$$t = \frac{2N}{N-1} \sum_{i=2}^{N} \sum_{j=2}^{N} (C_N(u_i, v_j) C_N(u_{i-1}, v_{j-1}) - C_N(u_i, v_{j-1}) C_N(u_{i-1}, v_j)). \quad (7)$$

A similar expression is known for Spearman's rho r; namely, we learn that

$$r = \frac{12}{N^2 - 1} \sum_{i=1}^{N} \sum_{j=1}^{N} (C_N(u_i, v_j) - u_i v_j).$$

3 Evolution of Copulas

3.1 Evolution of Copulas in Continuous Processes

In our previous study [3], we have introduced the time evolution of copulas. That is, we consider a time parameterized family of copulas $\{C(u, v, t)\}_{t \geq 0}$, which satisfy the heat equation:

$$\frac{\partial C}{\partial t}(u, v, t) = \left(\frac{\partial^2}{\partial u^2} + \frac{\partial^2}{\partial v^2} \right) C(u, v, t). \quad (8)$$

Here, by the definition of copula, we understand that $C(\cdot, \cdot, t)$ fulfills (1) and (2); to be precisely, we postulate that

(i) for every $(u, v, t) \in I^2 \times (0, \infty)$,

$$C(u, 0, t) = C(0, v, t) = 0, \quad and \quad C(u, 1, t) = u \quad and \quad C(1, v, t) = v. \quad (9)$$

(ii) for every $(u_i, v_i, t) \in I^2 \times (0, \infty)$ $(i = 1, 2)$ with $u_1 \leq u_2$ and $v_1 \leq v_2$,

$$C(u_1, v_1, t) - C(u_1, v_2, t) - C(u_2, v_1, t) + C(u_2, v_2, t) \geq 0. \qquad (10)$$

The stationary solution to (8), which is referred to as the harmonic copula, is uniquely determined to be $\Pi(u, v) := uv$, in view of the boundary condition (1). We note that the copula Π represents the independent structure between two respective random variables.

The unique existence and the convergence of solutions to (8) with satisfying the conditions (9) and (10) are established in [3, 4].

Theorem 2. *Let $C = C(u, v, t)$ $(t \geq 0)$ be a time-parametrized family of bivariate copulas, which satisfy (8), (9) and (10) for $(u, v, t) \in I^2 \times (0, \infty)$ and $C(u, v, 0) = C_0(u, v)$ on $(u, v) \in I^2$, where C_0 denotes a given arbitrary initial copula. Then it follows that*

$$C(u, v, t) = uv$$

$$+ 4 \sum_{m,n=1}^{\infty} e^{-\pi^2(m^2+n^2)t} \sin m\pi u \sin n\pi v \iint_{I^2} \sin m\pi\xi \sin n\pi\eta (C_0(\xi, \eta) - \xi\eta) d\xi d\eta,$$

and moreover

$$|\tau_{C_t}| + |\rho_{C_t}| \leq A e^{-Bt} \quad ast \to \infty,$$

where τ_{C_t}, ρ_{C_t} are defined through (4) and A, B are positive constants. In particular, $C_t \to \Pi$ exponentially as $t \to \infty$.

3.2   Evolution of Copulas in Discrete Processes

The evolution of copulas is able to be considered in discrete processes, which is indispensable to the numerical computation. Indeed, given a system of division (5), let a family of copulas $\{C^n(u, v)\}_{n=0,1,2,\cdots}$ be defined as follows: First,

$$C^0(u, v) := C_0(u, v),$$

where C_0 denotes given initial copula. Next, at $\{(u_i, v_j)\}_{i,j=0,1,\cdots,N}$, the value $C_{i,j}^n := C^n(u_i, v_j)$ is defined to be governed by the system of difference equations: For $i, j = 1, 2, \cdots, N - 1$,

$$\frac{C_{i,j}^{n+1} - C_{i,j}^n}{\Delta t} = \frac{C_{i+1,j}^n - 2C_{i,j}^n + C_{i-1,j}^n}{(\Delta u)^2} + \frac{C_{i,j+1}^n - 2C_{i,j}^n + C_{i,j-1}^n}{(\Delta v)^2}, \qquad (11)$$

combined with the boundary conditions

$$\begin{cases} C_{0,j}^n = 0 = C_{i,0}^n \\ C_{i,N}^n = u_i, \quad C_{N,j}^n = v_j \end{cases} \qquad \text{for } i, j = 0, 1, \cdots, N.$$

As to the point $(u, v) \in I^2$ other than $\{(u_i, v_j)\}_{i,j=0,1,\ldots,N}$, the value $C^n(u, v)$ is provided by interpolation. That is, if for instance

$$u_i \leq u \leq u_{i+1}, \quad v_j \leq v \leq v_{j+1},$$

then we put

$$C^n(u, v) := C^n_{i,j} + \frac{C^n_{i+1,j} - C^n_{i,j}}{u_{i+1} - u_i}(u - u_i) + \frac{C^n_{i,j+1} - C^n_{i,j}}{v_{j+1} - v_j}(v - v_j)$$
$$+ \frac{C^n_{i+1,j+1} - C^n_{i+1,j} - C^n_{i,j+1} + C^n_{i,j}}{(u_{i+1} - u_i)(v_{j+1} - v_j)}(u - u_i)(v - v_j).$$

Other parts are computed similarly.

It is also easy to check that a sequence of copulas $\{C^n(u, v)\}_{n=0,1,2,\cdots}$ formulated as above verify the boundary conditions (1) as well as the 2-increasing condition (2) provided $\lambda \leq 1/4$. Furthermore, we infer that in this range of λ, the difference scheme (11) is stable. We thus obtain the next theorem [10].

Theorem 3. *For any C_0, there exists a family of copulas $\{C^n(u, v)\}_{n=0,1,2,\cdots}$, which satisfy the system of difference equations (11) at every $\{(u_i, v_j)\}_{i.j=0,1,\cdots,N}$. As $n \to \infty$, it follows that*

$$C^n(u, v) \to \Pi(u, v) = uv \quad \text{uniformly on } I^2.$$

As expected, Kendall's tau and Spearman's rho in discrete setting also converge to zero as $n \to \infty$. See [10].

We remark that the difference equation (11) can be generalized as

$$C^{n+1}_{i,j} = (1 - 4\nu\lambda)C^n_{i,j} + \nu\lambda(C^n_{i+1,j} + C^n_{i-1,j} + C^n_{i,j+1} + C^n_{i,j-1}), \qquad (12)$$

where the diffusion coefficient $\nu > 0$ satisfies $\nu\lambda \leq 1/4$ (observe [2]).

It is further to be noted that, in [9], the evolution of copulas in discrete processes is proved to converge to the one in continuous processes as $N \to \infty$. Moreover, a multivariate version in discrete processes is constructed (see [2]).

4   Numerical Implementation with Empirical Data

This section is devoted to the numerical study on the dependence of euro-Japanese yen foreign exchange rates with those of the Swiss franc-Japanese yen. The evolution of copulas suits events whose dependence monotonically increases or decreases. We thus focus on rapidly changing events whose directions are almost stable. We treat foreign exchange rates on January 15, 2015, when Swiss franc endured a shock breakout after the announcement that the Swiss central bank had stopped monetary policy efforts to maintain the Swiss franc against the euro at more than 1.20. Furthermore, we collect the second time scale in order to capture its monotonic directivity. Most ingredients are reproduced from Sect. 3.5 of Yoshizawa [9].

First, we construct empirical copulas of the euro-Japanese yen rates and the Swiss franc-Japanese yen rates for every second of 40 min through the formula (6). We then calculate their Kendall's tau by applying the formula (7). Figure 1 charts the transition of the Kendall's tau as well as the euro-Swiss franc foreign exchange rates. In Fig. 1, Tau denotes Kendall's tau of the euro against the Japanese yen and the Swiss franc against the Japanese yen, while EURO/CHF denotes the euro against the Swiss franc. Data source is Bloomberg, exchange rate.

We collect the foreign exchange rates on the second scale interval from 18:20 to 19:00 on January 15, 2015. We calculate the euro-Japanese yen rates by multiply the Swiss franc-Japanese yen rates by the euro-Swiss franc rates. With reference to the formula (6), we construct empirical copulas for every second by using 60 datasets for both the euro-Japanese yen and the Swiss franc-Japanese yen, where we collect 60 datasets in 1 min. Then, we calculate Kendall's tau by applying the formula (7) to the empirical copulas to every second.

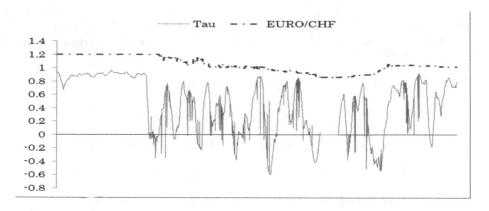

Fig. 1. Kendall's tau versus the euro against the Swiss franc

Second, we apply a moving average method to 600 datasets of Kendall's tau, which are shown in Fig. 2. This treatment is because that Kendall's tau fluctuate and may include some singular data. The average of 600 datasets for every second refers to the data average over 10 min. Thus, the start point of the average of 600 datasets is 18:30. In Fig. 2, Tau denotes the Kendall's tau of the euro against the Japanese yen and the Swiss franc against the Japanese yen, while TauAve600 denotes the moving averages of "Tau" for 600 datasets.

Finally, we compare Kendall's tau of the evolution of empirical copulas to the above mentioned moving averages of Kendall's tau of empirical copulas. We extract the empirical copula at 18:31:47, at which time the Kendall's tau of the empirical copulas is almost equal to the smoothed Kendall's tau. We evolve them 1,200 times, which means for 20 min. For this evolution, we set parameters for the recursion equation (12) as $\Delta u = \Delta v = 1/N = 1/(60\text{data})$, $\Delta t = 1\,\text{s}$, $\lambda = \Delta t/(\Delta u)^2 = \Delta t/(\Delta v)^2 = 3,600$, and $\nu = 720$. Thus, we derive the equation

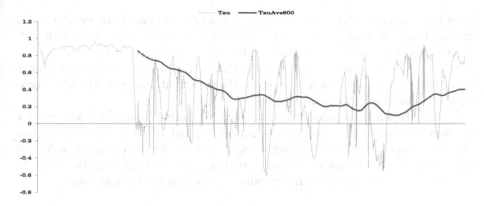

Fig. 2. Moving averages of Kendall's tau

$C_{i,j}^{n+1} = (1/5)(C_{i,j}^n + C_{i+1,j}^n + C_{i-1,j}^n + C_{i,j+1}^n + C_{i,j-1}^n)$, which is used to evolve the empirical copulas.

The results are plotted in Fig. 3, which shows that the evolution of empirical copulas approximate the smoothed transitions of empirical copulas from the viewpoint of Kendall's tau. In Fig. 3, TauAve 600 denotes the moving averages of "Tau" for 600 data, while Evolution of Tau denotes Kendall's tau of the evolution of copulas.

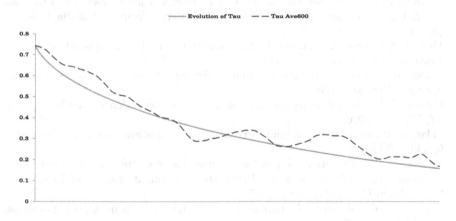

Fig. 3. Kendall's tau of evolution of copulas

5   Discussion

Copulas are flexible tools for the study on a nonlinear dependence structure. The dependence relation between exchange rates is a common theme for researches in the copula theory [7]. Using flexible discrete copula models, we are able to numerically analyze the movement of dependence relations between the Swiss franc and the euro against the Japanese yen approximately rather well. Although the event treated is restricted such that the dependence does not fluctuate drastically but changes monotonically, we have verified the practicality of some aspect on the theory of the evolution of copulas. We hope that our method will become much popular in relevant communities.

Various topics of copulas still have been an active area which is open for researches both from theoretical and practical point of view. We believe that a possible application of our theory seems much wider, since the dependence structure normally varies with time. Further project of investigations are in progress and we may report our outcomes somewhere in the near future.

References

1. Frees, E.W., Valdez, E.A.: Understanding relationships using copulas. North Am. Actuar. J. **2**, 1–25 (1998)
2. Ishimura, N.: Evolution of copulas and its applications. In: Vanmaele, M., Deelstra, G., De Schepper, A., Dhaene, J., Schoutens, W., Vanduffel, S., Vyncke, D. (Eds.) Proceedings of Actuarial and Financial Mathematics Conference 2014, pp. 85–89 (2014). http://www.afmathconf.ugent.be/
3. Ishimura, N., Yoshizawa, Y.: On time-dependent bivariate copulas. Theoret. Appl. Mech. Jpn. **59**, 303–307 (2011)
4. Ishimura, N., Yoshizawa, Y.: A note on the time evolution of bivariate copulas. In: Stoynov, P. (Ed.) Proceedings of the Fourth International Conference, Financial and Actuarial Mathematics, FAM-2011, pp. 3–7. Perun-Sprint Ltd., Sofia, Bulgaria (2011)
5. McNeil, A.J., Frey, R., Embrechts, P.: Quantitative Risk Management. Princeton University Press, Princeton (2005)
6. Nelsen, R.B.: An Introduction to Copulas. Springer Series in Statistics, 2nd edn. Springer, New York (2006)
7. Patton, A.J.: Modelling asymmetric exchange rate dependence. Int. Econ. Rev. **47**, 527–556 (2006)
8. Sklar, A.: Random variables, joint distribution functions, and copulas. Kybernetika **9**, 449–460 (1973)
9. Yoshizawa, Y.: Evolution of copulas: continuous, discrete, and its application to quantitative risk management. Ph.D. thesis, Graduate School of Economics, Hitotsubashi University, March 2015
10. Yoshizawa, Y., Ishimura, N.: Evolution of bivariate copulas in discrete processes. JSIAM Lett. **3**, 77–80 (2011)

Using ϵ-nets for Solving the Classification Problem

Maria A. Ivanchuk[1]($\boxtimes$) and Igor V. Malyk[2]

[1] Bukovinian State Medical University, Chernivtsi, Ukraine
mgracia@ukr.net
[2] Yuriy Fedkovych Chernivtsi National University, Chernivtsi, Ukraine
malyk.igor.v@gmail.com

Abstract. We propose a new approach for solving the classification problem, which is based on the using ϵ-nets theory. It is showed that for separating two sets one can use their ϵ-nets, which considerably reduce the complexity of the separating algorithm for large sets' sizes. The necessary and sufficient conditions of separable ϵ-nets of two sets are proved. The algorithm of building separable ϵ-nets is proposed. The ϵ-nets, constructed according to this algorithm, have size $O(1/\varepsilon)$, which does not depended on the size of set. The set of possible values of ϵ for ϵ-nets of both sets is considered. The properties of this set and the theorem of its convergence are proved. The proposed algorithm of solving the classification problem using the ϵ-nets has the same computational complexity as Support Vector Machine $O(n \ln n)$ and its accuracy is comparable with SVM results.

Keywords: Epsilon-nets · Sets' separation · VC-dimension

1 Introduction

The classification problem can be formalized as the problem of finding separation hyperplane of two training sets in multidimensional space. Given two finite sets $A \subset R^d$ and $B \subset R^d$, $|A| = n_A$, $|B| = n_B$. Let $A \not\subset conv_B$, $B \not\subset conv_A$ and sets A and B are not separable: $conv(A) \bigcap conv(B) \neq \emptyset$. We will find a way to separate sets A and B using ϵ-nets [1].

Definition 1. Sets A and B are called ϵ-**separable**, if there exist sets $A_1 \subset A$, $B_1 \subset B$, such that

$$conv(A \backslash A_1) \bigcap conv(B \backslash B_1) = \emptyset \tag{1}$$

and

$$|A_1| + |B_1| < \varepsilon(n_A + n_B). \tag{2}$$

Definition 2. Hyperplane L_ε is called ϵ-**separating** for the sets A and B if

$$\frac{|A \bigcap L_\varepsilon^+| + |B \bigcap L_\varepsilon^-|}{n_A + n_B} \geq 1 - \varepsilon.$$

© Springer International Publishing AG 2017
I. Dimov et al. (Eds.): NAA 2016, LNCS 10187, pp. 367–374, 2017.
DOI: 10.1007/978-3-319-57099-0_40

Let (R^d, H^d) is an infinite range space, where H^d are the closed halfspaces in R^d.

Theorem 1. A necessary and sufficient condition that two sets of points A and B are ϵ-separable is there exist $\varepsilon_A, \varepsilon_B$ and corresponding ϵ-nets $N_{\varepsilon_A}^A$, $N_{\varepsilon_B}^B$ in (R^d, H^d) such that

$$\varepsilon_A n_A + \varepsilon_B n_B < \varepsilon (n_A + n_B) \tag{3}$$

and

$$conv N_{\varepsilon_A}^A \bigcap conv N_{\varepsilon_B}^B = \emptyset. \tag{4}$$

Theorem 1 is proved in [2].

2 Separation Space

To find $\varepsilon_A, \varepsilon_B$ which satisfy the condition (3) consider the separation space.

Definition 3. The set

$$D_{A,B} = \left\{ (\varepsilon_1, \varepsilon_2) \in (0,1)^2 : \exists N_A^{\varepsilon_1}, N_B^{\varepsilon_2}, conv N_A^{\varepsilon_1} \bigcap conv N_B^{\varepsilon_2} = \emptyset \right\} \tag{5}$$

is called **the separation space for** A, B.

Let $\xi, \eta \in R^1$ are continuous random variables with distribution functions F_ξ, F_η.

Definition 4. The set D_l

$$D_l := \left\{ (x,y) \in (0,1)^2 : \exists h \in R^1, P\{\xi \in h_+\} \le x, P\{\eta \in h_-\} \le y \right\} \tag{6}$$

is called **the separation space for** ξ, η.

Lemma 2. Let the inverse function F_ξ exists. Then the sets D_l and $\bar{D}_l := (0,1)^2 \setminus D$ are separated by the line

$$y(x) = \min \left(F_\eta \left(F_\xi^{-1}(1-x) \right), 1 - F_\eta \left(F_\xi^{-1}(x) \right) \right). \tag{7}$$

Proof. Consider two possible cases.

1. Let the inequality $F_\xi(h) > F_\eta(h)$ is hold for $h \in (-\infty, \infty)$, then the set D_l is described by the system of inequality

$$\begin{cases} x \ge 1 - F_\xi(h) \\ y \ge F_\eta(h) \end{cases}.$$

Then the line that separates sets D_l and $\bar{D}_l$ is

$$y(x) = F_\eta \left(F_\xi^{-1}(1-x) \right)$$

2. Let the inequality $F_\xi(h) \le F_\eta(h)$ is hold for $h \in (-\infty, \infty)$, then the set D_l is described by the system of inequality

$$\begin{cases} x \ge F_\xi(h) \\ y \ge 1 - F_\eta(h) \end{cases}$$

In this case the line that separates sets D_l and $\bar{D}_l$ is $y(x) = 1 - F_\eta\left(F_\xi^{-1}(x)\right)$. So, in general, sets D_l and $\bar{D}_l$ are separated by the line $y(x) = \min\left(F_\eta\left(F_\xi^{-1}(1-x)\right), 1 - F_\eta\left(F_\xi^{-1}(x)\right)\right)$.

Lemma 2 is proved.

Theorem 3. Let the following conditions are exist:

1. The sets A, B of size n_A, n_B are generated by the independent continuous random variables ξ, η.
2. The sets $D_{A,B}$ and $\overline{D_{A,B}}$ are separated by the line $y_{A,B}(x)$.

Then there exist the following equality $\lim\limits_{n_A,n_B \to \infty} y_{A,B}(x) = y(x)$, where

$$y(x) = \min\left(F_\eta\left((F_\xi)_G^{-1}(1-x)\right), 1 - F_\eta\left((F_\xi)_G^{-1}(x)\right)\right)$$

Proof. To proof the theorem it is enough to show that the relations

$$F_{n_B}\left(F_{n_A}^{-1}(y)\right) \to F_\eta\left(F_\xi^{-1}(y)\right), \quad y \in (0,1) \tag{8}$$

and

$$F_{n_B}\left(1 - F_{n_A}^{-1}(y)\right) \to F_\eta\left(1 - F_\xi^{-1}(y)\right), \quad y \in (0,1) \tag{9}$$

hold. Let's show that the relation (8) hold.

$$\sup_{y \in [0,1]} \left| F_{n_B}\left(F_{n_A}^{-1}(y)\right) - F_\eta\left(F_\xi^{-1}(y)\right) \right|$$

$$= \sup_{y \in [0,1]} \left| F_{n_B}\left(F_{n_A}^{-1}(y)\right) + F_\eta\left(F_{n_A}^{-1}(y)\right) - F_\eta\left(F_{n_A}^{-1}(y)\right) - F_\eta\left(F_\xi^{-1}(y)\right) \right|$$

$$\le \sup_{y \in [0,1]} \left| F_{n_B}\left(F_{n_A}^{-1}(y)\right) - F_\eta\left(F_{n_A}^{-1}(y)\right) \right| + \sup_{y \in [0,1]} \left| F_\eta\left(F_{n_A}^{-1}(y)\right) - F_\eta\left(F_\xi^{-1}(y)\right) \right|$$

$$\le \sup_{x \in R^1} \left| F_{n_B}(x) - F_\eta(x) \right| + \sup_{y \in [0,1]} \left| F_\eta\left(F_{n_A}^{-1}(y)\right) - F_\eta\left(F_\xi^{-1}(y)\right) \right|.$$

According to the Glivenko-Cantelli theorem [3] the first term $\sup\limits_{x \in R^1} |F_{n_B}(x) - F_\eta(x)| \to 0$, $n_B \to \infty$.

Let η has density of distribution $f_\eta < K$. Then we have for a second term $\sup\limits_{y \in [0,1]} \left| F_\eta\left(F_{n_A}^{-1}(y)\right) - F_\eta\left(F_\xi^{-1}(y)\right) \right| \le K \sup\limits_{y \in [0,1]} \left| F_{n_A}^{-1}(y) - F_\xi^{-1}(y) \right|$. Let's show that $\sup\limits_{y\in[0,1]} \left|F_{n_A}^{-1}(y) - F_\xi^{-1}(y)\right| \to 0$. Suppose that $F_\xi(x_0) = y_0 \in (0,1)$ is fixed.

Assume that $\sup\limits_{y\in[0,1]} \left|F_{n_A}^{-1}(y) - F_\xi^{-1}(y)\right| \nrightarrow 0$, then $|F_\xi(x_0) - F_{n_A}(x_0)| \nrightarrow 0$. We arrive at contradiction. So, the relation (8) holds. By analogy, the relation (9) also holds.

Theorem 3 is proved.

Theorem 4. Let ξ_i, $i \geq 1$, η_j, $j \geq 1$ be independent random variables with distributions F_ξ, F_η and let functions F_ξ, F_η have inverse. Then the weak convergence

$$\zeta_{n,m} = \zeta_{n,m}(y) = \sqrt{n}\left(F_n\left(F_m^{-1}(y)\right) - F_\eta\left(F_\xi^{-1}(y)\right)\right) \to N(0,\sigma^2),$$

where $\sigma^2 = F_\eta\left(F_\xi^{-1}(y)\right)\left(1 - F_\eta\left(F_\xi^{-1}(y)\right)\right)$, as $n \to \infty$, $m = O(n^\alpha)$, $\alpha > 0$ holds.

Proof. Consider the mean of $\zeta_{n,m}$:

$$E\zeta_{n,m} = \sqrt{n}\left(EF_n\left(F_m^{-1}(y)\right) - F_\eta\left(F_\xi^{-1}(y)\right)\right).$$

Using definition of empirical distribution function we get:

$$|E\zeta_{n,m}| = \left|\sqrt{n}\left(EI_{\eta<F_m^{-1}(y)} - F_\eta\left(F_\xi^{-1}(y)\right)\right)\right|$$
$$= \left|\sqrt{n}\left(P\left\{\eta < F_m^{-1}(y)\right\} - P\left\{\eta < F_\xi^{-1}(y)\right\}\right)\right|$$
$$= \left|\sqrt{n}\left(P\left\{F_\xi^{-1}(y) \leq \eta < F_m^{-1}(y)\right\} + P\left\{F_m^{-1}(y) \leq \eta < F_\xi^{-1}(y)\right\}\right)\right|.$$

So, to proof convergence $E\zeta_{n,m} \to 0$ it's enough to show for any ϵ

$$\sqrt{n}P\left\{\left|F_\xi^{-1}(y) - F_m^{-1}(y)\right| > \varepsilon\right\} \to 0$$

According to definition of general inverse function we get

$$\sqrt{n}P\left\{\left|F_\xi^{-1}(y) - F_m^{-1}(y)\right| > \varepsilon\right\} = \sqrt{n}P\left\{\left|F_\xi(y) - F_m(y)\right| > \varepsilon\right\},$$

using Dvoretzky–Kiefer–Wolfowitz inequality [4] we get

$$\sqrt{n}P\left\{\left|F_\xi(y) - F_m(y)\right| > \varepsilon\right\} \leq \sqrt{n}2e^{-2m\varepsilon^2} = 2e^{0,5\ln n - 2m\varepsilon^2}$$

So, $\lim\limits_{n\to\infty} E\zeta_{n,m} = 0$ as $m = O(n^\alpha)$, $\alpha > 0$.

Consider the variance of $\zeta_{n,m}$: $D\zeta_{n,m} = E\zeta_{n,m}^2 - (E\zeta_{n,m})^2$.

Second term tends to 0, so it is important to consider the first term of the previous equality.

$$E\zeta_{n,m}^2 = nE\left(F_n\left(F_m^{-1}(y)\right) - F_\eta\left(F_\xi^{-1}(y)\right)\right)^2 = E\left(I_{\eta < F_m^{-1}(y)} - F_\eta\left(F_\xi^{-1}(y)\right)\right)^2$$
$$+\frac{1}{n}\sum_{i \neq j}\left(I_{\eta_i < F_m^{-1}(y)} - F_\eta\left(F_\xi^{-1}(y)\right)\right)\left(I_{\eta_j < F_m^{-1}(y)} - F_\eta\left(F_\xi^{-1}(y)\right)\right)$$

Ignoring the order of smallness $O(1)$, we get

$$D\zeta_{n,m} = E\left(I_{\eta < F_m^{-1}(y)} - F_\eta\left(F_\xi^{-1}(y)\right)\right)^2$$
$$+\frac{1}{n}\left(\sum_{i \neq j}\left(EI_{\eta_i < F_m^{-1}(y)}I_{\eta_j < F_m^{-1}(y)} - p^2\right)\right) = A_1 + A_2,$$

where $p = F_\eta\left(F_\xi^{-1}(y)\right)$.

Consider each term separately. Using Lebesgue's theorem [5], we get

$$A_1 = E\left(I_{\eta < F_m^{-1}(y)} - F_\eta\left(F_\xi^{-1}(y)\right)\right)^2 \rightarrow E\left(I_{\eta < F_\xi^{-1}(y)} - F_\eta\left(F_\xi^{-1}(y)\right)\right)^2 =$$
$p(1 - p)$ as $m \rightarrow \infty$

For the second term, using independent $\eta_i, i \geq 1$, we get

$$A_2 = (n-1)\left(EI_{\eta_1 < F_m^{-1}(y)}I_{\eta_2 < F_m^{-1}(y)} - p^2\right)$$
$$= (n-1)\left(P\left\{\eta_1 < F_m^{-1}(y), \eta_2 < F_m^{-1}(y)\right\} - p^2\right) = (n-1)$$
$$\cdot\left(P\left\{\eta_1 < F_m^{-1}(y), \eta_2 < F_m^{-1}(y)\right\} - p^2 \pm P\left\{\eta_1 < F_m^{-1}(y), \eta_2 < F_\xi^{-1}(y)\right\}\right)$$
$$= (n-1)\left(P\left\{\eta_1 < F_m^{-1}(y), \eta_2 < F_m^{-1}(y)\right\} - P\left\{\eta_1 < F_m^{-1}(y), \eta_2 < F_\xi^{-1}(y)\right\}\right)$$
$$+(n-1)\left(P\left\{\eta_1 < F_m^{-1}(y), \eta_2 < F_\xi^{-1}(y)\right\} - p^2\right) = A_{21} + A_{22}.$$

For the first term we get

$$|A_{21}| = (n-1)\left|P\left\{\eta_1 < F_m^{-1}(y)\right\}\right|$$
$$\cdot\left|P\left\{\eta_2 < F_m^{-1}(y)\middle| \eta_1 < F_m^{-1}(y)\right\} - P\left\{\eta_2 < F_\xi^{-1}(y)\middle| \eta_1 < F_m^{-1}(y)\right\}\right|$$
$$\leq (n-1)\left|P\left\{\eta_2 < F_m^{-1}(y)\middle| \eta_1 < F_m^{-1}(y)\right\} - P\left\{\eta_2 < F_\xi^{-1}(y)\middle| \eta_1 < F_m^{-1}(y)\right\}\right|$$

By analogues to previous, using Dvoretzky–Kiefer–Wolfowitz inequality we get $|A_{21}| \rightarrow 0$.

From the equality $A_{22} = (n-1)p\left(P\left\{\eta_1 < F_m^{-1}(y)\right\} - p\right)$, we get $|A_{22}| \rightarrow 0$. So, $D\zeta_{n,m} \rightarrow p(1-p)$.

The final step of the proof is using the central limit theorem for asymptotic independent random variables [6]

$$u_{i,m} = I_{\eta_i < F_m^{-1}(y) - p}, \ i \geq 1, \text{ as } m \rightarrow \infty$$

Theorem 4 is proved.

3   Algorithm of the ϵ-net Building

Using Lemma 2 and Theorem 3 find separation space $D_{A,B}$. Select $\varepsilon_A, \varepsilon_B \in D_{A,B}$.

Let set A contain the point with minimal y-coordinate. Let's denote the point of set A with minimal x-coordinate by a_{min}, and the point with maximal x-coordinate by a_{max}. Let's draw $k = \left[\frac{1}{\varepsilon_A}\right] + 1$ vertical lines from a_{min} to a_{max} in a manner that there are $\varepsilon_A n_A$ points in each of $\left[\frac{1}{\varepsilon_A}\right]$ bands. Vertical lines which separate the bands are described by the equations $x = C_i$, $i = \overline{1, k}$, where constant C_i can be founded from the equation $F(C_i) = i\varepsilon_A$.

For each i-th band, $1 \leq i \leq \left[\frac{1}{\varepsilon}\right]$ let's denote:

A^i is the set which contains points from the set A contained in the i-th band;
B^i is the set (may be empty) which contains points from the set B contained in the i-th band;
ay^i_{min}, ay^i_{max} are points from the set A^i with minimal and maximal y-coordinates;
by^i_{min}, by^i_{max} are points from the set B^i with minimal and maximal y-coordinates.

Denote $N_A^{\varepsilon_A}$ is the set of points which we will select in the ϵ-net of the set A. From the i-th band we will select two points in the set N_A. The first point is the point ay^i_{min}. According to assumption, set A is placed below the set B. The second point from the set A^i in the set $N_A^{\varepsilon_A}$ we will select according to the following rule.

If $B^i = \emptyset$ (it means that i-th band does not contain points from the set B),
  include point ay^i_{max} in the set $N_A^{\varepsilon_A}$
else
  if $ay^i_{max} < by^i_{min}$ (it means that in i-th band convex hulls of sets A, B are not intersected),
  include point ay^i_{max} in the set $N_A^{\varepsilon_A}$
  else (some points of the set B exist in the i-th band and they are placed bellow some points of set A) include in the set $N_A^{\varepsilon_A}$ point $a^i \in A$ such that point a^i is the nearest neighbor to the point by^i_{min}. We will called point a^i basis point of the set A.

The same way we build horizontal bands and select two points from each band to the set $N_A^{\varepsilon_A}$.

Lemma 5. The set $N_A^{\varepsilon_A}$ is ϵ-net for the set A.

Proof. Let's make an indirect proof. Let $N_A^{\varepsilon_A}$ is not an ϵ-net of the set A. It means that there exist a halfspace $H \subset R^2$ such that contain at least $\varepsilon_A n_A$ sets of point A, but each point does not belong to the set $N_A^{\varepsilon_A}$. Let's denote Z is the set of points from the set A, such that belong to the halfspace H and $|Z| \geq \varepsilon_A n_A$. Consider a point $z \in Z$. This point belongs to one horizontal and one vertical band. Together with point z one of extreme points of horizontal or vertical band or basis point is belong to the halfspace H. According to the building, set $N_A^{\varepsilon_A}$ consists of extreme and basis points of the set A, so $Z \bigcap N_A^{\varepsilon_A} \neq \emptyset$. This contradicts the assumption.

Lemma 5 is proved.

According to the algorithm, ϵ-net $N_A^{\varepsilon_A}$ consists of $[4/\varepsilon_A]$ points. ϵ-net $N_B^{\varepsilon_B}$ is built by the same algorithm. We propose to separate sets $N_A^{\varepsilon_A}$ and $N_B^{\varepsilon_B}$ using separation of the convex hulls algorithm, which is described in [7].

The complexity of the algorithm of separating to sets using ϵ-nets is $O(n \ln n)$ what is the same as the complexity of the most popular modern data mining algorithm SVM [8].

3.1 Example

Consider two sets A and B, which are generated from the normal distributions with parameters $n_A = 500$, $\mu_A = (1; 5)$, $\sigma^A = \begin{pmatrix} 1 & 0 \\ 0 & 3 \end{pmatrix}$, $n_B = 500$, $\mu_B = (6; 10)$, $\sigma^B = \begin{pmatrix} 3 & 0 \\ 0 & 1 \end{pmatrix}$ (Fig. 1). D_l and $\bar{D}_l$ are separated by the line which is illustrated in the (Fig. 2).

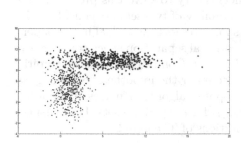

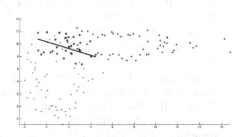

Fig. 1. Two sets generated by the normal distribution

Fig. 2. Bound line for the separation space

Since $\frac{|A \cap convB| + |B \cap convA|}{n_A + n_B} = 0,072$, let $\varepsilon_A = 0,07$; $\varepsilon_B = 0,07$. According to the Fig. 2, $(0,07; 0,07) \in D_l$. ϵ-nets of the sets A, B are illustrated in the Fig. 3.

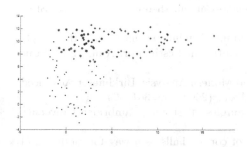

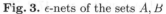

Fig. 3. ϵ-nets of the sets A, B

Fig. 4. The separating line for ϵ-nets

The separating line for ϵ-nets is illustrated in the Fig. 4. Classification using ϵ-nets gives 6,1% errors; classification by SVM [8] gives 6,2% errors (Fig. 5).

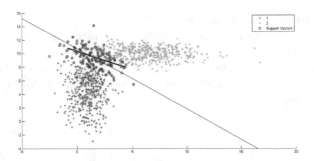

Fig. 5. Comparing classification by using ε-nets and by using SVM

4   Conclusions

The algorithm of two sets A and B in R^d separating is proposed in the paper. Unlike other similar works, we use the ε-nets theory to solve this problem. The necessary and sufficient conditions of ε-separability of two sets are considered in the paper. The algorithm of building ε-nets for two sets is described in the paper. The sufficient conditions of empiric and theoretical separation space convergence are proved. The asymptotic normality of the point T on the board of the empiric distribution is proved. It makes possible to consider the hypothesis of belonging of this point to the interval $[a, b]$. The proposed algorithm makes possible to build ε-nets of the size $O\left(1/\varepsilon_A\right)$ and $O\left(1/\varepsilon_B\right)$ for ε-separable sets. It intensifies and simplifies necessary and sufficient conditions of Theorem 1.

References

1. Haussler, D., Welzl, E.: Epsilon-nets and simplex range queries. Discrete Comput. Geom. **2**, 127–151 (1987)
2. Ivanchuk, M.A., Malyk, I.V.: Using-nets for linear separation of two sets in a Euclidean space R^d. Cybern. Syst. Anal. **51**(6), 965–968 (2015)
3. Tucker, H.G.: Generalization of the Glivenko-Cantelli theorem. Ann. Math. Stat. **30**(3), 828–830 (1959)
4. Dvoretzky, A., Kiefer, J., Wolfowitz, J.: Asymptotic minimax character of the sample distribution function and of the classical multinomial estimator. Ann. Math. Stat. **27**(3), 642–669 (1956)
5. Benedetto, J.J., Czaja, W.: Integration and Modern Analysis. Birkhäuser Advanced Texts Basler Lehrbücher. Springer, Heidelberg (2009). pp. 361–364
6. Durrett, R.: Probability: Theory and Examples, 4.1st edn. Cambridge University Press, Cambridge (2013). 386 p.
7. Ivanchuk, M.A., Malyk, I.V.: Separation of convex hulls as a way for modeling of systems of prediction of complications in patients. J. Autom. Inf. Sci. **47**(4), 78–84 (2015)
8. Christopher, J.C.: Tutorial on support vector machines for pattern recognition. Data Min. Knowl. Disc. **2**(2), 121–167 (1998)

Convergence of a Factorized Finite Difference Scheme for a Parabolic Transmission Problem

Zorica Milovanović Jeknić[1]($^{(\boxtimes)}$) and Boško Jovanović[2]

[1] Faculty of Construction Management, University "Union-Nikola Tesla",
Cara Dušana 62-64, Belgrade, Serbia
zorica.milovanovic@gmail.com
[2] Faculty of Mathematics, University of Belgrade, Studentski trg 16, Belgrade, Serbia
bosko@matf.bg.ac.rs

Abstract. In this paper, we consider a non-standard parabolic transmission problem in disjoint domains. A priori estimate for its weak solution in appropriate Sobolev-like space is proved. The convergence of a factorized finite difference scheme approximating this problem is analyzed.

1 Introduction

In applications, especially in engineering, composite or layered structures are often encountered, where the properties of individual layers can differ considerably from the properties of the surrounding material. Mathematical models of energy and mass transfer in domains with layers lead to so called transmission problems. In this paper a parabolic transmission problem in disjoint rectangular domains is investigated (see [4–6,8,10]). In each rectangle an initial-boundary problem of parabolic type is given, while the interaction between their solutions is described by means of nonlocal integral conjugation conditions.

2 Formulation of the Initial-Boundary-Value Problem

Let us consider two unit squares symmetrically situated in $\mathbb{R}^2$: $\Omega^1 = (-a - 1, -a) \times (0, 1)$ and $\Omega^2 = (a, a+1) \times (0, 1)$, where $a > 0$ is given real number. We denote $\Gamma^k = \partial \Omega^k = \cup_{i,j=1}^2 \Gamma^k{}_{ij}$, where $\Gamma^1{}_{11} = \{x = (x_1, x_2) \in \Gamma^1 \,|\, x_1 = -a-1\}$, $\Gamma^1{}_{12} = \{x \in \Gamma^1 \,|\, x_1 = -a\}$, $\Gamma^1{}_{21} = \{x \in \Gamma^1 \,|\, x_2 = 0\}$, $\Gamma^1{}_{22} = \{x \in \Gamma^1 \,|\, x_2 = 1\}$, while $\Gamma^2{}_{ij}$ are defined analogously. Further we denote $Q^k = \Omega^k \times (0, T)$, $\Sigma^k = \Gamma^k \times (0, T)$, $\Sigma^k{}_{ij} = \Gamma^k{}_{ij} \times (0, T)$ and $\Delta^k = \Gamma^k{}_{1,3-k} \times \Gamma^{3-k}{}_{1,k}$ $(k, i, j = 1, 2)$.

As a model example, we consider the following initial-boundary-value problem:

$$\frac{\partial u^k}{\partial t} + L^k u^k = f^k, \qquad (x, t) \in Q^k, \tag{1}$$

$$l^k u^k = \begin{cases} r^k u^{3-k}, & (x, t) \in \Sigma^k{}_{1,3-k}, \\ 0, & (x, t) \in \Sigma^k \setminus \Sigma^k{}_{1,3-k}, \end{cases} \tag{2}$$

© Springer International Publishing AG 2017
I. Dimov et al. (Eds.): NAA 2016, LNCS 10187, pp. 375–382, 2017.
DOI: 10.1007/978-3-319-57099-0_41

$$u^k(x,0) = u_0^k(x), \qquad x \in \Omega^k, \tag{3}$$

where

$$L^k u^k := - \sum_{i,j=1}^{2} \frac{\partial}{\partial x_i} \left(a_{ij}^k \frac{\partial u^k}{\partial x_j} \right), \tag{4}$$

$$l^k u^k := \sum_{i,j=1}^{2} a_{ij}^k \frac{\partial u^k}{\partial x_j} \cos(\nu, x_i) + \alpha^k u^k, \tag{5}$$

$$\left(r^k u^{3-k} \right)(x,t) := \int_{\Gamma_{1,k}^{3-k}} \beta^k(x,x') \, u^{3-k}(x',t) \, d\Gamma^{3-k}, \tag{6}$$

and ν is the unit outward normal to Γ^k and $k = 1,2$.

Notice that (2) on $\Sigma^k \setminus \Sigma^k_{1,3-k}$ reduces to a standard oblique derivative boundary condition, while on $\Sigma^k_{1,3-k}$ it can be considered as a conjugation condition of non-local Robin-Dirichlet type. Such conjugation condition models linearized radiation heat transfer (see [1]).

We assume that the standard conditions of regularity and ellipticity are satisfied:

$$a_{ij}^k = a_{ji}^k \in L^\infty(\Omega^k), \qquad \alpha^k \in L^\infty(\Gamma^k), \qquad \beta^k \in L^\infty(\Delta^k), \tag{7}$$

$$c_0^k \sum_{i=1}^{2} \xi_i^2 \le \sum_{i,j=1}^{2} a_{ij}^k \xi_i \xi_j \le c_1^k \sum_{i=1}^{2} \xi_i^2, \qquad \forall x \in \bar{\Omega}^k, \quad \forall \xi \in \mathbb{R}^2. \tag{8}$$

By C, c_i and c_i^k we denote positive constants, independent of the solution of the IBVP and the mesh-sizes. In particular, C may take different values in different formulas.

We also assume that the generalized solution of the problem (1)–(6) belongs to the Sobolev space $H^{s,s/2}$, $2 < s \le 3$, while the data satisfy the following smoothness conditions:

$$\begin{aligned} a_{ij}^k \in H^{s-1}(\Omega^k), \quad \alpha^k \in H^{s-3/2}(\Gamma^k_{ij}), \quad \alpha^k \in C(\Gamma^k), \quad \beta^k \in H^{s-1}(\Delta^k), \\ f^k \in H^{s-2,\,s/2-1}(Q^k), \quad u_0^k \in H^{s-1}(\Omega^k), \quad k,i,j = 1,2. \end{aligned} \tag{9}$$

3   Factorized Finite Difference Approximation

Let $n, m \in \mathbb{N}$, $n \ge 2$, $m \ge 1$, $h = 1/n$ and $\tau = T/m$. We consider the uniform spatial meshes $\bar{\omega}^k$ with mesh size h on $\bar{\Omega}^k$, $k = 1,2$, and the uniform temporal mesh $\bar{\omega}_\tau$ with mesh size τ on $[0,T]$. We also denote $\omega^k = \bar{\omega}^k \cap \Omega^k$, $\omega_\tau = \bar{\omega}_\tau \cap (0,T)$, $\omega_\tau^- = \bar{\omega}_\tau \cap [0,T)$, $\omega_\tau^+ = \bar{\omega}_\tau \cap (0,T]$, $\gamma^k = \bar{\omega}^k \cap \Gamma^k$, $\bar{\gamma}_{ij}^k = \bar{\omega}^k \cap \Gamma^k_{ij}$, $\gamma_{ij}^k = \{x \in \bar{\gamma}_{ij}^k : 0 < x_{3-i} < 1\}$, $\gamma_{ij}^{k-} = \{x \in \bar{\gamma}_{ij}^k : 0 \le x_{3-i} < 1\}$, $\gamma_{ij}^{k+} = \{x \in \bar{\gamma}_{ij}^k : 0 < x_{3-i} \le 1\}$, $\gamma_{ij}^{k\star} = \bar{\gamma}_{ij}^k \setminus \gamma_{ij}^k$, $\gamma_\star^k = \gamma^k \setminus \{\cup_{i,j} \gamma_{ij}^k\}$, $\sigma_{ij}^k = \gamma_{ij}^k \times \omega_\tau^-$, $\bar{\sigma}_{ij}^k = \bar{\gamma}_{ij}^k \times \omega_\tau^-$ and $\bar{Q}_{h\tau}^k = \bar{\omega}^k \times \bar{\omega}_\tau$, $i,j,k = 1,2$.

The finite difference operators are defined in the usual manner [11]:

$$v^k_{x_i} = \frac{(v^k)^{+i} - v^k}{h}, \quad v^k_{\bar{x}_i} = \frac{v^k - (v^k)^{-i}}{h}, \quad v^k_t = \frac{\hat{v}^k - v^k}{\tau}, \quad v^k_{\bar{t}} = \frac{v^k - \check{v}^k}{\tau},$$

where $(v^k)^{\pm i}(x,t) = v^k(x \pm he_i, t)$, e_i is the unit vector of the axis x_i, $\hat{v}^k(x,t) = v^k(x, t+\tau)$ and $\check{v}^k(x,t) = v^k(x, t-\tau)$, $i, k = 1, 2$.

We define the following discrete inner products and norms:

$$[v^k, w^k]_k = h^2 \sum_{x \in \omega^k} v^k(x)w^k(x) + \frac{h^2}{2} \sum_{x \in \gamma^k \backslash \gamma^k_*} v^k(x)w^k(x) + \frac{h^2}{4} \sum_{x \in \gamma^k_*} v^k(x)w^k(x),$$

$$[v^k, w^k]_{k,i} = h^2 \sum_{x \in \omega^k \cup \gamma^k_{i1}} v^k(x)w^k(x) + \frac{h^2}{2} \sum_{x \in \gamma^{k-}_{3-i,1} \cup \gamma^{k-}_{3-i,2}} v^k(x)w^k(x),$$

$$(v^k, w^k]_{k,i} = h^2 \sum_{x \in \omega^k \cup \gamma^k_{i2}} v^k(x)w^k(x) + \frac{h^2}{2} \sum_{x \in \gamma^{k+}_{3-i,1} \cup \gamma^{k+}_{3-i,2}} v^k(x)w^k(x),$$

$$\|[v^k]\|^2_k = [v^k, v^k]_k, \qquad \|[v^k]\|^2_{k,i} = [v^k, v^k]_{k,i}, \qquad \|v^k]\|^2_{k,i} = (v^k, v^k]_{k,i},$$

$$[v^k, w^k)_k = h^2 \sum_{x \in \omega^k \cup \gamma^{k-}_{11} \cup \gamma^{k-}_{21}} v^k(x)w^k(x), \qquad \|v^k]\|^2_k = [v^k, v^k)_k,$$

$$(v^k, w^k]_k = h^2 \sum_{x \in \omega^k \cup \gamma^{k+}_{12} \cup \gamma^{k+}_{22}} v^k(x)w^k(x), \qquad \|v^k]\|^2_k = (v^k, v^k]_k,$$

$$\|[v^k]\|^2_{H^1(\bar{\omega}^k)} = \|[v^k]\|^2_k + \|v^k_{x_1}]\|^2_{k,1} + \|v^k_{x_2}]\|^2_{k,2}, \qquad \|[v^k]\|_{C(\bar{\omega}^k)} = \max_{x \in \bar{\omega}^k} |v^k(x)|,$$

$$[v^k, w^k]_{\bar{\gamma}^k_{ij}} = h \sum_{x \in \gamma^k_{ij}} v^k(x)w^k(x) + \frac{h}{2} \sum_{x \in \gamma^{k*}_{ij}} v^k(x)w^k(x), \qquad \|[v^k]\|^2_{\bar{\gamma}^k_{ij}} = [v^k, v^k]_{\bar{\gamma}^k_{ij}},$$

$$[v^k, w^k)_{\gamma^{k-}_{ij}} = h \sum_{x \in \gamma^{k-}_{ij}} v^k(x)w^k(x), \qquad \|[v^k]\|^2_{\gamma^{k-}_{ij}} = [v^k, v^k)_{\gamma^{k-}_{ij}},$$

$$|v^k|^2_{H^{1/2}(\gamma^{k-}_{ij})} = h^2 \sum_{x, \, x' \in \gamma^{k-}_{ij}, \, x' \neq x} \left[\frac{v^k(x) - v^k(x')}{x_{3-i} - x'_{3-i}} \right]^2,$$

$$\|[v^k]\|^2_{H^{1/2}(\gamma^{k-}_{ij})} = |v^k|^2_{H^{1/2}(\gamma^{k-}_{ij})} + \|[v^k]\|^2_{\gamma^{k-}_{ij}},$$

$$\|[v^k]\|^2_{\ddot{H}^{1/2}(\gamma^{k-}_{ij})} = |v^k|^2_{H^{1/2}(\gamma^{k-}_{ij})} + h \sum_{x \in \gamma^{k-}_{ij}} \left(\frac{1}{x_{3-i} + h/2} + \frac{1}{1 - x_{3-i} - h/2} \right) |v^k(x)|^2,$$

$$\|v^k\|^2_\tau = \tau \sum_{t \in \omega^+_\tau} |v^k(t)|^2, \qquad \|[v^k]\|^2_{k,i,h\tau} = \tau \sum_{t \in \omega^+_\tau} \|[v^k(\cdot, t)]\|^2_{k,i},$$

$$\|v^k\|^2_{\sigma^k_{ij}} = \tau \sum_{t \in \omega^+_\tau} h \sum_{x \in \gamma^k_{ij}} |v^k(x,t)|^2, \qquad \|[v^k]\|^2_{\bar{\sigma}^k_{ij}} = \tau \sum_{t \in \omega^+_\tau} \|[v^k(\cdot, t)]\|^2_{\bar{\gamma}^k_{ij}},$$

$$\|[v^k\|^2_{L^2(\omega^+_\tau,\, \ddot{H}^{1/2}(\gamma^{k-}_{ij}))} = \tau \sum_{t\in\omega^+_\tau} \|[v^k(\cdot,t)\|^2_{\ddot{H}^{1/2}(\gamma^{k-}_{ij})},$$

$$|v^k|^2_{H^{1/2}(\bar\omega_\tau,\, L^2(\bar\omega^k))} = \tau^2 \sum_{t,\, t'\in\bar\omega_\tau,\ t'\neq t} \frac{\|[v^k(\cdot,t) - v^k(\cdot,t')\|^2_k}{(t-t')^2},$$

$$\|[v^k\|^2_{\ddot{H}^{1/2}(\bar\omega_\tau,\, L^2(\bar\omega^k))} = |v^k|^2_{H^{1/2}(\bar\omega_\tau,\, L^2(\bar\omega^k))} + \tau \sum_{t\in\bar\omega_\tau} \left(\frac{1}{t+\tau/2} + \frac{1}{T-t+\tau/2}\right)\|[v^k(\cdot,t)\|^2_k,$$

$$\|[v^k\|^2_{L^2(\omega^+_\tau,\, H^1(\bar\omega^k))} = \tau \sum_{t\in\omega^+_\tau} \|[v^k(\cdot,t)\|^2_{H^1(\bar\omega^k)},$$

$$\|[v^k\|^2_{H^{1,\,1/2}(Q^k_{h\tau})} = \|[v^k\|^2_{L^2(\omega^+_\tau,\, H^1(\bar\omega^k))} + |v^k|^2_{H^{1/2}(\bar\omega_\tau,\, L^2(\bar\omega^k))}.$$

For $v = (v^1, v^2)$ and $w = (w^1, w^2)$ we denote

$$[v, w] = [v^1, w^1]_1 + [v^2, w^2]_2, \qquad \|[v]\|^2 = [v, v],$$

$$\|[v]\|^2_{H^1_h} = \|[v^1]\|^2_{H^1(\bar\omega^1)} + \|[v^2]\|^2_{H^1(\bar\omega^2)},\ \|[v]\|^2_{H^{1,\,1/2}_{h\tau}} = \|[v^1]\|^2_{H^{1,\,1/2}(Q^1_{h\tau})} + \|[v^2]\|^2_{H^{1,\,1/2}(Q^2_{h\tau})}.$$

We also define the Steklov smoothing operators with the step sizes h and τ [12]:

$$T^+_i f^k(x,t) = \int_0^1 f^k(x + hx'_i e_i, t)\, \mathrm{d}x'_i = T^-_i f^k(x + he_i, t) = T_i f^k(x + 0.5he_i, t),$$

$$T^{2\pm}_i f^k(x,t) = 2\int_0^1 (1 - x'_i) f^k(x \pm hx'_i e_i, t)\, \mathrm{d}x'_i, \qquad i = 1, 2,$$

$$T^+_t f^k(x,t) = \int_0^1 f^k(x, t + \tau t')\, \mathrm{d}t' = T^-_t f^k(x, t + \tau) = T_t f^k(x, t + 0.5\tau).$$

We approximate the initial-boundary-value problem (1)–(6) with the following factorized finite difference scheme:

$$(I^1 + \sigma^1\tau\Lambda^{11})(I^1 + \sigma^1\tau\Lambda^{12})v^1_t + L^1_h v = \tilde f^1, \qquad v^1\big|_{t=0} = u^1_0, \tag{10}$$

$$(I^2 + \sigma^2\tau\Lambda^{21})(I^2 + \sigma^2\tau\Lambda^{22})v^2_t + L^2_h v = \tilde f^2, \qquad v^2\big|_{t=0} = u^2_0, \tag{11}$$

where

$$\Lambda^{k1}v^k = \begin{cases} -v^k_{x_1\bar x_1}, & x_1 \in \omega^k \cup \gamma^k_{21} \cup \gamma^k_{22}, \\ -\dfrac{2}{h}v^k_{x_1}, & x \in \bar\gamma^k_{11}, \\ \dfrac{2}{h}v^k_{\bar x_1}, & x \in \bar\gamma^k_{12}, \end{cases} \qquad \Lambda^{k2}v^k = \begin{cases} -v^k_{x_2\bar x_2}, & x_2 \in \omega^k \cup \gamma^k_{11} \cup \gamma^k_{12}, \\ -\dfrac{2}{h}v^k_{x_2}, & x \in \bar\gamma^k_{21}, \\ \dfrac{2}{h}v^k_{\bar x_2}, & x \in \bar\gamma^k_{22}, \end{cases}$$

$$L_h^1 v = \begin{cases} -\frac{1}{2}\sum\limits_{i,j=1}^{2}\left[\left(a_{ij}^1 v_{x_j}^1\right)_{\bar{x}_i} + \left(a_{ij}^1 v_{\bar{x}_j}^1\right)_{x_i}\right], & x \in \omega^1 \\[2mm] \frac{2}{h}\left[-\frac{a_{11}^1+(a_{11}^1)^{+1}}{2}v_{x_1}^1 - a_{12}^1\frac{v_{x_2}^1+v_{\bar{x}_2}^1}{2}+\tilde{\alpha}^1 v^1\right] - \left(a_{12}^1 v_{\bar{x}_2}^1\right)_{x_1} \\ \quad -\left(a_{21}^1 v_{x_1}^1\right)_{\bar{x}_2} - \frac{1}{2}\left(a_{22}^1 v_{x_2}^1\right)_{\bar{x}_2} - \frac{1}{2}\left(a_{22}^1 v_{\bar{x}_2}^1\right)_{x_2}, & x \in \gamma_{11}^1 \\[2mm] \frac{2}{h}\left[-\frac{a_{11}^1+(a_{11}^1)^{+1}}{2}v_{x_1}^1 - a_{12}^1 v_{x_2}^1 - a_{21}^1 v_{x_1}^1 - \frac{a_{22}^1+(a_{22}^1)^{+2}}{2}v_{x_2}^1\right. \\ \quad \left. +(\tilde{\alpha}_1^1+\tilde{\alpha}_2^1)v^1\right], & x = (-a-1,0) \\[2mm] \frac{2}{h}\left[\frac{a_{11}^1+(a_{11}^1)^{-1}}{2}v_{\bar{x}_1}^1 + a_{12}^1\frac{v_{x_2}^1+v_{\bar{x}_2}^1}{2}+\tilde{\alpha}^1 v^1\right. \\ \quad \left. -[\tilde{\beta}^1(x,\cdot),v^2(\cdot)]_{\bar{\gamma}_{11}^2}\right] - \left(a_{12}^1 v_{\bar{x}_2}^1\right)_{\bar{x}_1} \\ \quad -\left(a_{21}^1 v_{\bar{x}_1}^1\right)_{x_2} - \frac{1}{2}\left(a_{22}^1 v_{x_2}^1\right)_{\bar{x}_2} - \frac{1}{2}\left(a_{22}^1 v_{\bar{x}_2}^1\right)_{x_2}, & x \in \gamma_{12}^1 \\[2mm] \frac{2}{h}\left[\frac{a_{11}^1+(a_{11}^1)^{-1}}{2}v_{\bar{x}_1}^1 + a_{12}^1 v_{x_2}^1 - a_{21}^1 v_{\bar{x}_1}^1 - \frac{a_{22}^1+(a_{22}^1)^{+2}}{2}v_{x_2}^1\right. \\ \quad \left. +(\tilde{\alpha}_1^1+\tilde{\alpha}_2^1)v^1 - [\tilde{\beta}^1(x,\cdot),v^2(\cdot)]_{\bar{\gamma}_{11}^2}\right] \\ \quad -2\left(a_{12}^1 v_{x_2}^1\right)_{\bar{x}_1} - 2\left(a_{21}^1 v_{\bar{x}_1}^1\right)_{x_2}, & x = (-a,0) \end{cases}$$

and analogously at the other boundary nodes $\bar{\omega}^1$,

$L_h^2 v$ is defined in an analogous manner,

$$\tilde{f}^k = \begin{cases} T_1^2 T_2^2 T_t^+ f^k, & x \in \omega^k \\ T_i^{2\pm} T_{3-i}^2 T_t^+ f^k, & x \in \gamma_{i1}^k \,/\, x \in \gamma_{i2}^k \\ T_1^{2\pm} T_2^{2\pm} T_t^+ f^k, & x \in \gamma_\star^k \end{cases}$$

$$\tilde{\alpha}^k = T_{3-i}^2 \alpha^k, \qquad x \in \gamma_{i1}^k \cup \gamma_{i2}^k, \qquad \tilde{\alpha}_i^k = T_i^{2\pm}\alpha^k, \qquad x \in \gamma_\star^k, \qquad i = 1, 2$$

$$\tilde{\beta}^k = \begin{cases} T_2^2 \beta^k, & x \in \gamma_{1,3-k}^k, \\ T_2^{2\pm} \beta^k, & x \in \gamma_{1,3-k}^{k\star}. \end{cases}$$

I^k is identity operator, $I^k v^k = v^k$ and $\sigma^k \in [0,1]$ is a real parameter. For sufficiently large σ^k

$$\sigma^k \geq 2 \max_{i,j} \max_{x \in \bar{\Omega}^k} |a_{ij}^k(x)|, \quad i,j,k = 1,2$$

from the general theory of difference schemes (see [9,11,12]) follows that the scheme (10)–(11) is unconditionaly stable. This scheme is computationally efficient, because on each time level it can be resolved by two application of Thomas (Gauss) algorithm (algorithm requires $O(n^2)$ arithmetic operations).

4   Error Analysis and Convergence of Factorized Scheme

Let $u = (u^1, u^2)$ be the solution of the initial-boundary-value problem (1)–(6), and let $v = (v^1, v^2)$ denote the solution of the factorized scheme (10)–(11). The error $z = (z^1, z^2) = u - v$ is defined on $\bar{Q}^1_{h\tau} \times \bar{Q}^2_{h\tau}$ and satisfies the following conditions

$$(I^1 + \sigma^1 \tau \Lambda^{11})(I^1 + \sigma^1 \tau \Lambda^{12})z^1_t + L^1_h z = \tilde{\psi}^1 \tag{12}$$

$$(I^2 + \sigma^2 \tau \Lambda^{21})(I^2 + \sigma^2 \tau \Lambda^{22})z^2_t + L^2_h z = \tilde{\psi}^2 \tag{13}$$

where

$$
\tilde{\psi}^k = \begin{cases}
\psi^k + \vartheta^k_{1,\bar{x}_1} + \vartheta^k_{2,\bar{x}_2} + \vartheta^k_{3,\bar{x}_1\bar{x}_2}, & x \in \omega^k, \\[2mm]
\psi^k + \dfrac{2}{h}\vartheta^k_1 + \vartheta^k_{2,\bar{x}_2} + \dfrac{2}{h}\vartheta^k_{3,\bar{x}_2}, & x \in \gamma^k_{11}, \\[2mm]
\psi^k - \dfrac{2}{h}(\vartheta^k_1)^{-1} + \vartheta^k_{2,\bar{x}_2} - \dfrac{2}{h}(\vartheta^k_{3,\bar{x}_2})^{-1}, & x \in \gamma^k_{12}, \\[2mm]
\psi^k + \dfrac{2}{h}\vartheta^k_1 + \dfrac{2}{h}\vartheta^k_{2,\bar{x}_2} + \dfrac{4}{h^2}\vartheta^k_3, & x = \begin{cases} (-a-1,0), & k=1, \\ (a,0), & k=2, \end{cases}
\end{cases}
$$

and analogously at the other boundary nodes $\bar{\omega}^k$,

$$\vartheta^k_1 = \sigma^k \tau u^k_{x_1 t}, \quad \vartheta^k_2 = \sigma^k \tau u^k_{x_2 t}, \quad \vartheta^k_3 = -(\sigma^k)^2 \tau^2 u^k_{x_1 x_2 t} = -\sigma^k \tau \vartheta^k_{1,x_2} = -\sigma^k \tau \vartheta^k_{2,x_1}$$

and $k = 1, 2$.

$$
\psi^1 = \begin{cases}
\xi^1_t + \displaystyle\sum_{i,j=1}^{2} \eta^1_{ij,\bar{x}_i}, & x \in \omega^1, \\[3mm]
\tilde{\xi}^1_t + \dfrac{2}{h}\eta^1_{11} + \dfrac{2}{h}\eta^1_{12} + \tilde{\eta}^1_{21,\bar{x}_2} + \tilde{\eta}^1_{22,\bar{x}_2} + \dfrac{2}{h}\zeta^1, & x \in \gamma^1_{11}, \\[3mm]
\tilde{\tilde{\xi}}^1_t + \dfrac{2}{h}\left(\tilde{\eta}^1_{11} + \tilde{\eta}^1_{12} + \tilde{\eta}^1_{21} + \tilde{\eta}^1_{22} + \zeta^1_1 + \zeta^1_2\right), & x = (-a-1,0), \\[3mm]
\tilde{\xi}^1_t - \dfrac{2}{h}(\eta^1_{11})^{-1} - \dfrac{2}{h}(\eta^1_{12})^{-1} + \tilde{\eta}^1_{21,\bar{x}_2} + \tilde{\eta}^1_{22,\bar{x}_2} + \dfrac{2}{h}\zeta^1 + \dfrac{2}{h}\chi^1, & x \in \gamma^1_{12}, \\[3mm]
\tilde{\tilde{\xi}}^1_t + \dfrac{2}{h}\left[-(\tilde{\eta}^1_{11})^{-1} - (\tilde{\eta}^1_{12})^{-1} + \tilde{\eta}^1_{21} + \tilde{\eta}^1_{22} + \zeta^1_1 + \zeta^1_2 + \chi^1\right], & x = (-a,0), \\[3mm]
\end{cases}
$$
and analogously at the other boundary nodes $\bar{\omega}^1$

the term ψ^2 is defined in an analogous manner,

$$
\begin{aligned}
\xi^k &= u^k - T^2_1 T^2_2 u^k, & x \in \omega^k, \\
\tilde{\xi}^k &= u^k - T^{2\pm}_i T^2_{3-i} u^k, & x \in \gamma^k_{i1} \,/\, x \in \gamma^k_{i2}, \\
\tilde{\tilde{\xi}}^k &= u^k - T^{2\pm}_1 T^{2\pm}_2 u^k, & x \in \gamma^k_\star,
\end{aligned}
$$

$$\eta^k_{ij} = T^+_i T^2_{3-i} T^+_t \left(a^k_{ij} \frac{\partial u^k}{\partial x_j}\right) - \frac{1}{2}\left[a^k_{ij} u_{x_j} + (a^k_{ij})^{+i}(u^k_{\bar{x}_j})^{+i}\right], \qquad x \in \omega^k,$$

$$\tilde{\eta}^k_{ii} = T^+_i T^{2\pm}_{3-i} T^+_t \left(a^k_{ii} \frac{\partial u^k}{\partial x_i}\right) - \frac{a^k_{ii} + (a^k_{ii})^{+i}}{2} u^k_{x_i}, \qquad x \in \gamma^{k-}_{3-i,1} \,/\, x \in \gamma^{k-}_{3-i,2},$$

$$
\tilde{\eta}^k_{i,3-i} = \begin{cases}
T^+_i T^{2+}_{3-i} T^+_t \left(a^k_{i,3-i} \dfrac{\partial u^k}{\partial x_{3-i}}\right) - a^k_{i,3-i} u^k_{x_{3-i}}, & x \in \gamma^{k-}_{3-i,1}, \\[3mm]
T^+_i T^{2-}_{3-i} T^+_t \left(a^k_{i,3-i} \dfrac{\partial u^k}{\partial x_{3-i}}\right) - (a^k_{i,3-i})^{+i} (u^k_{\bar{x}_{3-i}})^{+i}, & x \in \gamma^{k-}_{3-i,2},
\end{cases}
$$

$$\zeta^k = (T_i^2 \alpha^k) u^k - T_i^2 T_t^+ (\alpha^k u^k), \qquad x \in \gamma_{3-i,1}^k \cup \gamma_{3-i,2}^k,$$
$$\zeta_i^k = (T_i^{2\pm} \alpha^k) u^k - T_i^{2\pm} T_t^+ (\alpha^k u^k), \qquad x \in \gamma_\star^k,$$
$$\chi^k = \int_{I_{1k}^{3-k}} T_2^2 \beta^k(x, x') T_t^+ u^{3-k}(x', t) \, d\Gamma_{1k}^{3-k}$$
$$\qquad - h \sum_{x' \in \bar{\gamma}_{1k}^{3-k}}' T_2^2 \beta^k(x, x') u^{3-k}(x', t), \qquad x \in \gamma_{1,3-k}^k,$$
$$\chi^k = \int_{\Gamma_{1k}^{3-k}} T_2^{2\pm} \beta^k(x, x') T_t^+ u^{3-k}(x', t) \, d\Gamma_{1k}^{3-k}$$
$$\qquad - h \sum_{x' \in \bar{\gamma}_{1k}^{3-k}}' T_2^{2\pm} \beta^k(x, x') u^{3-k}(x', t), \qquad x \in \gamma_{1,3-k}^{k\star}.$$

Under the assumption $\tau \asymp h^2$, factorized finite difference scheme (12)–(13) satisfies the following a priori estimate

$$\|[z]\|_{H_{h\tau}^{1,1/2}} \le C \sum_{k=1}^2 \left\{ \|[\xi^k]\|_{\dot{H}^{1/2}(\bar{\omega}_\tau, L_2(\bar{\omega}^k))} + \sum_{i,j=1}^2 \left(\|[\eta_{ij}^k\|_{k,i,h\tau} + \|\zeta^k\|_{\sigma_{ij}^k} \right) + \|\chi^k\|_{\sigma_{1,3-k}^k} \right.$$

$$+ h \sum_{i,j,l=1}^2 \|[\bar{\eta}_{ij}^k\|_{L_2(\omega_\tau^-, \dot{H}^{1/2}(\gamma_{3-i,l}^{k-}))} + h \sum_{i,j=1}^2 \|[\lambda_i^k\|_{L_2(\omega_\tau^-, \dot{H}^{1/2}(\gamma_{ij}^{k-}))}$$

$$+ h|[\kappa^k]\|_{\bar{\sigma}_{1,3-k}^k} + h \sum_{i,j=1}^2 \left(\|[\mu_i^k]\|_{\bar{\sigma}_{ij}^k} + \|[\nu_i^k]\|_{\bar{\sigma}_{ij}^k} \right) \qquad (14)$$

$$+ h \sqrt{\log \frac{1}{h}} \sum_{i=1}^2 \sum_{x \in \gamma_\star^k} \left(\|\zeta_i^k(x, \cdot)\|_\tau + \|\bar{\lambda}_i^k(x, \cdot)\|_\tau \right) + \sum_{i=1}^2 \|[\vartheta_i^k\|_{k,i,h\tau} \right\}.$$

The problem of deriving the convergence rate estimate for the factorized scheme (10)–(11) is reduced to estimating the right-hand side terms in the inequality (14).

When $s > 2$, ϑ_i^k is a bounded bilinear functional of $u^k \in W_2^{s,s/2}(\Omega^k)$, which vanishes when $u^k = 1, x_1, x_2, t, x_1^2, x_1 x_2, x_2^2$. Using Bramble-Hilbert lemma (see [2]), we obtain the following result

$$\|[\vartheta_i^k\|_{k,i,h\tau} \le C \|u^k\|_{W_2^{s,s/2}(\Omega^k)}, \qquad 2 < s \le 3.$$

The estimates for the other terms in the inequality (14) see in [3,7–9].

Theorem 1. *Let $\tau \asymp h^2$ and the condition (9) satisfied. Then the solution of factorized finite difference scheme (10)–(11) converges in the norm $H_{h\tau}^{1,1/2}$ to the solution of initial-boundary-value problem (1)–(6) and the convergence rate estimates*

$$\|[u - v]\|_{H_{h\tau}^{1,1/2}} \le Ch^{s-1} \sqrt{\log \frac{1}{h}} \left(1 + \max_{i,j,k} \|a_{ij}^k\|_{H^{s-1}(\Omega^k)} + \max_{i,j,k} \|\alpha^k\|_{H^{s-3/2}(\Gamma^k_{ij})} \right.$$
$$\left. + \max_k \|\beta^k\|_{H^{s-1}(\Delta^k)} \right) \|u\|_{H^{s,s/2}}, \qquad 2.5 < s < 3$$

and

$$\|[u-v]\|_{H^{1,1/2}_{h\tau}} \leq Ch^2 \left(\log \frac{1}{h}\right)^{3/2} \left(1 + \max_{i,j,k} \|a^k_{ij}\|_{H^2(\Omega^k)} + \max_{i,j,k} \|\alpha^k\|_{H^{3/2}(\Gamma^k_{ij})} \right. $$
$$\left. + \max_{k} \|\beta^k\|_{H^2(\Delta^k)}\right)\|u\|_{H^{3,3/2}}, \qquad s = 3$$

hold.

Acknowledgement. The research of authors was supported by Ministry of Education, Science and Technological Development of Republic of Serbia under project 174015.

References

1. Amosov, A.A.: Global solvability of a nonlinear nonstationary problem with a nonlocal boundary condition of radiation heat transfer type. Differ. Equ. **41**(1), 96–109 (2005)
2. Bramble, J.H., Hilbert, S.R.: Bounds for a class of linear functionals with application to the Hermite interpolation. Numer. Math. **16**, 362–369 (1971)
3. Jovanović, B.S.: Finite difference method for boundary value problems with weak solutions. Posebna izdanja Mat. Instituta 16, Belgrade (1993)
4. Jovanović, B.S., Koleva, M.N., Vulkov, L.G.: Convergence of a FEM and two-grid algorithms for elliptic problems on disjoint domains. J. Comput. Appl. Math. **236**, 364–374 (2011)
5. Jovanović, B.S., Vulkov, L.G.: Finite difference approximation of strong solutions of a parabolic interface problem on disconected domains. Publ. Inst. Math. **84**(98), 37–48 (2008)
6. Jovanović, B.S., Vulkov, L.G.: Numerical solution of a two-dimensional parabolic transmission problem. Int. J. Numer. Anal. Model. **7**(1), 156–172 (2010)
7. Jovanović, B.S., Milovanović, Z.: Finite difference approximation of a parabolic problem with variable coefficients. Publ. Inst. Math. **95**(109), 49–62 (2014)
8. Jovanović, B.S., Milovanović, Z.: Numerical approximation of 2D parabolic transmission problem in disjoint domains. Appl. Math. Comput. **228**, 508–519 (2014)
9. Jovanović, B.S., Süli, E.: Analysis of Finite Difference Schemes. Springer Series in Computational Mathematics, vol. 46. Springer, Heidelberg (2013)
10. Milovanović, Z.: Finite difference scheme for a parabolic transmission problem in disjoint domains. In: Dimov, I., Faragó, I., Vulkov, L. (eds.) NAA 2012. LNCS, vol. 8236, pp. 403–410. Springer, Heidelberg (2013). doi:10.1007/978-3-642-41515-9_45
11. Samarskii, A.A.: The Theory of Difference Schemes. Marcel Dekker, New York (2001)
12. Samarskii, A.A., Lazarov, R.D., Makarov, V.L.: Difference Schemes for Differential Equations with Generalized Solutions. Vyshaya Shkola, Moscow (1987). (in Russian)

Compound Log-Series Distribution with Negative Multinomial Summands

Pavlina Jordanova[1]($\boxtimes$), Monika P. Petkova[2], and Milan Stehlík[3,4]

[1] Faculty of Mathematics and Informatics, Shumen University,
115 Universitetska Str., 9712 Shumen, Bulgaria
pavlina_kj@abv.bg
[2] Faculty of Mathematics and Informatics, Sofia University,
5 "James Bourchier" Blvd., 1164 Sofia, Bulgaria
monikapetevapetkova@abv.bg
[3] Institute of Statistics, Universidad de Valparaíso, Valparaíso, Chile
[4] Department of Applied Statistics, Johannes Kepler University,
Altenbergerstrasse 69, 4040 Linz, Austria
mlnstehlik@gmail.com

Abstract. The paper presents the first full characterization of multivariate random sum with one and the same Logarithmic Series number of summands in each coordinate. The summands with equal indexes in any coordinate are Negative Multinomially distributed. We show that considered as a mixture, the resulting distribution coincides with Mixed Negative Multinomial distribution with scale changed Logarithmic Series distributed first parameter.

Keywords: Compound distributions · Mixed distributions · Negative multinomial distribution · Logarithmic series distributions

1 Introduction

Negative multinomial (NMn) distribution is a generalization of multivariate geometric distribution. It can be interpreted as the distribution of the numbers of outcomes A_i, $i = 1, 2, ..., k$ before the n-th realization of the event $B = \overline{A}_1 \cap \overline{A}_2 \cap ... \cap \overline{A}_k$ happen, in a series of independent repetitions of a trial. Here A_i, $i = 1, 2, ..., k$ and B are assumed to form a partition of the sample space. This distribution seems to be introduced by Bates and Neyman [1] and Wishart [15]. The best estimators of its parameters are obtained by Tweedie [14]. Sibuya et al. [12] makes clear the probability structure of the individual distributions and describe its relation with Binomial distib. The next definition and some properties of NMn distribution can be found e.g. in Johnson et al. [4].

Definition 1. Let $n \in \mathbb{N}$, $0 < p_i$, $i = 1, 2, ..., k$ and $p_1 + p_2 + ... + p_k < 1$. A random vector $(\xi_1, \xi_2, ..., \xi_k)$ is called Negative multinomially distributed with parameters $n, p_1, p_2, ..., p_k$, if its probability mass function (p.m.f.) is

© Springer International Publishing AG 2017
I. Dimov et al. (Eds.): NAA 2016, LNCS 10187, pp. 383–390, 2017.
DOI: 10.1007/978-3-319-57099-0_42

$$P(\xi_1 = i_1, \xi_2 = i_2, ..., \xi_k = i_k)$$

$$= \binom{n + i_1 + i_2 + ... i_k - 1}{i_1, i_2, ..., i_k, n - 1} p_1^{i_1} p_2^{i_2} ... p_k^{i_k} (1 - p_1 - p_2 - ... - p_k)^n,$$

$i_s = 0, 1, ..., s = 1, 2, ..., k$. Briefly $(\xi_1, \xi_2, ..., \xi_k) \sim NMn(n; p_1, p_2, ..., p_k)$.
For $m = 2, 3, ..., k - 1$, its finite dimensional distributions (f.d.ds) are,

$$(\xi_{i_1}, \xi_{i_2}, ..., \xi_{i_m}) \sim NMn(n; \rho_{i_1}, \rho_{i_2}, ..., \rho_{i_m}),$$

with $\rho_{i_s} = \frac{p_{i_s}}{1 - \sum_{j \notin \{i_1, i_2, ..., i_m\}} p_j}$, $s = 1, 2, ..., m$. For $k = 1$, we obtain a version of
Negative Binomial distribution that we will use along the paper. For $n = 1$ the
NMn distribution is Multivariate geometric distribution.

As we mentioned any coordinate in the vector that we consider can be presented as sum of M independent identically distributed (i.i.d.) random variables (r.v.) where M is Logarithmic Series (LS) distributed. Fisher et al. [2] and Kendall [7] seems to be the first who have investigated this distribution. This distribution is closely related with NMn distribution. As Fisher notes, the Compound Poisson distribution with LS i.i.d. summands coincides with Negative Binomial distribution. Philippou et al. [11] prove multivariate version of this property. Let us remind the definition about LS distribution.

Definition 2. A r.v. X is Logarithmic series distributed with parameter $p_L \in (0, 1)$, if it has p.m.f.

$$P(X = n) = \frac{p_L^n}{-n \log(1 - p_L)}, \quad n = 1, 2, ... \tag{1}$$

Briefly $X \sim LS(p_L)$.

If $X \sim LS(p_L)$, then $X - 1$ has Kemp Generalized Hypergeometric probability distribution. See [5]. We use the following multivariate version introduced by Patil [10] and Khatri [8] and considered by Johnson et al. [4].

Definition 3. Let the coefficients in the power series

$$A(x_1, x_2, ..., x_k) = \sum_{i_1=0}^{\infty} ... \sum_{i_k=0}^{\infty} a_{(i_1, i_2, ..., i_k)} x_1^{i_1} x_2^{i_2} ... x_k^{i_k}, \quad (x_1, x_2, ..., x_k) \in \mathfrak{A}_A$$

be non-negative, independent on $(x_1, x_2, ..., x_k)$ and for all $(\theta_1, \theta_2, ..., \theta_k) \in \Theta_k$, $A(\theta_1, \theta_2, ..., \theta_k) < \infty$. We call the distribution of a random vector $\boldsymbol{X} = (X_1, X_2, ..., X_k)$ with p.m.f.

$$P(X_1 = n_1, X_2 = n_2, ..., X_k = n_k)$$

$$= \frac{a_{(n_1, n_2, ..., n_k)} \theta_1^{n_1} \theta_2^{n_2} ... \theta_k^{n_k}}{A(\theta_1, \theta_2, ..., \theta_k)}, \quad n_1, n_2, ..., n_k = 0, 1, ...,$$

Multivariate Power Series Distribution w.r.t. $(\theta_1, \theta_2, ..., \theta_k) \in \Theta_k$ with parameters $A(\boldsymbol{x})$ and $\boldsymbol{\theta} = (\theta_1, \theta_2, ..., \theta_k)$. Briefly $\boldsymbol{X} \sim MPS(A(\boldsymbol{x}), \boldsymbol{\theta} \in \Theta_k)$.

Along the paper $k = 2, 3, ...$, is fixed. $\overset{d}{=}$ means coincidence in distribution, "$\sim$" is for the fact that a r.v. belongs to a given class of distributions, $G_{\xi_1, \xi_2, ..., \xi_k}(z_1, z_2, ..., z_k)$ is for the joint p.g.f. of a random vector $(\xi_1, \xi_2, ..., \xi_k)$ and $FI\xi$ means the index of dispersion of the r.v. ξ. Analogously to the well known Compound Poisson distribution, under "compounding" distribution we understand the one of the number of summand in a sum of i.i.d. r.vs. Here multivariate Compound Log-series Distribution with Negative Multinomial Summands (CLSNMn) is defined and its conditional distributions, main properties and numerical characteristics are obtained. We show that this distribution is a particular case of MPS distributions and find the explicit form of its parameters. Considered as a mixture, this distribution would be Mixed NMn distribution with possibly scale changed LS distributed first parameter. The paper finishes with some conclusive remarks.

With respect to the recursive formulae these distributions are considered in [13]. For the case with MPS distributed summands see [3] and for NMn summands see [6]. The case $n = 1$ and $k = 2$ is partially investigated in [9].

2   Definition and Main Properties

Along this section $n = 1, 2, ..., p_L, \pi_1, \pi_2, ..., \pi_k \in (0, 1)$ and $\pi_0 := 1 - \pi_1 - \pi_2 - ... - \pi_k \in (0, 1)$.

Definition 4. A random vector $X = (X_1, X_2, ..., X_k)$ is called Compound Log-series Distributed with Negative Multinomial summands and with parameters $p_L; n; \pi_1, ..., \pi_k$, if for $m_1, m_2, ..., m_k = 0, 1, 2, ...$,

$$P(X_1 = m_1, X_2 = m_2, ... X_k = m_k)$$
$$= \frac{\pi_1^{m_1} \pi_2^{m_2} ... \pi_k^{m_k}}{-log(1 - p_L)} \sum_{j=1}^{\infty} \frac{p_L^j}{j} \binom{jn + m_1 + m_2 + ... m_k - 1}{m_1, m_2, ..., m_k, jn - 1} \pi_0^{nj}, \quad (2)$$

Briefly $X \sim CLSNMn(p_L; n, \pi_1, \pi_2, ..., \pi_k)$.
In particular

$$P(X_1 = 0, X_2 = 0, ... X_k = 0) = \frac{log(1 - p_L \pi_0^n)}{log(1 - p_L)}.$$

This distribution possess the following properties.

Theorem 1. *If* $X \sim CLSNMn(p_L; n, \pi_1, \pi_2, ..., \pi_k)$, *then*

1. $X \sim MPSD(A(\boldsymbol{x}), \boldsymbol{\theta})$, *where* $\boldsymbol{\theta} = (\pi_1, \pi_2, ..., \pi_k)$, $a_{(0,....0)} = -log(1 - \theta \pi_0^n)$,

$$a_{(i_1, i_2, ..., i_k)}$$
$$= \sum_{j=1}^{\infty} \frac{p_L^j \pi_0^{nj}}{j} \binom{jn + i_1 + i_2 + ... i_k - 1}{i_1, i_2, ..., i_k, jn - 1}, \quad (i_1, i_2, ..., i_k) \neq (0, 0, ..., 0),$$

$$A(x_1, x_2, ..., x_k) = -log\left\{ 1 - p_L \frac{\pi_0^n}{[1 - (x_1 + x_2 + ... + x_k)]^n} \right\}.$$

2. For $|\pi_1 z_1 + \pi_2 z_2 + ... + \pi_k z_k| < 1$,

$$G_{X_1,X_2,...,X_k}(z_1, z_2, ..., z_k) = \frac{\log\left[1 - p_L\left(\frac{\pi_0}{1-(\pi_1 z_1+\pi_2 z_2+...+\pi_k z_k)}\right)^n\right]}{\log(1-p_L)}.$$

3. For all $r = 2, 3, ..., k$,

$$(X_{i_1}, X_{i_2}, ..., X_{i_r}) \sim CLSNMn(p_L; n, \frac{\pi_{i_1}}{\pi_0 + \pi_{i_1} + \pi_{i_2} + ... + \pi_{i_r}},$$

$$\frac{\pi_{i_2}}{\pi_0 + \pi_{i_1} + \pi_{i_2} + ... + \pi_{i_r}}, ..., \frac{\pi_{i_r}}{\pi_0 + \pi_{i_1} + \pi_{i_2} + ... + \pi_{i_r}}).$$

4. For $i = 1, 2, ..., k$, $X_i \sim CLSNBi(p_L; n, \frac{\pi_i}{\pi_0 + \pi_i})$,

$$G_{X_i}(z_i) = \frac{\log\left[1 - p_L\left(\frac{\pi_0}{\pi_0 + \pi_i - \pi_i z_i}\right)^n\right]}{\log(1-p_L)}, |\pi_i z_i| < \pi_0 + \pi_i,$$

$$EX_i = \frac{-np_L\pi_i}{\pi_0(1-p_L)\log(1-p_L)},$$

$$VarX_i = \frac{-n\,\pi_i p_L}{\pi_0(1-p_L)\log(1-p_L)}\left[\frac{\pi_i}{\pi_0}\left(n\frac{p_L+\log(1-p_L)}{(1-p_L)\log(1-p_L)}+1\right)+1\right].$$

$$FIX_i = 1 + \frac{\pi_i}{\pi_0}\left(n\frac{p_L+\log(1-p_L)}{(1-p_L)\log(1-p_L)}+1\right).$$

5. For $i \neq j = 1, 2, ..., k$, $(X_i, X_j) \sim CLSNMn(p_L; n, \frac{\pi_i}{\pi_0+\pi_i+\pi_j}, \frac{\pi_j}{\pi_0+\pi_i+\pi_j})$.
   For $|\pi_i z_i + \pi_j z_j| < \pi_0 + \pi_i + \pi_j$,

$$G_{X_i,X_j}(z_i, z_j) = \frac{\log\left[1 - p_L\left(\frac{\pi_0}{\pi_0+(1-z_i)\pi_i+(1-z_j)\pi_j}\right)^n\right]}{\log(1-p_L)},$$

$$cov(X_i, X_j) = \frac{-n\,\pi_i\pi_j p_L}{\pi_0^2(1-p_L)\log(1-p_L)}\left[n\frac{p_L+\log(1-p_L)}{(1-p_L)\log(1-p_L)}+1\right],$$

$$cor(X_i, X_j) = \sqrt{\frac{(FIX_i-1)(FIX_j-1)}{FIX_i\,FIX_j}}.$$

6. For $i, j = 1, 2, ..., k$ $j \neq i$,
   (a) For $m_j \neq 0$ and $m_i = 0, 1, ...$,

$$P(X_i = m_i | X_j = m_j) = \binom{m_i + m_j}{m_i}.$$

$$\frac{\pi_i^{m_i}(\pi_0+\pi_j)^{m_j}}{(\pi_0+\pi_i+\pi_j)^{m_i+m_j}}\frac{\sum_{s=1}^{\infty}\frac{p_L^s}{s}\binom{sn+m_i+m_j-1}{sn-1}\left(\frac{\pi_0}{\pi_0+\pi_i+\pi_j}\right)^{ns}}{\sum_{s=1}^{\infty}\frac{p_L^s}{s}\binom{sn+m_j-1}{sn-1}\left(\frac{\pi_0}{\pi_0+\pi_j}\right)^{ns}}.$$

In particular,

$$(X_i|X_j = 0) \sim CLSNMn \left[p_L \left(\frac{\pi_0}{\pi_0 + \pi_j} \right)^n ; n, \frac{\pi_i}{\pi_0 + \pi_i + \pi_j} \right]$$

$$P(X_i = m_i | X_j = 0)$$

$$= \frac{\pi_i^{m_i}}{(\pi_0 + \pi_i + \pi_j)^{m_i}} \frac{\sum_{s=1}^{\infty} \frac{p_L^s}{s} \binom{sn + m_i - 1}{sn - 1} \left(\frac{\pi_0}{\pi_0 + \pi_i + \pi_j} \right)^{ns}}{-log \left[1 - p_L \left(\frac{\pi_0}{\pi_0 + \pi_j} \right)^n \right]}.$$

$$P(X_i = 0 | X_j = 0) = \frac{log \left[1 - p_L \left(\frac{\pi_0}{\pi_0 + \pi_i + \pi_j} \right)^n \right]}{log \left[1 - p_L \left(\frac{\pi_0}{\pi_0 + \pi_j} \right)^n \right]}.$$

(b) For $i, j = 1, 2, ..., k$ $j \neq i$, $m_j = 0, 1, ...$

$$E(z_i^{X_i} | X_j = m_j) =$$

$$\frac{(\pi_0 + \pi_j)^{m_j}}{(\pi_0 + \pi_i + \pi_j - z_i \pi_i)^{m_j}} \frac{\sum_{s=1}^{\infty} \frac{p_L^s}{s} \frac{(sn + m_j - 1)!}{(sn - 1)!} \left(\frac{\pi_0}{\pi_0 + \pi_j + (1 - z_i) \pi_i} \right)^{ns}}{\sum_{s=1}^{\infty} \frac{p_L^s}{s} \frac{(sn + m_j - 1)!}{(sn - 1)!} \left(\frac{\pi_0}{\pi_0 + \pi_j} \right)^{ns}}.$$

In particular,

$$E(z_i^{X_i} | X_j = 0) = \frac{log \left[1 - p_L \left(\frac{\pi_0}{\pi_0 + \pi_j + (1 - z_i) \pi_i} \right)^n \right]}{log \left[1 - p_L \left(\frac{\pi_0}{\pi_0 + \pi_j} \right)^n \right]}.$$

(c)

$$E(X_i | X_j = z_j) = \frac{\pi_i}{\pi_0 + \pi_j} \frac{\sum_{m=1}^{\infty} \frac{p_L^m}{m} \frac{(mn + z_j)!}{(mn - 1)!} \left(\frac{\pi_0}{\pi_0 + \pi_j} \right)^{mn}}{\sum_{m=1}^{\infty} \frac{p_L^m}{m} \frac{(mn + z_j - 1)!}{(mn - 1)!} \left(\frac{\pi_0}{\pi_0 + \pi_j} \right)^{mn}}, \quad z_j = 1, 2, ...$$

$$E(X_i | X_j = z_j) = \frac{\pi_i}{\pi_j} \frac{P(X_j = z_j + 1)}{P(X_j = z_j)} (z_j + 1), \tag{3}$$

$$E(X_i | X_j = 0) = \frac{-n p_L \pi_0^n \pi_i}{(\pi_0 + \pi_j)^{n+1} \left[1 - p_L \left(\frac{\pi_0}{\pi_0 + \pi_j} \right)^n \right] log \left[1 - p_L \left(\frac{\pi_0}{\pi_0 + \pi_j} \right)^n \right]}.$$

7. $X_1 + X_2 + ... + X_k \sim CLSNMn(p_L; n, 1 - \pi_0)$

8. For $i = 1, 2, ..., k$

$$(X_i, X_1 + X_2 + ... + X_k - X_i) \sim CLSNMn(p_L; n, \pi_i, 1 - \pi_0 - \pi_i).$$

9. For $i = 1, 2, ..., k$, $m \in \mathbb{N}$, $(X_i | X_1 + X_2 + ... + X_k = m) \sim Bi(m, \frac{\pi_i}{1 - \pi_0})$.

10. *There exists a probability space* $(\Omega, \mathcal{A}, \mathbb{P})$, *a r.v.* $M \sim LS(p_L)$ *and a random vector* $\boldsymbol{Y} = (Y_1, Y_2, ..., Y_k)$ *defined on it, such that*

$$\boldsymbol{Y}|M = m \sim NMn(nm, \pi_1, \pi_2, ..., \pi_k), \ m = 1, 2, ..$$

and $\boldsymbol{X} \overset{d}{=} \boldsymbol{Y}$.

Note: The last statement in the previous theorem means that

$$CLSNMn(p_L; n, \pi_1, \pi_2, ..., \pi_k) \overset{d}{=} NMn(nM, \pi_1, \pi_2, ..., \pi_k) \underset{M}{\wedge} LS(p_L),$$

where we have used the notations introduced in Johnson et al. [4]. Therefore considered as a Mixture, CLSNMn distribution would be Mixed NMn distribution with scale changed LS first parameter.

The name of CLSNMn distribution comes from the following representation.

Theorem 2. *Let* $M \sim LS(p_L)$ *and* $(Y^{(1)}, ..., Y^{(k)}) \sim NMn(n; \pi_1, ..., \pi_k)$ *be independent. Define a random vector* $(T_M^{(1)}, T_M^{(2)}, ..., T_M^{(k)})$ *by*

$$T_M^{(j)} = \sum_{i=1}^{M} Y_i^{(j)}, j = 1, 2, ..., k.$$

Then

1. *For* $m \in \mathbb{N}$, $(T_M^{(1)}, T_M^{(2)}, ..., T_M^{(k)}|M = m) \sim NMn(nm; \pi_1, \pi_2, ..., \pi_k)$;
2. $(T_M^{(1)}, T_M^{(2)}, ..., T_M^{(k)}) \sim CLSNMn(p_L; n, \pi_1, \pi_2, ..., \pi_k)$.

If we would like to work with the corresponding conditional distribution given that no one of the coordinates is equal to 0, we need the following definition.

Definition 5. A random vector $\boldsymbol{X} = (X_1, X_2, ..., X_k)$ is called Compound Logarithmic series distributed with Negative Multinomial summands on $\mathbb{N}^k$ and with parameters $p_L; n, \pi_1, ..., \pi_k$, if for $(m_1, m_2, ..., m_k) \in \mathbb{N}^k$,

$$P(X_1 = m_1, X_2 = m_2, ...X_k = m_k)$$

$$= \frac{1}{\rho} \frac{\pi_1^{m_1} \pi_2^{m_2} ... \pi_k^{m_k}}{-log(1 - p_L \pi_0^n)} \sum_{j=1}^{\infty} \frac{(p_L \pi_0^n)^j}{j} \binom{jn + m_1 + m_2 + ...m_k - 1}{m_1, m_2, ..., m_k, jn - 1},$$

$$\rho = 1 - \sum_{m=1}^{k} (-1)^{m+1} \sum_{1 \leq i_1 < i_2 < ... < i_m \leq k} \frac{log\left[1 - p_L \pi_0^n (\pi_0 + \pi_{i_1} + ... + \pi_{i_m})^{-n}\right]}{log(1 - p_L)}.$$

Briefly $\boldsymbol{X} \sim CLSNMn_{\mathbb{N}^k}(p_L; n, \pi_1, \pi_2, ..., \pi_k)$.

Note: It is easy to see that, if $\boldsymbol{X} \sim CLSNMn(p_N; n, \pi_1, \pi_2, ..., \pi_k)$, then

$$(X_1, X_2, ..., X_k|X_1 \neq 0, X_2 \neq 0, ..., X_k \neq 0) \sim CLSNMn_{\mathbb{N}^k}(p_L; n, \pi_1, \pi_2, ..., \pi_k).$$

3   Conclusions

In this paper we have defined and have made the first full characterization of CLSNMn distribution.

✓ The name of the distribution comes from the fact that any random vector with such type can be presented as multivariate random sum with one and the same LS number of summands in any coordinate and NMn summands with equal indexes in different coordinates.

✓ This distribution is a particular case on MPS distributions, considered in [4].

✓ CLSNMn coincides with Mixed NMn with scale changed LS first parameter

$$CLSNMn(p_L; n, \pi_1, \pi_2, ..., \pi_k) \overset{d}{=} NMn(nM, \pi_1, \pi_2, ..., \pi_k) \underset{M}{\wedge} LS(p_L),$$

✓ Under our settings, it is easy to see that for any $CLSNMn(p_L; n, \pi_1, \pi_2, ..., \pi_k)$,

$$\frac{p_L + log(1 - p_L)}{(1 - p_L)log(1 - p_L)} > 0. \tag{4}$$

Therefore Theorem 1, 4) implies that the Fisher index of dispersion of anyone of the coordinates increases, when n increases. Now having in mind Theorem 1, 5), we can conclude that the bigger the n, the bigger the correlation between the coordinates is.

✓ The Fisher indexes of dispersion and the correlation between the coordinates increases with $p_L \in (0, 1)$.

✓ It can be proven that (3) is true also if the LS distribution is replaced by arbitrary univariate Power series distribution. Due to the fact that the fraction $\frac{P(X_j = z_j + 1)}{P(X_j = z_j)}$ very fast approaches 1, when $z_j = 1, 2, ...$ increases, the mean square regression of these distributions become linear. In order to visualize this relation we have simulated m realizations of $(X_1, X_2) \sim CLSMNn(p_L; n, \pi_1, \pi_2)$ and plotted their scatter plots. Two of them, for $m = 100$ and different parameters p_L, n, π_1 and π_2 are given on Fig. 1.

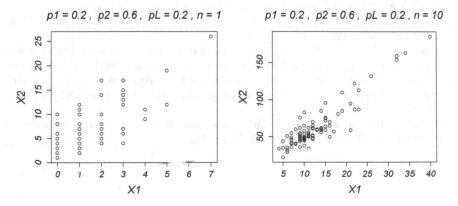

Fig. 1. Scatterplot of X_1 and X_2 for different parameters p_L, n, π_1 and π_2, $m = 100$.

Acknowledgements. This work is partially supported by project Fondecyt Proyecto Regular No. 1151441, the Project RD-08-69/02.02.2016 from the Scientific Research Fund in Konstantin Preslavsky University of Shumen, Bulgaria and by the financial funds allocated to the Sofia University St. Kliment Ohridski, Bulgaria, grant No. 197/13.04.2016.

References

1. Bates, G.E., Neyman, J.: Contributions to the Theory of Accident Proneness. 1. An Optimistic Model of the Correlation Between Light and Severe Accidents, vol. 132. California University, Berkeley (1952)
2. Fisher, R.A., Corbet, A.S., Williams, C.B.: The relation between the number of species and the number of individuals in a random sample of an animal population. J. Animal Ecol. 42–58 (1943)
3. Jose, K.K., Jacob, S.: Type II Bivariate Generalized Power Series Poisson Distribution and its Applications in Risk Analysis (2016)
4. Johnson, N.L., Kotz, S., Balakrishnan, N.: Discrete Multivariate Distributions, vol. 165. Wiley, New York (1997)
5. Johnson, N.L., Kotz, S., Kemp, A.W.: Univariate Discrete Distributions, vol. 444. Wiley, New York (2005)
6. Jordanova, P.K., Petkova, M.M., Stehlik, M.: Compound Power Series Distribution with Negative Multinomial Summands (2016). Submitted
7. Kendall, D.G.: On some modes of population growth leading to RA Fisher's logarithmic series distribution. Biometrika **35**(1/2), 6–15 (1948)
8. Khatri, C.G.: On certain properties of power-series distributions. Biometrika **46**(3/4), 486–490 (1959)
9. Kostadinova, K., Minkova, L.: Type II family of bivariate inflated-parameter generalized power series distributions. Serdica Math. J. **42**(1), 27–42 (2016)
10. Patil, G.P.: On multivariate generalized power series distribution and its applications to the multinomial and negative multinomial. In: Proceedings of International Symposium at McGill University on Classical and Contageous Discrete Distributions, 15–20 August 1963, pp. 183–194. Statistical Publishing Society/Pergamon Press, Calcutta/Oxford (1965)
11. Philippou, A.N., Roussas, G.G.: A note on the multivariate logarithmic series distribution. Commun. Stat. Theory Methods **3**(5), 469–472 (1974)
12. Sibuya, M., Yoshimura, I., Shimizu, R.: Negative multinomial distribution. Ann. Inst. Stat. Math. **16**(1), 409–426 (1964)
13. Sundt, B., Vernic, R.: Recursions for Convolutions and Compound Distributions with Insurance Applications. Springer Science and Business Media, Berlin (2009)
14. Tweedie, M.C.K.: The estimation of parameters from sequentially sampled data on a discrete distribution, series B (methodological). J. Roy. Stat. Soc. **14**(2), 238–245 (1952)
15. Wishart, J.: Cumulants of multivariate multinomial distribution. series B (methodological). Biometrika **36**, 47–58 (1949)

Inverse Problems of Determination of the Right-Hand Side Term in the Degenerate Higher-Order Parabolic Equation on a Plane

Vitaly L. Kamynin and Tatiana I. Bukharova[✉]

National Research Nuclear University MEPhI (Moscow Engineering Physics
Institute), 31, Kashirskoe shosse, 115409 Moscow, Russia
vlkamynin2008@yandex.ru, bukharova_t@mail.ru

Abstract. We establish existence and uniqueness theorems as well as
the theorem on stability under perturbations of the input data for the
solution of the inverse problem for a degenerate higher-order parabolic
equation on a plane with integral observation. We also obtain the estimates of the solution with constants explicitly written out in terms of
the input data of the problem.

1 Introduction

In present paper we study the existence, uniqueness and stability under perturbations of the input data for the solution $\{u(t,x), p(t)\}$ of the inverse problem
for nonuniformly parabolic equation

$$D_t u + (-1)^m a(t,x) D_x^{2m} u = p(t)g(t,x) + r(t,x), \quad (t,x) \in Q; \tag{1}$$

with initial and boundary conditions

$$u(0,x) = u_0(x), \ x \in [0,l]; \ D_x^j u|_{x=0,l} = 0, \ j = 0,1,...,m-1, \ t \in [0,T]; \tag{2}$$

and the additional condition of integral observation

$$\int_0^l u(t,x)\omega(x)dx = \varphi(t), \quad t \in [0,T]; \tag{3}$$

here $Q = [0,T] \times [0,l]$, where T and l are some numbers, $D_t u \equiv \frac{\partial u}{\partial t}$, $D_x^j u \equiv \frac{\partial^j u}{\partial x^j}$, $j = 1,2,...,2m$, $D_x^0 u \equiv u$. In the setting of the inverse problem under
consideration one allowed the degeneration ($a(t,x) \geq 0$) and unboundedness
of the coefficient $a(t,x)$ as well as unboundedness of the functions $g(t,x)$ and
$r(t,x)$.

We establish sufficient conditions under which the problem (1)–(3) is well
solved (we prove the existence and the uniqueness of generalized solution an its
stability with respect to perturbations of the input data).

© Springer International Publishing AG 2017
I. Dimov et al. (Eds.): NAA 2016, LNCS 10187, pp. 391–397, 2017.
DOI: 10.1007/978-3-319-57099-0_43

It should be noted that the proofs of the theorems are based on the proof of contractibility of a special operator that enters into the equation for finding the unknown function $p(t)$. This fact allows us to use the iteration method for approximate determination of this function which is important for numerical solution of the inverse problem (1)–(3).

A similar inverse problem for second order parabolic equation was studied in recent paper [1].

Inverse problems for parabolic equations admitting degeneracy in various settings other than in the present paper (and in paper [1]) were considered in [2–6], etc.

Note that inverse problem for degenerate parabolic equations are important in application in particular they arise in the study of models of price formation for options in financial markets (see, for example, [7–9], etc.). We also note that for nondegenerate higher-order parabolic equations the inverse problem of determination the right-hand side of the equation with additional condition (3) were considered in [10–12], etc.

In this paper we use Lebesgue and Sobolev spaces with corresponding norms in usual sense (see [13]). Set $Q(0,\tau) = [0,\tau] \times [0,l], 0 < \tau \leq T, Q(0,T) \equiv Q$.

We assume that the functions occurring in the input data of the problem (1)–(3) are measurable and satisfy the following conditions:

(A) $0 \leq a(t,x), (t,x) \in Q;$ $a(t,x), 1/a(t,x) \in L_q(Q), q > 1,$

$$\|a\|_{L_q(Q)} \leq a_1, \|1/a\|_{L_q(Q)} \leq a_2;$$

(B) $g(t,x), r(t,x) \in L_\infty(0,T; L_1([0,l])), \|r\|_{L_\infty(0,T;L_1([0,l]))} \leq K_r,$

$$g^2(t,x)/a(t,x) \in L_\infty(0,T; L_1([0,l])), \|g^2/a\|_{L_\infty(0,T;L_1([0,l]))} \leq K_g^*,$$

$$r^2(t,x)/a(t,x) \in L_1(Q), \|r^2/a\|_{L_1(Q)} \leq K_r^*;$$

(C) $u_0(x) \in \overset{0}{W}_2^m([0,l]);$

(D) $\varphi(t), \varphi'(t) \in L_\infty([0,T]),$ $\|\varphi'(t)\|_{L_\infty([0,T])} \leq K_\varphi^*;$

(E) $\omega(x) \in L_\infty(Q), D_x^j(a\omega) \in L_\infty(0,T; L_2([0,l])), j = 0,...,m;$

$$D_x^j(a\omega)|_{x=0,l} = 0, j = 0,...,m-1; \|D_x^m(a\omega)\|_{L_\infty(0,T;L_2([0,l]))} \leq K_a;$$

$$\|\omega\|_{L_\infty([0,l])} \leq K_\omega;$$

(F) $\left| \int\limits_{0}^{l} a(t,x)g(t,x)\omega(x)dx \right| \geq g_0 > 0;$

(G) $\varphi(0) = \int\limits_{0}^{l} u_0(x)\omega(x)dx.$

Here $a_1, a_2, K_g^*, K_\omega, K_a, g_0 = \text{const} > 0,\ \ K_r, K_\varphi^* = \text{const} \geq 0.$

Remark 1. Obviously the conditions imposed above are satisfied, for example, by the following functions:

$$a(t,x) = x^{\kappa_1}(l-x)^{-\nu_1}\hat{a}(t,x),\ g(t,x) = (l-x)^{-\nu_2/2}\hat{g}(t,x),$$

$$r(t,x) = (l-x)^{-\nu_3/2}\overset{0}{\hat{r}}(t,x),\ \omega(x) \in \overset{0}{C^\infty}((0,l)),\ \omega(x) \geq 0$$

and there exists $x_0 \in (0,l)$ such that $\omega(x_0) > 0$.

Here $\hat{a}(t,x) \in C^\infty(Q)$, and $\hat{a}(t,x) \geq \Lambda_0 > 0$, $\hat{g}(t,x)$ is bounded in Q and $\hat{g}(t,x) \geq \Lambda_1 > 0$, $\hat{r}(t,x)$ is bounded in Q; the constants $\kappa_1, \nu_1, \nu_2, \nu_3$ are satisfied the inequalities $0 < \kappa_1 < 1,\ 0 < \nu_1 < 1,\ 0 < \nu_2, \nu_3 \leq \nu_1 + 1$.

In what follows we shall use the notations

$$G(t) = \int\limits_{0}^{l} g(t,x)\omega(x)dx,\ R(t) = \int\limits_{0}^{l} r(t,x)\omega(x)dx. \tag{4}$$

Definition 1. By generalized solution of the inverse problem (1)–(3) we mean a pair of functions $\{u(t,x), p(t)\}$ such that

(1) $u(t,x) \in W_s^{1,2m}(Q) \bigcap L_\infty(0,T; \overset{0}{W_2^m}([0,l])), s > 1,\ a(t,x)(D_x^{2m}u)^2,$
    $(D_t u)^2/a(t,x) \in L_1(Q),\ \ p(t) \in L_\infty([0,T]);$
(2) this pair satisfies Eq. (1) almost everywhere in Q;
(3) $\lim_{t\to 0+} \int\limits_{0}^{l} |u(t,x) - u_0(x)|\, dx = 0;$
4) equality (3) is satisfied at each point $t \in [0,T]$.

2   A Priori Estimates and Solvability of the Direct Problem

Suppose that $p(t) \in L_\infty([0,T])$ is a known function and consider the direct problem (1)–(2) with this function $p(t)$ in the right-hand side of Eq. (1). For convenience we introduce a notation $f(t,x) = p(t)g(t,x) + r(t,x)$. Using the ideas of [13] we establish the following theorem.

Theorem 1. Let conditions (A)–(C) hold and $p(t) \in L_\infty([0,T])$. Set $q^* = 2q/(q+1)$. Then the solution $u(t,x)$ of direct problem (1)–(2) satisfies the following estimates:

$$\sup_{0 \leq t \leq \tau} \|D_x^m u(t,\cdot)\|_{L_2([0,l])}^2 + \|(D_t u)^2/a\|_{L_1(Q(0,\tau))} + \|a(D_x^{2m} u)^2\|_{L_1(Q(0,\tau))}$$

$$\leq \|D_x^m u_0\|_{L_2([0,l])}^2 + \int_{Q(0,\tau)} \frac{f^2(t,x)}{a(t,x)} \, dx dt, \quad \tau \in (0,T], \quad (5)$$

$$\||D_t u\|_{L_{q^*}(Q)}^2 + \||D_x^{2m} u\|_{L_{q^*}(Q)}^2$$

$$\leq (a_1 + a_2) \left(\|D_x^m u_0\|_{L_2([0,l])}^2 + \int_Q \frac{f^2(t,x)}{a(t,x)} \, dx dt \right), \quad (6)$$

Remark 2. Using the well-known Ehrling-Nirenberg-Gagliardo interpolation inequality (see [14], p. 236) we obtain from the estimates (5), (6) that

$$\|u\|_{W_{q^*}^{1,2m}(Q)}^2 \leq c_1 \left(\|D_x^m u_0\|_{L_2([0,l])}^2 + \int_Q \frac{f^2(t,x)}{a(t,x)} \, dx dt \right). \quad (7)$$

where $c_1 = \text{const} > 0$.

Remark 3. Having the a priori estimates (5), (6) and (7) by standard methods (for example by the method of continuation on a parameter) we prove the existence and the uniqueness of the generalized solution $u((t,x))$ of direct problem (1) and (2) which belong to the class of functions indicated in the Definition 1 with $s = q^*$. Moreover such a solution satisfies the estimates (5), (6) and (7).

3   Solvability of the Inverse Problem

Let us assume that the input data of the inverse problem (1)–(3) satisfy the conditions (A)–(G), and $q^* > 1$ is defined in Theorem 1. Let us derive the operator equation for the unknown coefficient $p(t)$.

Let $p(t)$ be an arbitrary function from $L_\infty([0,T])$, and let $u(t,x)$ be a generalized solution of direct problem (1) and (2) with the chosen function $p(t)$ in the right-hand side of the Eq. (1). In view of Remark 3 such a solution exists and is unique and also $u(t,x) \in W_{q^*}^{1,2}(Q)$.

Let us multiply Eq. (1) by $\omega(x)$ and integrate the resulting relation over the segment $[0,l]$. Taking into account conditions (2) notation (4) and integrating by parts using conditions (D)–(F) we obtain the relation

$$p(t) = \frac{1}{G(t)} \left[\int_0^l D_x^m(a\omega) D_x^m u \, dx + \varphi'(t) - R(t) \right]. \quad (8)$$

Let us introduce the operator $\mathcal{A} : L_\infty([0,T]) \to L_\infty([0,T])$ by the formula

$$\mathcal{A}(p) = \frac{1}{G(t)} \int_0^l D_x^m(a\omega) D_x^m u \, dx, \tag{9}$$

where $u(t,x)$ is a solution of problem (1) and (2) with a given function $p(t)$ in the right-hand side of Eq. (1). Then the relation (8) can be written as

$$p(t) = \mathcal{A}(p)(t) + \frac{\varphi'(t)}{G(t)} - \frac{R(t)}{G(t)}. \tag{10}$$

Lemma 1. Let conditions (A)–(G) hold. Then the pair of functions $\{u(t,x), p(t)\}$ is a generalized solution of problem (1)–(3) if and only if this pair of functions satisfies the relations (1), (2) and (10).

The proof of this Lemma is standard (see, for example [1]).

Using this lemma we prove the existence and the uniqueness of the solution of the inverse problem (1)–(3).

Theorem 2. Let conditions (A)–(G) hold and let q^* be defined in Theorem 1. Set

$$\beta = \frac{K_a}{g_0} \sqrt{K_g^*}, \ \gamma = 4\beta^2. \tag{11}$$

Then there exists a unique generalized solution $\{\hat{u}(t,x), \hat{p}(t)\}$ of the inverse problem (1)–(3), $\hat{u}(t,x) \in W_{q^*}^{1,2}(Q)$ and the following estimates hold:

$$\|\hat{p}\|_{L_\infty([0,T])} \leq \frac{2e^{\gamma T/2}}{g_0} \left\{ K_a \left(\|D_x^m u_0\|_{L_2([0,l])}^2 + K_r^* \right)^{1/2} + K_\varphi^* + K_\omega K_r \right\}, \tag{12}$$

$$\sup_{0 \leq t \leq T} \|D_x^m u(t,\cdot)\|_{L_2([0,l])}^2 + \|D_x^{2m} u\|_{L_{q^*}(Q)}^2 + \|D_t u\|_{L_{q^*}(Q)}^2$$

$$\leq (1 + a_1 + a_2) \left[\|D_x^m u_0\|_{L_2([0,l])}^2 + 2TK_g^* \|\hat{p}\|_{L_2([0,l])}^2 + 2K_r^* \right]. \tag{13}$$

The proof of this theorem is based on the proof that the operator $\mathcal{A}$ defined in (9) is a contraction operator in $L_\infty([0,T])$ where the norm is chosen in the following way: $\|z\|_\gamma = \|e^{-\gamma t/2} z\|_{L_\infty([0,T])}$, $z(t) \in L_\infty([0,T])$ and γ is defined in (11).

4 Stability of the Solution of the Inverse Problem

Using the fact that the operator $\mathcal{A}$ is a contraction operator in $L_\infty([0,T])$ with the norm $\|\cdot\|_\gamma$ we can obtain estimates for the stability of the solution of inverse problem (1)–(3) with respect to the perturbations of the functions $g(t,x)$, $r(t,x)$, $\varphi(t)$, $u_0(x)$ with constants that can be written out explicitly in terms of the input data of the problem.

Theorem 3. Consider the following two inverse problems in Q:

$$D_t u + (-1)^m a(t,x) D_x^{2m} u = p(t) g^{(i)}(t,x) + r^{(i)}(t,x),$$

$$u(0,x) = u_0^{(i)}(x), \quad D^j u|_{x=0,l} = 0, \quad j = 0,1,...,m-1,$$

$$\int_0^l u(t,x)\omega(x)\,dx = \varphi^{(i)}(t),$$

$i = 1,2.$

Suppose that for these problems conditions (A)–(G) hold. Let $\{u^{(i)}(t,x), p^{(i)}(t)\}$ be the corresponding solutions of these problems. (These solutions exist and are unique by Theorem 2). Then the following estimates hold:

$$\|p^{(1)}(t) - p^{(2)}(t)\|_{L_\infty([0,T])} \leq \frac{2e^{\gamma T/2}}{g_0} \Big\{ K_a + \Big[\|D_x^m u_0^{(1)} - D_x^m u_0^{(2)}\|_{L_2([0,l])}^2$$

$$+ 2T\|p^{(2)}\|_{L_\infty([0,T])}^2 \left\| (g^{(1)} - g^{(2)})^2/a \right\|_{L_\infty(0,T;L_1([0,l]))}$$

$$+ 2\left\| (r^{(1)} - r^{(2)})^2/a \right\|_{L_1(Q)} \Big]^{1/2} + \|\varphi^{(1)\prime} - \varphi^{(2)\prime}\|_{L_\infty([0,T])}$$

$$+ \|p^{(2)}\|_{L_\infty([0,T])} K_\omega \|g^{(1)} - g^{(2)}\|_{L_\infty(0,T;L_1([0,l]))}$$

$$+ K_\omega \|r^{(1)} - r^{(2)}\|_{L_\infty(0,T;L_1([0,l]))} \Big\}, \quad (14)$$

$$\sup_{0\leq t\leq T} \|D_x^m u^{(1)}(t,\cdot) - D_x^m u^{(2)}(t,\cdot)\|_{L_2([0,l])}^2 + \|D_x^{2m} u^{(1)} - D_x^{2m} u^{(2)}\|_{L_{q*}(Q)}^2$$

$$+ \|D_t u^{(1)} - D_t u^{(2)}\|_{L_{q*}(Q)}^2$$

$$\leq (1 + a_1 + a_2) \Big[\|D_x^m u_0^{(1)} - D_x^m u_0^{(2)}\|_{L_2([0,l])}^2 + 3T K_g^* \|p^{(1)} - p^{(2)}\|_{L_\infty([0,T])}^2$$

$$+ 3T\|p^{(2)}\|_{L_\infty([0,T])}^2 \left\| (g^{(1)} - g^{(2)})^2/a \right\|_{L_\infty(0,T;L_1([0,l]))}$$

$$+ 3\left\| (r^{(1)} - r^{(2)})^2/a \right\|_{L_1(Q)} \Big]. \quad (15)$$

In estimates (14), (15) the norm $\|p^{(2)}\|_{L_\infty([0,T])}^2$ satisfies estimate (12).

References

1. Kamynin, V.L.: On the solvability of the inverse problem for determining the right-hand side of a degenerate parabolic equation with integral observation. Math. Notes **98**(5), 765–777 (2015)
2. Ivanchov, M., Saldina, N.: An inverse problem for strongly degenerate heat equation. J. Inverse Ill-Posed Prob. **14**(5), 465–480 (2006)

3. Cannarsa, P., Tort, J., Yamamoto, M.: Determination of source terms in degenerate parabolic equation. Inverse Prob. **26**(10), 105003 (2010)
4. Deng, Z.C., Qian, K., Rao, X.B., Yang, L.: An inverse problem of identifying the source coefficient in degenerate heat equation. Inverse Prob. Sci. Eng. **23**(3), 498–517 (2014)
5. Huzyk, N.: Inverse problem of determining the coefficients in degenerate parabolic equation. Electron. J. Differ. Equ. **172**, 1–11 (2014)
6. Kawamoto, A.: Inverse problems for linear degenerate parabolic equations by "time-like" Carleman estimate. J. Inverse Ill-posed Prob. **23**(1), 1–21 (2015)
7. Bouchouev, I., Isakov, V.: Uniqueness, stability and numerical methods for the inverse problem that arises in financial markets. Inverse Prob. **15**(3), 95–116 (1999)
8. Lishang, J., Yourshan, T.: Identifying the volatibility of underlying assets from option prices. Inverse Prob. **17**(1), 137–155 (2001)
9. Lishang, J., Qihong, C., Lijun, W., Zhang, J.E.: A new well-posed algorithm to recover implied local volatility. Quant. Finance **3**(6), 451–457 (2003)
10. Prilepko, A.I., Orlovskii, D.G.: Determination of the parameter of an evolution equation and inverse problems of mathematical physics I. Differ. Equ. **21**(1), 96–104 (1985)
11. Prilepko, A.I., Orlovskii, D.G.: Determination of the parameter of an evolution equation and inverse problems of mathematical physics II. Differ. Equ. **21**(4), 472–477 (1985)
12. Kamynin, V.L., Francini, E.: An inverse problem for a higher-order parabolic equation. Math. Notes **64**(5–6), 590–599 (1999)
13. Kruzhkov, S.N.: Quasilinear parabolic equations and systems with two independent variables. Trudy Sem. im. I.G.Petrovskogo **5**, 217–272 (1979)
14. Besov, O.V., Il'in, V.P., Nikolskii, S.M.: Integral'nye predstavleniya funkcii i teoremy vlozheniya (Integral reprezentation of functions and embedding theorems). Nauka, Moscow (1975)

Numerical Methods of Solution of the Dirichlet Boundary Value Problem for the Fractional Allers' Equation

Fatimat A. Karova$^{(\boxtimes)}$

Institute of Applied Mathematics and Automation,
Russian Academy of Sciences, ul. Shortanova 89 a, Nalchik 360000, Russia
_timka_86_86@mail.ru

Abstract. Solution of Dirichlet boundary value problem for the Allers' equation in differential and difference settings are studied. By the method energy inequalities, a priori estimate is obtained for the solution of the differential problems.

1 Introduction

Moisture movement is in capillary porous environment is described by the equation of Aller [1]. Boundary value problems for classical Allers' equation is studied in [2]. However, it was found that the fractional derivatives are more effective in describing the properties of viscoelastic fluid. In this regard, there are models for fractional Allers' equation. The starting point of the fractional derivative model of viscoelastic fluid is usually a classical differential equation which is modified by replacing the time derivative of an integer order by the so-called Riemann-Louville fractional calculus operator. The method of energy inequalities has been applied for the numerical solution of boundary value problems for differential equations of fractional order with variable coefficient [3,4]. The stability and convergence of the numerical scheme for solving the boundary value problem for generalized Allers' equation are analysed [5] by the energy method.

2 The Dirichlet Boundary Value Problem in Differential Setting

In rectangle $\overline{Q}_T = \{(x,t) : \ 0 \leq x \leq l, 0 \leq t \leq T\}$ let us study the boundary value problem

$$\partial_{0t}^{\alpha} u = \frac{\partial}{\partial x}\left(k(x,t)\frac{\partial u}{\partial x}\right) + \partial_{0t}^{\beta}\frac{\partial}{\partial x}\left(\eta(x,t)\frac{\partial u}{\partial x}\right) - q(x,t)u + f(x,t), \quad (1)$$

$$u(0,t) = 0, \ u(l,t) = 0, \ 0 \leq t \leq T, \quad (2)$$

$$u(x,0) = u_0(x), \ 0 \leq x \leq l, \quad (3)$$

© Springer International Publishing AG 2017
I. Dimov et al. (Eds.): NAA 2016, LNCS 10187, pp. 398–405, 2017.
DOI: 10.1007/978-3-319-57099-0_44

where $\partial_{0t}^{\gamma} u(x,t) = \frac{1}{\Gamma(1-\gamma)} \int\limits_0^t u_s(x,s)(t-s)^{-\gamma}\,ds$ is a Caputo fractional derivative

of order $\gamma, 0 < \gamma < 1, 0 < c_1 \le k(x,t), \eta(x,t) \le c_2, \eta_t(x,t) \ge 0, q(x,t) \ge 0$
on $\overline{Q}_T$.

Suppose that the existence of solution $u(x,t) \in C^{2,1}\left(\overline{Q}_T\right)$ for the problem
(1)–(3), where $C^{m,n}\left(\overline{Q}_T\right)$ is the class of functions, continuous together with their
partial derivatives of the order m with respect to x and order n with respect to
t on $\overline{Q}_T$.

Lemma 1. [3] For any function $v(t)$ is absolutely continuous on $[0,T]$, we have
the equality

$$v(t)\partial_{0t}^{\alpha} v(t) \ge \frac{1}{2}\partial_{0t}^{\alpha} v^2(t), \quad 0 < \alpha < 1.$$

Let us use the following notation

$$\|u\|_0^2 = \int\limits_0^l u^2(x,t)dx, \quad \|u\|_{W_2^1(0,l)}^2 = \|u\|_0^2 + \|u_x\|_0^2,$$

$$D_{0t}^{-\alpha} u(x,t) = \frac{1}{\Gamma(\alpha)} \int\limits_0^l (t-\tau)^{\alpha-1} u(x,\tau)d\tau$$

is a fractional Riemann-Liouville integral of order α.

Theorem 1. If $k(x,t) \in C^{1,0}\left(\overline{Q}_T\right), \eta(x,t) \in C^{1,1}\left(\overline{Q}_T\right), q(x,t), f(x,t) \in \left(\overline{Q}_T\right), 0 < c_1 \le k(x,t), \eta(x,t) \le c_2, \eta_t(x,t) \ge 0$, then the solution $u(x,t)$ of
the problem (1)–(3) satisfies a priori estimate

$$D_{0t}^{\alpha-1}\|u\|_0^2 + \int\limits_0^t \|u_x\|_0^2 ds + D_{0t}^{\beta-1}\|u_x\|_0^2$$

$$\le M\left(\int\limits_0^t \|f\|_0^2 ds + \|u_0'\|_0^2 + \|u_0\|_0^2\right). \tag{4}$$

Proof. Let us multiply (1) by u and integrate the resulting relation over x from
0 to l:

$$\int\limits_0^l u\partial_{0t}^{\alpha} u\,dx - \int\limits_0^l u(ku_x)_x dx - \int\limits_0^l u\partial_{0t}^{\beta}(\eta u_x)_x dx + \int\limits_0^l qu^2 dx = \int\limits_0^l uf\,dx. \tag{5}$$

Then transform the terms in identity (5) as follow:

$$-\int\limits_0^l u(ku_x)_x dx = \int\limits_0^l ku_x^2 dx \ge c_1\|u_x\|_0^2, \tag{6}$$

$$\int_0^l uf dx \leq \varepsilon \|u\|_0^2 + \frac{1}{4\varepsilon}\|f\|_0^2, \quad \varepsilon > 0. \tag{7}$$

Using Lemma 1 one obtains

$$\int_0^l u \partial_{0t}^\alpha u \, dx \geq \frac{1}{2}\int_0^l \partial_{0t}^\alpha u^2 \, dx = \frac{1}{2}\partial_{0t}^\alpha \|u\|_0^2, \tag{8}$$

$$-\int_0^l u \partial_{0t}^\beta (\eta u_x)_x dx = \int_0^l u_x \partial_{0t}^\beta (\eta u_x) dx \geq \frac{1}{2}\int_0^l \frac{1}{\eta}\partial_{0t}^\beta (\eta u_x)^2 dx. \tag{9}$$

Taking into account the above transformations, from the identity (5) we arrive to the following inequality

$$\partial_{0t}^\alpha \|u\|_0^2 + 2c_1\|u_x\|_0^2 + \int_0^l \frac{1}{\eta}\partial_{0t}^\beta (\eta u_x)^2 dx \leq 2\varepsilon\|u\|_0^2 + \frac{1}{2\varepsilon}\|f\|_0^2. \tag{10}$$

Using the inequality $\|u\|_0^2 \leq \frac{l^2}{2}\|u_x\|_0^2$, from the inequality (10) at $\varepsilon = \frac{c_1}{l^2}$ we get

$$\partial_{0t}^\alpha \|u\|_0^2 + c_1\|u_x\|_0^2 + \int_0^l \frac{1}{\eta}\partial_{0t}^\beta (\eta u_x)^2 dx \leq \frac{l^2}{2c_1}\|f\|_0^2. \tag{11}$$

Replacing the variable t by s in the inequality (11) and integrating it over s from 0 to t, one obtains

$$D_{0t}^{\alpha-1}\|u\|_0^2 + c_1\int_0^t \|u_x\|_0^2 ds + \int_0^t \frac{1}{\eta}\int_0^l \partial_{0t}^\beta (\eta u_x)^2 dx ds$$

$$\leq \frac{l^2}{2c_1}\int_0^t \|f\|_0^2 ds + \frac{t^{1-\alpha}}{\Gamma(2-\alpha)}\|u_0(x)\|_0^2. \tag{12}$$

Evaluate expression $\int_0^t \frac{1}{\eta}\partial_{0t}^\beta v \, ds$, where $v = (\eta u_x)^2$:

$$\int_0^t \frac{1}{\eta}\partial_{0t}^\beta v \, d\tau = \frac{1}{\Gamma(1-\beta)}\int_0^t \frac{1}{\eta(x,\tau)}\int_0^\tau \frac{v'(\xi)d\xi}{(\tau-\xi)^\beta}d\tau$$

$$= \frac{1}{\Gamma(1-\beta)}\int_0^t v'(\xi)\int_\xi^t \frac{1}{\eta(x,\tau)}\frac{1}{(t-\xi)^\beta}d\tau d\xi$$

$$= \frac{1}{\Gamma(1-\beta)}\int_0^t (t-\xi)^{1-\beta}v'(\xi)\int_0^1 \frac{\theta^{-\beta}d\theta}{\eta(x,\xi+\theta(t-\xi))}d\xi$$

$$= \frac{(t-\xi)^{1-\beta}}{\Gamma(1-\beta)} \int_0^1 \frac{\theta^{-\beta}d\theta}{\eta(x,\xi+\theta(t-\xi))} v(\xi)\big|_0^t$$

$$-\frac{1}{\Gamma(1-\beta)} \int_0^t v(\xi)\frac{\partial}{\partial\xi}\left[(t-\xi)^{1-\beta}\int_0^1 \frac{\theta^{-\beta}d\theta}{\eta(x,\xi+\theta(t-\xi))}\right]d\xi$$

$$\geq -t^{1-\beta}\frac{v(0)}{\Gamma(1-\beta)}\int_0^1 \frac{\theta^{-\beta}d\theta}{\eta(x;\theta t)}$$

$$+\frac{1}{\Gamma(1-\beta)}\int_0^t v(\xi)\left[(1-\beta)(t-\xi)^{-\beta}\int_0^1 \frac{\theta^{-\beta}d\theta}{\eta(x,\xi+\theta(t-\xi))}\right]$$

$$\geq \frac{1}{c_2\Gamma(1-\beta)}\int_0^t (t-\xi)^{-\beta}v(\xi)d\xi - t^{1-\beta}\frac{v(0)}{\Gamma(1-\beta)}\int_0^1 \frac{\theta^{-\beta}d\theta}{\eta(x;\theta t)}$$

$$= \frac{1}{c_2}D_{0t}^{\beta-1}v(\xi) - \frac{t^{1-\beta}}{c_1\Gamma(2-\beta)}v(0)$$

$$= \frac{c_1^2}{c_2}D_{0t}^{\beta-1}u_x^2(x,t) - \frac{c_2^2}{c_1\Gamma(2-\beta)}\frac{t^{1-\beta}}{}u_x^2(x,0). \qquad (13)$$

Integrating the inequality (13) in respect to x from 0 to l and substituting it in (12), obtain a priori estimate (4).

The uniqueness and continuous dependence of the solution of the problem (1)–(3) on the input data follow from a priori estimate (4).

3   Difference Schemes for the Dirichlet Boundary Value Problem. Stability and Convergence

Suppose that solution $u(x,t) \in C^{4,3}(Q_T)$ of the problem (1)–(3) exists, and coefficients of the Eq. (1) and functions $f(x,t), u_0(x)$ satisfy the conditions, required for the construction of difference schemes with then order of approximation $O(\tau^{2-max\{\alpha,\beta\}} + h^2)$.

In this section we considered case, when $\eta(x,t) = \eta(x)$. In the rectangle $\overline{Q}_T$ we introduce the grid $\overline{\omega}_{h\tau} = \overline{\omega}_h \times \overline{\omega}_\tau$, where

$$\overline{\omega}_h = \{x_i = ih, \ i = 0,1,\ldots,N, \ hN = l\},$$

$$\overline{\omega}_\tau = \{t_j = j\tau, \ j = 0,1,\ldots,j_0, \ \tau j_0 = T\}.$$

To problem (1)–(3) we assign the difference scheme

$$\Delta_{0t_{j+1}}^\alpha y = \Lambda_1 y^{j+1} + \Delta_{0t_{j+1}}^\beta \Lambda_2 y + \varphi_i^{j+1}, \quad 1 \leq i \leq N-1, \quad 1 \leq j \leq j_0-1, \ (14)$$

$$y(0,t) = 0, \quad y(l,t) = 0, \quad t \in \overline{\omega}_\tau, \qquad (15)$$

$$y(x,0) = u_0(x), \quad x \in \bar{\omega}_h, \tag{16}$$

where

$$\Lambda_1 y = (ay_{\bar{x}})_x, \quad a_i^j = k(x_{i-1/2}, t_{j+1}),$$

$$\Lambda_2 y = (by_{\bar{x}})_x, \quad b_i = \eta(x_{i-1/2}),$$

$\varphi_i^{j+1} = f(x_i, t_{j+1})$, $\Delta_{0t_{j+1}}^\gamma y = \frac{1}{\Gamma(2-\gamma)} \sum_{s=0}^{j} \left(t_{j-s+1}^{1-\gamma} - t_{j-s}^{1-\gamma} \right) y_t^s$ is the difference analogue of the Caputo fractional derivative of order $\gamma, 0 < \gamma < 1$ [6].

The difference scheme (14)–(16) has the order of approximation $O(\tau^{2-max\{\alpha,\beta\}} + h^2)$ [7].

Lemma 2. [8] For any function $y(t)$ defined on the grid $\bar{\omega}_\tau$ we have the inequality

$$y^{j+1} \Delta_{0t_{j+1}}^\alpha y \geq \frac{1}{2} \Delta_{0t_{j+1}}^\alpha (y^2).$$

Theorem 2. *The difference scheme (14)–(16) is absolutely stable and its solution satisfies a priori estimate*

$$\sum_{s=0}^{j} \left(t_{j-s+1}^{1-\alpha} - t_{j-s}^{1-\alpha} \right) \|y^{s+1}\|_0^2$$

$$+ \sum_{s=0}^{j} \left(t_{j-s+1}^{1-\beta} - t_{j-s}^{1-\beta} \right) \|\sqrt{b} y_{\bar{x}}^{s+1}]\|_0^2$$

$$\leq M \left(\sum_{j'=0}^{j} \|\varphi^s\|_0^2 \tau + \|y^0\|_0^2 + \|\sqrt{b} y_{\bar{x}}^0]\|_0^2 \right), \tag{17}$$

where $(y,v) = \sum_{i=1}^{N-1} y_i v_i h, (y,v] = \sum_{i=1}^{N} y_i v_i h, \|y\|_0^2 = (y,y), \|y]\|_0^2 = (y,y].$

Proof. Let us multiply scalarly Eq. (17) by y^{j+1}:

$$\left(\Delta_{0t_{j+1}}^\alpha y, y^{j+1} \right) = (\Lambda_1 y^{j+1}, y^{j+1}) + \left(\Delta_{0t_{j+1}}^\beta \Lambda_2 y, y^{j+1} \right) + (\varphi, y^{j+1}). \tag{18}$$

Transform the terms in identity (18):

$$- (\Lambda_1 y^{j+1}, y^{j+1}) = (a, y_{\bar{x}}^2] \geq c_1 \|y_{\bar{x}}]\|_0^2 \geq \kappa \|y^{j+1}\|_0^2, \quad \kappa = \frac{2c_1}{l^2}, \tag{19}$$

$$|(\varphi, y^{j+1})| \leq \varepsilon \|y^{j+1}\|_0^2 + \frac{1}{4\varepsilon} \|\varphi\|_0^2, \quad \varepsilon > 0. \tag{20}$$

Using Lemma 2 obtain:

$$\left(\Delta_{0t_{j+1}}^\alpha y, y^{j+1} \right) \geq \frac{1}{2} \Delta_{0t_{j+1}}^\alpha \|y\|_0^2, \tag{21}$$

$$- \left(\Delta^{\beta}_{0t_{j+1}}(by_{\bar{x}})_x, y^{j+1} \right) = \left(\Delta^{\beta}_{0t_{j+1}}(by_{\bar{x}}), y^{j+1}_x \right] = \left(b, y^{j+1}_x \Delta^{\beta}_{0t_{j+1}} y_{\bar{x}} \right]$$

$$\geq \frac{1}{2} \left(b, \Delta^{\beta}_{0t_{j+1}} y^2_{\bar{x}} \right] = \frac{1}{2} \Delta^{\beta}_{0t_{j+1}} \| \sqrt{b} y_{\bar{x}} \|^2_0. \qquad (22)$$

Taking into account the above transformations, from identity (18) arrive at the inequality

$$\frac{1}{2} \Delta^{\alpha}_{0t_{j+1}} \| y \|^2_0 + \kappa \| y^{j+1} \|^2_0 + \frac{1}{2} \Delta^{\beta}_{0t_{j+1}} \| \sqrt{b} y_{\bar{x}} \|^2_0 \leq \varepsilon \| y^{j+1} \|^2_0 + \frac{1}{4\varepsilon} \| \varphi \|^2_0. \qquad (23)$$

From (23) at $\varepsilon = \kappa > 0$ obtain the inequality

$$\Delta^{\alpha}_{0t_{j+1}} \| y \|^2_0 + \Delta^{\beta}_{0t_{j+1}} \| \sqrt{b} y_{\bar{x}} \|^2_0 \leq \frac{l^2}{2} \| \varphi \|^2_0. \qquad (24)$$

Multiplying the inequality (24) by τ and summing over s from 0 to j, one obtains the a priori estimate (17). The stability and convergence of the difference scheme (14)–(16) follow from the a priori estimate (17).

Here the results are obtained for the homogeneous boundary conditions $u(0,t) = 0$, $u(l,t) = 0$. In the case of inhomogeneous boundary conditions $u(0,t) = \mu_1(t)$, $u(l,t) = \mu_2(t)$ the boundary conditions for the difference problem will have the following form:

$$y(0,t) = \mu_1(t), \quad y(l,t) = \mu_2(t). \qquad (25)$$

Convergence of the difference scheme (14), (16) and (25) follows from the results obtained above. Actually, let us introduce the notation $y = z + u$. Then, the error $z = y - u$ is a solution on the following problem

$$\Delta^{\alpha}_{0t_{j+1}} z = \Lambda_1 z^{j+1} + \Delta^{\beta}_{0t_{j+1}} \Lambda_2 z + \psi, \quad 1 \leq i \leq N - 1, 1 \leq j \leq j_0 - 1, \qquad (26)$$

$$z(0,t) = 0, \quad z(l,t) = 0, \quad j = 0, ..., j_0, \qquad (27)$$

$$z(x,0) = 0, \quad i = 0, ..., N, \qquad (28)$$

where

$$\psi \equiv \Lambda_1 u^{j+1} + \Delta^{\beta}_{0t_{j+1}} \Lambda_2 u - \Delta^{\alpha}_{0t_{j+1}} u + \varphi = O(\tau^{2-max\{\alpha,\beta\}} + h^2).$$

The solution of the problem (26)–(28) satisfies the estimation (17) so that the solution of the difference scheme (14), (16) and (25) converges to the solution of the corresponding differential problem with order $O(\tau^{2-max\{\alpha,\beta\}} + h^2)$.

4   Numerical Results

In this section, the following time fractional Allers' equation is considered:

$$\partial^{\alpha}_{0t} u = \frac{\partial}{\partial x} \left(k(x,t) \frac{\partial u}{\partial x} \right) + \partial^{\beta}_{0t} \frac{\partial}{\partial x} \left(\eta(x) \frac{\partial u}{\partial x} \right) - q(x,t)u + f(x,t), \qquad (29)$$

Table 1. L_2-norm and maximum norm error behaviour versus τ-grid size reduction when $h = 1/1000$

α	β	τ	$\max\limits_{0 \le j \le j_0} \|z^n\|_0$	CR in $\|\cdot\|_0$	$\|z\|_{C(\bar{\omega}_{h\tau})}$	CR in $\|\cdot\|_{C(\bar{\omega}_{h\tau})}$
0.5	0.5	1/20	3,6495e–3		5,2917e–3	
		1/40	1,3382e–3	1,447	1,9404e–3	1,447
		1/80	4,8525e–4	1,463	7,0359e–4	1,464
0.9	0.1	1/20	1,4076e–3		2,0706e–3	
		1/40	6,2902e–4	1,162	9,2673e–4	1,160
		1/80	2,8502e–4	1,142	4,2039e–4	1,140
0.4	0.7	1/20	9,3828e–3		1,3610e–2	
		1/40	3,8905e–3	1,270	5,6438e–3	1,270
		1/80	1,5989e–3	1,283	2,3195e–3	1,283

Table 2. L_2-norm and maximum norm error behaviour versus grid size reduction when $h^2 = \tau^{2-max\{\alpha,\beta\}}$

α	β	h	$\max\limits_{0 \le j \le j_0} \|z^n\|_0$	CR in $\|\cdot\|_0$	$\|z\|_{C(\bar{\omega}_{h\tau})}$	CR in $\|\cdot\|_{C(\bar{\omega}_{h\tau})}$
0.5	0.5	1/20	3,9757e–3		5,7482e–3	
		1/40	9,9936e–4	1,992	1,4448e–3	1,989
		1/80	2,5035e–4	1,997	3,6197e–4	2,000
0.8	0.4	1/20	3,2057e–3		4,6703e–3	
		1/40	7,9347e–4	2,014	1,1595e–3	2,010
		1/80	1,9704e–4	2,010	2,8802e–4	2,009
0.3	0.7	1/20	4,4687e–3		6,4354e–3	
		1/40	1,1184e–3	1,998	1,6106e–3	1,998
		1/80	2,7974e–4	1,999	4,0285e–4	1,999

$$u(0,t) = \mu_1(t), \ u(1,t) = \mu_2(t), \ 0 \le t \le 1, \tag{30}$$

$$u(x,0) = u_0(x), \ 0 \le x \le 1, \tag{31}$$

where $k(x,t) = (2 - sin(\pi x/2))e^t$, $\eta(x) = e^x$, $q(x,t) = 1 - sin(xt)$,

$$f(x,t) = (sin(\pi x) + cos(\pi x)) \left(\frac{6t^{3-\alpha}}{\Gamma(4-\alpha)} + \frac{t^{1-\alpha}}{\Gamma(2-\alpha)} \right)$$

$$+ \left(\frac{6t^{3-\beta}}{\Gamma(4-\beta)} + \frac{t^{1-\beta}}{\Gamma(2-\beta)} \right) (\pi^2(sin(\pi x) + cos(\pi x)) - \pi(cos(\pi x) - sin(\pi x)))e^x$$

$$+ (1 - sinxt)(sin(\pi x) + cos(\pi x))(t^3 + t + 1),$$

$$\mu_1(t) = t^3 + t + 1, \quad \mu_2(t) = -(t^3 + t + 1), \quad u_0(x) = sin(\pi x) + cos(\pi x).$$

The exact solution is $u(x,t) = (sin(\pi x) + cos(\pi x))(t^3 + t + 1)$.

Tables 1 and 2 present the accuracy and the convergence rate (CR) of the difference scheme. The convergence order is computed by the formula $\log_{\frac{h_1}{h_2}} \frac{e_1}{e_2}$.

References

1. Chudnovsky, A.F.: Thermophysics Soil. Nauka, Moscow (1976). p. 137. (in Russian)
2. Shkhanukov-Lafishev, M.Kh.: About boundary value problems for equation of the third order. Differ. Equ. **18**(4), 1785–1795 (1982)
3. Alikhanov, A.A.: Boundary value problems for the diffusion equation of the variable order in differential and difference settings. Appl. Math. Comput. **219**, 3938–3946 (2012)
4. Alikhanov, A.A.: A new difference scheme for the time fractional diffusion equation. J. Comput. Phys. **280**, 424–438 (2015)
5. Wu, Ch.: Numerical solution for Stokes' first problem for a heated generalized second grade fluid with fractional derivative. Appl. Num. Math. **59**, 2571–2583 (2009)
6. Shkhanukov-Lafishev, M.Kh., Taukenova, F.I.: Difference methods for solving boundary value problems for fractional differential equations. Comput. Math. Math. Phys. **46**(10), 1785–1795 (2006)
7. Alikhanov, A.A.: Numerical methods of solutions of boundary value problems for the multi-term variable-distributed order diffusion equation. Appl. Math. Comput. **268**, 12–22 (2015)
8. Alikhanov, A.A.: A priori estimates for solutions of boundary value problems for fractional-order equations. Differ. Equ. **46**(5), 660–666 (2010)

Slot Machine Base Game Evolutionary RTP Optimization

Delyan Keremedchiev$^{(\boxtimes)}$, Petar Tomov, and Maria Barova

Institute of Information and Communication Technologies,
Bulgarian Academy of Sciences, acad. Georgi Bonchev Str,
Block 2, 1113 Sofia, Bulgaria
d_keremedchiev@bas.bg
http://www.iict.bas.bg/

Abstract. Slot machines are casino gambling machines with three or more reels which spin when a button is pushed. The machine pays are based on patterns of symbols visible on the front of the machine when it stops. Most of the modern slots consist of a base game, free games and bonus games. The base game is the core of the playing process. Player's money are usually taken as bet in the base game and no bet is taken during free games or bonus games. Each slot machine has a parameter called return to player (RTP). RTP is the average amount of money which a player will get back, in average, after each spin of the reels. The total RTP (measured in percents) can be in the range between 75% and 98%. Its components are: base game RTP, free games RTP, bonus games RTP. The base game also controls how often free games will be activated and how often bonus games will be played. In this paper an evolutionary optimization algorithm for optimization of slot machine base game RTP by rearrangement of the symbols in the reels, is proposed. The problem itself is a combinatorial problem and the fitness function used checks all possible slot machine winning screens.

Keywords: Slot machine · Gambling · Genetic algorithms · Return to player · Optimization

1 Introduction

Slot machines are electronic gambling devices which are popular all over the world. The most popular slot machines consist of five reels. The reels start spinning when the button is pushed. Nowadays slot machines are computerized with PRNG embedded in them. In 1984 Inge Telnaes received a patent for a device titled, "Electronic Gaming Device Utilizing a Random Number Generator for Selecting the Reel Stop Positions" (US Patent 4448419) [1]. In the beginning slot machines were mechanical. They had a lever on one (the left) side of the machine (because of this lever, machines were known as one-armed bandits), which were used for reels spinning activation. The machine pays according to

I. Dimov et al. (Eds.): NAA 2016, LNCS 10187, pp. 406–413, 2017.
DOI: 10.1007/978-3-319-57099-0_45

symbol patterns, visible on the screen, when the reels stop. Slot machines are the most popular gambling method in casinos and constitute about 70 percent of the average US casino income [2].

A gambler playing a slot machine has credit inserted - either cash or by printed ticket or loaded by the attendant. The machine is activated by means of a lever (or a button), or by pressing a touchscreen. The objective of the game is to win money from the machine, which usually involves matching symbols on reels (mechanical or virtual) that spin and stop to reveal one or several symbols. Most games have a variety of winning combinations of symbols. If a player matches a combination according to the given patterns, the slot machine rewards the player [6].

Each machine has a table that lists the number of credits the player will receive if the symbols listed on the pay table line up on the pay line of the machine. Some symbols are wild and can represent many (or all) of the other symbols to complete a winning line [3]. Symbols are statistically distributed on the reels. Some symbols show up more often than others. Some symbols pay more than others, according to the pay table. Slot machines are usually adjusted to pay out winnings of 75 to 98% of the money that the players bet. It is known as a theoretical payout percentage or RTP (return to player). The minimum RTP varies among jurisdictions and is subject of law regulations [6].

$$RTP = SUM(bet)/SUM(win) * 100 \qquad (1.1)$$

This research is motivated by a previous work based on GA (genetic algorithm) optimization of slot machine RTP [5]. In this work RTP is optimized for the base game of the slot machine. Less important, free games and bonus game winnings frequency are also optimized. The source code used in this research is available as an open-source project in Github global repository [4]. The rest of this paper is organized as follows: Sect. 2 presents the model proposed. Section 3 is about some experiments and results. The final Sect. 4 is a conclusion and presents some ideas for further research.

2   The Model Proposed

The model proposed is based on Genetic Algorithms (GAs). GA is applied in RTP, free games and bonus game winning frequency, as parallel computing evolutionary optimization. Unlike [5,6], the fitness function is calculated by an exact numerical method instead of Monte-Carlo simulation.

2.1   RTP Optimization

Most of the modern slot machines are computerized. They have virtual reels with symbols distributed on them. Stops of the virtual reels are selected by randomly generated numbers. RTP of the game is directly dependant on symbols distribution on the reels. In most of the cases the total RTP is the sum of base game RTP, free games RTP and bonus game RTP. Usually symbols (and their positions) are selected manually by mathematicians. From mathematical point of view, slot machine reels are discrete distribution of symbols. Such distribution can be generated by discrete optimization, according given constraints like desired RTP, free games and bonus game winning frequency. Multi-criteria cost function is converted into a single criteria by Euclidean distance calculation. [4] Slot machine reels are presented as chromosomes in the GA optimization. The cost function is an exact numeric calculation of RTP in order to make solution space exploration more efficient.

2.2 Genetic Algorithms

Genetic algorithms (GAs) are search heuristic algorithms inspired by the process of natural selection [8,9]. GAs are routinely used to select points (candidate solutions) from the solutions space. By application of the techniques for inheritance (crossover), mutation and selection, the generated solutions get closer to the optimum. GAs are classified also as population based algorithms because each point in the solution space represents an individual inside the GA population. Each individual has a set of properties which are subject to mutation and modification (usually a crossover). The traditional representation of the properties is binary, as a sequence of zeros and ones, but other encodings are also possible (a binary tree for example) [5].

The optimization usually starts with a randomly generated population of individuals, but this may vary in different implementations. The optimization process is iterative and the population in each iteration is called generation. A fitness value is calculated for each individual of the generation. The fitness value usually represents the objective function which is subject to optimization. The fittest individuals in the population are selected (according to a selection rule) and recombined (crossover and/or mutation) to form a new generation. This new generation is used at the next iteration of the algorithm. The algorithm termination is usually achieved either by reaching the maximum number of generations or by reaching the desired level of the fitness value [5].

In order to run GAs, the researcher should provide: 1. Genetic representation of the solution space (the solution domain); 2. An appropriate fitness function to evaluate the solution domain. Once these two conditions are met, GAs can proceed with the population initialization and the iterative population improvement by repetitive application of the selection, crossover, mutation and individuals evaluation [5].

2.3 Implementation

GA chromosomes are two-dimensional arrays of symbols. All symbols are marked as integer numbers. Each GA chromosome is a point in discrete finite solution space. Valid reel should consist of integer numbers listed in Table 1. The slot machine model in this research consists of symbols from 3 to 12 as regular winning combination symbols. Symbols1 is used as a special symbols called wild (it substitutes all other symbols). Symbol16 is used as a special symbol called scatter. Scatter pays everywhere on the screen and it is not needed to form special line for combination. Scatter is also responsible for the game to play extra reels spins, called free games. Symbol15 is used as a special symbol called bonus symbol. The bonus symbol is responsible the game to enter bonus mode, which is extra game inside the slot game play. The optimization process runs on parallel CPUs. There is a central node with global GA population and group of calculating nodes. Each calculating node has a subset of the global population as described in [7]. Initialization of the global population is done by randomly generated reels. The size of the global population vary from several chromosomes to hundreds or thousands of chromosomes. The central node operation is as follows:

Table 1. Genetic algorithm parameters.

Parameter	Value
Generation gap	0.97
Crossover rate	0.98
Mutation rate	0.02
Maximum generations	100
Number of individuals	37
Number of variables	315
Inserted rate	100 %

1. Generate random global population;
2. Distribution for each local node:
   2.1 Random selection of local population replacement;
   2.2 Distribute local populations;
   2.3 Collect local optimization results;
   2.4 Stop if predefined number of distributions reached;
   2.5 Repeat from 2.1;
3. Finish;

Local population size is fixed on 37 or 57 usually (it is subject of empirical estimation). Local GA's parameters are described in Table 1.
Local node genetic algorithm is as follows:

1. Load subset of the global population into the memory;
2. Optimization:
   2.1 Select parents;
   2.2 Crossover;
   2.3 Mutation;
   2.4 Calculate RTP and frequencies;

2.5 Keep newly generated chromosome;
2.6 Stop if predefined number of generations reached;
2.7 Repeat from 2.1;
3. Report results;

The first step of local GA optimization is selecting parents and result vector. As a second step, binomial crossover is applied. In order mutation to be valid an element from randomly selected chromosome replaces an element into the result chromosome. In the third step, a mutation is done. The final fourth step is related to fitness value calculation. The main difference between this work and those in [5,6] is that in the fitness value calculation the Monte-Carlo simulation is not used. A full generation of the possible game screens is applied instead. In this approach only base game reels can be optimized which is a disadvantage. When there is a complex game play in free games mode and bonus game mode the only approach could be the Monte-Carlo simulation.

Multi-criteria problem is converted to a single-criteria problem by the formula of the Euclidean distance. Because there are only three criteria the problem is not so interesting from multi-criteria point of view. The dominant criteria is the RTP, when frequencies (winning, free spins mode and bonus mode) are not so important.

The maximum number of recombinations is used as an optimization termination. Manual observation/termination of the process is also possible. The final solution, found by GA, is an integer matrix. This matrix is directly applicable as slot machine reels strips. For example, if there is a slot game with 5 reels (visible on the screen as 5 columns and 3 rows), and each reel consists of 63 symbols, the final GA solution would be an integer matrix of 5×63 values (refer to [4] for more details).

3   Experiments and Results

All experiments have been done on an open-source slot machine simulator [4] (5×3 screen) with a particular pay Table (Table 1) and 50 winning lines (Fig. 1). All winning combinations are paid from left to right. The lowest winning is 2 - for a combination of 3 symbols SYM12 (Table 2). The highest winning is 500 -

Table 2. Slot machine pay table. Each column represents the winning of one particular symbol (10 possible symbols in this game). Each row shows the winning of the symbols when the combination is of 2, 3, 4 or 5 symbols.

	SYM03	SYM04	SYM05	SYM06	SYM07	SYM08	SYM09	SYM10	SYM11	SYM12
2 of	30	20	0	0	0	0	0	0	0	0
3 of	150	75	50	40	5	4	4	3	3	2
4 of	250	100	100	75	50	40	20	6	5	4
5 of	500	250	175	150	100	50	30	25	15	10

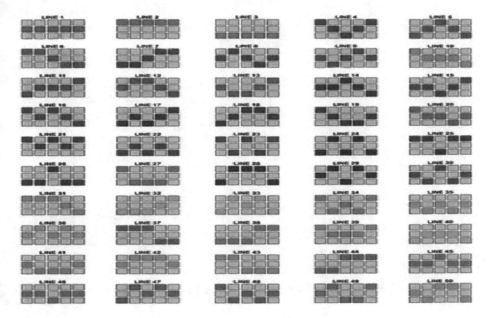

Fig. 1. Slot machine winning lines. There are 50 possible lines. On each of these lines (from left to right) win patterns can appear. Win patterns are formed by 2, 3, 4 or 5 symbols.

for a combination of 5 symbols SYM03 (Tab. 1). There are 10 regular symbols and 1 scatter symbol, which form the winning patterns on the screen.

All experiments are done by elitism rule keeping the best solution found. Global population expands, while local populations are fixed to 37 chromosomes. The maximum number of local recombinations is 100. All initial reels are randomly generated.

The target RTP was selected to 90, which is the lowest legal value for the Bulgarian gambling market. Thirty independent runs of the algorithm were done and it is visible that fast convergence was achieved in the beginning (Fig. 3). GA convergence is faster and smoother than DDE (Fig. 2), presented in [6].

The results were compared with those presented in [5,6], which is related to slot machine RTP optimization by GA and DDE, but with Monte-Carlo simulation as fitness value. The convergence is a little bit faster than that achieved in [5,6]. There are common differences in the models, but the general optimization idea is similar. In the case of this research the main disadvantage is the cost function. It is very time consuming and It is more time consuming and much slower than the cost functions used in both previous implementations.

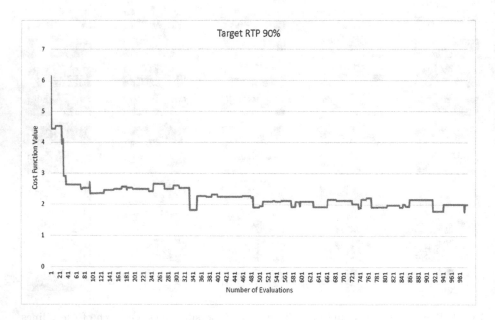

Fig. 2. Discrete differential evolution RTP optimization.

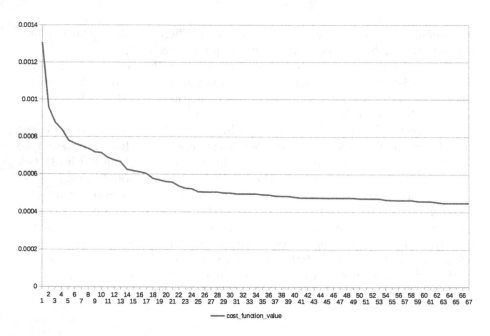

Fig. 3. Average convergence of thirty runs with mean value of 0.0005560259 and standard deviation of 0.0001492231.

4   Conclusions

The experiments show that the use of GA with exact numerical calculation of RTP as fitness function is very efficient and improves the slot game parameters by better adjustment of RTP, free games and bonus game frequencies. The optimization convergence depends on the probabilistic nature of GA. The biggest disadvantage is the exact RTP calculation which is time consuming and slows down the optimization process. The implementation of Discrete Diferential Evolution (DDE) instead of GA could be used for further research.

Acknowledgements. This work was supported by private funding of Velbazhd Software LLC.

References

1. Inge, S.: Electronic gaming device utilizing a random number generator for selecting the reel stop positions. US 4448419 A, Published 1984-05-15 (1984)
2. Cooper, M.: How slot machines give gamblers the business. The Atlantic Monthly Group (2005). Accessed 21 Apr 2008
3. Observer, C.: How to Play Slots. CasinoObserver.com (2013). http://casinoobserver.com/how-to-play-slots.htm. Accessed 06 Mar 2013
4. Balabanov, T.: Slot machine base game evolutionary RTP optimization as parallel implementation with MPI (2016). http://github.com/TodorBalabanov/SlotMachineBaseGameEvolutionaryOptimization/
5. Balabanov, T., Zankinski, I., Shumanov, B.: Slot machines RTP optimization with genetic algorithms. In: Dimov, I., Fidanova, S., Lirkov, I. (eds.) NMA 2014. LNCS, vol. 8962, pp. 55–61. Springer, Cham (2015). doi:10.1007/978-3-319-15585-2_6
6. Balabanov, T., Zankinski, I., Shumanov, B.: Slot machine RTP optimization and symbols wins equalization with discrete differential evolution. In: Lirkov, I., Margenov, S.D., Waśniewski, J. (eds.) LSSC 2015. LNCS, vol. 9374, pp. 210–217. Springer, Cham (2015). doi:10.1007/978-3-319-26520-9_22
7. Balabanov, T., Zankinski, I., Barova, M.: Distributed Evolutionary computing migration strategy by incident node participation. In: Lirkov, I., Margenov, S.D., Waśniewski, J. (eds.) LSSC 2015. LNCS, vol. 9374, pp. 203–209. Springer, Cham (2015). doi:10.1007/978-3-319-26520-9_21
8. Eiben, A.E., Raué, P.-E., Ruttkay, Z.: Genetic algorithms with multi-parent recombination. In: Davidor, Y., Schwefel, H.-P., Männer, R. (eds.) PPSN 1994. LNCS, vol. 866, pp. 78–87. Springer, Heidelberg (1994). doi:10.1007/3-540-58484-6_252
9. Ting, C.-K.: On the mean convergence time of multi-parent genetic algorithms without selection. In: Capcarrère, M.S., Freitas, A.A., Bentley, P.J., Johnson, C.G., Timmis, J. (eds.) ECAL 2005. LNCS (LNAI), vol. 3630, pp. 403–412. Springer, Heidelberg (2005). doi:10.1007/11553090_41

Numerical Analysis of Reinforced Concrete Deep Beams

Aleksandr E. Kolesov[1], Petr V. Sivtsev[1(✉)], Piotr Smarzewski[2],
and Petr N. Vabishchevich[3]

[1] North-Eastern Federal University, 58, Belinskogo, 677000 Yakutsk, Russia
sivkapetr@mail.ru
[2] Lublin University of Technology, Nadbystrzycka 40, 20-618 Lublin, Poland
[3] Nuclear Safety Institute, 52, B. Tulskaya, 115191 Moscow, Russia

Abstract. In this work we consider numerical analysis of elasticity problem for reinforced concrete deep beams. Main investigation is made to define the effect of presence of steel-polypropylene fibres in concrete mixture for different types of reinforcement. For numerical solution we use finite element method approximation. Numerical realization of method performed on collection of free software FEniCS. As model problem we consider computation of elastically-deformed state of reinforced concrete structure, consisting of concrete matrix and steel reinforcement, loaded in 3-point bending test. Numerical results of three-dimensional problem with complex geometry are presented.

1 Introduction

The development of concrete production technology and its properties together with notion to create safe and durable constructions contributed to the development of modern concrete composites [7,8]. In many cases they develop from the existing solutions through improvements of concrete properties. The methods of obtaining high strength concrete result from modifications of ordinary concrete by appropriate dosage of carefully selected ingredients and its proper curing. The benefits of additional fibres include increased resistance to failures, fatigue strength and the ability to absorb concrete energy. Another innovation is fibre hybridisation, which concerns rational use of various types of fibres in the composite and creating the concrete characterised by resistance to broad range of cracks. It is most advantageous to use large steel fibres, which ensure resistance to significant cracks, and small polypropylene fibres, which strengthen the mortar phase before and after cracking.

There are many applied mathematical problems related to calculation of stress-strain state of solid bodies [2–5]. In first approximation, one uses models of linear elasticity, which are described by Lame equations for displacement.

In this work we describe computational algorithm of solution for linear elasticity problem. This algorithm is based on finite-element approximation of displacement by space [1,9]. Numerical realization of method performed on collection of free software FEniCS [6].

© Springer International Publishing AG 2017
I. Dimov et al. (Eds.): NAA 2016, LNCS 10187, pp. 414–421, 2017.
DOI: 10.1007/978-3-319-57099-0_46

Features of computational algorithm are illustrated by calculation data for three-dimensional problem of reinforced concrete structure. Calculations are performed using the North-Eastern Federal University computational cluster *Arian Kuzmin*.

2   Problem Statement

Let us consider mathematical model that describe stress-strain state in computational domain Ω, which consist of reinforced deep beam, loading plate and supports

$$\operatorname{div} \boldsymbol{\sigma}(\boldsymbol{x}) = 0, \quad \boldsymbol{x} \in \Omega = \sum_{i=1}^{7} \Omega_i, \tag{1}$$

where $\boldsymbol{x} = (x_1, x_2, x_3)$ is coordinate vector.

Here Ω_1 is concrete subdomain of deep beam, Ω_2 is subdomain corresponding to loading plate, Ω_3 is subdomain related with support and other subdomains are introduced for different parts of inner reinforcement: Ω_4 for tension rebars, Ω_5 for compression rebars, Ω_6 for stirrups, Ω_7 for web mesh.

Next, we add relation between stress tensor $\boldsymbol{\sigma}$ and deformation tensor $\varepsilon_{i,j}$

$$\boldsymbol{\sigma}(\boldsymbol{x}) = \lambda \operatorname{div} \boldsymbol{u}(\boldsymbol{x}) E + 2\mu\varepsilon(\boldsymbol{x}),$$

where

$$\varepsilon_{i,j} = \frac{1}{2} \left(\frac{\partial u_i}{\partial x_j} + \frac{\partial u_j}{\partial x_i} \right),$$

E is unit tensor, $\boldsymbol{u} = (u_1, u_2, u_3)$ is displacement of body, λ, μ are Lame parameters, which are depend on subdomain and defined as

$$\mu = \mu_i, \quad \boldsymbol{x} \in \Omega_i,$$
$$\lambda = \lambda_i, \quad \boldsymbol{x} \in \Omega_i.$$

Then, we add boundary conditions, which are correspond to surface forces and support fixation. In particular, we define Dirichlet boundary condition

$$\boldsymbol{u}(\boldsymbol{x}) = \boldsymbol{u_0}, \quad \boldsymbol{x} \in \Gamma_D, \tag{2}$$

which defines fixation of displacement at Dirichlet boundary. We also add Neumann boundary condition

$$(\boldsymbol{\sigma} \cdot \boldsymbol{n})(\boldsymbol{x}) = \boldsymbol{f}(\boldsymbol{x}), \quad \boldsymbol{x} \in \Gamma_N. \tag{3}$$

This boundary condition related to pressing forces at Neumann boundary.

3   Finite-Element Discretisation

To get numerical solution for Eq. (1), we build finite-element approximation by space.

Let $H = L_2(\Omega)$ be Hilbert space with inner product and norm as

$$(u, v) = \int_\Omega u(\boldsymbol{x})\, v(\boldsymbol{x})\, dx, \quad ||u|| = (u, u)^{1/2},$$

and $\boldsymbol{H} = (L_2(\Omega))^d$ be space for displacement, where $\Omega \in \mathbb{R}^d$, $d = 2, 3$. Also let the test function $\boldsymbol{v}$ set to zero at Dirichlet boundary Γ_D, where solution is already known.

Then, using boundary conditions (2) and (3), we obtain the following variational formulation of problem: find $\boldsymbol{u} \in \hat{V}$ that

$$\int_\Omega \boldsymbol{\sigma}(\boldsymbol{u})\, \boldsymbol{\varepsilon}(\boldsymbol{v}) dx = \int_{\Gamma_N} (\boldsymbol{f}, \boldsymbol{v}) ds, \quad \forall \boldsymbol{v} \in \hat{V}, \tag{4}$$

where $\hat{V}$ is test function space, which is defined as

$$\hat{V} = \{\boldsymbol{v} \in \boldsymbol{H}(\Omega) : \boldsymbol{v}(\boldsymbol{x}) = 0, \quad \boldsymbol{x} \in \Gamma_D\},$$

and V is space of trial functions, shifted from test function space by the value of Dirichlet condition.

$$V = \{\boldsymbol{v} \in \boldsymbol{H}(\Omega) : \boldsymbol{v}(\boldsymbol{x}) = \boldsymbol{u}_0, \quad \boldsymbol{x} \in \Gamma_D\}.$$

Then, we define bilinear and linear forms as

$$a(\boldsymbol{u}, \boldsymbol{v}) = \int_\Omega \boldsymbol{\sigma}(\boldsymbol{u})\, \boldsymbol{\varepsilon}(\boldsymbol{v}) dx,$$

$$g(\boldsymbol{v}) = -\int_{\Gamma_N} (\boldsymbol{f}, \boldsymbol{v}) ds.$$

Then Eq. (4) switches to following variational formulation: find $\boldsymbol{u} \in V$ so, that

$$a(\boldsymbol{u}, \boldsymbol{v}) + g(\boldsymbol{v}) = 0, \quad \forall \boldsymbol{v} \in \hat{V}.$$

4   Investigation Object

Our objects of investigation are reinforced concrete deep beams, which is concrete block with inner steel reinforcement. There are three different samples, named as DB1, DB2 and DB3. Feature of first sample DB1 is different shape of reinforcement and concrete mixture without steel-polypropylene fibres. In this sample we have additional web mesh. Samples DB2, DB3 got the identical reinforcement and different concentration of steel-polypropylene fibres. Quantitative composition of concrete mixes are presented in Table 1. The investigation must

Table 1. Quantitative composition of concrete mixes

Sample	DB1		DB2		DB3	
Fibres concentration	kg/m^3	$\%, Vol.$	kg/m^3	$\%, Vol.$	kg/m^3	$\%, Vol.$
Polypropylene fibres	0.00	0.00	2.30	0.25	4.50	0.50
Steel fibres	0.00	0.00	78.00	1.00	156.00	2.00

define strength of samples depending on type of reinforcement and concentration of steel-polypropylene fibres.

Geometry of steel reinforcement in sample DB1 is presented in Fig. 1. Samples DB2 and DB3 got a similar steel reinforcement without web mesh, which is highlighted as dashed line with gray color. The reinforcements are consist of following parts: tension rebars with 22 mm diameter, compression rebars with 12 mm diameter, web mesh with 8 mm diameter and stirrups with 6 mm diameter.

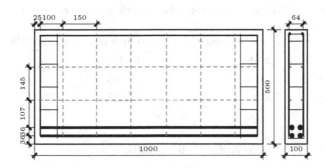

Fig. 1. Geometry of steel reinforcement in sample DB1

The samples are stand on cylindrical supports and affected to compressive force from hydraulic press. So complete geometry of problem, shown in Fig. 2, consist of deep beam set-up. In this figure there are boundaries, where we shown applied boundary condition. At lower surface of supports, which we modelled as rectangular, we have Dirichlet boundary condition that define fixed zero value for displacement. While, we model load as steel bar affected to vertical force at the top of the sample, which is defined as Neumann boundary condition. To achieve converging solution for stress and strain we used round form of intersection between sample and support/loading plate [10].

For all samples of concrete we got different concentration of steel-polypropylene fibre, so elastic properties of concrete are differs from each other. Table 2 shows elastic parameters for all concrete samples. In Table 3 we presented elastic parameters of steel for different parts of reinforcement. These parts can be seen in Fig. 3 as pieces of finished mesh.

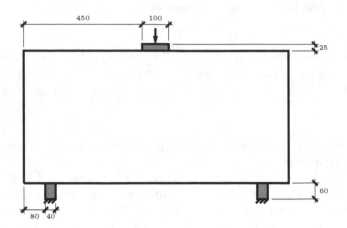

Fig. 2. Complete sample geometry with boundary conditions

Table 2. Concrete-fibre elastic parameters

Sample	Lame parameter λ, GPa	Lame parameter μ, GPa
DB1	7.2188	16.844
DB2	7.2294	16.869
DB3	7.3360	17.117

Table 3. Steel rebars elastic parameters

Reinforcement element	λ, GPa	μ, GPa
Tension rebars	117.1	78.1
Compression rebars	114.8	76.5
Web mesh	113.1	75.4
Stirrup	111.3	74.2

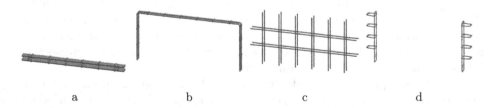

a   b   c   d

Fig. 3. Parts of steel reinforcement: a - tension rebars, b - compression rebars, c - web mesh, d - stirrups.

For comparative analysis of elastic-strain properties of samples we consider maximum values of displacement and Mizes's stress under fixed vertical pressure. The Mizes's stress is calculated as

$$\sigma_M = \frac{1}{2}[(\sigma_{11} - \sigma_{22})^2 + (\sigma_{22} - \sigma_{33})^2 + (\sigma_{33} - \sigma_{11})^2 + 6(\sigma_{23}^2 + \sigma_{31}^2 + \sigma_{12}^2)],$$

and used as scalar variable for arising stress σ.

5 Comparative Analysis

In order to guarantee accuracy of results the investigation for meshes with different size are performed [10]. And for example mesh for DB2 which provide proper accuracy of results consist of 1056625 cells and 5987417 vertices. So this analysis and solution of problem are performed on computational cluster.

To compare different reinforcement types and concrete structure, we solve linear elasticity model problem for three samples of reinforced concrete deep beams under vertical load force equal to 40 kN.

Table 4. Maximum values of displacement and Mizes stress for different samples

Sample	Maximum value of	
	$u_{max}, \mu m$	$\sigma_{M,max}, MPa$
DB1	32.07	19.83
DB2	32.52	19.85
DB3	32.10	19.88

Solutions of problem for different samples are shown in Table 4. In this table we have maximum values for displacement and Mizes stress. By analysing this data we conclude that by using additional web mesh in sample DB1 we get lesser value of displacement than in sample DB2. But high concentration of fibres in sample DB3 provide displacement close to value for sample DB1. And for all samples we have approximately same Mizes stress located in accuracy range.

For comparison of experimental data and numerical results. We solve same problem for deep beams under load equal to 150 kN in the measurement area for deep beam in Aramis system shown in Fig. 4. Comparison of experimental and numerical values of major strains are presented in Fig. 5.

According to pictures of major strains we can see that numerical strains are approximately less than experimental strains by 100 times. This can be explained by different nature of numerical and experimental data. The load equal to 150 kN generates strains which are beyond linear elasticity domain.

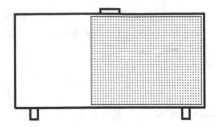

Fig. 4. The measurement area for the deep beam in Aramis system.

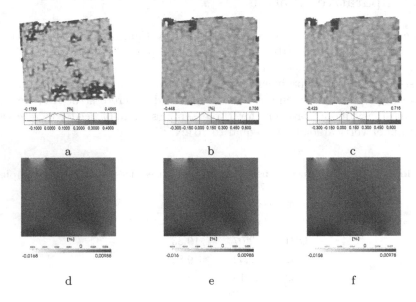

Fig. 5. Experimental: a - DB1, b - DB2, c - DB3 and numerical: d - DB1, e - DB2, f - DB3 major strain for load equal to 150 kN

6   Conclusion

We made numerical solution for linear elasticity problem for deep beam. As result of comparison of numerical solutions we find out positive impact of extra mesh web for reinforcement and mixing steel-polypropylene fibres in concrete by decreasing displacement.

In future we plan to use non-linear elasticity models to receive more appropriate results.

Acknowledgements. The research was supported by "Scientific and Educational Foundation for Young Scientists of Republic of Sakha (Yakutia)" 201604010196 and "Russian Foundation for Basic Research" (project N15-31-20856).

References

1. Afanas'eva, N.M., Vabishchevich, P.N., Vasil'eva, M.V.: Unconditionally stable schemes for convection-diffusion problems. Russ. Math. **57**(3), 1–11 (2013)
2. Vabishhevich, P.N., Vasil'eva, M.V., Kolesov, A.E.: Shema rasshheplenija dlja zadach porouprugosti i termouprugosti. Zhurnal vychislitel'noj matematiki i matematicheskoj fiziki **54**(8), 1345–1355 (2014)
3. Kolesov, A.E., Vabishchevich, P.N., Vasil'eva, M.V.: Splitting schemes for poroelasticity and thermoelasticity problems. Comput. Math. Appl. **67**(12), 2185–2198 (2014)
4. Kolesov, A.E., Vabishchevich, P.N., Vasilyeva, M.V., Gornov, V.F.: Splitting scheme for poroelasticity and thermoelasticity problems. In: Dimov, I., Faragó, I., Vulkov, L. (eds.) FDM 2014. LNCS, vol. 9045, pp. 241–248. Springer, Cham (2015). doi:10.1007/978-3-319-20239-6_25
5. Sivtsev, P.V., Vabishchevich, P.N., Vasilyeva, M.V.: Numerical simulation of thermoelasticity problems on high performance computing systems. In: Dimov, I., Faragó, I., Vulkov, L. (eds.) FDM 2014. LNCS, vol. 9045, pp. 364–370. Springer, Cham (2015). doi:10.1007/978-3-319-20239-6_40
6. Logg, A., Mardal, K.A., Wells, G.N.: Automated Solution of Differential Equations by the Finite Element Method. Springer, Heidelberg (2012)
7. Rombach, G.A.: Finite Element Design of Concrete Structures. Thomas Telford, London (2004)
8. Smarzewski, P., Poreba, J., Rentflejsz, A.: Badania doswiadczalne tarcz zelbetowych z betonu wysokowartosciowego z dodatkiem wlokien. Budownictwo i Architektura **10**, 15–26 (2012)
9. Lui, S.H.: Numerical Analysis of Partial Differential Equations. Wiley, Hoboken (2012)
10. Sivtsev, P.V.: Chislennoe modelirovanie zadachi uprugosti zhelezobetonnyh plit. Vestnik SVFU **4**(48), 64–74 (2015)

Numerical Solution of Thermoporoelasticity Problems

Alexandr E. Kolesov[1,2]([✉]) and Petr N. Vabishchevich[1,3]

[1] North-Eastern Federal University, Yakutsk, Russia
Kolesov.svfu@gmail.com
[2] University of North Carolina at Charlotte, Charlotte, NC 28223, USA
[3] Nuclear Safety Institute, RAS, Moscow, Russia

Abstract. We consider the numerical solution of thermoporoelasticity problems. The basic system of equations includes the Lame equation for the displacement and two nonstationary equations for the fluid pressure and temperature. The computational algorithm is based on the finite element approximation in space and the finite difference approximation in time. We construct standard implicit scheme and unconditionally stable splitting schemes with respect to physical processes, when the transition to a new time level is associated with solving separate sub-problems for the desired displacement, pressure, and temperature. The stability of the scheme is achieved by passing to three-level difference scheme and by choosing a weight used as a regularization parameter. We provide the stability condition of the splitting scheme and present numerical experiments supporting this condition.

1 Introduction

Thermoporoelasticity problems study the thermoelastic behaviour of porous medium saturated by fluid and arise in many areas of science and technology such as the construction of buildings in permafrost and the enhanced oil recovery by hot water. The mathematical model of this problems includes the Lame elliptic equation for the displacement vector and two nonstationary parabolic equations for the fluid pressure and temperature increment. The most important feature of mathematical models of thermoporoelasticity consists in that these three equations are strongly tied together. The equation for displacements contains the body forces proportional to the pressure and temperature gradients. On the other hand, the both equations for pressure and temperature include the divergence of displacement velocity and hydrothermal coupling between each other.

In general, the numerical solution of poroelasticity or thermoelasticity problems is mostly based on the finite element approximation in space [4,8] and the finite difference schemes for time discretization. In work [2] the stability and convergence of two-level schemes for thermoporoelasticity problem were investigated on the basis of the general theory of stability (correctness) for operator-difference schemes [11,12]. The solution of two-level schemes leads to a coupled system of equations, which requires the use of special algorithms [1,3] or additive (splitting) schemes [10,13].

© Springer International Publishing AG 2017
I. Dimov et al. (Eds.): NAA 2016, LNCS 10187, pp. 422–429, 2017.
DOI: 10.1007/978-3-319-57099-0_47

For thermoporoelasticity problems, it is essential to construct schemes with decoupling into physical processes, when the transition to a new time level is associated with the sequential solution of separate sub-problems for the displacement, pressure, and temperature. Various splitting schemes for poroelasticity an thermoelasticity problems are constructed and studied in [5–7,14].

In this work, on the basis of Samarskii's regularization principle for operator-difference schemes [11] we construct the splitting scheme for thermoporoelasticity problems. The stability of the scheme is achieved by passing to three-level difference scheme and by choosing a weight used as a regularization parameter. We provide the stability condition of the splitting scheme and present numerical experiments supporting this condition.

2 Mathematical Model

We consider thermoporoelasticity problems, which involves a strong coupling among displacement u, pore pressure p, and temperature increment θ. Neglecting the heat convection and some nonlinear terms, the mathematical model is given by the system of equations:

$$\operatorname{div}\boldsymbol{\sigma}(\boldsymbol{u}) - \alpha_p \operatorname{grad} p - \alpha_\theta \operatorname{grad}\theta = 0, \tag{1}$$

$$c_p \frac{\partial p}{\partial t} - 3\beta \frac{\partial \theta}{\partial t} + \alpha_p \frac{\partial \operatorname{div}\boldsymbol{u}}{\partial t} - \operatorname{div}(k_p \operatorname{grad} p) = 0, \tag{2}$$

$$c_\theta \frac{\partial \theta}{\partial t} - 3\beta T_0 \frac{\partial p}{\partial t} + \alpha_\theta T_0 \frac{\partial \operatorname{div}\boldsymbol{u}}{\partial t} - \operatorname{div}(k_\theta \operatorname{grad}\theta) = 0. \tag{3}$$

Here, $\boldsymbol{\sigma}$ is the stress tensor:

$$\boldsymbol{\sigma} = 2\mu\boldsymbol{\varepsilon}(\boldsymbol{u}) + \lambda \operatorname{div}\boldsymbol{u}\boldsymbol{I},$$

where μ is the shear modulus, λ are the Lame coefficient, $\boldsymbol{I}$ is the unit tensor, and $\boldsymbol{\varepsilon}$ is the strain tensor:

$$\boldsymbol{\varepsilon} = \frac{1}{2}(\operatorname{grad}\boldsymbol{u} + \operatorname{grad}\boldsymbol{u}^T).$$

Also, α_p is Biot coefficients, $\alpha_\theta = \alpha(3\lambda + 2\mu)$, α is the thermal expansion coefficient, $c_p = 1/M$, M is the Biot modulus, c_θ is the heat capacity, β is the hydrothermal coupling coefficient, k_p and k_θ are the hydraulic and thermal conductivity, respectively, T_0 is the initial temperature. In (1)–(3) there are no body forces, fluid and heat sources. Also, for simplicity, we set the initial temperature $T_0 = 1$.

The system (1)–(3) is considered in a bounded domain Ω with a boundary Γ, on which we set the following homogeneous conditions for displacements, pressure, and temperature increment, respectively:

$$\boldsymbol{u} = 0, \quad x \in \Gamma_D, \quad \boldsymbol{\sigma}\boldsymbol{n} = 0, \quad x \in \Gamma_N, \tag{4}$$

$$p = 0, \quad \boldsymbol{x} \in \Gamma_N, \quad \frac{\partial p}{\partial n} = 0, \quad \boldsymbol{x} \in \Gamma_D, \tag{5}$$

$$\theta = 0, \quad \boldsymbol{x} \in \Gamma_N, \quad \frac{\partial \theta}{\partial n} = 0, \quad \boldsymbol{x} \in \Gamma_D, \tag{6}$$

Here, $\boldsymbol{n}$ is the unit normal to the boundary, $\Gamma = \Gamma_D + \Gamma_N$.

In addition, we specify the zero initial conditions for pressure and temperature increment

$$p(\boldsymbol{x}, 0) = p_0, \quad \theta(\boldsymbol{x}, 0) = \theta_0, \quad \boldsymbol{x} \in \Omega. \tag{7}$$

3   Finite Element Discretization

For the numerical solution, we use finite element approximation in space, so we need obtain a variational formulation of problem (1)–(7). First, we define the subspaces of scalar and vector functions

$$Q = \{q \in H^1(\Omega) : q(\boldsymbol{x}) = 0, \, \boldsymbol{x} \in \Gamma_D\},$$

$$V = \{v \in \boldsymbol{H}^1(\Omega) : v(\boldsymbol{x}) = 0, \, \boldsymbol{x} \in \Gamma_D\}.$$

where $H^1(\Omega)$ and $\boldsymbol{H}^1(\Omega)$ are the Sobolev spaces. After multiplying (1), (2), and (3) by test functions $v \in V$, $q \in Q$, and $s \in Q$, respectively, and integrating by parts to eliminate the second order derivatives, we come to the following variational problem: find $\boldsymbol{u} \in V$, $p, \theta \in Q$ such that

$$a(\boldsymbol{u}, \boldsymbol{v}) + \alpha_p \, g(p, \boldsymbol{v}) + \alpha_\theta \, g(\theta, \boldsymbol{v}) = 0, \quad \boldsymbol{v} \in V, \tag{8}$$

$$c_p \left(\frac{dp}{dt}, q \right) + r \left(\frac{d\theta}{dt}, q \right) + \alpha_p \, d \left(\frac{d\boldsymbol{u}}{dt}, q \right) + b_p(p, q) = 0, \quad q_p \in Q, \tag{9}$$

$$c_\theta \left(\frac{d\theta}{dt}, s \right) + r \left(\frac{dp}{dt}, s \right) + \alpha_\theta \, d \left(\frac{d\boldsymbol{u}}{dt}, s \right) + b_\theta(\theta, s) = 0, \quad s \in Q. \tag{10}$$

Here, the bilinear forms are defined as

$$a(\boldsymbol{u}, \boldsymbol{v}) = \int_\Omega \boldsymbol{\sigma}(\boldsymbol{u}) \varepsilon(\boldsymbol{v}) \, d\boldsymbol{x}, \quad c_p(p, q) = \int_\Omega c_p p \, q \, d\boldsymbol{x}, \quad c_\theta(\theta, s) = \int_\Omega c_\theta \theta \, s \, d\boldsymbol{x},$$

$$b_p(p, q) = \int_\Omega k_p \operatorname{grad} p \, \operatorname{grad} q \, d\boldsymbol{x}, \quad b_\theta(\theta, s) = \int_\Omega k_\theta \operatorname{grad} \theta \, \operatorname{grad} s \, d\boldsymbol{x}.$$

$$g(p, \boldsymbol{v}) = \int_\Omega \operatorname{grad} p \, \boldsymbol{v} \, d\boldsymbol{x}, \quad d(\boldsymbol{u}, q) = \int_\Omega \operatorname{div} \boldsymbol{u} \, q \, d\boldsymbol{x}, \quad r(p, q) = -\int_\Omega 3\beta p \, q \, d\boldsymbol{x}.$$

The bilinear forms $a(\cdot, \cdot)$, $b_p(\cdot, \cdot)$, $b_\theta(\cdot, \cdot)$, $c_p(\cdot, \cdot)$, $c_\theta(\cdot, \cdot)$, and $r(\cdot, \cdot)$ are symmetric and positive definite. According to the Green's formula, we have

$$\int_\Omega \operatorname{div} \boldsymbol{u} \, p \, d\boldsymbol{x} = -\int_\Omega \operatorname{grad} p \, \boldsymbol{u} \, d\boldsymbol{x} + \int_\Gamma \boldsymbol{u} n \, p \, d\boldsymbol{x}, \quad p \in Q, \quad \boldsymbol{u} \in V.$$

Then, taking into account the boundary conditions (4)–(6), forms $g(\cdot, \cdot)$ and $d(\cdot, \cdot)$ are related as $d(\boldsymbol{u}, p) = -g(p, \boldsymbol{u})$.

The discrete problem is obtained by restricting the variational problem (8)–(10) to discrete spaces: find $\boldsymbol{u}_h \in V_h \subset V$, $p_h, \theta_h \in Q_h \subset Q$ such that

$$a(\boldsymbol{u}_h, \boldsymbol{v}) + \alpha_p g(p_h, \boldsymbol{v}) + \alpha_\theta g(\theta_h, \boldsymbol{v}) = 0, \quad \boldsymbol{v} \in V_h, \tag{11}$$

$$c_p \left(\frac{dp_h}{dt}, q \right) + r \left(\frac{d\theta_h}{dt}, q \right) + \alpha_p d \left(\frac{d\boldsymbol{u}_h}{dt}, q \right) + b_p(p_h, q) = 0, \; q \in Q_h, \tag{12}$$

$$c_\theta \left(\frac{d\theta_h}{dt}, s \right) + r \left(\frac{dp_h}{dt}, s \right) + \alpha_\theta d \left(\frac{d\boldsymbol{u}_h}{dt}, s \right) + b_\theta(\theta_h, s) = 0, \quad s \in Q_h. \tag{13}$$

For further consideration, it is convenient to use the operator formulation of problem (11)–(13). We define finite-dimensional operators A, B_p, B_θ, C_p, C_θ, D, G related to the corresponding bilinear forms by setting, for example,

$$(A\boldsymbol{u}, \boldsymbol{v}) = a(\boldsymbol{u}, \boldsymbol{v}), \quad \forall \boldsymbol{u}, \boldsymbol{v} \in V.$$

As a result, we come from problem (11)–(13) to the following Cauchy problem for a system of equations:

$$A\boldsymbol{u}_h + \alpha_p G p_h + \alpha_\theta G \theta_h = 0, \tag{14}$$

$$C_p \frac{dp_h}{dt} + R \frac{d\theta_h}{dt} + \alpha_p D \frac{d\boldsymbol{u}_h}{dt} + B_p p_h = 0, \tag{15}$$

$$C_\theta \frac{d\theta_h}{dt} + R \frac{dp_h}{dt} + \alpha_\theta D \frac{d\boldsymbol{u}_h}{dt} + B_\theta \theta_h = 0, \tag{16}$$

$$p_h(0) = 0, \quad \theta_h(0) = 0. \tag{17}$$

Here, the operators A, B_p, B_θ, C_p, C_θ, and R are self-adjoint and positive definite, while D and G are adjoint with to each other with an opposite sign $D = -G^*$.

4   Time Discretization

For the discretization in time we use a uniform grid with a step $\tau > 0$. Let $\boldsymbol{u}^n = \boldsymbol{u}_h(\boldsymbol{x}, t^n)$, $p^n = p_h(\boldsymbol{x}, t^n)$, and $\theta^n = \theta_h(\boldsymbol{x}, t^n)$, where $t^n = n\tau$, $n = 0, 1, \ldots$. To obtain an approximate solution of problem (14)–(17) we use the standard fully implicit two-level scheme:

$$A\boldsymbol{u}^{n+1} + \alpha_p G p^{n+1} + \alpha_\theta G \theta^{n+1} = 0, \tag{18}$$

$$C_p \frac{p^{n+1} - p^n}{\tau} + R \frac{\theta^{n+1} - \theta^n}{\tau} + \alpha_p D \frac{\boldsymbol{u}^{n+1} - \boldsymbol{u}^n}{\tau} + B_p p^{n+1} = 0, \tag{19}$$

$$C_\theta \frac{\theta^{n+1} - \theta^n}{\tau} + R \frac{p^{n+1} - p^n}{\tau} + \alpha_\theta D \frac{\boldsymbol{u}^{n+1} - \boldsymbol{u}^n}{\tau} + B_\theta \theta^{n+1} = 0, \tag{20}$$

$$p^0 = 0, \quad \theta^0 = 0. \tag{21}$$

The numerical implementation of the scheme (18)–(21) is associated with determining u^{n+1}, p^{n+1}, and θ^{n+1} at each time level by solving the coupled system. The solution of this system of equations is time consuming and requires special algorithms [9].

In this work, on the basis of Samarskii's regularization principle, we construct the three-level splitting scheme with weights [7]:

$$Au^{n+1} + \alpha_p G p^n + \alpha_\theta G \theta^n = 0, \tag{22}$$

$$C_p \left(\sigma \frac{p^{n+1} - p^n}{\tau} + (1-\sigma) \frac{p^n - p^{n-1}}{\tau} \right) + R \frac{\theta^n - \theta^{n-1}}{\tau} +$$
$$\alpha_p D \frac{u^{n+1} - u^n}{\tau} + B_p p^{n+1} = 0. \tag{23}$$

$$C_\theta \left(\sigma \frac{\theta^{n+1} - \theta^n}{\tau} + (1-\sigma) \frac{\theta^n - \theta^{n-1}}{\tau} \right) + R \frac{p^n - p^{n-1}}{\tau} +$$
$$\alpha_\theta D \frac{u^{n+1} - u^n}{\tau} + B_\theta \theta^{n+1} = 0. \tag{24}$$

where the solution is determined by separately solving equations for displacement, pressure, and temperature. In addition to the initial condition (21) we need p^1 and θ^1, which can be calculated using the two-level scheme (18)–(20). The splitting scheme (22)–(24) is stable for $\sigma \geq \sigma_0$, where σ_0 does not depend on mesh or time step.

5   Numerical Experiments

We consider a two-dimensional problem in a unit square domain Ω (Fig. 1). At the upper part of the domain Γ_1 the load $\sigma n = -\mu \sin((\lambda + 2\mu)k_p t)n$ is applied in strip of length 0.2 m. The remaining of the top boundary Γ_2 is traction free $\sigma n = 0$. On the vertical boundaries Γ_3 we set zero horizontal displacement $un = 0$ and vertical surface traction $(\sigma n) \times n = 0$. The bottom of the domain Γ_4 is assumed to be fixed, i.e. the displacement vector is taken as zero $u = 0$. For pressure and temperature increment, we prescribe them to be zero $p = \theta = 0$ at Γ_2, while the remaining boundaries are impermeable and there are no heat flux $\partial p/\partial n = \partial \theta/\partial n = 0$.

For the numerical solution we use two computational meshes: the coarse mesh with 3100 cells and the fine mesh with 12300 cells (Fig. 2). The time step is chosen as $\tau = 0.005$ s. The material parameters chosen for computations are listed in Table 1.

Figure 3 shows the pressure (left), temperature increment (middle), and displacement (right) at time $t = 0.5$, which are obtained using the implicit scheme (18)–(19). The displacement is presented on the deformed domain (overstated for visual contrast).

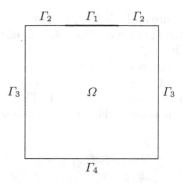

Fig. 1. Computational domain

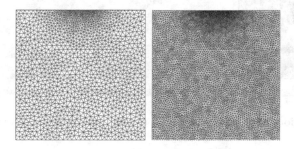

Fig. 2. Meshes

Table 1. Problem properties

Parameter	Unit	Value
μ	Pa	$6 \cdot 10^7$
λ	Pa	$6 \cdot 10^7$
c_p	Pa^{-1}	$1.25 \cdot 10^{-9}$
c_θ	$J/(m^3\ K)$	$2.18 \cdot 10^6$
k_p	$m^2/(Pa{\cdot}s)$	10^{-9}
k_θ	$W/(m{\cdot}\ K)$	1.51
α_p	–	0.9
α	$1/K$	$1.2 \cdot 10^{-5}$
β	K^{-1}	$1.84 \cdot 10^{-5}$

For analysis of the stability of the splitting scheme (22)–(24) we evaluate the pressure error:

$$\varepsilon_p = ||p_e - p||, \quad ||p_e - p||^2 - \int_\Omega (p_e - p)^2 d\varpi,$$

where p_e is the etalon pressure and p is the pressure, calculated using the scheme (22)–(24). As the etalon pressure we use the solution of the fully implicit scheme (18)–(19) on the fine mesh.

Figure 4 illustrates the errors of the pressure ε_p obtained using the splitting scheme for several values of weights σ on the coarse mesh. We see that the splitting scheme (22)–(24) is unstable for $\sigma = 1$. The increase in the value of σ improves the stability of the scheme. For $\sigma = 3.2$, the scheme is stable.

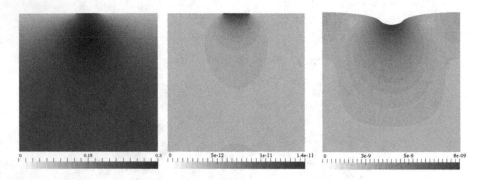

Fig. 3. Distributions of pressure (left), temperature (middle), and displacement (right)

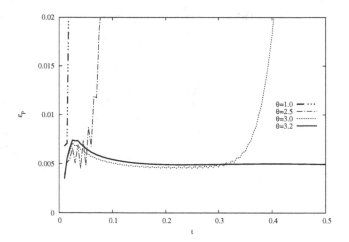

Fig. 4. Pressure errors for different weights σ

Acknowledgements. This research was supported by RFBR (project N14-01-00785A).

References

1. Alexsson, O., Blaheta, R., Byczanski, P.: Stable discretization of poroelastiicty problems and efficient preconditioners for arising saddle point type matrices. Comput. Vissualizat. Sci. **15**(4), 191–2007 (2013)
2. Boal, N., Gaspar, F., Vabishchevich, P.: Finite-difference analysis for the linear thermoporoelasticity problem and its numerical resolution by multigrid methods. Math. Model. Anal. **17**(2), 227–244 (2012)
3. Gaspar, F.J., Lisbona, F.J.: An efficient multigrid solver for a reformulated version of the poroelasticity system. Comput. Methods Appl. Mech. Eng. **196**(8), 1447–1457 (2007)
4. Haga, J.B., Osnes, H., Langtangen, H.P.: On the causes of pressure oscillations in low-permeable and low-compressible porous media. Int. J. Numer. Anal. Methods Geomech. **36**(12), 1507–1522 (2012)
5. Kim, J., Tchelepi, H., Juanes, R.: Stability and convergence of sequential methods for coupled flow and geomechanics: drained and undrained splits. Comput. Methods Appl. Mech. Eng. **200**(23–24), 2094–2116 (2011)
6. Kim, J., Tchelepi, H., Juanes, R.: Stability and convergence of sequential methods for coupled flow and geomechanics: fixed-stress and fixed-strain splits. Comput. Methods Appl. Mech. Eng. **200**(13–16), 1591–1606 (2011)
7. Kolesov, A.E., Vabishchevich, P.N., Vasil'eva, M.V.: Splitting schemes for poroelasticity and thermoelasticity problems. Comput. Math. Appl. **67**(12), 2185–2198 (2014)
8. Lewis, R., Schrefler, B.: The Finite Element Method in the Static and Dynamic Deformation and Consolidation of Porous Media. Wiley, Hoboken (1998)
9. Lisbona, F., Vabishchevich, P., Gaspar, F., Oosterlee, C.: An efficient multigrid solver for a reformulated version of the poroelasticity system. Comput. Methods Appl. Mech. Eng. **196**(8), 1447–1457 (2007)
10. Marchuk, G.I.: Splitting and alternating direction methods. In: Ciarlet, P.G., Lions, J.L. (eds.) Handbook of Numerical Analysis, I, North-Holland, pp. 197–462 (1990)
11. Samarskii, A.A.: The Theory of Difference Schemes. Marcel Dekker, New York (2001)
12. Samarskii, A.A., Matus, P.P., Vabishchevich, P.N.: Difference Schemes with Operator Factors. Kluwer Academic Publisher, Dordrecht (2002)
13. Vabishchevich, P.N.: Additive Operator-difference Schemes. Splitting schemes. de Gruyter, Berlin (2014)
14. Vabishchevich, P.N., Vasil'eva, M.V., Kolesov, A.E.: Splitting scheme for poroelasticity and thermoelasticity problems. Comput. Math. Math. Phys. **54**(8), 1305–1315 (2014)

Computation of Delta Greek for Non-linear Models in Mathematical Finance

Miglena N. Koleva[✉] and Lubin G. Vulkov

Faculty of Natural Science and Education,
University of Ruse, 8 Studentska str., 7017 Rousse, Bulgaria
{mkoleva,lvalkov}@uni-ruse.bg

Abstract. We consider a class of non-linear models in mathematical finance. The focus is on numerical study of Delta equation, where the unknown solution is the first spatial derivative of the option value. We also discuss the convergence to the viscosity solution. Newton's and Picard's iteration methods for solving the non-linear system of algebraic equations are proposed. Illustrative numerical examples are presented.

1 Introduction and Model Formulation

The pricing and hedging of European options is a long-standing problem of modern finance. Following the classical theory to Black, Sholes and Merton, see e.g. [13], an option value can be priced by solving linear equation, which has been derived under several restrictive assumptions, for example, zero transaction costs, perfectly replicated portfolio, frictionless, market completeness, etc. The motivation to study the non-linear modifications of Black-Sholes equation arises from more realistic option pricing models, in which one can take into account nontrivial transaction costs, market feedback, risk from unprotected portfolio and other effects.

We consider the following non-linear equation

$$\frac{\partial V}{\partial t} + \frac{1}{2}\sigma^2\left(S, t, \frac{\partial^2 V}{\partial S^2}\right) S^2 \frac{\partial^2 V}{\partial S^2} + (r-q)S\frac{\partial V}{\partial S} - rV = 0, \quad S > 0, \ 0 \le t < T, \quad (1)$$

where $V = V(S, t)$ is the option value, t is the time variable, S is the underlying asset price, $r > 0$ is the interest rate, $q \ge 0$ is the dividend yield rate and T is the maturity data.

For the purpose of computations it is necessary to restrict the underlying stock price in a finite region $I = (0, S_{max})$, where S_{max} denoted a sufficiently large positive number to ensure accuracy of the solution (cf. for example [14]). Then, we define the terminal and boundary conditions for (1) in this finite region:

$$V(S, T) = g_1(S), \quad S \in (0, S_{max}), \tag{2}$$
$$V(0, t) = g_2(t), \quad V(S_{max}, t) = g_3(t), \quad t \in (0, T],$$

satisfying the compatibility conditions $g_1(0) = g_2(0)$ and $g_1(S_{max}) = g_3(0)$, where g_1, g_2 and g_3 are given functions.

© Springer International Publishing AG 2017
I. Dimov et al. (Eds.): NAA 2016, LNCS 10187, pp. 430–438, 2017.
DOI: 10.1007/978-3-319-57099-0_48

The choices of g_1, g_2 and g_3 depend on the type of an option. Popular European options are Vanilla and Butterfly Spread for which the initial and boundary data are given by

$$g_1 = \begin{cases} \max(S-K,0) & \text{for Vanilla call,} \\ \max(K-S,0) & \text{for Vanilla put,} \\ \max(S-K_1,0) - 2\max(S-K,0) + \max(S+K_2,0), & \text{Butterfly Spread,} \end{cases}$$

$$g_2 = \begin{cases} 0 & \text{for Vanilla call,} \\ Ke^{-rt} & \text{for Vanilla put,} \\ 0 & \text{for Butterfly Spread,} \end{cases} \qquad g_3 = \begin{cases} S_{\max} - Ke^{-rt} & \text{for Vanilla call,} \\ 0 & \text{for Vanilla put,} \\ 0 & \text{Butterfly Spread,} \end{cases}$$

where K, K_1 and K_2 denote the strike prices of the options.

The partial derivatives of the solution - *Greeks*, are considered of major importance in finance, cf. e.g. [1,5,9,11,13]. In particular, the first spatial derivative, referred to as the *Delta Greek* in finance, is the key for the hedging process in time i.e. portfolio projection against market movements as it follows from the Black-Sholes hedging argument. Efficient techniques and fast computational methods for pricing derivative securities is a practical task in financial quotes markets. Thus, we have a good reason to find numerically Delta Greek more precisely, solving the corresponding equation, and then to restore the option value by numerical integration.

We invert the time $t := T - t$, introduce $W := V_S$ and formally differentiate (1) and (2) with respect to S in order to derive the equation

$$\frac{\partial W}{\partial t} = \frac{1}{2}\frac{\partial}{\partial S}\left[\sigma^2\left(S, t, \frac{\partial W}{\partial S}\right) S^2 \frac{\partial W}{\partial S}\right] + (r-q)S\frac{\partial W}{\partial S} - qW, \qquad (3)$$

for $S \in (0, S_{\max})$, $0 < t \le T$ and initial data (here $H(x) = 1_{[0,\infty)}(x)$ stands for Heviside functions):

$$g_1'(S) = \begin{cases} H(S-K) & \text{for Vanilla call,} \\ -H(K-S) & \text{for Vanilla put,} \\ H(S-K_1) - 2H(S-K) + H(S-K_2) & \text{for Butterfly Spread.} \end{cases} \qquad (4)$$

The boundary conditions for (3): g_2^W and g_3^W at $S = 0$ and $S = S_{\max}$ respectively, are compatible with the differential problem (3) and initial conditions (4), namely

$$g_2^W = \begin{cases} 0 \text{ for Vanilla call and },\\ -1 \text{ for Vanilla put,} \\ 0 \text{ for Butterfly Spread,} \end{cases} \qquad g_3^W = \begin{cases} 1 \text{ for Vanilla call,} \\ 0 \text{ for Vanilla put,} \\ 0 \text{ Butterfly Spread.} \end{cases} \qquad (5)$$

There are a number of papers on the numerical solution of (1) with different non-linear volatility terms, governing European option (cf., for example [5,7,9,17]). An explicit finite difference scheme, which requires a restrictive stability condition on the time and spatial mesh sizes, is developed in [3]. Ankudova and Ehrhardt [1] use a Grank-Nickolson method combined with a high order compact difference scheme, proposed in [12], to construct a numerical scheme for the linearized Black-Sholes equation using frozen values of nonlinear volatility. Lesmana and Wang [11] develop first order upwind finite difference method for a non-linear Black-Scholes equation.

The papers [15,16] present numerical approach for computing the Delta Greek and the option price of the Black-Sholes-Barenblatt equation.

The main goal in this paper is to construct and investigate efficient finite difference approximations for solving a class of non-linear modifications of the linear Black-Sholes equation. We are concentrated on Delta equation (3).

2   Finite Difference Approximations

We now introduce finite difference approximations for the discretization of (3). Let $I := (0, S_{\max})$ to be divided into M sub-intervals $I_i = (S_i, S_{i+1})$, $i = 0, 1, \ldots, M-1$ satisfying $0 = S_0 < S_1 < S_2 < \cdots < S_M = S_{\max}$. For $i = 0, 1, \ldots, M-1$ we denote $h_i = S_{i+1} - S_i$ and $\hbar_0 = h_0/2$, $\hbar_i = (h_{i-1} + h_i)/2$, $i = 1, 2, \ldots, M-1$, $\hbar_M = h_{M-1}/2$.

Similarly, we divide $[0, T]$ into sub-intervals with mesh nodes $\{t_n\}_{n=0}^N$ satisfying $0 = t_0 < t_1 < \cdots < t_N = T$ and let $\triangle t_n = t_{n+1} - t_n$, $n = 0, 1, \ldots, N-1$.

The numerical solution at point (S_i, t^n) is denoted by W_i^n. For the finite difference approximations of the first derivative we use the following notations

$$(W_S)_i^n = \frac{W_{i+1}^n - W_i^n}{h_i}, \quad (W_{\bar{S}})_i^n = \frac{W_i^n - W_{i-1}^n}{h_{i-1}}, \quad (W_{\mathring{S}})_i^n = \frac{h_{i-1}(W_S)_i^n + h_i(W_{\bar{S}})_i^n}{2\hbar_i}.$$

At each inner grid node S_i, $i = 1, 2, \ldots, M-1$, we approximate the non-linear term in (3) as follows:

$$
\begin{aligned}
&\frac{\partial}{\partial S} \left(\sigma^2 \left(S, t, \frac{\partial W}{\partial S}, \right) S^2 \frac{\partial W}{\partial S} \right)_i \\
&\approx \frac{1}{\hbar_i} \left[\left\{ \sigma^2 \left(S, t, \frac{\partial W}{\partial S} \right) S^2 \frac{\partial W}{\partial S} \right\}_{i+1/2} - \left\{ \sigma^2 \left(S, t, \frac{\partial W}{\partial S} \right) S^2 \frac{\partial W}{\partial S} \right\}_{i-1/2} \right] \\
&\approx \frac{1}{\hbar_i} \left(S_{i+1/2}^2 \sigma^2 \left(S_{i+1/2}, t, (W_S)_i \right) (W_S)_i - S_{i-1/2}^2 \sigma^2 \left(S_{i-1/2}, t, (W_{\bar{S}})_i \right) (W_{\bar{S}})_i \right).
\end{aligned}
\tag{6}
$$

Using (6) and upwind scheme for the convection term, we construct the full implicit discretization of (3)

$$
\begin{aligned}
&\frac{W_i^{n+1} - W_i^n}{\triangle t_n} - (r-q)^+ S_i (W_S)_i^{n+1} + (r-q)^- S_i (W_{\bar{S}})_i^{n+1} + q W_i^{n+1} \\
&- \frac{1}{2\hbar_i} \left[S_{i+1/2}^2 \widehat{\sigma}_{i+1/2}^{2,\,n+1} (W_S)_i^{n+1} - S_{i-1/2}^2 \widehat{\sigma}_{i-1/2}^{2,\,n+1} (W_{\bar{S}})_i^{n+1} \right] = 0,
\end{aligned}
\tag{7}
$$

where $v^+ = \max(0, v)$, $v^- = \max(0, -v)$ and

$$\widehat{\sigma}_{i+1/2}^{2,\,n+1} := \sigma^2 \left(S_{i+1/2}, t_{n+1}, (W_S)_i^{n+1} \right), \quad \widehat{\sigma}_{i-1/2}^{2,\,n+1} := \sigma^2 \left(S_{i-1/2}, t_{n+1}, (W_{\bar{S}})_i^{n+1} \right).$$

For any $n = 0, 1, \ldots, N - 1$, $i = 1, 2, \ldots, M - 1$, (7) can be written as

$$H_i^{n+1} := -\alpha_i^{n+1} W_{i-1}^{n+1} + \beta_i^{n+1} W_i^{n+1} - \gamma_i^{n+1} W_{i+1}^{n+1} - \frac{W_i^n}{\Delta t_n} = 0, \qquad (8)$$

where $H_i^{n+1} = H_i^{n+1}(W_i^{n+1}, W_{i-1}^{n+1}, W_{i+1}^{n+1}, W_i^n)$:

$$\alpha_i^{n+1} = \frac{1}{2\hbar_i h_{i-1}} S_{i-1/2}^2 \widehat{\sigma}_{i-1/2}^{2,\, n+1} + \frac{(r-q)^- S_i}{h_{i-1}},$$

$$\gamma_i^{n+1} = \frac{1}{2\hbar_i h_i} S_{i+1/2}^2 \widehat{\sigma}_{i+1/2}^{2,\, n+1} p_i + \frac{(r-q)^+ S_i}{h_i}, \qquad \beta_i^{n+1} = \frac{1}{\Delta t_n} + \alpha_i^{n+1} + \gamma_i^{n+1} + q.$$

To complete the finite difference scheme we define boundary and initial conditions in agreement with (5)

$$W_0^n = g_2^W(t_n), \ W_M^n = g_3^W(t_n), \ n = 1, \ldots, N, \ W_i^0 = g_1'(S_i), \ i = 0, \ldots, M. \quad (9)$$

The discretization (8) and (9) can be written in equivalent matrix-vector form

$$A^{n+1}(W^{n+1})W^{n+1} = F^n, \quad W^n = (W_0^n, W_1^n, \ldots, W_M^n)^T, \quad n = 0, 1, \ldots, N-1.$$

For the coefficient matrix A^n we establish the next property.

Theorem 1. *For any given W^n, $n = 1, \ldots, N$, the matrix $A^n = (A_{ij}^n)$ is an M-matrix.*

Proof (Outline). From the definition of the coefficients of (8) follows that for A^n both the sign condition and strictly diagonal domination are fulfilled. □

3   Convergence

Following the results in [2] for second order non-linear PDE, to guarantee the convergence of the fully implicit scheme (7) and (9) to the viscosity solution, it is sufficiently to prove consistency, stability and monotonicity of the discretization. Such questions are discussed in this section.

Suppose that $\sigma^2(S, t, U)\, U$, where $U = \frac{\partial W}{\partial S}$, is non-decreasing function with respect to U. Mostly of the models in financial mathematics satisfy this property. As an example we consider two well known models, written in terms of U:

- *Barles-Soner model* [3], where the non-linear volatility is given by

$$\sigma^2(S, t, U) = \sigma_0^2 \left(1 + \psi\left[e^{rt} a^2 S^2 U\right]\right).$$

Here σ_0^2 represents the historical volatility, $a = \kappa \sqrt{\widetilde{\gamma} \widetilde{N}}$ with κ being the transaction cost parameter, $\widetilde{\gamma}$ a risk aversion factor and $\widetilde{N}$ the number of options to be sold. The function ψ is the solution to the following non-linear initial value problem

$$\psi'(z) = \frac{\psi(z) + 1}{2\sqrt{z\psi(z)} - z} \quad \text{for } z \neq 0 \text{ and } \psi(0) = 0.$$

This model is derived on the base of the assumption that investor's preferences are characterized by an exponential utility function.

Function $\psi(z)$ is strictly increasing in z. Using the implicit representation of the solution (see [4]), in [11] is proved that the function $\left(1 + \psi\left[e^{rt}a^2 S^2 U\right]\right) U$ is increasing with respect to U.

- *Leland model* [10]:

$$\sigma^2\left(U\right) = \sigma_0^2\left(1 + \text{Le} \times \text{sign}\left(U\right)\right), \quad \text{Le} = \sqrt{\frac{2}{\pi}}\left(\frac{\kappa}{\sigma_0\sqrt{\delta t}}\right),$$

where $0 < \text{Le} < 1$ is the Leland number, δt denotes the transaction frequency and κ denotes transaction cost measure.

It is evidently that $\sigma^2\left(U\right) U$ is non-decreasing function with respect to U.

Lemma 1 (*Monotonicity*). *The discretization* (8), (9) *is monotone, i.e. for any* $\varepsilon > 0$ *and* $i = 1, 2, \ldots, M - 1$ *we have*

$$H_i^{n+1}(W_i^{n+1}, W_{i+1}^{n+1} + \varepsilon, W_{i-1}^{n+1} + \varepsilon, W_i^n + \varepsilon) \le H_i^{n+1}(W_i^{n+1}, W_{i+1}^{n+1}, W_{i-1}^{n+1}, W_i^n),$$
$$H_i^{n+1}(W_i^{n+1} + \varepsilon, W_{i+1}^{n+1}, W_{i-1}^{n+1}, W_i^n) \ge H_i^{n+1}(W_i^{n+1}, W_{i+1}^{n+1}, W_{i-1}^{n+1}, W_i^n).$$

Proof (Outline). We apply the property, that $\sigma^2(S, t, U)U$ is non-decreasing function with respect to U, taking into account that $U_{i-1/2} = (W_{\bar{S}})_i$ and $U_{i+1/2} = (W_S)_i$. □

Lemma 2 (*Stability*). *For the solution of the discretization* (8) *and* (9) *is fulfilled the following estimate*

$$\|W^{n+1}\|_\infty \le \max\{\|g_1'\|_\infty, \|g_3\|_\infty, \|g_3\|_\infty\}.$$

Proof (Outline). In view of Theorem 1, we apply maximum principle. □

Lemma 3 (*Consistency*). *The discretization* (8) *and* (9) *is consistent.*

Proof (Outline). We apply classical consistent notion, relaying on Taylor series expansion of the solution of the differential problem. □

The next statement follows straightforwardly from [2] and Lemmas 1–3.

Theorem 2. *The solution of* (8), (9) *converges to the viscosity solution as* $(|h|, \triangle t) \to (0^+, 0^+)$, *where* $|h| = \max\limits_{0 \le i \le M} h_i$ *and* $\triangle t = \max\limits_{0 \le n \le N-1} \triangle t_n$.

4   Solution of the Non-linear Algebraic Systems

The obtained, after discretization, non-linear systems of algebraic equations will be solved by classical Newton iteration process, if the function $\sigma^2(\cdot)$ is differentiable; and Picard-like iteration method, which can be applied in both cases - for differentiable or non-differentiable function $\sigma^2(\cdot)$.

Classical Newton Method.
We find W^{n+1} by initiating a Newton's iteration process with initial guess $W^{(0)} = W^n$, where the Newton increment on the $(k+1)$-th step $\triangle^{(k+1)} = W^{(k+1)} - W^{(k)}$, $k = 0, 1, \ldots$, is the solution of the following linear system:

$$\triangle_0^{(k+1)} = g_2(t_{n+1}) - W_0^{(k)},$$

$$-D_i^{(k)}\triangle_{i-1}^{(k+1)} + P_i^{(k)}\triangle_i^{(k+1)} - Q_i^{(k)}\triangle_{i+1}^{(k+1)} = \frac{1}{\Delta t_n}W_i^n$$

$$+ \alpha_i^{(k)}W_{i-1}^{(k)} - \beta_i^{(k)}W_i^{(k)} + \gamma_i^{(k)}W_{i+1}^{(k)}, \quad i = 1, \ldots, M-1, \tag{10}$$

$$\triangle_M^{(k+1)} = g_3(t_{n+1}) - W_M^{(k)},$$

where

$$D_i^{n+1} = \frac{(r-q)^- S_i}{h_{i-1}} + \frac{S_{i-1/2}^2}{2\hbar_i h_{i-1}}\frac{\partial(\hat{\sigma}_{i-1/2}^{2,\,n+1}U_{i-1/2}^{n+1})}{\partial U_{i-1/2}^{n+1}} \geq 0,$$

$$Q_i^{n+1} = \frac{(r-q)^+ S_i}{h_i} + \frac{S_{i+1/2}^2}{2\hbar_i h_i}\frac{\partial(\hat{\sigma}_{i+1/2}^{2,\,n+1}U_{i+1/2}^{n+1})}{\partial U_{i+1/2}^{n+1}} \geq 0,$$

$$P_i^{n+1} = \frac{1}{\Delta t_n} + q + D_i^{n+1} + Q_i^{n+1} > 0.$$

Note, that the coefficient matrix of the linearized system (10) is an M-matrix.

Picard-Like Method.
The solution W^{n+1} is a consequence of the next iteration process for $k = 0, 1, \ldots$
and $W^{(0)} = W^n$:

$$-\alpha_i^{(k)}W_{i-1}^{(k+1)} + \beta_i^{(k)}W_i^{(k+1)} - \gamma_i^{(k)}W_{i+1}^{(k+1)} = \frac{1}{\Delta t_n}W_i^n, \quad i = 1, \ldots, M-1,$$

$$W_M^{(k+1)} = g_3(t_{n+1}), \quad W_0^{(k+1)} = g_3(t_{n+1}).$$

5   Numerical Examples

In this section we verify the accuracy, the order of convergence and the efficiency of the proposed methods for Delta equation of the Leland model, in the case of call option and more challenging butterfly option.

We consider a smooth non-uniform grid, cf. [6].- uniform inside the region $[S_l, S_r]$, and non-uniform outside with stretching parameter $c = K/10$:

$$S_i := \phi(\xi_i) = \begin{cases} S_l + c\sinh(\xi_i), & \xi_{\min} \leq \xi_i < 0, \\ S_l + c\xi_i, & 0 \leq \xi_i \leq \xi_{\text{int}}, \\ S_r + c\sinh(\xi_i - \xi_{\text{int}}), & \xi_{\text{int}} \leq \xi_i < \xi_{\max}. \end{cases}$$

The uniform partition of $[\xi_{\min}, \xi_{\max}]$ is defined by $\xi_{\min} = \xi_0 < \cdots < \xi_M = \xi_{\max}$:

$$\xi_{\min} = \sinh^{-1}\left(\frac{-S_l}{c}\right), \quad \xi_{\text{int}} = \frac{S_r - S_l}{c}, \quad \xi_{\max} = \xi_{\text{int}} + \sinh^{-1}\left(\frac{S_{\max} - S_r}{c}\right).$$

The payoff is further averaged in the neighbouring nodes of the strikes K, K_1 and K_2 (depending on the option) in order to minimize oscillations in the spatial error and to obtain g_1' we use numerical differentiation.

Computations are performed for fixed time step Δt. The likely local consistency is $O(\Delta t + |h|)$. Thus, in order to verify this at strike nodes, where the mesh is uniform and $h = \min_i h_i$, we choose $\Delta t = h$. Further, we set $\Delta t = T/\lceil T/h \rceil$, where $\lceil u \rceil$ is the smallest integer greater than or equal to u in order to get exactly the desired final time T and avoiding time interpolation.

We give the values of the solution at strike point K at maturity $T = 1$. Model parameters are: $r = 0.1$, $q = 0.05$, $\sigma_0 = 0.2$, $Le = 0.5$, $S_{\max} = 100$, $K = 50$, $K_1 = 40$ and $K_2 = 60$. Thus, for computations we may set $[S_l, S_r] = [K/2, 3K/2]$. The notation diff stands for the error (absolute value of the difference) between the solution on two consecutive refinement grids. The order of convergence CR is calculated as $\log_2$ from the ratio between two consecutive values of diff. The stopping criterion for iteration processes is $\max_i |\triangle_i^{(k+1)}| \leq 1.e{-}6$.

In Table 1 we give the results from the computations - grid nodes in space, the solution at strike point K, diff, the corresponding order of convergence and computer time in seconds (CPU). For both options we observe first order convergence in space and therefore first order convergence in time, because the ratio $\Delta t/h \approx 1$ is fixed. The order of convergence for butterfly option stabilizes for more fine meshes. For this option, we obtain better order of convergence, also for the coarse space meshes, if we decrease the time step, for example $\Delta t = h/5$.

Table 1. Leland model, call and butterfly options

| M | Call option | | | | Butterfly option | | | |
	$W(K,1)$	diff	CR	CPU	$-W(K,1)$	diff	CR	CPU
800	0.595807			0.01	0.093981			0.02
1600	0.596634	8.27e-4		0.03	0.093997	1.62e-5		0.05
3200	0.597052	4.18e-4	0.9825	0.11	0.094041	4.41e-5	-1.4427	0.18
6400	0.597263	2.11e-4	0.9910	0.39	0.094079	3.80e-5	0.2134	0.65
12800	0.597368	1.06e-4	0.9955	1.62	0.094104	2.44e-5	0.6409	2.54
25600	0.597421	5.27e-5	1.0015	6.24	0.094117	1.38e-5	0.8179	9.56
51200	0.597447	2.65e-5	0.9931	29.87	0.094125	7.37e-6	0.9091	45.759
102400	0.597461	1.33e-5	0.9966	131.04	0.094129	3.81e-6	0.9526	193.74
204800	0.597467	6.65e-6	0.9983	681.33	0.094131	1.94e-6	0.9756	954.86
409600	0.597471	3.32e-6	0.9995	2667.84	0.094132	9.76e-7	0.9882	3178.24

6   Conclusions

In this paper we present numerical methods for computation Delta Greek of option pricing non-linear models. Essentially, Newton and Picard methods are efficient and robust, but significant improvement can be achieved by implementing two-grid idea [8]. We intend to accelerate the efficiency and to improve the order of convergence of the numerical method in our next work.

Acknowledgements. This research was supported by the European Union under Grant Agreement number 304617 (FP7 Marie Curie Action Project Multi-ITN STRIKE - Novel Methods in Computational Finance) and the Bulgarian National Fund of Science under Project I02/20-2014.

References

1. Ankudinova, J., Ehrhardt, M.: On the numerical solution of nonlinear Black-Scholes equation. Comp. Math Appl. **56**, 799–812 (2008)
2. Barles, G.: Convergence of numerical schemes for degenerate parabolic equations arising in finance. In: Rogers, L.C.G., Talay, D. (eds.) Numerical Methods in Finance. Cambridge University Press, Cambridge (1997)
3. Barles, G., Soner, M.-H.: Option pricing with transaction costs and a nonlinear Black-Scholes equation. Finance Stoch. **2**, 369–397 (1998)
4. Company, R., Navarro, E., Pintos, J.R., Ponsoda, E.: Numerical solution of linear and nonlinear Black-Scholes option pricing equation. Comput. Math. Appl. **56**, 813–821 (2008)
5. Forsyth, P., Vetzal, K.: Numerical methods for nonlinear PDEs in finance, handbook of computational finance. In: Duan, J.-C., Härdle, W.K., Gentle, J.E. (eds.) Part of the Series Springer Handbooks of Computational Statistics, pp. 503–528. Springer, Heidelberg (2011)
6. Haentjens, T., In't Hout, K.J.: Alternating direction implicit finite difference schemes for the Heston-Hull-White partial differential equation. J. Comp. Fin. **16**(1), 83–110 (2012)
7. Koleva, M.N.: Positivity preserving numerical method for non-linear Black-Scholes models. In: Dimov, I., Faragó, I., Vulkov, L. (eds.) NAA 2012. LNCS, vol. 8236, pp. 363–370. Springer, Heidelberg (2013). doi:10.1007/978-3-642-41515-9_40
8. Koleva, M.N., Valkov, R.L.: Two-grid algorithms for pricing American options by a penalty method, In: ALGORITMY 2016, Slovakia, Publishing House of Slovak University of Technology in Bratislava, pp. 275–284 (2016)
9. Koleva, M.N., Vulkov, L.G.: On splitting-based numerical methods for nonlinear models of European options. Int. J. Comput. Math. **3**(5), 781–796 (2016)
10. Leland, H.E.: Option pricing and replication with transaction costs. J. Fin. **40**, 1283–1301 (1985)
11. Lesmana, D.C., Wang, S.: An upwind finite difference method for a nonlinear Black-Scholes equation governing European option valuation under transaction costs. Appl. Math. Comput. **16**, 8811–8828 (2013)
12. Rigal, A.: High order difference schemes for unsteady one-dimensional diffusion-convection problems. J. Comput. Phys. **114**(1), 59–76 (1994)
13. Ševčovič, D., Stehlíková, B., Mikula, K.: Analytical and Numerical Methods for Pricing Financial Derivatives. Nova Science Publishers, New York (2011)
14. Ševčovič, D., Žitňanská, M.: Analysis of the nonlinear option pricing model under variable transaction costs. Asia-Pacific Finan. Markets. **23**(2), 153–174 (2016)

15. Valkov, R.: Fitted strong stability-preserving schemes for the Black-Scholes-Barenblatt equation. Int. J. Comput. Math. **92**(12), 2475–2497 (2015)
16. Valkov, R.: Predictor-Corrector balance method for the worst-case 1D option pricing. Comput. Methods Appl. Math. **16**(1), 175–186 (2015)
17. Wang, S.: Anovel fitted finite volume method for the Black-Scholes equation governing option pricing. IMA J. Numer. Anal. **24**, 699–720 (2004)

Research of Optimum Strategy for Semi-Markov Queueing Models at Control of CBSMAP-Flow. Algorithmization

Elizaveta V. Kondrashova[✉] and Victor A. Kashtanov

Department of Applied Mathematics,
National Research University Higher School of Economics (HSE),
Tallinskaya Street 34, Moscow, Russia
elizavetakondr@gmail.com

Abstract. The functioning of different systems can be described using queueing models. Optimization is used to increase the efficiency of the system functioning. The research is devoted to CBSMAP-flow. Note that it is very reasonable to change the characteristics of arrival flows and others characteristics in various queueing models for optimization of its functioning. In the research a process of system functioning is investigated at an controlled arrival process (flow).

Algorithmization of the problem is necessary because choice of optimum strategy at a large number of the controlled objects is very labor-intensive process. The research is very interesting and actual. The researched model and algorithm describe a lot of mathematical problems such as control of telecommunication networks, transport management and other subjects. . .

Keywords: Control · Semi-markov process · Arrival flow · CBSMAP-flow · Algorithmization · Strategy

1 Introduction

The functioning of different systems can be described using queueing models (semi-markov models) [2,6]. The controlled strategies are used to increase the efficiency of the system. In the present paper a process of system functioning is investigated at an controlled arrival process (flow).

This controlled process is a generalization of a BMAP-flow (Batch Markov Arrival Process) [3,4].

Define CBSMAP-flow (Controlled Batch Semi-Markov Arrival Process-flow). After holding in the state comes to an end, a controlled semi-markov process jumps to the other state and the batch of queries of CBSMAP-flow will be generated [8].

The purpose of the queueing model theory is to present recommendations on maintenance for high efficiency system functioning. Notice that the functioning

© Springer International Publishing AG 2017
I. Dimov et al. (Eds.): NAA 2016, LNCS 10187, pp. 439–447, 2017.
DOI: 10.1007/978-3-319-57099-0_49

increase can be reached using control parameters. There can be several control parameters in the model. There may be the interarrival time distribution, the service time distribution, the number of servers, the system capacity and others. We are interested in several control parameters, that means the control set expansion.

One of the main results of the controlled semi-markov processes theory is strategy determination that gives the maximum value for the functional.

It is important to note that we can obtain new results using investigation for not only stationary characteristics (stationary queue length, loss probability, etc.), but using structure model changing. Stochastic character of the interarrival time and the service time generates a stochastic process in the queueing model. For the controlled stochastic process the problem of a choice of optimum control strategy is given.

In the previous papers [12–14] control in different models was carried out. In the present research control of the model is complicated. The control of three (four) parameters at the same time is used. One of the main problems is algorithmization for creating a program for calculation. It is planned to use algorithmization with all necessary steps for programming.

2   Multiserver-Queueing Model

Describe the model. The given system consists of N subsystems.

The subsystem of k-th type $k = \overline{1, N}$, in designations of Kendall's notation can be described as follows:

$CBSMAP/M_k/n_k/N_k$, where

- $CBSMAP$ means, that the arrival flow is controlled flow;
- $Symbol_k$ means, that service duration of customer in a subsystem is exponentionally distributed with parameter μ_k;
- Symbols n_k and N_k define the quantity of the service buffers and the number of places in the queue n_k and N_k accordingly.

In classification of queueing systems, the system can be considered as controlled semi-markov system as its evolution is defined with controlled semi-markov process.

For construction of controlled semi-markov arrival process, describing the evolution of the system, it is necessary to realize the following algorithm:

- Define Markov moments - Define the states of semi-markov process - Define control set and control strategy - Define semi-markov kernel and a matrix of transition probabilities for embedded Markov chain - Construct income functional on the trajectories of CSMP - Define optimum strategy of control.

2.1 Markov Moments, System States and Control Measures

In the given model the Markov moments are the moments of arrivals of any type customers in system. In case of k-th type customers arrival, the given customers are taken on service to a subsystem of k-th type, and in other subsystems the batch of zero quantity "arrives".

The system states are defined using a vector $(i, l_1, l_2, \ldots, l_N)$, where i - a state of an arrival flow (at Markov moment the batch of i-th type customers arrives), l_k - a quantity of queries in a subsystem of k-th type, $M_k + N_k + n_k > l_k \geq 0, k \neq i, M_i + N_i + n_i > l_i > 0$, n_k and N_k- accordingly the quantity of service channels and the quantity of places in the queue in the System (k), $i \in E = \{1, 2, \ldots, N\}$. The quantity of customers in a subsystem is final and depends on the admission discipline and the structure of a queueing model. Therefore $l_k \in E_k = \{1, 2, \ldots, M_k + N_k + n_k - 1\}$ and $(i, l_1, l_2, \ldots, l_N) \in E \times E_1 \times \ldots \times E_N$.

Enter the following designations:

$$(l_1, l_2, \ldots, l_N) = l;$$
$$E \times E_1 \times \ldots \times E_N = \tilde{E}.$$

The transition from state $(i, l_1, l_2, \ldots, l_N)$ to state $(j, l'_1, l'_2, \ldots, l'_j, \ldots, l'_N)$ with positive probability occurs if $l'_k \leq l_k, k \neq j$, so in all subsystems, except for a subsystem of j-th type, there is only a service customers which can be presented as process of death process, accordingly the quantity of customers in these subsystems is not more than the quantity of customers in subsystems at previous Markov moment of batch arrivals.

Note, that the system control is carried out using a control of arrival-flow at the moments of SMP (semi-Markov process) states change, at the Markov moments.

Remind, that control Markov strategy $G = (G_{(i,l)}(u), (i, l) \in \tilde{E})$, depending only on a current state of controlled process, is a set of the probability measures, given for each state $(i, l) \in \tilde{E}$ on σ-algebra of subsets of decisions' set $U_{(i,l)}$.

Remind, that control Markov strategy $G = (G_{(i,l)}(u), (i, l) \in \tilde{E})$, depending only on a current state of controlled process, is a set of the probability measures, given for each state $(i, l) \in \tilde{E}$ on σ-algebra of subsets of decisions' set $U_{(i,l)}$.

2.2 Parameters of Control

As it was noted above, the problem of controlled arrival flow is investigated. The flow is set as semi-Markov process. The arrival flow is defined by three factors: the type of the next batch, the arrival time of the next batch and the number of queries in the batch.

Therefore, the following options for construction of control measures set are possible:

– control of the next batch type,
– control of the next batch type and the moment of the batch arrival,

– control of the next batch type, the moment of the batch arrival and the number of queries in the batch.

1. Construct a control set. At Markov moments (transition moments, state $(i, l) \in \tilde{E}$), the type of queries is chosen. Then equality $U_{(i,l)} = E = \{1, 2, \ldots, N\}$ is fair, and the probability measure on discrete set $E = \{1, 2, \ldots, N\}$ is defined with a set of probabilities

$$G_{(i,l)}(j) = P\{u(t) = j/\xi(t) = (i, l)\} = p_{(i,l),j}, \\ p_{(i,l),j} \geq 0, \sum_{j \in E} p_{(i,l),j} = 1, E = \{1, 2, \ldots, N\}, \tag{1}$$

where the decision accepted at Markov moment t is designated $u(t)$;

2. If depending on the state of semi-markov process $(i, l) \in \tilde{E}$ the next batch type and the moment of the batch arrival are chosen, then the equations are fair for control measers

$$U_{(i,l)} = E \times [0, \infty) = \{(j, u), j \in E = \{1, 2, \ldots, N\}, u \in R^+ = [0, \infty)\},$$

that is the control set consists of final number of half-lines. A probability measure on the set can be given as a set of probabilities (1) and conditional distributions of continuous component. If to designate θ as casual interval of time through which the next batch of queries arrives to system then it is possible continuous to determine conditional distribution components by equality

$$F_{(i,l)}(j, u) = P\{\theta < u/\xi(t) = (i, l), u(t) = j\}, j \in E = \{1, 2, \ldots, N\}, u \in R^+,$$

and the strategy is given by equalities

$$G_{(i,l)}(j, u) = P\{u(t) = j, \theta < u/\xi(t) = (i, l)\} = p_{(i,l),j} F_{(i,l),j}(u), \\ j \in E = \{1, 2, \ldots, N\}, u \in R^+. \tag{2}$$

3. If depending on the state of semi-markov process $(i, l) \in \tilde{E}$ the next batch type, the moment of the batch arrival and the number of queries in the batch are chosen, then the equations are fair for control measures

$$U_{(i,l)} = E \times [0, \infty) \times \tilde{M}_k \\ = \{(j, u, s), j \in E = \{1, 2, \ldots, N\}, u \in R^+ = [0, \infty), s \in \tilde{M}_k = \{1, \ldots, M_k\},$$

that is the control set consists of final number of half-lines. The probability measure on the set is given as a set of probabilities

$$P\{u(t) = (j, s)/\xi(t) = (i, l)\} = p_{(i,l),(j,s)} = p_{(i,l),j} p^{(j)}(s), \\ p_{(i,l),(j,s)} \geq 0, \sum_{(j,s) \in E \times E_k} p_{(i,l),(j,s)}, = 1, j \in E = \{1, 2, \ldots, N\}, s \in \tilde{M}_k \tag{3}$$

Conditional distribution of continuous components is set by equality:

$$F_{(i,l)}(j, u, n) = P\{\theta < u/\xi(t) = (i, l), u(t) = (j, n)\}, \\ j \in E = \{1, 2, \ldots, N\}, u \in R^+, n \in \tilde{M}_k,$$

The strategy is given by equalities:

$$G_{(i,l)}(j, u, n) = P\{u(t) = (j, n), \theta < u/\xi(t) = (i, l)\}$$
$$= p_{(i,l),j} p^{(j)}(n) F_{(i,l)}(j, u, n), \qquad (4)$$
$$j \in E = \{1, 2, \dots, N\}, u \in R^+, n \subset (1, \dots, M_i).$$

The next steps of typical algorithm is a definition of semi-markov kernel and a matrix of transition probabilities for embedded Markov chain and construction of income functional on the trajectories of controlled semi-markov process.

Note several types of constants describing the incomes and charges during system functioning: $c_1^{(k)}$ - an income received for the service of one query (customer); $c_2^{(k)}$ - a payment for a time unit during the working of one server during the service; $c_3^{(k)}$ - a payment for a time unit of idle time of one server; $c_4^{(k)}$ - a payment for a time unit for staying in the queue for one query; $c_5^{(k)}$ - a payment for one lost query of k-th type.

3 Problem of Functional Optimization and Search of Optimum Strategy

In the previous papers [12,13] the following fact was proved: the income functional for the subsystems and for the system (CBSMAP-model) is a fractional-linear functional concerning the distributions $G = \{G_{(i,m)}(u), (i, m) \in \tilde{E}\}$ defining the Markov homogeneous strategy.

The final stage of the research is construction of optimum control strategy. For the solution of the problem we will use the known fact [2,9]: if the fractional-linear functional has an extremum (a maximum or a minimum), then this extremum is reached in a class of the degenerate determined strategy. Therefore the set by which extreme value of functionality is determined is significantly reduced.

We will describe a set of degenerate strategies.

1. The type of queries is chosen. The strategy is defined by equality (1). The fixed determined measure in state (i, l) is defined by equality

$$G_{(i,l)}(j) = P\{u(t) = j/\xi(t) = (i, l)\} = p_{(i,l),j} = 1,$$
$$G_{(i,l)}(n) = P\{u(t) = n/\xi(t) = (i, l)\} = p_{(i,l),n} = 0. \qquad (5)$$

Therefore, the number of degenerate measures in state (i, l) is equal N. If the number of states $(i, l) \in \tilde{E}$ is equal $K = \sum_{s=1}^{N}(M_s + N_s + n_s - 1) \prod_{l=1}^{N} (M_l + N_l + n_l)$, then the number of degenerated strategies is equal $L = N^K$.

$l \neq s$

Thus, we have the algorithm for searching of optimum strategy:

– For the fixed strategy (5) the matrix of transition probabilities is calculated;

$$p_{(i,m),(j,l)}^{(k)} = \lim_{t\to\infty} Q_{(i,m),(j,l)}^{(k)}(t) = \int_{U_{(i,m)}} Q_{(i,m),(j,l)}^{(k)}(\infty, u) G_{(i,m)}(du)$$
$$= p_{(i,m),j} \int_0^\infty \sum_{s=0}^{m_j} p_{m_j,s}^{(j)}(x) \tilde{p}_{N_j+n_j-s}^{(j)}(l_j - s) \prod_{v\neq j} p_{m_v,l_v}^{(v)}(x) dF_{i,j}(x). \qquad (6)$$

Where the probability $Q^{(k)}_{(i,m),(j,l)}(t,u)$ is semi-markov kernel.

The following designations are used:

- The probability $p^{(j)}_{m_j,s}(x)$ is probability of that in k-th subsystem during time t (between the markov moments) $(m-s)$ queries are operated, provided that during the initial moment in System there were m queries (depending on subsystem structure);
- Probabilities $\tilde{p}^{(j)}_m(l)$ - probability to accept m queries from batch of j-th type at presence of empty seats depending on an admission discipline.
- $F_{i,j}(x)$ - probability of that the following batch of customers will arrive in system before the moment x, provided that it is a batch of j-th type and the previous batch was a batch of i-th type.
- For this matrix the system of the algebraic equations is solved and the stationary distribution of embedded Markov chain at chosen fixed strategy is determined;
- At chosen fixed strategy characteristics (7) and (8) are calculated:

$$m^{(k)}_{(i,m)} = \int_0^\infty [1 - \sum_{(j,l)\in\tilde{E}} Q^{(k)}_{(i,m)(j,l)}(t)]dt \tag{7}$$

- mathematical expectation for the time of continuous staying of the process $\xi(t)$ in state $(i, \boldsymbol{m})$;

$$s^{(k)}_{(i,m)} = \int_{u\in U_{(i,m)}} \sum_{(j,l)} [\int_0^\infty R^{(k)}_{(i,m)(j,l)}(x,u) dQ^{(k)}_{(i,m)(j,l)}(x,u)] G_{(i,m)}(du) \tag{8}$$

- mathematical expectation of the saved up income during the system (k) working during continuous staying of the process $\xi(t)$ in state $(i, \boldsymbol{m})$.
- the income functional (9) is calculated for the chosen fixed strategy:

$$S^{(k)} = \frac{\sum_{(i,\bar{l})\in\tilde{E}} s^{(k)}_{(i,\bar{l})} \pi^{(k)}_{(i,l)}}{\sum_{(i,\bar{l})\in\tilde{E}} m^{(k)}_{(i,\bar{l})} \pi^{(k)}_{(i,l)}} \tag{9}$$

where $S^{(k)}_{(j,l)}(t)$ - the mathematical expectation of the saved up income in subsystem k during time $t > 0$, provided that the process starts from the state (j, l).

- Using all fixed strategies and the values of the income functional for these strategies, we define the maximal income and optimum strategy.

2. The next batch type and the moment of the batch arrival are chosen. The strategy is defined by equality (2). The fixed degenerated measure in state (i, l) is defined by equality (5) for discrete component and equality

$$F_{(i,m),j}(x) = \begin{cases} 1, \tau_{(i,m)} < x, \\ 0, \tau_{(i,m)} > x. \end{cases} \tag{10}$$

Thus, we have the algorithm for searching of optimum strategy:

- For the fixed strategy (5) the matrix of transition probabilities (6) is calculated;
- For this matrix the system of the algebraic equations is solved and the stationary distribution of embedded Markov chain at chosen fixed strategy is determined;
- At chosen fixed strategy characteristics (7) and (8) are calculated;
- The income functional (9) is calculated for the chosen fixed strategy;
- The maximum of the specific income (9) is determined by variables $\tau_{(i,m)} \geq 0, (i, m) \in \tilde{E}$;
- Using all fixed strategies and the values of the income functional for these strategies, we define the maximal income and optimum strategy.

3. The next batch type, the moment of the batch arrival and the number of queries in the batch are chosen. The strategy is defined by equality (4). The fixed degenerated measure in state (i, l) is defined by equality

$$G_{(i,l)}(j, n) = P\{u(t) = (j, n)/\xi(t) = (i, l)\} = p_{(i,l),j}p^{(j)}(n) = 1,$$
$$G_{(i,l)}(s, m) = P\{u(t) = (s, m)/\xi(t) = (i, l)\} = p_{(i,l),s} = 0, (s, m) \neq (j, n) \quad (11)$$

for discrete component and by equality

$$F_{(i,m)}(j, x, n) = \begin{cases} 1, \tau_{(i,m)(j,n)} < x, \\ 0, \tau_{(i,m)(j,n)} > x. \end{cases} \quad (12)$$

In essence, for this case the previous algorithm is able, but the number of options for search discrete components increases.

Thus, the algorithms formulated above allow to construct compliance:

$$(i, m) \to j^{(0)}_{(i,m)}, (i, m) \to (j^{(0)}_{(i,m)}, \tau^{(0)}_{(i,m)}), (i, m) \to (j^{(0)}_{(i,m)}, \tau^{(0)}_{(i,m)}, n^{(0)}_{(i,m)}), \quad (13)$$

and every time at hit in a state (i, m) to make the relevant decision with probability 1.

3.1   Scheme of the Automated Solution

We will present the main units of the algorithm.

"Choice of control" unit: parameters of control for arrival flow are chosen.

"Data input" unit: the maximum number of possible places in the queue and the maximum number of servers in each subsystem, intensity of the arrival flow for each subsystem, the constants defining the income and expenses of system for each subsystem are given by user.

"The Fixed Strategy" unit: the fixed degenerate strategy depending on chosen control are set.

"Calculations" unit: for the fixed degenerate strategy a matrix of transitional probabilities is calculated - the system of the algebraic equations is solved and the rated decision is determined - stationary distribution of the enclosed Markov's chain at the chosen fixed degenerate strategy is defined - the specific income corresponding to the chosen degenerate strategy discrete components (functionality of accumulation) is calculated.

"Optimum Strategy" unit: the maximum of the specific income for degenerate strategy is defined. The specific income is compared to the specific income for the next degenerate strategy. The strategy with maximum value is chosen.

"Set of the optimum strategies for every system state" unit: set of optimum strategies is displayed.

4 Conclusion

Thus we obtained the algoritmization for search of optimum strategy for semi-Markov queueing models at control of CBSMAP-flow. The scheme of the automated solution is submitted. All necessary points and compliance between strategies are described.

It is possible to complicate the control of the model. We can add one more parameter of control for each of the control types. If depending on the state of semi-markov process $(i, l) \in \tilde{E}$ the number of additional places in the queue for each subsystem is chosen then we have an additional part for control measure: $U_k = \{0, \ldots, u_k\}$, where $0 \leq u_k \leq n_k - l_k$, u_k - the number of additional places in the queue for k-th subsystem. For the whole system $U_{(i,l)} = (\{0, \ldots, u_1\}, \{0, \ldots, u_2\}, \ldots, \{0, \ldots, u_N\})$. This parameter of control can be added to three previous types of control.

The researched model and algorithm describe a lot of mathematical problems such as control of telecommunication networks, transport management and other subjects.

References

1. Barbu, V., Limnios, N.: Semi-Markov Chains and Hidden Semi-Markov Models toward Applications. Springer, New York (2008)
2. Barzilovich, E., Belyaev, J., Kashtanov, V.A.: The Problems of Mathematical Reliability-Theory. Radio I Svyaz', Moscow (1983)
3. Lucantoni, D.M., Hellstem, K.S., Neuts, M.F.: A single-server queue with server vacations and a class of non-renewal arrival processes. Advant. Appl. Probab. **22**, 676–705 (1990)
4. Lucantoni, D.M.: New results on the single server queue with a batch Markovian arrival process. Commun. Stat. Stoch. Models **7**, 1–46 (1991)
5. Janssen, J., Manca, R.: Applied Semi-Markov Processes. Springer US, New York (2006)
6. Janssen, J., Limnios, N.: Semi-Markov Models and Applications. Springer US, New York (1999)
7. Jewell, W.S.: Markov-renewal programming. Oper. Res. **11**, 938–971 (1967)
8. Kashtanov, V.A., Kondrashova, E.V.: Analisys of arrival flow, controlled with Markov chain. Reliab. J. (1), 52–68, (2012)

9. Kashtanov, V.A.: Elements of Stochastic Processes Theory. MIEM, Moscow (2012)
10. Kashtanov, V.A., Kondrashova, E.V.: Controlled semi-markov queueing model, In: ALT 2010 (Accelerated Life testing, Reliability-Based Analysis and Design), pp. 243–249. Universite Blaise Pascal (2010)
11. Kashtanov, V.A., Kondrashova, E.V.: Controlled semi-markov queueing models. In: 2011 International Conference on Integrated Modeling and Analysis in Applied Control and Automation, IMAACA, pp. 10–18, Rome (2011)
12. Kondrashova, E.V.: Optimization of income functional in controlled queueing model. Large-Scale Control J. **36** (2012). Moscow, IPO
13. Kondrashova, E.V., Kashtanov, V.A.: Optimization of the CBSMAP queueing model. In: Proceedings of the World Congress on Engineering, WCE 2013, London, UK, 3–5 July 2013, vol. I, pp. 69–73 (2013)
14. Kondrashova, E.V.: Optimization of the CBSMAP queueing model. J. Math. Syst. Sci. DP **3**, 359–364 (2014)

On Some Piecewise Quadratic
Spline Functions

Oleg Kosogorov and Anton Makarov[✉]

St. Petersburg State University, 7/9, Universitetskaya nab.,
St. Petersburg 199034, Russia
{o.kosogorov,a.a.makarov}@spbu.ru

Abstract. We study some C^1 quadratic spline functions on bounded domain. The spline functions comprise polynomials, trigonometric functions, hyperbolic functions or their combinations. We show that some subset of minimal splines share most properties of the classical polynomial B-splines (positivity, compact support, smoothness, partition of unity). Some examples of polynomial and non-polynomial minimal splines are given.

Keywords: Minimal splines · B-splines · Trigonometric splines · Hyperbolic splines · Exponential splines · Polynomial splines · Non-polynomial splines

1 Introduction

Polynomial B-splines have gained widespread application, in particular, the non-uniform rational B-spline curves and surfaces have been widely applied in many CAD/CAM systems. However, it is well known that they have several drawbacks that limit their applications. A great interest in the design of curves in spaces mixing algebraic, trigonometric and hyperbolic functions has arisen. Thus various methods for constructing splines sharing properties of B-splines are developed (for more details see [1–14] and reference therein).

In this paper we continue the study of properties of minimal splines initiated in [15–17]. The purpose of the paper is to construct the minimal splines of second order and to derive a number of properties of splines similar to familiar properties of the polynomial case (positivity, compact support, smoothness, partition of unity). First, in Sect. 1 we give basic notation. Then, in Sect. 2 we study spline functions on uniform and non-uniform partitions of a bounded interval of the real line. We regard approximation relations as a system of equations which leads to (polynomial or non-polynomial) minimal splines of maximal smoothness. Here we discuss some properties of minimal splines: compact support, smoothness, partition of unity. Finally, in Sect. 3 we give examples of polynomial, trigonometric and hyperbolic splines. We complete the study of properties of spline functions and discuss positivity of some subset of minimal splines.

© Springer International Publishing AG 2017
I. Dimov et al. (Eds.): NAA 2016, LNCS 10187, pp. 448–455, 2017.
DOI: 10.1007/978-3-319-57099-0_50

2 Notation and Auxiliaries

Let $\mathbb{Z}$ be the set of integers, and let $\mathbb{R}$ be the set of real numbers. We identify vectors of the space $\mathbb{R}^{m+1}$ with one-column matrices and apply the usual matrix operations, in particular, for two vectors $\mathbf{a}, \mathbf{b} \in \mathbb{R}^{m+1}$ the expression $\mathbf{a}^T \mathbf{b}$ is the Euclidean inner product of these vectors.

Components of vectors are written in the square brackets and enumerated by $0, 1, \ldots, m$, for example, $\mathbf{a} = ([\mathbf{a}]_0, [\mathbf{a}]_1, \ldots, [\mathbf{a}]_m)^T$. Reasons to use such a notation for components of vectors will be clear from the context.

The quadratic matrix with columns $\mathbf{a}_0, \mathbf{a}_1, \ldots, \mathbf{a}_m \in \mathbb{R}^{m+1}$ (in the indicated order) is denoted by $(\mathbf{a}_0, \mathbf{a}_1, \ldots, \mathbf{a}_m)$, and $\det(\mathbf{a}_0, \mathbf{a}_1, \ldots, \mathbf{a}_m)$ denotes its determinant.

An ordered set $\mathbf{A} = \{\mathbf{a}_j\}_{j \in \mathbb{Z}}$ of vectors $\mathbf{a}_j \in \mathbb{R}^{m+1}$ is called a *chain*. A chain is *complete* if $\det(\mathbf{a}_{j-m}, \mathbf{a}_{j-m+1}, \ldots, \mathbf{a}_j) \neq 0$ for all $j \in \mathbb{Z}$.

The set of all functions continuous on (α, β) is denoted by $C(\alpha, \beta)$. If the components of a vector-valued function $\mathbf{u} \in \mathbb{R}^{m+1}$ belong to $C(\alpha, \beta)$ we write $\mathbf{u} \in \mathbf{C}(\alpha, \beta)$. We use similar notation $C[a, b]$ and $\mathbf{C}[a, b]$ for the corresponding spaces on a segment $[a, b]$.

3 Space of Piecewise Functions

On an interval $(\alpha, \beta) \subset \mathbb{R}$ we consider a partition $X = \{x_j\}_{j \in \mathbb{Z}}$,

$$X : \ldots < x_{-1} < x_0 < x_1 < \ldots, \tag{1}$$

where $\alpha = \lim_{j \to -\infty} x_j$ and $\beta = \lim_{j \to +\infty} x_j$ (the cases $\alpha = -\infty$ and $\beta = +\infty$ are not excluded).

For $K_0 \geqslant 1$, $K_0 \in \mathbb{R}$, we denote by $\mathcal{X}(K_0, \alpha, \beta)$ the class of partitions of the form (1) possessing the *local quasiuniformity* property

$$K_0^{-1} \leqslant \frac{x_{j+1} - x_j}{x_j - x_{j-1}} \leqslant K_0.$$

We introduce the *fineness characteristics* h_X of a partition X by the formula

$$h_X = \sup_{j \in \mathbb{Z}} (x_{j+1} - x_j).$$

For $k, j \in \mathbb{Z}$ we use the notation

$$M = \cup_{j \in \mathbb{Z}} (x_j, x_{j+1}), \ S_j = [x_j, x_{j+3}], \ J_k = \{k - 2, k - 1, k\}.$$

Let $\mathbb{X}(M)$ be the linear space of real-valued functions on the set M. We consider a vector-valued function $\boldsymbol{\varphi} : (\alpha, \beta) \mapsto \mathbb{R}^3$ with components in $\mathbb{X}(M)$. If a chain of vectors $\{\mathbf{a}_j\}$ is complete, then the relations

$$\sum_{j' \in J_k} \mathbf{a}_{j'} \, \omega_{j'}(t) \equiv \boldsymbol{\varphi}(t) \quad \forall t \in (x_k, x_{k+1}), \ \forall k \in \mathbb{Z},$$

$$\omega_j(t) \equiv 0 \qquad \forall t \notin S_j \cap M, \tag{2}$$

uniquely determine the functions $\omega_j(t)$, $t \in M$, $j \in \mathbb{Z}$. It is clear from relations (2) that $\operatorname{supp} \omega_j \subset S_j$. The conditions (2) are called the *approximation relations*.

By the Cramer formula, from the system of linear algebraic equations (2) we find

$$\omega_j(t) = \frac{\det(\{\mathbf{a}_{j'}\}_{j' \in J_k, j' \neq j} \|'^j \varphi(t))}{\det(\mathbf{a}_{k-2}, \mathbf{a}_{k-1}, \mathbf{a}_k)} \quad \forall t \in (x_k, x_{k+1}), \ \forall j \in J_k,$$

where $\|'^j$ means that the determinant in the numerator is obtained from the determinant in the denominator by replacing $\mathbf{a}_j$ with $\varphi(t)$ (preserving the column order).

The linear span of functions $\{\omega_j\}_{j \in \mathbb{Z}}$ is denoted by

$$\mathbb{S}(X, \mathbf{A}, \varphi) = \left\{ u \mid u = \sum_{j \in \mathbb{Z}} c_j \omega_j \ \forall c_j \in \mathbb{R} \right\}.$$

For a vector-valued function φ we set $\varphi_j = \varphi(x_j)$, $j \in \mathbb{Z}$. For $\varphi \in \mathbf{C}^1(\alpha, \beta)$ we consider the vectors

$$\mathbf{d}_j = \varphi_j \times \varphi'_j, \tag{3}$$

where by $\times$ we denote the vector product in the space $\mathbb{R}^3$.

Let $\varphi \in \mathbf{C}^2(\alpha, \beta)$. We introduce the Wronskian determinant

$$W(t) = \det\big(\varphi(t), \varphi'(t), \varphi''(t)\big).$$

We define the following vector chain $\mathbf{A}^* = \{\mathbf{a}_j^*\}$:

$$\mathbf{a}_j^* = -\mathbf{d}_{j+1} \times \mathbf{d}_{j+2}. \tag{4}$$

Theorem 1. *Let $\varphi \in \mathbf{C}^2(\alpha, \beta)$. If*

$$|W(t)| \geqslant c = \operatorname{const} > 0 \ \forall t \in (\alpha, \beta), \tag{5}$$

and $X \in \mathcal{X}(K_0, \alpha, \beta)$ for some $K_0 \geqslant 1$, then there exists h_X such that the space $\mathbb{S}(X, \mathbf{A}^*, \varphi)$ lies in the space $C^1(\alpha, \beta)$.

Proof. The proof of this theorem can be found in [16]. □

Corollary 1. *Under the assumptions of Theorem 1, the chain $\{\mathbf{d}_j\}_{j \in \mathbb{Z}}$ is complete, and $\mathbf{d}_{j+p}^T \mathbf{a}_j^* \neq 0$ for $p = 0, 3$. For $p = 1, 2$ the equalities $\mathbf{d}_{j+p}^T \mathbf{a}_j^* = 0$ hold in view of the properties of the vector product.*

The functions $\{\omega_j^*\}$ of the space $\mathbb{S}(X, \mathbf{A}^*, \varphi)$ are referred to as *minimal splines of maximal smoothness* of the third order. We called them minimal because they have minimal support. The difference between the order of the spline and the order of the highest continuous derivative is called the *defect* of the spline. So splines $\{\omega_j^*\}$ have minimal defect.

Theorem 2. *If* $\varphi \in \mathbf{C}^2(\alpha, \beta)$ *and the condition (5) is fulfilled, then functions* $\omega_j^* \in C^1(\alpha, \beta)$ *and*

$$
\omega_j^*(t) = \begin{cases}
\dfrac{\mathbf{d}_j^T \varphi(t)}{\mathbf{d}_j^T \mathbf{a}_j^*}, & t \subset [x_j, x_{j+1}), \\[2ex]
\dfrac{\mathbf{d}_j^T \varphi(t)}{\mathbf{d}_j^T \mathbf{a}_j^*} - \dfrac{\mathbf{d}_j^T \mathbf{a}_{j+1}^*}{\mathbf{d}_j^T \mathbf{a}_j^*} \dfrac{\mathbf{d}_{j+1}^T \varphi(t)}{\mathbf{d}_{j+1}^T \mathbf{a}_{j+1}^*}, & t \in [x_{j+1}, x_{j+2}), \\[2ex]
\dfrac{\mathbf{d}_{j+3}^T \varphi(t)}{\mathbf{d}_{j+3}^T \mathbf{a}_j^*}, & t \in [x_{j+2}, x_{j+3}).
\end{cases}
\tag{6}
$$

Proof. The equality (6) is obtained by substituting the required formulas into approximation relations (2) and using the definition of $\mathbf{d}_j$ and $\mathbf{a}_j^*$ (cf. (3) and (4)). The continuity of the function ω_j^* and its first derivative is checked at the grid points x_{j+i}, $i = 0, 1, 2, 3$ by applying the formulas which we prove. At the remaining points of (α, β) the continuity is obvious. $\quad\square$

Theorem 3. *Let* $[\varphi(t)]_0 \equiv 1$ *for all* $t \in (\alpha, \beta)$. *If a vector chain* $\mathbf{A}^N = \{\mathbf{a}_j^N\}$ *is defined by the formula*

$$
\mathbf{a}_j^N = \frac{\mathbf{d}_{j+1} \times \mathbf{d}_{j+2}}{[\mathbf{d}_{j+1} \times \mathbf{d}_{j+2}]_0},
\tag{7}
$$

then

$$
\sum_{j \in \mathbb{Z}} \omega_j^N(t) \equiv 1 \quad \forall t \in (\alpha, \beta).
$$

Proof. The required equality is obtained from the approximation relations (2) with the vector $\mathbf{a}_j^N$ written in the componentwise form. $\quad\square$

The functions $\{\omega_j^N\}$ of the space $\mathbb{S}(X, \mathbf{A}^N, \varphi)$ is *normalized* minimal splines of maximal smoothness of the third order.

We consider finite-dimensional spaces of splines. We set

$$
a = x_0, \quad b = x_n, \quad J_{2,n} = \{-2, \ldots, n-1, n\}, n \in \mathbb{N}, n \geqslant 3.
$$

From the infinite partition X we extract a finite partition. It is convenient to define the extended knot sequence

$$
X_n = \{x_{-2} < \ldots < a = x_0 < x_1 < \ldots < x_{n-1} < x_n = b < \ldots < x_{n+2}\}.
$$

From the complete infinite chain $\mathbf{A}^N$ we extract a finite chain

$$
\mathbf{A}_n^N = \{\mathbf{a}_{-3}^N, \ldots, \mathbf{a}_{n-1}^N, \mathbf{a}_n^N\}.
$$

We restrict all functions in the space $\mathbb{S}(X, \mathbf{A}^N, \varphi)$ onto the segment $[a, b]$. The set of these restrictions is the finite-dimensional linear space

$$
\mathbb{S}(X_n, \mathbf{A}_n^N, \varphi) = \left\{ u \mid u = \sum_{j \in J_{2,n-1}} c_j \, \omega_j^N \quad \forall c_j \in \mathbb{R} \right\} \subset C^1[a, b].
$$

Remark 1. The restrictions of ω_j^N form a linearly independent system on the segment $[a, b]$. Moreover, $\dim \mathbb{S}(X_n, \mathbf{A}_n^N, \boldsymbol{\varphi}) = n + 2$.

Theorem 4. *For $j \in J_{2,n-1}$ the following assertions hold:*

1. $\{\omega_j^N\}$ is a linearly independent system of functions.
2. $\operatorname{supp} \omega_j^N = [x_j, x_{j+3}]$.
3. $\omega_j^N \in C^1[a, b]$.
4. $\sum_j \omega_j^N = 1$.

Proof. We obtain the required assertion from the definition of spline functions $\{\omega_j^N\}$ by formulas (6), (7), and the condition (5) that the Wronskian $W(t)$ does not vanish on $[a, b]$. $\qquad\square$

The splines $\{\omega_j^N\}$ share many of the properties of classical polynomial B-splines of the third order (the second degree). The property of positivity will be discussed in the next section.

4   Examples of Nonnegative Normalized Splines

Let $\boldsymbol{\varphi}(t) = (1, t, t^2)^T$. Then the Wronskian $W(t) = 2$ is different from zero. For all $j \in \mathbb{Z}$ from (3) we obtain a vector $\mathbf{d}_j = (x_j^2, -2\,x_j, 1)^T$, using representation (7), we find a vector

$$\mathbf{a}_j^N = \left(1, \frac{x_{j+1} + x_{j+2}}{2}, x_{j+1}x_{j+2}\right)^T.$$

Now for $j, k \in \mathbb{Z}$ easily derive equalities

$$\mathbf{d}_j^T \boldsymbol{\varphi}(t) = (t - x_j)^2, \qquad \mathbf{d}_j^T \mathbf{a}_k^N = (x_j - x_{k+1})(x_j - x_{k+2}).$$

From mentioned above equalities and formula (6) follow representation of *polynomial spline* denoted by ω_j^B:

$$\omega_j^B(t) = \frac{(t-x_j)^2}{(x_j-x_{j+1})(x_j-x_{j+2})}, \quad t \in [x_j, x_{j+1}),$$

$$\omega_j^B(t) = \frac{1}{x_{j+1}-x_j}\left[\frac{(t-x_j)^2}{x_{j+2}-x_j} - \frac{(t-x_{j+1})^2(x_{j+3}-x_j)}{(x_{j+2}-x_{j+1})(x_{j+3}-x_{j+1})}\right],$$

$$t \in [x_{j+1}, x_{j+2}),$$

$$\omega_j^B(t) = \frac{(t-x_{j+3})^2}{(x_{j+3}-x_{j+1})(x_{j+3}-x_{j+2})}, \quad t \in [x_{j+2}, x_{j+3}).$$

We note that ω_j^B coincide with the known polynomial third order B-splines (second degree).

Let $\varphi(t) = (1, \sin t, \cos t)^T$. Then the Wronskian $W(t) = -1$ is different from zero. For all $j \in \mathbb{Z}$ from (3) we obtain a vector $\mathbf{d}_j = (-1, \sin x_j, \cos x_j)^T$, using representation (7), we find a vector

$$\mathbf{a}_j^N = \left(1, \frac{\sin \frac{x_{j+1}+x_{j+2}}{2}}{\cos \frac{x_{j+1}-x_{j+2}}{2}}, \frac{\cos \frac{x_{j+1}+x_{j+2}}{2}}{\cos \frac{x_{j+1}-x_{j+2}}{2}}\right)^T.$$

Then for all $j, k \in \mathbb{Z}$ yields

$$\mathbf{d}_j^T \varphi(t) = -2 \sin^2\left(\frac{x_j - t}{2}\right), \qquad \mathbf{d}_j^T \mathbf{a}_k^N = -\frac{2 \sin \frac{x_j - x_{k+1}}{2} \sin \frac{x_j + - x_{k+2}}{2}}{\cos \frac{x_{k+1} - x_{k+2}}{2}}.$$

From mentioned above equalities and formula (6) follow representation of *trigonometrical spline*, denoted by w_j^{trig}:

$$w_j^{trig}(t) = \frac{\cos \frac{x_{j+2} - x_{j+1}}{2} \sin^2\left(\frac{t - x_j}{2}\right)}{\sin \frac{x_{j+1} - x_j}{2} \sin \frac{x_{j+2} - x_j}{2}}, \quad t \in [x_j, x_{j+1}),$$

$$w_j^{trig}(t) = \frac{\cos \frac{x_{j+2} - x_{j+1}}{2}}{\sin \frac{x_{j+1} - x_j}{2}}\left(\frac{\sin^2\left(\frac{t - x_j}{2}\right)}{\sin \frac{x_{j+2} - x_j}{2}} - \frac{\sin \frac{x_{j+3} - x_j}{2} \sin^2\left(\frac{t - x_{j+1}}{2}\right)}{\sin \frac{x_{j+3} - x_{j+1}}{2} \sin \frac{x_{j+2} - x_{j+1}}{2}}\right), \quad (8)$$

$$t \in [x_{j+1}, x_{j+2}),$$

$$w_j^{trig}(t) = \frac{\cos \frac{x_{j+2} - x_{j+1}}{2} \sin^2\left(\frac{x_{j+3} - t}{2}\right)}{\sin \frac{x_{j+3} - x_{j+1}}{2} \sin \frac{x_{j+3} - x_{j+2}}{2}}, \quad t \in [x_{j+2}, x_{j+3}).$$

Theorem 5. *For* $h_X < \pi/2$ *and for* $t \in (x_j, x_{j+3})$ *trigonometrical spline*

$$w_j^{trig}(t) > 0.$$

Proof. By the representation of the sine and cosine functions it is easily verified that the function $w_j^{trig}(t) > 0$ for $t \in (x_j, x_{j+1}]$ and $t \in [x_{j+2}, x_{j+3})$, and it is concave function on mentioned intervals.

By the representation (8) we have the sign of function w_j^{trig} for $t \in [x_{j+1}, x_{j+2})$ depends on the sign of expression is denoted by $f(t)$,

$$f(t) = \sin \frac{x_{j+3} - x_{j+1}}{2} \sin \frac{x_{j+2} - x_{j+1}}{2} \sin^2\left(\frac{t - x_j}{2}\right)$$

$$- \sin \frac{x_{j+2} - x_j}{2} \sin \frac{x_{j+3} - x_j}{2} \sin^2\left(\frac{t - x_{j+1}}{2}\right). \qquad (9)$$

Using the formula

$$\sin^2(a) - \sin^2(b) = \sin(a + b) \sin(a - b), \quad \forall a, b \in \mathbb{R},$$

for substitution $a = \frac{t - x_j}{2}, b = \frac{t - x_{j+1}}{2}$, from equality (9) we obtain

$$f(t) = \sin^2\left(\frac{t - x_{j+1}}{2}\right)\left[\sin \frac{\beta + \gamma}{2} \sin \frac{\beta}{2} - \sin \frac{\alpha + \beta}{2} \sin \frac{\alpha + \beta + \gamma}{2}\right]$$

$$+ \sin\left(t - \frac{x_j + x_{j+1}}{2}\right) \sin \frac{\beta + \gamma}{2} \sin \frac{\beta}{2} \sin \frac{\alpha}{2}, \qquad (10)$$

where $\alpha = x_{j+1} - x_j, \beta = x_{j+2} - x_{j+1}, \gamma = x_{j+3} - x_{j+2}$.

Using the formula

$$2\sin(a)\sin(b) = \cos(a-b) - \cos(a+b), \quad \forall\, a, b \in \mathbb{R},$$

in the square bracket in (10) for passing from the product to the sum of the trigonometric functions, in view of equalities

$$\frac{d^2}{dt^2}\sin^2\left(\frac{t - x_{j+1}}{2}\right) = \frac{\cos(\delta)}{2}, \quad \frac{d^2}{dt^2}\sin\left(t - \frac{x_j + x_{j+1}}{2}\right) = -\sin\left(\frac{\alpha}{2} + \delta\right),$$

where $\delta = t - x_{j+1}$, from (10) we obtain

$$\frac{d^2}{dt^2} f(t) = -\sin\frac{\alpha}{2}\left(\frac{\cos(\delta)}{2}\sin\frac{\alpha + 2\beta + \gamma}{2} + \sin\left(\frac{\alpha}{2} + \delta\right)\sin\frac{\beta + \gamma}{2}\sin\frac{\beta}{2}\right). \quad (11)$$

For $h_X < \pi/2$ all sine and cosine functions in (11) are positive. Therefore function f is convex, and hence function w_j^{trig} is convex on the interval $[x_{j+1}, x_{j+2})$. For $t = x_{j+1}$ and $t = x_{j+2}$ function $w_j^{trig} \in C^1$ is positive. Hence it is positive on the interval (x_{j+1}, x_{j+2}). $\qquad\square$

Let $\varphi(t) = (1, \sinh t, \cosh t)^T$. Then the Wronskian $W(t) = 1$ is different from zero. From the formula (6) follow representation of *hyperbolical spline*, denoted by w_j^{hyp}:

$$w_j^{hyp}(t) = \frac{\cosh\frac{x_{j+1} - x_{j+2}}{2}\sinh^2\left(\frac{x_j - t}{2}\right)}{\sinh\frac{x_j - x_{j+1}}{2}\sinh\frac{x_j - x_{j+2}}{2}}, \quad t \in [x_j, x_{j+1}),$$

$$w_j^{hyp}(t) = \frac{\cosh\frac{x_{j+1} - x_{j+2}}{2}}{\sinh\frac{x_j - x_{j+1}}{2}}\left(\frac{\sinh^2\left(\frac{x_j - t}{2}\right)}{\sinh\frac{x_j - x_{j+2}}{2}} - \frac{\sinh\frac{x_j - x_{j+3}}{2}\sinh^2\left(\frac{x_{j+1} - t}{2}\right)}{\sinh\frac{x_{j+1} - x_{j+3}}{2}\sinh\frac{x_{j+1} - x_{j+2}}{2}}\right),$$

$$t \in [x_{j+1}, x_{j+2}),$$

$$w_j^{hyp}(t) = \frac{\cosh\frac{x_{j+1} - x_{j+2}}{2}\sinh^2\left(\frac{x_{j+3} - t}{2}\right)}{\sinh\frac{x_{j+3} - x_{j+1}}{2}\sinh\frac{x_{j+3} - x_{j+2}}{2}}, \quad t \in [x_{j+2}, x_{j+3}).$$

The positivity of hyperbolical spline $w_j^{hyp}(t) > 0$ for $t \in (x_j, x_{j+3})$ does not depend on the fineness characteristics h_X of a partition X. The exact proof will be published elsewhere.

Acknowledgments. The reported study was funded by a grant of the President of the Russian Federation (MD-6766.2015.9) and by RFBR, according to the research project No. 16-31-60060 mol_a_dk.

References

1. Lyche, T.: A recurrence relation for Chebyshevian B-splines. Constr. Approx. **1**, 155–173 (1985)
2. Schumaker, L., Trass, C.: Fitting scattered data on spherelike surfaces using tensor products of trigonometric and polynomial splines. J. Numer. Math. **60**, 133–144 (1991)
3. Koch, P.E., Lyche, T.: Interpolation with exponential B-splines in tension. Comput. Suppl. **8**, 173–190 (1993)
4. Pottman, H., Wagner, M.G.: Helix splines as an example of affine Tchebycheffian splines. Adv. Comput. Math. **2**, 123–142 (1994)
5. Kvasov, B.I.: GB-splines and their properties. Ann. Numer. Math. **3**, 139–149 (1996)
6. Mazure, M.L.: Chebyshev-Bernstein bases. Comput. Aided Geom. Des. **16**, 649–669 (1999)
7. Mainar, E., Peña, J.M., Sânchez-Reyes, J.: Shape preserving alternatives to the rational Bezier model. Comput. Aided Geom. Des. **18**, 37–60 (2001)
8. Burova, I.G., Dem'yanovich, Y.K.: On the smoothness of splines. Mat. Model. **16**(12), 40–43 (2004)
9. Dem'yanovich, Y.K.: Embedded spaces of trigonometric splines and their wavelet expansion. Math. Notes **78**(5), 615–630 (2005)
10. Makarov, A.A.: Normalized trigonometric splines of Lagrange type. Vestn. St. Petersbg. Univ. Math. **41**(3), 266–272 (2008)
11. Costantini, P., Manni, C., Pelosi, F., Sampoli, M.L.: Quasi-interpolation in isogeometric analysis based on generalized B-splines. Comput. Aided Geom. Des. **27**, 656–668 (2010)
12. Makarov, A.A.: Reconstruction matrices and calibration relations for minimal splines. J. Math. Sci. **178**(6), 605–621 (2011)
13. Dem'yanovich, Y.K., Lebedinskii, D.M., Lebedinskaya, N.A.: Two-sided estimates of some coordinate splines. J. Math. Sci. **216**(6), 770–782 (2016)
14. Makarov, A.A.: Biorthogonal systems of functionals and decomposition matrices for minimal splines. J. Math. Sci. **187**(1), 57–69 (2012)
15. Dem'yanovich, Y.K., Kosogorov, O.M.: Splines and biorthogonal systems. J. Math. Sci. **165**(5), 501–510 (2010)
16. Makarov, A.A.: Construction of splines of maximal smoothmess. J. Math. Sci. **178**(6), 589–603 (2011)
17. Makarov, A.A.: Knot insertion and knot removal matrices for nonpolynomial splines. Numer. Methods Program. **13**, 74–86 (2012). [in Russian]

Moving Limit Cycles Model
of an Economic System

Vladimir Kryuchkov[1], Mark Solomonovich[2(✉)], and Cristina Anton[2]

[1] Financial University Under the Government of the Russian Federation, Dmitrov,
141802 Moscow Region, Russia
[2] MacEwan University, Edmonton, AB T5J 4S2, Canada
solomonovichm@macewan.ca

Abstract. We consider a model explaining the dependence between the
productivity of labor (PL) and the fixed capital per worker (FC) in an
economic system. The core of the model is a nonlinear oscillator with a
limit cycle as an attractor. We run numerical simulations of the dynamics
specific to this non-autonomous model to compare with the actual data
recorded for the years 1987–2001 for the enterprise Omsk Bacon. Based
on the numerical analysis we can conclude that the interior dynamics
is not affected by exterior perturbations. The numerical simulations can
help the managers of the enterprise to take the right steps to avoid stagnation.

Keywords: Dynamical system · Limit cycle · Complex systems
modeling · Fiber bundle · Numerical scheme

1   Introduction

We consider a model explaining a curve illustrating the dependence between the
productivity of labor (hereafter PL) and the fixed capital per worker (hereafter
FC) in an economic system. The curve has been obtained from statistical data
collected at the modern industrialized enterprise Omsk Bacon for the years 1973–
1987 (the dotted curve in Fig. 1) and the years 1987–2001 (the plain curve in
Fig. 1). Such a curve contradicts the traditional economic predictions because
according to the latter, the PL must gradually grow following the FC growth
until it reaches saturation level, hence the curve must look like a sigmoid - a
branch of the solution of some logistic equation.

In order to explain the curve observed from the experimental data for the
years 1973–1987, we have proposed in [3] a dynamical model assuming that the
curve is a phase portrait of a nonlinear oscillator with a stable limit cycle as
the attractor. The equilibrium of the oscillator can move along a certain sigmoid
(probably the one economists would love to see instead of the curve observed)
and the equilibrium moves only if the external impulse applied to the system is
high enough to overcome a certain threshold. From the economic viewpoint the
novelty of the model consists in the assumption that the system does not ever

© Springer International Publishing AG 2017
I. Dimov et al. (Eds.): NAA 2016, LNCS 10187, pp. 456–463, 2017.
DOI: 10.1007/978-3-319-57099-0_51

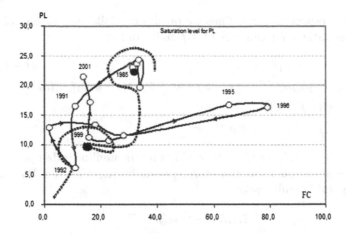

Fig. 1. The curves of the PL-FC interdependencies: data for 1973–1987 (dotted line), and for 1987–2001 (plain line). The painted dots denote the locations of the supposed equilibria.

stay in the equilibrium position, but rather oscillates about it. The oscillations are caused by the existence of the nonlinear feedback between the PL and FC variables.

The goal of the current paper is to extend the model proposed in [3] to represent the dependence between the PL and FC in the years 1987–2001. We consider a non-autonomous system of differential equations and we run numerical simulations to study the oscillations.

2   The Model

The core of the model is a nonlinear oscillator for the variables f and p, representing the FC and PL measurements respectively. The choice of the nonlinearity and the parameters of the system correspond to the well-known Van der Pole oscillator [1] with a limit cycle as an attractor. The system oscillates about an equilibrium with the coordinates F and P respectively in the $f - p$ plane, and thus the coordinates of the equilibrium F and P are also dynamic variables. We assume that they are connected by a constraint equation, which determines a sigmoidal curve in the $f - p$ plane:

$$P = \frac{\sqrt{\Gamma}}{\sqrt{1 + \omega \exp(-\lambda F)}} \tag{1}$$

where parameters $\Gamma = 840$, $\omega = 72$, $\lambda = 0.19$ have been introduced and estimated in [3].

In order to describe the dynamics of the system, we have introduced in [3] an auxiliary variable v, which can be interpreted as the rate of change of the F

variable with respect to time. Thus we obtain the following system of ordinary differential equations for the variables f, p, F, and v:

$$\dot{f} = p - P + \sigma_1\delta(t - t_1) + \sigma_2\delta(t - t_2) + \sigma_3\delta(t - t_3) + \sigma_5\delta(t - t_5) \qquad (2)$$

$$\dot{p} = -K(p)(p - P) - (f - F) + \sigma_4\delta(t - t_4) + \sigma_6\delta(t - t_6) \qquad (3)$$

$$\dot{F} = v\Theta(\rho|f - F|^\alpha - \epsilon) \qquad (4)$$

$$\dot{v} = (\rho|f - F|^\alpha - \epsilon)\Theta(\rho|f - F|^\alpha - \epsilon), \qquad (5)$$

where Θ denotes the Heaviside function [3], and the nonlinear damping coefficient $K(p)$ is chosen in the form that corresponds to the Van der Pole oscillator, exhibiting a stable limit cycle:

$$K(p) = -\gamma + (p - P)^2. \qquad (6)$$

The external impulses acting on the oscillator are represented by the terms of type $\sigma_i\delta(t - t_i), i = 1, \ldots, 6$, where δ is the Dirac's delta function [3] of the variable $t - t_i$. Thus the action of the ith impulse occurs at the moment t_i. The amplitudes of the impulses are determined by the values of the parameters σ_i. In order to avoid strong singularities in numerical experiments, and make the model more realistic (real-life impulses are never instantaneous), we replaced in simulations the δ- function with its regularization δ_a defined as

$$\delta_a(t) = \frac{a}{\pi(a^2t^2 + 1)} \qquad (7)$$

with the value of the regularization parameter $a = 20$.

3   Numerical Simulations

The model presented in [3] was composed in order to explain the dynamics of the PL and FC interdependence during the time interval between 1973 and 1987, a period of stability in the Soviet economy. The next period, from 1987 to 2001 corresponds to a time when the economy experienced great disturbances, not to say cataclysms.

Each disturbance has been modeled by a respective impulse - a term of type $\sigma_i\delta(t - t_i)$ occurring at the corresponding moment t_i. An impulsive term is added to the right hand side of the corresponding differential Eqs. (2) or (3). The value of the parameter σ_i is determined individually for each impulse; the values of the rest of the parameters are chosen as follows: $\alpha = 5$, $\rho = 1.5$, $\epsilon = 200$, $\gamma = 1.5$, $\Gamma = 840$, $\omega = 72$, $\lambda = 0.19$. For the numerical solutions we use a Runge-Kutta 4th order numerical scheme with time step $h = 10^{-4}$.

The first numerical experiment describes the curve shown in Fig. 2(a) by a dotted line, the behavior of the system in the years 1977–1987 (it was already described in [3]). The first plant of the enterprise started working at its full capacity in 1977, and it had immediately fallen into the first limit cycle; then with the introduction of the second plant in 1981 (moment t_1), a sharp increase of FC occurred. This sharp increase was modeled by the first impulsive term $\sigma_1\delta(t-t_1)$ acting on the FC (f variable). The jump in the f-value was followed by the respective increase in the p-value, and the system has entered into the second limit cycle, in which it was oscillating until 1987 (Fig. 2(a)).

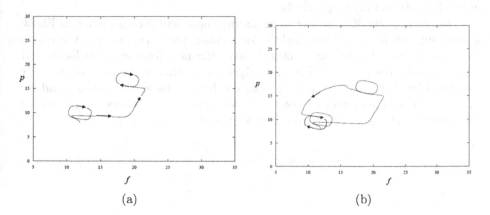

$$(a) \qquad\qquad\qquad\qquad\qquad (b)$$

Fig. 2. (a) $\sigma_1 = 13, \sigma_2 = \ldots = \sigma_6 = 0$ and (b) $\sigma_1 = 13, \sigma_2 = -10, \sigma_3 = \ldots = \sigma_6 = 0$

The next impulse $\sigma_2\delta(t-t_2)$ represents the fall of both FC and PL in 1990–92. In 1990 prices increased dramatically, which resulted in the impoverishment of the population. This resulted in the sharp decrease of the purchases of Omsk Bacon's products, and a simultaneous re-estimation of FC in the direction of decrease. The amount of the income of the enterprise (the numerator in the estimator for the productivity) went down, and so did the PL. The fall of the variables is modeled by adding the next perturbation $\sigma_2\delta(t-t_2)$ with $\sigma_2 = -10$. The resulting phase portrait is illustrated in Fig. 2(b). As a result of the second impulse, the system returns to the oscillation at a lower level, close to the position it had been before the first disturbance. The third limit cycle has been formed. On the experimental curve only the beginning of the oscillation can be seen, whereas in simulation we let the system perform a complete oscillation for the sake of showing that this is the same limit cycle.

As one can see from Fig. 1, in the years 1994–96 the FC of Omsk Bacon had quickly grown to a value approximately twice as large as the value of FC in 1985. It happened because by 1994 the centralized federal system of supplying the forage and selling the products had been lost, and in order to overcome the difficulties with supplies, Omsk Bacon purchased eight plant-growing agricultural enterprises. This surge in the value of capital funds has been modeled by adding the impulse $\sigma_3\delta(t-t_3)$ with a huge amplitude $\sigma_3 = 25$.

The above surge in the FC value was followed by an immediate fall in this value in 1997–98. The sharp decrease was caused by the introduction of the new taxation system. In order to avoid the deadly consequences of the latter, the owners of the firm had to re-evaluate the capital funds so as to decrease the taxable value. The nation-wide Russian financial crisis of 1998 also led to the sharp reduction in the income of the enterprise with the ensuing fall of the PL variable.

The fall of both variables in 1997–98 has been modeled by adding the impulses $\sigma_4\delta(t-t_4)$ with $\sigma_4 = -50$, and $\sigma_5\delta(t-t_5)$ with $\sigma_5 = -25$ to the right hand sides of the Eqs. (3) and (2) respectively.

The phase portrait resulting from adding these impulses is shown in Fig. 3: a new limit cycle, at a level slightly lower than the very first limit cycle, is about to emerge. In the experimental curve the new limit cycle is located at approximately the same level as the original one; this could have been easily simulated by slightly reducing the absolute values of the amplitudes σ_4, and σ_5; however, the authors preferred not to impose a cycle onto another cycle, just for the readers convenience, to have a better view of each of them.

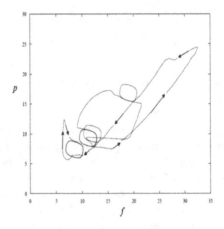

Fig. 3. $\sigma_1 = 13$, $\sigma_2 = -10$, $\sigma_3 = 25$, $\sigma_4 = -50$, $\sigma_5 = -25$, $\sigma_6 = 30$

As one can see from both the experimental curve in Fig. 1 and from the recent diagram in Fig. 3, the most recent limit cycle did not have enough time to emerge, since the system got hit by another impulse in 1999. The last impulse $\sigma_6\delta(t-t_6)$ with $\sigma_6 = 30$ characterizes the sharp growth of the PL in 1999–2001. This increase was an exhibition of recovery from the preceding crisis of 1998: due to the high cost of the US dollar, the import of meat products had been reduced, and the demand for the products of Omsk Bacon became really high. This allowed the owners of the enterprise to increase the prices. Thus, the income of the enterprise went up rapidly, and so did the PL-value.

The experimental curve (Fig. 1) ends in the year 2001, at the peak of the latter surge. In Fig. 3, this surge is followed by a fall back to the attracting limit cycle that started emerging in 1999 but did not form then because of the impulse $\sigma_6\delta(t-t_6)$. Now this limit cycle is shown in the diagram as the eventual attractor of the trajectory. This part illustrates the forecasting: if the internal parameters of the system do not change, it will inevitably fall down to the low level oscillation until the next impulse is applied.

It should be noted that the phase portraits obtained are not sensitive to the changes in the parameters of the dynamical system, which shows that the model is robust.

4   Geometric Interpretation and Its Generalization

In this section we would like to visualize the model, and respectively the system, as a geometrical structure, and generalize this structure to acquire a new perspective of the system, and create more opportunities for the management.

It is easy to see that the first two equations of the system, (2) and (3), describe the dynamics of the deviations of the FC and PL values from the equilibrium values. We shall introduce notations for these deviations: $x = f - F$, $y = p - P$. Thus, for any fixed location of the equilibrium (F, P), the (x, y) -variables oscillate about the equilibrium as prescribed by the equations obtained from (2) and (3) in the absence of impulses:

$$\dot{x} = v^1(x, y, \zeta), \qquad \dot{y} = v^2(x, y, \zeta), \tag{8}$$

where ζ denotes the parameters participating in these equations.

All possible equilibria are located on a smooth (sigmoidal) curve described by the constraint Eq. (1). We can view this equilibrium curve as a manifold, which has in this particular case a trivial structure: since it is a smooth non-closed curve, it is homeomorphic to R, and it can be covered by a single coordinate chart [2]. At each point of this manifold, a 2-dimensional (x, y)-plane is attached, which leads to the construction of a fiber bundle E with the equilibrium curve as the base manifold B, and the (x, y)-planes attached to the points of B as typical fibers [2].

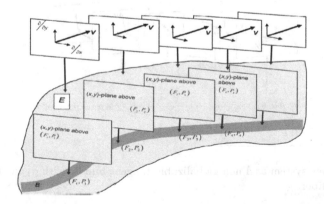

Fig. 4. The system viewed as the tangent bundle above the manifold B of equilibria.

E can be viewed as the tangent bundle $T_{(x,y)}B$ to the base manifold B created by assigning the tangent space $T_{(x,y)}$ to each fiber and defining the transition functions between the tangent spaces (these transition functions are elements of the structural group of the bundle [2]). The dynamical system (8) can be seen as a vector field above B (or a section of the bundle $T_{(x,y)}B$), obtained by selecting a vector $\mathbf{V} = (\dot{x}, \dot{y}) = \left(v^1(x, y, \zeta), v^2(x, y, \zeta)\right)$ from each tangent space from the corresponding fiber.

The geometric structure described above is illustrated in Fig. 4. The base manifold B of E is the space of the equilibrium values of the FC and PL variables. The dynamical system for the (x, y)-variables is determined in every fiber of the tangent bundle by the same vector $\mathbf{V}$, shown in the diagram in the tangent planes. The coordinates of $\mathbf{V}$ with respect to the coordinate basis $(\partial/\partial x, \partial/\partial y)$ do not change when moving between the fibers, and the dynamics inherent in the system do not change either.

The tangent bundle constructed above B is also globally trivializable, and vectors $\mathbf{V}$ above different fibers are exact copies of one another, since the dynamical system for the (x, y)-variables does not change. One can say that the structural group G of this tangent bundle consists of a single element - the identity. Then the following question emerges naturally: what may happen if the structural group G of the bundle is nontrivial? Can we assume some other structural group, which will change the vectors $\mathbf{V}$, and therefore the dynamics of the system when moving between the fibers? Obviously this is the goal of the management: to change the dynamics of the (x, y)-variables, so as to get rid of the stable limit cycle, which does not allow the growth of the productivity of labor.

The answer probably lies in changing the structure of B. If it is a manifold that cannot be covered with a single chart, the nontrivial transition functions between the charts will generate nontrivial transition functions between the fibers in the tangent bundle of B. Thus, the dynamics of the system, determined by vector $\mathbf{V}$, may change while moving from a fiber to another fiber.

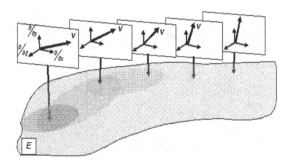

Fig. 5. The new system as a non-globalizable tangent bundle with dynamics changing from fiber to fiber.

Therefore it seems necessary to supply the base B with a nontrivial manifold structure, which can be done by adding more dimensions to the local patches covering B. In a way, we did that by introducing the impulses, and thus making the system non-autonomous. It did not help, however, with changing the dynamics of the (x, y)-system.

The new system, which can be viewed as a non-globalizable tangent bundle, with dynamics changing from fiber to fiber, is illustrated in Fig. 5. We have to introduce some other variables, denoted ξ, that can influence the transitions between the equilibrium states, and generate the changes in the transition functions between the fibers in E, and consequently in the fiber bundle. Such variables (usually called latent variables) may be disguised in the current system as some of its parameters. By changing them one can hope to change the dynamics inherent in the system, and, in particular, to break the limit cycle, or at least to increase its amplitude. For example, parameter γ from (6) may be related to latent variables (see the discussion in [3]).

5 Conclusions

Based on the numerical analysis of the model and the comparison of the results obtained with the two sets of statistical data from the intervals 1973–87, and 1987–2001, we notice that the interior dynamics inherent in the enterprise (Omsk Bacon) have not changed in spite of serious structural changes in exterior conditions, such as the transition of the model of action of the Russian economy from socialistic to capitalistic, an imposed privatization of the enterprise, three changes in ownership, and the deprivation of the enterprise of its forage supplies and distribution system.

Due to their failure to understand the dynamics of the enterprise, the management resorts to applying impulses, which shake the system, but do not generate its development. The knowledge of the model proposed and its possible generalizations can help the managers of the enterprise to take the right steps that will allow them to change the dynamics inherent in the enterprise and thus to avoid stagnation.

References

1. Andronov, A.A., Witt, A.A., Haikin, S.E.: Oscillation Theory. Physmatgiz, Moscow (1959)
2. Choquet-Bruhat, Y., DeWitt-Morette, C.: Analysis, Manifolds and Physics, Part 1: Basics. Elsevier Science B.V., North Holland (1982)
3. Kriuchkov, V.N., Solomonovich, M.: Self-sustained oscillation in economics. Appl. Math. Comput. **49**, 63–77 (1992)

The Use of Contrast Structures Theory for the Mathematical Modelling of the Wind Field in Spatially Heterogeneous Vegetation Cover

Natalia Levashova[✉], Julia Muhartova, and Marina Davydova

Lomonosov Moscow State University, bld. 2, Leninskiye Gory, 1, Moscow, Russia
natasha@npanalytica.ru

Abstract. In this work a three-dimensional model of turbulent exchange between the land surface and heterogeneous vegetation cover is presented. The model is based on so-called family of $E - \varphi$ models. For the description of the air flow interaction with vegetation elements on the land surface the theory of dissipative contrast structures is used by analogy with the active environments considered in biophysics.

Keywords: Wind field · Contrast structure · Navier-Stokes equation

1 Introduction

The problem of description of greenhouse gases transfer processes between the land surface and the atmosphere demands the development of the universal model approaches allowing to simulate the wind field of the atmospheric ground layer taking into account a non-uniform land surface with difficult relief and mosaic vegetation cover. The majority of modern models describing the transfer processes in a ground layer are based on the system of Navier-Stokes equation and the continuity equation [1]. In this work a three-dimensional model of turbulent exchange between the land surface and heterogeneous vegetation cover is presented. The model is based on so-called family of $E - \varphi$ models [2–4]. For the description of the air flow interaction with vegetation elements on the land surface the theory of dissipative contrast structures is used by analogy with the active environments considered in biophysics [5]. The contrast structure is referred to some function the domain of which has an area, where the function values undergo sharp changes. This area is called an interior transition layer. The idea to use the contrast structures theory came from the fact that in the vicinity of the two environments with various kinematic viscosity border (for example, a wood edge) big gradients of wind speed occur and therefore the graphs of the functions corresponding to the wind speed components have inner transition layers.

© Springer International Publishing AG 2017
I. Dimov et al. (Eds.): NAA 2016, LNCS 10187, pp. 464–472, 2017.
DOI: 10.1007/978-3-319-57099-0_52

2   Description of the Model

The developed model consists of system of six equations: three equations for a wind speed components, the equations for turbulent kinetic energy E and the specific dissipation of turbulent kinetic energy φ and the continuity equation.

For the three components $u(x,y,z,t)$, $v(x,y,z,t)$, $w(x,y,z,t)$ of the wind speed we have the equations

$$\frac{\partial u}{\partial t} + u\frac{\partial u}{\partial x} + v\frac{\partial u}{\partial y} + w\frac{\partial u}{\partial z} = F_x$$

$$-\frac{1}{\rho_0}\frac{\partial(\delta P)}{\partial x} - \frac{2}{3}\frac{\partial E}{\partial x} + 2\frac{\partial}{\partial x}\left(K\frac{\partial u}{\partial x}\right) + \frac{\partial}{\partial y}\left(K\frac{\partial u}{\partial y}\right) + \frac{\partial}{\partial z}\left(K\frac{\partial u}{\partial z}\right)$$

$$+\frac{\partial}{\partial y}\left(K\frac{\partial v}{\partial x}\right) + \frac{\partial}{\partial z}\left(K\frac{\partial w}{\partial x}\right) + 2\Omega v\sin\psi;$$

$$\frac{\partial v}{\partial t} + u\frac{\partial v}{\partial x} + v\frac{\partial v}{\partial y} + w\frac{\partial v}{\partial z} = F_y$$

$$-\frac{1}{\rho_0}\frac{\partial(\delta P)}{\partial y} - \frac{2}{3}\frac{\partial E}{\partial y} + \frac{\partial}{\partial x}\left(K\frac{\partial v}{\partial x}\right) + 2\frac{\partial}{\partial y}\left(K\frac{\partial v}{\partial y}\right) + \frac{\partial}{\partial z}\left(K\frac{\partial v}{\partial z}\right) \qquad (1)$$

$$+\frac{\partial}{\partial x}\left(K\frac{\partial u}{\partial y}\right) + \frac{\partial}{\partial z}\left(K\frac{\partial w}{\partial y}\right) - 2\Omega u\sin\psi;$$

$$\frac{\partial w}{\partial t} + u\frac{\partial w}{\partial x} + v\frac{\partial w}{\partial y} + w\frac{\partial w}{\partial z} = F_z$$

$$-\frac{1}{\rho_0}\frac{\partial(\delta P)}{\partial z} - \frac{2}{3}\frac{\partial E}{\partial z} + \frac{\partial}{\partial x}\left(K\frac{\partial w}{\partial x}\right) + \frac{\partial}{\partial y}\left(K\frac{\partial w}{\partial y}\right) + 2\frac{\partial}{\partial z}\left(K\frac{\partial w}{\partial z}\right)$$

$$+\frac{\partial}{\partial x}\left(K\frac{\partial u}{\partial z}\right) + \frac{\partial}{\partial y}\left(K\frac{\partial v}{\partial z}\right).$$

Here δP is an average deviation of pressure from standard p_0, determined by hydrostatic distribution

$$\frac{dp_0}{dz} = -\rho_0 g,$$

ρ_0 is the air density that we consider to be constant, g is a free fall acceleration, K is the turbulent diffusion coefficient equal in size to the main components of turbulent viscosity tensor. The functions F_x, F_y, F_z describe the interaction of the air flow with vegetation elements. The terms $2\Omega v\sin\psi$ it the first equation and $-2\Omega u\sin\psi$ in the second equation describe the influence of the rejecting force of the Earth rotation on driving of some volume of the air (Coriolis's force), here Ω is an angular speed of the earth rotation, ψ is geographic latitude.

The equations for the values E and φ are the following:

$$\frac{\partial E}{\partial t} + u\frac{\partial E}{\partial x} + v\frac{\partial E}{\partial y} + w\frac{\partial E}{\partial z}$$

$$= \frac{\partial}{\partial x}\left(\frac{K}{\sigma_E}\frac{\partial E}{\partial x}\right) + \frac{\partial}{\partial y}\left(\frac{K}{\sigma_E}\frac{\partial E}{\partial y}\right) + \frac{\partial}{\partial z}\left(\frac{K}{\sigma_E}\frac{\partial E}{\partial z}\right) + P - E\varphi;$$

$$\frac{\partial \varphi}{\partial t} + u\frac{\partial \varphi}{\partial x} + v\frac{\partial \varphi}{\partial y} + w\frac{\partial \varphi}{\partial z}$$

$$= \frac{C_{1\varphi}P - C_{2\varphi}\varphi E}{E}\varphi + \frac{\partial}{\partial x}\left(\frac{K}{\sigma_\varphi}\frac{\partial \varphi}{\partial x}\right) + \frac{\partial}{\partial y}\left(\frac{K}{\sigma_\varphi}\frac{\partial \varphi}{\partial y}\right) + \frac{\partial}{\partial z}\left(\frac{K}{\sigma_\varphi}\frac{\partial \varphi}{\partial z}\right) - \Delta_\varphi,$$

$$(2)$$

where P describes the generation of turbulent kinetic energy; it is expressed through the wind speed components as follows:

$$P = 2K\left(\left(\frac{\partial u}{\partial x}\right)^2 + \left(\frac{\partial v}{\partial y}\right)^2 + \left(\frac{\partial w}{\partial z}\right)^2\right)$$

$$+K\left(\left(\frac{\partial u}{\partial y} + \frac{\partial v}{\partial x}\right)^2 + \left(\frac{\partial u}{\partial z} + \frac{\partial w}{\partial x}\right)^2 + \left(\frac{\partial v}{\partial z} + \frac{\partial w}{\partial y}\right)^2\right).$$

The values of the parameters in the equations are taken in accordance with [3]: $\sigma_E = 2$, $\sigma_\varphi = 2$, $C_{1\varphi} = 0.52$, $C_{2\varphi} = 0.8$. The term Δ_φ characterizes the increase in dissipation of the turbulent kinetic energy due to vegetation. Whether the air density is considered to be constant the continuity equation may be written as following:

$$\frac{\partial u}{\partial x} + \frac{\partial v}{\partial y} + \frac{\partial w}{\partial z} = 0. \tag{3}$$

For closuring of the system (1)–(3) an additional condition expressing the turbulent diffusion coefficient through values E and φ is used:

$$K = C_\mu E\varphi^{-1}, \tag{4}$$

where $C_\mu = 0.09$ is dimensionless coefficient of proportionality (see [3]). The system of Eqs. (1)–(3) is considered in a some closed area of space representing a rectangular parallelepiped:

$$\Pi = \{-a \leq x \leq a,\ -a \leq y \leq a,\ 0 \leq z \leq b\}. \tag{5}$$

On the lower bound of the modelled area the "sticking" conditions are set for the wind velocity components and the turbulent kinetic energy:

$$u|_{z=0} = v|_{z=0} = w|_{z=0} = 0; \quad E|_{z=0} = 0. \tag{6}$$

We assume that out of the area considered there are no external sources or sinks of energy and substance affecting the regarded object. Thus it is possible to accept that flows of substance and energy through the sides of the parallelepiped are equal to zero from where it follows that derivatives of the E, ε δP functions in the direction normal to sides of the parallelepiped are also equal to zero:

$$\frac{\partial E}{\partial n}\bigg|_{\substack{x = \pm a, \\ y = \pm a, \\ z = b}} = 0; \qquad \frac{\partial \varphi}{\partial n}\bigg|_{\substack{x = \pm a, \\ y = \pm a, \\ z = b, \ z = 0}} = 0; \qquad \frac{\partial(\delta P)}{\partial n}\bigg|_{\substack{x = \pm a, \\ y = \pm a, \\ z = 0}} = 0. \quad (7)$$

The same conditions are set for the wind velocity vector components – u, v, w – on the top and sides of the parallelepiped:

$$\frac{\partial u}{\partial n}\bigg|_{\substack{x = \pm a, \\ y = \pm a, \\ z = b}} = 0; \qquad \frac{\partial v}{\partial n}\bigg|_{\substack{x = \pm a, \\ y = \pm a, \\ z = b}} = 0; \qquad \frac{\partial w}{\partial n}\bigg|_{\substack{x = \pm a, \\ y = \pm a, \\ z = b}} = 0. \quad (8)$$

Whether the top border $z = b$ of the considered parallelepiped is high enough we can assume that the average deviation of pressure from standard on this border is equal to zero:

$$\delta P|_{z=b} = 0. \quad (9)$$

For the description of vegetation structure the data on density and height of the forest, $LAD(x, y, z)$ is used. For carrying out the numerical model experiments the rectangular area of the forest $\{x_1 \leq x \leq x_2; \ y_1 \leq x \leq y_2\}$, where $x_1 = y_1 = -40\,\text{m}$, $x_2 = y_2 = 40\,\text{m}$ has been chosen. Height of the wood has been accepted equal to $H_0 = 20\,\text{m}$. For LAD within the forest area we chose the following function:

$$LAD = c_d \cdot h(z) \cdot l(x, y),$$

c_d is the dimensionless drag coefficient of vegetation elements (in the present work the value c_d has been accepted equal to 0.4), the functions $h(z)$ and $l(x, y)$ are the following:

$$h(z) = \max(1. - \exp((z - H_0)/2), 0),$$

$$l(x, y) = 2 \cdot \min\left(\exp\left(40\frac{(x_1 - x)(x - x_2)}{(x_1 - x_2)^2}\right) \exp\left(40\frac{(y_1 - y)(y - y_2)}{(y_1 - y_2)^2}\right) - 1, 1\right).$$

Outside the forest LAD is equal to zero. The graphs of LAD distribution are shown on Fig. 1.

The value Δ_φ in the second Eq. (2) is defined by the equality

$$\Delta_\varphi = 12\sqrt{C_\mu c_d LAD}\sqrt{u^2 + v^2 + w^2}\,(C_{1\varphi} - C_{2\varphi})\,\varphi.$$

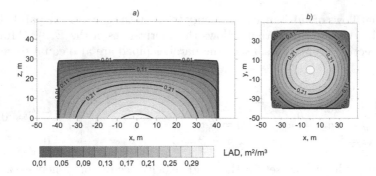

Fig. 1. LAD distribution (a) along the plane which is carried out through the central part of the forest area, (b) horizontal LAD distribution inside the forest at the height of 10 m

3 The Choice of the Function Describing the Interaction of the Air Flow with Vegetation

For the description of the air flow interaction with the inhomogeneous vegetation in the present model we used the theory of contrast structures by analogy with the active environments considered in biophysics [5]. The active environment is characterized by the physical properties called the excitable elements. Some of the excitable elements possess the feature of multistability, that means an opportunity to be in one of the possible states determined by external factors. The change of the stable state is a result of external influence or is due to spatial heterogeneity. Monostable elements of the active environment can be brought out of an equilibrium state due to the feedforwards and feedbacks occurring in the environment.

In this model we assume the multistable element is the horizontal wind speed component along the direction of an air flow in case of lack of vegetation. We will consider that it is u, the wind speed component along X axis. Under the certain physical conditions this parameter has two stable states: u_{open}, the wind speed on the open district and u_{forest}, the wind speed in area with the forest vegetation located at sufficient distance from the wood edge. Other unknown sizes in system (1)–(3) play a role of monostable elements and have a stable zero state. The continuity equation plays a feedback role for wind speed components, and E and φ parameters are linked with the value u by means of Eq. (2).

Multistable elements with two steady states are often described by means of parabolic equation with cubic nonlinearity [5]. Besides, in paper [6] the conditions under which the formation of stable in time contrast structure is possible are formulated. Therefore we have chosen the function F_x in the Eq. (1) for u component as follows:

$$F_x(u, x, y, z) = \frac{1}{u_*} \cdot LAD \cdot (u - u_{open})(u - f(x, y, z))(u - u_{forest}),$$

the factor $1/u_*$ is added to coordinate dimensions. Far from borders between vegetation communities with different structures the horizontal speed component u_{open} is given by the expression [7].

$$u_{open} = u_* \frac{1}{\kappa} \ln \frac{z + z_g}{z_g}, \tag{10}$$

where $u_* \simeq 0.3\,\text{m/s}$ is a friction velocity $\kappa = 0.4$ is a von Karman constant, $z_g = 0.1\,\text{m}$ is the roughness length for grass. The wind field over the high forest vegetation and also inside the forest is described by means of sectionally continuous function [7]

$$u_{forest} = u_h(H_0) \cdot \exp\left(\alpha\left(\frac{z}{H_0} - 1\right)\right), \text{ if } z < H_0, \quad u_{forest} = u_h(z), \text{ if } z \geq H_0,$$

where $\quad u_h(z) = u_* \frac{1}{\kappa} \ln \frac{z - d_f}{z_f},$

$z_f \simeq 2\,\text{m}$ is the value of roughness length for forest received by means of the model offered by Raupach [8]; d_f is the height of the displacement layer calculated according to the equation [9]

$$1 - \frac{d}{H_0} = \frac{1}{\sqrt{7.5 \cdot LAI}} \left(1 - \exp\left(-\sqrt{7.5 \cdot LAI}\right)\right),$$

$\alpha = H_0 \cdot LAD/c_d$. The non-dimensional parameter LAI (leaf area index), defined as the relation of the total area of plant leaves surface to the land surface is calculated as $\quad LAI = \int\limits_0^{H_0} LAD\,dz.$

The function $f(x, y, z)$ is chosen in a way that for the first Eq. (1) the conditions of the stable state solution existence specified in paper [6] were fulfilled:

$$f(x, y, z) = A \cdot LAD \cdot (u_{open} + u_{forest}) \cdot e^{-B(z - H_0/C)}.$$

The coefficients A, B and C are chosen during the numerical experiments. In the present work it was taken $A = 2$; $B = 0.3$; $C = 3$.

For the functions F_y and F_z we will use the expression for the aerodynamic friction [3]:

$$F_y = -LAD \cdot v \cdot \sqrt{u^2 + v^2 + w^2}, \quad F_z = -LAD \cdot w \cdot \sqrt{u^2 + v^2 + w^2}.$$

4   The Results of the Numerical Experiments

The system (1)–(3) with boundary conditions (6)–(9) was solved by method of evolutionary factorization [10]. The initial conditions were the following:

$$u|_{t=0} = u_{open}(z) \cdot (1 - LAD); \quad v|_{t=0} = w|_{t=0} = 0;$$

$$\delta P|_{t=0} = 0; \quad E|_{t=0} = E_0; \quad \varepsilon|_{t=0} = \varepsilon_0,$$

the constants E_0 and ε_0 were chosen in a wy that the turbulent diffusion coefficient K, determined by expression (4), in an initial time point was equal to 1. In the considered area (5) the uniform mesh was set. For solving the continuity equation the method of splitting on physical processes [11] was used.

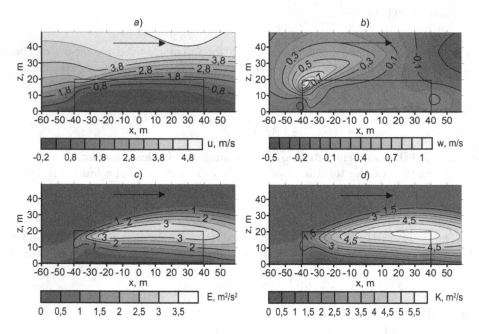

Fig. 2. The distribution along the plane which is carried out through the central part of the forest area: (a) u wind component, (b) w wind component, (c) turbulent kinetic energy, (d) turbulent diffusion coefficient. The wind direction is marked by an arrow

The results of numerical experiments are shown on Figs. 2 and 3.

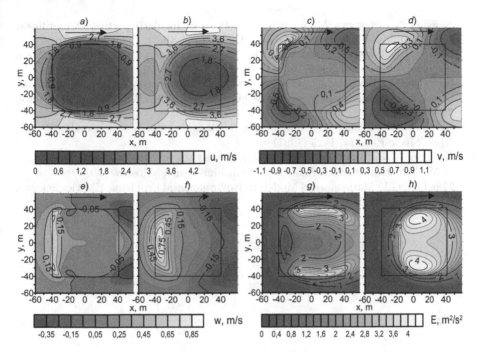

Fig. 3. The horizontal distribution: (a) u wind component in crowns ($z = 10\,\mathrm{m}$), (b) u wind component at the top border of the forest ($z = 20\,\mathrm{m}$), (c) v wind component in crowns ($z = 10\,\mathrm{m}$), (d) v wind component at the top border of the forest ($z = 20\,\mathrm{m}$), (e) w wind component in crowns ($z = 10\,\mathrm{m}$), (f) w wind component at the top border of the forest ($z = 20\,\mathrm{m}$), (g) turbulent kinetic energy in crowns ($z = 10\,\mathrm{m}$), (h) turbulent kinetic energy at the top border of the forest ($z = 20\,\mathrm{m}$). The wind direction is marked by an arrow

Acknowledgements. The study was supported by grant of Russian Science Foundation 14-14-00956.

References

1. Garrat, J.: The Atmospheric Boundary Layer. Cambridge University Press, Cambridge (1992)
2. Sogachev, A., Panferov, O.: Modification of two-equation models to account for plant drag. Bound.-Layer Meteorol. **121**, 229–266 (2006)
3. Sogachev, A.: A note on two-equation closure modelling of canopy flow. Bound.-Layer Meteorol. **130**, 423–436 (2009)
4. Olchev, A., Radler, K., Sogachev, A., Panferov, O., Gravenhorst, G.: Application of a three-dimensional model for assessing effects of small clear-cuttings on radiation and soil temperature. Ecol. Model. **220**, 3046–3056 (2009)
5. Vasilev, V., Romanovskii, Y., Chernavskii, D., Yakhno, V.: Autowave Processes in Kinetic Systems. Spatial and Temporal Self-Organization in Physics, Chemistry, Biology and Medicine. VEB Deutscher Verlag der Wissenschaften, Berlin (1986)

6. Nefedov, N., Davydova, M.: Contrast structures in singularly perturbed quasilinear reaction-diffusion-advection equations. Differ. Equ. **49**(6), 715–733 (2013)
7. Cionco, R.: A mathematical model for air flow in a vegetative canopy. J. Appl. Meteorol. **4**, 517–522 (1965)
8. Raupach, M.: Simplified expressions for vegetation roughness length and zeroplane displacement as functions of canopy height and area index. Bound.-Layer Meteorol. **71**, 211–216 (1994)
9. Raupach, M., Antonia, R., Rajagopalan, S.: Rough wall turbulent boundary layers. Appl. Mech. Rev. **44**, 1–25 (1991)
10. Kalitkin, N.N., Koryakin, P.V.: Numerical Methods in 2 Books. Book 2: Methods of Mathematical Physics. Academia Publishing Centre, Moscow (2013)
11. Marchuk, G.I.: Mathematical Models in Environmental Problems. Elsevier Science, Amsterdam (1986)

An Approximate Solution of Optimization Problems for Elliptic Interface Problems with Variable Coefficients and Imperfect Contact

Aigul Manapova[✉]

Bashkir State University, Zaki Validi Street, 32,
Republic of Bashkortostan, Russia
aygulrm@yahoo.com

Abstract. The present work is devoted to computational aspects of solving the optimization problems for semi-linear elliptic interface problems. We develop numerical algorithms for minimizing a cost functional, depending on a state of the system and a control. Numerical experiments are included. The results from computer experiment showed the effectiveness of the approximate method of solution.

Keywords: Optimal control problem · Semi-linear elliptic equations · Numerical method

1 Introduction and Setting of the Problem

In the paper we study an optimal control problem (OCP) of processes described by elliptic equations in heterogeneous anisotropic media with discontinuous coefficients and solutions (states) (DCS) subject to the boundary interface conditions of imperfect type [1,2]. The coefficients in the right-hand side of the state equation are used as a control function. Such problems arise often in the mathematical modeling and optimization of heat transfer, diffusion, filtration, elasticity, etc., in a study of inverse problems and of optimal control problems for equations of mathematical physics (EMP) in multilayered media. The discontinuity of the coefficients and solution occurs when the media is heterogeneous and consists in several parts with different properties or when the domain contains thin layers S with physical properties sharply different from the main media.

Nonlinear optimization problems is one of the most complicated class of problems in the optimization theory, and they require application of numerical methods and their computer realizations. The situation becomes even more complicated when the states in optimal control problems are described by nonlinear EMP with discontinuous coefficients and the solutions can be discontinuous due to the character of the physical process under study.

The numerical solving of optimal control problems (NSOCP) is in a wide sense related with studying the following issues:

© Springer International Publishing AG 2017
I. Dimov et al. (Eds.): NAA 2016, LNCS 10187, pp. 473–481, 2017.
DOI: 10.1007/978-3-319-57099-0_53

1. The formulation of optimization problem ensuring the existence of solution on the set of admissible controls being a set of some infinite-dimensional vector space;
2. The reduction of optimal control problem to a sequence of finite-dimensional problem guaranteeing the convergence in some sense of solutions to the finite-dimensional problems to the solutions of original optimal control problems;
3. Numerical solving of the finite-dimensional problems.

This work is a natural continuation of [3–6]. Particularly, in works [3,4] for studying optimization problems described by nonlinear EMP with DCS there were developed methods, based on constructing and studying difference approximations of extremal problems, establishing the estimates for the approximation accuracy w.r.t. the state and functional, and on the regularization of approximations. In papers [5,6] an iterative method for solving boundary value problems of contact for EMP of elliptic type with DCS is developed and validated.

The present paper is aimed on solving the third step of NSOCP, namely, on developing effective numerical methods of solving constructed finite-dimensional grid optimal control problems. We develop numerical algorithms for minimizing a grid cost functional of the approximating grid problems, depending on a state of the system and a control. The minimizing algorithms are based on difference analogues of gradient methods such as gradient projection method, conjugate gradient projection method, conditional gradient method. The calculation of the gradients uses the numerical solutions of direct problems for the state and adjoint problems. Numerical experiments are included. Computational experiments, illustrating application of developed methods, are relatively simple to implement and enable efficiently solving numerically of large classes of nonlinear optimal control problems within a reasonable time. We note that such issues were not considered before.

Now consider the following problem. Let $\Omega = \{r = (r_1, r_2) \in \mathbf{R}^2 : 0 \leq r_\alpha \leq l_\alpha, \alpha = 1, 2\}$ be a rectangle in $\mathbf{R}^2$ with a boundary $\partial \Omega = \Gamma$. Suppose that the domain Ω is splitted by an internal interface $\overline{S} = \{r_1 = \xi, \ 0 \leq r_2 \leq l_2\}$, where $0 < \xi < l_1$, into the left $\Omega_1 \equiv \Omega^- = \{0 < r_1 < \xi, \ \ 0 < r_2 < l_2\}$ and right $\Omega_2 \equiv \Omega^+ = \{\xi < r_1 < l_1, \ \ 0 < r_2 < l_2\}$ subdomains with boundaries $\partial \Omega_1 \equiv \partial \Omega^-$ and $\partial \Omega_2 \equiv \partial \Omega^+$. Thus, $\Omega = \Omega_1 \cup \Omega_2 \cup S$, while $\partial \Omega$ is the outer boundary of Ω. Let $\overline{\Gamma}_k$ denote the boundaries of Ω_k without S, $k = 1, 2$. Therefore $\partial \Omega_k = \overline{\Gamma}_k \cup S$, where Γ_k, $k = 1, 2$ are open nonempty subsets of $\partial \Omega_k$, $k = 1, 2$; and $\overline{\Gamma}_1 \cup \overline{\Gamma}_2 = \partial \Omega = \Gamma$. Let n_α, $\alpha = 1, 2$ denote the outward normal to the boundary $\partial \Omega_\alpha$ of Ω_α, $\alpha = 1, 2$. Let $n = n(x)$ be a unit normal to S at a point $x \in S$, directed, for example, so that n is the outward normal on S with respect to Ω_1; i.e., n is directed inside Ω_2. While formulating boundary value problems for states of control processes below, we assume that S is a straight line across which the coefficients and solutions of the problems are discontinuous, while being smooth within Ω_1 and Ω_2.

It is required to minimize the cost functional $J : U \to \mathbf{R}^1$, defined as

$$g \to J(g) = \int_{\Omega_1} \left| u(r_1, r_2; g) - u_0^{(1)}(r) \right|^2 d\Omega_1, \tag{1}$$

where $u_0^{(1)} \in W_2^1(\Omega_1)$ is a given function, over the solutions $u(r; g)$ of the Dirichlet problem for a semi-linear elliptic equation with DCS:

$$-\sum_{\alpha=1}^{2} \frac{\partial}{\partial x_\alpha} \left(k_\alpha(x) \frac{\partial u}{\partial x_\alpha} \right) + d(x)q(u) = f(x), \ x \in \Omega_1 \cup \Omega_2,$$

and the conditions $\quad u(x) = 0, \quad x \in \partial\Omega = \overline{\Gamma}_1 \cup \overline{\Gamma}_2,$ $\tag{2}$

$$\left[k_1(x) \frac{\partial u}{\partial x_1} \right] = 0, \ G(x) = \left(k_1(x) \frac{\partial u}{\partial x_1} \right) = \theta(x_2)[u], \ x \in S,$$

corresponding to all admissible controls $g = (f_1, f_2) \in U$:

$$U = \prod_{\alpha=1}^{2} U_\alpha \subset H = L_2(\Omega_1) \times L_2(\Omega_2), \tag{3}$$

$$U_\alpha = \{ g_\alpha = f_\alpha \in L_2(\Omega_\alpha) : g_0 \le g_\alpha(x) \le \overline{g}_0 \text{ a.e. on } \Omega_\alpha \}, \ \alpha = 1, 2.$$

Here

$$k_\alpha(x), d(x), f(x) = \begin{cases} k_\alpha^{(1)}(x), d_1(x), f_1(x), x \in \Omega_1; \\ k_\alpha^{(2)}(x), d_2(x), f_2(x), x \in \Omega_2, \end{cases} \quad u(x) = \begin{cases} u_1(x), x \in \Omega_1; \\ u_2(x), x \in \Omega_2, \end{cases}$$

$q(\xi) = \begin{cases} q_1(\xi_1), \xi_1 \in \mathbf{R}; \\ q_2(\xi_2), \xi_2 \in \mathbf{R}, \end{cases}$ $[u] = u_2(x) - u_1(x)$ is the jump in $u(x)$ across S; $k_\alpha(x)$,
$\alpha = 1, 2$, $d(x)$ are given functions that are defined independently in Ω_1 and Ω_2, and having a first kind jump at S; $q_\alpha(\xi_\alpha)$, $\alpha = 1, 2$, are given functions defined for $\xi_\alpha \in \mathbf{R}$, $\alpha = 1, 2$; $\theta(x_2)$, $x_2 \in S$, is a given function. The given functions are assumed to satisfy the following conditions: $k_\alpha(x) \in W_\infty^1(\Omega_1) \times W_\infty^1(\Omega_2)$, $\alpha = 1, 2$, $d(x) \in L_\infty(\Omega_1) \times L_\infty(\Omega_2)$, $\theta(x_2) \in L_\infty(S)$, $0 < \nu \le k_\alpha(x) \le \overline{\nu}$, $\alpha = 1, 2$, $0 \le d_0 \le d(x) \le \overline{d}_0$ for $x \in \Omega_1 \cup \Omega_2$ and $0 < \theta_0 \le \theta(x_2) \le \overline{\theta}_0$ for $x \in S$, where $\nu, \overline{\nu}, d_0, \overline{d}_0, \theta_0, \overline{\theta}_0, g_0$ and $\overline{g}_0$ are given constants; and the functions $q_\alpha(\zeta_\alpha)$ satisfy the conditions: $q_\alpha(0) = 0$, $0 < q_0 \le (q_\alpha(\zeta_\alpha) - q_\alpha(\overline{\zeta}_\alpha))/(\zeta_\alpha - \overline{\zeta}_\alpha) \le L_q < \infty$, for all $\zeta_\alpha, \overline{\zeta}_\alpha \in \mathbf{R}$, $\zeta_\alpha \ne \overline{\zeta}_\alpha$, $L_q = \text{Const}$.

In what follows we shall need some spaces introduced in work [3]. In particular, let $\overset{\circ}{\Gamma}_k$ be a portion of the boundary $\partial\Omega_k$. Denote by $W_2^1\left(\Omega_k; \overset{\circ}{\Gamma}_k\right)$ the closed subspace of $W_2^1(\Omega_k)$ in which the set of all functions from $C^1(\overline{\Omega}_k)$ vanishing near $\overline{\Gamma}_k \subset \partial\Omega_k$, $k = 1, 2$ in a dense set. We introduce the space $\overset{\circ}{V}_{\Gamma_1, \Gamma_2}(\Omega^{(1,2)})$ of pairs $u = (u_1, u_2)$: $\overset{\circ}{V}_{\Gamma_1, \Gamma_2}(\Omega^{(1,2)}) = \{ u = (u_1, u_2) \in W_2^1(\Omega_1; \Gamma_1) \times W_2^1(\Omega_2; \Gamma_2) \}$ with

the norm $\|u\|_{\overset{\circ}{V}_{\Gamma_1, \Gamma_2}}^2 = \sum_{k=1}^{2} \int_{\Omega_k} \sum_{\alpha=1}^{2} \left(\frac{\partial u_k}{\partial x_\alpha} \right)^2 d\Omega_k + \int_S [u]^2 \, dS.$

The solution of direct problem (2) with a fixed control $g \in U$ is a function $u \in \overset{\circ}{V}_{\Gamma_1,\Gamma_2}(\Omega^{(1,2)})$ satisfying for $\forall \vartheta \in \overset{\circ}{V}_{\Gamma_1,\Gamma_2}(\Omega^{(1,2)})$ the identity:

$$\int_{\Omega_1 \cup \Omega_2} \left[\sum_{\alpha=1}^{2} k_\alpha \frac{\partial u}{\partial x_\alpha} \frac{\partial \vartheta}{\partial x_\alpha} + d\,q(u)\,\vartheta \right] d\Omega_0 + \int_S \theta\,[u][\vartheta]dS = \int_{\Omega_1 \cup \Omega_2} f\,\vartheta\,d\Omega_0.$$

2 Difference Approximation of the OCPs

Optimal control problems (1)–(3) are associated with the following difference approximations: minimize the grid functional

$$J_h(\Phi_h) = \sum_{\overline{\omega}^{(1)}} |y(\Phi_h) - u_{0h}^{(1)}|^2 \hbar_1 \hbar_2 = \|y(\Phi_h) - u_{0h}^{(1)}\|_{L_2(\overline{\omega}^{(1)})}^2, \tag{4}$$

provided that the grid function $y \in \overset{\circ}{V}_{\gamma^{(1)}\gamma^{(2)}}(\overline{\omega}^{(1,2)})$, which is the solution of the difference boundary value problem for problem (2), satisfies, for any grid function $v \in \overset{\circ}{V}_{\gamma^{(1)}\gamma^{(2)}}(\overline{\omega}^{(1,2)})$, the summation identity

$$\sum_{\alpha=1}^{2} \Bigg(\sum_{\omega_1^{(\alpha)+} \times \omega_2} a_{1h}^{(\alpha)} y_{\alpha \overline{x}_1} v_{\alpha \overline{x}_1} h_1 h_2 + \sum_{\omega_1^{(\alpha)} \times \omega_2^+} a_{2h}^{(\alpha)} y_{\alpha \overline{x}_2} v_{\alpha \overline{x}_2} h_1 h_2$$
$$+ \frac{1}{2} \sum_{\gamma_S} a_{2h}^{(\alpha)} y_{\alpha \overline{x}_2} v_{\alpha \overline{x}_2} h_1 h_2 + \sum_{\omega^{(\alpha)}} d_{\alpha h}\, q_\alpha(y_\alpha)\, v_\alpha\, h_1 h_2$$
$$+ \frac{1}{2} \sum_{\gamma_S} d_{\alpha h}\, q_\alpha(y_\alpha)\, v_\alpha\, h_1 h_2 \Bigg) + \sum_{\gamma_S} \theta_h(x_2)\,[y]\,[v]\,h_2 \tag{5}$$
$$= \sum_{\alpha=1}^{2} \Bigg(\sum_{\omega^{(\alpha)}} \Phi_{\alpha h}\, v_\alpha\, h_1 h_2 + \frac{1}{2} \sum_{\gamma_S} \Phi_{\alpha h}\, v_\alpha\, h_1 h_2 \Bigg),$$

while the grid controls Φ_h are such that $\Phi_h = (\Phi_{1h}, \Phi_{2h}) \in \prod_{\alpha=1}^{2} U_{\alpha h} = U_h \subset H_h$,

$$H_h = \prod_{\alpha=1}^{2} L_2(\omega^{(\alpha)} \cup \gamma_S), \quad U_{\alpha h} = \{\Phi_{\alpha h} : g_0 \le \Phi_{\alpha h}(x) \le \overline{g}_0, x \in \omega^{(\alpha)} \cup \gamma_S\},$$

$\alpha = 1, 2$. Here, $a_{\alpha h}^{(1)}(x)$, $a_{\alpha h}^{(2)}(x)$, and $d_{\alpha h}(x)$, $\alpha = 1, 2$, $\theta_h(x_2)$, and $u_{0h}^{(1)}(x)$ are grid approximations of the functions $k_\alpha^{(1)}(r)$, $k_\alpha^{(2)}(r)$, and $d_\alpha(r)$, $\alpha = 1, 2$, $\theta(r_2)$, and $u_0^{(1)}(r)$ defined via Steklov averages. For the definition of grids $\overline{\omega}^{(1,2)}$, $\omega^{(\alpha)} \cup \gamma_S$, $\alpha = 1, 2$ and grid spaces $\overset{\circ}{V}_{\gamma^{(1)}\gamma^{(2)}}(\overline{\omega}^{(1,2)})$ see work [4].

Problem (5) is a grid analogue of the original problem for state with DCS (2).

3   Numerical Solution of Finite-Dimensional Optimization Problem

Generally speaking, the problem (4)–(5) is not a well posed minimization problem in the sense of Tikhonov (see [3], pp. 1714). Therefore there is no ground to expect that any minimizing sequence for the functional $J_h(\Phi_h)$ on U_h converges strongly in the norm of H_h to the set $U_{h*} = \{\Phi_{h*} \in U_h : J(\Phi_{h*}) = J_h*\} \neq \emptyset$ of minimizers of $J_h(\Phi_h)$. Note in practice the grid functionals $J_h(\Phi_h)$ are computed approximately because of the inaccurate input data and rounding errors, so that the functional $J_h(\Phi_h)$ is, in fact, replaced by an approximate one $J_{h\delta_h}(\Phi_h)$ that is related to $J_h(\Phi_h)$ as follows:

$$J_{h\delta_h}(\Phi_h) = J_h(\Phi_h) + \chi_{\delta_h}(\Phi_h), \quad |\chi_{\delta_h}(\Phi_h)| \leq \delta_h, \forall \Phi_h \in U_h, \delta_h \to +0 \text{ as } |h| \to 0.$$

To regularize the family of grid optimization problems (4)–(5) we apply Tikhonov regularization method. For every $h = (h_1, h_2)$ the grid Tikhonov functional of problem (4)–(5) on U_h is defined as

$$T_{h\alpha_h}(\Phi_h) = J_{h\delta_h}(\Phi_h) + \alpha_h \Omega_h(\Phi_h), \quad \Phi_h \in U_h, \tag{6}$$

where $\Omega_h(\Phi_h) = \|\Phi_h\|^2_{H_h}$, $\Phi_h \in U_h$, is a grid stabilizing functional and $\{\alpha_h\}$ is an arbitrary sequence of positive numbers converging to zero as $|h| \to 0$.

Consider the minimization of the functional (6) under the conditions (4)–(5): for every $h > 0$ find a grid control $\widehat{\Phi}_h \equiv \Phi_{h\alpha_h\nu_h}(x) \in U_h$, satisfying the conditions

$$T_{h\alpha_h*} = \inf_{\Phi_h \in U_h} T_{h\alpha_h}(\Phi_h) \leq T_{h\alpha_h}(\widehat{\Phi}_h) \leq T_{h\alpha_h*} + \nu_h, \tag{7}$$

where the sequence $\{\nu_h\}$ is such that $\nu_h > 0$ and $\nu_h \to 0$ as $|h| \to 0$. Here, the sequence $\{\nu_h\}$ characterizes the accuracy of the solution to the minimization problem for $T_{h\alpha_h}$ on U_h. By the definition of an infimum, such a point $\widehat{\Phi}_h$ exists, since ν_h is positive and $T_{h\alpha_h}(\Phi_h)$ is bounded below on U_h. Since there is at least one point $\widehat{\Phi}_h$ at which the functional $T_{h\alpha_h}(\Phi_h)$ attains an infimum $T_{h\alpha_h*}$ on U_h, it is possible (can be assumed) that $\nu_h = 0$ in (7). The point $\widehat{\Phi}_h$ can be determined from condition (7) by applying difference gradient methods for solving OCP (7), (4)–(5), [7] (see also the calculation of the functional gradients and adjoint problems construction in [5], pp. 187–193).

The gradient of the functional (6) is given by

$$\frac{\partial T_{h\alpha_h}}{\partial \Phi_h} = \frac{\partial J_h}{\partial \Phi_h} + 2\, \alpha_h \Phi_h(x), \tag{8}$$

using, correspondingly, the numerical solutions $y(x) = y(x; \Phi_h)$ and $\psi(x) = \psi(x; \Phi_h)$ of direct problems for the state (5) and adjoint problems:

$$- \left(a_{1h}^{(1)}(x)\psi_{1\overline{x}_1} \right)_{x_1} - \left(a_{2h}^{(1)}(x)\psi_{1\overline{x}_2} \right)_{x_2} + d_{1h}(x)\, q_{1y_1}\psi_1(x) = -2\left(y(x) - u_{0h}^{(1)}(x) \right),$$
$$x \in \omega^{(1)},$$
$$\psi_1(x) = 0, \quad \gamma^{(1)} = \partial\omega^{(1)} \setminus \gamma_S;$$
$$- \left(a_{1h}^{(2)}(x)\psi_{2\overline{x}_1} \right)_{x_1} - \left(a_{2h}^{(2)}(x)\psi_{2\overline{x}_2} \right)_{x_2} + d_{2h}(x)\, q_{2y_2}\psi_2(x) = 0, \quad x \in \omega^{(2)}, \tag{9}$$
$$\psi_2(x) = 0, \quad x \in \gamma^{(2)} = \partial\omega^{(2)} \setminus \gamma_S;$$
$$\frac{2}{h_1}\left[a_{1h}^{(1)}(\xi,x_2)\psi_{1\overline{x}_1}(\xi,x_2) + \theta_h(x_2)\psi_1(\xi,x_2) \right] + d_{1h}(\xi,x_2)\, q_{1y_1}\psi(\xi,x_2))$$
$$- \left(a_{2h}^{(1)}(\xi,x_2)\psi_{1\overline{x}_2}(\xi,x_2) \right)_{x_2} = -2\left(y(\xi,x_2) - u_{0h}^{(1)}(\xi,x_2) \right) + \frac{2}{h_1}\theta_h(x_2)\psi_2(\xi,x_2),$$
$$x \in \gamma_S = \{x_1 = \xi, x_2 \in \omega_2\},$$
$$- \frac{2}{h_1}\left[a_{1h}^{(2)}(\xi + h_1, x_2)\psi_{2x_1}(\xi,x_2) - \theta_h(x_2)\psi_2(\xi,x_2) \right] + d_{2h}(\xi,x_2)\, q_{2y_2}(\xi,x_2))$$
$$- \left(a_{2h}^{(2)}(\xi,x_2)\psi_{2\overline{x}_2}(\xi,x_2) \right)_{x_2} = \frac{2}{h_1}\theta_h(x_2)\psi_1(\xi,x_2), \ x \in \gamma_S = \{x_1 = \xi, x_2 \in \omega_2\}.$$

where $\dfrac{\partial J_h}{\partial \Phi_h}$ is the gradient of the functional $J_h(\Phi_h)$ (see [5], pp. 193):

$$\frac{\partial J_h}{\partial \Phi_h} = \left(\frac{\partial J_h}{\partial \Phi_{1h}}, \frac{\partial J_h}{\partial \Phi_{2h}} \right),$$
$$\frac{\partial J_h}{\partial \Phi_{1h}} = -\psi_1(x), \quad x \in \omega^{(1)} \cup \gamma_S, \qquad \frac{\partial J_h}{\partial \Phi_{2h}} = -\psi_2(x), \quad x \in \omega^{(2)} \cup \gamma_S. \tag{10}$$

Minimizing sequences for the functional (6) can be constructed on the base of difference analogues of gradient projection method, conjugate gradient projection method, conditional gradient method. Moreover, to find a projection on the set of restrictions U_h is not difficult. Here we confine ourselves to the construction of the conjugate gradient projection method algorithm for the original finite-dimensional problem (5).

Starting from initial approximation $\widehat{\Phi}_h^{(0)}(x)$ for every $h > 0$, we construct a sequence of approximations as follows:

$$\widehat{\Phi}_{\alpha h}^{(n+1)}(x) = \begin{cases} \widehat{\Phi}_{\alpha h}^{(n)}(x) - \gamma_h^{(n)} p_{\alpha h}^{(n)} = G_{\alpha h}^{(n)}, & \text{if } g_0 \le G_{\alpha h}^{(n)} \le \overline{g}_0; \\ g_0, & \text{if } G_{\alpha h}^{(n)} < g_0; \\ \overline{g}_0, & \text{if } G_{\alpha h}^{(n)} > \overline{g}_0. \end{cases}$$

$$= P_{U_h}\left(\widehat{\Phi}_{\alpha h}^{(n)}(x) - \gamma_h^{(n)} p_{\alpha h}^{(n)} \right), \ n = 0, 1, 2, \ldots, \ \alpha = 1, 2;$$

$$p_{\alpha h}^{(0)} = \frac{\partial T_{h\alpha h}(\widehat{\Phi}_h^{(0)})}{\partial \Phi_{\alpha h}}, \quad p_{\alpha h}^{(n)} = \frac{\partial T_{h\alpha h}(\widehat{\Phi}_{\alpha h}^{(n)})}{\partial \Phi_{\alpha h}} - \beta_h^{(n)} p_{\alpha h}^{(n-1)}, \quad n = 1, 2, \ldots, \ \alpha = 1, 2,$$

$$\beta_h^{(n)} = \frac{\left(\dfrac{\partial T_{h\alpha h}(\widehat{\Phi}_{\alpha h}^{(n)})}{\partial \Phi_{\alpha h}}, \dfrac{\partial T_{h\alpha h}(\widehat{\Phi}_{\alpha h}^{(n-1)})}{\partial \Phi_{\alpha h}} - \dfrac{\partial T_{h\alpha h}(\widehat{\Phi}_{\alpha h}^{(n)})}{\partial \Phi_{\alpha h}} \right)_{H_h}}{\left\| \dfrac{\partial T_{h\alpha h}(\widehat{\Phi}_{\alpha h}^{(n-1)})}{\partial \Phi_{\alpha h}} \right\|_{H_h}^2},$$

where $\dfrac{\partial T_h(\Phi_h)}{\partial \Phi_{\alpha h}}$, $\alpha = 1, 2$, are defined in (8)–(10), H_h is the linear space, equipped with the inner product

$$\left(\Phi_h, \widetilde{\Phi}_h\right)_{H_h} = \sum_{\omega^{(1)} \cup \gamma_S} \Phi_{1h} \widetilde{\Phi}_{1h} h_1 h_2 + \sum_{\omega^{(2)} \cup \gamma_S} \Phi_{2h} \widetilde{\Phi}_{2h} h_1 h_2,$$

and a method parameter (a descent step) $\gamma_h^{(n)}$ is chosen from monotone decreasing condition of the functional $T_{h\alpha_h}(\widehat{\Phi}_h)$:

$$T_{h\alpha_h}(\widehat{\Phi}_h^{(n+1)}) < T_{h\alpha_h}(\widehat{\Phi}_h^{(n)}), \qquad n = 0, 1, 2, \dots .$$

To stop the algorithm, determining $\widehat{\Phi}_h^{(n)}(x)$, where n is an iteration number, we select one of the following conditions:

$$\theta^h(n) = \left| T_{h\alpha_h}(\widehat{\Phi}_h^{(n)}) - T_{h\alpha_h}(\widehat{\Phi}_h^{(n-1)}) \right| \le \delta_{1h}, \tag{11}$$

$$\chi_2^h(n) = \left\| \widehat{\Phi}_h^{(n)}(x) - \widehat{\Phi}_h^{(n-1)}(x) \right\|_{L_2(\omega^{(1)} \cup \gamma_S) \times L_2(\omega^{(2)} \cup \gamma_S)} \le \delta_{2h}, \tag{12}$$

Besides, to assess approximation accuracy of the algorithms rms and uniform relative errors (estimates)

$$\sigma_2^h(n) = \left\| \widehat{\Phi}_h^{(n)}(x) - g(x) \right\|_{L_2(\omega^{(1)} \cup \gamma_S) \times L_2(\omega^{(2)} \cup \gamma_S)}, \tag{13}$$

$$\sigma_\infty^h(n) = \left\| \widehat{\Phi}_h^{(n)}(x) - g(x) \right\|_{L_\infty(\omega^{(1)} \cup \gamma_S) \times L_\infty(\omega^{(2)} \cup \gamma_S)}, \tag{14}$$

are used, where $g(x) = g_*(x) = (f_{1*}(x), f_{2*}(x))$ is the exact solution of the initial optimization problem (1)–(3).

We apply the developed iterative method (see [5,6]) to solve the problems (5) and (9) under the fixed control $\Phi_h = (\Phi_{1h}, \Phi_{2h})$.

4   Numerical Experiments

We present numerical experiments corresponding to the application of the previous section. Computer experiments are included, using IDE Embarcadero Delphi. We present here one test example.

Example. Nonlinear optimization problem for imperfect contact problem in a unit square $\overline{\Omega} = \overline{\Omega}_1 \times \overline{\Omega}_2$, $\overline{\Omega}_1 = \Omega_1 \cup \partial\Omega_1 = [0, \xi] \times [0, 1] = [0, 1/3] \times [0, 1]$, $\overline{\Omega}_2 = \Omega_2 \cup \partial\Omega_2 = [\xi, 1] \times [0, 1] = [1/3, 1] \times [0, 1]$. In this example interface is at $\xi = \dfrac{1}{3}$. The coefficients k_1, k_2 are different for different subdomains: $k_1(x) = $
$\begin{cases} k_1^{(1)}(x) = \sqrt{3}/4, \\ k_1^{(2)}(x) = 1, \end{cases}$ $k_2^{(1)}(x) = 1$, $k_2^{(2)}(x) = 1$, $\theta(x_2) = \dfrac{4\sqrt{3}\pi}{8\sqrt{3} - 3}$, $d_1 = 0.4x_1 + x_2$,
$q_1(u_1) = u_1^3$, $d_2 = 0.4x_1 + x_2$, $q_2(u_2) = u_2^3$, $u_0^{(1)}(x) = x_2(2 - 2x_2)\sin(2\pi x_1)$. And

the controls $g(x) = (f_1(x), f_2(x))$ are such that $f_\alpha \in U_\alpha = \{g_\alpha = f_\alpha \in L_2(\Omega_\alpha) : g_0 \leq g_\alpha(x) \leq \overline{g}_0 \text{ a.e. on } \Omega_\alpha\}$, $\alpha = 1, 2$, where $g_0 = -10$ and $\overline{g}_0 = 12.7$.

It is clear that $J(g) \geq 0$. Moreover, since the exact solution of (5) is given by

$$u(x) = \begin{cases} u_1(x) = x_2(2 - 2x_2)\sin(2\pi x_1), \\ u_2(x) = 0.25x_2(2 - 2x_2)\sin(2\pi x_1)\sin(2\pi x_1), \end{cases} \text{ when } g_1(x) = f_{1*}(x) =$$

$\sqrt{3}x_2(2 - 2x_2)\pi^2 \sin(2\pi x_1) + 4\sin(2\pi x_1) + (0.4x_1 + x_2)(x_2(2 - 2x_2)\sin(2\pi x_1))^3$,
$g_2(x) = f_{2*}(x) = -2\pi^2 x_2(2 - 2x_2)\cos(4\pi x_1) + \sin(2\pi x_1)\sin(2\pi x_1) + (0.4x_1 + x_2)(0.25x_2(2 - 2x_2)\sin(2\pi x_1)\sin(2\pi x_1))^3$, then the minimum functional value $J(g)$ on U is achieved on the control-function $g_* = (f_{1*}, f_{2*}) \in U : J_* = \inf_{g \in U} J(g) = J(g_*) = 0$, $g \in U$.

Numerical experiments are carried out by choosing different initial approximations and variations of parameters $h_1^{(1)}$, $h_1^{(2)}$, h_2 and α_h. They showed that functional values $T_{h\alpha_h}(\widehat{\Phi}_h^{(n)})$ decrease rapidly in a few iterations, and the optimal control is determined with quite satisfactory accuracy. Subsequently, while increasing iterations number the process of approximation refinement in the neighborhood of the exact solution is sharply slowed down.

Computational experiments, based on agreed parameters change in the implementation of computational algorithms have led to a substantial increase in the rate of convergence. For example, for one of the initial approximations the value of the functional decreases from $T_{h\alpha_h}(\widehat{\Phi}_h^{(0)}) \approx 1,7639058258$ to $T_{h\alpha_h}(\widehat{\Phi}_h^{(n)}) \approx 0,508590187$. At the same time two subsequent iteration have coincided with a given accuracy $\delta_{1h} = 10^{-4}$. Note that the Tikhonov functional value on the exact solution is equal to $T_{h\alpha_h}(g_*) \approx 0,514962611$. The initial cost functional value $J(g_*) \approx 9,617247945E - 5$.

Acknowledgements. This work was supported by a grant of the President of the Russian Federation for state support of young Russian scientists and PhDs, number MK 4147.2015.1.

References

1. Samarskii, A.A., Andreev, V.B.: Difference Methods for Elliptic Equations. Nauka, Moscow (1976). [in Russian]
2. Samarskii, A.A.: The Theory of Difference Schemes. (Nauka, Moscow, 1989; Marcel Dekker, New York, 2001)
3. Lubyshev, F.V., Manapova, A.R., Fairuzov, M.E.: Approximations of optimal control problems for semilinear elliptic equations with discontinuous coefficients and solutions and with control in matching boundary conditions. Comput. Math. Math. Phys. **54**(11), 1700–1724 (2014)
4. Manapova, A.R., Lubyshev, F.V.: Accuracy estimate with respect to state of finite-dimensional approximations for optimization problems for semi-linear elliptic equations with discontinuous coefficients and solutions. Ufim. Mat. Zh. **6**(3), 72–87 (2014)
5. Manapova, A.R., Lubyshev, F.V.: Numerical solution of optimization problems for semi-linear elliptic equations with discontinuous coefficients and solutions. Appl. Numer. Math. **104**, 182–203 (2016). doi:10.1016/j.apnum.2015.12.006

6. Manapova, A.: An approximate solution to state problem in coefficient-optimal-control problems. IFAC-PapersOnLine **48**(25), 011–015 (2015). doi:10.1016/j.ifacol.2015.11.051. http://www.sciencedirect.com/science/article/pii/S2405896315023095
7. Vasilev, F.P.: Optimization Methods. Faktorial, Moscow (2002). [in Russian]

An Antithetic Approach of Multilevel Richardson-Romberg Extrapolation Estimator for Multidimensional SDES

Cheikh Mbaye[1]([✉]), Gilles Pagès[2], and Frédéric Vrins[1]

[1] Université Catholique de Louvain, Louvain-la-Neuve, Belgium
{cheikh.mbaye,frederic.vrins}@uclouvain.be
[2] Université Pierre et Marie Curie, Paris, France
gilles.pages@upmc.fr

Abstract. The Multilevel Richardson-Romberg (*ML2R*) estimator was introduced by Pagès & Lemaire in [1] in order to remove the bias of the standard Multilevel Monte Carlo (*MLMC*) estimator in the 1D Euler scheme. Milstein scheme is however preferable to Euler scheme as it allows to reach the optimal complexity $O(\varepsilon^{-2})$ for each of these estimators. Unfortunately, Milstein scheme requires the simulation of Lévy areas when the SDE is driven by a multidimensional Brownian motion, and no efficient method is currently available to this purpose so far (except in dimension 2). Giles and Szpruch [2] recently introduced an antithetic multilevel correction estimator avoiding the simulation of these areas without affecting the second order complexity. In this work, we revisit the *ML2R* and *MLMC* estimators in the framework of the antithetic approach, thereby allowing us to remove the bias whilst preserving the optimal complexity when using Milstein scheme.

Keywords: Multilevel Monte Carlo · Antithetic Multilevel Monte Carlo · Richardson-Romberg extrapolation · Milstein scheme · Option pricing

1 Introduction

Monte Carlo methods provide a flexible and elegant tool for pricing derivatives securities that do not require the calculation of optimal exercise decision, like European options, exotic options (Asian, barrier, lookback, etc.). Unfortunately, these methods suffer from chronic deficit in performance. Various techniques have been implemented to accelerate their convergence: antithetic prints, control variate, importance sampling, just to name a few. Except the first, all of these techniques rely on the specificities of the product we want to price and hence fail by lack of genericness. Multilevel Monte Carlo (*MLMC*) aims at providing a universal solution applicable to this convergence problem. In the case of Euler scheme, Lemaire and Pagès introduced a more efficient method called Multilevel Richardson-Romberg (*ML2R*). This leads to a reduction of the computational

© Springer International Publishing AG 2017
I. Dimov et al. (Eds.): NAA 2016, LNCS 10187, pp. 482–491, 2017.
DOI: 10.1007/978-3-319-57099-0_54

cost from $O(\varepsilon^{-2}(\log(1/\varepsilon))^2)$ to $O(\varepsilon^{-2}(\log(1/\varepsilon)))$ (see [1]), for a given accuracy $\varepsilon > 0$. The Milstein scheme allows to reach the optimal complexity $O(\varepsilon^{-2})$ for each of these estimators. In the multidimensional case, an antithetic approach avoiding the simulation of Lévy area allows to achieve this optimal complexity for MLMC estimator [2].

In this paper, we introduce a suitable antithetic correction for *ML2R* estimator without affecting the optimal complexity. We show that this complexity is achieved by both *ML2R* and *MLMC* but our new *antithetic ML2R* is better in term of performance than the standard *antithetic MLMC* (see Sect. 4).

This paper is organized as follows: in Sect. 2, we introduce a general framework to formalize the optimization of a biased Monte Carlo simulation as described in [1]. In Sect. 3, we introduce our new antithetic *ML2R* estimmtor (see Eq. (8)) with its optimal allocation given by Theorem 3 and the standard antithetic *MLMC* after a brief background on the Multilevel estimators (*MLMC* & *ML2R*). In the last section, we present the numerical experiments.

2   General Framework

In this section, we will revisit in detail the preliminaries we need before introducing our Multilevel estimators.

2.1   Milstein Scheme

Let $T > 0$ and $(\Omega, \mathcal{F}, \{\mathcal{F}_t\}_{t \in [0,T]}, \mathbb{P})$ be a complete probability space with a filtration $\{\mathcal{F}_t\}_{t \in [0,T]}$ satisfying the usual conditions. We consider the multidimensional $\mathbb{R}^d$-valued SDEs of the form

$$dX_t = b(X_t)dt + \sigma(X_t)dW_t, \quad X_0 \in \mathbb{R}^d, \ t \in [0,T] \tag{1}$$

where $b : \mathbb{R}^d \to \mathbb{R}^d$ and $\sigma : \mathbb{R}^d \to \mathrm{M}(d,q)$ are smooth functions and W is a q-dimensional Brownian motion. Define the tensor $g_{ijk}(x)$ as

$$g_{ijk}(x) = \frac{1}{2} \sum_{m=1}^{d} \frac{\partial \sigma_{ij}(x)}{\partial x_m} \sigma_{mk}(x), \quad i = 1, \ldots, d, \ j = 1, \ldots, q.$$

Then X can be approximated by the Milstein scheme with uniform step $h = T/n$

$$\Delta \tilde{X}_{i,\ell} = b_i(\tilde{X}_\ell)h + \sum_{j=1}^{q} \sigma_{ij}(\tilde{X}_\ell)\Delta W_{j,\ell} + \sum_{j,k=1}^{q} g_{ijk}(\tilde{X}_\ell)(\Delta W_{j,\ell}\Delta W_{k,\ell} - hC_{i,k} - A_{jk,\ell})$$

where C is the correlation matrix for the driving Brownian paths $\Delta W_{j,\ell} = W_{t_{\ell+1}}^j - W_{t_\ell}^j$, $\Delta \tilde{X}_{i,\ell} = \tilde{X}_{i,\ell+1} - \tilde{X}_{i,\ell}$ with $t_\ell = \ell h$, and $A_{jk,\ell}$ is the Lévy area of (W^j, W^k) between t_ℓ and $t_{\ell+1}$ defined as (see [2])

$$A_{jk,\ell} = \int_{t_\ell}^{t_{\ell+1}} (W_t^j - W_{t_\ell}^j)dW_t^k - \int_{t_\ell}^{t_{\ell+1}} (W_t^k - W_{t_\ell}^k)\,dW_t^j.$$

The Milstein scheme without Lévy area, or *approximate Milstein scheme*, is obtained by setting $A \equiv 0$, namely:

$$\Delta \bar{X}_{i,\ell} = b_i(\bar{X}_\ell)h + \sum_{j=1}^{q} \sigma_{ij}(\bar{X}_\ell)\Delta W_{j,\ell} + \sum_{j,k=1}^{q} g_{ijk}(\bar{X}_\ell)(\Delta W_{j,\ell}\Delta W_{k,\ell} - hC_{i,k}). \quad (2)$$

2.2   Variance and Complexity (Effort)

We consider a family of linear statistical estimators $(I_\pi^N)_{N \geq 1}$ of $I_0 \in \mathbb{R}$ where π lies in a parameter set $\Pi \subset \mathbb{R}^d$. By linear, we mean, on the one hand, that

$$\mathbb{E}[I_\pi^N] = \mathbb{E}[I_\pi^1], \quad N \geq 1,$$

and, on the other hand, that the numerical cost, $\text{Cost}(I_\pi^N)$ induced by the simulation of I_π^N is given by

$$\text{Cost}(I_\pi^N) = N\kappa(\pi)$$

where $\kappa(\pi) = \text{Cost}(I_\pi^1)$ is the cost of a single simulation or *unitary complexity*. We also assume that our estimator is of *Monte Carlo type* in the sense that its variance is *inverse linear* in the size N of the simulation:

$$\text{var}(I_\pi^N) = \frac{\nu(\pi)}{N}$$

where $\nu(\pi) = \text{var}(I_\pi^1)$ denotes the variance of one simulation. We are looking for the "best" estimator in the family $\{(I_\pi^N), \pi \in \Pi\}$ *i.e.* the estimator minimizing the computational cost for a given $\varepsilon > 0$. This generic problem reads:

$$(\pi(\varepsilon), N(\varepsilon)) = \underset{\|I_\pi^N - I_0\|_2 \leq \varepsilon}{\text{argmin}} \; \text{Cost}(I_\pi^N). \quad (3)$$

In order to solve this minimization problem, we introduce the notion of *effort* $\phi(\pi)$ of a linear Monte Carlo type estimator I_π^N.

Definition 1. *The effort of the estimator I_π^N is defined for every $\pi \in \Pi$ by*

$$\phi(\pi) = \nu(\pi)\kappa(\pi) \quad \text{so that} \quad \text{Cost}(I_\pi^N) = N\frac{\phi(\pi)}{\nu(\pi)}. \quad (4)$$

When the estimators $(I_\pi^N)_{N \geq 1}$ are *biased*, the $\mathbb{L}^2$-error becomes

$$\mathbb{E}[(I_\pi^N - I_0)^2] = \mu^2(\pi) + \frac{\nu(\pi)}{N} \quad \text{where} \quad \mu(\pi) = \mathbb{E}[I_\pi^N] - I_0 = \mathbb{E}[I_\pi^1] - I_0$$

denotes the bias (which does not depend on N).

2.3   Assumptions on Weak and Strong Approximation Errors

We consider a family of real-valued random variables $(Y_h)_{h\in\mathcal{H}}$ associated to a random variable $Y_0 \in \mathbb{L}^2$. The index set $\mathcal{H}$ is a *consistent set of step* parameters in the sense that $\mathcal{H} = \{\mathbf{h}/n, n \geq 1\}$ for a fixed $\mathbf{h} \in (0, +\infty)$. The family satisfies two assumptions which formalize the strong and weak rates of approximation of Y_0 by Y_h when the bias parameter $h \to 0$ in $\mathcal{H}$.

Bias Error Expansion (Weak Error Rate):

$$\exists \alpha > 0, \ \bar{R} \in \mathbb{N}, \ \mathbb{E}[Y_h] = \mathbb{E}[Y_0] + \sum_{k=1}^{\bar{R}} c_k h^{\alpha k} + h^{\alpha \bar{R}} \eta_{\bar{R}(h)}, \ \lim_{h \to 0} \eta_{\bar{R}(h)} = 0, \ (WE_{\alpha, \bar{R}})$$

where c_k, $k = 1, \ldots, \bar{R}$ are real coefficients and $\eta_{\bar{R}}$ is a real function defined on $\mathcal{H}$.

Strong Approximation Error Assumption:

$$\exists \beta > 0, \ V_1 \in \mathbb{R}_+, \quad \|Y_h - Y_0\|_2^2 = \mathbb{E}\left[|Y_h - Y_0|^2\right] \leq V_1 h^\beta. \tag{SE_β}$$

The parameters α, β and $\bar{R}$ are structural parameters which depend on the family $(Y_h)_{h\in\mathcal{H}}$. In the sequel, we will consider a free parameter $R \in \{2, \ldots, \bar{R}\}$ for which $(WE_{\alpha,R})$ is always satisfied.

We associate to the family $(Y_h)_{h\in\mathcal{H}}$, the $\mathbb{R}^R$-valued random vector

$$Y_{h,\underline{n}} = (Y_h, Y_{\frac{h}{n_2}}, \ldots, Y_{\frac{h}{n_R}}) \tag{5}$$

where the R-tuple of integers $\underline{n} := (n_1, n_2, \ldots, n_R) \in \mathbb{N}^R$, called *refiners*, satisfy

$$n_1 = 1 < n_2 < \cdots < n_R. \tag{6}$$

3   Antithetic Multilevel Monte Carlo

Let $R \geq 2$ and let $(Y_{h,\underline{n}}^{(j),k})_{k\geq 1}$ be an i.i.d sequence of copies of $Y_{h,\underline{n}}$.

3.1   Brief Recall of Multilevel Monte Carlo Estimators

A *Multilevel Richardson-Romberg (ML2R) estimator* of depth R attached to an *allocation strategy* $q = (q_1, \ldots, q_R)$ with $q_j > 0$, $j = 1, \ldots, R$, $\sum_j q_j = 1$ and a weight vector $\mathbf{W}$, is defined for every integer $N \geq 1$ and $h \in \mathcal{H}$ by (see [1])

$$\bar{Y}_{h,\underline{n}}^{N,q} = \frac{1}{N_1} \sum_{k=1}^{N_1} Y_h^{(1),k} + \sum_{j=2}^{R} \frac{\mathbf{W}_j}{N_j} \sum_{k=1}^{N_j} \left(Y_{\frac{h}{n_j}}^{(j),k} - Y_{\frac{h}{n_{j-1}}}^{(j),k}\right)$$

where for all $j \in 1, \ldots, R$, $N_j = \lceil q_j N \rceil$, $\mathbf{W}_j = \mathbf{w}_j + \ldots + \mathbf{w}_R$, and

$$\mathbf{w}_i = \frac{(-1)^{R-i} n_i^{-\alpha(R-1)}}{\prod_{1 \leq j < i} (n_i^\alpha - n_j^\alpha) \prod_{i < j \leq R} (n_j^\alpha - n_i^\alpha)}.$$

The vector $\mathbf{w} = (\mathbf{w}_1, \ldots, \mathbf{w}_R)$ is defined as the unique solution of a Vandermonde system (see [1] for more details).

Remark 1. If the R-level weight vector $\mathbf{W}$ satisfies $\mathbf{W} = (1, \ldots, 1)$, the estimator is called a Multilevel Monte Carlo *MLMC* estimator.

3.2 Bias Parameter and Depth R Optimization for Geometric Refiners

In this approach, we consider geometric refiners with *root* $M \geq 2$ of the form

$$n_i = M^{i-1}, \ i = 1, \ldots, R. \tag{7}$$

Theorem 1 (see[1]). *The ML2R estimator satisfies*

$$\limsup_{\varepsilon \to 0} v(\beta, \varepsilon) \times \inf_{\substack{h \in \mathcal{H}, R \geq 2 \\ |\mu(h, R, q^*)| < \varepsilon}} \mathrm{Cost}\left(\bar{Y}_{h,\underline{n}}^{N,q^*}\right) \leq K(\alpha, \beta, M) < +\infty$$

$$\text{with} \quad v(\beta, \varepsilon) = \begin{cases} \varepsilon^2 (\log(1/\varepsilon))^{-1} & \text{if } \beta = 1, \\ \varepsilon^2 & \text{if } \beta > 1, \\ \varepsilon^2 e^{-\frac{1-\beta}{\sqrt{\alpha}}\sqrt{2\log(1/\varepsilon)\log(M)}} & \text{if } \beta < 1. \end{cases}$$

The MLMC *estimator satisfies (see also Giles in [3])*

$$\limsup_{\varepsilon \to 0} v(\beta, \varepsilon) \times \inf_{\substack{h \in \mathcal{H}, R \geq 2 \\ |\mu(h, R, q^*)| < \varepsilon}} \mathrm{Cost}\left(\bar{Y}_{h,\underline{n}}^{N,q^*}\right) \leq K(\alpha, \beta, M) < +\infty$$

$$\text{with} \quad v(\beta, \varepsilon) = \begin{cases} \varepsilon^2 (\log(1/\varepsilon))^{-2} & \text{if } \beta = 1, \\ \varepsilon^2 & \text{if } \beta > 1, \\ \varepsilon^{2+\frac{1-\beta}{\alpha}} & \text{if } \beta < 1. \end{cases}$$

We refer to [1] for explicit formulas of $K(\alpha, \beta, M)$.

3.3 Antithetic Multilevel Estimators

In this section we set $M = 2$. Let $Y = f(\bar{X})$ where f is a smooth payoff function and $\bar{X}$ the approximation of the Milstein scheme without Lévy area.
First we define $\bar{X}^f$ the approximate Milstein scheme with the fine time steps and its antithetic "twin" $\bar{X}^a$. This second approximate Milstein scheme is defined (or simulated) by simply inverting two by two the successive Brownian increments used to simulate $\bar{X}^f$. Finally, the approximate Milstein scheme on the coarse time step $\bar{X}^c$, associated to the payoff Y^c, is still simulated using the above Brownian increments, but grouped two by two. Hence $\bar{X}^f$ and $\bar{X}^a$ have the same conditional distribution, given $\bar{X}^c$. In particular $\mathbb{E}[Y^f] = \mathbb{E}[Y^a]$ where Y^f

and Y^a are the payoffs "written" on $\bar{X}^f$ and $\bar{X}^a$ respectively. At the same time, one has

$$(\bar{X}^f - \bar{X}^a) \approx -(\bar{X}^a - \bar{X}^c) \quad \text{and} \quad (Y^f - Y^c) \approx -(Y^a - Y^c)$$

so that $\frac{1}{2}(Y^f + Y^a) \approx Y^c$. This suggests that $\frac{1}{2}(Y^f + Y^a) - Y^c$ has a much smaller variance than the standard estimator $Y^f - Y^c$ (see [2]).

In this case, an *antithetic ML2R estimator* with weight vector $\mathbf{W}$ is given by

$$\bar{Y}_{h,\underline{n}}^{N,q,a} = \frac{1}{N_1} \sum_{k=1}^{N_1} Y_h^{(1),k,c} + \sum_{j=2}^{R} \frac{\mathbf{W}_j}{N_j} \sum_{k=1}^{N_j} \left(\frac{1}{2} \left(Y_{\frac{h}{n_j}}^{(j),k,f} + Y_{\frac{h}{n_j}}^{(j),k,a} \right) - Y_{\frac{h}{n_{j-1}}}^{(j),k,c} \right) \quad (8)$$

Remark 2. If the weight vector $\mathbf{W}$ satisfies $\mathbf{W} = (1, \ldots, 1)$, we obtain the original antithetic *MLMC* estimator devised in [2].

Cost, Complexity and Effort of an Antithetic Multilevel Estimator.
For a simulation size N, the numerical cost induced by the antithetic estimators $Y_{h,\underline{n}}^{N,q,a}$, $N \geq 1$, with the convention $n_0 = (n_0)^{-1} = 0$, is

$$\text{Cost}(\bar{Y}_{n,\underline{n}}^{N,q,a}) = \sum_{j=1}^{R} \frac{N_j}{h}(n_j + n_{j-1}) = N\kappa(\pi) \quad (9)$$

where $\kappa(\pi)$ is the unitary complexity given by

$$\kappa(\pi) = \frac{1}{h} \sum_{j=1}^{R} q_j(n_j + n_{j-1}). \quad (10)$$

The effort (product of variance by complexity) of such an estimator reads

$$\phi(\pi) = \left(\sum_{j=1}^{R} \frac{1}{q_j} \text{var} \left[\mathbf{W}_j \left(\frac{1}{2} \left(Y_{\frac{h}{n_j}}^{(j),f} + Y_{\frac{h}{n_j}}^{(j),a} \right) - Y_{\frac{h}{n_{j-1}}}^{(j),c} \right) \right] \right) \kappa(\pi) \quad (11)$$

with the convention $Y_{\frac{h}{n_0}}^{(1),c} \equiv 0$. The proof of Theorem 3 below relies on the following lemma which is a straightforward consequence of Schwartz's Inequality.

Lemma 1. *For all* $j = 1, \ldots, R$, *let* $a_j > 0$, $b_j > 0$ *and* $q_j > 0$ *such that* $\sum_{j=1}^{R} q_j = 1$. *Then*

$$\left(\sum_{j=1}^{R} \frac{a_j}{q_j} \right) \left(\sum_{j=1}^{R} b_j q_j \right) \geq \left(\sum_{i=1}^{R} \sqrt{a_j b_j} \right)^2 \quad (12)$$

and equality holds iff $q_j = \sqrt{a_j b_j^{-1}} \left(\sum_{k=1}^{R} \sqrt{a_k b_k^{-1}} \right)^{-1}$, $j = 1, \ldots, R$.

3.4   Optimization of the Effort for Antithetic Multilevel Estimators

In this section, we first present a lemma and theorem which give an upper-bound of the antithetic Multilevel estimator and then allow us to prove the last theorem of this section which gives the optimal allocation of simulated paths across the levels of our new antithetic *ML2R* estimator. This allows to achieve the optimal complexity without simulating the Lévy areas.

Theorem 2 (see [2]).

(a) For $p \geq 2$, there exists a constant K_p, independent of step $h = \frac{T}{n}$, such that the approximate Milstein scheme $\bar{X}$ satisfies

$$\forall \ell \geq 1 \qquad \mathbb{E}\left[\max_{0 \leq \ell \leq n} \|\bar{X}_\ell\|^p\right] \leq K_p. \tag{13}$$

(b) For all $p \geq 2$, there exists a constant K'_p such that

$$\forall \ell \geq 1 \qquad \mathbb{E}\left[\max_{0 \leq \ell \leq n} \left\|\frac{1}{2}(\bar{X}^f_\ell + \bar{X}^a_\ell) - \bar{X}^c_\ell\right\|^p\right] \leq K'_p h^p \tag{14}$$

with $h = \frac{T}{n}$ the coarse time step. Hence one has $\beta = 2$, for $p = 2$.

The next theorem is the main result for the antithetic estimators.

Theorem 3 (Allocation optimization). *The minimal effort ϕ^* of an antithetic multilevel estimator satisfies*

$$\phi^*(\pi_0) \leq \bar{\phi}(\pi_0, q^*) = \frac{1}{h}\left(\sqrt{K_p} + h\sqrt{K'_p}\sum_{j=2}^{R}|\mathbf{W}_j|\frac{\sqrt{n_j + n_{j-1}}}{n_{j-1}}\right)^2$$

where $q^* = q^*(\pi_0)$ is the allocation strategy given by

$$\begin{cases} q^*_1(\pi_0) = \mu^*\sqrt{K_p} \\ q^*_j(\pi_0) = \mu^* h\sqrt{K'_p}\left(|\mathbf{W}_j|\frac{n_{j-1}^{-1}}{\sqrt{n_j + n_{j-1}}}\right), \quad j = 2, \ldots, R \end{cases}$$

and μ^* is the normalizing constant such that $\sum_{j=1}^{R} q^*_j = 1$ (since $\beta = 2$, see [1]). The constants K_p and K'_p are respectively defined in (13) and (14).

Proof. Consider Theorem 2 with $(p = 2)$. Using the independence of the random variables, one has, $\forall j \geq 2$,

$$\mathrm{var}(\bar{Y}^{N,q,a}_{h,\underline{n}}) = \frac{\mathrm{var}(Y^{(1),f}_h)}{N_1} + \sum_{j=2}^{R}\frac{(\mathbf{W}_j)^2}{N_j}\mathrm{var}\left[\frac{1}{2}\left(Y^{(j),f}_{\frac{h}{n_j}} + Y^{(j),a}_{\frac{h}{n_j}}\right) - Y^{(j),c}_{\frac{h}{n_{j-1}}}\right]$$

$$\leq \frac{K_p}{N_1} + \sum_{j=2}^{R}\frac{(\mathbf{W}_j)^2}{N_j}K'_p(h/n_{j-1})^2 = \frac{K_p}{N_1} + \sum_{j=2}^{R}\frac{(\mathbf{W}_j)^2}{N_j}K'_p h^2 n_{j-1}^{-2}.$$

Using (10), one derives an upper-bound for the effort $\phi(\pi) \le \bar{\phi}(\pi)$, with

$$\bar{\phi}(\pi) = \frac{1}{h} \left(\frac{K_p}{q_1} + \sum_{j=2}^{R} \frac{(\mathbf{W}_j)^2}{q_j} K_p' h^2 n_{j-1}^{-2} \right) \left(\sum_{j=1}^{R} q_j (n_j + n_{j-1}) \right). \qquad (15)$$

Using Lemma 1 with $a_1 = K_p$, $b_1 = 1$, $a_j = K_p' h^2 (\mathbf{W}_j)^2 n_{j-1}^{-2}$ et $b_j = (n_j + n_{j-1})$, $j = 2, \ldots, R$, the proof is complete.

Remark 3. If $\mathbf{W}_j = 1$, $j = 1, \ldots, R$ then, one recovers Giles and Szpruch' antithetic *MLMC* estimator.

4 Numerical Experiments

Practitioner's Corner. To implement the antithetic Multilevel estimators (8), we need to know both the weak and strong rates of convergence of the biased estimator Y_h toward Y_0. The Milstein scheme is a second order scheme which satisfies (SE_β) with $\beta = 2$ and $(WE_{\alpha,\bar{R}})$ for $\alpha = 1$, at least when dealing with "vanilla" Lipschitz poyoffs.

Note that in the antithetic case for multidimensional SDEs, the root M of Eq. (7) is set to $M = 2$ to allow the permutation of the two successive pair of Brownian increments coming from the fine time step to simulate the antithetic paths. A natural approximation of K_p and K_p' respectively from (13) and (14) is
$K_p \sim \operatorname{var}(Y_h^{(1)})$ and $K_p' \sim \operatorname{var}\left[\frac{1}{2}(Y_{\frac{h}{2}}^{(1),f} + Y_{\frac{h}{2}}^{(1),a}) - Y_h^{(1),c} \right] h^2$.

Best-of-Call option by 2D Milstein scheme, $\alpha = 1$, $\beta = 2$. We consider the diffusion Eq. (1) with $d = 2$, $b(X_t) = rX_t$, $\sigma(X_t) = \sigma X_t$ (2D Black-Scholes model), $X_0^1 = 100$, $X_0^2 = 100$, $r = 0.15$, $\sigma_1 = 0.3$, $\sigma_1 = 0.4$ and the Milstein scheme without Lévy area defined by (2) with $q = 2$ and $C = \begin{pmatrix} 1 & 0 \\ 0 & 1 \end{pmatrix}$. The payoff of a Best-of-Call option is given by

$$f(x_1, x_2) = e^{-rT} (\max(x_1(T), x_2(T)) - K)_+ \quad (x_1, x_2) \in \mathcal{C}([0, T], \mathbb{R}^2)$$

with maturity $T = 1$ and strike $K = 80$ and the biased estimator $Y_h = f(\bar{X}_T^1, \bar{X}_T^2)$. For the numerical simulation, we apply the antithetic procedure clearly detailed in [2]. The constants K_p and K_p' have been estimated by $K_p = 7.55$ and $K_p' = 965.04$ and the bias is computed using the exact price $I_0 = 51.0867$. The results are summarized in Table 1 for *ML2R* estimator and in Table 2 for *MLMC* estimator (see the Appendix). Note that the depth R required to achieve the optimal complexity is more important for the *MLMC* estimator (Tables 1 and 2).

This leads to a reduction of the computational cost with our new *ML2R* estimator which is more efficient than the *MLMC*. Figure 2 shows that *MLMC* takes 50 to 60% more time than *ML2R*. Other experiments carried out on Margrabe model lead to the same conclusions.

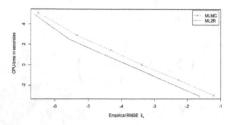

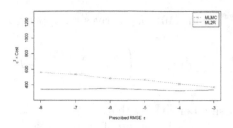

Fig. 1. CPU-time (y-axis log scale) as a function of $\tilde{\varepsilon}_L$ (x-axis $\log_2$ scale)

Fig. 2. $\varepsilon^2 \times$ Cost as a function of ε ($\log_2$ scale for the x-axis)

Acknowledgements. The work of Cheikh Mbaye is supported by the *Fédération Wallonie-Bruxelles* and the *National Bank of Belgium* via an *FSR* grant.

The opinions expressed in this paper are those of the authors and do not necessarily reflect the views of the *National Bank of Belgium*.

Appendix: Main Results for Antithetic Estimators

Table 1. Best-of-Call option: parameters and results of *ML2R* estimator.

k	$\varepsilon = 2^{-k}$	L^2-error	Time (s)	Bias	R	M	h^{-1}	N	Cost
3	$1.25 \cdot 10^{-01}$	$3.23 \cdot 10^{-01}$	$4.69 \cdot 10^{-02}$	$-1.85 \cdot 10^{-02}$	4	2	1	$1.64 \cdot 10^{+04}$	$2.18 \cdot 10^{+04}$
4	$6.25 \cdot 10^{-02}$	$1.64 \cdot 10^{-02}$	$1.84 \cdot 10^{-01}$	$-2.07 \cdot 10^{-03}$	4	2	1	$6.38 \cdot 10^{+04}$	$8.46 \cdot 10^{+04}$
5	$3.12 \cdot 10^{-02}$	$8.33 \cdot 10^{-02}$	$7.50 \cdot 10^{-01}$	$8.36 \cdot 10^{-03}$	5	2	1	$2.62 \cdot 10^{+05}$	$3.53 \cdot 10^{+05}$
6	$1.56 \cdot 10^{-02}$	$4.05 \cdot 10^{-02}$	$3.15 \cdot 10^{+00}$	$1.99 \cdot 10^{-03}$	5	2	1	$1.09 \cdot 10^{+06}$	$1.48 \cdot 10^{+06}$
7	$7.81 \cdot 10^{-03}$	$2.07 \cdot 10^{-02}$	$1.24 \cdot 10^{+01}$	$-1.03 \cdot 10^{-03}$	5	2	1	$4.20 \cdot 10^{+06}$	$5.67 \cdot 10^{+06}$
8	$3.91 \cdot 10^{-03}$	$1.04 \cdot 10^{-02}$	$1.40 \cdot 10^{+02}$	$9.00 \cdot 10^{-04}$	5	2	1	$1.67 \cdot 10^{+07}$	$2.25 \cdot 10^{+07}$

Table 2. Best-of-Call option: parameters and results of *MLMC* estimator.

k	$\varepsilon = 2^{-k}$	L^2-error	Time (s)	Bias	R	M	h^{-1}	N	Cost
3	$1.25 \cdot 10^{-01}$	$4.33 \cdot 10^{-01}$	$5.08 \cdot 10^{-02}$	$-3.18 \cdot 10^{-01}$	5	2	1	$1.86 \cdot 10^{+04}$	$2.41 \cdot 10^{+04}$
4	$6.25 \cdot 10^{-02}$	$2.07 \cdot 10^{-01}$	$2.30 \cdot 10^{-01}$	$-1.47 \cdot 10^{-01}$	6	2	1	$8.08 \cdot 10^{+04}$	$1.06 \cdot 10^{+05}$
5	$3.12 \cdot 10^{-02}$	$9.59 \cdot 10^{-02}$	$1.02 \cdot 10^{+00}$	$-6.49 \cdot 10^{-02}$	7	2	1	$3.57 \cdot 10^{+05}$	$4.80 \cdot 10^{+05}$
6	$1.56 \cdot 10^{-02}$	$4.99 \cdot 10^{-02}$	$4.23 \cdot 10^{+00}$	$-3.53 \cdot 10^{-02}$	8	2	1	$1.47 \cdot 10^{+06}$	$2.00 \cdot 10^{+06}$
7	$7.81 \cdot 10^{-03}$	$2.45 \cdot 10^{-02}$	$1.80 \cdot 10^{+01}$	$-1.75 \cdot 10^{-02}$	9	2	1	$6.37 \cdot 10^{+06}$	$8.78 \cdot 10^{+06}$
8	$3.91 \cdot 10^{-03}$	$1.12 \cdot 10^{-03}$	$1.65 \cdot 10^{+02}$	$-7.31 \cdot 10^{-03}$	10	2	1	$2.64 \cdot 10^{+07}$	$3.67 \cdot 10^{+07}$

Optimal Parameters for the Antithetic Estimators. In all the numerical experiments presented in the paper, we set $\mathbf{h} = T$ in the definition of the parameters of the simulation for both *MLMC* and *ML2R* estimators. In the table below, keep in mind that $n_0 = 0$ by convention (Table 3).

Table 3. Optimal parameters for antithetic multilevel estimators.

	ML2R	MLMC
$q(\varepsilon)$	$q_1 = \mu^* \sqrt{K_p}$	$q_1 = \mu^* \sqrt{K_p}$
	$q_j = \mu^* h \sqrt{K_p'} \left(|\mathbf{W}_j| \frac{n_j^{-1}}{\sqrt{n_j + n_{j-1}}} \right), \ j = 2 \ldots, R$	$q_j = \mu^* h \sqrt{K_p'} \left(\frac{n_j^{-1}}{\sqrt{n_j + n_{j-1}}} \right) \ j = 2 \ldots, R$
$N(\varepsilon)$	$\left(1 + \frac{1}{2\alpha R}\right) \dfrac{\left(1 + h\sqrt{K_p'} \sum_{j=1}^{R} |\mathbf{W}_j|(n_j^{-1})\sqrt{n_j + n_{j-1}}\right)^2}{\varepsilon^2 \sum_{j=1}^{R} q_j(n_j + n_{j-1})}$	$\left(1 + \frac{1}{2\alpha}\right) \dfrac{\left(1 + h\sqrt{K_p'} \sum_{j=1}^{R} (n_j^{-1})\sqrt{n_j + n_{j-1}}\right)^2}{\varepsilon^2 \sum_{j=1}^{R} q_j(n_j + n_{j-1})}$

Note that in the table above, μ^* is defined such that $\sum_{j=1}^{R} q_j = 1$. For the optimal parameters R and h, their formulas remain the same as in [1].

References

1. Lemaire, V., Pagès, G.: Multilevel Richardson-Romberg extrapolation. pre-pub. LPMA 1603 arXiv:1401.1177 (2014). (forthcoming in Bernoulli)
2. Giles, M.B., Szpruch, L.: Antithetic multilevel Monte Carlo for multi-dimensional SDEs without Lévy area simulation. Ann. Appl. Prob. **24**(4), 1585–1620 (2014)
3. Giles, M.B.: Multilevel Monte Carlo path simulation. Oper. Res. **56**(3), 607–617 (2008)

Front Dynamics in an Activator-Inhibitor System of Equations

Alina Melnikova[✉], Natalia Levashova, and Dmitry Lukyanenko

Faculty of Physics, Department of Mathematics,
Lomonosov Moscow State University, Moscow 119991, Russia
melnikova@physics.msu.ru

Abstract. We consider the construction of formal asymptotic approximation for solution of the singularly perturbed boundary value problem
of an activator-inhibitor type with a solution in a form of moving front.
Corresponding asymptotic analysis provides *a priori* information about
the localization of the transition point for moving front that is further
used for constructing of dynamic adapted mesh. This mesh significantly
improves numerical stability of numerical calculations for the considered
system.

1 Statement of the Problem

We consider the singularly perturbed initial-boundary value problem with Neu-
man's boundary conditions:

$$
\begin{cases}
\varepsilon^4 u_{xx} - \varepsilon^2 u_t = u\big(u - \alpha(x)\big)(u - 1) + uv, \\
\varepsilon^2 v_{xx} - \varepsilon^2 v_t = \gamma v - \beta u, \qquad x \in (a, b), \quad t \in (0, T], \\
u_x(a, t) = u_x(b, t) = 0, \quad v_x(a, t) = v_x(b, t) = 0, \quad t \in [0, T], \\
u(x, 0) = u_{init}(x), \quad v(x, 0) = v_{init}(x), \quad x \in [a, b].
\end{cases}
\tag{1}
$$

This system is considered in the area $\Omega : \{u \in I_u \subset \mathbb{R}^+ \cup v \in I_v \subset \mathbb{R}^+ \cup x \in [a; b] \cup t \in [0; T]\}$, the values $0 < \alpha < 1$, $\beta > 0$ and $\gamma > 0$ are parameters of the system, $\varepsilon > 0$ is a small parameter. This is a problem for an activator-inhibitor system of equations of Fitz-Hugh-Nagumo type [5]. The investigation of the solutions of such boundary value problems having the form of contrast structures is quite important as they play significant role in mathematical modelling in biophysics, particulary in simulating processes of excitement in a neuro-muscular tissue [2,5]. The above–mentioned contrast structure is some function which domain has an area, where the function values undergo sharp changes. This area is called an inner transition layer. Carrying out the numerical calculations for the problems with solutions having inner transition layers is a complicated problem as it demands using of detailed meshes and using of stiff numerical

Electronic supplementary material The online version of this chapter (doi:10.
1007/978-3-319-57099-0_55) contains supplementary material, which is available to
authorized users.

© Springer International Publishing AG 2017
I. Dimov et al. (Eds.): NAA 2016, LNCS 10187, pp. 492–499, 2017.
DOI: 10.1007/978-3-319-57099-0_55

methods. For the purpose of creating efficient numerical algorithms the analytical investigations can be very useful. In the present paper the construction of the zero-order formal asymptotic approximation for the solution of the system (1), which have moving inner transition layer (moving front) is held. That provides a priori information about the localization and speed of the front that is further used to create a dynamic adapted mesh (DAM) appropriate for carrying out numerical calculations for the system (1).

2   Asymptotic Approximation

We constructed the zero-order asymptotic approximation according to Vasil'eva algorithm [3,6,7] in a following form:

$$u = \bar{u} + Q_0 u\,(\xi, t) + M_0 u\,(\sigma, t)\,, \quad v = \bar{v} + Q_0 v\,(\xi, t) + M_0 v\,(\sigma, t)\,,$$

where $\bar{u}$, $\bar{v}$ are the regular part; $Q_0 u\,(\xi, t)$, $Q_0 v\,(\xi, t)$, $M_0 u\,(\sigma, t)$, $M_0 v\,(\sigma, t)$ are the functions describing the two scaled inner transition layer, $\xi = (x - x_{tr}(t))/\varepsilon$, $\sigma = (x - x_{tr}(t))/\varepsilon^2$. Function $x_{tr}(t)$ shows where the inner layer is localized at each time point.

2.1   Regular Part

The regular part, functions $\bar{u}$, $\bar{v}$ are the solution of so-called degenerated system that turns out from the system of equations (1) if ε is equal to zero:

$$f(u,v) := u\big(u - \alpha(x)\big)(u - 1) + uv = 0, \quad g(u,v) := \gamma v - \beta u = 0. \qquad (2)$$

Let us consider that the first equation of system (2) has three roots: $u = \varphi^i(v)$, $i = 1, 2, 3 : \varphi^1(v) = 0$, $\varphi^{2,3}(v) = 0.5(\alpha + 1 \mp \sqrt{(\alpha + 1)^2 - 4v})$. The roots $\bar{u} = \varphi^1(v)$ and $\bar{u} = \varphi^3(v)$ satisfy the stability condition $f_u(\varphi^{1,3}(v), v) > 0$ for positive values α and v. We consider the parameters α, β and γ such that we can substitute $\varphi^1 = 0$ and $\varphi^3(v)$ in the second equation of system (2) and obtain the regular part for v–component of the asymptotic approximation:

$$v^1 = 0, \quad v^3 = 0.5\beta\gamma^{-1}\left(\alpha + 1 - \beta\gamma^{-1} + \sqrt{(\alpha + 1 - \beta\gamma^{-1})^2 - 4\alpha}\right).$$

Solution $v = v^1$ is a regular part at the left side of the inner layer and $v = v^3$ is the regular part at the right side. Function $\varphi^1 = 0$ is a regular part of the u–component of the asymptotic at the left side of the inner layer and the function $\varphi^3(v^3)$ is the regular part at the right side.

2.2   Zero-Order Terms of the Inner Layer Functions

For zero-order terms of the inner layer functions we have the following system of equations [6]:

$$f\left(\varphi^{1,3}(v^{1,3}) + Q_0^{1,3}u,\ v^{1,3} + Q_0^{1,3}v\right) = 0,$$

$$\frac{\partial^2 Q_0^{1,3} v}{\partial \xi^2} = \gamma\left(v^{1,3} + Q_0^{1,3}v\right) - \beta\left(\varphi^{1,3}(v^{1,3}) + Q_0^{1,3}u\right),$$

$$\frac{\partial^2 M_0^{1,3} u}{\partial \sigma^2} + W_0 \frac{\partial M_0^{1,3} u}{\partial \sigma}$$

$$= f\left(\varphi^{1,3}(v^{1,3}) + Q_0^{1,3}u(0,t) + M_0^{1,3}u,\ v^{1,3} + Q_0^{1,3}v(0,t) + M_0^{1,3}v\right),$$

$$\frac{\partial^2 M_0^{1,3} v}{\partial \sigma^2} = 0. \tag{3}$$

Here functions $Q_0^1 u(\xi,t)$ and $Q_0^3 u(\xi,t)$ denotes function $Q_0 u(\xi,t)$ at the left and right side of the point $\xi = 0$ respectively, the notations $Q_0^{1,3}v$, $M_0^{1,3}u$, $M_0^{1,3}v$ have the similar meaning and $W_0 = dx_{tr}/dt$ is the front velocity in zero order approximation of the parameter ε.

For the inner layer functions we demand the decreasing when $\xi \to \mp\infty$ or $\sigma \to \mp\infty$. From this demand it turns out that $M_0^{1,3}v(\sigma,t) = 0$. The first Eq. (3) gives

$$\varphi^{1,3}(v^{1,3}) + Q_0^{1,3}u = \varphi^{1,3}\left(v^{1,3} + Q_0^{1,3}v\right). \tag{4}$$

Let us introduce the designation

$$\tilde{v}(\xi,t) = \begin{cases} v^1 + Q_0^1 v(\xi,t), & \xi \le 0; \\ v^3 + Q_0^3 v(\xi,t), & \xi \ge 0. \end{cases} \tag{5}$$

Let's demand the function $\tilde{v}$ to be smooth for all $\xi \in \mathbb{R}$ and define $\tilde{v}(0,t) = v_0$, where the time-independent value v_0 is to be determined from smooth matching conditions. From the first two equations Eq. (3) and designation (5) it comes that function $\tilde{v}(\xi,t)$ is the solution of the following problem:

$$\frac{\partial^2 \tilde{v}}{\partial \xi^2} = h^1(\tilde{v}), \quad \tilde{v}(0,t) = v_0, \quad \tilde{v}(-\infty,t) = v^1, \quad \xi \le 0;$$

$$\frac{\partial^2 \tilde{v}}{\partial \xi^2} = h^3(\tilde{v}), \quad \tilde{v}(0,t) = v_0, \quad \tilde{v}(+\infty,t) = v^3, \quad \xi \ge 0, \tag{6}$$

where $h^1(\tilde{v}) = \gamma\tilde{v}$, $h^3(\tilde{v}) = \gamma\tilde{v} - 0.5\beta\left(\alpha + 1 + \sqrt{(\alpha+1)^2 - 4\tilde{v}}\right)$.

The second-order equations in (6) are equal to systems of the two first–order differential equations:

$$\frac{d\tilde{v}}{d\xi} = \varPhi, \quad \frac{d\varPhi}{d\xi} = h^{1,3}(\tilde{v}); \tag{7}$$

The points $(v^1,0)$ and $(v^3,0)$ are respectively the saddle-type rest points of system (7) on the phase plane $(\tilde{v},\varPhi)$ as inequalities $h_{\tilde{v}}^i(v^i) > 0,\ i = 1,3$ are valid. As $\tilde{v}$ is smooth $\varPhi$ is continuous, that means that the outcoming separatrix of the saddle point $(v^1,0)$ intersects the incoming separatrix of the saddle point $(v^3,0)$ on the phase plane $(\tilde{v},\varPhi)$. The explicit expressions $\varPhi = \varPhi(\tilde{v})$ for the separatrices can be obtained from systems (7). The condition of their intersection yields to the following equality that is the equation for the value v_0.

$$6\beta(\alpha+1)v_0 - \beta\left((\alpha-1)^2 - 4v_0\right)^{3/2} = 6\beta(\alpha+1)v^3 - \left((\alpha-1)^2 - 4v^3\right)^{3/2} - 6\gamma(v^3)^2.$$

Let us introduce the smooth function $\tilde{u}$

$$\tilde{u}(\sigma, t) = \begin{cases} \varphi^1(v_0) + M_0^1 u(\sigma, t), & \sigma \leq 0; \\ \varphi^3(v_0) + M_0^3 u(\sigma, t), & \sigma \geq 0. \end{cases}$$

From the third Eq. (3), designations (4) and (5), relations $M_0^{1,3} v(\sigma, t) = 0$ and $\tilde{v}(0, t) = v_0$ and the demand of the decreasing of functions $M_0^{1,3} u$ when $\sigma \to \mp\infty$ it comes that function $\tilde{u}(\sigma, t)$ is the solution of the following problem:

$$\frac{\partial^2 \tilde{u}}{\partial\sigma^2} + W_0 \frac{\partial \tilde{u}}{\partial\sigma} = f(\tilde{u}, v_0), \quad \tilde{u}(0, t) = \varphi^2(v_0), \quad \tilde{u}(\mp\infty, t) = \varphi^{1,3}(v_0). \quad (8)$$

The existence of smooth solution of the problem (8) for each value $v_0 \in I_v$ was proved in [4].

The second-order equation in (8) is equal to system of the two first–order differential equations:

$$\frac{d\tilde{u}}{d\sigma} = \Psi, \quad \frac{d\Psi}{d\sigma} = -W_0\Psi + f(\tilde{u}, v_0). \quad (9)$$

The points $(\varphi^1(v_0), 0)$ and $(\varphi^3(v_0), 0)$ are respectively the saddle-type rest points of system (9) on the phase plane $(\tilde{u}, \Psi)$ due to condition $f_u(\varphi^{1,3}(v), v) > 0$. The existence of the smooth solution $\tilde{u}$ of the problem (9) means that there exists a connecting separatrix $\Psi(\tilde{u})$ that links these two saddle points on the phase plane. In case of cubic dependence of function $f(u, v)$ from u it is possible to express the function $\Psi(\tilde{u})$ in an explicit form. For this purpose we will divide the second equation of system (9) to the first and then multiply at Ψ:

$$\Psi \frac{d\Psi}{d\tilde{u}} = -W_0\Psi + f(\tilde{u}, v_0). \quad (10)$$

We will look for the solution of (10) in a form of a square parabola connecting the saddle points: $\Psi(\tilde{u}) = C(\tilde{u} - \varphi^1(v_0))(\tilde{u} - \varphi^3(v_0))$. Substituting the function $\Psi(\tilde{u})$ in this form in the Eq. (10) we can find C and W_0:

$$C = -1/\sqrt{2}, \quad W_0 = \sqrt{2}\left(\varphi^2(v_0) - 0.5(\varphi^1(v_0) + \varphi^3(v_0))\right).$$

Therefore we can write the formula for the position of the inner transition layer:

$$x_{tr}(t) = x_{tr}(0) + W_0 t. \quad (11)$$

Solving the differential equation $\dfrac{\partial \tilde{u}}{\partial\sigma} = -\dfrac{1}{\sqrt{2}}(\tilde{u} - \varphi^1(v_0))(\tilde{u} - \varphi^3(v_0))$ with an additional condition $\tilde{u}(0, t) = \varphi^2(v_0)$ we get the expression for $\tilde{u}(\sigma, t)$:

$$\tilde{u}(\sigma, t) = \left(\varphi^3(v_0) + \varphi^1(v_0)\frac{\varphi^3(v_0) - \varphi^2(v_0)}{\varphi^2(v_0) - \varphi^1(v_0)}\exp\left(-(\varphi^3(v_0) - \varphi^1(v_0))\sigma/\sqrt{2}\right)\right)$$
$$\times \left(1 + \frac{\varphi^3(v_0) - \varphi^2(v_0)}{\varphi^2(v_0) - \varphi^1(v_0)}\exp\left(-(\varphi^3(v_0) - \varphi^1(v_0))\sigma/\sqrt{2}\right)\right)^{-1}.$$

Thus we obtain the zero-order asymptotic approximation for the solution of the system (1):

$$v(x, t) = \bar{v}\big((x - x_{tr}(t))/\varepsilon, t\big),$$

$$u(x, t) = \begin{cases} \tilde{u}\big((x - x_{tr}(t))/\varepsilon^2, t\big) - \varphi^1(v_0) + \varphi^1\big(\bar{v}((x - x_{tr}(t))/\varepsilon, t)\big), & x \leq x_{tr}(t), \\ \tilde{u}\big((x - x_{tr}(t))/\varepsilon^2, t\big) - \varphi^3(v_0) + \varphi^3\big(\bar{v}((x - x_{tr}(t))/\varepsilon, t)\big), & x \geq x_{tr}(t). \end{cases} \quad (12)$$

3   Dynamic Adapted Mesh Construction

Using as *a priori* information the result from Sect. 2 about dependence of the transition point position on time (11) we are able to construct so–called *dynamic adapted mesh* [11] (which is ε–independent on spatial variable x). This mesh will have in each time moment the same (or proportional) number of nodes inside and outside of the interior layer and this number will not depend on the small parameter ε.

At first, we introduce uniform mesh T_M only on t–dimension (it is also possible to use quasiuniform mesh [8] without any changes at the further algorithm) that has number of nodes $M + 1$ (that equals to M intervals): $T_M = \{t_m, \ 0 \leq m \leq M : \ 0 = t_0 < t_1 < t_2 < \ldots < t_{M-1} < t_M = T\}$. The choice of the time step τ (in the case of uniform mesh) is very important. In process of constructing of dynamic adapted mesh it is possible to obtain a situation in which local refining of the mesh on the new time layer does not have intersection (or this intersection is small enough) with local refining of the mesh on the previous time layer. In order to avoid such situation we have to check the condition on the maximal allowed time step τ — the transition point position does not have to leave the current local refining of the mesh in one time step:

$$\tau = t_{m+1} - t_m, \quad \text{where} \quad t_{m+1} = \text{argmax}_{t_m \in T_M} |x'_{tr}(t_m)|. \quad (13)$$

In the case of the considered example the time step τ must be less than $(T - 0)/|W_0\, 2\varepsilon \ln \varepsilon|$.

After that, for each time node m of the mesh T_M we are able to construct a piecewise uniform mesh XT_m on x–dimension that has some uniform refining in $\varepsilon \ln \varepsilon$–neighborhood of the transition point $x_{tr}(t_m)$:

$$\begin{aligned} XT_m = \{\, x_n, \ & 0 \leq n \leq N + KN : \\ & x_n = a \quad \text{for} \quad n = 0, \\ & x_n = x_{n-1} + h_{left} \quad \text{for} \quad n = \overline{1, N_{left}}, \\ & x_n = x_{n-1} + h_{int} \quad \text{for} \quad n = \overline{N_{left} + 1, N_{left} + KN}, \\ & x_n = x_{n-1} + h_{right} \quad \text{for} \quad n = \overline{N_{left} + KN + 1, N + KN}\}, \end{aligned}$$

where N is a control parameter that determine number of intervals allocated inside and outside of the interior layer (intervals outside of the layer rationed out, so $h_{left} \cong h_{right}$),

$$N_{left} = \left[\frac{\left(x_{tr}(t_m) - C|\varepsilon \ln \varepsilon| \right) - a}{b - a - 2C|\varepsilon \ln \varepsilon|} \, N \right] , \; N_{right} = N - N_{left}, \; h_{int} = \frac{2C|\varepsilon \ln \varepsilon|}{KN},$$

$$h_{left} = \frac{\left(x_{tr}(t_m) - C|\varepsilon \ln \varepsilon| \right) - a}{N_{left}}, \quad h_{right} = \frac{b - \left(x_{tr}(t_m) + C|\varepsilon \ln \varepsilon| \right)}{N_{right}},$$

C is a control parameters that is able to adjust thickness of the interior layer ($C = 1$ on default); K is a control parameters that denote relative density of the refining inside the interior layer ($K = 1$ on default).

As a result we have constructed the *dynamic adapted mesh* (DAM) XT_M (see for example Fig. 1), which has on each time layer T_m number of intervals equals to $N + KN$. Also we want to remind that this mesh is ε-independent on spatial variable x.

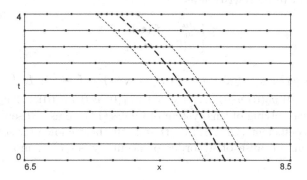

Fig. 1. Some example of the *dynamic adapted mesh* XT_M ($a = 6.5$, $b = 8.5$).

In the formulas of the following Sect. 4 we will use the redefining $N := N + KN$.

4   Numerical Example

Let us consider the initial-boundary value problem (1) for which we have performed asymptotic investigations in Sect. 2.

For numerical solving of the system (1) we apply the stiff method of lines (SMOL) [10] in order to reduce the system of PDEs to the system of ODEs that can be solved by the Rosenbrock scheme with complex coefficient [9] which is the most suitable and very efficient scheme for stiff systems of ODEs [1].

Approximating the derivatives with second order accuracy by finite-differences in (1) we obtain the following system of ODEs from which we should determine $N - 1$ unknown functions $u_n \equiv u_n(t) \equiv u(x_n, t)$ ($n = \overline{1, N-1}$, u_0 and u_N we know from the boundary conditions: $u_0 = \frac{4}{3}u_1 - \frac{1}{3}u_2$, $u_N = \frac{4}{3}u_{N-1} - \frac{1}{3}u_{N-2}$) and $N - 1$ unknown functions

$v_n \equiv v_n(t) \equiv v(x_n, t)$ $(n = \overline{1, N-1}$, v_0 and v_N we also know from the boundary conditions: $v_0 = \frac{4}{3}v_1 - \frac{1}{3}v_2$, $v_N = \frac{4}{3}v_{N-1} - \frac{1}{3}v_{N-2})$:

$$\frac{du_n}{dt} = \frac{2\varepsilon^2}{x_{n+1} - x_{n-1}} \left(\frac{u_{n+1} - u_n}{x_{n+1} - x_n} - \frac{u_n - u_{n-1}}{x_n - x_{n-1}} \right) - \frac{u_n \left(u_n - \alpha_n \right) \left(u_n - 1 \right) + u_n v_n}{\varepsilon^2}$$

$$\frac{dv_n}{dt} = \frac{2}{x_{n+1} - x_{n-1}} \left(\frac{v_{n+1} - v_n}{x_{n+1} - x_n} - \frac{v_n - v_{n-1}}{x_n - x_{n-1}} \right) - \frac{1}{\varepsilon^2} \left(\gamma v_n - \beta u_n \right),$$

plus initial condition: $u_n(0) = u_{init}(x_n)$, $v_n(0) = v_{init}(x_n)$.

Note, that we have used approximations of the first derivatives in boundary conditions with asymmetric formula with the second order of approximation on an uniform mesh. This assumption has to do with the fact that we suppose that the moving front never achieve the boundaries (so in a neighborhood of boundary the mesh is always uniform).

This system can be rewritten as

$$\begin{cases} \frac{d\boldsymbol{u}}{dt} = \boldsymbol{f}(\boldsymbol{u}), \\ \boldsymbol{u}(0) = \boldsymbol{u}_{init}, \end{cases} \tag{14}$$

where $\boldsymbol{u} = \begin{pmatrix} u_1 \ u_2 \ u_3 \ \dots \ u_{N-1} \ v_1 \ v_2 \ v_3 \ \dots \ {}_{N-1} \end{pmatrix}^T$, $\boldsymbol{f} = \begin{pmatrix} f_1 \ f_2 \ f_3 \ \dots \ f_{2N-2} \end{pmatrix}^T$ and $\boldsymbol{u}_{init} = \begin{pmatrix} u_1(0) \ u_2(0) \ u_3(0) \ \dots \ u_{N-1}(0) \ v_1(0) \ v_2(0) \ v_3(0) \ \dots \ v_{N-1}(0) \end{pmatrix}^T$.

For numerical solving of this system of ODEs (14) we use Rosenbrock scheme with complex coefficient (CROS1) [9] that is the best choice for solving of such kind of problems because of its order of accuracy ($O(\tau^2)$), monotonicity and stability (L_2) [1]

$$\boldsymbol{u}(t_{m+1}) = \boldsymbol{u}(t_m) + (t_{m+1} - t_m)\,Re\,\boldsymbol{w},$$
where $\boldsymbol{w}$ is the solution of the SLAE
$$\left[E - \frac{1+i}{2}\,(t_{m+1} - t_m)\,\boldsymbol{f}_u\left(\boldsymbol{u}(t_m) \right) \right] \boldsymbol{w} = \boldsymbol{f}\left(\boldsymbol{u}(t_m) \right). \tag{15}$$

Here, E is identity matrix, $\boldsymbol{f}_u$ is the Jacobian matrix.

So the matrix of the SLAE (15) consists of four $(N-1) \times (N-1)$ blocks which inner structure is allows to implement an algorithm that gets the solution of the SLAE (15) in $O(N)$ operations.

Note, that in the numerical scheme (15) we use for calculations $\boldsymbol{u}(t_m)$ that has been determined on the XT_m mesh. So, after applying (15), we have to interpolate $\boldsymbol{u}(t_{m+1})$ on XT_{m+1} mesh.

Some example of numerical calculations is represented on the Fig. 2 and in the video-file that is attached to this paper. The calculations were performed for $\varepsilon = 0.1$, $\alpha = 0.2$, $\beta = 0.2$, $\gamma = 1$, $a = 6.5$, $b = 8.5$, $N = 20$, $K = 2$, $C = 0.8$, $T = 4$, $M = 100$. The initial functions $u_{init}(x)$, $v_{init}(x)$ selected as zero-order asymptotic approximation (12) with $x_{tr}(t) \equiv 8$.

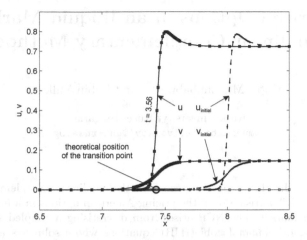

Fig. 2. The example of numerical calculations (some refining of the mesh in $\varepsilon \ln \varepsilon$– neighborhood of the transition point has been performed at the current time moment).

Acknowledgements. This study was supported by grants of the Russian Foundation for Basic Research projects No. 16-01-00437, 15-01-04619 and 16-01-00755.

References

1. Alshin, A.B., Alshina, E.A., Kalitkin, N.N., Koryagina, A.: Rosenbrock schemes with complex coefficients for stiff and differential algebraic systems. Comput. Math. Math. Phys. **46**, 1320–1340 (2006)
2. Barkley, D.: A model for fast computer simulation of waves in excitable media. Phys. D: Nonlinear Phenom. **49**, 445–466 (1991)
3. Butuzov, V.F., Levashova, N.T., Mel'nikova, A.A.: Steplike contrast structure in a singularly perturbed system of equations with different powers of small parameter. Comput. Math. Math. Phys. **52**, 1526–1546 (2012)
4. Fife, P., McLeod, J.: The approach of solutions of nonlinear diffusion equations to travelling front solutions. Arch. Ration. Mech. Anal. **65**, 335–361 (1977)
5. FitzHugh, R.: Impulses and physiological states in theoretical model of nerve membrane. Biophys. J. **1**, 445–466 (1961)
6. Levashova, N.T., Mel'nikova, A.A.: Step-like contrast sructure in a singularly perturbed system of parabolic equations. Differ. Equ. **51**, 342–367 (2015)
7. Vasil'eva, A.B., Butuzov, V.F., Kalachev, L.V.: The Boundary Function Method for Singular Perturbation Problems. SIAM, Bangkok (1995)
8. Kalitkin, N.N., Alshin, A.B., Alshina, E.A., Rogov, B.V.: Computations on Quasi-Uniform Grids. Fizmatlit, Moscow (2005). (in Russian)
9. Rosenbrock, H.H.: Some general implicit processes for the numerical solution of differential equations. Comput. J. **5**(4), 329–330 (1963)
10. Hairer, E., Wanner, G.: Solving of Ordinary Differential Equations. Stiff and Differential-Algebraic Problems. Springer, Heidelberg (2002)
11. Lukyanenko, D.V., Volkov, V.T., Nefedov, N.N., Recke, L., Schneider, K.: Analytic-numerical approach to solving singularly perturbed parabolic equations with the use of dynamic adapted meshes. Model. Anal. of Inf. Syst. **23**(3), 334–341 (2016)

American Options in an Illiquid Market: Nonlinear Complementary Method

Walter Mudzimbabwe[✉] and Lubin Vulkov

Ruse University, Ruse, Bulgaria
{wmudzimbabwe,lvalkov}@uni-ruse.bg

Abstract. In this paper, we consider the nonlinear complementary problem (NCP) arising from the pricing American options in a liquidity switching market. The NCP arises from discretising a coupled system of Hamilton-Bellman-Jacobi (HJB) equations whose solutions are the American option buyer indifference prices. In order to price American options, we derive a complementary problem. Due to the form of liquidity assumptions, the system of (HJB) equations are nonlinear which when discretised give rise to a NCP. We apply two Newton-like methods and perform various numerical experiments to illustrate the method.

Keywords: Nonlinear complementary problem · Nonsmooth equations · American option · Liquidity · Regime switching · Finite differences · *theta*-method · B-derivative · Semismooth Newton method

1 Introduction

On most financial markets, American options are the most traded compared with European options partly because of the early exercise feature which enable the holder of this option to exerice the right to buy or sell the underlying stock for a prescribed price at any time between entering the contract and preset maturity, whereas European options can only exercise the option at maturity.

This feature however complicates the pring of the option becuase the holder must decide at each time whether or not to exercise. The price of the underlying for which it's favourable to exerice the right to buy of sell is called the optimal exerice price which is not known apriori. Mathematically, the problem becomes a free boundary problem. It is possible however to formulate the problem as a obstacle one and treat the free boundary implicitly by posing the problem by a complementary approach.

Recently Ludkovski and Shen have proposed a nonlinear parabolic-ordinary differential system to price European options in a market that switches liquidity. It is possible using techniches in [6] to deduce the following nonlinear complementary (NLCP) problem

© Springer International Publishing AG 2017
I. Dimov et al. (Eds.): NAA 2016, LNCS 10187, pp. 500–507, 2017.
DOI: 10.1007/978-3-319-57099-0_56

$$\begin{cases} p_\tau - \frac{1}{2}\sigma^2 S^2 p_{SS} - \frac{\nu_{01}}{\gamma}\frac{F_1}{F_0}(1 - e^{-\gamma(q-p)}) \geq 0, \\ q_\tau - \frac{\nu_{10}}{\gamma}\frac{F_0}{F_1}(1 - e^{-\gamma(p-q)}) \geq 0, \\ \left(p_\tau - \frac{1}{2}\sigma^2 S^2 p_{SS} - \frac{\nu_{01}}{\gamma}\frac{F_1}{F_0}(1 - e^{-\gamma(q-p)})\right)(p - h) = 0 \\ \left(q_\tau - \frac{\nu_{10}}{\gamma}\frac{F_0}{F_1}(1 - e^{-\gamma(p-q)})\right)(q - h) = 0 \\ p \geq h, q \geq h \\ p(0, S) = q(0, S) = h(S). \end{cases} \quad (1)$$

2   Finite Difference Approximation

The NCP can be written more compactly as follows

$$\begin{cases} p_{i,\tau} - \frac{1}{2}\sigma^2 S^2 p_{i,SS} - \frac{\nu_{i1-i}}{\gamma}\frac{F_{1-i}}{F_i}(1 - e^{-\gamma(p_{1-i}-p_i)}) \geq 0, \\ p_i \geq h \\ \left(p_{i,\tau} - \frac{1}{2}\sigma^2 S^2 p_{i,SS} - \frac{\nu_{i1-i}}{\gamma}\frac{F_{1-i}}{F_i}(1 - e^{-\gamma(p_{1-i}-p_i)})\right)(p_i - h) = 0, \end{cases} \quad (2)$$

where $i \in \{0, 1\}, \sigma_0 = \sigma, \sigma_1 = 0, p_0 = p, p_1 = q$ and $p_i(T, S) = h(S)$.

Let $S_j = (j - 1)\Delta S, j = 1, \cdots, M, \tau_n = (n - 1)\Delta\tau, n = 1, \cdots, N$, where $\Delta S = S_{max}/(M - 1), \Delta\tau = T/(N - 1)$. We make the following approximation $P_i^{j,n} \approx p_i(n\Delta\tau, j\Delta S)$.

Let

$$\nu_{i1-i}^n = \Delta t\frac{\nu_{i1-i}}{\gamma}\frac{F_{1-i}^n}{F_i^n}, \quad g_i^j(P_i^{j,n}, P_{1-i}^{j,n}) = e^{-\gamma(P_{1-i}^{j,n}-P_i^{j,n})}$$

then using a simple *theta* method the scheme can be written as

$$\begin{aligned} f_i^j = &-\theta L_i^j P_i^{j-1,n+1} + (1 + \theta D_i^j)P_i^{j,n+1} - \theta L_i^j P_i^{j+1,n+1} \\ &- \theta\nu_{i1-i}^{n+1}(1 - g_i^j(P_i^{j,n+1}, P_{1-i}^{j,n+1})) - (1 - \theta)L_i^j P_i^{j-1,n} - (1 - (1 - \theta)D_i^j)P_i^{j,n} \\ &- (1 - \theta)L_i^j P_i^{j+1,n} - (1 - \theta)\nu_{i1-i}^n(1 - g_i^j(P_i^{j,n}, P_{1-i}^{j,n})) \end{aligned} \quad (3)$$

where

$$L_i^j = \alpha\sigma_i^2 S_j^2, \quad D_i^j = 2\alpha\sigma_i^2 S_j^2, \quad \alpha = \frac{1}{2}\frac{\Delta t}{(\Delta S)^2}.$$

We introduce the following vectors

$$\mathbf{x} = (P_0^{2,n+1}, \cdots, P_0^{M-1,n+1}, P_1^{2,n+1}, \cdots, P_1^{M-1,n+1})$$

$$\mathbf{f} = (f_0^2, \cdots, f_0^{M-1}, f_1^2, \cdots, f_1^{M-1}).$$

Note that function $\mathbf{f}^n$ can be alternatively written as

$$\begin{aligned} \mathbf{f}^n(\mathbf{x}) = &(\mathbf{I} + \theta\mathbf{A})\mathbf{x} - (\mathbf{I} - (1 - \theta)\mathbf{A})\mathbf{P}^n - \theta\nu^{n+1}(1 - \mathbf{g}(\mathbf{x})) \\ &- (1 - \theta)\nu^n(1 - \mathbf{g}(\mathbf{P}^n)) - \mathbf{BC}^{n+1}, \end{aligned} \quad (4)$$

where

$$\mathbf{g}(\mathbf{x}) = (g_0^2(\mathbf{x}), \cdots, g_0^{M-1}(\mathbf{x}), g_1^2(\mathbf{x}), \cdots, g_1^{M-1}(\mathbf{x}))^T$$
$$\mathbf{P}^n = (P_0^{2,n+1}, \cdots, P_0^{M-1,n}, P_1^{2,n}, \cdots, P_1^{M-1,n})^T$$
$$\boldsymbol{\nu}^n = (\overbrace{\nu_{01}^n, \cdots, \nu_{01}^n}^{M-2}, \overbrace{\nu_{10}^n, \cdots, \nu_{10}^n}^{M-2})^T$$

In (4) the matrix $\mathbf{A} \in \mathbf{R}^{2(M-2),2(M-2)}$ is given by

$$\mathbf{A} = \begin{pmatrix} \mathbf{A}_0 & \mathbf{0} \\ \mathbf{0} & \mathbf{A}_1 \end{pmatrix},$$

where the tridiagonal matrices $\mathbf{A}_i \in \mathbf{R}^{M-2,M-2}$, $i = 1, 2$ are given by

$$\mathbf{A}_i = \begin{bmatrix} D_i^2 & -L_i^2 & & \\ -L_i^3 & D_i^3 & -L_i^3 & \\ & & \ddots & \\ & & -L_i^{M-1} & D_i^{M-1} \end{bmatrix}$$

To summarise, the discrete NLCP takes the form (cf. [9])

$$\mathbf{x} - \mathbf{h} \geq \mathbf{0}, \quad \mathbf{f}(\mathbf{x}) \geq \mathbf{0} \text{ and } (\mathbf{x} - \mathbf{h})^T \mathbf{f}(\mathbf{x}) = \mathbf{0} \tag{5}$$

This formulation can be written as (cf. [9])

$$\min\{\mathbf{x} - \mathbf{h}, \mathbf{f}(\mathbf{x})\} = 0, \tag{6}$$

where the min is taken componentwise.

3   Existence of a Solution

In this section we discuss the existence of a solution to (5). The result mainly follows from diagonal dorminance of the associated matrix. Define

$$\mathbf{z} = \mathbf{x} - \mathbf{h}, \quad \mathbf{F}(\mathbf{z}) = \mathbf{q} + \mathbf{M}\mathbf{z} + \boldsymbol{\Psi}(\mathbf{z}) \tag{7}$$

where

$$\mathbf{q} = ((1 - \theta)\mathbf{A} - \mathbf{I})\mathbf{P}^n - \theta\boldsymbol{\nu}^{n+1} - (1 - \theta)\boldsymbol{\nu}^n(1 - \mathbf{g}(\mathbf{P}^n)) - \mathbf{B}\mathbf{C}^{n+1},$$
$$\mathbf{M} = (\mathbf{I} + \theta\mathbf{A}),$$
$$\boldsymbol{\Psi}(\mathbf{z}) = \theta\boldsymbol{\nu}^{n+1}\mathbf{g}(\mathbf{z}),$$

see [11]. Since the matrix $\mathbf{M}$ is strictly diagonally dorminant and the nonlinear function $\boldsymbol{\Psi}(\mathbf{z})$ is nonnegative and continuous, the following theorem follows from results obtained by Huang and Pang [11].

Theorem 1 (Existence of solution). *There exists a solution to the problem*

$$\mathbf{z} \geq \mathbf{0}, \quad \mathbf{F}(\mathbf{z}) \geq \mathbf{0} \quad and \quad \mathbf{z}^T \mathbf{F}(\mathbf{z}) = \mathbf{0} \tag{8}$$

4   Newton Methods

4.1   Semismooth Approach

The semismooth approach is based on the Fischer function [3,4] $\phi(a,b) = \sqrt{a^2 + b^2} - (a + b)$. The following property makes it usefull in complementarity problem: $\phi(a,b) = 0 \Leftrightarrow a \geq 0, b \geq 0$ and $a \cdot b = 0$. Other complementary functions can be used e.g.,

$$\phi_P(a,b) = \min\{a - h, b\}$$

[9,10] etc. It is important to note that ϕ is continuously differentiable everywhere but in the origin.

The idea to apply Newton to nonsmooth system $\Phi(x) = 0$ where

$$\Phi(x) = \begin{pmatrix} \phi(x_1 - h_1, F_1(x)) \\ \vdots \\ \phi(x_{2M-4} - h_{2M-4}, F_{2M-4}(x)) \end{pmatrix}.$$

In order to ensure that a Newton method can gives a global solution a merit function is introduced $\Psi(x) = \frac{1}{2}\Phi(x)^T\Phi(x)$. Since the system is nonsmooth a more general form of differentiability is used and a generalised Jacobian - Clarke derivative should be found. The following algorithm by De Luca et al. [3] is used to calculate the generalised Jacobian of the system.

Semismooth Newton.

Step 0: Let $\rho > 0, p > 2, \beta \in (0, 1/2)$ and choose x^0
Step 1: (Stopping criteria). If $\|\nabla\Psi(x^k)\| \leq \epsilon$ stop.
Step 2: (Direction calculation). Select $H^k \in \partial\Phi(x^k)$ and solve $H^k d^k = -\Phi(x^k)$. If H^k is singular or if d^k is not good enough set $d^k = -\nabla\Psi(x^k)$.
Step 3: (Linear search). Find the smallest $i^k = 0, 1, 2, \cdots$, such that

$$\Psi(x^k + 2^{-i^k}d^k) \leq \Psi(x^k) + \beta 2^{-i^k}\nabla\Psi(x^k)^T d^k.$$

Set $x^{k+1} = x^k + 2^{-i^k}d^k, k \leftarrow k + 1$. Go to Step 1.

In Luca et al. [3] they show how to calculate the generalised Jacobian $H(x) \in \partial\Phi(x)$.

4.2   B-Derivative Method

This approach was derived by Pang in [9] as an application Newton method for B-derivative systems. In Harker and Pang [5] the method is applied for linear complementary problems and several numerical results are given. The technique is based on the associated function $H : \mathbb{R}^{2M-4} \to \mathbb{R}^{2M-4}$

$$\mathbf{H(x)} = \min\{\mathbf{x} - \mathbf{h}, \mathbf{f(x)}\}. \tag{9}$$

The idea being to solve $\mathbf{H}(\mathbf{x}) = 0$. The function $\mathbf{H}$ is not F-differentiable and so the classical Newton method does not apply [9]. It is however Boulivard differentiable (B-differentiable) a much more general derivative whose properties are given in Pang [9]. Usually a merit function such as

$$\mathbf{g}(\mathbf{x}) = \frac{1}{2}\mathbf{H}(\mathbf{x})^T\mathbf{H}(\mathbf{x}), \tag{10}$$

is used to ensure that the Newton method converges globally.

The B-derivative of $\mathbf{H}$ is given by [5, 9, 10]:

Theorem 2. *Let* $\mathbf{f} : \mathbb{R}^{2M-4} \to \mathbb{R}^{2M-4}$ *be* $F-differentiable$. *Let* $\mathbf{H}$ *be defined in (9) and let* $\mathbf{g}(\mathbf{x})$ *be a merit function. Then* $\mathbf{H}$ *is everywhere B-differentiable with B-derivative given by*

$$\mathbf{BH}_i(\mathbf{x}) = \begin{cases} \nabla\mathbf{f}_i(\mathbf{x})^T, & if\ i \in \alpha(x), \\ \min\{\nabla\mathbf{f}_i(\mathbf{x})^T, \mathbf{e}_i^T\}, & if\ i \in \beta(x), where \\ \mathbf{e}_i^T, & if\ i \in \gamma(x), \end{cases} \tag{11}$$

$\alpha(\mathbf{x}) = \{i : \mathbf{f}_i(\mathbf{x}) < \mathbf{x}_i - \mathbf{h}_i\}$, $\beta(\mathbf{x}) = \{i : \mathbf{f}_i(\mathbf{x}) = \mathbf{x}_i - \mathbf{h}_i\}$, $\gamma(\mathbf{x}) = \{i : \mathbf{f}_i(\mathbf{x}) > \mathbf{x}_i - \mathbf{h}_i\}$

A similar derivative called *slant derivative* is given in Sun and Zheng [13]. Using the B-derivative $\mathbf{BH}$ one may apply a Newton-like method of the form

$$\mathbf{BH}(\mathbf{x}^k)\mathbf{d}^k = -\mathbf{H}(\mathbf{x}^k), \qquad \mathbf{x}^{k+1} = \mathbf{x}^k + \mathbf{d}^k \tag{12}$$

Together with a Amijo linear search to the merit function to adjust the calculated direction $\mathbf{d}^k$ so that it stays in the feasible region.

To summarise the Newton method with line search:

Newton for B-differentiable systems.

Step 0: Let $s > 0, \beta \in (0,1), \sigma \in (0,1/2)$ and choose x^0
Step 1: (Direction calculation). Solve $\mathbf{BH}(\mathbf{x}^k)\mathbf{d}^k = -\mathbf{H}(\mathbf{x}^k)$.
Step 2: (Linear search). Find the smallest $m_k = 0, 1, \cdots$ such that

$$\mathbf{g}(\mathbf{x}^k) - \mathbf{g}(\mathbf{x}^k + \beta^m s\mathbf{d}^k) \geq 2\sigma\beta^m s\mathbf{g}(\mathbf{x}^k).$$

Step 3: $\mathbf{x}^{k+1} = \mathbf{x}^k + \beta^{m_k}s\mathbf{d}^k$.
Step 4: (Stopping criteria). If $||\mathbf{x}^{k+1} - \mathbf{x}^k|| \leq \epsilon$ stop, else $\to$ Step 1.

5   Numerical Examples

In this section we present several numerical examples to illustrate the performance of the method for the pricing different vanilla. For the domain truncation we also take $S_\infty = 20$ and take $\mu_0 = 0.06, \sigma_0 = 0.3, \nu_{01} = 1, \nu_{10} = 12, K = 10, T = 1, \gamma = 1$ [7]. For the algorithm parameters we take the following values: for the Semismooth Newton $\beta = 0.4, p = 2.1, \rho = 0.5$ and for B-Newton $\beta = 0.5, s = 1.0, \sigma = 0.1$. We choose different initial gueses such as $\mathbf{x}_1^0 = \mathbf{P}^n$, $\mathbf{x}_2^0 = (0, \cdots, 0)^T$, $\mathbf{x}_3^0 = (-1, \cdots, -1)^T$, $\mathbf{x}_4^0 = (10, \cdots, 10)^T$ and $\mathbf{x}_5^0 = \mathbf{h}$.

5.1   Recovering Black-Scholes Prices

To compare Ludkovski and Shen model with the classical Black-Scholes model, we take $\nu_{01} = \nu_{10} = 0, q \equiv 0$ adjusting μ to be $\mu - \sigma^2/2$ in (1) and use $r = 0.0$ for classical Black-Scholes model. In Fig. 1, we compare the values from Semi-Newton and B-Newton algorithms and the Black-Scholes prices. The price of American vanilla options is a still an open question even in the classical Black-Scholes model. There are however several analytical approximations that have been published over the years. One example is the pricing using formulae developed by Bjerksund and Stensland [1]. We denote by BJS the Black-Scholes values when $t = 0$ using Bjerksund-Stensland model. Figure 1 shows that our Newton methods give very close prices to Bjerksund-Stensland prices.

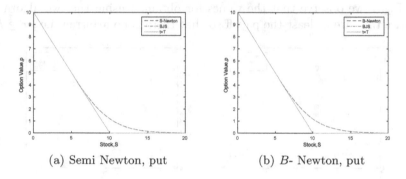

(a) Semi Newton, put   (b) B- Newton, put

Fig. 1. Comparison with Bjerksund-Stensland option pricing model

Figure 2 depicts the typical solution curves for vanilla calls and puts for both p and q.

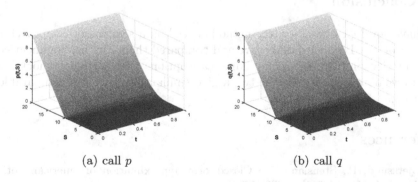

(a) call p (b) call q

Fig. 2. Solution curves

5.2   Comparing the Two Newton Schemes

In Table 1 we compare the the the at the money American call option prices using the two Newton approaches, Semismooth Newton (SN) and B-Newton (BN) for varying space sizes for constant time step $N = 2^7$.

Table 1. Starting point $\mathbf{x}^0 = \mathbf{P}^n$

M	Semismooth Newton (SN)	B-Newton (BN)	$|SN - BN|$
128	1.144057343741911	1.144057155169366	1.8857E-007
256	1.143725176628988	1.143722531817765	2.6448E-006
512	1.131656237869542	1.143631569988881	1.1975E-002

In Fig. 3 we observe that the values for obtained using the two algorithms give price that are at least the payoff so that there is no arbitrag, i.e., $x \geq h$.

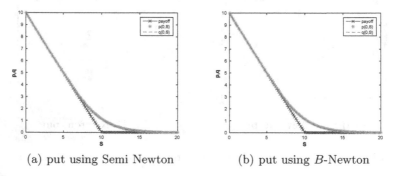

(a) put using Semi Newton   (b) put using B-Newton

Fig. 3. Comparing the two Newton methods

6   Conclusion

We have developed two Newton methods based on a semismooth approach and another on the Boulivard derivative and compared them for the Nonlinear complementarity problem of pricing American options with liquidity shocks. The results we obtain show that the two a give similar results and a comparable to other approaches.

References

1. Bjerksund, P., Stensland, G.: Closed-form approximation of american options. Scand. J. Manag. **9**, S88–S99 (1993)
2. Carmona, R.: Indifference Pricing: Theory and Applications. Princeton University Press, Princeton (2009)
3. De Luca, T., Facchinei, F., Kanzow, C.: A semismooth equation approach to the solution of nonlinear complementarity problems. Math. Program. **75**, 407–439 (1996)

4. De Luca, T., Facchinei, F., Kanzow, C.: A theoretic and numerical comparison of some semismooth algorithms for complementary problems. Comput. Optim. Appl. **16**, 173–205 (1996)
5. Harker, P.T., Pang, J.S.: A damped-Newton method for the linear complementarity problem. In: Lectures in Applied Mathematics, vol. 26, pp. 265–284. American Mathematical Society (1996)
6. Leung, T.S.T.: A Markov-modulated stochastic control problem with optimal multiple stopping with application to finance. In: IEEE Conference on Decision and Control (CDC), pp. 559–566 (2010)
7. Ludkovski, M., Shen, Q.: European option pricing with liquidity shocks. Int. J. Theor. Appl. Finan. **16**, 219–249 (2013)
8. Øksendal, B.: Stochastic Differential Equations. Springer, Heidelberg (2010)
9. Pang, J.S.: Newton's method for B-differential equations. Math. Oper. Res. **15**(2), 311–341 (1990)
10. Pang, J.S.: A B-differentiable equation-based, globally and locally quadratically convergent algorithm for nonlinear programs, complementarity and variational inequality problems. Math. Program. **51**, 101–131 (1991)
11. Pang, J.S., Huang, J.: Pricing American options with transaction costs by complementarity methods. In: Avellaneda, M. (ed.) Quantitative Analysis in Financial Markets, pp. 172–198 (2002)
12. Seydel, R.U.: Tools for Computational Finance. Springer, Heidelberg (2006)
13. Sun, Z., Zheng, J.: A monotone semismooth Newton type method for a class of complementarity problems. J. Comput. Appl. Math. **235**, 1261–1274 (2011)
14. Wilmott, P., Howison, S., Dewynne, J.: The Mathematics of Financial Derivatives: A Student Introduction. Cambridge University Press, Cambridge (1995)

Inverse Problem for Paleo-Temperature Reconstruction Based on the Tree-Ring Width and Glacier-Borehole Data

Oleg V. Nagornov$^{(\boxtimes)}$ and Sergey A. Tyuflin

National Research Nuclear University MEPhI,
Kashirskoe Shosse, 31, 115409 Moscow, Russia
nagornov@yandex.ru

Abstract. There is studied the inverse problem to determine solution $\{T(z,t), \mu(t)\}$ of the equation $\rho(z)C(z)T_t = (k(z)T_z)_z - \rho(z)C(z)w(z)T_z$, $(t,z) \in Q \equiv [0,t_f] \times [0,H]$, with initial condition $T(z,0) = U(z), z \in [0,H]$, and boundary conditions $T(0,t) = U_s + \mu(t), -k(H)T_z(H,t) = q, t \in [0,t_f]$, and redetermination condition $T(z,t_f) = \chi(z), z \in [0,H]$, where $w(z)$ is the advection rate of glacier layers, $\rho(z)$, $C(z)$, and $k(z)$ are the density, specific heat, and thermal conductivity of ice, respectively, q is the geothermal heat flux, U_s is the steady-state surface temperature in the past, t_f is the present time, H is the borehole depth, $\chi(z)$ is the measured temperature-depth profile. The solution of the inverse problem $\mu(t)$ is looking for in the finite Fourier series form where periods correspond to the climatic signals retrieved by the annual tree-ring width index. It is derived that the solution is unique and stable and can be found out by the Tikhonov's regularization method. This method is applied for the Kamchatka region.

Keywords: Boreholes · Heat and mass transfer · Climate reconstruction · Inverse problems

1 Introduction

Development of methods for the past surface temperature reconstruction is important problem on the global change. It is due to relative short period of the instrumental temperature measurements that started about two centuries ago. Information on the previous temperature changes can be provided by indirect sources of climate. These sources can be the Earth temperature distribution, the annual tree-ring widths, and the ice core data. The tree-ring widths correlate with air temperature changes. In glaciers the paleotemperatures are recorded as a disturbance of the steady-state temperature connected with penetration of the temperature changes from the surface to deep layers while the seasonal temperature variations are noticeable near 10–20 m at the surface. The most reliable data for the temperature reconstruction are in the temperature-depth profiles measured in the boreholes because this temperature distribution is response to the

© Springer International Publishing AG 2017
I. Dimov et al. (Eds.): NAA 2016, LNCS 10187, pp. 508–516, 2017.
DOI: 10.1007/978-3-319-57099-0_57

surface temperature history. Unfortunately, due to fast attenuation of the surface temperature high-frequency oscillations, the borehole information is degraded for large time scale. It can be improved by usage of additional data based on the dominant periods retrieved from the high resolution climate indicators.

2   Region Under Study

The Kamchatka peninsula (56° N 160° E) is situated in the sea mild climate, its length is about 1600 km and 500 km width. The Sredinniy and Easten crests support the continental climate. There are many glaciers and coniferous trees, spruces and lurches, on the peninsula. The most powerful glacier is the Ushkovsky glacier. The ice core (56°04′ N 160°28′ E) was extracted from this glacier and analyzed [1,2]. The ice core length was 212 m, and the borehole was upper by 28 m than the bedrock. The accuracy of temperature measurements was 0.01 °C. The measured data are shown in Fig. 1. The annual temperature of the surface was −15.8 °C [1].

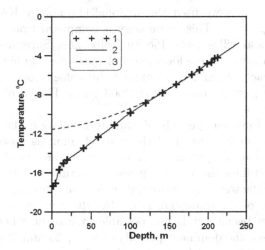

Fig. 1. Borehole temperature profiles (1 - asured temperature, 2 - proximated temperature, 3 - steady state temperature).

The isotope measurements of the ice core derived the ice core dating of the first hundred meters. The advection rate of the annual layers was calculated by this dating. These data were extrapolated up to the bedrock base. The approximation of the advection rate of the annual layers was done by the quadratic function for the first 40 m and by the linear function for the other part (Fig. 2).

The density of glacier was approximated by function $\rho = \rho_{ice}(1 - c_0 \exp(-\gamma z))$ where $c_0 = 0.54$, $\gamma \approx 0.034$ m^{-1} that corresponds to the density at the surface $\rho \approx 423$ kg/m^3. This approximation is shown in Fig. 3.

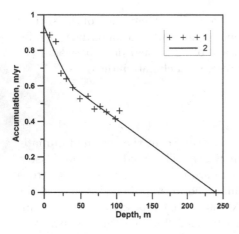

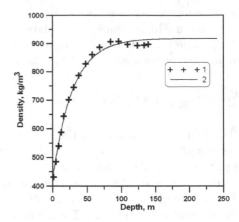

Fig. 2. The advection rate (1 - values based on the dating, 2 - approximated curve).

Fig. 3. The density of glacier (1 - measured density profile, 2 - approximated density profile).

Except ice core data we used the regional chronology KAML for lurches of this region (Fig. 4a) [3]. This chronology is composed from local ones based on lurches from Esso village, the Ploskii Tolbachek, Shiveluch volcanos, and the Kronockiy peninsula. These local chronologies were combined in one due to good correlation each other. The summary chronology consists from 144 tree cores and contains periods from 1622 to 2003 years. The correlation coefficient between series is 0.65.

To derive the dominant periods of climate oscillations based on the high resolution lurch chronology we used the wavelet analysis based on the Morle wavelet that reflects most appropriate frequency parameters of signals [4].

Figure 4(b,c) exhibits the wavelet power spectrum and the global wavelet power spectrum of the wavelet transformation for the summary regional chronology of lurches. Out of the triangle of reliability the wavelet coefficients are calculated with errors because it is not possible to use the whole length of the wavelet. One can see the dominant periods of 8, 16, 25 and 53 years. Probably there is period of 200 years and more but this information is out of the triangle of reliability.

Also we carried out the wavelet analysis of the Pacific Ocean decade oscillations that occur influence on the climate change in the studied region. These results are shown in Fig. 5. The most sharp period is 25 years. Also there is period inside 50–60 years. These data are in a good agreement with data noted by [5]. Our data contain periods of 6 and 9 years. Thus, the periods for the summary regional chronology are in agreement with data on the Pacific Ocean decade oscillations.

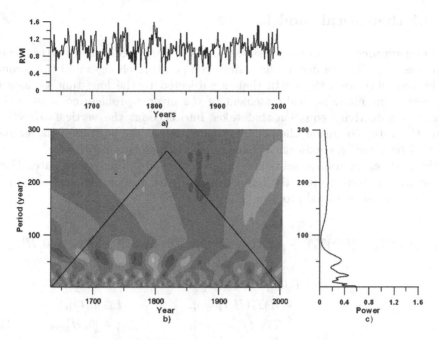

Fig. 4. Tree-ring chronology (a); wavelet power spectrum (b) and global wavelet power spectrum (c).

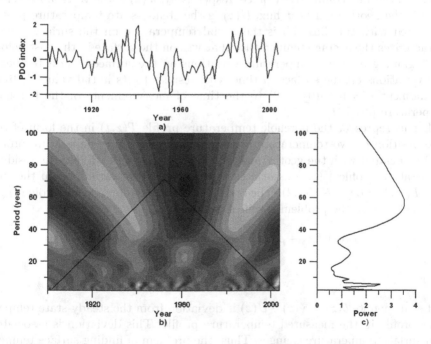

Fig. 5. PDO index (a); wavelet power spectrum (b) and global wavelet power spectrum (c).

3   Mathematical Model

The underground temperature distribution is mainly determined by two types
of processes [6,7]. The first is the surface temperature changes and the second
is the heat flux from the Earth that is subjected to the long-time geological
processes. The mathematical statement of the inverse problem consists of the
thermal conductivity equation that takes into account the vertical advection
term, the initial condition, the boundary condition at the bottom of glacier and
the re-determination condition. The measured temperature-depth profile is used
as the re-determination condition, $\chi(z)$, where z is vertical coordinate. Then
the inverse problem to find the temperature in the past is the solution of the
following one-dimensional problem [9]:

$$
\begin{aligned}
\rho(z)C(z)T_t = (k(z)T_z)_z - \rho(z)C(z)w(z)T_z, \quad & (t,z) \in Q \equiv [0,t_f] \times [0,H], \\
T(z,0) = U(z), \quad & z \in [0,H], \\
T(0,t) = U_s + \mu(t), \quad & t \in [0,t_f], \\
-k(H)T_z(H,t) = q, \quad & t \in [0,t_f], \\
T(z,t_f) = \chi(z), \quad & z \in [0,H], \quad (1)
\end{aligned}
$$

where H is the ice sheet thickness, $\rho(z)$, $C(z)$, and $k(z)$ are the density, specific
heat, and thermal conductivity of ice, respectively, $w(z)$ is the vertical ice velocity, q is the geothermal heat flux, $U(z)$ is the steady-state temperature profile
associated with this flux. U_s is the initial temperature on the surface, which
characterizes the average temperature that was on the surface in the past before
the beginning of sharp temperature variations on the surface, $\mu(t)$ is temperature variations on the surface in time with respect to its initial value U_s from
the moment $t = 0$ ($\mu(0) = 0$) to the time of measurements of the borehole
temperature profile t_f.

Let us represent the borehole temperature profile $T(z,t)$ in the form of the
superposition of two temperature profiles: the steady-state temperature profile
$U(z)$ associated with the geothermal heat flow from the Earth and the residual
temperature profile $V(z,t)$ associated with temperature variations on the surface: $T(z,t) = U(z) + V(z,t)$. Then, the steady-state temperature profile $U(z)$
is the solution of the problem specified as

$$
\begin{aligned}
(k(z)U_z)_z - \rho(z)C(z)w(z)u_z = 0, \quad & z \in [0,H], \\
U(0) = U_s, & \\
-k(H)U_z(H,t) = q, & \quad (2)
\end{aligned}
$$

Let us denote $\theta(z) = \chi(z) - U(z)$ is deviation from the steady-state temperature profile in the measured temperature profile. This deviation is associated
with surface temperature changes. Thus, the problem of finding surface temperature history is reduced to the solution of the problem

$$\rho(z)C(z)V_t = (k(z)V_z)_z - \rho(z)C(z)w(z)V_z, \quad (t,z) \in Q \equiv [0,t_f] \times [0,H],$$
$$V(z,0) = 0, \qquad\qquad z \in [0,H],$$
$$V(0,t) = \mu(t), \qquad\qquad t \in [0,t_f],$$
$$V_z(H,t) = 0, \qquad\qquad t \in [0,t_f],$$
$$V(z,t_f) = \theta(z), \qquad\qquad z \in [0,H]. \qquad (3)$$

4   Method of Past Temperature Reconstruction

The Tikhonov method is applied to determine the past surface temperatures [8]. The Tikhonov regularization method is the determination of the boundary temperature $\mu(t)$ minimizing a smoothing functional consisting of the difference and stabilizer:

$$\Psi = \frac{1}{2}\int_0^H [R\{\mu(t)\} - \theta(z)]^2 dz + \alpha\Omega\{\mu(t)\} \qquad (4)$$

where $R\{\mu(t)\}$ is the solution of the direct problem (3) represented in the form of the operator relation, α is the regularization parameter matched with the accuracy of the input data. The functional $\Omega\{\mu(t)\}$ is called the stabilizing functional or stabilizer

$$\Omega\{\mu(t)\} = \int_0^{t_f} \sum_{j=0}^r q_j \left(\frac{d^j\mu(t)}{dt^j}\right)^2 dt \qquad (5)$$

where r is the stabilizer order, $q_j \geq 0$, and $q_r > 0$. The procedure of the minimization of the smoothing functional Ψ can be performed by means of the gradient method and is an iteration procedure. The iteration procedure is carried out until the functional Ψ reaches the minimum with a given accuracy, which corresponds to the optimal solution of the inverse problem.

Let us write the surface temperature in the form of finite set of the Fourier series:

$$\mu(t) = \frac{a_0}{2}\sum_{m=1}^M a_m cos(\frac{2\pi T}{T_m}) + b_m sin(\frac{2\pi T}{T_m}) \qquad (6)$$

The initial Fourier coefficients are given in the first iteration step while the next n-th iterations are determinated by the following equations:

$$a_0^{n+1} = a_0^n - \gamma^n \frac{\partial \Psi^n}{\partial a_0^n},$$
$$a_m^{n+1} = a_m^n - \gamma^n \frac{\partial \Psi^n}{\partial a_m^n},$$
$$b_m^{n+1} = b_m^n - \gamma^n \frac{\partial \Psi^n}{\partial b_m^n}, \quad m = 1, 2, \ldots M, \qquad (7)$$

where $\gamma^n > 0$ is the gradient step. The derivatives of the functional in (7) with respect to the corresponding Fourier coefficients are given by the expressions:

$$\frac{\partial \Psi^n}{\partial a_0^n} = \int_0^H W_{a_0}(z)[R\{\mu^n(t)\} - \theta(z)]dz + \alpha \frac{\partial \Omega^n}{\partial a_0^n},$$

$$\frac{\partial \Psi^n}{\partial a_m^n} = \int_0^H W_{a_m}(z)[R\{\mu^n(t)\} - \theta(z)]dz + \alpha \frac{\partial \Omega^n}{\partial a_m^n},$$

$$\frac{\partial \Psi^n}{\partial b_m^n} = \int_0^H W_{b_m}(z)[R\{\mu^n(t)\} - \theta(z)]dz + \alpha \frac{\partial \Omega^n}{\partial b_m^n}, \quad m = 1, 2, \ldots M, \qquad (8)$$

Here, the profiles $W_{a_0}(z)$, $W_{a_m}(z)$, and $W_{b_m}(z)$ are the solutions of the problem specified by (3) with the boundary conditions on the surface $\mu(t) = 1/2$, $\mu(t) = cos(2\pi t/T_m)$ and $\mu(t) = sin(2\pi t/T_m)$, respectively. It is easy to show that the term with the stabilizer in (4) when the boundary condition on the surface $\mu(t)$ has the form of the segment of the trigonometric Fourier series has the form:

$\alpha\Omega = \alpha_0 \frac{a_0^2}{2} + \sum\limits_{m=1}^{M}(a_m^2 + b_m^2)\xi_m$, where $\xi_m = \alpha_0 + \alpha_1(2\pi/T_m)^2 + \ldots + \alpha_r(2\pi/T_m)^2 r$, $\alpha_i = \alpha q_i t_f/2$, $i = 0, 1, \ldots r$. In this case

$$\alpha \frac{\partial \Omega^n}{\partial a_0^n} = \alpha_0 a_0^n,$$

$$\alpha \frac{\partial \Omega^n}{\partial a_m^n} = 2a_m^n \xi_m,$$

$$\alpha \frac{\partial \Omega^n}{\partial b_m^n} = 2b_m^n \xi_m, \quad m = 1, 2, \ldots M. \qquad (9)$$

Thus, to determine the Fourier coefficients of the boundary condition given by (6), the iteration procedure specified by (7) is performed with the use of (8) and (9).

5   Reconstruction of the Past Temperature

We reconstruct the surface temperature based on the Ushkovsky glacier data jointly with the summary regional lurch chronology. The steady-state temperature and the geothermal flow we determine by (2). The bottom of the glacier is in the steady-state regime. The geothermal flow is $Q = 0.11\,W/m^2$. The calculated steady-state temperature is shown in Fig. 1. In the range from 210 to 240 m the measured data are absent. That is why they were extrapolated.

We look for the surface temperature as the finite set of the Fourier series. The periods of the trigonometric series are taken from the wavelet analysis (KAML).

The temperature reconstruction is shown in Fig. 6. It correlates with annual temperature for the Arctic region [10]. The correlation coefficient for time from 1918 to 1990 years is 0.47 while it equals to 0.76 for 1950–1990.

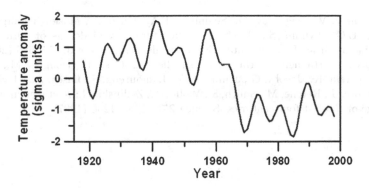

Fig. 6. Past surface temperature reconstruction.

6   Conclusions

We developed approach for the past surface temperature reconstruction based on the measured temperature profile in glacier borehole. The surface temperature is looking for the finite set of the Fourier series. The inverse problem is solved by usage of indirect data on the dominant climatic oscillations. The periods of this set are determined by the wavelet analysis. It is proved that the problem has the unique and stable solution. As example there was considered the temperature reconstruction for the Kamchatka region. It was derived the dominant periods of 8, 16, 25 and 53 years, and amplitudes of these oscillations.

References

1. Shiraiwa, T., Nishio, F., Kameda, T., Takahashi, A., Toyama, Y., Muraviev, Y., Ovsyannikov, A.: Ice core drilling at Ushkovsky ice cap, Kamchatka, Russia. Seppyo **61**, 25–40 (1999)
2. Shiraiwa, T., Muravyev, Y.D., Kameda, T., Nishio, F., Toyama, Y., Takahashi, A., Ovsyannikov, A.A., Salamatin, A.N., Yamagata, K.: Characteristics of a crater glacier at Ushkovsky volcano as revealed by the physical properties of ice cores and borehole thermometry. J. Glaciol. **47**, 423–432 (2001)
3. Solomina, O.N., Wiles, G., Shiraiwa, T.: Multiproxy records of climate variability for Kamchatka for the past 400 years. Clim. Past **3**, 119–128 (2007)
4. Daubechies, I.: Ten Lectures on Wavelets. SIAM, Philadelphia (1992)
5. Mantua, N.J., Hare, S.R.: The pacific decadal oscillation. J. Oceanogr. **58**, 35–44 (2002)
6. Paterson, W.S.B.: The Physics of Glaciers, 3rd edn. Butterworth-Heinemann, Birlington (1994)
7. Kotlyakov, V.M., Arkhipov, S.M., Henderson, K.A., Nagornov, O.V.: Deep drilling of glaciers in Eurasian Arctic as a source of paleoclimatic records. Quat. Sci. Rev. **23**, 1371–1390 (2004)
8. Nagornov, O.V., Konovalov, Y.V., Zagorodnov, V.S., Thompson, L.G.: Reconstruction of the surface temperature of Arctic glaciers from the data of temperature measurements in wells. J. Eng. Phys. Thermophys. **74**, 253–265 (2001)

9. Zagorodnov, V., Nagornov, O., Scambos, T.A., Muto, A., Mosley-Thompson, E., Pettit, E.C., Tyuflin, S.: Borehole temperatures reveal details of 20th century warming at Bruce Plateau Antarctic Peninsula. Cryosphere **6**, 675–686 (2012)
10. Overpeck, J., Hughen, K., Hardy, D., Bradley, R., Case, R., Douglas, M., Finney, B., Gajewski, K., Jacoby, G., Jennings, A., Lamoureux, S., Lasca, A., MacDonald, G., Moore, J., Retelle, M., Smith, S., Wolfe, A., Zielinski, G.: Arctic environmental change of the last four centuries. Science **278**, 1251–1256 (1997)

Collision of Solitons for a Non-homogenous Version of the KdV Equation: Asymptotics and Numerical Simulation

G. Omel'yanov[(✉)]

Department of Mathematics, University of Sonora,
Rosales y Blvd. Encinas s/n, 83000 Hermosillo, Mexico
omel@mat.uson.mx

Abstract. We consider a generalized KdV equation with a small dispersion and C^1-nonlinearity $g'(u)$. We present sufficient conditions for $g'(u)$ under which a soliton type solution exists and, moreover, pairs of solitary waves collide preserving in an asymptotic sense the KdV type scenario of interaction. Furthermore, we create a finite difference scheme to simulate the solution of the Cauchy problem and present some numerical results for the interaction problem.

Keywords: Generalized Korteweg-de Vries equation · Soliton · Interaction · Weak asymptotics method · Finite difference scheme

1 Introduction

We consider a generalization of the KdV equation of the form:

$$\frac{\partial u}{\partial t} + \frac{\partial g'(u)}{\partial x} + \varepsilon^2 \frac{\partial^3 u}{\partial x^3} = 0, \; x \in \mathbb{R}^1, \; t > 0, \tag{1}$$

where $g'(u) \overset{\text{def}}{=} \partial g / \partial u \in C^1$ is a real function (for more detail see below) and $\varepsilon \ll 1$ is a small parameter. Such equations describe nonlinear wave phenomena in plasma physics. In particular, for some specific plasma states, the ion-acoustic or dust-acoustic phenomena can be described by the KdV-type equation (1) with non-linearities $g'(u) = \alpha u^{3/2} + \beta u^2$ or $g'(u) = \alpha u^2 + \beta u^3, \alpha, \beta = \text{const}$ (see e.g. [1]). To simplify the situation we restrict ourselves by non-negative u. Moreover, we assume that uniformly in $u \geq 0$

$$c_1 u^{1+\delta_1} \leq g'(u) \leq c_2 u^{5-\delta_2}, \tag{2}$$

where c_i, δ_i are positive constants. These restrictions imply for $\varepsilon = \text{const}$ both the solvability of the Cauchy problem for (1) and the solution stability with respect to initial data (see [2,3]). For homogenous case $g'(u) = u^\kappa$, $\kappa > 1$, it is easy to find explicit solitary wave solutions. Moreover, as it is well known nowadays, the solitons interact elastically in the integrable case ($\kappa = 2$ and 3).

© Springer International Publishing AG 2017
I. Dimov et al. (Eds.): NAA 2016, LNCS 10187, pp. 517–524, 2017.
DOI: 10.1007/978-3-319-57099-0_58

More in detail, N solitons collide and form after that the sequence of N solitons again with the same amplitudes and velocities. Some shifts of trajectories appear as the unique result of the wave interaction.

Almost the same is true for non-integrable homogenous case: the solitary waves interact elastically in the principal term in an asymptotic sense, whereas the non-integrability implies the appearance of small radiation-type corrections [4–10]. At the same time, the character of the solitary wave collision remains unknown for arbitrary non-linearity. Our aim is to consider this open problem.

2   Solitary Wave Solution

First of all, we should determine what type of solitary waves will be under consideration.

Definition 1. *A function*

$$u = A\omega(\beta(x - Vt)/\varepsilon, A) \tag{3}$$

is called a soliton type solitary wave if $\beta = \beta(A)$, $V = V(A)$, *and* $\omega = \omega(\eta, A)$ *are smooth functions uniformly* $\eta \in \mathbb{R}^1$ *and* $A > 0$; *and* ω *is an even function,* $\omega(-\eta, A) = \omega(\eta, A)$, *such that* $\omega(0, \cdot) = 1$, $\omega'(0, \cdot) = 0$, *and* $\omega''(0, \cdot) < 0$, *where the prime denotes the derivative with respect to* η. *Moreover, we assume that*

$$\omega(0, \cdot) \to 0 \quad as \quad \eta \to \pm\infty \tag{4}$$

with an exponential rate and $\omega(\eta, \cdot) < 1$ *for* $\eta \neq 0$.

Theorem 1. *Let* $g(u) \in C^2(u \geq 0) \bigcap C^\infty(u > 0)$ *be such that*

$$g(u) = u^2 g_1(u), \tag{5}$$

where the Hölder continuous function g_1 *satisfies the conditions:*

$$g_1(0) = 0, \quad g_1(u) > 0 \quad and \quad g_1'(u) > 0 \quad for \quad u > 0. \tag{6}$$

Then the Eq. (1) has a soliton type solitary wave solution such that

$$V = 2g_1(A), \quad \beta^2 = V, \quad \partial\omega(\eta, A)/\partial A \to 0 \quad as \quad \eta \to 0 \text{ or } \eta \to \pm\infty. \tag{7}$$

3   Asymptotic Solution

For essentially nonintegrable problems of interaction it is impossible to construct either explicit solutions (classical or weak) or asymptotics in the classical sense. However, it is possible to construct an asymptotic solution in the weak sense (see e.g. [4–10] and references therein).

Definition 2. *A sequence $u(t, x, \varepsilon)$, belonging to $\mathcal{C}^\infty(0, T; \mathcal{C}^\infty(\mathbb{R}^1_x))$ for $\varepsilon = \text{const} > 0$ and belonging to $\mathcal{C}(0, T; \mathcal{D}'(\mathbb{R}^1_x))$ uniformly in ε, is called a weak asymptotic mod $O_{\mathcal{D}'}(\varepsilon^2)$ solution of (1) if the relations*

$$\frac{d}{dt} \int_{-\infty}^{\infty} u\psi\, dx - \int_{-\infty}^{\infty} g'(u)\frac{\partial \psi}{\partial x} dx = O(\varepsilon^2), \tag{8}$$

$$\frac{d}{dt} \int_{-\infty}^{\infty} u^2 \psi\, dx + \int_{-\infty}^{\infty} \left\{ 2(g(u) - ug'(u)) + 3\left(\varepsilon\frac{\partial u}{\partial x}\right)^2 \right\}\frac{\partial \psi}{\partial x} dx = O(\varepsilon^2) \tag{9}$$

hold uniformly in t for any test function $\psi = \psi(x) \in \mathcal{D}(\mathbb{R}^1)$.

Here the right-hand sides are $\mathcal{C}^\infty$-functions for $\varepsilon = \text{const} > 0$ and piecewise continuous functions uniformly in $\varepsilon \geq 0$. The estimates are understood in the $\mathcal{C}(0, T)$ sense:

$$z(t, \varepsilon) = O(\varepsilon^k) \leftrightarrow \max_{t \in [0,T]} |z(t, \varepsilon)| \leq c\varepsilon^k.$$

Definition 3. *A function $v(t, x, \varepsilon)$ is said to be of the value $O_{\mathcal{D}'}(\varepsilon^k)$ if the relation*

$$\int_{-\infty}^{\infty} v(t, x, \varepsilon)\psi(x)\, dx = O(\varepsilon^k)$$

holds uniformly in t for any test function $\psi \in \mathcal{D}(\mathbb{R}^1_x)$.

Let us consider the interaction of two solitary waves for the model (1) with the initial data

$$u|_{t=0} = \sum_{i=1}^{2} A_i \omega\left(\beta_i \frac{x - x_i^0}{\varepsilon}, A_i\right), \tag{10}$$

where $A_2 > A_1 > 0$, $x_1^0 - x_2^0 = \text{const} > 0$ and we assume the same relations between A_i, β_i and V_i as in (7). Obviously, the trajectories $x = V_i t + x_i^0$ have a joint point $x = x^*$ at a time instant $t = t^*$.

Let

$$A_i \gg 1, \quad i = 1, 2, \qquad \theta \overset{\text{def}}{=} \beta_1/\beta_2 \ll 1. \tag{11}$$

Moreover, let

$$g_1(z) = \sum_{k=1}^{n} c_k z^{q_k}, \quad \delta_1 \leq q_1 < q_2 < \cdots < q_n \leq 4 - \delta_2, \quad c_k = \text{const}_k. \tag{12}$$

The main result, which is known for the problem (1), (10), is the following:

Theorem 2. [10] *Let the assumptions (11), (12) be satisfied. Then the solitary wave collision in the problem (1), (10) preserves the elastic scenario with accuracy $O_{\mathcal{D}'}(\varepsilon^2)$ in the sense of Definition 3.*

4   Finite Differences Scheme

Obviously, it is impossible to create any finite difference scheme for the problem (1), (10) which remains stable uniformly in $\varepsilon \to 0$ and $t \in (0, T]$, $T = \text{const}$. So we will treat ε as a small but fixed constant.

Concerning the original KdV-type equation (1), we simulate firstly the Cauchy problem by a mixed problem over a $\Omega_T = \{x \in (0, L), t \in (0, T], L = \text{const}\}$ with zero boundary value. Next we define a mesh $\Omega_{T,\tau,h} = \{(x_i, t_j) \overset{\text{def}}{=} (ih, j\tau), i = 0, \ldots, N, j = 0, \ldots, J\}$. To simulate the interaction phenomena, we assume that L, T, and the initial front positions x_i^0 of solitons are such that the solitary wave trajectories have an intersection point which belongs to $(0, L) \times (0, T/2)$. Furthermore, we assume that, uniformly in $t \leq T$,

$$\left| u(x, t) \right|_{x \in [0, \delta]} \leq c\varepsilon^2, \quad \left| u(x, t) \right|_{x \in [L-\delta, L]} \leq c\varepsilon^2 \tag{13}$$

for some $c > 0$ and sufficiently small $\delta > 0$.

To create a finite difference scheme for the Eq. (1) we should choose appropriate approximations for the differential terms and for the nonlinear term. Let us do it separately. We write firstly a preliminary nonlinear "scheme" of the local accuracy $O(\tau + h^2)$:

$$y_{\bar{t}} + Q(y) + \varepsilon^2 \gamma y_{x\bar{x}\dot{x}} + \varepsilon^2 \nu h y_{x\bar{x}\bar{x}} = 0, \tag{14}$$

where $y \overset{\text{def}}{=} y_i^j \overset{\text{def}}{=} u(x_i, t_j)$, $\gamma = 1 - h\nu$, $\nu > 0$ is a constant,

$$y_{ix}^j \overset{\text{def}}{=} \frac{y_{i+1}^j - y_i^j}{h} \overset{\text{def}}{=} y_x, \quad y_{i\bar{x}}^j \overset{\text{def}}{=} \frac{y_i^j - y_{i-1}^j}{h} \overset{\text{def}}{=} y_{\bar{x}}, \quad y_{i\bar{t}}^j \overset{\text{def}}{=} \frac{y_i^j - y_i^{j-1}}{\tau} \overset{\text{def}}{=} y_{\bar{t}},$$

and $y_{\dot{x}} \overset{\text{def}}{=} (y_x + y_{\bar{x}})/2$.

For the nonlinearity of the form (5), (12) we define

$$Q(y) = \sum_{k=1}^{n} c_k (q_k + 1) Q_k(y), \quad Q_k(y) = \left\{ \left(y^{q_k+1} \right)_{\dot{x}} + y^{q_k} y_{\dot{x}} \right\}. \tag{15}$$

The last term in the left-hand side of (14) means the parabolic regularization of the value $O(\varepsilon^2 h^2)$. This and the choice of Q of the form (15) guarantees the equality

$$h \sum_{i=1}^{N-1} y_i^j Q(y_i^j) = 0$$

for all set functions y with zero boundary value. This implies the identity

$$\partial_{\bar{t}} \|y^j\|^2 + \tau \|y_{\bar{t}}^j\|^2 + \nu h^2 \|\varepsilon y_{x\bar{x}}^j\|^2 = 0, \tag{16}$$

where $\| \cdot \|$ is the discrete version of the $L^2(0, L)$ norm. Obviously, the equality (16) is the discrete version of the conservation law

$$\frac{d}{dt} \int_{-\infty}^{\infty} u^2 dx = 0,$$

which contains two regularized terms. On the other hand, (16) implies a weak convergence of y_i^j as $\tau, h \to 0$ to the solution of the corresponded mixed problem for (1).

Theorem 3. *Let the system (14), (15) supplemented by initial and boundary conditions have a solution $y = y_{\tau,h}$ and let $y_{\tau,h}(x,t)$ be its continuation over Ω_T. Then there exists a subsequence $y_{\bar{\tau},\bar{h}}(x,t)$ such that*

$$y_{\bar{\tau},\bar{h}} \to u \quad *\text{-weakly in} \quad L^\infty((0,T); L^2(0,L)) \cap L^2((0,T); H_0^1(0,L)) \quad (17)$$

as $\tau, h \to 0$.

Furthermore we note that for polynomial g', that is for integer q_k,

$$g'(u) = \sum_{k=1}^{3} c_k(k+2)u^{k+1}, \qquad (18)$$

it should be used a more effective formula [6]:

$$Q(y) = 2\sum_{k=1}^{3} c_k Q_k(y), \quad Q_k(y) = \sum_{l=1}^{k+1} y^{k+1-l}(y^l)_{\hat{x}}. \qquad (19)$$

Such approximation implies that the solution of (14) satisfies both (16) and the mean value conservation law:

$$\partial_{\bar{t}} \sum_{i=0}^{N} y_i^j = 0, \quad j = 1, 2, \dots. \qquad (20)$$

Moreover, for the KdV case, $g' = u^2$, and sufficiently smooth initial data it can be proved the following *a-priori* estimate:

$$\|(\varepsilon\partial_x)^k y^j\|^2 + \tau\||(\varepsilon\partial_x)^k y_{\bar{t}}\||^2(j) + h^2\||(\varepsilon\partial_x)^k \varepsilon y_{x\bar{x}}\||^2(j) \le C_k, \qquad (21)$$

where $k \ge 1$, $\||\cdot\||(j)$ is the discrete version of the $L^2(\Omega_{t_j})$ norm, and $C_k = C_k(\|y^0\|, \dots, \|(\varepsilon\partial_x)^k y^0\|, \varepsilon)$ does not depend on τ and h. Obviously, the estimate (21) implies the convergence of a subsequence of y.

Theorem 4. *Let the assumptions of Theorem 3 be satisfied and $g' = u^2$. Then there exists a subsequence $y_{\bar{\tau},\bar{h}}(x,t)$ such that*

$$y_{\bar{\tau},\bar{h}} \to u \quad *\text{-weakly in} \quad L^\infty((0,T); H_0^4(0,L)),$$

$$\frac{\partial y_{\bar{\tau},\bar{h}}}{\partial t} \to \frac{\partial u}{\partial t} \quad *\text{-weakly in} \quad L^\infty((0,T); H_0^1(0,L)) \qquad (22)$$

as $\tau, h \to 0$.

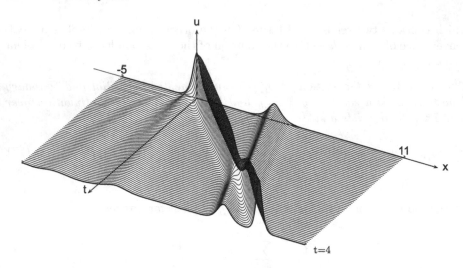

Fig. 1. Interaction of the solitary waves for $g' = u^{3/2} + u^2$ and $\varepsilon = 0.1$.

To solve the Eq. (14) for any fixed $j \geq 1$ we apply an iterative procedure. Namely, we construct a sequence of functions $\varphi(s) \overset{\text{def}}{=} \{\varphi_0(s), \ldots, \varphi_N(s)\}$, $s \geq 0$, where $\varphi(0) = \check{y} \overset{\text{def}}{=} y^{j-1}$ and the consequent terms $\varphi(s)$ we define depending on the nonlinearity. If $q_1 < 1$, then to find $\varphi(s)$ we solve the linear system:

$$\varphi + \tau R(\bar{\varphi}, \varphi) + \tau \varepsilon^2 \gamma \varphi_{x\bar{x}\dot{x}} + \tau \varepsilon^2 \nu h \varphi_{x\bar{x}x} = \check{y}, \tag{23}$$

where $\varphi = \varphi(s)$, $\bar{\varphi} = \varphi(s - 1)$,

$$R(\bar{\varphi}, \varphi) = \sum_{k=1}^{n} c_k (q_k + 1) R_k(\bar{\varphi}, \varphi), \quad R_k(\bar{\varphi}, \varphi) = (\bar{\varphi}^{q_k} \varphi)_{\dot{x}} + \bar{\varphi}^{q_k} \varphi_{\dot{x}}. \tag{24}$$

If $q_1 \geq 1$, then $\varphi(s)$ should satisfy the following system:

$$\varphi + \tau \sum_{k=1}^{n} c_k (q_k + 1) R_k(\bar{\varphi}, w) + \tau \varepsilon^2 \gamma \varphi_{x\bar{x}\dot{x}} + \tau \varepsilon^2 \nu h \varphi_{x\bar{x}x} = \check{y} - \tau Q(\bar{\varphi}), \tag{25}$$

where $w = \varphi - \bar{\varphi}$, $Q(y)$ has been defined in (15), and

$$R_k(\bar{\varphi}, w) = (q_k + 1)(\bar{\varphi}^{q_k} w)_{\dot{x}} + \bar{\varphi}^{q_k} w_{\dot{x}} + q_k \bar{\varphi}^{q_k - 1} \bar{\varphi}_{\dot{x}} w. \tag{26}$$

Finally, for polynomial nonlinearity of the form (18) we consider again the equations similar to (25) but with others Q and R_k:

$$\varphi + 2\tau \sum_{k=1}^{n} c_k R_k(\bar{\varphi}, w) + \tau \varepsilon^2 \gamma \varphi_{x\bar{x}\dot{x}} + \tau \varepsilon^2 \nu h \varphi_{x\bar{x}x} = \check{y} - \tau Q(\bar{\varphi}), \tag{27}$$

where $Q(y)$ has been defined in (19), and

$$R_k(\bar{\varphi}, w) = \sum_{l=1}^{k+1} l\bar{\varphi}^{k+1-l}\left(\bar{\varphi}^{l-1}w\right)_{\hat{x}} + \sum_{l=1}^{k}(k+1-l)\bar{\varphi}^{k-l}\left(\bar{\varphi}^l\right)_{\hat{x}}w. \qquad (28)$$

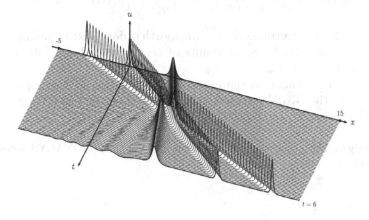

Fig. 2. Evolution of the soliton triplet for $g' = u^4$ and $\varepsilon = 0.1$

The solvability of the described above algebraic systems is obvious for sufficiently small τ/h^3. Next let us note that to estimate $\|\varphi\|$ we should have uniform in j estimates for $\|y^j_{xx}\|$. For this reason we consider the KdV case and use the estimate (21). A detailed analysis [6] concludes that the φ-sequence converges very rapidly,

$$\frac{c}{\varepsilon}\|\varphi(s+1) - \varphi(s)\|^2_{(2,\varepsilon)} \le \left(\sqrt{\frac{c}{\varepsilon}}\,\tau\right)^{2^s}, \quad s \ge 1, \qquad (29)$$

where $\|f\|^2_{(r,\varepsilon)} \stackrel{\text{def}}{=} \|f\|^2 + \|(\varepsilon\partial_x)^r f\|^2$. This implies the statement:

Theorem 5. *Let the assumption the Theorem 4 be satisfied. Then the sequence $\varphi(s)$ converges in the H^2_0 sense to the solution of the Eqs. (14), (19). Moreover,*

$$\|y - \varphi(2)\| \le c\tau^2/\sqrt{\varepsilon}, \qquad (30)$$

where $c > 0$ dos not depend on h, τ, and ε.

The convergence $\varphi(s) \to y^j$ as $s \to \infty$ for the scheme (23), (24) has been proved numerically. Moreover it turns out that $\varphi(2)$ approximates the solution y^j of (14) sufficiently well. This implies that we set $y^j = \varphi(2)$ in the general case also.

To solve the systems of linear equations of the form (23), (25), (27) we use the Gauss method adapted to systems with five non-zero diagonals. This implies the efficiency of the schemes in the sense that it executes $O(N)$ arithmetic operations to pass to the next time-level.

To define the function $\omega = \omega(\eta, A)$ and subsequently the initial data (10), we consider the problem

$$\frac{d\omega}{d\eta} = -\omega\sqrt{1 - g_1(A\omega)/g_1(A)}, \quad \eta > h,$$

$$\omega(h, A) = 1 - A\frac{h^2}{4}\frac{g_1'(A)}{g_1(A)} + A^2\frac{h^4}{4!}\frac{g_1'(A)}{g_1^2(A)}\left\{g_1'(A) + \frac{A}{4}g_1''(A)\right\},$$

and apply the Runge-Kutta method of the fourth order. Next we define $\omega(0, A) = 1$ and $\omega(-\eta, A) = \omega(\eta, A)$. Some results of the numerical simulations are presented in the Figs. 1 and 2.

This numerical scheme is the natural generalization of the algorithm suggested in [6] for the GKdV equation with homogeneous nonlinearity $g'(u) = u^m$. The reader can find there all detail of the scheme analysis.

Acknowledgments. The research was supported by SEP-CONACYT under grant 178690 (Mexico).

References

1. Schamel, H.: A modified Korteweg-de Vries equation for ion acoustic waves due to resonant electrons. J. Plasma Phys. **9**, 377–387 (1973)
2. Faminskii, A.V.: On an initial boundary-value problem in a bounded domain for the generalized Korteweg-de Vries equation. Funct. Differ. Equ. **8**(1–2), 183–194 (2001)
3. Bona, J.L., Souganidis, P.E., Strauss, W.: Stability and instability of solitary waves of Korteweg-de Vries type. Proc. R. Soc. London Ser. A **411**(1841), 395–412 (1987)
4. Danilov, V.G., Omel'yanov, G.A.: Weak asymptotics method and the interaction of infinitely narrow delta-solitons. Nonlinear Anal.: Theory Methods Appl. **54**, 773–799 (2003)
5. Danilov, V.G., Omel'yanov, G.A., Shelkovich, V.M.: Weak asymptotics method and interaction of nonlinear waves. In: Karasev, M.V. (ed.) Asymptotic Methods for Wave and Quantum Problems. AMS Translations: Series 2, vol. 208, pp. 33–164. AMS, Providence (2003)
6. Garcia, M., Omel'yanov, G.: Interaction of solitary waves for the generalized KdV equation. Commun. Nonlinear Sci. Numer. Simul. **17**(8), 3204–3218 (2012)
7. Garcia, M., Omel'yanov, G.: Interaction of solitons and the effect of radiation for the generalized KdV equation. Commun. Nonlinear Sci. Numer. Simul. **19**(8), 2724–2733 (2014)
8. Omel'yanov, G., Valdez-Grijalva, M.: Asymptotics for a C^1-version of the KdV equation. Nonlinear Phenom. Complex Syst. **17**(2), 106–115 (2014)
9. Omel'yanov, G.: Soliton-type asymptotics for non-integrable equations: a survey. Math. Methods Appl. Sci. **38**(10), 2062–2071 (2015)
10. Omel'yanov, G.: Collision of solitons for a non-homogenous version of the KdV equation, pp. 1–18 (2015). http://arxiv.org/abs/1511.09063

On Construction of Combined Shock-Capturing Finite-Difference Schemes of High Accuracy

Vladimir Ostapenko[1,2] and Olyana Kovyrkina[1(✉)]

[1] Lavrentyev Institute of Hydrodynamics SB RAS, Novosibirsk, Russia
{Ostapenko_VV,olyana}@ngs.ru
[2] Novosibirsk State University, Novosibirsk, Russia

Abstract. We show that compact scheme of the third order of weak approximation (unlike the TVD scheme) allows to obtain the second order of integral convergence in intervals crossing the front line of the shock wave and, as consequence, to conserve the high order of local convergence in the domain of shock influences. It allows to use the compact scheme as a basis scheme in construction of combined shock-capturing finite-difference schemes of high accuracy.

Keywords: Shock-capturing difference schemes · Monotonicity and high accuracy of finite-difference schemes · Discontinuous solutions · Integral order of convergence

1 Introduction

In the classic work [1], widely known due to discontinuity decay scheme, the concept of the monotonicity of finite-difference schemes was introduced and it was shown that there are no monotone high-order accurate schemes among the linear finite-difference schemes. Further development of shock-capturing finite-difference schemes for hyperbolic systems of conservation laws in the main was directed to overcoming this "Godunov's taboo". As a result, there were developed different classes of shock-capturing schemes, in which high order of approximation on smooth solutions and monotonicity (in the approximation of linear system and scalar conservation law) were achieved by nonlinear correction of fluxes, leading to non-linearity of these schemes even in the approximation of linear transport equation. Here are the main classes of such schemes, which we will shortly call NFC (Nonlinear Flux Correction) schemes: FCT [2], MUSCL [3], TVD [4], ENO [5], NED [6], WENO [7,8], KABARET [9–11] schemes. The main advantage of these schemes is that they localize shock waves with high accuracy without significant non-physical oscillations.

However, in most studies dealing with the construction of NFC schemes [2–11], the accuracy of a scheme is understood as the order of its Taylor expansion for smooth solutions, which does not guarantee a similar increase in the accuracy in the computation of discontinuous solutions. Nevertheless, it has been a longstanding misconception that these schemes preserve a high order of convergence

© Springer International Publishing AG 2017
I. Dimov et al. (Eds.): NAA 2016, LNCS 10187, pp. 525–532, 2017.
DOI: 10.1007/978-3-319-57099-0_59

in all smooth parts of weak solutions. This belief was supported by the fact that, in the overwhelming majority of works, the testing of difference schemes was based on various versions of the Riemann problem, whose exact solution is a set of simple waves (steady shocks and centered rarefaction waves) joined by constant flow regions. Such tests effectively evaluate the scheme resolvability of strong and weak discontinuities (the width of their smearing) and the presence or absence of shock front oscillations. However, they fail to provide information on the actual accuracy of the scheme in shock influence regions, since the exact solution behind the front is a constant. Moreover, this accuracy cannot be estimated from the computation of unsteady shock waves developing in the solution of a scalar conservation law, since in this case the influence region of a stable shock wave coincides with its front.

To determine the accuracy of schemes in shock influence regions, discontinuous solutions have to involve unsteady shock waves, i.e., variable velocity shock waves with a nonconstant solution forming behind their fronts. As a rule, such a solution of systems of conservation laws is not described by exact formulas and, to determine the order of convergence of a discrete solution, we need a series of (at least) three runs on a sequence of refining grids. In this case, we can use the Runge method for the approximate determination the order of convergence. Following this approach, it was shown in [12–14] that NFC schemes have at most the first order of local convergence in unsteady shock influence regions; i.e., in fact, they are not high order accurate schemes. One of the main reasons for this is that the mini-max correction of fluxes, which is typical for NFC-schemes, leads to reduce the smoothness of these difference fluxes, which in turn adduce to reduction of the approximation order of the Rankine-Hugoniot conditions in the shock wave fronts [15].

The accuracy of transmitting the Rankine-Hugoniot conditions by a scheme can be directly estimated by analyzing the convergence of integrals of the discrete solution over domains containing a shock front. Moreover, these integrals have to admit the potential possibility of obtaining high (at least the second) order of convergence for shock capturing schemes. As a result, this convergence cannot be strong, for example, in the L_1 or L_2 norm. This is associated with the fact that, in shock capturing schemes, the discrete solution does not locally converge to the exact one at several nodes near the shock front. As a result, the order of convergence of the discrete solution in a strong norm containing a discontinuity line cannot be higher than the first.

In the paper [16] it was proposed to estimate the scheme's accuracy of translation of the Rankine-Hugoniot conditions across the shock front by numerical calculation the order of the integral convergence of the finite-difference solution (rather than of its absolute value, as in the L_1 norm). The basic idea of this approach, belonging to Godunov and Ryaben'kii, is that the error arising ahead of a shock wave front because of its "smearing" can be compensated with a similar error of the opposite sign behind the shock wave front in case of convergence in such negative norm. Applying this approach to the second order TVD-schemes [4] and the fifth order WENO-schemes [8] showed that in

NFC-schemes the indicated order of integral convergence was reduced to the first in the areas containing non-stationary shock wave front [16,17].

At the same time explicit nonmonotonic high order schemes, such as the second order McCormack scheme [18] and the third order Rusanov scheme [19] with the analytical functions of numerical fluxes approximating the ε-Rankine-Hugoniot conditions with the high accuracy [15], allow to obtain a higher (the second) order of convergence in the negative norm for the integration in the regions, containing strong discontinuities [16,20]. As a result, these non-monotonic schemes, unlike NFC schemes, allow to obtain high order of convergence in the areas of non-stationary shock wave influence, in spite of significant scheme oscillations on their fronts.

Thus, at present time in the theory of shock-capturing finite-difference schemes there is the following alternative: it is impossible to localize strong discontinuities with high accuracy and at the same time keep high order of convergence in areas of their influence. Meanwhile, in practice NFC-schemes (and particularly WENO-schemes) are widely used in numerical modeling of complex gas-dynamics and hydrodynamics flows with a large number of shock waves with different amplitude, therefore all such calculations have only the first order of accuracy. So the main purpose of this paper is to formulate essentially new approach for the construction of shock-capturing finite-difference schemes, which (like NFC-schemes) will localize shock fronts with high accuracy and at the same time (like classic non-monotonic high-order schemes) will conserve high order of convergence in the all areas of smoothness of calculated discontinuous solutions.

In this paper we show that compact scheme [21] of the third order of weak approximation (unlike the TVD scheme [4]) allows to obtain the second order of integral convergence in intervals crossing the front line of the shock wave and, as consequence, to conserve the high order of local convergence in the domain of shock influences. It allows to use the compact scheme [21] as a basis scheme in construction of combined shock-capturing finite-difference schemes of high accuracy.

2   Cauchy Problem for the Shallow Water Equations

We consider a system of the shallow water equations [22] in the case of a rectangular horizontal bed less the bottom friction which in vector form is

$$\mathbf{v}_t + \mathbf{f}(\mathbf{v})_x = 0 \tag{1}$$

$$\mathbf{v} = \begin{pmatrix} h \\ u \end{pmatrix}, \quad \mathbf{f}(\mathbf{v}) = \begin{pmatrix} q \\ qu + gh^2/2 \end{pmatrix} \tag{2}$$

where h, q and $u = q/h$ are the depth, flux and horizontal velocity of the fluid, g is the acceleration of gravity. Let's set for system (1) and (2) the Cauchy problem with periodic initial data

$$u(x,0) = a \sin\left(\frac{2\pi x}{X} + \frac{\pi}{4}\right), \tag{3}$$

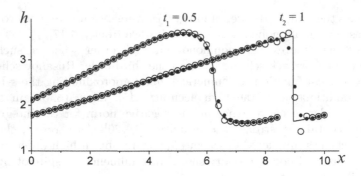

Fig. 1. Depth profiles obtained in the calculation by the Harten TVD scheme (dots) and by the compact scheme (circles) on a coarse grid with $\Delta = 0.2$ step compared to the solution obtained by the Harten TVD scheme on a fine grid with a step of $\Delta = 0.001$ (solid line)

$$h(x,0) = \frac{(u(x,0) + b)^2}{4g} = \frac{1}{4g}\left(a\sin\left(\frac{2\pi x}{X} + \frac{\pi}{4}\right) + b\right)^2 \qquad (4)$$

to which the following initial values of invariants correspond

$$w_1(x,0) = -b = \text{const}, \quad w_2(x,0) = 2u + b,$$

where $w_1 = u - 2c$, $w_2 = u + 2c$, $c = \sqrt{gh}$, $X = 10$, $a = 2$, $b = 10$.

In Fig. 1, at time moments $t_1 = 0.5$ and $t_2 = 1$ the profiles of depth h for problem (1)–(4) are given by solid lines. This solution is modeled by the numerical calculation according to the TVD-scheme on a fine grid with the space step $\Delta = 0.001$. From these calculations, it follows that at $t \approx 0.5$, as a result of a gradient catastrophe at point $x \approx 6$ of interval $[0, X]$, a shock wave is formed propagating in the positive direction of axis x. At $t_2 = 1$, this shock wave reaches the point $x \approx 9$ and behind its front a domain of influence (wave track) is formed lying inside interval $(4, 9)$. Similar shock waves are formed in all the intervals $[iX, (i+1)X]$, $i \in \mathbb{Z}$.

3   Results of Numerical Calculations

Figures 1–3 present the calculation results of problem (1)–(4) by two difference schemes: the Harten TVD-scheme [4] and the compact scheme [21]. The TVD-scheme is explicit, obtained by monotonization of the Lax-Wendroff scheme [23], has a TVD feature in the scalar case, is monotonic at approximation of linear systems, five-point in space, and has the second order of approximation on smooth solutions. The compact scheme is implicit, nonmonotonic, three-point in space and three layer in time, and has the third order of classical approximation on smooth solutions and the third order of weak approximation on discontinuous solutions. For both the schemes, the initial conditions are set by accurate approximation of the initial data (3) and (4). The difference solution on the first time

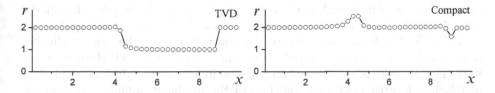

Fig. 2. Orders of integral convergence at time $t_2 = 1$

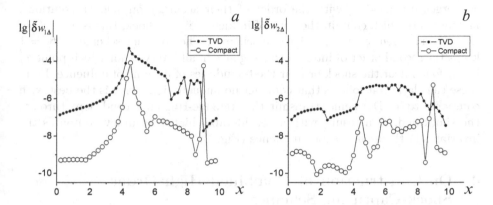

Fig. 3. Relative errors of the calculation of invariants $w_1 = u - 2c$ and $w_2 = u + 2c$ at time moment $t_2 = 1$

layer for the compact scheme is calculated by the second order Lax-Wendroff scheme [23].

The circles and dots in Fig. 1 show the results of the calculation on a coarse grid with the space step $\Delta = 0.2$. The circles correspond to calculations by compact scheme, and the dots correspond to calculations by TVD-scheme. Unlike the TVD-scheme, the nonmonotonic compact scheme has noticeable oscillations in the neighborhood of the shock front (Fig. 1, time moment $t_2 = 1$). Figure 2 gives at the time moment $t_2 = 1$ the orders of convergence $r = r(t_2, x_j, X)$ of the integral $\mathbf{V}_\Delta(t, x_j, X)$ of difference solution to the integral $\mathbf{V}(t, x_j, X)$ of exact solution, such as

$$\mathbf{V}(t, x_j, X) = \int_{x_j}^{X} \mathbf{v}(t, y) dy, \quad \mathbf{V}_\Delta(t, x_j, X) = \int_{x_j}^{X} \mathbf{v}_\Delta(t, y) dy$$

where continuous difference solution $\mathbf{v}_\Delta(t, x)$ is the result of spatial linear interpolation of the grid difference solution $\mathbf{v}_j^n = \mathbf{v}(t^n, x_j)$, $t^n = n\tau$, $x_j = j\Delta$, and $\tau = \tau(\Delta)$ is the time step which is chosen from the Courant stability condition. Figure 3 shows relative errors $\tilde{\delta}w_{k\Delta}$ in the computation of invariants w_k, $k = 1, 2$. The orders of integral convergence r and relative errors $\tilde{\delta}w_{k\Delta}$ are calculated by the Runge method which expounded in details in [16,20]. The results of the calculations given in Figs. 2 and 3 were carried out on a basis grid with spatial step $\Delta = 0.004$. They are shown for each 50th node $j = 50m$ of this grid.

From Fig. 2, it follows that at time moment $t_2 = 1$, when the zone of influence of the shock occupies only a part of the calculation area, the TVD Harten scheme have the first order of integral convergence on intervals the right-hand boundary of which is located before the shock, and the left-hand boundary behind the shock in the area of its influence. On intervals whose boundaries lie outside the zone of influence of the shock, this scheme keeps the second order of integral convergence coinciding with the order of their accuracy on smooth solutions. For intervals which contain the zone of influence of the shock, this is a result of the conservativeness of the difference schemes. The compact scheme at $t_2 = 1$ keeps the second order of integral convergence in all intervals for which point x_j is not found near the shock or near the boundaries of its zone of influence. From these calculations it follows that compact nonmonotonic scheme is the best with compared to TVD-scheme, transmit the Rankine-Hugoniot conditions through the shock and, as a result, with a considerably higher accuracy computes the invariants in the area of shock influence (Fig. 3).

4   On Construction of Combined High Order Shock-Capturing Schemes

The method of construction of combined high order shock-capturing schemes is due to that the compact scheme [21] has the third order of weak approximation on discontinuous solutions and, as a result, this scheme transmits the Rankine-Hugoniot conditions through the smeared shock with high accuracy. It allows to use the compact scheme as basic at construction of combined high order scheme.

Let's assume that difference solution $\mathbf{v}_j^n$, containing a unique smeared shock, is known on time layers $(n-1)$ and n. Let's consider that difference solution $\mathbf{v}_j^n$ on these time layers is monotonous (i.e. has no nonphysical oscillation on the smeared shock) and has the high accuracy in the shock influence domain. For definition of the difference solution on the $(n+1)$th time layer we at first use the basic compact scheme. The difference solution $\mathbf{u}_j^{(n+1)}$ obtained by this scheme, we correct in the mesh interval $[j_1, j_2]$ of the large gradients, containing the smeared shock. For this correction it is possible to use one of monotonous two layer in time NFC-schemes of high accuracy at smooth solutions. For this NFC-scheme we solve on the mesh interval $[j_1, j_2]$ and in one time step the following initial boundary value problem with next initial conditions

$$\mathbf{w}_j^n = \mathbf{u}_j^n, \quad j \in [j_1, j_2];$$

and boundary conditions

$$j_1 : \quad \varphi_k(\mathbf{w}_{j_1}^{n+1}) = \varphi_k(\mathbf{u}_{j_1}^{n+1}), \quad k = \overline{1, K},$$
$$j_2 : \quad \psi_l(\mathbf{w}_{j_2}^{n+1}) = \psi_l(\mathbf{u}_{j_2}^{n+1}), \quad l = \overline{1, L},$$

where $\mathbf{w}_j^{n+1}$ is the required difference solution of the NFC-scheme on $(n+1)$th time layer; φ, ψ is set functions to ensure the correctness of the given problem. In particular, this functions may be the values of the hyperbolic system invariants coming in the boundaries of the interval $[j_1, j_2]$ from outside of it. The final solution of the combined scheme on $(n+1)$th time layer is set by the formula

$$\mathbf{v}_j^{n+1} = \begin{cases} \mathbf{u}_j^{n+1}, & j \le j_1, \\ \mathbf{w}_j^{n+1}, & j_1 < j < j_2, \\ \mathbf{u}_j^{n+1}, & j \ge j_2. \end{cases}$$

As a result the combined scheme will conserve a high order of convergence in all smooth pats of the calculated exact solution, that the basic compact scheme provides, and also will ensure the high monotonous resolution of the shock fronts, that the chosen NFC-scheme provides.

5   Conclusion

In the present work the general methodological principles of the construction of combined high order shock-capturing schemes are stated. Further it is planned to develop technics of construction of such schemes in details and to apply these schemes to calculation of different weak solutions of various quasilinear hyperbolic systems of conservation laws, in particular conservation laws of hydraulics and gas dynamics, both in one-dimensional, and in multidimensional cases.

Acknowledgments. The work was supported by the Russian Science Foundation (grant No. 16-11-10033).

References

1. Godunov, S.K.: A difference method for numerical calculation of discontinuous solutions of the equations of hydrodynamics. Mat. Sb. **47**, 271–306 (1959)
2. Boris, J.P., Book, D.L., Hain, K.: Flux-corrected transport. II. Generalizations of the method. Comput. Math. Math. Phys. **18**, 248–283 (1975)
3. Van Leer, B.: Toward the ultimate conservative difference scheme. V. A second-order sequel to Godunov's method. J. Comput. Phys. **32**, 101–136 (1979)
4. Harten, A.: High resolution schemes for hyperbolic conservation laws. J. Comput. Phys. **49**, 357–393 (1983)
5. Harten, A., Osher, S.: Uniformly high-order accurate nonoscillatory schemes. SIAM J. Numer. Anal. **24**, 279–309 (1987)
6. Nessyahu, H., Tadmor, E.: Non-oscillatory central differencing for hyperbolic conservation laws. J. Comput. Phys. **87**, 408–463 (1990)
7. Liu, X.-D., Osher, S., Chan, T.: Weighted essentially non-oscillatory schemes. J. Comput. Phys. **115**, 200–212 (1994)
8. Jiang, G.-S., Shu, C.-W.: Efficient implementation of weighted ENO schemes. J. Comput. Phys. **126**, 202–228 (1996)
9. Iserles, A.: Generalized leapfrog methods. IMA J. Numer. Anal. **6**, 381–392 (1986)

10. Karabasov, S.A., Goloviznin, V.M.: New efficient high-resolution method for non-linear problems in aeroacoustics. AIAA J. **45**, 2861–2871 (2007)
11. Karabasov, S.A., Goloviznin, V.M.: Compact accurately boundary-adjusting high-resolution technique for fluid dynamics. J. Comput. Phys. **228**, 7426–7451 (2009)
12. Ostapenko, V.V.: The convergence of difference schemes for non-stationary shock wave front. Comput. Math. Math. Phys. **37**, 622–625 (1997)
13. Casper, J., Carpenter, M.H.: Computational consideration for the simulation of shock-induced sound. SIAM J. Sci. Comput. **19**, 813–828 (1998)
14. Engquist, B., Sjogreen, B.: The convergence rate of finite difference schemes in the presence of shocks. SIAM J. Numer. Anal. **35**, 2464–2485 (1998)
15. Ostapenko, V.V.: On the finite-difference approximation of the Hugoniot conditions on a shock wave front propagating with variable velocity. Comput. Math. Math. Phys. **38**, 1299–1311 (1998)
16. Kovyrkina, O.A., Ostapenko, V.V.: On the convergence of shock-capturing difference schemes. Dokl. Math. **82**, 599–603 (2010)
17. Mikhailov, N.A.: On convergence rate of WENO schemes behind a shock front. Matem. Mod. **27**, 129–138 (2015)
18. MacCormack, R.W.: The effect of viscosity in hypervelocity impact cratering. In: AIAA Paper 69-354 (1969)
19. Rusanov, V.V.: Difference schemes of third order accuracy for the through calculation of discontinuous solutions. Dokl. Akad. Nauk SSSR **180**, 1303–1305 (1968)
20. Kovyrkina, O.A., Ostapenko, V.V.: On the practical accuracy of shock-capturing schemes. Math. Model. Comput. Simul. **6**, 183–191 (2014)
21. Ostapenko, V.V.: Construction of high-order accurate shock-capturing finite-difference schemes for unsteady shock waves. Comput. Math. Math. Phys. **40**, 1784–1800 (2000)
22. Stoker, J.J.: Water Waves: The Mathematical Theory with Applications. Wiley, Hoboken (1957)
23. Lax, P., Wendroff, B.: Systems of conservation laws. Commun. Pure Appl. Math. **13**, 217–237 (1960)

Numerical Methods for a Class of Fractional Advection-Diffusion Models with Functional Delay

Vladimir Pimenov[1,2]($\boxtimes$) and Ahmed Hendy[1]

[1] Department of Computational Mathematics, Ural Federal University,
Ekaterinburg, Russia
v.g.pimenov@urfu.ru
[2] Institute of Mathematics and Mechanics, Ekaterinburg, Russia

Abstract. In this paper, we consider a technique of creation of difference schemes for time and space fractional partial differential equations with effect of delay on time. For two sided space fractional diffusion equation and fractional advection equations with time functional after-effect, an implicit numerical method is constructed. We use shifted Grunwald-Letnikov formulae to approximate space fractional derivatives and L1-algorithm to approximate time fractional derivatives. We also use piecewise constant interpolation and extrapolation by continuation for the prehistory of model with respect to time. The algorithm is a fractional analogue of the pure implicit numerical method in which the model is reduced on each time step to the solution of linear algebraic system. The order of convergence is obtained. Numerical experiments are carried out to support the obtained theoretical results.

Keywords: Fractional partial differential equation · Functional delay · Grunwald-Letnikov approximations · Grid schemes · Interpolation · Extrapolation · Convergence order

1 Introduction

Fractional differential equations [1–3] gained a great interest in the past decades due to their accuracy in modeling a lot of problems in many fields of science. The equations in partial derivatives of a fractional order were subdivided into two different classes: with a fractional derivative on space and with a fractional derivative on time. Now there are many papers in which numerical methods for such equations are discussed [4,5]. In this paper, we depend on results of work [6] in which implicit numerical methods for the solution of the advection-dispersion models with fractional derivative on time and two sided space fractional derivative were investigated.

Advection and diffusion equations with delay of general type, constant or variable, concentrated or distributed appeared in the simulation of dynamics processes. There are two effects often combined with these equations: distribution

© Springer International Publishing AG 2017
I. Dimov et al. (Eds.): NAA 2016, LNCS 10187, pp. 533–541, 2017.
DOI: 10.1007/978-3-319-57099-0_60

of parameters in space and heredity in time [7,8]. Numerical methods for solving such equations were considered in many papers, for example [9–12]. In paper [13], a technique of study on stability and convergence of numerical algorithms using the general theory of differential schemes [14] and using the theory of numerical methods of the solution of the functional and differential equations [15] were discussed for the heat conduction equation with delay. After that, this technique was applied to research of numerical methods of the solution of the equations of hyperbolic type with delay [16], various equations of parabolic type [17,18] and other types of the equations in partial derivatives with effect of heredity [19,20]. In this work, this technique is applied to the equations with partial derivatives of a fractional order effected with functional delay. It is noted that this technique was announced in [21] for the numerical method of the solution of the diffusion equation contains a left hand side fractional derivative on space and delay on time.

We consider the class of the fractional advection-diffusion equations [6] with functional delay [13]

$$\beta_1 \frac{\partial u}{\partial t} + \beta_2 \frac{\partial^\gamma u}{\partial t^\gamma} = -V \frac{\partial u}{\partial x} + D(\frac{1}{2} + \frac{q}{2}) \frac{\partial^\alpha u}{\partial x^\alpha} + D(\frac{1}{2} - \frac{q}{2}) \frac{\partial^\alpha u}{\partial(-x)^\alpha} + f(x, t, u(x, t), u_t(x, \cdot)), \tag{1}$$

where $x \in [a, b]$, $t \in [t_0, T]$, $u_t(x, \cdot) = \{u(x, t+s), -\tau \leq s < 0\}$ is the prehistory function, $\tau > 0$ is the value of delay, with initial conditions: $u(x, t) = \varphi(x\,t)$, $x \in [a, b]$, $t \in [t_0 - \tau, t_0]$, and boundary conditions: $u(a, t) = \varphi_1(t)$, $u(b, t) = \varphi_2(t)$, $t \in [t_0, T]$. We assume that $\beta_1 \geq 0$, $\beta_2 \geq 0$, $|\beta_1| + |\beta_2| \neq 0$, $-1 \leq q \leq 1$, $V > 0$, $D > 0$, $1 < \alpha \leq 2$, $0 < \gamma \leq 2$.

The time fractional derivative $\frac{\partial^\gamma u}{\partial t^\gamma}$ is defined in Caputo sense, the left hand side $\frac{\partial^\alpha u}{\partial x^\alpha}$ and the right hand side $\frac{\partial^\alpha u}{\partial(-x)^\alpha}$ space fractional derivatives are defined in Riemann-Liouville sense.

We denote by $Q = Q[-\tau, 0)$ the set of functions $u(s)$ that are piecewise continuous on $[-\tau, 0]$ with a finite number of points of discontinuity of first kind and right continuous at the points of discontinuity, $\|u(\cdot)\|_Q = \sup_{s \in [-\tau, 0)} |u(s)|$. We assume that the functional $f(x, t, u, v(\cdot))$ is given on $[a, b] \times [t_0, T] \times R \times Q$. We assume that the functional f, the functions $\varphi(x, t)$, $\varphi_1(t)$, $\varphi_2(t)$, and the coefficients β_1, β_2, q, V, D are such that this problem has a unique solution $u(x, t)$. We additionally assume that the functional $f(x, t, u, v(\cdot))$ is Lipschitz in the last two arguments, i.e., there exists a constants L and K such that, for all $x \in [a, b]$, $t \in [t_0, T]$, $u^1 \in R$, $u^2 \in R$, $v^1(\cdot) \in Q$ and $v^2(\cdot) \in Q$ the following inequality holds:

$$|f(x, t, u^1, v^1(\cdot)) - f(x, t, u^2, v^2(\cdot))| \leq L_f(|u^1 - u^2| + \| v^1(\cdot) - v^2(\cdot) \|_Q). \tag{2}$$

Further for simplicity, we consider the case when $\beta_1 = 1$, $\beta_2 = \beta$, $0 < \gamma \leq 1$. The cases $\beta_1 = 0$ and $1 < \gamma \leq 2$ can be considered similarly.

2   Derivation of the Difference Scheme

Let $h = (b - a)/N$, $\Delta = (T - t_0)/M$, where N, M are a positive integers, we assume that $\tau/\Delta = m$ is a positive integer, we introduce $x_i = a + ih$, $i = 0, \ldots, N$, and $t_j = t_0 + j\Delta$, $j = 0, \ldots, M$. Denote by u_j^i approximations of functions $u(x_i, t_j)$ at the nodes.

For every fixed $i = 0, \ldots, N$ let us introduce a discrete prehistory for the time points t_j, $j = 0, \ldots, M$: $\{u_k^i\}_j = \{u_k^i, j - m \le k \le j\}$. The mapping I : $\{u_k^i\}_j \to v^i(t)$, $t \in [t_j - \tau, t_j + \Delta]$ will be called the interpolation-extrapolation operator of discrete prehistory. As we will construct implicit method of the first order on time, we will use piecewise-constant interpolation with extrapolation by continuation

$$v^i(t) = \begin{cases} u_{l-1}^i, & t_{l-1} \le t < t_l, \ 1 \le l \le j, \\ u_{j-1}^i, & t_j \le t \le t_j + \Delta, \\ \varphi(x_i, t), & t_0 - \tau \le t \le t_0. \end{cases}$$

We discrete Riemann-Liouville space fractional derivatives by the shifted Grünwald-Letnikov formulae

$$\frac{\partial^\alpha}{\partial x^\alpha} u(x_i, t_{k+1}) = \frac{1}{h^\alpha} \sum_{j=0}^{i+1} \omega_j^\alpha u(x_{i+1-j}, t_{k+1}) + O(h^p),$$

$$\frac{\partial^\alpha}{\partial(-x)^\alpha} u(x_i, t_{k+1}) = \frac{1}{h^\alpha} \sum_{j=0}^{N-i+1} \omega_j^\alpha u(x_{i-1+j}, t_{k+1}) + O(h^p).$$

The coefficients ω_j^α are defined not unique, taking

$$\omega_j^\alpha = (-1)^j \frac{\alpha(\alpha - 1) \cdots (\alpha - j + 1)}{j!}, \ j = 0, 1, \ldots,$$

then the Grünwald-Letnikov approximations provide order $p = 1$.

We discrete the Caputo time fractional derivative using the $L1$-algorithm [6]

$$\frac{\partial^\gamma}{\partial t^\gamma} u(x, t_{k+1}) = \frac{1}{\Delta^\gamma \Gamma(2 - \gamma)} \sum_{j=0}^{k} b_j^\gamma [u(x, t_{k+1-j}) - u(x, t_{k-j})] + O(\Delta^{2-\gamma}),$$

where $b_j^\gamma = (j + 1)^{1-\gamma} - j^{1-\gamma}$, $j = 0, 1, \ldots, M$.

Also we will use formulas of backward difference scheme

$$\frac{\partial}{\partial x} u(x_i, t_{k+1}) = \frac{u(x_i, t_{k+1}) - u(x_{i-1}, t_{k+1})}{h} + O(h),$$

$$\frac{\partial}{\partial t} u(x_i, t_{k+1}) = \frac{u(x_i, t_{k+1}) - u(x_i, t_k)}{\Delta} + O(\Delta).$$

As result, we receive the following implicit difference scheme

$$\frac{u_{k+1}^i - u_k^i}{\Delta} + \frac{\beta}{\Delta^\gamma \Gamma(2-\gamma)} \sum_{j=0}^{k} b_j^\gamma [u_{k+1-j}^i - u_{k-j}^i] = -V \frac{u_{k+1}^i - u_{k+1}^{i-1}}{h} +$$

$$(\frac{1}{2} + \frac{q}{2}) \frac{D}{h^\alpha} \sum_{j=0}^{i+1} \omega_j^\alpha u_{k+1}^{i+1-j} + (\frac{1}{2} - \frac{q}{2}) \frac{D}{h^\alpha} \sum_{j=0}^{N-i+1} \omega_j^\alpha u_{k+1}^{i-1+j} + f_{j+1}^i, \qquad (3)$$

$f_{j+1}^i = f(x_i, t_{j+1}, v^i(t_j + \Delta), v_{t_j+\Delta}^i(\cdot))$.

The scheme is supplemented with initial conditions $u_0^i = \varphi(t_i)$, $i = 0, 1,$ $\dots, N$, and boundary conditions $u_k^0 = \varphi_1(t_k)$, $u_k^M = \varphi_2(t_k)$, $k = 1, 2, \dots, M$.

The scheme can be written as follows

$$\mu u_{k+1}^i - r_2 u_{k+1}^{i-1} - \frac{1+q}{2} r_3 \sum_{j=0, j\neq 1}^{i+1} \omega_j^\alpha u_{k+1}^{i+1-j} - \frac{1-q}{2} r_3 \sum_{j=0, j\neq 1}^{N-i+1} \omega_j^\alpha u_{k+1}^{i-1+j}$$

$$= u_{k+1}^i + \beta r_1 [b_k^\gamma u_0^i + \sum_{j=0}^{k-1} (b_j^\gamma - b_{j+1}^\gamma) u_{k-j}^i] + \Delta f_{j+1}^i. \qquad (4)$$

where $r_1 = \Delta^{1-\gamma}/\Gamma(2-\gamma)$, $r_2 = V\Delta/h$, $r_3 = D\Delta/h^\alpha$, $\mu = 1 + \beta r_1 + r_2 + \alpha r_3$.

Lemma 1 [6]. *The coefficients matrix A of the systems (4) are strictly diagonally dominant, hence the system is solvable and has a unique solution.*

3   Error of Approximation

Now, we investigate the error of approximation (residual) of method (3).

For every fixed $i = 0, \dots, N$ let's introduce the discrete prehistory of exact solution for the time points t_j, $j = 0, \dots, M$: $\{u(x_i, t_k)\}_j = \{u(x_i, t_k), j - m \le k \le j\}$. We will use piecewise-constant interpolation with extrapolation by continuation of exact solution

$$w^i(t) = \begin{cases} u(x_i, t_{l-1}), & t_{l-1} \le t \le t_l, \ 1 \le l \le j, \\ u(x_i, t_{j-1}), & t_j \le t \le t_j + \Delta, \\ \varphi(x_i, t), & t_0 - \tau \le t \le t_0. \end{cases}$$

The residual (with interpolation) of methods (3) can be defined using the following grid function

$$\nu_k^i = \frac{u(x_i, t_{k+1}) - u(x_i, t_k)}{\Delta} + \frac{\beta}{\Delta^\gamma \Gamma(2-\gamma)} \sum_{j=0}^{k} b_j^\gamma [u(x_i, t_{k+1-j}) - u(x_i, t_{k-j})]$$

$$+V \frac{u(x_i, t_{k+1}) - u(x_{i-1}, t_{k+1})}{h} - (\frac{1}{2} + \frac{q}{2}) \frac{D}{h^\alpha} \sum_{j=0}^{i+1} \omega_j^\alpha u(x_{i+1-j}, t_{k+1}) - \check{f}_{k+1}^i$$

$$-(\frac{1}{2} - \frac{q}{2}) \frac{D}{h^\alpha} \sum_{j=0}^{N-i+1} \omega_j^\alpha u(x_{i-1+j}, t_{k+1}), \quad \check{f}_{k+1}^i = f(x_i, t_{k+1}, w^i(t_k + \Delta), w_{t_k+\Delta}^i(\cdot)).$$

Lemma 2. *Suppose that the exact solution $u(x,t)$ is twice continuously differentiable in t and in x and also the fractional derivatives $\frac{\partial^\alpha u}{\partial_+ x^\alpha}$ and $\frac{\partial^\alpha u(x,t)}{\partial(-x)^\alpha}$ are continuously differentiable in t and in x. Then, the residual with piecewise-constant interpolation and extrapolation by continuation of the method has order $\Delta + h$, i.e. there is such constant C_2, that $|\nu_j^i| \leq C_2(h + \Delta)$ for all $i = 0, 1, \cdots, N$ and $j = 0, 1, \cdots, M$.*

Proof of lemma based on Taylor series expansions for the residual with the use of a condition (2) and properties of a piecewise-constant interpolation with extrapolation by continuation.

4   Convergence for the Difference Scheme

Denote by $\varepsilon_j^i = u(x_i, t_j) - u_j^i$ the error of the method at the nodes. We will say that the method converges with order $h^p + \Delta^q$ if there exists a constant C independent of h and Δ such that $|\varepsilon_j^i| \leq C(h^p + \Delta^q)$ for all $i = 0, 1, \cdots, N$ and $j = 0, 1, \cdots, M$.

We will define for any temporary layer of $j = 0, 1, \cdots, M$ a layer-by-layer error $\varepsilon_j = (\varepsilon_j^1, \varepsilon_j^2, \cdots, \varepsilon_j^{N-1})$ with norm $\|\varepsilon_j\| = \max_{1 \leq i \leq N-1} |\varepsilon_j^i|$.

Also, we will define collecting prehistory of a layer-by-layer error for the time t_j, $j = 0, 1, \cdots, M$:

$$\{\varepsilon_k\}_j = \{\varepsilon_k, 0 \leq k \leq j\} \text{ with norm } \|\{\varepsilon_k\}_j\| = \max_{0 \leq k \leq j} \|\varepsilon_k\|.$$

Lemma 3. *Let $|\varepsilon_{k+1}^{i_0}| = \max_{1 \leq i \leq N-1} |\varepsilon_{k+1}^i|$, than*

$$(1 + \beta r_1)\|\varepsilon_{k+1}\| \leq |\varepsilon_k^{i_0} + \beta r_1[b_k^\gamma \varepsilon_0^{i_0} + \sum_{j=0}^{k-i}(b_j^\gamma - b_{j+1}^\gamma)\varepsilon_{k-j}^{i_0}] + \Delta \check{f}_{k+1}^{i_0} - \Delta f_{k+1}^{i_0} + \Delta \nu_k^{i_0}|.$$

Proof. We will rewrite definition for the residual with interpolation in the form

$$\mu u(x_i, t_{k+1}) - r_2 u(x_{i-1}, t_{k+1}) - \frac{1+q}{2} r_3 \sum_{j=0, j\neq 1}^{i+1} \omega_j^\alpha u(x_{i+1-j}, t_{k+1})$$

$$-\frac{1-q}{2} r_3 \sum_{j=0, j\neq 1}^{N-i+1} \omega_j^\alpha u(x_{i-1+j}, t_{k+1})$$

$$= u(x_i, t_{k+1}) + \beta r_1[b_k^\gamma u(x_i, t_0) + \sum_{j=0}^{k-1}(b_j^\gamma - b_{j+1}^\gamma)u(x_i, t_{k-J})] + \Delta \check{f}_{j+1}^i + \Delta \nu_k^i.$$

From here and from definition of a method (4), we will derive the equation of error

$$\mu \varepsilon_{k+1}^i - r_2 \varepsilon_{k+1}^{i-1} - \frac{1+q}{2} r_3 \sum_{j=0, j\neq 1}^{i+1} \omega_j^\alpha \varepsilon_{k+1}^{i+1-j} - \frac{1-q}{2} r_3 \sum_{j=0, j\neq 1}^{N-i+1} \omega_j^\alpha \varepsilon_{k+1}^{i-1+j}$$

$$= \varepsilon_{k+1}^i + \beta r_1[b_k^\gamma \varepsilon_0^i + \sum_{j=0}^{k-1}(b_j^\gamma - b_{j+1}^\gamma)\varepsilon_{k-j}^i] + \Delta \check{f}_{j+1}^i + \Delta \nu_k^i - \Delta f_{j+1}^i.$$

As $1 + \beta r_1 = \mu - r_2 - \alpha r_3$, using properties of coefficients $\omega_1^\alpha = -\alpha$, $\omega_j^\alpha > 0$, $j = 2, 3, \cdots$, $\sum_{j=0}^{i_0+1} \omega_j^\alpha < 0$, $\sum_{j=0}^{N-i_0+1} \omega_j^\alpha < 0$ [6], we receive

$$(1 + \beta r_1)\|\varepsilon_{k+1}\| \le (\mu - r_2 - \frac{1+q}{2} r_3 \sum_{j=0, j\neq 1}^{i_0+1} \omega_j^\alpha - \frac{1-q}{2} r_3 \sum_{j=0, j\neq 1}^{N-i_0+1} \omega_j^\alpha)|\varepsilon_{k+1}^{i_0}|$$

$$\le \mu|\varepsilon_{k+1}^{i_0}| - r_2|\varepsilon_{k+1}^{i_0-1}| - \frac{1+q}{2} r_3 \sum_{j=0, j\neq 1}^{i_0+1} \omega_j^\alpha |\varepsilon_{k+1}^{i_0+1-j}| - \frac{1-q}{2} r_3 \sum_{j=0, j\neq 1}^{N-i_0+1} \omega_j^\alpha |\varepsilon_{k+1}^{i_0-1+j}|$$

$$\le |\mu\varepsilon_{k+1}^{i_0} - r_2\varepsilon_{k+1}^{i_0-1} - \frac{1+q}{2} r_3 \sum_{j=0, j\neq 1}^{i_0+1} \omega_j^\alpha \varepsilon_{k+1}^{i_0+1-j} - \frac{1-q}{2} r_3 \sum_{j=0, j\neq 1}^{N-i_0+1} \omega_j^\alpha \varepsilon_{k+1}^{i_0-1+j}|$$

$$= |\varepsilon_k^{i_0} + \beta r_1 [b_k^\gamma \varepsilon_0^{i_0} + \sum_{j=0}^{k-i} (b_j^\gamma - b_{j+1}^\gamma)\varepsilon_{k-j}^{i_0}] + \Delta \check{f}_{k+1}^{i_0} - \Delta f_{k+1}^{i_0} + \Delta \nu_k^{i_0}|.$$

Lemma 4. *Assume that the condition in Lemma 2 is satisfied, then*

$$\|\{\varepsilon_j\}_{k+1}\| \le (1 + (L + K)\Delta)\|\{\varepsilon_j\}_k\| + C_2\Delta(h + \Delta).$$

Proof. Let $|\varepsilon_{k+1}^{i_0}| = \max_{1 \le i \le N-1} |\varepsilon_{k+1}^i|$, then from Lemma 4, we receive

$$(1 + \beta r_1)\|\varepsilon_{k+1}\| \le |\varepsilon_k^{i_0} + \beta r_1 [b_k^\gamma \varepsilon_0^{i_0} + \sum_{j=0}^{k-i} (b_j^\gamma - b_{j+1}^\gamma)\varepsilon_{k-j}^{i_0}] + \Delta \check{f}_{k+1}^{i_0} - \Delta f_{k+1}^{i_0} + \Delta \nu_k^{i_0}|$$

$$\le [1 + \beta r_1(b_k^+ \sum_{j=0}^{k-i} (b_j^\gamma - b_{j+1}^\gamma))]\|\{\varepsilon_j\}_k\| + \Delta|\check{f}_{k+1}^{i_0} - f_{k+1}^{i_0}| + \Delta|\nu_k^{i_0}|$$

$$\le (1 + \beta r_1)\{\varepsilon_j\}_k\| + \Delta|\check{f}_{k+1}^{i_0} - f_{k+1}^{i_0}| + \Delta|\nu_k^{i_0}|.$$

Divide both sides of the previous inequality by $(1 + \beta r_1)$, we used that f is Lipschitz in the last two arguments, the definition of piecewise continuous interpolation with extrapolation by continuation, and also the statement of a Lemma 3, we receive the statement of a Lemma 4.

Theorem 1. *Assume that the condition in Lemma 2 is satisfied, then method (4) converges with order $h + \Delta$.*

The theorem is output from lemma 5 by a standard way [13,20].

5   Numerical Experiments

Let us consider the following equation with varying delay in the variable t

$$\frac{\partial u(x,t)}{\partial t} + \frac{\partial^\gamma u(x,t)}{\partial t^\gamma} = \frac{1}{2}\frac{\partial^\alpha u(x,t)}{\partial x_+^\alpha} + \frac{1}{2}\frac{\partial^\alpha u(x,t)}{\partial x_-^\alpha} + f, \tag{5}$$

$$f = \frac{1}{\ln((t^2/4)(x-1/2)^3(3/2-x)^3)}\left(2(t+\frac{t^{2-\gamma}}{\Gamma(3-\gamma)})(x-1/2)^3(3/2-x)^3 - \frac{1}{2}\frac{\Gamma(4)}{\Gamma(4-\alpha)}\right)$$

$$((x-1/2)^{3-\alpha} + (3/2-x)^{3-\alpha}) + \frac{1}{2}\frac{3\Gamma(5)}{\Gamma(5-\alpha)}((x-1/2)^{4-\alpha} + (3/2-x)^{4-\alpha}) - \frac{1}{2}\frac{3\Gamma(6)}{\Gamma(6-\alpha)}$$

$$((x-1/2)^{5-\alpha} + (3/2-x)^{5-\alpha}) + \frac{1}{2}\frac{\Gamma(7)}{\Gamma(7-\alpha)}((x-1/2)^{6-\alpha} + (3/2-x)^{6-\alpha})\Big)\ln(u(x,t-$$

$t/2))$, such that $1/2 \le x \le 3/2$, $1 \le t \le 5$,

with initial and boundary conditions on the form

$$u(x,r) = r^2(x-1/2)^3(3/2-x)^3, \quad 1/2 \le r \le 1, \quad 1/2 \le x \le 3/2,$$
$$u(1/2,t) = 0, \quad u(3/2,t) = 0, \quad 1 \le t \le 5.$$

The exact solution is $u(x,t) = t^2(x-1/2)^3(3/2-x)^3$.

Define the maximum error by $E(\Delta,h) = \max_{0 \le j \le M,\, 0 \le i \le N} | u(x_i,t_j) - u_j^i |$, such that $u(x_i,t_j)$ is the exact solution at the grid point (x_i,t_j). Let us test the spatial errors and their convergence orders by letting h varies from $\frac{1}{20}$ to $\frac{1}{320}$, fix the time step $\Delta = \frac{4}{1024}$. The convergence order with respect to the spatial step size is given by $order_s = \log_2\left(\frac{E(\Delta,2h)}{E(\Delta,h)}\right)$ which is clarified in Table 1. For the sake of temporal errors, let Δ varies from $\frac{1}{16}$ to $\frac{1}{256}$ and fix $h = \frac{1}{4000}$. The convergence order with respect to the time step size is given by $order_t = \log_2\left(\frac{E(2\Delta,h)}{E(\Delta,h)}\right)$ and this is illustrated in Table 2. From these tables, we can notice that the numerical results have a good agreement with the theoretical results.

Table 1. The spatial maximum norm errors and their convergence orders.

h	$\gamma = 0.15, \alpha = 1.1$		$\gamma = 0.85, \alpha = 1.9$	
	$E(\Delta,h)$	$order_s$	$E(\Delta,h)$	$order_s$
1/20	0.00568		0.00437	
1/40	0.00293	0.9567	0.00222	0.9765
1/80	0.00149	0.9687	0.00112	0.9884
1/160	0.00075	0.9854	0.00056	0.9996
1/320	0.00037	0.9975	0.00028	1.0004

Table 2. The temporal maximum norm errors and their convergence orders.

Δ	$\gamma = 0.15, \alpha = 1.1$		$\gamma = 0.85, \alpha = 1.9$	
	$E(\Delta,h)$	$order_t$	$E(\Delta,h)$	$order_t$
1/16	0.00074		0.000032	
1/32	0.00038	0.9432	1.625×10^{-5}	0.97743
1/64	0.00019	0.9632	8.19×10^{-6}	0.9886
1/128	0.00009	0.9778	4.098×10^{-6}	0.9988
1/256	0.00004	0.9965	2.05×10^{-6}	0.9999

Acknowledgements. This work was supported by Government of the Russian Federation program 02.A03.21.0006 on 27.08.2013 and by Russian Science Foundation 14-35-00005.

References

1. Oldham, S.G., Spanier, J.: The Fractional Calculus. Academic Press, New York (1974)
2. Samko, S.G., Kilbas, A.A., Marichev, O.I.: Fractional Integrals and Derivatives: Theory and Applications. CRC Press, Boca Raton (1993)
3. Podlubny, I.: Fractional Differential Equations. Academic Press, San Diego (1999)
4. Meerschaert, M.M., Tadjeran, C.: Finite difference approximations for fractional advection-dispersion flow equations. J. Comput. Appl. Math. **172**, 65–77 (2004)
5. Liu, F., Zhuang, P., Anh, V., Turner, I., Burrage, K.: Stability and convergence of the difference methods for the space-time fractional advection-diffusion equation. Appl. Math. Comput. **191**, 12–20 (2007)
6. Liu, F., Zhuang, P., Burrage, K.: Numerical methods and analysis for a class of fractional advection-dispersion models. Comput. Math. Appl. **64**, 2990–3007 (2012)
7. Wu, J.: Theory and Applications of Partial Functional Differential Equations. Springer-Verlag, New York (1996)
8. Zhang, B., Zhou, Y.: Qualitative Analysis of Delay Partial Difference Equations. Hindawi Publishing Corporation, New York (2007)
9. Tavernini, L.: Finite difference approximations for a class of semilinear volterra evolution problems. SIAM J. Numer. Anal. **14**(5), 931–949 (1977)
10. Van Der Houwen, P.J., Sommeijer, B.P., Baker, C.T.H.: On the stability of predictor-corrector methods for parabolic equations with delay. IMA J. Numer. Anal. **6**, 1–23 (1986)
11. Zubik-Kowal, B.: The method of lines for parabolic differential-functional equations. IMA J. Numer. Anal. **17**, 103–123 (1997)
12. Kropielnicka, K.: Convergence of implicit difference methods for parabolic functional differential equations. Int. J. Mat. Anal. **1**(6), 257–277 (2007)
13. Pimenov, V.G., Lozhnikov, A.B.: Difference schemes for the numerical solution of the heat conduction equation with aftereffect. Proc. Steklov Inst. Math. **275**(Suppl. 1), 137–148 (2011)
14. Samarskii, A.A.: The Theory of Difference Schemes. Nauka, Moscow (1977)
15. Pimenov, V.G.: General linear methods for numerical solving functional differential equations. Differ. Equ. **37**(1), 105–114 (2001)
16. Pimenov, V.G., Tashirova, E.E.: Numerical methods for solving a hereditary equation of hyperbolic type. Proc. Steklov Inst. Math. **281**(Suppl. 1), 126–136 (2013)
17. Lekomtsev, A.V., Pimenov, V.G.: Convergence of the alternating direction methods for the numerical solution of a heat conduction equation with delay. Proc. Steklov Inst. Math. **272**(Suppl. 1), 101–118 (2011)
18. Lekomtsev, A., Pimenov, V.: Convergence of the scheme with weights for the numerical solution of a heat conduction equation with delay for the case of variable coefficient of heat conductivity. Appl. Math. Comput. **256**, 83–93 (2015)
19. Pimenov, V., Lozhnikov, A.: Numerical methods for evolutionary equations with delay and software package PDDE. In: Dimov, I., Faragó, I., Vulkov, L. (eds.) NAA 2012. LNCS, vol. 8236, pp. 437–444. Springer, Heidelberg (2013). doi:10.1007/978-3-642-41515-9_49

20. Pimenov, V.G.: Difference methods for the solution of the equations in partial derivatives with heredity. Ural University, Ekaterinburg (2014). (in Russian)
21. Pimenov, V.G., Hendy, A.S.: Numerical methods for the equation with fractional derivative on state and with functional delay on time. Bull. Tambov Univ. Ser.: Nat. Tech. Sci. **20**(5), 1358–1361 (2015)

Calculation of Kinetic Coefficients for Real Gases on Example of Nitrogen

Viktoriia O. Podryga[(✉)]

Keldysh Institute of Applied Mathematics, 4 Miusskaya sq., 125047 Moscow, Russia
pvictoria@list.ru

Abstract. Problem of obtaining data on the properties of gaseous media is considered. Gases of interest are the gases used as transport systems in technical facilities. The focus is on calculating the kinetic coefficients of gaseous medium considering the molecular processes that take place in the gas flow. Molecular dynamics method is selected as the method of modeling. Various techniques for determining the kinetic coefficients of gases are described in detail and compared. The problem is considered on the example of nitrogen flow. For this goal calculating the coefficients of self-diffusion, shear viscosity and thermal conductivity for nitrogen is made. The obtained numerical results are in good agreement with known theoretical estimates and experimental data.

Keywords: Molecular dynamics · Transport coefficients · Diffusion · Shear viscosity · Thermal conductivity · Nitrogen

1 Introduction

With development of nanotechnologies the gas mixture flows in micro- and nanochannels are of particular interest. Investigation of gas microflows assumes good knowledge of materials and medium properties which are used in technological processes. Some gas properties are well studied experimentally in particular temperature and pressure ranges, and are described in literature. However there are properties which can be predicted only theoretically on the basis of the kinetic theory of gases [1]. At the same time the obtained theoretical data correspond to very limited temperature and pressure ranges and can differ significantly from actual properties of gas. In this situation one way of obtaining the information on real properties of a gaseous medium is molecular dynamic modeling [2].

In this paper the main attention is paid to calculations of kinetic coefficients of the gases taking into account the molecular processes that occur in a gas flow. As an example of gas dynamic system the system of nitrogen molecules is considered. As a method of simulation the molecular dynamics method (MMD) is selected, which numerical implementation is based on application of Verlet integration [3]. The paper describes three techniques for determining the transport coefficients of gases. Using the MMD calculation of shear viscosity, self-diffusion and thermal conductivity coefficients of real gas is made.

© Springer International Publishing AG 2017
I. Dimov et al. (Eds.): NAA 2016, LNCS 10187, pp. 542–549, 2017.
DOI: 10.1007/978-3-319-57099-0_61

2   Mathematical Statement

In MMD investigated system is a set of the particles moving under laws of classical mechanics [2]:

$$m\frac{d\mathbf{v}_i}{dt} = \mathbf{F}_i, \quad \mathbf{v}_i = \frac{d\mathbf{r}_i}{dt}, \quad i = 1, ..., N, \tag{1}$$

where i is particle number, N is amount of particles, m is mass of single particle, $\mathbf{r}_i = (r_{x,i}, r_{y,i}, r_{z,i})$ and $\mathbf{v}_i = (v_{x,i}, v_{y,i}, v_{z,i})$ are coordinate and velocity vectors of i-th particle, $\mathbf{F}_i = (F_{x,i}, F_{y,i}, F_{z,i})$ is total force acted on specific particle.

Equations system (1) is solved by Velocity Verlet algorithm:

$$\mathbf{r}^{n+1} = \mathbf{r}^n + \mathbf{v}^n \triangle t + \frac{\mathbf{F}^n}{m}\frac{\triangle t^2}{2}, \quad \mathbf{v}^{n+1} = \mathbf{v}^n + \frac{\mathbf{F}^{n+1} + \mathbf{F}^n}{2m}\triangle t,$$

here $\triangle t$ is integration step, n is step number, $\mathbf{F}^{n+1}$ is force on concrete step.

The forces represent the sum of two components. The first component is force of interacting the i-th particle with surrounded particles, it depends on the potential energy. The second component is responsible for the external influence.

$$\mathbf{F}_i = -\frac{\partial U(\mathbf{r}_1, ..., \mathbf{r}_N)}{\partial \mathbf{r}_i} + \mathbf{F}_i^{ext}, \quad i = 1, ..., N,$$

here U is total potential energy, $\mathbf{F}_i^{ext}$ is external force.

The potential energy of the system depends on the particle coordinates and describes the interaction between the particles of the system [4]:

$$U = \sum_{i=1}^{N} \varphi_i, \quad \varphi_i = \sum_{j>i} \varepsilon \frac{n}{n-6} \left(\frac{n}{6}\right)^{\frac{6}{n-6}} \left[\left(\frac{\sigma}{r_{ij}}\right)^n - \left(\frac{\sigma}{r_{ij}}\right)^6\right], \quad i, j = 1, ..., N, \tag{2}$$

where $r_{ij} = |\mathbf{r}_i - \mathbf{r}_j| = |\mathbf{r}_{ij}|$ is a distance between the i-th and the j-th particles, ε is an energy of molecular interaction that is equal to the deep of potential well, σ is an effective distance of interaction.

Initial conditions include distributions of coordinates and velocities of particles and are defined by an equilibrium thermodynamic state of system [5]. Equilibrium distributions of coordinates and velocities of the gas molecules in the selected microvolume are taken uniform on coordinates and maxwellian on velocities. At the same time average velocity of the particles must correspond to the given temperature of gas. However such artificial state can be far from the true state of thermodynamic equilibrium. So for achieving the true state at a given temperature a corrective molecular dynamic calculation using the thermostatic control method is carried out [2]. This calculation is made up to the value of microsystem temperature and pressure will be permanent.

Boundary conditions at the molecular level are selected depending on the modeled situation. When calculations are aimed on determining the medium properties, some selected its volume is considered, out of which periodic continuation of the medium on unlimited distances in all coordinates is assumed. In this case, in all three coordinates periodic boundary conditions (PBC) are used.

3   Transport Coefficients Models

MMD allows tracking the full trajectories of system particles, to measure numerically time-dependent equilibrium and nonequilibrium properties. This makes it possible to determine different parameters associated with the correlation functions, such as coefficients of diffusion, shear (dynamic) and kinematic viscosity, thermal conductivity. For definition of this coefficients the Einstein relation [2], the Green-Kubo formulas [2] and the mixed approach are presented.

3.1   Einstein Relation

The diffusion coefficient is determined as follows:

$$D = \frac{1}{6N} \frac{1}{t} \left\langle \sum_i [\mathbf{r}_i(t_0 + t) - \mathbf{r}_i(t_0)]^2 \right\rangle, \quad t = NSTEPS \cdot \Delta t, \quad i = 1, ..., N, \quad (3)$$

where $NSTEPS$ is a number of steps from the beginning of calculation, t_0 is an initial time in calculating coefficients, $\langle \rangle$ is an average value on system states.

For PBC formula (3) will provide correct result, but it is necessary to enter an additional array of coordinates and calculate the diffusion using it. This array should contain atomic displacements without cyclic return, i.e., excluding PBC.

Shear (dynamic) viscosity coefficient:

$$\mu = \frac{m^2}{6k_b TV} \frac{1}{t} \left\langle \sum_{\alpha\beta} \left[\sum_i (r_{i,\alpha}(t_0 + t) \cdot v_{i,\beta}(t_0 + t) - r_{i,\alpha}(t_0) \cdot v_{i,\beta}(t_0)) \right]^2 \right\rangle, \quad (4)$$

$$\alpha\beta = xy, yz, zx,$$

where V is a volume of the investigated system, k_b is a Boltzmann constant, T is a temperature.

Thermal conductivity coefficient is defined as follows:

$$\chi = \frac{1}{6k_b T^2 V} \frac{1}{t} \left\langle \sum_\alpha [\delta E_\alpha(t_0 + t) - \delta E_\alpha(t_0)]^2 \right\rangle, \quad \alpha = x, y, z, \quad (5)$$

$$\delta E_\alpha = \sum_i r_{\alpha,i} E_i, \quad E_i = \frac{m\mathbf{v}_i^2}{2} + \varphi_i,$$

where E_i is an instantaneous total energy of the molecule i.

Under PBC formulas (4) and (5) will give incorrect results, i.e., they can not be used [6].

Calculation of transport coefficients (3)–(5) is made until values don't quit on some constant. These constant values correspond to the result. But for Einstein's relation time of functions growth to this value can be quite large therefore follow functions are calculated:

$$D'(t) = D \cdot t, \quad \mu'(t) = \mu \cdot t, \quad \chi'(t) = \chi \cdot t \quad (6)$$

These functions are built depending on time, resulting in a line and the coefficients are determined according to their inclination. Coefficients calculation time is significantly reduced.

3.2   Green-Kubo Formulas

Alternatives to Einstein relations are Green-Kubo formulas which are based on the integration of autocorrelation functions. These formulas are applicable in case of PBC, for their use additional arrays aren't needed.

Diffusion coefficient depends on the velocity autocorrelation function C^{DIFF}:

$$D = \frac{1}{N} \int_0^\infty \langle C^{DIFF} \rangle \, dt, \quad C^{DIFF} = \frac{1}{3} \sum_\alpha C_{\alpha\alpha}^{DIFF}, \tag{7}$$

$$C_{\alpha\alpha}^{DIFF} = \sum_i [v_{\alpha,i}(t_0+t) \cdot v_{\alpha,i}(t_0)], \quad \alpha = x,y,z, \quad i = 1,...,N.$$

Viscosity depends on the autocorrelation function of pressure tensor C^{VISC}:

$$\mu = \frac{1}{k_b T V} \int_0^\infty \langle C^{VISC} \rangle \, dt, \quad C^{VISC} = \frac{1}{3} \sum_{\alpha\beta} C_{\alpha\beta}^{VISC}, \tag{8}$$

$$C_{\alpha\beta}^{VISC} = p_{\alpha\beta}(t_0+t) \cdot p_{\alpha\beta}(t_0), \quad \alpha\beta = xy, yz, zx, \quad i,j = 1,...,N.$$

$$p_{\alpha\beta} = \sum_i m \cdot v_{\alpha,i} \cdot v_{\beta,i} + \sum_i \sum_{j>i} (r_{\alpha,i} - r_{\alpha,j}) \cdot (F_{\beta,i} - F_{\beta,j}),$$

where $p_{\alpha\beta}$ are the off-diagonal components of pressure tensor.

Thermal conductivity coefficient depends on the autocorrelation function of the heat flow C^{TCOND}:

$$\chi = \frac{1}{k_b T^2 V} \int_0^\infty \langle C^{TCOND} \rangle \, dt, \quad C^{TCOND} = \frac{1}{3} \sum_\alpha C_{\alpha\alpha}^{TCOND}, \tag{9}$$

$$C_{\alpha\alpha}^{TCOND} = q_\alpha(t_0+t) \cdot q_\alpha(t_0), \quad q_\alpha = \frac{d\delta E_\alpha}{dt} = \sum_i E_i v_i + \frac{1}{2} \sum_i \sum_{j \neq i} r_{ij}(F_{ij} \cdot v_i),$$

$$F_{ij} = F_i - F_j, \quad \alpha = x,y,z, \quad i,j = 1,...,N.$$

Computing time of transport coefficients (7)–(9) is defined by the relaxation time of the autocorrelation functions. The autocorrelation functions will decrease, converge to zero and after some time come to stationary state close to zero. Transport coefficients should be computed at least this time. Calculation of coefficients according to Green-Kubo formulas has no restrictions on the use of boundary conditions, but is more expensive on the performed operations.

3.3   Mixed Approach

Computations of viscosity and thermal conductivity coefficients are costly on number and complexity of operations, and therefore on necessary machine time and memory capacity. The diffusion coefficient is calculated much easier, and there are no additional restrictions associated with the use of PBC.

It is offered to use mixed approach based on determining viscosity and thermal conductivity coefficients according to the gas dynamics formulas [7] and using diffusion values computed by MMD on Einstein or Green-Kubo formulas:

$$\mu = \frac{\rho \cdot D}{a}, \quad a = 1.3, \tag{10}$$

$$\chi = b \cdot C_V \cdot \mu, \quad b = \frac{1}{4}(9\gamma - 5), \quad \gamma = 1 + \frac{R}{C_V}, \tag{11}$$

where $\gamma = \gamma(T)$ is an adiabatic index, ρ is a density of matter, R is an universal gas constant, C_V is a specific heat capacity at constant volume, which can also be calculated using the MD [8] or can be taken from the experimental data [9].

Value of constant a in (10) is taken in the generalized case [7]. For temperature range of 100–700 K it leads to an error of 1–10% depending on type of gas and value of temperature. If you need higher accuracy, it is possible to specify constant [10]: $a = 3A(\Omega)$, where Ω is a collision integral [1,11].

4   Calculation Results

System of nitrogen molecules concluded in microvolume under normal conditions ($T = 273.15$ K, $p = 101325$ Pa) was considered. Microvolume is a cube with side Lx, which volume is $V = Lx^3$. Distribution of particles corresponds to primitive cubic crystal lattice with cell length r_{cr} and cell numbers in each direction nc. Particle number corresponds to $N = nc^3$, side length of cube is $Lx = nc \cdot r_{cr}$.

Density of nitrogen corresponds to $\rho = 1.25 \, \text{kg/m}^3$ [9], mass of the nitrogen molecule is $m = 46.517 \cdot 10^{-27}$ kg. Then total volume is $V = m \cdot N/\rho$, and the length of the unit cell can be defined as $r_{cr} = \sqrt[3]{m/\rho}$.

The number of particles should be selected so that it was not really much for easier computing, but was enough that the selection was representative and the obtained results had the smallest deviations. The smallest representative system size for calculation of transport coefficients corresponds to $nc = 30$. Also the sizes of $10, 20, 60, 90, 150$ were considered, which showed that for smaller number of particles there were big deviations, and for bigger - accuracy was higher. For $nc = 90, 150$ it is possible not use the ensemble averaging, but the capacity of necessary computer memory and transaction number increase by several times.

The Berendsen thermostat [12] was used to achieve the desired temperature of gas in thermodynamic equilibrium; it was realized via rescaling the velocities.

The technique of obtaining the average characteristics within the chosen strategy was as follows. At first the start point on P-T diagram corresponding to normal conditions was calculated. Calculation of a start point was carried out for 27000 molecules of nitrogen in volume 1 004 936 nm^3.

Calculation was executed in 3 stages. At first the system was balanced under conditions of the switched on thermostat. Thermostating time was 6 ns that corresponds to number of steps $NSTEPS = 3\,000\,000$, 1 step is $\Delta t = 2$ fs $= 0.002$ ps. Then also under conditions of thermostatic control during 0.5 ns

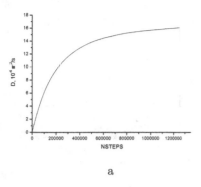

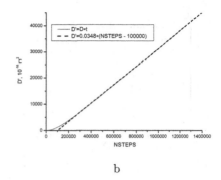

a   b

Fig. 1. Calculations of self-diffusion coefficient of nitrogen based on Einstein relation. On (a) diffusion is defined by direct use of a ratio. On (b) an inclination of function (6) is used for determination of diffusion.

the first calculation of average macroparameters of gas was carried out. Further in the conditions of the switched-off thermostat during 0.5 ns more the second calculation of average macroparameters of gas was carried out. Comparing the macroparameters of gas in the first and second calculations confirmed an achievement of dynamic equilibrium by gas system. Output data of the carried-out calculations are coordinates and velocities distributions of particle system that is in thermodynamic equilibrium at selected temperature and pressure. The obtained arrays are used in further calculations as the starting data.

Main computation step of transport coefficients starts from the equilibrium coordinates and velocities. Duration of the calculation is 2–3 ns.

Direct calculation of diffusion based on the Einstein relation showed that function was increasing for a long time. On Fig. 1a it is possible to see that time of 2.6 ns isn't sufficient for determination of coefficient. Figure 1b shows the result of determining the coefficient through the inclination of function (6). For this purpose, function (6) depending on number of step is built, the linear approximation is superimposed on it and its inclination is determined. For this calculation the inclination $\alpha_D = 0.0348$ is obtained, then diffusion is defined as $D \sim \alpha_D / \triangle t = 0.0348/0.002 = 17.4$, $D = 17.4 \cdot 10^{-6}$ m^2/s.

The calculation results of self-diffusion coefficient of nitrogen according to the formulas of the Green-Kubo are presented in Fig. 2a. Apparently from the figure function increases until it reaches a certain value. Figure 2b shows the graph of the velocity autocorrelation function which attenuation and tendency to zero confirm the settling time of the diffusion value. Thus the diffusion convergence on the correct value occurs from 1.2 ns, which is sufficient calculation time. This result is consistent with the previous computation $D = 17.4 \cdot 10^{-6}$ m^2/s.

Calculation results of nitrogen diffusion coefficient obtained numerically in work are in agreement with high accuracy with the values obtained experimentally $D = (17.2 \pm 0.2) \cdot 10^{-6}$ m^2/s [13] and numerically $D = 17.4 \cdot 10^{-6}$ m^2/s [1].

Results of viscosity calculations for nitrogen on Green-Kubo's formulas are provided in Fig. 3a. Function grows to an output on some value; further function fluctuates around this value. Figure 3b shows the pressure autocorrelation

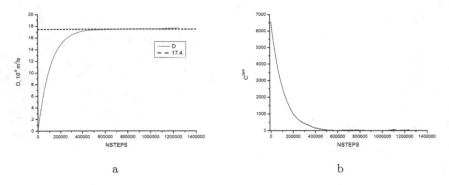

a   b

Fig. 2. Calculations of self-diffusion coefficient of nitrogen on Green-Kubo formulas. On (a) diffusion coefficient is presented. On (b) velocity autocorrelation function is shown, which convergence to zero value means the necessary calculation time.

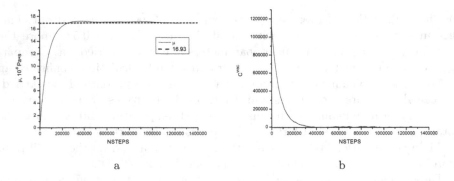

a   b

Fig. 3. Calculations of shear viscosity coefficient of nitrogen on Green-Kubo formulas. On (a) viscosity coefficient is presented. On (b) pressure autocorrelation function is shown, which convergence to zero value means the necessary calculation time.

function which attenuation and tendency to zero similarly to Fig. 2 confirm the settling time of needed state. Calculation was also made to 2.6 ns, which is sufficient calculation time. The obtained numerical result $\mu = 16.93 \cdot 10^{-6}$ Pa·s is in agreement with experimental data $\mu = (16.65 \pm 0.01) \cdot 10^{-6}$ Pa·s [7,9].

For calculating the viscosity and thermal conductivity coefficients by the mixed approach the results of earlier simulation and also the value of specific heat capacity (for example, [8,9]) can be used. The results are: $\mu = 16.73 \cdot 10^{-6}$ Pa·s and $\chi = 0.024$ J/(s·m·K) $= 0.024$ W/(m·K), that is in agreement with experimental data [9] and data obtained through Green-Kubo formulas.

5   Conclusions

Work is devoted to the problem of calculating the kinetic coefficients of the gaseous media by means of molecular dynamics methods. This problem is relevant for many technical applications related to the use of nanotechnologies. The outcomes are the descriptions of techniques for calculating the diffusion, viscosity and thermal conductivity coefficients of gas, and also the simulation results of transport properties for nitric medium. The obtained numerical results are in agreement with the known theoretical and experimental data and thus confirm efficiency of the provided techniques and the developed algorithms.

Acknowledgements. This work was supported by Russian Foundation for Basic Research (projects no. 15-07-06082-a, 15-01-04620-a, 16-37-00417-mol_a).

References

1. Hirschfelder, J.O., Curtis, C.F., Bird, R.B.: Molecular Theory of Gases and Liquids. Wiley, New York (1964)
2. Rapaport, D.C.: The Art of Molecular Dynamics Simulation. Cambridge University Press, Cambridge (2004)
3. Verlet, L.: Computer experiments on classical fluids. I. Thermodynamical properties of Lennard-Jones molecules. Phys. Rev. **159**, 98–103 (1967)
4. Fokin, L.R., Kalashnikov, A.N.: The transport properties of an N2–H2 mixture of rarefied gases in the EPIDIF database. High Temp. **5**, 643–655 (2009)
5. Podryga, V.O.: Molecular dynamics method for simulation of thermodynamic equilibrium. Math. Models Comput. Simul. **3**, 381–388 (2011)
6. Allen, M.P., Masters, A.J.: Some notes on einstein relationships. Mol. Phys. **79**, 435–443 (1993)
7. Golubev, I.F.: Viscosity of gases and gas mixtures. Handbook. Fizmatgiz, Moscow (1959) (in Russian). (Translated in English by Israel Program for Scientific Translations 1970)
8. Podryga, V.O., Polyakov, S.V.: Correction of the gas flow parameters by molecular dynamics. In: 4th International Conference on Particle-Based Methods Fundamentals and Applications, pp. 779–788. CIMNE, Barcelona (2015)
9. Lemmon, E.W., McLinden, M.O., Friend, D.G.: Thermophysical properties of fluid systems. In: Linstrom, P.J., Mallard, W.G. (eds.) NIST Chemistry WebBook, NIST Standard Reference Database Number 69, National Institute of Standards and Technology, Gaithersburg MD, 20899. http://webbook.nist.gov. Accessed 13 June 2016
10. Winter, E.R.S.: Diffusion properties of gases. Trans. Faraday Soc. **46**, 81–92 (1950)
11. Chapman, S., Cowling, T.G.: The Mathematical Theory of Non-uniform Gases. Cambridge University Press, Cambridge (1939)
12. Berendsen, H.J.C., Postma, J.P.M., Gunsteren, W.F., et al.: Molecular dynamics with coupling to an external bath. J. Chem. Phys. **81**, 3684–3690 (1984)
13. Winter, E.R.S.: Diffusion properties of gases. Part IV. The self-diffusion coefficients of nitrogen, oxygen and carbon dioxide. Trans. Faraday Soc. **47**, 342–347 (1951)

New Grid Approach for Solution of Boundary Problems for Convection-Diffusion Equations

Sergey V. Polyakov[1,2(✉)], Yuri N. Karamzin[1],
Tatiana A. Kudryashova[1,2], and Viktoriia O. Podryga[1,2]

[1] Keldysh Institute of Applied Mathematics,
4 Miusskaya sq., 125047 Moscow, Russia
polyakov@imamod.ru
[2] National Research Nuclear University MEPhI,
31 Kashirskoe sh., 115409 Moscow, Russia

Abstract. The numerical solution of boundary value problems is considered for multidimensional equations of convection-diffusion (CDE). These equations are used for many physical processes in solids, liquids and gases. A new approach to the spatial approximation for such equations is proposed. This approach is based on the integral transformation of second order differential operators. A linear version of CDE was selected to simplify analysis. For this variant, a new exponential finite difference scheme was offered, algorithms of its implementation were developed, and brief analysis of the stability and convergence was fulfilled.

Keywords: Convection-diffusion equation (CDE) · Integral transformation · Finite-difference schemes · Non-monotonic sweep procedure

1 Introduction

The equations of convection-diffusion are the basis for many mathematical models [1] and the main component of computational fluid dynamics problems. The methods for solving these equations have been discussed in [2,3] etc. However, the solution of this type equations still generates some difficulties. In our earlier studies by analogy with [4], class of schemes has been introduced. These schemes use the integral transformation of the flux vector with allocation of the exponential factor. This class of schemes were called as an exponential schemes.

Initially the exponential schemes were used in [5] for the calculation of nonlinear processes in semiconductors for a spatially one-dimensional case. Later, these schemes were extended to the two-dimensional case and irregular triangular grids [6,7]. Recently, the idea of applying a double integral transformation was appeared [8,9]. The conversion has been proposed and tested numerically for 1D case in the works [10,11]. In the present study, these results are generalized to the multidimensional case and used for construction of approximations in the center of mass of rectangular cells. Convergence study of schemes to differential solution has also been carried out.

© Springer International Publishing AG 2017
I. Dimov et al. (Eds.): NAA 2016, LNCS 10187, pp. 550–558, 2017.
DOI: 10.1007/978-3-319-57099-0_62

2   Formulation and Transformation of the Problem

Consider a stationary multidimensional linear convection-diffusion equation of general view with real coefficients in a bounded domain Ω of the real Euclidean space R^m:

$$Lu \equiv \sum_{\alpha=1}^{m} L_\alpha u - qu = -f, \quad \mathbf{x} = (x_1, ..., x_m) \in \Omega \subset R^m, \qquad (1)$$

where $u = u(\mathbf{x})$ is unknown function, L is multi-dimensional differential operator, $f = f(\mathbf{x})$ is given right part, $q = q(\mathbf{x})$ is given scalar coefficient, L_α are one-dimensional differential operators defined by the formulas:

$$L_\alpha u = \frac{\partial w_\alpha}{\partial x_\alpha} + r_\alpha \frac{\partial u}{\partial x_\alpha}, \quad w_\alpha = \sum_{\beta=1}^{m} k_{\alpha\beta} \frac{\partial u}{\partial x_\beta} + p_\alpha u, \quad \alpha = 1, ..., m, \qquad (2)$$

$w_\alpha = w_\alpha(\mathbf{x}, u)$ are components of the flux vector $\mathbf{w} = \mathbf{k}\nabla u + \mathbf{p}u$, $k_{\alpha\beta} = k_{\alpha\beta}(\mathbf{x})$ are elements of diffusion matrix $\mathbf{k}$, $p_\alpha = p_\alpha(\mathbf{x})$ and $r_\alpha = r_\alpha(\mathbf{x})$ are elements of the vectors of internal and external convections $\mathbf{p}$ and $\mathbf{r}$.

Equation (1) at the boundary of the area $\partial\Omega$ is closed by the conditions:

$$u = \mu, \quad \text{or} \quad (\mathbf{w}, \mathbf{n}) = -\lambda u + \mu, \quad \mathbf{x} \in \partial\Omega, \qquad (3)$$

where $\lambda = \lambda(\mathbf{x})$ and $\mu = \mu(\mathbf{x})$ are given functions, (,) is scalar product in the space R^m, $\mathbf{n}$ is vector of the external normal to the boundary $\partial\Omega$.

About initial data of the problem (1)–(3) we assume the following:

P1. Area Ω is unclosed unit m-dimensional cube: $\Omega = (0, 1)^m$.

P2. Functions $k_{\alpha\beta}$, p_α, r_α, q, f, λ, μ are piecewise smooth.

P3. Matrix $\mathbf{k}$ is symmetric and strictly positive definite at every point of closure $\bar{\Omega}$ of area Ω (then the inverse matrix $\mathbf{k}^{-1} = \{\tilde{k}_{\beta\alpha}\}$ exists).

P4. The necessary and sufficient conditions for the existence of a classical solution of the problem (1)–(3) were performed.

When constructing the algorithm for solving the problem (1)–(3) formulas (2) are transformed:

$$Lu = \operatorname{div} \mathbf{w} + (\mathbf{r}, \nabla u) - qu = (\nabla, \mathbf{w}) + (\mathbf{r}, \mathbf{k}^{-1}(\mathbf{k}\nabla u \pm \mathbf{p}u))$$
$$= \left(\nabla + (\mathbf{k}^{-1})^T \mathbf{r}, \mathbf{w}\right) - \left(q + \left((\mathbf{k}^{-1})^T \mathbf{r}, \mathbf{p}\right)\right) u \equiv (\nabla + \tilde{\mathbf{r}}, \mathbf{w}) - \tilde{q}u; \qquad (4)$$
$$\mathbf{w} = \mathbf{k}\nabla u + \mathbf{p}u = \mathbf{k}\left(\nabla u + \mathbf{k}^{-1}\mathbf{p}u\right) \equiv \mathbf{k}\left(\nabla u + \tilde{\mathbf{p}}u\right).$$

Consider component-wise records (4) and use the equivalent transformations:

$$Lu = \sum_{\alpha=1}^{m} \tilde{L}_\alpha u - \tilde{q}u, \quad \tilde{L}_\alpha u = \frac{\partial w_\alpha}{\partial x_\alpha} + \tilde{r}_\alpha \frac{\partial u}{\partial x_\alpha} = \frac{1}{g_\alpha} \frac{\partial}{\partial x_\alpha}(g_\alpha w_\alpha),$$

$$g_\alpha = \exp\left[\int\limits_0^{x_\alpha} \tilde{r}_\alpha \, dx_\alpha\right], \quad \tilde{r}_\alpha = \sum_{\beta=1}^m \tilde{k}_{\alpha\beta} r_\beta, \quad \alpha = 1, ..., m;$$

$$w_\alpha = \sum_{\beta=1}^m k_{\alpha\beta}\left(\frac{\partial u}{\partial x_\beta} + \tilde{p}_\beta u\right) = \sum_{\beta=1}^m k_{\alpha\beta}\frac{1}{e_\beta}\frac{\partial}{\partial x_\beta}(e_\beta u), \quad \alpha = 1, ..., m,$$

$$e_\beta = \exp\left[\int\limits_0^{x_\beta} \tilde{p}_\beta \, dx_\beta\right], \quad \tilde{p}_\beta = \sum_{\alpha=1}^m \tilde{k}_{\beta\alpha} p_\alpha, \quad \beta = 1, ..., m.$$

As a result, formulas (2) take the following equivalent view:

$$\tilde{L}_\alpha u = \frac{1}{g_\alpha}\frac{\partial}{\partial x_\alpha}(g_\alpha \, w_\alpha), \quad w_\alpha = \sum_{\beta=1}^m k_{\alpha\beta}\frac{1}{e_\beta}\frac{\partial}{\partial x_\beta}(e_\beta u), \quad \alpha = 1, ..., m. \quad (2')$$

The boundary conditions (3) can be rewritten as:

$$u = \mu, \quad \text{or} \quad w_\alpha = n_\alpha\left(-\lambda u + \mu\right), \quad x_\alpha = 0, 1, \quad \alpha = 1, ..., m, \quad (3')$$

where n_α is component of corresponding normal vector, it is equal to -1 or $+1$.

3   Construction of Finite Difference Schemes

A uniform conservative finite difference scheme is constructed for numerical solution of the problem (1), (2'), (3'). To do this, we introduce two grids Ω_h and Ω_C in domain $\bar{\Omega}$. The grid $\Omega_h = \prod\limits_{\alpha=1}^m \omega_{\hat{x}_\alpha}$. Each grid $\omega_{\hat{x}_\alpha} = \{0 = x_{\alpha,0} < ... < x_{\alpha,N_\alpha} = 1\}$ is 1D irregular grid on coordinate x_α with nodes x_{α,i_α} ($i_\alpha = 0, ..., N_\alpha$), and steps $h_{\alpha,i_\alpha} = x_{\alpha,i_\alpha} - x_{\alpha,i_\alpha-1}$ ($i_\alpha = 1, ..., N_\alpha$), $\hbar_{\alpha,i_\alpha} = 0.5\left(x_{\alpha,i_\alpha^{(+)}} - x_{\alpha,i_\alpha^{(-)}}\right)$; $i_\alpha^{(+)} = \min(i_\alpha + 1, N_\alpha)$, $i_\alpha^{(-)} = \max(i_\alpha - 1, 0)$, $i_\alpha = 0, ..., N_\alpha$. The grid of cells $\Omega_C = \{C_{i_1,...,i_m}, i_\alpha = 1, ..., N_\alpha, \alpha = 1, ..., m\}$, it contains m-dimensional parallelepipeds formed by neighboring nodes of grid Ω_h. Further, set of indexes $i_1, ..., i_m$ will be replaced by the vector index $\mathbf{i}$ so that $C_{\mathbf{i}} \equiv C_{i_1,...,i_m}$.

All continuous functions of the problem (1), (2'), (3') are defined in the centers of the grid cells Ω_C: $M_{\mathbf{i}} \equiv M_{i_1,...,i_m} = \left(x_{1,i_1-1/2}, ..., x_{m,i_m-1/2}\right)$, where $x_{\alpha,i_\alpha-1/2} = 0.5\left(x_{\alpha,i_\alpha} + x_{\alpha,i_\alpha-1}\right)$, $i_\alpha = 1, ..., N_\alpha$, $\alpha = 1, ..., m$. We use the notations u_h, L_h, f_h, $k_{h,\alpha\beta}$, $p_{h,\alpha}$, $r_{h,\alpha}$, q_h and etc. for exact values of continuous functions and operators in the centers of the grid cells Ω_C, and U_h, Λ_h, F_h, $K_{h,\alpha\beta}$, $P_{h,\alpha}$, $R_{h,\alpha}$, Q_h and etc. for their approximate values.

Construction of the finite difference scheme will be carried out by means of integral-interpolation method [2,3]. We integrate the Eq. (1) on cell $C_{\mathbf{i}}$ with the weight $g = \prod\limits_{\alpha=1}^m g_\alpha$ and we divide obtained equality by $|C_{\mathbf{i}}| = \int\limits_{C_{\mathbf{i}}} g \, d\mathbf{x}$:

$$\frac{1}{|C_{\mathbf{i}}|}\int\limits_{C_{\mathbf{i}}}(Lu \cdot g)\,d\mathbf{x} = -\frac{1}{|C_{\mathbf{i}}|}\int\limits_{C_{\mathbf{i}}}(f \cdot g)\,d\mathbf{x} \quad (d\mathbf{x} \equiv dx_1...dx_m). \quad (5)$$

The right side of (5) is denoted as $-\varphi_{h,\mathbf{i}}$ and it is approximate by formula

$$- \frac{1}{|C_\mathbf{i}|} \int\limits_{C_\mathbf{i}} (f \cdot g)\, d\mathbf{x} = -\varphi_{h,\mathbf{i}} \approx -f_{h,\mathbf{i}}. \tag{6}$$

The left side of (5) is transformed as formula:

$$\frac{1}{|C_\mathbf{i}|} \int\limits_{C_\mathbf{i}} (Lu \cdot g)\, d\mathbf{x} = \frac{1}{|C_\mathbf{i}|} \int\limits_{C_\mathbf{i}} \left(\sum_{\alpha=1}^{m} \frac{g}{g_\alpha} \frac{\partial}{\partial x_\alpha} (g_\alpha w_\alpha) \right) d\mathbf{x} - \frac{1}{|C_\mathbf{i}|} \int\limits_{C_\mathbf{i}} (\tilde{q}\, u \cdot g)\, d\mathbf{x}. \tag{7}$$

The first term on the right side (7) is transformed using the formula of average values and integrating along one of coordinates:

$$\frac{1}{|C_\mathbf{i}|} \int\limits_{C_\mathbf{i}} \left(\sum_{\alpha=1}^{m} \frac{g}{g_\alpha} \frac{\partial}{\partial x_\alpha} (g_\alpha w_\alpha) \right) d\mathbf{x} \approx \sum_{\alpha=1}^{m} \frac{(g_\alpha w_\alpha)_{h,i_\alpha} - (g_\alpha w_\alpha)_{h,i_\alpha-1}}{(g_\alpha)_{h,i_\alpha-1/2} h_{\alpha,i_\alpha}}. \tag{8}$$

In this formula the expressions $(g_\alpha w_\alpha)_{h,i_\alpha}$, $(g_\alpha w_\alpha)_{h,i_\alpha-1}$ are taken in points x_{α,i_α} and $x_{\alpha,i_\alpha-1}$ of grid Ω_h along coordinate x_α, and for others coordinates its are taken in the center of the cell $C_\mathbf{i}$. If it is necessary to emphasize approximation on coordinate x_α with index i_α, we will change the expression $(\cdot)_{h,\mathbf{i}}$ on $(\cdot)_{h,i_\alpha-1/2}$.

The second term on the right side of (7) is denoted as $-Q_{h,\mathbf{i}}$ and also approximated by the formula of average values:

$$- \frac{1}{|C_\mathbf{i}|} \int\limits_{C_\mathbf{i}} (\tilde{q}\, u \cdot g)\, d\mathbf{x} = -Q_{h,\mathbf{i}} \approx -\tilde{q}_{h,\mathbf{i}} \cdot u_{h,\mathbf{i}}. \tag{9}$$

The future scheme is obtained by substituting of the (6)–(9) in (5):

$$\sum_{\alpha=1}^{m} \frac{g}{g_\alpha} \frac{\partial}{\partial x_\alpha} (g_\alpha w_\alpha)\, d\mathbf{x} - \frac{1}{|C_\mathbf{i}|} \int\limits_{C_\mathbf{i}} (\tilde{q}\, u \cdot g)\, d\mathbf{x} - \tilde{q}_{h,\mathbf{i}} \cdot u_{h,\mathbf{i}} \approx -f_{h,\mathbf{i}}. \tag{10}$$

Consider approximations of functions g_α and $g_\alpha w_\alpha$. For g_α are true the following forms and approximations:

$$(g_\alpha)_{h,i_\alpha} = \exp\left[\int\limits_{0}^{x_{\alpha,i_\alpha}} \tilde{r}_\alpha\, dx_\alpha \right] \approx \exp\left[\sum_{j_\alpha=1}^{i_\alpha} (\tilde{r}_\alpha)_{h,j_\alpha-1/2} h_{\alpha,j_\alpha} \right], \tag{11}$$

$$(g_\alpha)_{h,i_\alpha-1/2} \approx \exp\left[\sum_{j_\alpha=1}^{i_\alpha-1} (\tilde{r}_\alpha)_{h,j_\alpha-1/2} h_{\alpha,j_\alpha} + 0.5\, (\tilde{r}_\alpha)_{h,i_\alpha-1/2} h_{\alpha,i_\alpha} \right]. \tag{12}$$

Together with (12) the approximation $0.5 \left[(g_\alpha)_{h,i_\alpha-1} + (g_\alpha)_{h,i_\alpha} \right]$ can be used.

For approximation $(g_\alpha w_\alpha)_{h,i_\alpha}$ we use formula $(g_\alpha w_\alpha)_{h,i_\alpha} = (g_\alpha)_{h,i_\alpha} \cdot (w_\alpha)_{h,i_\alpha}$. To obtain approximations of flow components $(w_\alpha)_{h,i_\alpha-1}$ and $(w_\alpha)_{h,i_\alpha}$ on faces ∂C_{i_u-1} and ∂C_{i_u} of cell $C_\mathbf{i}$ relations are used:

$$(\mathbf{k}^{-1}\mathbf{w})_\gamma = \sum_{\beta=1}^{m} \tilde{k}_{\beta\gamma} w_\beta = \frac{1}{e_\gamma} \frac{\partial}{\partial x_\gamma} (e^\gamma u), \quad e^\gamma = \exp\left[\int\limits_{0}^{x_\gamma} \tilde{p}_\gamma\, dx_\gamma \right]. \tag{13}$$

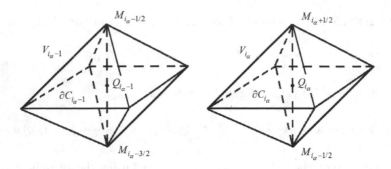

Fig. 1. Determination of the flow through the faces $\partial C_{i_\alpha-1}$ and ∂C_{i_α} of the cell C_i in three-dimensional space ($m = 3$).

The (13) are integrated by volumes $V_{\alpha,i_\alpha-1}$, V_{α,i_α} for $\gamma = 1, ..., m$. Each volume is formed of two pyramids, which have the bases $\partial C_{i_\alpha-1}$, ∂C_{i_α} with centers $Q_{i_\alpha-1}$, Q_{i_α} and apexes $M_{i_\alpha-3/2}$, $M_{i_\alpha-1/2}$, and $M_{i_\alpha-1/2}$, $M_{i_\alpha+1/2}$ (see Fig. 1).

As values of streams through a fixed face of two neighboring grid cells differ only in sign, it is sufficient to consider a single case. For example, a definition of the flow $(w_\alpha)_{h,i_\alpha}$ through the face ∂C_{i_α} (where $i_\alpha = 1, ..., N_\alpha - 1$). Integrating the first of equalities (13) we obtain for each $\alpha, \gamma = 1, ..., m$:

$$\frac{1}{|V_{\alpha,i_\alpha}|} \int\limits_{V_{\alpha,i_\alpha}} \left[\sum_{\beta=1}^{m} \tilde{k}_{\beta\gamma} w_\beta \right] d\mathbf{x} = \frac{1}{|V_{\alpha,i_\alpha}|} \int\limits_{V_{\alpha,i_\alpha}} \frac{1}{e_\gamma} \frac{\partial}{\partial x_\gamma} (e^\gamma u) \, d\mathbf{x}. \qquad (14)$$

In the left side of (14) we take out the summation behind the integral symbol, apply the formula of average values for w_β and replace the integral coefficients by their approximations:

$$\frac{1}{|V_{\alpha,i_\alpha}|} \int\limits_{V_{\alpha,i_\alpha}} \left[\sum_{\beta=1}^{m} \tilde{k}_{\beta\gamma} w_\beta \right] d\mathbf{x} \approx \sum_{\beta=1}^{m} I_{\alpha\beta\gamma} (w_\beta)_{h,i_\alpha},$$

$$I_{\alpha\beta\gamma} = \frac{1}{|V_{\alpha,i_\alpha}|} \int\limits_{V_{\alpha,i_\alpha}} \tilde{k}_{\beta\gamma} \, d\mathbf{x} \approx \frac{h_{\alpha,i_\alpha}}{2\hbar_{\alpha,i_\alpha}} \left(\tilde{k}_{\beta\gamma} \right)_{h,i_\alpha-1/2} + \frac{h_{\alpha,i_\alpha+1}}{2\hbar_{\alpha,i_\alpha}} \left(\tilde{k}_{\beta\gamma} \right)_{h,i_\alpha+1/2}.$$

The right side of (14) we use the approximations:

$$\frac{1}{e_\gamma} \frac{\partial}{\partial x_\gamma} (e^\gamma u) \approx \delta_{\alpha\gamma} \frac{(e_\gamma u)_{h,i_\alpha+1/2} - (e_\gamma u)_{h,i_\alpha-1/2}}{0.5 \left[(e_\gamma)_{h,i_\alpha+1/2} + (e_\gamma)_{h,i_\alpha-1/2} \right] \hbar_{\alpha,i_\alpha}}, \qquad \mathbf{x} \in V_{\alpha,i_\alpha}.$$

and took out them outside the integral symbol ($\delta_{\alpha\gamma}$ is the Kronecker symbol).

As a result from (14) the following approximate relations are obtained:

$$\sum_{\beta=1}^{m} (A_{\alpha,\beta,\gamma})_{h,i_\alpha} (w_\beta)_{h,i_\alpha} \approx (B_{\alpha,\gamma})_{h,i_\alpha}, \qquad \alpha, \gamma = 1, ..., m, \qquad (15)$$

$$(A_{\alpha,\beta,\gamma})_{h,i_\alpha} = \frac{h_{\alpha,i_\alpha}}{2\hbar_{\alpha,i_\alpha}} \left(\tilde{k}_{\beta\gamma}\right)_{h,i_\alpha-1/2} + \frac{h_{\alpha,i_\alpha+1}}{2\hbar_{\alpha,i_\alpha}} \left(\tilde{k}_{\beta\gamma}\right)_{h,i_\alpha+1/2}, \qquad (16)$$

$$(B_{\alpha,\gamma})_{h,i_\alpha} = \frac{(e_\gamma u)_{h,i_\alpha+1/2} - (e_\gamma u)_{h,i_\alpha-1/2}}{0.5\left[(e_\gamma)_{h,i_\alpha+1/2} + (e_\gamma)_{h,i_\alpha-1/2}\right]\hbar_{\alpha,i_\alpha}}, \qquad \alpha,\beta,\gamma = 1,...,m. \quad (17)$$

In (17) it is necessary to calculate the values $(e_\gamma)_{h,i_\alpha\pm1/2}$ only when $\gamma = \alpha$. By analogy with (11), (12) these values are approximated by expressions:

$$(e_\alpha)_{h,i_\alpha-1/2} \approx \exp\left[\sum_{j_\alpha=1}^{i_\alpha-1}(\tilde{p}_\alpha)_{h,j_\alpha-1/2}\,h_{\alpha,j_\alpha} + 0.5\,(\tilde{p}_\alpha)_{h,i_\alpha-1/2}\,h_{\alpha,i_\alpha}\right]. \qquad (18)$$

The relations (15) can be solved for $(w_\beta)_{h,i_\alpha}$ (it is obviously that they are m independent linear systems of equations, each with m unknowns). Without performing the systems (15) solutions procedure, we write its general form by introducing elements $\left(A_{\alpha,\beta,\gamma}^{-1}\right)_{h,i_\alpha}$ of the corresponding inverse matrix:

$$(w_\beta)_{h,i_\alpha} \approx \sum_{\gamma=1}^{m}\left(A_{\alpha,\beta,\gamma}^{-1}\right)_{h,i_\alpha}(B_{\alpha,\gamma})_{h,i_\alpha}, \quad i_\alpha = 1,...,N_\alpha, \quad \alpha,\beta = 1,...,m.$$
$$(19)$$

Among all values in (19) the values $(w_\alpha)_{h,i_\alpha}$ are needed only.

Write the final form of the scheme for an approximate solution U_h:

$$\Lambda_h U_{h,\mathbf{i}} \equiv \sum_{\alpha=1}^{m}\Lambda_{h,\alpha}U_{h,\mathbf{i}} - \tilde{q}_{h,\mathbf{i}}\,U_{h,\mathbf{i}} = -f_{h,\mathbf{i}}, \qquad (20)$$

$$\Lambda_{h,\alpha}U_{h,\mathbf{i}} = \frac{(g_\alpha)_{h,i_\alpha}(w_\alpha)_{h,i_\alpha} - (g_\alpha)_{h,i_\alpha-1}(w_\alpha)_{h,i_\alpha-1}}{0.5\left[(g_\alpha)_{h,i_\alpha} + (g_\alpha)_{h,i_\alpha-1}\right]h_{\alpha,i_\alpha}}, \qquad (21)$$
$$i_\alpha = 1,...,N_\alpha, \quad \alpha = 1,...,m;$$

$$(g_\alpha)_{h,i_\alpha} = \exp\left[\sum_{j_\alpha=1}^{i_\alpha}(\tilde{r}_\alpha)_{h,j_\alpha-1/2}\,h_{\alpha,j_\alpha}\right], \quad i_\alpha = 0,...,N_\alpha, \quad \alpha = 1,...,m; \quad (22)$$

$$(w_\alpha)_{h,i_\alpha} = \left(A_{\alpha,\alpha,\alpha}^{-1}\right)_{h,i_\alpha}(B_{\alpha,\alpha})_{h,i_\alpha}, \quad i_\alpha = 1,...,N_\alpha-1, \quad \alpha = 1,...,m; \quad (23)$$

$$(A_{\alpha,\alpha,\gamma})_{h,i_\alpha} = \frac{h_{\alpha,i_\alpha}}{2\hbar_{\alpha,i_\alpha}}\left(\tilde{k}_{\alpha\gamma}\right)_{h,i_\alpha-1/2} + \frac{h_{\alpha,i_\alpha+1}}{2\hbar_{\alpha,i_\alpha}}\left(\tilde{k}_{\alpha\gamma}\right)_{h,i_\alpha+1/2},$$

$$(B_{\alpha,\alpha})_{h,i_\alpha} = \frac{(e_\alpha u)_{h,i_\alpha+1/2} - (e_\alpha u)_{h,i_\alpha-1/2}}{0.5\left[(e_\alpha)_{h,i_\alpha+1/2} + (e_\alpha)_{h,i_\alpha-1/2}\right]\hbar_{\alpha,i_\alpha}}, \qquad (24)$$

$$(e_\alpha)_{h,i_\alpha-1/2} = \exp\left[\sum_{j_\alpha=1}^{i_\alpha-1}(\tilde{p}_\alpha)_{h,j_\alpha-1/2}\,h_{\alpha,j_\alpha} + 0.5\,(\tilde{p}_\alpha)_{h,i_\alpha-1/2}\,h_{\alpha,i_\alpha}\right],$$

$$i_\alpha = 1,...,N_\alpha-1, \quad \alpha,\gamma = 1,...,m.$$

To finally get the scheme, we need determinate fluxes $(w_\alpha)_{h,i_\alpha}$ on the boundary of computational domain (for $i_\alpha = 0$ and $i_\alpha = N_\alpha$). It is connected with the assignment of boundary conditions (3'). If the first-type boundary conditions are given, the flows are determined from the expansion of the function in a Taylor series in the corresponding boundary point. If the boundary conditions are the conditions of second- or third-types, the fluxes are known to us, but boundary values of function are unknown. In this case, the Taylor expansion is needed too.

As a result, it can be assumed that the flow values are always calculated according to the formulas:

$$(w_\alpha)_{h,i_\alpha} = (n_\alpha)_{h,i_\alpha} \left[-\bar{\lambda}_{h,i_\alpha} U_{h,i_\alpha \pm 1/2} + \bar{\mu}_{h,i_\alpha} \right], \quad i_\alpha = 0, N_\alpha. \tag{25}$$

Here and below the index $i_\alpha \pm 1/2$ is $i_\alpha + 1/2$ for $i_\alpha = 0$, and $i_\alpha - 1/2$ for $i_\alpha = N_\alpha$. The values $\bar{\lambda}_{h,i_\alpha}$ and $\bar{\mu}_{h,i_\alpha}$ in (25) differ from their boundary values λ_{h,i_α} and μ_{h,i_α} in (3') as the function $U_{h,i_\alpha \pm 1/2}$ approximates the value in the cell center, and is moved from the boundary by half step towards α.

To correct this error in the case of a diagonal matrix $\mathbf{k}$ and conditions of the first-type, the formulas can be used:

$$\bar{\lambda}_{h,i_\alpha} = (k_{\alpha\alpha})_{h,i_\alpha} \frac{(e_\alpha)_{h,i_\alpha \pm 1/2}}{(e_\alpha)_{h,i_\alpha} \hbar_{\alpha,i_\alpha}}, \quad \bar{\mu}_{h,i_\alpha} = (k_{\alpha\alpha})_{h,i_\alpha} \frac{\mu_{h,i_\alpha}}{\hbar_{\alpha,i_\alpha}}, \quad i_\alpha = 0, N_\alpha. \tag{26}$$

For second- and third-types conditions the following formulas can be used

$$\bar{\lambda}_{h,i_\alpha} = \frac{\lambda_{h,i_\alpha} \phi_{h,i_\alpha}}{1 + \delta_{h,i_\alpha}}, \quad \bar{\mu}_{h,i_\alpha} = \frac{\mu_{h,i_\alpha}}{1 + \delta_{h,i_\alpha}}, \quad \delta_{h,i_\alpha} = \frac{\hbar_{\alpha,i_\alpha} \lambda_{h,i_\alpha} \phi_{h,i_\alpha}}{2 (k_{\alpha\alpha})_{h,i_\alpha}},$$

$$\phi_{h,i_\alpha} = \left[1 - 0.5 \hbar_{\alpha,i_\alpha} (k_{\alpha\alpha})_{h,i_\alpha}^{-1} (p_\alpha)_{h,i_\alpha} \right]^{-1}, \quad i_\alpha = 0, N_\alpha. \tag{27}$$

In the case of a not diagonal matrix $\mathbf{k}$ more complex calculations are needed.

4   Brief Analysis

1. The maximum order of approximation is equal to 2, that is $\Psi_h = O\left(\hbar^2\right)$, where $\hbar$— minimum or average grid step on all directions. This order is achieved by using uniform or quasi-uniform grids.
2. The solution of schemes exists by the reason of the weak maximum principle fulfillment.
3. The using of the method of energy inequalities [3] for case of sufficiently smooth coefficients of task we can prove the convergence of schemes to the differential solution in the norm $L_2\left(\Omega_C\right)$ of the same order, what order the approximation error Ψ_h has.
4. The implementation of the presented exponential schemes is based on the one-dimensional algorithms of non-monotonic sweep proposed in [10,11]. For one-dimensional variant these formulas are enough to generalize to the case of cells, it means actually to use the stream non-monotonic sweep. For multidimensional case you should use an iterative process based on the method of alternating directions [12]. Flux sweep is used at each step of the process consecutively on each spatial direction.

5. The conditions for the stability of non-monotonic sweep formulas and stability of the schemes are:

$$1 - 0.5h_{\alpha,i_\alpha}|\tilde{p}_{\alpha,i_\alpha - 1/2}| > 0, \quad i_\alpha - 1, ..., N_\alpha;$$
$$1 - 0.5\hbar_{\alpha,i_\alpha}|\tilde{r}_{\alpha,i_\alpha}| > 0, \quad i_\alpha = 0, ..., N_\alpha; \quad \alpha = 1, ..., m.$$

5   Conclusion

The paper is devoted to the numerical solution of boundary value problems for multidimensional equations of convection-diffusion. A new approach is proposed for spatial approximation of such type equations. It is based on the double integral transforms of one-dimensional differential operators of the second order. For the linear version of convection-diffusion equations the exponential finite difference schemes were constructed, algorithms for their implementation were developed, a brief analysis of stability and convergence was performed.

This project has received funding from Russian Fund for Basic Researches (projects No. 15-01-04620-a, 16-29-15095-ofi-m).

References

1. Samarskii, A.A., Mikhailov, A.P.: Principles of Mathematical Modelling: Ideas, Methods, Examples. Tailor & Francis, London (2002)
2. Samarskii, A.A., Andreev, V.B.: Difference Methods for Elliptic Equations. Mir, Moscow (1978). (Russian)
3. Samarskii, A.A.: The Theory of Difference Schemes. Marcel Dekker Inc., New-York (2001)
4. Karetkina, N.V.: An unconditionally stable difference scheme for parabolic equations containing first derivatives. USSR Comput. Math. Math. Phys. **20**(1), 257–262 (1980)
5. Polyakov, S.V., Sablikov, V.A.: Light-induced charge carriers lateral transfer in heterostructures with 2D electron gas. Mat. Model. **9**(12), 76–86 (1997). (Russian)
6. Polyakov, S.V.: Exponential finite volume schemes for solving of evolutional equations on irregular grids. Phys. Math. Kazan. Gos. Univ. Uchen. Zap. Ser. Fiz.-Mat. Nauki **149**(4), 121–131 (2007). (in Russian)
7. Karamzin, Y.N., Polyakov, S.V.: Exponential finite volume schemes for solution of elliptic and parabolic equations of general type on irregular grids. In: Materials of VIII Russian Conference on Grids Methods for Boundary Problems and Applications, pp. 234–248. Kazan University, Kazan (2010). (Russian)
8. Samarskii, A.A.: Monotonic difference schemes for elliptic and parabolic equations in the case of a non-selfadjoint elliptic operator. Zh. Vychisl. Mat. Mat. Fiz. **5**(3), 548–551 (1965)
9. Golant, E.I.: Conjugate families of difference schemes for equations of parabolic type with lowest terms. Zh. Vychisl. Mat. Mat. Fiz. **18**(5), 1162–1169 (1978). (Russian)
10. Polyakov, S.V.: Exponential difference schemes for convection-diffusion equation. Math. Montisnigri **25**, 1–16 (2012). (Russian)

11. Polyakov, S.V.: Exponential difference schemes with double integral transformation for solving convection-diffusion equations. Math. Models Comput. Simul. **5**(4), 338–340 (2013)
12. Samarskii, A.A., Nikolaev, E.S.: Numerical Methods for Grid Equations. Direct Methods, vol. I, Iterative Methods, vol. II. Birkhauser Verlag, Basel-Boston-Berlin (1989)

Finite-Difference Method for Solution of Advection Equation by Unstable Schemes

Igor V. Popov[1,2(✉)]

[1] Keldysh Institute of Applied Mathematics,
4 Miusskaya sq., 125047 Moscow, Russia
piv2964@mail.ru
[2] National Research Nuclear University MEPhI,
31 Kashirskoe sh., 115409 Moscow, Russia

Abstract. In this work the analysis of two-layer differential schemes for the one-dimensional uniform linear equation of transfer is carried out: (a) the first finite-difference scheme is the UPWIND evolution scheme; (b) the second finite-difference scheme has central differential approximation of the convective term (Explicit Central Space evolution scheme); (c) the third scheme represents a downstream finite difference with a regularization term (Regularized Downwind evolution scheme). The special attention in this work is paid to a question of stability enhancement of differential schemes.

Keywords: Advection equation · Finite-difference schemes · Stability · Regularization

1 Introduction

Let's consider the one-dimensional uniform linear equation of transfer:

$$\frac{\partial u}{\partial t} + a\frac{\partial u}{\partial x} = 0, \tag{1}$$

in computation area $0 \leq x \leq L$, $0 \leq t \leq T$, and $a = \text{const} > 0$. For a correct Cauchy problem definition it is necessary to set the following initial and boundary conditions

$$u(x,0) = u_0(x), \quad u(0,t) = u_1(t). \tag{2}$$

The solution of the problem (1), (2) is function

$$u(x,t) = f(x - at),$$

where f is an arbitrary function.

Let us solve a problem (1), (2) numerically on uniform grid $\omega_x \times \omega_t$ with steps $h > 0$ and $\tau > 0$. For this purpose we will construct three difference schemes and carry out their comparison.

© Springer International Publishing AG 2017
I. Dimov et al. (Eds.): NAA 2016, LNCS 10187, pp. 559–567, 2017.
DOI: 10.1007/978-3-319-57099-0_63

2   UPWIND Scheme

Let us consider the well known UPWIND evolution scheme:

$$\frac{y_k^{n+1} - y_k^n}{\tau} + a\frac{y_k^n - y_{k-1}^n}{h} = 0, \quad \text{or} \quad y_k^{n+1} = (1 - \gamma)\, y_k^n + \gamma y_{k-1}^n,$$

where y_k^n is grid analogue of the function u, $\gamma = a\tau/h$ is the Courant number. As it is known, this scheme is monotonous and stable when the conditions $0 < \gamma \leq 1$ are fulfilled (see, for example, [1–3]). This scheme has the first order with respect to time and space.

3   Forward Time Central Space (FTCS) Scheme

Let's consider now FTCS evolution scheme:

$$\frac{y_k^{n+1} - y_k^n}{\tau} + a\frac{y_{k+1}^n - y_{k-1}^n}{2h} = 0, \quad \text{or} \quad y_k^{n+1} = 0.5\gamma y_{k-1}^n + y_k^n - 0.5\gamma y_{k+1}^n.$$

This scheme is unstable and is non-monotonic, has the first order on time and the second order on space. Non-monotonicity of this scheme is that the coefficient at grid function y_{k+1}^n is negative, therefore the nonmonotonicity results also in instability.

Let's regularize this scheme. For this we will add to the right side of the Eq. (1) the term $\frac{\partial}{\partial x}\left(\mu\left(x, t, u\right)\frac{\partial u}{\partial x}\right)$, where μ is the artificial viscosity (some function). In this case the difference scheme will take the following form

$$\frac{y_k^{n+1} - y_k^n}{\tau} + a\frac{y_{k+1}^n - y_{k-1}^n}{2h} = \frac{1}{h}\left\{\mu_{k+1/2}^n \frac{y_{k+1}^n - y_k^n}{h} - \mu_{k-1/2}^n \frac{y_k^n - y_{k-1}^n}{h}\right\},$$

or in the three-point form under the condition of $\mu = \text{const} > 0$:

$$y_k^{n+1} = (\beta + 0.5\gamma)\, y_{k-1}^n + (1 - 2\beta)\, y_k^n + (\beta - 0.5\gamma)\, y_{k+1}^n, \tag{3}$$

where $\beta = \mu\tau/h^2 = \gamma\mu/(ah)$.

Lemma 1. The scheme (3) is monotonous when following conditions are satisfied:

$$0 < \gamma < 1, \quad \mu_{\min} \leq \mu < \mu_{\max}, \tag{4}$$

where $\mu_{\min} = 0.5ah$, $\mu_{\max} = 0.5ah/\gamma$.

Proof. The difference scheme (3) will be monotonous if all coefficients under terms u_{k-1}^n, u_k^n, u_{k+1}^n in right part of (3) will be positive:

$$\beta + 0.5\gamma \geq 0, \quad 1 - 2\beta > 0, \quad \beta - 0.5\gamma \geq 0.$$

The first inequality is right because β and γ are positive.

The second inequality gives the up limit for the μ values:

$$\beta < \frac{1}{2}, \ or \ \mu < \frac{ah}{2\gamma}.$$

The third inequality gives the down limit for the μ values:

$$\beta \geq 0.5\gamma, \ or \ \mu \geq \frac{ah}{2}.$$

The range of μ values will not empty if γ will less than 1. For definiteness we will demand $\gamma = 1/(1+\varepsilon)$, where $\varepsilon > 0$. Then $1/\gamma = 1+\varepsilon$. If to give μ from range

$$\frac{ah}{2} \leq \mu < \frac{ah}{2}(1+\varepsilon),$$

we can to prove the Lemma 1 with constants $\mu_{\min} = 0.5ah$, $\mu_{\max} = 0.5ah/\gamma$.

Lemma 2. The scheme (3) is stable when the following conditions are satisfied:

$$\frac{h^2}{2\tau}A_- = \mu_- \leq \mu \leq \mu_+ = \frac{h^2}{2\tau}A_+, \quad A_{\pm} = \frac{1 \pm \sqrt{1 - \gamma^2 \sin^2 \varphi}}{1 - \cos \varphi}, \tag{5}$$

$$0 < \gamma < 1, \quad 0 < \varphi < 0.5\pi. \tag{6}$$

Here $\varphi = wh$, w is positive character parameter of differential solution.

Proof. Let us prove this lemma using the separation of variables (see, e.g., [2]). We seek a solution of the difference scheme in the form $y_k^n = \rho^n \exp(ik\varphi)$ where $\varphi = wh$, $i = \sqrt{-1}$ is the imaginary unit. Substituting these expressions into (3) and taking into account $\exp(ik\varphi) = \cos(k\varphi) + i\sin(k\varphi)$, we obtain

$$\rho = 1 + 2\beta(\cos \varphi - 1) - i\gamma \sin \varphi.$$

Let us denote $A = 2\beta$ and taking into account $\mathrm{Re}\,\rho = 1 - A(1 - \cos\varphi)$, $\mathrm{Im}\,\rho = -\gamma \sin \varphi$, we receive

$$|\rho|^2 = [1 - A(1 - \cos \varphi)]^2 + \gamma^2 \sin^2 \varphi = [1 - A(1 - \cos \varphi)]^2 + \gamma^2 (1 - \cos^2 \varphi).$$

Let us find the condition for the artificial viscosity μ which supports the inequality $|\rho|^2 \leq 1$. In solving this quadratic inequality with respect to A, we find the values $A_{\pm}$:

$$A_{\pm} = \frac{1 \pm \sqrt{1 - \gamma^2 \sin^2 \varphi}}{1 - \cos \varphi}.$$

These values $A_{\pm}$ will real if the conditions (6) are true. Choosing the values of viscosity μ from a range

$$\frac{h^2}{2\tau}A_- \leq \mu \leq \frac{h^2}{2\tau}A_+,$$

we obtain the approval of the Lemma 2.

Theorem 1. The finite-difference scheme (3) is monotonous and stable if the following conditions are satisfied:

$$0 < \gamma < 0.8, \quad 0 < wh < 0.5\pi, \quad \mu_{min} \leq \mu \leq \mu_{max}. \tag{7}$$

Proof. Let us bring together the estimates obtained in Lemmas 1 and 2.

Let us consider the following Taylor series and trivial inequalities

$$\sqrt{1 - \gamma^2} = 1 - \frac{1}{2}\gamma^2 - \frac{1}{8}\gamma^4 - \frac{1}{16}\gamma^6 - \frac{5}{128}\gamma^8 - \ldots = 1 - \frac{1}{2}\gamma^2 R(\gamma),$$

$$R(\gamma) = 1 + \frac{1}{4}\gamma^2 + \frac{1}{8}\gamma^4 + \frac{5}{64}\gamma^6 + \frac{7}{128}\gamma^8 + \ldots \leq \exp(\gamma^2),$$

$$1 - \sqrt{1 - 2\gamma^2} = 0.5\gamma^2 R(\gamma) \leq 0.5\gamma^2 \exp(\gamma^2),$$

$$1 + \sqrt{1 - 2\gamma^2} = 2 - 0.5\gamma^2 R(\gamma) \geq 2 - 0.5\gamma^2 \exp(\gamma^2).$$

These expressions we will use in the following estimates:

$$\mu_- = \frac{h^2}{2\tau}A_- \equiv \frac{ah}{2\gamma}A_- = \frac{ah}{2}\frac{1 - \sqrt{1 - \gamma^2 \sin^2\varphi}}{\gamma(1 - \cos\varphi)}$$

$$\leq \mu_{min}\frac{1 - \sqrt{1 - \gamma^2}}{\gamma} \leq \mu_{min}0.5\gamma \exp(\gamma^2);$$

$$\mu_+ = \frac{h^2}{2\tau}A_+ \equiv \frac{ah}{2\gamma}A_+ = \frac{ah}{2\gamma}\frac{1 + \sqrt{1 - \gamma^2 \sin^2\varphi}}{1 - \cos\varphi}$$

$$\geq \mu_{max}\frac{1 + \sqrt{1 - \gamma^2}}{1 - \cos\varphi} \geq \mu_{max}\frac{2 - 0.5\gamma^2 \exp(\gamma^2)}{1 - \cos\varphi}.$$

Now we require the following conditions:

$$0.5\gamma \exp(\gamma^2) \leq 1, \quad \frac{2 - 0.5\gamma^2 \exp(\gamma^2)}{1 - \cos\varphi} \geq 1.$$

These conditions are exactly satisfied if

$$\gamma \in (0, 0.8), \quad \varphi \in (0, 0.5\pi).$$

For these values of γ and φ are valid the following final estimates

$$\mu_- \leq \mu_{min}, \quad \mu_+ \geq \mu_{max}.$$

So the Theorem 1 is proved.

Lemma 3. The scheme (3), recorded for the case of constant viscosity coefficient, approximates the problem (1), (2) with an error

$$\Psi = \begin{cases} O\left(h^2 + \tau^2\right), & \mu = \dfrac{a^2\tau}{2} = \dfrac{ah}{2}\gamma; \\ O\left(h + \tau\right), & \mu = O\left(h\right). \end{cases}$$

Proof. We will write the approximation error of the FTCS for the exact solution on grid u_k^n

$$\Psi = \frac{u_k^{n+1} - u_k^n}{\tau} + a\frac{u_{k+1}^n - u_{k-1}^n}{2h} - \mu\frac{u_{k+1}^n - 2u_k^n + u_{k-1}^n}{h^2},$$

and we will take into to account the following Taylor series

$$\frac{u_k^{n+1} - u_k^n}{\tau} = \left(\frac{\partial u}{\partial t}\right)_k^n + \left(\frac{\partial^2 u}{\partial t^2}\right)_k^n \frac{\tau}{2} + \left(\frac{\partial^3 u}{\partial t^3}\right)_k^n \frac{\tau^2}{6} + O\left(\tau^3\right),$$

$$\frac{u_{k+1}^n - u_{k-1}^n}{2h} = \left(\frac{\partial u}{\partial x}\right)_k^n + \left(\frac{\partial^3 u}{\partial x^3}\right)_k^n \frac{h^2}{6} + O\left(h^4\right),$$

$$\frac{u_{k+1}^n - 2u_k^n + u_{k-1}^n}{h^2} = \left(\frac{\partial^2 u}{\partial x^2}\right)_k^n + \left(\frac{\partial^4 u}{\partial x^4}\right)_k^n \frac{h^2}{12} + O\left(h^4\right),$$

and the relations

$$\frac{\partial^m u}{\partial t^m} = (-1)^m a^m \frac{\partial^m u}{\partial x^m}.$$

If we substitute these expressions in the formula for Ψ we obtain

$$\Psi = \left(\frac{a^2\tau}{2} - \mu\right)\left(\frac{\partial^2 u}{\partial x^2}\right)_k^n + O\left(h^2 + \tau^2\right).$$

This fact proves the Lemma 3.

The stability and approximation imply the convergence of the proposed difference scheme (3).

Theorem 2. Under the conditions (7) the scheme (3) with a constant viscosity coefficient converges to the exact solution of the problem (1), (2) in the norm $C\left(\omega_x \times \omega_t\right)$ at a rate proportional to the norm $C\left(\omega_x \times \omega_t\right)$ of the approximation error Ψ. The order of convergence corresponds to the order of approximation.

Proof. Let us consider the error function $z_k^n = y_k^n - u_k^n$ for the scheme (3). This grid function can be determinate from the scheme

$$\frac{z_k^{n+1} - z_k^n}{\tau} + a\frac{z_{k+1}^n - z_{k-1}^n}{2h} - \mu\frac{z_{k+1}^n - 2z_k^n + z_{k-1}^n}{h^2} = -\Psi_k^n, \quad z_k^0 = 0, \quad z_0^n = 0.$$

When the conditions (7) are valid, this scheme is also monotonous, and satisfies the estimate arising from the maximum principle:

$$||z^{n+1}||_{C(\omega_x)} \le ||z^n||_{C(\omega_x)} + \tau||\Psi^n||_{C(\omega_x)}, \quad ||z^0||_{C(\omega_x)} = 0.$$

It follows that

$$||z||_{C(\omega_x \times \omega_t)} \le T||\Psi||_{C(\omega_x \times \omega_t)}.$$

This evaluation proves the theorem.

4   The Regularized Downwind (RDWIN) Scheme

Let us consider the third finite-difference scheme for problem (1), (2). This scheme uses downstream difference with regularization and has the form

$$\frac{y_k^{n+1} - y_k^n}{\tau} + a\frac{y_{k+1}^n - y_k^n}{h} = \frac{1}{h}\left\{\mu_{k+1/2}^n\frac{y_{k+1}^n - y_k^n}{h} - \mu_{k-1/2}^n\frac{y_k^n - y_{k-1}^n}{h}\right\}.$$

For the case of constant viscosity we has the following scheme

$$\frac{y_k^{n+1} - y_k^n}{\tau} + a\frac{y_{k+1}^n - y_k^n}{h} = \mu\frac{y_{k+1}^n - 2y_k^n + y_{k-1}^n}{h^2},$$

or in three-point form

$$y_k^{n+1} = \beta y_{k-1}^n + (1 + \gamma - 2\beta) y_k^n + (\beta - \gamma) y_{k+1}^n. \tag{8}$$

By analogy with the previous one can prove the following assertions.

Lemma 1'. The scheme (8) is monotonous when following conditions are satisfied:

$$0 < \gamma < 1, \quad \mu_{min} \equiv \frac{h^2}{\tau}\gamma = ah \le \mu < ah\frac{1+\gamma}{2\gamma} = \frac{h^2}{\tau}\frac{1+\gamma}{2} \equiv \mu_{max}. \tag{9}$$

Lemma 2'. The scheme (8) is stable when the conditions (6) and

$$\frac{h^2}{2\tau}(\gamma + A_-) = \mu_- \le \mu \le \mu_+ = \frac{h^2}{2\tau}(\gamma + A_+), \tag{10}$$

are satisfied. Here $A_\pm$ are given in (5).

Theorem 1'. The finite-difference scheme (8) is monotonous and stable if the following conditions

$$0 < \gamma < 1, \quad 0 < \omega h < 0.5\pi, \quad \mu_{min} \le \mu \le \mu_{max}, \tag{11}$$

are satisfied.

Lemma 3'. The scheme (8), recorded for the case of constant viscosity coefficient, approximates the problem (1), (2) with an error

$$\Psi = \begin{cases} O\left(h^2 + \tau^2\right), & \mu = \dfrac{ah}{2} + \dfrac{a^2\tau}{2} = \dfrac{ah}{2}\left(1 + \gamma\right); \\ O\left(h + \tau\right), & \mu = O\left(h\right). \end{cases}$$

Theorem 2'. Under the conditions (11) the scheme (8) with a constant viscosity coefficient converges to the exact solution of the problem (1), (2) in the norm $C\left(\omega_x \times \omega_t\right)$ at a rate proportional to the norm $C\left(\omega_x \times \omega_t\right)$ of the approximation error Ψ. The order of convergence corresponds to the order of approximation.

5 Stages of Calculations of the Regularized Schemes

The computations by the scheme (8) on each time step are divided into three stages. On the first stage, the scheme equation solved with the zero artificial viscosity, i.e. $\mu \equiv 0$. On the second stage for the obtained values $\tilde{y}_k^{n+1}$ we determine the areas of computational domain where the solution oscillations occur. In these areas the inequality takes place:

$$\left(\tilde{y}_k^{n+1} - \tilde{y}_{k-1}^{n+1}\right)\left(\tilde{y}_{k+1}^{n+1} - \tilde{y}_k^{n+1}\right) < 0.$$

If this condition is satisfied, on the segments $[x_{k-1}, x_k]$ and $[x_k, x_{k+1}]$ non-monotonicity should be suppressed. To do this in the right-hand side of the equation with numbers $k-1, k$ and $k+1$ we add the term with the artificial viscosity, the value of artificial viscosity is equal

$$\mu = \frac{1}{2}\left(\mu_{\min} + \mu_{\max}\right).$$

On the third stage we solve the all regularized equations.

6 Numerical Experiments

Let us perform the numerical experiments with the finite-difference schemes and compare them with each other. For the first test we consider a function with the initial profile

$$u\left(x, 0\right) = \begin{cases} 1, & x \in [l_1, l_2], \\ 0, & x \notin [l_1, l_2], \end{cases}$$

and the second test with continuous initial profile "cosine"

$$u\left(x, 0\right) = \begin{cases} 0.5\left(1 - \cos\varphi\right), & \varphi = 2\pi\left(x - l_1\right)/\left(l_2 - l_1\right), & x \in [l_1, l_2], \\ 0, & x \notin [l_1, l_2], \end{cases}$$

with parameters of problem $a = 1$, $L = 400$, $T = 300$, $l_1 = 10$, $l_2 = 30$, and the Courant number $\gamma = 0.3$. In all calculations the adaptive artificial viscosity was taken equal to the value $\mu = 0.5\left(\mu_{\min} + \mu_{\max}\right)$.

The results of numerical solution using the proposed three difference schemes and the exact solution are shown in Figs. 1 and 2. The isolines of transition amplitude multiplier function module according to the Courant number and wave number for the RDWIN and FTCS schemes are shown in Fig. 3. These results demonstrate that RWIND scheme is more effective for numerical solution of considered problem.

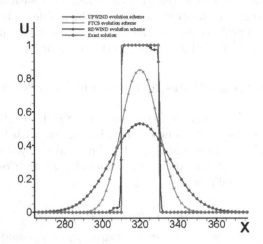

Fig. 1. The results of numerical solution of the rectangle's transfer using the proposed three difference schemes and the exact solution.

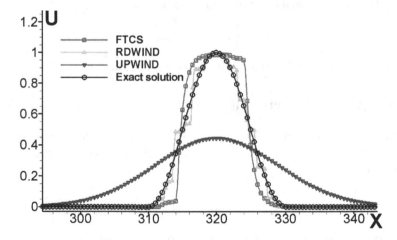

Fig. 2. The results of numerical solution of the transport of a continuous function "cosine" using the three proposed difference schemes and the exact solution.

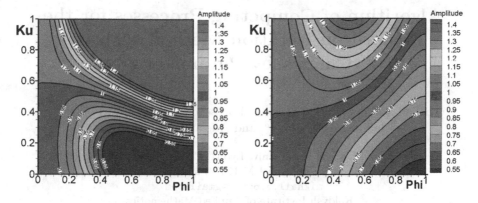

Fig. 3. The isolines of transition amplitude multiplier function module according to the Courant number and wave number for RDWIN scheme (on a left) and FTCS scheme (on a right).

Acknowledgements. This project has received funding from Russian Fund for Basic Researches (projects no. 15-01-04620-a, 16-07-00519-a).

References

1. Samarskii, A.A.: The Theory of Difference Schemes. Marcel Dekker Inc., New York (2001)
2. Kalitkin, N.N.: Numerical Methods: Tutorial. BKhV-Peterburg Publ., St. Petersburg (2011). (in Russian)
3. Galanin, M.P., Savenkov, E.B.: Methods of Numerical Analysis of the Mathematical Models. Bauman MSTU Publ, Moscow (2010). (in Russian)

Algorithm of Competing Processes for the Richardson Iteration Method with the Chebyshev Parameters

Mikhail V. Popov[1,2]([✉]), Yuriy A. Poveschenko[2,3], Igor V. Popov[2,3], Vladimir A. Gasilov[2,3], and Alexander V. Koldoba[4]

[1] Univ. Lyon, ENS de Lyon, Univ. Lyon1, CNRS, Centre de Recherche Astrophysique de Lyon UMR5574, 69007 Lyon, France
mikhail.v.popov@gmail.com
[2] Keldysh Institute of Applied Mathematics,
4 Miusskaya sq., 125047 Moscow, Russia
hecon@mail.ru
[3] National Research Nuclear University MEPhI,
31 Kashirskoe sh., 115409 Moscow, Russia
[4] Moscow Institute of Physics and Technology,
Dolgoprudnyy 141700, Moscow Region, Russia

Abstract. A method for solving linear systems of equations with sign-definite self-adjoint operator matrix by the Richardson iteration method in case of the absence of information about the lower spectral bound of a problem is presented. The algorithm is based on the simultaneous operation of two competing iterative processes, the effectiveness of which is constantly analyzed. The method is explained on an example of one-dimensional steady-state heat equation.

Keywords: Systems of linear algebraic equations · Matrix inversion · Iterative methods · Richardson iteration method

1   Introduction

The Richardson iteration method is one of the well-known explicit two-layer iterative schemes for constructing the solution of linear algebraic equations with sign-definite self-adjoint operator. The advantage of the scheme is its simplicity and the possibility of effective parallelization. At the same time application of Chebyshev accelerator significantly increases the speed of convergence. However, in the classic formulation the Richardson method assumes knowledge of the bounds of the spectral radius of a matrix. Estimation of the upper bound is not a problem since the Gershgorin circle theorem could be applied [1]. But the lower bound is known analytically only for special cases of invertible operators, e.g. for model problems with the Laplacian of the required function in a rectangular computational domain. To estimate lower spectrum bound by a strictly positive value, approaching to the true minimum eigenvalue, is particularly difficult for complex geometry of a computational domain, for complex-structured or irregular grids, which in practice reduces the efficiency of the iterative process.

© Springer International Publishing AG 2017
I. Dimov et al. (Eds.): NAA 2016, LNCS 10187, pp. 568–575, 2017.
DOI: 10.1007/978-3-319-57099-0_64

In this paper present the shorten description of the method, which will be published elsewhere [2]. We propose an algorithm for setting the lower bound of the spectrum for any computational domain geometry and mesh type. The algorithm is based on the simultaneous operation of two competing iterative processes, the effectiveness of which is constantly analyzed. The method is explained on the example of one-dimensional steady-state heat equation. In this case the lower bound of the matrix spectrum is known exactly what allows to check how well algorithm approaches to its value in the contex of the convergence.

2 Iterative Method for Solving a System of Linear Algebraic Equations

Let us consider the algebraic system of equations in the form

$$Ay = f, \tag{1}$$

where A is a sign-definite self-adjoint operator (with the corresponding matrix), denoting a scalar product; y is a vector of unknowns; f is a right part. The solution is $y = A^{-1}f$, however, in practice the matrix A can be difficult invertible. To solve (1) an iterative process can be applied

$$B\frac{y^{k+1} - y^k}{\tau} + Ay^k = f \tag{2}$$

with easily inversible precondition matrix B, which is a self-adjoint positive definite operator. Here y^k is a vector of unknowns, computed on the k-iteration.

Left multiplying (2) by τB^{-1}, we get

$$y^{k+1} - y^k + \tau B^{-1}Ay^k = \tau B^{-1}f,$$

$$\delta y^{k+1} = \left(I - \tau B^{-1}A\right)\delta y^k, \tag{3}$$

where I is the identity matrix.

Changing variables by $x^k = B^{1/2}\delta y^k$, (3) is transformed to

$$x^{k+1} = (I - \tau C)\, x^k. \tag{4}$$

The operators A, B and $C = B^{-1/2}AB^{1/2}$ are self-adjoint and positive definite. In particular for $C = C^* > 0$ there is a complete set of orthogonal eigenvectors r_n with positive eigenvalues λ_n:

$$Cr_n = \lambda_n r_n.$$

The spectrum of C is unknown but it is assumed it belongs to the range

$$0 < \gamma_1 \leq \lambda_n \leq \gamma_2, \tag{5}$$

where $\gamma_1 = \min_n \lambda_n$, $\gamma_2 = \max_n \lambda_n$. From $\gamma_1 B \leq A \leq \gamma_2 B$ it apparently follows $\gamma_1 I \leq C \leq \gamma_2 I$. Decomposition of the initial residual $\mathbf{x}^0$ by the eigenvectors $\mathbf{r}_n$ is

$$\mathbf{x}^0 = \sum_n C_n^0 \mathbf{r}_n,$$

where C_n^0 are the expansion coefficients. According to (4) the coefficients C_n^s on the s-iteration for each λ_n have the form

$$C_n^s = \underbrace{(1 - \tau\lambda_n)(1 - \tau\lambda_n)\ldots(1 - \tau\lambda_n)}_{s} C_n^0 = (1 - \tau\lambda_n)^s C_n^0.$$

Thus for the chosen matrix B, the iterative process (2) converges the faster the closer the factor $(1 - \tau\lambda_n)$ to zero. Moreover, the spectral harmonics of the iterative error converge independently. In other words the convergence of a particular harmonic is not affected by other harmonics and the only issue is to optimize the speed of convergence of the iterative process in a spectral integral norm in full range of the spectrum of the operator $C = C^* > 0$. For the convergence analysis we apply the mesh analog of the uniform norm C. As it is required to minimize the factor for all possible λ_n of (5), then the best possibility is

$$\tau = \tau_0 = \frac{2}{\gamma_1 + \gamma_2}. \tag{6}$$

For this τ the maximum of $|1 - \tau\lambda|$ is reached at $\lambda = \gamma_1$ or $\lambda = \gamma_2$ and equals to

$$\max_\lambda |1 - \tau\lambda| = \frac{\gamma_2 - \gamma_1}{\gamma_2 + \gamma_1}. \tag{7}$$

From (7) it follows that if $\gamma_2 \gg \gamma_1$, then the $\max_\lambda |1 - \tau\lambda| \approx 1$ and the iterative process (2) converges slowly. But from (7) it also follows that the iterations (2) converge also for $0 < \min_n \lambda_n < \gamma_1$. To create an iterative algorithm with an unknown lower spectral bound it is important to take into account that a similar result holds also for the Richardson method with variable iteration parameters.

By choosing the parameter τ at each iteration k, we get a set of $\{\tau_k\}$, $k = 1, \ldots, s$, where s is the preset number of iterations. Then instead of (2) a more complex iterative process holds:

$$B\frac{\mathbf{y}^{k+1} - \mathbf{y}^k}{\tau_k} + A\mathbf{y}^k = \mathbf{f}. \tag{8}$$

Instead of (4), considering the changing τ, we have [2]

$$x^s = P_s(C)x^0, \quad P_s(C) = \prod_{k=1}^{s}(I - \tau_k C).$$

Norm estimation gives

$$\|x^s\| \le \|P_s(C)\| \cdot \|x^0\|.$$

The norm of the resolving polynomial operator $P_s(C)$ with the argument $C = C^* > 0$, $\gamma_1 I \le C \le \gamma_2 I$ could be estimated via a standard polynomial $P_s(t)$ as [3,4]

$$\|P_s(C)\| \le \max_{t \in [\gamma_1, \gamma_2]} |P_s(t)|, \quad P_s(t) = \prod_{k=1}^{s} (1 - \tau_k t).$$

The parameters $\tau_1, \ldots, \tau_s$ are obtained from a condition for minimum $\|P_s(C)\|$, i.e. $\min\limits_{\tau_j} \left\{ \max\limits_{t \in [\gamma_1, \gamma_2]} |P_s(t)| \right\}$ and are expressed via Chebyshev polynomial zeros as [5]

$$T_s(x) = \cos(s \arccos x), \quad -1 \le x \le 1, \quad s = 0, 1, 2, \ldots, \quad T_0(x) = 1, \quad T_1(x) = 1.$$

The recursive formula is valid:

$$2T_{s+1}(x) = 2x T_s(x) - T_{s-1}(x),$$

as well as the supplement relation for $|x| > 1$:

$$2T_s(x) = \left(x + \sqrt{x^2 - 1}\right)^s + \left(x - \sqrt{x^2 - 1}\right)^s.$$

The roots of $T_s(x)$ are

$$x_k = \cos \frac{2k-1}{2s} \pi, \quad k = 1, 2, \ldots, s.$$

The determination of a polynomial $P_s(t)$ on the interval $t \in [\gamma_1, \gamma_2]$ with minimum deviation from zero and normalized to unity by the condition $P_s(0) = 1$ gives the following Chebyshev parameters for (8) [2,5]:

$$\tau_k = \frac{\tau_0}{1 + \rho_0 x_k}, \quad x_k = \cos\left(\frac{2k-1}{2s}\pi\right), \quad k = 1, 2, \ldots, s, \tag{9}$$

with

$$\rho_0 = \frac{1 - \xi}{1 + \xi}, \quad \xi = \gamma_1/\gamma_2$$

and τ_0 defined by (6).

It could be proved that in the case of overestimation of the lower spectral bound the iterative process (8) also converges, although with increased, but still less than 1, norm [2].

3   Grid Solution of 1D Steady-State Heat Transfer Equation

Let us consider 1D steady-state heat transfer equation in the form

$$-\partial^2 y/\partial x^2 = f(x), \quad y(a) = y_a, \; y(b) = y_b, \tag{10}$$

where $x \in [a, b]$ is a spatial coordinate, $y(x)$ is an unknown, $f(x)$ is the right-hand side; y_a, y_b are the known values on the boundaries of domain.

We consider a uniform grid with a spatial step $h = (b-a)/N$, where N is the number of cells. Approximating (10) in a standard way we obtain a system of algebraic equations of the form (1) with a positive definite matrix A. We apply the process (8), where for simplicity we assume $B = I$:

$$\mathbf{y}^{k+1} = \mathbf{y}^k + \tau_k \left(\mathbf{f} - A\mathbf{y}^k\right). \tag{11}$$

As initial approximation any vector is suited, as well as the zero one:

$$\mathbf{y}^0 = (y_1, \dots, y_N)^T = (0, \dots, 0)^T.$$

Application of (11) gives a sequential set of the vectors $\mathbf{y}^k$. Iterations should be stopped if e.g. $\|\mathbf{y}^k - \mathbf{y}^{k-1}\|_C \equiv \max_i |y_i^k - y_i^{k-1}| < \varepsilon$, where ε is the given accuracy of the solution.

For numerical treatment of (10) we set $a = 0$, $b = 1$, $y(0) = 0$, $y(1) = 1$ and zero source $f(x) = 0$. As τ_k, $k = 1, \dots, s$ the Chebyshev set of parameters should be taken. We stress that the number of iterations must be a power of two:

$$s = 2^n, n = 3, 4, 5, \dots.$$

For the stability of the algorithm it is required to order the Chebyshev parameters from 1 to s [3,6,7]. A FORTRAN procedure for such ordering is presented in [2]. After the values x_k (see (9)) are grouped into quadruples, as

$$x_{1,s} = -\cos\mu, \quad x_{2,s} = \cos\mu, \quad x_{3,s} = -\sin\mu, \quad x_{4,s} = \sin\mu.$$

In this the quantity

$$\Delta_k = \|y^k - y^{k-1}\|_C \equiv \max_i |y_i^k - y_i^{k-1}| \tag{12}$$

has the tendency to decrease with each iteration within the quadruple, but it increases in the transition from one quadruple to the next one:

$$\Delta_4 < \Delta_3 < \Delta_2 < \Delta_1, \quad \Delta_5 > \Delta_4.$$

Still the iteration error at the beginning of every quadruple is generally smaller than the one at the beginning of the previous quadruple, which makes the algorithm converging.

After choosing s and corresponding set $x_{k,s}$, $k = 1, \dots, s$, other required parameters are defined by (9). The spectrum bounds for the Laplacian on a uniform grid are known [3]:

$$\gamma_1 = \frac{4}{h^2} \sin^2 \frac{\pi h}{2l} > \frac{8}{l^2}, \quad \gamma_2 = \frac{4}{h^2} \cos^2 \frac{\pi h}{2l} < \frac{4}{h^2}, \quad l = b - a. \tag{13}$$

4   Algorithm of Competing Processes

The Gershgorin circle theorem does not give a correct estimation for the lower spectral bound for the strictly positive operator. At the same time, the Richardson iteration method requires $\gamma_1 > 0$. Thus we suggest an algorithm of competing processes for tuning to an unknown positive lower spectral bound on the basis of competing iterative processes with different γ_1:

$$\gamma_1^* = q\gamma_2, \quad \gamma_1^{**} = p\gamma_2,$$

where the upper bound γ_2 is defined by the Gershgorin theorem (see (13)); $q < 1$, $p < 1$ are some factors. We have chosen $q = 1/10$ and $p = q/5$. For the domain $[0,1]$ with the grid of $N = 100$ cells this choice gives $\gamma_1^* = 4000$, which far exceeds the analytical estimation $\gamma_1 = 8$. This process rapidly converges on initial iterations, but the rate of convergence in a given norm will fall quickly due to the fact that the large part of the lower part of the spectrum is excluded. At some point, the competing process $\gamma_1^{**} = 800$, which initially converges slower due to the optimization in more representative coverage of the entire width of the spectrum, overtakes the process γ_1^*, which is still convergent, but not optimal in sense of Richardson method particularly in the lower part of the spectrum.

Figure 1-A represents the relative errors (12) in two competing processes γ_1^* and γ_1^{**} with $s = 64$. Starting from the iteration $k = 27$ the process with the smaller value of the lower spectral bound γ_1^{**} becomes more rapid.

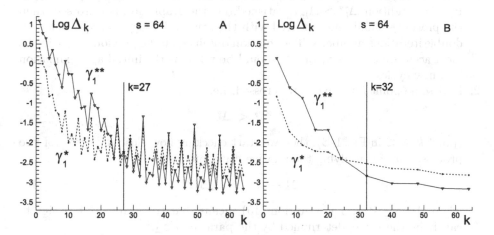

Fig. 1. Convergence of the competing processes γ_1^* and γ_1^{**} with $q = 1/10$, $p = q/5$, $s = 64$. Case A shows the relative errors in each iteration, case B shows monotone graphs, built on selected points.

However, it is convenient to compare monotone graphs, presented in Fig. 1-B. Since the error values are grouped by quadruples, to construct a monotonic curve we place starting points Δ_4^* and Δ_4^{**} for the both processes γ_1^* and γ_1^{**}, and subsequently select only the point at the end of each quadruple, in which the following conditions are satisfied in both processes:

$$\Delta_k^* < \Delta_{k-4}^*, \quad \Delta_k^{**} < \Delta_{k-4}^{**}. \tag{14}$$

If (14) is not fulfilled at the end of some quadruple, then such a point is skipped, and at the end of the next quadruple we check $\Delta_k^* < \Delta_{k-8}^*$ and $\Delta_k^{**} < \Delta_{k-8}^{**}$, etc.

Comparison of the monotone graphs (Fig. 1-B) gives a critical iteration number $k = 32$, where the process γ_1^{**} becomes more effective than γ_1^*. At the beginning the value γ_1^* presumably fits better the unknown true value $\gamma_1 > 0$ (in the sense of error estimated for all spectral components), but in the following the value γ_1^* is dropped as useless. Hereafter its role is taken by the γ_1^{**}. In this way the algorithm of processes switching works.

On the other hand, if the value $\gamma_1^* > 0$ is already smaller than the true unknown value $\gamma_1 > 0$, then the Richardson process γ_1^*, which suppresses spectral errors, is more effective than the process γ_1^{**} ($\gamma_1^{**} < \gamma_1^*$). To prevent changing of the processes in the latter case, it is necessary that in the described above graphs intersection (if it takes place) the gap between the graphs be large enough and exceeds some positive threshold. The changing of processes will also be allowed only after full s-iterative cycle.

From all has been said the algorithm consisting of checking several conditions follows.

1. If at all the points of the s-cycle, which are built the monotone graphs (Fig. 1-B), the condition $\Delta_k^{**} > \Delta_k^*$ is satisfied, i.e. the graphs are not crossed, then the processes should be restarted with the same values γ_1^* and γ_1^{**}, but with double iterations number s. The solution obtained on the previous stage in the most accurate process γ_1^* or γ_1^{**} should be taken as the initial approximation in a new cycle.

2. If at some point k the plots are crossed, i.e.

$$\Delta_k^{**} < \Delta_k^*$$

(point $k = 32$ in Fig. 1-B), then we start checking how close the errors of two processes are by applying the relative criterion

$$\Delta_k^* - \Delta_k^{**} > \varepsilon_\Delta \Delta_k^*, \tag{15}$$

where $0 \le \varepsilon_\Delta < 1$. The criterion (15) means that the graphs differs significantly by the value, determined by the parameter ε_Δ.

3. If (15) is not valid at all the points during s-cycle, then the processes should be restarted with $s \to 2s$. The value γ_1^* should be kept the same, the value γ_1^{**} should be decreased by a factor of 4 ($\gamma_1^{**} \to \gamma_1^{**}/4$). For details see [2].

4. If (15) is valid at some point during s-cycle, then the process γ_1^* should be stopped, only γ_1^{**} process should be running till the end of the cycle. After both processers are restarted with $s \to 2s$. Here γ_1^* takes the value of γ_1^{**} from the previous cycle; γ_1^{**} should be decreased by a factor of 4 ($\gamma_1^{**} \to \gamma_1^{**}/4$) in the same way as in the p. 3 of the algorithm.

If in (15) we set $\varepsilon_\Delta = 0$, then p. 3 of the algorithm is always skipped. At each iteration an absolute ε-criterion of convergence should be under constant check. If in any of two processes $\Delta_k^* < \varepsilon$ (or $\Delta_k^{**} < \varepsilon$), where ε is the given accuracy of the solution, then the problem is considered solved.

More details about an application example of such algorithm are presented in [2].

5   Conclusions

A possibility to develop a class of convergent iterative algorithms, tuning on the unknown lower bound of the problem spectrum without reference to the type of a computational domain or a grid, was justified and proved. The important fact that is the basis of such algorithms is the property of the Richardson scheme to converge in the case of overestimation of the lower spectral bound.

The algorithm is built on the basis of two competing simultaneous iterative processes, the effectiveness of which is analyzed in terms of convergence of iterative cycles. The proposed method is tested on a simple 1D steady-state heat transfer equation. In this case the lower boundary of the matrix spectrum is known exactly what gives an indication of how well the iterative algorithm approaches to the exact value. It was also demonstrated that the convergence rate increases significantly in spectral integral norm while approaching the exact lower bound of the spectrum.

Acknowledgements. This project has received funding from Russian Science Foundation (project no. 16-11-00100).

References

1. Strang, G.: Linear Algebra and its Applications, 4th edn. Brooks/Cole/Cengage, Pacific Grove (2006)
2. Popov, M.V., Poveschenko, Y., Gasilov, V.A., Koldoba, A.V., Poveschenko, T.S.: Application of the Richardson iteration method with the Chebyshev parameters in case of the absence of information about the lower spectral bound. Math. Models Comput. Simul. **9**(4) (2017)
3. Samarskii, A.A.: The Theory of Difference Schemes. Marcel Dekker Inc., New York City (2001)
4. Goncharov, V.L.: The Theory of Interpolation and Approximation of Functions. GosTekhIzdat (1954). (in Russian)
5. Kantorovich, L.V., Akilov, G.P.: Functional Analysis in Normed Spaces. Pergamon Press, Oxford (1964)
6. Samarskii, A.A., Nikolaev, E.S.: Numerical Methods for Grid Equations. Birkhauser Verlag AG, Basel (1988)
7. Lebedev, V.I., Finogenov, S.A.: Solution of the parameter ordering problem in Chebyshev iterative methods. USSR Comput. Math. Math. Phys. **13**, 21–41 (1974)

Unstable Flow Modes of the Non-isothermal Liquid Film

Ludmila A. Prokudina$^{(\boxtimes)}$

The South Ural State University, Chelyabinsk, Russia
prokudina-la@mail.ru

Abstract. In industrial gas-liquid systems, the processes of heat and mass transfer occur under conditions of instability of the gas-liquid interface. Volatility is associated with various nonlinear phenomena in distributed open systems of different nature. In open systems there are energy, matter, information exchanges with external environment. Obviously, studying non-linear phenomena in such systems has a significant impact on important features of realized chemical processes. Examples include intensive convective motion of the liquid near the gas-liquid interface, contributing to a more rapid renewal of the surface rupture of the liquid film in plastic film and packed apparatuses.

Keywords: Liquid film · Unstability · Wave packet

1 Introduction

The importance of the study of flowing thin layers of viscous liquids (liquid films) is related to their wide distribution in various heat and mass transfer apparatuses. The investigation of thin layers of viscous fluids, that combines thin and large contacted surfaces, is connected with the implementation of the trends in heat and mass transfer apparatuses (i.e. thermal power, chemical, food, pharmaceutical industry [1,2]). The study of the unstable flows of the liquid films was started by Kapitsa and Kapitsa [3].

Numerous physical and chemical factors influence on flow modes of liquid films [4]. Temperature and concentration gradients cause surface tention heterogeneity and lead to the appearance of surface tension gradients resulting in the thermo-capillary forces on the interphase surface.

Technological processes based on the liquid film flows are complex due to their non-stationarity and non-linearity. As result, development of nonlinear mathematical models is required. The paper presents a nonlinear mathematical model as a nonlinear parabolic equation for a wave packet amplitude envelope.

The work's novelty is studying the influence of temperature gradients on the instability and on the nonlinear development and interaction of the perturbation wave packet initiated on the free surface of the liquid film.

© Springer International Publishing AG 2017
I. Dimov et al. (Eds.): NAA 2016, LNCS 10187, pp. 576–582, 2017.
DOI: 10.1007/978-3-319-57099-0_65

2   Reduction of the Problem

As an open system, we consider a model of liquid film (a thin layer of viscous fluid) which flowing modes in heat-mass transfer apparatuses are affected by various physical and chemical factors [1,2]. The mathematical model of the liquid film, flowing down on the solid impermeable base under action of gravity, is the system of the Navier-Stokes equations and the continuity equation with boundary conditions [4]:

$$
\begin{cases}
\dfrac{\partial u_+}{\partial t_+} + u_+ \dfrac{\partial u_+}{\partial x_+} + v_+ \dfrac{\partial u_+}{\partial y_+} = -\dfrac{\partial P_+}{\partial x_+} + Fr_x + \dfrac{1}{Re} \cdot \left(\dfrac{\partial^2 u_+}{\partial x_+^2} + \dfrac{\partial^2 u_+}{\partial y_+^2} \right), \\[2mm]
\dfrac{\partial v_+}{\partial t_+} + u_+ \dfrac{\partial v_+}{\partial x_+} + v_+ \dfrac{\partial v_+}{\partial y_+} = -\dfrac{\partial P_+}{\partial y_+} + Fr_y + \dfrac{1}{Re} \cdot \left(\dfrac{\partial^2 v_+}{\partial x_+^2} + \dfrac{\partial^2 v_+}{\partial y_+^2} \right), \quad (1) \\[2mm]
\dfrac{\partial u_+}{\partial x_+} + \dfrac{\partial v_+}{\partial y_+} = 0,
\end{cases}
$$

$$
y = 0: \quad u = 0, \quad v = V_0; \tag{2}
$$

$y = \delta$:

$$
\frac{1}{Re} \left(2 \frac{\partial u_+}{\partial x_+} \frac{\partial \delta_+}{\partial x_+} - 2 \frac{\partial v_+}{\partial y_+} \frac{\partial \delta_+}{\partial x_+} - \frac{\partial v_+}{\partial x_+} - \frac{\partial u_+}{\partial y_+} \right) + M \frac{\partial \delta_+}{\partial x_+} + N \frac{\partial^2 u_+}{\partial x_+^2} = 0, \tag{3}
$$

$$
P_+ = \frac{2}{Re} \left(\frac{\partial v_+}{\partial y_+} - \frac{\partial \delta_+}{\partial x_+} \left(\frac{\partial u_+}{\partial y_+} + \frac{\partial v_+}{\partial x_+} \right) \right) + \sigma_+ \frac{\partial^2 \delta_+}{\partial x_+^2} + P_0, \tag{4}
$$

$$
\frac{\partial \delta_+}{\partial t_+} = v_+ - u_+ \frac{\partial \delta_+}{\partial x_+}, \tag{5}
$$

Here u and v are projections of the velocity on the coordinate axes OX and OY respectively, Re is the Reynolds number, Fr_x and Fr_y are projections of the Froude number, M is the Marangoni parameter, σ is the surface tension, N is the surface viscosity parameter. Index $+$ marks dimentionless quantities.

The system (1) with its boundary conditions (2)–(5) is reduced to a nonlinear parabolic equation [4,5]. Its one-dimentional form is written below:

$$
\frac{\partial A}{\partial t} + i \frac{\partial \omega_i}{\partial k_x} \frac{\partial A}{\partial x} - \frac{i}{2} \left(\frac{\partial^2 \omega_r}{\partial k_x^2} + i \frac{\partial^2 \omega_i}{\partial k_x^2} \right) \frac{\partial^2 A}{\partial x^2} = \gamma A - (\beta_1 + i\beta_2) |A|^2 A. \tag{6}
$$

Here A is the complex amplitude envelop of the wave packet, t is time, x is spatial coordinate, ω_r is the frequency, ω_i is the increment.

$$\omega_r = \frac{Y - X \cdot Z}{1 + Z^2}, \qquad \omega_i = X + \omega_r \cdot Z,$$

where $X = a_1 k_x^4 - a_6 k_z^2$, $Y = a_4 k_x^3 - a_{11} k_x$, $Z = a_7 k_x$, $a_1 = -\frac{1}{3} Re\sigma$, $a_4 = -\frac{1}{2} Re^2 Fr_x N$, $a_6 = -\frac{1}{3} ReFr_y - \frac{1}{2} ReM + \frac{3}{40} Re^3 Fr_x^2$, $a_7 = \frac{5}{24} Re^2 Fr$, $a_{11} = -ReFr_x^2$.

The coefficients of the Eq. (6) are expressed through the frequency ω_r, the increment ω_i and their first and second-order derivatives and describe the next: γ is the increment from the non-linearity, β_1 is the damp of non-linear perturbations, β_2 is the nonlinear dependence of phase velocity on the amplitude A.

3   Numerical Method

Let us consider the nonlinear parabolic equation (6). We denote $d = \frac{\partial \omega_i}{\partial k_x}$, $\alpha_1 = \frac{1}{2} \frac{\partial^2 \omega_r}{\partial k_x^2}$, $\alpha_2 = \frac{1}{2} \frac{\partial \omega_i}{\partial k_x}$ and rewrite the Eq. (6) as follows

$$\frac{\partial A}{\partial t} + id\frac{\partial A}{\partial x} - (\alpha_1 + i\alpha_2)\frac{\partial^2 A}{\partial x^2} = \gamma A - (\beta_1 + i\beta_2)|A|^2 A. \tag{7}$$

With substitution $A = U(x,t) + iV(x,t)$, the Eq. (7) is led to the system of differential equations:

$$\begin{cases} \dfrac{\partial U}{\partial t} = d\dfrac{\partial V}{\partial x} + \alpha_1 \dfrac{\partial^2 U}{\partial x^2} - \alpha_2 \dfrac{\partial^2 V}{\partial x^2} + \gamma U - \left(U^2 + V^2\right)(\beta_1 U - \beta_2 V), \\[3mm] \dfrac{\partial V}{\partial t} = -d\dfrac{\partial U}{\partial x} + \alpha_1 \dfrac{\partial^2 V}{\partial x^2} + \alpha_2 \dfrac{\partial^2 U}{\partial x^2} + \gamma V - \left(U^2 + V^2\right)(\beta_1 V + \beta_2 U). \end{cases} \tag{8}$$

In (8), by finite differences we replace derivatives with respect to spacial coordinate

$$U_j' = \frac{U_{j+1} - U_{j-1}}{2\Delta x}, \qquad U_j'' = \frac{U_{j+1} - 2U_j + U_{j-1}}{\Delta x^2},$$

$$V_j' = \frac{V_{j+1} - V_{j-1}}{2\Delta x}, \qquad V_j'' = \frac{V_{j+1} - 2V_j + V_{j-1}}{\Delta x^2},$$

and obtain the system of the ordinary differential equations:

$$\begin{cases} \dfrac{dU_j}{dt} = d \cdot V_j' + \alpha_1 U_j'' - \alpha_2 V_j'' + \gamma U_j - \left(U_j^2 + V_j^2\right)(\beta_1 U_j - \beta_2 V_j), \\[3mm] \dfrac{dV_j}{dt} = -d \cdot U_j' + \alpha_1 V_j'' + \alpha_2 U_j'' + \gamma V_j - \left(U_j^2 + V_j^2\right)(\beta_1 V_j + \beta_2 U_j). \end{cases} \tag{9}$$

4   Computational Experiments

The Wave Characteristics. The unstable flow modes of liquid films for the Reynolds numbers $Re \leqslant 20$ are investigated. The criterion of liquid film instability is a positive increment $\omega_i > 0$. We consider the influence of temperature gradients (the Marangoni parameter $M > 0$) and insoluble surface-active substances (oils, fats) on wave characteristics of the liquid film and the coefficients of the nonlinear parabolic equation. For example on Figs. 1, 2 and 3 we show how wave characteristics of vertical water film change for the Reynolds number $Re = 5$.

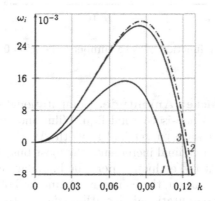

Fig. 1. Increment ω_i as a function of wave number k. $1 - M = 0$; $2 - M = 1$; $3 - M = 1$, $N = 1$

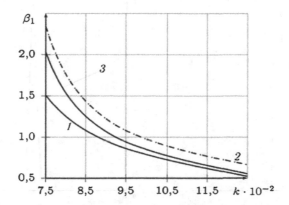

Fig. 2. Coefficient β_1 as a function of wave number k. $1 - M = 0$; $2 - M = 1$; $3 - M = 1$, $N = 1$

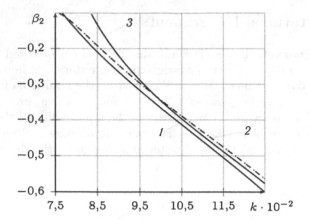

Fig. 3. Coefficient β_2 as a function of wave number k. $1-M=0$; $2-M=1$; $3-M=1$, $N=1$

Envelop of Wave Packet Amplitude. Computational experiment has been conducted for a vertical non-isothermal liquid film and the Reynolds numbers $Re \leq 20$. A considered wave packet had a width $\Delta k = o(\varepsilon)$ and was centered on the harmonic of the maximal increment (the most busy excited wave packet) for length of $L = 25$. At the endpoints periodic boundary conditions were put.

For the system (9) solution, the Runge-Kutta method of the 4th order, implemented as a separate computational algorithm adapted to solve a specific problem, is used. Modelling of processes in the system under study has been made with an error not exceeding 0.1%. Initial amplitude perturbation is presented on Fig. 4.

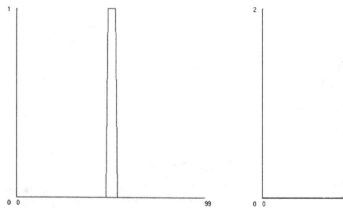
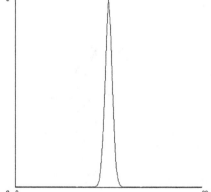

Fig. 4. The amplitude $A(x,0)$ **Fig. 5.** The amplitude $A(x,0.3)$

In the subsequent moments of time growth of the Central harmonic amplitude begins in the system (Fig. 5). Since the time $t = 0.43$, the Central harmonic

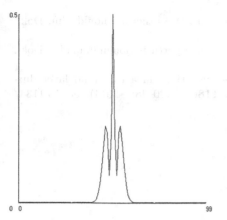

Fig. 6. The amplitude $A(x, 0.75)$

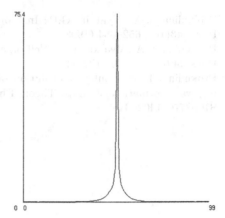

Fig. 7. The amplitude $A(x, 4.035)$

amplitude ceases growing and starts decreasing. This process is accompanied by increasing in amplitudes of lateral harmonics of the wave packet (Fig. 6).

After side harmonics gain sufficient energy, energy transfer to the Central one takes back. As a result, since time $t = 1.81$, the Central harmonic amplitude rises rapidly (Fig. 7), exceeding many times the Central harmonic amplitude of the initial perturbation.

There is a narrow spectrum of wave packets, which leads to a monocromatic waves. In Kapitsa's experiments [3] monochromatic wave modes with wave number from the vicinity of the maximal harmonic increment were realized.

5   Conclusions

1. We have presented the mathematical model of a non-isothermal liquid film for the Reynolds numbers $Re \leqslant 20$.
2. We have presented the nonlinear parabolic equation for the complex amplitude of the wave packet developing on surface of a non-isothermal liquid film.
3. We have calculated the wave characteristics and the coefficients of the nonlinear parabolic equation and shown the impact of temperature gradients.
4. We have calculated the forms for the amplitude of the wave packet of the free surface of a liquid film under inhomogeneous surface tension.

References

1. Filippov, A.G., Saltanov, G.A., Kukushkin, A.N.: Fluid Flow and Heat and Mass Transfer in the Presence of Surfactants. Energoizdat, Moscow (1988). (in Russian)
2. Kholpanov, L.P., Shkadov, V.Y.: Hydrodynamic and Heat and Mass Transfer with Free Surface. Nauka, Moscow (1990). (in Russian)
3. Kapitsa, P.L., Kapitsa, S.P.: Wave flow of thin layers of viscous fluid. Zh. Eksper. Teor. Fiz. **19**, 105–120 (1949)

4. Prokudina, L.A., Vyatkin, G.P.: Instability of a non-isothermal liquid film. Dokl. Phys. **43**(10), 652–654 (1998)
5. Prokudina, L.A., Vyatkin, G.P.: Self-organization of perturbation in fluid film. Dokl. Phys. **56**(8), 444–447 (2011)
6. Prokudina, L.A.: Nonlinear evolution of perturbations in a thin fluid layer during wave formation. J. Exp. Theor. Phys. **118**(3), 480–488 (2014). doi:10.1134/S1063776114020046

Discrete Modeling of Oscillatory Processes in a Blocky Medium

Vladimir M. Sadovskii and Evgenii P. Chentsov[✉]

Institute of Computational Modeling SB RAS,
Akademgorodok 50/44, 660036 Krasnoyarsk, Russia
sadov@icm.krasn.ru, chencov.evg@gmail.com

Abstract. Resonance phenomena in inhomogeneous layered and blocky media are investigated on the basis of the discrete model of transverse-rotational oscillations in linear monoatomic chain. To analyze the behavior of the system in near-resonance region, spectral portraits are built. It allows to visualize resonant frequencies and to analyze alteration of an amplitude vector. Using mentioned model, a set of resonant frequencies are found. It is shown that there exists one specific frequency, which does not depend on the number of particles in the chain. In passing to the limit as the chain length tends to infinity, this frequency, related to rotational particles vibration, is the unique resonant frequency. The discussed property is associated with the previously studied property of the Cosserat continuum, where the resonant frequency of rotational motion exists, which does not depend on the specimen size and on the conditions of loading at its boundary.

Keywords: Microstructure · Elasticity · Resonance · Discrete chain · Cosserat continuum · Blocky medium · Rotational motion

1 Introduction

To analyze wave motion in a structurally inhomogeneous deformable medium, discrete and continuous mathematical models are applied. In the simplest model of a multilayer medium with compliant interlayers a linear particle chain is considered. Particles in the chain are consequently connected with each other by elastic springs. In a monoatomic chain the masses of all particles and spring rates are equal; such approximation is possible if interlayers are thin enough, therefore their masses may be neglected. In a diatomic chain the particles with two different masses are introduced, so one of them is considered as the mass of a block and the other one – as an interlayer's mass. Wave processes in chains caused by the external periodic perturbations (in particular, resonances) were investigated in linear [1–3] and nonlinear [4–6] approximations. Considering viscosity forces, it was proved that the resonance amplitude becomes finite [7], and the spectrum of resonant frequencies rebuilds because of connection defects [8,9]. 2D waves in lattices (rectangular and triangular) were studied in [10–12].

© Springer International Publishing AG 2017
I. Dimov et al. (Eds.): NAA 2016, LNCS 10187, pp. 583–590, 2017.
DOI: 10.1007/978-3-319-57099-0_66

The simplest continuous model of a deformable medium with microstructure is formulated in terms of 1D wave equation that can be acquired from the discrete chain model by the passage to the limit as a number of particles tends to infinity. In such case it is assumed that the velocities of elastic waves and the materials densities are equal in blocks and interlayers. Resonant solutions of the wave equation for different types of nonlinearity was deduced in [13–15]. In [16,17] 1D model equations that describe resonance were obtained.

Common matters of oscillations theory and resonance phenomena were represented in [18,19]. Many authors analyzed various applied problems related to resonant excitation of mechanical and physical systems. Resonances in complex systems, such as quantum dots and photon crystals, were examined using discrete models in [8]. In [20] a way to apply resonant method of the ice cover destruction using flexure-gravitational superficial waves was proposed.

In this paper the study of resonances is based on a discrete modeling, resonances are caused by transverse and rotational oscillations in layered and blocky media.

2 General Case

Discrete modeling of the elastic wave propagation with an infinitely small amplitude leads to the system of ordinary differential equations:

$$A\ddot{U} + BU = F. \tag{1}$$

Here U is a vector of generalized coordinates, F is a vector of external forces, A is a symmetric positive definite matrix of generalized masses, B is a positive semidefinite stiffness matrix. Dots above symbols denote time derivatives.

For the system (1) the following equation is satisfied: $\dot{E} = F\dot{U}$, which characterizes the change in full energy $E = (\dot{U}A\dot{U} + UBU)/2$. If vector F depends on time periodically: $F = \hat{F}e^{i\omega t}$ with the frequency ω, then vector $U = \hat{U}e^{i\omega t}$ is periodic, too, and $(B - \omega^2 A)\hat{U} = \hat{F}$. Resonant frequencies (eigenfrequencies) are an exception, so periodicity condition for U is not performed. Squared frequencies $\lambda = \omega^2$ are the roots of characteristic equation: $\det(B - \lambda A) = 0$. Due to the fact that matrices A and B are symmetric and that A is sign-definite, the roots of the equation are real, and their number equals to the system order. The roots correspond to linearly independent vectors, each of which is the solution of the degenerate homogeneous system: $(B - \lambda A)Z = 0$.

For non-resonant frequencies the equation $\hat{U} = -R(\omega^2)\,A^{-1}\hat{F}$ is fulfilled, where $R(\lambda) = (\lambda E - A^{-1}B)^{-1}$ is a resolvent and E is the unit matrix. To analyze the behaviour of system in a neighborhood of resonant frequencies, the matrix spectral portrait $A^{-1}B$ is constructed. According to one of the technologies [21], on the first step a set of values

$$s(\lambda) = \frac{1}{||A^{-1}B|| \cdot ||R(\lambda)||} \tag{2}$$

is calculated. Represented spectral matrix norm is used in the nodes of some defined lattice on the complex plane. On the second step contours for $s(\lambda)$ are built; as a result, a "portrait" of contours is obtained. Spectral portrait allows to segregate eigenfrequencies visually and to analyze the amplitude vector $\ddot{U}$ alteration in their neighborhood. Rapid reduction of $s(\lambda)$ while approaching resonant frequency indicates that the system is strongly sensitive to perturbations. Such sensitivity corresponds to general perceptions of resonance.

If the interlayers viscosity is taken into account, the system (1) is supplemented with addends that depend on $\dot{U}$:

$$A\ddot{U} + M\dot{U} + BU = F. \tag{3}$$

Here M is a positive semidefinite (generally speaking, non-symmetric) matrix with small coefficients. Dissipative addend $\dot{U}M\dot{U} \geqslant 0$ appears in the left side of this equation. The eigenfrequencies determination problem leads to an algebraic equation:

$$\det(B + i\omega M - \omega^2 A) = 0.$$

This equation indicates that spectral portraits should be represented on a complex plane. For nonlinear models the coefficients of the system (3) and vector located on the right side of equation can depend on U and $\dot{U}$. In such case authors believe that there is no universal algorithm to analyze resonant processes.

3   Transverse and Rotational Oscillations

Let us consider a linear discrete chain of n particles, each one of them is of mass m. Particles are linked to each other by springs of stiffness k. Distance between material points equals h, so the total chain length is $l = (n-1)h$. Every particle is affected by transverse forces Q_j and angular momentum R_j; as a result, they rotate by a small angle φ_j and move in transverse direction for w_j (Fig. 1). It is assumed that chain boundaries are free to rotate but unable to move in transverse direction:

$$w_1 + w_0 = 0, \quad w_{n+1} + w_n = 0, \quad \varphi_1 - \varphi_0 = 0, \quad \varphi_{n+1} - \varphi_n = 0. \tag{4}$$

Here particles with indices $j = 0$ and $j = n + 1$ are added to find resonant frequencies in closed form easily.

The strain state of a chain is described by parameters

$$\Lambda_{j-1/2} = \frac{w_j - w_{j-1}}{h} - \frac{\varphi_j + \varphi_{j-1}}{2}, \quad M_{j-1/2} = \frac{\varphi_j - \varphi_{j-1}}{h},$$

where $h = l/n, j = 1, 2, \ldots, n{+}1$. First attribute characterizes shear deformation, while the second one describes curvature. Such selection of kinematic characteristics is explained in the following way: both of them equal zero when motion of particles is absolutely rigid, i.e. when the whole chain with free boundaries rotates for an arbitrary infinitely small angle.

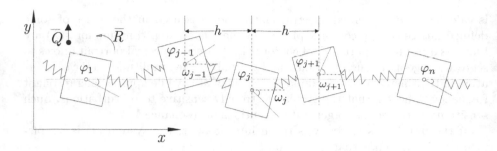

Fig. 1. The scheme of a chain with rotating particles

Elastic stiffness of the spring in a chain is modeled using the equations: $\tau_{j-1/2} = a\,\Lambda_{j-1/2}$, $\mu_{j-1/2} = b\,M_{j-1/2}$, that bind shear forces $\tau_{j-1/2}$ and angular moments $\mu_{j-1/2}$ to deformations and curvatures. Kinetic and potential energies in mentioned mechanical system are calculated using the equations:

$$T = \frac{m}{2}\sum_{j=0}^{n+1} \dot{w}_j^2 + \frac{J}{2}\sum_{j=0}^{n+1}\dot{\varphi}_j^2, \quad \Pi = \frac{a}{2}\sum_{j=1}^{n+1}\Lambda_{j-1/2}^2 + \frac{b}{2}\sum_{j=1}^{n+1}M_{j-1/2}^2,$$

where J is the moment of inertia of particles. Lagrange equations take the next form:

$$m\,\ddot{w}_j = a\,\frac{w_{j+1} - 2\,w_j + w_{j-1}}{h^2} - a\,\frac{\varphi_{j+1} - \varphi_{j-1}}{2\,h} + Q_j,$$
$$J\ddot{\varphi}_j = a\,\frac{w_{j+1} - w_{j-1}}{2h} - a\,\frac{\varphi_{j+1} + 2\varphi_j + \varphi_{j-1}}{4} + b\,\frac{\varphi_{j+1} - 2\varphi_j + \varphi_{j-1}}{h^2} + R_j. \tag{5}$$

Equations (5) can be rewritten as the system (1) with symmetric and definite matrices. One can substitute terms in homogeneous equations:

$$w_j = \hat{w}\,\mathrm{e}^{i\omega t}\sin\left(j - \frac{1}{2}\right)\alpha, \quad \varphi_j = \hat{\varphi}\,\mathrm{e}^{i\omega t}\cos\left(j - \frac{1}{2}\right)\alpha, \quad \alpha = \frac{\pi s}{n+1}.$$

Here $s = 1, 2, \ldots, n$. Mentioned terms automatically satisfy the boundary conditions (4), so the substitution leads to the system of equations for the amplitudes $\hat{w}$ and $\hat{\varphi}$.

Condition whereby determinant of the system equals zero (which is equivalent to the existence of nontrivial solutions) allows to acquire biquadratic equation for finding eigenfrequencies of the chain:

$$m\,J\,\omega^4 - C\,\omega^2 + 16\,\frac{ab}{h^4}\sin^4\frac{\alpha}{2} = 0, \quad C = ma\cos^2\frac{\alpha}{2} + 4\,\frac{Ja + mb}{h^2}\sin^2\frac{\alpha}{2}.$$

Hence,

$$\omega^2 = \frac{C \pm \sqrt{D}}{2\,m\,J},$$
$$D = (ma)^2\cos^4\frac{\alpha}{2} + 2\,ma\,\frac{Ja + mb}{h^2}\sin^2\alpha + 16\,\frac{(Ja - mb)^2}{h^4}\sin^4\frac{\alpha}{2}. \tag{6}$$

Analysis of formulas (6) shows that eigenfrequencies are always real and vary for different $s = 1, 2, \ldots, n$ in the case of essential limitations on the chain mechanical parameters (m, J, a, $b > 0$). Moreover, a specific eigenfrequency ω_0 exudes as the chain length increases. One can detect it on Fig. 2, where squares of eigenfrequencies are represented for the chain of 9 elements with various length $0.5\,\text{m} < l < 2.5\,\text{m}$. Mass of the elements $m = 112.5 \cdot 10^{-3}\,\text{kg}$, moment of inertia $J = 46.68 \cdot 10^{-6}\,\text{kg} \cdot \text{m}^2$ and elasticity parameters $a = 1300\,\text{N} \cdot \text{m}$, $b = 3.3\,\text{N} \cdot \text{m}^3$ are calculated for a blocky medium that is made of ice cubes with edges of $h_0 = 0.05\,\text{m}$. Blocks interact through compliant interlayers that have different thickness.

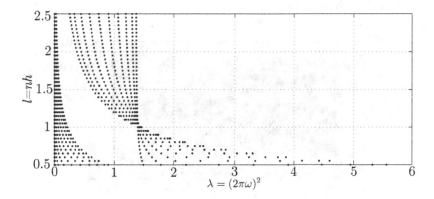

Fig. 2. Squares of eigenfrequencies for the chains of different length

If number s is fixed, then for the current length l confined limits exist and have the following form:

$$\lim_{h \to 0} \frac{4}{h^2} \sin^2 \frac{\alpha}{2} = \nu, \quad \lim_{h \to 0} \frac{1}{h^2} \sin^2 \alpha = \nu, \quad \nu = \frac{\pi^2 s^2}{l^2}.$$

These limits simplify coefficients of the formula (6) for the square frequency of an infinite chain:

$$C = ma + (Ja + mb)\nu, \quad D = (ma)^2 + 2ma(Ja + mb)\nu + (Ja - mb)^2 \nu^2.$$

Tending $l \to \infty$ (i.e. $\nu \to 0$) one may determine that $\omega_0 = \sqrt{a/J}$. This is the only resonant frequency of an infinite chain (that has infinite length), related to the particles rotations.

Value ω_0 may be determined alternatively by considering the model of a chain consisting of three elements, i.e. $n = 1$. Substituting the boundary conditions (4) and the equality $Q_1 = R_1 = 0$, the equations (5) transform to the next form:

$$m\ddot{w}_1 = -\frac{4a}{h^2} w_1, \quad J\ddot{\varphi}_1 = -a\varphi_1.$$

First equation describes transverse oscillations, while the second one – independent rotational vibrations of an element at a frequency ω_0.

Figure 3 represents spectral portrait for the discrete chain of 9 particles with the chain length $l = 0.5\,\mathrm{m}$. Frequencies are divided by two "groups", which is obvious because of (6). According to the spot sizes on the figure, one may say that oscillations amplitude increases while approaching ω_0 in the same way as approaching other eigenfrequencies. Figure 4 represents spectral portrait for the same chain of length $l = 1\,\mathrm{m}$. Both large spots correspond to the frequencies from two "groups", including specific frequency (right spot); spot size shows attainability of the frequency. Further increase of the chain length (Fig. 5) to $l = 1.5\,\mathrm{m}$ shows that frequency "groups" diverge and the only motionless frequency is ω_0; its spot has comparable size.

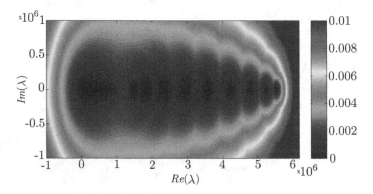

Fig. 3. Spectral portrait of the matrix ($n = 9$, $l = 0.5\,\mathrm{m}$)

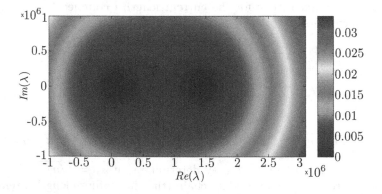

Fig. 4. Spectral portrait of the matrix ($n = 9$, $l = 1\,\mathrm{m}$)

In the passage to the limit as the chain density tends to infinity with fixed length l, Eqs. (5) transform to the 1D Cosserat differential continuum equations:

$$\rho_0\,\ddot{w} = a_0\,(w_{xx} - \varphi_x) + q(x,t), \quad J_0\,\ddot{\varphi} = a_0\,(w_x - \varphi) + b_0\,\varphi_{xx} + r(x,t) \quad (7)$$

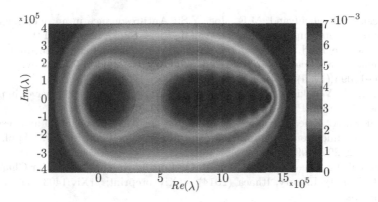

Fig. 5. Spectral portrait of the matrix $(n = 9,\ l = 1.5)$

with boundary conditions $w(0) = w(l) = 0$, $\varphi_x(0) = \varphi_x(l) = 0$. Equation coefficients are recalculated via mechanical parameters of the discrete model. It is considered that the conservation of full mass and joint inertia of the system is performed:

$$\rho_0 = \lim_{n \to \infty} \frac{m}{h}, \quad J_0 = \lim_{n \to \infty} \frac{J}{h}, \quad a_0 = \lim_{n \to \infty} \frac{a}{h}, \quad b_0 = \lim_{n \to \infty} \frac{b}{h}.$$

The continuum eigenfrequencies are calculated using formulas that can be acquired by the passage to the limit in the formulas for calculating eigenfrequencies of the limited discrete chain. Resonant properties of the Cosserat continuum based on the stress-strain state model were studied in [22–24]. Authors established that there is a special resonant frequency in a couple-stress medium, that is related to the particle rotations and does not depend on the size of domain and on the type of boundary conditions.

Under transverse and rotational oscillations of a discrete chain there is a set of resonant frequencies, that depend on the number of particles and on the chain length. At the same time there is a specific rotational eigenfrequency, that is determined only by means of mechanical parameters of the system.

Acknowledgements. This work was supported by the Russian Foundation for Basic Research (grant no. 14-01-00130), Krasnoyarsk Region Science and Technology Support Fund and joint grant of RFBR and Government of Krasnoyarsk Territory (grant no. 16-41-240415).

References

1. Kosevich, A.M., Kovalev, A.S.: Self-localization of vibrations in an one-dimensional anharmonic chain. Sov. Phys. J. Exp. Theor. Phys. **40**(5), 891–896 (1975)
2. Grundmann, M.: The Physics of Semiconductors: An Introduction Including Nanophysics and Applications. Springer, Heidelberg (2010)

3. Belbasi, S., Foulaadvand, M.E., Joe, Y.S.: Anti-resonance in an one-dimensional chain of driven coupled oscillators. Am. J. Phys. **82**(1), 32–38 (2014)
4. Coste, J., Peyraud, J.: Stationary waves in a nonlinear periodic medium: strong resonances and localized structures. I. The discrete model. Phys. Rev. B **39**(18), 13086–13095 (1989)
5. Filip, A.-M., Venakides, S.: Existence and modulation of traveling waves in particle chains. Commun. Pure Appl. Math. **52**(6), 693–735 (1999)
6. Georgieva, A., Kriecherbauer, T., Venakides, S.: Wave propagation and resonance in an one-dimensional nonlinear discrete periodic medium. SIAM J. Appl. Math. **60**(1), 272–294 (1999)
7. Bonanomi, L., Theocharis, G., Daraio, C.: Locally Resonant Granular Chain. Cornell University Library, Ithaca (2014). arXiv preprint: arXiv:1403.1052v1 [cond-mat.mtrl-sci]
8. Cun-Xi, Z., Xiu-Huan, D., Rui, W., Yun-Qing, Z., Ling-Min, K.: Fano resonance and wave transmission through a chain structure with an isolated ring composed of defects. Chin. Phys. B **21**(3), 034202 (2012)
9. Man, Y., Boechler, N., Theocharis, G., Kevrekidis, P.G., Darairo, C.: Defect modes in one-dimensional granular crystals. Phys. Rev. E **85**, 037601 (2012)
10. Mishuris, G.S., Movchan, A.B., Slepyan, L.I.: Waves and fracture in an inhomogeneous lattice structure. Waves Random Complex Media **17**(4), 409–428 (2007)
11. Feckan, M., Rothos, V.M.: Travelling waves in hamiltonian systems on 2D lattices with nearest neighbor interactions. Nonlinearity **20**(2), 319–341 (2007)
12. Ayzenberg-Stepanenko, M.V., Slepyan, L.I.: Resonant-frequency primitive waveforms and star waves in lattices. J. Sound Vib. **313**(35), 812–821 (2008)
13. Jeffrey, A., Taniuti, T.: Nonlinear Wave Propagation with Applications to Physics and Magnetohydrodynamics. Academic, New York (1964)
14. Collins, W.D.: Forced oscillations of systems governed by one-dimensional nonlinear wave equations. Q. J. Mech. Appl. Math. **24**(2), 129–153 (1971)
15. Manevich, A.I., Manevich, L.I.: The Mechanics of Nonlinear Systems with Internal Resonances. Imperial College Press, London (2005)
16. Bretherton, F.P.: Resonant interactions between waves. The case of discrete oscillations. J. Fluid Mech. **20**(3), 457–479 (1964)
17. Shipman, S.P., Venakides, S.: An exactly solvable model for nonlinear resonant scattering. Nonlinearity **25**(9), 2473–2501 (2012)
18. Bogolyubov, N.N., Mitropolsky, Y.A.: Asymptotic Methods in the Theory of Nonlinear Oscillations. Gordon and Breach, New York (1968)
19. Karlov, N.V., Kirichenko, N.A.: Oscillations, Waves, and Structures. Fizmatlit, Moscow (2003). (in Russian)
20. Kozin, V.M.: Resonant Method of Breaking Ice Cover: Inventions and Experiments. Akad. Estestvoz, Moscow (2007). (in Russian)
21. Godunov, S.K.: Modern Aspects of Linear Algebra. American Mathematical Society, Providence (1998)
22. Sadovskaya, O.V., Sadovskii, V.M.: Analysis of rotational motion of material microstructure particles by equations of the Cosserat elasticity theory. Acoust. Phys. **56**(6), 942–950 (2010)
23. Sadovskii, V.M., Sadovskaya, O.V., Varygina, M.P.: Numerical solution of dynamic problems in couple-stressed continuum on multiprocessor computer systems. Int. J. Numer. Anal. Mod. Ser. B **2**(2–3), 215–230 (2011)
24. Sadovskaya, O., Sadovskii, V.: Mathematical Modeling in Mechanics of Granular Materials. Advanced Structured Materials, vol. 21. Springer, Heidelberg (2012)

An Overlapping Domain Decomposition Method for the Helmholtz Exterior Problem

Alexander Savchenko$^{(\boxtimes)}$ and Artem Petukhov

Institute of Computational Mathematics and Mathematical Geophysics,
pr.Lavrentieva 6, Novosibirsk 630090, Russia
savch@ommfao1.sscc.ru, petukhov@lapasrv.sscc.ru
http://www.icmmg.nsc.ru

Abstract. The new method for solving the three-dimensional boundary value problem for the Helmholtz equation, based on decomposition of an exterior domain, is proposed. The approach proposed is based on using of Swartz alternating method with solving sequentially interior and exterior boundary value problems in overlapping subdomains with iterated interface conditions on their adjacent borders. The sufficient conditions of the method convergence in the case of a negative coefficient in the Helmholtz equation are found. The approach in question is implemented to solve a problem with a complex configuration of boundaries with using the finite volume method for solving the interior boundary value problems and Greens formula for solving the exterior boundary value problems. The rate of convergence of iterations and achievable accuracy of calculations are illustrated on a series of numerical experiments.

Keywords: Helmholtz equation · Exterior boundary value problem · Domain decomposition · Green formula

1 Introduction

In this paper we study the numerical solution to the Dirichlet problem for the Helmholtz equation in an exterior domain. The approach proposed consists in finding approximate values for a sought-for function and its normal derivative on the surface of an artificial sphere that includes the original boundary. Then one can find a sought-for function at any point of the domain using Green formula. The introduction of an artificial boundary is due to the fact that the original boundary can be of an arbitrary shape. Then the calculation of a normal derivative on it which can be in the general case a discontinuous function, is a complicated calculation problem. To solve the problem we apply the alternating Schwartz method in which we consistently recalculate the boundary conditions on an artificial sphere ∂S and on an outer artificial boundary $\partial \Omega$. The solution of the interior boundary value problem in the domain bounded by the outer boundary $\partial \Omega$ and the original boundary ∂D is sought for only at the artificial surface ∂S. To solve the exterior boundary value problem in the domain outside

© Springer International Publishing AG 2017
I. Dimov et al. (Eds.): NAA 2016, LNCS 10187, pp. 591–598, 2017.
DOI: 10.1007/978-3-319-57099-0_67

the sphere ∂S we use the Green formula. We seek the solution of the exterior problem only on the outer surface $\partial \Omega$.

We have obtained sufficient conditions for the convergence of iterative method and the estimate for a derivative of the solution to the Helmholtz equation with a negative coefficient. The method proposed is illustrated by numerical experiments to solve the Dirichlet problem for a representative selection of test functions.

2   The Statement of the Problem and the Method of Solution

Let us consider the unbounded domain D in the space $\mathbf{R}^3$, bounded by the boundary ∂D. The exterior boundary value problem for the Helmholtz equation is: find the function $u \in C^1(\mathbf{R}^3 \backslash D) \cap C^2(\mathbf{R}^3 \backslash \bar{D})$ that satisfies the equation

$$\Delta u(\mathbf{r}) + \kappa^2 u(\mathbf{r}) = 0, \quad \mathbf{r} \in \mathbf{R}^3 \backslash \bar{D}, \tag{1}$$

the boundary condition

$$u(\mathbf{r}) = f(\mathbf{r}), \quad \mathbf{r} \in \partial D, \tag{2}$$

and the radiation condition at infinity $\lim_{r \to \infty} \left(\frac{\partial u}{\partial r} - i\kappa u \right) = 0$.

Problem (1)–(2) is solved if one finds the sought-for function u_S and its normal derivative u_n at the surface of the artificial sphere ∂S that is the boundary of the ball $\bar{S} = S \cup \partial S$, $\bar{D} \subset S$. Really, in this case one can find the function at any point $\mathbf{r} \in \mathbf{R}^3 \backslash \bar{D}$ by the Green formula

$$u(\mathbf{r}) = \frac{1}{4\pi} \int\limits_{\partial S} \left[u_n \frac{e^{i\kappa R}}{R} - u_S \frac{\partial}{\partial n} \left(\frac{e^{i\kappa R}}{R} \right) \right] ds, \tag{3}$$

where r_0 is the radius of the sphere ∂S, $R = |\mathbf{r} - \mathbf{r}_0| = \sqrt{r^2 - 2r_0 r \cos \gamma + r_0^2}$, $\mathbf{r}_0 \in \partial S$, γ is the angle between the vectors $\mathbf{r}$ and $\mathbf{r}_0$, $\cos \gamma = \cos \theta \cos \theta_0 + \sin \theta \sin \theta_0 \cos(\varphi - \varphi_0)$, $\frac{\partial}{\partial n}$ is the normal derivative to the sphere at the point $\mathbf{r}_0$.

Formula (3) is the base for the proposed iterative method to solve problem (1)–(2), that is a subsequent solution to auxiliary boundary value problems in bounded and unbounded domains.

We introduce an extra bounded domain Ω_0 with the boundary $\partial \Omega_0$ such that $\bar{S} \subset \Omega_0$. We will solve the interior boundary value problem by the finite volume method in the domain $\Omega_0 \backslash \bar{D}$ under given boundary conditions on ∂D and recalculated under every iteration boundary condition on $\partial \Omega_0$. On the first method iteration we set the zero boundary conditions on the surface $\partial \Omega_0$ and solve the interior boundary value problem in the domain $\Omega_0 \backslash \bar{D}$. Wherein we are interested only in the values of the function and its normal derivative on the artificial sphere ∂S. For these values we define the new boundary conditions on the outer surface of the calculated domain $\partial \Omega_0$ by formula (3), and carry out the second and the following iterations in a similar manner. Note, that on different iterations of the method the calculated domains can be chosen by the dynamic way, i.e. on the j-th iteration we can define the domain Ω_{j+1} with the boundary $\partial\Omega_{j+1}$, $j = 0, 1, \ldots, J$ instead of Ω_0. The number of iterations J is chosen according the condition of given accuracy $\varepsilon \ll 1$. The choice of the domain Ω_j that includes the ball S, on the second and subsequent iterations is taken for the sake of economy of calculating resources and the possibility to obtain a more precise solution on the sphere ∂S. Wherein the condition of inclusion $S \subset \Omega_j$ is valid. On all the next iterations we solve the interior boundary value problem in the domain $\Omega_j \backslash \bar{D}$ and rebuild the new boundary values of the Dirichlet problem at the surface $\partial\Omega_j$ by formula (3).

3   The Iterative Method Convergence

Let us prove the convergence of the method proposed for the equation

$$\Delta u(\mathbf{r}) - \kappa^2 u(\mathbf{r}) = 0, \tag{4}$$

where κ is a real number. Rewrite formula (3) in the form

$$u(\mathbf{r}) = \int_{\partial S} \left[G(\mathbf{r}, \mathbf{r}_0) \frac{\partial u}{\partial n}(\mathbf{r}_0) - u(\mathbf{r}_0) \frac{\partial}{\partial n}(G(\mathbf{r}, \mathbf{r}_0)) \right] ds,$$

where $G(\mathbf{r}, \mathbf{r}_0) = \frac{1}{4\pi} \frac{e^{-\kappa R}}{R}$. Then the iterative process is defined by the following way:

$$\begin{cases} \Delta u^{j+1}(\mathbf{r}) - \kappa^2 u^{j+1}(\mathbf{r}) = 0, & \mathbf{r} \in \Omega_j \backslash \bar{D} \\ u^{j+1}(\mathbf{r}) = \Phi^j(\mathbf{r}), & \mathbf{r} \in \partial\Omega_j, \\ u^{j+1}(\mathbf{r}) = f(\mathbf{r}), & \mathbf{r} \in \partial D, \quad j = 0, 1, \ldots \end{cases} \tag{5}$$

where

$$\Phi^0(\mathbf{r}) = 0,$$

$$\Phi^j(\mathbf{r}) = \int_{\partial S} \left[G(\mathbf{r}, \mathbf{r}_0) \frac{\partial u^j}{\partial n}(\mathbf{r}_0) - u^j(\mathbf{r}_0) \frac{\partial}{\partial n}(G(\mathbf{r}, \mathbf{r}_0)) \right] ds, \tag{6}$$

$$\mathbf{r} \in \partial\Omega_j, \ \mathbf{r}_0 \in \partial S, \ j = 1, 2, \ldots$$

We define the error of the method as

$$w^j(\mathbf{r}) = u^j(\mathbf{r}) - u(\mathbf{r}), \quad \mathbf{r} \in \Omega_{j-1} \backslash \bar{D},$$

$$w_n^j(\mathbf{r}_0) = \frac{\partial u^j}{\partial n}(\mathbf{r}_0) - \frac{\partial u}{\partial n}(\mathbf{r}_0), \quad \mathbf{r}_0 \in \partial S. \tag{7}$$

Then

$$w^{j+1}(\mathbf{r}) = \int_{\partial S} \left[G(\mathbf{r}, \mathbf{r}_0) w_n^j(\mathbf{r}_0) - w^j(\mathbf{r}_0) \frac{\partial}{\partial n}(G(\mathbf{r}, \mathbf{r}_0)) \right] ds \equiv \varphi^j(\mathbf{r}), \tag{8}$$

and the error will satisfy the equations

$$\begin{cases} \Delta w^{j+1}(\mathbf{r}) - \kappa^2 w^{j+1}(\mathbf{r}) = 0, & \mathbf{r} \in \Omega_j \backslash \bar{D} \\ w^{j+1}(\mathbf{r}) = \varphi^j(\mathbf{r}), & \mathbf{r} \in \partial\Omega_j, \\ w^{j+1}(\mathbf{r}) = 0, & \mathbf{r} \in \partial D. \end{cases} \tag{9}$$

To determine the condition for the method convergence we have to find the estimation for the normal derivative of the solution to the Helmholtz equation on the sphere ∂S. We define ρ_0 as the distance between the sphere and the outer surface, $\rho_0 = \min\limits_{\mathbf{r},\mathbf{r}_0} |\mathbf{r} - \mathbf{r}_0|$, $\mathbf{r} \in \partial\Omega_j \cup \partial D$, $\mathbf{r}_0 \in \partial S$; and u_l as a derivative from the function u in the direction l, $u_l = \frac{\partial u}{\partial l}$. If the function u satisfies the Helmholtz equation, then the function u_l will satisfy the same equation, and for any δ, $\delta \leq \rho_0$ the mean value theorem for solution of the Helmholtz equation is valid [1]

$$\int\limits_{\partial S_\delta} u_l \, ds_\delta = u_l(\mathbf{r}_0) \frac{4\pi\delta}{\kappa} \sin(\kappa\delta), \tag{10}$$

where ∂S_δ is a sphere with the radius δ with the center at the point $\mathbf{r}_0$. We integrate Eq. (10) from 0 to ρ in variable δ, $0 \leq \delta \leq \rho \leq \rho_0$. Then we obtain

$$I_\rho = \int\limits_{S_\rho} u_l \, dv_\rho = 4\pi u_l(\mathbf{r}_0) \alpha(\kappa, \rho), \tag{11}$$

where S_ρ is a ball with the radius ρ, and

$$\alpha(\kappa, \rho) = \frac{\sin(\kappa\rho)}{\kappa^3} - \frac{\rho\cos(\kappa\rho)}{\kappa^2}. \tag{12}$$

Integrating equality (11) by parts we obtain

$$I_\rho = \int\limits_{\partial S_\rho} u \frac{\partial l}{\partial n_s} ds_\rho,$$

where n_s is the normal at a current point on the sphere ∂S_ρ.

We estimate the integral I_ρ, which will allow us not only to obtain the sought-for estimation of the derivative at the point $\mathbf{r}_0$, but to define more precisely the estimation of the derivative of the harmonic function that was obtained in [1]. We choose the axis $\mathbf{Z}$ in the Cartesian coordinates coinciding with the direction l. Then

$$I_\rho = \rho^2 \int\limits_0^{2\pi} \int\limits_0^{\pi} u(\theta, \varphi) \sin\theta \cos\varphi \, d\theta d\varphi.$$

To estimate $|I_\rho|$, we divide the interval of integration by the variable φ into subintervals $\Phi_1 = [\pi/2, 3\pi/2]$ and $\Phi_2 = [0, 2\pi] \backslash \Phi_1$, on which the function $\cos\varphi$ has a constant sign. Then

$$|I_\rho| \leq \rho^2 \max_{\mathbf{r}\in\partial S_\rho} |u(\mathbf{r})| \int\limits_0^{\pi} \sin\theta d\theta \left\{ \int\limits_{\Phi_1} |\cos\varphi| \, d\varphi + \int\limits_{\Phi_2} \cos\varphi \, d\varphi \right\} = 8\rho^2 \max_{\mathbf{r}\in\partial S_\rho} |u(\mathbf{r})|.$$

Hence, taking into account (11), we obtain the estimation

$$|u_l(\mathbf{r}_0)| \leq \frac{2\rho^2}{\pi\alpha(\kappa, \rho)} \max_{\mathbf{r}\in\partial S_\rho} |u(\mathbf{r})|, \tag{13}$$

where $\alpha(\kappa, \rho)$ is defined by formula (12).

Note, that the right-hand side of inequality (13) is the function that depends on an arbitrary radius of the ball ρ, $\rho \leq \rho_0$, so is naturally to choose this radius such that the right-hand side of (13) has a minimum value. One can do this simply with relatively big values of the parameter κ. We define $x = \kappa\rho$. Then, taking into account (12), we have

$$\frac{\rho^2}{\alpha(\kappa, \rho)} = \frac{\kappa x^2}{\sin(x) - x\cos(x)}. \tag{14}$$

The function in the right-hand side of formula (14) has a minimum at $x_* \approx 2.08$ and at the point x_* the value $\rho_*^2/\alpha(\kappa, \rho_*) \approx 2.3\kappa$, where $\rho_* = x_*/\kappa$. Then inequality (13) takes the form

$$|u_l(\mathbf{r}_0)| \leq 1.465\kappa \max_{\mathbf{r} \in \partial S_{\rho_*}} |u(\mathbf{r})|. \tag{15}$$

If $\rho_* > \rho_0$, then due to decreasing the function in the right-hand side of formula (14) on the interval $(0, x_*]$, we choose $\rho = \rho_0$ and inequality (13) takes the form

$$|u_l(\mathbf{r}_0)| \leq \frac{2\rho_0^2}{\pi\alpha(\kappa, \rho_0)} \max_{\mathbf{r} \in \partial S_{\rho_0}} |u(\mathbf{r})| = \frac{2\kappa x_0^2}{\pi(\sin(x_0) - x_0\cos(x_0))} \max_{\mathbf{r} \in \partial S_{\rho_0}} |u(\mathbf{r})|, \tag{16}$$

where $x_0 = \kappa\rho_0$.

It is not difficult to show that $\lim_{\kappa \to 0}(\rho_0^2/\alpha(\kappa, \rho_0)) = 3/\rho_0$. From here it follows that on $\kappa = 0$ the estimation of a derivative for the Laplace equation in (16) is $\pi/2$ times better than the estimation $|u_x(\mathbf{r}_0)| \leq 3M/\rho_0$, given in [1], where M is a maximum modulus value of the function $u(\mathbf{r})$ in the given domain.

We now turn to the study of the method convergence. We will find the explicit form for a normal derivative from the fundamental solution, and integrals from the fundamental solution and its normal derivative at the surface of the sphere

$$\frac{\partial G}{\partial n} = -\frac{\partial G}{\partial R}\frac{\partial R}{\partial n} = e^{-\kappa R}\frac{1 + \kappa R}{R^2}\cos(R, n) = e^{-\kappa R}\frac{(1 + \kappa R)}{R^3}\frac{(r_0^2 + R^2 - r^2)}{2r_0},$$

$$I_0(\mathbf{r}) = \int_{\partial S} G(\mathbf{r}, \mathbf{r}_0)\, ds = \frac{r_0}{2\kappa r}e^{-\kappa r}\left[e^{\kappa r_0} - e^{-\kappa r_0}\right], \tag{17}$$

$$I_1(\mathbf{r}) = \int_{\partial S} \frac{\partial}{\partial n}G(\mathbf{r}, \mathbf{r}_0)\, ds = \frac{1}{2\kappa r}e^{-\kappa r}\left[(1 - \kappa r_0)e^{\kappa r_0} - (1 + \kappa r_0)e^{-\kappa r_0}\right], \tag{18}$$

where $r = |\mathbf{r}|$. It is easy to see that $I_0(\mathbf{r}) > 0$, $I_1(\mathbf{r}) < 0$, and both functions decrease in modulus with increasing the argument r.

We estimate the norm of the method error on the $j + 1$-th iteration by the norm of the error on the j-th iteration. Due to the fact that the maximum principle is valid for the Helmholtz equation in the form of (4), from (8) follows the inequality

$$\max_{\mathbf{r} \in \bar{\Omega}_j \backslash D}|\omega^{j+1}(\mathbf{r})| \leq \max_{\mathbf{r} \in \partial S}|\omega_n^j(\mathbf{r})| \max_{\mathbf{r} \in \partial\Omega_j} I_0(\mathbf{r}) + \max_{\mathbf{r} \in \partial S}|\omega^j(\mathbf{r})| \max_{\mathbf{r} \in \partial\Omega_j}|I_1(\mathbf{r})|,$$

where the integrals $I_0(\mathbf{r})$ and $I_1(\mathbf{r})$ are defined by formulas (17), (18). Hence, taking into account formulas (15) and (16), we obtain the final estimation

$$\max_{\mathbf{r}\in\bar{\Omega}_j\backslash D}\left|\omega^{j+1}(\mathbf{r})\right| \leq M(\kappa,d,r_0,x_0)\max_{\mathbf{r}\in\bar{\Omega}_{j-1}\backslash D}\left|\omega^j(\mathbf{r})\right|,$$

where

$$M(\kappa,d,r_0,x_0) = \frac{e^{-\kappa d}}{2\kappa d}\left[e^{\kappa r_0}(\kappa r_0 + r_0\beta - 1) + e^{-\kappa r_0}(\kappa r_0 - r_0\beta + 1)\right], \quad (19)$$

$$d = \min_{\mathbf{r}\in\partial\Omega_j}|\mathbf{r}|, x_0 = \kappa\rho_0,$$

$$\beta = \beta(\kappa,x_0) = \begin{cases} \frac{2\kappa x_0^2}{\pi(\sin(x_0) - x_0\cos(x_0))}, & x_0 < 2.08 \\ 1.465\kappa, & x_0 \geq 2.08 \end{cases}.$$

Therefore the sufficient condition for the convergence of the iteration process to solve Eq. (4) will be

$$M(\kappa,d,r_0,x_0) < 1. \quad (20)$$

4   Numerical Experiments and Discussion

We seek for the numerical solution of the exterior boundary value problem that has the following accurate solution

$$u(x,y,z) = e^{-\kappa r_1}/r_1 + e^{-\kappa r_2}/r_2, \quad (21)$$

or

$$u(x,y,z) = ze^{-\kappa r}(\kappa + 1/r)/r^2, \quad (22)$$

$r = \sqrt{x^2 + y^2 + z^2}$, $r_1 = \sqrt{(x - x_0)^2 + y^2 + z^2}$, $r_2 = \sqrt{(x + x_0)^2 + y^2 + z^2}$, $x_0 = 0.1$. Let us set the boundary conditions in accord with the precise solution to (21) or (22) at the surface of the cube with edges equal to 0.4. We suppose that the cube and all domains to be chosen, have the center of symmetry at the origin. We chose the boundary of the exterior domain $\partial\Omega_0$ that is permanent on every iteration, as the boundary of the cube with edges equal to 2, and define the zero conditions at its surface. We solve the interior Dirichlet boundary value problem in the domain $\Omega_0\backslash\bar{D}$, that is bounded by surfaces of two cubes, and find values of the function and its normal derivative on the sphere that is inside the domain $\Omega_0\backslash\bar{D}$. Taking into account these values, the new boundary values at the surface $\partial\Omega_0$ are determined by formula (3). Further, the Dirichlet problem is solved again in the domain $\Omega_0\backslash\bar{D}$, and new approximate values of the function and its normal derivative are determined on the sphere. The sphere where we find an approximate solution is permanent in every numerical experiment, but may vary in different experiments. An approximate solution of the interior boundary value problem had been obtained by the finite volume method. Information about constructing the finite volumes, forming local matrices of balance and the right-hand side of the equation is discussed in [2] and references listed there.

The finite-volume approximation of the interior boundary value problem was realized on the series of three uniform condensing meshes. The number of points and finite elements for the meshes are

$$
\begin{aligned}
Sparse\ Mesh\ (SM): &\quad L = 29666, &\quad Ltet = 160704; \\
Intermediate\ Mesh\ (IM): &\quad L = 225650, &\quad Ltet = 1285632; \\
Dense\ Mesh\ (DM): &\quad L = 1759794, &\quad Ltet = 10285056.
\end{aligned}
$$

We denote as ε_v, ε_d and δ_v, δ_d the mean quadratic and maximum errors of deviation from the precise solution and from its normal derivative on the sphere ∂S calculated by formulas

$$
\varepsilon_v = \sqrt{\sum_{i,j} \left(u\left(\theta_i, \varphi_j\right) - \tilde{u}\left(\theta_i, \varphi_j\right) \right)^2 / \sum_{i,j} u^2\left(\theta_i, \varphi_j\right)},
$$

$$
\delta_v = \max_{i,j} \left| u\left(\theta_i, \varphi_j\right) - \tilde{u}\left(\theta_i, \varphi_j\right) \right| / \max_{i,j} \left| u\left(\theta_i, \varphi_j\right) \right|,
$$

$$
\varepsilon_d = \sqrt{\sum_{i,j} \left(\frac{\partial u}{\partial n}\left(\theta_i, \varphi_j\right) - \frac{\partial \tilde{u}}{\partial n}\left(\theta_i, \varphi_j\right) \right)^2 / \sum_{i,j} \left(\frac{\partial u}{\partial n} \right)^2 \left(\theta_i, \varphi_j\right)},
$$

$$
\delta_d = \max_{i,j} \left| \frac{\partial u}{\partial n}\left(\theta_i, \varphi_j\right) - \frac{\partial \tilde{u}}{\partial n}\left(\theta_i, \varphi_j\right) \right| / \max_{i,j} \left| \frac{\partial u}{\partial n}\left(\theta_i, \varphi_j\right) \right|,
$$

where $u\left(\theta_i, \varphi_j\right)$, $\frac{\partial u}{\partial n}\left(\theta_i, \varphi_j\right)$ and $\tilde{u}\left(\theta_i, \varphi_j\right)$, $\frac{\partial \tilde{u}}{\partial n}\left(\theta_i, \varphi_j\right)$ are precise and approximate values of the sought-for function in the nodes θ_i, φ_j in the spherical coordinates on sphere.

In Table 1 present the errors of the function and its normal derivative on the sphere with the radius $r_0 = 0.5$ when $\kappa^2 = 1$, $N_\varphi = N_\theta = 17$ after 5 iterations of the method. The precise solution was defined by formula (22). Calculations were carried out on sparse, intermediate and dense meshes, where the length of an elementary tetrahedron in the finite volume method was decreased approximately by the factor of two when passing to a finer mesh. The calculations have shown that further increasing the number of iterations does not lead to decreasing errors. Note, that increasing the number of nodes on the sphere does not result in a essential decrease in the calculated errors, but lead to an extra calculation time. This is because the total error of the solution of the problem is mainly caused by the error of its interior part.

Table 1. The dependence of errors of solution and normal derivative on the form of the mesh

Error	ε_v	δ_v	ε_d	δ_d
SM	0.0192	0.0383	0.1782	0.3721
IM	0.0055	0.0077	0.0761	0.1310
DM	0.0011	0.0018	0.0389	0.0596

The results shown in the table illustrate the quadratic convergence for approximate solution and linear convergence for approximate values of the normal derivative. In the next numerical experiment we determined the error for different radii of the sphere and for different coefficients in the Helmholtz equation. The calculations were done on a dense mesh; the precise solution was defined by formula (21). The maximum errors of approximate values of the function and its normal derivative are shown in Table 2.

Table 2. The dependence of errors of solution and normal derivative on the radius of the sphere and the coefficient in the Helmholtz equation

r_0	0.4	0.5	0.6	0.7	0.8	0.9	0.95
$\delta_v, \kappa^2 = 1$	0.0020	0.0008	0.0011	0.0015	0.0095	0.0566	0.2090
$\delta_v, \kappa^2 = 10$	0.0032	0.0025	0.0019	0.0016	0.0086	0.0444	0.1420
$\delta_d, \kappa^2 = 1$	0.0564	0.0492	0.0440	0.0413	0.0561	0.1389	1.1966
$\delta_d, \kappa^2 = 10$	0.0710	0.0619	0.0608	0.0585	0.0586	0.0892	0.7608

To analyze the values obtained we make use of formula (19) for the reduction coefficient $M(\kappa, d, r_0, x_0)$ before passing to the next iteration of the method. This coefficient for the given values of parameters will be less a unit when $r_0 < 0.658$ for $\kappa^2 = 1$, and when $r_0 < 0.83$ for $\kappa^2 = 10$. The obtained values for errors allow us to conclude that the convergence of the method takes place for a larger radius of the sphere, up to $r_0 < 0.95$. This circumstance does not contradict the condition of convergence (20), because it is sufficient. The errors for large radii of a sphere decrease when the parameter κ^2 is increasing. This fact corresponds to decreasing the coefficient $M(\kappa, d, r_0, x_0)$ when the parameter κ^2 is increasing for given values of other parameters.

Acknowledgments. This work was supported by the Russian Foundation for Basic Research, project no. 14-11-00485.

References

1. Courant, R., Hilbert, D.: Methods of Mathematical Physics, vol. 2. Interscience, New York (1962)
2. Petukhov, A.V.: The barycentric finite volume method for 3D Helmholtz complex equation. Optoelectron. Instrum. Data Process. **43**(2), 182–191 (2007)

A Semi-Lagrangian Numerical Method for the Three-Dimensional Advection Problem with an Isoparametric Transformation of Subdomains

Vladimir Shaydurov[1], Alexander Vyatkin[1(✉)], and Elena Kuchunova[2]

[1] Institute of Computational Modelling of SB RAS,
Akademgorodok, 660036 Krasnoyarsk, Russia
shaidurov04@mail.ru, vyatkin@icm.krasn.ru
[2] Siberian Federal University, 79 Svobodny pr., 660041 Krasnoyarsk, Russia
hkuchunova@sfu-kras.ru

Abstract. We develop a semi-Lagrangian algorithm for solving the three-dimensional advection problem. A numerical solution is determined on a uniform cubic grid as a piecewise trilinear function. The method is based on the integral balance equation between two neighboring time levels. The domain of integration at the previous time level is a curved cuboid. To compute an integral over this domain numerically, we approximate this cuboid by another one with the same 8 vertices. The latter cuboid is obtained by a trilinear (isoparametric) transformation of the unit cube. This leads to the integration over the unit cube with the help of the composite midpoint rule. Such a technique provides the validity of the local balance equation and does not involve computational and algorithmic complexity for solving the three-dimensional problem. The numerical experiments confirm the first-order convergence.

Keywords: Semi-Lagrangian method · Advection equation · Isoparametric transformation · Local conservation

1 Introduction

Initially, the idea of applying the characteristic trajectories was developed for the applications in weather prediction [1]. The evolution of this approach leads to wide diversity of algorithms from the family of semi-Lagrangian methods [2–5]. Nowadays, these methods are a powerful tool for solving time-dependent hyperbolic conservation laws. In contrast to the traditional Eulerian schemes, semi-Lagrangian algorithms do not involve the time step restriction [6]. Modern versions of semi-Lagrangian methods provide the validity of the balance equation [7–9]. These schemes usually are based on the integral balance equation between two neighboring time levels. The procedure algorithmically consists of three main steps: calculation of the integral of an approximate solution at the current time

© Springer International Publishing AG 2017
I. Dimov et al. (Eds.): NAA 2016, LNCS 10187, pp. 599–607, 2017.
DOI: 10.1007/978-3-319-57099-0_68

level; computation of characteristic trajectories backward in time to the previous time level; calculation of the integral at the previous time level. There are many ways to implement each step. The improvement of properties of a numerical method usually leads to computational and algorithmic complications. This is especially valid for three-dimensional problems [10,11].

In this paper, we determine a numerical solution as a piecewise trilinear function. Our approach is also based on the integral balance equation between two neighboring time levels. To simplify the computations, we use a uniform cubic grid. To calculate an integral at the current time level, we consider a node of the grid and take the integral over its neighborhood. Since the integrand is a piecewise trilinear function, we perform the integration directly. This yields the linear combination of values of a numerical solution at the node mentioned above and at its neighboring nodes. Thus, we get the left-hand side of an algebraic equation. To construct the right-hand side of this equation, we project the rectangular neighborhood of the node onto the previous time level along characteristic trajectories and get a curved cuboid. It is of irregular shape with respect to a uniform cubic grid. To compute it numerically, we make several steps. First we approximate the curved cuboid by a new one which has the same 8 vertices and is obtained by a trilinear (isoparametric) transformation of the unit cube. This leads to the integration over the unit cube with the help of the composite midpoint rule. The Jacobian of the transformation is computed directly. This local integration for all nodes of the grid results in a system of linear algebraic equations. To solve it, we use the Jacobi iterative method. As an initial guess for the iteration, we use a numerical solution from the previous time level. The proposed algorithm is of first-order convergence and is stable regardless of the ratio between mesh sizes in time and space.

2   The Formulation of the Problem

Let $D = [0,1] \times [0,1] \times [0,1]$ be the unit cube and ∂D be its boundary. In the cylinder $[0,T] \times D$ consider the three-dimensional advection equation

$$\frac{\partial \rho}{\partial t} + \frac{\partial(u\rho)}{\partial x} + \frac{\partial(v\rho)}{\partial y} + \frac{\partial(w\rho)}{\partial z} = f. \tag{1}$$

The function $\rho(t, \mathbf{x})$ is unknown, $\mathbf{x} = (x, y, z)$. The components $u(t, \mathbf{x})$, $v(t, \mathbf{x})$, $w(t, \mathbf{x})$ of the velocity vector $\mathbf{u} = (u, v, w)$ and the function $f(t, \mathbf{x})$ are known and sufficiently smooth in $[0, T] \times D$. Denote the inflow boundary by $\Gamma_{\text{input}} = \partial D\big|_{x=0}$, the outflow boundary by $\Gamma_{\text{output}} = \partial D\big|_{x=1}$ and the rigid boundary by $\Gamma_{\text{chan}} = \partial D \setminus (\Gamma_{\text{input}} \cup \Gamma_{\text{output}})$. Impose the following conditions for the velocity vector $\mathbf{u}$:

$$\mathbf{u}(t, \mathbf{x})\big|_{\Gamma_{\text{chan}}} = 0, \quad \mathbf{u}(t, \mathbf{x})\big|_{\Gamma_{\text{input}}} \geq 0, \quad \mathbf{u}(t, \mathbf{x})\big|_{\Gamma_{\text{output}}} \geq 0 \qquad \forall\, t \in [0, T]. \tag{2}$$

For the unknown function $\rho(t, \mathbf{x})$ the following boundary and initial conditions are specified:

$$\rho(t, \mathbf{x})\big|_{\Gamma_{\text{input}}} = \rho_{\text{input}}(t, y, z) \quad \forall\, t \in [0, T],\ y \in [0, 1],\ z \in [0, 1], \tag{3}$$

$$\rho(0, \mathbf{x}) = \rho_{\text{init}}(\mathbf{x}) \quad \forall\, \mathbf{x} \in D. \tag{4}$$

Here we suppose that the functions $\rho_{\text{input}}(t, y, z)$ and $\rho_{\text{init}}(\mathbf{x})$ are known and sufficiently smooth.

3 The Numerical Method

3.1 The Local Integral Balance Equation

To solve problem (1)–(4), we subdivide the time segment $[0, T]$ by $M + 1$ time nodes $t_m = m\tau$, $m = 0, 1, \dots, M$, for step $\tau = T/M$. In D we construct a uniform grid D^h for mesh-size $h = 1/N$, $N \geq 2$,

$$D^h = \{(x_i, y_j, z_k) : x_i = ih, y_j = jh, z_k = kh;\ i, j, k = 0, 1, \dots, N\}.$$

Denote $\mathbf{x}_{i,j,k} = (x_i, y_j, z_k), i, j, k = 0, \dots, N$, and define the neighborhood

$$\Omega_{i,j,k} = ([x_i - h/2, x_i + h/2] \times [y_j - h/2, y_j + h/2] \times [z_k - h/2, z_k + h/2]) \cap D.$$

For convenience we introduce the notations $g_{i,j,k}^m = g(t_m, \mathbf{x}_{i,j,k})$ and $g_{i,j,k}^{h,m} = g^h(t_m, \mathbf{x}_{i,j,k})$ for a function g and a grid function g^h defined in D^h, respectively.

To compute the numerical solution ρ^h at the time level $t_m, m = 1, \dots, M$, we suppose that it is already defined at all nodes $\mathbf{x}_{i,j,k} \in D^h$ at the time level t_{m-1}. To compute $\rho_{i,j,k}^{h,m}$, $i = 1, \dots, N;\ j, k = 0, \dots, N$, we consider the neighbourhood $\Omega_{i,j,k}$. For each point $P = (P_x, P_y, P_z)$ at the time level t_m we construct the characteristic trajectory which issues out of this point and passes to the previous time level t_{m-1}. The trajectory is defined as a solution of the Cauchy problem for the system of ordinary differential equations

$$\hat{x}'(t) = u(t, \hat{x}, \hat{y}, \hat{z}),\quad \hat{y}'(t) = v(t, \hat{x}, \hat{y}, \hat{z}),\quad \hat{z}'(t) = w(t, \hat{x}, \hat{y}, \hat{z}) \tag{5}$$

backward in time on $[t_{m-1}, t_m]$ with the 'initial' condition

$$\hat{x}(t_m) = P_x,\quad \hat{y}(t_m) = P_y,\quad \hat{z}(t_m) = P_z. \tag{6}$$

Denote its solution by $\bar{\mathbf{x}}(t; P) = (\bar{x}(t; P), \bar{y}(t; P), \bar{z}(t; P))$. Considering the solutions for all points $P = (P_x, P_y, P_z) \in \Omega_{i,j,k}$, at each instant of time t we get the set $V_{i,j,k}^m(t) = \{\bar{\mathbf{x}}(t; P) : P \in \Omega_{i,j,k}\}$. It is clear that $V_{i,j,k}^m(t_m) = \Omega_{i,j,k}$.

Remark 1. Generally speaking, for some points $P \in \Omega_{i,j,k}$ near the inflow boundary Γ_{input}, a trajectory may be calculated only up to the boundary Γ_{input} at a point $P^{*,m-1} = (0, P_y^{*,m-1}, P_z^{*,m-1})$ at the instant $t_{m-1}^* > t_{m-1}$. Combining these points on Γ_{input}, we get a boundary polyhedron $\Gamma_{i,j,k} \subset [0, 1] \times \Gamma_{\text{input}}$. It is obvious that $V_{i,j,k}^m(t_{m-1})$ is empty provided that all trajectories issued out of $\Omega_{i,j,k}$ reach the boundary Γ_{input} before the instant t_{m-1}.

The following statement about the integral balance equation [7, 10, 11] is valid.

Statement 1. For a smooth solution of problem (1)–(4) we have the equality

$$\int\limits_{\Omega_{i,j,k}} \rho(t_m, \mathbf{x}) \, d\Omega = \int\limits_{V_{i,j,k}^m(t_{m-1})} \rho(t_{m-1}, \mathbf{x}) \, dV$$

$$+ \int\limits_{\Gamma_{i,j,k}} u(t, 0, y, z) \rho_{\text{input}}(t, y, z) \, d\Gamma \, dt + \int\limits_{t_{m-1}}^{t_m} \int\limits_{V_{i,j,k}^m(t)} f(t, \mathbf{x}) \, dV \, dt. \qquad (7)$$

If $V_{i,j,k}^m(t_{m-1})$ or $\Gamma_{i,j,k}$ is empty, the corresponding integral is supposed to vanish.

To construct the numerical scheme, we substitute an approximate solution $\rho^h(t, \mathbf{x})$ instead of $\rho(t, \mathbf{x})$ in two first integrals and use the trilinear interpolants $u_I(t, 0, y, z)$, $\rho_I(t, y, z)$ in two last ones:

$$\int\limits_{\Omega_{i,j,k}} \rho^h(t_m, \mathbf{x}) \, d\Omega = \int\limits_{V_{i,j,k}^m(t_{m-1})} \rho^h(t_{m-1}, \mathbf{x}) \, dV$$

$$+ \int\limits_{\Gamma_{i,j,k}} u_I(t, 0, y, z) \rho_I(t, y, z) \, d\Gamma \, dt + \int\limits_{t_{m-1}}^{t_m} \int\limits_{V_{i,j,k}^m(t)} f(t, \mathbf{x}) \, dV \, dt. \qquad (8)$$

3.2   The Integration at the Current Time Level

We calculate integral over $\Omega_{i,j,k}$ in (8) directly:

$$\int\limits_{\Omega_{i,j,k}} \rho^h(t_m, \mathbf{x}) \, d\Omega = h^3 \Big[\rho_{i-1,j-1,k-1}^{m,h} + 6\rho_{i-1,j-1,k}^{m,h} + \rho_{i-1,j-1,k+1}^{m,h}$$

$$+6\rho_{i-1,j,k-1}^{m,h} + 36\rho_{i-1,j,k}^{m,h} + 6\rho_{i-1,j,k+1}^{m,h} + \rho_{i-1,j+1,k-1}^{m,h} + 6\rho_{i-1,j+1,k}^{m,h}$$

$$+\rho_{i-1,j+1,k+1}^{m,h} + 6\rho_{i,j-1,k-1}^{m,h} + 36\rho_{i,j-1,k}^{m,h} + 6\rho_{i,j-1,k+1}^{m,h} + 36\rho_{i,j,k-1}^{m,h}$$

$$+216\rho_{i,j,k}^{m,h} + 36\rho_{i,j,k+1}^{m,h} + 6\rho_{i,j+1,k-1}^{m,h} + 36\rho_{i,j+1,k}^{m,h} + 6\rho_{i,j+1,k+1}^{m,h}$$

$$+\rho_{i+1,j-1,k-1}^{m,h} + 6\rho_{i+1,j-1,k}^{m,h} + \rho_{i+1,j-1,k+1}^{m,h} + 6\rho_{i+1,j,k-1}^{m,h} + 36\rho_{i+1,j,k}^{m,h}$$

$$+6\rho_{i+1,j,k+1}^{m,h} + \rho_{i+1,j+1,k-1}^{m,h} + 6\rho_{i+1,j+1,k}^{m,h} + \rho_{i+1,j+1,k+1}^{m,h} \Big] \Big/ 512. \qquad (9)$$

For $k = 0$ and $i, j = 1, \ldots, N - 1$, we have

$$\int\limits_{\Omega_{i,j,k}} \rho^h(t_m, \mathbf{x}) \, d\Omega = h^3 \Big[108\rho_{i,j,0}^{m,h}$$

$$+18 \left(\rho_{i,j+1,0}^{m,h} + \rho_{i,j-1,0}^{m,h} + \rho_{i+1,j,0}^{m,h} + \rho_{i-1,j,0}^{m,h} \right)$$

$$+3 \left(\rho_{i-1,j-1,0}^{m,h} + \rho_{i+1,j-1,0}^{m,h} + \rho_{i+1,j+1,0}^{m,h} + \rho_{i-1,j+1,0}^{m,h} \right)$$

$$+36\rho_{i,j,1}^{m,h} + 6 \left(\rho_{i,j+1,1}^{m,h} + \rho_{i,j-1,1}^{m,h} + \rho_{i+1,j,1}^{m,h} + \rho_{i-1,j,1}^{m,h} \right)$$

$$+\rho_{i-1,j-1,1}^{m,h} + \rho_{i+1,j-1,1}^{m,h} + \rho_{i+1,j+1,1}^{m,h} + \rho_{i-1,j+1,1}^{m,h} \Big] \Big/ 512. \qquad (10)$$

For $j = k = 0$ and $i = 1, \ldots, N - 1$, the following relation is valid:

$$\int_{\Omega_{i,j,k}} \rho^h(t_m, \mathbf{x}) \, d\Omega = h^3 \left[54\rho_{i,0,0}^{m,h} + 18 \left(\rho_{i,1,0}^{m,h} + \rho_{i,0,1}^{m,h} \right) \right.$$

$$+9 \left(\rho_{i-1,0,0}^{m,h} + \rho_{i+1,0,0}^{m,h} \right) + 6\rho_{i,1,1}^{m,h}$$

$$\left. +3 \left(\rho_{i-1,0,1}^{m,h} + \rho_{i-1,1,0}^{m,h} + \rho_{i+1,0,1}^{m,h} + \rho_{i+1,1,0}^{m,h} \right) + \rho_{i-1,1,1}^{m,h} + \rho_{i+1,1,1}^{m,h} \right] \Big/ 512. \quad (11)$$

Finally, for $i = j = k = N$ we obtain

$$\int_{\Omega_{i,j,k}} \rho^h(t_m, \mathbf{x}) \, d\Omega = h^3 \left[27\rho_{N,N,N}^{m,h} + 9 \left(\rho_{N-1,N,N}^{m,h} + \rho_{N,N-1,N}^{m,h} + \rho_{N,N,N-1}^{m,h} \right) \right.$$

$$\left. +3 \left(\rho_{N,N-1,N-1}^{m,h} + \rho_{N-1,N,N-1}^{m,h} + \rho_{N-1,N-1,N}^{m,h} \right) + \rho_{N-1,N-1,N-1}^{m,h} \right] \Big/ 512. \quad (12)$$

All other combinations of indices i, j, and k lead to formulae similar to (10)–(12).

To calculate the integral for the integrand $f(t, \mathbf{x})$ in (8), we use the rectangle rule in the t-direction:

$$\int_{t_{m-1}}^{t_m} \int_{V_{i,j,k}^m(t)} f(t, \mathbf{x}) \, dV \, dt \approx \tau \int_{V_{i,j,k}^m(t_m)} f(t_m, \mathbf{x}) \, dV. \quad (13)$$

Since $V_{i,j,k}^m(t_m) = \Omega_{i,j,k}$, we use suitable formulae similar to (9)–(12) to compute the integral in the right-hand side of (13).

3.3   The Integration at the Previous Time Level

To compute approximately the integral over $V_{i,j,k}^m(t_{m-1})$ in (8), we consider 8 vertices $\mathbf{A}_{p,q,r} = (A_{p,q,r,x}, A_{p,q,r,y}, A_{p,q,r,z})$, $p, q, r = 0, 1$, of the cube $\Omega_{i,j,k}$ and use them for the initial conditions for problem (5) and (6). Denote by $\mathbf{B}_{p,q,r} = (B_{p,q,r,x}, B_{p,q,r,y}, B_{p,q,r,z})$ the values of a numerical solution of this problem at the time level t_{m-1} calculated by one step of the Euler method:

$$B_{p,q,r,x}^h = A_{p,q,r,x} - \tau u(t_m, \mathbf{A}_{p,q,r}), \quad B_{p,q,r,y}^h = A_{p,q,r,y} - \tau v(t_m, \mathbf{A}_{p,q,r}),$$

$$B_{p,q,r,z}^h = A_{p,q,r,z} - \tau w(t_m, \mathbf{A}_{p,q,r}).$$

In this section we suppose that $\mathbf{B}_{p,q,r}^h \in D$ for all $p, q, r = 0, 1$. If some points $\mathbf{B}_{p,q,r}^h$ do not belong to D, according to Remark 1 they form two polyhedrons on the planes $t = t_{m-1}$ and $x = 0$.

We construct the trilinear isoparametric transformation $G_{i,j,k}^I = \left(G_{i,j,k,x}^I, \right.$ $\left. G_{i,j,k,y}^I, G_{i,j,k,z}^I \right)$ of the unit cube $[0, 1] \times [0, 1] \times [0, 1]$ in the following way:

$$G_{i,j,k,d}^I = \sum_{p=0}^{1} \sum_{q=0}^{1} \sum_{r=0}^{1} \varphi_p(\xi)\psi_q(\eta)\chi_r(\theta)B_{p,q,r,d}^h \quad \xi, \eta, \theta \in [0, 1], \quad d = x, y, z. \quad (14)$$

Here

$$\varphi_p(\xi) = \begin{cases} 1 - \xi, & \text{if } p = 0, \\ \xi, & \text{if } p = 1, \end{cases} \quad \psi_q(\eta) = \begin{cases} 1 - \eta, & \text{if } q = 0, \\ \eta, & \text{if } q = 1, \end{cases} \quad \chi_r(\theta) = \begin{cases} 1 - \theta, & \text{if } r = 0, \\ \theta, & \text{if } r = 1. \end{cases}$$

Define the cuboid $V_{i,j,k}^{m,h}(t_{m-1})$ as a closed domain obtained from $[0,1] \times [0,1] \times [0,1]$ by the transformation $G_{i,j,k}^I$. It is clear that $V_{i,j,k}^{m,h}(t_{m-1})$ has the same vertices as $V_{i,j,k}^m(t_{m-1})$. Since the faces of cuboids are not flat, we need to prove the absence of voids or excess intersections between the cuboids. For instance, consider two neighboring cuboids $V_{i,j,k}^{m,h}(t_{m-1})$ and $V_{i,j+1,k}^{m,h}(t_{m-1})$. They have 4 common vertices $\mathbf{B}_{p,r,d}^{h,\text{com}} = \left(B_{p,r,x}^{h,\text{com}}, B_{p,r,y}^{h,\text{com}}, B_{p,r,z}^{h,\text{com}} \right)$, $p, r = 0, 1$. Denote by $S_{i,j,k}^{\text{right}}$ the curved face of $V_{i,j,k}^{m,h}(t_{m-1})$ obtained from the face $y = 1$ of the unit cube by the transformation $G_{i,j,k}^I$. Besides, denote by $S_{i,j,k}^{\text{left}}$ the part of the boundary of $V_{i,j+1,k}^{m,h}(t_{m-1})$ obtained from the face $y = 0$ by the transformation $G_{i,j+1,k}^I$. Then the following statement is valid.

Statement 2. The surfaces $S_{i,j,k}^{\text{right}}$ and $S_{i,j,k}^{\text{left}}$ coincide.

Proof. The coordinates of a point $S_{i,j,k}^{\text{right}} = \left(S_{i,j,k,x}^{\text{right}}, S_{i,j,k,y}^{\text{right}}, S_{i,j,k,z}^{\text{right}} \right)$ are computed in the following way:

$$S_{i,j,k,d}^{\text{right}} = \sum_{p=0}^{1} \sum_{q=0}^{1} \sum_{r=0}^{1} \varphi_p(\xi) \psi_q(1) \chi_r(\theta) B_{p,q,r,d}^h$$

$$= \sum_{p=0}^{1} \sum_{r=0}^{1} \varphi_p(\xi) \chi_r(\theta) B_{p,r,d}^{h,\text{com}} \quad \xi, \theta \in [0,1], \quad d = x, y, z.$$

Similarly

$$S_{i,j+1,k,d}^{\text{left}} = \sum_{p=0}^{1} \sum_{r=0}^{1} \varphi_p(\xi) \chi_r(\theta) B_{p,r,d}^{h,\text{com}} \quad \xi, \theta \in [0,1], \quad d = x, y, z.$$

□

Thus, to compute the integral over $V_{i,j,k}^m(t_{m-1})$ in (8), we approximate it by $V_{i,j,k}^{m,h}(t_{m-1})$, use the isoparametric transformation $G_{i,j,k}^I$ defined by (14), and transform the integral over $V_{i,j,k}^{m,h}(t_{m-1})$ to the integral over the unit cube

$$\int_{V_{i,j,k}^{h,m}(t_{m-1})} \rho^h(t_{m-1}, \mathbf{x}) \, dV = \int_0^1 \int_0^1 \int_0^1 \rho^h \left(t_{m-1}, G_{i,j,k}^I(\xi, \eta, \theta) \right) J \, d\xi \, d\eta \, d\theta, \quad (15)$$

where the Jacobian J has the following form:

$$J(\xi, \eta, \theta) = \begin{vmatrix} \partial G_{i,j,k,x}^I/\partial \xi & \partial G_{i,j,k,x}^I/\partial \eta & \partial G_{i,j,k,x}^I/\partial \theta \\ \partial G_{i,j,k,y}^I/\partial \xi & \partial G_{i,j,k,y}^I/\partial \eta & \partial G_{i,j,k,y}^I/\partial \theta \\ \partial G_{i,j,k,z}^I/\partial \xi & \partial G_{i,j,k,z}^I/\partial \eta & \partial G_{i,j,k,z}^I/\partial \theta \end{vmatrix}. \quad (16)$$

Each component of the Jacobian J is computed directly from (14). For instance,

$$\frac{\partial G^I_{i,j,k,d}}{\partial \xi} = \sum_{q=0}^{1} \sum_{r=0}^{1} (B^h_{1,q,r,d} - B^h_{0,q,r,d}) \psi_q(\eta) \chi_r(\theta), \quad \eta, \theta \in [0,1], \quad d = x, y, z.$$

To compute the multiple integral in the right-hand side of (15), we use the composite midpoint rule with a number N_{loc} for each of 3 variables:

$$\int_0^1 g(\lambda)\, d\lambda \approx \frac{1}{N_{\text{loc}}} \sum_{\tilde{k}=0}^{N_{\text{loc}}-1} g\left((\lambda_{\tilde{k}} + \lambda_{\tilde{k}+1})/2\right), \quad \lambda = \xi, \eta, \theta, \tag{17}$$

where $\lambda_i = i/N_{\text{loc}}$. In the numerical experiments we use the parameter N_{loc} in such a way that the approximation of (17) be achieved within good accuracy.

To compute the multiple integral in the right-hand side of (8) for the integrand $u_I \rho_I$, we use the algorithm which is similar to that for computation of the integral over $V^m_{i,j,k}(t_{m-1})$ in (8).

Thus, combining the previous considerations we get the system of linear algebraic equations at each time level $t = t_m$, $m = 1, \ldots, M$. To solve it, we use Jacobi iterative method. As an initial guess, we use the numerical solution from the previous time level.

4   A Numerical Experiment

To test the proposed algorithm, several numerical experiments were carried out for model problems. The main purpose was to study the order of convergence. We take the following components of velocity

$$u = 10y(1-y)z(1-z)[\pi/2 - \operatorname{arctg}(x)],$$
$$v = w = \operatorname{arctg}\left(x(1-x)y(1-y)z(1-z)(1+t)/10\right)$$

and consider the function $\rho = 1.1 + \sin(t\,x\,y\,z)$, $t \in [0,1]$. We substituted these functions into (1) and get the function $f(t,x,y,z)$. To study the convergence of proposed scheme, we use the discrete analogue of the L_1-norm

$$\left\|\rho^{m,h}\right\|_{L_1^h} = \sum_{i,j,k=0}^{N} \left|\rho^{m,h}_{i,j,k}\right| \operatorname{mes}(\Omega_{i,j,k})$$

where $\operatorname{mes}(\Omega_{i,j,k})$ is the square of $\Omega_{i,j,k}$. We perform the computation of a numerical solution $\rho^{m,h}$ on the set of regular grids consisting of $(N+1)^3$ nodes, $N = 10 \cdot 2^n$ for $n = 0, 1, \ldots, 4$. Since the order of convergence depends on h and τ, we put $\tau = c \cdot h$ for the constant $c = 0.1$. We use $N_{\text{loc}} = 4$. Denote by ρ^h_n a numerical solution $\rho^{M,h}$ which is computed on the grid D^h consisting of $(N+1)^3$ nodes. To evaluate the order of convergence, we use the quantity $\log_2\left(\left\|\rho - \rho^h_n\right\|_{L_1^h} / \left\|\rho - \rho^h_{n+1}\right\|_{L_1^h}\right)$. As shown in Fig. 1, the scheme has the first order of convergence.

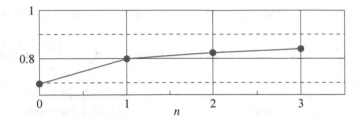

Fig. 1. The order of convergence

5 Conclusion

We propose a new numerical scheme from the family of semi-Lagrangian methods for the three-dimensional advection equation. The distinctive feature of the scheme consists in the technique of the computation of an integral at the previous time level. The proposed approach helps one to avoid algorithmic complexity and to decrease resource-intensive computations. The numerical experiments show the first-order convergence.

Acknowledgments. The work was supported by Project 14-01-00147 of Russian Scientific Foundation.

References

1. Wiin-Nielson, A.: On the application of trajectory methods in numerical forecasting. Tellus **11**, 180–186 (1959)
2. Morton, K.: Numerical Solution of Convection-Diffusion Problems. Chapman and Hall, London (1996)
3. Russell, T., Celia, M.: An over view of research on Eulerian-Lagrangian localized adjoint methods. Adv. Water Resour. **25**, 1215–1231 (2002)
4. Andreeva, E., Vyatkin, A., Shaidurov, V.: The semi-Lagrangian approximation in the finite element method for Navier-Stokes equations for a viscous incompressible fluid. In: AIP Conference Proceedings, vol. 1611 (2014). doi:10.1063/1.4893794
5. Celledoni, E., Kometa, B., Verdier, O.: High order semi-Lagrangian methods for the incompressible Navier-Stokes equations. J. Sci. Comput. **66**(1), 91–115 (2016)
6. Lentine, M., Gretarsson, J., Fedkiw, R.: An unconditionally stable fully conservative semi-Lagrangian method. J. Comput. Phys. **230**, 2857–2879 (2011)
7. Arbogast, T., Wang, W.: Convergence of a fully conservative volume corrected characteristic method for transport problems. SIAM J. Numer. Anal. **48**(3), 797–823 (2010)
8. Efremov, A., Karepova, E., Shaydurov, V., Vyatkin, A.: A computational realization of a semi-Lagrangian method for solving the advection equation. J. Appl. Math. **2014** (2014). ID 610398. doi:10.1155/2014/610398
9. Phillips, T.N., Williams, A.J.: Conservative semi-Lagrangian finite volume schemes. Numer. Meth. Part. Diff. Eq. **17**, 403–425 (2001)

10. Behrens, J., Mentrup, L.: A conservative scheme for 2D and 3D adaptive semi-Lagrangian advection. Contemp. Math. **383**, 175–189 (2005)
11. Vyatkin, A.: A semi-Lagrangian algorithm based on the integral transformation for the three dimensional advection problem. In: AIP Conference Proceedings, vol. 1684, p. 090012 (2015). doi:10.1063/1.4934337

Some Quadrature-Based Versions of the Generalized Newton Method for Solving Unconstrained Optimization Problems

Marek J. Śmietański[(⊠)]

Faculty of Mathematics and Computer Science, University of Lodz, Łódź, Poland
smietan@math.uni.lodz.pl

Abstract. Many practical problems require solving system of nonlinear equations. A lot of optimization problems can be transformed to the equation $F(x) = 0$, where the function F is nonsmooth, i.e. nonlinear complementarity problem or variational inequality problem. We propose a modifications of a generalized Newton method based on some rules of quadrature. We consider these algorithms for solving unconstrained optimization problems, in which the objective function is only LC^1, i.e. has not differentiable gradient. Such problems often appear in nonlinear programming, usually as subproblems. The methods considered are Newton-like iterative schemes, however they use combination of elements of some subdifferential. The methods are locally and at least superlinearly convergent under mild conditions imposed on the gradient of the objective function. Finally, results of numerical tests are presented.

Keywords: Unconstrained optimization problems · Generalized Newton method · B-differential · Superlinear convergence

1 Introduction

We consider nonlinear unconstrained programming problem

$$\min f(x), x \in D \subset R^n \qquad (1)$$

where $f : D \to R$ (D is an open convex set) is assumed to be only LC^1 but not C^2. The LC^1 property means that the objective function f is continuously differentiable and its gradient ∇f is locally Lipschitzian. If f is LC^1, then (1) is called an unconstrained LC^1 optimization problem.

In many problems functions may not possess a sufficient degree of smoothness (here, the second-order differentiability), so algorithms for solving nondifferentiable unconstrained problems can be important tools. In fact, application of proposed methods is much wider than solving equations which arise from the reformulation of nonlinear complementarity problems or variational inequalities. The augmented Lagrangian of a twice smooth nonliear programming problem is a LC^1 function (examples were discussed in [13,15]). The effective Newton-like

© Springer International Publishing AG 2017
I. Dimov et al. (Eds.): NAA 2016, LNCS 10187, pp. 608–616, 2017.
DOI: 10.1007/978-3-319-57099-0_69

methods for solving some optimization problems with nonsmooth gradient were also introduced for LC^1 optimization problem, e.g. in [5,14]. Some constructions of the unconstrained optimization problem with the objective function, which may not be twice differentiable, by using penalty functions or Lagrange multiplier functions was presented in [3,4].

Obviously, finding the solution of problem (1) is equivalent to solving of nonsmooth equation $F(x) = 0$, where function F is a gradient of f ($F = \nabla f$). One of the efficient methods for solving nonsmooth equations is the generalized Jacobian-based Newton method. Such superlinearly convergent methods were introduced i.e. in [6,12,15,17]. The general form of iteration is defined by

$$x^{(k+1)} = x^{(k)} - (V^{(k)})^{-1} F(x^{(k)}), \ k = 0, 1, \dots \tag{2}$$

where $V^{(k)}$ is taken from some subdifferential of F at $x^{(k)}$. It was assumed to be an element of the Clarke generalized Jacobian [15], the B-differential [12], the b-differential [17] and the *-differential [6]. New variants of Newton's method for solving smooth equations, based on various rules of quadrature, were proposed in [2]. These methods have local and quadratic convergence for a continuously differentiable function F. Some of them were extended in [18] to nonsmooth case.

The paper is organized as follows. In Sect. 2, we recall the well-known results for the generalized Jacobian, various subdifferentials and semismoothness. Quadrature-based versions of the generalized Newton method for solving unconstrained LC^1 optimization problem are studied in Sect. 3. In Sect. 4, we give some numerical examples.

2 Preliminaries

Let $F : R^n \rightarrow R^n$ be a locally Lipschitz function and $f_i : R^n \rightarrow R$ be the ith component of F for $i = 1, \dots, n$. According to Rademacher's theorem, the local Lipschitz continuity of F implies that F is differentiable almost everywhere. The set

$$\partial_B F(x) = \{ \lim_{x_i \to x} JF(x_i), x_i \in D_F \},$$

where D_F denotes the set where F is differentiable and $JF(x)$ - the usual Jacobian matrix of partial derivatives of F at x, is called *the B-differential* of F at x. If all elements of $\partial_B F(x)$ are nonsingular, then F is called *BD-regular* at x^*. *The generalized Jacobian* of F at x [1] is $\partial F(x) = \operatorname{conv} \partial_B F(x)$. The set $\partial_b F(x) = \partial_B f_1(x) \times \dots \times \partial_B f_n(x)$ is called *the b-differential* of F at x [17]. In turn, *the *-differential* $\partial_* F(x)$ introduced in [6] is a non-empty bounded set for each x such that $\partial_* F(x) \subset \partial f_1(x) \times \dots \times \partial f_n(x)$, where $\partial f_i(x)$ is the Clarke generalized gradient of f_i at x.

It is well-known that if $n = 1$, then $\partial F(x)$ reduces to the Clarke generalized gradient of F at x and $\partial_b F(x) = \partial_B F(x)$. Moreover, $\partial F(x), \partial_B F(x)$ and $\partial_b F(x)$ are *-differentials for a locally Lipschitz function [6].

The notion of semismoothness was originally introduced for functionals by Mifflin [8]. The following definition is taken from [15]. A function F is *semismooth* at a point x if F is locally Lipschitzian at x and

$$\lim_{V \in \partial F(x+th'),h' \to h, t \downarrow 0} Vh'.$$

exists for any $h \in R^n$.

Semismoothness implies some important properties for the convergence analysis of methods in nonsmooth optimization.

Proposition 1 *(Theorem 2.3, [15]). The following statements are equivalent:*

(i) F *is semismooth at* x;
(ii) $Vh - F'(x; h) = o(\|h\|)$ *for* $V \in \partial F(x + h)$, $h \to 0$;
(iii) $\lim_{h \to 0} \frac{F'(x+h;h)-F'(x;h)}{\|h\|} = 0$.

If for any $V \in \partial F(x + h)$, as $h \to 0$,

$$Vh - F'(x, h) = O(\|h\|^{1+p}), \tag{3}$$

where $0 < p \leq 1$, then we say that F is *p-order semismooth* at x. Clearly, p-order semismoothness implies semismoothness.

Qi and Sun [15] remarked that if F is semismooth at x, then for any $h \to 0$,

$$F(x + h) - F(x) - F'(x; h) = o(\|h\|),$$

and if F is p-order semismooth at x, then for any $h \to 0$,

$$F(x + h) - F(x) - F'(x; h) = O(\|h\|^{1+p}).$$

If $p = 1$ then the function F is called *strongly semismooth* [10].

Lemma 1. *(Lemma 2.6, [12]) If F is BD-regular at x, then there are a neighborhood N of x and a constant $C > 0$ such that for any $y \in N$ and $V \in \partial_B F(y)$, V is nonsingular and*

$$\|V^{-1}\| \leq C. \tag{4}$$

If F is also semismooth at $y \in N$, then, for any $h \in R^n$,

$$\|h\| \leq C \|F'(y; h)\|.$$

We can use the following more general assumption on the function F instead of semismoothness:

Assumption A. *Assume that function F is Lipschitz continuous. We say that F satisfies A at x if for any $y \in R^n$ and any $V_y \in \partial_B F(y)$, the following equality holds:*

$$F(y) - F(x) = V_y (y - x) + o (\|y - x\|).$$

Moreover, we say that F satisfies A at x with degree p if F is Lipschitz continuous and the following equality holds:

$$F(y) - F(x) = V_y (y - x) + O \left(\|y - x\|^{1+p} \right).$$

There are at least three classes of functions that satisfied Assumption A [11]. Beside semismoothness, the second-order C-differentiability (see [13]) and the H-differentiability (see [7]) are properties that imply A. Moreover, it is easy to verify that the strongly semismoothness implies A with degree 1.

3   Quadrature-Based Generalized Newton Methods

Now, we present the new versions of the generalized Jacobian-based Newton method, which are obtained using some rules deriving from numerical quadratures. Using methods from [18] based on ideas from [2], we apply similar methods for solving unconstrained optimization problem (1). We prove a local convergence theorem for the method using the B-differential of F.

Let $x^{(0)} \in R^n$ be some approximation to solution of (1). *The trapezoidal generalized Newton method* provides to obtain $x^{(k+1)}$ from $x^{(k)}$ by means of formula

$$x^{(k+1)} = x^{(k)} - 2[V_x^{(k)} + V_z^{(k)}]^{-1} F(x^{(k)}), \ k = 0, 1, ... \tag{5}$$

where $V_x^{(k)}$ and $V_z^{(k)}$ are taken form some subdifferentials of F at $x^{(k)}$ and $z^{(k)}$, respectively, and $z^{(k)} = x^{(k)} - (V_x^{(k)})^{-1} F(x^{(k)})$.

Using the midpoint rule as a source one, we obtain *the midpoint generalized Newton method* in the form

$$x^{(k+1)} = x^{(k)} - (V_{xz}^{(k)})^{-1} F(x^{(k)}), \ k = 0, 1, ... \tag{6}$$

where $V_{xz}^{(k)}$ is an element of some subdifferential of F at $\frac{1}{2}(x^{(k)} + z^{(k)})$ and as previously $z^{(k)} = x^{(k)} - (V_x^{(k)})^{-1} F(x^{(k)})$.

Moreover, Gao suggested in [6] that different $*$-differentials $\partial_* F(x)$ generate different superlinearly convergent Newton methods based on the iteration (2). In this way we can obtain other methods from (5) and (6).

To prove convergence results for method (6) with the B-differential, we need some helpful lemma.

Lemma 2. *Let F be BD-regular at x^*. Then the function*

$$G(x) = x - (\bar{V}_x)^{-1} F(x),$$

where $\bar{V}_x \in \partial_B F(x - \frac{1}{2}(V_x)^{-1} F(x))$ and $V_x \in \partial_B F(x)$, is well-defined in a neighborhood of x^*.

Proof. Since F is BD-regular at x^*, then from Lemma 1 there exists a constant $\beta > 0$ and a neighborhood N of x^* such that V_x is nonsingular and $\|V_x^{-1}\| \le \beta$ for any $x \in N$ and $V_x \in \partial_B F(x)$.

Let ε be such that $0 < \varepsilon < 1/2\beta$. First, we claim that there is $V_{x^*} \in \partial_B F(x^*)$ such that

$$\|V_x - V_{x^*}\| < \varepsilon \text{ for any } x \in S(x^*, \delta) \text{ and } V_x \in \partial_B F(x). \tag{7}$$

If the above inequality is not true, then there is a sequence $\{y^{(k)} : y^{(k)} \in D_F\}$ convergent to x^* such that

$$\|JF(y^{(k)}) - V_{x^*}\| \ge \varepsilon \text{ for all } V_{x^*} \in \partial_B F(x^*).$$

By passing to a subsequence, we may assume that $\{JF(y^{(k)})\}$ converges to $V_{x^*} \in \partial_B F(x^*)$, which contradicts the above inequality.

Now, we consider $z = x - \frac{1}{2}(V_x)^{-1} F(x)$, where $V_x \in \partial_B F(x)$. By the local convergence of the generalized Newton method (see [12]), it can be guaranteed that $z \in S(x^*, \delta)$ and (7) holds. So, using the Banach perturbation lemma [9], we obtain that $\bar{V}_x$ is nonsingular and

$$\|(\bar{V}_x)^{-1}\| = \|(V_z)^{-1}\| = \|[V_{x^*} + (V_z - V_{x^*})]^{-1}\|$$

$$\le \frac{\|(V_{x^*})^{-1}\|}{1 - \|(V_{x^*})^{-1}\|\|V_z - V_{x^*}\|} \le \frac{\beta}{1 - \beta\varepsilon} < 2\beta \text{ for } x \in S(x^*, \delta).$$

So, the function $G(x)$ is well-defined in $S(x^*, \delta)$. □

Theorem 1. *Suppose that F satisfies Assumption A at x^*, F is BD-regular at x^* and $\|V_{x^*}\| \le \gamma$ for all $V_{x^*} \in \partial_B F(x^*)$. Then there exists a neighborhood of x^* such that for any starting point $x^{(0)}$ belonging to this neighborhood, the sequence $\{x^{(k)}\}$ generated by the method (6) with the B-differential converges superlinearly to x^*. Moreover, if $F(x^{(k)}) \ne 0$ for all k, then the norm of F decreases superlinearly in a neighborhood of x^*, i.e.*

$$\lim_{k \to \infty} \frac{\|F(x^{(k+1)})\|}{\|F(x^{(k)})\|} = 0. \tag{8}$$

If F satisfies Assumption A with degree 1 at x^*, then the convergence is quadratic.

Proof. By Lemmas 1 and 2, the iterative schema (6) is well-defined in a neighborhood of x^* for the first step $k = 0$. Further, if we consider $z = x - \frac{1}{2}(V_x)^{-1} F(x)$, where $V_x \in \partial_B F(x)$, then it can be guaranteed that $z \in S(x^*, \delta)$ (like in Lemma 2) and

$$\|x^{(k+1)} - x^*\| = \|x^{(k)} - (V_{xz}^{(k)})^{-1} F(x^{(k)}) - x^*\|$$
$$= \|(V_{xz}^{(k)})^{-1}\|\|F(x^{(k)}) - F(x^*) - V_{xz}^{(k)}(x^{(k)} - x^*)\| = o(\|x^{(k)} - x^*\|). \tag{9}$$

The last equality is due to Lemma 1 and Assumption A. This shows that the sequence $\{x^{(k)}\}$ is superlinearly convergent to x^*.

Now, we prove (8). By Lemma 1, there are a scalar C and $\delta_1 > 0$ such that if $x \in S(x^*, \delta_1)$ and $V \in \partial_B F(x)$ then V is nonsingular and (4) holds. By Assumption A, for any $\alpha \in (0, 1)$, there is a $\delta_2 \in (0, \delta_1)$ such that if $x \in S(x^*, \delta_2)$,

$$\|F(x) - V(x - x^*)\| \leq \alpha \|x - x^*\|. \tag{10}$$

By (9) there is a $\delta \in (0, \delta_2)$ such that if $x^{(k)} \in S(x^*, \delta)$, then

$$\|x^{(k+1)} - x^*\| \leq \alpha \|x^{(k)} - x^*\|. \tag{11}$$

Since $\{x^{(k)}\}$ converges to x^*, there is an integer k_δ such that $\|x^{(k)} - x^*\| \leq \delta$ for all $k \geq k_\delta$. By (11), $\|x^{(k+1)} - x^*\| \leq \delta \leq \delta_2$. Furthermore, (7) implies $\|V_{xz}^{(k+1)}\| \leq \varepsilon + \|V_{x^*}\| \leq \varepsilon + \gamma$. By (10) and (11) we have

$$\|F(x^{(k+1)})\| \leq \|V_{xz}^{(k+1)}(x^{(k+1)} - x^*)\| + \alpha \|x^{(k+1)} - x^*\|$$
$$\leq (\varepsilon + \gamma + \alpha) \|x^{(k+1)} - x^*\| \leq \alpha(\varepsilon + \gamma + \alpha) \|x^{(k)} - x^*\|. \tag{12}$$

By (6), (11) and (4) we obtain

$$\|x^{(k)} - x^*\| \leq \|x^{(k+1)} - x^{(k)}\| + \|x^{(k+1)} - x^*\|$$
$$\leq \|(V_{xz}^{(k)})^{-1} F(x^{(k)})\| + \alpha \|x^{(k)} - x^*\| \leq C \|F(x^{(k)})\| + \alpha \|x^{(k)} - x^*\|.$$

So,

$$\|x^{(k)} - x^*\| \leq \frac{C}{1 - \alpha} \|F(x^{(k)})\|. \tag{13}$$

By (12) and (13),

$$\|F(x^{(k+1)})\| \leq \alpha(\varepsilon + \gamma + \alpha) \|x^{(k)} - x^*\| \leq \frac{C\alpha(\varepsilon + \gamma + \alpha)}{1 - \alpha} \|F(x^{(k)})\|.$$

Since $F(x^{(k)}) \neq 0$ for all k and α may be arbitrarily small as $k \to \infty$, we have (8).

Moreover, if the function F satisfies A with degree 1, then we have that $F(x) - F(x^*) - V(x - x^*) = O(\|x - x^*\|^2)$. Hence the sequence $\{x^{(k)}\}$ defined by (6) with the B-differential is quadratically convergent to x^*.   □

Remark: The convergence of the variants of the generalized Newton method with other subdifferentials may be proved in a similar way.

4   Numerical Tests and Conclusions

The methods proposed here can be applied to solve the following constrained optimization problem

$$\begin{cases} \min_{x \in R^n} h(x) \\ g_i(x) \leq 0, i = 1, ..., m, \end{cases} \tag{14}$$

where h and g_i are twice differentiable. As in [4], we can construct an unconstrained optimization problem (1) with use the Di Pillo-Grippo type Lagrange multiplier function $f : R^{n+m} \to R$ in the form

$$f(x, \mu, C) = h(x) + \sum_{i=1}^{m} \frac{(\max\{0, \mu_i + Cg_i(x)\})^2 - \mu_i^2}{2C} + C\|P(x)[\nabla h(x)$$
$$+ \sum_{i=1}^{m} \mu_i \nabla g_i(x)]\|^2, \tag{15}$$

where $\mu = (\mu_1, ..., \mu_m)$ are Lagrange multipliers, $P(x)$ is a $m \times n$ matrix and $C > 0$ is a constant. If h and g_i are twice continuously differentiable, f may not be twice differentiable at (x, μ, C) with $\mu_i + Cg_i(x) = 0$, but it is LC^1 and satisfies Assumption A at such points. Problems (14) and (15) are equivalent for large enough parameter C (Table 1).

To illustrate a performance of the both established methods, we present numerical results. Computations were performed in double-precision arithmetic (C++). We declare a failure of algorithm when the condition $\|x^{(k+1)} - x^{(k)}\|_2 \leq 10^{-10}$ is not satisfied after 1000 iterations. Tables summarize the results in terms of the number of iterations N for miscellaneous starting points $x^{(0)}$ and constants C. A symbol "×" denotes that the test failed (Table 2).

We consider some problems (14) (taken from [16]) with the following functions:
Problem 1.

$$h(x) = x_2 \text{ and } g_1(x) = -(x_2 - x_1^2).$$

Problem 2.

$$h(x) = -x_1 x_2 x_3 \text{ and } g_1(x) = -(x_1 + 2x_2 + 2x_3), g_2(x) = -(72 - x_1 - 2x_2 - 2x_3).$$

Table 1. Numerical results for Problem 1.

Initial data			Method		
$x^{(0)}$	μ	C	Midpoint	Trapezoid	Classic
$(1, 1)$	1	10	7	10	15
$(1, 1)$	5	10	8	11	15
$(5, 5)$	5	10	9	14	21
$(5, 5)$	5	20	10	16	21
$(5, 5)$	5	50	×	19	21
$(10, 10)$	1	10	12	17	23
$(10, 10)$	5	10	×	18	25
$(10, 10)$	5	50	×	21	×

Table 2. Numerical results for Problem 2.

Initial data			Method		
$x^{(0)}$	μ	C	Midpoint	Trapezoid	Classic
$(5,5,5)$	$(1,1)$	10	5	8	12
$(-5,-5,-5)$	$(1,1)$	10	7	11	17
$(10,10,10)$	$(1,1)$	10	7	9	19
$(-10,-10,-10)$	$(1,1)$	10	9	14	31
$(5,5,5)$	$(2,2)$	20	4	7	12
$(-5,-5,-5)$	$(2,2)$	20	8	23	17
$(10,10,10)$	$(2,2)$	20	$\times$	19	19
$(-10,-10,-10)$	$(2,2)$	20	$\times$	26	31

We presented some versions of the generalized Newton algorithms for solving unconstrained LC^1 optimization problem (1), proving that under mild assumptions, the sequence generated by the midpoint variant of the method with the B-differential is locally and superlinearly convergent to the solution. Under some stronger condition imposed on objective function, the method has quadratic convergence.

The important conclusion is that the studied versions of the generalized Newton method allow us to find the solutions of optimization problem usually more efficiently than the fundamental version. However, it should be remembered that the computational cost of one step of the quadrature-based method is double that for one step of the classic method, because two matrices are required. Moreover, examples show that the midpoint version is faster than the trapezoid one (but it is not always convergent). This is not surprising if we recall that we use the arithmetic mean of $V_x^{(k)}$ and $V_z^{(k)}$ in the trapezoidal version and the arithmetic mean of $x^{(k)}$ and $z^{(k)}$ in the midpoint one. So, the efficiency of the generalized Newton method depends on the successive iteration points rather than on matrices taken from the subdifferential.

References

1. Clarke, F.H.: Optimization and Nonsmooth Analysis. Wiley, New York (1983)
2. Cordero, A., Torregrosa, J.R.: Variants of Newton's method for functions of several variables. Appl. Math. Comput. **183**, 199–208 (2006)
3. Di Pillo, G.: Exact penalty method. In: Spedicato, E. (ed.) Algorithms for Continuous Optimization: The State of Art, pp. 209–253. Kluwer, Dordrecht (1994)
4. Di Pillo, G., Grippo, L.: A new class of lagrangians in nonlinear optimization. SIAM J. Control Optim. **17**, 616–628 (1979)
5. Djuranović-Milicić, N.I.: On an optimization algorithm for LC^1 unconstrained optimization. Yugoslav J. Oper. Res. **15**, 301–306 (2005)
6. Gao, Y.: Newton methods for solving nonsmooth equations via a new subdifferential. Math. Methods Oper. Res. **54**, 239–257 (2001)

7. Gowda, M.S., Ravindran, G., Song, Y.: On the characterizations of P- and P_0-properties in nonsmooth functions. Math. Oper. Res. **25**, 400–408 (2000)

8. Mifflin, R.: Semismooth and semiconvex functions in constrained optimization. SIAM J. Control Optim. **15**, 959–972 (1977)

9. Ortega, J.M., Rheinboldt, W.C.: Iterative Solution of Nonlinear Equations in Several Variables. Academic Press, New York (1970)

10. Potra, F.A., Qi, L., Sun, D.: Secant methods for semismooth equations. Numer. Math. **80**, 305–324 (1998)

11. Pu, D., Tian, W.: Globally convergent inexact generalized Newton's methods for nonsmooth equations. J. Comput. Appl. Math. **138**, 37–49 (2002)

12. Qi, L.: Convergence analysis of some algorithms for solving nonsmooth equations. Math. Oper. Res. **18**, 227–244 (1993)

13. Qi, L.: Superlinearly convergent approximate Newton methods for LC^1 optimization problems. Math. Program. **64**, 277–295 (1994)

14. Qi, L., Chen, X.: A globally convergent successive approximation method for severely nonsmooth equations. SIAM J. Control Optim. **33**, 402–418 (1995)

15. Qi, L., Sun, J.: A nonsmooth version of Newton's method. Math. Program. **58**, 353–367 (1993)

16. Schittkowski, K.: More Test Examples for Nonlinear Programming Codes. Springer, Berlin (1988)

17. Sun, J., Han, J.: Newton and Quasi-Newton methods for a class of nonsmooth equations an related problems. SIAM J. Optim. **7**, 463–480 (1997)

18. Śmietanski, M.J.: Some Quadrature-based versions of the generalized Newton method for solving nonsmooth equations. J. Comput. Appl. Math. **235**, 5131–5139 (2011)

One Parallel Method for Solving the Multidimensional Transfer Equation with Aftereffect

Svyatoslav I. Solodushkin[1,2]($\boxtimes$), Arsen A. Sagoyan[2], and Irina F. Yumanova[1]

[1] Ural Federal University, Turgeneva st. 4, Ekaterinburg, Russia
s.i.solodushkin@urfu.ru
[2] Institute of Mathematics and Mechanics, Ural Branch of the RAS,
S. Kovalevskoy st. 16, Ekaterinburg, Russia

Abstract. We describe a finite difference scheme for a multidimensional advection equation with time delay. The difference scheme has the second order in space and the first order in time and is unconditionally stable. The difference scheme lead to a big system of linear algebraic equations which could be solved in parallel. The performance of a sequential algorithm and several parallel implementations with the MPI technology in the C/C++ language has been studied in test examples, strong scalability is closed to ideal one.

Keywords: Parallel numerical method · Difference scheme · Multidimensional transfer equation · Time delay

1 Introduction

Let $\overline{G} = \prod_{\alpha=1}^{p}[0, X_\alpha]$ be a p-dimensional bar with boundary Γ. We have to find a sufficiently smooth function $u(x, t)$ which satisfies the transfer equation (also known as advection equation) with aftereffect

$$\frac{\partial u}{\partial t} + a \sum_{\alpha=1}^{p} \frac{\partial u}{\partial x_\alpha} = f(x, t, u(x, t), u_t(x, \cdot)), \tag{1}$$

in the cylinder $\overline{G} \times [t_0, \theta]$. Here $x = (x_1, \ldots, x_p)$, $x_\alpha \in [0, X_\alpha]$ are state variables and $t \in [t_0, \theta]$ represents time; $u(x, t)$ is an unknown function; $u_t(x, \cdot) = \{u(x, t + \xi), -\tau \le \xi < 0\}$ is a prehistory-function of the unknown function up to the moment t, and $\tau > 0$ is a value of time delay, $a > 0$ is a constant.

Together with the advection equation we have the following initial and boundary conditions

$$u(x, t) = \varphi(x, t), \quad x \in \overline{G}, \; t \in [t_0 - \tau, t_0], \tag{2}$$

$$u(x, t) = g(x, t), \quad x \in \Gamma, \; t \in [t_0, \theta]. \tag{3}$$

© Springer International Publishing AG 2017
I. Dimov et al. (Eds.): NAA 2016, LNCS 10187, pp. 617–624, 2017.
DOI: 10.1007/978-3-319-57099-0_70

Questions of the existence and uniqueness of a solution to the stated initial-boundary value problem (1)–(3) were considered in [1] and we assume that the functional f and functions φ and g are such that problem has a unique solution.

We denote by $\mathcal{Q} = \mathcal{Q}[-\tau, 0)$ the set of functions $v(\xi)$ that are piecewise continuous on $[-\tau, 0)$ with a finite number of points of discontinuity of the first kind and right continuous at the points of discontinuity. We define a norm on $\mathcal{Q}$ by $\|v\|_{\mathcal{Q}} = \sup_{\xi \in [-\tau, 0)} |v(\xi)|$. We additionally assume that the functional $f(x, t, u, v(\cdot))$ is given on $\overline{G} \times [t_0, \theta] \times \mathbb{R} \times \mathcal{Q}$ and is Lipschitz in the last two arguments: $\exists\, L_f \in \mathbb{R}\ \forall x \in \overline{G},\ t \in [t_0; \theta],\ u^1 \in \mathbb{R},\ u^2 \in \mathbb{R},\ v^1 \in \mathcal{Q},\ v^2 \in \mathcal{Q}:$

$$\left| f(x, t, u^1, v^1(\cdot)) - f(x, t, u^2, v^2(\cdot)) \right| \le L_f \left(|u^1 - u^2| + \|v^1(\cdot) - v^2(\cdot)\|_{\mathcal{Q}} \right).$$

2   The Difference Scheme

Let us consider an equidistant partition of $[0, X_\alpha]$ into parts with step size $h_\alpha = X_\alpha / N_\alpha$. On the set $\overline{G}$ we introduce a rectangular grid $\overline{\omega}_h = \{(i_1 h_1, \ldots, i_p h_p),\ i_\alpha = 0, 1, \ldots\ldots, N_\alpha\}$ which is uniform with respect to each direction. For brevity we denote $x_{(i)} = (x_{i_1}, \ldots, x_{i_p}) = (i_1 h_1, \ldots, i_p h_p),\ i_\alpha = 0, 1, \ldots, N_\alpha$, where the index i is a p-dimensional vector. To avoid any confusion, note that $x_\alpha \in \mathbb{R}$ is an α-coordinate of the vector x and $x_{(i)} \in \mathbb{R}^p$ is a particular node of the grid $\overline{\omega}_h$. Let us split the time interval $[t_0, \theta]$ into M parts with step size Δ and define the grid $\overline{\omega}_\Delta = \{t_j = t_0 + j\Delta,\ j = 0, 1, \ldots, M\}$. Without loss of generality and to simplify the narration we assume that the value $\tau/\Delta = m$ is a integer. In order to replace the differential Eq. (1) for a finite difference equation we define a grid on the cylinder $\overline{G} \times [t_0, \theta]$ and consider the inner product $\overline{\omega}_{h\Delta} = \overline{\omega}_h \times \overline{\omega}_\Delta = \{(x_i, t_j),\ x_i \in \overline{\omega}_h,\ t_j \in \overline{\omega}_\Delta\}$.

Denote by u_j^i the approximation of the function value $u(x_{(i)}, t_j)$ at the respective node, here $i = (i_1, \ldots, i_p),\ i_\alpha = 0, 1, \ldots, N_\alpha,\ j = 0, \ldots, M$.

Functional $f(x_{(i)}, t_j, u(x_{(i)}, t_j), u_{t_j}(x_{(i)}, \cdot))$ may depend on values of the function u between grid nodes, so interpolation may be needed for it calculation. For every fixed t_j and time delay $\xi \in [-\tau, 0)$ there are only two possibilities: if $t_j + \xi \le t_0$, we use the initial condition, $u(x_{(i)}, t_j + \xi) = \varphi(x_{(i)}, t_j + \xi)$, interpolation is not needed, otherwise we use the interpolation as described below.

For every fixed node $(x_{(i)}, t_j)$ we introduce its *discrete prehistory* as

$$\{u_l\}_j^i = \{u_l \mid \max\{0, j - m\} \le l \le j\}.$$

Piecewise linear interpolation operator has the properties required for the numerical method that we are going to construct. For the discrete prehistory $\{u_l\}_j^i$ we define piecewise linear interpolation operator as follow

$$v^{i,j}(t_j + \xi) = \frac{1}{\Delta}\left((t_{l+1} - t_j - \xi)\, u_l^i + (t_j + \xi - t_l)\, u_{l+1}^i \right),\ t_l \le t_j + \xi \le t_{l+1}. \tag{4}$$

Piecewise linear interpolation operator has the second order of error on the exact solution, i.e. there exist constants C_1 and C_2 such that, for all

$i = (i_1, \ldots, i_p)$, $i_\alpha = 0, 1, \ldots, N_\alpha$, $j = 1, \ldots, M$, and $t \in [\max\{0, t_j - \tau\}, t_j]$ the following inequality holds:

$$\left| v^{i,j}(t) - u(x_{(i)}, t) \right| \le C_1 \max_{\max\{0, j-m\} \le l \le j} \left| u_l^i - u(x_{(i)}, t_l) \right| + C_2 \Delta^2.$$

Let us introduce into consideration difference operators which approximate first derivatives with respect to state variables

$$\Omega_\alpha u_j^i = \begin{cases} \dfrac{-4u_j^{i[-1_\alpha]} - 2h_\alpha \dot{u}_j^{i[-1_\alpha]} + 4u_j^i}{2h_\alpha}, & i_\alpha = 1, \\[3mm] \dfrac{u_j^{i[-2_\alpha]} - 4u_j^{i[-1_\alpha]} + 3u_j^i}{2h_\alpha}, & i_\alpha \ge 2; \end{cases}$$

here $u_j^{i[-1_\alpha]}$ is a grid approximation of the function $u(x, t)$ which is evaluated in the node $(x_{(i)[-1_\alpha]}, t_j)$, and $\dot{u}_j^{i[-1_\alpha]}$ is a value of thee first derivative with respect to x_α. Based on the Eq. (1) and boundary condition (3) it is easy to check that

$$\dot{u}_j^{i[-1_\alpha]} = \frac{1}{a} \left(f_j^{i[-1_\alpha]} - \sum_{\beta=1, \beta \ne \alpha}^{p} \frac{\partial g(x_{(i)[-1_\alpha]}, t_j)}{\partial x_\beta} - \frac{\partial g(x_{(i)[-1_\alpha]}, t_j)}{\partial t} \right).$$

We use the notation $x_{(i)[-1_\alpha]} = (i_1 h_1, \ldots, i_{\alpha-1} h_{\alpha-1}, (i_\alpha - 1) h_\alpha, i_{\alpha+1} h_{\alpha+1}, \ldots, i_p h_p)$, that means $x_{(i)[-1_\alpha]}$ is a left neighbor of the node $x_{(i)}$ which is shifted by one step h_α in a corresponding coordinate. The way how these operators are constructed for the simplest case $p = 1$ is comprehensively described in [7].

We consider the following family of difference schemes (parametrized by s, $0 \le s \le 1$), with $j = 0, \ldots, M - 1$, and $i = (i_1, \ldots, i_p)$, $i_\alpha = 1, \ldots, N_\alpha$:

$$\frac{u_{j+1}^i - u_j^i}{\Delta} + a \sum_{\alpha=1}^{p} \left(s \, \Omega_\alpha u_{j+1}^i + (1-s) \, \Omega_\alpha u_j^i \right) = f_j^i, \tag{5}$$

with the initial and boundary conditions

$$u_0^i = \varphi(x_{(i)}, t_0), \quad v^{i,0}(t) = \varphi(x_{(i)}, t), \ t < t_0, \ \text{for all possible } i, \tag{6}$$

$$u_j^i = g(x_{(i)}, t_j), \quad x_{(i)} \in \Gamma, \ j = 0, \ldots, M. \tag{7}$$

Here $f_j^i = f\big(x_{(i)}, t_j, u_j^i, v^{i,j}(\cdot)\big)$ is the value of the functional f, calculated on an approximate solution and $v^{i,j}(\cdot)$ is the result of an interpolation. For constructing a numerical method we additionally assume that $g(x, t)$ is a differentiable function.

We call the mesh function

$$\Psi_j^i = \frac{u(x_{(i)}, t_{j+1}) - u(x_{(i)}, t_j)}{\Delta}$$

$$+ a \sum_{\alpha=1}^{p} \left(s \, \Omega_\alpha u(x_{(i)}, t_{j+1}) + (1-s) \, \Omega_\alpha u(x_{(i)}, t_j) \right) - \bar{f}_j^i, \tag{8}$$

the residual of method (5). Here $\bar{f}_j^i = f(x_{(i)}, t_j, u(x_{(i)}, t_j), u_{t_j}(x_{(i)}, \cdot))$ is the value of the functional f calculated on the exact solution.

Theorem 1. *Let the exact solution* $u(x, t)$ *of problem* (1)–(3) *have continuous derivatives with respect to state variables* x_α *up to third order, continuous derivatives with respect to time* t *up to second order and all first derivatives of the solution with respect to* x_α *are continuously differentiable in* t. *Then the residual of method* (5) *has order* $\sum_{\alpha=1}^{p} h_\alpha^2 + \Delta$.

3   Stability and Convergence of the Method

Due to the nonlinear dependence of the functional f on the state and its prehistory we apply the technique of abstract schemes with aftereffect, which was developed earlier [2] in the case of function-differential equations with ordinary derivatives and was generalized for the partial differential equations in [3–5].

In this section we consider problems with the homogeneous boundary condition $u(0, t) = 0$, $t \in [t_0, \theta]$. The replacement $\tilde{u}(x, t) = u(x, t) - \hat{g}(x, t)$ turns the initial problem into the mentioned one, here $\hat{g}(x, t)$ is a sufficiently smooth function such that $\hat{g}(x, t) = g(x, t)$, for $x \in \Gamma$, $t \in [t_0, \theta]$. We embed the schemes from family (5) into the general difference scheme with aftereffect [2]. The idea is close to [7,8] and is based on the method of dimension increase.

Let p be fixed. For every $t_j \in \overline{\omega}_\Delta$ we define the values of the discrete prehistory by the vector $\boldsymbol{y}_j$, for this we order components of u_j^i in the lexicographical order

$$\boldsymbol{y}_j = \left(u_j^{(1,\ldots,1)}, u_j^{(2,\ldots,1)}, \ldots, u_j^{(N_1,\ldots,1)}, \ldots, u_j^{(N_1,\ldots,N_p)} \right)^\top \in Y_p,$$

here $j = 0, \ldots, M$, the sign $\cdot^\top$ means transposition, Y_p is a linear space, $\dim Y_p = N_1 \times N_2 \times \ldots \times N_p$. For example, when $p = 1$, we have $\boldsymbol{y}_j = \left(u_j^1, u_j^2, \ldots, u_j^{N_1} \right)^\top$.

To build the difference operator $A : Y_p \to Y_p$ that correspond to $\sum_{\alpha=1}^{p} \Omega_\alpha$ we consider a sequence of matrices D_α, $\alpha = 1, \ldots, p$, where each next matrix is recursively defined.

We start with the $N_1 \times N_1$-matrix D_1 which corresponds to the difference operator Ω_1 as follows

$$\mathsf{D}_1 = \frac{a}{2h_1} \begin{pmatrix} 4 & 0 & 0 & 0 & \cdots & \cdots & \cdots & 0 \\ -4 & 3 & 0 & 0 & \cdots & \cdots & \cdots & 0 \\ 1 & -4 & 3 & 0 & \cdots & \cdots & \cdots & 0 \\ 0 & 1 & -4 & 3 & \cdots & \cdots & \cdots & 0 \\ \vdots & & \ddots & \ddots & \ddots & & & \vdots \\ 0 & \cdots & \cdots & 1 & -4 & 3 & 0 & 0 \\ 0 & \cdots & \cdots & 0 & 1 & -4 & 3 & 0 \\ 0 & \cdots & \cdots & 0 & 0 & 1 & -4 & 3 \end{pmatrix}.$$

We define the matrix $E_1 = \frac{a}{2h_2}I_1$, where I_1 is identity $N_1 \times N_1$-matrix. Next, we can define the matrix D_2 as a matrix which has a block-3-banded form: each block has size $N_1 \times N_1$ and there are N_2 blocks in the line:

$$D_2 = \begin{pmatrix} D_1 + 4E_1 & 0 & 0 & 0 & \cdots & 0 & 0 & 0 \\ -4E_1 & D_1 + 3E_1 & 0 & 0 & \cdots & 0 & 0 & 0 \\ E_1 & -4E_1 & D_1 + 3E_1 & 0 & \cdots & 0 & 0 & 0 \\ 0 & E_1 & -4E_1 & D_1 + 3E_1 & \cdots & 0 & 0 & 0 \\ \vdots & & \ddots & \ddots & \ddots & & & \vdots \\ 0 & 0 & 0 & 0 & \cdots & -4E_1 & D_1 + 3E_1 & 0 \\ 0 & 0 & 0 & 0 & \cdots & E_1 & -4E_1 & D_1 + 3E_1 \end{pmatrix}$$

and so on. . . In such a way we can define a linear operator A by the square matrix of size $N_1 \times N_2 \times \cdots \times N_p$:

$$A = \begin{pmatrix} D_{p-1} + 4E_{p-1} & 0 & \cdots & 0 & 0 & 0 \\ -4E_{p-1} & D_{p-1} + 3E_{p-1} & \cdots & 0 & 0 & 0 \\ \vdots & & \ddots & & & \vdots \\ 0 & 0 & \cdots & -4E_{p-1} & D_{p-1} + 3E_{p-1} & 0 \\ 0 & 0 & \cdots & E_{p-1} & -4E_{p-1} & D_{p-1} + 3E_{p-1} \end{pmatrix},$$

where $E_{p-1} = \frac{a}{2h_p}I_{p-1}$ and I_{p-1} is the identity matrix of dimension $N_1 \times \cdots \times N_{p-1}$. •

Proposition 3.1. *For each p, the matrix D_p is positive definite.*

Now we can rewrite system (5) in the form

$$\frac{y_{j+1} - y_j}{\Delta} + sA\, y_{j+1} + (1 - s)A\, y_j = F_j. \tag{9}$$

Let us use the obvious identity

$$y_{j+1} = y_j + \Delta \frac{y_{j+1} - y_j}{\Delta}$$

and define the linear operator $B = I + s\Delta A$, (I is the identity operator of the appropriate dimension) to rewrite (9) as a two-layer difference scheme in the canonical form [6]

$$B\frac{y_{j+1} - y_j}{\Delta} + A\, y_j = F_j. \tag{10}$$

The operator A is positive definite therefore B is a positive definite operator. Since B is invertible, we can rewrite (10) in the form

$$y_{j+1} = S\, y_j + \Delta B^{-1}\, F_j,$$

where $S = (I - \Delta B^{-1} A)$ is the transition operator.

In the space Y_p we introduce scalar product and the energy norm

$$(y, u) = h_1 \, h_2 \cdots h_p \sum_{i=1}^{N_1 \times \cdots \times N_p} y^i u^i, \qquad \|y\|_{Y_p} = \sqrt{(Ay, y)},$$

thereafter we define the corresponding induced operator norm.

Definition 1. *The difference scheme* (10) *is said to be stable, if* $\|S\|_{Y_p} < 1$.

Note that the equivalent formalization of stability of two-layer difference scheme is given in [6, pp. 324–330].

Theorem 2. *If the condition* $s \geq 1/2$ *is fulfilled then the difference scheme* (10) *is stable.*

Proof. Let us consider (10) from the point of view of operator-difference equations and apply methods of the stability verification for a two-layer difference scheme [6] and the separation of finite-dimensional and infinite-dimensional components [3].

We symmetrize 10 by multiplying through by A^{-1}

$$(A^{-1} + s\Delta E) \frac{y_{j+1} - y_j}{\Delta} + E \, y_j = A^{-1} F_j.$$

Denoting $\hat{B} = A^{-1} + s\Delta E$, $\hat{A} = E$, and $\hat{F}_j = A^{-1} F_j$, we obtain

$$\hat{B} \frac{y_{j+1} - y_j}{\Delta} + \hat{A} \, y_j = \hat{F}_j. \tag{11}$$

Method (11) is stable in the energy norm if and only if $2\hat{B} \geq \hat{A}$, see [6, p. 333 Theorem 1]. This requirement is equivalent to $A^{-1} + \Delta E(s - 0.5) \geq 0$. Since A^{-1} is a positive definite operator, the latter inequality is fulfilled for any Δ, as soon as $s \geq 0.5$.

By embedding the method (5) into the general scheme [2] we can prove the following

Theorem 3. *Let the conditions of Theorem 2 be satisfied, the interpolation operator* (4) *is used and* $s \geq 0.5$, *then method* (5) *converges with the second order in space and the first order in time, i.e. there exists such constant C that* $\|u(x_{(i)}, t_j) - u_j^i\| \leq C \left(\sum_{\alpha=1}^{p} h_\alpha^2 + \Delta \right)$, $i = (i_1, \ldots, i_p)$, $i_\alpha = 1, \ldots, N_\alpha$, $j = 0, \ldots, M$.

For one dimensional case, i.e. $p = 1$, the way of embedding described in [7].

4   The Parallel Algorithm and Numerical Experiments

To simplify narration let us consider the case $p = 2$; for this reason the results of numerical experiments are provided for the case $p = 2$.

To find the solutions $u_j^{i_1,i_2}$, $i_1 \geq 2$, $i_2 \geq 2$, it is required to know the solution in 9 grid points. This type of recurrent information dependencies allow ones to find solutions sequentially. Therefore the method (5) may be programmed in three nested loops: one external in the time and two internal in space.

We organize calculation in such way that the solution in some grid points could be calculated in parallel. We split the grid into $K_1 \times K_2$ *spacial wells*. We name as a spacial well (k_1, k_2), $k_1 = 1, \ldots, K_1$, $k_2 = 1, \ldots, K_2$, the following set of mesh nodes $\{(i_1, i_2, j) : i_1 = \frac{(k_1-1)N_1}{K_1}, \ldots, \frac{k_1 N_1}{K_1} - 1, i_2 = \frac{(k_2-1)N_2}{K_2}, \ldots, \frac{k_2 N_2}{K_2} - 1,$ $j = 1, \ldots, M\}$. The problem of finding the solution in the grid nodes shall be divided between processes. Namely, let the process (k_1, k_2) calculates solution in the spatial wells (k_1, k_2), $k_1 = 1, \ldots, K_1$, $k_2 = 1, \ldots, K_2$, i.e. it is required to organize $K_1 \times K_2$ processes.

The process $(1, 1)$ has no information dependencies on other processes. To find the solution on the j time layer, process (k_1, k_2), $k_1 \geq 2$, $k_2 \geq 2$, has to receive values $u_j^{i_1,i_2}$, where $i_1 = (k_1 - 1)N_1/K_1 - 2, (k_1 - 1)N_1/K_1 - 1$, $i_2 = (k_2 - 1)N_2/K_2, \ldots, k_2 N_2/K_2 - 1$, from the process $(k_1 - 1, k_2)$ and values $u_j^{i_1,i_2}$, where $i_1 = (k_1 - 1)N_1/K_1, \ldots, k_1 N_1/K_1 - 1$, $i_2 = (k_2 - 1)N_2/K_2 - 2$, $(k_2 - 1)N_2/K_2 - 1$, from the process $(k_1, k_2 - 1)$. Notice, that other required values were received or claculated on the previous time step.

For all fixed k_1, k_2, j we call as a block $B_j^{k_1,k_2}$ the following set $\{(i_1, i_2, j) : i_1 = (k_1 - 1)N_1/K_1, \ldots, k_1 N_1/K_1 - 1, i_2 = (k_2 - 1)N_2/K_2, \ldots, k_2 N_2/K_2 - 1,$ $j = 1, \ldots, M\}$. To find the solution in the block $B_j^{k_1,k_2}$ it is necessary to complete the calculation in the blocks $B_j^{k_1-1,k_2}$ and $B_j^{k_1,k_2-1}$.

Proposition 4.1. *Calculation could organized in such way that solutions in blocks $B_j^{k_1,k_2}$ and $B_{j'}^{k_1',k_2'}$ could be searched in parallel if $k_1 + k_2 + j = k_1' + k_2' + j'$.*

Table 1. Numerical experiments: time and speedup

$N_1 \times N_2 \times M$	Number of processors							
	1	4	9	16	25	36	49	64
$64 \times 64 \times 100$	0.039	0.022	0.02	0.014	0.015	0.015	0.018	0.017
$128 \times 128 \times 100$	0.152	0.043	0.023	0.018	0.018	0.019	0.02	0.019
$256 \times 256 \times 100$	0.478	0.16	0.079	0.049	0.038	0.031	0.031	0.028
$512 \times 512 \times 100$	2.103	0.503	0.274	0.174	0.12	0.088	0.074	0.068
$1024 \times 1024 \times 100$	6.934	2.267	0.879	0.595	0.399	0.299	0.238	0.201
Speedup		3.058	7.889	11.647	17.399	23.165	29.179	34.471

To present the results of the parallel implementation let us consider the test equation

$$\frac{\partial u}{\partial x} + \frac{\partial u}{\partial y} + \frac{\partial u}{\partial t} = 2u(x,y,t) - u(x,y,t-1) - x - y - t + 2,$$

where $0 \leq x \leq 2$, $0 \leq y \leq 2$, $0 \leq t \leq 4$, with the following initial and the boundary conditions

$$u(x,y,t) = x + y + t, \quad 0 \leq x \leq 2, \quad 0 \leq y \leq 2, \quad -1 \leq t \leq 0,$$
$$u(0,y,t) = y + t, \quad u(x,0,t) = x + t, \quad 0 \leq x \leq 2, 0 \leq y \leq 2, 0 \leq t \leq 4.$$

This initial-boundary problem has the exact solution $u(x,y,t) = x + y + t$. Grid parameters $N_1 = N_2$ ranges from 64 to 1024 while $M = 100$ through the experiments. first line. Time in seconds is reported in the cells. Speedup is calculated for multi processes variants (compared to the single process variant).

These calculations were performed on a cluster of Ural Branch of RAS "Uran". The following hardware and software configuration were used: CPU INTEL XEON E5450, 24 cores, 3 GHz; Cache memory 26 MB Level 2 cache (5400 Sequence); RAM 16 GB DDR2; OS Linux 2.6.32; language: C/C++ with Intel C++ compiler (ICC) v14.0.0; MPI library MVAPICH2 Intel 13.0.

Strong scaling is defined as how the solving time varies with the number of processors for a fixed total problem size. Table 1 allows one to check that the experimental estimation of strong scaling is extremely close to perfect one.

Acknowledgements. This research is supported by RFBR 14-01-00065, Russian Science Foundation (RSF) 14-35-00005 and Program 02.A03.21.0006 on 27.08.2013.

References

1. Wu, J.: Theory and Applications of Partial Functional Differential Equations. Springer, New York (1996)
2. Pimenov, V.G.: General linear methods for the numerical solution of functional-differential equations. Differ. Eq. **37**(1), 116–127 (2001)
3. Pimenov, V.G., Lozhnikov, A.B.: Difference schemes for the numerical solution of the heat conduction equation with aftereffect. Proc. Steklov Inst. Math. **275**, 137–148 (2011)
4. Pimenov, V.G., Lekomtsev, A.V.: Convergence of the alternating direction methods for the numerical solution of a heat conduction equation with delay. Proc. Steklov Inst. Math. **272**, 101–118 (2011)
5. Pimenov, V.G., Sviridov, S.: Numerical methods for advection equations with delay. AIP CP **1631**, 114–121 (2014)
6. Samarskii, A.A.: Theory of Difference Schemes. Nauka, Moscow (1989). (in Russian)
7. Solodushkin, S.I.: A difference scheme for the numerical solution of an advection equation with aftereffect. Rus. Math. **57**, 65–70 (2013). Allerton Press
8. Solodushkin, S.I., Yumanova, I.F., Staelen, R.H.: First order partial differential equations with time delay and retardation of a state variable. J. Comput. Appl. Math. **289**, 322–330 (2015)

Numerical Simulation of Heat Transfer of the Pile Foundations with Permafrost

Sergei P. Stepanov[1,4(✉)], Ivan K. Sirditov[1], Petr N. Vabishchevich[2], Maria V. Vasilyeva[1], Vasiliy I. Vasilyev[1], and Anastasiya N. Tceeva[3]

[1] North-Eastern Federal University, Yakutsk, Russia
Cepe2a@inbox.ru
[2] Nuclear Safety Institute, RAS, Moscow, Russia
[3] Yakut State Project Scientific-Research Institute of Building, Yakutsk, Russia
[4] Melnikov Permafrost Institute SB RAS, Yakutsk, Russia

Abstract. In this work we consider the numerical simulation of the dynamics of soil temperature in a permafrost area. The mathematical formulation of the problem with appropriate initial and boundary conditions is presented. A computational algorithm is based on the finite element approximation in space. To approximate in time we use the standard fully implicit scheme with linearisation from previous time layer. We present results of research of temperature stabilization time and the impact of presence of piles on the temperature of the surrounding soil. Numerical comparisons of two-dimensional and three-dimensional model problems are presented.

1 Introduction

Permafrost in Yakutia is different from some other, because of extreme complexity of frozen soil distribution. Thereby, during engineering exploration of territory of Yakutia, it is necessary to develop measures to manage the interaction between permafrost and engineering constructions to ensure the effective exploitation of already constructed objects.

For the solution of this complex problem it is necessary to consider heat transfer with taking into account phase transitions as in classical setting of Stefan. Frozen soil are considered as unsaturated porous media and they consist of several phases: solid skeleton, ice, water, and air [1–7].

The computational algorithm is based on the finite element approximation in space [8,9]. To approximate in time we use the standard implicit scheme with linearisation from previous time layer [10]. The numerical implementation of this problem has several difficulties with geometry, since piles are small compared to whole domain, which leads to large unstructured computational meshes. The numerical solution of such problems is currently impossible without using of the high performance computing systems.

In this paper we consider a numerical solution of two-dimensional and three-dimensional model problems. For the calculation of heat transfer in the

© Springer International Publishing AG 2017
I. Dimov et al. (Eds.): NAA 2016, LNCS 10187, pp. 625–632, 2017.
DOI: 10.1007/978-3-319-57099-0_71

ground we take into account the installation of piles and the seasonal fluctuations in ambient temperature. Numerical simulations of the problem in three-dimensional case on the computational cluster of NEFU "Arian Kuzmin" are presented [5].

2 Mathematical Model and Computational Implementation

We present a mathematical model describing the heat transfer processes in the frozen and thawed soils. This model takes into account the presence of phase transitions of pore water in the soil at a given temperature phase transition T^* in domain $\Omega = \Omega^- \cup \Omega^+$:

$$\Omega^+(t) = \{x|x \in \Omega, \quad T(x,t) > T^*\}, \quad \Omega^-(t) = \{x|x \in \Omega, \quad T(x,t) < T^*\}.$$

Here Ω^+ is a domain occupied by the liquid phase, where the temperature is higher than the phase transition temperature and Ω^- is a domain occupied by the solid phase, phase transition occurs at the interface S.

For the simulation of heat transfer processes with phase transitions, we use the classic Stefan model [1,2]

$$\left(\alpha(\phi) + \rho^+ L\phi'\right)\frac{\partial T}{\partial t} - \mathrm{div}\left(\lambda(\phi)\,\mathrm{grad}\,T\right) = 0, \tag{1}$$

where L is the specific heat of the phase transition, m is porosity, ρ^+, c^+, λ^+ and ρ^-, c^-, λ^- are density, specific heat capacity and the thermal conductivity of melted and frozen zones, respectively. We have the following coefficients [3]

$$\alpha(\phi) = \rho^- c^- + \phi(\rho^+ c^+ - \rho^- c^-), \quad \lambda(\phi) = \lambda^- + \phi(\lambda^+ - \lambda^-),$$

$$c^- \rho^- = (1-m)c_{sc}\rho_{sc} + mc_i\rho_i, \quad \lambda^- = (1-m)\lambda_{sc} + m\lambda_i,$$

$$c^+ \rho^+ = (1-m)c_{sc}\rho_{sc} + mc_w\rho_w, \quad \lambda^+ = (1-m)\lambda_{sc} + m\lambda_w.$$

and

$$\phi = \begin{cases} 0, \text{ when } T < T^*, \\ 1, \text{ when } T > T^*, \end{cases}$$

The indexes sc, w, i denote the solid skeleton, water, and ice, respectively.

In practice, the phase transitions occur in a small temperature range $[T^* - \Delta, T^* + \Delta]$. As the function ϕ we take ϕ_Δ:

$$\phi_\Delta = \begin{cases} 0, & T \leq T^* - \Delta, \\ \dfrac{T - T^* + \Delta}{2\Delta}, & T^* - \Delta < T < T^* + \Delta, \\ 1, & T \geq T^* + \Delta, \end{cases} \quad \phi'_\Delta = \begin{cases} 0, & T \leq T^* - \Delta, \\ \dfrac{1}{2\Delta}, & T^* - \Delta < T < T^* + \Delta, \\ 0, & T \geq T^* + \Delta. \end{cases}$$

Therefore, we get the following equation for nonlinear parabolic temperature in the domain Ω:

$$(\alpha(\phi_\Delta) + \rho_l L \phi'_\Delta) \frac{\partial T}{\partial t} - \operatorname{div}(\lambda(\phi_\Delta) \operatorname{grad} T) = 0. \tag{2}$$

The Eq. (2) is supplemented with the initial condition:

$$T(x, 0) = T_0(x), \quad x \in \Omega, \tag{3}$$

and boundary conditions

$$-k \frac{\partial T}{\partial n} = \frac{Q(1 - A) + I - \alpha(T - T_{air})}{\alpha R + 1}, \quad x \in \Gamma_1, \tag{4}$$

$$-k \frac{\partial T}{\partial n} = 0, \quad x \in \Gamma_2, \tag{5}$$

$$T = T_0, \quad x \in \Gamma_3, \tag{6}$$

where Q is summary short-wave radiation, A is albedo, α is heat transfer coefficient, I is low-frequency radiation, T_{air} is outdoor temperature, R is thermal resistance of the ground cover (winter is snow).

The Eq. (2) with appropriate boundary and initial conditions can be approximated using the finite element method and the standard implicit difference scheme to approximate by time. For the linearisation of the equation we use take the coefficients for the temperature from previous time layer [8–10].

We write the variational formulation of the problem for each time layer: find $T \in H^1_D$, $H^1_D = \{T \in H^1(\Omega) : T(x) = T_0 \quad x = \Gamma_3\}$ which is

$$a(T^{n+1}, v) = L(v), \quad \forall v \in H^1_0,$$

where

$$a(T^{n+1}, v) = \frac{1}{\tau} \int_\Omega (\alpha(\phi_\Delta^n) + \rho_l L \phi_\Delta'^n) T^{n+1} v \, dx + \int_\Omega (\lambda(\phi_\Delta^n) \operatorname{grad} T^{n+1}, \operatorname{grad} v) dx$$

$$+ \int_{\Gamma_1} \frac{\alpha}{\alpha R + 1} T^{n+1} v ds$$

$$L(v) = \frac{1}{\tau} \int_\Omega (\alpha(\phi_\Delta^n) + \rho_l L \phi_\Delta'^n) T^n v \, dx$$

$$+ \int_{\Gamma_1} \frac{Q(1 - A) + I + \alpha T_{air}}{\alpha R + 1} v ds, \tag{7}$$

For the numerical solution of the problem it is necessary to move from the continuous variational problem to discrete one. For this we introduce the finite-dimensional space $V_h, \in H^1_D$, $\hat{V}_h \in H^1_0$ and define the following problem: find $T_h \in V_h$ which is

$$a(T_h^{n+1}, v) = L(v), \quad \forall v \in \hat{V}_h,$$

$$a(T_h^{n+1}, v) = \frac{1}{\tau} \int_{\Omega} (\alpha(\phi_{\Delta}^n) + \rho_l L \phi_{\Delta}'^n) T_h^{n+1} v_h dx + \int_{\Omega} (\lambda(\phi_{\Delta}^n) \operatorname{grad} T_h^{n+1}, \operatorname{grad} v_h) dx$$
$$+ \int_{\Gamma_1} \frac{\alpha}{\alpha R + 1} T_h^{n+1} v ds$$

$$L(v) = \frac{1}{\tau} \int_{\Omega} (\alpha(\phi_{\Delta}^n) + \rho_l L \phi_{\Delta}'^n) T_h^n v_h dx$$
$$+ \int_{\Gamma_1} \frac{Q(1 - A) + I + \alpha T_{air}}{\alpha R + 1} v ds, \tag{8}$$

As a basis functions we use standard linear basis functions.

3 Numerical Results

For the calculations in a complex three-dimensional domain we use high-performance computing system of North-Eastern Federal University. In the simulation of problem for three-dimensional statement the system of linear equations is very large. The dimension of the matrix depends on the number of nodes of the computational mesh.

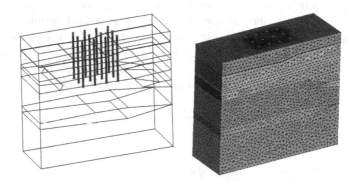

Fig. 1. The calculated geometric domain in three-dimensional statement and settlement unstructured tetrahedral mesh.

Table 1. Solution time and total number of iterations for the different number of processes for three-dimensional problem.

Number of processes	$iter_{sum}$	Time, sec
4	783931	4514.7
8	835445	3340.4
16	877862	2752.6
32	982994	2331.2

To solve large system of linear equations it is necessary to use iterative methods with preconditioner. Because direct methods use a much more resources, and may be out of memory.

We will conduct a numerical study of the temperature field of soil around a group pile. The computational domain consists of several layers of soil (Fig. 1). Initial temperature of soil $T_0 = -3$. The thermal characteristics are taken from the construction norms and regulations (SNiP) 2.02.04-88. We assume that the

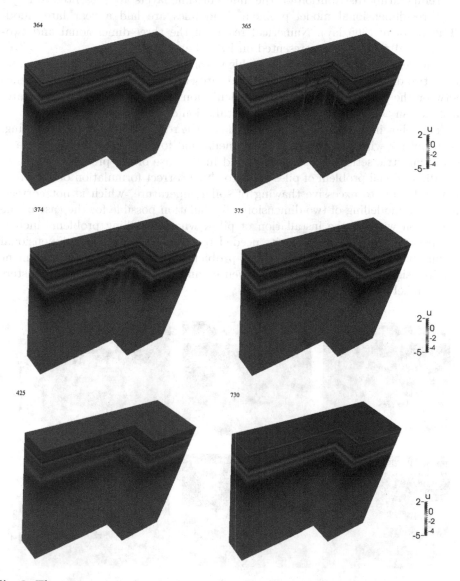

Fig. 2. The temperature distribution for $T = 364, 365, 374, 375, 425$ and 725 with the existence of the installation of piles.

phase transition temperature is $T^* = 0°C$. For calculation we use $T_{max} = 2$ years and time step $\tau = 1$ day. For numerical simulations we set zero Neumann boundary conditions for the side boundaries and on the bottom boundary.

Computational mesh consisting of 201 553 nodes and 1 171 690 cells (pic. 1). As iterative solver we use the conjugate gradient method (CG) with SOR preconditioner. Solution time (sec.) and total number of iterations are presented in Table 1 for different number of processes.

Temperature distribution for the different time layers are presented in Fig. 2 for three-dimensional model problem. The piles are laid a year later modelling problem (365 day). Numerical result of the three-dimensional and two-dimensional problems are presented on Figs. 3 and 4.

The results of comparison of problem solution between three-dimensional with two-dimensional formulations are presented on Fig. 5. The difference between the results show necessity of calculation in the three-dimensional statement for an adequate account of the installation of piles.

The following conclusions can be made on the results of numerical modelling: for problems with one pile at a two-dimensional formulation can be used axially symmetric setting heat equation and in the case of the presence of several two-dimensional problem of piles leads to the incorrect formulation of the problem and leads to excessive thawing of soil temperature, which is not correct; numerical modelling of two-dimensional formulation possible for the case of the simulation without the installation of piles; when modelling problems including installation of several piles its needed to calculate using three-dimensional formulation; numerical solution of the problem in three-dimensional statement requires significant computational expenses and use of computational clusters and parallel computing are necessary.

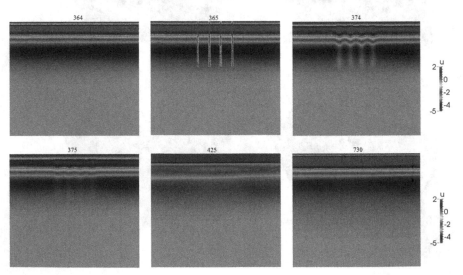

Fig. 3. The temperature distribution in the entire region, taking into account the installation of the pile for the case of three-dimensional formulation.

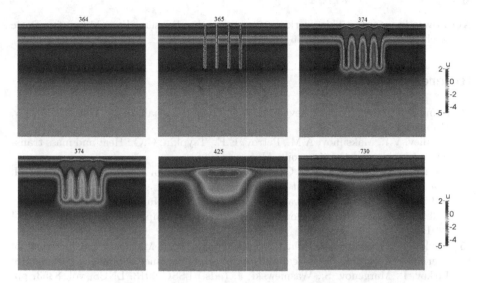

Fig. 4. The temperature distribution in the entire region, taking into account the installation of the pile for the case of two-dimensional formulation.

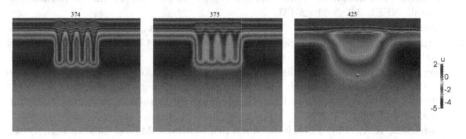

Fig. 5. Comparison of isotherms $T = 0$, -1 and -2 degrees: red is grid 2D and white is 3D. On the left its after 9 days. In the middle its after 10 days. On the right its after 2 month.

4   Conclusion

The studies allow to evaluate the temperature variation of the soil around the pile in a permafrost zone. We investigated halo thawing soils during installation of piles subject to seasonal fluctuations in ambient temperature. A numerical comparison of two-dimensional and three-dimensional models of calculation of heat transfer in soil is presented.

It should be noted the need for three-dimensional domains to simulate heat distribution around the pile.

Thus, to maintenance of accuracy of the calculation, along with a choice of a numerical method for solving the Stefan's problem is necessary to analyze the geometric domain.

Acknowledgements. This work is supported by Russian Foundation for Basic Research (project no. 15-31-20856)

References

1. Samarskii, A.A., Vabishchevich, P.N.: Computational Heat Transfer. Wiley, Hoboken (1995)
2. Vasiliev, V.I., Maksimov, A.M., Petrov, E.E., Tsypkin, G.G.: Heat and mass transfer in freezing and thawing soils. Moscow (1996)
3. Pavlov, A.V., Melnikov, P.I.: Calculation and regulation regime of permafrost soil. Science, Siberian Department (1980)
4. Vabishchevich, P.N., Vasilyeva, M.V., Pavlova, N.V.: Numerical modeling of thermal stabilization of filter ground. Math. Models Comput. Simul. **7**(2), 154–164 (2015)
5. Pavlova, N.V., Vabishchevich, P.N., Vasilyeva, M.V.: Mathematical modeling of thermal stabilization of vertical wells on high performance computing systems. In: Lirkov, I., Margenov, S., Waśniewski, J. (eds.) LSSC 2013. LNCS, vol. 8353, pp. 636–643. Springer, Heidelberg (2014). doi:10.1007/978-3-662-43880-0_73
6. Vabishhevich, P.N., Varlamov, S.P., Vasil'ev, V.I., Vasil'eva, M.V., Stepanov, S.P.: Mathematical modeling of the thermal regime of the railway line in permafrost. Vestnik NEFU **10**, 5–11 (2013)
7. Gornov, V.F., Stepanov, S.P., Vasilyeva, M.V., Vasilyev, V.I.: Mathematical modeling of heat transfer problems in the permafrost. AIP Conf. Proc. **1629**, 424–431 (2014)
8. Logg, A., Mardal, K.-A., Wells, G.N.: Automated solution of differential equations by the finite element method: The FEniCS book. Springer, Heidelberg (2012)
9. Thomee, V.: Galerkin Finite Element Methods for Parabolic Problems. Springer, Heidelberg (2006)
10. Samarskii, A.A.: The Theory of Difference Schemes. Marcel Dekker Inc., Basel, New York (2001). 761 p.

The Inverse Problem of the Simultaneous Determination of the Right-Hand Side and the Lowest Coefficients in Parabolic Equations

LingDe Su[1], P.N. Vabishchevish[2]([✉]), and V.I. Vasil'ev[1]

[1] Institute of Mathematics and Information Science,
North-Eastern Federal University, Yakutsk, Russia
[2] Nuclear Safety Institute, Russian Academy of Science, Moscow, Russia
vabishchevich@gmail.com

Abstract. In this paper, we propose a numerical scheme to solve the inverse problem of determining two lower coefficients that depends on time only in the parabolic equation. The time dependence of the right-hand side of a parabolic equation is determined using additional solution values at points of the computational domain. For solving the nonlinear inverse problem, linearized approximations in time are constructed using the fully implicit scheme, and standard finite difference procedures are used in space. The results of numerical experiments are presented, confirming the capabilities of the proposed computational algorithms for solving the coefficients inverse problem.

Keywords: Inverse problem · Finite difference method · Parabolic partial differential equation · Identification of the coefficients

1 Introduction

Recently, much attention is paid to the problem of the identification of coefficients from some additional information. The existence and uniqueness of the solution to such an inverse problem and well-posed of this problem in various functional classes are examined [1,2]. Numerical methods for solving the problem of the identification of the lower coefficient of parabolic equations are also considered in many works [3–5].

Different from direct problem, in inverse problems, the master equation, initial conditions and boundary conditions are not fully specified, instead, some additional information is available. So, many inverse problems are formulated as non-classical problems for PDEs. In other words, most standard numerical methods cannot achieve good accuracy in solving this problems. To solve approximately these problems, emphasis is on the development of stable computational algorithms that take into account peculiarities of inverse problems [6–8].

In this paper, we consider the inverse problem to find the two coefficients $\gamma(t)$ and $\psi(t)$ in the following two dimensional equation,

$$u_t - \mathscr{L}u - \gamma(t)u = \psi(t)g(\mathbf{x}, t), \tag{1}$$

© Springer International Publishing AG 2017
I. Dimov et al. (Eds.): NAA 2016, LNCS 10187, pp. 633–639, 2017.
DOI: 10.1007/978-3-319-57099-0_72

where $\mathbf{x} = (x_1, x_2) \in \overline{\Omega} \subseteq \mathbb{R}^2$, the operator $\mathscr{L}$ is Laplacian, $\mathscr{L}u = \frac{\partial^2 u}{\partial x_1^2} + \frac{\partial^2 u}{\partial x_2^2}$, $g(\mathbf{x}, t)$ is known analytic function, the solution $u(\mathbf{x}, t)$ and the two coefficients $\gamma(t)$, $\psi(t)$ are unknown.

To solve this inverse problem, the standard finite difference method is used for the approximation in space. The finite difference method was one of the first numerical methods used to solve partial differential equations (PDEs). It replaces differential operators by finite differences and the PDE becomes a (finite) system of equations.

The main features of the nonlinear inverse problem are taken into account via a proper choice of the linearized approximation in time. Linear problems at a particular time level are solved on the basis of a special decomposition into three standard elliptic problems.

The paper is organized as follows: In Sect. 2, we briefly introduce the formulation of the problem. In Sect. 3, we introduce the algorithm and apply on the inverse problems. The results of numerical experiments are presented in Sect. 4. Section 5 is dedicated to a brief conclusion. Finally, some references are introduced at the end.

2   The Problem Formulation

Let $\mathbf{x} = (x_1, x_2)$ and $\Omega = \{\mathbf{x} | \mathbf{x} = (x_1, x_2),\ 0 < x_i < l_i,\ i = 1, 2\}$ with the boundary $\partial\Omega = \{\mathbf{x} | \mathbf{x} = (x_1, x_2),\ x_i = 0,\ l_i,\ i = 1, 2\}$. The direct problem is formulated as follows,

$$u_t - \mathscr{L}u - \gamma(t)u = \psi(t)g(\mathbf{x}, t),\ \mathbf{x} \in \Omega,\ 0 < t \leqslant T, \tag{2}$$

where $\mathscr{L}u = \frac{\partial^2 u}{\partial x_1^2} + \frac{\partial^2 u}{\partial x_2^2}$, with initial conditions,

$$u(\mathbf{x}, 0) = u_0(\mathbf{x}),\ \mathbf{x} \in \overline{\Omega}, \tag{3}$$

and the Dirichlet boundary conditions,

$$u(\mathbf{x}, t) = 0,\ \mathbf{x} \in \partial\Omega,\ 0 < t \leqslant T, \tag{4}$$

where $\gamma(t)$, $\psi(t)$, $g(\mathbf{x}, t)$ and $\varphi(t)$ are known functions, $u(\mathbf{x}, t)$ is the solution of the Eq. (2) of second order, the initial condition $u_0(\mathbf{x})$ is known.

In this work, we consider the coefficients inverse problem instead of the directly. In this inverse problem, the coefficients $\gamma(t)$ and $\psi(t)$ in (2) are unknown, instead, some additional conditions are available as follows,

$$\begin{aligned} \int_{\Omega} u(\mathbf{x}, t)\lambda(\mathbf{x})d\mathbf{x} &= \Phi(t),\ 0 < t \leqslant T, \\ \int_{\Omega} u(\mathbf{x}, t)\chi(\mathbf{x})d\mathbf{x} &= \Psi(t),\ 0 < t \leqslant T, \end{aligned} \tag{5}$$

where $\lambda(\mathbf{x})$, $\chi(\mathbf{x})$ are weight functions. Choosing $\lambda(\mathbf{x}) = \delta(\mathbf{x} - \mathbf{x}^*)$ and $\chi(\mathbf{x}) = \delta(\mathbf{x} - \mathbf{x}^{**})$, $(\mathbf{x}^*, \mathbf{x}^{**} \in \Omega)$, where $\delta(\mathbf{x})$ is the Dirac δ-function, from (5), we get,

$$u(\mathbf{x}^*, t) = \Phi(t), \ 0 < t \leqslant T,$$
$$u(\mathbf{x}^{**}, t) = \Psi(t), \ 0 < t \leqslant T. \tag{6}$$

The inverse problem that we consider is to find $u(\mathbf{x}, t)$, $\gamma(t)$ and $\psi(t)$ from problems (2)–(4) and additional conditions (5) or (6). The corresponding conditions for existence and uniqueness of the solution are available in many works [9–11]. In this paper, we consider only the numerical solution of these inverse problems omitting theoretical issues of the convergence of an approximate solution to the exact one.

3   The Computational Algorithm

We consider the inverse problem (2)–(4), (6) with $\overline{\Omega} = [0, l]^2$. To numerically solve the inverse problem, we introduce the uniform grid of size $h = l/M$ in the domain Ω,

$$\omega_h = \{\mathbf{x}_{ij} | \mathbf{x}_{ij} = (ih_1, jh_2), \ i, j = 1, 2, \ \ldots, \ M - 1\},$$

where M is positive integers, $h_1 = h_2 = h$. For the time we have,

$$\omega_\tau = \{t^n | t^n = n\tau, \ n = 0, 1, \ \ldots, \ N, \ N\tau = T\}.$$

Using the fully implicit scheme for approximation in time, and the notation $u^n(\mathbf{x}) = u(\mathbf{x}, t^n)$,

$$u_t(\mathbf{x}, t) \approx \frac{u^{n+1}(\mathbf{x}) - u^n(\mathbf{x})}{\tau},$$

the finite difference approximation at each interior grid point, and the notation $u_{ij}(t) = u(\mathbf{x}_{ij}, t)$,

$$\mathcal{L} u_{ij} \approx \frac{u_{(i+1)j} - 2u_{ij} + u_{(i-1)j}}{h^2} + \frac{u_{i(j+1)} - 2u_{ij} + u_{i(j-1)}}{h^2},$$

For all grid nodes, except boundary ones, introducing the operator $\mathcal{D}$,

$$\mathcal{D} u_{ij} = -\frac{u_{(i+1)j} - 2u_{ij} + u_{(i-1)j}}{h^2} - \frac{u_{i(j+1)} - 2u_{ij} + u_{i(j-1)}}{h^2}.$$

Using the notations and the operator $\mathcal{D}$, we obtain the following variational problem,

$$\frac{u^{n+1} - u^n}{\tau} + \mathcal{D} u^{n+1} - \gamma^{n+1} u^n = \psi^{n+1} g^{n+1}, \ \mathbf{x}_{ij} \in \omega_h, \tag{7}$$

with initial conditions,

$$u(\mathbf{x}, 0) = u_0(\mathbf{x}), \ \mathbf{x} \in \Omega, \tag{8}$$

and the Dirichlet boundary conditions,

$$u(\mathbf{x}, t) = 0, \ \mathbf{x} \in \partial\Omega, \ t \in \omega_\tau, \tag{9}$$

the addition conditions (6) take the form,

$$u^n(\mathbf{x}^*) = \Phi^n, \; n = 0, \, 1, \, 2, \, \ldots, \, N,$$
$$u^n(\mathbf{x}^{**}) = \Psi^n, \; n = 0, \, 1, \, 2, \, \ldots, \, N. \tag{10}$$

We use the following decomposition for the solution u^{n+1},

$$u^{n+1}(\mathbf{x}) = y^{n+1}(\mathbf{x}) + \gamma^{n+1}v^{n+1}(\mathbf{x}) + \psi^{n+1}w^{n+1}(\mathbf{x}). \tag{11}$$

Applying (11) at every point $\mathbf{x}_{ij} \in \omega_h$, and substituted (11) into (7), we obtain three standard elliptic problems,

$$\left(\frac{1}{\tau} + \mathscr{D}\right)y_{ij}^{n+1} = \frac{u_{ij}^n}{\tau}, \; n = 0, \, 1, \, \ldots, \, N - 1, \, \mathbf{x}_{ij} \in \omega_h, \tag{12}$$

$$\left(\frac{1}{\tau} + \mathscr{D}\right)v_{ij}^{n+1} = u_{ij}^n, \; n = 0, \, 1, \, 2, \, \ldots, \, N - 1, \, \mathbf{x}_{ij} \in \omega_h, \tag{13}$$

$$\left(\frac{1}{\tau} + \mathscr{D}\right)w_{ij}^{n+1} = g_{ij}^{n+1}, \; n = 0, \, 1, \, 2, \, \ldots, \, N - 1, \, \mathbf{x}_{ij} \in \omega_h, \tag{14}$$

Written (12)–(14) in a matrix form, we have,

$$\left(\frac{1}{\tau}E + A\right)Y^{n+1} = \frac{1}{\tau}U^n,$$
$$\left(\frac{1}{\tau}E + A\right)V^{n+1} = U^n, \tag{15}$$
$$\left(\frac{1}{\tau}E + A\right)W^{n+1} = G^{n+1},$$

where

$$Y^{n+1} = [y_{11}^{n+1}, \ldots, y_{(M-1)1}^{n+1}, y_{12}^{n+1}, \ldots, y_{(M-1)2}^{n+1}, \ldots, y_{(M-1)(M-1)}^{n+1}]^T,$$
$$V^{n+1} = [v_{11}^{n+1}, \ldots, v_{(M-1)1}^{n+1}, v_{12}^{n+1}, \ldots, v_{(M-1)2}^{n+1}, \ldots, v_{(M-1)(M-1)}^{n+1}]^T,$$
$$W^{n+1} = [w_{11}^{n+1}, \ldots, w_{(M-1)1}^{n+1}, w_{12}^{n+1}, \ldots, w_{(M-1)2}^{n+1}, \ldots, w_{(M-1)(M-1)}^{n+1}]^T,$$
$$G^{n+1} = [g_{11}^{n+1}, \ldots, g_{(M-1)1}^{n+1}, g_{12}^{n+1}, \ldots, g_{(M-1)2}^{n+1}, \ldots, g_{(M-1)(M-1)}^{n+1}]^T,$$
$$U^n = [u_{11}^n, \ldots, u_{(M-1)1}^n, u_{12}^n, \ldots, u_{(M-1)2}^n, \ldots, u_{(M-1)(M-1)}^n]^T,$$

matrix A have the following forms,

$$A = \frac{1}{h^2}\begin{pmatrix} T & -I & & & \\ -I & T & -I & & \\ & \ddots & \ddots & \ddots & \\ & & -I & T & -I \\ & & & -I & T \end{pmatrix}, \quad T = \begin{pmatrix} 4 & -1 & & & \\ -1 & 4 & -1 & & \\ & \ddots & \ddots & \ddots & \\ & & -1 & 4 & -1 \\ & & & -1 & 4 \end{pmatrix}.$$

both T and I above are $(M-1) \times (M-1)$ matrices with I is the identity matrix, E is an identity matrix with the same size of A.

To evaluate γ^{n+1} and ψ^{n+1}, the addition conditions (10) are used, substituted (11) into (10), we get

$$\gamma^{n+1}v^{n+1}(\mathbf{x}^*) + \psi^{n+1}w^{n+1}(\mathbf{x}^*) = \Phi^{n+1} - y^{n+1}(\mathbf{x}^*),$$
$$\gamma^{n+1}v^{n+1}(\mathbf{x}^{**}) + \psi^{n+1}w^{n+1}(\mathbf{x}^{**}) = \Psi^{n+1} - y^{n+1}(\mathbf{x}^{**}), \tag{16}$$

where $\mathbf{x}^*, \mathbf{x}^{**} \in \omega_h$ and $\mathbf{x}^* \neq \mathbf{x}^{**}$. To solve γ^{n+1} and ψ^{n+1} from (16), we assume $v^{n+1}(\mathbf{x}^*)w^{n+1}(\mathbf{x}^{**}) - v^{n+1}(\mathbf{x}^{**})w^{n+1}(\mathbf{x}^*) \neq 0$, where y^{n+1}, v^{n+1} and w^{n+1} determined from (15).

Thus, the computational algorithm for solving the inverse problem (2)–(4), (6) based on the linearized scheme (7) and (10) involves the solution of three standard grid elliptic equations for the auxiliary functions $y^{n+1}(\mathbf{x})$ from Eq. (12), $v^{n+1}(\mathbf{x})$ from Eq. (13) and $w^{n+1}(\mathbf{x})$ from Eq. (14), the further evaluation of γ^{n+1} and ψ^{n+1} from (16), and the final calculation $u^{n+1}(\mathbf{x})$ from the relation (11).

4   Numerical Examples

In this section we present numerical results to test the efficiency of the new scheme for solving the coefficients inverse problems, in the example, we put $\mathbf{x} = (x_1, x_2) \in [0,1] \times [0,1]$ with the conditions,

$$u_0(\mathbf{x}) = -5 \exp\left(-100(x_1 - 0.5)^2 - 100(x_2 - 0.5)^2\right),$$

$$g(\mathbf{x}, t) = 8(1 - x_1)x_2 \exp(t), \ 0 < t \leqslant T.$$

The coefficients $\gamma(t)$ and $\psi(t)$ are taken in the forms,

$$\gamma(t) = -\frac{t^3}{1 + \exp\left(\zeta_1(t - 0.7T)\right)},$$

$$\psi(t) = \frac{100(T - t)}{1 + \exp\left(\zeta_2\left((T - t) - 0.8T\right)\right)}.$$

We consider the inverse problem with $h = \frac{1}{M}, \tau = \frac{T}{N}, M = 50, N = 100$. The observation points are $\mathbf{x}^* = (0.4, 0.4)$ and $\mathbf{x}^{**} = (0.7, 0.7)$. The solution of the direct problem at the observation point are depicted in Fig. 1(a), the solution at the finial time moment $(T = 1)$ is presented in Fig. 1(b), with $\zeta_1 = 50, \zeta_2 = 100$.

We also give the solution of $\gamma(t)$ (Fig. 2(a)) and $\psi(t)$ (Fig. 2(b)) of the inverse problem with $\zeta_1 = 50, \zeta_2 = 100$. The graphs show the very good accuracy and efficiency of the new approximate scheme.

The solution $\gamma(t)$ of the inverse problem with different ζ_1 are present in Fig. 3(a). For large ζ_1 (see Fig. 3(a)), $\gamma(t)$ approach discontinuous functions with a discontinuity point at $t = 0.7T$.

The solution $\psi(t)$ of the inverse problem with different ζ_2 are present in Fig. 3(b). For large ζ_2 (see Fig. 3(b)), $\psi(t)$ also approach discontinuous functions with a discontinuity point at $t = 0.2T$.

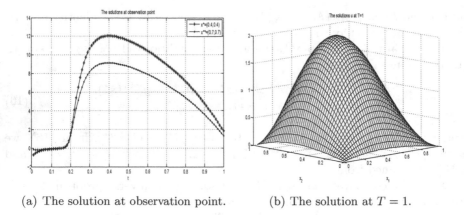

(a) The solution at observation point.   (b) The solution at $T = 1$.

Fig. 1. The solution at the observation point and the solution at $T = 1$.

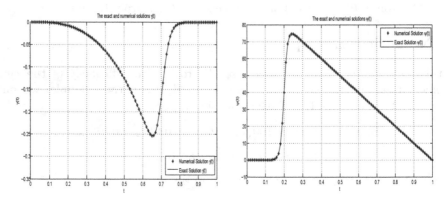

(a) The exact and numerical solutions $\gamma(t)$.   (b) The exact and numerical solutions $\psi(t)$.

Fig. 2. The exact and numerical solutions of the inverse problem.

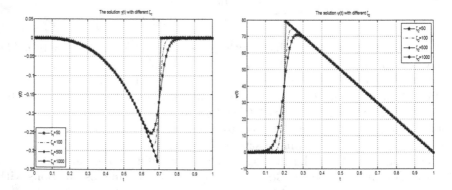

(a) The solution $\gamma(t)$ with different ζ_1.   (b) The solution $\psi(t)$ with different ζ_2.

Fig. 3. The solutions of the inverse problem with different variables.

5   Conclusions

In this paper, we proposed a numerical scheme to solve the inverse problems using the finite difference method. The numerical implementation using the linearized approximations in time and second-order approximation for the operator Laplacian in space, based on a decomposition of the approximate solution, where the transition to a new time level involves the solutions of three standard elliptic problems. Numerical solutions of the model problem demonstrate the convergence of the approximate solution of the inverse problem.

References

1. Isakov, V.: Inverse Problems for Partial Differential Equations. Applied Mathematical Sciences, vol. 127. Springer, Berlin (2006)
2. Prilepko, A.I., Orlovsky, D.G., Vasin, I.A.: Methods for Solving Inverse Problems in Mathematical Physics. Marcel Dekker, Berlin (2000)
3. Dehghan, M.: An inverse problem of finding a source parameter in a semilinear parabolic equation. Appl. Math. Model. **25**(9), 743–754 (2001)
4. Chan, T.F., Tai, X.C.: Level set and total variation regularization for elliptic inverse problems with discontinuous coefficients. J. Comput. Phys. **193**(1), 40–66 (2004)
5. Vabishchevicha, P.N., Vasilév, V.I., Vasiléva, M.V.: Computational identification of the right-hand side of a parabolic equation. Comput. Math. Math. Phys. **55**(9), 1015–1021 (2015)
6. Hansen, P.C.: The L-curve and it's use in the numerical treatment of inverse problems. In: Johnston, P. (ed.) Computational Inverse Problems in Electrocardiology. Advances in Computational Bioengineering Series, pp. 119–142. WIT Press, Southampton (2000)
7. Samarskii, A.A., Vabishchevich, P.N.: Numeircal Methods for Solving Inverse Problems of Mathematical Physics. VSP, Utrecht (2007)
8. Hansen, P.C.: Regularization tools: a Matlab package for analysis and solution of discrete ill-posed problems. Numer. Algorithms **6**(1), 1–35 (1994)
9. Prilepko, A., Soloviev, V.: On the solvability of an inverse boundary value problem of determining a coefficient at a lower order in a parabolic equation. Differ. Equ. **23**(1), 136–143 (1987)
10. Kamynin, V.L.: The inverse problem of the simultaneous determination of the right-hand side and the lowest coefficient in parabolic equations with many space variables. Math. Notes **97**(3), 349–361 (2015)
11. Kamynin, V.L.: The inverse problem of determining the lower-order coefficient in parabolic equations with integral observation. Math. Notes **94**(1), 205–213 (2013)

The GPU Solvers for High-Frequency Induction Logging

Irina Surodina[1,2](✉)

[1] Institute of Computational Mathematics and Mathematical Geophysics,
pr. Lavrentieva 6, Novosibirsk 630090, Russia
sur@ommfao1.sscc.ru
http://www.icmmg.nsc.ru
[2] A.A. Trofimuk Institute of Petroleum Geology and Geophysics SB RAS, 3,
Akademika Koptyuga, Novosibirsk 630090, Russia

Abstract. The results of numerical 2D and 3D simulations of high-frequency induction logging on a GPU platform are presentid. The conjugate A-orthogonal conjugate residual method with our full parallel preconditioner derived from the Hotelling-Schulz algorithm is used. Comparative analysis of the calculation time of the GPU implementation and CPU implementation is done. Numerical experiments show that the speed-up, obtained on GPU as compared to the CPU implementation for the 3D problems, is as great as 15–20 fold.

Keywords: Helmgoltz equation · Logging · Simulation · Krylov subspase methods · Preconditioner

1 Introduction

The fast solutions of direct problems can serve as a basis for solutions of certain inverse geophysical problems. One of the promising current directions for the speed-up of solutions to these problems is the use of parallel calculations on graphical processing units (GPU). Numerical algorithms can be significantly accelerated if the algorithms fit well to the characteristics of the GPU.

A numerical solution of the direct problem of the electromagnetic logging, through finite element or finite difference discretization, bring about large sparse linear systems usually solved by iterative methods of conjugate directions. With a suitable preconditioner, the perfomance can bedrastically increased. However, constructing preconditioners suitable for GPU architecture still remains a challenging task. Modern parallel implementations of the preconditioned conjugate directions on GPUs uses sparse approximate inverse preconditioners (AINV) due to their attractive features. First, columns or rows of the approximate inverse matrix can be generated in parallel. Second, the preconditioner matrix is used in preconditioned conjugate directions through matrix-vector multiplications, which are easy to parallelize. In published work [1], we have presented an algorithm for building a series of AINV preconditioners for the conjugate gradient

© Springer International Publishing AG 2017
I. Dimov et al. (Eds.): NAA 2016, LNCS 10187, pp. 640–647, 2017.
DOI: 10.1007/978-3-319-57099-0_73

method (CG). The computation of the AINV preconditioners is inherently parallel. The algorithm presented is derived from the Hotelling-Schulz algorithm for improving inverse matrix entries. This algorithm has been successfully applied to solve 2D and 3D potential field problems arising in mathematical geophysics from well resistivity logging applications.

In this paper we focus on the implementation of this approach for the numerical solution of the Helmgoltz equation, arising in the simulation of high-frequency induction logging. A numerical solution of the Helmgoltz equation through finite difference discretization leads to a large sparse linear system with an non-Hermitian symmetric matrix. For solving such systems efficiently, Tomohiro Sogabe and Shao-Liang Zhang proposed the conjugate A-orthogonal conjugate residual method (COCR) [2]. We use this method with our preconditioner.

This paper is organized as follows. In Sects. 2 and 3, we formulate the direct 2D and 3D problems. The implementation of the PCOCR algorithm on GPU is presented in Sect. 4. In Sect. 5, we present some numerical examples.

2   The 2D Numerical Simulation of High-Frequency Induction Logging

2D numerical simulation of high-frequency induction logging in an axially symmetric medium reduces to the solution of the Dirichlet problem for the Helmgoltz equation [3].

$$\frac{\partial^2}{\partial z^2} E_\varphi + \frac{\partial}{\partial r}\left(\frac{1}{r}\frac{\partial}{\partial r}(rE_\varphi)\right) + k^2 E_\varphi = i\omega\mu j\varphi_0. \tag{1}$$

Where E_φ is the tangential component of the quasi-stationary electric field, j_{φ_0} is the tangential component of the current density, $k^2 = \omega^2\mu\epsilon - i\omega\mu\sigma$ is the square wave number; ω is the circular frequency, $\sigma(r,z)$ is the electrical conductivity; $\epsilon(r,z)$ is the dielectric permittivity, $\mu = 4\pi * 10^{-7}$ H/m is the magnetic permeability.

Let us introduce a tangential component of the anomalous electrical field $E_\varphi^a = E_\varphi - E_\varphi^0$, where E_φ^0 is the tangential component of the quasi-stationary electric field in a uniform conductor with the conductivity $\sigma = \sigma_0$. Then

$$\frac{\partial^2}{\partial z^2} E_\varphi^a + \frac{\partial}{\partial r}\left(\frac{1}{r}\frac{\partial}{\partial r}(rE_\varphi^a)\right) + k^2 E_\varphi^a = (k_0^2 - k^2)E_\varphi^0 \tag{2}$$

where $k_0^2 = -i\omega\mu\sigma_0$.

After a carryng out finite difference approximation of Eq. (2) [4] and its subsequent symmetrization [5,6] we obtain the system of linear algebraic equation

$$Ay = f \tag{3}$$

where A is the block five-diagonal symmetric non-Hermitian matrix.

3   The 3D Numerical Simulation of High-Frequency Induction Logging

The 3D numerical simulation of high-frequency induction logging in any arbitrary medium reduces to the solution of the Dirichlet problem for partial differential equations of the second order with respect to the vector of the anomalous electric field [7]

$$rotrot\boldsymbol{E}^a + \boldsymbol{E}^a(i\omega\epsilon - \sigma)i\omega\mu = \boldsymbol{E}^0(\sigma - \sigma_0 - i\omega\epsilon)i\omega\mu \tag{4}$$

where $\sigma(x, y, z)$ is the electrical conductivity; $\epsilon(x, y, z)$ is the dielectric permittivity, $\boldsymbol{E}^a = (E_x, E_y, E_z)$ is the vector of the anomalous electric field in the Cartesian coordinate system. (The axis Z coincides with the axis of the probe and is directed downward).

We consider the magnetic dipole as a source of electromagnetic field. The magnetic dipole has the moment $M = (0, 0, M_z)$ directed along the axis Z. The components of the electric field E_x^0, E_y^0, E_z^0 are described in a homogeneous medium:

$$E_x^0 = -\frac{i\omega\mu M_z}{4\pi R^2}\frac{y}{R}(1 + k_0 R)e^{-k_0 R}, \quad E_y^0 = \frac{i\omega\mu M_z}{4\pi R^2}\frac{x}{R}(1 + k_0 R)e^{-k_0 R}, \quad E_z = 0$$

where $R = \sqrt{x^2 + y^2 + z^2}$. Discretization of (3) and its subsequent symmetrization [6] results in the linear system

$$Ay = f \tag{5}$$

where A is an $N \times N$ complex symmetric non-Hermitian matrix ($A \neq \overline{A}$, $A = A^T$). The matrix has a block-diagonal dramatically sparse structure.

4   The Method of Solution

Usually to solve systems (3), (5) direct and iterative methods are used. In general, the matrix A is large and sparse so that the direct methods become unsuitable. Although for small 2D - problems we often use the direct method (PARDISO from the Intel MKL library).

The conjugate A-orthogonal conjugate residual method (COCR) [2] is one of the best iterative methods for solving sparse complex symmetric linear systems of the form of (3), (5). The method is flexible, easy to implement and converges (theoretically) in a finite number of steps. The algorithm is as follows: Let X_n be the nth approximate solution in the method. Then, the corresponding nth residual vector r_n and search direction P_n are given by the following coupled two-term recurrences:

$$r_0 = b - Ax_0, \quad p_0 = r_0, \tag{6}$$

$$r_n = r_{n-1} - \alpha_{n-1} A P_{n-1}, \tag{7}$$

$$p_n = r_n + \beta_{n-1} A P_{n-1}, \quad n = 1, 2, \ldots, \tag{8}$$

The coefficients α_n and β_n are determined by the following orthogonality conditions

$$r_n \perp W \quad and \quad A p_n \perp W$$

If A is real symmetric and positive definite, then the choice of $W = K_n(A, r_0)$, where $K_n = span(r_0, Ar_0, \ldots, A^{n-1}r_0)$ leads to the conjugate gradient method (CG).

If A is complex symmetric, then the choice of $W = \overline{A} K_n(\overline{A}, \overline{r_0})$ leads to the COCR method. Then

$$\alpha_n = \frac{(\overline{r}_n, Ar_n)}{(\overline{Ap}_n, Ap_n)}, \beta_n = \frac{\overline{r}_{n+1}, Ar_{n+1}}{(\overline{r}_n, Ar_n)}. \tag{9}$$

All the necessary calculations in formulas (6)–(9) are the matrix-vector operations and can be well parallelized on the GPU. For the NVIDIA GPU implementation, all the required operations can be found in the standard CUBLAS Library. However, in this case, the rate of convergence will be low, because the matrix has a large condition number, i.e. $\mathrm{cond}\,(A) \gg 1$. The efficiency of the COCR method can be increased by applying a preconditioning technique that improves the properties of the matrix. Suppose that M is a symmetric matrix that approximates A, but it is easier to invert. We can solve the equation $Ax = b$ indirectly by solving $M^{-1}Ax = M^{-1}b$. If $\mathrm{cond}\,(M^{-1}A) \ll \mathrm{cond},(A)$ or if the eigenvalues of $M^{-1}A$ are better clustered than those of A, we can iteratively solve (5) faster than the initial one [8]. In the case of the GPU realization we impose the requirement of high parallelization on onstructing the matrix M^{-1}.

In this paper we use the original approach [1] to constructing the preconditioning matrix relying on an approximation of the inverse matrix. Based on the Hotelling-Schulz algorithm [9,10], this approach is fully parallel. The main idea of the approach is the following. Let us assume that D_0 is an initial approximation of A^{-1}. If

$$\|E - AD_0\| \le k < 1, \tag{10}$$

then we can build an iterative process:

$$D_1 = D_0 + D_0(E - AD_0). \tag{11}$$

$$D_2 = D_1 + D_1(E - AD_1) = 2D_1 - D_1 AD_1 \tag{12}$$

$$\begin{aligned} D_3 &= D_2 + D_2(E - AD_2) = \\ &= (2D_1 - D_1 AD_1)(2E - A(2D_1 - D_1 AD_1)) \\ &= 2(2D_1 - D_1 AD_1) - (2D_1 - D_1 AD_1)A(2D_1 - D_1 AD_1) \end{aligned} \tag{13}$$

$$D_{m+1} = D_m + D_m(E - AD_m)$$

This process converges to A^{-1} [11] and in this case the error is

$$\|D_n - A^{-1}\| \le \|D_0\| \frac{k^{2^n}}{1 - k}. \tag{14}$$

In addition, the algorithm preserves the symmetry of all matrices A: If $A = A^T$ and $D_0 = D_0^T$, them $D_m = D_m^T, m = 1, 2, \ldots$. With a reasonable initial approximation D_0, we can use D_m as the preconditioning matrix M in the PCOCR algorithm. The iteration scheme of preconditioned conjugate A-orthogonal conjugate residual method (PCOCR):

$$
\begin{aligned}
&k = 0 : Initialization : &&x_0, r_0 = b - Ax_0, z_0 = D_i r_0, p_0 = z_0 \\
&k \geq 0 : while \|r_k\| / \|r_0\| > \epsilon 1. &&q_k = Ap_k, \alpha_k = \frac{z_k, Az_k}{q_k, D_i q_k} \\
&2. &&x_{k+1} = x_k + \alpha_k p_k, r_{k+1} = r_k - \alpha_k q_k && (15) \\
&3. &&z_{k+1} = D_i r_{k+1} \\
&4. &&\beta_i = \frac{z_{k+1}, Az_{k+1}}{z_k, Az_k}, p_{k+1} = r_{k+1} + \beta_k p_k.
\end{aligned}
$$

We use a well-known Jacobi preconditioner $D_0 = diag(a_{11}^{-1}, a_{22}^{-1}, \ldots, a_{nn}^{-1})$ as an initial approximation to A^{-1}. In our case this is possible since in Eqs. (3) and (5) we have the matrices with a weak diagonal domination. The matrix D_1 is easy to calculate

$$
\begin{aligned}
d_{ii} &= 1/a_{ii} \\
d_{i,i+1} &= a_{i,i+1}/(a_{i+1,i+1}a_{ii}) && (16) \\
d_{i,i+m} &= a_{i,i+m}/(a_{i+m,i+m}a_{ii})
\end{aligned}
$$

Formulas (16) are given for the 2D problem. Similar trivial calculations can be done in the 3D case. We can see that D_1 has the same structure as the matrix A. So it means that the same matrix-vector product implementation can be used for the multiplication of the original matrix and the preconditioner in the PCOCR algorithm. The factorized forms (12), (13) are properly suitable for computations, because we know the structure of the matrices A and D_1. It is difficult to use the explicit form of the matrix with index 2 and higher, because the number of fill-ins accordingly increases. For example, the matrix D_1 contains 5 diagonals for a 2D finite difference grid, D_2 contains 25 diagonals, and D_3 contains 113 diagonals.

As a criterion for choosing the optimal preconditioner we took a minimum time for solving the problem with the prescribed accuracy.

As compared to the PCG method, which we used for the potential field problems [1,13], the PCOCR method requires the multiplication by the preconditioner at two steps. If we use the preconditioner D_1, we will need two additional matrix-vector products (one for step 1 and one for step 3). If we use D_2, we will need 6 additional multiplications (3 per each step). If we use D_3, we will need 14 additional multiplications. Also, additional operations are required such as the vector additions, the multiplications by a constant. In view of the somewhat larger (as compared to PCG method) computer cost of implementing the PCOCR method, we should expect the more modest results than in [13]. This is confirmed by the calculations for real models.

5 Simulations

Let us consider a typical model with axisymmetrical electrical conductivity (Fig. 1). The medium is divided into laterally inhomogeneous strata by a system of parallel flat boundaries. There is a well of radius 0.108 m with resistance 2 Ohm·m. Some strata could include the drilling fluid zone and the surrounding zone. The resistance of beds varies from 3 to 100 Ohm ·m. The dielectric permittivity in this model is assumed to be 0.

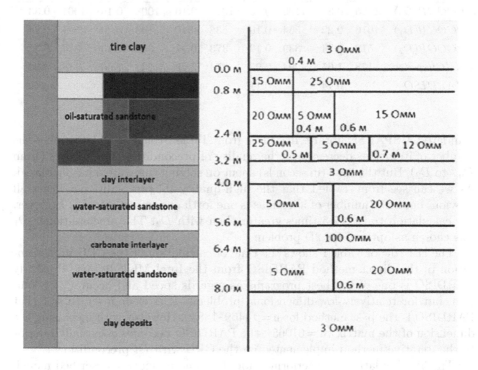

Fig. 1. Model of medium

Table 1 shows the results of simulations for one probe position in the above described model for the 2D case. The results of numerical calculations for 5 different probes VIKIZ (with frequencies ranging from 14 MHz to 0.875 MHz) are shown. For the numerical experiments, we used the following processors

CPU: Intel Xeon X5670 2.93 GHz (using Intel Fortran 11.0),
GPU: NVIDIA Tesla M2090, 512 Core, 4 GB RAM (using CUDA).

A calculated grid for each probe was constructed. The non-uniform grid was built into according to the probe configuration and the model in question. In all cases the iteration was started with $x_0 = 0$, and the stopping criterion was $\|r_k\|\|r_0\|^{-1} \leq 10^{-9}$. It is this value of the residual that we needed for our simulation to solve the problem with a given accuracy ~3%

In Table 1, we compare the GPU implementations of the PCOCR method with D_0, D_1 and D_2 preconditioners to the GPU implementation of the PCOCR method with IC preconditioner (Incomplete Cholesky factorization) from the

Table 1. The 2D problem. Results of calculations by the COCR method with various preconditioners. Results of PARDISO.

Method	n = 14469		n = 17819		n = 18795		n = 18849		n = 61065	
	Iter.	Time	Iter.	Time	Iter.	Time	Iter.	Time	Iter.	Time
$PCOCR(D_0)$	1925	0.3	1100	0.15	1078	0.16	1008	0.15	1061	0.33
$PCOCR(D_1)$	1016	0.22	554	0.10	538	0.10	499	0.10	521	0.24
$PCOCR(D_2)$	512	0.24	339	0.12	373	0.13	366	0.13	376	0.31
$IC(Cusparse)$	178	1.65	294	2.95	254	2.65	233	2.66	188	4.22
$PARDISO$		0.14		0.11		0.10		0.10		0.40

standard CUSPARSE library. It is clear from Table 1 that for all the probes the number of iterations decreases as the quality of preconditioning improves (from D_0 - to D_2). But the time (in seconds) spent on solving the system has increased. As we can see from Table 1 that the high-quality IC preconditioner beats all previous ones: the number of iterations is one forth as compared to D_2. However, the calculation time is 7.5 times greater than with D_2. The preconditioner D_1 was chosen as optimal for 2D problem.

The last row of Table 1 shows the time of calculation of the CPU implementation of the direct method PARDISO from the Intel MKL library. Presently PARDISO is one of the best programs as regards speed and accuracy of solution, but for relatively low-dimensional problems. It is clear from Table 1 that PARDISO is the best method for n = 14469–18849. However with increasing the dimension of the matrix (n = 61065), the PARDISO becomes essentially inferior to the iterative method implemented on the GPU with Di preconditioner.

The 3D simulation was performed for the same previously described model, but the well had a slope of 75°. The preconditioner D_1 was optimal in this case. In Table 2, the time (in seconds) of the PCOCR calculation for one probe position is show. The calculations are performed for 5 probes VIKIZ. We compare the GPU implementations of the PCOCR method with D_1 preconditioner and a CPU implementation of the PCOCR method with SSOR (Symmetric Successive Over-Relaxation) preconditioner. We have performed the calculations on the GPU M2090 and K40 (NVIDIA Tesla K40, 2880 Core, 12 GB RAM).

Table 2. 3D problem. The results of calculations by the PCOCR method

n	1260681	1976475	2032344	2414475	2546154
CPU	111.80	167.11	167.78	219.60	346.70
$GPU(M2090)$	7.36	11.02	11.25	14.54	23.24
$GPU(K40)$	5.50	8.04	8.08	10.54	16.32

As one can see from Table 2, the speed-up is 15 - times for M2090 and 20.5 - times for k40 in comparison with the running time of the sequential software versions (for CPU) for all the probes VIKIZ.

6   Conclusion

The algorithms and programs for the fast GPU implementation of high-frequency induction logging have been created. The proposed preconditioner for the conjugate gradient method in [1] has proved this applicability for the conjugate A-orthogonal conjugate residual metod. The direct method (PARDISO) for the small-size grids remains preferable ($N = 14{,}000 - 19{,}000$). But with an increase in the size of the matrix, the GPU implementation of the iterative method PCOCR with the preconditioner D_1 works more efficiently. Basing on this two-dimensional program, we created an inversion program [12].

References

1. Labutin, I.B., Surodina, I.V.: Algorithm for sparse approximate inverse preconditioners in conjugate gradient method. Reliab. Comput. (Interval Comput.) J. **18** (2013). http://interval.louisiana.edu/reliable-computing-journal/tables-of-contents.html
2. Sogabe, T., Zhang, S.: A COCR method for solving complex symmetric linear systems. J. Comp. App. Math. **199**, 297–303 (2007)
3. Kaufman, A.A.: Theory of Induction Logging. Nauka Novosibirsk (1965). (in Russian)
4. Martakov, S.V., Epov, M.I.: Direct two-dimensional problems of electromagnetic logging. Geologiya i Geofizika (Rus. Geol. Geoph.) **40**(2), 250–255 (1999)
5. Surodina, I.V., Epov, M.I.: High-frequency induction data affected by biopolymer-based drilling fluids. Geologiya i Geofizika (Rus. Geol. Geoph.) **53**(8), 817–822 (2012)
6. Kuznetsov, Y.I., Agapitova, N.S.: Computer Modeling: Mathematical Background. YuSIEPI, Yuzhno-Sakhalinsk (2003). (in Russian)
7. Surodina, I.V., Martakov, S.V., Epov, M.I.: 3D mathematical simulations of the harmonic electromagnetic fields in the problems logging in inclined horizontal wells. In: International Conference on Computational Mathematics, ICCM 2004, vol. 2, pp. 699–703 (2004). (in Russian)
8. Saad, Y.: Iterative Methods for Sparse Linear Systems, 2nd edn. Society for Industrial and Applied Mathematics, Philadelphia (2003)
9. Hotelling, H.: Analysis of a complex of statistical variables into principal components. J. Educ. Psych. **417–441**, 498–520 (1933)
10. Schulz, G.: Iterative Berechnung der reziproken Matrix. Z. Angew. Math. Mech. **13**, 57–59 (1933)
11. Faddeev, D.K., Faddeeva, V.N.: Computational Methods of Linear Algebra. Fizmatgiz, Moscow (1963). (in Russian)
12. Nikitenko, M.N., Surodina, I.V., Mikhaylov, I.V., Glinskikh, V.N., Suhorukova, C.V.: Formation evaluation via 2D processing of induction and galvanic logging data using high-performance Computing. In: Abstract of 77th EAGE Conference and Exhibition, 2015 Madrid, p. Tu N107 15 (2015)
13. Surodina, I.V.: Parallel algorithms for direct electrical logging problems. Yakut. Math. J. **22**(2) (2016)

Mathematical Modeling of Fan-Structure Shear Ruptures Generated in Hard Rocks

Boris G. Tarasov[1] and Vladimir M. Sadovskii[2(✉)]

[1] University of Western Australia, Stirling Highway 35, 6009 Perth, WA, Australia
boris.tarasov@uwa.edu.au
[2] Institute of Computational Modeling SB RAS,
Akademgorodok 50/44, 660036 Krasnoyarsk, Russia
sadov@icm.krasn.ru

Abstract. The main goal of this paper is to analyze the fan-mechanism of rotational motion transmission in a system of elastically bonded slabs on flat surface, simulating growth of shear ruptures in super brittle rocks. A physical model recently designed demonstrates that the fan-structure formation can be stable at the absence of distributed shear stress applied. The action of distributed shear stress causes the fan propagation as a wave representing the rupture head. The developed mathematical model of a fan-structure as a continuous system establishes the relation between the fan velocity and the fan length. It is shown that in the absence of friction the fan velocity may be arbitrary, but not greater than the limit velocity which is determined by the moment of inertia of slabs, the initial angle of their orientation and the elastic coefficient of bonds. In a system with friction the velocity of traveling fan is solely determined by the opening angle. The action of distributed shear stress leads to the instability start before the fan-structure completion. The fan length decreases with increasing velocity.

Keywords: Shear rupture · Super brittle rock · Fan-structure · Continuous model · Euler equation · Traveling fan

1 Introduction

Shear rupture is a typical mode of rock failure at confined compression including earthquakes. A proper understanding of rupture mechanisms provides a basis for better understanding of dynamic events in the Earth's crust. Recently a new rupture mechanism of extreme dynamics has been identified, which differs fundamentally from the generally accepted mechanism. Figure 1 illustrates the difference between the frictional (conventional) and fan-hinged (new) rupture mechanisms and in rock properties at failure caused by these mechanisms. It shows two specimens loaded by axial and confining compressional stresses $\sigma_1 > \sigma_2 = \sigma_3$. The induced shear σ_τ and normal σ_n stresses are related to shear ruptures propagating from points A towards points C. Any rupture is capable to propagate in its own plane owing to the creation of an echelon of microtensile cracks at

© Springer International Publishing AG 2017
I. Dimov et al. (Eds.): NAA 2016, LNCS 10187, pp. 648–656, 2017.
DOI: 10.1007/978-3-319-57099-0_74

the rupture tip directed along σ_1, which form intercrack slabs (or domino-like blocks) [1]. The slabs are subjected to rotation due to the relative displacement of the rupture faces [2,3]. According to the frictional mechanism, the slabs collapse at rotation creating friction within the rupture head, which provides the variation of the specimen strength according to the stress–displacement curve in Fig. 1a. When the rupture has crossed the specimen at point C, the specimen strength is determined by friction along the completed rupture. It is believed today that the frictional strength represents the lower limit on rock shear strength [4].

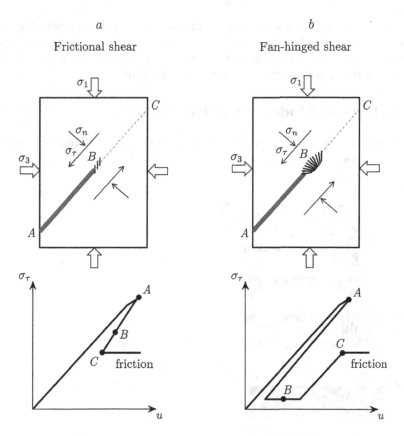

Fig. 1. Fundamental difference in post-peak properties of hard rocks associated with shear rupture propagation, governed by the frictional (conventional) rupture mechanism (a) and the fan-hinged (new) rupture mechanism (b)

Recent studies [5–8] showed that in hard rocks at confining stresses corresponding to the seismogenic depths the domino-blocks can withstand rotation between the shearing rupture faces. Because the relative shear displacement of the rupture faces increases with distance from the rupture tip, the domino-blocks form a fan-structure that represents the rupture head. In this case, the domino blocks behave as hinges between the moving rupture faces that dramatically decreases friction within the rupture head (up to an order of magnitude compared with the conventional frictional strength). The fan-head determines the abnormally low transient strength of the material during the failure process and can propagate through the material as a wave with extreme dynamics. The fan-mechanism can provide intersonic rupture velocity. The stress–displacement curve in Fig. 1b illustrates the variation in rock strength during the post-peak failure governed by the fan-hinged mechanism. When the fan-head has crossed the specimen, the specimen strength becomes equal to the conventional frictional strength.

In this paper, within the framework of the continuum approach, the mathematical model of the fan-structure shear rupture is constructed, based on the assumption that a large number of thin slabs is involved in a fan and that the fan length is much greater than the length of the slabs.

2   Continuous Model of the Fan System

Photographs of a laboratory physical model illustrating different stages of the fan-structure formation and propagation are represented in Fig. 2 [8].

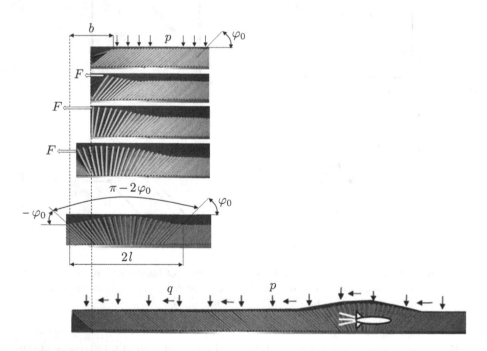

Fig. 2. Physical model of fan-mechanism of the hard rock failure under high stresses

In case that the length of a fan is more than the characteristic size a of slabs and the fan consists of many slabs, the fan travel can be described using a continuous model. In this model, the strain ε of elastic bonds is calculated in terms of current coordinates of the upper ends of the slabs $X = x + u$ and $Y = \sqrt{a^2 - u^2}$ by the formula:

$$\varepsilon = \frac{ds^2 - ds_0^2}{2\,ds_0^2} = \frac{dX^2 + dY^2 - dx^2}{2\,dx^2}$$

$$= \frac{1}{2}\left((a\,\varphi' - \sin\varphi)^2 + \cos^2\varphi - 1\right) \approx -a\,\varphi' \sin\varphi = u'.$$

Here x is the coordinate of the bottom ends, $u = a\cos\varphi$ is the longitudinal displacement, φ is the angle of rotation, ds and ds_0 are the actual and initial lengths of the element of a bond, the prime marks a derivative with respect to the variable x. Under the stable equilibrium of the fan, given there are no tangential forces, the minimum potential energy principle holds true:

$$I = \int_{x_e}^{x_b} W(u, u')\,dx, \quad W = \frac{k}{2}\varepsilon^2 - pY = \frac{k}{2}(u')^2 - p\sqrt{a^2 - u^2},$$

where x_b and x_e are the beginning and end coordinates of the fan, p is the normal force per unit length, k is the elastic modulus of bonds. The related Euler's equation

$$\frac{d}{dx}\frac{\partial W}{\partial u'} - \frac{\partial W}{\partial u} = 0 : \quad k\,u'' + p\frac{u}{\sqrt{a^2 - u^2}} = 0$$

describes configuration of a quiescent fan. Insertion of $u' = z(u)$ integrates the equation:

$$k\,z^2 = 2p\sqrt{a^2 - u^2} + C_1.$$

The integration constant C_1 is determined from the condition that the tension force in the head of the fan is zero: $z(a\cos\varphi_0) = 0$, where φ_0 is the initial angle. The obtained solution is a first-order ordinary differential equation for displacement:

$$u' = \lambda\sqrt{\sqrt{1 - \frac{u^2}{a^2}} - \sin\varphi_0}, \quad \lambda^2 = \frac{2p\,a}{k}.$$

Separation of the variables yields the formula for the half-length $l = (x_b - x_e)/2$ of the fan:

$$l = \frac{a}{\lambda}\int_0^{\cos\varphi_0} \frac{dw}{\sqrt{\sqrt{1 - w^2} - \sin\varphi_0}} = \frac{a}{\lambda}\int_{\varphi_0}^{\pi/2} \frac{\sin\varphi\,d\varphi}{\sqrt{\sin\varphi - \sin\varphi_0}}. \tag{1}$$

The quiescent fan is symmetric relative to the peak point and the fan is complete at the point x_e, where $\varphi = \pi - \varphi_0$ and the connection is free of tension.

For a traveling fan, the equation of motion can be derived based on the principle of possible displacements in combination with D'Alembert's principle:

$$-j\left(\ddot\varphi + \frac{\dot\varphi}{\tau}\right)\delta\varphi + \left(k\,u'' + p\frac{u}{\sqrt{a^2 - u^2}} - q\right)\delta u = 0,$$

where j and q are the moment of inertia and tangential force per unit length, τ is the relaxation time, δ means the variation, dot over a symbol denotes the derivative with respect to time. Since

$$\delta\varphi = -\frac{\delta u}{\sqrt{a^2 - u^2}}, \quad \dot\varphi = -\frac{\dot u}{\sqrt{a^2 - u^2}}, \quad \ddot\varphi = -\frac{\ddot u}{\sqrt{a^2 - u^2}} - \frac{u\,\dot u^2}{\left(\sqrt{a^2 - u^2}\right)^3},$$

the above equation takes the form:

$$j\,\frac{\ddot u}{a^2 - u^2} + j\left(\frac{u\,\dot u}{a^2 - u^2} + \frac{1}{\tau}\right)\frac{\dot u}{a^2 - u^2} - k\,u'' - p\,\frac{u}{\sqrt{a^2 - u^2}} + q = 0. \quad (2)$$

For Eq. (2) the energy balance equation

$$\frac{j}{2}\frac{d}{dt}\frac{\dot u^2}{a^2 - u^2} - k\,u''\,\dot u + p\frac{d}{dt}\sqrt{a^2 - u^2} + q\,\dot u = -\frac{j}{\tau}\frac{\dot u^2}{a^2 - u^2}$$

is satisfied with allowance for dissipative processes.

3   Traveling Fan

Insertion of $u = u(x - Vt)$ for the fan traveling at a constant velocity $V > 0$ leads to an equation that, after integrating, transforms as follows:

$$\left(k - \frac{j\,V^2}{a^2 - u^2}\right)\frac{(u')^2}{2} - p\sqrt{a^2 - u^2} - q\,u = C_2 - \frac{j\,V}{\tau}\int_x^{x_b}\frac{(u')^2\,dx}{a^2 - u^2}. \quad (3)$$

Here $x_e < x < x_b$, the integration constant $C_2 = -p\,a\,\sin\varphi_0 - q\,a\,\cos\varphi_0$ is determined by virtue of the boundary conditions $u = a\,\cos\varphi_0$, $u' = 0$ in the head of the fan.

Given there is no friction and tangential forces ($\tau \to \infty$, $q = 0$), Eq. (3) yields an equation of free fan traveling in horizontal plane. Such fan is symmetric relative to the peak point, can move at any velocity below the critical value $V_* = a\,\sin\varphi_0\,\sqrt{k/j}$ and its half-length is given by:

$$l = \frac{a}{\lambda}\int_{\varphi_0}^{\pi/2}\sqrt{\frac{\sin^2\varphi - \nu^2}{\sin\varphi - \sin\varphi_0}}\,d\varphi, \quad \lambda^2 = \frac{2\,p\,a}{k}, \quad \nu^2 = \frac{j\,V^2}{k\,a^2}.$$

The simple analysis shows that the traveling fan is in this case always shorter than the quiescent fan, the half-length of which is found from the formula (1), and that the fan shortens with the higher velocity. In a system with friction ($\tau < \infty$), the horizontal fan velocity V equals zero. This follows from Eq. (3) taken at $x = x_e$, with regard to the boundary conditions at the end point of the fan. Figure 3 shows the plots of the dimensionless half-length $\bar{l} = \lambda l/a$

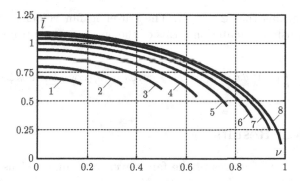

Fig. 3. Half-length versus velocity of the fan for different initial angles

versus the dimensionless velocity ν of the fan for the initial displacement angles $\varphi_0 = 10°, 20°, \ldots, 80°$ (curves $1, 2, \ldots, 8$, respectively). On the right, the curves terminate at the critical velocity ν_* dependent on the initial angle.

Under the supposition that the tangential force is non-zero and the friction in a system is neglectable, (3) produces the equation

$$\left(k - \frac{jV^2}{a^2 - u^2}\right)\frac{(u')^2}{2} = p\left(\sqrt{a^2 - u^2} - a\sin\varphi_0\right) + q\left(u - a\cos\varphi_0\right),$$

showing incomplete opening of the fan. As judged by this equation, the bond of the fan end slab vanishes at the rotation angle φ_*, for which the right-hand side of this equation turns out to equal zero:

$$p\sin\varphi_* + q\cos\varphi_* = p\sin\varphi_0 + q\sin\varphi_0.$$

Let ψ be the angle between the external force vector and the axis x. Then

$$p = \sqrt{p^2 + q^2}\sin\psi, \quad q = \sqrt{p^2 + q^2}\cos\psi,$$
$$\cos(\psi - \varphi_*) = \cos(\psi - \varphi_0) \quad \Rightarrow \quad \varphi_* = 2\psi - \varphi_0.$$

The angle ψ ranges from φ_0 to $\pi/2$, inasmuch as the violation of the condition that $\psi \geqslant \varphi_0$ would result in the concurrent overturn of all slabs, whereas the condition $\psi < \pi/2$ fits with the accepted orientation of the tangential force. Consequently, the equation of motion describes in this case a traveling fan with the incomplete opening angle $\varphi_* < \pi - \varphi_0$. The velocity of such fan may be an arbitrary value, not to exceed the critical velocity V_*.

It turns out that the opening angle of a traveling fan narrows because of the dissipative forces. With the friction at the fan end ($\varphi = \varphi^*$) taken into account, on the strength of (2), the strict inequality is satisfied:

$$p\left(\sin\varphi^* - \sin\varphi_0\right) + q\left(\cos\varphi^* - \cos\varphi_0\right),$$

which exactly means that $\varphi^* < \varphi_*$. The traveling fan velocity is in the given case unambiguously calculated in terms of the opening angle from the equation:

$$\frac{jV}{a\tau}\int_{x_e}^{x_b}(\varphi')^2\,dx = p(\sin\varphi^* - \sin\varphi_0) + q(\cos\varphi^* - \cos\varphi_0).$$

4   Numerical Results

The above described features of the fan travel were analyzed using a discrete model, represented in [9]. In particular, the comparison of the computations of half-lengths using the formula (1), by which $l = 0.107\,\mathrm{m}$ at k selected at a level $120 - 150\,\%$, and the computations using the discrete model showed complete conformity of the results.

Figure 4 illustrates computations of incomplete opening of a fan at the inclination angle $\alpha = 20°$. The calculations have shown that initiation of a fan naturally takes place when the first slab is forced to rotate up to a critical angle close to φ_*. Moreover, setting the fan initiation velocity, i.e. the angular velocity of the first slab, allows a fan traveling at a certain velocity. Under fast rotation of the first slab for $T = 1\,\mathrm{s}$, the fan opening up to the moment of initiation has a wider angle (see Fig. 4a) than in the case of comparatively slow opening ($T = 6\,\mathrm{s}$ in Fig. 4b). When this happens, the fan velocity grows, which qualitatively agrees with Eq. (3).

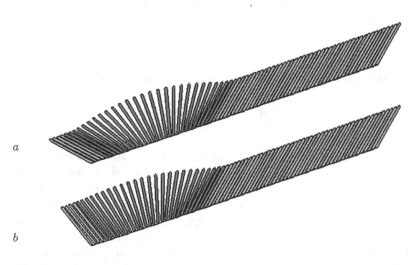

Fig. 4. Fan initiation in a system of inclined slabs at various opening times: (a) $T = 1\,\mathrm{s}$ ($t = 0.82\,\mathrm{s}$), (b) $T = 6\,\mathrm{s}$ ($t = 4\,\mathrm{s}$)

On the whole, the correlation of the results has shown that the equations of the continuous model under the long fan approximation describe the basic quantitative and qualitative characteristics of the Tarasov model. Using the data on physical parameters of hard rocks at seismogenic depth, the model will enable investigation in the process of dynamic propagation of shear ruptures with the fan-mechanism. This will allow deeper insight into the nature of the dynamic events in the Earth's crust.

5 Conclusions

We have derived the equation describing the equilibrium of the fan-structure under the long fan approximation, and have calculated the fan length depending on the weight of the slabs and elasticity coefficient of the bonds. It is shown that when friction is absent, a stable fan may travel at any velocity not to exceed a critical value governed by the size and moment of inertia of slabs, initial angle and elasticity coefficient of bonds, and that the fan length shortens with increasing velocity. When there are no tangential forces and the system of slabs is in a horizontal position, the fan travel is interrupted by friction. The action of tangential forces results in incomplete fan, and the opening angle of fan decreases with increasing friction. In the system with friction, the traveling fan velocity is unambiguously determined in terms of the opening angle, or it may possess any value within a permitted range if friction is neglected.

The comparisons of computations of the fan length and velocity using the analytical formulas and using the discrete model and laboratory observations has yielded good compliance of the results. The developed continuous model will be used to study the nature of initiation of dynamic shear ruptures in rocks at seismogenic depths.

Acknowledgements. This work was supported by the Centre for Offshore Foundation Systems (The University of Western Australia), the Complex Fundamental Research Program no. II.2P "Integration and Development"of SB RAS (project no. 0361-2015-0023) and the Russian Foundation for Basic Research (grant no. 14-01-00130).

References

1. Reches, Z., Lockner, D.A.: Nucleation and growth of faults in brittle rocks. J. Geophys. Res. Solid Earth **99**(B9), 18159–18173 (1994)
2. Ortlepp, W.D.: Rock Fracture And Rockbursts: An Illustrative Study. Monograph Series M9. South African Institute of Mining and Metallurgy, Johannesburg (1997)
3. King, G.C.P., Sammis, C.G.: The mechanisms of finite brittle strain. Pure Appl. Geophys. **138**(4), 611–640 (1992)
4. Scholz, C.H.: The Mechanics of Earthquakes and Faulting. Cambridge University Press, Cambridge (2002)
5. Stavrogin, A.N., Tarasov, B.G.: Experimental physics and rock mechanics: results of laboratory studies. A.A. Balkema Publishers, Lisse - Abingdon - Exton - Tokio (2001)

6. Tarasov, B.G., Randolph, M.F.: Frictionless shear at great depth and other paradoxes of hard rocks. Int. J. Rock Mech. Min. Sci. **45**(3), 316–328 (2008)
7. Tarasov, B.G.: Intersonic shear rupture mechanism. Int. J. Rock Mech. Min. Sci. **45**(6), 914–928 (2008)
8. Tarasov, B.G.: Hitherto unknown shear rupture mechanism as a source of instability in intact hard rocks at highly confined compression. Tectonophysics **621**, 69–84 (2014)
9. Tarasov, B.G., Sadovskii, V.M., Sadovskaya, O.V.: Analysis of fan waves in a laboratory model, simulating the propagation of shear ruptures in rocks. Comput. Contin. Mech. **9**(1), 38–51 (2016). [in Russian]

On the Numerical Analysis of Fan-Shaped Waves

Boris G. Tarasov[1]($\boxtimes$), Vladimir M. Sadovskii[2], and Oxana V. Sadovskaya[2]

[1] Centre for Offshore Foundation Systems, University of Western Australia,
Stirling Highway 35, Perth, WA 6009, Australia
boris.tarasov@uwa.edu.au
[2] Institute of Computational Modeling SB RAS,
Akademgorodok 50/44, 660036 Krasnoyarsk, Russia
{sadov,o_sadov}@icm.krasn.ru

Abstract. The fan-shaped mechanism of rotational motion transmission in a system of elastically bonded slabs on flat surface is studied. This mechanism governs the propagation of shear ruptures in super brittle rocks at stress conditions of seismogenic depths. The current paper analyzes a built laboratory physical model, which demonstrates the process of fan waves propagation. Equations of the dynamics of the fan-structure as a mechanical system with a finite number of degrees of freedom are obtained. Computational algorithm, taking into account contact interaction of slabs, is worked out. The computations, showing the incomplete disclosure of fans with different opening angles due to fast or slow change in the velocity of rotation of the first slab, are performed. Comparison of the results of computations of length and velocity of a fan by means of a discrete model with laboratory measurements and observations shows good correspondence between the results.

Keywords: Shear rupture · Fan-shaped wave · Discrete model · Lagrange equations · Computational algorithm

1 Introduction

Natural observations and laboratory studies demonstrated that the propagation of shear ruptures in rocks under confined compression is associated with consecutive creation of inclined slabs (domino-like blocks) in the fracture tip which, due to rotation caused by shear displacement of the fracture interfaces, collapse creating friction in the fracture head [1,2]. Recent studies [3–5] showed that in hard rocks at stress conditions corresponding to seismogenic depths the domino-blocks can withstand rotation and form a fan-structure in the rupture head. The fan-head is characterised by very low shear resistance and can propagate as a wave of extreme dynamics through intact rock mass at shear stresses below the frictional strength. The fan-mechanism has been proposed as a possible mechanism of earthquakes [6]. Unique features of the fan-mechanism associated with extremely low shear resistance of propagating fan waves have been studied on a laboratory model [5]. The main goal of this paper is to analyze this model numerically on the basis of the equations of motion.

© Springer International Publishing AG 2017
I. Dimov et al. (Eds.): NAA 2016, LNCS 10187, pp. 657–664, 2017.
DOI: 10.1007/978-3-319-57099-0_75

2   Mathematical Model

The model of fan-shaped mechanism represents a structure composed of elastically bonded slabs on a flat surface (Fig. 1). From the viewpoint of the classical mechanics, this is a system with n degrees of freedom, characterized by the rotational angles φ_1, φ_2, ..., φ_n of the slabs.

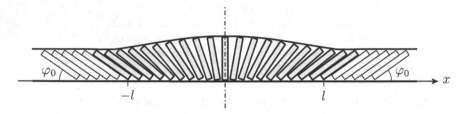

Fig. 1. Fan-shaped structure of slabs

Motion of this system is described using Lagrange's equations:

$$\frac{d}{dt}\frac{\partial L}{\partial \dot{\varphi}_i} - \frac{\partial L}{\partial \varphi_i} = F_i \qquad (i = 1, 2, ..., n), \tag{1}$$

where $\dot{\varphi}_i = d\varphi_i/dt$ are the angular velocities, $L = T - \Pi$ is Lagrange's function equal to the difference between the kinetic energy $T = \frac{J}{2}\sum_{i=1}^{n}\dot{\varphi}_i^2$ and potential energy $\Pi = \sum_{i=1}^{n-1} U(\varepsilon_{i+1/2})$ of the fan-structure. Here J is the moment of inertia of a slab relative to the axis of rotation, U is the elastic potential of the bond, $\varepsilon_{i+1/2} = (s_{i+1/2}^2 - h^2)/(2\,h^2)$ is the strain of the elastic bond between neighbor slabs with the numbers i and $i+1$, $h = h_0/\sin\varphi_0$ is the distance between the axes of rotation, h_0 and a are the thickness and height of a slab, φ_0 is the initial angle, $s_{i+1/2}^2 = \left(h + a\left(\cos\varphi_{i+1} - \cos\varphi_i\right)\right)^2 + a^2(\sin\varphi_{i+1} - \sin\varphi_i)^2$ is the squared distance between the top ends of the slabs. The constraint reaction as a function of strains is given by:

$$R_{i+1/2} = \frac{dU}{ds_{i+1/2}} = U'\frac{d\varepsilon_{i+1/2}}{ds_{i+1/2}} = \frac{s_{i+1/2}}{h^2}U'(\varepsilon_{i+1/2}). \tag{2}$$

The generalized forces F_i on the right-hand sides of Eq. (1), induced in the fan-structure by the vertical P_i and horizontal Q_i forces, and by the rotary moments M_i of the forces of contact interaction between the slabs, are calculated in conformity with the expression derived for the virtual work:

$$\delta A = \sum_{i=1}^{n} F_i\,\delta\varphi_i = \sum_{i=1}^{n}\left(-Q_i\,\delta X_i - P_i\,\delta Y_i + M_i\,\delta\varphi_i - \eta\,\dot{\varphi}_i\,\delta\varphi_i\right),$$

where $X_i = (i-1)h_0 + a \cos\varphi_i$, $Y_i = a \sin\varphi_i$ are the coordinates of the top ends of slabs, η is the coefficient of viscous rotation resistance. The forces equal: $F_i = Q_i\,a\,\sin\varphi_i - P_i\,a\,\cos\varphi_i + M_i - \eta\,\dot\varphi_i$. When the external effect on the fan-structure, disposed horizontally, is only conditioned by the own weight of the slabs, $2\,P_i = m\,g$ and $Q_i = 0$. In case that the slab foundation is inclined at an angle α to the horizontal plane: $2\,P_i = m\,g\,\cos\alpha$, $2\,Q_i = m\,g\,\sin\alpha$.

Below, the system (1) is given in the expanded form:

$$
J\,\ddot\varphi_i = F_i - R_{i+1/2}\,\frac{\partial s_{i+1/2}}{\partial\varphi_i} - R_{i-1/2}\,\frac{\partial s_{i-1/2}}{\partial\varphi_i},
$$

$$
s_{i+1/2}\,\frac{\partial s_{i+1/2}}{\partial\varphi_i} = \big(h + a\,(\cos\varphi_{i+1} - \cos\varphi_i)\big)a\,\sin\varphi_i -
$$
$$
- (\sin\varphi_{i+1} - \sin\varphi_i)\,a^2\,\cos\varphi_i, \tag{3}
$$

$$
s_{i-1/2}\,\frac{\partial s_{i-1/2}}{\partial\varphi_i} = -\big(h + a\,(\cos\varphi_i - \cos\varphi_{i-1})\big)a\,\sin\varphi_i +
$$
$$
+ (\sin\varphi_i - \sin\varphi_{i-1})\,a^2\,\cos\varphi_i.
$$

The system (3) is composed of $(n-1)$ second-order differential equations for $i = 2, 3, \ldots, n$. For $i = 1$, it is set that $\varphi_1 = \varphi_0 + \omega_0\,t$ for the forced rotation of the first slab at a constant angular velocity $\omega_0 = (\pi - 2\varphi_0)/T$, which ensures completed fan during the time T. It is assumed that the end slab in the fan-structure is only connected with the previous slab and is, therefore, free from the stretching force from the right $(R_{n+1/2} = 0)$. The moments M_i are set as the functions of rotational angles and angular velocities of the slabs. They ensure the fulfillment of one-sided constraints for the rotational angles $\varphi_0 \leqslant \varphi_i \leqslant \pi - \varphi_0$, implemented in the course of the computation using a special algorithm for solution correction. For the system (3), the Cauchy problem is set: $\varphi_i = \varphi_0$ and $\dot\varphi_i = 0$ at $t = 0$.

3   Algorithm for Solution Correction

For numerical implementation of the physical model of contact interaction between the slabs in the fan, a special algorithm has been developed based on variational formulation of contact conditions. The moment of reaction forces M_i applied to the slab with the number i equals zero if the slab is in the free state, is above zero if the slab rests on the $(i+1)$-th slab, and is below zero if the slab bears on the $(i-1)$-th slab. Under simultaneous contact of neighbor slabs, M_i may have any sign. In the general case, the variational inequality holds true

$$
\delta A' = \sum_{i=1}^{n} M_i\,\delta\varphi_i \geqslant 0, \tag{4}
$$

that is a mathematical formulation of the principle of minimum work spent by the reaction force for any possible displacements in the fan-structure. The possible displacements obey the kinematic constraints

$$
\pi - \varphi_0 \geqslant \varphi_1 \geqslant \varphi_2 \geqslant \ldots \geqslant \varphi_n \geqslant \varphi_0. \tag{5}
$$

If there is no contact interaction in the system of three neighbor slabs, i.e. when $\varphi_{i-1} > \varphi_i > \varphi_{i+1}$, then, considering arbitrariness of the variation $\delta\varphi_i$, it follows from (4) that $M_i = 0$. Under the contact with the $(i+1)$-th slab: $\varphi_{i-1} > \varphi_i = \varphi_{i+1}$, then $\delta\varphi_i \geqslant 0$ and $M_i \geqslant 0$; and under the contact with the $(i-1)$-th slab: $\varphi_{i-1} = \varphi_i > \varphi_{i+1}$, then $\delta\varphi_i \leqslant 0$ and $M_i \leqslant 0$. When three slabs contact simultaneously at $\varphi_{i-1} = \varphi_i = \varphi_{i+1}$, the permissible variation $\delta\varphi_i$ is zero, and the sign of M_i is therefore undetermined. This reasoning shows that the variational inequality (4) adequately describes contact conditions for slabs without regard to their rebound from each other due to elastic forces.

The moment M_i is additively included on the right-hand side of the equation of rotational motion (3) for an i-th slab; for this reason, after approximation of the equations at each time step, the linear relation $M_i = \mu\left(\varphi_i - \bar{\varphi}_i\right)$ arises between the reaction force moment, the current rotational angle φ_i and the angle $\bar{\varphi}_i$ calculated from (3) with no regard for contact interaction, i.e. at $M_i = 0$. In the relation above, μ is a positive constant proportional to the moment of inertia of a slab. Exclusion of M_i transforms the variational inequality (4) to the form:

$$\sum_{i=1}^{n}(\varphi_i - \bar{\varphi}_i)\,\delta\varphi_i \geqslant 0. \tag{6}$$

The constraints (5) assign a convex and closed set in the space of generalized coordinates. Using the conditions of the convexity and closedness, it is possible to show that the solution of the inequality (6) under the constraints (5) exists, is unique and is the projection of a vector with the coordinates $\bar{\varphi}_i$ onto the given set. A rigorous proof of this statement, more generally formulated, can be found, for example, in the monograph [7], where constitutive relationships of media having different resistance to compression and tension under deformation are reduced to variational inequalities.

When finding projections, the computations use a version of the Uzawa iteration algorithm [8] for saddle points. The application of the Kuhn–Tucker theorem reduces the problem to searching a saddle point of the Lagrangian:

$$\min_{\varphi_i} \max_{\lambda_i \geqslant 0} \left(\frac{1}{2}\sum_{i=1}^{n}\left(\varphi_i - \bar{\varphi}_i\right)^2 + \sum_{i=1}^{n}\lambda_{i+1/2}\left(\varphi_{i+1} - \varphi_i\right)\right),$$

where $\lambda_{i+1/2}$ are the Lagrangian multipliers conforming with the constraints, that $\varphi_{i+1} - \varphi_i \leqslant 0$, and $\varphi_{n+1} = \varphi_0$ for the compactness of notation. At the initial stage of the algorithm, it is assumed that $\varphi_i^{(0)} = \bar{\varphi}_i$, $\lambda_{i+1/2}^{(0)} = 0$. Then recomputation is performed using the recurrence formulas:

$$\lambda_{i+1/2}^{(k+1)} = \left(\lambda_{i+1/2}^{(k)} + \gamma\left(\varphi_{i+1}^{(k)} - \varphi_i^{(k)}\right)\right)_+, \quad \varphi_i^{(k+1)} = \bar{\varphi}_i + \lambda_{i+1/2}^{(k+1)} - \lambda_{i-1/2}^{(k+1)}. \tag{7}$$

Here $\gamma > 0$ is an iteration parameter, the subscript "+" denotes the positive part of a number: $z_+ = \max(z, 0)$. In accordance with the boundary conditions, $\varphi_1^{(k+1)}$ remains equal to $\bar{\varphi}_1$, and $\varphi_{n+1}^{(k+1)}$ equal φ_0. The formulas (7) are suitable for calculating Lagrangian multipliers in the range of $i = 1, 2, \ldots, n$. The first formula in (7) provides a gradient approximation of the solution to the Lagrangian maximum point of $\lambda_{i+1/2}$, and the second formula is the condition of the minimum for the variables φ_i. The computations by recurrence formulas are terminated, when the preset accuracy is achieved: for the sufficiently small ε the condition $\left| \varphi_i^{(k+1)} - \varphi_i^{(k)} \right| \leqslant \varepsilon$ is fulfilled. It is possible to prove that the algorithm converges, when $k \to \infty$, if $\gamma < 1/2$. The optimal value of the iteration parameter to ensure the quickest convergence of the algorithm is set by the formula $\gamma = -2/(\nu_1 + \nu_n)$, where $\nu_k = -4 \sin^2 \pi i/(2n)$ are the eigenvalues of a tridiagonal symmetric matrix with the coefficients -2 on the main diagonal and 1 on the secondary diagonals (refer to, e.g., [9]).

4 Computational Results

The system of Eq. (3) was solved numerically using the Merson method possessing good safety factor for solving nonlinear problems and having a simple procedure of automatic selection of integration step with the control of computational accuracy [10]. Software implementation of the Merson method was performed in Matlab.

Figure 2a shows the plot of $R = R(\varepsilon)$, and Fig. 2b illustrates distribution of strains of the bonds in the complete fan, based on the computational data. The points mark the laboratory measurements in Fig. 2a and the positions of the slabs in the fan in Fig. 2b. As follows from Fig. 2b, the complete fan is a symmetric self-balanced (equilibrium) structure.

In computations, the fan opening time T was varied in a range from 2 to 15 s. The parameters of the problem, consistent with Tarasov's laboratory model, were accepted: $m = 0.11$ kg, $a = 0.1$, $h = 0.005$ m, $\varphi_0 = 40°$. The viscous resistance coefficient was calculated in terms of the rotational motion relaxation time τ as $\eta = J/\tau$. The relaxation time was ranged between 0.1 and 5 s.

For the estimation of the effect of friction, comparative computations of slow opening of the fan in the time $T = 6$ s were performed at the short relaxation time $\tau = 0.5$ s and at the relatively long time $\tau = 2$ s. The computations showed

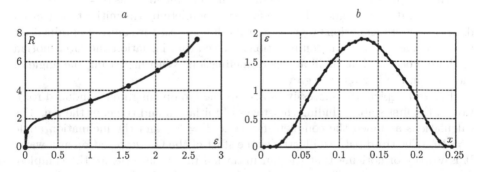

Fig. 2. Tension–strain curve (a) and distribution of strains of bonds in a fan (b)

that viscous friction contributed to rapid cessation of motion. In the first case, the fan completely inhibited at a small distance from the first slab (Fig. 3a). In the second case, the distance of the fan slowdown reached the length of the fan (Fig. 3b).

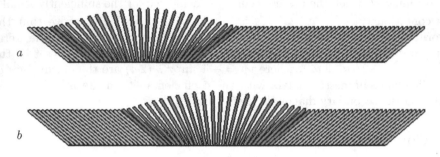

Fig. 3. The fan-structure after cessation of motion at different times of relaxation: a) $\tau = 0.5\,\mathrm{s}$, b) $\tau = 2\,\mathrm{s}$

The travel distance of the fan up to the moment of cessation grows with the velocity of the fan opening. For instance, at $T = 2$ s, the distance of slowdown covers the half-length of the fan at the short relaxation time and increases, when the opening time is shorter. This exhibits essential influence of the rate of loading.

The equilibrium of the fan is unbalanced when the domino-system, including the fan, is subjected to supplementary horizontal forces. In a physical model, such forces are induced by inclining the domino-system at an angle α. The non-equilibrium of the fan invokes the fan motion as a traveling wave along the domino-system. In this case, the fan-mechanism ensures successive displacement of the upper interface of the shear rupture relative its lower interface as the traveling wave propagates. The unique feature of the fan-mechanism is that the fan provides displacement of the shear rupture interfaces at very low shear stresses, that may be an order of magnitude lower than friction between the interfaces of the rupture at the absence of the fan.

The computations were carried out at different inclinations $\alpha = 5°$, $10°$, $20°$ and higher. It was assumed that the fan was complete by the initial time (initial distribution of inclinations was set in accordance with the solution obtained for the system at $\alpha = 0$). The computations showed that initiation and later motion of the fan occurred at small angles of inclination. With increasing the angle of inclination, the fan velocity grows.

Figure 4 depicts the successive stages of the fan opening under induced rotation of the first slab, including the stages of initiation and travel of the fan. The calculations assumed the complete fan time $T = 3$ s and the inclination angle $\alpha = 10°$. The third pattern illustrates the start of the fan propagation as a wave. It is worthy of noticing that the fan initiation takes place before the complete fan moment corresponding to rotation of the first slab by an angle $\pi - \varphi_0$.

Fig. 4. The stages of the formation and travel of the fan: $t = 0.67$, 1.71, 2.38, 3.12, and $3.63\,\mathrm{s}$

The series of computational experiments included modeling the process of motion of incomplete fan. In case that the fan equilibrium is disturbed at the initial time, there are two possible scenarios. In the first scenario, when the angle of inclination is smaller than the critical value, the fan closure to the initial state takes place. When the inclinations exceed the critical value, the fan initiates and speeds up along the system of slabs.

5   Conclusions

Based on the Lagrange equations, the discrete mathematical model has been constructed to simulate rotational motion of a system of elastically bonded slabs on flat surface. The algorithm proposed for the model implementation takes into account contact interaction between the slabs. The processes of initiation and travel of a fan under forced rotation of the first slab are calculated with

Matlab. The computations performed in the framework of the discrete model have demonstrated the incomplete disclosure of fans with different opening angles owing to fast or slow change in rotation velocity of the first slab. Comparison of the computations and laboratory observations has yielded good compliance of the results. The developed model will be used to study the nature of initiation of dynamic shear ruptures in rocks at seismogenic depths.

Acknowledgements. This work was supported by the Centre for Offshore Foundation Systems (The University of Western Australia), the Complex Fundamental Research Program no. II.2P "Integration and Development" of SB RAS (project no. 0361-2015-0023) and the Russian Foundation for Basic Research (grant no. 14–01–00130).

References

1. Reches, Z., Lockner, D.A.: Nucleation and growth of faults in brittle rocks. J. Geophys. Res.: Solid Earth **99**(B9), 18159–18173 (1994)
2. Ortlepp, W.D.: Rock Fracture and Rockbursts An Illustrative Study. Monograph Series M9. South African Institute of Mining and Metallurgy, Johannesburg (1997)
3. Tarasov, B.G., Randolph, M.F.: Frictionless shear at great depth and other paradoxes of hard rocks. Int. J. Rock Mech. Min. Sci. **45**(3), 316–328 (2008)
4. Tarasov, B.G.: Intersonic shear rupture mechanism. Int. J. Rock Mech. Min. Sci. **45**(6), 914–928 (2008)
5. Tarasov, B.G.: Hitherto unknown shear rupture mechanism as a source of instability in intact hard rocks at highly confined compression. Tectonophysics **621**, 69–84 (2014)
6. Tarasov, B.G., Randolph, M.F.: Superbrittleness of rocks and earthquake activity. Int. J. Rock Mech. Min. Sci. **48**(6), 888–898 (2011)
7. Sadovskaya, O., Sadovskii, V.: Mathematical Modeling in Mechanics of Granular Materials. Ser.: Advanced Structured Materials, vol. 21. Springer, Heidelberg (2012)
8. Glowinski, R., Lions, J.L., Trémolières, R.: Numerical Analysis of Variational Inequalities, vol. 1-2. Dunod, Paris (1976). [in French]
9. Marchuk, G.I.: Methods of Numerical Mathematics. Springer, New York (1982)
10. Novikov, E.A.: Explicit Methods for Stiff Systems. Nauka, Novosibirsk (1997). [in Russian]

High Performance Computations
for Study the Stability of a Numerical Procedure
for Crossbar Switch Node

Tasho D. Tashev[1(✉)], Vladimir V. Monov[1], and Radostina P. Tasheva[2]

[1] Institute of Information and Communication Technologies - B.A.S.,
Acad. G. Bonchev St., Block 2, 1113 Sofia, Bulgaria
{ttashev,vmonov}@iit.bas.bg
[2] Technical University-Sofia, 8 Kl.Ohridski Blvd., 1000 Sofia, Bulgaria
rpt@tu-sofia.bg

Abstract. In the present paper the problem concerning computation of the upper bound of the throughput (THR) of crossbar switch node is formulated. The task is solved using an appropriate numerical procedure. The input data for this procedure are the results of high performance simulations executed on the grid-structure of the IICT-BAS (www.grid. bas.bg). The modeling of the THR utilizes MiMa-algorithm for a node with $(n \times n)$ commutation field, specified by the Generalized Nets. For studying the stability of the numerical procedure we use a modified family of patterns for hotspot load traffic simulation. The obtained results give an upper bound of the THR for $n \in [3, 97]$ which enables us to estimate the limit of the THR for $n \to \infty$. This estimate is obtained to be 100% for the applied family of patterns. The numerical procedure is stable (small input perturbations lead to small changes in the output).

Keywords: Numerical methods · Generalized nets · Switch node · Modeling

1 Introduction

Crossbar switch node maximizes the speed of data transfer using parallel paths between his input and output lines. The switch must send packets without delay and without losses in the ideal case. This is obtained by means of a conflict-free commutation schedule calculated by the control block of the node [9].

The calculation a conflict-free schedule from a mathematical point of view is NP-complete task [8]. The existing solutions partly solve the problem by using different formalisms [6]. The observed increase in the volume of telecommunications traffic [4] requires new algorithms, which have to be checked for efficiency. Similar problems arise in the development of efficient wireless sensor networks [1] and distributed evolutionary algorithms [5]. The efficiency of the switch is firstly evaluated by the throughput (THR) submitted by the node [6].

© Springer International Publishing AG 2017
I. Dimov et al. (Eds.): NAA 2016, LNCS 10187, pp. 665–673, 2017.
DOI: 10.1007/978-3-319-57099-0_76

At the stage of switch design the THR obtained by algorithms for conflict-free schedule is initially assessed. For a chosen algorithm THR will depend on the incoming traffic. For a given traffic model, THR of a switch node depends on the load intensity ρ of input lines [6].

In the present paper the task for computation of the upper bound of the THR is formulated and a solution of this task is presented. In the calculations we use a numerical procedure [13] with input data from high performance computations for computer simulation of the THR executed on the grid-structure of IICT-BAS. Our modeling of the THR utilizes the MiMa-algorithm [12] for switch with (n x n) commutation field, specified by the Generalized Nets (GN) [3]. We will study whether the numerical procedure is stable in the sense that small values of intensity perturbation ρ lead to small changes in the output. For this aim we use a modified family of patterns for hotspot load traffic simulation [7] with a introduced asymptotically vanishing perturbation.

2   The Problem of Upper Bound of the Throughput

For a chosen algorithm, traffic model and load intensity ρ of the input lines, THR depends on the dimension of its commutation field $n \times n$ (n input lines, n output lines) and the dimension of the input buffer i. Usually researchers obtain THR for discrete values of n $(2, 3, 4, 8, 16, 32, 64, \dots)$ and the buffer is supposed to be large. Then conclusions are drawn about the upper bound of THR for $n \to \infty$ and $i \to \infty$ [9]. In our computer simulations we examine THR for every integer value in the chosen intervals for n and i. This is possible due to the usage of the grid-structure for the neccessary high performance computations. As a consequence we can make a more precise assessment of the upper bound.

We shall use the following definitions.

Definition 1. Function f is defined as

$$f(n, i) = THR(n, i) \tag{1}$$

where $n = 2, 3, 4, \dots$ is the dimension of the input (output) lines of the node, $i = 1, 2, 3, \dots$ is the dimension of the input buffer of each input line and $0 \leq THR(n, i) \leq 1$.

Here, THR with value 1 corresponds to 100% - normalized throughput with respect to the maximum throughput of the output lines of the switch.

Definition 2. Function V is defined as

$$V(n) = \lim_{\substack{i \to \infty, \\ n \in [n1, n2]}} f(n, i) \tag{2}$$

where $i \to \infty$ means infinitely large input buffer and n belongs to the interval $[n_1.n_2]$ which is determined by working range of simulations.

Thus, $V(n)$ is the upper bound in the working range of the switch.

Definition 3. The absolute upper bound U of THR is defined as

$$U = \lim_{\substack{i \to \infty, \\ n \to \infty}} f(n,i) \tag{3}$$

where $i \to \infty$ means infinitely large input buffer and $n \to \infty$ means infinitely large commutation field.

If $V(n)$ is known calculation of the U is standart mathemathical problem. Therefore we focus on the following task.

Task: Determine $V(n)$ in working interval $[n_1, n_2]$ from known results $f(n,i)$ for THR of computer simulations for $n \in [n_1, n_2], i \in [i_1, i_2]$.

In our investigations we solved the task in two steps: existence of solution; calculating of the solution. The description of the first step is given below.

3 The Existence of a Solution for $V(n)$

Let us chose values for i as : $i = 1, 2, 3, \ldots, q$, where q is the apriori selected integer constant ($q \geq 3$ for standart simulation sequence, $n \in [2, n_2]$).

We shall perform q simulations in order to obtain q curves for THR. The obtained curves will be denoted as follows:

$$f_1(n) = f(n,1), \ f_2(n) = f(n,2), \ \ldots, f_q(n) = f(n,q)$$

Typical curves obtained in our previous works are given in [10, 11].
Denote the difference between two successive curves f_j and f_{j+1} by res_j:

$$res_j(n) = f_{j+1}(n) - f_j(n) = f(n, j+1) - f(n, j)$$

We will compute last $(q-1)$-th difference as:

$$res_{q-1}(n) = f_q(n) - f_{q-1}(n) = f(n,q) - f(n, q-1) \tag{4}$$

Denote the ratio of the values of two curves res_j and res_{j+1} through δ_j:

$$\delta_j(n) = \frac{res_{j+1}(n)}{res_j(n)} = \frac{f_{j+2}(n) - f_{j+1}(n)}{f_{j+1}(n) - f_j(n)} = \frac{f(n, j+2) - f(n, j+1)}{f(n, j+1) - f(n, j)}$$

We will compute last $(q-2)$-th ratio as :

$$\delta_{q-2}(n) = \frac{res_{q-1}(n)}{res_{q-2}(n)} = \frac{f_q(n) - f_{q-1}(n)}{f_{q-1}(n) - f_{q-2}(n)} = \frac{f(n,q) - f(n, q-1)}{f(n, q-1) - f(n, q-2)} \tag{5}$$

Simulation data allow us to calculate $\delta_1, \delta_2, \ldots, \delta_{q-2}$. From the last formula (5) we obtain:

$$f(n,q) = f(n, q-1) + \delta_{q-2}(n).[f(n, q-1) - f(n, q-2)]$$

or

$$f_q(n) = f_{q-1}(n) + \delta_{q-2}(n).res_{q-2}(n) \tag{6}$$

Then we can extrapolate the result for the following future values

$$f_{q+1}(n) = f_q(n) + \delta_{q-1}(n).res_{q-1}(n) \tag{7}$$
$$f_{q+2}(n) = f_{q+1}(n) + \delta_q(n).res_q(n) \tag{8}$$
$$f_{q+3}(n) = f_{q+2}(n) + \delta_{q+1}(n).res_{q+1}(n) \tag{9}$$

Substituting (7) in (8), we obtain

$$f_{q+2}(n) = f_q(n) + [\delta_{q-1}(n). + \delta_{q-1}(n).\delta_q(n)].res_{q-1}(n) \tag{10}$$

Substituting (10) in (9), we obtain

$$f_{q+3}(n) = f_q(n) + [\delta_{q-1}(n). + \delta_{q-1}(n).\delta_q(n)+$$
$$+\delta_{q-1}(n).\delta_q(n).\delta_{q+1}(n)].res_{q-1}(n) \tag{11}$$

By induction we obtain accordingly for p-th extrapolation ($p \geq 1, q \geq 3$)

$$f_{q+p}(n) = f_q(n) + [\delta_{q-1}(n) + \delta_{q-1}(n).\delta_q(n) + \dots$$
$$+\delta_{q-1}(n).\delta_q(n))\dots\delta_{q+p-2}(n)].res_{q-1}(n) \tag{12}$$

From this we can derive the following:

Statement. When $p \to \infty$ in (12) then $f_{(q+p \to \infty)}(n)$ is the seeking $V(n)$.

Thus, if there is an upper bound of the throughput of a switch node, then there exists the bound

$$\lim_{\substack{p \to \infty, \\ q = const}} [\delta_{q-1}(n) + \delta_{q-1}(n).\delta_q(n) + \dots \delta_{q-1}(n).\delta_q(n))\dots\delta_{q+p-2}(n)]$$

Corollary 1. Under the above assumption, the sum

$$\delta_{q-1}(n) + \delta_{q-1}(n).\delta_q(n) + \dots + \delta_{q-1}(n).\delta_q(n))\dots\delta_{q+p-2}(n) \tag{13}$$

for $p \to \infty$ is convergent and has a boundary S.

Corollary 2. If the upper bound of THR exists then the solution for $V(n)$ is unique :

$$V(n) = f_q(n) + S.res_{q-1}(n) \tag{14}$$

If we can calculate S from (13) knowing $\delta_1(n), \delta_2(n), \dots \delta_{q-2}(n)$ this allows us to obtain $V(n)$. For this aim we need the relation $\delta_{j+1} = \phi(\delta_j)$.

4 Calculating the Solution for $V(n)$

Finding dependence $\delta_{j+1} = \phi(\delta_j)$ in output data from simulations like those in [10,11] proved to be a difficult task. We have found a relation in the case of the special choice of the sequence parameter i. The short description is shown below according to [13].

We have found a relation for the model of PIM-algorithm [2] specified by means of Generalized nets [3] with Chao-model for hotspot load traffic [7]. For this aim we define a family of patterns $Chao_i$ for traffic matrices (shown below - traffic matrix T^i for $Chao_i$ (15)) [10].

$$T^i_{(2\times2)} = \begin{bmatrix} i & i \\ i & i \end{bmatrix} \quad T^i_{(3\times3)} = \begin{bmatrix} 2i & i & i \\ i & 2i & i \\ i & i & 2i \end{bmatrix} \quad \cdots \quad T^i_{(k\times k)} = \begin{bmatrix} (k-1)i & \cdots & i \\ \vdots & \ddots & \vdots \\ i & \cdots & (k-1)i \end{bmatrix} \quad (15)$$

Let us choose for a simulation with this family of patterns the sequences for i : $i = 1, 10, 100, 1000$. In this case we find the dependence $\delta_1 = \delta_2 = 1/(3, 15\pm0.01)$. Is this dependence a constant? The answer is not because when we choose the sequences for i : $i = 1, 4, 16, \ldots, 256$, i.e. $4^0, 4^1, 4^2, \ldots, 4^5$ then the dependence is $1/\delta_{j+1} = 1/\delta_j = 2$ where $j \in [1, 3]$.

Using high performance computations for large-scale simulations we have reached more precise result:

Corollary. The dependance $\delta_{j+1} = \phi(\delta_j)$ is $\delta_{j+1} = \delta_j = m^{-1/2}$ with an accuracy within the uncertainty of simulations (when $i \in [1, i2), n \in [n1, n2], m \in [2, 3, 4, \ldots), i = 1, m^1, \ldots, m^p, \ldots$).

The validity of this assertion we test by simulations with a different algorithm. The utilized algorithm is the MiMa-algorithm [12], working with the same model of load traffic (and $m = 2$). The results of the simulations confirm dependence $\delta_{j+1} = \delta_j = 2^{-1/2}$ for choise i : $i = 1, 2, 4, 8, \ldots, 256$ where $j \in [1, 4]$ [14].

A numerical procedure for calculation of $V(n)$ is described in [13,14] and the uncertainty of simulation can be assessed. The next question for investigation is whether the procedure is stable. Here we will research the case of a small, asymptotically vanishing perturbations.

5 Computer Simulations for Studying the Stability of the Numerical Procedure

We introduce perturbations in Chao model as follows: first, we modify the family template according to [14] (see (16) bellow for model $Chao_1$).

$$T^1_{(3\times3)} = \begin{bmatrix} 2 & 1 & 1 \\ 1 & 2 & 1 \\ 1 & 1 & 2 \end{bmatrix} \Rightarrow \begin{bmatrix} 2 & 1 & 1 \\ 1 & 2 & 1 \\ 1 & 1 & 2 \end{bmatrix}, \begin{bmatrix} 1 & 2 & 1 \\ 1 & 1 & 2 \\ 2 & 1 & 1 \end{bmatrix}, \begin{bmatrix} 1 & 1 & 2 \\ 2 & 1 & 1 \\ 1 & 2 & 1 \end{bmatrix} \quad (16)$$

The resulting throughput is the average for n runs for each size $n \times n$. This family has $\rho = 100\%$ load intensity of each input (i.i.d. Bernoulli).

Second, we reduce the number of requests in a chosen (e.g., the first) input line by one (i.e. minus one request). Then the general element T_{per}^i for $(k \times k)$ dimension of the pattern $Chao_p^i$ for Chao-model with pertubations has a view

$$T_{per}^{\ i}{}_{(k \times k)} \Rightarrow$$

$$\begin{bmatrix} (k-1)i-1 & i & \cdots & i \\ i & (k-1)i & \cdots & i \\ \vdots & & \ddots & \vdots \\ i & \cdots & i & (k-1)i \end{bmatrix}, \cdots, \begin{bmatrix} i & i & \cdots & (k-1)i-1 \\ (k-1)i & i & \cdots & i \\ \vdots & & \ddots & \vdots \\ i & \cdots & (k-1)i & i \end{bmatrix}$$

(17)

This is equivalence of perturbation of load intensity ρ for one (first) input line. For this input the load intensity is $\rho = (100 - \theta)\%$ where θ is a asymptotically vanishing perturbation. The curves for θ and ρ are shown in Fig. 1 where $\theta = (2.i.(n-1)-1)/(2.i.(n-1))$ is normalized with respect to 100% ($\rho = 1$).

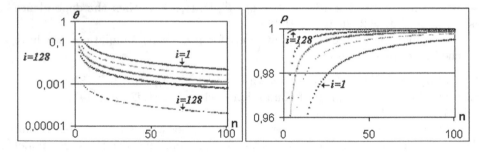

Fig. 1. The curves for perturbation θ and intensity ρ

In this computer simulation we are using GN-model of the MiMa-algorithm [12]. The transition from a GN-model to executive program is performed as in [8] using the program package VFort [15]. The source code has been compiled by means of the grid-structure BG01-IPP of the Institute of information and communication technologies - Bulgarian Academy of Sciences (www.grid.bas.bg). The resulting code is executed in the grid-structure locally. The operation system is Scientific Linux release 6.5. We used the following grid-resources: up to 16 CPU, 32 threads, 12 GB RAM. The main restriction is the time for execution ≤ 72 h.

We choose value $m = 2$ (for comparison with the results from [14]). When $m = 2$, then $i = 1, 2, 4, 8, \ldots, 2^p, \ldots$. The initial estimate of the required number of curves for THR is at least 4 (from Pattern $Chao_p^1$). In our example, we have eight curves (patterns). In the figures below, $Chao_p^i$ is denoted as C_p^i for $i = 1, 2, \ldots$ We get results for $C_p^1, C_p^2, \ldots, C_p^{64}, C_p^{128}$ which are shown in Fig. 2.

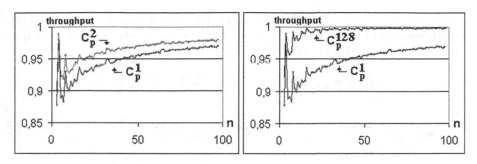

Fig. 2. Throughput for $Chao_p^1$, $Chao_p^2$, $Chao_p^{128}$

The dimension n varies from 3×3 to 97×97 and n runs (simulations) for each size ($n \times n$) of pattern $Chao_p^i$ are executed. The resulting throughput is the average for n runs. Then we calculate the difference between throughput for neighboring patterns δ_j and the obtained curves are shown in Fig. 3. The values of δ_1 vary around $(1,5)^{-1}$ but the values of δ_6 tend to $(1,4)^{-1}$.

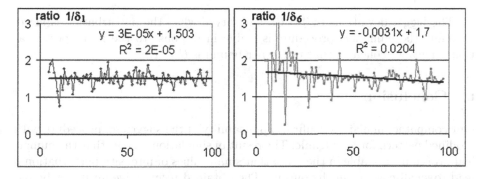

Fig. 3. Ratio $1/\delta_1$, $1/\delta_6$ between differences

From our simulations in the case $m = 2$, we have drawn the following:

Conclusion: The dependence $\delta_{j+1} = \phi(\delta_j)$ is not a constant but tend to constant $2^{-1/2}$. These differences arise due to the input disturbance.

As a consequence, the upper boundary in case $m = 2$ can be calculated according to (14) as:

$$V(n) = = f(n, m^{p-1}) + [\delta(m) + \delta^2(m) + \cdots + \delta^p(m) + \ldots].res(m^{p-2})$$
$$= f(n, m^{p-1}) + [(m^{1/2} - 1)^{-1}].(f(n, m^{p-1}) - f(n, m^{p-2}))$$

In this simulation ($m = 2$) we calculate the boundary by

$$V(n) = f(n, 128) + [(2^{1/2} - 1)^{-1}].(f(n, 128) - f(n, 64))$$

This result is obtained for δ_6 which has the least deviation from $m^{-1/2}$. The result is shown in Fig. 4 (right). For comparison in Fig. 4 (left) we have shown a boundary which is calculated for δ_1 :

$$f_{i\to\infty}(n,i) = f(n,4) + [(2^{1/2} - 1)^{-1}].(f(n,4) - f(n,2))$$

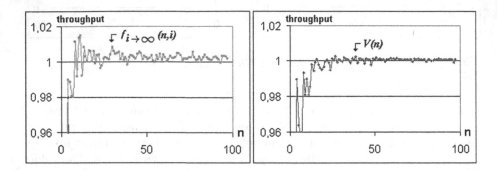

Fig. 4. Upper boundary $V(n)$ for δ_1 and δ_6

The bigger deviations in THR are observed when the perturbation is large, therefore the numerical procedure is stable in response to the input perturbations. Thus we conclude that the upper bound is $U = 1$.

6   Conclusion

Our computer simulation confirms applicability of the suggested procedure with modified pattern for load traffic. The result of simulations shows that the numerical procedure is stable in the sense that small values of intensity perturbation ρ lead to small changes in the output. The obtained results give an upper bound of the THR for $n \in [3, 97]$ which enables us to estimate the limit of the THR of MiMa-algorithm for $n \to \infty$. This estimate is obtained to be 100%.

In a future study, the numerical procedure will be tested using other models of the incoming traffic that include perturbations.

References

1. Alexandrov, A.: Ad-hoc Kalman filter based fusion algorithm for real-time wireless sensor data integration. Flexible Query Answering Systems 2015. AISC, vol. 400, pp. 151–159. Springer, Cham (2016). doi:10.1007/978-3-319-26154-6_12
2. Anderson, T., Owicki, S., Saxe, J., Thacker, C.: High speed switch scheduling for local area networks. ACM Trans. Comput. Syst. **11**(4), 319–352 (1993)
3. Atanassov, K.: Generalized Nets and System Theory. Prof. M. Drinov Academic Publishing House, Sofia (1997)

4. Atanasova, T., Mishina, A.: Multiservice networks in digital houses. J. Probl. Eng. Cybern. Robot. **65**, 14–21 (2012). Prof. M. Drinov Academic Publishing House, Sofia, Bulgaria
5. Balabanov, T., Zankinski, I., Barova, M.: Distributed evolutionary computing migration strategy by incident node participation. In: Lirkov, I., Margenov, S.D., Waśniewski, J. (eds.) LSSC 2015. LNCS, vol. 9374, pp. 203–209. Springer, Cham (2015). doi:10.1007/978-3-319-26520-9_21
6. Chao, H., Lui, B.: High Performance Switches and Routers. Wiley, Hoboken (2007)
7. Chao-Lin, Y., Chang, C.-S., Lee, D.-S.: CR switch: a load-balanced switch with contention and reservation. IEEE/ACM Trans. Netw. **17**(5), 1659–1671 (2007)
8. Chen, T., Mavor, J., Denyer, P., Renshaw, D.: Traffic routing algorithm for serial superchip system customisation. IEE Proc. **137**(1), 65–73 (1990)
9. Kang, K., Park, K., Sha, L., Wang, Q.: Design of a crossbar VOQ real-time switch with clock-driven scheduling for a guaranteed delay bound. Real-Time Syst. **49**(1), 117–135 (2013)
10. Tashev, T.: Modelling throughput crossbar switch node with nonuniform load traffic. In: Proceedings of the International Conference on DCCN, October 26–28 2011, Moscow, Russia, pp. 96–102. R&D Company, Mosco (2011). (in Russian)
11. Tashev, T.: MiMa algorithm throughput modelling for crossbar switch with hotspot load traffic. In: Proceedings of the International Conference on DCCN 2013, October 7–10 2013, Moscow, Russia, pp. 257–264. JSC TECHNOSPHERA, Moscow (2013). (in Russian)
12. Tashev, T., Atanasova, T.: Computer simulation of MiMa algorithm for input buffered crossbar switch. Int. J. Inf. Technol. Knowl. **5**(2), 183–189 (2011)
13. Tashev, T., Bakanov, A., Tasheva, R.: Determination of the value of convergence parameter in a procedure of calculating the upper boundary of throughput for packet switch. In: Proceedings of International Conference on RAM 2013, October 8–10 2013, Bankya, Bulgaria, pp. 34–37. Prof. M. Drinov Academic Publishing House, Sofia (2013)
14. Tashev, T., Monov, V.: A numerical study of the upper bound of the throughput of a crossbar switch utilizing MiMa-algorithm. In: Dimov, I., Fidanova, S., Lirkov, I. (eds.) NMA 2014. LNCS, vol. 8962, pp. 295–303. Springer, Cham (2015). doi:10.1007/978-3-319-15585-2_33
15. Vabishchevich, P., Fort, V.: http://www.nomoz.org/site/629615/vfort.html. Accessed 15 Apr 2016

Solving a Singularly Perturbed Elliptic Problem by a Multigrid Algorithm with Richardson Extrapolation

Svetlana Tikhovskaya[✉]

Sobolev Institute of Mathematics SB RAS, Omsk Branch,
Pevtsova str., 13, 644043 Omsk, Russia
s.tihovskaya@yandex.ru

Abstract. A two-dimensional linear elliptic equation with regular boundary layers is considered. It is solved by using an upwind difference scheme on the Shishkin mesh with the property of uniform convergence with respect to small parameter ε. It is known that the application of multigrid methods leads to essential reduction of the number of arithmetical operations. Earlier the two-grid method with the application Richardson extrapolation to increase the ε-uniform accuracy of the difference scheme is investigated. In this paper the multigrid algorithm of the same structure is considered. The application of the Richardson extrapolation with the usage of numerical solutions on all of meshes leads to increase the ε-uniform accuracy of the difference scheme by two orders. Also we construct a better initial guess on refined mesh, using numerical solutions on coarse meshes. The results of some numerical experiments are discussed.

Keywords: Singularly perturbed elliptic problem · Regular boundary layers · Difference scheme · Shishkin mesh · ε-uniform accuracy · Multigrid method · Richardson extrapolation

1 Introduction

The two-dimensional linear elliptic problem with regular boundary layers in the unit square is considered. It is well known that the application of classical difference schemes for a singularly perturbed problem leads to large errors for small values of the perturbation parameter [1–5]. The uniform convergence of a difference scheme for such problem can be provided by fitting the scheme to a boundary layer component [1] or by using a mesh which is dense in a boundary layer [2,3]. The multigrid method [6–9] and the two-grid method [10,11] as a special case of multigrid method leads to essential reduction of the number of arithmetical operations. These methods can be effective applied for singularly perturbed problems, see [12–19] and the references therein. To increase the accuracy of the difference scheme in multigrid method the Richardson extrapolation [5,20] can be applied. The two-grid method with the application Richardson

© Springer International Publishing AG 2017
I. Dimov et al. (Eds.): NAA 2016, LNCS 10187, pp. 674–681, 2017.
DOI: 10.1007/978-3-319-57099-0_77

extrapolation to increase the ε-uniform accuracy of the difference scheme on the Shishkin mesh is investigated in [17–19]. In this paper the multigrid algorithm of the same structure based on three meshes is considered. To reduce the number of iteration the idea of the extrapolation of numerical solutions on coarse meshes [21, 22] is investigated.

2 Preliminaries

We consider the following boundary value problem:

$$
\varepsilon u_{xx} + \varepsilon u_{yy} + a(x)u_x + b(y)u_y - c(x,y)u = f(x,y), \ (x,y) \in \Omega = (0,1)^2, \\
u(x,y) = g(x,y), \quad (x,y) \in \Gamma = \overline{\Omega} \setminus \Omega, \tag{1}
$$

where the coefficients a, b, c are bounded and satisfy the condition:

$$
a(x) \geqslant \alpha > 0, \quad b(y) \geqslant \beta > 0, \quad c(x,y) \geqslant 0, \tag{2}
$$

the perturbation parameter ε takes arbitrary values in the open-closed interval $(0, 1]$. The coefficients, the right-hand side f and the boundary function g are sufficiently smooth. We also assume that the sufficient compatibility conditions are satisfied.

It is known [3–5] that the solution of problem (1) under condition (2) is uniformly bounded and has two regular boundary layers near $x = 0$ and $y = 0$.

Define the piecewise-uniform mesh [3] in the domain $\overline{\Omega}$:

$$
\Omega_N = \{(x_i, y_j), \ i,j = 0,1,\ldots,N, \ h_i = x_i - x_{i-1}, \ \tau_j = y_j - y_{j-1}\}, \\
h_i = 2\sigma_x/N, \quad 1 \leqslant i \leqslant N/2; \qquad h_i = 2(1-\sigma_x)/N, \quad N/2 < i \leqslant N, \\
\tau_j = 2\sigma_y/N, \quad 1 \leqslant j \leqslant N/2; \qquad \tau_j = 2(1-\sigma_y)/N, \quad N/2 < j \leqslant N, \tag{3} \\
\sigma_x = \min\left\{\frac{1}{2}, \frac{2\varepsilon}{\alpha} \ln N\right\}, \quad \sigma_y = \min\left\{\frac{1}{2}, \frac{2\varepsilon}{\beta} \ln N\right\}.
$$

where α and β are given in (2).

We consider the upwind difference scheme for the problem (1) on the mesh Ω_N:

$$
\frac{2\varepsilon}{h_i + h_{i+1}} \left(\frac{u_{i+1,j}^N - u_{i,j}^N}{h_{i+1}} - \frac{u_{i,j}^N - u_{i-1,j}^N}{h_i} \right) +
$$

$$
+ \frac{2\varepsilon}{\tau_j + \tau_{j+1}} \left(\frac{u_{i,j+1}^N - u_{i,j}^N}{\tau_{j+1}} - \frac{u_{i,j}^N - u_{i,j-1}^N}{\tau_j} \right) + a(x_i)\frac{u_{i+1,j}^N - u_{i,j}^N}{h_{i+1}} \tag{4}
$$

$$
+ b(y_j)\frac{u_{i,j+1}^N - u_{i,j}^N}{\tau_{j+1}} - c(x_i, y_j)u_{i,j}^N = f(x_i, y_j), \quad (x_i, y_j) \in \Omega_N,
$$

$$
u_{i,j}^N = g(x_i, y_j), \qquad (x_i, y_j) \in \Gamma_N = \Gamma \cap \Omega_N.
$$

Let $[u]_\Omega$ be the projection of a function $u(x,y)$ on a mesh Ω. According to [3, 4] the difference scheme (4) on the mesh (3) converges ε-uniformly and the following accuracy estimate is satisfied:

$$
\max_{0 \leqslant i,j \leqslant N} |u_{i,j}^N - u(x_i, y_j)| = \|u^N - [u]_{\Omega_N}\| \leqslant C\Delta_N, \qquad \Delta_N = \ln N/N, \tag{5}
$$

where C is positive constant that is independent of the perturbation parameter ε and the parameters of the mesh (3).

Notice that the matrix of this system (4) is M-matrix and a number of iterative methods for its resolving converge [23, 24].

3 Multigrid Method

We investigate a multigrid method based on three meshes with structure like in the articles [17–19]. To improve the accuracy of the difference scheme (4) in the multigrid method we investigate the application of Richardson extrapolation [5, 20] using all of three numerical solutions.

Let u^N be the solution of the difference scheme (4) on the mesh Ω_N. For the application of Richardson extrapolation we use the solutions of the difference scheme (4) on the meshes $\Omega_{N/2}$ and $\Omega_{N/4}$ which ones have the same value of the parameters σ_x and σ_y as the refined mesh Ω_N. Thus these meshes are nested that is $\Omega_{N/4} = \{(X_l, Y_m)\} \subset \Omega_{N/2} = \{(X_p, Y_q)\} \subset \Omega_N = \{(x_i, y_j)\}$.

Let us define the solution of the difference scheme (4) on the mesh $\Omega_{N/2}$ as $u^{N/2}$ and on the mesh $\Omega_{N/4}$ as $u^{N/4}$. According to the Richardson extrapolation we introduce the function u^{nN} on the mesh Ω_N. At first we define the function u^{nN} at the nodes of the coarse mesh $(X_l, Y_m) \in \Omega_{N/4}$ the formula:

$$u^{nN}(X_l, Y_m) = 8/3\, u^{N/4}(X_l, Y_m) - 2\, u^{N/2}(X_l, Y_m) + 1/3\, u^N(X_l, Y_m).$$

Now let us specify $u^{nN}(x_i, y_j)$ at the nodes of refined mesh $\Omega_N \backslash \Omega_{N/4}$ using the interpolation function. Then we define for each of nodes $(x_i, y_j) \in \Omega_N$ from some cell $S_{l,m} = [X_{l-2}, X_{l+2}] \times [Y_{m-2}, Y_{m+2}]$ as

$$u^{nN}(x_i, y_j) = I([u^{nN}]_{\Omega_{N/4}}, x_i, y_j),\tag{6}$$

where we use ε-uniform interpolation, see [18, 19] and reference therein.

Using the similar idea of articles [21, 22] we construct a better initial guess on refined mesh Ω_N. According to the error estimate (5) we have:

$$\begin{aligned}
u(X_l, Y_m) - u^N(X_l, Y_m) &= A(X_l, Y_m)\ln N/N + O(\Delta_N^2),\\
u(X_l, Y_m) - u^{N/2}(X_l, Y_m) &= A(X_l, Y_m)\ln N/(N/2) + O(\Delta_{N/2}^2),\\
u(X_l, Y_m) - u^{N/4}(X_l, Y_m) &= A(X_l, Y_m)\ln N/(N/4) + O(\Delta_{N/4}^2),
\end{aligned}\tag{7}$$

where the function A is ε-uniformly bounded. Therefore, eliminating the u in (7), we obtain

$$\begin{aligned}
u^N(X_l, Y_m) - u^{N/2}(X_l, Y_m) &= A(X_l, Y_m)\ln N/N,\\
u^{N/2}(X_l, Y_m) - u^{N/4}(X_l, Y_m) &= 2A(X_l, Y_m)\ln N/N.
\end{aligned}\tag{8}$$

We can eliminate the term with $A(X_l, Y_m)$ in (8), then we have

$$Eu_{l,m}^N = \frac{3}{2}u_{l,m}^{N/2} - \frac{1}{2}u_{l,m}^{N/4}, \qquad l = i,\, i+1,\ m = j,\, j+1,\tag{9}$$

where $u^{N/2}(X_l, Y_m) = u_{l,m}^{N/2}$, $u^{N/4}(X_l, Y_m) = u_{l,m}^{N/4}$. It follows from (8) that

$$A_{l,m} = (u_{l,m}^{N/2} - u_{l,m}^{N/4})/(2\ln N/N), \qquad A_{l,m} = A(X_l, Y_m). \tag{10}$$

To calculate the averaging value $A_{i,j+1/2}$, where $(X_i, Y_{j+1/2}) \in \Omega_{N/2}$ denotes the midpoint $(X_i, Y_j) \in \Omega_{N/4}$ and $(X_i, Y_{j+1}) \in \Omega_{N/4}$, we use the formula

$$A_{i,j+1/2} = (A_{i,j} + A_{i,j+1})/2. \tag{11}$$

We calculate the averaging values $A_{i+1/2,j}$, $A_{i+1,j+1/2}$, $A_{i+1/2,j+1}$, $A_{i+1/2,j+1/2}$ analogically. Taking into account (10) and substituting (11) into (8) we obtain

$$Eu_{i,j+1/2}^N = u_{i,j+1/2}^N + \left((u^{N/2} - u^{N/4})_{i,j} + (u^{N/2} - u^{N/4})_{i,j+1}\right)/4, \tag{12}$$

and similarly we have

$$Eu_{i+1/2,j}^N = u_{i+1/2,j}^N + \left((u^{\frac{N}{2}} - u^{\frac{N}{4}})_{i,j} + (u^{\frac{N}{2}} - u^{\frac{N}{4}})_{i+1,j}\right)/4,$$
$$Eu_{i+1,j+1/2}^N = u_{i+1,j+1/2}^N + \left((u^{\frac{N}{2}} - u^{\frac{N}{4}})_{i+1,j} + (u^{\frac{N}{2}} - u^{\frac{N}{4}})_{i+1,j+1}\right)/4,$$
$$Eu_{i+1/2,j+1}^N = u_{i+1/2,j+1}^N + \left((u^{\frac{N}{2}} - u^{\frac{N}{4}})_{i,j+1} + (u^{\frac{N}{2}} - u^{\frac{N}{4}})_{i+1,j+1}\right)/4, \tag{13}$$
$$Eu_{i+1/2,j+1/2}^N = u_{i+1/2,j+1/2}^N + \frac{1}{8} \sum_{\substack{l=i,i+1 \\ m=j,j+1}} (u^{\frac{N}{2}} - u^{\frac{N}{4}})_{l,m}.$$

We can get the values at other nodes on the mesh $\Omega_N \setminus \Omega_{N/2}$, using an appropriate interpolation.

4   Results of Numerical Experiments

We consider the following boundary layer problem:

$$\begin{aligned} \varepsilon u_{xx} + \varepsilon u_{yy} + u_x + u_y &= f(x,y), \qquad (x,y) \in \Omega, \\ u(x,y) &= 0, \qquad (x,y) \in \Gamma, \end{aligned} \tag{14}$$

where f corresponds to the exact solution:

$$u(x,y) = (1-x)\left(1 - e^{-x/\varepsilon}\right)(1-y)\left(1 - e^{-y/\varepsilon}\right).$$

The solution of the problem (14) is computed based on the difference scheme (4). We define the initial guess as

$$u^{(0)}(x_i, y_j) = 0, \qquad (x_i, y_j) \in \Omega_N.$$

We investigate the realization of the difference scheme (4) based on methods of Gauss-Seidel and successive over relaxation [23, 24]. The difference scheme (4) can be written as the five-point scheme:

$$a_{i,j}u_{i-1,j}^N + b_{i,j}u_{i,j-1}^N + c_{i,j}u_{i+1,j}^N + d_{i,j}u_{i,j+1}^N - e_{i,j}u_{i,j}^N = -f_{i,j}^N, \quad 0 < i,j < N.$$

Then vector-matrix form of the Gauss-Seidel method can be presented as:

$$u^{(m)} = D^{-1}\left(f + Lu^{(m)} + Uu^{(m-1)}\right),$$
$$(Lv)_{i,j} = a_{i,j}v_{i-1,j} + b_{i,j}v_{i,j-1}, (Dv)_{i,j} = e_{i,j}v_{i,j}, (Uv)_{i,j} = c_{i,j}v_{i+1,j} + d_{i,j}v_{i,j+1},$$

and successive over relaxation method [24] can be presented as:

$$u^{(m)} = \omega D^{-1}\left(f + Lu^{(m)} + Uu^{(m-1)}\right) + (1-\omega)u^{(m-1)},$$

where ω is the iterative parameter.

Notice that the efficiency of Gauss-Seidel method in the case of the problem (1) depends on the ordering of equations and unknowns [23,25]. Then vector-matrix form of the Gauss-Seidel method with localization of boundary layers can be written as:

$$u^{(m)} = D^{-1}(f + Lu^{(m-1)} + Uu^{(m)}).$$

There is more stable symmetric point Gauss-Seidel method [23] and it can be presented as

$$u^{(m)} = e_{i,j}^{-1}(a_{i,j}^{N}u_{i-1,j}^{(m)} + b_{i,j}^{N}u_{i,j-1}^{(m)} + c_{i,j}^{N}u_{i+1,j}^{(m-1)} + d_{i,j}^{N}u_{i,j+1}^{(m-1)} + f_{i,j}^{N}),$$
$$u^{(m)} = e_{i,j}^{-1}(a_{i,j}^{N}u_{i-1,j}^{(m-1)} + b_{i,j}^{N}u_{i,j-1}^{(m)} + c_{i,j}^{N}u_{i+1,j}^{(m)} + d_{i,j}^{N}u_{i,j+1}^{(m-1)} + f_{i,j}^{N}),$$
$$u^{(m)} = e_{i,j}^{-1}(a_{i,j}^{N}u_{i-1,j}^{(m)} + b_{i,j}^{N}u_{i,j-1}^{(m-1)} + c_{i,j}^{N}u_{i+1,j}^{(m)} + d_{i,j}^{N}u_{i,j+1}^{(m-1)} + f_{i,j}^{N}),$$
$$u^{(m)} = e_{i,j}^{-1}(a_{i,j}^{N}u_{i-1,j}^{(m-1)} + b_{i,j}^{N}u_{i,j-1}^{(m-1)} + c_{i,j}^{N}u_{i+1,j}^{(m)} + d_{i,j}^{N}u_{i,j+1}^{(m)} + f_{i,j}^{N}).$$

We can apply the same idea for the order of unknowns for successive over relaxation method.

Table 1 contains the number of iterations for the successive over relaxation (SOR) method, where the iterative parameter $\omega = 2/(1 + \sqrt{0.9})$, the Gauss-Seidel (GS) method with localization of boundary layers, the symmetric point successive over relaxation and the symmetric point Gauss-Seidel methods in the case of $\varepsilon = 10^{-4}$. For each of methods a bottom values correspond to the implementation by a one-grid method, a middle values correspond to the implementation by a multigrid method, a top values correspond to the implementation by a multigrid method with extrapolation (9), (12), (13). The number of iterations on the coarse meshes is given in brackets. The number of iterations on the refined is given without brackets.

Table 2 contains the time of computing for the successive over relaxation method and the Gauss-Seidel method with localization of boundary layers, the symmetric point successive over relaxation and the symmetric point Gauss-Seidel methods in the case of $\varepsilon = 10^{-4}$. For each of methods a bottom values correspond to the implementation by a one-grid method, a middle values correspond to the implementation by a multigrid method, a top values correspond to the implementation by a multigrid method with extrapolation (9), (12), (13).

It follows from Tables 1, 2 that the application of extrapolation (9), (12), (13) decreases the number of iteration and the time of computing the iterative methods.

Table 1. The number of iteration in the case of $\varepsilon = 10^{-4}$

Method	N			
	64	128	256	512
GS with localization	497(171)(72)	1398(467)(214)	3831(1356)(662)	10750(4210)(2160)
	602(171)(72)	1709(467)(214)	5147(1356)(662)	16414(4210)(2160)
	975	3081	10141	34465
GS symmetric point	544(192)(88)	1472(500)(244)	3964(1408)(712)	10988(4296)(2248)
	636(192)(88)	1764(500)(244)	5240(1408)(712)	16584(4296)(2248)
	1032	3180	10324	34812
SOR	470(162)(68)	1323(442)(201)	3628(1284)(625)	10187(3989)(2044)
	570(162)(68)	1619(442)(201)	4878(1284)(625)	15563(3989)(2044)
	923	2918	9612	32678
SOR symmetric point	516(180)(84)	1400(472)(228)	3760(1336)(672)	10424(4076)(2132)
	604(180)(84)	1676(472)(228)	4972(1336)(672)	15732(4076)(2132)
	976	3016	9788	33016

Table 2. The time of computing in the case of $\varepsilon = 10^{-4}$

Method	N				Method	N			
	64	128	256	512		64	128	256	512
GS with	0.129	1.379	19.088	333.560	SOR	0.133	1.392	18.767	326.091
localization	0.162	1.658	25.163	496.402		0.152	1.672	24.745	487.200
	0.220	2.718	45.077	990.065		0.217	2.738	44.627	979.968
GS	0.086	0.888	12.777	258.200	SOR	0.089	0.869	13.125	258.598
symmetric	0.096	0.948	16.545	381.998	symmetric	0.096	1.018	16.404	383.680
point	0.129	1.543	31.288	742.529	point	0.134	1.660	31.545	736.203

Table 3 contains the error norm for a one-grid method (left table) and a multigrid method with Richardson extrapolation (right table) for various values of ε and N.

Table 4 contains the numerical order of convergence for a one-grid method (left table) and a multigrid method with Richardson extrapolation (right table) for various values of N and ε.

The theoretical order of convergence CR_t depending on N is given for comparison in the bottom line of the tables. We calculated the numerical order of convergence CR by the formula:

$$CR = \log_2 \frac{D_N}{D_{2N}}, \qquad D_N = \|u^N - [u]_{\Omega_N}\|,$$

and the theoretical order of convergence CR_t that corresponds to the error estimate (5) for a one-grid method and corresponds to the error estimate of $O(\ln^3 N/N^3)$ for a multigrid method with Richardson extrapolation respectively.

It follows from Tables 3, 4 that the application of Richardson extrapolation (6) increases the accuracy of the difference scheme (4) to the order $O(\ln^3 N/N^3)$.

Table 3. The error norm for a one-grid Gauss-Seidel method (left) and a multigrid Gauss-Seidel method with Richardson extrapolation (right)

ε	N				ε	N			
	64	128	256	512		64	128	256	512
10^{-2}	$6.05e{-}2$	$3.65e{-}2$	$2.13e{-}2$	$1.22e{-}2$	10^{-2}	$4.03e{-}3$	$1.10e{-}3$	$2.64e{-}4$	$4.93e{-}5$
10^{-3}	$6.17e{-}2$	$3.72e{-}2$	$2.18e{-}2$	$1.24e{-}2$	10^{-3}	$4.44e{-}3$	$1.16e{-}3$	$2.88e{-}4$	$6.07e{-}5$
10^{-4}	$6.19e{-}2$	$3.73e{-}2$	$2.18e{-}2$	$1.24e{-}2$	10^{-4}	$4.49e{-}3$	$1.18e{-}3$	$2.94e{-}4$	$6.24e{-}5$
10^{-5}	$6.19e{-}2$	$3.73e{-}2$	$2.18e{-}2$	$1.24e{-}2$	10^{-5}	$4.50e{-}3$	$1.18e{-}3$	$2.95e{-}4$	$6.26e{-}5$
10^{-6}	$6.19e{-}2$	$3.73e{-}2$	$2.18e{-}2$	$1.24e{-}2$	10^{-6}	$4.50e{-}3$	$1.18e{-}3$	$2.95e{-}4$	$6.27e{-}5$

Table 4. The order of convergence for a one-grid Gauss-Seidel method (left) and a multigrid Gauss-Seidel method with Richardson extrapolation (right)

ε	N			ε	N		
	64	128	256		64	128	256
10^{-2}	0.729	0.775	0.810	10^{-2}	1.878	2.052	2.424
10^{-3}	0.730	0.774	0.809	10^{-3}	1.932	2.012	2.247
10^{-4}	0.730	0.774	0.809	10^{-4}	1.928	2.005	2.237
10^{-5}	0.730	0.774	0.809	10^{-5}	1.927	2.004	2.235
10^{-6}	0.730	0.774	0.809	10^{-6}	1.927	2.004	2.234
CR_t	0.778	0.807	0.830	CR_t	2.333	2.422	2.490

Acknowledgments. The research has been partially supported by the Russian Foundation for Basic Research grant 15-01-06584 and 16-01-00727.

References

1. Ilyin, A.M.: A difference scheme for a differential equation with a small parameter at the highest derivative. Mat. Zametki. **6**, 237–248 (1969). (in Russian)
2. Bakhvalov, N.S.: On optimization of methods to solve boundary value problems in the presence of a boundary layer. Zh. Vych. Mat. Mat. Fiz. **9**, 841–859 (1969). (in Russian)
3. Shishkin, G.I.: Grid Approximations of Singular Perturbation Elliptic and Parabolic Equations. UB RAS, Yekaterinburg (1992). (in Russian)
4. Roos, H.-G., Stynes, M., Tobiska, L.: Robust Numerical Methods for Singularly Perturbed Differential Equations. Springer, Berlin (2008)
5. Shishkin, G.I., Shishkina, L.P.: Difference Methods for Singular Perturbation Problems. Chapman & Hall/CRC, Boca Raton (2009)
6. Fedorenko, R.P.: The speed of convergence of one iterative process. Zh. Vychisl. Mat. Mat. Fiz. **4**, 559–564 (1964). (in Russian)
7. Hackbusch, W.: Multigrid convergence for a singular perturbation problem. Linear Algebra Appl. **58**, 125–145 (1984)
8. Shaidurov, V.V.: Multigrid Methods for Finite Elements. Springer, Berlin (1995)
9. Trottenberg, U., Oosterlee, C.W., Schuller, A.: Multigrid. Academic Press Inc, San Diego (2001)

10. Axelsson, O., Layton, W.: A two-level discretization of nonlinear boundary value problems. SIAM J. Numer. Anal. **33**, 2359–2374 (1996)
11. Xu, J.: A novel two-grid method for semilinear elliptic equation. SIAM J. Sci. Comput. **15**, 231–237 (1994)
12. Gaspar, F.J., Clavero, C., Lisbona, F.: Some numerical experiments with multigrid methods on Shishkin meshes. J. Comput. Appl. Math. **138**, 21–35 (2002)
13. Olshanskii, M.A.: Analysis of a multigrid method for convection-diffusion equations with the Dirichlet boundary conditions. Comput. Math. Math. Phys. **44**, 1374–1403 (2004)
14. Angelova, I.T., Vulkov, L.G.: Comparison of the two-grid method on different meshes for singularly perturbed semilinear problems. In: Applications of Mathematics in Engineering and Economics, pp. 305–312. American Institute of Physics (2008)
15. Vulkov, L.G., Zadorin, A.I.: Two-grid algorithms for the solution of 2d semilinear singularly perturbed convection-diffusion equations using an exponential finite difference scheme. In: AIP Conference Proceedings Application of Mathematics in Technical and Natural Sciences, pp. 371–379 (2009)
16. MacLachlan, S., Madden, N.: Robust solution of singularly perturbed problems using multigrid methods. SIAM J. Sci. Comput. **35**, A2225–A2254 (2013)
17. Tikhovskaya, S.V.: A two-grid method for an elliptic equation with boundary layers on a Shishkin mesh. Lobachevskii J. Math. **35**, 409–415 (2014)
18. Zadorin, A.I., Tikhovskaya, S.V., Zadorin, N.A.: A two-grid method for elliptic problem with boundary layers. Appl. Numer. Math. **93**, 270–278 (2015)
19. Tikhovskaya, S.V.: Investigation of a two-grid method of improved accuracy for elliptic reaction-diffusion equation with boundary layers. Kazan. Gos. Univ. Uchen. Zap. Ser. Fiz.-Mat. Nauki. **157**, 60–74 (2015). (in Russian)
20. Shishkin, G.I., Shishkina, L.P.: A higher-order Richardson method for a quasilinear singularly perturbed elliptic reaction-diffusion equation. Differ. Equ. **41**, 1030–1039 (2005)
21. Chen, C.M., Hu, H.L., Xie, Z.Q., Li, C.L.: Analysis of extrapolation cascadic multigrid method (EXCMG). Sci. China Ser. A: Math. **51**, 1349–1360 (2008)
22. Li, M., Li, C.L., Cui, X., Zhao, J.: Cascadic multigrid methods combined with sixth order compact scheme for Poisson equation. Numer. Algorithms **71**, 715–727 (2016)
23. Wessiling, P.: An Introduction to Multigrid Methods. Wiley, Chichester (1992)
24. Ilin V.P.: Finite Difference and Finite Volume Methods for Elliptic Equations. ICMMG Publ., Novosibirsk (2001). (in Russian)
25. Han, H., Il'in, V.P., Kellogg, R.B.: Flow directed iterations for convection dominated flow. In: Proceeding of the Fifth International Conference on Boundary and Interior Layers, pp. 7–17 (1988)

Conservative Finite-Difference Scheme for Computer Simulation of Field Optical Bistability

Vyacheslav A. Trofimov, Maria M. Loginova$^{(\boxtimes)}$, and Vladimir A. Egorenkov

Lomonosov Moscow State University, Moscow 119992, Russian Federation
mmloginova@gmail.com

Abstract. We investigate 2D switching wave of nonlinear absorption in a semiconduntor under the high intensive laser pulse action. A laser pulse interaction with semiconductor is described by the set of 2D nonlinear differential equations. To solve these equations numerically we have developed the conservative finite-difference scheme. It's realization is based on the original two-stage iteration process. It is very important, that the finite-difference scheme is conservative one on each of iterations because we have to provide a simulation on big time interval.

Keywords: Conservative finite-difference scheme · Iteration process · Optical bistability

1 Introduction

Investigation of a laser radiation interaction with semiconductor is very modern problem because semiconductors are widely used in many applications. Among them we would stress an optically bistable element, based on using of various nonlinear responses of semiconductor, exposed by a laser radiation. Optical bistability (OB) occurs because of a nonlinear dependence of semiconductor absorption coefficient on semiconductor characteristics. This phenomenon characterizes by existence of the hysteresis loop [1] for semiconductor characteristics. That means an existence of two stable states of the system (upper state and low one) for the same value of the incident optical pulse intensity, and switching from the low state to the upper one and vice versa occurs at different intensities. To investigate the switching wave, a computer simulation is widely used. One of the well-known approaches for solving of multi-dimensional equations is the split-step method [2]. In [3] we had shown that this method possessed some disadvantages for the problem under consideration and proposed a new finite-difference scheme for computer simulation of the problem.

2 Statement of 2D Problem

The process under consideration is described by the following set of 2D dimensionless differential equations [1,4]:

© Springer International Publishing AG 2017
I. Dimov et al. (Eds.): NAA 2016, LNCS 10187, pp. 682–689, 2017.
DOI: 10.1007/978-3-319-57099-0_78

$$\frac{\partial^2 \varphi}{\partial x^2} + \frac{\partial^2 \varphi}{\partial y^2} = \gamma(n - N), \quad 0 < x < L_x, \quad 0 < y < L_y, \quad t > 0, \qquad (1)$$

$$\frac{\partial n}{\partial t} = D_x \frac{\partial}{\partial x}\left(\frac{\partial n}{\partial x} - \mu_x n \frac{\partial \varphi}{\partial x}\right) + D_y \frac{\partial}{\partial y}\left(\frac{\partial n}{\partial y} - \mu_y n \frac{\partial \varphi}{\partial y}\right) + G(N, n) - R(N, n),$$

$$\frac{\partial N}{\partial t} = G(N, n) - R(N, n), \quad \frac{\partial I}{\partial y} + \delta_o \delta(N, \varphi)I = 0.$$

Above the following notations are introduced. Function n denotes a free electron concentration in the conductivity zone of a semiconductor; N is a concentration of ionized donors. Function φ describes a dimensionless electric field potential. I is the intensity of laser radiation propagating along the y axis. The coordinate x is a coordinate that is transverse to the laser pulse propagation direction. Variables x, y are dimensionless spatial coordinates and L_x, L_y denote their maximal values, correspondingly. Variable t denotes dimensionless time, its maximal value is equal to L_t. Coefficients of electron diffusion D_x, D_y and coefficients of electron mobility μ_x, μ_y are non-negative constants. Parameter γ depends, in particular, on the maximal achieving concentration of free charged particles, δ_0 denotes a maximal semiconductor absorption coefficient of laser energy.

Light energy absorption coefficient $\delta(N, \varphi)$ can be approximated by different ways in dependence of physical experiment conditions. Below we consider its following approximation

$$\delta(N, \varphi) = (1 - N)e^{-\alpha|\varphi|}, \alpha > 0, \qquad (2)$$

which is close to one of the experimental dependencies corresponding to the concentration OB existence. This dependence takes into account the Burstein-Moss effect [4]. The functions G and R, describing generation and recombination of free charged particles in the semiconductor, are given by the formulas

$$G(N, n) = q_0 I \delta(N, \varphi), \quad R(N, n) = \frac{nN - n_0^2}{\tau_p}, \qquad (3)$$

where n_0 is an equilibrium value of the free electron concentration and ionized donor concentration, τ_p characterizes a recombination time of free electron. q_0 is a maximal intensity of the incident laser pulse, its profile is Gaussian one along the x-coordinate

$$I|_{y=0} = \exp\left(-\left(\frac{x - 0.5L_x}{0.1L_x}\right)^2\right)(1 - \exp(-10t)). \qquad (4)$$

Boundary conditions (BC) for the set of Eq. (1) are written below if an electric current is absent through the semiconductor faces and if a semiconductor is placed in the external electric field:

$$\left(\frac{\partial n}{\partial x} - \mu_x n \frac{\partial \varphi}{\partial x}\right)\Bigg|_{x=0, L_x} = \left(\frac{\partial n}{\partial y} - \mu_y n \frac{\partial \varphi}{\partial y}\right)\Bigg|_{y=0, L_y} = 0, \qquad (5)$$

$$\left.\frac{\partial\varphi}{\partial x}\right|_{x=0,L_x} = -E_x, \quad \left.\frac{\partial\varphi}{\partial y}\right|_{y=0,L_y} = -E_y$$

But it should be stressed, that in the present paper we provide computer simulation with $E_x = E_y = 0$ and $\mu_x = \mu_y = \mu$. In this case, the problem solution is symmetrical concerning the laser beam centre. So, it is important feature for an accuracy estimate of the finite-difference scheme.

Initial conditions for the charged particle concentrations depend on the BC for the set of equations. If external electric field exist, then we have to solve an additional stationary problem for semiconductor characteristics initial distribution. If the external electric field is absent then the initial conditions are written in the following manner:

$$n|_{t=0} = N|_{t=0} = n_0, \quad \varphi|_{t=0} = 0, \quad I|_{t=0} = 0. \tag{6}$$

For the problem (1)–(6) the law of charge preservation takes place:

$$Q(t) = \int_0^{L_y} \int_0^{L_x} (n(x,y,t) - N(x,y,t))\, dx dy = 0. \tag{7}$$

If the difference analogue for the invariant (7) preserves, then our finite-difference scheme is a conservative one. This property shouldn't be loosen due to the accumulation of a computing error even for computation on long time interval. So, our aim is to construct the conservative finite-difference scheme with the asymptotic stability property.

3   Finite-Difference Scheme

To solve the problem (1)–(6) we have developed a finite-difference scheme (FDS). Below we describe it briefly. With this aim let us introduce in the domain

$$\bar{G} = \{0 \le x \le L_x\} \times \{0 \le y \le L_y\} \times \{0 \le t \le L_t\}$$

the uniform grids in time and space

$$\Omega = \omega_x \times \omega_y \times \omega_t, \quad \Omega' = \omega_x \times \omega_y \times \omega'_t, \quad \Omega" = \omega_x \times \omega'_y \times \omega'_t,$$

$$\omega_x = \left\{ x_i = ih_x, i = \overline{0, N_x}, h_x = L_x/N_x \right\},$$

$$\omega_y = \left\{ y_j = jh_y, j = \overline{0, N_y}, h_y = L_y/N_y \right\},$$

$$\omega'_y = \left\{ y'_j = (j - 0.5)h_y, j = \overline{0, N_y + 1}, h_y = L_y/N_y \right\},$$

$$\omega_t = \left\{ t_k = k\tau, k = \overline{0, N_t}, \tau = L_t/N_t \right\},$$

$$\omega'_t = \left\{ t'_k = (k + 0.5)\tau, k = \overline{0, N_t - 1}, \tau = L_t/N_t \right\}.$$

Let's define grid functions n_h, N_h, φ_h on Ω by the following way:

$$n_{ijk} = n\left(x_i, y_j, t_k\right), \quad N_{ijk} = N\left(x_i, y_j, t_k\right), \quad \varphi_{ijk} = \varphi\left(x_i, y_j, t_k\right).$$

Function I_h we define on the grid Ω'' shifted additionally on spatial coordinate y:
$$I_{ijk} = I\left(x_i, y_j', t_k'\right).$$
For brevity, below we used the following index-free notations:

$$f = f_{ijk}, \quad f_{i\pm1} = f_{i\pm1jk}, \quad f_{j\pm1} = f_{ij\pm1k}, \quad f_{i\pm0.5} = 0.5\left(f_{ijk} + f_{i\pm1jk}\right),$$

$$f_{j\pm0.5} = 0.5\left(f_{ijk} + f_{ij\pm1k}\right), \quad \hat{f} = f_{ijk+1}, \quad \overset{0.5}{f} = 0.5\left(f + \hat{f}\right),$$

$$I = I_{ijk}, \quad I_{i\pm1} = I_{i\pm1jk}, \quad \hat{I} = I_{ij+1k}, \quad \overset{0.5}{I} = 0.5\left(I + \hat{I}\right),$$

where f is one of the grid functions n_h, N_h, φ_h.

The first and the second differential derivatives are defined in standard way and notated as follows: f_x, $f_{\bar{x}}$, $f_{\bar{x}x}$, f_y, $f_{\bar{y}}$, $f_{\bar{y}y}$, f_t. We also use the standard designation of Laplace difference operator: $\Lambda f = f_{\bar{x}x} + f_{\bar{y}y}$.

Let's notice that the FDS conservatism means validity of a difference analogue for the conservation law (7). For numerical calculation of the integral (7) we used a trapezoid rule. We follow the invariant validity accuracy at carrying out the computer simulation, because it is an important characteristic of a FDS efficiency. For the problem, considered in this paper, violation of asymptotic stability property can cause violation of the conservatism property (invariant (8) values increase in time) or in violation of the problem solution symmetry.

For numerical solution of the problem (1)–(6) we approximate the initial-boundary problem by the set of finite-difference equations. For their resolvability we use various iterative processes.

3.1 FDS on the Base of Two-Stage Iteration Process

For our research we constructed conservative FDS. We solve the problem of difference equations non-linearity by means of two-stage iteration process. Below the first stage of the iteration process with BC is written:

$$\Lambda \overset{s+1}{\hat{\varphi}} = \gamma \left(\overset{s+1}{\hat{n}} - \overset{s+1}{\hat{N}} \right), \quad \frac{\overset{s+1}{\hat{N}} - N}{\tau} = \overset{s}{G}^{0.5} - \overset{s}{R}^{0.5}, \quad \frac{\overset{s+1}{\hat{I}} - I}{h_y} + \delta_0 \overset{s+1}{\delta} \overset{s+1}{I}^{0.5} = 0, \quad (8)$$

$$\frac{\overset{s+1}{\hat{n}} - n}{\tau} = \frac{D_x}{2}\left(n_{\bar{x}x} + \overset{s+1}{\hat{n}}_{\bar{x}x} \right) + \frac{D_y}{2}\left(n_{\bar{y}y} + \overset{s}{\hat{n}}_{\bar{y}y} \right) + \overset{s}{G}^{0.5} - \overset{s}{R}^{0.5} -$$

$$- \frac{D_x \mu}{2h_x}\left(n_{i+0.5}\varphi_x - n_{i-0.5}\varphi_{\bar{x}} + \overset{s}{\hat{n}}_{i+0.5}\overset{s}{\hat{\varphi}}_x - \overset{s}{\hat{n}}_{i-0.5}\overset{s}{\hat{\varphi}}_{\bar{x}} \right) -$$

$$-\frac{D_y\mu}{2h_y}\left(n_{j+0.5}\varphi_y - n_{j-0.5}\varphi_{\bar{y}} + \overset{s}{\hat{n}}_{j+0.5}\overset{s}{\hat{\varphi}}_y - \overset{s}{\hat{n}}_{j-0.5}\overset{s}{\hat{\varphi}}_{\bar{y}}\right),$$

$$\frac{\overset{s+1}{\hat{\varphi}}_{i1} - \overset{s+1}{\hat{\varphi}}_{i0}}{h_y} = \frac{\overset{s+1}{\hat{\varphi}}_{iN_y} - \overset{s+1}{\hat{\varphi}}_{iN_y-1}}{h_y} = 0, \quad i = 0,...,N_x, \tag{9}$$

$$\frac{\overset{s+1}{\hat{\varphi}}_{1j} - \overset{s+1}{\hat{\varphi}}_{0j}}{h_x} = \frac{\overset{s+1}{\hat{\varphi}}_{N_xj} - \overset{s+1}{\hat{\varphi}}_{N_x-1j}}{h_x} = 0, \quad j = 0,...,N_y,$$

$$\frac{\overset{s+1}{\hat{n}}_{1j} - \overset{s+1}{\hat{n}}_{0j}}{h_x} = \frac{\overset{s+1}{\hat{n}}_{N_xj} - \overset{s+1}{\hat{n}}_{N_x-1j}}{h_x} = 0, \quad j = 0,...,N_y,$$

$$I_{i0k} = \exp\left(-\left(\frac{i - N_x/2}{0.1N_x}\right)^2\right)(1 - \exp(-10k\tau)), \quad k = 1,...,N_t, \quad i = 0,...,N_x,$$

The second stage of the iteration process is:

$$\Lambda\overset{s+2}{\hat{\varphi}} = \gamma\left(\overset{s+2}{\hat{n}} - \overset{s+2}{\hat{N}}\right), \quad \frac{\overset{s+2}{\hat{N}} - N}{\tau} = \overset{s+1}{G}_{0.5} - \overset{s+1}{R}_{0.5}, \quad \frac{\overset{s+2}{\hat{I}} - I}{h_y} + \delta_0\overset{s}{\delta}\overset{s+2}{I}_{0.5\,0.5} = 0, \tag{10}$$

$$\frac{\overset{s+2}{\hat{n}} - n}{\tau} = \frac{D_x}{2}\left(n_{\bar{x}x} + \overset{s+1}{\hat{n}}_{\bar{x}x}\right) + \frac{D_y}{2}\left(n_{\bar{y}y} + \overset{s+2}{\hat{n}}_{\bar{y}y}\right) + \overset{s+1}{G}_{0.5} - \overset{s+1}{R}_{0.5} -$$

$$-\frac{D_x\mu}{2h_x}\left(n_{i+0.5}\varphi_x - n_{i-0.5}\varphi_{\bar{x}} + \overset{s+1}{\hat{n}}_{i+0.5}\overset{s+1}{\hat{\varphi}}_x - \overset{s+1}{\hat{n}}_{i-0.5}\overset{s+1}{\hat{\varphi}}_{\bar{x}}\right) -$$

$$-\frac{D_y\mu}{2h_y}\left(n_{j+0.5}\varphi_y - n_{j-0.5}\varphi_{\bar{y}} + \overset{s+1}{\hat{n}}_{j+0.5}\overset{s+1}{\hat{\varphi}}_y - \overset{s+1}{\hat{n}}_{j-0.5}\overset{s+1}{\hat{\varphi}}_{\bar{y}}\right),$$

BC for the second stage of iteration process approximated in the same way as for the first stage. The FDS has the second order of an approximation on spatial coordinates and on time in the inner grid nodes. The BC have the first order of approximation. This is a consequence of the conservatism property validity. Necessity of such BC approximation was substantiated in [5] for 1D case, and could be easily generalized for 2D case. It is very important to pay attention that for our finite-difference scheme we check the criterion of iteration convergence only after we make both iteration stages:

$$\left|\overset{s+2}{\hat{n}} - \overset{s}{\hat{n}}\right| \leq \varepsilon_1\left|\overset{s}{\hat{n}}\right| + \varepsilon_2, \quad \left|\overset{s+2}{\hat{N}} - \overset{s}{\hat{N}}\right| \leq \varepsilon_1\left|\overset{s}{\hat{N}}\right| + \varepsilon_2,$$

$$\left|\overset{s+2}{\hat{\varphi}} - \overset{s}{\hat{\varphi}}\right| \leq \varepsilon_1\left|\overset{s}{\hat{\varphi}}\right| + \varepsilon_2, \quad \left|\overset{s+2}{\hat{I}} - \overset{s}{\hat{I}}\right| \leq \varepsilon_1\left|\overset{s}{\hat{I}}\right| + \varepsilon_2, \quad \varepsilon_1, \varepsilon_2 > 0. \tag{11}$$

The values of functions, calculated on the previous time layer, are undertaken as an initial approach for the iterative process:

$$\overset{s=0}{\hat{n}} = n, \quad \overset{s=0}{\hat{N}} = N, \quad \overset{s=0}{\hat{\varphi}} = \varphi, \quad \overset{s=0}{\hat{I}} = I. \tag{12}$$

3.2   Additional iteration process for the Poison equation.

One more complexity, appearing at solving the problem (1)–(6), is caused by the solution of the 2D Poisson equation concerning semiconductor electric field potential. Obviously, if this equation has zero-value BC, then for its solution one can apply the method of Fast Fourier Transform. However, in more general case this method cannot be used. In connection with this we arrange additional iteration process for the Poisson equation. For this purpose we introduce additional an auxiliary grid function F on the grid $\bar{\Omega} = \omega_x \times \omega_y$, which is governed by the problem:

$$F^0 = \overset{s}{\hat{\varphi}}, \quad \frac{F^{p+1} - F^p}{\bar{\tau}} = F^{p+1}_{\bar{x}x} + F^p_{\bar{y}y} - \gamma \left(\overset{s+1}{\hat{n}} - \overset{s+1}{\hat{N}} \right),$$

$$\frac{F^{p+2} - F^{p+1}}{\bar{\tau}} = F^{p+1}_{\bar{x}x} + F^{p+2}_{\bar{y}y} - \gamma \left(\overset{s+1}{\hat{n}} - \overset{s+1}{\hat{N}} \right), i = 1, ..., N_x - 1, j = 1, ..., N_y - 1.$$

$$(13)$$

Here p is an iteration number, parameter $\bar{\tau}$ is an iteration step, which is not equal to time step τ. The functions n and N are taken on upper layer in time coordinate (here we consider the additional iteration process for the first stage (8) of two-stage iteration process, as for the second stage (10) this additional process is constructed in the same way). Computer simulation showed, that an accuracy of the Poisson equation solution influences significantly on the problem (1)–(6) solution. To achieve a hight accuracy of electric field potential calculation and to decrease the number of iterations we use convergence criterion based on the discrepancy assessment:

$$|\Psi(\varphi)| = \left| F^{p+1}_{\bar{x}x} + F^{p+2}_{\bar{y}y} - \gamma \left(\overset{s+1}{\hat{n}} - \overset{s+1}{\hat{N}} \right) \right| \leq \varepsilon_3, \quad \varepsilon_3 > 0. \quad (14)$$

If the solution, obtained on the $p+2$ iteration, satisfies to this criterion, then it is the Poisson equation solution for $s+1$ iteration with respect to the concentration of free electrons and ionized donors: $\overset{s+1}{\hat{\varphi}} = F^{p+2}$.

4   Computer Simulation Results

For the problem under consideration the OB point-wise model can't be written below because of the distributed electric field. So, we estimate OB realization by analysing the presence of explosive absorption of light energy in semiconductor and hysteresis loop occurrence. As it is known, the explosive absorption consists in sharp increasing of free electron concentration under the insignificant increase of input pulse intensity at the certain set of interaction parameters. In this case the system (semiconductor - optical radiation) switches from the low state to upper one. When the system is in the upper state the distribution of free electrons concentration represents domain of high concentration with sharp boundaries - switching waves.

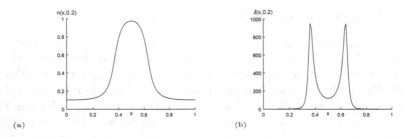

Fig. 1. Stationary distribution of free electron concentration n (a) and absorption coefficient δ (b), obtained at time moment t=100.

Below we presented results of computer modelling for parameters: $\delta_0 = 0.01$, $q_0 = 0.01$, $D_x = D_y = 10^{-3}$, $\gamma = 10^3$, $n_0 = 0.1$, $\mu_x = \mu_y = 1$, $\alpha = 5$. For illustration in Fig. 1 the stationary distributions of free electron concentration and absorption coefficient are shown. Form of these domains confirms an OB existence. To obtain a hysteresis loop we start our calculations with small input intensity value. In this case the system (optical pulse-semiconductor) is in low state. With further discrete increasing of input intensity

$$q_{max} = q_{previous} + \triangle q \left(1 - e^{-10t}\right), \quad \triangle q = const \tag{15}$$

the system gradually tends to the point of switching. Then in the vicinity of it, an insignificant increasing in input intensity q_{max} tending in sharp free electron concentration growth and the system switches to the upper state. Reverse switching to the low state realizes at q_{max} value, witch is smaller than the intensity corresponding to the upper state switching. Therefore, OB takes place. Switching intensities and contrast of switching for free electron concentration and width of bistable loop are important characteristics of the hysteresis. For hysteresis loop demonstration we show free electron concentration stationary distribution at the centre of laser beam ($x = 0.5$) and in different sections of a semiconductor (Fig. 2). As one can see, the hysteresis dependence exists not only on front surface of semiconductor (Fig. 2a), but also in other sections (Fig. 2b). In this case the width of bistable region doesn't change, but a contrast of the switching decreasing because of the light energy absorption in the semiconductor.

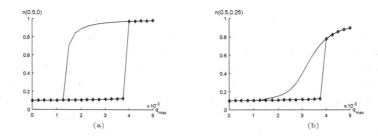

Fig. 2. Dependence of electrons concentration value at light beam axis from the maximum input optical pulse intensity.

5   Conclusions

In this paper we developed conservative FDS on the base of two-stage iteration process for the problem of field OB realization in semiconductor under the action of high intensive laser pulse. One of its main advantages consists in the asymptotic stability property. Thus, it is possible to provide a computation on long time interval without round error accumulation. For the Poisson equation solution we apply additional iteration process. For an assessment of this iteration process convergence we use the criterion based on the discrepancy assessment. This allows to calculate the electric field potential with high accuracy and with good high-speed performance. We demonstrated that the developed FDS is applicable for calculating of nonlinear complicated processes, which occur in a semiconductor under the action of high intensive laser pulse. Computer experiments, providing on the base of proposed finite-difference scheme, confirmed possibility of field optical bistability realization in the system: semiconductor-laser radiation.

Acknowledgments. The investigation was made using support of the Russian Science Foundation (Grant 14-21-00081).

References

1. Gibbs, H.M.: Optical Bistability: Controlling Light with Light. Academic Press, NY (1985)
2. Strang, G.: On the construction and comparison of difference schemes. SIAM J. Numer. Anal. **5**(N3), 506–517 (1968)
3. Trofimov, V.A., Loginova, M.M., Egorenkov, V.A.: New two-step iteration process for solution of semiconductor plasma generation problem with arbitrary Boundary Conditions in 2D case. WIT Trans. Model. Simul. **59**, 85–96 (2015)
4. Bonch-Bruevich, V.L., Kalashnikov, S.G.: Physics of Semiconductors. Nauka, Moscow (1990). (in Russian)
5. Trofimov, V.A., Loginova, M.M.: Difference scheme for the problem of femtosecond pulse interaction with semiconductor in the case of nonlinear electron mobility. J. Comput. Math. Math. Phys. **45**(N12), 2185–2196 (2005)

Numerical Modeling of Micropolar Thin Elastic Plates

Maria Varygina[✉]

Institute of Computational Modeling, Siberian Branch of Russian Academy
of Sciences, Akademgorodok, 660036 Krasnoyarsk, Russia
vmp@icm.krasn.ru

Abstract. System of equations describing micropolar elastic thin plates
is written in symmetric hyperbolic form. For the solution of dynamic
problems in the framework of micropolar thin plates the numerical
algorithm is proposed. The algorithm based on the two-cyclic splitting method in combination with monotone finite-difference 1D schemes
is proposed. The results of computations of problems on distributed
impulse action loads are shown.

Keywords: Cosserat continuum · Micropolar plates · Elasticity · Finite
-difference scheme · Dynamic problems

1 Introduction

Thin-walled structures such as rods, plates and shells are widely used in civil
engineering, aero-space industry, medical and biological fields as basic structural elements. Structure is one of the most important indicator of the quality
of materials directly influencing on theirs strength characteristics. Depending
on type of the material and the scope of research in practical problems it is
necessary to take into account the sctructure on nano-, micro- or mesolevel. To
describe complex inner structure of a material the construction of new models of
micropolar media is required. In micropolar or Cosserat continuum in addition
to translational motion characterized with the velocity vector independent small
rotations of particles are considered [1,2]. And together with the antisymmetric
stress tensor antisymmetric couple stress tensor is introduced.

In papers [3–6] numerical solution of three-dimensional dynamic problems of
Cosserat continuum was presented. In particular, it is shown that in Cosserat
continuum there is a resonant frequency depending only on the inertial properties
of particles and the elasticity parameters of the material. The present paper gives
the results of numerical modeling of micropolar thin elastic plates.

There are some approaches of constructing two-dimensional mathematical
models of micropoar plates. Within the framework of the direct approach the
plate is modeled as a deformable surface with material points; see for example [7–9] and references therein. Another approach is based on the reduction

I. Dimov et al. (Eds.): NAA 2016, LNCS 10187, pp. 690–697, 2017.
DOI: 10.1007/978-3-319-57099-0_79

of three-dimensional micropolar continuum equations. Various averaging procedures together with the approximation of the displacements and rotations in the
thickness direction are applied; see for example [9–13] and references therein. In
present paper assumptions on linear approximation of translation and rotation
together with the through-the thickness integration procedure are made.

2 Mathematical Model

2.1 Equations of Three-Dimensional Cosserat Continuum

In Cosserat continuum translational motion denoted by u and independent rotations of particles denoted by φ are considered. The stress state of the material
is characterized by the antisymmetric stress tensor σ and antisymmetric couple
stress tensor m. The complete system of equations in three-dimensional Cosserat
model consists of the motion equations, the kinematic equations and the generalised law of linear elasticity theory [2]

$$\rho \ddot{u} = \nabla \cdot \sigma + \rho g, \quad j \ddot{\varphi} = \nabla \cdot m - 2\sigma_x + jq,$$
$$\Lambda = \nabla u + \varphi, \quad M = \nabla \varphi,$$
$$\sigma = \lambda I I \cdot \cdot \Lambda^S + 2\mu \Lambda^S + 2\alpha \Lambda^A,$$
$$m = \beta I I \cdot \cdot M^S + 2\gamma M^s + 2\varepsilon M^A. \tag{1}$$

Here g and q are the mass force and couple vectors, ρ is the material density,
j is the inertial parameter equal to the product of the inertia moment of a particle about the axis through its center of gravity and the numbers of particles
in unit volume. The formula $r = \sqrt{5j/(2\rho)}$ is valid to estimate the linear parameter of material microstructure. Constants λ and μ are the Lame parameters,
and $\alpha, \beta, \gamma, \varepsilon$ are the phenomenological elasticity coefficients for an isotropic
material. Λ and M are the strain and wryness tensors, the superscripts S and A
correspond to the symmetric and antisymmetric tensor components respectively.
The antisymmetric component is identified with its corresponding vector. A dot
above a symbol denotes the derivative with respect to time t.

Boundary conditions have the following form in terms of translations and
rotations

$$u = u^0, \quad \varphi = \varphi^0$$

or stresses

$$n \cdot \sigma = p^0, \quad n \cdot m = q^0,$$

where u^0, φ^0 are given functions, p^0 and q^0 are the surfaces forces and surface
couples acting on a part of a boundary of micropolar body.

2.2   Equations of Two-Dimensional Micropolar Plates

The transition to the two-dimensional equations is based on the linear in x_3 approximation of the displacement and rotation with independent integration of motion equations in (1) through the plate thickness. Let the isotropic plate-like body occupy the volume $V = \{(x_1, x_2, x_3) \in R^3 : (x_1, x_2) \in S \subset R^2, x_3 \in [-h, h]\}$. Here h is a half of the plate thickness, $h = \text{const}$. Let us assume that the plate thickness $2h$ is small compared to characteristic sizes of a plate. For translations and rotations of the plate-like body the following aproximation is made

$$
\begin{aligned}
u_i(t, x_1, x_2, x_3) &= x_3 \psi_i(t, x_1, x_2), \quad i = 1, 2 \\
u_3(t, x_1, x_2, x_3) &= w(t, x_1, x_2), \\
\varphi_i(t, x_1, x_2, x_3) &= \omega_i(t, x_1, x_2), \\
\varphi_3(t, x_1, x_2, x_3) &= \omega_3(t, x_1, x_2) + x_3 \theta(t, x_1, x_2).
\end{aligned} \tag{2}
$$

Normal to the midplane displacement u_3 and rotations φ_1, φ_2 are independent on coordinate x_3. Hence, there are 7 kinematically independent scalar fields: ψ_1, ψ_2, w, φ_1, φ_2, φ_3, θ. In static case when ω_3 is equal to zero kinematical assumptions (2) coincide with hypotheses in [12].

On the front planes of the plate $x_3 = \pm h$ boundary conditions are assumed to be homogeneous, stress and couple stress tensors are zero.

Stress and couple stress tensor in constitutive Eq. (1) after the integration through the thickness are written in terms of stress and couple stress resultants

$$
N_{ij} = <\sigma_{ij}>, \quad T_{ij} = <x_3\,\sigma_{ij}>, \quad L_{ij} = <m_{ij}>, \quad K_{ij} = <x_3\,m_{ij}>,
$$

where $<(\dots)> = \int\limits_{-h}^{h} (\dots)\, dx_3$. Here and below indices $i, j = 1, 2$, $i \neq j$.

In terms of velocities and angular velocities

$$
\Psi_i = \dot{\psi}_i, \quad W = \dot{w}, \quad \Omega_i = \dot{\omega}_i, \quad \Omega_3 = \dot{\omega}_3, \quad \Theta = \dot{\theta},
$$

integration of the motion equations in (1) leads to the following motion equations:

$$
\begin{aligned}
\frac{2h^3}{3}\rho\dot{\Psi}_i &= T_{1i,1} + T_{2i,2} - N_{3i}, \quad 2h\rho\dot{W} = N_{13,1} + N_{23,2}, \\
2hj\dot{\Omega}_i &= L_{1i,1} + L_{2i,2} + (-1)^j(N_{j3} - N_{3j}), \\
2hj\dot{\Omega}_3 &= L_{13,1} + L_{23,2} + N_{12} - N_{21}, \\
\frac{2h^3}{3}j\dot{\Theta} &= K_{13,1} + K_{23,2} + T_{12} - T_{21} - L_{33}.
\end{aligned} \tag{3}
$$

The subscripts after a comma denote the partial derivatives with respect to the corresponding coordinate.

Elasticity relations after the integration are as follows

$$\dot{T}_{ii} = \frac{2h^3}{3} \left(\lambda \left(\Psi_{1,1} + \Psi_{2,2}\right) + 2\mu\Psi_{i,i}\right), \quad \dot{T}_{33} = \frac{2h^3}{3}\lambda \left(\Psi_{1,1} + \Psi_{2,2}\right),$$

$$\dot{T}_{ij} = \frac{2h^3}{3} \left((\mu + \alpha)\Psi_{j,i} + (\mu - \alpha)\Psi_{i,j} + (-1)^i \cdot 2\alpha\Theta\right),$$

$$\dot{N}_{i3} = 2h \left((\mu + \alpha)W_{,i} + (\mu - \alpha)\Psi_i + (-1)^j \cdot 2\alpha\Omega_j\right),$$

$$\dot{N}_{3i} = 2h \left((\mu - \alpha)W_{,i} + (\mu + \alpha)\Psi_i + (-1)^i \cdot 2\alpha\Omega_j\right),$$

$$\dot{L}_{ii} = 2h \left(\beta(\Omega_{1,1} + \Omega_{2,2} + \Theta) + 2\gamma\Omega_{i,i}\right), \tag{4}$$

$$\dot{L}_{33} = 2h \left(\beta(\Omega_{1,1} + \Omega_{2,2} + \Theta) + 2\gamma\Theta\right),$$

$$\dot{L}_{ij} = 2h \left((\gamma + \varepsilon)\Omega_{j,i} + (\gamma - \varepsilon)\Omega_{i,j}\right),$$

$$\dot{K}_{i3} = \frac{2h^3}{3}(\gamma + \varepsilon)\Theta_{,i}, \quad \dot{K}_{3i} = \frac{2h^3}{3}(\gamma - \varepsilon)\Theta_{,i},$$

$$\dot{L}_{i3} = 2h(\gamma + \varepsilon)\Omega_{3,i}, \quad \dot{L}_{3i} = 2h(\gamma - \varepsilon)\Omega_{3,i},$$

$$\dot{N}_{ij} = (-1)^i 4h\alpha\Omega_3.$$

From the system (3)–(4) two independent systems are derived. One of them contains equations for Ω_3, N_{ij}, L_{i3}, L_{3i}:

$$2hj\dot{\Omega}_3 = L_{13,1} + L_{23,2} + N_{12} - N_{21},$$

$$\dot{L}_{i3} = 2h(\gamma + \varepsilon)\Omega_{3,i}, \quad \dot{L}_{3i} = 2h(\gamma - \varepsilon)\Omega_{3,i}, \tag{5}$$

$$\dot{N}_{ij} = (-1)^i 4h\alpha\Omega_3.$$

System (5) may be written as two-dimensional Klein-Gordon equation for angular velocity Ω_3:

$$\ddot{\Omega}_3 = \frac{\gamma + \varepsilon}{j} \left(\Omega_{3,11} + \Omega_{3,22}\right) - 4\frac{\alpha}{j}\Omega_3.$$

The second independent system from (3)–(4) includes 24 equations for 24 unknowns and can be written in matrix form

$$A\frac{\partial U}{\partial t} = B^1\frac{\partial U}{\partial x_1} + B^2\frac{\partial U}{\partial x_2} + QU + G, \tag{6}$$

where U is the vector-function

$$U = (\Psi_i, W, \Omega_i, \Theta, T_{ii}, T_{33}, T_{ij}, N_{i3}, N_{3i}, L_{ii}, L_{33}, L_{ij}, K_{i3}, K_{3i}).$$

The matrix coefficients A, B_1, B_2 containing elasticity parameters of a material are symmetric, and Q is antisymmetric. G is the given vector of mass forces and couples. The matrix A is positive definite if its diagonal blocks are positive definite. According to the Sylvester criterion this condition restricts the admissible values of the material parameters:

$$3\lambda + 2\mu > 0, \quad \mu, \alpha > 0, \quad 3\beta + 2\gamma > 0, \quad \gamma, \varepsilon > 0. \tag{7}$$

If inequalities (7) are fulfilled system (6) is hyperbolic in the sense of Friedrichs. The potential energy of elastic deformation is a positive-definite quadratic form and conservation law is fulfilled:

$$\frac{\partial(UAU)}{\partial t} = \frac{\partial(UB_1U)}{\partial x_1} + \frac{\partial(UB_2U)}{\partial x_2}.$$

The characteristic properties of this system are described by the equation

$$\det(cA + n_1B_1 + n_2B_2) = 0, \quad n_1^2 + n_2^2 = 1.$$

Its positive roots are the velocities of longitudinal waves c_p, transverse waves c_s, torsional waves c_m, and rotational waves c_ω. These roots are

$$c_p = \sqrt{\frac{\lambda + 2\mu}{\rho}}, \quad c_s = \sqrt{\frac{\mu + \alpha}{\rho}}, \quad c_m = \sqrt{\frac{\beta + 2\gamma}{j}}, \quad c_\omega = \sqrt{\frac{\gamma + \varepsilon}{j}}.$$

For hyperbolic system (6) the boundary-value problem with initial conditions $U(0,x) = U^0(x)$ and dissipative boundary conditions is well-posed. In particular among the dissipative conditions there are the conditions in terms of velocities

$$\Psi_i = \Psi_i^0, \quad W = w^0, \quad \Omega_i = \Omega_i^0, \quad \Omega_3 = \Omega_3^0, \quad \Theta = \Theta^0,$$

or stresses

$$n_1 T_{1i} + n_2 T_{2i} = p_i, \quad n_1 N_{13} + n_2 N_{23} = p_3,$$
$$n_1 L_{1i} + n_2 L_{2i} = q_i, \quad n_1 L_{13} + n_2 L_{23} = q_3, \quad n_1 K_{13} + n_2 K_{23} = k,$$

where Ψ_i^0, W^0, Ω_i^0, Ω_3^0 and Θ^0 are given functions, p_i, p_3, q_i, q_3, k are given surface forces and couples acting on part of the boundary of micropolar body.

3   Numerical Modeling

3.1   Numerical Algorithm

The algorithm of numerical solution of linear system (6) is based on two-cyclic splitting method with respect to spatial variables and time. On time interval $(t, t + \triangle t)$ the method consists of five stages: the solution of a one-dimensional problem in the x_1 direction on time interval $(t, t + \triangle t/2)$; similar stage in the x_2 direction; the stage of solution of a system of linear ordinary differential equations with matrix Q with full time-step; and two stages of repeated recalculations of a problem in the x_2 and x_1 directions respectively on time interval $(t + \triangle t/2, t + \triangle t)$. As applied to system (6) the splitting method has the following form:

$$A\dot{U}^1 = B^1 U_{,1}^1 + G^1, \quad U^1(t,x) = U(t,x),$$
$$A\dot{U}^2 = B^2 U_{,2}^2 + G^2, \quad U^2(t,x) = U^1(t + \triangle t/2, x),$$
$$A\dot{U}^3 = QU^3, \quad U^3(t,x) = U^2(t + \triangle t/2, x),$$
$$A\dot{U}^4 = B^2 U_{,2}^4 + G^2, \quad U^4(t + \triangle t/2, x) = U^3(t + \triangle t, x),$$
$$A\dot{U}^5 = B^1 U_{,1}^5 + G^1, \quad U^5(t + \triangle t/2, x) = U^4(t + \triangle t, x).$$

Vector-function U^1 is taken from the prevous time step. At $t = 0$ it is taken from the initial conditions. The unknown value $U(t + \triangle t, x)$ is $U^5(t + \triangle t, x)$, $G^1 + G^2 = G$.

At the third stage Crank-Nickolson finite-difference scheme with full time step is used. Each of four remaining one-dimensional problems are solved with the help of explicit monotone ENO-scheme of "predictor-corrector" type. This scheme is a generalization of Godunov collapse of the gap scheme.

The two-cyclic splitting method ensures the stability of a numerical solution provided Courant-Friedrichs-Levy stability condition for one-dimensional systems is fulfilled. It has second order of accuracy if second-order schemes at its stages are used.

The verification of the algorithm is performed by comparig the results of numerical computations with the exact solution describing wave propagation in micropolar plate.

3.2   Numerical Results

The results of numerical computations of the elastic waves propagation in a rectangular thin plate are presented in Figs. 1 and 2. Top and bottom sides of the plate are nonreflecting boundaries. The right side of the plate is fixed. On the left side distributed periodic load of Λ-impulses of T_{11} is given. The area of action load is one third of a side in the central part. As a result of impulse load a sequence of loading and unloading waves propagates over the material. Waves are generated at the points of the boundary of the area of action load on the left side of the plate. In the first case (Fig. 1) a single wave induced by the loading impulse is observed, in the second case (Fig. 2) three waves caused by three impulses propagate.

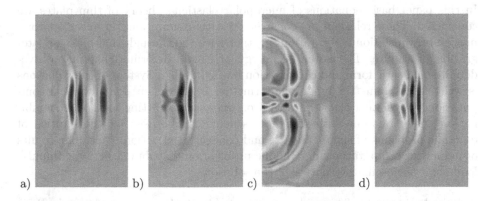

a)   b)   c)   d)

Fig. 1. The action of Λ-shaped impulse of T_{11}: level curves of the couple stress resultant T_{11} (a), velocity W (b), angular velocities Ω_1 (c) and Ω_2 (d)

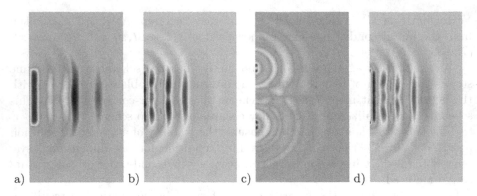

Fig. 2. The action of Λ-shaped impulses of T_{11}: level curves of the couple stress resultant T_{11} (a), velocity W (b), angular velocities Ω_1 (c) and Ω_2 (d)

The level curves of the couple stress resultant T_{11}, velocity W, angular velocities Ω_1 and Ω_2 are shown in Figs. 1 and 2 for the time moment $t = 18$ μs. The calculations are performed on a rectangular plate of sides 0.05×0.1 m for synthetic polyurethane. Material parameters are taken from [14]. Velocities of elastic waves are $c_p = 2687$, $c_s = 1394$, $c_\omega = 893$ m/s. The plate thickness $2h = 5$ mm. Characteristic scale of the microstructure of a material is $r = 0.15$ mm. The uniform difference grid used in computations consists of 1000×1000 cells with a mesh size of 0.1 mm less than r. On coarser grids calculations with satisfactory accur acy may not be performed.

4 Conclusions

In this paper basic equations of micropolar elasticity theory of thin plates are considered. These relations are constructed with assumptions on linear approximation of translation and rotation together with the through-the thickness integration procedure. The potential energy of elastic deformation is a positive-definite quadratic form and conservation law is fulfilled. System of the equations is written in symmetric hyperbolic form that is convenient for numerical computations. Numerical algorithm based on two-cyclic splitting method in combination with monotone finite-difference scheme can be used for the solution of dynamic problems of shock, impulse and concentrated action loads. The results of numerical computations of dynamic problem on disrtibuted periodic impulse load show the oscillatory nature of the solution.

Acknowledgements. This work was supported by the Russian Foundation for Basic Research (project no. 16-31-00078).

References

1. Cosserat, E., Cosserat, F.: Theorie des Corps Deformables. Librairie Scientifique A. Hermann et Fils, Paris (1909)
2. Palmov, V.A.: Governing equations of the of nonsymmetrical elasticity theory. Prikl. Mat. Mekh. **28**(3), 401–408 (1964). (in Russian)
3. Varygina, M.P., Kireev, I.V., Sadovskaya, O.V., Sadovskii, V.M.: Software for analysis of wave motions in Cosserat media on multiprocessor computer systems. Vest. SibSAU **2**(23), 104–108 (2009). (in Russian)
4. Varygina, M.P., Sadovskaya, O.V., Sadovskii, V.M.: Resonant properties of moment Cosserat continuum. J. Appl. Mech. Tech. Phys. **51**(3), 405–413 (2010)
5. Sadovskii, V., Sadovskaya, O., Varygina, M.: Numerical solution of dynamic problems in couple-stressed continuum on multiprocessor computer systems. Int. J. Numer. Anal. Modeling, Ser. B **2**(2–3), 215–230 (2011)
6. Sadovskaya, O., Sadovskii, V.: Mathematical Modeling in Mechanics of Granular Materials. Advanced Structured Materials, vol. 21. Springer, Heidelberg (2012)
7. Zhilin, P.A.: Mechanics of deformable directed surfaces. Int. J. Solids Struct. **12**, 635–648 (1976)
8. Altenbach, J., Altenbach, H., Eremeyev, V.: On generalized Cosserat-type theories of plates and shells: a short review and bibliography. Arch. Appl. Mech. **80**(1), 73–92 (2010)
9. Altenbach, H., Eremeyev, V.: On the linear theory of micropolar plates. ZAMM. **89**(4), 242–256 (2009)
10. Eringen, A.C.: Theory of micropolar plates. ZAMP. **18**(1), 12–30 (1967)
11. Eringen, A.C.: Microcontinuum Field Theory. I. Foundations and Solids. Springer, Heidelberg (1999)
12. Sarkisyan, S.O.: Mathematical model of micropolar elastic thin plates and their strength and stiffness characteristics. J. Appl. Mech. Tech. Phys. **53**(2), 275–282 (2012)
13. Sargsyan, S.O.: The theory of micropolar thin elastic shells. J. Appl. Math. Mech. **76**, 235–249 (2012)
14. Lakes, R.: Experimental methods for study of Cosserat elastic solids and other generalized elastic continua. In: Continuum Models for Materials with Micro-Structure, pp. 1–22 (1995). (Chap. 1)

Iterative Solution of the Retrospective Inverse Problem for a Parabolic Equation Using the Conjugate Gradient Method

V.I. Vasil'ev[✉] and A.M. Kardashevsky[✉]

North-Eastern Federal University, Yakutsk, Russia
vasvasil@mail.ru, kardam123@gmail.com

Abstract. In theory and practice of the inverse problems for unsteady partial differential equations, significant attention is paid to the problems of determination of the initial condition based on the values of the initial function in a finite time.

In this paper, we propose a numerical method for the solution of a retrospective inverse problem for a multidimensional parabolic equation by a rapidly convergent conjugate gradient method. The results of computational experiments on model problems with quasi-solutions are being discussed. Besides, the results of computations obtained at specifying a perturbated additional condition with random errors are presented.

Keywords: Parabolic equations · Inverse problem · Retrospective problem · Finite-difference method · Iterative method · Conjugate gradient method

1 Introduction

Inverse problems represent a quite large class of partial differential equations that are widely discussed, because these equations are most often used in mathematical descriptions of various processes. On the other hand, when the structure of a mathematical model of a process under research is known, we can pose the problem of identification of a mathematical model, for example, the definition of a coefficient of differential equations [1,2], of the right-hand side [1,2], domain boundaries [3,4], boundary conditions [5,6], or initial conditions [7], etc.

Problems with reverse time (retrospective reverse problems) [8–10] at a fixed point of time belong to a class of incorrect problems in the classical sense [7]. Several methods were developed for solution of this problem: the quasi-inversion method [3,4], A.N. Tikhonov regularization method and its modifications [8]. As indicated in [7] the iterative algorithms carry a great potential in solving evolutionary inverse problems, which take into consideration the specifics of problems at task with iteratively defined primary condition, that is at each iteration an ordinary direct correct primary boundary problem for parabolic equations is solved [9].

© Springer International Publishing AG 2017
I. Dimov et al. (Eds.): NAA 2016, LNCS 10187, pp. 698–705, 2017.
DOI: 10.1007/978-3-319-57099-0_80

Most of these works use the parameters of regularization. It is well-known that incorrect problems, are sensitive to the parameters of regularization. In this study we use an iterative method where the parameter of regularization is a number of iterations that is in line with the error of the input data. Let us bring the calculations that demonstrate the possibilities of this method.

2   Problem Formulation

We consider a parabolic equation

$$\frac{\partial u}{\partial t} = \sum_{\alpha=1}^{p} \frac{\partial}{\partial x_\alpha} \left(k_\alpha(x) \frac{\partial u}{\partial x_\alpha} \right), \quad x \in \Omega, \quad 0 \leqslant t < T, \tag{2.1}$$

where $\Omega \in \mathcal{R}^p$ is computational domain

$$\Omega = \{ x \mid x = (x_1, \ldots, x_p), \quad 0 < x_\alpha < l_\alpha, \quad \alpha = 1, \ldots, p \}$$

and coefficient $k_\alpha(x)$, $\alpha = 1, \ldots, p$ satisfies the following condition

$$0 < \kappa_1 \leqslant k_\alpha(x,t) \leqslant \kappa_2 < \infty. \tag{2.2}$$

On the boundary of domain homogeneous Dirichlet boundary conditions are applied

$$u(x,t) = 0, \quad x \in \partial\Omega, \quad 0 \leqslant t < T. \tag{2.3}$$

We formulate a retrospective inverse problem in the following way: we set the values of the desired function at the final time $u(x,T)$, and the initial condition $u(x,0)$ to be determined [7]

$$u(x,T) = \phi(x), \quad x \in \overline{\Omega}. \tag{2.4}$$

It should be noted that direct equation (2.1) in the case of $k = 1$ has an exact analytical solution

$$u(x_1, x_2, \ldots, x_p, t) = \frac{1}{\sqrt{(\beta + 4t)^p}} e^{-\frac{\|x - x_0\|^2}{\beta + 4t}}, \quad \beta > 0, \tag{2.5}$$

where $x_o = (l_1/2, \ldots, l_p/2)$. Function (2.5) is bounded, sufficiently smooth and can be used to establish the accuracy of the proposed iterative method.

3   Discretization of the Retrospective Problem

For numerical solution of the non-classical problem with reverse time for the parabolic equation (2.1)–(2.4) we use the finite-difference method. In domain Ω, we define uniform computational grid

$$\omega_\alpha = \left\{ x_\alpha = i_\alpha h_\alpha, i_\alpha = 1, 2, \ldots, N_\alpha - 1, N_\alpha h_\alpha = l_\alpha \right\}, \quad \alpha = 1, \ldots, p; \quad \omega = \omega_1 \times \ldots \times \omega_p$$

with constant mesh size in each direction, $h_\alpha, \alpha = 1, \ldots, p$.

Let H be the Hilbert space of grid functions. For $y, v \in H = L_2(\omega)$ we define inner product and norm as follows

$$(y, w) \equiv \sum_{x \in \omega} y(x) w(x) \prod_{\alpha=1}^{p} h_\alpha, \quad \|y\| \equiv \sqrt{(y, y)},$$

where $y = 0$, $x \in \partial \omega$.

For elliptic operator A, we write approximation as follows

$$A = \sum_{\alpha=1}^{p} A_\alpha, \quad x \in \omega, \tag{3.1}$$

where $A_\alpha(t)$ is the discrete analogue of differential operator in problem (2.1)–(2.3) on α-th direction

$$A_\alpha y = -(a_\alpha(x) y_{\bar{x}_\alpha})_{x_\alpha}, \quad \alpha = 1, \ldots, p, \quad x \in \omega. \tag{3.2}$$

where for the coefficients we have [11]

$$a_\alpha(x) = k_\alpha(x_1, \ldots, x_{\alpha-1}, x_\alpha + 0.5 h_\alpha, x_{\alpha+1}, \ldots, x_p), \quad \alpha = 1, \ldots, p, \quad x \in \omega.$$

The operator A is self-adjoint and positively defined: $A = A^* > 0$.

For the retrospective problem (2.1)–(2.4), we define the Cauchy problem with reverse time

$$\frac{dy}{dt} + Ay = 0, \quad x \in \omega, \quad 0 < t \leqslant T, \tag{3.3}$$

with additional condition

$$y(x, T) = \phi(x), \quad x \in \bar{\omega}. \tag{3.4}$$

For the numerical solution of the non-classical problem (3.3)–(3.4), we use conjugate gradient method associated with the iterative updating of the initial condition. Then instead of solution of the inverse problem (3.3)–(3.4), we solve the direct problem for Eq. (3.3) with following initial condition

$$y(x, 0) = \nu(x), \quad x \in \bar{\omega}. \tag{3.5}$$

For approximation of the Eq. (3.3) with initial condition (3.5) by time we use two-layered weighted scheme

$$\frac{y^{n+1} - y^n}{\tau} + A\left(\sigma y^{n+1} + (1 - \sigma) y^n\right) = 0, \quad x \in \omega, \quad n = 0, 1, \ldots M - 1, \tag{3.6}$$

with

$$y_0 = \nu, \quad x \in \bar{\omega}, \tag{3.7}$$

where $\sigma \in [0, 1]$ is the weight parameter, y^n is the solution for $t^n = n\tau$, $\tau > 0$ is the time step where $M\tau = T$.

It is well known that the weighted scheme (3.6)–(3.7) is stable when

$$\sigma \geq \frac{1}{2} - \frac{1}{\tau \|A\|} \tag{3.8}$$

and for the solution, we have following a priori estimate

$$\|y^{n+1}\| \leq \|y^n\| \leq \dots \|y_0\| = \|\nu\|. \tag{3.9}$$

Therefore, difference scheme (3.6)–(3.7) is stable for $\sigma \geq \frac{1}{2}$. When $\sigma < \frac{1}{2}$ we have following time step limit for guaranteed stability

$$\tau \leq \frac{|h|^2}{\gamma p \kappa_2}, \quad \gamma = \frac{1}{2} - \sigma > 0, \quad |h| = min(h_1, \dots, h_p). \tag{3.10}$$

Then when condition (3.8) is hold, the norm of solution is not grow in time.

4   Computational Algorithm for the Inverse Problem

For numerical implementation of the solution of the retrospective inverse problem (3.1)–(3.2) we use an iterative conjugate gradient method based on successive iterative refinement of the initial condition by the solution of the direct problem in each iteration. Let us give this problem in the corresponding operator formulation. We can write Eqs. (3.6)–(3.7) with defined y^0 as follows

$$y^M = A y^0, \quad x \in \overline{\omega}, \quad A = S^n, \tag{4.1}$$

where S is the transition operator from one time layer to another:

$$S = (I + \sigma \tau A)^{-1}(I + (\sigma - 1)\tau A). \tag{4.2}$$

Then the approximate solution of the inverse problem is naturally associated with the solution of the following grid operator equation:

$$A\nu = \phi(x), \quad x \in \overline{\omega}. \tag{4.3}$$

Due to the self-adjoint nature of the operator A, the transition operator S and operator A in the Eq. (3.9) are both self-adjoint. Unique solvability of the operator Eq. (3.9) will take place, for example, with the positivity of the operator A. This condition will be satisfied for the positive transition operator S. Taking into account the representation (4.1), we get $S > 0$ when the stability conditions are performed

$$\sigma \geq \frac{1}{2} - \frac{1}{\tau \|A\|}. \tag{4.4}$$

Condition (4.1) for weighted scheme (3.4)–(3.5) is more strict condition than standard condition of stability, when estimation (3.6) is connected with inequality $0 < S \leq I$ for any $\tau > 0$. So for (3.9) with limitation (4.1) we have

$$0 < A = A^* < I.$$

For the numerical solution of Eq. (4.3), it is expedient to use the rapidly convergent iterative method of conjugate gradients [12].

5   Numerical Experiments

First, to demonstrate the efficiency of the proposed computational algorithm for solution of the retrospective inverse problem for a parabolic equation, we consider a one-dimensional model problem

$$\frac{\partial u}{\partial t} = \frac{\partial^2 u}{\partial x^2}, \quad 0 < x < l, \quad 0 \leqslant t < T, \tag{5.1}$$

$$u(0,t) = u(l,t) = \frac{1}{\sqrt{1+4t}} e^{-\frac{(l/2)^2}{1+4t}}, \quad 0 \leqslant t < T, \tag{5.2}$$

$$u(x,T) = \phi(x), \quad 0 \leqslant x \leqslant l, \tag{5.3}$$

where the additional condition is given in the form of a function

$$\phi(x) = \frac{1}{\sqrt{1+4T}} e^{-\frac{(x-l/2)^2}{1+4T}}. \tag{5.4}$$

In this case there is an exact analytical solution [8] (function (2.2) in one-dimensional case at $\beta = 1$), given by the formula

$$u(x,t) = \frac{1}{\sqrt{\beta+4t}} e^{-\frac{(x-l/2)^2}{\beta+4t}}. \tag{5.5}$$

Calculations were done for $l = 24$, $N = 128$, $\varepsilon = 10^{-7}$, $T = 1, 3$, respectively, $M = 50$.

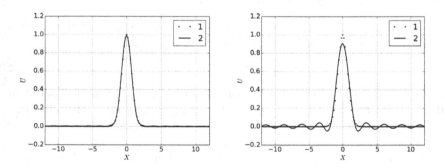

Fig. 1. The exact and computed initial condition at $T = 1$ (left) and at $T = 3$ (right)

In the Fig. 1 on the left chart, we show the graphs of the desired function $\nu(x)$ at $T = 1$ that found using proposed iterative method. The number of iterations equal to $k_{max} = 31$ and error equal to $R = 0.001334$, where the relative error is in the l_2 norm, is defined by the following formula

$$R = \frac{\|y^0 - \nu\|}{\|\nu\|}.$$

Similar results at $T = 3$ are shown in the right graph. Here, the number of iterations equal to $k_{max} = 18$, and error equal to $R = 0.1328$. These results show a better approximation of the initial condition, which are given in the work of [8], obtained by the modified method of the Tikhonov regularization.

Next, we present the results for the solution of this problem when function $\phi(x)$ is defined with perturbation by some error. In the experiments for function $\varphi(x), x \in \omega$ we use:

$$\varphi_\delta(x) = \varphi + \delta F(x), x \in \omega,$$

where $F(x)$ are random variables uniformly distributed on the interval $[-1, 1]$. To select a smooth solution as a smoothing operator we use K times repeating of three-point formula:

$$\varphi_i^{k+1} = (\varphi_{i+1}^k + 4\varphi_i^k + \varphi_{i-1}^k)/6, \quad i = 1, \ldots, M-1, \ k = 0, \ldots, K-1;$$

$$\tilde{\varphi}_i = \varphi_i^K, \quad i = 1, \ldots, M-1.$$

grid functions vanishes at the boundary nodes.

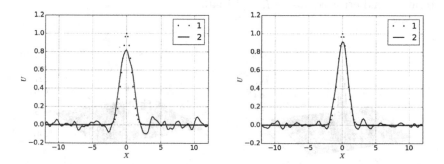

Fig. 2. The computed initial condition for $\delta = 0.01$ (left) and $\delta = 0.001$ (right)

In Fig. 2 we present a numerical results for desired function $\nu(x)$ at $K = 6$, $T = 1$, $N = 128$, $M = 50$, $l = 24$, $k_{max} = 5$, $\delta = 0.01\,(left)$, and $k_{max} = 10$, $\delta = 0.001\,(right)$ that is accurate and found through a proposed iterative method.

Next, we present the computational experiment with an additional quasi-real condition derived from the solution of the direct problem for the problem (5.2) – (5.3) with the initial condition

$$\nu(x) = \frac{x}{0.5 - e^{5(x-l/2)}}, \quad x \in [0, l]. \tag{5.6}$$

As additional condition (5.1), (5.2), (5.6), we take the solution of the direct problem at the final time:

$$\phi(x_i) = y_i^M, \quad i = 0, \ldots, N.$$

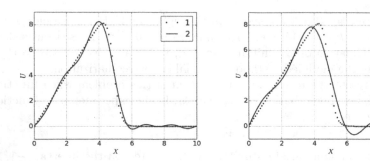

Fig. 3. The quasi-exact initial condition founded through an iterative method at $T = 1$ (left) and at $T = 2$ (right)

The results of the calculations for input data equal to $l = 10$, $N = 100$, $M = 100$ are presented in Fig. 3. On the left graph, we show numerical results at $T = 1$, the number of iterations is equal to 36 and error equal to $R = 0.0726$. On the left graphs, we show numerical results at $T = 2$, the number of iterations is equal to 15 iteration and error equals to $R = 0.1346$.

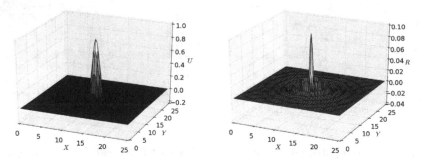

Fig. 4. Quasi-exact approximation of the initial condition found through an iterative method at $T = 1$ (left) with an error (right)

Finally, we present the results of the computational experiment for the two-dimensional parabolic equation

$$\frac{\partial u}{\partial t} = \frac{\partial^2 u}{\partial x^2} + \frac{\partial^2 u}{\partial y^2}, \quad 0 < x < l, \quad 0 < y < l, \quad 0 < t \leqslant T. \tag{5.7}$$

Solution of a retrospective inverse problem (5.7) looks like (see (2.2), $\beta = 1$):

$$u(x, y, t) = \frac{1}{1 + 4t} e^{-\frac{((x-l/2)^2 + (y-l/2)^2)}{1+4t}}.$$

Calculations were made for $l = 25$, $N_1 = N_2 = 125$, $\varepsilon = 10^{-9}$, $T = 1$ and $M = 60$. In Fig. 4 on the left graph, we present the initial condition that obtained using conjugate gradients method. On the right graph, we present a deviation from the exact solution: number of iterations is 70 and error is equal to $R = 0.08975$.

6   Conclusion

In this paper, we considered an iterative conjugate gradient method for the numerical solution of a retrospective inverse problem for a parabolic equation. A theoretical study and numerical simulation on the model of one-dimensional and two-dimensional problems are presented. In addition, the proposed method can be used for other inverse problems, and these studies will find further development in the future.

Acknowledgement. The authors express their sincere gratitude to Professor P.N. Vabischevich for constructive comments and fruitful discussions.

References

1. Aster, R.C.: Parameter Estimation and Inverse Problems. Elsevier Science, Amsterdam (2011)
2. Kabanikhin, S.I.: Inverse and Ill-Posed Problems. Theory and Applications. De Gruyter, Germany (2011)
3. Lavrent'ev, M.M., Romanov, V.G., Shishatskii, S.P.: Ill-Posed Problems of Mathematical Physics and Analysis. American Mathematical Society, Providence (1986)
4. Isakov, V.: Inverse Problems for Partial Differential Equations. Springer, New York (2006)
5. Latte's, R., Lions, J.L.: The Method of Quasi-Reversibility. Applications to Partial Differential Equations. American Elsevier Publishing Company, Amsterdam (1969)
6. Prilepko, A.I., Orlovsky, D.G., Vasin, I.A.: Methods for Solving Inverse Problems in Mathematical Physics. Marcel Dekker, New York (2000)
7. Samarskii, A.A., Vabishchevich, P.N.: Numerical Methods for Solving Inverse Problems of Mathematical Physics. De Gruyter, Germany (2007)
8. Zhao, Z., Meng, Z.: A modified Tikhonov regularization method for a backward heat equation. Inverse Prob. Sci. Eng. **19**(8), 1175–1182 (2011)
9. Samarskii, A.A., Vabishchevich, P.N., Vasil'ev, V.I.: Iterative solution of a retrospective inverse problem of heat conduction. Mat. Model. **9**(5), 119–127 (1997)
10. Vasil'ev, V.I., Popov, V.V., Eremeeva, M.S., Kardashevsky, A.M.: Iterative solution of a nonclassical problem for the equation of string vibrations. Vest. Mosk. Gos. Univ. im. N.E.Baumana, Estest. Nauki **3**, 77–87 (2015)
11. Samarskii, A.A.: The Theory of Difference Schemes. Marcel Dekker, New York (2001)
12. Saad, Y.: Iterative Methods for Sparse Linear Systems, 2nd edn. SIAM, Philadelphia (2003)

Discrete Approximations for Multidimensional Singular Integral Operators

Alexander Vasilyev[1,2] and Vladimir Vasilyev[1,2]($\boxtimes$)

[1] Department of Mathematical Analysis, National Research Belgorod State University, Studencheskaya 14/1, 308007 Belgorod, Russia
alexvassel@gmail.com, vbv57@inbox.ru
[2] Chair of Pure Mathematics, Lipetsk State Technical University, Moskovskaya 30, 398600 Lipetsk, Russia

Abstract. For discrete operator generated by singular kernel of Calderon–Zygmund one introduces a finite dimensional approximation which is a cyclic convolution. Using properties of a discrete Fourier transform and a finite discrete Fourier transform we prove a solvability for approximating equation in corresponding discrete space. For comparison discrete and finite discrete solution we obtain an estimate for a speed of convergence for a certain right-hand side of considered equation.

Keywords: Discrete Calderon–Zygmund operator · Discrete fourier transform · Cyclic convolution · Approximation rate

1 Introduction

A basic object of this paper is a multidimensional singular integral

$$v.p. \int_D K(x, x - y)u(y)dy, \quad x \in D,$$

which generates the Calderon–Zygmund operator with variable kernel [1,2], where D is a domain in $\mathbb{R}^m$ unbounded as a rule.

Taking into account forthcoming studies of such a general operator using a local principle here we consider a case of constant coefficients, i.e. when the kernel does not depend on a first variable and the Calderon–Zygmund operator looks as follows

$$(Ku)(x) \equiv v.p. \int_D K(x - y)u(y)dy \equiv \tag{1}$$

$$\lim_{\varepsilon \to 0} \int_{D \cap \{|x-y|>\varepsilon\}} K(x - y)u(y)dy, \ x \in D.$$

We assume here that the kernel $K(x)$ of the integral (1) satisfy the following conditions: (1) $K(x)$ is a homogeneous function of order $-m$, $K(tx) - t^{-m}K(x), \forall t > 0$; (2) $K(x)$ is a differentiable function on a unit sphere S^{m-1}; (3) the function $K(x)$ has zero mean value on the S^{m-1} [1,2].

© Springer International Publishing AG 2017
I. Dimov et al. (Eds.): NAA 2016, LNCS 10187, pp. 706–712, 2017.
DOI: 10.1007/978-3-319-57099-0_81

1.1   Canonical Domains

We will consider different types of domains D because the theory is essentially dependend on this type. So for example cases $D = \mathbb{R}^m$ and $D - \mathbb{R}^m_+ - \{x \in \mathbb{R}^m : x = (x_1, \cdots, x_m, x_m > 0\}$ are essentially distinct. Invertibility conditions for the operator K do not coincide for these cases.

1.2   Infinite Discrete Case–Series

To obtain a good approximation for the integral (1) we will use the following reduction. First instead of the integral (1) we introduce the series

$$\sum_{\tilde{y} \in D \cap h\mathbb{Z}^m} K(\tilde{x} - \tilde{y}) u_d(\tilde{y}) h^m, \tag{2}$$

which generates a discrete operator K_d defined on functions u_d of discrete variable $\tilde{x} \in D \cap h\mathbb{Z}^m$. Since the Calderon–Zygmund kernel has a strong singularity at the origin we mean $K(0) = 0$. Convergence for the series (2) means that the following limit

$$\lim_{N \to +\infty} \sum_{\tilde{y} \in h\mathbb{Z}^m \cap Q_N \cap D} K(\tilde{x} - \tilde{y}) u_d(\tilde{y}) h^m$$

exists, where $Q_N = \{x \in \mathbb{R}^m : \max_{1 \le k \le m} |x_k| < N\}$. It was shown earlier that a norm of the operator $K_d : L_2(h\mathbb{Z}^m \cap D) \to L_2(h\mathbb{Z}^m \cap D)$ does not depend on h [9]. But although the operator is a discrete object it is an infinite one, and to solve equations with a such operator one needs to replace it by a certain finite object.

1.3   Finite Discrete Case–Systems of Linear Algebraic Equations

It is natural to consider a system of linear algebraic equation instead of an infinite system generated by the series (2). This method is called projection method [3]. It proposes following actions. Let $P_N : L_2(D \cap \mathbb{Z}^m) \to L_2(D \cap Q_N \cap \mathbb{Z}^m)$ be a projector on a finite dimensional space. One needs to solve the equation

$$P_N K_d P_N u_d = P_N v_d \tag{3}$$

in the space $L_2(D \cap Q_N \cap \mathbb{Z}^m)$ instead of the infinite system of linear algebraic equations

$$K_d u_d = v_d \tag{4}$$

in the space $L_2(D \cap \mathbb{Z}^m)$.

The following result plays a key role in the theory of projection methods, and it was proved for some special domains D and integrable kernel $K(x)$ [3].

Proposition 1. *If the Eq. (4) is uniquely solvable in the space $L_2(D \cap \mathbb{Z}^m)$ for arbitrary right-hand side $v_d \in L_2(D \cap \mathbb{Z}^m)$ then the Eq. (3) is uniquely solvable in the space $L_2(D \cap Q_N \cap \mathbb{Z}^m)$ for enough large N.*

Here we consider the case $D = \mathbb{R}^m$ and prove this proposition 1 with some novelties because the Eq.(4) is a system of linear algebraic equations and for large N one needs much time to solve it.

Our considerations are based on two steps: continual infinite object (1) $\longrightarrow$ discrete infinite object (2) $\longrightarrow$ discrete finite object (3) with justification and error estimates. Some pieces of this programm were realized in authors' papers [4–9].

2   Discrete Fourier Transform and Symbols

Let us define the **discrete Fourier transform** for functions u_d of a discrete variable $\tilde{x} \in h\mathbb{Z}^m$

$$(F_d u_d)(\xi) = \sum_{\tilde{x} \in h\mathbb{Z}^m} u_d(\tilde{x}) e^{i\tilde{x}\cdot\xi} h^m, \quad \xi \in \hbar\mathbb{T}^m, \hbar = \frac{h^{-1}}{2\pi},$$

where $\mathbb{T}^m$ is m-dimensional cube$[-\pi, \pi]^m$.

Such discrete Fourier transform preserves all basic properties of the classical fourier transform, particularly for a discrete convolution of two discrete functions u_d, v_d

$$(u_d * v_d)(\tilde{x}) \equiv \sum_{\tilde{y} \in h\mathbb{Z}^m} u_d(\tilde{x} - \tilde{y}) v_d(\tilde{y}) h^m$$

we have the well known multiplication property

$$(F_d(u_d * v_d))(\xi) = (F_d u_d)(\xi) \cdot (F_d v_d)(\xi).$$

If we apply this property to the operator K_d we obtain

$$(F_d(K_d u_d))(\xi) = (F_d K_d)(\xi) \cdot (F_d u_d)(\xi).$$

Let us denote $(F_d K_d)(\xi) \equiv \sigma_d(\xi)$ and give the following

Definition 1. *The function* $\sigma_d(\xi), \xi \in \hbar\mathbb{T}^m$, *is called a periodic symbol of the operator* K_d.

We will assume below that the symbol $\sigma_d(\xi) \in C(\hbar\mathbb{T}^m)$ therefore we have immediately the following

Property 1. The operator K_d is invertible in the space $L_2(h\mathbb{Z}^m)$ iff $\sigma_d(\xi) \neq 0, \forall \xi \in \hbar\mathbb{T}^m$.

Definition 2. *A continuous periodic symbol is called an elliptic symbol if* $\sigma_d(\xi) \neq 0, \forall \xi \in \hbar\mathbb{T}^m$.

So we see that an arbitrary elliptic periodic symbol $\sigma_d(\xi)$ corresponds to an invertible operator K_d in the space $L_2(h\mathbb{Z}^m)$.

Remark 1. It was proved earlier that operators (1) and (2) for cases $D = \mathbb{R}^m, D = \mathbb{R}^m_+$ are invertible or non-invertible in spaces $L_2(\mathbb{R}^m), L_2(\mathbb{R}^m_+)$ and $L_2(h\mathbb{Z}^m), L_2(h\mathbb{Z}^m_+)$ simultaneously [6,8].

3   Periodic Approximation and Cyclic Convolutions

Here we will introduce a special discrete periodic kernel $K_{d,N}(\tilde{x})$ which is defined as follows. We take a restriction of the discrete kernel $K_d(\tilde{x})$ on the set $Q_N \cap \mathbb{Z}^m \equiv Q_N^d$ and periodically continue it to a whole $\mathbb{Z}^m$. Further we consider discrete periodic functions $u_{d,N}$ with discrete cube of periods Q_N^d. We can define a cyclic convolution for a pair of such functions $u_{d,N}, v_{d,N}$ by the formula

$$(u_{d,N} * v_{d,N})(\tilde{x}) = \sum_{\tilde{y} \in Q_N^d} u_{d,N}(\tilde{x} - \tilde{y}) v_{d,N}(\tilde{y}) h^m. \tag{5}$$

(We would like to remind that such convolutions are used in digital signal processing [10]). Further we introduce **finite discrete Fourier transform** by the formula

$$(F_{d,N} u_{d,N})(\tilde{\xi}) = \sum_{\tilde{x} \in Q_N^d} u_{d,N}(\tilde{x}) e^{i\tilde{x} \cdot \tilde{\xi}} h^m, \quad \tilde{\xi} \in R_N^d,$$

where $R_N^d = \hbar \mathbb{T}^m \cap \hbar \mathbb{Z}^m$. Let us note that here $\tilde{\xi}$ is a discrete variable.

According to the formula (5) one can introduce the operator

$$K_{d,N} u_{d,N}(\tilde{x}) = \sum_{\tilde{y} \in Q_N^d} K_{d,N}(\tilde{x} - \tilde{y}) u_{d,N}(\tilde{y}) h^m$$

on periodic discrete functions $u_{d,N}$ and a finite discrete Fourier transform for its kernel

$$\sigma_{d,N}(\tilde{\xi}) = \sum_{\tilde{x} \in Q_N^d} K_{d,N}(\tilde{x}) e^{i\tilde{x} \cdot \tilde{\xi}} h^m, \quad \tilde{\xi} \in R_N^d.$$

Definition 3. *A function $\sigma_{d,N}(\tilde{\xi}), \tilde{\xi} \in R_N^d$, is called s symbol of the operator $K_{d,N}$. This symbol is called an elliptic symbol if $\sigma_{d,N}(\tilde{\xi}) \neq 0, \forall \tilde{\xi} \in R_N^d$.*

Theorem 1. *Let $\sigma_d(\xi)$ be an elliptic symbol. Then for enough large N the symbol $\sigma_{d,N}(\tilde{\xi})$ is elliptic symbol also.*

Proof. The function

$$\sum_{\tilde{x} \in Q_N^d} K_{d,N}(\tilde{x}) e^{i\tilde{x} \cdot \xi} h^m, \quad \xi \in \hbar \mathbb{T}^m,$$

is a segment of the Fourier series

$$\sum_{\tilde{x} \in \hbar \mathbb{Z}^m} K_d(\tilde{x}) e^{i\tilde{x} \cdot \xi} h^m, \quad \xi \in \hbar \mathbb{T}^m,$$

and according our assumptions this is continuous function on $\hbar \mathbb{T}^m$. Therefore values of the partial sum coincide with values of $\sigma_{d,N}$ in points $\tilde{\xi} \in R_N^d$. Besides these partial sums are continuous functions on $\hbar \mathbb{T}^m$. $\square$

As before an elliptic symbol $\sigma_{d,N}(\tilde{\xi})$ corresponds to the invertible operator $K_{d,N}$ in the space $L_2(Q_N^d)$.

4   Approximation Rate

Let $A : B \to B$ be a linear bounded operator acting in a Banach space B, $B_N \subset B$ be its finite dimensional subspace, $P_N : B \to B_N$ be a projector, $A_N : B_N \to B_N$ linear finite-dimensional operator [5].

Definition 4. *Approximation rate for operators A and A_N is called the following operator norm*

$$||P_N A - A_N P_N||_{B \to B_N}$$

We will obtain a "weak estimate" for approximation rate but enough for our purposes. We assume additionally that a function u_d is a restriction on $h\mathbb{Z}^m$ of continuous function with certain estimates [4,5]. Let's define the discrete space $C_h(\alpha, \beta)$ as a functional space of discrete variable $\tilde{x} \in h\mathbb{Z}^m$ with finite norm

$$||u_d||_{C_h(\alpha,\beta)} = ||u_d||_{C_h} + \sup_{\tilde{x},\tilde{y}\in h\mathbb{Z}^m} \frac{|\tilde{x} - \tilde{y}|^\alpha}{(\max\{1 + |\tilde{x}|, 1 + |\tilde{y}|\})^\beta}.$$

It means that the function $u_d \in C_h(\alpha, \beta)$ satisfies the following estimates

$$|u_d(\tilde{x}) - u_d(\tilde{y})| \le c \frac{|\tilde{x} - \tilde{y}|^\alpha}{(\max\{1 + |\tilde{x}|, 1 + |\tilde{y}|\})^\beta},$$

$$|u_d(\tilde{x})| \le \frac{c}{(1 + |\tilde{x}|)^{\beta-\alpha}}, \qquad \forall \tilde{x}, \tilde{y} \in h\mathbb{Z}^m, \; \alpha, \beta > 0, \; 0 < \alpha < 1. \qquad (6)$$

Let us note that under required assumptions $C_h(\alpha, \beta) \subset L_2(h\mathbb{Z}^m)$.

Theorem 2. *For operators K_d and $K_{d,N}$ we have the following estimate*

$$||(P_N K_d - K_{d,N} P_N)u_d||_{L_2(Q_N^d)} \le C N^{m+2(\alpha-\beta)}$$

for arbitrary $u_d \in C_h(\alpha, \beta), \beta > \alpha + m/2$.

Proof. Let us write

$$(P_N K_d - K_{d,N} P_N)u_d = P_N K_d P_N u_d - K_{d,N} P_N u_d + P_N K_d (I - P_N)u_d,$$

where I is an identity operator in $L_2(h\mathbb{Z}^m)$.

First two summands have annihilated, and we need to estimate only the last summand. We have

$$||P_N K_d (I - P_N)u_d|| \le C ||(I - P_N)u_d||$$

because norms of operators K_d are uniformly bounded, and for the last norm taking into account (6) we can write

$$||(I - P_N)u_d||^2 \le C \sum_{\tilde{x}\in h\mathbb{Z}^m \setminus Q_N} |u_d(\tilde{x})|^2 h^m \le C \sum_{\tilde{x}\in h\mathbb{Z}^m \setminus Q_N} \frac{h^m}{(1 + |\tilde{x}|)^{2(\beta-\alpha)}} \le$$

and further

$$C \int_{\mathbb{R}^m \setminus Q_N} |x|^{2(\alpha-\beta)} dx$$

The last integral using spherical coordinates gives the estimate $N^{m+2(\alpha-\beta)}$ which tends to 0 under $n \to \infty$ if $\beta > \alpha + m/2$. $\qquad \square$

5 Main Theorem on Approximation

Here we consider the equation

$$K_{d,N}u_{d,N} = P_N v_d \tag{7}$$

instead of the Eq. (4) and give a comparison for these two solutions.

Below we assume that operator K_d is invertible in $L_2(h\mathbb{Z}^m)$.

Theorem 3. *If $v_d \in C_h(\alpha, \beta), \beta > \alpha + m/2, u_d$ is a solution of the Eq. (4), $u_{d,N}$ is a solution of (7) then the estimate*

$$\|u_d - u_{d,N}\|_{L_2(h\mathbb{Z}^m)} \leq CN^{m+2(\alpha-\beta)}$$

is valid, and C is a constant non-depending on N.

Proof. Let us write

$$u_d - u_{d,N} = K_d^{-1}v_d - K_{d,N}^{-1}P_N v_d =$$

$$(I - P_N)K_d^{-1}v_d + P_N K_d^{-1}v_d - K_{d,N}^{-1}P_N v_d$$

For the summand $P_N K_d^{-1}v_d - K_{d,N}^{-1}P_N v_d$ we have a corresponding estimate by the Theorem 2 because the operators K_d^{-1} and $K_{d,N}^{-1}$ are constructed similar initial operators K_d and $K_{d,N}$.

The first summand is estimated like the proof of the Theorem 2 and using the property that operator K_d is uniformly on h is bounded in the space $C_h(\alpha, \beta)$ [9] and the operator K_d^{-1} has a symbol with required properties [8]. □

6 Conclusion

We have introduced such a finite approximation for original integral (1) because there are a lot of algorithms for calculating a finite discrete Fourier transform, these are so called fast Fourier transform algorithms [10]. On the other hand this step-by-step approximation permits to justify mathematically without additional difficulties results on a solvability for a corresponding approximate equation.

Acknowledgments. The work was supported by Russian Foundation for Basic Research and Lipetsk regional government of Russia, project No. 14-41-03595-r-center-a.

References

1. Selected papers of Alberto P. Calderon with commentary. In: Bellow, A., Kenig, C.E., Malliavin, P. (eds.) AMS, Providence (2008)
2. Mikhlin, S.G., Prößdorf, S.: Singular Integral Operators. Akademie-Verlag, Berlin (1986)

3. Gohberg, I.C., Feldman, I.A.: Convolution Equations and Projection Methods for Their Solution. AMS, Providence (1974)
4. Vasilyev, A.V., Vasilyev, V.B.: Numerical analysis for some singular integral equations. Neural, Parallel, Sci. Comput. **20**, 313–326 (2012)
5. Vasilyev, A.V., Vasilyev, V.B.: Approximation rate and invertibility for some singular integral operators. Proc. Appl. Math. Mech. **13**, 373–374 (2013)
6. Vasilyev, A.V., Vasilyev, V.B.: Discrete singular operators and equations in a half-space. Azerb. J. Math. **3**, 84–93 (2013)
7. Vasil'ev, A.V., Vasil'ev, V.B.: Periodic Riemann problem and discrete convolution equations. Diff. Equat. **51**, 652–660 (2015)
8. Vasilyev, A.V., Vasilyev, V.B.: Discrete singular integrals in a half-space. In: Mityushev, V., Ruzhansky, M. (eds.) Current Trends in Analysis and Its Applications. Trends in Mathematics. Research Perspectives, pp. 663–670. Birkhäuser, Basel (2015)
9. Vasilyev, A.V., Vasilyev, V.B.: On the solvability of certain discrete equations and related estimates of discrete operators. Dokl. Math. **92**, 585–589 (2015)
10. Oppenheim, A.V., Schafer, R.W.: Digital Signal Processing. Prentice Hall, Englewood Cliffs (1975)

A Generalized Multiscale Finite Element Method for Thermoelasticity Problems

Maria Vasilyeva[1,2]([✉]) and Denis Stalnov[1]

[1] North-Eastern Federal University, Yakutsk, Russia
vasilyevadotmdotv@gmail.com
[2] Texas A&M University, College Station, TX, USA

Abstract. In this work, we consider the coupled systems of a partial differential equations, which arise in the modeling of thermoelasticity processes in heterogeneous domains. Heterogeneity of the properties requires a high resolution solve that adds many degrees of freedom that can be computationally costly. For the numerical solution, we use a Generalized Multiscale Finite Element Method (GMsFEM) that solves problem on a coarse grid by constructing local multiscale basis functions [1–3]. We construct multiscale basis functions for the temperature and for the displacements on the offline stage in each coarse block using local spectral problems [4–7]. On the online stage we construct coarse scale system using precalculated multiscale basis functions and solve problem with any forcing and boundary conditions. The numerical results are presented for heterogeneous and perforated domains.

1 Problem Formulation and Fine Scale Approximation

We consider linear thermoelasticity problem for temperature, T, and for displacement, u [8–10]

$$
\begin{aligned}
-\operatorname{div}\sigma(u) + \beta\operatorname{grad}T &= 0 \text{ in } \Omega, \\
\beta\operatorname{div}\frac{\partial u}{\partial t} + c\frac{\partial T}{\partial t} - \operatorname{div}(k\operatorname{grad}T) &= f \text{ in } \Omega,
\end{aligned}
\tag{1}
$$

where f is a source term, c is a heat capacity, k is a thermal conductivity and β is the coupling coefficient.

The stress and strain tensors are given by

$$
\sigma(u) = 2\mu\varepsilon(u) + \lambda\operatorname{div}(u)\mathcal{I}, \quad \varepsilon(u) = \frac{1}{2}\left(\operatorname{grad}u + \operatorname{grad}u^{T}\right),
$$

where μ, λ are Lame parameters, $\mathcal{I}$ is the identity tensor.

We consider (1) with initial condition $T(x,0) = T_0$ and boundary conditions for displacement and for temperature

$$
\sigma n = 0, \quad x \in \Gamma_N^u, \quad u = u_1, \quad x \in \Gamma_D^u,
$$

© Springer International Publishing AG 2017
I. Dimov et al. (Eds.): NAA 2016, LNCS 10187, pp. 713–720, 2017.
DOI: 10.1007/978-3-319-57099-0_82

$$-k\frac{\partial T}{\partial n} = 0, \quad x \in \Gamma_N^T, \quad T = T_1, \quad x \in \Gamma_D^T,$$

where n is the unit normal to the boundary.

For numerical solution on fine grid, we use a standard finite element method and implicit scheme for approximation by time [4, 5, 8]

$$a_u(u^{n+1}, v) + b(T^{n+1}, v) = 0,$$
$$b(u^{n+1} - u^n, q) + m(T^{n+1} - T^n, q) + \tau a_T(T^{n+1}, q) = \tau(f, q), \tag{2}$$

for $(u, T) \in W = (V, Q)$ and $(v, q) \in \hat{W} = (\hat{V}, \hat{Q})$ where

$$V = \{v \in [H^1(\Omega)]^d : v(x) = u_1, x \in \Gamma_D^u\}, \quad Q = \{q \in H^1(\Omega) : q(x) = T_1, x \in \Gamma_D^T\},$$
$$\hat{V} = \{v \in [H^1(\Omega)]^d : v(x) = 0, x \in \Gamma_D^u\}, \quad \hat{Q} = \{q \in H^1(\Omega) : q(x) = 0, x \in \Gamma_D^T\}.$$

Here for bilinear and linear forms we have

$$a_u(u, v) = \int_\Omega (\sigma(u), \varepsilon(v)) dx, \quad a_T(T, q) = \int_\Omega (k \operatorname{grad} T, \operatorname{grad} q) \, dx,$$
$$b(T, u) = \int_\Omega \beta(\operatorname{grad} T, u) dx, \quad m(T, q) = \int_\Omega cT q \, dx, \quad (f, q) = \int_\Omega f q \, dx.$$

where as basis functions on fine grid we use standard linear basis functions for both temperature and displacement.

2   Coarse-Scale Approximaiton Using GMsFEM

Let $\mathcal{T}^H$ be a standard conforming partition of the computational domain Ω into finite elements. We refer to this partition as the coarse-grid and assume that each coarse element is partitioned into a connected union of fine grid blocks. The fine grid partition will be denoted by $\mathcal{T}^h$. Let $\{x_i\}_{i=1}^N$ is the vertices of the coarse mesh $\mathcal{T}^H$, where N is the number of coarse nodes. We define the neighborhood (local) domain of the node x_i by

$$\omega_i = \bigcup_j \{K_j \in \mathcal{T}^H \mid x_i \in \overline{K}_j\},$$

where K_j to denote a coarse element.

In the GMsFEM algorithm, we have three steps [1–3]:

Step 1: Generate the coarse-grid, $\mathcal{T}^H$ and local domains ω_i, $i = 1, 2, \dots, N$;
Step 2: The construction of the multiscale basis functions in local domains, ω_i, $i = 1, 2, \dots, N$ (offline space);
Step 3: Use offline space to find the solution of a coarse-grid problem for any force term and/or boundary conditions.

We construct multiscale basis functions for temperature and displacements separately.

Multiscale Basis Functions for Pressure. To construct the offline space Q_{off} for temperature, we solve following the eigenvalue problem in the local domain ω:

$$A_T \Psi_k^{\text{off}} = \lambda_k^{\text{off}} S_T \Psi_k^{\text{off}},$$

$$A_T = [a_{ij}], \quad a_{ij} = \int_\Omega (k \operatorname{grad} \phi_i, \operatorname{grad} \phi_j)\, dx, \quad M_T = [s_{ij}], \quad s_{ij} = \int_\Omega k \phi_i \phi_j\, dx,$$

$$(3)$$

and choose the eigenvectors ψ_k^{off} that corresponds to the smallest $M_{\text{off}}^{\omega,T}$ eigenvalues in Eq. (3) and denote the span of this reduced space as Q_{off}^ω.

For construction of the offline space, to ensure the functions we construct form an conforming basis, we define multiscale partition of unity functions χ_i

$$a_T(\chi_i, q) = 0 \quad \text{in } K, \quad \chi_i = g_i \quad \text{on } \partial K, \tag{4}$$

for all $K \in \omega$. Here g_i is a continuous on K and is linear on each edge of ∂K.

Finally, we multiply the partition of unity functions by the eigenfunctions in the offline space $Q_{\text{off}}^{\omega_i}$ to construct the resulting basis functions $\psi_{i,k} = \chi_i \psi_k^{\omega,\text{off}}$, for $1 \leq i \leq N$ and $\leq k \leq M_{\text{off}}^{\omega_i,T}$, where $M_{\text{off}}^{\omega_i,T}$ denotes the number of offline eigenvectors that are chosen for each coarse node i.

We define the multiscale space using a single index notation as

$$Q_{\text{off}} = \operatorname{span}\{\psi_i\}_{i=1}^{M_T^{\text{off}}}, \quad \text{and} \quad R_T = \left[\psi_1, \ldots, \psi_{M_T^{\text{off}}}\right]^T, \tag{5}$$

where $M_T^{\text{off}} = \sum_{i=1}^N M_{\text{off}}^{\omega_i,T}$ denotes the total number of basis functions.

Multiscale Basis Functions for Displacement. For construction of multiscale basis functions for displacements we use similar algorithm that we used for the temperature. We solve the following eigenvalue problem in $V_h(\omega)$ [3–5]

$$A_u \Phi_k^{\text{off}} = \lambda_k^{\text{off}} S_u \Phi_k^{\text{off}},$$

$$A_u = [a_{ij}], \quad a_{ij} = \int_\Omega \left(2\mu\varepsilon(\varphi_m) : \varepsilon(\varphi_n) + \lambda \operatorname{div}(\varphi_m) \cdot \operatorname{div}(\varphi_n)\right), \tag{6}$$

$$S_u = [s_{ij}], \quad s_{ij} = \int_\Omega (\lambda + 2\mu)\varphi_m \cdot \varphi_n.$$

We then choose the eigenvectors that corresponds to the smallest $M_{\text{off}}^{\omega,u}$ eigenvalues from Eq. (6) and denote the span of this reduced space as V_{off}^ω.

For construction of multiscale partition of unity functions for the mechanics solve, we proceed as before and solve for all $K \in \omega$

$$a_u(\xi_i, v) = 0 \quad \text{in } K, \quad \xi_i = g_i \quad \text{on } \partial K, \tag{7}$$

where g_i is a continuous function on K and is linear on each edge of ∂K. Finally, we multiply the partition of unity functions by the eigenfunctions in the offline space $V_{\text{off}}^{\omega_i}$ to construct the resulting basis functions $\varphi_{i,k} = \xi_i \varphi_k^{\omega_i,\text{off}}$ for $1 \leq i \leq N$ and $1 \leq k \leq M_{\text{off}}^{\omega_i,u}$, where $M_{\text{off}}^{\omega_i,u}$ denotes the number of offline eigenvectors that are chosen for each coarse node i.

Next, we define the multiscale space as

$$V_{\text{off}} = \text{span}\{\varphi_i\}_{i=1}^{M_u^{\text{off}}}, \quad \text{and} \quad R_u = [\psi_1, \ldots, \varphi_{M_u^{\text{off}}}]^T, \tag{8}$$

where $M_u^{\text{off}} = \sum_{i=1}^{N} M_{\text{off}}^{\omega_i, u}$ denotes the total number of basis functions.

Coarse-Scale System. The variational form in (2) yields the following linear algebraic system

$$\begin{pmatrix} A_u^c & (B_c)^T \\ B^c & (M_c + \tau A_T^c) \end{pmatrix} \begin{pmatrix} u_H^{n+1} \\ T_H^{n+1} \end{pmatrix} = \begin{pmatrix} 0 \\ Q_c \end{pmatrix}, \tag{9}$$

where

$$A_u^c = R_u A_u R_u^T, \quad A_T^c = R_T A_T R_T^T \quad B_c = R_T B R_u^T, \quad M_c = R_T M R_T^T$$

and $Q_c = R_T \tau F + M_c T_H^n + B_c u_H^n$. Here u_H and T_H denotes the coarse-scale solutions that we can project into the fine-grid $u_h^{n+1} = R_u^T u_H^{n+1}$ and $T_h^{n+1} = R_T^T T_H^{n+1}$.

3   Numerical Examples

In this section, we present numerical examples to demonstrate the performance of the GMsFEM for computing the solution of the thermoelasticity problem in heterogeneous and perforated domains where the inclusions can have different size (see Figs. 1 and 2).

We present results for perforated and heterogeneous domains with random distribution of the inclusions (Fig. 1). For high-constrast domain, we consider case with one type of particles (Fig. 2). For numerical simulations we use following thermomechanical coefficients: $c_1 = 1000$, $c_2 = 100$, $k_1 = 1$, $k_2 = 100$, $E_1 = 100$, $E_2 = 10$, $\nu = 0.3$ and $\beta = 1.0$.

We consider three test cases:

Case 1a. Perforated domain with homogeneous backround with source term $f = 100$ and zero Dirichlet boundary conditions for temperature and displacement on perforations;

Case 1b. Perforated domain with heterogeneous backround with source term $f = 100$ and zero Dirichlet boundary conditions for temperature and displacement on perforations;

Case 2. Heterogeneous domain with circle particles with zero source term $f = 0$ and boundary conditions: a fixed temperature $T = 1.0$ on cavity, a fixed displacements $u_x = 0$ for left boundary and $u_y = 0$ on top boundary.

For numerical comparison, we calculate a weighted relative errors using L^2 norm and H^1 semi-norm for temperature

$$\|\epsilon_T\|_{L^2} = \left(\int_\Omega k \epsilon_T^2 dx \right)^{1/2}, \quad |\epsilon_T|_{H^1} = \left(\int_\Omega (k \, \text{grad} \, \epsilon_T, \text{grad} \, \epsilon_T) \, dx \right)^{1/2},$$

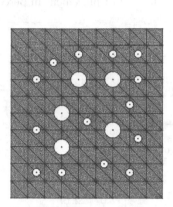
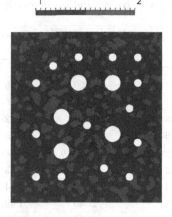

Fig. 1. Coarse and fine computational grids (left) and heterogeneous backround (right). Blue color is the subdomain 2 and red is the subdomain 1. Fine grid contains 12426 vertices and 24124 cells. Coarse grid have 110 vertices and 180 cells. (Color figure online)

and for displacement

$$||\epsilon_u||_{L^2} = \left(\int_\Omega (\lambda + 2\mu)(\epsilon_u, \epsilon_u) dx \right)^{1/2}, \quad |\epsilon_u|_{H^1} = \left(\int_\Omega (\sigma(\epsilon_u), \varepsilon(\epsilon_u)) dx \right)^{1/2},$$

where $\epsilon_T = T_f - T_{ms}$, $\epsilon_u = u_f - u_{ms}$. Here (u_f, T_f) and (u_{ms}, T_{ms}) are fine-scale and coarse-scale (multiscale) solutions, respectively for displacement and temperature.

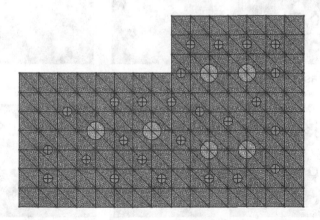

Fig. 2. Coarse and fine computational grids (left) for domain with circle particles. Orange color is the subdomain 2 and blue is the subdomain 1. Fine grid contains 18378 vertices and 36254 cells. Coarse grid have 152 vertices and 252 cells. (Color figure online)

Table 1. Relative L_2 and H_1 errors for temperature and displacement in percentage

$M_T^{off} = M_u^{off}$	$dim(W^{off})$	Temperature errors, ϵ^T		Displacement errors, ϵ^u	
		$\epsilon_{L_2}^T$	$\epsilon_{H_1}^T$	$\epsilon_{L_2}^u$	$\epsilon_{H_1}^u$
Perforated domain, *Case 1a*					
2	660	7.08	27.07	18.55	45.05
4	1320	2.48	15.22	6.63	27.32
8	2640	0.72	7.60	1.94	14.53
16	5280	0.20	3.65	0.43	6.01
Perforated domain with heterogeneous backround, *Case 1b*					
2	660	47.03	71.60	61.57	70.29
4	1320	17.74	45.50	26.16	43.28
8	2640	1.42	16.73	5.07	22.19
16	5280	0.20	6.37	0.98	10.14
20	6600	0.13	4.76	0.61	7.83
Heterogeneous domain with circle particles, *Case 2*					
2	912	6.085	43.48	11.33	32.25
4	1824	4.81	19.05	7.17	25.76
8	3648	3.06	12.55	1.23	14.86
16	7296	1.47	7.41	0.43	9.05
20	9120	1.15	6.53	0.82	7.53

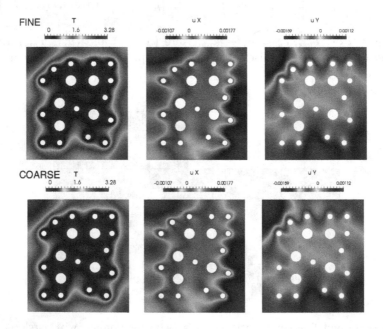

Fig. 3. Fine-scale solution (top) and coarse-scale solution using 8 basis functions for temperature and 8 for displacement (bottom) for the *Case 1a*. Left: temperature. Middle: displacement u_x. Right: displacement u_y.

In Fig. 3, we show the fine-scale and coarse-scale solutions for the *Case 1a* and in Fig. 4 for the *Case 1b*. For multiscale solution we used 8 multiscale basis functions for temperature and 8 multiscale basis functions for displacement. Comparing the fine-scale and coarse-scale solutions in Figs. 3 and 4, we can

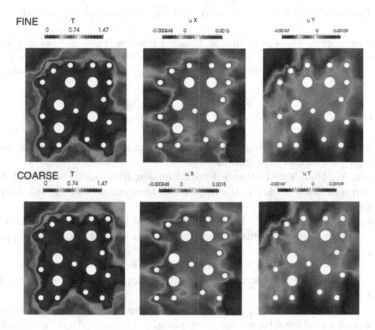

Fig. 4. Fine-scale solution (top) and coarse-scale solution using 8 basis functions for temperature and 8 for displacement (bottom) for the *Case 1b*. Left: temperature. Middle: displacement u_x. Right: displacement u_y.

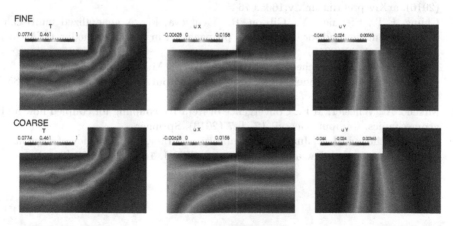

Fig. 5. Fine-scale solution (top) and coarse-scale solution using 8 basis functions for temperature and 8 for displacement (bottom) for the *Case 2*. Left: temperature. Middle: displacement u_x. Right: displacement u_y.

observe a good accuracy of the proposed multiscale method for both homogeneous and heterogeneous backround coefficients for perforated domain. In Fig. 5 we show solutions for the *Case 2* for heterogeneous domain with circle particles. In Table 1 we present relative errors for the coarse-scale solutions with different number of the multiscale basis functions. We observe a good accuracy for all cases for multiscale solution using only $\approx 0.2\%$ of fine-scale system size.

Acknowledgement. We would like to thank Professor Yalchin Efendiev for many interesting discussions. This work is partially supported by the grant of the President of the Russian Federation MK-9613.2016.1 and RFBR (project N 15-31-20856).

References

1. Efendiev, Y., Galvis, J., Hou, T.: Generalized multiscale finite element methods. J. Comput. Phys. **251**, 116–135 (2013)
2. Efendiev, Y., Hou, T.: Multiscale Finite Element Methods: Theory and Applications. Surveys and Tutorials in the Applied Mathematical Sciences, vol. 4. Springer, New York (2009)
3. Chung, E.T., Efendiev, Y., Li, G., Vasilyeva, M.: Generalized multiscale finite element method for problems in perforated heterogeneous domains. Appl. Anal. **255**, 1–15 (2015)
4. Brown, D.L., Vasilyeva, M.: A generalized multiscale finite element method for poroelasticity problems I: linear problems. J. Comput. Appl. Math. **294**, 372–388 (2016)
5. Brown, D.L., Vasilyeva, M.: A generalized multiscale finite element method for poroelasticity problems II: nonlinear coupling. J. Comput. Appl. Math. **297**, 132–146 (2016)
6. Chung, E.T., Efendiev, Y., Leung, W.T., Vasilyeva, M., Wang, Y.: Online adaptive local multiscale model reduction for heterogeneous problems in perforated domains (2016). arXiv preprint arXiv:1605.07645
7. Chung, E.T., Efendiev, Y., Gibson, R., Vasilyeva, M.: A generalized multiscale finite element method for elastic wave propagation in fractured media. GEM-Int. J. Geomath. 1–20 (2015)
8. Kolesov, A.E., Vabishchevich, P.N., Vasilyeva, M.V.: Splitting schemes for poroelasticity and thermoelasticity problems. Comput. Math. Appl. **67**(12), 2185–2198 (2014)
9. Mikelic, A., Wheeler, M.F.: Convergence of iterative coupling for coupled flow and geomechanics. Comput. Geosci. **17**, 1–7 (2013). Springer
10. Kim, J., Tchelepi, H.A., Juanes, R.: Stability, accuracy, and efficiency of sequential methods for coupled flow and geomechanics. SPE J. **16**(2), 249–262 (2011)

Asymptotic-Numerical Method for the Location and Dynamics of Internal Layers in Singular Perturbed Parabolic Problems

Vladimir Volkov[✉], Dmitry Lukyanenko, and Nikolay Nefedov

Department of Mathematics, Faculty of Physics,
Lomonosov Moscow State University, 119991 Moscow, Russia
volkovvt@mail.ru

Abstract. A singularly perturbed initial-boundary value problem for the parabolic reaction-diffusion-advection (RDA) equation is considered. Some effective asymptotic-numerical approach for the description of internal layers location and moving fronts dynamics is proposed.

Asymptotic analysis allows to reduce the spatial dimension of the numerical problem and highlight a priori information to optimize numerical calculations and save computational resources. But for some classes of RDA problems, featuring the internal layers or moving fronts, the layers location or fronts speed could not be found explicitly and asymptotic algorithm needs to be supplemented by the appropriate numerical calculations. In this paper we show the main ideas of the asymptotic algorithm for this type of solutions and outline some problems which need to use numerical calculations on some steps of the asymptotic procedure. The main features of numerical algorithm are presented.

Keywords: Singularly perturbed parabolic problems · Reaction-diffusion-advection equation · Internal layers · Moving fronts · Asymptotic methods

1 Introduction

We consider a singularly perturbed RDA problem featuring the solutions with internal layers or moving fronts. Important that layers location (fronts speed) are not known a priori and could be determined from the asymptotic procedure by smooth joining of asymptotic expansions. In quite general cases this procedure can be done explicitly and explicit asymptotic formulas for the layers location or the fronts speed can be written. But for some classes of RDA problems it could not be done explicitly and asymptotic algorithm needs to be supplemented by the appropriate numerical calculations. The main purpose of this paper is, on the one hand, to show the ideas of the asymptotic algorithm for the solutions with internal layers or moving fronts and, on the other hand, to outline some problems which need to use numerical calculations on some steps of the asymptotic procedure.

© Springer International Publishing AG 2017
I. Dimov et al. (Eds.): NAA 2016, LNCS 10187, pp. 721–729, 2017.
DOI: 10.1007/978-3-319-57099-0_83

We demonstrate our approach by the following problems:

(a) stationary problem

$$
\begin{cases}
\varepsilon^2 \dfrac{d^2 y}{dx^2} = f(y, x, \varepsilon), & x \in (0,1) \\
y(0) = y^0, \quad y(1) = y^1;
\end{cases}
\tag{1}
$$

(b) nonstationary problem

$$
\begin{cases}
\varepsilon^2 \dfrac{\partial^2 y}{\partial x^2} - \varepsilon \dfrac{\partial y}{\partial t} = f(y, x, \varepsilon), & x \in (0,1) \\
y(0,t) = y^0, \quad y(1,t) = y^1; & y(x,0) = y_{init}(x),
\end{cases}
\tag{2}
$$

where function $f(y, x, \varepsilon)$ is sufficiently smooth; $0 < \varepsilon \ll 1$ is a small parameter.

It is well known [1,2] that under certain conditions (1) and (2) can feature solutions of the contrast structure type what means, that solution is close to two different levels on the opposite sides of some (unknown, may be moving) point $\hat{x}$, and contains sharp transition internal layer located near this point (Fig. 1).

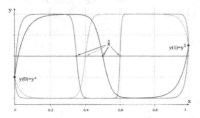

Fig. 1. Contrast structure

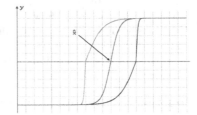

Fig. 2. $C^{(1)}$ – matching

Asymptotics solution of the problems (1) or (2) could be constructed as a combination of two solutions of boundary value problems in domains $[0; \hat{x}]$ and $[\hat{x}; 1]$. The point $\hat{x}$ of the transition layer location is not known *a priori* and can be determined by the asymptotic procedure from the condition of the smooth joining of these two solutions ($C^{(1)}$ – matching condition).

If we introduce some operator $\hat{H}(x)$, which defines the difference of limits of first derivatives of the asymptotic solutions on the opposite sides from the point $x \in (0;1)$ (see Fig. 2), we can write the problem as $\hat{H}(x) = 0$. The solution $\hat{x} = \hat{H}^{-1}(0)$ defines the internal layer location.

In some cases the inverse operator $\hat{H}^{-1}$ can be built explicitly by the asymptotic procedure (for example, as power series of ε) and explicit formulas for the internal layer location or the moving front speed also can be written explicitly.

But, as a rule, this procedure can't be done explicitly, and numerical algorithm for $C^{(1)}$ – matching condition must be developed. In this work we propose some asymptotic-numerical approach to the description of the internal layers location and moving fronts dynamics.

2   Stationary Problem with Internal Layer

Let consider the problem (1) with the following main conditions for $f(y, x, \varepsilon)$.

C 1. *The reduced problem* $f(y, x, 0) = 0$ *has 3 roots in* $x \in (0; 1)$ *that satisfy*

$$(1) \qquad \varphi^{(-)}(x) < \varphi(x) < \varphi^{(+)}(x);$$

$$(2) \qquad f_y(\varphi^{(\pm)}(x), x, 0) > 0, \qquad f_y(\varphi(x), x, 0) < 0.$$

C 2. *Assume that the equation*

$$(a) \qquad I(x) \equiv \int_{\varphi^{(-)}(x)}^{\varphi^{(+)}(x)} f(y, x, 0)dy = 0$$

has a solution $x = x_0 \in (0; 1)$, *that satisfy*

$$(b) \qquad \frac{dI}{dx}(x_0) \equiv \int_{\varphi^{(-)}(x_0)}^{\varphi^{(+)}(x_0)} f_x(y, x_0, 0)dy \neq 0.$$

We construct the asymptotic solution of (1) as a combination of the solutions of two boundary value problems

$$\begin{cases} \varepsilon^2 \dfrac{d^2 y^{(\pm)}}{dx^2} = f(y^{(\pm)}, x, \varepsilon), \\ y^{(-)}(0, \varepsilon) = y^0, \quad y^{(+)}(1, \varepsilon) = y^1; \qquad y^{(\pm)}(\hat{x}, \varepsilon) = \varphi(\hat{x}), \end{cases} \tag{3}$$

where indexes $(-)$ and $(+)$ corresponds to domains $[0, \hat{x}]$ and $[\hat{x}, 1]$ respectively.

Note that the transition point $\hat{x}$ is not known *a priori* and we define it as the intersection of the solution $y(x, \varepsilon)$ and the root $\varphi(x)$ of the reduced equation $f(y, x, 0) = 0$. Its location can be determined from the $C^{(1)}$ – matching condition: smooth joining of the solutions of (3) with indexes $(-)$ and $(+)$ at the point $\hat{x}$.

Asymptotic of the solutions $y^{(\pm)}(x, \varepsilon)$ of the problems (3) can be built according scheme [1, 2] in the form

$$y^{(\pm)} = \bar{y}^{(\pm)}(x, \varepsilon) + \Pi^{(\pm)} + Q^{(\pm)}(\xi, \varepsilon), \qquad \xi = \frac{x - \hat{x}}{\varepsilon}. \tag{4}$$

Functions $\Pi^{(\pm)}$ describe the boundary layers near the points $x = 0$, $x = 1$ and exponentially decrease. So we eliminate them from (4) for further considerations and focus only on the description of the internal layer, representing regular part $\bar{y}^{(\pm)}(x, \varepsilon)$ of (3), transition layer functions $Q^{(\pm)}(\xi, \varepsilon)$ and layer location $\hat{x}$ as

$$\bar{y}^{(\pm)}(x, \varepsilon) = \bar{y}_0^{(\pm)}(x) + \varepsilon \bar{y}_1^{(\pm)}(x) + \varepsilon^2 \bar{y}_2^{(\pm)}(x) + \dots \tag{5}$$

$$Q^{(\pm)}(\xi, \varepsilon) = Q_0^{(\pm)}(\xi) + \varepsilon Q_1^{(\pm)}(\xi) + \varepsilon^2 Q_2^{(\pm)}(\xi) + \dots. \tag{6}$$

$$\hat{x}(\varepsilon) = x_0 + \varepsilon x_1 + \varepsilon^2 x_2 + \dots. \tag{7}$$

The solutions of (3) are joint continuously at the point $\hat{x}$. If denote the difference of the limits of their first derivatives at this point as $H(\hat{x}, \varepsilon)$, the $C^{(1)}$ – matching condition takes the form $H(\hat{x}, \varepsilon) = 0$.

Using (5), (6), (7) we have

$$H(\hat{x}, \varepsilon) = H_0(\hat{x}) + \varepsilon H_1(\hat{x}) + \dots = H_0(x_0) + \varepsilon \left[H_1(x_0) + \frac{dH_0}{d\hat{x}}(x_0) \cdot x_1 \right] + \dots, \tag{8}$$

and the $C^{(1)}$ – matching condition means that all term of (8) are equal to zero. Particularly, at higher order of ε this condition is

$$H_0(x_0) = \left. \frac{dQ_0^{(+)}}{d\xi} \right|_{\substack{\xi=0 \\ \hat{x}=x_0}} - \left. \frac{dQ_0^{(-)}}{d\xi} \right|_{\substack{\xi=0 \\ \hat{x}=x_0}} = 0. \tag{9}$$

According to [1, 2] we put $\bar{y}_0^{(-)}(x) = \varphi^{(-)}(x)$ for $x \in [0; \hat{x})$ and $\bar{y}_0^{(+)}(x) = \varphi^{(+)}(x)$ for $x \in (\hat{x}; 1]$ and get the following equations for the transition layer functions $Q_0(\xi)$:

$$\frac{d^2 Q_0^{(\pm)}}{d\xi^2} = f\left(\varphi^{(\pm)}(\hat{x}) + Q_0^{(\pm)}(\xi), \hat{x}, 0 \right) \qquad (\hat{x} - \text{parameter}),$$

$$Q_0^{(\pm)}(0) = \varphi(\hat{x}) - \varphi^{(\pm)}(\hat{x}), \qquad Q_0^{(\pm)}(\pm\infty) = 0. \tag{10}$$

In (10) indexes $(-)$ and $(+)$ corresponds to $\xi < 0$ and $\xi > 0$ respectively.

If define continuous function

$$\tilde{Q}(\xi) = \begin{cases} \varphi^{(-)}(\hat{x}) + Q_0^{(-)}(\xi), & \xi < 0 \\ \varphi^{(+)}(\hat{x}) + Q_0^{(+)}(\xi), & \xi > 0 \end{cases}$$

the problems (10) can be formulated as the following:

$$\frac{d^2 \tilde{Q}}{d\xi^2} = f(\tilde{Q}, \hat{x}, 0); \qquad \tilde{Q}(0) = \varphi(\hat{x}), \quad \tilde{Q}(\pm\infty) = \varphi^{(\pm)}(\hat{x}). \tag{11}$$

In this case the $C^{(1)}$ – matching condition (9) for higher order of ε at the point $\xi = 0$ can be written explicitly. If substitute x_0 instead of $\hat{x}$ and once integrate (11) with the conditions at $\xi \to \pm\infty$, the equation for x_0 – zero order term of the internal layer location – takes the form

$$\int\limits_{\varphi^{(-)}(x_0)}^{\varphi^{(+)}(x_0)} f(u, x_0, 0)du \equiv I(x_0) = 0, \tag{12}$$

which is equal to the cell condition at the phase plane of (11) (Fig. 3a).

For example, if $f(y, x, \varepsilon) = (y^2 - 1)(y - h(x))$ the Eq. (12) is $h(x) = 0$. In this case function $f(x, y, \varepsilon)$ and related phase diagrams of (11) for two different values of $h(x)$ are presented on Fig. 3. The phase plane picture depends on parameter x and contains two saddle type points ($\varphi^{(\pm)}(x) = \pm 1; 0$) and the center type point ($h(x); 0$) between them. The $C^{(1)}$ – matching condition means that for some $x \in (0; 1)$ there exists the joining separatrix between two saddle points (the cell on the phase plane, Fig. 3a). According to the condition C 3 it satisfies when $x = x_0 \in (0; 1)$.

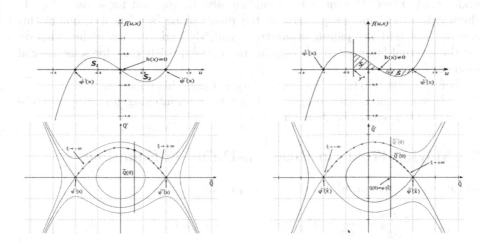

Fig. 3. (a) The cell at the phase plane. (b) No cell at the phase plane

Remark. If we consider the problem (1) with small advection

$$\begin{cases} \varepsilon^2 \dfrac{d^2 y}{dx^2} + \varepsilon A \dfrac{dy}{dx} = f(y, x, \varepsilon), & x \in (0, 1) \\ y(0) = y^0, \quad y(1) = y^1, \end{cases}$$

the equation for the main term of the internal layer location takes the form

$$\frac{d^2 \tilde{Q}}{d\xi^2} + A \cdot \frac{d\tilde{Q}}{d\xi} = f(\tilde{Q}, \hat{x}, 0) \qquad (\hat{x} \text{ - parameter}),$$

$$\tilde{Q}(0, \hat{x}) = \varphi(\hat{x}), \qquad \tilde{Q}(\pm \infty, \hat{x}) = \varphi^{(\pm)}(\hat{x}). \tag{13}$$

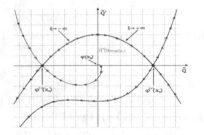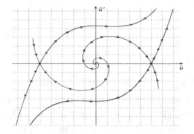

Fig. 4. (a) The joining separatrix. (b) No joining separatrix

Related phase diagrams of (13) are presented on Fig. 4. Here stationary points $(\varphi^{(\pm)}(x), 0)$ of (13) are the saddle points, but $(\varphi(x), 0)$ is not the center type point (as previously), and now it is a point of nonstable focus type (or nonstable node type). The $C^{(1)}$ – matching condition also means that for some $x \in (0; 1)$ there exists the joining separatrix at the phase plane (see Fig. 4a), but in this case we can find the joining separatrix explicitly and also write the equation for the internal layer's location (similar to (12)) explicitly only for some special kinds of function $f(y, x, \varepsilon)$.

For example, if $f(y, x, \varepsilon) = (y^2 - 1)(y - h(x))$ this equation is $A = \pm\sqrt{2} \cdot h(x)$. But in general case of $f(y, x, \varepsilon)$ the $C^{(1)}$ – matching procedure for the problem (13) needs to be done numerically.

3   Moving Fronts in Reaction-Diffusion Equations

In this section we consider the following problem

$$
\begin{aligned}
&\varepsilon^2 \frac{\partial^2 u}{\partial x^2} - \varepsilon \frac{\partial u}{\partial t} = f(u, x, \varepsilon), \qquad x \in (0; 1), \qquad t \in (0, T], \\
&u_x(0, t, \varepsilon) = 0, \qquad u_x(1, t, \varepsilon) = 0, \\
&u(x, 0, \varepsilon) = u_{init}(x, \varepsilon), \qquad x \in [0, 1].
\end{aligned}
\tag{14}
$$

We construct the solution of moving front type that is close to different levels on the left and right side from some (moving) point $\hat{x}(t, \varepsilon)$ with sharp transition layer near this point (Fig. 5).

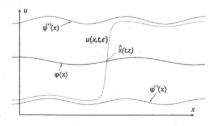

Fig. 5. .

<u>Main conditions</u> for $f(u, x, \varepsilon)$.

C 1. *Reduced problem* $f(u, x, 0) = 0$ *has 3 roots in* $x \in (0; 1)$ *that satisfy*

$$(1) \qquad \varphi^{(-)}(x) < \varphi(x) < \varphi^{(+)}(x);$$

$$(2) \qquad f_u(\varphi^{(\pm)}(x), x, 0) > 0, \qquad f_u(\varphi(x), x, 0) < 0.$$

C 2. *Assume, that at moment* $t = 0$ *front is already exists near the point* $x_{00} \in (0; 1)$, *so* $\hat{x}(0, \varepsilon) = x_{00}$.

Process of the front generation was studied at [3,4].

C 3. *Let*

$$\int_{\varphi^{(-)}(x)}^{\varphi^{(+)}(x)} f(u, x, 0) du > 0 \quad \text{for all} \quad x \in [0; 1].$$

Condition C 3. means that there are no stationary solutions of the problem (14) on $x \in [0; 1]$.

Writing two problems similar to (3) for the transition layer functions $Q(\xi, t)$ at zero order of ε we have [see [1,2]]:

$$\frac{d^2 Q_0^{(\pm)}}{d\xi^2} + \frac{d\hat{x}}{dt} \cdot \frac{dQ_0^{(\pm)}}{d\xi} = f(\varphi^{(\pm)}(\hat{x}(t, \varepsilon)) + Q_0^{(\pm)}, \hat{x}(t, \varepsilon), 0),$$

$$Q_0^{(\pm)}(0, t) = \varphi(\hat{x}(t, \varepsilon)) - \varphi^{(\pm)}(\hat{x}(t, \varepsilon)), \quad Q_0^{(\pm)}(\pm\infty, t) = 0,$$

$$(15)$$

where index $(-)$ corresponds to $\xi < 0$ and index $(+)$ to $\xi > 0$.

Defining continuous function

$$\tilde{u}(\xi, \hat{x}) = \begin{cases} \varphi^{(-)}(\hat{x}) + Q_0^{(-)}(\xi, t), & \xi < 0 \\ \varphi^{(+)}(\hat{x}) + Q_0^{(+)}(\xi, t), & \xi > 0 \end{cases}$$

and introducing the designation $V = \dfrac{d\hat{x}}{dt}$ we can rewrite problems (15) as

$$\frac{d^2 \tilde{u}}{d\xi^2} + V \cdot \frac{d\tilde{u}}{d\xi} = f(\tilde{u}, \hat{x}, 0) \qquad (\hat{x} \text{ - parameter}),$$

$$\tilde{u}(0, \hat{x}) = \varphi(\hat{x}), \qquad \tilde{u}(\pm\infty, \hat{x}) = \varphi^{(\pm)}(\hat{x}).$$

$$(16)$$

To get the main term of the front location we must use (9) and substitute x_0 instead of $\hat{x}$ to (16). It follows from the condition C 3 that if $V = 0$, the phase plane of (16) for all $\hat{x} \in [0; 1]$ looks like at Fig. 6a and there is no $\hat{x} \in (0; 1)$ when the cell on the phase plane exists. So the problem

$$\frac{d^2 \tilde{u}}{d\xi^2} = f(\tilde{u}, x, 0), \qquad \tilde{u}(0, x) = \varphi(x), \qquad \tilde{u}(\pm\infty, x) = \varphi^{(\pm)}(x) \qquad (x \text{ - parameter})$$

has no solution and (14) has no *stationary* solutions with internal layers.

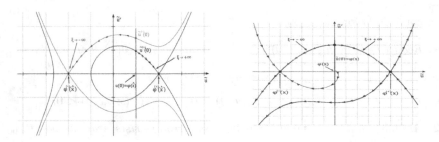

Fig. 6. (a) Phase plane for $V = 0$. (b) Phase plane for $V \neq 0$

The phase plane of (16) if $V \neq 0$ is presented at Figs. 4b and 6b. Stationary points $(\varphi^{(\pm)}(x), 0)$ of (16) are the saddle points, but $(\varphi(x), 0)$ is a point of nonstable focus or nonstable node type. In this case we can implement the $C^{(1)}$ – matching procedure (finding the separatrix, which join two saddles at the phase plane, Fig. 6b), by variation of the parameter V. It is known [5] that for all $\hat{x} \in (0; 1)$ there exists the unique V when also exists the unique solution of (16), which exponentially tends to $\varphi^{(\pm)}(\hat{x})$ when $\xi \to \pm\infty$.

So, the $C^{(1)}$ – matching procedure allows to define $V(x)$ for all $x \in (0; 1)$ and write ODE for the moving front's location with initial condition from C 2:

$$\frac{dx}{dt} = V(x), \qquad x(0, \varepsilon) = x_{00}. \tag{17}$$

For example, if $f(y, x, \varepsilon) = (y^2 - 1)(y - h(x))$ the parameter $V(x)$ can be determined explicitly and the equation (17) is $\frac{dx}{dt} = \pm\sqrt{2} \cdot h(x)$. In general case of $f(y, x, \varepsilon)$ the $C^{(1)}$ – matching procedure for the problem (16) needs to be done numerically, and the problem (17) also must be solved numerically.

4   The Numerical Approach

Summarizing the main ideas, we can say that to determine the location of the internal layer or speed of the moving front we must develop the $C^{(1)}$ – matching procedure for the problems (13) or (16). Note, that it can be done explicitly only for some special cases but, in general, requires to implement some numerical procedure. In this section we briefly describe the main features of the numerical calculations for (13) and (16).

1. We consider two boundary-value problems on half-lines for $\xi < 0$ and $\xi > 0$.
2. The both problems are solved using the relaxation count method. In this case we have to solve two PDE equations. We use the stiff method of lines (SMOL) in order to reduce these PDEs to stiff systems of ODEs. The finite-difference approximation of spatial derivatives are performed on quasi-uniform meshes. Each of these systems is solved by Rosenbrock scheme with complex coefficient (CROS1) because of its stability (L_2) and monotonicity.

3. Then we calculate the difference between left and right derivatives at the different joining points $x \in (0; 1)$.
4. So we are able to construct the function of this difference depending on the location of joining point. The right position of the joining point could be found by minimization of modulus of this function by method of golden section.

Acknowledgements. This work partially supported by RFBR, projects No. 16-01-00437 and 16-01-00755.

References

1. Vasilieva, A.B., Butuzov, V.F., Nefedov, N.N.: Contrast structures in singularly perturbed problems. J. Fund. Prikl. Math. **4**(3), 799–851 (1998)
2. Volkov, V., Nefedov, N., Antipov, E.: Asymptotic-numerical method for moving fronts in two-dimensional R-D-A problems. In: Dimov, I., Faragó, I., Vulkov, L. (eds.) FDM 2014. LNCS, vol. 9045, pp. 408–416. Springer, Cham (2015). doi:10.1007/978-3-319-20239-6_46
3. Volkov, V.T., Grachev, N.E., Nefedov, N.N., Nikolaev, A.N.: On the formation of sharp transition layers in two-dimensional reaction-diffusion models. J. Comput. Math. Math. Phys. **47**(8), 1301–1309 (2007)
4. Volkov, V., Nefedov, N.: Asymptotic-numerical investigation of generation and motion of fronts in phase transition models. In: Dimov, I., Faragó, I., Vulkov, L. (eds.) NAA 2012. LNCS, vol. 8236, pp. 524–531. Springer, Heidelberg (2013). doi:10.1007/978-3-642-41515-9_60
5. Fife, P.C., Hsiao, L.: The generation and propagation of internal layers. Nonlinear Anal. Theory Meth. Appl. **12**(1), 19–41 (1988)

Influence of Snow Cover on the Seismic Waves Propagation

Gyulnara Voskoboynikova[1](✉), Kholmatzhon Imomnazarov[1],
Aleksander Mikhailov[1], and Jian-Gang Tang[2]

[1] Institute of Computational Mathematics and Mathematical Geophysics,
Siberian Branch of Russian Academy of Sciences, Akademik Lavrentjev Avenue 6,
630090 Novosibirsk, Russia
gulya@opg.sscc.ru, {imom,alex_mikh}@omzg.sscc.ru
[2] YiLi Normal University, 448 Jiefang Road, Yinning, Xinjiang,
People's Republic of China
tjg@ylsy.edu.cn
http://www.sscc.ru

Abstract. In this paper the mathematical modeling of seismic waves propagation in the snow based on the theory of dynamic poroelasticity is considered. The snow cover is approximated as porous medium, saturated with liquid or air, where the three elastic parameters are expressed via three elastic wave velocities. These velocities are recalculated using the elastic wave's velocities via the Biot theory, which are expressed through elastic parameters of the snow. The obtained solutions allow the study the peculiarities of the seismic wave's propagation in the liquid or air, which saturating snow cover. In this case, the obtained equations are used for simulation the displacement velocity of the porous frame and the saturating fluid in it, as well as the pore pressure and the stress tensor components with given elastic parameters of the medium and the velocities of propagation of transverse and longitudinal waves in a porous medium.

Keywords: Snow cover · Porous medium · Seismic waves · Elastic velocities · Mathematical model · Transverse and longitudinal waves

1 Introduction

The main important geoecological problem is the problem of studying the impact from industrial such as quarry and test ground explosions to environment. It is known that different factors influence on seismic and acoustic waves propagation from explosion source. Among these factors there are weather factors, related with atmosphere state (wind, humidity, temperature), relief, state of the Earth's surface and also presence of snow cover and forest.

In the given paper, we consider the snow cover influences on the seismic and acoustic waves propagation using mathematical simulation for two-layer model. The model consist of two homogenous layers. Upper layer is snow and

© Springer International Publishing AG 2017
I. Dimov et al. (Eds.): NAA 2016, LNCS 10187, pp. 730–736, 2017.
DOI: 10.1007/978-3-319-57099-0_84

lower layer is elastic semispace. In mathematical simulation the snow cover is presented by elastic-saturated medium which described via equations of motions and state. There is fundamental property for elastic-porous saturated medium as two longitudinal waves of the first and second kind and one rotational wave in two-layer model. The waves of the second kind are highly attenuated.

Mathematical modeling of seismic waves propagation in snow is based on the theory of dynamic poroelasticity. Three elastic parameters are expressed via three elastic wave velocities per Biot theory taking into account parameters of the snow medium. Note that obtained solutions allow studying features of the seismic waves propagation in the snow medium, saturated liquid or air. The obtained equations allow to simulate the displacement velocities of porous frame saturated fluid or air, and also porous pressure and stress tensor components by given elastic parameters of the medium and also by given velocities of the longitudinal and transverse waves propagation in the porous medium.

This problem is the subject of theoretical and experimental studies, including the using acoustic methods. Effective use of the acoustic methods involves knowledge of the processes occurring in saturated porous media, including the elastic wave's propagation as well as non-destructive methods of the snow structure classification; determination of snow physical features; developing effective methods of explosive control for snow slopes; and monitoring acoustic influence. The development of structure classification techniques and determination of the appropriate physical parameters of snow require an accurate acoustic propagation model. Pre-existing models of the seismic wave propagation made use either porous media, which has a rigid ice frame, or continuum elastic or inelastic models [1–6]. These models do not adequately explain the wave propagation phenomena observed in the snow. The air pressure waves, propagating in the interstitial pore space, and dilatational and shear stress waves, propagating in the ice frame, were detected in the snow [7–10]. However, neither the porous medium or the continuum models can explain all the three wave propagation modes. In [11], based on the Biot model the propagation of seismic waves in the snow as a porous material with an elastic frame saturated with a compressible viscous fluid (air) was discussed. The characteristics of the solutions describing acoustic waves in the snow are discussed in detail and compared with the experimental results [8,9,12,13]. In this paper, we described three elastic parameters [14–16] for the propagation of seismic waves in the snow using dynamic theory poroelasticity.

2 The Problem Statement

In poroelasticity theory [14], the stress-strain relations for a porous aggregate including the effects of fluid's pressure and dilatation are considered. As well as in [11,17] we study the dynamics of a material and the coupling between a fluid and a solid provided that the material is statistically homogeneous and isotropic in the region of interest, behaves in a linearly elastic manner and that thermoelastic effects are negligible. The macroscopic stress-strain relation for the medium was derived by assuming the isotropic medium. The coupling effect between the elastic frame and the compressible fluid was taken into account by introducing a kinetic coefficient into the dissipation function of the system [14]. Dissipation of energy by the viscous fluid was expressed in terms of a relative velocity between the fluid and the solid.

The constitutive relations describing the porous material by the tree parameters are the following [16]:

$$\sigma_{ij} = 2\mu\varepsilon_{ij} + \tilde{\lambda}\varepsilon_{kk}\delta_{ij} - \left(1 - \frac{K}{\alpha p^2}\right)p\delta_{ij} \tag{1}$$

$$p = (K - \alpha\rho\rho_s)\varepsilon_{kk} - \alpha\rho\rho_l e_{kk} \tag{2}$$

$$\varepsilon_{ij} = \frac{1}{2}\left(\frac{\partial u_i}{\partial x_j} + \frac{\partial u_j}{\partial x_i}\right), \ e_{ij} = \frac{1}{2}\left(\frac{\partial U_i}{\partial x_j} + \frac{\partial U_j}{\partial x_i}\right),$$

where σ_{ij} is the stress in the solid framework, p is the fluid pore pressure, ε_{kk} and e_{kk} are the dilatations of the solid and the fluid, ε_{ij} is the strain in the solid, and δ_{ij} is the Kronecker delta, $u = (u_1, u_2, u_3)$ and $U = (U_1, U_2, U_3)$ are the displacement vectors of the elastic matrix and the saturating fluid with the respective partial densities ρ_s and ρ_l, $\rho = \rho_l + \rho_s$, $\tilde{\lambda} = \lambda - (\alpha p^2)^{-1}K^2$, $K = \lambda + \frac{2}{3}\mu$, $\lambda, \mu, \alpha p^2$ are elastic parameters of the porous medium [14].

The dynamic poroelasticity theory for the porous media is derived using the method of conservation laws, which include the motion and state equations [18]

$$\frac{\partial u_i}{\partial t} + \frac{1}{\rho_s}\frac{\partial\sigma_{ik}}{\partial x_k} + \frac{\rho_l}{\rho_s\rho}\frac{\partial p}{\partial x_i} + \frac{\rho_l}{\rho_s}\chi\rho_l(u_i - v_i) = F_i, \tag{3}$$

$$\frac{\partial v_i}{\partial t} + \frac{1}{\rho}\frac{\partial p}{\partial x_i} - \chi\rho_i(u_i - v_i) = F_i, \tag{4}$$

$$\frac{\partial\sigma_{ik}}{\partial t} + \mu\left(\frac{\partial u_k}{\partial x_i} + \frac{\partial u_i}{\partial x_k}\right) + \left(\frac{\rho_s}{\rho}K - \frac{2}{3}\mu\right)\delta_{ij}\mathbf{div u} - \frac{\rho_s}{\rho}K\delta_{ik}\mathbf{div v} = 0, \tag{5}$$

$$\frac{\partial p}{\partial t} - (K - \alpha\rho\rho_s)\mathbf{div u} - \alpha\rho\rho_l\mathbf{div v} = 0. \tag{6}$$

$$u_i\mid_{t=0} = v_i\mid_{t=0} = \sigma_{ik}\mid_{t=0} = p\mid_{t=0} = 0,$$

$$\sigma_{22} + p\mid_{x_2=0} = \sigma_{12}\mid_{x_2=0} = \frac{\rho_l}{\rho}p\mid_{x_2=0} = 0,$$

where χ is the friction coefficient.

Determining the Physical Parameters for the Snow. The dynamic nature of a porous material is determined by the three elastic parameters K, μ, and α. These parameters are expressed via velocity of propagation of the shear wave c_s and the velocities of the longitudinal waves c_{p_1}, c_{p_2} [12,13].

$$\mu = \rho_s c_s^2, K = \frac{\rho}{2}\frac{\rho_s}{\rho_l}\left(c_{p_1}^2 + c_{p_2}^2 - \frac{8}{3}\frac{\rho_l}{\rho}c_s^2 - \sqrt{(c_{p_1}^2 - c_{p_2}^2)^2 - \frac{64}{9}\frac{\rho_l\rho_s}{\rho^2}c_s^4}\right),$$

$$\alpha_3 = \frac{1}{2\rho^2}\left(c_{p_1}^2 + c_{p_2}^2 - \frac{8}{3}\frac{\rho_s}{\rho}c_s^2 + \sqrt{(c_{p_1}^2 - c_{p_2}^2)^2 - \frac{64}{9}\frac{\rho_l\rho_s}{\rho^2}c_s^4}\right).$$

The velocities c_s and c_{p_1}, c_{p_2} are determined by the Biot parameters A, N, R, and Q. These parameters are calculated from the four measurable coefficients for the snow [11]. The friction coefficient is expressed by the dissipation coefficient from [11,17].

3   Algorithm of Numerical Solving

We have been used the algorithm of numerical solving of the 2D dynamic problem of seismic wave propagation in the porous medium with allowance for the energy dissipation. To solve numerically this problem, we used the method for combining the Laguerre integral transform with respect to time with a finite-difference approximation along the spatial coordinates. The proposed method of the solution can be considered as analog to the known spectral-difference method based on Fourier transform, only instead of frequency we have a parameter m, i.e. the degree of the Laguerre polynomials. However, unlike Fourier transform, application of the Laguerre integral transform with respect to time allows us to reduce the initial problem to solving a system of equations in which the parameter of division is present only in the right side of the equations and has a recurrent dependence. The algorithm used for the solution makes it possible to perform efficient calculations when simulating a complicated porous medium and studying wave effects emerging in such media.

As of result of this Laguerre transform the initial problem (3)–(6) is reduced to two-dimensional spatial differential problem in spectral domain [19]

$$\frac{h}{2}u_i^m + \frac{1}{\rho_s}\frac{\partial \sigma_{ik}^m}{\partial x_k} + \frac{1}{\rho}\frac{\partial P^m}{\partial x_i} = f_i^m + h\sum_{n=0}^{m-1}u_i^n, \tag{7}$$

$$\frac{h}{2}v_i^m + \frac{1}{\rho}\frac{\partial P^m}{\partial x_i} = f_i^m + h\sum_{n=0}^{m-1}v_i^n, \tag{8}$$

$$\frac{h}{2}\sigma_{ik}^m + \mu\left(\frac{\partial u_k^m}{\partial x_i} + \frac{\partial u_i^m}{\partial x_k}\right) + \left(\lambda - \frac{\rho_s}{\rho}K\right)\delta_{ij}\mathbf{div}\boldsymbol{u}^m - \frac{\rho_s}{\rho}K\delta_{ik}\mathbf{div}\boldsymbol{v}^m = -h\sum_{n=0}^{m-1}\sigma_{ik}^n, \tag{9}$$

$$\frac{h}{2}P^m - (K - \alpha\rho\rho_s)\mathbf{div}\boldsymbol{u}^m - \alpha\rho\rho_l\mathbf{div}\boldsymbol{v}_m = -h\sum_{n=0}^{m-1}P^n.$$

$$\sigma_{22}^m + P^m\,|_{x_2=0} = \sigma_{12}^m\,|_{x_2=0} = \frac{\rho_l}{\rho}P^m\,|_{x_2=0} = 0, \tag{10}$$

For solution of the problem (7)–(10) we used finite-difference approximation of 4th order along the spatial coordinates on the shifted grids [19]. As result of the finite-difference approximation for the problem (7)–(10) we obtain the system of linear algebraic equations in the vector form

$$\left(A_\Delta + \frac{1}{2}E\right)\boldsymbol{W}(m) = \boldsymbol{F}_\Delta(m-1) \tag{11}$$

where $W(m) = (V_0(m), ..., V_{M+N}(m))$ is solution vector with the wave field components.

Finding the solution of this system of linear algebraic equations (11), we can define spectral values for all wave field components. Then we obtain the solution of the initial problem (3)–(6) using Laguerre transform.

4   Results of Numerical Simulation

In given section we present the numerical results of simulation of seismic waves fields for test model of the porous medium. The given medium is consist of two homogenous layers, namely snow (upper layer) and elastic semispace (lower layer). The physical medium's features were specified as

– snow layer with parameters $\rho_0 = 0.4\,\text{g/cm}^3$ is density, $\rho^f_{0,l} = 0.01\,\text{g/cm}^3$ is density, $c_{p_1} = 1.1\,\text{km/s}$, $c_{p_2} = 0.25\,\text{km/s}$, $c_s = 0.7\,\text{km/s}$, $d = 0.5$ is porosity coefficient, $\chi = 100\,\text{cm}^3/(\text{g} \cdot \text{sec})$ is friction coefficient;
– lower elastic semispace with parameters $\rho = 1.5\,\text{g/cm}^3$, $c_p = 1.2\,\text{km/s}$, $c_s = 0.8\,\text{km/s}$.

Wave field is modeled from point source like expansion center with coordinates $x_0 = 200\,\text{m}$, $z_0 = 4\,\text{m}$, situated in upper layer. Time signal in the source was given as Puzyrev's pulse with main frequency equaled 100 Hz.

$$f(t) = \exp\left(-\frac{2\pi f_0(t - t_0)^2}{\gamma^2}\right)$$

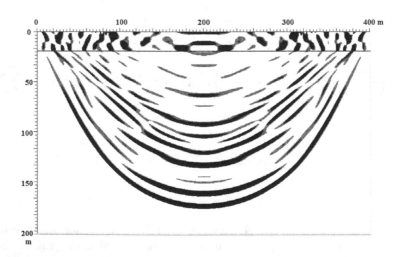

Fig. 1. Snapshot of the wave's field for vertical component of the displacement velocity $u_z(x, z)$ in the time moment $t = 0.15\,\text{s}$. The main frequency of signal in the source is 100 Hz

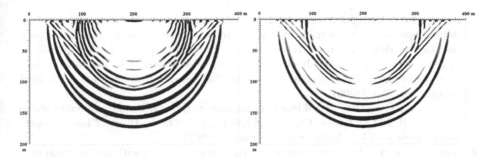

Fig. 2. Snapshots of the wave field for vertical component of the displacements velocity $u_z(x, z)$ in the moment time t = 0.15 s. On the left side without energy dissipation, on the right side with energy dissipation. The main frequency of signal in the source is 100 Hz.

The example of calculation wave's field results for different models is represented on the Figs. 1 and 2. The figure is shown that in snow's layer the multiply reflected waves arise. They generates different longitudinal and transverse waves in elastic semispace.

The Fig. 1 demonstrates snapshot of wave's field for vertical velocity component of displacement u_z in the fixed moment of time T = 0.15 s when thickness of the upper snow layer is 20 m. The Fig. 2 represents the snapshot of wave's field for vertical velocity component of displacement when thickness of the snow layer is 3 m. The right snapshot shows the wave field with energy dissipation in the snow's layer and the left snapshot presents wave field without energy dissipation. From analysis of Figs. 1 and 2. It is shown that if thickness of the snow layer less spatial length transverse and quick longitudinal wave, then presence of energy dissipation in medium significantly influences to wave field. Since the presence of energy dissipation firstly significantly influences to attenuation of the slow longitudinal waves and decreases the number of reflections in thin layers.

Acknowledgments. This work was supported by the Russian Foundation for Basic Research (grant 16-07-01052).

References

1. Nakaya, U.: Visco-elastic properties of snow and ice in the Greenland ice cap. U.S. Snow, Ice and Permafrost Research Establishment, Research report 46, 33 p. (1959)
2. Nakaya, U.: Visco-elastic properties of processed snow. U.S. Snow, Ice and Permafrost Establishment, Research report 58, 22 p. (1959)
3. Nakaya, U.: Elastic properties of processed snow with reference to its internal structure. U.S. Cold Regions Research and Engineering Laboratory, Hanover, N.H., Research report 82, 25 p. (1961)
4. Ishida, T.: Acoustic properties of snow. Contrib. Inst. Low Temp. Sci. Ser. A **20**, 23–68 (1965)

5. Smith, J.L.: The elastic constants, strength and density of Greenland snow as determined from measurements of sonic wave velocity. U.S. Cold Regions Research and Engineering Laboratory, Hanover, N.H. Technical report 167, 18 p. (1965)
6. Smith, N.: Determining the dynamic properties of snow and ice by forced vibration. U.S. Cold Regions Research and Engineering Laboratory, Hanover, N.H., Technical report 216, 17 p. (1969)
7. Chae, Y.S.: Frequency dependence of dynamic modulus of, and damping in snow. In: Oura, H. (ed.) Physics of Snow and Ice, pp. 827–842. Institute of Low Temperature Science, Hokkaido University, Sapporo (1967)
8. Yen, Y.C., Fan, S.S.T.: Pressure wave propagation in snow with non-uniform permeability. U.S. Cold Regions Research and Engineering Laboratory, Hanover, N.H., Research report 210, 9 p. (1966)
9. Oura, H.: Sound velocity in the snow cover. Low Temp. Sci. **9**, 171–178 (1952)
10. Smith, J.L.: The elastic constants, strength and density of Greenland snow as determined from measurements of sonic wave velocity. U.S. Cold Regions Research and Engineering Laboratory, Hanover, N.H., Technical report 167, 18 p. (1965)
11. Yamada, T., Hasemi, T., Izumi, K., Sata, A.: On the dependencies of the velocities of P and S waves and thermal conductivity of snow upon the texture of snow. Contrib. Inst. Low Temp. Sci. Ser. A **32**, 71–80 (1974)
12. Johnson, J.B.: On the application of Biot's theory to acoustic propagation in snow. Cold Reg. Sci. Technol. **6**, 49–60 (1982)
13. Gubler, H.: Artificial release of avalanches by explosives. J. Glaciol. **19**, 419–429 (1977)
14. Bogorodskii, V.V., Gavrilo, V.P., Nikitin, V.A.: Characteristics of sound propagation in snow. Akust. Zh. **20**, 195–198 (1974). (Sov. Phys. Acoust. **20**(2), 121–122)
15. Blokhin, A.M., Dorovsky, V.N.: Mathematical Modelling in the Theory of Multivelocity Continuum. Nova Science, New York (1995)
16. Imomnazarov, K.K.: Some remarks on the Biot system of equations describing wave propagation in a porous medium. Appl. Math. Lett. **13**(3), 33–35 (2000)
17. Zhabborov, N.M., Imomnazarov, K.K.: Some Initial Boundary Value Problems of Mechanics of Two-Velocity Media. National University of Uzbekistan Named After Mirzo Ulugbek, Tashkent (2012). (in Russian)
18. Biot, M.A.: Theory of propagation of elastic waves in a fluid-saturated porous solid. I. Low frequency range. J. Acoust. Soc. Am. **28**(2), 168–178 (1956)
19. Imomnazarov, Kh.Kh., Mikhailov, A.A.: Laguerre spectral method as applied to numerical solution of a two dimensional linear dynamic seismic problem for porous media. Sib. J. Ind. Math. **11**(3), 86–95 (2008)

Hybrid Model of Particle Acceleration on a Shock Wave Front

Lyudmila Vshivkova[1](✉) and Galina Dudnikova[2,3]

[1] Institute of Computational Mathematics and Mathematical
Geophysics SB RAS, Novosibirsk, Russia
lyudmila.vshivkova@parbz.sscc.ru
[2] Institute of Computational Technologies SB RAS, Novosibirsk, Russia
dudn@ict.nsc.ru
[3] University of Maryland, College Park, MD 20742, USA
http://www.icmmg.nsc.ru, http://www.ict.nsc.ru, http://www.umd.edu

Abstract. A new 2D hybrid numerical plasma model to investigate the
processes of particle acceleration on a shock wave front is presented.
This problem has a fundamental interest for astrophysics, plasma physics
and charged particle accelerators. The model is based on the hybrid (or
combined) approach where an electron component of plasma is described
by the MHD-equations, while ions are treated kinetically via the Vlasov
equation. One of the advantages of this approach is that it allows reduce
the requirements for computing resources essentially comparing to a fully
kinetic model. Another important advantage of it is the possibility to
study the important instabilities on the ion time scale, neglecting high-
frequency modes associated with electrons.

Keywords: Hybrid model · Magnetohydrodynamics (MHD) · Vlasov
kinetic equation · Particle-in-cell (PIC) method

1 Introduction

The paper deals with the research based on the numerical modeling referring to
a fundamental research problem – the study of mechanism of cosmic ray generation and acceleration of their particles on a shock wave front. Regardless to the
fact that cosmic rays, which are a high-energetic particle flow (electrons, protons
and atomic nuclei), were discovered almost a hundred years ago, their nature has
not been still studied well. The interest to study the cosmic rays deals not only
with the astrophysical direction, but also with the study of the fundamental
properties of elementary particles of high energies. The last new direct observations of galactic remnants of supernovae approve the hypothesis that exactly
the supernovae are the places where the cosmic rays are generated. Mechanisms
of their acceleration are not very clear but, however, it is clear that strong electric and magnetic fields play a fundamental role in the process. Nowadays, many
researchers study this matter and try to find out the physical acceleration mechanism of the cosmic rays. The detailed review of it is given in [1]. One of the

© Springer International Publishing AG 2017
I. Dimov et al. (Eds.): NAA 2016, LNCS 10187, pp. 737–743, 2017.
DOI: 10.1007/978-3-319-57099-0_85

possible acceleration mechanisms is the mechanism of acceleration by a shock wave which has been widely discussed since 70s [2,3]. The numerical models, which are used to study the problem of generation and dynamics of cosmic rays, are divided into three types and deal with the kinetic or magnetohydrodynamics (MHD) approach [4]. The most fully description is possible when using the kinetic Vlasov equation and the Maxwell system of equations. However, the difficulties of numerical implementation of such model, associated with a great difference in space and time scales for electron and ion components of plasma, complicate their implementation even when carrying out computations on modern computational systems [5]. Also, the MHD-models [6–8] are used very often which, nevertheless, do not permit to describe the violation of flow one-streaming when particles are reflected from a shock wave front. The research based on the hybrid models, where electrons are described by the MHD-approach and ions – via the Vlasov kinetic equation, allows reduce requirements to computational resources, considerably, and in our days are perspective [9–15].

The distinction of the numerical model presented in this paper from the other ones (e.g., [16]) is as follows: an electron mass ($m_e \neq 0$) and a plasma heat conductivity are taken into account. Also, an implicit scheme is used for computing the magnetic field intensity $\mathbf{B}$ to increase the stability.

2   Problem Statement

Let us consider a 2D problem (in Cartesian coordinates) of shock wave generation and ion acceleration caused by their reflection from the wave front. The plasma flow, consisting of hydrogen ions and electrons, comes into a uniform plasma background. At initial time $t = 0$ the plasma flow of a density n, temperature T_e, charge q and constant initial velocity $\mathbf{v} = \mathbf{v_0}$ enters the plasma rectangular box (x, y) of the uniform magnetic field $\mathbf{B}$, constant density n, temperature T_e, charge q and velocity $\mathbf{v} = \mathbf{0}$. Entering the box from the left (at $x = x_0$) it forms a shock wave which accelerates charged particles of the background plasma towards the right boundary. It is supposed that plasma is quasi-neutral ($n_i = n_e = n$).

3   System of Equations

Let us write the initial system of equations of the proposed hybrid model. The Vlasov kinetic equation for ions in the problem is solved by the particle-in-cell (PIC) method [17–19], therefore, the following characteristic equations of this kinetic equation are used

$$\frac{d\mathbf{r}}{dt} = \mathbf{v}_\alpha, \quad m_\alpha \frac{d\mathbf{v}_\alpha}{dt} = Z_\alpha e \left(\mathbf{E} + \frac{1}{c} \mathbf{v}_\alpha \times \mathbf{B} \right) + \mathbf{R}_\alpha. \tag{1}$$

Here Z_α is a degree of ionization of ions of a sort α, $\mathbf{E} = \{\mathbf{E_x}, \mathbf{E_y}, \mathbf{E_z}\}$ and $\mathbf{B} = \{\mathbf{B_x}, \mathbf{B_y}, \mathbf{B_z}\}$ are intensities of electric and magnetic fields, $\mathbf{R}_\alpha = -eZ_\alpha \mathbf{j}/\sigma$

is the force of friction between the ions of the sort α and electrons [20, 21]. The density n_α and average velocity of the ions $\mathbf{V}_\alpha$ of the sort α are determined by the ion distribution function f_α by velocities

$$n_\alpha = \int f_\alpha d\mathbf{v}, \quad \mathbf{V}_\alpha = \frac{1}{n_\alpha} \int f_\alpha \mathbf{v}_\alpha d\mathbf{v}. \tag{2}$$

The motion of an electron component of plasma is described by the magnetic hydrodynamics (MHD) approach

$$m_e \left(\frac{\partial \mathbf{V}_e}{\partial t} + (\mathbf{V}_e \cdot \nabla) \mathbf{V}_e \right) = -e \left(\mathbf{E} + \frac{1}{c} \mathbf{V_e} \times \mathbf{B} \right) - \frac{\nabla p_e}{n_e} + \mathbf{R}_e, \tag{3}$$

$$n_e \left(\frac{\partial T_e}{\partial t} + (\mathbf{V}_e \cdot \nabla) T_e \right) + (\gamma - 1) p_e \nabla \cdot \mathbf{V}_e = (\gamma - 1) \left(Q_e - \nabla \cdot \mathbf{q_e} \right), \tag{4}$$

where n_e, $\mathbf{V_e}$ are a density and an electron velocity, $p_e = n_e T_e$ is an electron pressure, $\mathbf{R}_e = - \sum_\alpha \mathbf{R}_\alpha$ is a force of friction between electrons and ions, T_e is an electron temperature, $Q_e = j^2/\sigma$ is an electron heating which is the result of collisions of electrons with ions (here $\sigma = n_e e^2/(m_e \nu)$, ν – frequency of collisions), $\mathbf{q}_e = -\kappa_1 \nabla T_e$ is a heat flow, where κ_1 is the coefficient of heat conductivity. Also, the following Maxwell equations are added to the system of equations

$$\nabla \times \mathbf{B} = \frac{4\pi}{c} \mathbf{j}, \quad \nabla \times \mathbf{E} = -\frac{1}{c} \frac{\partial \mathbf{B}}{\partial t}, \quad \nabla \cdot \mathbf{B} = 0. \tag{5}$$

Here $\mathbf{j}$ is a current density which in case of multicomponent plasma is defined as $\mathbf{j} = e \left(\sum_\alpha Z_\alpha n_\alpha \mathbf{V}_\alpha - n_e \mathbf{V}_e \right)$. Plasma is quasi-neutral, i.e. $n_e = \sum_\alpha Z_\alpha n_\alpha$, consequently, we do not consider the equation $\nabla \cdot \mathbf{E} = 4\pi\rho$ in the system, where $\rho = e \left(\sum_\alpha Z_\alpha n_\alpha - n_e \right)$ is a volume charge, $\mathbf{V}_\alpha = (V_{\alpha x}, V_{\alpha y}, V_{\alpha z})$ is an average ion velocity.

4 Algorithm

The problem is considered in the area $0 \leq x \leq x_{\max}, 0 \leq y \leq y_{\max}$. In compliance with the stated problem the initial conditions of the background plasma at $t = 0$ are as follows: $n(x, y) = n_0 = const$, $q(x, y) = q_0 = const$, $B_x(x, y) = B_y(x, y) = 0$, $B_z(x, y) = B_0 = const$, $T_e(x, y) = T_0$, $v_x(x, y) = 0$, $v_y(x, y) = v_z(x, y) = 0$ and $E_x(x, y) = E_y(x, y) = E_z(x, y) = 0$. The initial conditions of the incoming flow are given as: $n(x, y) = n_0 = const$, $q(x, y) = q_0 = const$, $T_e(x, y) = T_0$, $v_x(x, y) = v_0$, $v_y(x, y) = v_z(x, y) = 0$. The boundary conditions are as follows: for particles (ions) – the reflection conditions by x (at $x = x_{max}$) and the periodic conditions by y; for grid functions – $E_x = 0$, $\partial E_y/\partial x = \partial E_z/\partial x = 0$, $\partial n/\partial x = 0$, $\partial T_e \partial x = 0$ by x and the periodic conditions by y.

The uniform grid with steps h_1, h_2 by the axis x and y, respectively, has been introduced in the computation area. The grid function B_z is defined in the grid nodes $x_i = ih_1$, $y_k = kh_2$, the functions V_x, V_{ex}, E_x, B_y – in the points x_i, $y_{k-1/2} = (k - 0.5)h_2$, the functions V_y, V_{ey}, E_y, B_x – in the points $x_{i-1/2} = (i - 0.5)h_1$, y_k, and the functions n, T_e, V_z, V_{ez}, E_z – in the cell centers $x_{i-1/2}$, $y_{k-1/2}$.

Let us write the scheme in dimensionless vector form in the order of computation. As the normalizations for the density, magnetic field intensity, temperature, velocity, coordinate and time the following values have been taken: n_0, B_0, $B_0^2/(8\pi n_0)$, $V_A = B_0/\sqrt{4\pi n_0 m_i}$, $c/\omega_{pi} = c\sqrt{m_i}/\sqrt{4\pi n_0 e^2}$ and $1/\omega_{ci} = m_i c/(eB_0)$, respectively. Below all variables are written in the dimensionless form.

The motion equations for model particles are solved on the first stage of computation:

$$\frac{\mathbf{v}^{m+1} - \mathbf{v}^m}{\tau} = \hat{E}^m + \mathbf{v}^m \times \mathbf{B}^m, \qquad \frac{\mathbf{r}^{m+1} - \mathbf{r}^m}{\tau} = \mathbf{v}^{m+1}. \qquad (6)$$

The density and the average ion velocity are found on the second stage using the PIC-method:

$$n^{m+1}(\mathbf{r}) = \sum_j q_j \overline{P}\left(\mathbf{r} - \mathbf{r}_j^{m+1}\right), \qquad (7)$$

$$\mathbf{V}^{m+1}(\mathbf{r}) = \sum_j q_j \mathbf{v}_j^{m+1} \overline{P}\left(\mathbf{r} - \mathbf{r}_j^{m+1}\right) / n^{m+1}(\mathbf{r}), \qquad (8)$$

where $\overline{P}(\mathbf{r}) = \overline{P}(x)\overline{P}(y)$

$$\overline{P}(l) = \begin{cases} 1 - |l|/h, & \text{where } |l| \le h, \\ 0, & \text{where } |l| > h \end{cases} \qquad (9)$$

where $l = \{x, y\}$ and $h = \{h_1, h_2\}$, respectively.

Then, using the average ion velocities, the electron velocities are determined (the third stage):

$$\mathbf{V}_e^{m+1} = \mathbf{V}^{m+1} - \nabla \times \mathbf{B}^m/n^{m+1}. \qquad (10)$$

To compute the magnetic field, the following equation is solved on the fourth stage:

$$\frac{\mathbf{B}^{m+1} - \mathbf{B}^m}{\tau} = -\nabla \times \left(\hat{\mathbf{E}}^{m+1} + \mathbf{R}_e\right), \qquad (11)$$

where $\mathbf{R}_e = \text{æ}\nabla \times \mathbf{B}^{m+1}/n^{m+1}$, $\text{æ} = m_e \nu c/(B_0 e)$,

$$\hat{\mathbf{E}}^{m+1} = -\mathbf{V}_e^{m+1} \times \mathbf{B}^{m+1} - \frac{1}{n^{m+1}}\nabla\left(n^{m+1}T_e^m\right) - \beta \mathbf{s}^{m+1}, \qquad (12)$$

$\beta = m_e/m_i$, $\mathbf{s}^{m+1} = \left(\mathbf{V}_e^{m+1} - \mathbf{V}_e^m\right)/\tau + \left(\mathbf{V}_e^{m+1} \cdot \nabla\right)\mathbf{V}_e^{m+1}$.

The implicit scheme Eq. (11) is solved by the directional splitting method and is implemented using the scalar sweep method. The difference of the present scheme from the previous one [22] is in implicit computation of the magnetic field intensity which permits to increase the stability. Then, using Eq. (12)

for $\hat{\mathbf{E}}^{m+1}$, one can compute the final value of the electric field intensity (the fifth stage):

$$\mathbf{E}^{m+1} = \hat{\mathbf{E}}^{m+1} + \mathbf{R}_e. \tag{13}$$

Afterwards, on the final sixth stage, the temperature equation is solved:

$$\frac{T_e^{m+1} - T_e^m}{\tau} + (\mathbf{V} \cdot \nabla)\, T_e^{m+1} + (\gamma - 1)\, n^{m+1} T_e^{m+1} \nabla \cdot \mathbf{V}_e^{m+1}$$
$$= (\gamma - 1) \left(Q_e^{m+1} - \nabla \cdot q_e^{m+1} \right). \tag{14}$$

The implicit scheme for the temperature is solved, also, by the splitting method and realized by the sweep method.

5 Computational Results

Let us consider some results of the computations. The convergence of the results of computer simulation depending on the number of particles per cell, space and time steps has been examined. It was found that for good reproducing of shock wave structure the number of particles per cell, N, should be greater or equal to 100. Therefore, when carrying out the computations, the particle number per cell was varying from 100 to 500. There were chosen ten cells per c/ω_{pi}, i.e. the space steps were taken $h_x = h_y = 0.1$.

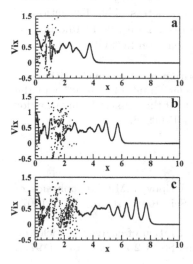

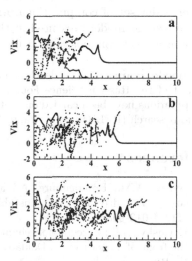

Fig. 1. Phase space of ions at times $t = 3.0$ (a), 4.5 (b), 6.0 (c). Here $N = 100$, $v_0 = 1.0$, $M_A = 1.3$, $T_e = 1.0$ and æ $= 0.01$.

Fig. 2. Phase space of ions at times $t = 2.0$ (a), 2.5 (b), 3.0 (c). Here $N = 100$, $v_0 = 3.0$, $M_A = 2.2$, $T_e = 1.0$ and æ $= 0.01$.

The distribution of plasma ions on the phase space (x, V_{ix}) for consistent moments of time $t = 3.0, 4.5, 6.0$ is given on Fig. 1. Here the following parameters had been chosen: the particle number per cell $N = 100$, the initial particle velocity $v_0 = 1.0$, the electron temperature $T_e = 1.0$ and the collision frequency $\ae = 0.01$. From this figure one can see that the shock wave, propagating across the magnetic field $\mathbf{B}$, has the oscillation structure with the characteristic oscillation size of about $0.3c/\omega_{pi}$. The Mach number $M_A = v^*/V_A = 1.3$, where v^* is the shock wave velocity. This Mach number is the lower critical value (the critical Mach number $M_A^* > 2$ [23]) and, as the result, there is no particle reflection from the wave front.

Another regime of formation of the shock wave structure at times $t = 2.0$, 2.5, 3.0 is represented on Fig. 2. Here the initial ion velocity $v_0 = 3.0$ and other parameters are the same as in the previous case. For these parameters the increased initial ion velocity of the incoming flow results in the Mach number of the shock wave to be $M_A = 2.2$, which is greater than the critical one. Consequently, the shock wave propagation is accompanied by the reflection of particles from its front. This mechanism of acceleration of ions on the front of the shock wave is the one of the possible acceleration mechanisms of cosmic rays.

6   Conclusion

The feature of the new scheme is in implicit computation of the magnetic field intensity and the electric field intensity associated with it. This procedure increases the scheme stability essentially, and permits to carry out calculations under wider set of computational and physical parameters. The distinction of the numerical model presented in this paper from other ones is in taking into account an electron mass ($m_e \neq 0$) and a plasma heat conductivity.

Acknowledgements. The development of the algorithm has been made within the project of the Russian Science Foundation (RSF) under Grant 14-11-00485 and the computations have been carried out within the projects of the Russian Foundation for Basic Research (RFBR) under Grants 16-01-00209, 16-07-00916.

References

1. Bykov, A.M., Brandenburg, A., Malkov, M.A., Osipov, S.M.: Microphysics of cosmic ray driven plasma instabilities. Space Sci. Rev. (2013). doi:10.1007/s11214-013-9988-3
2. Krymsky, G.F.: A regular mechanism for the acceleration of charged particles on the front of a shock wave. Doklady Akademii Nauk SSSR **234**, 1306–1308 (1977). (in Russian)
3. Bell, A.R.: The acceleration of cosmic rays in shock fronts - I. Mon. Not. R. Astron. Soc. **182**, 147–156 (1978)
4. Malkov, M.A., Drury, L.O.: Nonlinear theory of diffusive acceleration of particles by shock waves. Rep. Prog. Phys. **64**, 429–481 (2001)
5. Riquelme, M.A., Spitkovsky, A.: Magnetic amplification by magnetized cosmic rays in supernova remnant shocks. Astrophys. J. **717**, 1054–1066 (2010)
6. Bell, A.R., Lucek, S.G.: Cosmic ray acceleration to very high energy through the non-linear amplification by cosmic rays of the seed magnetic field. Mon. Not. R. Astron. Soc. **321**, 433–438 (2001)

7. Zirakashvili, V.N., Ptuskin, V.S.: Diffusive shock acceleration with magnetic amplification by nonresonant streaming instability in supernova remnants. Astrophys. J. **678**, 939–949 (2008)
8. Bykov, A.M., Osipov, S.M., Ellison, D.C.: Cosmic ray current driven turbulence in shocks with efficient particle acceleration: the oblique, long-wavelength mode instability. Mon. Not. R. Astron. Soc. **410**, 39–52 (2011)
9. Berezin, Y.A., Dudnikova, G.I.: Numerical Plasma Models and Reconnection Processes. Nauka, Moskva (1985). (in Russian)
10. Lipatov, A.S.: The Hybrid Multiscale Simulation Technology. An Introduction with Application to Astrophysical and Laboratory Plasmas. Springer, Heidelberg (2002)
11. Damiano, P.A., Sydora, R.D., Samson, J.C.: Hybrid magnetohydrodynamic-kinetic model of standing shear Alfven waves. J. Plasm. Phys. **69**(4), 277–304 (2003)
12. Caprioli, D., Spitkovsky, A.: Cosmic-ray-induced filamentation instability in collisionless shocks. J. Lett. Astrophys. **765**, 8 (2013)
13. Vshivkova, L.V.: Numerical simulation of plasma using the hybrid MHD-kinetic model. Bull. Novosibirsk Comput. Center, Ser.: Numer. Anal. **14**, 95–114 (2009)
14. Vshivkova, L.V., Dudnikova, G.I.: Numerical modeling of plasma phenomena using the PIC-method. In: 2012 25th International Symposium on IEEE Conference Electrical Insulation in Vacuum (ISDEIV), pp. 398–400 (2012)
15. Vshivkova, L.V., Dudnikova, G.I.: Dispersion analysis of the hybrid plasma model. Bull. Novosibirsk Comput. Center, Ser.: Numer. Anal. **16**, 101–106 (2013)
16. Gargate, L., Bingham, R., Fonseca, R.A., Silva, L.O.: dHybrid: a massively parallel code for hybrid simulations of space plasmas. Comput. Phys. Commun. **176**, 419–425 (2007)
17. Berezin, Y.A., Vshivkov, V.A.: Pirticle-in-Cell Method in the Dynamics of Low-density Plasma. Science, Siberian Branch, Novosibirsk (1980). (in Russian)
18. Hockney, R.W., Eastwood, J.W.: Computer Simulation Using Particles. CRC Press, Boca Raton (1988)
19. Grigoryev, Y.N., Vshivkov, V.A., Fedoruk, M.P.: Numerical Particle-In-Cell Methods. Theory and Applications, pp. 1–260. de Gruyter, Berlin (2002)
20. Braginski, S.I.: Protsessy perenosa v plasme. Voprosy Teorii Plazmy Gosatomizdat, Moskva **1**, 183–272 (1963). (in Russian); transl. 'Transport processes in plasma'. Reviews of Plasma Physics, **1**, 205–311 (1965)
21. Vshivkova, L.V.: Numerical modeling of the multi-component plasma dynamics. Bull. Novosibirsk State Univ. **3**(2), 3–20 (2003). (in Russian)
22. Vshivkova, L.V., Dudnikova, G.I., Liseykina, T.V., Mesyats, E.A.: Hybrid simulation of collisionless shock waves using the PIC-method. Bull. Novosibirsk Comput. Center, Numer. Anal. **17**, 79–88 (2015)
23. Sagdeev, R.Z.: Cooperative phenomena and shock waves in collisionless plasma. Rev. Plasma Phys. **4**, 23–91 (1966)

Solution of the Stochastic Differential Equations Equivalent to the Non-stationary Parker Transport Equation by the Strong Order Numerical Methods

Anna Wawrzynczak[1]($\boxtimes$) and Renata Modzelewska[2]

[1] Institute of Computer Sciences, Siedlce University, Siedlce, Poland
awawrzynczak@uph.edu.pl
[2] Institute of Mathematics and Physics, Siedlce University, Siedlce, Poland
renatam@uph.edu.pl

Abstract. We present the newly developed stochastic model of the galactic cosmic ray (GCR) particles transport in the heliosphere. The model is based on the numerical solution of the Parker transport equation (PTE) describing the non-stationary transport of charged particles in the turbulent medium. We present the method of deriving from PTE the equivalent stochastic differential equations (SDEs) in the heliocentric spherical coordinate system for the backward approach. We present the formulas for the numerical solution of the obtained set of SDEs driven by a Wiener process in the case of the full three-dimensional diffusion tensor. We introduce the solution applying the strong order Euler-Maruyama, Milstein, and stochastic Runge-Kutta methods. We compare the convergence and stability of the solution for the listed methods. We also discuss the advantages and disadvantages of the presented numerical methods in the context of increasing the accuracy of the solution of the PTE. We present the comparison of the stochastic model of the Forbush decrease (Fd) of the GCR intensity with the experimental data.

Keywords: Parker transport equation · Fokker-Planck equation · Stochastic differential equations · Numerical approximation

1 Introduction

Nowadays in the era of easy access to the high-performance computing clusters, efficient parallel algorithms to realize large-scale simulations are crucial in solving realistic problems inspired by physical sciences and engineering. Especially the stochastic approach which from its nature is highly parallelized becomes very popular in the simulation of different random physical processes. One of the still open tasks is the simulation of the non-stationary stochastic motion of charged particles in a multidimensional-dimensional space, described by diffusion equation of Fokker-Planck type. The difficulty of the numerical solution of this type of equations increases with the problem dimension. The reason is the instability of the numerical schemes like finite-differences and finite-volume

© Springer International Publishing AG 2017
I. Dimov et al. (Eds.): NAA 2016, LNCS 10187, pp. 744–751, 2017.
DOI: 10.1007/978-3-319-57099-0_86

in the higher dimensions. To ensure the scheme stability and convergence the density of numerical grid must be improved, with the computational complexity enlargement, at the same time. To overcome this problem, the stochastic methods can be applied (e.g. [1,2]). In this approach, the individual particle motion is described as a Markov stochastic process, and the system evolves probabilistically. Employment of probabilistic description with Monte Carlo simulations allow reducing the solution of the partial differential equation (PDE) describing the analyzed phenomena to the integration of stochastic differential equations (SDEs).

The aim of this paper is to propose an efficient model of the galactic cosmic ray (GCR) particles transport in the heliosphere based on the stochastic solution of the five-dimensional Parker transport equation (PTE). This paper is an extension of our previous results. In [3] we have presented that the GCR particles transport can be effectively modeled based on the solution of the set of the SDEs corresponding to the PTE (Eq. 1). In [3,4] the Euler-Maruyama method of the SDE's numerical solution was applied. In this paper, we continue our work [5] and implement the strong order integration Euler-Maruyama, Milstein and stochastic Runge-Kutta numerical methods of SDEs solution to simulate the Forbush decrease (Fd) of the GCR intensity being in agreement with the experimental data.

2   Fokker-Planck Equation and Ito Stochastic Differential Equation

Fokker-Planck equation (FPE) plays a prominent role in stochastic modeling. The PDEs of Fokker-Planck type in general describe the transport of the distribution function $f = f(r, p, t)$ being a function of the coordinates r, momentum p and time t. In this paper, we will focus on the numerical solution of the general form of the four-dimensional FPE i.e. the PTE describing the propagation of the GCR particles in the heliosphere. PTE has a form [6]:

$$\frac{\partial f}{\partial t} = \boldsymbol{\nabla} \cdot (K_{ij}^S \cdot \boldsymbol{\nabla} f) - (\boldsymbol{v}_d + \boldsymbol{U}) \cdot \boldsymbol{\nabla} f + \frac{R}{3}(\boldsymbol{\nabla} \cdot \boldsymbol{U})\frac{\partial f}{\partial R}, \tag{1}$$

where $f = f(r, R, t)$ is an omnidirectional distribution function of three spatial coordinates (r, θ, φ), r - radial distance, θ - heliolatitudes, φ - heliolongitudes; particles rigidity R and time t; $\boldsymbol{U}$ is the solar wind velocity, $\boldsymbol{v}_d$ the drift velocity, and K_{ij}^S is the symmetric part of the diffusion tensor of the GCR particles.

Equation 1 pictures the modulation of the GCR particles as a interplay between four core processes: convection by the solar wind, diffusion on irregularities of the heliospheric magnetic field, particles drifts in the non-uniform magnetic field and adiabatic cooling (e.g. [7]). PTE is unsolvable analytically, so numerical methods must be applied.

Depending on the direction of integration FPE can be expressed in two forms [8], the time-forward:

$$\frac{\partial F}{\partial t} = -\sum_i \frac{\partial}{\partial x_i}(A_i \cdot F) + \frac{1}{2}\sum_{i,j} \frac{\partial^2}{\partial x_i \partial x_j}(B_{ij}B_{ij}^T \cdot F), \tag{2}$$

and the time-backward:

$$\frac{\partial F}{\partial t} = \sum_i A_i \frac{\partial F}{\partial x_i} + \frac{1}{2} \sum_{i,j} B_{ij} B_{ij}^T \frac{\partial^2 F}{\partial x_i \partial x_j}. \tag{3}$$

SDE equivalent to Eqs. 2 and 3 has the form (e.g. [8]):

$$d\boldsymbol{r} = \boldsymbol{A}_i \cdot dt + B_{ij} \cdot d\boldsymbol{W}, \tag{4}$$

where $\boldsymbol{r}$ is the individual pseudoparticle trajectory in the phase space and $d\boldsymbol{W}$ is the Wiener process.

Solving the Eq. 4 we recognize the path of the pseudoparticles as a Markov process and define a transition density $f(r_{old}, t_{old}; r_{new}, t_{new},)$ describing the probability density for a transition from the 'old' to the 'new' point. Both time-forward and time-backward integration of Eq. 4, having specified an initial and final state, compute the transition density for a particle to reach current state. However, it imposes different understanding of the initial and final state. Choosing between the forward or backward in time approach should be adjusted to the problem that we want to solve.

3   Stochastic Differential Equations Corresponding to the Parker Transport Equation

In the case of the modeling of the GCR transport in the heliosphere, the time-backward approach is much more computationally effective (see [3,5]). Thus, to solve PTE by stochastic approach we must bring it to the form of the time-backward FPE diffusion equation, as:

$$\frac{\partial f}{\partial t} = a_1 \frac{\partial^2 f}{\partial r^2} + a_2 \frac{\partial^2 f}{\partial \theta^2} + a_3 \frac{\partial^2 f}{\partial \varphi^2} + a_4 \frac{\partial^2 f}{\partial r \partial \theta} + a_5 \frac{\partial^2 f}{\partial r \partial \varphi} + a_6 \frac{\partial^2 f}{\partial \theta \partial \varphi} \tag{5}$$
$$+ a_7 \frac{\partial f}{\partial r} + a_8 \frac{\partial f}{\partial \theta} + a_9 \frac{\partial f}{\partial \varphi} + a_{10} \frac{\partial f}{\partial R}$$

with following coefficients given in the 3-D spherical coordinate system (r, θ, φ):

$$a_1 = K_{rr}^S, a_2 = \frac{K_{\theta\theta}^S}{r^2}, a_3 = \frac{K_{\varphi\varphi}^S}{r^2 sin^2\theta}, a_4 = \frac{2K_{r\theta}^S}{r}, a_5 = \frac{2K_{r\varphi}^S}{rsin\theta}, a_6 = \frac{2K_{\theta\varphi}^S}{r^2 sin\theta}$$

$$a_7 = \frac{2}{r}K_{rr}^S + \frac{\partial K_{rr}^S}{\partial r} + \frac{ctg\theta}{r}K_{\theta r}^S + \frac{1}{r}\frac{\partial K_{\theta r}^S}{\partial \theta} + \frac{1}{rsin\theta}\frac{\partial K_{\varphi r}^S}{\partial \varphi} - U - v_{d,r}$$

$$a_8 = \frac{K_{r\theta}^S}{r^2} + \frac{1}{r}\frac{\partial K_{r\theta}^S}{\partial r} + \frac{1}{r^2}\frac{\partial K_{\theta\theta}^S}{\partial \theta} + \frac{ctg\theta}{r^2}K_{\theta\theta}^S + \frac{1}{r^2 sin\theta}\frac{\partial K_{\varphi\theta}^S}{\partial \varphi} - \frac{1}{r}v_{d,\theta}$$

$$a_9 = \frac{K_{r\varphi}^S}{r^2 sin\theta} + \frac{1}{rsin\theta}\frac{\partial K_{r\varphi}^S}{\partial r} + \frac{1}{r^2 sin\theta}\frac{\partial K_{\theta\varphi}^S}{\partial \theta} + \frac{1}{r^2 sin^2\theta}\frac{\partial K_{\varphi\varphi}^S}{\partial \varphi} - \frac{1}{rsin\theta}v_{d,\varphi}$$

$$a_{10} = \frac{R}{3}\nabla \cdot U.$$

The drift velocity is calculated as: $v_{d,i} = \frac{\partial K_{ij}^{(A)}}{\partial x_j}$, where $K_{ij}^{(A)}$ is the antisymmetric part of the full 3D anisotropic diffusion tensor of the GCR particles $K_{ij} = K_{ij}^{(S)} + K_{ij}^{(A)}$ containing the symmetric $K_{ij}^{(S)}$ and antisymmetric $K_{ij}^{(A)}$ parts presented in [9].

The equivalent to Eq. 5 set of SDEs has the following form (the same can be found in [3,4]):

$$
\begin{aligned}
dr &= a_7 \cdot dt + [B \cdot dW]_r \\
d\theta &= a_8 \cdot dt + [B \cdot dW]_\theta \\
d\varphi &= a_9 \cdot dt + [B \cdot dW]_\varphi \\
dR &= a_{10} \cdot dt;
\end{aligned}
\quad
B_{i,j} =
\begin{bmatrix}
\sqrt{2a_1} & 0 & 0 \\
\frac{a_4}{\sqrt{2a_1}} & \sqrt{2a_2 - \frac{a_4^2}{2a_1}} & 0 \\
\frac{a_5}{\sqrt{2a_1}} & \frac{a_6 - \frac{a_4 a_5}{2a_1}}{B_{\theta\theta}} & \sqrt{2a_3 - B_{\varphi r}^2 - B_{\varphi\theta}^2}
\end{bmatrix}.
\tag{6}
$$

Solving Eq. 4 by the time-backward aproach pseudoparticles are iniciated from the point of interest (e.g. Earth orbit) and are traced backward in time until crossing the heliospheric boundary (in this paper this boundary is assumed at 100 AU). During theirs travel throughout the heliosphere, the pseudoparticles gain/lose their energy/rigidity proportionally to their travel time. In the result the value of the particle distribution function, $f(\boldsymbol{r}, R)$, for the starting point can be found as an average value of $f_{LIS}(R)$ for pseudoparticles characteristics at the entry positions $f(\boldsymbol{r}, R) = \frac{1}{N} \sum_{n=1}^{N} f_{LIS}(R)$ where $f_{LIS}(R) = 21.1/(T^{2.8} + 5.85 \cdot T^{1.58} + T^{0.26})$ is the cosmic ray local interstellar spectrum (LIS) [10] for rigidity R of the nth particle at the exit/entrance point, $T = \sqrt{R^2 + 0.938^2} - 0.938$ is the kinetic energy in GeV.

4   Numerical Solution of the Stochastic Differential Equations

The most commonly applied in the literature method of solution of the SDEs is the unconditionally stable Euler-Maruyama scheme with the convergence of order $\gamma = 0.5$. However, it involves the high statistical error, which can be reduced by increasing the number of simulated pseudoparticles or by applying the higher order extension of the SDEs solution in the Ito - Taylor series e.g. [11]. Therefore, we have solved the set of Eq. 4 also by the Milstein method and by the stochastic Runge-Kutta method. Both methods increase the computational complexity, but in return, the convergence order increases to $\gamma = 1$ in the case of the Milstein method and $\gamma = 1.5$ for stochastic Runge-Kutta method.

The numerical approximation of the solution of the Eq. 6 has a form:

$$
dr = \underbrace{\underbrace{\underbrace{a_7 \cdot dt + B_{rr} \cdot dW_r}_{Euler-Maruyama} + \frac{1}{2} B_{rr} \frac{\partial B_{rr}}{\partial r}(dW_r^2 - dt)}_{Milstein} + \Phi_1}_{stochastic\ Runge-Kutta}
\tag{7}
$$

$$
d\theta = \underbrace{\underbrace{\underbrace{a_8 \cdot dt + B_{\theta r} \cdot dW_r + B_{\theta\theta} \cdot dW_\theta}_{Euler-Maruyama} + \frac{B_{\theta r}}{2} \frac{\partial B_{\theta r}}{\partial \theta}(dW_r^2 - dt) + \frac{B_{\theta\theta}}{2} \frac{\partial B_{\theta\theta}}{\partial \theta}(dW_\theta^2 - dt)}_{Milstein} + \Phi_2}_{stochastic\ Runge-Kutta}
$$

$$
d\varphi = \underbrace{\underbrace{a_9 \cdot dt + B_{\varphi r} \cdot dW_r + B_{\varphi\theta} \cdot dW_\theta + B_{\varphi\varphi} \cdot dW_\varphi}_{Euler-Maruyama} + \Phi_3 + \Phi_4}_{stochastic\ Runge-Kutta}
$$
$$
\underbrace{\phantom{d\varphi = a_9 \cdot dt + B_{\varphi r} \cdot dW_r + B_{\varphi\theta} \cdot dW_\theta + B_{\varphi\varphi} \cdot dW_\varphi}}_{Milstein}
$$

$$
dR = a_{10} \cdot dt.
$$

The coefficients Φ_1, Φ_2, Φ_3 and Φ_4 have the form:

$\Phi_1 = B_{rr} \cdot dZ_r \frac{\partial a_7}{\partial r} + \frac{1}{2}(a_7 \frac{\partial a_7}{\partial r} + \frac{1}{2}B_{rr}^2 \frac{\partial^2 a_7}{\partial r^2})dt^2 + (a_7 \frac{\partial B_{rr}}{\partial r} + \frac{1}{2}B_{rr}^2 \frac{\partial^2 B_{rr}}{\partial r^2})(dW_r \cdot dt - dZ_r) + \frac{1}{2}B_{rr}(B_{rr}\frac{\partial^2 B_{rr}}{\partial r^2} + (\frac{\partial B_{rr}}{\partial r})^2)(\frac{1}{3}dW_r^2 - dt)dW_r$;

$\Phi_2 = B_{\theta r} \cdot dZ_r \frac{\partial a_8}{\partial \theta} + \frac{1}{2}(a_8 \frac{\partial a_8}{\partial \theta} + \frac{1}{2}B_{\theta r}^2 \frac{\partial^2 a_8}{\partial \theta^2})dt^2 + (a_8 \frac{\partial B_{\theta r}}{\partial \theta} + \frac{1}{2}B_{\theta r}^2 \frac{\partial^2 B_{\theta r}}{\partial \theta^2})(dW_r \cdot dt - dZ_r) + \frac{1}{2}B_{\theta r}(B_{\theta r}\frac{\partial^2 B_{\theta r}}{\partial \theta^2} + (\frac{\partial B_{\theta r}}{\partial \theta})^2)(\frac{1}{3}dW_r^2 - dt)dW_r + B_{\theta\theta} \cdot dZ_\theta \frac{\partial a_8}{\partial \theta} + \frac{1}{2}(a_8 \frac{\partial a_8}{\partial \theta} + \frac{1}{2}B_{\theta\theta}^2 \frac{\partial^2 a_8}{\partial \theta^2})dt^2 + (a_8\frac{\partial B_{\theta\theta}}{\partial \theta} + \frac{1}{2}B_{\theta\theta}^2 \frac{\partial^2 B_{\theta\theta}}{\partial \theta^2})(dW_\theta \cdot dt - dZ_\theta) + \frac{1}{2}B_{\theta\theta}(B_{\theta\theta}\frac{\partial^2 B_{\theta\theta}}{\partial \theta^2} + (\frac{\partial B_{\theta\theta}}{\partial \theta})^2)(\frac{1}{3}dW_\theta^2 - dt)dW_\theta$;

$\Phi_3 = \frac{1}{2}B_{\varphi r}\frac{\partial B_{\varphi r}}{\partial \varphi}(dW_r^2 - dt) + \frac{1}{2}B_{\varphi\theta}\frac{\partial B_{\varphi\theta}}{\partial \varphi}(dW_\theta^2 - dt) + \frac{1}{2}B_{\varphi\varphi}\frac{\partial B_{\varphi\varphi}}{\partial \varphi}(dW_\varphi^2 - dt)$;

$\Phi_4 = B_{\varphi r} \cdot dZ_r \frac{\partial a_9}{\partial \varphi} + \frac{1}{2}(a_9 \frac{\partial a_9}{\partial \varphi} + \frac{1}{2}B_{\varphi r}^2 \frac{\partial^2 a_9}{\partial \varphi^2})dt^2 + (a_9 \frac{\partial B_{\varphi r}}{\partial \varphi} + \frac{1}{2}B_{\varphi r}^2 \frac{\partial^2 B_{\varphi r}}{\partial \varphi^2})(dW_r \cdot dt - dZ_r) + \frac{1}{2}B_{\varphi r}(B_{\varphi r}\frac{\partial^2 B_{\varphi r}}{\partial \varphi^2} + (\frac{\partial B_{\varphi r}}{\partial \varphi})^2)(\frac{1}{3}dW_r^2 - dt)dW_r + B_{\varphi\theta} \cdot dZ_\theta \frac{\partial a_9}{\partial \varphi} + \frac{1}{2}(a_9 \frac{\partial a_9}{\partial \varphi} + \frac{1}{2}B_{\varphi\theta}^2 \frac{\partial^2 a_9}{\partial \varphi^2})dt^2 + (a_9\frac{\partial B_{\varphi\theta}}{\partial \varphi} + \frac{1}{2}B_{\varphi\theta}^2 \frac{\partial^2 B_{\varphi\theta}}{\partial \varphi^2})(dW_\theta \cdot dt - dZ_\theta) + \frac{1}{2}B_{\varphi\theta}(B_{\varphi\theta}\frac{\partial^2 B_{\varphi\theta}}{\partial \varphi^2} + (\frac{\partial B_{\varphi\theta}}{\partial \varphi})^2)(\frac{1}{3}dW_\theta^2 - dt)dW_\theta + B_{\varphi\varphi} \cdot dZ_\varphi \frac{\partial a_9}{\partial \varphi} + \frac{1}{2}(a_9 \frac{\partial a_9}{\partial \varphi} + \frac{1}{2}B_{\varphi\varphi}^2 \frac{\partial^2 a_9}{\partial \varphi^2})dt^2 + (a_9\frac{\partial B_{\varphi\varphi}}{\partial \varphi} + \frac{1}{2}B_{\varphi\varphi}^2 \frac{\partial^2 B_{\varphi\varphi}}{\partial \varphi^2})(dW_\varphi \cdot dt - dZ_\varphi) + \frac{1}{2}B_{\varphi\varphi}(B_{\varphi\varphi}\frac{\partial^2 B_{\varphi\varphi}}{\partial \varphi^2} + (\frac{\partial B_{\varphi\varphi}}{\partial \varphi})^2)(\frac{1}{3}dW_\varphi^2 - dt)dW_\varphi$.

The numerical code for the numerical solution of the set Eq. 6 is realized in MatLab environment. The code is easy to parallelize versus the number od simulated pseudoparticles using Matlab Parallel Toolbox. This approach fallouts from the assumption that any random process is independent of the other realization, accordingly each pseudoparticle's trajectory is independent on another. We performed the simulations applying all three numerical methods given by Eq. 7. The pseudoparticles were initialized in the point representing the Earth's orbit ($r = 1AU$, $\theta = 90°$, $\varphi = 250°$) and traced backward in time until crossing the heliosphere boundary assumed at 100 AU. To solve Eq. 7 in spherical coordinates system we used the boundary conditions given in [3].

Figure 1 presents the trajectories of the pseudoparticles with the rigidity of 10 GV for the $A > 0$ solar magnetic cycle with respect to all coordinates. For the reliable comparison, all simulations were based on the same Wiener process. One can see that there arises some subtle difference: the trajectories of pseudoparticles vs. the heliolongitude are the widest for Euler-Maruyama and more narrow for the stochastic Runge-Kutta method. At the same time, the opposite distribution is observed vs. the heliolatitude.

In the case, when we know the analytical solution of the equation that we want to solve numerically, it is easy to compare the efficiency of each applied numerical method. However, in the case of GCR particles transport in the heliosphere, we do not have such a possibility. Implementation of the higher order numerical method should result in the increase of the accuracy of the solution. Particularly, the smaller number of simulated particles for higher order method should give better or comparable results to those provided by a weaker method employing more pseudoparticles. Thus, we have analyzed the changes of the standard deviations of the particle distribution function, $f(r, R)$ calculated for the Euler-Maruyama, Milstein, and stochastic Runge-Kutta methods vs. the number of simulated pseudoparticles with the rigidity of 10 GV. We have also

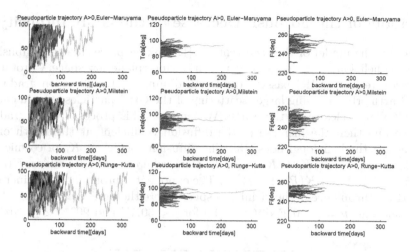

Fig. 1. Trajectories of the 500 pseudoparticles with the rigidity of 10 GV for the A > 0 solar magnetic cycle obtained by applying the Euler-Maruyama, Milstein and stochastic Runge-Kutta methods. The specific colors highlight the trajectories of the sample pseudoparticles, based on the same Wiener process, traced backward in time from the heliosphere boundary until they reach the position $r = 1AU$, $\theta = 90°$, $\varphi = 250°$.

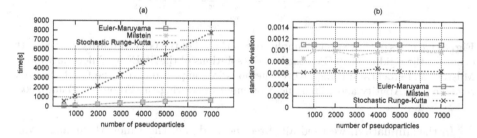

Fig. 2. (a) The average time required to compute the particle distribution function $f(\mathbf{r}, R)$ for a single space point; and (b) the standard deviations of the particle distribution function $f(\mathbf{r}, R)$ calculated for the Euler-Maruyama, Milstein and stochastic Runge-Kutta methods vs. a number of simulated pseudoparticles with the rigidity of 10 GV.

analyzed the computational time required by each numerical method. Figure 2 shows that the standard deviation of the numerical solution of the Eq. 7 is the largest for Euler-Maruyama method and the lowest for stochastic Runge-Kutta. So, by applying the higher order method (especially stochastic Runge-Kutta), the statistical accuracy of the numerical solution is significantly increased. From the other side, the computational time required by the stochastic Runge-Kutta method is nine times greater than for other two methods.

5   Model of the Forbush Decrease of the GCR Intensity

We present the model of the recurrent Fd [12] taking place due to established corotating heliolongitudinal disturbances in the interplanetary space. Corotating interaction regions (CIR) passing the Earth diminishes the diffusion gradually at the Earth orbit, causing larger scattering of the GCR particles, and in effect, fewer GCR particles reach the Earth. We simulate this process by the gradual decrease and then the increase of the diffusion coefficient at the Earth orbit with respect the heliolongitudes [13]. The diffusion coefficient $K_\parallel$ of cosmic ray particles was taken as: $K_\parallel = K_0 \cdot K(r) \cdot K(R, \nu)$, where $K_0 = 10^{21}\,\mathrm{cm}^2/\mathrm{s}$, $K(r) = 1 + 0.5 \cdot (r/1AU)$ and $K(R, \nu) = R^{2-\nu}$. The exponent ν pronounces the increase of the HMF turbulence in the vicinity of space where the Fd is created (e.g. [14]), and is taken as: $\nu = 1 + 0.25 sin(\varphi - 90°)$ for $r < 30AU$ and $90° \leq \varphi \leq 270°$.

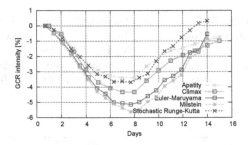

Fig. 3. Changes of the expected amplitudes of the Fd of the GCR intensity at the Earth orbit, for the rigidity of 10 GV based on the solutions of the SDEs by Euler-Maruyama, Milstein and stochastic Runge-Kutta methods in comparison with the GCR intensity registered by Apatity and Climax neutron monitors during the Fd on 16–30 Jun 2003.

In the simulation, we used all three methods i.e. the Euler-Maruyama, Milstein, and stochastic Runge-Kutta method. For comparison, all model assumptions were the same and all methods were based on the same Wiener process. The expected relative changes in the GCR intensity for the rigidity of 10 GV during the simulated Fd in comparison with the profiles of the daily GCR intensities recorded by the neutron monitors with different cut-off rigidities on 16–30 Jun 2003 presents Fig. 3. One can see that the proposed models are in a good coincidence with the experimental data. However, there are the some differences. The Euler-Maruyama and Milstein methods results are very similar, but the amplitude of the Fd obtained by the stochastic Runge-Kutta method is a bit smaller and the recovery time is shorter. This type of profile is more expected as far in the model we did not assume the prolonged recovery of the diffusion coefficient.

6   Summary

We presented the numerical solution of the PTE using a numerical solution of the set of SDEs driven by a Wiener process with the strong order Euler-Maruyama,

Milstein, and stochastic Runge-Kutta methods. We showed that application of the higher order methods (especially stochastic Runge-Kutta) significantly increased the statistical accuracy of the numerical solution. However, choosing between these methods we have to take into account the substantial increase in the computational time. The presented model of the Forbush decrease of the GCR intensity is in a good agreement with the experimental data.

Acknowledgments. This work is supported by The Polish National Science Centre grant awarded by decision number DEC-2012/07/D/ST6/02488.

We thank the principal investigators of Apatity and Climax neutron monitors for the ability to use their data.

References

1. Zhang, M.: Astrophys. J. **513**, 409–420 (1999)
2. Gervasi, M., Rancoita, P.G., Usoskin, I.G., Kovaltsov, G.A.: Nucl. Phys. B Proc. Suppl. **78**, 26–31 (1999)
3. Wawrzynczak, A., Modzelewska, R., Gil, A.: J. Phys.: Conf. Ser. **574**, 012078 (2015)
4. Kopp, A., Busching, I., Strauss, R.D., Potgieter, M.S.: Comput. Phys. Commun. **183**, 530–542 (2012)
5. Wawrzynczak, A., Modzelewska, R., Kluczek, M.: J. Phys.: Conf. Ser. **633**, 1742–6596 (2015)
6. Parker, E.N.: Planet. Space Sci. **13**, 9–49 (1965)
7. Moraal, H.: Space Sci. Rev. **176**, 299–319 (2013)
8. Gardiner, C.W.: Handbook of Stochastic Methods. For Physics, Chemistry and the Natural Sciences. Springer Series in Synergetics. Springer, Berlin (2009)
9. Alania, M.V.: Acta Phys. Pol. B **33**(4), 1149–1166 (2002)
10. Webber, W.R., Lockwood, J.A.: J. Geophys. Res. **106**, 29323–29332 (2001)
11. Kloeden, P.E., Platen, E., Schurz, H.: Numerical solution of SDE through computer experiments **24**, 140–146 (2003)
12. Forbush, S.E.: Phys. Rev. **51**, 1108–1109 (1937)
13. Wawrzynczak, A., Alania, M.V.: Adv. Space Res. **41**(2), 325–334 (2008)
14. Wawrzynczak, A., Alania, M.V.: Adv. Space Res. **45**, 622–631 (2010)

Numerical Method for Solving an Inverse Boundary Problem with Unknown Initial Conditions for Parabolic PDE Using Discrete Regularization

Natalia M. Yaparova[✉]

Department of Computational Mathematics and High-Performance Computing,
South Ural State University (National Research University),
pr. Lenin 76, 454080 Chelyabinsk, Russia
ddjy@math.susu.ac.ru

Abstract. We consider an inverse boundary values problem for parabolic PDE with unknown initial conditions. In this problem both Dirichlet and Neumann boundary conditions are given on a part of the boundary and it is required to determine the corresponding function on the remaining part of the boundary. To solve this problem, the numerical method based on finite difference schemes and regularization technique is proposed. The computing scheme involves solving the equation for each spatial step that allows to obtain the numerical solution in internal points of the domain and on the boundary. We prove a conditional stability of the method. The reliability and the efficiency of the method were confirmed by computational results.

Keywords: Heat conduction problem · Inverse problem · Regularization method · Computational scheme · Conditional stability

1 Introduction

Inverse boundary value problems for parabolic PDEs with unknown initial conditions arise in a wide variety of physical and engineering applications such as, material science, thermal monitoring, heat and mass transfer, diffusion processes, combustion processes in the operating engines. As some applied examples, we refer to heat conduction problems, which were considered in, e.g. [1,2].

The numerical methods for solving inverse problems for PDEs are widely developed, see e.g. [2–6]. Existing methods as a rule, involve two parts. First, the unknown boundary function is calculated via different regularization methods. Next, the function that satisfies the partial differential equation in internal points of the domain is determined. The essential feature of these methods is that their applications are impossible without initial condition. On other hand, quite often it is impossible to formulate the initial conditions. For example, we have unknown conditions in situation when we can't measure the initial temperature in the entire medium or inside object. It leads to inapplicability of the existing

© Springer International Publishing AG 2017
I. Dimov et al. (Eds.): NAA 2016, LNCS 10187, pp. 752–759, 2017.
DOI: 10.1007/978-3-319-57099-0_87

methods. Thus, it is required to create numerical method for solving problem with unknown initial conditions.

The development of numerical method for solving the inverse problems with unknown initial conditions is of great interest. For example, Lavrentiev et al. in [7] proved the existence of solution to inverse boundary problem with unknown initial conditions and its uniqueness in some subdomain. Calderon in [8] formulated the some first results to solve the inverse problem of electrical impedance tomography. Uhlmann [9] has generalized some of the results on this inverse boundary problems in multidimensional case. In regard with the Cauchy problem for elliptic equations we also mention the works [10,11]. The inverse boundary problems for parabolic PDEs with specific initial conditions are consider, for example in [12,13].

In this paper, we propose the method for numerical solving a multidimensional inverse boundary problem with unknown initial conditions. Our method is based on the finite-difference explicit scheme and discrete regularization method. The application of regularization technique ensures the steady of the scheme in some subdomain and allows simplify the computational procedure significantly. The reliability of the method were verified by comparing the numerical results with the exact solution functions. The computational results for the some test data are presented in this paper.

2   Setting of the Problem

Let $Q \subset R^n$ be bounded domain with smooth boundary and $Q_T = Q \times (0,T)$ for $T > 0$. We consider the equation

$$\frac{\partial u}{\partial t} = \sum_{r,s=1}^{n} a_{rs}(x,t)\frac{\partial^2 u}{\partial x_r \partial x_s} + \sum_{s=1}^{n} b_s(x,t)\frac{\partial u}{\partial x_s} + c(x,t)u + f(x,t), \quad (x,t) \in Q_T, \quad (1)$$

where $x = (x_1, x_2, \ldots, x_n) \in Q$ and $a_{rs}(x,t), b_s(x,t) \in H^{2,1+\beta}(\overline{Q_T})$, with $\beta \in (0,1)$, $r,s = \overline{1,n}$ and $c(x,t), f(x,t) \in C(\overline{Q_T})$. Assume that the coefficients $a_{rs}(x,t)$ satisfy the conditions $\eta_1|\xi|^2 \leq \sum_{r,s=1}^{n} a_{rs}(x,t)\xi_r\xi_s \leq \eta_2|\xi|^2$ for any $\xi = (\xi_1, \xi_2, \ldots, \xi_n)$ with fixed $\eta_1, \eta_2 > 0$.

Given functions $p(x,t), q(x,t) \in H^{2,1+\beta}(\overline{Q_T})$ on the part of boundary $\Gamma \subset \partial Q$ the induced function $u(x,t)$ solves the Eq. (1) with boundary conditions

$$u(x,t)\,|_\Gamma = p(x,t), \quad \frac{\partial u}{\partial \nu}\,|_\Gamma = q(x,t), \tag{2}$$

where ν denotes the unit outer normal to Γ. The inverse problem is to determine $u(x,t)$ knowing $p(x,t), q(x,t)$ and then to find the boundary function $\varphi(x,t)$

$$u(x,t)\,|_{\partial Q \backslash \Gamma} = \varphi(x,t). \tag{3}$$

The existence of the solution $u(x,t) \in H^{2,1+\beta}(Q_T) \cap H^{1,\beta}(\overline{Q_T})$, $\beta \in (0,1)$ to problem (1)–(3) for some p_0, g_0 and its uniqueness in some domain $D \subset \overline{Q_T}$ have been proved in [7]. However, instead of p_0, g_0 we are given some approximations p_δ, q_δ and an error level $\delta > 0$ such that $\max\{\|p_\delta - p_0\|, \|q_\delta - q_0\|\} \leqslant \delta$. If the physical meaning of the problem allows to assume that function $u(x, t)$ satisfying to (1)–(3) is smooth function and there exist constants $\Phi, \gamma, R > 0$ such that

$$\max_{x \in Q} |u(x, t)| \leq \Phi e^{\gamma t}, \max\left\{\max_{(x,t) \in \overline{Q_T}} |\partial_t^2 u|, \max_{(x,t) \in \overline{Q_T}} |\partial_x^3 u|\right\} \leq R. \tag{4}$$

for any $(x, y, t) \in \overline{Q_T}$ we can use the computational scheme based on discrete regularization method (DRM).

3   Discrete Regularization Method

We consider a computational scheme only for PDE with two spatial variables. Nevertheless, it is easy to extend our results to 1-d and 3-d cases and for a more general parabolic PDE of the second order.

Let $Q = \{(x, y) : (x, y) \in (0, X) \times (0, Y)\}$. We take the following representation to the problem (1)–(3)

$$u_t = a(x, y, t)(u_{xx} + u_{yy}) + f(x, y, t), \quad (x, y, t) \in Q_T, \tag{5}$$

$$u(x, 0, t) = p(x, t), u_y(x, 0, t) = q(x, t), \quad (x, t) \in M_x, \tag{6}$$

$$u(0, y, t) = g(y, t), u(X, y, t) = h(y, t) \quad (y, t) \in M_y, \tag{7}$$

where $M_x = [0, X] \times [0, T]$ and $M_y = [0, Y] \times [0, T]$. In additional, instead of the exact p, q, g and h we know the noised values $p_\delta, q_\delta, g_\delta$ and h_δ and error label δ such that $\max\{\|p - p_\delta\|, \|q - q_\delta\|, \|g - g_\delta\|, \|h - h_\delta\|\} \leq \delta$. In this problem, it is required to find the function $u_\delta^\alpha(x, y, t)$ satisfying (5) and (6) and then to obtain the boundary functions $u_\delta^\alpha(x, Y, t) = \varphi_\delta(x, t)$ and $u_\delta^\alpha(X, Y, t) = \phi_\delta(t)$. We introduce a finite difference grid G in $\overline{Q_T}$, such that:

$$G = \begin{cases} (x_i, y_j, t_k) : x = (i-1)h_x, y = (j-1)h_y, t = (k-1)\tau, \\ h_x = X/N_x; h_y = Y/N_y; \tau = T/N_t; \\ i = \overline{1, N_x + 1}; j = \overline{1, N_y + 1}; k = \overline{1, N_t + 1}. \end{cases}$$

Let $V_h = \{v(x_i, y_j, t_k)\} = \{v_{i,j,k}\}$ be the set of discrete functions defined on G. Following Samarskii [14], we approximate the partial derivatives in each point of G as

$$v_{xx}^{i,j,k} = \frac{v_{i+1,j,k} - 2v_{i,j,k} + v_{i-1,j,k}}{h_x^2}, \quad v_{yy}^{i,j,k} = \frac{v_{i,j+1,k} - 2v_{i,j,k} + v_{i,j-1,k}}{h_y^2},$$

$$v_t^{i,j,k} = \frac{v_{i,j,k+1} - v_{i,j,k}}{\tau}. \tag{8}$$

Next, using the finite difference analog (8), we replace the differential Eq. (5) on a finite-difference equations:

$$v_t^{i,j,k} = a_{i,j,k}\left(v_{xx}^{i,j,k} + v_{yy}^{i,j,k}\right) + f_{i,j,k}, \quad k = \overline{1, N_t}, \tag{9}$$

$$v_t^{i,j,N_t+1} = \vartheta a_{i,j,N_t+1} \left[v_{xx}^{i,j,N_t+1} + v_{yy}^{i,j,N_t+1} \right]$$
$$+ (1-\vartheta)a_{i,j,N_t} \left[v_{xx}^{i,j,N_t} + v_{yy}^{i,j,N_t} \right] + f_{i,j,N_t+1},$$

where $a(x_i, y_j, t_k) = a_{i,j,k}$, $f(x_i, y_j, t_k) = f_{i,j,k}$ and $\vartheta \in (0,1)$.

Setting $Av = v_t^{i,j,k} - a_{i,j,k} \left(v_{xx}^{i,j,k} + v_{yy}^{i,j,k} \right)$, we represent the Eq. (9) as $Av = f$. It is well known that the explicit computational scheme based on (9) is unsteady. Nevertheless, we can improve it. To do this, we use the discrete regularization method proposed in [15].

The basic idea of the DRM is that we enhance the Eq. (9) via additional stabilizing functional and then we obtain the approximate solution u_δ^α to the problem (1)–(3) by moving from Γ on spatial variable. Thus, according this approach, the Eq. (9) reduced to following

$$Av + \alpha v = f, \tag{10}$$

where α is regularization parameter. Further, solving it, we obtain the values $v_{i,j+1,k}$. The purpose of the our further analytical effort is to investigate the stability question of the numerical procedure based on (10).

4 Stability Analysis

The important aspect of numerical solving to the inverse problem with unknown initial conditions lies in the fact, that it is impossible to obtain unique solution in $\overline{Q_T}$ and different uncontrollable errors inevitably occur. This feature to the problem have been proved in [7]. According this result, we can obtain exact error estimates only in $D \subset \overline{Q_T}$, such that $D = \left\{ (x,y,t) : 0 < \frac{x}{\Lambda} < \left(1 - \frac{y^2}{2(\mu\Lambda)^2} - \frac{t^2}{2(\rho T)^2} \right) \right\}$ where $\mu, \rho \in (0,1)$ and $\Lambda^2 = \frac{x^2}{X^2} + \frac{y^2}{Y^2}$. Therefore, we evaluate the stability of the regularized solutions with some assumptions.

The stability of DRM is examined proceeding as it is always done in the regularization theory, see, e.g. [5]. Namely, we define parameters W, m, F and functions w_j as follows:

$$W = \max_{i,k} |q_{i,k}|, \quad m = \min_{i,j,k} |a_{i,j,k}|, \quad F = \max_{i,j,k} |f_{i,j,k}|,$$
$$w_j = \max_{i,k} |v_{i,j+1,k} - v_{i,j,k}|,$$

where $v_{i,j,k}$ satisfy (10) with conditions (6) and (7). Taking into account the condition (4) we obtain

$$w_{i+1} \leq w_i + \left[4\frac{h_y^2}{h_x^2} + 2\frac{h_y^2}{m\tau} + \frac{\alpha h_y^2}{m} \right] \Phi e^{\beta t} + \frac{h_y^2}{m} F \tag{11}$$

Define parameter C as $C = 4\frac{h_y^2}{h_x^2} + 2\frac{h_y^2}{m\tau} + \frac{\alpha h_y^2}{m}$. From (11) it is follows that

$$w_{N_y+1} \leq (W + h_y)\,\delta + (h_y Y - 1)\left(C\Phi e^{\beta t} + \frac{h_y^2}{m}F\right). \qquad (12)$$

By linking the regularization parameter α and spatial steps h_y h_x with τ and noise level δ, we obtain that $C < 1/(Y\tau)$. Next, using the estimate (12) and taking into consideration the relationship between discretization steps, we get

$$w_{N_y+1} \leq \frac{\sqrt{m}}{2\sqrt{Y}}\delta^2 + \left(W + \frac{F}{4} + \Phi e^{\beta t}\right)\delta. \qquad (13)$$

From the relations (12) and (13) it follows, that the proposed scheme is conditionally stable.

5   Computational Results

In order to evaluate the reliability and efficiency of the proposed scheme, computational experiments were carried out. In the experiment we consider $\overline{Q_T} = [0, 1.5] \times [0, 1] \times [0, 2]$. The accuracy of the presented method was verified by comparing the numerical results for calculating the boundary functions $\varphi_\delta(x, t)$ with the test functions $u(x, Y, t) = \varphi_0(x, t)$. In order to make a comparative analysis numerical solutions to the problem (5)–(7) with exact solution, we solve the following direct problem on the first stage of experiment.

$$u_t = a(x, y, t)\,(u_{xx} + u_{yy}) + f(x, y, t), \quad (x, y, t) \in Q_T, \qquad (14)$$

$$u(x, 0, t) = p(x, t),\, u(x, Y, t) = \varphi(x, t), \quad (x, t) \in M_x, \qquad (15)$$

$$u(0, y, t) = g(y, t),\, u(X, y, t) = h(y, t), \quad (y, t) \in M_y, \qquad (16)$$

$$u(x, y, 0) = v(x, y), \quad (x, y) \in \overline{Q}. \qquad (17)$$

where functions p, φ, g, h, v are known. We obtain the exact solution $u_0(x, y, t)$ which further we use as test function. Next, the values of $p_\delta, q_\delta g_\delta, h_\delta$ are calculated by formulas:

$$p_\delta = u_0(x_i, y_1, t_k) + erp^{i,k}, \quad q_\delta = \frac{u_0(x_i, y_2, t_k) - u_0(x_i, y_1, t_k)}{y_2 - y_1} + erq^{i,k},$$

$$g_\delta = u_0(x_1, y_j, t_k) + erg^{j,k}, \quad h_\delta = u_0(X, y_j, t_k) + erh^{j,k},$$

where values $erp^{i,k}, erq^{i,k}, erg^{j,k}, erh^{j,k}$ are simulated as evenly distributed random variable in $[-\delta, \delta]$ at each respective point. Further, we choose the discretization steps in keeping with the conditions that guarantee the stability of the method. Then, the numerical solution to the problem (5)–(7) is calculated via proposed method.

The computational results for some numerical examples are illustrated in Figs. 1, 2 and 3. The same notations were used in all the figures. The one-dimensional figures illustrate the graphs of calculated boundary function $u_\delta^\alpha = \phi_\delta(t)$ and exact function $u_0 = \phi_0(t)$ in points $(0.75, 1, t)$. We considered these graphs of the boundary function as illustrations because the points $(0.75, 1, t)$ are most distant from points with known functions. Thus, it reasonable to expect, that the numerical solution has a highest error level in $(0.75, 1, t)$, $t \in [0, T]$.

The notation u_0 corresponds to the exact function $u_0(0.75, 1, t)$ and the designation u_δ corresponds to the numerical solution $u_\delta^\alpha(0.75, 1, t)$ to the problem (5)–(7) obtained via discrete regularization method. The noise level is denoted as δ.

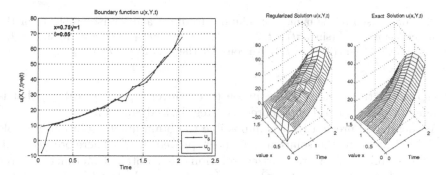

Fig. 1. Comparison of the numerical solutions to direct and inverse problems for the test functions $p = 8.66e^t \cos(x - \pi/12)$, $\varphi = 5e^t \cos(x - \pi/12)$, $g = e^t \cos(y - \pi/6)$, $h = 0.33e^t \cos(y - \pi/6)$. We consider them as *Example 1*.

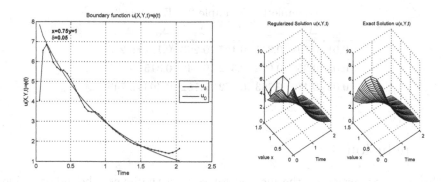

Fig. 2. Comparison of the numerical solutions to direct and inverse problems for the test functions $p = 8.66e^{-t} \cos(x - \pi/12)$, $\varphi = 5e^{-t} \cos(x - \pi/12)$, $g = e^{-t} \cos(y - \pi/6)$ and $h = 0.33e^{-t} \cos(y - \pi/6)$. We consider them as *Example 2*.

The two-dimensional figures illustrate the surfaces that corresponded to the test functions $\varphi_0(x, t)$ and to the numerical solutions $\varphi_\delta(x, t)$ to inverse problem. The designation "Exact Solution" used for indicating the surface of the solution to the direct problem (14)–(17) for the test boundary functions. The surface

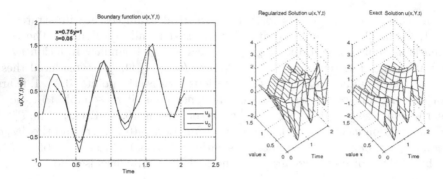

Fig. 3. Comparison of the numerical solutions to direct and inverse problems for the test functions $p = e^{-x} (\sin(3\pi t) + t/2)$, $\varphi = e^{-x} (\sin(3\pi t) + t/2)$, $g = \sin(3\pi t) + t/2$ and $h = 1/3e^{y-1.5} (\sin(3\pi t) + t/2)$. We consider them as *Example 3*.

corresponding to the numerical solution to the inverse problem (5)–(7) that is obtained in points (x, Y, t) is denoted as "Regularized Solution".

As mentioned previously, it is impossible to obtain unique solution in whole $\overline{Q_T}$. This feature to the solution have been proved in [7]. Therefore, we evaluate the stability of the regularized solutions and obtain experimental error estimates of deviations in $D \subset \overline{Q_T}$. To do this, we calculated both quantities $\Delta_\phi = \|\phi_\delta - \phi_0\|_{C(\varepsilon, T-\varepsilon)}$ and $\Delta_\varphi = \|\varphi_\delta(x, t) - \varphi_0(x, t)\|_{C(D)}$. In Table 1, we give some average values of these quantities.

Table 1. Experimental estimates

δ	Example 1		Example 2		Example 3	
	Δ_ϕ	Δ_φ	Δ_ϕ	Δ_φ	Δ_ϕ	Δ_φ
0.01	0.191	0.341	0.187	0.293	0.181	0.243
0.03	0.358	0.603	0.342	0.649	0.315	0.597
0.05	0.470	0.793	0.482	0.782	0.403	0.743

6   Conclusion

In this work, we proposed the method for solving the inverse boundary values problem with the unknown initial conditions based on the Discrete Regularization Method and the finite-difference schemes. The application of explicit scheme allows to simplify the computational procedure and to reduce the impact of unknown initial conditions on the solvability of the problem. A significant feature of the approach is the possibility to obtain the numerical solution inside of the domain and on the boundary. We prove the conditional stability of the method. The results of the experiments show the reasonable stability of the offered scheme in respective subset of the domain.

References

1. Alifanov, O.M.: Inverse Heat Transfer Problems. Springer, New York (2011)
2. Samarskii, A.A., Vabishchevich, P.N.: Numerical Methods for Solving Inverse Problems of Mathematical Physics. Walter de Gruyter, Berlin (2007)
3. Kabanikhin, S.I.: Inverse and Ill-Posed Problems. Theory and Applications. Walter de Gruyter, Berlin (2011)
4. Romanov, V.G., Kabanikhin, S.I., Bukhgeim, A.L.: Ill-Posed and Inverse Problems. VSP, Utrecht (2002)
5. Tikhonov, A.N., Goncharsky, A.V., Stepanov, V.V., Yagola, A.G.: Numerical Methods for the Solution of Ill-Posed Problems. Kluwer, London (1995)
6. Dorofeev, K.Y., Nikolaeva, N.N., Titarenko, V.N., Yagola, A.G.: New approaches to error estimation to Ill-posed problems with application to inverse problems of heat conductivity. J. Inverse Ill-Posed Probl. **10**(2), 155–169 (2002)
7. Lavrentiev, M.M., Romanov, V.G., Shishatskii, S.P.: Ill-Posed Problems of Mathematical Physics and Analysis. AMS, Providence (1986)
8. Calderon, A.: On an inverse boundary value problem, In: Seminar on Numerical Analysis and Its Applications to Continuum Physics (Rio de Janeiro, 1980), pp. 65–73. Sociedade Brasileira de Pesquisa em Materiais, Rio de Janeiro (1980)
9. Uhlmann, G.: Inverse boundary problems in two dimensions. In: Haroske, D., Tunst, T., Schmeisser, H.-J. (eds.) Function Spaces, Differential Operators and Nonlinear Analysis - The Hans Triebel Anniversary Volume, pp. 183–203. Birkhaeuser, Boston (2003)
10. Avdonin, S., Kozlov, V., Maxwell, D., Truffer, M.: Iterative methods for solving a nonlinear boundary inverse problem in glaciology. J. Inverse Ill-Posed Probl. **17**, 239–258 (2009)
11. Ben Belgacem, F., Du, D.T., Jelassi, F.: Local convergence of the Lavrentiev method for the Cauchy problem via a Carleman inequality. J. Sci. Comput. **53**(2), 320–341 (2012)
12. Glazyrina, O.V., Pavlova, M.F.: Study of the convergence of the finite-element method for parabolic equations with a nonlinear nonlocal spatial operator. Differ. Eq. **51**(7), 872–885 (2015)
13. Hào, D.N., Thanh, P.X., Lesnic, D., Johansson, T.B.: A boundary element method for a multi-dimensional inverse heat conduction problem. Int. J. Comput. Math. **89**(11), 1540–1554 (2012)
14. Samarskii, A.A.: The Theory of Difference Schemes. Marcel Dekker Inc., New York (2001)
15. Yaparova, N.M.: Method for solving some multidimensional inverse boundary value problems for parabolic PDEs without conditions. Bull. South Ural State Univ. Ser.: Comput. Technol. Autom. Control Radioelectron. **3**, 97–108 (2015)

Two-Dimensional Interpolation of Functions with Large Gradients in Boundary Layers

Alexander Zadorin[✉]

Sobolev Mathematics Institute SB RAS,
Omsk Branch, Pevtsova 13, Omsk 644043, Russia
zadorin@ofim.oscsbras.ru

Abstract. Question of two-dimensional interpolation of functions with large gradients in the boundary layers is considered. The problem is that an application of polynomial interpolation on an uniform mesh to functions with large gradients leads to significant errors. We consider two approaches for increase of accuracy of interpolation: a fitting of the interpolation formula to a boundary layer component and the application of polynomial interpolation on Shishkin mesh. Numerical results are discussed.

Keywords: Function of two variables · Boundary layer · Polynomial interpolation · Shishkin mesh · Nonpolynomial interpolation

1 Introduction

It is known that an application of the polynomial interpolation to functions with large gradients in the boundary layers leads to the errors of the order $O(1)$ [1,2]. Such function corresponds to the solution of elliptic problem with boundary layers. We consider two approaches to the interpolation of a function of two variables with large gradients in the boundary layers. Approaches consist in the fitting of interpolation formula to boundary layer components and in the using of polynomial interpolation on Shishkin mesh [3].

Through the paper C and C_j denote generic positive constants independent of ε and mesh size.

2 Interpolation Formula Exact on the Boundary Layer Components

Let us a function $u(x, y)$ be smooth enough with the following representation:

$$u(x, y) = p(x, y) + d_1(y)\Phi(x) + d_2(x)\Theta(y) + d_3\Phi(x)\Theta(y), \tag{1}$$

where $(x, y) \in \bar{\Omega}$, $\bar{\Omega} = [0, 1]^2$. We suppose that functions p, d_1, d_2 have bounded derivatives up to some order and are not known, d_3 is not given. Boundary layer

© Springer International Publishing AG 2017
I. Dimov et al. (Eds.): NAA 2016, LNCS 10187, pp. 760–768, 2017.
DOI: 10.1007/978-3-319-57099-0_88

functions $\Phi(x), \Theta(y)$ are known, but have large gradients. The representation (1) takes a place for the solution of a singular perturbed elliptic problem:

$$\varepsilon u_{xx} + \varepsilon u_{yy} + a(x)u_x + b(y)u_y - c(x,y)u = f(x,y), \ (x,y) \in \Omega;$$
$$u(x,y) = g(x,y), \ (x,y) \in \Gamma, \tag{2}$$

where $\Gamma = \overline{\Omega} \backslash \Omega$, functions a, b, c, f, g are smooth enough,

$$a(x) \geq \alpha > 0, \quad b(y) \geq \beta > 0, \quad c(x,y) \geq 0, \quad \varepsilon \in (0,1].$$

According to [4], the solution of the problem (2) can be written in the form (1) with

$$\Phi(x) = \exp(-a(0)\varepsilon^{-1}x), \quad \Theta(y) = \exp(-b(0)\varepsilon^{-1}y).$$

We will notice that derivatives of functions $\Phi(x), \Theta(y)$ are not ε-uniformly bounded.

We are going to construct the interpolation formula for the function of the form (1), which is exact on boundary layer components $\Phi(x), \Theta(y)$. For this purpose, we use such formula, constructed in [5] for one-dimensional case.

Let us $\bar{\Delta}^h$ be the uniform grid of the region $\bar{\Omega}$:

$$\bar{\Delta}^h = \{(x_i, y_j), \ x_i = ih_1, y_j = jh_2, \ i = 0,1,\ldots,N_1, \ j = 0,1,\ldots,N_2\},$$

where $h_1 = 1/N_1$, $h_2 = 1/N_2$.

We suppose that region $\bar{\Omega}$ is divided into not crossed cells $\Omega_{i,j}$:

$$\bar{\Omega} = \bigcup_{\substack{i=0,\,k_1-1,\ldots,N_1-k_1+1 \\ j=0,\,k_2-1,\ldots,N_2-k_2+1}} \Omega_{i,j}, \quad \Omega_{i,j} = [x_i, x_{i+k_1-1}] \times [y_j, y_{j+k_2-1}]. \tag{3}$$

Every cell $\Omega_{i,j}$ contains k_1 nodes on x and k_2 nodes on y. We construct the interpolation formula in every cell $\Omega_{i,j}$. We use one-dimensional formula from [5] and obtain two-dimensional interpolation formula:

$$L_x(u,x,y) = L_{k_1-1}(u,x,y) + \frac{[x_i, x_{i+1}, \ldots, x_{i+k_1-1}]u}{[x_i, x_{i+1}, \ldots, x_{i+k_1-1}]\Phi}\left[\Phi(x) - L_{k_1-1}(\Phi,x)\right], \tag{4}$$

$$L_{\Phi,\Theta,k_1,k_2}(u,x,y) = L_{k_2-1}(L_x(u,x,y),x,y)$$
$$+ \frac{[y_j, y_{j+1}, \ldots, y_{j+k_2-1}]L_x(u,x,y)}{[y_j, y_{j+1}, \ldots, y_{j+k_2-1}]\Theta}\left[\Theta(y) - L_{k_2-1}(\Theta,y)\right], \tag{5}$$

where $[x_i, x_{i+1}, \ldots, x_{i+k_1-1}]u$ is the divided difference for function $u(x,y)$ [6] and $L_{k_1}(u,x,y)$ is Lagrange polynomial on x with fixed y:

$$L_{k_1}(u,x,y) = \sum_{n=i}^{i+k_1-1} u(x_n,y) \prod_{\substack{j=i \\ j\neq n}}^{i+k_1-1} \frac{x-x_j}{x_n-x_j}. \tag{6}$$

It is similarly defined $L_{k_2}(u, x, y)$.

We will notice that in the case $k_1 = k_2 = 2$ the formula (5) was constructed in [7], the case $k_1 = k_2 = 3$ was considered in [2].

According to [8] the following lemma takes a place.

Lemma 1. *For some constant C independent from derivatives of functions $\Phi(x)$ and $\Theta(y)$ and for each cell $\Omega_{i,j}$ the following estimate is true*

$$|u(x, y) - L_{\Phi, \Theta, k_1, k_2}(u, x, y)| \leq$$
$$C\left(1 + \max_x |M_{k_1}(\Phi, x)|\right)\left(1 + \max_y |M_{k_2}(\Theta, y)|\right) \times \left[h_1^{k_1-1} + h_2^{k_2-1}\right], \quad (7)$$

where $(x, y) \in \Omega_{i,j}$, $M_{k_1}(\Phi, x)$ is defined by formula

$$M_{k_1}(\Phi, x) = \frac{\Phi(x) - L_{k_1-1}(\Phi, x)}{\Phi(x_{i+k_1-1}) - L_{k_1-1}(\Phi, x_{i+k_1-1})}. \quad (8)$$

Now we will obtain more exact estimate of the interpolation error.

Lemma 2. *Let us*

$$\Phi^{(m-1)}(x) > 0, \Phi^{(m)}(x) \geq 0 \text{ or } \Phi^{(m-1)}(x) < 0, \Phi^{(m)}(x) \leq 0, x_i < x < x_{i+k_1-1},$$

$$\Theta^{(m-1)}(y) > 0, \Theta^{(m)}(y) \geq 0 \text{ or } \Theta^{(m-1)}(y) < 0, \Theta^{(m)}(y) \leq 0, y_j < y < y_{j+k_2-1}.$$

Then for some constant C independent from derivatives of functions $\Phi(x), \Theta(y)$ the following estimate is fullfiled

$$|u(x, y) - L_{\Phi, \Theta, k_1, k_2}(u, x, y)| \leq C\left[h_1^{k_1-1} + h_2^{k_2-1}\right], \quad (x, y) \in \Omega_{i,j}. \quad (9)$$

Proof. We define $w_{k_1}(x) = (x - x_i)(x - x_{i+1}) \cdots (x - x_{i+k_1-1})$. Then we use the relation [6, p. 44] $\Phi(x) - L_{k_1-1}(\Phi, x) = w_{k_1-1}(x)[x_i, x_{i+1}, \ldots, x_{i+k_1-2}, x]\Phi$ and from (8) obtain

$$M_{k_1}(\Phi, x) = \frac{w_{k_1-1}(x)[x_i, x_{i+1}, \ldots, x_{i+k_1-2}, x]\Phi}{w_{k_1-1}(x_{i+k_1-1})[x_i, x_{i+1}, \ldots, x_{i+k_1-1}]\Phi}.$$

We define $g(x) = [x_i, x_{i+1}, \ldots, x_{i+k_1-2}, x]\Phi$. Using conditions $\Phi^{(m-1)}(x) > 0$, $\Phi^{(m)}(x) \geq 0$ and relations [6]:

$$[x_i, x_{i+1}, \ldots, x_{i+k_1-1}]\Phi = \Phi^{(k_1-1)}(s)/(k - 1)!, \quad \exists s \in (x_i, x_{i+k_1-1}),$$

$$g'(x) = [x_i, x_{i+1}, \ldots, x_{i+k_1-2}, x, x]\Phi,$$

we obtain that the function $g(x)$ is positive and increasing. Using the inequality $|w_{k_1-1}(x)| \leq w_{k_1-1}(x_{i+k_1-1})$, $x \in [x_i, x_{i+k_1-1}]$, we obtain $|M_{k_1}(\Phi, x)| \leq 1$. Similarly we obtain $|M_{k_2}(\Theta, y)| \leq 1$. Using these inequalities in (7), we obtain (9). The lemma is proved. $\diamond$

According to Lemmas 1 and 2 the interpolation formula (4)–(5) has the accuracy of the order $O(h_1^{k_1-1} + h_2^{k_2-1})$ uniformly in gradients of the function $u(x, y)$ in the boundary layers.

It is easy to obtain the following estimate of stability for constructed interpolant (5):

$$|L_{\Phi,\Theta,k_1,k_2}(u - \tilde{u}, x, y)| \leq \max_{n=1,2,\ldots,k_1} \max_{m=1,2,\ldots,k_2} |u(x_n, y_m) - \tilde{u}(x_n, y_m)| \Lambda_{\Phi,k_1} \Lambda_{\Theta,k_2},$$

where $(x, y) \in \Omega_{i,j}$,

$$\Lambda_{\Phi,k_1} = 2^{k_1-2} + \left(1 + 2^{k_1-2}\right) \max_x |M_{k_1}(\Phi, x)|,$$

the value of Λ_{Θ,k_2} is defined similarly.

3   Polynomial Interpolation on Shishkin Mesh

Now we consider other approach to interpolation of the function with large gradients in boundary layers. We investigate the accuracy of polynomial interpolation on Shishkin mesh which is dense in the boundary layers.

In this section we suppose that the function $u(x, y)$ has the representation:

$$u(x, y) = p(x, y) + E_1(x, y) + E_2(x, y) + E_{1,2}(x, y), \tag{10}$$

where $(x, y) \in \bar{\Omega}$, $\bar{\Omega} = [0, 1]^2$. We suppose that $p(x, y)$ is the regular component with bounded derivatives up to some order and $E_1, E_2, E_{1,2}$ are boundary layer functions with large gradients. Functions in (10) aren't given in an explicit form. We suppose that for some constant C independent from ε the following estimates are fullfiled:

$$\left|\frac{\partial^i p(x, y)}{\partial x^i}\right| \leq C, \quad \left|\frac{\partial^j p(x, y)}{\partial y^j}\right| \leq C, \tag{11a}$$

$$\left|\frac{\partial^i E_1(x, y)}{\partial x^i}\right| \leq \frac{C}{\varepsilon^i} e^{-\alpha x/\varepsilon}, \quad \left|\frac{\partial^j E_1(x, y)}{\partial y^j}\right| \leq C, \tag{11b}$$

$$\left|\frac{\partial^i E_2(x, y)}{\partial x^i}\right| \leq C, \quad \left|\frac{\partial^j E_2(x, y)}{\partial y^j}\right| \leq \frac{C}{\varepsilon^j} e^{-\beta y/\varepsilon}, \tag{11c}$$

$$\left|\frac{\partial^i E_{1,2}(x, y)}{\partial x^i}\right| \leq \frac{C}{\varepsilon^i} e^{-\alpha x/\varepsilon}, \quad \left|\frac{\partial^j E_{1,2}(x, y)}{\partial y^j}\right| \leq \frac{C}{\varepsilon^j} e^{-\beta y/\varepsilon}, \tag{11d}$$

where $0 \leq i \leq k_1$, $0 \leq j \leq k_2$, α, β are positive and separated from zero, $\varepsilon \in (0, 1]$. According to (11b)–(11d) the derivatives of functions $E_1(x, y), E_2(x, y), E_{1,2}(x, y)$ are not ε-uniformly bounded.

According to [3, 9], the solution of the problem (2) can be represented in the form (10) with restrictions (11) for some values of k_1, k_2.

Define Shishkin mesh [3] in the region $\bar{\Omega}$:

$$\bar{\Omega}^h = \{(x_i, y_j), \ i = 0, 1, \ldots, N_1, \ j = 0, 1, \ldots, N_2\},$$

$$h_i = x_i - x_{i-1}, \ \tau_j = y_j - y_{j-1}, \ x_0 = 0, x_{N_1} = 1, \ y_0 = 0, y_{N_2} = 1, \qquad (12)$$

where

$$h_i = \frac{2\sigma_1}{N_1}, \ 1 \le i \le \frac{N_1}{2}; \ h_i = \frac{2(1-\sigma_1)}{N_1}, \ \frac{N_1}{2} < i \le N_1,$$

$$\tau_j = \frac{2\sigma_2}{N_2}, \ 1 \le j \le \frac{N_2}{2}; \ \tau_j = \frac{2(1-\sigma_2)}{N_2}, \ \frac{N_2}{2} < j \le N_2,$$

$$\sigma_1 = \min\left\{\frac{1}{2}, \frac{k_1\varepsilon}{\alpha}\ln N_1\right\}, \quad \sigma_2 = \min\left\{\frac{1}{2}, \frac{k_2\varepsilon}{\beta}\ln N_2\right\}. \qquad (13)$$

We will carry out the interpolation in cells $\Omega_{i,j}$ of the mesh $\bar{\Omega}^h$, where $\Omega_{i,j}$ is defined just as in (3), but for mesh $\bar{\Omega}^h$. We suppose that N_1 is multiple $2(k_1 - 1)$ and N_2 is multiple $2(k_2 - 1)$. Then the cell $\Omega_{i,j}$ has the uniform mesh steps on each direction. We suppose that $h_{i,x}$ and $h_{j,y}$ are mesh steps in the cell $\Omega_{i,j}$.

Now we define two-dimensional polynomial interpolation formula for the cell $\Omega_{i,j}$:

$$L_{k_1,k_2}(u, x, y) = L_{k_2}(L_{k_1}(u, x, y), x, y), \qquad (14)$$

where $L_{k_1}(u, x, y)$ is defined by (6) and $L_{k_2}(u, x, y)$ is defined similarly.

Using transition from (14) to one-dimensional interpolation formulas and accounting the accuracy of Lagrange interpolation, we proved the following lemma.

Lemma 3. *For some constant C_1 the following estimate is fullfiled*

$$\left| L_{k_1,k_2}(u, x, y) - u(x, y) \right| \le C_1 \left[M_{k_1,x} h_{i,x}^{k_1} + M_{k_2,y} h_{j,y}^{k_2} \right], \ (x, y) \in \Omega_{i,j}, \qquad (15)$$

where

$$M_{k_1,x} = \max_{(x,y)\in\Omega_{i,j}} \left| \frac{\partial^{k_1} u(x,y)}{\partial x^{k_1}} \right|, \ M_{k_2,y} = \max_{(x,y)\in\Omega_{i,j}} \left| \frac{\partial^{k_2} u(x,y)}{\partial y^{k_2}} \right|.$$

The estimate (15) of the interpolation error depends on gradients of function $u(x, y)$. We will notice that in Lemma 3 the cell $\Omega_{i,j}$ can correspond to both a uniform and piecewise uniform grid.

Lemma 4. *Let the cell $\Omega_{i,j}$ corresponds to Shishkin mesh (12). Then for some constant C_2 the following estimate is fullfiled*

$$|L_{k_1,k_2}(u, x, y) - u(x, y)| \le C_2 \left[\left(\frac{\ln N_1}{N_1}\right)^{k_1} + \left(\frac{\ln N_2}{N_2}\right)^{k_2} \right], (x, y) \in \Omega_{i,j}. \qquad (16)$$

Proof. We use the representation (10) and obtain the estimate (16) for each component of decomposition of function $u(x, y)$.

We estimate the interpolation error for the function $E_1(x, y)$.

Consider the case $\sigma_1 < 1/2$, $\sigma_2 < 1/2$.

Let us $x_{i+k_1-1} \leq \sigma_1$. We use relations (11b), (13) in (15) and for some constant C_3 obtain

$$\left| L_{k_1,k_2}(E_1, x, y) - E_1(x, y) \right| \leq C_3 \left[\left(\frac{\ln N_1}{N_1} \right)^{k_1} + \frac{1}{N_2^{k_2}} \right], \quad (x, y) \in \Omega_{i,j}.$$

Let us $x_i \geq \sigma_1$. Then according to (11b), (13)

$$|E_1(x, y)| \leq \frac{C}{N_1^{k_1}}. \tag{17}$$

We use known stability estimate for Lagrange polynomial

$$\left| L_{k_1}(u, x, y) \right| \leq \max_{x,y} |u(x, y)| 2^{k_1-1}$$

and obtain

$$|L_{k_1,k_2}(E_1, x, y)| \leq \max |E_1(x, y)| 2^{k_1+k_2-2}, \quad (x, y) \in \Omega_{i,j}. \tag{18}$$

Now we use estimates (17), (18) and obtain

$$\left| L_{k_1,k_2}(E_1, x, y) - E_1(x, y) \right| \leq \frac{C_4}{N^{k_1}}, \quad (x, y) \in \Omega_{i,j}.$$

Other cases can be considered similarly.   ◇

4   Numerical Results

We consider the function

$$u(x, y) = \left(1 - e^{-x/\varepsilon} \right) \left(1 - e^{-2y/\varepsilon} \right)(1 - x)(1 - y) + \cos \frac{\pi x}{2} e^{-y},$$

$$x, y \in [0, 1], \quad \varepsilon \in (0, 1]$$

with boundary layer components $\Phi(x) = e^{-x/\varepsilon}$, $\Theta(y) = e^{-2y/\varepsilon}$.

Let us $N_1 = N_2 = N$. Define $\tilde{x}_i = (x_{i-1} + x_i)/2$, $\tilde{y}_j = (y_{j-1} + y_j)/2$, $i, j = 1, 2, \ldots, N$. Tables contain an interpolation error

$$\Delta_{N,\varepsilon} = \max_{i,j} \left| Int(u, \tilde{x}_i, \tilde{y}_j) - u(\tilde{x}_i, \tilde{y}_j) \right|,$$

where $Int(u, x, y)$ is studied interpolant. In tables $e \pm m$ means $10^{\pm m}$.

Table 1 lists the error $\Delta_{N,\varepsilon}$ of polynomial interpolant $L_{3,3}(u, x, y)$ on the uniform mesh for various values of N and ε. It follows from the Table 1 that the error of polynomial interpolation is of order $O(1)$ as $\varepsilon \leq h$.

Table 2 lists the error $\Delta_{N,\varepsilon}$ of the interpolant $L_{\Phi,\Theta,3,3}(u, x, y)$ from (5), fitted to boundary layer components. For $\varepsilon = 1$ there are no boundary layers and

Table 1. The error of piecewise polynomial interpolation with $k_1 = k_2 = 3$ on the uniform mesh

ε	N				
	2^3	2^4	2^5	2^6	2^7
1	$6.72e-4$	$8.90e-5$	$1.14e-5$	$1.45e-6$	$1.82e-7$
2^{-3}	$7.98e-2$	$2.07e-2$	$3.88e-3$	$6.02e-4$	$8.40e-5$
2^{-4}	$2.19e-1$	$8.84e-2$	$2.29e-2$	$4.30e-3$	$6.65e-4$
2^{-5}	$4.37e-1$	$2.29e-1$	$9.48e-2$	$2.46e-2$	$4.60e-3$
2^{-6}	$5.46e-1$	$4.52e-1$	$2.40e-1$	$9.94e-2$	$2.57e-2$
2^{-11}	$5.63e-1$	$5.86e-1$	$5.98e-1$	$6.04e-1$	$6.06e-1$
2^{-12}	$5.63e-1$	$5.86e-1$	$5.98e-1$	$6.04e-1$	$6.06e-1$

Table 2. The interpolation error of fitted formula (5) with $k_1 = k_2 = 3$ on the uniform mesh

ε	N				
	2^3	2^4	2^5	2^6	2^7
1	$4.57e-4$	$5.90e-5$	$7.41e-6$	$9.26e-7$	$1.16e-7$
2^{-3}	$4.67e-3$	$1.20e-3$	$2.16e-4$	$3.24e-5$	$4.44e-6$
2^{-4}	$7.27e-3$	$2.66e-3$	$7.07e-4$	$1.23e-4$	$1.95e-5$
2^{-5}	$3.40e-3$	$4.14e-3$	$1.47e-3$	$3.94e-4$	$7.21e-5$
2^{-6}	$6.83e-3$	$1.96e-3$	$2.20e-3$	$7.77e-4$	$2.10e-4$
2^{-11}	$8.08e-3$	$2.11e-3$	$5.35e-4$	$1.34e-4$	$5.33e-5$
2^{-12}	$8.08e-3$	$2.11e-3$	$5.35e-4$	$1.34e-4$	$3.07e-5$

Table 3. The error of piecewise polynomial interpolation with $k_1 = k_2 = 3$ on Shishkin mesh

ε	N				
	2^3	2^4	2^5	2^6	2^7
1	$6.72e-4$	$8.90e-5$	$1.14e-5$	$1.45e-6$	$1.82e-7$
2^{-3}	$5.36e-2$	$2.06e-2$	$3.88e-3$	$6.02e-4$	$8.40e-5$
2^{-4}	$7.24e-2$	$2.79e-2$	$8.42e-3$	$2.24e-3$	$5.10e-4$
2^{-5}	$7.11e-2$	$2.99e-2$	$9.92e-3$	$2.65e-3$	$6.06e-4$
2^{-6}	$7.04e-2$	$2.98e-2$	$9.88e-3$	$2.64e-3$	$6.02e-4$
2^{-11}	$6.97e-2$	$2.99e-2$	$9.92e-3$	$2.65e-3$	$6.04e-4$
2^{-12}	$6.96e-2$	$2.99e-2$	$9.93e-3$	$2.65e-3$	$6.04e-4$

formula (5) has error of order $O(h^3)$; the interpolation error increases up to the order $O(h^2)$ with decreasing ε. The error estimate (9) with $k_1 = k_2 = 3$ is supported by numerical results.

Table 4. Numerical and theoretical orders of accuracy of polynomial interpolation on Shishkin mesh, $k_1 = k_2 = 3$

ε	N				
	2^3	2^4	2^5	2^6	2^7
1	2.9	3.0	3.0	3.0	2.9
2^{-3}	1.4	2.4	2.7	2.8	2.9
2^{-7}	1.2	1.6	1.9	2.2	2.3
2^{-12}	1.2	1.6	1.9	2.1	2.3
CR_N	1.7	2.0	2.2	2.3	2.4

Table 3 lists the error $\Delta_{N,\varepsilon}$ of polynomial interpolation $L_{3,3}(u, x, y)$ on Shishkin mesh.

Now we define the numerical order of accuracy $M_{N,\varepsilon} = \log_2(\Delta_{N,\varepsilon}/\Delta_{2N,\varepsilon})$ and theoretical order of accuracy $CR_N = 3\log_2(2\ln N/\ln(2N))$, corresponding to the estimate (16) with $N_1 = N_2 = N$, $k_1 = k_2 = 3$. Table 4 contains numerical and theoretical orders of accuracy of interpolation formula $L_{3,3}(u, x, y)$ on Shishkin mesh. For small values of ε the numerical order of accuracy is close to the theoretical order of accuracy. Therefore, the error estimate (16) is supported by numerical experiments.

5 Conclusion

The error of polynomial interpolation of functions with large gradients can be of order O(1). For such functions we have compared two approaches to interpolation, which are based on the application of interpolation exact on boundary layer components and on the application of polynomial interpolation on a mesh which is dense in boundary layers. The numerical experiments are carried out.

Acknowledgements. Supported in part by Russian Foundation for Basic Research under Grants 15-01-06584, 16-01-00727.

References

1. Zadorin, A.I.: Method of interpolation for a boundary layer problem. Sib. J. Numer. Math. **10**(3), 267–275 (2007). (in Russian)
2. Zadorin, A.I., Zadorin, N.A.: Interpolation of functions with boundary layer components and its application to the two-grid method. Sib. Elektron. Math. Rep. **8**, 247–267 (2011). (in Russian)
3. Shishkin, G.I.: Grid Approximations of Singular Perturbation Elliptic and Parabolic Equations. UB RAS, Yekaterinburg (1992). (in Russian)
4. Roos, H.-G., Stynes, M., Tobiska, L.: Numerical Methods for Singularly Perturbed Differential Equations. Convection-Diffusion and Flow Problems, vol. 24. Springer, Berlin (2008)

5. Zadorin, A.I., Zadorin, N.A.: Interpolation formula for functions with a boundary layer component and its application to derivatives calculation. Siberian Electron. Math. Rep. **9**, 445–455 (2012)
6. Bakhvalov, N.S.: Numerical Methods. Nauka, Moskow (1975). (in Russian)
7. Vulkov, L.G., Zadorin, A.I.: Two-grid algorithms for the solution of 2D semilinear singularly perturbed convection-diffusion equations using an exponential finite difference scheme. In: American Institute of Physics Conference Proceedings, vol. 1186, pp. 371–379 (2009)
8. Zadorin, A.I.: Interpolation of a function of two variables with large gradients in boundary layers. Lobachevskii J. Math. **37**(3), 349–359 (2016)
9. Lins, T., Stynes, M.: Asymptotic analysis and Shishkin-type decomposition for an elliptic convection-diffusion problem. J. Math. Anal. Appl. **261**, 604–632 (2001)

A Volunteer-Computing-Based Grid Architecture Incorporating Idle Resources of Computational Clusters

Oleg Zaikin[1](✉), Maxim Manzyuk[2], Stepan Kochemazov[1], Igor Bychkov[1], and Alexander Semenov[1]

[1] Matrosov Institute for System Dynamics and Control Theory SB RAS, Irkutsk, Russia
zaikin.icc@gmail.com, veinamond@gmail.com, bychkov@icc.ru, biclop.rambler@yandex.ru
[2] Internet-portal BOINC.ru, Moscow, Russia
hoarfrost@rambler.ru

Abstract. In this paper, we suggest a new architecture of a computational grid that involves resources of BOINC-based volunteer computing projects and idle resources of computational clusters. We constructed a computational grid of the proposed kind, based on several computational clusters and the volunteer computing project SAT@home. This project, launched and maintained by us, is aimed at solving hard computational problems, which can be effectively reduced to Boolean satisfiability problem. In the constructed grid several new combinatorial designs based on diagonal Latin squares of order 10 were found, and also several weakened cryptanalysis problems for the Bivium cipher were solved.

Keywords: Grid · Volunteer computing · BOINC · SAT

1 Introduction

Volunteer computing [1] is a type of distributed computing [11] which uses computational resources of personal computers (PCs) of private persons called volunteers. Each volunteer computing project is designed to solve one or several hard problems. When PC is connected to the project, all the calculations are performed automatically and do not inconvenience user since only idle resources of PC are used. Nowadays the most popular platform for organizing volunteer computing projects is Berkeley Open Infrastructure for Network Computing (BOINC) [1], that was developed in Berkeley in 2002.

A volunteer computing project consists of the following basic parts: server daemons, database, web site and client applications. Daemons include work generator (generates tasks to be processed), validator (checks the correctness of the results received from volunteer's PCs) and assimilator (processes correct results). Each client application, which is being launched by BOINC manager (common

© Springer International Publishing AG 2017
I. Dimov et al. (Eds.): NAA 2016, LNCS 10187, pp. 769–776, 2017.
DOI: 10.1007/978-3-319-57099-0_89

for all projects) on a volunteer's PC, should have versions for the widespread computing platforms.

One of the attractive features of volunteer computing is its low cost—to maintain a project one needs only a dedicated server working 24/7. Main difficulty here lies in the development of software and in database administration. In addition, it is crucial to provide the feedback to volunteers using the web site of the project and special forums.

There are some restrictions on problems that are to be solved by volunteer computing—such problems should be decomposed into separated subproblems (i.e. the embarrassing parallelism [11] should be used). However, there are many scientific areas (astronomy, biology, mathematics, etc.) where such problems occur and, consequently, there is a number of volunteer computing projects (most of which are based on BOINC) with good results from the aforementioned areas.

So, BOINC-based volunteer computing projects can provide large amount of computational tasks from various scientific areas. Meanwhile, a lot of computational clusters are systematically underutilized. We propose a new architecture of computational grids incorporating resources of volunteer computing projects and idle resources of computational clusters.

Below we present a brief outline of our paper. In the second section we propose the aforementioned architecture of computational grids. In this section we also describe CLUBORUN—the tool developed for constructing grids of the proposed kind. In the third section the computational grid, aimed at solving hard instances of Boolean satisfiability problem (SAT), is described. This grid, constructed via CLUBORUN, is based on the volunteer computing project SAT@home and several clusters. In the fourth section we consider some results obtained in the constructed grid. In particular, several new pairs of orthogonal diagonal Latin squares of order 10 were found, and also several weakened cryptanalysis problems of the Bivium stream cipher were solved.

2   Grid Architecture Incorporating Resources of Computational Clusters and Volunteer Computing Projects

We propose a new architecture of computational grids. The main idea of this architecture is to increase the performance of a volunteer computing project desktop grid by incorporating idle resources of computing clusters. Below the main features of the proposed architecture are listed.

- In one computational grid exactly one volunteer computing project and several computational clusters are used.
- Only idle computational resources of clusters are utilized.
- Only ordinary user rights must be used for launching calculations on clusters.
- Finally, calculations on a particular cluster node are performed by the standard BOINC manager.

– Any involved cluster node is perceived by a volunteer computing project as an individual host, just like an ordinary volunteer's PC.

For constructing grids of the proposed kind we have implemented a CLUBORUN (Cluster for BOINC Run) tool. This tool consists of several shell scripts and a C++ MPI program. Several scripts periodically monitor the current condition of a cluster queue. If idle resources are available, then CLUBORUN launches BOINC calculations as MPI tasks, which are processed by cluster scheduling system (as all other tasks of other users of a cluster). After being launched on a cluster node the MPI program starts BOINC manager. The BOINC manager connects to a volunteer computing project server and performs calculations using standard client applications of a project. Thus, from the BOINC manager point of view, cluster node is just another PC. When tasks from another user appear in a cluster queue, CLUBORUN stops BOINC tasks in a queue if new tasks can be launched on freed resources. It should be noted that CLUBORUN doesn't violate any rules of cluster use since all ordinary user's restrictions apply (limit on total CPU hours or on a number of simultaneously active tasks in queue, etc.).

One of the main difficulties in development of CLUBORUN is a necessity to make a separate version for each specific job scheduling system (because each system has its own commands to work with cluster queue). At the moment CLUBORUN can work with the Cleo, SUPPZ, SLURM and PBS Torque job scheduler systems.

3   Computational Grid Aimed at Solving Hard SAT Instances

With the help of CLUBORUN tool we constructed a computational grid of the proposed kind. This grid is based on the volunteer computing project SAT@home [14], which was launched and is being maintained by us. This project is aimed at solving hard instances of Boolean satisfiability problem (SAT) [2]. The CLUBORUN tool has been successfully working on three computational clusters for more than two years. In particular, it utilized the idle resources of the "Academician V.M. Matrosov" cluster (Irkutsk supercomputing center of SB RAS) with PBS Torque job scheduling system. In total, SAT@home desktop grid, augmented by idle resources of computational clusters forms a computational grid for solving hard SAT instances. The cluster resources made it possible to significantly improve the performance of the project (peak performance boost was 40% or about 2 teraflops).

Let us describe SAT@home in more detail. Problems from various areas (verification, cryptography, combinatorics, bioinformatics, etc.) can be effectively reduced to SAT. SAT problems are usually considered as the problems of search for solutions of Boolean equations in the form of CNF = 1, where CNF is a conjunctive normal form. All known SAT solving algorithms are exponential in the worst case since SAT itself is NP-hard. Nevertheless, modern SAT solvers successfully cope with many classes of tests based on the problems from the areas mentioned above. Improvement of the effectiveness of SAT solving algorithms, including the development of algorithms that are able to work in parallel and distributed computing environments, is a very important direction of research.

The SAT@home project has been actively functioning since September 2011. The project server employs a number of standard BOINC daemons responsible for sending and processing tasks (transitioner, feeder, scheduler, etc.). For other daemons, such as work generator, validator and assimilator, we implemented our specific versions. The work generator decomposes the original SAT problem to subproblems based on the previously found decomposition parameters. In SAT@home we use the method for finding such parameters that was proposed in [14]. The work generator creates 2 copies of each task in accordance with the concept of redundant calculations used in BOINC. The validator checks the correctness of the results, and the assimilator processes correct results. When the assimilator finds satisfying assignment in the obtained results, it checks its correctness. If the assignment is correct, then the original problem is marked as solved and the generation of tasks for this problem stops. The SAT@home client application is based on the SAT solver MINISAT [6], which was slightly modified to use less RAM.

The characteristics of the SAT@home project as of 5 of July 2016 are listed below. It should be noted, that these characteristics in fact describe the computational grid, constructed with the help of CLUBORUN on the base of SAT@home.

- 3509 active PCs (active PC in volunteer computing is a PC that sent at least one result in last 30 days).
- 1348 active users (active user is a user that has at least one active PC).
- Versions of the client application for CPU: Windows x86, Windows x86-64, Linux x86, Linux x86-64.
- Average real performance: 8.5 teraflops, maximal performance: 15 teraflops.

On the first stage, SAT@home was used to solve several cryptanalysis problems of the A5/1 keystream generator [14]. On the second stage new pairs of orthogonal diagonal Latin squares of order 10 were found [15]. On the third stage several weakened problems of cryptanalysis of the Bivium cipher were solved [16]. In the next section we describe some of the obtained results.

4   Computational Problems Solved in the Constructed Grid

The main goal of constructing grids of the proposed in Sect. 2 kind is to solve some hard problems faster. Here we consider new results, obtained in the grid, described in Sect. 3. With the help of SAT approach we solved hard problems from two areas: combinatorics and cryptography.

4.1   The Search for Systems of Mutually Orthogonal Diagonal Latin Squares

One of the most promising areas of application of SAT approach is the search for combinatorial designs [18]. In particular, combinatorial problems related to

Latin squares [5] are very interesting. Latin square of order n is a square $n \times n$ table filled with elements from some set M, $|M| = n$ in such a way that each element from M appears in each row and each column exactly once. Initially Leonard Euler used the set of Latin letters as M, therefore the corresponding combinatorial designs were named Latin squares. In this paper for convenience we will use as M the set $\{0, \ldots, n-1\}$. The Latin square is called diagonal if both its main diagonal and main antidiagonal contain all numbers from 0 to $n-1$. In other words, the constraint on the uniqueness is extended from rows and columns to two diagonals.

A pair of Latin squares of the same order is called orthogonal if all ordered pairs of the kind (a, b) are different, where a is the number in some cell of the first Latin square and b is the number from the same cell in the second Latin square. If there are m different orthogonal Latin squares, from which each pair is orthogonal, then it is called the system of m mutually orthogonal Latin squares (MOLS). One of the most well-known unsolved problems in this area is the following: to answer the question whether there exists a triple of MOLS of order 10.

The existence of a pair of mutually orthogonal diagonal Latin squares (MODLS) of order 10 was proved in 1992—in the paper [3] three such pairs were presented. In 2012 we started in SAT@home the computational experiment aimed at finding new pairs of MODLS of order 10. The experiment was finished in 2013 and its results were published in [15]. First we constructed a propositional encoding for this problem. The obtained CNF had 2000 Boolean variables and 434440 clauses. The size of the corresponding SAT instance in the DIMACS format was 10 Mb. We used the so-called "naive" encoding (for example, see [12]). We decomposed the obtained SAT instance as follows. The first row of the first diagonal Latin square was fixed to be equal to "0 1 2 3 4 5 6 7 8 9". It does not lead to the loss of generality because of the properties of Latin squares. After this we processed all possible values of the first 8 cells of the second and the third rows of the first square. As a result we obtained a family of subproblems, in which for each subproblem for the first diagonal Latin square the values were fixed for 26 out of 100 cells (10 from the first row and 8 from the second and the third rows). In terms of SAT encoding, it translated into assigning values to 260 out of 2000 variables in each SAT instance. In SAT@home experiment each job batch contained 20 such SAT instances. For each instance MINISAT had a limit of 2600 restarts that is more or less equal to 4 min on one core of state-of-the-art CPU. To process 20 million subproblems generated for the experiment it took about 9 months of work of the SAT@home project (from September 2012 to May 2013). As a result we found 17 new pairs of MODLS of order 10 (in addition to three previously known pairs from [3]). From April 2015 to February 2016 we tried other decompositions engaging the cells of the first rows—as a result new 32 pairs were found [17].

In March 2016 we started in SAT@home another experiment, aimed at searching for new pairs of MODLS of order 10. In this experiment we decided to use another approach to decomposition—to vary values of 18 cells from the main diagonal and antidiagonal of the first square (the first row is fixed in the same manner as in the previous experiment). As a result, in total from March 3, 2016 to July 5, 2016 we managed to find 22 new previously unknown pairs of the considered kind (compared to 3 pairs from [3], 17 pairs from [15]) and 32 pairs from [17]. All found solutions are available online at the web site of the SAT@home project[1]. In this experiment the limit on the amount of restarts in the client application was increased from 2600 to 5000. While the performance of SAT@home increased (compared to 2012–2013), it is easy to see that the new approach to decomposition is more effective than that from the previous experiment.

4.2   SAT-Based Cryptanalysis of the Bivium Keystream Generator

Usually if the cryptanalysis is considered as a SAT problem then it is called a SAT-based cryptanalysis. In this case to find a secret key it is sufficient to find a solution of corresponding satisfiable SAT instance. Here we consider the SAT-based cryptanalysis of the Bivium keystream generator. This generator [4] uses two shift registers of a special kind. The first register contains 93 cells and the second contains 84 cells. To initialize the cipher, a secret key of length 80 bit is put to the first register, and a fixed (known) initialization vector (IV) of length 80 bit is put to the second register. All remaining cells are filled with zeros. An initialization phase consists of 708 rounds during which keystream output is not released.

In accordance with [13] we considered cryptanalysis problems for Bivium in the following formulation. Based on the known fragment of keystream we search for the values of all registers cells (177 bits) at the end of the initialization phase. Therefore, in our experiments we used CNF encodings where the initialization phase was omitted. Usually it is believed that to uniquely identify the secret key it is sufficient to consider keystream fragment of length comparable to the total length of shift registers. Here we followed [7] and set the keystream fragment length to 200 bits.

The SAT-based cryptanalysis of Bivium turned out to be very hard, that is why we decided to solve several weakened cryptanalysis instances for this generator. Below we use the notation *BiviumK* to denote a weakened problem for Bivium with known values of K variables encoding the last K cells of the second shift register. In [16] we described how 5 instances of *Bivium9* were solved in SAT@home in 2014. In the corresponding experiment all values of several chosen variables were checked in the form of computational tasks in SAT@home.

We also tried another approach to solving weakened Bivim instances, which was not mentioned in [14]. Let us describe it below. On the first stage a SAT instance is being processed on a computational cluster by running PDSAT [14] in the solving mode (here all values of several variables are checked too). During this process, the time limit equal to 0.1 s (this value was selected according to experiments) for every subproblem is used. PDSAT collects (by writing to a file)

---

[1] http://sat.isa.ru/pdsat/solutions.php.

all subproblems which could not be solved within the time limit. It turned out, that this approach allowed to solve 2 out of 3 instances *Bivium10* on a cluster (i.e., despite the time limit, PDSAT found a satisfying assignments for these 2 instances). It should be noted, that during processing of these 2 instances the new approach was about 2 times faster than the approach without time limits. Solving of the remaining cryptanalysis instance was performed in SAT@home with the help of the file with data about the hard subproblems (on which the solving was interrupted due to the time limit), collected by PDSAT. So, we can conclude that with the help of the proposed approach some instances can be quickly processed on a computational cluster, and a volunteer computing project suits well for processing the remaining instances. We hope that this approach will help us to solve non-weakened instances of cryptanalysis of Bivium in the nearest future.

5 Related Work

With the help of BNB-GRID [9] a service grid based on several clusters can be constructed. In such grid all calculations are performed by MPI programs, launched via standard cluster job scheduler, so only ordinary user rights are required (CLUBORUN has the similar feature). However, BNB-GRID cannot operate with a volunteer computing project, and it also cannot utilize only idle cluster resources (like CLUBORUN does).

The tools 3G BRIDGE [10] and HTCONDOR [8] are aimed at combining resources of several grid systems (for example, of a service grid based on clusters and a BOINC-based desktop grid). However, 3G BRIDGE requires administrator rights of a cluster, and it also cannot operate with its idle resources. HTCONDOR can be used for utilizing only idle resources of a cluster, but it requires administrator rights too.

We can conclude that, compared to all aforementioned tools, only CLUBORUN can launch calculations of a BOINC-based volunteer computing project on a cluster with ordinary user rights, and utilize only idle cluster resources.

Acknowledgements. The research was funded by Russian Science Foundation (project no. 16-11-10046). We thank all SAT@home participants for their resources and fruitful feedback.

References

1. Anderson, D.P., Fedak, G.: The computational and storage potential of volunteer computing. In: Sixth IEEE International Symposium on Cluster Computing and the Grid (CCGrid 2006), Singapore, 16–19 May 2006, pp. 73–80. IEEE Computer Society (2006)
2. Biere, A., Heule, M.J.H., van Maaren, H., Walsh, T. (eds.): Handbook of Satisfiability, Frontiers in Artificial Intelligence and Applications, vol. 185. IOS Press, Amsterdam (2009)

3. Brown, J., Cherry, F., Most, L., Parker, E., Wallis, W.: Completion of the spectrum of orthogonal diagonal latin squares. In: Lecture Notes in Pure and Applied Mathematics, vol. 139, pp. 43–49 (1992)
4. Cannière, C.: TRIVIUM: a stream cipher construction inspired by block cipher design principles. In: Katsikas, S.K., López, J., Backes, M., Gritzalis, S., Preneel, B. (eds.) ISC 2006. LNCS, vol. 4176, pp. 171–186. Springer, Heidelberg (2006). doi:10. 1007/11836810_13
5. Colbourn, C.J., Dinitz, J.H.: The CRC Handbook of Combinatorial Designs. CRC Press, Inc., Boca Raton (1996)
6. Eén, N., Sörensson, N.: An extensible SAT-solver. In: Giunchiglia, E., Tacchella, A. (eds.) SAT 2003. LNCS, vol. 2919, pp. 502–518. Springer, Heidelberg (2004). doi:10. 1007/978-3-540-24605-3_37
7. Eibach, T., Pilz, E., Völkel, G.: Attacking bivium using SAT solvers. In: Kleine Büning, H., Zhao, X. (eds.) SAT 2008. LNCS, vol. 4996, pp. 63–76. Springer, Heidelberg (2008). doi:10.1007/978-3-540-79719-7_7
8. Epema, D., Livny, M., van Dantzig, R., Evers, X., Pruyne, J.: A worldwide flock of condors: load sharing among workstation clusters. Future Gener. Comput. Syst. **12**, 53–65 (1996)
9. Evtushenko, Y., Posypkin, M., Sigal, I.: A framework for parallel large-scale global optimization. Comput. Sci. Res. Dev. **23**(3), 211–215 (2009)
10. Farkas, Z., Kacsuk, P., Balaton, Z., Gombás, G.: Interoperability of BOINC and EGEE. Future Gener. Comput. Syst. **26**(8), 1092–1103 (2010)
11. Foster, I.: Designing and Building Parallel Programs: Concepts and Tools for Parallel Software Engineering. Addison-Wesley Longman Publishing Co., Inc., Boston (1995)
12. Lynce, I., Ouaknine, J.: Sudoku as a SAT problem. In: International Symposium on Artificial Intelligence and Mathematics (ISAIM 2006), Fort Lauderdale, Florida, USA, 4–6 January 2006
13. Maximov, A., Biryukov, A.: Two trivial attacks on TRIVIUM. In: Adams, C., Miri, A., Wiener, M. (eds.) SAC 2007. LNCS, vol. 4876, pp. 36–55. Springer, Heidelberg (2007). doi:10.1007/978-3-540-77360-3_3
14. Semenov, A., Zaikin, O.: Algorithm for finding partitionings of hard variants of boolean satisfiability problem with application to inversion of some cryptographic functions. SpringerPlus **5**(1), 1–16 (2016)
15. Zaikin, O., Kochemazov, S.: The search for systems of diagonal Latin squares using the SAT@home project. In: Second International Conference BOINC-Based High Performance Computing: Fundamental Research and Development (BOINC: FAST 2015), Petrozavodsk, Russia, 14–18 September 2015, vol. 1502, pp. 52–63. CEUR-WS (2015)
16. Zaikin, O., Semenov, A., Otpuschennikov, I.: Solving weakened cryptanalysis problems for the Bivium cipher in the volunteer computing project SAT@home. In: Second International Conference BOINC-Based High Performance Computing: Fundamental Research and Development (BOINC: FAST 2015), Petrozavodsk, Russia, 14–18 September 2015, vol. 1502, pp. 22–30. CEUR-WS (2015)
17. Zaikin, O., Vatutin, E., Zhuravlev, A., Manzyuk, M.: Applying high-performance computing to searching for triples of partially orthogonal Latin squares of order 10. In: 10th Annual International Scientific Conference on Parallel Computing Technologies, Arkhangelsk, Russia, 29–31 March 2016, vol. 1576, pp. 155–166. CEUR-WS (2016)
18. Zhang, H.: Combinatorial Designs by SAT Solvers, pp. 533–568. In: Biere et al. [2], vol. 185, February 2009

Effects of the Neuron Permutation Problem on Training Artificial Neural Networks with Genetic Algorithms

Iliyan Zankinski[✉]

Institute of Information and Communication Technologies,
Bulgarian Academy of Sciences,
acad. Georgi Bonchev Str, Block 2, 1113 Sofia, Bulgaria
iict@bas.bg
http://www.iict.bas.bg/

Abstract. A method of investigation of numerical schemes deriving from the variational formulation of the problem (variational-difference method and FEM) is discusses. The method is based on the reduction of the numerical schemes to the canonical finite difference form. The resulting numerical scheme standard notation in the form of a grid operator equality is used for analyzing its approximation, stability and other properties. The application of this approach to a wider classes of finite elements (from the simplest ones to the Hermitian elements and serendipities) is discussed. These opportunities are illustrated by the analysis of FEM schemes for Timoshenko shells and elasticity dynamic problems.

1 Introduction

Genetic algorithms find applications in neural networks decades after their early development. If preparing a training set is inefficient or even impossible (reinforcement learning), gradient methods such as Error backpropagation are unsuitable. In these cases a heuristic approach could be the only option and genetic algorithms (neuroevolution) have a history of quality and robustness, successfully overcoming problems with local optima and valleys in the search space. A limiting factor for the algorithm is the destructive nature of the crossover operator. With the increase of the number of free parameters, the likelihood that a crossover of two fit individuals leads to a new fit individual is reduced.

The problem has been extensively investigated, both theoretically and practically, with inconclusive or mixed results. Some are shown here:

- The effect is masked by successful recombinations - Building Block Hypothesis [1,2];
- This is a positive effect which helps with more thorough exploration of the search space and is viewed as a macromutation;
- It has a negative effect but is still acceptable;
- It has a severe negative impact on results and crossover probability should be limited towards later stages of the search.

© Springer International Publishing AG 2017
I. Dimov et al. (Eds.): NAA 2016, LNCS 10187, pp. 777–782, 2017.
DOI: 10.1007/978-3-319-57099-0_90

When applied to neural networks training, genetic algorithms face an additional limitation. The same network can be achieved using different positions of the weights or even different weights altogether. In the first case, networks with equal weights but different arrangement of neurons in the hidden layer always give the same result. The most often used names are the Neuron Permutation Problem and Competing Conventions.

The case where different weights give the same results is harder to illustrate but intuitively comes to saturation of the activation function when it is nonlinear (step function or sigmoid function). This paper focuses on studying the effects the Neuron Permutation Problem has on the genetic algorithm by artificially simulating permutations when solving a real example of time series prediction.

2   Problem Analysis

2.1   Search Space

Given that the number of neurons in the hidden layer is Nh, the possible internal representations for a network with every hidden neuron being different is Nh!. Consequently the search space increases vastly with increasing the number of hidden neurons. It may seem that this form of symmetry should not have any effect on the search time as the number of maxima increase just as much as the minima. This is true for a random search as the probability of finding the optimum stays the same. For a genetic algorithm the increase in search space should have a negative effect. The increase in the number of local optima leads to separation of the population and exploration of the same hills in different locations. Combining individuals that are very different, possibly located on separate hills in search space, is much more likely to lead to destructive interference and have negative impact on the offspring.

2.2   Neuron Position

The current encoding of neurons in neural networks strictly ties these neurons to their positions. In the situation that two parents evolved to have different advantageous neurons in the same position, a crossover operator can either preserve one of them or a combination of them.

One process that may hide these effects in practice is the tendency for one population to contain very similar individuals (close to each other in search space). This is especially the case for small populations where successful individuals quickly spread their genes and the whole population is closely related. Permutations and coinciding advantageous neurons are harder to arise this way. This may be one of the reasons behind the mixed results of other research and it deserves additional attention.

A method that is more vulnerable to these problems is the separate evolution of populations with periodic exchange of individuals, suitable for parallel implementation [3]. The separation allows for similar patterns to evolve with different positions of the neurons. Transferred individuals are not as suitable for crossover as the ones that evolved together.

An encoding scheme for neural networks that is invariant to position would be much more efficient by keeping the search space small and allowing advantageous neurons to survive more often even when multiple populations are evolved.

Some suggestions for solving the problem exist in the literature (some may also evolve network topology):

- Identification of functionally similar neurons by the similarity in activation patterns and keeping these neurons with higher probability [4]
- Sorting neurons based on similarity prior to crossover [5]
- Identification of functionally similar neurons by the identity of weights and keeping these neurons with higher probability [6]
- A population of neurons with individual solutions using these neurons in different combinations, (Symbiotic, Adaptive Neuro-Evolution, SANE) [7]
- Improvement to SANE with separate population for every neuron position, (Enforced SubPopulations, ESP) [8]

3 Experiments

The effects of the Permutation Problem are studied here by introducing a single permutation to half of the population. The permutation is chosen at random for every new generation. Six versions of the algorithm are compared with and without permutation:

- No crossover
- Uniform crossover - individual weights have equal probability to be chosen from one of the parents or the other
- Point crossover - a random crossover point is chosen for the chromosome
- Whole neurons crossover - whole neurons have equal probability to be chosen from one of the parents or the other
- Single neuron crossover, location invariant - the new chromosome is the same as the first parent with the exception of one neuron chosen at random from the second parent and copied to a new random location
- Single neuron uniform crossover, location invariant - similar but the weights of the chosen neuron have equal probability to be copied or not.

All the experiments use truncation selection where the best quarter of the population survives and the mutation probability is 1/128 for every weight. Results are obtained by averaging ten trials for every test set up.

If neuron permutations and population convergence are present, location invariance would reduce convergence towards a single best individual without sacrificing the quality of the solution.

The diversity of the population is measured using the geometric distance between the most distinct chromosomes.

4   Results

When no artificial permutations are introduced, the more widely used versions of the genetic algorithm show better performance than the location invariant ones on Fig. 1. In these cases the whole population quickly converges except for the location invariant versions as seen on Fig. 2.

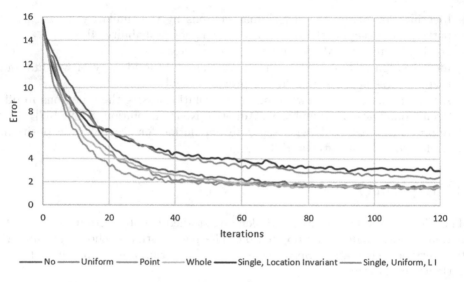

Fig. 1. Shows the results without artificial permutations.

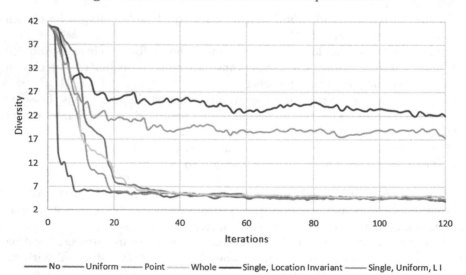

Fig. 2. Shows the population diversity without artificial permutations.

With the introduction of artificial permutations, the general versions of the genetic algorithm get worse results because of not being able to converge on one best solution as seen on Figs. 3 and 4.

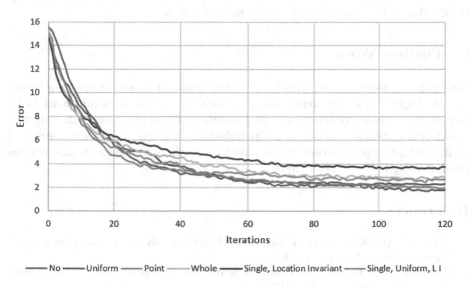

Fig. 3. Shows the results with artificial permutations.

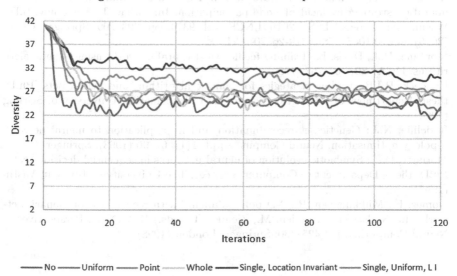

Fig. 4. Shows the population diversity with artificial permutations.

5   Conclusion

The application of a location invariant encoding scheme for artificial neural networks shows positive results when crossover is a limiting factor for success due to population convergence and neuron permutations.

6   Further Work

This paper presents a basic idea for location invariant crossover operator where only a single neuron gets relocated. More advanced versions would need to be implemented, possibly receiving equal number of neurons from both parents. Additional evaluation using actual populations evolving separately instead of artificially introduced permutations would prove the significance of this method for distributed genetic algorithms.

References

1. Holland, J.H.: Adaptation in Natural and Artificial Systems. The University of Michigan Press, Ann Arbor (1975)
2. Goldberg, D.E.: Genetic Algorithms in Search, Optimization, and Machine Learning. Addison Wesley, Reading (1989)
3. Balabanov, T., Zankinski, I., Barova, M.: Distributed evolutionary computing migration strategy by incident node participation. In: Lirkov, I., Margenov, S.D., Waśniewski, J. (eds.) LSSC 2015. LNCS, vol. 9374, pp. 203–209. Springer, Cham (2015). doi:10.1007/978-3-319-26520-9_21
4. Montana, D.J., Davis, L.: Training feedforward neural networks using genetic algorithms. IJCAI **89**, 762–767 (1989)
5. Hancock, P.J.B.: Coding strategies for genetic algorithms and neural nets. Ph.D. thesis Department of Computing Science and Mathematics, University of Stirling (1992)
6. Radcliffe, N.J.: Genetic set recombination and its application to neural network topology optimisation. Neural Comput. Appl. **1**(1), 67–90 (1993). Springer
7. Moriarty, D.E.: Symbiotic evolution of neural networks in sequential decision tasks. Ph.D. thesis Department of Computer Sciences, The University of Texas at Austin (1997)
8. Gomez, F., Miikkulainen, R.: 2-D pole balancing with recurrent evolutionary networks. In: Niklasson, L., Bodén, M., Ziemke, T. (eds.) ICANN 98. Perspectives in Neural Computing, pp. 425–430. Springer, London (1998)

Author Index

Printed in the United States
By Bookmasters

Printed in the United States
By Bookmasters